4

THEORETICAL AND COMPUTATIONAL CHEMISTRY

Recent Developments and Applications of Modern Density Functional Theory

THEORETICAL AND COMPUTATIONAL CHEMISTRY

SERIES EDITORS

Professor P. Politzer

Department of Chemistry
University of New Orleans
New Orleans, LA 70148, U.S.A.

Professor Z.B. Maksić

Ruder Bošković Institute
P.O. Box 1016,
10001 Zagreb, Croatia

VOLUME 1

Quantative Treatments of Solute/Solvent Interactions

P. Politzer and J.S. Murray (Editors)

VOLUME 2

Modern Density Functional Theory: A Tool for Chemistry

J.M. Seminario and P. Politzer (Editors)

VOLUME 3

Molecular Electrostatic Potentials: Concepts and Applications

J.S. Murray and K. Sen (Editors)

VOLUME 4

Recent Developments and Applications of Modern Density Functional Theory

J.M. Seminario (Editor)

4

THEORETICAL AND COMPUTATIONAL CHEMISTRY

Recent Developments and Applications of Modern Density Functional Theory

Edited by

J.M. Seminario

Department of Chemistry
University of New Orleans
New Orleans, LA 70148, USA

1996

ELSEVIER

Amsterdam – Lausanne – New York – Oxford – Shannon – Tokyo

ELSEVIER SCIENCE B.V.
Sara Burgerhartstraat 25
P.O. Box 211, 1000 AE Amsterdam, The Netherlands

ISBN: 0 444 82404 9

This book is printed on acid-free paper.

Printed in The Netherlands.

Preface to Density Functional Theory

Jorge M. Seminario
Department of Chemistry, University of New Orleans, New Orleans, Louisiana 70148, USA

THIS BOOK describes the present status of Density Functional Theory (DFT), which now has turned out to be the main technique for the study of matter at the atomistic level. Knowing the behavior of atoms and molecules provides a sure avenue for the design of new materials with specific features and properties in many areas of science and technology. The specialist or designer of new materials would benefit from a technique based on purely first principles that allows large savings in time and money.

The motivation and encouragement for preparing this book come from the experience and satisfaction obtained in editing a previous volume two years ago, to which this can be considered a sequel. There we tried to put in perspective and in rational terms the ideas behind modern DFT, which can be considered now a new and first principles (ab initio) technique. We have been witnessing for more than three decades how DFT evolved from a mere simplification of the Hartree-Fock (HF) method to what it is now: a fully formal and exact approach to the many-body problem. Comments and reviews have helped in the decision to undertake this second volume on DFT. Especial mention deserves the review by Robert Parr [J. Am. Chem. Soc., **117**, 12904 (1995)]. There were also words of concern and critiques by others. They were considered in this volume. Some criticisms were directly or indirectly connected to the premise of not having presented DFT in a historical fashion. In my personal opinion, such an approach would be confusing and misleading to the novice since DFT can be seen and analyzed from a quantum mechanical perspective.

The range of areas where DFT is applied has expanded and continues to do so. Any area where a molecular system is the center of attention can be studied using DFT. The scope of the 22 chapters of this book is an example of this.

Why do we use DFT? The exact answer to this question has been evolving since the inception of pre-DFT methods more than 60 years ago. The best answer is that DFT is nowadays the most precise tool available for dealing with the microscopic nature of matter. Perhaps this idea requires further elaboration. In its earlier times, DFT was used because it was faster than ab initio (initially just HF and simple corrections for correlation); however it was not as precise as ab initio. When only local functionals were available, DFT had the advantage over standard ab initio (HF) because DFT formally scaled as N^3 (provided a fitting of the density was performed) while HF is N^4. (In practice both methods scale lower than N^3.) What happened is that techniques for the standard ab initio methods have developed in such a way that the advantages of fitting the electron density are practically irrelevant.

It could also be argued that DFT is preferred over ab initio because the former uses the electron density with only three variables and the latter uses the wavefunction with 3N variables. However, those ab initio methods never actually deal with a 3N wavefunction, but also with 3-variable one-electron orbitals. The problem is in the fact that ab initio methods necessarily need to do 4-center integrals to compute the direct and exchange terms. This was avoided in old DFT because the density was fitted to a set of gaussians so the direct term was reduced to a product of gaussians and the exchange term was calculated using a functional of the density (as well as the correlation term). However modern DFT methods make use of practically the same ab initio techniques and carry out those four-electron integrals. Therefore, in this regard, there is no advantage to using DFT over a wavefunction-based method. In fact, the most precise DFT methods are as expensive computationally as the HF, and they use an auxiliary noninteracting wavefunction identical in structure to the HF wavefunction. The new so-called hybrid methods, where part of the exchange is calculated in the same fashion as in the HF method, are no more than an extension of the semilocal Generalized Gradient Aproximation (GGA) to incorporate fully nonlocal character in the exchange-correlation functional.

Present DFT results have surpassed in quality those from standard ab initio techniques like HF, MP, CC, CI, etc. These standard ab initio techniques were used to validate earlier DFT results. However, now they have become too imprecise when compared to modern DFT techniques. To make this point clear, let us see what happens to the atomization energy of water using several ab initio and DFT techniques shown in Table 1. Atomization energies are the most difficult energy differences to obtain since they imply the breaking of all the bonds in a molecule, and therefore most error cancellations are disturbed. These results for the atomization energy of water show a very interesting and common situation. Techniques like the MP4 or CC cannot compete any more with present DFT gradient-corrected ones. Results of this kind are common for many molecular systems that can be treated by conventional ab initio techniques.

Although a few studies have reported a "total failure of DFT", the use of DFT methods continues to increase because the cause of the failure is due, in several cases, to a particular functional, usually a local one as in the case of water in Table 1. In other cases, the use of a small basis set suitable for less sophisticated functionals but not for modern ones like the hybrid functionals can lead to wrong results. Failures were also reported in cases where standard ab initio methods were used as references. This is really preposterous in view of the results of Table I. If we would not know the experimental atomization energy of water and, as is frequently suggested, one used the CCSD(T) results as reference (similarly with any of the other N^7 methods), we would have obtained the errors shown in the column under "Error". In that case, Table 1 could be erroneously used to demonstrate the total failure of DFT in calculating the atomization energy of water. Other evaluations of DFT have been made comparing quantities that are not quantum mechanical expectation values.

Table 1
Water atomization energy errors (kcal/mol). The column labeled error (exp.) shows the errors with respect to the experimental value of 219.3 kcal/mol at 0 K. The column labeled as "error"(ab initio) is with respect to the CCSD(T)/6-311G** value, which is sometimes considered as a reference when no experimental value exists.

Method	Error (exp.)	"Error" (ab initio)
HF/6-311G(2df,p)	-79.9	62.7
MP2/6-311G(2df,p)	-11.7	5.5
MP4/6-311G**	-16.9	0.3
MP4/6-311+G	-12.5	4.7
MP4/6-311G(2df,p)	-13.2	4.0
QCISD(T)/6-311G**	-17.3	-0.1
CCSD(T)/6-311G**	-17.2	0.0
LSD(VWN)/DZVP	32.9	50.1
BLYP/6-31G**	-5.7	11.5
BLYP/6-311G(3df,3pd)	1.2	18.4
PW86/TZVP	3.7	20.5
PW91/TZVP	1.9	19.1
BP86/6-31G**	1.9	19.1
B3PW91/6-311G(2df,p)	-3.0	14.2

Explanations can be given for other apparent failures. For example, there is the incorrect comparison of calculated and experimental energies. Usually experimental values are given in the form of heats of formation at 298.15 K; however, calculations yield total energies at 0 K (provided the zero-point energy correction is also computed). Also, regarding the accuracy of the calculation of weak forces like van der Waals interactions, it has been argued that DFT cannot bind two He or Ne atoms. This is not completely true; in some cases it does (for instance using BLYP/6-31G**) and in others not. The reason is that these interactions are too small, and if they are below the precision of present functionals (3-5 kcal/mol) certainly the barrier will be missed by the calculation. This is a case of crossing-zero error rather than a failure of DFT. Most of the above "failures" have been widely discussed in the Computational Chemistry List (CCL).

DFT as currently used by theoretical chemists solves one of the fundamental equations of quantum mechanics. This equation evolved from an enigmatic theory developed at the beginning of the present century by many outstanding scientists like Planck, Einstein, Bohr, Heisenberg, Schrödinger, de Broglie, Born, Pauli and Dirac. The equation named after Schrödinger represents the most accurate of the theories ever developed. If solved precisely, it will tell us with certainty the energy or state of a particular atomic or molecular system. In very general form, for most situations of interest, this equation can be written as

$$\hat{H}\Psi = E\Psi$$

where $\hat{H}$ is the Hamiltonian operator for a nonrelativistic, Born-Oppenheimer, time-independent system of electrons. E is the electronic energy of the system and Ψ is the wavefunction of the system. The wavefunction Ψ is of paramount importance to the interpretation of quantum mechanics. The most conventional interpretation, the so-called Copenhagen interpretation, states that the square of the wavefunction yields the probability distribution of finding a set of electrons in a given place. What modern DFT does, based on the Hohenberg-Kohn-Sham theorems and procedure as amended by Levy, is to bypass the use of the interacting many-body wavefunction Ψ, making use of a much simpler function called the electron density ρ together with an auxiliary wavefunction of noninteracting electrons.

The electron density ρ is sometimes considered a physical observable, an experimentally obtainable function, although this is not quite true for isolated atoms or molecules where the uncertainty principle hinders its visualization. On the other hand, ρ is very well known in classical electrodynamics. It is an important quantity in the 19th century Maxwell Equations. So how the density appears in quantum mechanics is a rather interesting question. If we calculate the interaction of an electronic system with wavefunction Ψ with its external potential $\hat{\upsilon}_{ext}$ the interaction energy is precisely

$$V_{ne}^{QM} = \langle \Psi | \hat{\upsilon}_{ext} | \Psi \rangle = \int dr_1 \hat{\upsilon}_{ext}(\vec{r}_1) N \int d\vec{r}_2 \cdots d\vec{r}_N \left| \Psi(\vec{r}_1, \cdots, \vec{r}_N) \right|^2,$$

On the other hand, for a classical (class) charge density system $\rho(\vec{r})$, the interaction with the external potential $\hat{\upsilon}_{ext}$ is given by

$$V_{ne}^{class} = \int d\vec{r} \hat{\upsilon}_{ext}(\vec{r}) \rho(\vec{r})$$

Therefore, the energy of interaction with the external potential $\hat{\upsilon}_{ext}$ of the quantum mechanical (QM) electrons with wave function $\Psi(\vec{r}_1,\cdots,\vec{r}_N)$ is equivalent to the interaction of a classical charge distribution $\rho(\vec{r})$ (i.e., $V_{ne}^{QM} = V_{ne}^{class}$) when

$$\rho(\vec{r}) = N\int d\vec{r}_2\cdots d\vec{r}_N \left|\Psi(\vec{r},\vec{r}_2\cdots,\vec{r}_N)\right|^2$$

This is simply an association to a classical interpretation, so that what we call $\rho(\vec{r})$ in quantum mechanics is a function that if it acts classically on a potential $\hat{\upsilon}_{ext}$ will yield the same energy as the quantum mechanical system. If we try to write $\rho(\vec{r})$ as an expectation value we obtain,

$$\rho(\vec{r}) = N\left\langle\Psi(\vec{r}_1,\vec{r}_2\cdots,\vec{r}_N)\middle|\delta(\vec{r}-\vec{r}_1)\middle|\Psi(\vec{r}_1,\vec{r}_2\cdots,\vec{r}_N)\right\rangle$$

where $\delta(\vec{r}-\vec{r}_1)$ is the Dirac delta function in 3-dimensions. This last equation implies that $\rho(\vec{r})$ is N times the expectation value of finding one electron at position $\vec{r}$; or in other way, it is the sum of probabilities of finding all electrons at position $\vec{r}$. Certainly the probability of finding a particular electron at $\vec{r}$ have physical meaning; however N times such a probability has only a mathematical sense. Nevertheless, this function $\rho(\vec{r})$ is of great importance since it allows the interacting wavefunction to be bypassed in order to obtain the energy eigenvalue in a much simpler way.

This book has been divided into four parts:

I Basic Theory
II Functionals and their Problems
III Approaches and Methods
IV Applications

This partition is subjective since all contributions contain several portions belonging to each of the major divisions.

In part I, the most basic and general principles were accommodated. In addition, two chapters were added, one related to very precise solutions of the Schrodinger equation and another related to classical (as opposed to quantum) DFT. In the first chapter, Levy shows his amended version of the Hohenberg-Kohn-Sham DFT. In chapter 2, Ludeña, López-Boada and Pino present DFT from another perspective, the local scaling transformation to DFT. In chapter 3, March takes us to a modernization of a historical review of DFT. In Chapter 4, Sanders makes an overview of the most precise methods which provide energies and other values that represent benchmarks against which very precise functionals should be tested in the future. To end part I and for the sake of completeness, Percus offers the basics of classical DFT in Chapter 5.

In part II, we have collected very interesting papers dealing with the central problem of practical DFT applications, i.e., the study, design and test of exchange-correlation functionals. In Chapter 6, Ernzerhof, Burke and Perdew make an analysis of the exchange hole energy and functional, studying the PW91 functional to end with the behavior related to the breaking of a bond, including a review of the so-called hybrid functionals. In Chapter 7, Alonso and Cordero present their nonlocal weighted density aproximation to exchange, correlation and kinetic energy. In Chapter 8, Filippi, Gonze, and Umrigar make a critical evaluation of the GGA functionals comparing results with those obtained by some of the techniques explained in Chapter 4. In Chapter 9, Savin talks about degeneracy problems in DFT, and Chapter 10 by Ovchinnikov, Bender and Labanowski deals with the spin contamination present in a system of noninteracting electrons.

Part III treats approaches and methods. Chapter 11 by Casida introduces a review of time-dependent density functional response theory. In Chapter 12, Johnson et al illustrate the linear scaling techniques in DFT. In Chapter 13, St-Amant Goh and Gallant describe an implementation of the divide-and-conquer method. In Chapter 14, Rösch, Krüger, Mayer and Nasluzov present a review of relativistic DFT followed by its application to metal complexes and clusters.

Part IV contains several applications of DFT techniques. Chapter 15, by Neyman, Pacchioni and Rösch, deals with adsorption complexes on oxides. Chapter 16 by Broclawik, Vetrivel and Miyamoto shows a study of catalytic processes. In Chapter 17, Balbuena and Seminario find a DFT force field to study the nickel-water interface using molecular dynamics techniques. Chapter 18 by Frisch, Trucks and Cheeseman tells us about the model chemistries. In Chapter 19, Jursic shows several examples of transition state calculations. In Chapter 20 Belcastro et al study molecules with biological and pharmacological interest. Chapter 21 by Geerlings, De Proft and Martin, deals with the properties of charge distributions. Finally, in Chapter 22, Politzer, Wiener and Seminario show that DFT can be used for computing heats of reaction.

Last but not least, I would like to thank several people who have helped me in putting together this human adventure: Peter Politzer for his patient and tedious job as series co-editor; Perla Balbuena for reading, reviewing and suggestions regarding the preparation of the whole volume; John Perdew for several suggestions regarding the book; Mel Levy for reading this preface and his comments and suggestions; and mainly to all contributors, since without their efforts this book would not had been possible.

TABLE OF CONTENTS

TABLE OF CONTENTS

PART I:

BASICS

J.M. Seminario (Editor)
Recent Developments and Applications of Modern Density Functional Theory
Theoretical and Computational Chemistry, Vol. 4

Elementary Concepts in Density Functional Theory

Mel Levy

Department of Chemistry and Quantum Theory Group, Tulane University, New Orleans, Louisiana 70118

A few of the essential concepts in ground-state density functional theory are studied, and several of the basic theorems are proved in a simple manner. Then it is reviewed that coordinate scaling and the constrained-search approach provide dimensional requirements for exchange and correlation, provide formulas for the kinetic and potential components of the correlation energy, and provide a convenient path for a line integration that generates the correlation energy functional from the correlation potential. Recent research is reviewed concerning the adiabatic connection and hybrid functionals, and it is pointed out that the correction term to the two-point adiabatic connection formula consists of half the derivative of the correlation energy with respect to a coordinate scale factor in the density. This correction term is analyzed, and properties of the adiabatic connection integrand are listed for approximation purposes.

1. INTRODUCTION

Density-Functional Theory (DFT) is presently enormously popular. The electron density is quite attractive to work with because it contains only three dimensions, no matter how many electrons are being considered.

Superb books have recently been published that have carefully and accurately reviewed, in a systematic manner, the basic principles, theorems, and derivations in DFT. See, for instance, references [1-4]. Nevertheless, new users of DFT computer programs, who typically have little or no experience with the calculus of variations, often request to be provided, on just a few digestible pages, with quick and simple proofs of the very basic ground-state theorems in DFT, including derivations of the resultant Euler equations. Consequently, I have written the first part of this chapter with these requests in mind. For instance, I shall actually show how the corresponding Euler equation follows from the marvelous Hohenberg-Kohn variational principle [5,6], instead of just presenting the Euler equation as a consequence of the minimization process. In a sense, the first section of this contribution supplements part of the valuable earlier chapter, by Jorge Seminario [4], in a companion volume.

The constrained-search formulation shall be employed throughout, even for the adiabatic connection derivation, because the constrained-search approach is simple, explicit, automatically encompasses degeneracies, and circumvents v-representability considerations.

The reader might observe that certain technical points are going to be oversimplified, even those that I myself have spent a good bit of time studying, and that I shall obviously omit many important and fascinating features of the theory in an attempt to keep the focus well-localized. Also, the reader is going to have to infer the spin-density results from the density results.

In the second part of this chapter, coordinate scaling is first discussed. The coordinate scaling is then used to help quickly reveal aspects of the form of the exchange-correlation functional. Then, in the third part of the chapter, coordinate scaling is employed to display the exact formal scaling correction to the two-point approximation [7-9] in the adiabatic connection method [10-13]. Then a hybrid scheme is discussed in terms of a generalized Kohn-Sham formulation [14-16]. Finally, for approximation purposes, relations are explored [17,18] that connect the exchange-correlation potential (functional derivative) to the corresponding exchange-correlation functional, and a line integral is investigated. In essence, the second and third parts of this chapter concern powerful techniques that are now in use for approximating exchange-correlation and its functional derivative.

2. HOHENBERG-KOHN VARIATIONAL THEOREM

Suppose we want the ground-state energy and density for the following Hamiltonian of N electrons:

$$\hat{H} = \sum_{i=1}^{N} v(\vec{r}_i) + \hat{T} + \hat{V}_{ee} \tag{1}$$

where

$$\hat{T} = \sum_{i=1}^{N} -\frac{1}{2}\nabla_i^2, \tag{2}$$

where

$$\hat{V}_{ee} = \sum_{i=1}^{N-1}\sum_{j=i+1}^{N} |\vec{r}_i-\vec{r}_j|^{-1}, \tag{3}$$

and where $v(\vec{r})$ is the electron-nuclear attraction operator of interest. Then according to the Hohenberg-Kohn theorem [5], there exists a universal variational functional, $F[\rho]$, of trial electron densities, ρ, such that

$$E_o = \min_{\rho}\left\{\int d^3r v(\vec{r})\,\rho(\vec{r}) + F[\rho]\right\} \tag{4}$$

and

$$E_o = \int d^3r\; v(\vec{r})\,\rho_o(\vec{r}) + F[\rho_o] \tag{5}$$

where E_0 and ρ_0 are, correspondingly, the ground-state energy E_0 and the ground-state density of v. What makes $F[\rho]$ attractive is that ρ is only 3-dimensional, independent of the size of the system of interest.

2.1. Constrained-Search Proof Of Theorem

It is my purpose here to review how simple the proof of equations (4) and (5) is by using the constrained-search formulation. We simply start with the familiar variational theorem:

$$E_o = \min_{\Psi} <\Psi|\hat{H}|\Psi> . \tag{6}$$

Next, the key point in the proof is the utilization of the fact that Eq. (6) may be formulated in terms of a two-step minimization. Namely [19,20]

$$E_o = \min_{\rho} \min_{\Psi \to \rho} <\Psi|\hat{H}|\Psi> = \min_{\rho} <\Psi_{\rho}^{min}|\hat{H}|\Psi_{\rho}^{min}> \tag{7}$$

or

$$E_o = \min_{\rho} \left\{ <\Psi_{\rho}^{min}|\sum_{i=1}^{N} v(\vec{r}_i)|\Psi_{\rho}^{min}> + <\Psi_{\rho}^{min}|\hat{T} + \hat{V}_{ee}|\Psi_{\rho}^{min}> \right\} \tag{8}$$

where Ψ_{ρ}^{min} is the antisymmetric wavefunction that yields ρ and simultaneously minimizes the expectation value of $\hat{H}$. Equation (8) concludes the proof of the theorem [expressions (4) and (5)], with

$$\int d^3r \, v(\vec{r}) \, \rho(\vec{r}) = <\Psi_{\rho}^{min}|\sum_{i=1}^{N} v(\vec{r}_i)|\Psi_{\rho}^{min}> \tag{9}$$

with

$$F[\rho] = <\Psi_{\rho}^{min}|\hat{T} + \hat{V}_{ee}|\Psi_{\rho}^{min}> \tag{10}$$

and with the realization that $\Psi_{\rho_o}^{min}$ has to be the ground-state wavefunction of $\hat{H}$. The proof is that simple! But, the following very important fact still has to be utilized. Since the left-hand-side of Eq. (9) is obtained as the expectation value of $\sum_{i=1}^{N} v(\vec{r}_i)$ with any Ψ that yields ρ, it follows that Ψ_{ρ}^{min} is identified as the wavefunction that yields ρ and minimizes the expectation value of <u>just</u> $\hat{T} + \hat{V}_{ee}$; the $v(\vec{r})$ term need not be included in the constrained search! Consequently, the F[ρ] given in expression (10) is <u>truly</u> <u>universal</u> because the particular electron-nuclear attraction operator of interest does not appear on the right-hand-side of expression (10). The functional in Eq. (10) follows closely the one first put down by Percus for the non-interacting problem [21].

In actual calculations, F[ρ] must be approximated, and good approximations to it have to take into consideration, directly or indirectly, its definition in Eq. (10). This universality is important for approximating F[ρ] because we could study its properties, independent of the different v's under study; the value of F depends only upon the trial ρ within F[ρ].

2.2 Functional Derivative and Euler Equation

Functional derivatives are very important entities in density-functional theory (DFT). They appear as central characters for the minimization process in Eqs. (4-5). There are different equivalent ways to define the first functional derivative. Given some functional called G[ρ], a particularly useful definition for its first functional derivative, for application in DFT, is given through

$$\left.\frac{\partial G[\bar{\rho} + \varepsilon g]}{\partial \varepsilon}\right|_{\varepsilon=0} = \int d^3r \left.\frac{\delta G[\rho]}{\delta \rho(\vec{r})}\right|_{\rho=\bar{\rho}} g(\vec{r}) \,. \tag{11}$$

Here $\left.\frac{\delta G[\rho]}{\delta \rho(\vec{r})}\right|_{\rho=\bar{\rho}}$ is the functional derivative of G[ρ] at the density $\rho=\bar{\rho}$, ε is a scalar, and g is an arbitrary function. By setting F = G, we are now able to derive the Euler equation corresponding to our minimization problem, expressions (4) and (5).

Expressions (4) and (5) dictate that

$$E_0 = \min_{\varepsilon}\left\{\int d^3r v(\vec{r})\,[\rho_o(\vec{r}) + \varepsilon g(\vec{r})] + F[\rho_o + \varepsilon g]\right\} \tag{12}$$

for all g such that $\int g(\vec{r}) d^3r = 0$. Since the minimizing ε is ε = 0, it follows that

$$\left.\frac{\partial}{\partial \varepsilon}\left\{\int d^3r\, v(\vec{r})[\rho_o(\vec{r}) + \varepsilon g(\vec{r})] + F[\rho_o + \varepsilon g]\right\}\right|_{\varepsilon=0} = 0 \tag{13}$$

or

$$\int d^3r\, v(\vec{r})\, g(\vec{r}) + \left.\frac{\partial F[\rho_o + \varepsilon g]}{\partial \varepsilon}\right|_{\varepsilon=0} = 0 \,. \tag{14}$$

By expression (11), Eq. (14) becomes

$$\int d^3r\, g(\vec{r}) \left\{v(\vec{r}) + \left.\frac{\delta F[\rho]}{\delta \rho}\right|_{\rho=\rho_o}\right\} = 0 . \tag{15}$$

This, in turn, means that

$$v(\vec{r}) = -\left.\frac{\delta F[\rho]}{\delta \rho}\right|_{\rho=\rho_o} \tag{16}$$

because identity (15) applies for arbitrary g. Expression (16) is a fundamental Euler equation in DFT.

Observe that Euler equation (16) implies that $v(\vec{r})$ is a unique functional of the density, as was first shown by Hohenberg and Kohn via a different proof. In other words, note that Eq. (16) dictates that, except for the possible addition of a trivial constant, ρ_0 determines the one-body multiplicative attractive potential, v. Also observe that Eq. (12) requires that [22]

$$\left.\frac{\partial^2 F[\rho_o + \varepsilon g]}{\partial \varepsilon^2}\right|_{\varepsilon=0} \geq 0 \tag{17}$$

for arbitrary g. It is anticipated that the above inequality, the convexity constraint, shall prove useful for generating approximations to F in the future.

3. DERIVATION OF KOHN-SHAM THEORY

Assume that ρ_o is simultaneously the ground-state density of the noninteracting Hamiltonian

$$\hat{H}_s = \hat{T} + \sum_{i=1}^{N} v_s(\vec{r}_i) \tag{18}$$

It then follows from the Hohenberg-Kohn proof for the above corresponding interacting situation that

$$E_s = \min_{\rho}\left\{\int d^3r\, v_s(\vec{r})\,\rho(\vec{r}) + T_s[\rho]\right\} = \int d^3r\, v_s(\vec{r})\rho_o(\vec{r}) + T_s[\rho_o] \tag{19}$$

where E_s is the ground-state energy of $\hat{H}_s$. Here

$$T_s[\rho] = \langle \Phi_\rho^{min} | \hat{T} | \Phi_\rho^{min} \rangle, \tag{20}$$

and Φ_ρ^{min} is the antisymmetric wavefunction that yields ρ and simultaneously minimizes [21,23,19,20] just the expectation value of $\hat{T}$. By the previous section, the Euler equation corresponding to expression (19) is then

$$v_s(\vec{r}) = -\left.\frac{\delta T_s[\rho]}{\delta \rho(\vec{r})}\right|_{\rho=\rho_o} \tag{21}$$

(Note that $T_s[\rho]$ is often called the non-interacting kinetic energy functional or the Kohn-Sham kinetic energy functional. Also, Φ_ρ^{min} most often turns out to be a single determinant.)

Now, partition $F[\rho]$ as

$$F[\rho] = T_s[\rho] + U[\rho] + E_{xc}[\rho], \tag{22}$$

where

$$U[\rho] = \frac{1}{2}\int\int \frac{\rho(\vec{r}_1)\,\rho(\vec{r}_2)}{|\vec{r}_1 - \vec{r}_2|}\, d^3r_1 d^3r_2 \tag{23}$$

and where $E_{xc}[\rho]$ is the unknown exchange-correlation functional. The partitioning in Eq. (22) is a convenient one because of uniform density considerations (see, for instance, the discussion on page 18 in reference [4]), and because T_s, which helps to build in the Pauli principle, can often be computed exactly. Next combine Eq. (22) with Euler equations (16) and (21), to obtain

$$v_s(\vec{r}) = v(\vec{r}) + u([\rho_o]; \vec{r}) + v_{xc}([\rho_o]; \vec{r}), \tag{24}$$

where

$$u([\rho_o]; \vec{r}) = \left.\frac{\delta U[\rho]}{\delta \rho(\vec{r})}\right|_{\rho=\rho_0} = \int d^3r_2 \frac{\rho_o(\vec{r}_2)}{|\vec{r} - \vec{r}_2|} \tag{25}$$

and where the exchange-correlation potential, v_{xc}, is defined by

$$v_{xc}([\rho_o];\vec{r}) = \left.\frac{\delta E_{xc}[\rho]}{\delta \rho(\vec{r})}\right|_{\rho=\rho_0} \quad . \tag{26}$$

This all leads to the following iterative Kohn-Sham procedure [6]. One begins with a starting guess for ρ_o. Call this guess n_1. This, in turn, leads to a starting guess for $u + v_{xc}$, and hence for $\hat{H}_s$. That is, the starting $\hat{H}_s$ is

$$\hat{H}_s[n_1] = \hat{T} + \sum_{i=1}^{N} v_s([n_1];\vec{r}_i) \tag{27}$$

with

$$v_s([n_1];\vec{r}) = v(\vec{r}) + u([n_1];\vec{r}) + v_{xc}([n_1];\vec{r}) \tag{28}$$

where

$$u([n_1];\vec{r}) = \int \frac{n_1(\vec{r}_2)}{|\vec{r}-\vec{r}_2|} d^3r_2 \tag{29}$$

and where

$$v_{xc}([n_1];\vec{r}) = \left.\frac{\delta E_{xc}[\rho]}{\delta \rho}\right|_{\rho=n_1} \quad . \tag{30}$$

Next, solve for the ground-state of $\hat{H}_s[n_1]$. Call this density n_2. Use n_2 for the next iteration. The process continues until $n_{\ell+1} = n_\ell$. In other words, the process continues until self-consistency is achieved, upon which $n_{\ell+1} = \rho_o$.

It is often the case that the ground-state of $\hat{H}_s[n_k]$ is non-degenerate. In this situation, the ground-state wavefunction of $\hat{H}_s[n_k]$, $\Phi^{min}_{n_{k+1}}$, is a single determinant. This single determinant consists of Kohn Sham orbitals φ_i^{k+1} that are the lowest eigenfunctions of $-1/2\ \nabla^2 + v_s([n_k];\vec{r})$. That is

$$\left\{-\frac{1}{2}\nabla^2 + v_s([n_k];\vec{r})\right\}\varphi_i^{k+1}(\vec{r}) = \varepsilon_i^{k+1}\,\varphi_i^{k+1}(\vec{r})\,; i = 1,2,\cdots,N \tag{31}$$

where

$$n_{k+1} = \sum_{i=1}^{N} |\varphi_i^{k+1}|^2 \quad . \tag{32}$$

Note that in our notation, $v_s([n_k];\vec{r}) \neq -\left.\frac{\delta T_s[\rho]}{\delta \rho}\right|_{\rho=n_k}$, unless self-consistency is reached and ρ_o is obtained. At self-consistency, $n_k = \rho_o$, and thus in our notation we have $v_s([\rho_o];\vec{r}) = v_s(\vec{r}) = -\left.\frac{\delta T_s[\rho]}{\delta \rho(\vec{r})}\right|_{\rho=\rho_o}$.

4. COORDINATE SCALING

The exchange-correlation functional must be approximated in density-functional theory. In order to help one understand its form and in order to aid with our discussion later about the adiabatic connection method, some key coordinate scaling properties of $E_{xc}[\rho]$ shall now be briefly reviewed.

4.1 Definition Of Exchange And Correlation

To discuss coordinate scaling, it is first convenient and useful to partition $E_{xc}[\rho]$ into its exchange and correlation components

$$E_{xc}[\rho] = E_x[\rho] + E_c[\rho] \tag{33}$$

with

$$E_x[\rho] = \langle \Phi_\rho^{min} | \hat{V}_{ee} | \Phi_\rho^{min} \rangle - U[\rho] \tag{34}$$

so that

$$E_c[\rho] = \langle \Psi_\rho^{min} | \hat{T} + \hat{V}_{ee} | \Psi_\rho^{min} \rangle - \langle \Phi_\rho^{min} | \hat{T} + \hat{V}_{ee} | \Phi_\rho^{min} \rangle \tag{35}$$

4.2 Coordinate Scaling Requirements For $E_x[\rho]$ And $E_c[\rho]$

The study of coordinate scaling is the study of the exact dimensional properties in density-functional theory. Coordinate scaling provides key requirements of E_x and E_c for approximation purposes. Moreover, as we shall see in section 5, comprehension of the adiabatic connection method is greatly enhanced through coordinate scaling. With these thoughts in mind, define the uniformly scaled electron density, $\rho_\lambda(\vec{r})$, by

$$\rho_\lambda(\vec{r}) \equiv \rho_\lambda(x,y,z) \equiv \lambda^3 \rho(\lambda x, \lambda y, \lambda z) \tag{36}$$

With definition (36), we know that E_x satisfies the following simple homogeneous scaling relationship [17]:

$$E_x[\rho_\lambda] = \lambda E_x[\rho] \tag{37}$$

Requirement (37) dictates that a dimensionless function must multiply $\rho^{4/3}(\vec{r})$ in the integrand for a viable approximation to $E_x[\rho]$. The 4/3 results from (1+3)/3, where the number 1 stems from the fact that λ is raised to the power of unity in Eq. (37), and where the number 3 stems from the fact that space is 3 dimensional [24].

The scaling for E_c is much more complicated than for exchange. This fact is reflected in the forms of present approximate functionals.

For instance [17],

$$E_c[\rho_\lambda] > \lambda E_c[\rho]; \ \lambda > 1, \tag{38}$$

and [14,25,26]

$$E_c[\rho_\lambda] = \varepsilon_{c,2}[\rho] + \lambda^{-1}\varepsilon_{c,3}[\rho] + \lambda^{-2}\varepsilon_{c,4}[\rho] + \ldots \tag{39}$$

for sufficiently high λ, so that [25]

$$\lim_{\lambda \to \infty} E_c[\rho_\lambda] > -\infty \quad . \tag{40}$$

While [25]

$$E_c[\rho_\lambda] = \lambda D[\rho] + \cdots \tag{41}$$

for λ close enough to zero, where $D[\rho]$ is an unknown functional. In other words, E_c transforms differently for high and low λ. This means that unlike in the E_x case, the integrand, for a viable approximation to E_c, does not simply consist of a dimensionless function which multiplies ρ^m, where m is some constant.

It has been shown [17] that the simple scaling for E_x arises from the fact that the constrained-search definition for Φ_ρ^{min}, in Eq. (34), involves just the operator $\hat{T}$, in the search for Φ_ρ^{min}, and from the fact that $\hat{V}_{ee}$ is homogeneous of degree -1, through $|\vec{r}_i - \vec{r}_j|^{-1}$. Consequently λ is raised to the power of unity in expression (37). On the other hand, the complicated scaling for E_c arises from the fact that the constrained-search definition of Ψ_ρ^{min} in Eq. (35) involves two operators with <u>different</u> homogeneous properties; $\hat{T}$ is homogeneous of degree -2, because it contains second derivatives, while $\hat{V}_{ee}$ is homogeneous of degree -1 as stated above. (Eq (38) shall be derived later to illustrate a scaling derivation.)

Equation (39) is equivalent, at $\lambda = 1$, to

$$E_c[\rho] = \sum_{i=2} \varepsilon_{c,i}[\rho] \tag{42}$$

where $\varepsilon_{c,i}$ is homogeneous of degree i-2. That is [27-29]

$$\varepsilon_{c,i}[\rho_\lambda] = \lambda^{i-2}\, \varepsilon_{c,i}[\rho]. \tag{43}$$

Moreover, $\varepsilon_{c,2}[\rho]$ is known exactly as the second-order contribution [14] in a recently developed DFT perturbation theory [14,25]. In this perturbation theory, the non-interacting Kohn-Sham single determinant of ρ is the zero-order wavefunction and

$$\hat{V}_{ee} - \sum_{i=1}^{N} v_x([\rho];\vec{r}_i) - \sum_{i=1}^{N} u([\rho];\vec{r}_i) \tag{44}$$

is the perturbation [14] for the second-order term.

4.3 Kinetic And Potential Contributions To The Correlation Energy

Coordinate scaling allows one to generate the kinetic energy, T_c, and the potential energy, V_c, contributions to the correlation energy. The definitions of these contributions are, respectively,

$$T_c[\rho] = \langle \Psi_\rho^{min} | \hat{T} | \Psi_\rho^{min} \rangle - \langle \Phi_\rho^{min} | \hat{T} | \Phi_\rho^{min} \rangle \tag{45}$$

and

$$V_c[\rho] = \langle \Psi_\rho^{min} | \hat{V}_{ee} | \Psi_\rho^{min} \rangle - \langle \Phi_\rho^{min} | \hat{V}_{ee} | \Phi_\rho^{min} \rangle \quad . \tag{46}$$

They are obtained through coordinate scaling by [17]

$$T_c[\rho] = -E_c[\rho] + \left.\frac{\partial E_c[\rho_\lambda]}{\partial\lambda}\right|_{\lambda=1} \tag{47}$$

and [17]

$$V_c[\rho] = 2E_c[\rho] - \left.\frac{\partial E_c[\rho_\lambda]}{\partial\lambda}\right|_{\lambda=1} \tag{48}$$

Expressions (47) and (48) shall be derived in section 7.1.

5. ADIABATIC CONNECTION METHOD

The adiabatic connection formula is indeed a powerful one because it allows one to just focus upon the electron-electron repulsion operator when approximating $E_{xc}[\rho]$. In this section, the adiabatic connection formula [10-13] and method shall be derived and discussed. The object of the adiabatic connection method is to approximate $E_{xc}[\rho]$ by approximating the integrand in the adiabatic connection formula. The study of the adiabatic connection method is presently a very active area of research and is now beginning to provide exchange-correlation functionals that are among the best available.

5.1 Derivation Of Adiabatic Connection Formula

For the purpose of quickly deriving the adiabatic connection formula [10-12] through the simple constrained-search orientation [13], let's now define $\Psi_\rho^{min,\lambda}$ as the antisymmetric wavefunction that yields ρ and simultaneously minimizes $<\hat{T} + \lambda\hat{V}_{ee}>$. By recognizing that

$$E_{xc}[\rho] + U[\rho] = F[\rho] - T_s[\rho], \tag{49}$$

it follows that

$$E_{xc}[\rho] + U[\rho] = \int_0^1 d\lambda \frac{\partial}{\partial\lambda} < \Psi_\rho^{min,\lambda} |\hat{T} + \lambda\hat{V}_{ee}| \Psi_\rho^{min,\lambda} >, \tag{50}$$

where utilization, in Eq. (50), has been made of the identities

$$\Psi_\rho^{min} = \Psi_\rho^{min,1} \text{ and } \Phi_\rho^{min} = \Psi_\rho^{min,0} . \tag{51}$$

Next, we take advantage of the minimizing (stationary) quality embodied in the definition of $\Psi_\rho^{min,\lambda}$, concerning the variation of λ at fixed ρ, to ascertain that in Eq. (50) we only have to differentiate with respect to the operator, $\hat{T} + \lambda\hat{V}_{ee}$; the derivative with respect to $\Psi_\rho^{min,\lambda}$ at fixed operator, is zero. Consequently

$$E_{xc}[\rho] + U[\rho] = \int_0^1 d\lambda < \Psi_\rho^{min,\lambda} | \hat{V}_{ee} | \Psi_\rho^{min,\lambda} > \tag{52}$$

or

$$E_{xc}[\rho] = \int_0^1 V_{xc}^{\lambda}[\rho]\, d\lambda \tag{53}$$

where

$$V_{xc}^{\lambda}[\rho] = < \Psi_\rho^{min,\lambda} | \hat{V}_{ee} | \Psi_\rho^{min,\lambda} > - U[\rho] . \tag{54}$$

The partitioning of E_{xc} into E_x and E_c in expressions (33)-(35) enables a separate adiabatic connection formula for $E_c[\rho]$. Namely,

$$E_c[\rho] = \int_0^1 V_c^{\lambda}[\rho] d\lambda \tag{55}$$

where

$$V_c^{\lambda}[\rho] = <\Psi_\rho^{min,\lambda} | \hat{V}_{ee} | \Psi_\rho^{min,\lambda}> - <\Phi_\rho^{min} | \hat{V}_{ee} | \Phi_\rho^{min}> . \tag{56}$$

5.2 Exact Properties Of Adiabatic Connection Integrand

The object in the adiabatic connection method is to approximate its integrand V_{xc}^{λ} as a function of λ. See Eq. (53). Knowledge of V_{xc}^{λ} as a function of λ is equivalent to having exact knowledge of $E_{xc}[\rho]$. Accordingly, for approximation purposes in mind, I now list exact properties of $V_{xc}^{\lambda}[\rho]$. Note that the superscript λ represents an adiabatic connection parameter, while λ is a scale factor when it is a subscript with ρ.

Let's commence by partitioning $V_{xc}^{\lambda}[\rho]$ as

$$V_{xc}^{\lambda}[\rho] = E_x[\rho] + V_c^{\lambda}[\rho] . \tag{57}$$

where definitions (34), (54), and (56) have been employed. Hence, we need only focus upon how V_c^{λ} behaves in order to understand how V_{xc}^{λ} behaves, because $E_x[\rho]$ separates out of V_{xc}^{λ}. With this in mind the basic uniform scaling requirement of V_c^{λ} is [17,25]

$$V_c^{\lambda}[\rho] = \lambda V_c^{1}[\rho_{1/\lambda}] = \lambda V_c[\rho_{1/\lambda}] ,$$

where (58)

$$\rho_{1/\lambda}(x,y,z) = \lambda^{-3}\rho(\lambda^{-1}x,\lambda^{-1}y,\lambda^{-1}z).$$

Next, let's follow V_c^{λ} along the λ path. Begin at λ=0. We know the result exactly at this point. It is

$$V_c^{\lambda=0}[\rho] = 0 . \tag{59}$$

Now let's look at λ slightly greater than zero. In this region V_c^λ is linear [9] in λ. To see this begin with [17,14]

$$V_c^\lambda[\rho] = 2\lambda E_c[\rho_{1/\lambda}] + \lambda^2 \frac{\partial E_c[\rho_{1/\lambda}]}{\partial \lambda} \tag{60}$$

and for $E_c[\rho_{1/\lambda}]$ substitute [14,25]

$$E_c[\rho_{1/\lambda}] = \varepsilon_{c,2}[\rho] + \lambda \varepsilon_{c,3}[\rho] + \ldots \tag{61}$$

which applies when λ is sufficiently small (See Eq. (39)). The result is [14,16,30]

$$V_c^\lambda[\rho] = 2\lambda\, \varepsilon_{c,2}[\rho] + 3\lambda^2 \varepsilon_{c,3}[\rho] + \ldots \tag{62}$$

Now let's investigate V_c^λ for λ greater than the region near λ=0. For this purpose, an important general property is the fact that V_c^λ is a monotonically decreasing function of λ. That is [17]

$$\frac{\partial}{\partial \lambda} V_c^\lambda[\rho] < 0, \text{ all } \lambda \geq 0 . \tag{63}$$

Next, although one integrates only up to λ=1 in expressions (53) and (55), it is useful to give a result for $\lambda \to \infty$. Accordingly, we exploit the fact that

$$< \Psi_\rho^{\min,\lambda} | \hat{V}_{ee} | \Psi_\rho^{\min,\lambda} > \geq 0 \tag{64}$$

to imply [25]

$$V_c^\lambda[\rho] \geq -U[\rho] - E_x[\rho] \tag{65}$$

which means that V_c^λ is bounded as $\lambda \to \infty$. In fact, through implementation of the Lieb-Oxford bound [31] it has been proven that [32,22]

$$\lim_{\lambda \to \infty} V_c^\lambda \geq -a \int \rho^{4/3}(\vec{r})\, d^3r - E_x[\rho] . \tag{66}$$

Observe that the coefficient of the linear term, $2\varepsilon_{c,2}[\rho]$, is twice the second-order contribution to the correlation energy [14,30]. That $V_c^\lambda[\rho]$ is linear in the region of small enough λ was utilized in reference [9].

Finally, expressions (59), (62), (63), and (64) suggest that perhaps the derivative of V_c^λ is a monotonically increasing function of λ. In other words, perhaps

$$\frac{\partial^2 V_c^\lambda[\rho]}{\partial \lambda^2} > 0, \text{ all } \lambda \geq 0 \tag{67}$$

It would be worthwhile to investigate the validity of expression (67).

5.3 Exact Correction Term To Linear Approximation In Adiabatic Connection Method:

It is the purpose here to deduce the exact expression for $E_{xc}[\rho]$ in terms of the linear approximation to the adiabatic connection integrand, V_{xc}^λ, and a scaling correction. Start with

$$E_c[\rho] = \frac{1}{2}V_c[\rho] + \frac{1}{2}\frac{\partial E_c[\rho_\lambda]}{\partial\lambda}\bigg|_{\lambda=1} \tag{68}$$

which is a rearrangement of Eq. (48). Then simply add $E_x[\rho]$ to both sides of Eq. (68) to obtain the desired result which is [16]

$$E_{xc}[\rho] = \frac{1}{2}V_{xc}^{o}[\rho] + \frac{1}{2}V_{xc}^{1}[\rho] + \frac{1}{2}\frac{\partial E_c[\rho_\lambda]}{\partial\lambda}\bigg|_{\lambda=1} \tag{69}$$

or, equivalently,

$$E_{xc}[\rho] = \frac{1}{2}E_x[\rho] + \frac{1}{2}V_{xc}^{1}[\rho] + \frac{1}{2}\frac{\partial E_c[\rho_\lambda]}{\partial\lambda}\bigg|_{\lambda=1} \tag{70}$$

because

$$V_{xc}^{o}[\rho] = E_x[\rho] \tag{71}$$

by Eq. (54), and

$$V_{xc}^{1}[\rho] = E_x[\rho] + V_c[\rho] \tag{72}$$

by Eq. (57). The first two terms on the right-hand-side of Eq. (70) constitute the linear approximation for E_{xc} within the adiabatic connection method. Finally, an alternative way of arriving at expressions (69) and (70) is given in section 9.

The last (scaling) term in equations (69-70) constitutes the <u>correction</u> to the linear approximation (the assumption that V_{xc}^{λ} is linear in λ.). Since we have derived Eqs. (69-70) by starting with relation 68, it follows that the linear approximation would be exact if the density-functional definition of the correlation energy, $E_c[\rho]$, and the density-functional definition of the potential energy contribution to the correlation energy, $V_c[\rho]$, were to satisfy the same virial relation, for atoms, as are satisfied by their more familiar standard quantum chemistry counterparts. In other words, the error in the linear approximation would be zero <u>if</u> $E_c[\rho] = \frac{1}{2}V_c[\rho]$. (See also references [7] and [9].)

Alternatively, the linear approximation would be exact if the perturbation expansion for $E_c[\rho_\lambda]$, Eq. (39), where to contain only the second order term, because

$$\frac{\partial E_c[\rho_\lambda]}{\partial\lambda}\bigg|_{\lambda=1} = -\varepsilon_{c,3}[\rho] - 2\varepsilon_{c,4}[\rho] - \cdots \tag{73}$$

by expressions (39), (42), and (43). Consequently, the two-point (linear) formula with the added higher-order perturbation correction is [16]

$$E_{xc}[\rho] = \frac{1}{2}E_x[\rho] + \frac{1}{2}V^1_{xc}[\rho] - \frac{1}{2}\varepsilon_{c,3}[\rho] - \varepsilon_{c,4}[\rho] - \cdots \tag{74}$$

5.4 An Analysis Of The Adiabatic Connection Method And Future Considerations

With expression (70) we can understand why immediately the linear (two-point) approximation offers improvement over the local density approximation (LDA) for the exchange-correlation energy when the LDA repulsion contribution to exchange-correlation is employed at a coupling-constant of unity but not at the coupling-constant of zero. Although Eq. (70) was not published at the time of Becke's work [8], Becke, in effect approximates the right-hand-side of Eq. (70) by neglecting $\frac{1}{2}\left.\frac{\partial E_c[\rho_\lambda]}{\partial\lambda}\right|_{\lambda=1}$ and by using

$$\frac{1}{2}E_x[\rho] + \frac{1}{2}V^{1,LDA}_{xc}[\rho]\ , \tag{75}$$

where E_x is, of course, the exact exchange energy and where $V^{1,LDA}_{xc}$ is the local density approximation for V^1_{xc}. Analogous to Eqs. (72) and (48), $V^{1,LDA}_{xc}$ is obtained from the LDA exchange and correlation functionals by

$$V^{1,LDA}_{xc}[\rho] = E^{LDA}_x[\rho] + 2E^{LDA}_c[\rho] - \left.\frac{\partial E^{LDA}_c[\rho_\lambda]}{\partial\lambda}\right|_{\lambda=1} \tag{76}$$

Equation (76) is equivalent to the LDA expression in reference [33]. Expression (75) typically improves upon the E^{LDA}_{xc} because, relative to the other terms, the neglected exact value of $\frac{1}{2}\left.\frac{\partial E_c[\rho_\lambda]}{\partial\lambda}\right|_{\lambda=1}$ is generally small in magnitude and because $\frac{1}{2}V^{1,LDA}_{xc}$ is generally closer to $\frac{1}{2}V^1_{xc}$ than is E^{LDA}_{xc} to E_{xc}. Indeed, it has recently been argued that the LDA describes the exchange-correlation hole better at the coupling-constant of unity than at the coupling-constant of zero.

Exact values for $\frac{1}{2}\left.\frac{\partial E_c[\rho_\lambda]}{\partial\lambda}\right|_{\lambda=1}$ have recently been obtained. The obtained values, in atomic units, for this derivative correction to the two-point formula are small in magnitude. They are -0.003, -0.012, and -0.032, for the He, Be, and Ne atoms [34-36], respectively, while the corresponding total values for $E_{xc}[\rho]$ are approximately -1.068, -2.76, and -12.50. (Incidentally, the corresponding LDA values for 1/2 the above scaling derivative are much too negative. They are, respectively, -0.045, -0.086, and -0.247. The Perdew-Wang (91) values [37], however, are very good here. They are, respectively, -0.004, -0.011, and -0.034).

When the exact V_{xc}^{1} and E_x are employed, the two-point approximation,

$$\frac{1}{2}E_x + \frac{1}{2}V_{xc}^{1} \tag{77}$$

will yield atomic estimates for the exchange-correlation energies that are generally too positive because the signs of $\left.\frac{1}{2}\frac{\partial E_c[\rho_\lambda]}{\partial \lambda}\right|_{\lambda=1}$ have always been found to be negative. That is

$$E_{xc}[\rho] < \frac{1}{2}E_x[\rho] + \frac{1}{2}V_{xc}^{1}[\rho] \tag{78}$$

is to be expected. Expression (78) is consistent with inequality in conjecture (67).

Even if one were to know the exact $V_{xc}^{1}[\rho]$, then the above numbers for the exact scaling derivative correction indicate that the linear (two-point) approximation might not really be quite accurate enough. So, it looks like one has to employ, directly or indirectly, the scaling derivative term. With this in mind, I think that one of many possibilities to try would be to use

$$\frac{1}{2}E_x + \frac{1}{2}V_{xc}^{\text{mod},1} + \left.\frac{1}{2}\frac{\partial E_c^{\text{mod}}[\rho_\lambda]}{\partial \lambda}\right|_{\lambda=1} \tag{79a}$$

where the superscript "mod" indicates that an approximate (model) functional is used for the term. It would even be worthwhile to try an approximation for the scaling derivative term which comes from a model that is different from the one used for $V_{xc}^{\text{mod},1}$. Along these lines, it has been found that the accuracy becomes spectacular when the exact functional is used for the third term as well as for the first term in Eq. (79a), and the PW-91 is used for the second term [38]. There are obviously countless possiblities and opportunities for investigation [9,38]. Finally, it is interesting that if the model functionals used in the last two terms of expression (79a) are the same, then the expression becomes

$$\frac{1}{2}E_x[\rho] + \frac{1}{2}E_x^{\text{mod}}[\rho] + E_c^{\text{mod}}[\rho] \tag{79b}$$

and a scaling derivative thus disappears. It is consequently noteworth that until quite recently, all the hybrid functionals constructed after the publication of reference [8] have not contained a scaling derivative term.

5.5 Generating Approximations For V_{xc}^{λ}

The object of course is to appproximate as best as possible the adiabatic connection integrand, V_{xc}^{λ}, from λ=0 to λ=1, in expression (53). For this purpose the exact properties listed in section 5.2, via equations (57) through (67), should be of great help. For instance, we know that V_{xc}^{λ} is E_x at λ=0 and we know through Eq. (62) that its slope is $2\varepsilon_{c,2}$ at λ=0. Both E_x and $\varepsilon_{c,2}$ can be evaluated through the Kohn-Sham orbitals and orbital energies. (In fact encouraging implementation of this knowledge of $\varepsilon_{c,2}$ has recently been made by Ernzerhof [30]). The rest of the constraints listed in the preceding section deal with the shape of the exact V_{xc}^{λ} curve and its value at $\lambda=\infty$. By employing their knowledge of the known properties of the shape of V_{xc}^{λ}, Burke, Ernzerhof, and Perdew have fruitfully been exploring new approximate functionals [30, 39-41].

In approximating the V_{xc}^{λ} integrand one needs, of course, to approximate V_{xc}^{λ} for at least one point other than for $\lambda=0$. Accordingly, it is emphasized here that one may generate a model V_{xc}^{λ} from a model E_{xc} through use of expressions analogous to (57) and (60). That is [17,14,9]

$$V_{xc}^{mod,\lambda} = E_x^{mod} + 2\lambda E_c^{mod}[\rho_{1/\lambda}] + \lambda^2 \frac{\partial E_c^{mod}[\rho_{1/\lambda}]}{\partial \lambda} \tag{80}$$

In fact, for mod = LDA, Becke [8] employed an expression from the literature [33] that is equivalent to the right-hand-side in the above equation. In Becke's second paper, however, a $V_{xc}^{mod,\lambda}$ was not employed when he investigated gradient functionals, perhaps because expression (80) is not all that well known. I feel that it is worthwhile to pursue the study of V_{xc}^{λ} with respect to gradient functionals beyond the LDA. In fact, one could actually employ a different approximate functional (model) for each term on the right-hand-side of Eq. (80).

Essentially all of the published prevalent hybrid schemes employ neither a discussion of V_{xc}^{λ} nor an approximation to it in the hybrid formula. Again, perhaps this is because expression (80) is not well known. Instead, a hybrid functional of the following form is usually presented [42]:

$$E_{xc}^{hybrid}[\rho] = \alpha E_x[\rho] + (1-\alpha)\bar{E}_x^{mod} + \bar{E}_c^{mod} \tag{81}$$

where, consistent with our notation throughout this chapter, E_x is the exact exchange energy, but both $\bar{E}_x^{mod}$ and $\bar{E}_c^{mod}$ are approximations. The bar on each functional indicates that it may consist of a linear combination of present approximations.

6. HYBRID METHODS AS GENERALIZED KOHN-SHAM THEORIES

In a class of present hybrid schemes, a part of the exact Fock potential is employed, and the rest of the functional derivative is approximated by some linear combination of a variety of present approximations for exchange and correlation. In order to improve these hybrid approximations, it is necessary, through formal construction, to define what has to be approximated. With this in mind, I now review here the proof of this formal construction [14-16].

Define F_{HF}^{α} by

$$F_{HF}^{\alpha}[\rho] = \langle \Phi_{HF}^{\alpha}[\rho] | \hat{T} + \alpha \hat{V}_{ee} | \Phi_{HF}^{\alpha}[\rho] \rangle \tag{82}$$

where $\Phi_{HF}^{\alpha}[\rho]$ is the single determinant that yields ρ and minimizes $\langle \hat{T} + \alpha\hat{V}_{ee} \rangle$. Here HF signifies Hartree-Fock. Observe that $\Phi_{HF}^{1}[\rho_{HF}]$ is the familiar Hartree-Fock single determinant for the Hamiltonian in Eq. (1), where ρ_{HF} is the corresponding Hartree-Fock density. Next define v_{α} such that

$$\min_{\rho} \left\{ \int d^3r v_{\alpha}(\vec{r})\rho(\vec{r}) + F_{HF}^{\alpha}[\rho] \right\} = \int d^3r v_{\alpha}(\vec{r})\rho_o(\vec{r}) + F_{HF}^{\alpha}[\rho_o] \tag{83}$$

where ρ_o, our physical interacting density of interest, is simultaneously the ground-state density of $\hat{H}$ in Eq. (1). (Notice that $\rho_o \neq \rho_{HF}$) The Euler equation corresponding to (83) is

$$v_\alpha(\vec{r}) = -\left.\frac{\delta F_{HF}^{\alpha}[\rho]}{\delta\rho(\vec{r})}\right|_{\rho=\rho_o} \tag{84}$$

Now utilize F_{HF}^{α} as a component of F. In other words, define an unknown functional G_α such that

$$F[\rho] = F_{HF}^{\alpha}[\rho] + G_\alpha[\rho]. \tag{85}$$

That is,

$$G_\alpha[\rho] = <\Psi_\rho^{min}|\hat{T} + \hat{V}_{ee}|\Psi_\rho^{min}> - <\Phi_{HF}^{\alpha}[\rho]|\hat{T} + \alpha\hat{V}_{ee}|\Phi_{HF}^{\alpha}[\rho]> \tag{86}$$

Euler expression (16) thereby becomes

$$v(\vec{r}) = -\left.\frac{\delta F_{HF}^{\alpha}[\rho]}{\delta\rho(\vec{r})}\right|_{\rho=\rho_o} - g_\alpha([\rho_o];\vec{r}), \tag{87}$$

where

$$g_\alpha([\rho_o];\vec{r}) = \left.\frac{\delta G_\alpha[\rho]}{\delta\rho(\vec{r})}\right|_{\rho=\rho_o}. \tag{88}$$

By comparing (84) and (87), we have

$$v_\alpha(\vec{r}) = v(\vec{r}) + g_\alpha([\rho_o];\vec{r}). \tag{89}$$

Hence, ρ_o, the physical interacting ground-state density of $v(\vec{r})$, may be obtained through the following Hartree-Fock calculation:

$$\min_{\Phi_{SD}} <\Phi_{SD}|\sum_{i=1}^{N}\{v(\vec{r}_i) + g_\alpha([\rho_o];\vec{r}_i)\} + \hat{T} + \alpha\hat{V}_{ee}|\Phi_{SD}>, \tag{90}$$

where the Φ_{SD} are single determinants. In other words, the ground-state density is obtained as the Hartree-Fock density for the attractive potential $v(\vec{r}) + g_\alpha([\rho_o];\vec{r})$ and the repulsion operator $\alpha\hat{V}_{ee}$. The minimizing single determinant, $\Phi_{HF}^{\alpha}[\rho_o]$, is thus obtained from those orbitals, φ_i^{α}, that solve the following corresponding system of Hartree-Fock equations:

$$\left\{-\frac{1}{2}\nabla^2 + v(\vec{r}) + g_\alpha([\rho_o];\vec{r}) + \alpha u([\rho_o];\vec{r}) + \alpha\hat{v}_x^{HF}(\{\varphi_i^{\alpha}\};\vec{r})\right\}\varphi_k^{\alpha}(\vec{r}) = \epsilon_k^{\alpha}\varphi_k^{\alpha}(\vec{r});\ k=1,2,\ldots,N \tag{91}$$

The operator in expression (91) is the usual Hartree-Fock operator, except that it contains the coupling parameter α and the multiplication potential g_α. The operator $\hat{v}_x^{HF}$ is the familiar non-local Fock exchange operator. Here it is composed of the orbitals φ_i^α. The density in $g_\alpha([\rho];\vec{r})$ and in $u([\rho];\vec{r})$, and the orbtials in $\hat{v}_x^{HF}$, are changed during each interation until self-consistency is reached. Upon self-consistency, $\Phi_{HF}^\alpha[\rho_o]$ is generated. The physical ground-state energy is then given by

$$E_o = \langle\Phi_{HF}^\alpha[\rho_o]\,|\hat{H} + (\alpha-1)\hat{V}_{ee}|\Phi_{HF}^\alpha[\rho_o]\rangle + G_\alpha[\rho_o]. \tag{92}$$

or

$$E_o = \int d^3r\, v(\vec{r})\rho_o(\vec{r}) + \langle\Phi_{HF}^\alpha[\rho_o]\,|\hat{T} + \alpha\hat{V}_{ee}|\Phi_{HF}^\alpha[\rho_o]\rangle + G_\alpha[\rho_o] \tag{93}$$

In a sense expression (93) is a reasonably direct justification for the class of approximations commonly employed on the right-hand-side of Eq. (81).

To help best appreciate this section, let's investigate the α=0 and α=1 situations. At $\alpha = 0$, $v + g_\alpha$ becomes the Kohn-Sham effective potential v_s and Φ_{HF}^α becomes the Kohn-Sham single determinant. At α=1, $G_\alpha[\rho_o]$ becomes a correlation functional that should be typically close in value to $E_c[\rho_o]$, with the inequality

$$G_1[\rho_o] \geq E_c[\rho_o] \tag{94}$$

dictated by the definitions of E_c and G_1.

For <u>all</u> α, we have

$$\varepsilon_N^\alpha = -I\ , \tag{95}$$

where ε_N^α is the highest-occupied orbital energy and I is the <u>exact</u> ionization energy. The value of ε_N^α is independent of the value of α because the fact that the ground-state density, ρ_o, is independent of α implies that the asymptotic decay of ρ_o is independent of α, and both I and ε_N^α depend only upon the asymptotic decay of the density (see, for instance, references [1] and [2] and the works cited within).

7. COORDINATE SCALING PROOFS

The reader will gain some understanding of the simplicity of the constrained-search orientation and coordinate scaling if we now go through some of the scaling proofs [17]. They are indeed straightforward. Let's begin by immediately observing that

$$F[\rho_\lambda] \leq \langle\lambda^{3N/2}\Psi_\rho^{min}(\lambda\vec{r}_1,...\lambda\vec{r}_N)|\hat{T} + \hat{V}_{ee}|\lambda^{3N/2}\Psi_\rho^{min}(\lambda\vec{r}_1,...\lambda\vec{r}_N)\rangle \tag{96}$$

because

$$\lambda^{3N/2}\Psi_\rho^{min}(\lambda\vec{r}_1,...,\lambda\vec{r}_N) \rightarrow \rho_\lambda(\vec{r}) \tag{97}$$

and because of the $\langle\hat{T} + \hat{V}_{ee}\rangle$ minimizing quality of $\Psi_{\rho_\lambda}^{min}$. Since $\hat{T}$ is homogeneous of degree -2, since $\hat{V}_{ee}$ is homogeneous of degree -1, and since it is known that

$$T_s[\rho_\lambda] = \lambda^2 T_s[\rho] \tag{98a}$$

and

$$U[\rho_\lambda] = \lambda U[\rho], \tag{98b}$$

Eq. (96) becomes

$$\lambda^2 T_s[\rho] + \lambda U[\rho] + \lambda E_x[\rho] + E_c[\rho_\lambda] \leq \lambda^2 T[\rho] + \lambda V_{ee}[\rho] \, , \tag{99}$$

where

$$T[\rho] = \langle \Psi_\rho^{\min} | \hat{T} | \Psi_\rho^{\min} \rangle = T_s[\rho] + T_c[\rho], \tag{100}$$

and where

$$V_{ee}[\rho] = \langle \Psi_\rho^{\min} | \hat{V}_{ee} | \Psi_\rho^{\min} \rangle = U[\rho] + E_x[\rho] + E_c[\rho] - T_c[\rho] \, . \tag{101}$$

Next, employ expressions (100) and (101) in combination with Eq. (99). After a number of cancellations, the result is

$$E_c[\rho_\lambda] + \lambda(1-\lambda)\, T_c[\rho] - \lambda E_c[\rho] \leq 0, \tag{102}$$

which implies inequality (38) because $T_c[\rho] > 0$.

7.1 Scaling Derivation Of T_c And V_{xc}^{λ}

To get expression (47) for T_c, utilize the fact that the left-hand-side of Eq. (102) achieves its maximum at $\lambda = 1$. Consequently, Eq. (47) is generated [17] by setting the derivative equal to zero, at $\lambda = 1$, of the left-hand-side of Eq. (102). Further, since the T_c expression implies the expression for V_c, Eq. (48), we have also derived the latter. Finally, combine Eq. (48) with

$$V_c^{\lambda}[\rho] = \lambda V_c[\rho_{1/\lambda}] \tag{103}$$

to arrive at [17,9]

$$V_c^{\lambda}[\rho] = 2\lambda E_c[\rho_{1/\lambda}] + \lambda^2 \frac{\partial E_c[\rho_{1/\lambda}]}{\partial \lambda} \tag{104}$$

or, by adding E_x to both sides,

$$V_{xc}^{\lambda}[\rho] = 2\lambda E_{xc}[\rho_{1/\lambda}] + \lambda^2 \frac{\partial E_{xc}[\rho_{1/\lambda}]}{\partial \lambda} \tag{105}$$

7.2 Derivation $V_c^{\lambda}[\rho] = \lambda V_c[\rho_{1/\lambda}]$

Let's now prove the important identity (103), which is a restatement of relation (58). To achieve this, let's first review that it has been shown that

$\lambda^{3N/2} \Psi_\rho^{\min}(\lambda \vec{r}_1 \ldots, \lambda \vec{r}_N)$ yields $\rho_\lambda(\vec{r})$ and minimizes [17]

$$\langle \hat{T}(\vec{r}_1, \ldots, \vec{r}_N) + \lambda \hat{V}_{ee}(\vec{r}_1, \ldots, \vec{r}_N) \rangle \, . \tag{106}$$

Consequently, it follows that

$$\Psi_\rho^{\min,\lambda}(\vec{r}_1,\ldots,\vec{r}_N) = \lambda^{3N/2}\Psi_{\rho_{1/\lambda}}^{\min,1}(\lambda\vec{r}_1,\ldots,\lambda\vec{r}_N)\,. \tag{107}$$

In other words, the right-hand-side of Eq. (107) yields

$$\rho(x,y,z) = \lambda^3\rho_{1/\lambda}(\lambda x,\lambda y,\lambda z) \tag{108}$$

and minimizes the operator in expression (106). This infers that

$$V_c^\lambda[\rho] = <\lambda^{3N/2}\Psi_{\rho_{1/\lambda}}^{\min,1}(\lambda\vec{r}_1,...,\lambda\vec{r}_N)\,|\hat{V}_{ee}(\vec{r}_1,...,\vec{r}_N)|\,\lambda^{3N/2}\Psi_{\rho_{1/\lambda}}^{\min,1}(\lambda\vec{r}_1,...,\lambda\vec{r}_N)> - U[\rho] - E_x[\rho] \tag{109}$$

or

$$V_c^\lambda[\rho] = \lambda V_{ee}[\rho_{1/\lambda}] - U[\rho] - E_x[\rho] \tag{110}$$

where

$$V_{ee}[\rho_{1/\lambda}] = <\Psi_{\rho_{1/\lambda}}(\vec{r}_1,\ldots,\vec{r}_N)\,|\hat{V}_{ee}(\vec{r}_1,\ldots,\vec{r}_N)|\,\Psi_{\rho_{1/\lambda}}(\vec{r}_1,\ldots,\vec{r}_N)>. \tag{111}$$

As earlier in this chapter, use has been made here that $\hat{V}_{ee}$ is homogeneous of degree -1. Finally, identity (58) is arrived at [17,25] when $U[\rho_\lambda] = \lambda U[\rho]$ is employed, etc.

8. E_{xc} FROM v_{xc} BY LINE INTEGRATION THROUGH SCALING PATH

Knowledge of the exact relationship between E_{xc} and v_{xc} is especially important now that an active area of research involves the direct approximation of v_{xc}, in order to obtain more accurate orbital energies and densitites. The relation for exchange is simply [17]

$$E_x[\rho] = -\int d^3r\rho(\vec{r})\vec{r}\cdot\nabla v_x([\rho];\vec{r}) \tag{112}$$

The situation is more complicated for correlation. Instead, for correlation Eq. (112) transforms into the following inequality

$$E_c[\rho] < -\int\rho(\vec{r})\vec{r}\cdot\nabla v_c([\rho];\vec{r}) \tag{113}$$

However, all is not lost. We could get E_c from v_c, provided that we take a line integral along some specified path. For instance [18]

$$E_c[\rho] = \int_0^1 d\lambda\int d^3r\,v_c([\rho_\lambda];\vec{r})\,\frac{\partial\rho_\lambda(\vec{r})}{\partial\lambda} \tag{114}$$

where

$$v_c([\rho_\lambda];\vec{r}) = \left.\frac{\delta E_c[\rho]}{\delta\rho}\right|_{\rho=\rho_\lambda} \tag{115}$$

Eq. (114) above, of van Leeuwen and Baerends, results from a line integration of a formula of Ghosh and Parr [43] or a formula of Levy and Perdew [17]. It is instructive to now derive the formula. Commence with

$$E_c[\rho_{\lambda'}] - E_c[\rho_\lambda] = \int d^3r\, v_c([\rho_\lambda]; \vec{r})[\rho_{\lambda'}(\vec{r}) - \rho_\lambda(\vec{r})] \\ + \text{higher order terms in } \rho_{\lambda'} - \rho_\lambda \,, \tag{116}$$

where λ' is very close to λ. Next divide both sides by $\lambda' - \lambda$ and let $\lambda' \rightarrow \lambda$. Obtain

$$\frac{\partial E_c[\rho_\lambda]}{\partial \lambda} = \int d^3r\, v_c([\rho_\lambda]; \vec{r}) \frac{\partial \rho_\lambda(\vec{r})}{\partial \lambda} \,, \tag{117}$$

which resembles more closely the representation of the corresponding expression in reference [43] than in reference [17]. Next integrate both sides of Eq. (117) from λ=0 to λ=1 to generate our desired result, expression (114), by incorporating the crucial end point identity [25]

$$E_c[\rho_{\lambda=0}] = 0 \tag{118}$$

which is consistent with Eq. (41).

I feel that relations (112) and (114) will see ever increasing roles in the future. The object is to model v_x and v_c directly by utilizing important aspects of their known properties. Then one uses relations (112) and (114) to obtain, respectively, E_x and E_c from v_x and v_c.

9. CLOSING REMARKS

Various parts of the development within are closely connected to other studies. See, for instance, references [44] and [45]. Moreover, the exact correction term (the scaling derivative) to the linear approximation in the adiabatic connection method can be obtained by a straightforward manipulation of Eq. (3) in reference [38]. This latter equation is

$$E_{xc}[\rho] = \frac{1}{2}\left[V_{xc}^1[\rho] - \int d^3r\, \rho(\vec{r})\vec{r}\cdot\nabla v_{xc}([\rho];\vec{r})\right] \tag{119}$$

where our present notation is used. Specifically, to obtain equations (69) and (70), in Eq. (119) partition v_{xc} as

$$v_{xc}([\rho];\vec{r}) = v_x([\rho];\vec{r}) + v_c([\rho];\vec{r}) \,, \tag{120}$$

utilize Eq. (112), which relates E_x to v_x, and utilize [17]

$$\left.\frac{\partial E_c[\rho_\lambda]}{\partial \lambda}\right|_{\lambda=1} = -\int d^3r\, \rho(\vec{r})\,\vec{r}\cdot\nabla v_c([\rho];\vec{r}) \tag{121}$$

which is an alternative way of writing the corresponding expression in reference [43].

Equation (119) was employed to obtain extremely accurate values for $E_{xc}[\rho]$ from experimental densities [38] with the generalized gradient approximation [32,37]. Advantage was taken of the fact that the exact exchange-correlation energy functional is more local for full-coupling strength than for the coupling-constant average [46].

A number of valuable scaling relations are known for E_c, as reviewed in this chapter. Nevertheless, it would be absolutely wonderfull if one were to find a general scaling identity for E_c that is comparable to the one for E_x, but would of course have to be more complicated. Such an identity would indeed impose a powerful constraint for approximating E_c. Towards this goal, combine expressions (39), (40), and (42), to obtain

$$E_c[\rho] = \lim_{\lambda\to\infty} E_c[\rho_\lambda] - \left.\frac{\partial E_c[\rho_\lambda]}{\partial\lambda}\right|_{\lambda=1} - 2\,\varepsilon_{c,4}[\rho] - \cdots \tag{122}$$

Thus, Eq. (122) provides this kind of identity, through third order, for those densities where the Taylor series is valid. Further study is of course needed.

REFERENCES

1. R. G. Parr and W. Yang, *Density-Functional Theory of Atoms and Molecules* (Oxford University Press, Oxford, 1989).
2. R. M. Dreizler and E. K. U. Gross, *Density Functional Theory* (Springer Verlag, Berlin, 1990).
3. *Density Functional Theory*, edited by E. K. U. Gross and R. M. Dreizler (Plenum, New York, 1995).
4. *Modern Density Functional Theory: A Tool for Chemistry*, edited by J. M. Seminario and P. Politzer (Elsevier, Amsterdam, 1995).
5. P. Hohenberg and W. Kohn, Phys. Rev. 136, B864 (1964).
6. W. Kohn and L. J. Sham, Phys. Rev. 140, A 1133 (1965).
7. M. Levy in *Density Functional Theory*, edited by J. Keller and J. L. Gasquez (Springer, New York, 1983).
8. A. D. Becke, J. Chem. Phys. 98, 1372 (1993).
9. M. Levy, N. H. March, and N. C. Handy, J. Chem. Phys. 104, 1989 (1996).
10. D. C. Langreth and J. P. Perdew, Solid State Commun. 17, 1425 (1975).
11. O. Gunnarsson and B. I. Lundqvist, Phys. Rev. B13, 4274 (1976).
12. D. C. Langreth and J. P. Perdew, Phys. Rev. B15, 2884 (1976).
13. A. Görling, M. Levy, and J. P. Perdew, Phys. Rev. B47, 1167 (1993).
14. A. Görling and M. Levy, Phys. Rev. B47, 13105 (1993).
15. A. Seidl, A. Görling, P. Vogl, and J. A. Majewski, and M. Levy, Phys. Rev. B53, 3764 (1996).
16. A. Görling and M. Levy, J. Chem. Phys., in press.
17. M. Levy and J. P. Perdew, Phys. Rev. A32, 2010 (1985).
18. R. van Leeuwen and E. J. Baerends, Phys. Rev. A51, 170 (1995).
19. M. Levy, Proc. Natl. Acad. Sci. USA 76, 6062 (1979).
20. M. Levy, Phys. Rev. A26, 1200 (1982).
21. J. K. Percus, Int. J. Quantum Chem. 13, 89 (1978).
22. M. Levy and J. P. Perdew, Phys. Rev. B48, 11638 (1993).
23. M. Levy, Bull. Am. Phys. Soc. 24, 626 (1979).
24. L. J. Sham, Phys. Rev. A1, 969 (1970).
25. M. Levy, Phys. Rev. A43, 4637 (1991).
26. A. Görling and M. Levy, Phys. Rev. A45, 1509 (1992).
27. A. Görling and M. Levy, Int. J. Quantum Chem. Symposium 29, 93 (1995).

28. A. Görling and M. Levy, Phys. Rev. A52, 4493 (1995).
29. S. Liu and R. G. Parr, Phys. Rev. A, 53, 2211 (1996).
30. M. Ernzerhof, submitted to Chem. Phys. Letters.
31. E. H. Lieb and S. Oxford, Int. J. Quantum Chem. 19, 427 (1981).
32. J. P. Perdew, in *Electronic Structure of Solids* '91, edited by P. Ziesche and H. Eschrig (Akademie Verlag, Berlin, 1991).
33. J. P. Perdew and W. Yang, Phys. Rev. B45, 13244 (1992), and references within.
34. C. J. Umrigar and X. Gonze, in *High Performance Computing and its Application to the Physical Sciences*, Proceedings of the Mardi Gras '93 Conferences, edited by D. A. Browne et.al. (World Scientific, Singapore, 1993).
35. C. J. Umrigar, private communication.
36. C. J. Umrigar and X. Gonze, Phys. Rev. A50, 3827 (1994).
37. J. P. Perdew, J. A. Chevary, S. H. Vosko, K. A. Jackson, M. R. Pederson, D. J. Singh, and C. Fiolhais, Phys. Rev. B 46, 6671 (1992); 48, 4978(E) 1993.
38. K. Burke, J. P. Perdew, and M. Levy, Phys. Rev. A53, R2915 (1996).
39. M. Ernzerhof, K. Burke, and J. P. Perdew, submitted to the International Journal of Quantum Chemistry Symposium Issue.
40. K. Burke, M. Ernzerhof, and J. P. Perdew, submitted to Chem. Phys. Letters.
41. J. P. Perdew, K. Burke, and M. Ernzerhof, submitted to J. Chem. Phys.
42. A. D. Becke, J. Chem. Phys. 98, 5648 (1993).
43. S. K. Ghosh and R. G. Parr, J. Chem. Phys. 82, 3307 (1985).
44. E. I. Proynov, E. Ruiz, A. Vela, and D. R. Salahub, Int. J. Quantum Chem. S29, 61 (1995).
45. O. V. Gritsenko, R. van Leeuwen, and E. J. Baerends, Int. J. Quantum Chem., to be published.
46. K. Burke, J. P. Perdew, and M. Ernzerhof, Int. J. Quantum Chem., to be published.

J.M. Seminario (Editor)
Recent Developments and Applications of Modern Density Functional Theory
Theoretical and Computational Chemistry, Vol. 4

Explicit Density Functionals for the Energy by Means of Padé Approximants to Local-Scaling Transformations

Eduardo V. Ludeña,*[a] Roberto López-Boada[a], and Ramiro Pino[a]

[a]Centro de Química, Instituto Venezolano de Investigaciones Científicas, IVIC, Apartado 21827, Caracas 1020–A, Venezuela

We review in the present work the explicit construction of energy density functionals within the *local-scaling transformation version of density functional theory*, LS-DFT, as accomplished through the use of Padé approximants. The various stages in the development of density functional theory are briefly surveyed. A general way for constructing explicit Hartree-Fock energy density functionals for atoms is advanced and specific functionals for selected first-row atoms are displayed. It is found that these functionals contain universal factors whose leading terms are $\rho^{5/3}$ for the kinetic energy and $\rho^{4/3}$ for the exchange energy. But, in addition, they contain specific modulating factors that depend on the number of particles N, on the angular momentum and on the spin symmetry of the particular atomic ground states under consideration. Numerical examples attesting to the accuracy of the proposed functionals are given. Also, some problems pertaining to the extension of these ideas to the electronic correlation problem in atoms are discussed and explicit density functionals are obtained for the dominant electron correlation term.

1. INTRODUCTION

Density functional theory is becoming more and more the choice theory for dealing with the properties of quantum mechanical many-particle systems. [1–4]. The reason for its success is that it bypasses the complexities appearing in the usual treatments based on perturbation expansions and variational approaches. However, its purported simplicity has a price, at least in the case of formulations of this theory originating in the Hohenberg-Kohn-Sham context: it does not allow for the development of a systematic procedure whereby properties can be calculated with an increasing degree of accuracy. In perturbation theory, the question of the convergence of the series plays a preponderant role. Similarly, the upper-bound character of variational methods guarantees that one approaches the true result by successively increasing the variational domain. In the Hohenberg-Kohn-Sham version of density functional theory [5,7,6], on the other hand, once an approximate energy functional is prescribed, there is no way to improve it by following some strategy based on fundamental principles. All that can be done is to switch

*Author to whom all correspondence should be addressed

to another approximate functional, with the hope that this new functional is sufficiently taylored so as to include improvements.

The formulation of a density functional theory fully equivalent to the Schrödinger equation requires an adequate inclusion of the functional N-representability condition on the energy. This condition simply demands that there be a one to one correspondence between the energy written as a functional of the N-particle wavefunction and any other realization of this functional in terms of the reduced 1- or 2-density matrices or of the density. When the many-body problem is rigorously rewritten using the reduced second-order density matrix, it takes the form of the N-representability condition for the reduced 2-matrix [8,10,12,13]. Although some requirements pertaining to this condition are known, this still remains as an unsolved problem.

In the case when the energy is rewritten as a functional of the one-particle density, a general solution to the functional N-representability problem does not exist in the sense that there is no specification of how one should proceed in order to write down the *exact energy functional* for a given problem. Nevertheless, the closest approximation to this aim is provided by the *local-scaling transformation version of density functional theory* [14–48]. This is a constructive theory that includes from the outset the functional N-representability condition. This is done by starting from some guessed "generating" wavefunction (which does not have to be calculated) and by distorting it by means of density-dependent coordinate transformations. The whole gist of LS-DFT is that the energy functional for this deformed wavefunction generates an energy functional written entirely in terms of the one-particle density.

An important difference that arises in LS-DFT as opposed to the usual versions based on the Hohenberg-Kohn-Sham approach is that if one wishes to insure a one to one correspondence between the energy density functionals and the Schrödinger equation, (for the purpose of fulfilling the functional N-representability condition) then one has to drop the claim that these functionals are "universal". The question concerning the existence of a universal energy density functional is discussed in Section 2.5.

In the case of ordinary quantum mechanics, one defines the energy functional $E[\Phi] = \langle\Phi|\widehat{H}|\Phi\rangle$, where Φ is the N-particle wavefunction and $\widehat{H}$ the Hamiltonian and varies this functional until its extremum is reached. The functional $E[\Phi]$ clearly depends upon the symmetry properties of the wavefunction characterizing a given state, namely, on the quantum numbers associated with a specific spectroscopic term. Similarly, because of the constructive nature of LS-DFT, we also obtain in its context energy functionals that depend on N as well as on these quantum numbers. The explicit development of these functionals is carried out in Section 4.1 for the kinetic energy, in Section 4.2 for the exchange energy and in Section 6.1 for the dominant correlation energy term. We present here these basic results:

$$\begin{aligned} T[\rho] &= T_W[\rho] + \int_0^\infty dr\, r^2 \rho^{5/3}(r)\Big(1 + \vec{r}\cdot\nabla\ln\lambda(r)\Big)^{4/3}\tau_N \\ &+ \int_0^\infty dr\, r^2 \rho^{5/3}(r)\Big(1 + \vec{r}\cdot\nabla\ln\lambda(r)\Big)^{-2/3}\kappa_N, \end{aligned}$$

$$E_x[\rho] = -\frac{1}{2}\int_0^\infty dr_1 r_1^2 \sum_{k=0} \rho(r_1)^{(4+k)/3}\Big(\frac{1}{1 + \vec{r}_1\cdot\nabla_{\vec{r}}\ln\lambda(r_1)}\Big)^{(k+1)/3} \chi_N^k$$

and

$$
\begin{aligned}
E_C^{(1)}[\tilde{\Phi}_\rho] &= \int dr_1 r_1^2 \rho^{5/3}(r)\Big(r + \vec{r}_1 \cdot \nabla_{\vec{r}_1}\lambda(r)\Big)^{4/3} C_N^{(1)} \\
&+ \int dr_1 r_1^2 \rho^{4/3}(r)\Big(r + \vec{r}_1 \cdot \nabla_{\vec{r}_1}\lambda(r)\Big)^{2/3} C_N^{(2)} \\
&+ \int dr_1 r_1^2 \rho^{5/3}(r)\Big(r + \vec{r}_1 \cdot \nabla_{\vec{r}_1}\lambda(r)\Big)^{-2/3} C_N^{(3)}.
\end{aligned}
$$

The modulating factors τ_N, κ_N, $\{\chi_N^k\}$, and $\{C_N^{(i)}\}$ are specific to the N-particle system under consideration. In particular, the factors $\{\chi_N^k\}$ include the angular contributions arising in the exchange terms. All these factors are functions of the local-scaling transformation function. This function also appears explicitly as $\lambda(r)$ in the above expressions. In Section 5.1, we present an approximate analytic form for this transformation function where it is written in terms of the one-particle density. Hence, the density functionals listed above become explicit functionals of ρ.

The present work has been elaborated with the view of obtaining functionals for atoms. Nevertheless, the above formulas are general and can be applied as well to molecules. For molecular systems, however, the modulating factors would acquire particular forms. Work is being carried out in this laboratory for the purpose of obtaining the modulating factors for diatomic molecules [49].

2. NON-CONSTRUCTIVE AND CONSTRUCTIVE APPROACHES TO DENSITY FUNCTIONAL THEORY

Density functional theory, DFT, is reaching maturity. Ever since its inception in 1927 (in the form of the Thomas-Fermi theory for atoms) it has undergone a number of changes. Basically, one may characterize its beginning in terms of the heuristic attemps to improve the Thomas-Fermi formulation. These were followed by the formal attemps to express the many-body problem by means of reduced density matrices; this approach led to what is now called "density matrix theory". It was in density matrix theory that the functional N-representability problem first appeared in the form of conditions on the 2-matrix. This theory, which would have seemed to be the logical starting point for the development of density functional theory, did not serve, however, this purpose, and has followed, by and large, a rather separate course.

An important stage in the development of DFT was reached with the Hohenberg-Kohn theorem. This theorem provided the basis for what might be described as the formal non-constructive stage of DFT. The reformulation of this theorem by Levy and the subsequent mathematical analysis by Lieb, gave rise to an implicit constructive stage. But it was only with the introduction of density transformations that an explicit constructive stage was achieved. It is not surprising that in this constructive stage some of the preoccupations of density matrix theory, such as the functional N-representability conditions, found their way back into DFT.

The purpose of this Section is to briefly review these various stages in the development of DFT. Necessarily, a coherent review of these matters lies beyond the intention of this article. The reader is referred, however, for a more detailed account of this history, to Ref. [1] (Chapters 4, 5 and 6).

2.1. The Thomas-Fermi model and the heuristic stage

The classical works of Thomas [50] and Fermi [51] laid the bases for the "statistical approximation" to quantum mechanical many-particle systems. The primary idea in this approach was to express the energy of a many-particle system as a functional of the one-particle density and to obtain a differential equation for the density as the dynamical equation for the system. Clearly, the advantage was that one could get rid of wavefunctions – i.e., objects which depend on $3N$ coordinates – and use in their place much simpler objects such as the one-particle density, which is a function of just three spatial variables.

The Thomas-Fermi model of an atom is based on a semiclassical assumption, namely, on the premise that a Wigner distribution can be used for the phase-space representation of quantum mechanics. This model, therefore, standing at the semi-classical level, cannot provide a well-founded description of quantum phenomena. Nevertheless, the lack of rigor is compensated by its simplicity; it is precisely because of this characteristic that this theory is so attractive. Clearly, the Thomas-Fermi theory showed that it was possible to write the energy, although not very accurately, as a functional of the one-particle density and thus estimulated work directed at its improvement.

The early work of Thomas and Fermi was refined during the thirties by Dirac [52], Wigner [53] and von Weizsäcker [54], and it was further developed in the following decades by Kirzhnitz [55], Hodges [56] and others [57–59]. The improvements on the initial work aimed at including inhomogeneous corrections as well as many-particle correlation effects into the energy functionals.

The "heuristic" functionals that resulted from these attempts are all signed by a distinctive characteristic: they are not functionally N-representable. That is, there does not exist an energy functional, constructed as the expectation value of the Hamiltonian with respect to a wavefunction in Hilbert space, that can be reduced to these "heuristic" expressions for the energy.

2.2. Density matrix theory

Because the many-electron Hamiltonian has only two-particle interactions, it can be written as a sum of two-electron operators:

$$\widehat{H} = \sum_{i=1}^{N-1} \sum_{j=i+1}^{N} \widehat{K}(\vec{r}_i, \vec{r}_j) \tag{1}$$

where

$$\widehat{K}(\vec{r}_i, \vec{r}_j) = \frac{1}{N-1}(h(\vec{r}_i) + h(\vec{r}_j)) + \frac{1}{|\vec{r}_i - \vec{r}_j|} \tag{2}$$

with $h(\vec{r}_i) = -\frac{1}{2}\nabla_{\vec{r}_i} + v(\vec{r}_i)$. The expectation value of the Hamiltonian with respect to an N-particle wavefunction Φ becomes:

$$E[\Phi] = Tr_2[\widehat{K} D^2_\Phi] \tag{3}$$

where D^2_Φ is the 2-reduced matrix corresponding to the function Φ. The energy, as expressed in Eq. (3), depends only upon the diagonal part of the 2-matrix $D^2_\Phi(\vec{r}_1, \vec{r}_2; \vec{r}_1, \vec{r}_2)$, namely, on the pair-distribution function, PDF, (this is the unknown) and on the known two-particle Hamiltonian $\widehat{K}(\vec{r}_1, \vec{r}_2)$. Note that the PDF depends only on six spatial coordinates (and two spin ones). Thus, this is a very simple object in comparison with the N-particle wavefunction which depends on $3N$ spatial coordinates (and N spin ones). Clearly, the simplicity of the energy expression (3) can be readily appreciated when we notice that only six spatial coordinates enter into its description, regardless of the number of particles N.

This expression is, however, deceptively simple. In order to compute the exact ground-state energy, one sets up the variational principle

$$E_o = \inf_{D^2 \in \mathcal{P}_N^2} \left\{ Tr_2[\widehat{K} D^2] \right\} \tag{4}$$

In this equation, $\mathcal{P}_N^2$ is the subdomain of all reduced 2-matrices whose pre-images are in Hilbert space $\mathcal{L}_N$ under the mapping $\Phi \longrightarrow D^2$. The requirement that $D^2 \in \mathcal{P}_N^2$ is essential if one wishes to guarantee that at all steps of the variation $D^2 = D_\Phi^2$, namely, that D^2 is N-representable (i.e., that it comes from an N-particle wavefunction $\Phi \in \mathcal{L}_N$).

Let us comment on the relation between the N-representability of the 2-matrix and the *functional* N-representability of the energy functional. The latter is defined as the one to one correspondence between the energy as a functional of the N-particle wavefunction and the energy as a functional of the 2-matrix:

$$E[\Phi] \longleftrightarrow \mathcal{E}[D^2] \equiv Tr_2[\widehat{K} D^2], \qquad D^2 \in \mathcal{P}_N^2 \tag{5}$$

It so happens that in Eq. (5), since the two-particle operator is known and fixed, this correspondence can be fulfilled just by demanding that $D^2 \in \mathcal{P}_N^2$, namely the 2-matrix, be N-representable.

A solution to this problem demands that the necessary and sufficient conditions for $\mathcal{P}_N^2$ be given, without having to resort to the N-particle wavefunction. Some, but not all these conditions are known. In this sense, the N-representability problem for the 2-matrix, remains unsolved (for a review see Section 4.4 of Ref. [1]).

The above problem cannot be skewed by resorting to approximate forms of the 2-matrix, namely, D^2_{appr}, containing parameters which have been fitted so as to match the energy of some model systems and which are introduced then into the energy functional $Tr_2[\widehat{K} D^2_{appr}]$. Clearly, when the energies of other systems are computed using such *ad-hoc* functionals, it is not possible to claim that the resulting energy is either an upper-bound or a lower-bound. Thus, the goodness of such approximations with respect to experimental values cannot be taken as a measure of accuracy. They just reflect sound physical intuition in the selection of the approximate functional forms and care in parameter fitting. Let us also mention that the functional N-representability conditions are also pertinent to some recent attempts to derive Kohn-Sham-like equations for pair distribution functions.

An approach based on idempotent 1-matrices was advanced by March and Young in 1959. This approach has been recently generalized by Holas and March to interacting systems of electrons in three dimensions [60] (see also [62,61,63]). Closely related to this work is that put forward by Cohen and Frishberg [64] and Nakatsuji [65] in 1976 where an expression for the energy eigenstate using low-order reduced density is obtained (for other related work see [66–69]). There have been, in addition, several attempts to establish direct connections between density matrix theory and density functional theory. Among such works, let us mention those of Harbola and Sahni [70], Valdemoro [71] and Süle et al. [72].

A proper treatment of the many-body problem based on the 1-matrix requires an adequate handling of the functional N-representability problem. Attempts to formulate a 1-matrix theory with built-in N-representability conditions have been made by Nguyen-Dang, Ludeña and Tal [73] (see also [74]). The difficulty along this line stem from the fact that whenever the energy is expressed as a functional of the 1-matrix, it is not sufficient to require that the latter be N-representable (a task which can be fulfilled rather easily) but that the energy as a functional of the 1-matrix be N-representable.

2.3. The Hohenberg-Kohn theorem and the formal constructive stage

Whereas in the Thomas-Fermi theory, the contention that the one-electron density could determine the total energy, was a desideratum rather than a sound fact, in the considerations that preceded the formulation of the Hohenberg-Kohn theorem, the motivation was precisely to determine the truth of this contention.

For the systems initially considered by Hohenberg and Kohn the one-particle density determined the energy through the term $\int d^3\vec{r}\rho(\vec{r})v(\vec{r})$. Thus, the hypothesis that "knowledge of $\rho(\vec{r})$ alone determined – at least in principle – the energy" could be turned into another one, namely, "that a unique ground-state density is produced by a unique $v(\vec{r})$" [6].

The very simple *reductio ad absurdum* proof of this last conjecture was of momentous importance for the future development of DFT (more elaborate proofs designed so as to avoid some mathematical difficulties present in the original Hohenberg-Kohn version have been advanced by Lieb [91] and more recently by Pino et al. [75]). Apparently, one of the added conclusions was that this one to one correspondence between $\rho(\vec{r})$ and $v(\vec{r})$ "clearly does not depend on the details of the many-electron wave function". In fact, according to these authors 'since a knowledge of $v(\vec{r})$ completes the definition of the Hamiltonian $\widehat{H}$, it means that through $\widehat{H}$, a knowledge of $\rho(\vec{r})$, a function of three variables, implicitly defines the entire $3N$-dimensional ground state wave function $\Psi(\vec{r}_1,..,\vec{r}_N)$ and its energy E" [6].

The theory is further developed by noticing that since Ψ is determined by ρ, then the functional $F[\rho] \equiv \langle\Psi|\widehat{H}|\Psi\rangle$ is a unique functional of ρ and hence one can define an energy functional $E_v[\rho'(\vec{r})] \equiv \int d^3\vec{r}v(\vec{r})\rho'(\vec{r}) + F[\rho'(\vec{r})]$ satisfying the variational condition $E \leq E_v[\rho'(\vec{r})]$. Thus, the Hohenberg-Kohn theorem [5] proved the existence of an energy-density functional that yields the exact ground-state energy when evaluated at its extremum, i.e., when $\rho'(\vec{r})$ is the exact ground-state density.

Clearly, this theorem provided support to the earlier attempts devoted to finding heuristic expressions for the energy as a functional of the density. But, it did not give the means for constructing this functional as it simply was an existence theorem. Without trying to minimized the importance of this theorem, one may mention, however, that a widespread *non sequitur* is the notion concerning the existence of a "universal" functional which, purportedly has the same form for all many-electron system regardless of their specific characteristics (this point is discussed in Subsection 2.5).

Another aspect worth noticing in the Hohenberg-Kohn based density functional theory is the flagrant absence of the functional N-representability conditions. Apparently, these conditions were thought to be fulfilled by having N-representable (or v-representable) one-particle densities. Without entering deeply into this problem, let us quote the disconcerted statement of Löwdin [8]: "It is intriguing to think that if there is no representability problem in the electron-density functional method based on the Hohenberg-Kohn theorem, there can be hardly any representability problem in the approach based on reduced density matrices. Personally, I do not believe this is the case".

2.4. Levy's constrained variation and the implicit constructive stage

The Hohenberg-Kohn theory required in addition that the trial densities be in correspondence with ground-state wavefunctions. This "v-representability" condition was removed in the reformulation due to Levy [76] (see also Lieb, Payne and the early work of Percus [77–79]). Levy's work is based on the idea of a density-constrained minimization. As in the Hohenberg-Kohn approach, the energy functional is defined by $E_v[\rho] = F[\rho] + \int v\rho$. The auxiliary functional $F[\rho]$, however, is found by minimizing the functional

$$\begin{aligned} F[\rho] &= \min_{\Phi_\rho \in \mathcal{L}_N} \left\{ \langle \Phi_\rho \,|\, \widehat{T} + \widehat{U} \,|\, \Phi_\rho \rangle \right\} \\ &= \langle \Phi_\rho^{min} \,|\, \widehat{T} + \widehat{U} \,|\, \Phi_\rho^{min} \rangle \end{aligned} \tag{6}$$

i.e., by minimizing the internal energy corresponding to the kinetic energy and electron-electron interaction operators subject to the constraint of *fixed* ρ. Thus, in Levy's formulation the auxiliary wavefunctions Φ_ρ are trial wavefunctions (and not exact ground-state wavefunctions) in Hilbert space ($\mathcal{L}_N$) endowed with the characteristic that all of them yield the same fixed density ρ. Because these wavefunctions need not be ground-state wavefunctions, the "v-representability" condition is bypassed.

In Levy's reformulation, wavefunctions in Hilbert space are conspicuously present as the internal energy must be evaluated scanning over all wavefunctions $\Phi_\rho \in \mathcal{L}_N$. Clearly, there exists a minimizing wavefunction Φ_ρ^{min} that yields $F[\rho]$. The absolute minimum is attained when the fixed density $\rho = \rho_o^v$ is the exact ground-state density as in that case $F[\rho_o^v] = \langle \Psi_o^v \,|\, \widehat{T} + \widehat{U} \,|\, \Psi_o^v \rangle$, and the minimizing wavefunction is the exact ground-state wavefunction Ψ_o^v.

The *implicit* dependence of the auxiliary functional $F[\rho]$ on wavefunctions coupled to the *explicit* dependence of these wavefunctions on the number of particles N and the quantum numbers arising from operators that commute with the Hamiltonian, makes improbable the existence of universal functionals. Moroever, the question of whether it is possible to construct the exact functional $F[\rho]$ without resorting to wavefunctions is left open in this *implicit constructive* formulation of density functional theory [80].

There has been a vigorous work directed at constructing approximate functionals. The limitations of some incipient attempts based on local-density approximation have been amply understood and corrections have been suggested in order to include non-locality. Of importance in this respect have been the use of the generalized gradient approximation. In addition, one should mention the conditions– based on scaling properties – determined by Levy and co-workers for the exchange and correlation functionals. But as things stand, these functionals cannot lead to exact results as they do not satisfy the functional N-representability requirement.

2.5. Does there exist a universal functional for the energy?

The contention pertaining the existence of these universal functionals in the framework of the Hohenberg-Kohn-Sham approach appears to be based on a misconception. Briefly, the Hohenberg-Kohn theorem establishes a one to one correspondence between the exact ground-state one-particle density $\rho_o^v(\vec{r})$ and the external potential $v(\vec{r})$. It follows from this theorem that the density also determines the Hamiltonian $\widehat{H}_v = \widehat{T} + \widehat{U} + \sum_{i=1}^{N} v(\vec{r}_i)$ as well as the exact ground-state wavefunction Ψ_o^v. However, the latter is determined through the solution of the Schrödinger equation $\widehat{H}_v \Psi_o^v = E_o^v \Psi_o^v$. It is through the one to one correspondence between the exact one-particle density and the exact ground-state wavefunction that a *non-constructive proof* [80] of the existence of a density functional $E[\rho]$ is given. In Hohenberg and Kohn's work, however, since nothing is said about the underlying presence of the wavefunction and of its need in constructing the energy density functional, the false impression is given that wavefunctions (and hence N and the quantum numbers defining a state) can be dispensed with. As a result, it is concluded that because the functional apparently does not depend on the wavefunction, then it must be universal.

The presence of the wavefunction in the energy density functional is made more explicit in Levy's constrained-search formulation of density functional theory [76]. Certainly, if the exact auxiliary functional $F[\rho_o^v]$ depends upon the exact wavefunction Ψ_o^v for a particular N-particle system (corresponding to a given external potential $v(\vec{r})$), then one must bear in mind the fact that the exact wavefunction is an eigenfunction of the set of operators that commute with the Hamiltonian. This implies that exact wavefunctions for different systems exhibit different symmetries. Thus, for example, the ground-state wavefunction of the carbon atom is a triplet, whereas that of the neon atom is a singlet. These symmetry conditions require that Levy's constrained search be carried in specific subspaces (i.e., in the subspace of triplet wavefunctions for the carbon atom and of singlets for the neon atom). In other words, the symmetry of the wavefunction must be reflected in the density functional $F[\rho]$. One should design, therefore, approximate functionals for singlets and triplets (in the case at hand) in order to incorporate into the functional the symmetry of the wavefunction, or to speak with more propriety, the symmetry of the N-particle system one is considering (a very pointed analysis of the spin-symmetry difficulties that besiege the Hohenberg-Kohn-Levy formulation of density functional theory has been recently advanced by McWeeny [81]). It follows from the above considerations that the energy functional $F[\rho]$ cannot be "universal", namely, that it cannot be the same for all N-particle systems. It is not surprising that an example of the non-universality of the exchange functional for the hydrogen atom (which is a case where this functional can be calculated exactly) has been found by Gill and Pople [82].

The fact that the functional $F[\rho]$ is not universal for any fixed density ρ and not only for ρ_o^v is a consequence of the specificity of the minimizing wavefunction defining $F[\rho]$ for a particular N-particle system. Moreover, the fact that the external potential does not appear explicitly in $F[\rho]$ does not mean that it does not affect the wavefunction. The reason is that the external potential determines the density and the latter is in a one to one correspondence with the wavefunction.

In contrast with what happens in the context of the Hohenberg-Kohn-Levy formulation of density functional theory, within the constructive approach discussed in the present work, we obtain highly accurate energy density functionals that contain universal factors (i.e, general expressions containing only ρ) but also other modulating factors that are specific to a given system. These modulating factors, of course, become constants when the system considered is a homogeneous electron gas. It is perhaps possible that the widespread belief concerning the existence of universal functionals may have been the result of this particular choice of a reference system in the traditional approaches to density functional theory.

3. EXPLICIT CONSTRUCTIVE DENSITY FUNCTIONAL THEORY

3.1. Density transformations and the explicit constructive stage

Density transformations have been used for a long time both in quantum chemistry and in physics. Already in 1955, Macke [83,84] applied them to the many-body problem in quantum mechanics. This work was continued in different perspectives by March and Young in 1957 [85], Hall in 1960 [86], Moser in 1965 [87], Gilbert in 1975 [88], Percus in 1978 [79], Harriman in 1981 [89], Lieb in 1982 [90,77,91], Ludeña in 1983 [92], Zumbach and Maschke in 1983 [93], Nyden and Parr [94], Ghosh and Parr in 1985 [95], Ten Hoor in 1988 [96], etc.

Quite early it was realized that these transformations had much bearing on density functional theory. Thus, Zumbach and Maschke [93] used a particular type of these transformations for the purpose of generating modified plane-waves which could be used to construct Levy's functional. Attempts based on these transformations for the purpose of creating approximate energy density functionals were carried out by Ludeña in 1983 [92], Nyden and Parr [94] and Gosh and Parr [95]. In addition, the pioneer work of March and Young [85] was later developed by Kozlowski and March [97]).

A novel version of density functional theory based on density transformations was advanced in 1986 by Petkov, Stoitsov and Kryachko [16]. This theory, namely, the local-scaling transformation version of density functional theory (LS-DFT), was further developed in the works of Kryachko and Ludeña [14] and of Kryachko, Ludeña, Koga and others (see [1,48,98]). Essentially this formulation gave rise to what has been called the *explicit constructive stage* of density functional theory. A similar formulation – known as the density-driven approach – was developed independently by Cioslowski in 1988 [99–101,80]. As shown by Kryachko and Ludeña in 1991 [27], the density-driven approach corresponds to a finite-basis realization of LS-DFT.

3.2. Density transformations in density functional theory

A historical review of the influence of density transformations in the development of density functional theory has been given elsewhere [102]. For the sake of conciseness we devote the present Subsection to the discussion of some of the basic ideas underlying density transformations.

In order to illustrate the meaning of density transformations, consider the simple transformation discussed by Hall in 1960 [86]. Given the following orthonormal set $\{\phi_i(r, \theta, \phi)\}$, the orbital transformation

$$\psi_i(r,\theta,\phi) = \left(\frac{df(r)}{dr}\right)^{1/2} \frac{f(r)}{r} \phi_i(f(r),\theta,\phi) \tag{7}$$

has the property of keeping the overlap invariant. This can be easily shown as follows:

$$\begin{aligned} S_{ij}(\psi) &= \int drr^2 \int d\theta \sin\theta \int d\phi \psi_i^*(r,\theta,\phi)\psi_j(r,\theta,\phi) \\ &= \int drr^2 \int d\theta \sin\theta \int d\phi \frac{df(r)}{dr}\frac{f(r)^2}{r^2}\phi_i^*(f(r),\theta,\phi)\phi_j(f(r),\theta,\phi) \\ &= \int df(r) f^2(r) \int d\theta \sin\theta \int d\phi \phi_i^*(f(r),\theta,\phi)\phi_j(f(r),\theta,\phi) = S_{ij}(\phi). \end{aligned} \tag{8}$$

The overlap-conserving property guarantees the orthonormality of the transformed set. Moreover, if we define an untransformed density by

$$\rho_\psi(\vec{r}) = \sum_{i=1}^{N} |\psi_i(\vec{r})|^2, \tag{9}$$

then, the density generated by the transformed orbitals is:

$$\rho_\phi(\vec{r}) = \sum_{i=1}^{N} |\phi_i(\vec{r})|^2. \tag{10}$$

It follows from Eq. (7) that these densities are related by the equation

$$\rho_\psi(\vec{r}) = \frac{df(r)}{dr}\left(\frac{f(r)}{r}\right)^2 \rho_\phi(\vec{r}). \tag{11}$$

Thus, it is evident that the transformation proposed by Hall is in effect a density transformation.

The mathematical problem associated with the transformation of a density $\rho_1(\vec{r})$ into a density $\rho_2(\vec{r})$ when neither of them is constant (i.e, when neither is the density of a homogeneous electron gas, for example) was solved in 1965 by Moser [87]. This solution, however, remained unknown to the physics and chemistry community as it was written for mathematicians. We employ here the more accessible rendition given by Bokanowski and Grebert (1995) [103].

Let us consider the N-particle function Φ_0 given as a linear combination of Slater determinants constructed from plane waves:

$$\Phi_0 = \sum_K C_K \chi_K, \tag{12}$$

where

$$\chi_K = (N)^{-1/2} \det[\phi_{\vec{k}_1}, \cdots, \phi_{\vec{k}_N}], \tag{13}$$

and

$$\phi_{\vec{k}_1} = \exp i\vec{k}_j \cdot \vec{r}. \tag{14}$$

It is clear that the N-particle function Φ_0 belongs to the space $L_A^2(V^N)$ where V is defined as the unit cube $V = [0,1]^3$. The one-particle density $\rho_0(\vec{r})$ associated with the N-particle function $\Phi_0 \in L_A^2(V^N)$ comes from the following integration in the unit cube $V = [0,1]^3$:

$$\rho_0(\vec{r}) = N \int_{V^{N-1}} d^3\vec{r}_2 \cdots d^3\vec{r}_N |\Phi_0(\vec{r}, \vec{r}_2, \ldots, \vec{r}_N)|^2. \tag{15}$$

It is easy to show that the resulting density is $\rho_0(\vec{r}) = N/V$ provided that the expansion coefficients satisfy

$$\sum_{K=1} \sum_{L=1} C_K \, C_L \, \Delta_{KL}^{k_i k_j} = 0 \quad \text{for} \quad K \neq L \quad \text{and} \quad k_i \neq k_j \tag{16}$$

where [104]

$$\Delta_{KL}^{k_i k_j} = (-1)^{i+j} \quad \text{when} \quad \{K - k_i\} = \{L - k_j\} \quad \text{and zero otherwise.} \tag{17}$$

Let us consider, in addition, the action of the transformation operator $\widehat{R}_\rho$ on the vector $\vec{r} = (r, \theta, \phi)$:

$$\widehat{R}_\rho o(r, \theta, \phi) = (R_{\rho,r}(r, \theta, \phi), R_{\rho,\theta}(r, \theta, \phi), R_{\rho,\phi}(r, \theta, \phi)). \tag{18}$$

Clearly, this transformation carries the original plane-wave into the new density-deformed plane-wave:

$$\psi_{\vec{k}_j}(\vec{r}) = \sqrt{\frac{\rho(\vec{r})}{N}} \exp i\vec{k}_j \cdot \vec{R}_\rho(\vec{r}). \tag{19}$$

In terms of the Jacobian, this orbital transformation is given by

$$\psi_{\vec{k}_j}(\vec{r}) = \sqrt{J(\vec{R}_\rho; \vec{r})} \phi_{\vec{k}_j}(\vec{R}_\rho(\vec{r})). \tag{20}$$

Similarly, we have for the density transformation:

$$\rho(\vec{r}) = J(\vec{R}_\rho; \vec{r}) \rho_0(\vec{R}_\rho(\vec{r})). \tag{21}$$

Let us consider an arbitrary N-electron wavefunction $\Phi_1 \in \mathcal{L}_N$ from which we obtain the density $\rho_1 \equiv \rho_{\Phi_1}$ as follows:

$$\rho_1(\vec{r}) = N \int d^3\vec{r}_2 \cdots d^3\vec{r}_N |\Phi_1(\vec{r}, \vec{r}_2, \ldots, \vec{r}_N)|^2, \tag{22}$$

and, another arbitrary wavefunction $\Phi_2 \in \mathcal{L}_N$ associated with the density $\rho_2 \equiv \rho_{\Phi_2}$.

Let us assume that the transformation connecting the density ρ_1 to the density ρ_2 can be performed by means of the operator $\widehat{f}$. Then, the Jacobian problem associated with these densities is that of finding the diffeomorphism $\vec{f}(\vec{r})$ for the density transformation [87,103]:

$$\forall \vec{r}, \quad \rho_2(\vec{r}) \; = \; J(\vec{f}; \vec{r}) \rho_1(\vec{f}(\vec{r})). \tag{23}$$

In order to obtain the transformation vector $\vec{f}(\vec{r})$, Moser [87] (see also Bokanowski and Grébert [103,105]) considered the density transformations

$$\forall \vec{r}, \quad \rho_2(\vec{r}) \; = \; J(\vec{R}_{\rho_2}; \vec{r}) \rho_0(\vec{R}_{\rho_2}(\vec{r})), \tag{24}$$

and

$$\forall \vec{r}, \quad \rho_1(\vec{r}) = J(\vec{R}_{\rho_1};\vec{r})\rho_0(\vec{R}_{\rho_1}(\vec{r})). \tag{25}$$

The general transformation connecting densities ρ_1 and ρ_2 can be written, therefore, as [87,103]:

$$\hat{f} = (\hat{R}_{\rho_1})^{-1} \, o \, \hat{R}_{\rho_2} \tag{26}$$

which corresponds to the Jacobian

$$J(\vec{f}(\vec{r});\vec{r}) = \frac{\rho_2(\vec{r})}{\rho_1(\vec{f}(\vec{r}))}. \tag{27}$$

From Eq. (26) we obtain:

$$\hat{R}_{\rho_1} \, o \, \hat{f} = \hat{R}_{\rho_2}. \tag{28}$$

Using Eq. (18) we can now determine the action of the operators appearing on both sides of Eq. (28) on the vector $\vec{r} = (r, \theta, \phi)$. The result of the action of operator $\hat{R}_{\rho_2}$ on the vector $\vec{r} = (r, \theta, \phi)$, is:

$$\hat{R}_{\rho_2} \, o \, \vec{r} \equiv \hat{R}_{\rho_2} \, o \, (r, \theta, \phi) = (R_{\rho_2,r}(r,\theta,\phi), R_{\rho_2,\theta}(r,\theta,\phi), R_{\rho_2,\phi}(r,\theta,\phi)). \tag{29}$$

Similarly, the effect of the operators $\hat{R}_{\rho_1} \, o \, \hat{f}$ on the vector $\vec{r}$ is:

$$\begin{aligned}\hat{R}_{\rho_1} \, o \, \hat{f} \, o \, \vec{r} \equiv \hat{R}_{\rho_1} \, o \, \hat{f} \, o \, (r,\theta,\phi) &= \hat{R}_{\rho_1}(f_r(r,\theta,\phi), f_\theta(r,\theta,\phi), f_\phi(r,\theta,\phi)) \\ &= (R_{\rho_1,r}(f_r, f_\theta, f_\phi), R_{\rho_1,\theta}(f_r, f_\theta, f_\phi), R_{\rho_1,\phi}(f_r, f_\theta, f_\phi)).\end{aligned} \tag{30}$$

Since the components in the right-hand side must equal those of the left-hand side we obtain the following set of equations:

$$\begin{aligned}R_{\rho_2,r}(r,\theta,\phi) &= R_{\rho_1,r}(f_r, f_\theta, f_\phi), \\ R_{\rho_2,\theta}(r,\theta,\phi) &= R_{\rho_1,\theta}(f_r, f_\theta, f_\phi), \\ R_{\rho_2,\phi}(r,\theta,\phi) &= R_{\rho_1,\phi}(f_r, f_\theta, f_\phi).\end{aligned} \tag{31}$$

In order to express these relations in integrated form, we must select a particular functional form for the components. This is done by setting (Ghosh and Parr [95], Zumbach and Maschke [93]):

$$R_{\rho,r}(r,\theta,\phi) = \frac{\int_0^r \int_0^\pi \int_0^{2\pi} dr' d\theta' d\phi' \mu(r', \theta', \phi')}{\int_0^{+\infty} \int_0^\pi \int_0^{2\pi} dr' d\theta' d\phi' \mu(r', \theta', \phi')}, \tag{32}$$

$$R_{\rho,\theta}(r,\theta,\phi) = \frac{\int_0^\theta \int_0^{2\pi} d\theta' d\phi' \mu(r, \theta', \phi')}{\int_0^\pi \int_0^{2\pi} d\theta' d\phi' \mu(r, \theta', \phi')}, \tag{33}$$

$$R_{\rho,\phi}(r,\theta,\phi) = \frac{\int_0^\phi d\phi' \mu(r, \theta, \phi')}{\int_0^{2\pi} d\phi' \mu(r, \theta, \phi')}, \tag{34}$$

where $\mu_\rho(r,\theta,\phi)$ is the measure

$$\mu_\rho(r,\theta,\phi) = \rho(r,\theta,\phi) r^2 \sin\theta. \tag{35}$$

The above choice allows us to write Eq. (31) explicitly. The general expressions for these transformations are given therefore by

$$\frac{\int_0^{f_r}\int_0^{\pi}\int_0^{2\pi} df_r' df_\theta' df_\phi' \mu_{\rho_1}(f_r', f_\theta', f_\phi')}{\int_0^{+\infty}\int_0^{\pi}\int_0^{2\pi} df_r' df_\theta' df_\phi' \mu_{\rho_1}(f_r', f_\theta', f_\phi')} = \frac{\int_0^{r}\int_0^{\pi}\int_0^{2\pi} dr' d\theta' d\phi' \mu_{\rho_2}(r', \theta', \phi')}{\int_0^{+\infty}\int_0^{\pi}\int_0^{2\pi} dr' d\theta' d\phi' \mu_{\rho_2}(r', \theta', \phi')}, \tag{36}$$

$$\frac{\int_0^{f_\theta}\int_0^{2\pi} df_\theta' df_\phi' \mu_{\rho_1}(f_r, f_\theta', f_\phi')}{\int_0^{\pi}\int_0^{2\pi} df_\theta' df_\phi' \mu_{\rho_1}(f_r, f_\theta', f_\phi')} = \frac{\int_0^{\theta}\int_0^{2\pi} d\theta' d\phi' \mu_{\rho_2}(r, \theta', \phi')}{\int_0^{\pi}\int_0^{2\pi} d\theta' d\phi' \mu_{\rho_2}(r, \theta', \phi')}, \tag{37}$$

$$\frac{\int_0^{f_\phi} df_\phi' \mu_{\rho_1}(f_r, f_\theta, f_\phi')}{\int_0^{2\pi} df_\phi' \mu_{\rho_1}(f_r, f_\theta, f_\phi')} = \frac{\int_0^{\phi} d\phi' \mu_{\rho_2}(r, \theta, \phi')}{\int_0^{2\pi} d\phi' \mu_{\rho_2}(r, \theta, \phi')}. \tag{38}$$

An important simplification occurs in the case of spherical symmetries:

$$\frac{4\pi}{N}\int_0^{f_r} df_r' f_r'^2 \rho_1(f_r') = \frac{4\pi}{N}\int_0^{r} dr' r'^2 \rho_2(r'), \tag{39}$$

$$\frac{1-\cos f_\theta}{2} = \frac{1-\cos\theta}{2}, \tag{40}$$

and

$$\frac{f_\phi}{2\pi} = \frac{\phi}{2\pi}. \tag{41}$$

3.3. Relationship between local-scaling and density transformations

Scaling transformation are well known in many physical and mathematical applications. These transformations carry a vector $\vec{r} \in \mathcal{R}^3$ into $\vec{f}(\vec{r}) \equiv \lambda\vec{r}$ where λ is a constant. Local-scaling transformations, on the other hand are characterized by the fact that λ is a function (i.e., $\lambda \equiv \lambda(\vec{r})$). In addition, they are direction-conserving transformations.

These transformations can be described by means of an operator $\hat{f}$ whose action on a vector $\vec{r} \in \mathcal{R}^3$ produces a transformed vector $\vec{f}(\vec{r}) \in \mathcal{R}^3$:

$$\vec{r} \xrightarrow{\hat{f}} \vec{f}(\vec{r}) = \lambda(\vec{r})\vec{r}. \tag{42}$$

It is easy to see that these transformations are particular instances of the density transformations discussed above. In order to see the connection, consider the action of the operator $\hat{f}$ on each one of the coordinates of the wavefunction $\Phi_1(\vec{r}_1, ..., \vec{r}_N) \in \mathcal{L}_N$ associated with a one-particle density $\rho_1(\vec{r})$. This action leads to the following transformed wavefunction

$$\Phi_2(\vec{r}_1, ..., \vec{r}_N) \equiv \underbrace{\hat{f}\ldots\hat{f}}_{\text{N-times}}\Phi_1(\vec{r}_1, ..., \vec{r}_N) = \prod_{i=1}^{N}\left[J(\vec{f}(\vec{r}_i); \vec{r}_i)\right]^{1/2}\Phi_1(\vec{f}(\vec{r}_1), ..., \vec{f}(\vec{r}_N)) \tag{43}$$

associated with the transformed one-particle density $\rho_2(\vec{r})$. It is easy to show that the Jacobian of this transformation satisfies Eq. (27). Uniqueness of the transformation requires that the Jacobians for both the forward and backward transformations be positive definite. This implies in turn that the densities must be positive definite. However, in the case of local-scaling transformations, we can also express the Jacobian as an explicit function of the local-scaling function $\lambda(\vec{r})$. This is done by computing the determinant:

$$J(\lambda(\vec{r})\vec{r};\vec{r}) = \begin{vmatrix} \frac{\partial \lambda(\vec{r})x}{\partial x} & \frac{\partial \lambda(\vec{r})y}{\partial x} & \frac{\partial \lambda(\vec{r})z}{\partial x} \\ \frac{\partial \lambda(\vec{r})x}{\partial y} & \frac{\partial \lambda(\vec{r})y}{\partial y} & \frac{\partial \lambda(\vec{r})z}{\partial y} \\ \frac{\partial \lambda(\vec{r})x}{\partial z} & \frac{\partial \lambda(\vec{r})y}{\partial z} & \frac{\partial \lambda(\vec{r})z}{\partial z} \end{vmatrix} = \lambda^3(\vec{r})[1 + \vec{r} \cdot \nabla_{\vec{r}} \ln \lambda(\vec{r})]. \tag{44}$$

It follows from Eqs. (27) and (44) that

$$\lambda^3(\vec{r})[1 + \vec{r} \cdot \nabla_{\vec{r}} \ln \lambda(\vec{r})] = \frac{\rho_2(\vec{r})}{\rho_1(\vec{f}(\vec{r}))}. \tag{45}$$

Rewriting $\lambda(\vec{r})$ as $\lambda(\vec{r}) = f(\vec{r})/r$, the Jacobian can be put as a function of $f(\vec{r})$. In that case we have:

$$J(\vec{f}(\vec{r});\vec{r}) = \frac{\vec{r} \cdot \nabla_{\vec{r}} f^3(\vec{r})}{r^3}. \tag{46}$$

Bearing in mind Eq. (27) and using polar coordinates Eq. (46) becomes

$$\frac{\partial f(r,\theta_o,\phi_o)}{\partial r} = \frac{r^2}{f^2(r,\theta_o,\phi_o)} \frac{\rho_2(r,\theta_o,\phi_o)}{\rho_1(f(r,\theta_o,\phi_o),\theta_o,\phi_o)}, \tag{47}$$

where θ_o and ϕ_o, describe fixed angles. For the spherically-symmetric case (or equivalently for spherically-averaged densities) Eq. (47) becomes:

$$\frac{df(r)}{dr} = \frac{r^2}{f^2(r)} \frac{\rho_2(r)}{\rho_1(f(r))}, \tag{48}$$

or in integrated form

$$\int_0^{f(r)} dx x^2 \rho_1(x) = \int_0^r dy y^2 \rho_2(y). \tag{49}$$

Note that Eq. (39) is exactly equal to Eq. (49). This establishes the equivalence between local-scaling and density transformations for the spherically symmetric case.

4. THE HARTREE-FOCK ENERGY AS AN EXPLICIT FUNCTIONAL OF BOTH THE ONE-PARTICLE DENSITY AND THE LOCAL-SCALING TRANSFORMATION

Let us first consider the general expression for the variational principle which one obtains in LS-DFT for the Hartree-Fock problem

$$E_o^{HF} = \inf_{\substack{\text{over all orbits} \\ \mathcal{O}_S^{[i]} \subset \mathcal{S}_N}} \left\{ \inf_{\substack{\rho(\vec{r}) \in \mathcal{N}_\Phi \\ \Phi_\rho^{[i]} \in \mathcal{O}_S^{[i]}}} \left\{ E[\Phi_\rho^{[i]}] \right\} \right\}. \tag{50}$$

This principle simply states a two-step minimization of the Hartree-Fock functional (i.e., the minimization of the expectation value of the Hamiltonian with respect to single Slater determinants). Notice, however, that this minimization is carried out by searching over all Slater determinants that are related to each other by a local-scaling transformation, namely, over wavefunctions which belong to an "orbit" $\mathcal{O}_S^{[i]}$. This is followed by a search over all orbits making up this single-determinant subspace $\mathcal{S}_N$ of Hilbert space. All wavefunctions in an orbit are generated by applying local-scaling transformations to an arbitrary "orbit-generating" determinant.

The reason why this variational principle is given in this form stems from the fact that within each orbit there exists a one to one correspondence between locally-scaled determinants and one-particle densities. This fact allows us to write the energy $E[\Phi_\rho^{[i]}]$ as a functional of the one-particle density[14]:

$$E[\Phi_\rho^{[i]}] = \mathcal{E}^{[i]}[\rho] \quad \text{for} \quad \Phi_\rho^{[i]} \in \mathcal{O}_S^{[i]} \subset \mathcal{S}_N, \tag{51}$$

where $\Phi_\rho \in \mathcal{S}_N$ is a locally-scaled single Slater determinant associated uniquely (within orbit $\mathcal{O}_S^{[i]}$) with the one-particle density $\rho(\vec{r}) \in \mathcal{N}_\Phi$. (Note that the energy functional $\mathcal{E}^{[i]}[\rho]$ is orbit-dependent.) As will be shown below, this characteristic makes it possible in this constructive procedure to write down explicit density functionals for the Hartree-Fock energy.

The orbit generating single Slater determinant is formed from an arbitrary set $\{\phi_{g,k}^{[i]}(\vec{r})\}_{k=1}^N$ of orthonormal single particle functions. We discuss in Subsections 4.1 and 4.2 some particular realizations for the case of the beryllium atom in order to show that the choice of these functions is not in any way limiting. The point is that these sets can be chosen as arbitrary polynomials. As will be shown in Subsection 5.4, one can obtain very accurate results using very simple forms for these functions.

Using these arbitrary single particle orbitals, we obtain the "orbit-generating" density

$$\rho_g^{[i]}(\vec{r}) = \sum_{k=1}^N \left|\phi_{g,k}^{[i]}(\vec{r})\right|^2 \tag{52}$$

associated with the single Slater determinant. We now consider the problem of finding a local-scaling transformation that connects this density to a final density $\rho^{[i]}(\vec{r})$ which, of course, is the basic entity in this whole procedure and which is to be determined variationally. The local-scaling transformation is then determined by the first-order differential equation (48) or by the integrated equation (49).

The transformation connecting these densities can then be used in order to generate locally-scaled single-particle functions. In view of the fact that the local-scaling transformation is a function of the density $\rho(\vec{r})$, it is clear that these tranformed orbitals $\left\{\phi_{\rho,k}^{[i]}(\vec{r})\right\}$ are also density dependent. Explicitly, these orbitals are given by:

$$\phi_{\rho,k}^{[i]}(\vec{r}) = \hat{f}\phi_{g,k}^{[i]}(\vec{r}) = [J(\vec{f}(\vec{r});\vec{r})]^{1/2}\phi_{g,k}^{[i]}(\vec{f}(\vec{r})) = \left[\frac{\rho^{[i]}(\vec{r})}{\rho_g^{[i]}(\vec{f}(\vec{r}))}\right]^{1/2} \phi_{g,k}^{[i]}(\vec{f}(\vec{r})). \tag{53}$$

Again, these orbitals are defined within a given "orbit".

Because local-scaling transformations maintain orthogonality, the density $\rho^{[i]}(\vec{r})$ acquires the simple form:

$$\rho^{[i]}(\vec{r}) = \sum_{k=1}^{N} |\phi_{\rho,k}^{[i]}(\vec{r})|^2. \tag{54}$$

As a result of these transformations, the locally-scaled single determinant that associates with the density $\rho^{[i]}(\vec{r})$ becomes:

$$\Phi_{\rho}^{[i]}(\vec{r}_1, \cdots, \vec{r}_N) \equiv \frac{\det}{\sqrt{N!}}[\phi_{\rho,1}^{[i]}(\vec{r}_1) \cdots \phi_{\rho,N}^{[i]}(\vec{r}_N)]. \tag{55}$$

Within each particular orbit, we can then write the Hartree-Fock energy functional corresponding to the single Slater determinant given by Eq. (55). Clearly, since all orbitals are density dependent, the following identity holds $E[\Phi_{\rho}^{[i]}] = \mathcal{E}^{[i]}[\rho]$ where in the right-hand side we are emphazising the fact that within a given orbit $\mathcal{O}_{S}^{[i]}$ the energy is an explicit functional of the one-particle density.

After these preliminary considerations, let us deal with the energy functional expressed in terms of the 1-matrix and of the one-particle density (we have in mind an ulterior transformation of the terms depending on this 1-matrix into terms depending solely on the density):

$$\begin{aligned} E[\Phi_{\rho}^{[i]}] &= \frac{1}{2}\int d^3\vec{r}_1 \nabla_{\vec{r}_1} \nabla_{\vec{r}_1'} D_{\rho}^{1[i]}(\vec{r}_1, \vec{r}_1')|_{\vec{r}_1'=\vec{r}_1} + \int d^3\vec{r} \rho^{[i]}(\vec{r}) v(\vec{r}) \\ &+ \frac{1}{2}\int d^3\vec{r}_1 \int d^3\vec{r}_2 \frac{\rho^{[i]}(\vec{r}_1)\rho^{[i]}(\vec{r}_2)}{|\vec{r}_1 - \vec{r}_2|} - \frac{1}{2}\int d^3\vec{r}_1 \int d^3\vec{r}_2 \frac{D_{\rho}^{1[i]}(\vec{r}_1, \vec{r}_2) D_{\rho}^{1[i]}(\vec{r}_2, \vec{r}_1)}{|\vec{r}_1 - \vec{r}_2|}. \end{aligned} \tag{56}$$

It follows from this expression that only the kinetic and the exchange energy terms depend on the locally-scaled 1-matrix. This transformed 1-matrix arising from the single Slater determinant of Eq. (55) is:

$$D_{\rho}^{1[i]}(\vec{r}_1, \vec{r}_1') = \sum_{k=1}^{N} \phi_{\rho,k}^{*}(\vec{r}) \phi_{\rho,k}(\vec{r}'). \tag{57}$$

Thus, we see that the effect of the local-scaling transformation is to modify the single-particle orbitals entering into the 1-matrix. This fact is used below for the purpose of writing energy functionals for the kinetic and exchange energies.

4.1. The kinetic energy as a functional of the one-particle density $\rho(\vec{r})$ and of the transformation function $f([\rho]; \vec{r})$

We review here the generation of the kinetic energy functional [102,106,107]. The kinetic energy for the case when Φ_{ρ} is a single Slater determinant is given by (we skip for simplicity the orbit superindex):

$$T[\Phi_{\rho}] = \frac{1}{2}\int d^3\vec{r} \nabla_{\vec{r}} \nabla_{\vec{r}'} D_{\rho}^{1}(\vec{r}, \vec{r}')|_{\vec{r}'=\vec{r}} \tag{58}$$

$$= \frac{1}{2}\sum_{i=1}^{N} \int d^3\vec{r} \nabla_{\vec{r}} \phi_{\rho,i}^{*}(\vec{r}) \nabla_{\vec{r}} \phi_{\rho,i}(\vec{r}) \tag{59}$$

$$= -\frac{1}{2}\sum_{i=1}^{N} \int d^3\vec{r} \phi_{\rho,i}^{*}(\vec{r}) \nabla_{\vec{r}}^2 \phi_{\rho,i}(\vec{r}). \tag{60}$$

For atoms, the transformed orbitals are given by

$$\phi_{\rho,i}(\vec{r}) = R_{\rho,n_il_i}(r)Y_{l_i,m_{l_i}}(\theta,\phi) \tag{61}$$

so that the effect of a Laplacian on an orbital yields

$$\nabla^2_{\vec{r}}\phi_{\rho,i}(\vec{r}) = \left(\frac{1}{r^2}\frac{\partial}{\partial r}\left(r^2\frac{\partial}{\partial r}\right) - \frac{l_i(l_i+1)}{r^2}\right)R_{\rho,n_il_i}(r)Y_{l_i,m_{l_i}}(\theta,\phi). \tag{62}$$

Substituting this expression into Eq. (60), we obtain:

$$\begin{aligned} T[\Phi_\rho] &= -\frac{1}{2}\sum_{i=1}^{N}\int_0^\infty drr^2\int_0^\pi d\theta\sin\theta\int_0^{2\pi} d\phi R_{\rho,n_il_i}(r)Y^*_{l_i,m_{l_i}}(\theta,\phi) \\ &\times\left(\frac{1}{r^2}\frac{\partial}{\partial r}\left(r^2\frac{\partial}{\partial r}\right) - \frac{l_i(l_i+1)}{r^2}\right)R_{\rho,n_il_i}(r)Y_{l_i,m_{l_i}}(\theta,\phi). \end{aligned} \tag{63}$$

Integrating over the angular part, the kinetic energy becomes

$$\begin{aligned} T[\Phi_\rho] &= -\frac{1}{2}\sum_{i=1}^{N}\int_0^\infty drr^2 R_{\rho,n_il_i}(r)\frac{1}{r^2}\frac{\partial}{\partial r}\left(r^2\frac{\partial R_{\rho,n_il_i}(r)}{\partial r}\right) \\ &+ \sum_{i=1}^{N}\frac{l_i(l_i+1)}{2}\int_0^\infty drr^2\left(R_{\rho,n_il_i}(r)\right)^2\frac{1}{r^2}. \end{aligned} \tag{64}$$

Making use of the identity

$$\int_0^\infty drr^2 R_{\rho,n_il_i}(r)\frac{1}{r^2}\frac{\partial}{\partial r}\left(r^2\frac{\partial R_{\rho,n_il_i}(r)}{\partial r}\right) = \int_0^\infty drr^2\left(\frac{\partial R_{\rho,n_il_i}(r)}{\partial r}\right)^2, \tag{65}$$

we can rewrite the kinetic energy as follows:

$$\begin{aligned} T[\Phi_\rho] &= \frac{1}{2}\sum_{i=1}^{N}\int_0^\infty drr^2\left(\frac{dR_{\rho,n_il_i}(r)}{dr}\right)^2 \\ &+ \sum_{i=1}^{N}\frac{l_i(l_i+1)}{2}\int_0^\infty drr^2\left(R_{\rho,n_il_i}(r)\right)^2\frac{1}{r^2}. \end{aligned} \tag{66}$$

In order to express the kinetic energy as a functional of ρ and $f(r)$, we now introduce the explicit form of the locally-scaled radial orbital:

$$R_{\rho,n_il_i}(r) = \sqrt{\frac{\rho(r)}{\rho_g(f)}}R_{g,n_il_i}(f), \tag{67}$$

where $\rho(r)$ and $\rho_g(f)$ are the spherically-averaged one-particle densities (see Eq. (39)):

$$\rho(r) = \int_0^\pi d\theta\sin\theta\int_0^{2\pi} d\phi\rho(r,\theta,\phi), \tag{68}$$

$$\rho_g(f) = \int_0^\pi d\theta\sin\theta\int_0^{2\pi} d\phi\rho_g(f,\theta,\phi). \tag{69}$$

We calculate $dR_{\rho,n_il_i}(r)/dr$ taking care to employ the chain rule

$$\frac{d}{dr} = \left(\frac{df}{dr}\right)\frac{d}{df} \tag{70}$$

which in view of Eq. (48) becomes:

$$\frac{d}{dr} = \frac{r^2}{f^2}\frac{\rho(r)}{\rho_g(f)}\frac{d}{df}. \tag{71}$$

Introducing Eqs. (67) and (71) into Eq. (66) and carrying out the algebra we obtain:

$$\begin{aligned} T[\Phi_\rho] &= \frac{1}{8}\int_0^\infty dr\, r^2 \frac{(d\rho(r)/dr)^2}{\rho(r)} + \int_0^\infty dr\, r^2 \frac{\rho^3(r)}{\rho_g^3(f)}\frac{r^4}{f^4}\Big[-\frac{1}{8}\frac{(d\rho_g(f)/df)^2}{\rho_g(f)} \\ &+ \frac{1}{2}\sum_{i=1}^{N}\Big(\frac{dR_{g,n_il_i}(f)}{df}\Big)^2\Big] + \int_0^\infty dr\, r^2 \frac{\rho(r)}{\rho_g(f)}\frac{1}{r^2}\sum_{i=1}^{N}\frac{l_i(l_i+1)}{2}\big(R_{g,n_il_i}(f)\big)^2. \end{aligned} \tag{72}$$

We now make use of the relation

$$\lambda(r) = \frac{f}{r} = \left(\frac{\rho(r)}{\rho_g(f)}\right)^{1/3}\left\{1 + \vec{r}\cdot\nabla_{\vec{r}}\ln\lambda(r)\right\}^{-1/3} \tag{73}$$

and rewrite Eq. (72) in the following form:

$$\begin{aligned} T[\Phi_\rho] &= T_W[\rho] + \int_0^\infty dr\, r^2\rho^{5/3}(r)\big(1 + \vec{r}\cdot\nabla\ln\lambda(r)\big)^{4/3}\tau_N \\ &+ \int_0^\infty dr\, r^2\rho^{5/3}(r)\big(1 + \vec{r}\cdot\nabla\ln\lambda(r)\big)^{-2/3}\kappa_N, \end{aligned} \tag{74}$$

where

$$T_W[\rho] = \frac{1}{8}\int_0^\infty dr\, r^2\frac{(\nabla\rho(r))^2}{\rho(r)} \tag{75}$$

is the well-known Weiszäcker kinetic energy term [54]. The radial kinetic energy modulating factor τ_N is defined as:

$$\tau_N = \frac{1}{\rho_g^{5/3}(f)}\Big[-\frac{1}{8}\frac{(d\rho_g(f)/df)^2}{\rho_g(f)} + \frac{1}{2}\sum_{i=1}^{N}\Big(\frac{dR_{g,n_il_i}(f)}{df}\Big)^2\Big], \tag{76}$$

and the angular kinetic energy modulating factor κ_N is, similarly,

$$\kappa_N = \frac{1}{\rho_g(f)}\sum_{i=1}^{N}\frac{l_i(l_i+1)}{2}\Big(\frac{R_{g,n_il_i}(f)}{f}\Big)^2. \tag{77}$$

A simpler way to express the radial modulating factor τ_N is the following:

$$\tau_N = \frac{1}{\rho_g^{8/3}(f)}\frac{1}{2}\sum_{i<j}\Big[R_{g,n_il_i}(f)\frac{dR_{g,n_jl_j}(f)}{df} - R_{g,n_jl_j}(f)\frac{dR_{g,n_il_i}(f)}{df}\Big]^2. \tag{78}$$

The aim of this Subsection has been to express the kinetic energy as an explicit functional of both the one-particle density $\rho(r)$ and the local-scaling transformation function

$f(r)$. This is accomplished in Eq. (74). It is gratifying to see that the Thomas-Fermi kinetic energy term $\rho^{5/3}(r)$ appears in a natural way as a "universal factor" in the present derivation. Notice that in view of the fact that $\lambda(r) = f(r)/r$, both factors $\left(1+\vec{r}\cdot\nabla \ln \frac{f}{r}\right)^{4/3}$ and $\left(1+\vec{r}\cdot\nabla \ln \frac{f}{r}\right)^{-2/3}$ are explicit functions of $f(r)$. Since their form is the same for all N-particle systems, they may also be regarded as "universal" (although any particular realization of the function $f(r)$ depends on the number of particles N and on the particular choice of generating functions).

The modulating factors τ_N and κ_N, on the other hand, are not universal as they explicitly depend on the set of arbitrary but otherwise known orbit-generating orbitals. In this sense, for any particular choice of these orbit generating one-particle functions, these factors become explicit functions of the the local-scaling transformation function $f(r)$. It is to be expected that a great deal of the particular physical effects that characterize a given system are reflected in these kinetic energy modulating factors. We show below, how we can obtain very simple (but quite accurate) approximations to these factors with a proper choice of generating functions.

In illustration, we present in what follows the explicit kinetic energy functionals for the ground states of the beryllium and boron atoms. We have selected these examples because they can be used to highlight the differences between a closed-shell and an open-shell atom.

Since the differences between the functionals lie in the radial and angular modulating factors, we write below their explicit form. For the ^1S state of the berylium atom, we have

$$\tau_4 = \frac{1}{\rho_g^{8/3}(f)} 2\left[R_{g,1s}(f)\frac{dR_{g,2s}(f)}{df} - R_{g,2s}(f)\frac{dR_{g,1s}(f)}{df}\right]^2. \tag{79}$$

Clearly, for beryllium, the factor $\kappa_4 = 0$. For the $^2P^o$ ground state of boron, we have:

$$\begin{aligned}\tau_5 &= \frac{1}{\rho_g^{8/3}(f)}\frac{1}{2}\left[2\left(R_{g,1s}(f)\frac{dR_{g,2s}(f)}{df} - R_{g,2s}(f)\frac{dR_{g,1s}(f)}{df}\right)^2 + \left(R_{g,1s}(f)\frac{dR_{g,2p}(f)}{df}\right.\right. \\ &- \left.\left. R_{g,2p}(f)\frac{dR_{g,1s}(f)}{df}\right)^2 + \left(R_{g,2s}(f)\frac{dR_{g,2p}(f)}{df} - R_{g,2p}(f)\frac{dR_{g,2s}(f)}{df}\right)^2\right], \end{aligned} \tag{80}$$

and

$$\kappa_5 = \frac{1}{\rho_g(f)}\frac{\left(R_{g,2p}(f)\right)^2}{f^2}. \tag{81}$$

4.2. The exchange-only energy as a functional of the one-particle density $\rho(\vec{r})$ and of the transformation function $f([\rho];\vec{r})$

The exchange energy for the Hartree-Fock case is a functional of the spin-dependent locally-scaled 1-matrix D_ρ^1 [102,106,107] (again, for simplicity, we drop the orbit superindex):

$$E_x[\Phi_\rho] = -\frac{1}{2}\sum_{s_1}\int d^3\vec{r}_1 \sum_{s_2}\int d^3\vec{r}_2 \frac{D_\rho^1(\vec{r}_1, s_1; \vec{r}_2, s_2) D_\rho^1(\vec{r}_1, s_1; \vec{r}_2, s_2)}{|\vec{r}_1 - \vec{r}_2|}. \tag{82}$$

Expressing the 1-matrix in its transformed orbital expansion and integrating over spin, we have

$$E_x[\Phi_\rho] = -\frac{1}{2}\sum_{i=1}^{N}\sum_{j=1}^{N}\delta(m_{s_i}, m_{s_j})(ij/ji), \tag{83}$$

where we define

$$(ij/ji) = \int d^3\vec{r}_1 \int d^3\vec{r}_2 \frac{\phi^*_{\rho,i}(\vec{r}_1)\phi_{\rho,j}(\vec{r}_1)\phi^*_{\rho,j}(\vec{r}_2)\phi_{\rho,i}(\vec{r}_2)}{|\vec{r}_1 - \vec{r}_2|}. \tag{84}$$

In order to proceed further, let us use the relation $\phi_{\rho,i}(\vec{r}) = \sqrt{\frac{\rho(r)}{\rho_g(f)}}\phi_{g,i}(\vec{f})$ given by Eq. (53) and assume that, for atoms, the generating orbitals can be written as:

$$\phi_{g,i}(\vec{f}) = R_{g,n_i l_i}(f) Y_{l_i, m_{l_i}}(\theta, \phi). \tag{85}$$

Introducing these functions into Eq. (84) we can rewrite (ij/ji) as

$$(ij/ji) = \sum_{k=0} b^k(l_i, m_{l_i}; l_j, m_{l_j}) \tilde{R}^k(ij/ji) \tag{86}$$

where $b^k(l_i, m_{l_i}; l_j, m_{l_j})$ is the well-known angular coefficient [108]

$$b^k(l_i, m_{l_i}; l_j, m_{l_j}) = [c^k(l_i, m_{l_i}; l_j, m_{l_j})]^2 \tag{87}$$

and where [108]

$$\begin{aligned} c^k(l, m; l', m') &= (-1)^{[m+|m|+m'+|m'|+(m-m')+|m-m'|]/2}\sqrt{\frac{(k-|m-m'|)!}{(k+|m-m'|)!}} \\ &\times \sqrt{\frac{(2l+1)(l-|m|)!}{(l+|m|)!}}\sqrt{\frac{(2l'+1)(l-|m'|)!}{(l'+|m'|)!}} \\ &\times \frac{1}{2}\int_{-1}^{1} P_l^{|m|}(\mu) P_{l'}^{|m'|}(\mu) P_k^{|m-m'|}(\mu) d\mu. \end{aligned} \tag{88}$$

$\tilde{R}^k(ij/ji)$ is the radial integral (the tilde is employed to emphasize the difference with the usual atomic integrals [108]):

$$\begin{aligned} \tilde{R}^k(ij/ji) &= \int_0^\infty dr_1\, r_1^2 \frac{\rho(r_1)}{\rho_g(f_1)} \frac{A_{ij}(f_1)}{r_1^{k+1}} \int_0^{r_1} dr_2\, r_2^2 \frac{\rho(r_2)}{\rho_g(f_2)} r_2^k A_{ji}(f_2) \\ &+ \int_0^\infty dr_1\, r_1^2 \frac{\rho(r_1)}{\rho_g(f_1)} r_1^k A_{ij}(f_1) \int_{r_1}^\infty dr_2\, r_2^2 \frac{\rho(r_2)}{\rho_g(f_2)} \frac{A_{ji}(f_2)}{r_2^{k+1}} \end{aligned} \tag{89}$$

The terms $A_{ij}(f_1)$ and $A_{ji}(f_2)$ appearing in the above equation are defined by

$$A_{ij}(f_1) \equiv R_{g,n_i l_i}(f_1) R_{g,n_j l_j}(f_1) \tag{90}$$

and

$$A_{ji}(f_2) \equiv R_{g,n_j l_j}(f_2) R_{g,n_i l_i}(f_2) \tag{91}$$

Using the Dirichlet formula [109]

$$\int_0^\infty dr_1 \int_{r_1}^\infty F(r_1,r_2)dr_2 = \int_0^\infty dr_2 \int_0^{r_2} F(r_1,r_2)dr_1 \tag{92}$$

we can rewrite Eq.(89) as

$$\begin{aligned}\tilde{R}^k(ij/ji) &= \int_0^\infty dr_1\, r_1^2 \frac{\rho(r_1)}{\rho_g(f_1)} \frac{A_{ij}(f_1)}{r_1^{k+1}} \int_0^{r_1} dr_2\, r_2^2 \frac{\rho(r_2)}{\rho_g(f_2)} r_2^k A_{ji}(f_2) \\ &+ \int_0^\infty dr_1\, r_1^2 \frac{\rho(r_1)}{\rho_g(f_1)} \frac{A_{ji}(f_1)}{r_1^{k+1}} \int_0^{r_1} dr_2\, r_2^2 \frac{\rho(r_2)}{\rho_g(f_2)} r_2^k A_{ij}(f_2).\end{aligned} \tag{93}$$

Making use of the identity

$$df_2 f_2^2 = dr_2 r_2^2 \frac{\rho(r_2)}{\rho_g(f_2)} \tag{94}$$

and introducing the auxiliary functions

$$P_{ji}^k(f_1) = \int_0^{f_1} df_2\, f_2^2 r_2^k(f_2) A_{ji}(f_2) \tag{95}$$

and

$$P_{ij}^k(f_1) = \int_0^{f_1} df_2\, f_2^2 r_2^k(f_2) A_{ij}(f_2) \tag{96}$$

we obtain

$$\tilde{R}^k(ij/ji) = \int_0^\infty dr_1\, r_1^2 \frac{\rho(r_1)}{\rho_g(f_1)} \frac{1}{r_1^{k+1}} (A_{ij}(f_1)P_{ji}^k(f_1) + A_{ji}(f_1)P_{ij}^k(f_1)) \tag{97}$$

$$= \int_0^\infty dr_1\, r_1^2 \frac{\rho(r_1)}{\rho_g(f_1)} \left(\frac{f_1}{r_1}\right)^{k+1} \frac{(A_{ij}(f_1)P_{ji}^k(f_1) + A_{ji}(f_1)P_{ij}^k(f_1))}{f_1^{k+1}}. \tag{98}$$

Noting that

$$\lambda^{k+1}(r_1) = \left(\frac{f_1}{r_1}\right)^{k+1} = \left(\frac{\rho(r_1)}{\rho_g(f_1)}\right)^{(k+1)/3} \left(\frac{1}{1+\vec{r}\cdot\nabla_{\vec{r}} \ln\lambda(r_1)}\right)^{(k+1)/3} \tag{99}$$

we finally obtain for the exchange-energy functional the following expression:

$$E_x[\Phi_\rho] = -\frac{1}{2}\int_0^\infty dr_1 r_1^2 \sum_{k=0} \rho(r_1)^{(4+k)/3} \left(\frac{1}{1+\vec{r}_1\cdot\nabla_{\vec{r}} \ln\lambda(r_1)}\right)^{(k+1)/3} \chi_N^k(f_1), \tag{100}$$

where the angular-momentum-dependent exchange-modulating factor is given by

$$\chi_N^k(f_1) = \frac{1}{f^{k+1}\rho_g(f_1)^{(4+k)/3}} \sum_{i=1}^N \sum_{j=1}^N \delta(m_{s_i}, m_{s_j}) b^k(l_i, m_{l_i}; l_j, m_{l_j}) G_{ij}^k(f_1). \tag{101}$$

In the above expresion we have introduced (in analogy with the usual exchange integrals [108]) the function $G_{ij}^k(f_1)$ which is defined by:

$$G_{ij}^k(f_1) = A_{ij}(f_1)P_{ji}^k(f_1) + A_{ji}(f_1)P_{ij}^k(f_1). \tag{102}$$

Note that the summation over k is finite and is controlled by the so-called triangular condition.

In illustration of the specificity of the exchange-modulating factor, we give the explicit exchange energy functionals for the beryllium and boron atoms. For ^1S state of beryllium the only permissible value is $k = 0$. Hence, we have:

$$E_x = -\frac{1}{2}\int_0^\infty dr_1 r_1^2 \rho(r_1)^{4/3}\Big(\frac{1}{1+\vec{r}_1\cdot\nabla_{\vec{r}}\ln\lambda(r_1)}\Big)^{1/3}\chi_4^0(f_1), \tag{103}$$

where the exchange modulating function is

$$\chi_4^0(f_1) = \frac{1}{f_1\rho_g(f_1)^{4/3}}\big(2G^0_{1s1s}(f_1) + 2G^0_{2s2s}(f_1) + 4G^0_{1s2s}(f_1)\big). \tag{104}$$

In turn, for the boron atom (in the $^2P^o$ ground state) the exchange energy functional is:

$$\begin{aligned} E_x = -\frac{1}{2}\int_0^\infty dr_1 r_1^2 \quad & \Big[\rho(r_1)^{4/3}\Big(\frac{1}{1+\vec{r}_1\cdot\nabla_{\vec{r}}\ln\lambda(r_1)}\Big)^{1/3}\chi_5^0(f_1) \\ & + \rho(r_1)^{5/3}\Big(\frac{1}{1+\vec{r}_1\cdot\nabla_{\vec{r}}\ln\lambda(r_1)}\Big)^{2/3}\chi_5^1(f_1) \\ & + \rho(r_1)^{2}\Big(\frac{1}{1+\vec{r}_1\cdot\nabla_{\vec{r}}\ln\lambda(r_1)}\Big)^{4/3}\chi_5^2(f_1)\Big], \end{aligned} \tag{105}$$

where the exchange modulating factors are

$$\chi_5^0(f_1) = \frac{1}{f_1\rho_g(f_1)^{4/3}}\big(2G^0_{1s1s}(f_1) + 2G^0_{2s2s}(f_1) + 4G^0_{1s2s}(f_1) + 2G^0_{2p_o2p_o}(f_1)\big), \tag{106}$$

$$\chi_5^1(f_1) = \frac{1}{f_1^2\rho_g(f_1)^{5/3}}\Big(\frac{2}{3}G^1_{1s2p_o}(f_1) + \frac{2}{3}G^1_{2s2p_o}(f_1)\Big), \tag{107}$$

$$\chi_5^2(f_1) = \frac{1}{f_1^3\rho_g(f_1)^{2}}\Big(\frac{4}{25}G^2_{2p_o2p_o}(f_1)\Big). \tag{108}$$

5. THE HARTREE-FOCK ENERGY AS AN EXPLICIT FUNCTIONAL OF THE ONE-PARTICLE DENSITY

In the previous Section, we have obtained expressions for the energy of a single determinant as a functional of both the one-particle density $\rho(r)$ and of the local-scaling transformation function $f(r)$. Of course, these expressions contain the set of orbit generating orbitals and thus they cannot be put in an explicit form unless some particular choice of these functions is made. In addition, it is necessary to obtain an analytic expression for the local-scaling transformation function $f([\rho]; r)$ as an explicit function of the one-particle density ρ. Now, the whole purpose of density functional theory is to avoid wavefunctions and orbitals. Does it mean that in some way by resorting to these orbit generating functions one is losing the primary advantage of density functional theory, namely, that of calculating the energy in terms solely of the one-particle density?

As we show below, this is not the case. To substantiate our claim, we select orbit generating functions which are generalizations of Slater orbitals. We do not use them as primitive functions for expanding the atomic functions. We simply take them as reasonable approximations to the true atomic orbitals (which, we do not have to calculate). What really matters is that the *form* of these orbitals be simple enough so as to allow us to express the energy of a single determinant as an explicit functional of the one-particle density in an uncomplicated albeit accurate manner. Moreover, we resort to a an explicit analytic representation of the local-scaling transformation function obtained by means of Padé approximants.

5.1. An analytic representation of the local-scaling transformation function $f([\rho];r)$ via Padé approximants

Let us consider the integrated form of Eq. (49) for the determination of the local-scaling transformation function $f(r)$. Rewriting $f(r)$ as [110]

$$f(r) = r + \Delta r \tag{109}$$

and substituting Eq. (109) in Eq. (49), we obtain

$$\int_0^r dt\, t^2 \rho_g(t) + \int_r^{r+\Delta r} dt\, t^2 \rho_g(t) = \int_0^r dt\, t^2 \rho(t). \tag{110}$$

From Eq. (110) we have:

$$\int_r^{r+\Delta r} dt\, t^2 \rho_g(t) = \int_0^r dt\, t^2 \rho(t) - \int_0^r dt\, t^2 \rho_g(t) = \int_0^r dt\, t^2(\rho(t) - \rho_g(t)) = \Delta Q(r). \tag{111}$$

We can now employ the Padé approximant of order [1, 1] [111]

$$\int_{x_i}^{x_{i+1}} dt\, f(t) \simeq h \frac{2f^2(x_i)}{2f(x_i) - h\, f'(x_i)} \tag{112}$$

and apply it to the left-hand side of Eq. (111), with the identifications $f(t) = \rho_g(t)\, t^2$ and $h = \Delta r$. In the present case, Eq. (111) becomes

$$\Delta r \frac{2\rho_g^2(r)\, r^4}{2\rho_g(r)\, r^2 - \Delta r \left[\rho_g'(r)\, r^2 + 2\, r\, \rho_g(r)\right]} = \Delta Q(r). \tag{113}$$

Solving for Δr in Eq. (113), we obtain:

$$\Delta r = \frac{2\rho_g(r)\, r \Delta Q(r)}{2\rho_g^2(r)\, r^3 + \left[\rho_g'(r)\, r + 2\rho_g(r)\right] \Delta Q(r)}. \tag{114}$$

In terms of the Padé approximant [1, 1], the local-scaling transformation function $f(r)$ becomes

$$f(r) = r + \frac{2\rho_g(r)\, r \Delta Q(r)}{2\rho_g^2(r)\, r^3 + \left[\rho_g'(r)\, r + 2\rho_g(r)\right] \Delta Q(r)}. \tag{115}$$

Equation (115) gives $f(r)$ as an explicit function of the one-particle density $\rho(r)$, namely $f(r) \equiv f([\rho(r)]; r)$. It is interesting to notice, however, that $\rho(r)$ appears under the integral sign, and that it gives rise to the charge difference

$$\Delta Q(r) = Q(r) - Q_g(r), \tag{116}$$

where we have indentified

$$Q(r) = \int_0^r dt\, t^2 \rho(t), \tag{117}$$

$$Q_g(r) = \int_0^r dt\, t^2 \rho_g(t). \tag{118}$$

5.2. A simple choice of orbit-generating functions

Let us consider the following choice of orbit generating functions for first-row atoms

$$R_{g,1s}([\alpha_1, \beta_1]; r) \equiv N_1(\alpha_1, \beta_1) \exp\left(-\alpha_1 r^{\beta_1}\right), \tag{119}$$

$$R_{g,2s}([\alpha_2, \beta_2]; r) \equiv N_2(\alpha_2, \beta_2)(1 + B(\alpha_1, \beta_1; \alpha_2, \beta_2)\, r) \exp\left(-\alpha_2 r^{\beta_2}\right), \tag{120}$$

$$R_{g,2p}([\alpha_3, \beta_3]; r) \equiv N_3(\alpha_3, \beta_3)\, r \exp\left(-\alpha_3 r^{\beta_3}\right), \tag{121}$$

where $N_1(\alpha_1, \beta_1)$, $N_2(\alpha_2, \beta_2)$ and $N_3(\alpha_3, \beta_3)$ are the normalization constants

$$N_1(\alpha_1, \beta_1) = (\beta_1)^{1/2} (2\alpha_1)^{3/(2\beta_1)} \Gamma(3/\beta_1)^{-1/2}, \tag{122}$$

$$N_2(\alpha_2, \beta_2) = (\beta_2)^{1/2} (2\alpha_2)^{3/(2\beta_2)} \left(\Gamma(3/\beta_2) + \frac{2B\Gamma(4/\beta_2)}{(2\alpha_2)^{1/\beta_2}} + \frac{B^2\Gamma(5/\beta_2)}{(2\alpha_2)^{2/\beta_2}}\right)^{-1/2}, \tag{123}$$

$$N_3(\alpha_3, \beta_3) = (\beta_3)^{1/2} (2\alpha_3)^{5/(2\beta_3)} \Gamma(5/\beta_3)^{-1/2}. \tag{124}$$

The important fact we wish to stress is that these orbit generating functions depend only on two parameters each, α_i and β_i and do not contain any linear coefficients (i.e., they are not linear combinations of primitive functions).

Calculations for first-row atoms using these orbitals are reported elsewhere [113]. In the present work we just present results for the beryllium and boron atoms. Moreover, we deal in this work with a particular instance of these functions, namely, when all β's are equal: $\beta_1 = \beta_2 = \beta_3 = \beta$. The practical advantage is that in this case, all integrals can be reduced to very simple forms leading to analytic expressions for the functionals. For completeness, we also present for these atoms results of calculations when $\beta_1 \neq \beta_2 \neq \beta_3$. In addition, we consider the case when all β's are equal to 1. The latter, is of course, the well-known case of single-ζ Slater orbitals.

The parameter $B(\alpha_1, \beta_1; \alpha_2, \beta_2)$ guarantees the orthogonality between $R_{g,1s}([\alpha_1, \beta_1]; r)$ and $R_{g,2s}([\alpha_2, \beta_2]; r)$. In general it is given by

$$B(\alpha_1, \beta_1; \alpha_2, \beta_2) = -\frac{\int_0^\infty dr\, r^2 \exp(-\alpha_1 r^{\beta_1} - \alpha_2 r^{\beta_2})}{\int_0^\infty dr\, r^3 \exp(-\alpha_1 r^{\beta_1} - \alpha_2 r^{\beta_2})}. \tag{125}$$

For the case when $\beta_1 = \beta_2$ this magnitude acquires the very simple form:

$$B(\alpha_1, \beta_1; \alpha_2, \beta_2) = -(\alpha_1 + \alpha_2)^{1/\beta} \frac{\Gamma(3/\beta)}{\Gamma(4/\beta)}. \tag{126}$$

In the above expressions, $\Gamma(a/\beta)$ is the Gamma function

$$\Gamma(a/\beta) = \beta k^{a/\beta} \int_0^\infty r^{a-1} e^{-k r^\beta} dr. \tag{127}$$

5.3. The kinetic energy as an explicit functional of $\rho(\vec{r})$

We present in this Subsection, the explicit expansion for the kinetic energy modulating factor τ_4 for the ^1S state of the beryllium atom. This explicit form is obtained by substituting the orbitals described by Eq. (119) and (120) into Eq. (79). For the case when $\beta_1 = \beta_2 = \beta$ this factor becomes:

$$\tau_4(f) = \frac{2}{\rho_g(f)^{8/3}} N_1^2 N_2^2 \exp -(2\alpha_1 + 2\alpha_2) f^\beta \Big[(\alpha_1 - \alpha_2)\beta f^{\beta-1}\big(1 + Bf\big) + B \Big]^2, \tag{128}$$

where the generating density is given by

$$\rho_g(f) = 2\Big(N_1^2 \exp -2\alpha_1 f^\beta + N_2^2 \big(1 + Bf\big)^2 \exp -2\alpha_1 f^\beta \Big). \tag{129}$$

It is clear that when the analytic expression for the local-scaling transformation function given by Eq. (115) is substituted in Eqs. (128) and (129) we obtain $\tau_4(f)$ and $\rho_g(f)$ as explicit functions of the one-particle density $\rho(r)$. In addition, in view of Eq. (74), the kinetic energy term for the beryllium atom is completely transformed into a functional of the one-particle density $\rho(r)$. Moreover, it is evident that by means of trivial algebraic manipulations, the kinetic energy terms for all first-row atoms can be generated by following the procedure exemplified above. In fact we have developed computational codes for handling this kind of problem for the general case when $\beta_1 \neq \beta_2 \neq \beta_3$ [112,113]. For larger atoms, we can apply the same strategy and construct also explicit kinetic energy functionals.

Let us remark at this point that the procedure discussed here applies strictly to a minimization procedure within an orbit, which is characterized by fixed values of the parameters $\alpha_1, \alpha_2, \beta_1$ and β_2. However, the search among all the possible orbits in Hilbert space (compatible with the type of generating functions) can be realized by optimizing these parameters. Thus, the present calculations involve parameter optimization, or equivalently, inter-orbit jumping. For each set of parameters, the local-scaling transformation was carried out from the initial (or generating) density $\rho_g(r)$ defined by Eq. (129) to the final density $\rho_{HF}(r)$, where the latter is the Clementi-Roetti Hartree-Fock density [114]. In this sense, the results presented in this work are not optimal as the optimizing density within the orbits considered need not be the Hartree-Fock one. Results where both orbital parameters and final density parameters have been optimized for first-row atoms are presented elsewhere [112,113].

In Table 1 and 2 we list the orbital parameters (Eqs. (119), (120) and (121)) for the beryllium and boron atoms, respectively. For $\beta_i \neq \beta_j$ as well as for $\beta_i = \beta_j \neq 1$. the parameters presented result of total energy optimizations. In Table 3, we list the kinetic energy for the beryllium and boron atoms for the various cases discussed above. For comparison purposes we also include the Hartree-Fock kinetic energy of [114]. The present results indicate that for $\beta_i \neq \beta_j$ and for $\beta_i = \beta_j \neq 1.$, the calculated kinetic energies differ by 10^{-4} hartrees from the corresponding Hartree-Fock values. When $\beta_i = \beta_j$ (i.e., for single-ζ orbitals), the calculated values differ by 10^{-2} hartrees. This shows that although the single-ζ orbitals fail to give an accurate value for the kinetic energy, the more general orbitals employed here yield quite accurate functionals. In Fig. 1, we present graphs of the kinetic energy modulating factor $\tau_4(f)$. In order to assess the effect of the local-scaling

Table 1
Orbital parameters for the beryllium atom .

Case	α_1	α_2	β_1	β_2
$\beta_i \neq \beta_j$	4.4602008	1.8826842	1.0328776	0.8812299
$\beta_i = \beta_j$	3.2832469	0.9869175	1.2829691	1.2829691
$\beta_i = \beta_j = 1$	3.4000124	1.2556198	1.	1.

Table 2
Orbital parameters for the boron atom .

Case	α_1	α_2	α_3	β_1	β_2	β_3
$\beta_i \neq \beta_j$	4.8561290	2.0820714	1.6227175	1.0268214	0.8913619	0.8954187
$\beta_i = \beta_j$	4.7940493	1.5595078	1.1444439	1.2366608	1.2366608	1.2366608
$\beta_i = \beta_j = 1$	5.2845460	2.0395871	1.5738737	1.	1.	1.

transformation on this modulating factor, the curve corresponding to the untransformed value $f = r$ is also included. In addition, we have drawn the graph of $r^2\rho(r)$ in order to show the location of the maxima of these modulating factors in relation to the shell-boundary region.

It is important to notice that the kinetic energy modulating factor $\tau_4(f)$ is by and large a bell-shaped curve located at the intershell boundary. Thus, it is clear that this modulating factor plays an essential role in shell formation. In the case of a homogeneous electron gas, because the generating orbitals are plane waves, this modulating factor becomes a constant [102]. In such a situation, we are led to a kinetic energy density functional that depends on $\rho^{5/3}$ and which – as is well known – fails to produce shell structure in atoms. The effect of local-scaling on the modulating factor is to displace the bell-shaped curve toward the intershell radius.

In Fig. 2, the corresponding graphs for the kinetic energy modulating factor $\tau_5(f)$ and one-particle density for boron are depicted. In Fig. 3, the angular modulating factor κ_5 for boron is shown. Again we observe that the modulating factors are bell-shaped functions whose maximum lies about the value of r marking the intershell region. The tendency of τ_5 to increase monotonously with increasing r is a result of the presence of $\rho_g^{8/3}(f)$ in the denominator (Eq. 78)). The presence of the denominator is also responsible for the great magnification of these modulating factors with respect to the density. By contrast, the angular kinetic energy factor κ_5 of boron is roughly of the same size as the

Table 3
Comparative study of the local-scaling transformation calculations of the kinetic energy for the ground-state of some first-row atoms with those obtained by the Hartree-Fock method (energy in hartrees).

Atom	Hartree-Fock	$\beta_i \neq \beta_j$	$\beta_i = \beta_j$	$\beta_i = 1$
Be	14.57304	14.57317	14.57349	14.58076
B	24.52896	24.52931	24.52942	24.53973

density. Again, its maximum is located about the intershell region.

5.4. The exchange-only energy as an explicit functional of $\rho(\vec{r})$

As it is shown in Eq. (104), the modulating factor χ_N^0 appearing in the exchange energy functional reduces to the case of Be to a sum of terms with $k = 0$. Introducing the form of the generating orbitals given by Eq. (119) and (120) we obtain the following expressions for the terms appearing in Eq. (102):

$$A_{1s1s}(f) = N_1^2 \exp(-2\alpha_1 f^\beta), \tag{130}$$

$$A_{2s2s}(f) = N_2^2 (1 + Bf)^2 \exp(-2\alpha_2 f^\beta), \tag{131}$$

$$A_{1s2s}(f) = N_1 N_2 (1 + Bf) \exp -(\alpha_1 + \alpha_2) f^\beta, \tag{132}$$

$$P^0_{1s1s}(f) = N_1^2 \frac{1}{\beta(2\alpha_1)^{3/\beta}} \gamma(3/\beta, f), \tag{133}$$

$$\begin{aligned} P^0_{2s2s}(f) &= N_2^2 \frac{1}{\beta}\left(\frac{1}{(2\alpha_2)^{3/\beta}}\gamma(3/\beta, f) + \frac{2B}{(2\alpha_2)^{4/\beta}}\gamma(4/\beta, f)\right. \\ &+ \left.\frac{B^2}{(2\alpha_2)^{5/\beta}}\gamma(5/\beta, f)\right), \end{aligned} \tag{134}$$

$$P^0_{1s2s}(f) = N_1 N_2 \frac{1}{\beta}\left(\frac{1}{(\alpha_1+\alpha_2)^{3/\beta}}\gamma(3/\beta, f) + \frac{B}{(\alpha_1+\alpha_2)^{4/\beta}}\gamma(4/\beta, f)\right). \tag{135}$$

Thus, the explicit form of the exchange modulating factor for the beryllium atom as a function of the local-scaling transformation function $f(r)$ is:

$$\begin{aligned} \chi_4^0(f) &= \frac{1}{f\rho_g^{4/3}(f)}\left[4N_1^4 \frac{1}{\beta}\exp(-2\alpha_1 f^\beta)\frac{1}{(2\alpha_1)^{3/\beta}}\gamma(3/\beta, f)\right. \\ &+ 4N_2^4 \frac{(1+Bf)^2}{\beta}\exp(-2\alpha_2 f^\beta)\left(\frac{1}{(2\alpha_2)^{3/\beta}}\gamma(3/\beta, f)\right. \\ &+ \left.\frac{2B}{(2\alpha_2)^{4/\beta}}\gamma(4/\beta, f) + \frac{B^2}{(2\alpha_2)^{5/\beta}}\gamma(5/\beta, f)\right) \\ &+ 8N_1^2 N_2^2 \frac{(1+Bf)}{\beta}\exp -(\alpha_1+\alpha_2) f^\beta \left(\frac{1}{(\alpha_1+\alpha_2)^{3/\beta}}\gamma(3/\beta, f)\right. \end{aligned}$$

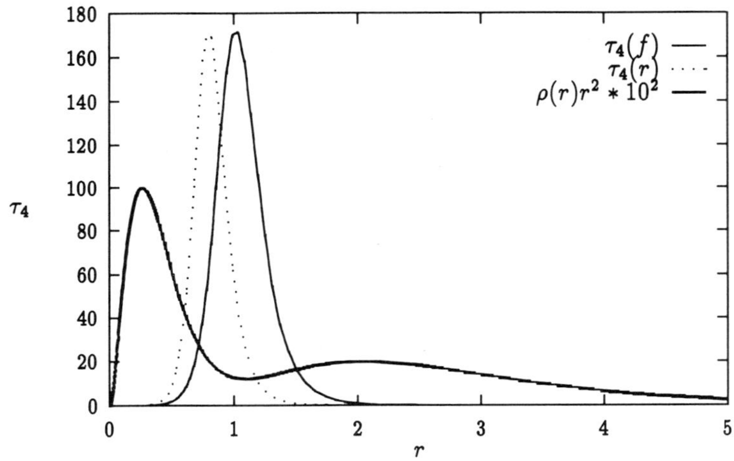

Figure 1. Behavior of the kinetic energy modulating factor τ_4 for the ^1S ground state of the beryllium atom evaluated both at f (full line) and at $f = r$ (dotted line). The density times r^2 (magnified by 10^2) is depicted in order to highlight the location of the maximum of τ_4.

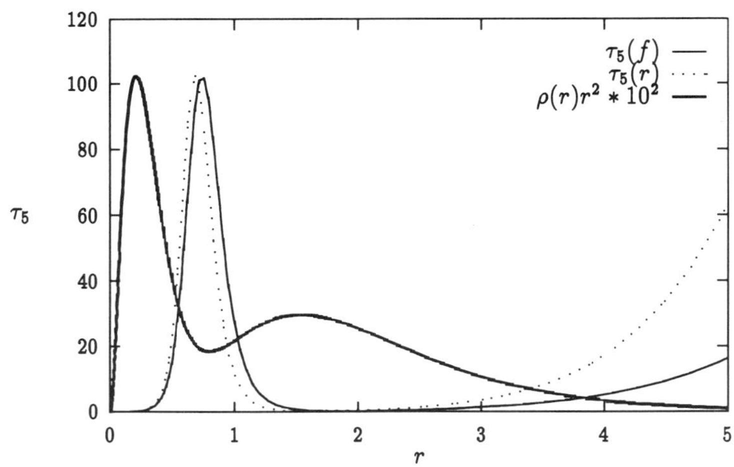

Figure 2. Behavior of the kinetic energy modulating factor τ_5 for the $^2P^o$ ground state of the boron atom evaluated both at f (full line) and at $f = r$ (dotted line). The density times r^2 (magnified by 10^2) is depicted in order to highlight the location of the maximum of τ_5.

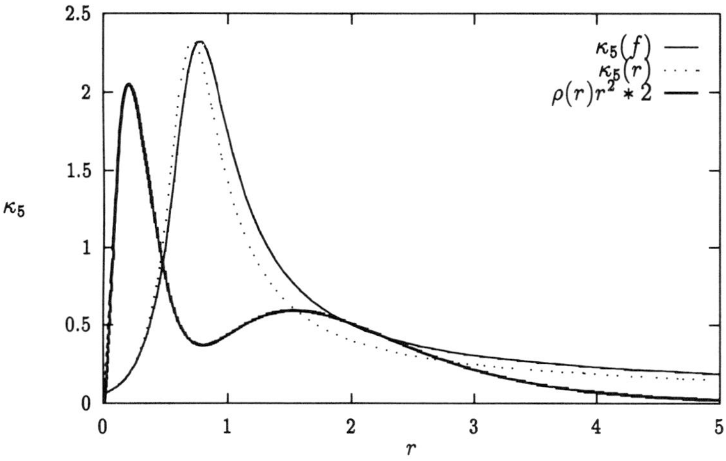

Figure 3. Behavior of the kinetic energy angular modulating factor κ_5 for the $^2P^o$ ground state of the boron atom evaluated both at f (full line) and at $f = r$ (dotted line). The density times r^2 (magnified by a factor of 2) is included for the purpose of indicating the relative position of the maximum of κ_5.

$$+ \quad \frac{B}{(\alpha_1 + \alpha_2)^{4/\beta}} \gamma(4/\beta, f) \Big) \Big] . \tag{136}$$

In the above expressions, we have made use of the incomplete gamma function:

$$\gamma(a/\beta, f) = \beta\,(k)^{a/\beta} \int_0^f \mu^{a-1} e^{-k\mu^\beta} d\mu \tag{137}$$

In Table 4, we list the values of the exchange energy for the various cases considered in this work. For completeness, we compare them to some values reported in the literature. The values labeled as Hartree-Fock, Perdew and Becke are taken from the Table appearing in Appendix 4.1 of March [4]. As we see from Table 4. our calculated results for Be coincide with the Hartree-Fock ones. For B, however, although our present values are in all cases closer to the Hartree-Fock exchange energies than those repoerted by Becke and Perdew, there is a small discrepancy of the order of 10^{-3} hartrees with respect to the Hartree-Fock results. Let us remark, however, that this is an outcome of the particular choice of orbit-generating functions. As has been shown elsewhere [113], within the context of LS-DFT, one attains a perfect coincidence with Hartree-Fock values when more flexible orbitals are adopted.

In Table 5, we list the total energies and compare them with the Hartree-Fock values of Clementi and Roetti [114] and of Harbola and Sahni [115,116]. For for $\beta_i \neq \beta_j$ the calculated total energies differ by 10^{-4} hartrees from the corresponding Hartree-Fock values. For $\beta_i = \beta_j \neq 1.$, the difference is of 10^{-4} hartrees for Be and of 10^{-3} hartrees for B. These values are considerably better than those of Harbola and Sahni. When $\beta_i = \beta_j$

Table 4
Exchange energies for beryllium and boron atoms from local-scaling method compared with others aproximations

Atom	Hartree-Fock	Perdew	Becke	$\beta_i \neq \beta_j$	$\beta_i = \beta_j$	$\beta_i = 1$
Be	-2.6669	-2.6810	-2.6692	-2.6669	-2.6669	-2.6673
B	-3.7587	-3.7864	-3.7679	-3.7513	-3.7505	-3.7515

Table 5
Comparative study of the local-scaling transformation calculations of the energy for the ground-state of some first-row atoms with those obtained by others methods (energy in hartrees with negative sign).

Atom	Hartree-Fock	Harbola-Sahni	$\beta_i \neq \beta_j$	$\beta_i = \beta_j$	$\beta_i = 1$
Be	14.57302	14.571	14.57279	14.57246	14.56560
B	24.52906	24.526	24.52859	24.52778	24.51829

(i.e., for single-ζ orbitals), the difference is increased to 10^{-2} hartrees. Again, we would like to emphasize that when more flexible bases are used, the Clementi-Roetti values are reproduced [113].

The upper-bound character of the total energies obtained in the the present calculations is an important characteristic of the local-scaling version of the version of density functional theory expounded here. Let us emphasize that LS-DFT is a strictly variational method that yields upper bounds to the total energy. Thus, when more flexible basis are used, the energy approaches from above the Hartree-Fock energy to any prescribed degree of accuracy. There is, hence, a substantial difference between the present approach and those based on the Hohenberg-Kohn-Levy version of DFT as in the latter cases there is no way to device a systematic improvement since the functionals are fixed.

In Fig. 4, the exchange modulating factor χ_4^0 for beryllium is shown. Again, we notice that this factor is a bell-shaped curve placed about the intershell region.

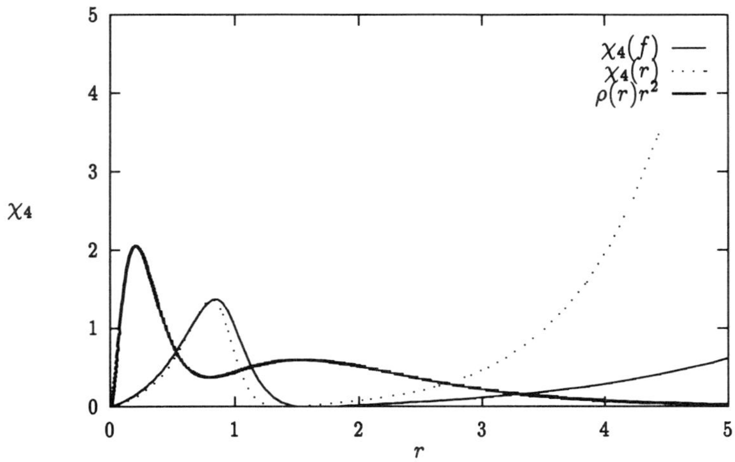

Figure 4. Behavior of the exchange energy modulating factor χ_4 for the ^1S ground state of the beryllium atom evaluated both at f (full line) and at $f = r$ (dotted line). The density times r^2 is included for the purpose of indicating the relative position of the maximum of χ_4.

6. DENSITY FUNCTIONALS FOR THE ELECTRON CORRELATION ENERGY

We sketch in the present Section a way to handle the electron correlation problem in the context of the local-scaling transformation version of density functional theory. The development takes advantage of the fact that both the kinetic-energy and the exchange-energy functionals corresponding to a single Slater determinant can be expressed, as it has been shown above, as explicit functionals of the one-particle density. The idea is, therefore, to device a way of expressing the correlation energy as the expectation value of a modified Hamiltonian with respect to a single Slater determinant and then to use this representation for the purpose of obtaining the electron correlation energy as an explicit functional of the one-particle density. A more thorough presentation including applications is given elsewhere [117].

The present work draws its inspiration from the method of transcorrelated functions introduced about thirty years ago by Boys and Handy [118]. This method, however, is recast as an strictly variational approximation.

Let us assume that the orbit-generating correlated wavefunction $\Psi_g^{[i]}$ can be written as the product of a single Slater determinant $\Phi_g^{[i]}$ times a correlation function C:

$$\Psi_g^{[i]}(\vec{r}_1, \cdots, \vec{r}_N) = C(\vec{r}_1, \cdots, \vec{r}_N)\Phi_g^{[i]}(\vec{r}_1, \cdots, \vec{r}_N), \tag{138}$$

where the correlation function is defined by

$$C(\vec{r}_1, \cdots, \vec{r}_N) = \prod_{i>j} g(\vec{r}_i, \vec{r}_j). \tag{139}$$

Note that the correlation function satisfies

$$C^{-1}C = C\,C^{-1} = 1. \tag{140}$$

As in Section 5, we also denote here by $\rho_g(\vec{r})$ the one-particle density associated with the single Slater determinant $\Phi_g^{[i]}$. Similarly, we assume that the wavefunction $\Psi_g^{[i]}$ is associated with the one-particle density $\tilde{\rho}_g(\vec{r})$.

Let us consider the application of local-scaling transformations from a starting density $\tilde{\rho}_g(\vec{r})$ to a final density $\rho(\vec{r})$ to each one of the coordinates of the wavefunction $\Psi_g^{[i]}$. As a result we obtain the transformed wavefunction:

$$\Psi_\rho^{[i]}(\vec{r}_1, \cdots, \vec{r}_N) = C(\vec{f}_1, \cdots, \vec{f}_N) \prod_{i=1}^{N} \sqrt{\frac{\rho(\vec{r}_i)}{\tilde{\rho}_g(\vec{f}_i)}}\,\Phi_g^{[i]}(\vec{f}_1, \cdots, \vec{f}_N). \tag{141}$$

Because the transformed single Slater determinant (from a density ρ_g to a final density ρ) is given by

$$\Phi_\rho^{[i]}(\vec{r}_1, \cdots, \vec{r}_N) = \prod_{i=1}^{N} \sqrt{\frac{\rho(\vec{r}_i)}{\rho_g(\vec{f}_i)}}\,\Phi_g^{[i]}(\vec{f}_1, \cdots, \vec{f}_N), \tag{142}$$

we can rewrite the transformed correlated wavefunction as

$$\Psi_\rho^{[i]}(\vec{r}_1, \cdots, \vec{r}_N) = C(\vec{f}_1, \cdots, \vec{f}_N) \prod_{i=1}^{N} \sqrt{\frac{\rho_g(\vec{f}_i)}{\tilde{\rho}_g(\vec{f}_i)}}\,\Phi_\rho^{[i]}(\vec{r}_1, \cdots, \vec{r}_N), \tag{143}$$

or more simply as

$$\Psi_\rho^{[i]}(\vec{r}_1, \cdots, \vec{r}_N) = C(\vec{f}_1, \cdots, \vec{f}_N)\tilde{\Phi}_\rho^{[i]}(\vec{r}_1, \cdots, \vec{r}_N), \tag{144}$$

where we have defined

$$\tilde{\Phi}_\rho^{[i]}(\vec{r}_1, \cdots, \vec{r}_N) = \prod_{i=1}^{N} \sqrt{\frac{\rho_g(\vec{f}_i)}{\tilde{\rho}_g(\vec{f}_i)}}\,\Phi_\rho^{[i]}(\vec{r}_1, \cdots, \vec{r}_N). \tag{145}$$

In general, we can approximate the exact wavefunction of an N-particle system by the correlated function $\Psi_\rho^{[i]}$. For this reason, the correlated wavefunctions satisfies the variational principle:

$$\langle \Psi_\rho^{[i]} | \widehat{H} | \Psi_\rho^{[i]} \rangle \geq E_o. \tag{146}$$

Using Eq. (140), we can rewrite Eq. (146) as follows (where for simplicity we have dropped the orbit superindex and where for clarity we explicitly indicate the dependence of C on the transformed particle coordinates):

$$\langle C^2(\vec{f}_1, \cdots, \vec{f}_N)\tilde{\Phi}_\rho \Big[C^{-1}(\vec{f}_1, \cdots, \vec{f}_N)\widehat{H}(\vec{r}_1, \cdots, \vec{r}_N)C(\vec{f}_1, \cdots, \vec{f}_N) \Big] \tilde{\Phi}_\rho^{[i]} \rangle \geq E_o. \tag{147}$$

The reason why we have rewritten the energy expectation value as above is that – as has been shown by Boys and Handy [118] – the modified Hamiltonian operator $C^{-1}\widehat{H}\,C$ can be expressed in terms of one-, two- and three-particle operators as follows:

$$C^{-1}\widehat{H}\,C = \frac{1}{2}\sum_{IJ}\frac{Z_I Z_J}{R_{IJ}} + \sum_{i=1}^{N}\widehat{F}_i + \sum_{i=1}^{N}\sum_{j\neq i}^{N}\widehat{F}_{ij} + \sum_{i=1}^{N}\sum_{j\neq i}^{N}\sum_{k\neq i,j}^{N}\widehat{F}_{ijk}. \tag{148}$$

In Eq. (148) the one-, two- and three-particle operators are:

$$\widehat{F}_i = -\frac{1}{2}\nabla^2_{\vec{r}_i} - \sum_I \frac{Z_I}{|\vec{R}_I - \vec{r}_i|}, \tag{149}$$

$$\widehat{F}_{ij} = -\frac{1}{2}\frac{1}{|\vec{r}_i - \vec{r}_j|} - \frac{1}{2}\frac{\nabla^2_{\vec{r}_i} g(\vec{f}_i, \vec{f}_j)}{g(\vec{f}_i, \vec{f}_j)} - \frac{\nabla_{\vec{r}_i} g(\vec{f}_i, \vec{f}_j)\cdot\nabla_{\vec{r}_i}}{g(\vec{f}_i, \vec{f}_j)}, \tag{150}$$

and

$$\widehat{F}_{ijk} = -\frac{1}{2}\frac{\nabla^2_{\vec{r}_i} g(\vec{f}_i, \vec{f}_j)}{g(\vec{f}_i, \vec{f}_j)} - \frac{\nabla_{\vec{r}_i} g(\vec{f}_i, \vec{f}_j)\cdot\nabla_{\vec{r}_i} g(\vec{f}_i, \vec{f}_k)}{g(\vec{f}_i, \vec{f}_j)\,g(\vec{f}_i, \vec{f}_k)}. \tag{151}$$

Noticing that

$$\sum_{i=1}^{N}\sum_{j\neq i}^{N}\widehat{F}_{ij} = \sum_{i<j}\Big(\widehat{F}_{ij} + \widehat{F}_{ji}\Big), \tag{152}$$

we can rewrite the two-electron term as

$$\sum_{i=1}^{N}\sum_{j\neq i}^{N}\widehat{F}_{ij} = \sum_{i<j}\Big(\frac{1}{|\vec{r}_i - \vec{r}_j|} + \widehat{u}(\vec{f}_i, \vec{f}_j)\Big), \tag{153}$$

where $\widehat{u}(\vec{f}_i, \vec{f}_j)$ is defined as

$$\widehat{u}(\vec{f}_i, \vec{f}_j) = -\frac{1}{2}\frac{\Big(\nabla^2_{\vec{r}_i} + \nabla^2_{\vec{r}_j}\Big) g(\vec{f}_i, \vec{f}_j)}{g(\vec{f}_i, \vec{f}_j)} - \frac{\nabla_{\vec{r}_i} g(\vec{f}_i, \vec{f}_j)\cdot\nabla_{\vec{r}_i} + \nabla_{\vec{r}_j} g(\vec{f}_i, \vec{f}_j)\cdot\nabla_{\vec{r}_j}}{g(\vec{f}_i, \vec{f}_j)}. \tag{154}$$

In view of Eqs. (149), (150) and (153), it is clear that Eq. (148) can be put in the form:

$$\begin{aligned} C^{-1}\widehat{H}C &= \widehat{H} + \sum_{i<j}\widehat{u}(\vec{f}_i, \vec{f}_j) + \sum_{i=1}^{N}\sum_{j\neq i}^{N}\sum_{k\neq i,j}^{N}\widehat{F}_{ijk} \\ &= \widehat{H} + \widehat{H}^{(1)} + \widehat{H}^{(2)}. \end{aligned} \tag{155}$$

In order to proceed further, let us assume that the pair correlation factor $g(\vec{f}_i, \vec{f}_j)$ is written as follows:

$$g(\vec{f}_i, \vec{f}_j) = 1 + \phi(\vec{f}_i, \vec{f}_j) \tag{156}$$

In particular, this form is quite useful for expanding the function C^2 appearing in Eq. (147). Using Eq. (156), we have

$$C = \prod_{i>j}(1 + \phi_{ij}) = 1 + \sum_{i>j}\phi_{ij} + \sum_{i>j}\phi_{ij}\sum_{k>l}\phi_{kl} + \cdots \tag{157}$$

which leads to

$$C^2 = 1 + 2\sum_{i>j}\phi_{ij} + 2(\sum_{i>j}\phi_{ij})(\sum_{k>l}\phi_{kl}) + \cdots \tag{158}$$
$$= 1 + c_{12} + c_{123} + c_{1234} + \cdots + c_{12\ldots N} \tag{159}$$

In this last expression, c_{12}, c_{123}, up to $c_{12\ldots N}$ are the cluster functions arising from products of the correlation functions. The cluster expansion of Jastrow-type correlated functions has been analyzed in another context by Clark and coworkers [119] and their results could be useful for the present case of correlated electrons.

Keeping in mind Eqs. (155) and (159), we can put the energy functional in the following form:

$$E[\Psi_\rho] = \langle(1 + c_{12} + c_{123} + c_{1234} + \cdots + c_{12\ldots N})\tilde{\Phi}_\rho|\widehat{H} + \widehat{H}^{(1)} + \widehat{H}^{(2)}|\tilde{\Phi}_\rho\rangle \tag{160}$$

$$\begin{aligned} &= \langle\tilde{\Phi}_\rho|\widehat{H}|\tilde{\Phi}_\rho\rangle + \langle c_{12}\tilde{\Phi}_\rho|\widehat{H}|\tilde{\Phi}_\rho\rangle + \cdots + \langle c_{12\cdots N}\tilde{\Phi}_\rho|\widehat{H}|\tilde{\Phi}_\rho\rangle \\ &+ \langle\tilde{\Phi}_\rho|\widehat{H}^{(1)}|\tilde{\Phi}_\rho\rangle + \langle c_{12}\tilde{\Phi}_\rho|\widehat{H}^{(1)}|\tilde{\Phi}_\rho\rangle + \cdots + \langle c_{12\cdots N}\tilde{\Phi}_\rho|\widehat{H}^{(1)}|\tilde{\Phi}_\rho\rangle \\ &+ \langle\tilde{\Phi}_\rho|\widehat{H}^{(2)}|\tilde{\Phi}_\rho\rangle + \langle c_{12}\tilde{\Phi}_\rho|\widehat{H}^{(2)}|\tilde{\Phi}_\rho\rangle + \cdots + \langle c_{12\cdots N}\tilde{\Phi}_\rho|\widehat{H}^{(1)}|\tilde{\Phi}_\rho\rangle, \end{aligned} \tag{161}$$

or equivalently as

$$\begin{aligned} E[\Psi_\rho] &= E[\tilde{\Phi}_\rho] + E[c_{12}\tilde{\Phi}_\rho] + \cdots E[c_{12\cdots N}\tilde{\Phi}_\rho] \\ &+ E_C^{(1)}[\tilde{\Phi}_\rho] + E_C^{(1)}[c_{12}\tilde{\Phi}_\rho] + \cdots E_C^{(1)}[c_{12\cdots N}\tilde{\Phi}_\rho] \\ &+ E_C^{(2)}[\tilde{\Phi}_\rho] + E_C^{(2)}[c_{12}\tilde{\Phi}_\rho] + \cdots E_C^{(2)}[c_{12\cdots N}\tilde{\Phi}_\rho]. \end{aligned} \tag{162}$$

Essentially in Sections 4 and 5 (which were devoted to the kinetic and exchange energies), we have advanced explicit energy functionals for the first term of Eq. (161), namely, for

$$E[\tilde{\Phi}_\rho] = \langle\tilde{\Phi}_\rho|\widehat{H}|\tilde{\Phi}_\rho\rangle. \tag{163}$$

Let us notice, however, that in the present case we deal with a slightly different single Slater determinant $\tilde{\Phi}_\rho$ instead of Φ_ρ (see Eq. (145)). This correction, however, can be trivially incorporated into the formulas developed in the previous sections.

6.1. Density-dependent correlation energy functionals

We consider in this section the problem of writing

$$E_C^{(1)}[\tilde{\Phi}_\rho] = \langle\tilde{\Phi}_\rho|\widehat{H}^{(1)}|\tilde{\Phi}_\rho\rangle \tag{164}$$

as a functional of the one-particle density. For this purpose we make use of Eq. (154) and rewrite

$$E_C^{(1)}[\tilde{\Phi}_\rho] = \int d^3\vec{r}_1 \int d^3\vec{r}_2 \hat{u}(\vec{f}_1, \vec{f}_2) D_{\tilde{\rho}}^2(\vec{r}_1, \vec{r}_2; \vec{r}_1, \vec{r}_2), \tag{165}$$

where $D_{\tilde{\rho}}^2$ is the two matrix corresponding to the single Slater determinant $\tilde{\Phi}_\rho$:

$$D_{\tilde{\rho}}^2(\vec{r}_1, \vec{r}_2; \vec{r}_1, \vec{r}_2) = \frac{\rho(r_1)}{\tilde{\rho}_g(f_1)}\frac{\rho_g(r_2)}{\tilde{\rho}_g(f_2)} D_g^2(\vec{f}_1, \vec{f}_2; \vec{f}_1, \vec{f}_2). \tag{166}$$

Expanding the orbit-generating 2-matrix

$$D_g^2(\vec{f}_1,\vec{f}_2;\vec{f}_1,\vec{f}_2)=\frac{1}{2}\Big(\rho_g(\vec{f}_1)\rho_g(\vec{f}_2)-D_g^1(\vec{f}_1,\vec{f}_2)D_g^1(\vec{f}_2,\vec{f}_1)\Big) \tag{167}$$

we can rewrite

$$\begin{aligned} E_C^{(1)}[\tilde{\Phi}_\rho] &= \frac{1}{2}\int d^3\vec{r}_1\frac{\rho(r_1)}{\tilde{\rho}_g(f_1)}\int d^3\vec{r}_2\frac{\rho_g(r_2)}{\tilde{\rho}_g(f_2)}\rho_g(\vec{f}_1)\hat{u}(\vec{f}_1,\vec{f}_2)\rho_g(\vec{f}_2) \\ &- \frac{1}{2}\int d^3\vec{r}_1\frac{\rho(r_1)}{\tilde{\rho}_g(f_1)}\int d^3\vec{r}_2\frac{\rho_g(r_2)}{\tilde{\rho}_g(f_2)}D_g^1(\vec{f}_1,\vec{f}_2)\hat{u}(\vec{f}_1,\vec{f}_2)D_g^1(\vec{f}_2,\vec{f}_1)\Big). \end{aligned} \tag{168}$$

It is convenient at this point to rewrite the two-particle operator as

$$\hat{u}(\vec{f}_1,\vec{f}_2)=\hat{u}_1(\vec{f}_1,\vec{f}_2)+\hat{u}_2(\vec{f}_1,\vec{f}_2), \tag{169}$$

where we have defined

$$\hat{u}_1(\vec{f}_i,\vec{f}_j)=-\frac{1}{2}\frac{\nabla^2_{\vec{r}_i}g(\vec{f}_i,\vec{f}_j)}{g(\vec{f}_i,\vec{f}_j)}-\frac{\nabla_{\vec{r}_i}g(\vec{f}_i,\vec{f}_j)\cdot\nabla_{\vec{r}_i}}{g(\vec{f}_i,\vec{f}_j)}, \tag{170}$$

with a similar definition for $\hat{u}_2(\vec{f}_1,\vec{f}_2)$. Using the explicit expressions for the Laplacian and gradient operators:

$$\nabla^2_{\vec{r}_1}=\left(\frac{2}{r_1}\frac{\partial}{\partial r_1}+\frac{\partial^2}{\partial r_1^2}\right)+\frac{1}{r_1^2}\nabla^2_{\Omega_1} \tag{171}$$

and

$$\nabla_{\vec{r}_1}g(\vec{f}_1,\vec{f}_2)\cdot\nabla_{\vec{r}_1}=\frac{\partial g(\vec{f}_1,\vec{f}_2)}{\partial r_1}\frac{\partial}{\partial r_1}+\frac{1}{r_1^2}\nabla_{\Omega_1}g(\vec{f}_1,\vec{f}_2)\cdot\nabla_{\Omega_1}, \tag{172}$$

where

$$\nabla^2_{\Omega_1}=\frac{1}{\sin\theta_1}\frac{\partial}{\partial\theta_1}\left(\sin\theta_1\frac{\partial}{\partial\theta_1}\right)+\frac{1}{\sin^2\theta_1}\frac{\partial^2}{\partial\phi_1^2}, \tag{173}$$

and

$$\nabla_{\Omega_1}g(\vec{f}_1,\vec{f}_2)\cdot\nabla_{\Omega_1}=\frac{\partial g(\vec{f}_1,\vec{f}_2)}{\partial\theta_1}\frac{\partial}{\partial\theta_1}+\frac{1}{\sin^2\theta_1}\frac{\partial g(\vec{f}_1,\vec{f}_2)}{\partial\phi_1}\frac{\partial}{\partial\phi_1}. \tag{174}$$

Introducing the chain rule given in Eq. (70) we are able to express the operator $\hat{\mu}_1(\vec{f}_1,\vec{f}_2)$ as follows:

$$\begin{aligned} \hat{\mu}_1(\vec{f}_1,\vec{f}_2) &= \left(\frac{df_1}{dr_1}\right)^2\frac{1}{g(\vec{f}_1,\vec{f}_2)}\left(2\frac{\partial g(\vec{f}_1,\vec{f}_2)}{\partial f_1}\frac{\partial}{\partial f_1}-\frac{1}{2}\frac{\partial^2 g(\vec{f}_1,\vec{f}_2)}{\partial f_1^2}\right) \\ &+ \left(\frac{df_1}{dr_1}\right)\frac{1}{g(\vec{f}_1,\vec{f}_2)}\left(-\frac{1}{r_1}\frac{\partial g(\vec{f}_1,\vec{f}_2)}{\partial f_1}\right) \\ &+ \frac{1}{r_1^2 g(\vec{f}_1,\vec{f}_2)}\left(-\frac{1}{2}\nabla^2_{\Omega_1}g(\vec{f}_1,\vec{f}_2)+2\nabla_{\Omega_1}g(\vec{f}_1,\vec{f}_2)\cdot\nabla_{\Omega_1}\right) \\ &= \left(\frac{df_1}{dr_1}\right)^2\frac{a(\vec{f}_1,\vec{f}_2)}{g(\vec{f}_1,\vec{f}_2)}+\left(\frac{df_1}{dr_1}\right)\frac{b(\vec{f}_1,\vec{f}_2)}{g(\vec{f}_1,\vec{f}_2)}+\frac{c(\vec{f}_1,\vec{f}_2)}{r_1^2 g(\vec{f}_1,\vec{f}_2)}. \end{aligned} \tag{175}$$

Since the operator $\widehat{\mu}_2(\vec{f}_1, \vec{f}_2)$ leads to a similar term, we can rewrite $E_C^{(1)}[\tilde{\Phi}_\rho]$ as:

$$\begin{aligned} E_C^{(1)}[\tilde{\Phi}_\rho] &= \int dr_1 r_1^2 \rho^{5/3}(r)\Big(r + \vec{r}_1 \cdot \nabla_{\vec{r}_1}\lambda(r)\Big)^{4/3} \mathcal{C}_N^{(1)}(f_1) \\ &+ \int dr_1 r_1^2 \rho^{4/3}(r)\Big(r + \vec{r}_1 \cdot \nabla_{\vec{r}_1}\lambda(r)\Big)^{2/3} \mathcal{C}_N^{(2)}(f_1) \\ &+ \int dr_1 r_1^2 \rho^{5/3}(r)\Big(r + \vec{r}_1 \cdot \nabla_{\vec{r}_1}\lambda(r)\Big)^{-2/3} \mathcal{C}_N^{(3)}(f_1), \end{aligned} \tag{176}$$

where we have defined

$$\mathcal{C}_1(f_1) = \eta_1(f_1) - \nu_1(f_1), \tag{177}$$

$$\mathcal{C}_2(f_1) = \eta_2(f_1) - \nu_2(f_1), \tag{178}$$

$$\mathcal{C}_3(f_1) = \eta_3(f_1) - \nu_3(f_1), \tag{179}$$

with

$$\eta_1(f_1) = \frac{1}{\tilde{\rho}_g^{5/3}(f_1)} \int d\Omega_1 \int df_2 f_2^2 \int d\Omega_2 \rho_g(\vec{f}_1) \frac{a(\vec{f}_1, \vec{f}_2)}{g(\vec{f}_1, \vec{f}_2)} \rho_g(\vec{f}_2), \tag{180}$$

$$\eta_2(f_1) = \frac{1}{\tilde{\rho}_g^{4/3}(f_1)} \int d\Omega_1 \int df_2 f_2^2 \int d\Omega_2 \rho_g(\vec{f}_1) \frac{b(\vec{f}_1, \vec{f}_2)}{g(\vec{f}_1, \vec{f}_2)} \rho_g(\vec{f}_2), \tag{181}$$

$$\eta_3(f_1) = \frac{1}{f_1^2 \tilde{\rho}_g^{5/3}(f_1)} \int d\Omega_1 \int df_2 f_2^2 \int d\Omega_2 \rho_g(\vec{f}_1) \frac{c(\vec{f}_1, \vec{f}_2)}{g(\vec{f}_1, \vec{f}_2)} \rho_g(\vec{f}_2), \tag{182}$$

and

$$\nu_1(f_1) = \frac{1}{\tilde{\rho}_g^{5/3}(f_1)} \int d\Omega_1 \int df_2 f_2^2 \int d\Omega_2 D_g^1(\vec{f}_1, \vec{f}_2)) \frac{a(\vec{f}_1, \vec{f}_2)}{g(\vec{f}_1, \vec{f}_2)} D_g^1(\vec{f}_2, \vec{f}_1)), \tag{183}$$

$$\eta_2(f_1) = \frac{1}{\tilde{\rho}_g^{4/3}(f_1)} \int d\Omega_1 \int df_2 f_2^2 \int d\Omega_2 D_g^1(\vec{f}_1, \vec{f}_2)) \frac{b(\vec{f}_1, \vec{f}_2)}{g(\vec{f}_1, \vec{f}_2)} D_g^1(\vec{f}_2, \vec{f}_1)), \tag{184}$$

$$\eta_3(f_1) = \frac{1}{f_1^2 \tilde{\rho}_g^{5/3}(f_1)} \int d\Omega_1 \int df_2 f_2^2 \int d\Omega_2 D_g^1(\vec{f}_1, \vec{f}_2)) \frac{c(\vec{f}_1, \vec{f}_2)}{g(\vec{f}_1, \vec{f}_2)} D_g^1(\vec{f}_2, \vec{f}_1)). \tag{185}$$

The main result of this Section is given in Eq. (176) where the dominant electron correlation-energy term is displayed as an explicit functional of the one-particle density and of the modulating functions $\mathcal{C}_N^{(i)}(f_1)$ which depend on the local-scaling transformation function f_1. It is again gratifying to observe that the terms appearing in this functional are quite similar to those obtained for the kinetic and exchange energy functionals. Again, we see that universal functionals behaving as $\rho^{5/3}(r)$ and $\rho^{4/3}(r)$ appear as the basic terms. Thus, we may conclude that the electron correlation functionals provide corrections to the contributions arising from the kinetic and exchange energies.

Let us finish this Section by mentioning that the rigorous treatment of electron correlation requires the inclusion of all terms appearing in Eq. (162). In the present discussion, we just have treated $E_C^{(1)}$ and have shown that it can be converted into a functional of the one-particle density. Although all the remaining terms of Eq. (162) are, in principle, amenable to the same treatment, their relative importance in terms of the magnitude of their contribution to the correlation energy, would make it unnecessary to deal with all of them with the same rigor. Thus, we believe that the present systematic approach could lay the foundations for selective treatment of these corrections. Clearly, if approximations have to be made, it would be advisable to derive them based on a general scheme such as the one advanced by the local-scaling transformation version of density functional theory and not on heuristic considerations. Applications of the present ideas for the purpose of generating electron correlation functionals akin to that of Colle and Salvetti are under under way in our laboratory [117].

ACKNOWLEDGMENT

E.V.L. would like to express his gratitude to the Consejo Nacional de Investigaciones Científicas y Tecnológicas, CONICIT, of Venezuela for its support of the present work through Project S1–95000702. He also gratefully acknowledges support of this work by the Commission of European Communities through Contract No. CI1*-CT93-0333.

REFERENCES

1. E.S. Kryachko and E.V. Ludeña, Energy Density Functional Theory of Many-Electron Systems, Kluwer Academic Publishers, Dordrecht, 1990.
2. R.G. Parr and W. Yang, Density Functional Theory of Atoms and Molecules, Oxford University Press, Oxford, 1989.
3. R.M. Dreizler and E.K.U. Gross, Density Functional Theory, Springer-Verlag, Berlin, 1990.
4. N.H. March, Electron Density Theory of Atoms and Molecules, Academic Press, New York, 1992.
5. P.C. Hohenberg and W. Kohn, Phys. Rev. **136B**, (1964) 864.
6. P.C. Hohenberg, W. Kohn and L.J. Sham, Adv. Quantum Chem. **21**, (1990) 7.
7. W. Kohn and L.J. Sham, Phys. Rev. **140A** (1965) 1133.
8. P.-O. Löwdin, in: Density Matrices and Density Functionals, R. Erdahl and V.H. Smith (eds.), D. Reidel, Dordrecht, 1987, p. 21.
9. N.H. March, Adv. in Phys. **6** (1957) 1.
10. A.J. Coleman, Revs. Mod. Phys. **35** (1963) 668.
11. E.H. Lieb, Rev. Mod. Phys. **53** (1981) 603.
12. A.J. Coleman, in: Density Matrices and Density Functionals, R. Erdahl and V.H. Smith (eds.), D. Reidel, Dordrecht, 1987, p. 5.
13. See also, Reference [1], Section 4.4.
14. E.S. Kryachko and E.V. Ludeña, Phys. Rev. **A 43** (1991) 2179.
15. See also, Reference [1], Chapters 7 and 8.
16. I.Zh. Petkov, M.V. Stoitsov and E.S. Kryachko, Int. J. Quantum Chem. **29** (1986) 149.
17. E.S. Kryachko, I.Zh. Petkov and M.V. Stoitsov, Int. J. Quantum Chem. **32** (1987) 467.
18. E.S. Kryachko, I.Zh. Petkov and M.V. Stoitsov, Int. J. Quantum Chem. **32** (1987) 473.
19. E.S. Kryachko and E.V. Ludeña, Phys. Rev. **A 35** (1987) 957.

20. E.S. Kryachko, I.Zh. Petkov and M.V. Stoitsov, Int. J. Quantum Chem. **34** (1988) 305(E).
21. E.S. Kryachko and E.V. Ludeña, Phys. Rev. **A 43** (1991) 2194.
22. I.Zh. Petkov and M.V. Stoitsov, Nuclear Density Functional Theory, Oxford University Press, New York, 1991.
23. E.S. Kryachko and E.V. Ludeña, in: Computational Chemistry: Structure, Interactions and Reactivity, S. Fraga (ed.), Elsevier, Amsterdam, 1992, p. 136.
24. E.V. Ludeña and E.S. Kryachko, Rev. Mex. Astr. y Astrofis., **23** (1992) 95.
25. E.S. Kryachko and E.V. Ludeña, in: Condensed Matter Theories, A.N. Proto and J.L. Aliaga, (eds.), Plenum, New York, 1992, Vol. 7, p. 101.
26. E.S. Kryachko and E.V. Ludeña, New J. Chem. **16** (1992) 1089.
27. E.S. Kryachko and E.V. Ludeña, J. Chem. Phys. **95** (1991) 9054.
28. E.S. Kryachko and T. Koga, J. Chem. Phys. **91** (1989) 1108.
29. T. Koga, Y. Yamamoto and E.S. Kryachko, J. Chem. Phys. **91** (1989) 4785.
30. T. Koga, Phys. Rev. **A 41** (1990) 1274.
31. T. Koga, Phys. Rev. **A 42** (1990) 3763.
32. T. Koga, J. Chem. Phys. **93** (1990) 5856.
33. T. Koga and Y. Yamamoto, Phys. Rev. **A 42** (1990) 6336.
34. T. Koga and E.S. Kryachko, J. Chem. Phys. **94** (1991) 2910.
35. T. Koga, Y. Yamamoto and E.V. Ludeña, J. Chem. Phys. **94** (1991) 3805.
36. T. Koga, Y. Yamamoto and E.V. Ludeña, Phys. Rev. **A 43** (1991) 5814.
37. T. Koga, J. Chem. Phys. **95** (1991) 4306.
38. E.S. Kryachko and T. Koga, Int. J. Quantum Chem. **42** (1992) 591.
39. T. Koga and H. Matsuyama, Phys. Rev. **A 45** (1992) 5266.
40. E.S. Kryachko, E.V. Ludeña and T. Koga, in: Methods and Techniques in Computational Chemistry, METECC-94, Vol. B: Medium Size Systems, E. Clementi (ed.), STEF, Cagliari, 1993, p. 23.
41. E.S. Kryachko, E.V. Ludeña and T. Koga, J. Math. Chem. **11** (1992) 325.
42. E.S. Kryachko and E.V. Ludeña, Int. J. Quantum Chem. **43** (1992) 769.
43. E.S. Kryachko, E.V. Ludeña, R. López-Boada and J. Maldonado, in: Condensed Matter Theories, L. Blum and F.B. Malik (eds.), Plenum, New York, 1993, Vol. 8, p. 373.
44. E.V. Ludeña, R. López-Boada, J. Maldonado, T. Koga and E.S. Kryachko, Phys. Rev. **A 48** (1993) 1937.
45. E.V. Ludeña, R. López-Boada and J. Maldonado, in: Condensed Matter Theories, J.W. Clark, A. Sadiq and K.A. Shoaib (eds.), Nova Science Publishers, Commack, N.Y., 1994, Vol. 9, p. 97.
46. E.V. Ludeña, R. López-Boada and J. Maldonado, J. Mol. Struct. (Theochem) **330**, 33.
47. E.V. Ludeña, J. Maldonado, R. López-Boada, T. Koga and E.S. Kryachko, J. Chem. Phys. **102** (1995) 318.
48. E.V. Ludeña, E.S. Kryachko, T. Koga, R. López-Boada, J. Hinze, J. Maldonado and E. Valderrama, in "Density Functional Theory: A Tool for Chemistry", J.M. Seminario and P. Politzer (eds.), Elsevier, Amsterdam (1995). p. 75
49. J. Maldonado, E.V. Ludeña, R. López-Boada, V. Karasiev, O. Bokanowski. Ap-

plication of the local-scaling transformation version of density functional theory to diatomic molecules. Int. J. Quantum Chem. (to be submitted).
50. L.H. Thomas, Proc. Cambridge Phil. Soc. **23** (1927) 542.
51. E. Fermi, Rend. Accad. Naz. Lincei **6** (1927) 602.
52. P.A.M. Dirac, Proc. Cambridge Phil. Soc. **26** (1930) 376.
53. E. Wigner, Phys. Rev. **A46** (1934) 1002.
54. C.F.von Weizsäcker, Z. Phys. **96** (1935) 431.
55. D.A. Kirzhnitz, Zh. Eksp. Teor. Fiz. **32** (1957) 115.
56. C.H. Hodges, Can. J. Phys. **51** (1973) 1428.
57. B. Grammaticos and A. Voros, Ann. Phys. **129** (1979) 153.
58. D.R. Murphy, Phys. Rev. **A24** (1981) 1682.
59. W. Yang, Phys. Rev. **A34** (1986) 4575.
60. A. Holas, and N.H. March, Phys. Rev. **A 51** (1993) 2040.
61. A. Holas, N.H. March, Y. Takahashi and C. Zhang, Phys. Rev. **A 48** (1993) 2708.
62. K.A. Dawson and N.H. March, J. Chem. Phys. **81** (1984) 5850.
63. N.H. March, in: Condensed Matter Theories, E.V. Ludeña, P. Vashishta and R.F. Bishop (eds.), Nova Science Publishers, Commack, N.Y., 1996, Vol. 11, p. 23.
64. L. Cohen and C. Frishberg, Phys. Rev. **A 13** (1976) 927.
65. H. Nakatsuji, Phys. Rev. **A 14** (1976) 41.
66. G. Hunter, in: Density Matrices and Density Functionals, R. Erdahl and V.H. Smith (eds.), Reidel, Dordrecht, 1985, p. 583.
67. G. Hunter, Int. J. Quantum Chem. **29** (1986) 197.
68. G. Hunter, Int. J. Quantum Chem. **19** (1981) 755.
69. G. Hunter, Int. J. Quantum Chem. **9** (1975) 237.
70. M.K. Harbola and V. Sahni, Phys. Rev. Lett. **62** (1989) 489.
71. C. Valdemoro, Phys. Rev. **A 45** (1992) 4462.
72. P. Sule, O.V. Gritsenko, A. Nagy and E.V. Baerends, J. Chem. Phys. **103** (1995) 10085.
73. T.T. Nguyen-Dang, E.V. Ludeña and Y. Tal, J. Mol. Struct. (Theochem) **120** (1985) 247.
74. E.V. Ludeña and J. Keller, Adv. Quantum Chem. **21** (1990) 47.
75. Pino, R., E.V. Ludeña, R. López-Boada (1995) In: *Condensed Matter Theories* **10**. J. Navarro and M. de Llano (eds.), Nova Science Publishers, Commack, N.Y., p. 189
76. M. Levy, Proc. Natl. Acad. Sci. USA **76** (1979) 6062.
77. E.H. Lieb, Int. J. Quantum Chem. **24** (1983) 243.
78. P.W. Payne, J. Chem. Phys. **71** (1979) 490.
79. J.K. Percus, Int. J. Quantum Chem. **13** (1978) 89.
80. J. Cioslowski, Adv. Quantum Chem. **21** (1990) 303.
81. R. McWeeny, Phil. Mag. **B 69** (1994) 727.
82. P.M.W. Gill and J.A. Pople, Phys. Rev. **A 47** (1993) 2383. It may be argued, however, that this example only proves that if such universal functional exists, it cannot be expressed in GGA form (we thank J.M. Seminario for bringing up this point).
83. W. Macke, Phys. Rev. **100** (1955) 992.
84. W. Macke, Ann. Phys. (Leipzig) **17** (1955) 1.

85. N.H. March and W.H. Young, Proc. Roy. Soc. London **72** (1958) 182.
86. G.G. Hall, Proc. Roy. Soc. London **75** (1960) 575.
87. J. Moser, Trans. Amer. Math. Soc. **120** (1965) 286.
88. T.L. Gilbert, Phys. Rev. **B12** (1975) 2111.
89. J.E. Harriman, Phys. Rev. **A 24** (1981) 680.
90. E.H. Lieb, in: Physics as Natural Philosophy: Essays in Honor of Laszlo Tisza on his 75th Birthday, H. Feshback and A. Shimony (eds.), MIT Press, Cambridge, MA, 1982, p. 111.
91. E.H. Lieb, in: Density Functional Methods in Physics, R.M. Dreizler and J. da Providencia (eds.), Plenum, New York, 1985, p. 31.
92. E.V. Ludeña, J. Chem. Phys. **79** (1983) 6174.
93. G. Zumbach and K. Maschke, Phys. Rev. **A 28** (1983) 544; Ibid, **A 29** (1984) 1585(E).
94. M.R. Nyden and R.G. Parr, J. Chem. Phys. **78** (1983) 4044.
95. S.K. Ghosh and R.G. Parr, J. Chem. Phys. **82** (1985) 3307.
96. M.J. Ten Hoor, Int. J. Quantum Chem. **33** (1988) 563; J. Chem. Ed. **66** (1989) 633.
97. P.M. Kozlowski and N.H. March, Int. J. Quantum Chem. **36** (1989) 741.
98. E.V. Ludeña, E.V., R. López-Boada, J. Maldonado, E. Valderrama, T. Koga, E.S. Kryachko, J. Hinze, Int. J. Quantum Chem. **56** (1995) 285.
99. J. Cioslowski, Phys. Rev. Lett. **60** (1988) 2141.
100. J. Cioslowski, J. Chem. Phys. **89** (1988) 4871.
101. J. Cioslowski, Int. J. Quantum Chem. **S23** (1989)
102. E.V. Ludeña and R. López-Boada, Top. Curr. Chem. (1996) (in press)
103. O. Bokanowski and B. Grébert, Math. Models and Meth. Appl. Sci **6** (1996) 0000.
104. E.R. Davidson, Reduced Density Matrices in Quantum Chemistry, Academic Press, New York, 1976, p. 135.
105. O. Bokanowski and B. Grébert, Int. Rep. MIP No. 95-18 UFR-MIG, Univ. P. Sabatier, Toulouse, France (1995)
106. E.V. Ludeña, R. López-Boada and R. Pino, Can. J. Chem. (1996) (in press).
107. E.V. Ludeña, R. López-Boada and R. Pino, in: Condensed Matter Theories, E.V. Ludeña, P. Vashishta and R.F. Bishop (eds.), Nova Science Publishers, Commack, N.Y., 1996, Vol. 11, p. 51.
108. J.C. Slater, *Quantum Theory of Atomic Structure*, MacGraw-Hill, New York (1960), v. 1, p. 311.
109. J.L. Calais and P.-O. Löwdin, J. Mol. Spectrosc. **8** (1962) 203.
110. R. López-Boada, R. Pino and E.V. Ludeña. Explicit expressions for $T_s[\rho]$ and $E_x[\rho]$ by means of Padé approximants to local-scaling transformations. Int. J. Quantum Chem. (submitted).
111. A. Cuyt and L. Wuytack, *Nonlinear Methods in Numerical Analysis, North Holland Mathematical Studies 136*, North Holland, Amsterdam (1988).
112. R. López-Boada and E.V. Ludeña. Hartree-Fock calculations in the context of the local-scaling transformation version of density functional theory. Applications to the lithium and beryllium atoms. Phys. Rev. A (submitted).
113. R. López-Boada, R. Pino and E.V. Ludeña. Hartree-Fock energy density functionals generated by local-scaling transformations: Applications to first-row atoms. J. Chem.

Phys. (submitted).
114. E. Clementi and C. Roetti, At. Data and Nucl. Data Tables **14** (1974) 177.
115. V. Sahni, Int. J. Quantum Chem. **53** (1995) 591.
116. V. Sahni, Y. Li, M.K. Harbola, Phys. Rev. **A 45** (1992) 1434.
117. R. López-Boada, E.V. Ludeña, V. Karasiev and R. Colle. Generation of explicit electron correlation functionals by means of local-scaling transformations. J. Chem. Phys. (to be submitted).
118. S.F. Boys and N.C. Handy, Proc. Roy. Soc. **A 310** (1969) 43.
119. J.W. Clark, R.L. Mead, E. Krotscheck, K.E. Kürten and M.L. Ristig, Nucl. Phys. **A328** (1979) 45.

J.M. Seminario (Editor)
Recent Developments and Applications of Modern Density Functional Theory
Theoretical and Computational Chemistry, Vol. 4

Inhomogeneous Electron Gas: Transcending Semiclassical Thomas-Fermi-Dirac Method

N. H. March

Oxford University, Oxford, ENGLAND

1. BACKGROUND

Following Hartree's pioneering work on the self-consistent field in atoms, Thomas [1] and Fermi [2] independently constructed a semiclassical theory of heavy atoms solely in terms of the ground-state electron density $\rho(\mathbf{r})$. Their theory was generalized by Dirac [3] to include exchange. This so called Thomas-Fermi-Dirac theory is the forerunner of modern density functional theory. The Thomas-Fermi-Dirac theory was formally completed by Hohenberg and Kohn [4] who proved that the ground state energy of a many-electron assembly was a unique functional of the ground-state electron density, just as Thomas, Fermi and Dirac asserted.

The basic idea underlying the Thomas-Fermi semiclassical theory was to start out from the homogeneous electron gas and to switch on a potential energy $V(\vec{r})$ to create from an initially uniform density an inhomogeneous electron gas. Thus, one took free electron relations, and employed them locally at position $\vec{r}$. To be quite specific, in a free electron gas the uniform density, ρ_0 say, is related to the maximum or Fermi momentum p_f by the phase space relation (see, for example, March, [5,6])

$$\rho_o = \frac{8\pi}{3h^3} p_f^3 \quad . \tag{1}$$

This homogeneous gas relation (1) is now taken over into the inhomogeneous charge cloud existing in an atom or molecule to read

$$\rho(\vec{r}) = \frac{8\pi}{3h^3} p_f^3(\vec{r}) \tag{2}$$

which relates the electron density $\rho(\vec{r})$ in the molecule to the maximum momentum $p_f(\vec{r})$ at position $\vec{r}$. Evidently, since relation (1) is exact in a homogeneous electron cloud, Eq. (2) will work best in an electronic assembly where the density varies relatively slowly in space. This condition of validity of Eq.

(2) will be made somewhat precise later on. But, as already mentioned, the inhomogeneity in the electron density has been created by a potential energy $V(\vec{r})$ switched on to the homogeneous electron gas. Thomas [1] and Fermi [2] therefore invoked the classical energy equation of the fastest electron in the assembly with momentum $p_f(\vec{r})$ at position $\vec{r}$, to write

$$\mu = \frac{p_f^2(\vec{r})}{2m} + V(\vec{r}) \tag{3}$$

where m is the mass of the electron. In Eq. (3) we have denoted the maximum energy by μ, as it will be identified subsequently with the chemical potential of the electronic charge cloud. Eliminating $p_f(\vec{r})$ from Eq. (3) using Eq. (2) leads to the basic density-potential relation of the Thomas-Fermi semiclassical method as

$$\mu = \frac{5}{3} c_k \rho(\vec{r})^{2/3} + V(\vec{r}) \tag{4}$$

where

$$c_k = \frac{3h^2}{10m}\left(\frac{3}{8\pi}\right)^{2/3},$$

the factor (5/3) being introduced for convenience in determining total kinetic energy in terms of the constant c_k.

This equation is already remarkable in its physical and chemical content. First of all, the left-hand side is a constant, independent of position. However, it is made up of two terms displayed on the right-hand side (RHS) which are individually dependent on the position $\vec{r}$. The first term is evidently from kinetic energy while the second is simply the potential energy $V(\vec{r})$ in which the fastest electron moves. The total energy μ of this electron being constant throughout the charge cloud in a molecule is essentially the requirement that charge redistribution is complete as atoms are brought from infinite separation to form the molecule. For if, at one point, μ were higher than at another point, electrons could evidently rearrange themselves spatially to lower the energy.

Eq. (4), though of course approximate due to (a) the use of Eq. (2), which is only quantitative for slowly varying density $\rho(\vec{r})$ and potential energy $V(\vec{r})$ and (b) application of the classical energy equation (3) is conceptually important. In a fully quantum-mechanical many-electron theory there is a equation analogous to Eq. (4), though the meaning of the kinetic plus potential energy terms has to be clarified. This will be a focus of the discussion below.

However, of prime interest for recent advances in the foundations of density functional theory is the construction of the force corresponding to the external (that is, nuclear) potential energy. ($-Ze^2/r$ in an atom with nuclear charge Ze).

In the simple Thomas-Fermi method, $V(\vec{r})$ in Eq. (4) is the sum of the nuclear potential energy $V_n(\vec{r})$ and the potential energy $V_e(\vec{r})$ created by the electronic density distribution $\rho(\vec{r})$. Thus, we form the external force $\vec{F}$ from Eq. (4) as

$$\vec{F} = -\frac{\partial V_n(\vec{r})}{\partial \vec{r}} = \frac{10}{9} c_k \rho(\vec{r})^{-1/3} \frac{\partial \rho(\vec{r})}{\partial \vec{r}} + \frac{\partial V_e(\vec{r})}{\partial \vec{r}}. \tag{5}$$

Eq (4) is then seen to have the content that the external force $\vec{F}$ can be expressed explicitly in terms of the electron density since $V_e(\vec{r})$ is given by

$$V_e(\vec{r}) = e^2 \int \frac{\rho(\vec{r}')}{|\vec{r} - \vec{r}'|} d\vec{r}'. \tag{6}$$

Just as the total energy E is determined uniquely by the ground-state density (see also below), so the external force is likewise a function(al) of the density. Already implicit in the original theory of Thomas and Fermi, this result was formally proved in an exact existence theorem by Hohenberg and Kohn [4] though their argument gave no route for the construction of a relation between the external force and the density. Such a (formally exact) construction will be sought below, but will require a generalization of the density $\rho(\vec{r})$ to density matrices. These quantities will provide a further focus of the present Chapter.

2. OFF-DIAGONAL DENSITY, OR DIRAC DENSITY MATRIX

Given a potential energy $V(\vec{r})$ in which electrons move, March and Murray [7-8] constructed a general perturbation theory, taking literally the idea of Thomas and Fermi of *switching on* the potential $V(\vec{r})$ in an initially uniform electron gas. These workers noted that appropriate tool was an *off-diagonal density*, or Dirac density matrix $\gamma(\vec{r}, \vec{r}')$. Suppose the potential $V(\vec{r})$, inserted in a one-electron Schrödinger equation, generates one-electron wave functions $\psi_i(\vec{r})$. Then the electron density $\rho(\vec{r})$ is obtained by summing the squares of these wave functions over the lowest occupied electronic states, consistent with the Pauli exclusion Principle. Thus the density $\rho(\vec{r})$ is given by

$$\rho(\vec{r}) = \sum_{occupied\, i} \psi_i(\vec{r}) \psi_i^*(\vec{r}). \tag{7}$$

The desired off-diagonal generalization of Eq. (7) was given by Dirac [3]; and is termed the Dirac density matrix. This is defined again in terms of the wave functions and is indeed an (elementary) generalization of Eq. (7) off the diagonal, namely

$$\gamma(\vec{r},\vec{r}') = \sum_{occupied\, i} \psi_i(\vec{r})\psi_i^*(\vec{r}'). \tag{8}$$

It is well known that the normalized wave functions $\psi_i(\vec{r})$ in Eqs. (7) and (8) are also orthogonal, as they are all generated by the common potential $V(\vec{r})$ in which the electrons move. This means in matrix language that γ is idempotent, satisfying $\gamma^2 = \gamma$, or in coordinate representation

$$\gamma(\vec{r},\vec{r}') = \int \gamma(\vec{r},\vec{r}'')\gamma(\vec{r}'',\vec{r}')d\vec{r}'' \tag{9}$$

for singly occupied levels, with an appropriate constant factor (see below) introduced for doubly filled levels. This matrix is the proper tool for dealing with single-particle Fermion problems, as emphasized especially in the early studies of Löwdin [9] and of McWeeny [10]. March and Murray [8] therefore focussed on calculating the Dirac matrix by switching on the potential energy $V(\vec{r})$ to a free electron gas, using perturbation theory to all orders. We shall merely state their main findings below, and then discuss the consequences which flow from them. To date, however, it has to be said that their perturbation series has only been summed to all orders in the case when the potential energy $V(\vec{r})$ is slowly varying in space. Then, not surprisingly, the relation between density and potential expressed through the chemical potential Eq. (4) is regained when $\vec{r}'$ in the Dirac density matrix is subsequently put equal to $\vec{r}$. Thus, it is fair to say that the March-Murray perturbation series transcends the Thomas-Fermi method, removing all semiclassical assumptions from its foundations. Because of the difficulty of carrying out the summation of the perturbation series: modern workers generally return to the one-electron Schrödinger equation and calculate directly the one-electron wave functions $\psi_i(\vec{r})$ from a suitably chosen potential energy $V(\vec{r})$. Much of the present Chapter will therefore be concerned with the proper choice of $V(\vec{r})$. That there was great merit in working with a common *local* potential $V(\vec{r})$ in which all electrons in a molecule (or indeed condensed phase) were assumed to move was emphasized in the pioneering work of Slater [11]. His work was formally completed by the study of Kohn and Sham [12]. We shall return to the way Slater proposed to construct the potential $V(\vec{r})$ in the following section. However, let us first state the main results of the March-Murray perturbation expansion. We shall give both (diagonal) density $\rho(\vec{r})$ and the kinetic energy density $t(\vec{r})$. But first let us write down the Dirac density matrix for free electrons, as this is the unperturbed problem onto which the potential energy $V(\vec{r})$ is switched.

2.1. Dirac density matrix for free electrons

When the potential $V(\vec{r}) = 0$, the wave functions $\psi_i(\vec{r})$ in Eq. (8) become plane waves, classified by wave vector $\vec{k}$, or momentum $p = \hbar\vec{k}$. The Dirac density matrix, for a normalization volume ϑ, is therefore

$$\gamma(\vec{r},\vec{r}') = \frac{1}{\vartheta} \sum_{|\vec{k}|<k_f} e^{i\vec{k}.(\vec{r}-\vec{r}')} \tag{10}$$

where k_f is the Fermi wave number, related to the Fermi momentum p_f in Eq. (1) by $p_f = \hbar k_f$. The summation can be replaced by an integration, with a density of states in $\vec{k}$ space given by $1/8\pi^3$ per unit volume of coordinate space. Completing the integration then yields the Dirac density matrix for free electrons as

$$\gamma_o(\vec{r},\vec{r}') = \frac{k_f^3}{\pi^2} \frac{j_1(k_f|\vec{r}-\vec{r}'|)}{k_f|\vec{r}-\vec{r}'|} \ : \tag{11}$$

$$j_1(x) = \frac{\sin x - x\cos x}{x^2}$$

which, on the diagonal $\vec{r} = \vec{r}'$ merely leads back to Eq. (1).

2.2. Exact Thomas-Fermi method in perturbation theory

This density matrix is the basis of the March-Murray [8] perturbation series for $\gamma(\vec{r},\vec{r}')$ in the presence of the potential $V(\vec{r})$. Before writing the complete series, let us summarize the shape of the first order result in V, namely

$$\gamma(\vec{r},\vec{r}') = \gamma_o(\vec{r},\vec{r}') - \int V(\vec{r}'')R(\vec{r},\vec{r}';\vec{r}'')d\vec{r}'' \tag{12}$$

where R is known explicitly in terms of the first-order spherical Bessel functions j_1. On the diagonal $\vec{r} = \vec{r}'$, this is a first-order theory of the density, namely

$$\rho(\vec{r}) = \rho_o - \frac{k_f^2}{\pi^2} \int V(\vec{r}'') \frac{j_1(k_f|\vec{r}-\vec{r}''|)}{|\vec{r}-\vec{r}''|^2} d\vec{r}''. \tag{13}$$

This can be solved for V by Fourier transform, due to the appearance of the convolution on the right-hand side. The shape of the result is then

$$\tilde{V}(\vec{k}) = \frac{\Delta\tilde{\rho}(k)}{F(k)} \tag{14}$$

where $\Delta\tilde{\rho}(k)$ denotes the Fourier transform (FT) of the *displaced* charge $\rho(\vec{r}) - \rho_o$ while $F(k)$, apart from a multiplying factor, is the FT of $R(\vec{r},\vec{r}';\vec{r}'')$ in Eq (12). In Ref. [8] the general jth term of $O(V^j)$ in the expansion is given, to read

$$\rho(\vec{r})-\rho_o=\sum_{j=1}^{\infty}\rho_j(\vec{r}) \tag{15}$$

where $\rho_j(\vec{r})$ is known explicitly in terms of the spherical Bessel function j_1, and the potential V, the quantity $\rho_j(\vec{r})$ being of order V^j. As already mentioned for slowly varying $V(\vec{r})$, the series (15) can be summed to infinity [8] and leads back to the Thomas-Fermi density-potential relation in Eq. (4). Equation (5) therefore constitutes an exact Thomas-Fermi method in perturbation theory, all semiclassical assumptions having been removed from the foundations of the original method. Stoddart and March [12] subsequently showed that the inverse relation (14) for potential in terms of density could be inverted term by term, only the first term being displayed there.

Since the analogue of Eq. (15) is known off the diagonal, the term of $O(V)$ in $\gamma(\vec{r},\vec{r}')$ being given in Eq. (12), the potential $V(\vec{r})$ can be replaced by the density (compare the first order Eq. (13)) and hence $\gamma(\vec{r},\vec{r}')$ can be got, term by term, as a functional of the diagonal displaced charge $\Delta\rho$. This result, that the Dirac density matrix is a unique functional of its diagonal density was exploited in a non-perturbative, but now variational framework in the early work of March and Young [13], see also [14]. This prompts us next to return to the variational basis of the original Thomas-Fermi method and to utilize that to introduce exchange into the theory, following Dirac [3].

3. VARIATIONAL BASIS OF ELECTRON DENSITY THEORY

It is a pillar of the Thomas-Fermi method that Eq. (4) can be derived from a variational principle in which the ground-state energy E is minimized with respect to the density $\rho(\vec{r})$, subject to the normalization constraint that

$$\int\rho(\vec{r})d\vec{r}=N \tag{16}$$

where N is the total number of electrons. This variational principle is explicitly

$$\delta[E-N\mu]=0 \tag{17}$$

where the quantity μ in Eq. (4) is evidently playing the role of the Lagrange multiplier taking care of the normalization condition (16).

In fact, using Eqs (4) and (17), one can construct the total energy in the Thomas-Fermi (TF) method as

$$E_{TF}=c_k\int\rho^{5/3}(\vec{r})d\vec{r}+\int\rho(\vec{r})V_n(\vec{r})d\vec{r}+\tfrac{1}{2}e^2\iint\frac{\rho(\vec{r})\rho(\vec{r}\,')}{|\vec{r}-\vec{r}\,'|}d\vec{r}d\vec{r}\,'+U_{nn}. \tag{18}$$

One can readily verify Eq. (18) using Eq. (17), Eq. (4) being regained. In Thomas-Fermi theory, from Eq. (17) one can write $\mu = \frac{\partial E}{\partial N}$ taken at constant external potential, completing the identification of μ with the chemical potential of the electronic charge cloud in the molecule. Evidently the terms on the right-hand-side of Eq. (18) are, successively, kinetic energy, electron-nuclear potential energy, the classical electrostatic energy of the charge cloud $\rho(\vec{r})$ and, in a molecule, the nuclear-nuclear potential energy U_{nn}.

Later, when we use the device of scaling the electron-electron interactions from e^2/r_{ij} to $\lambda e^2/r_{ij}$, we shall return to the explicit form the ground-state energy E_{TF} takes for positive atomic ions. However, let us turn next to introduce exchange into the TF method: to reach the Thomas-Fermi-Dirac (TFD) theory.

3.1 Introduction of exchange

It is easy to show from the Dirac density matrix, Eq. (10), for free electrons, using the analogue of the THF $(\nabla\Psi)^2$ form for the kinetic energy associated with wave function Ψ that the kinetic energy per unit volume, t_o say, of a uniform free electron gas is simply $t_o = c_k \rho_o^{5/3}$ and the kinetic energy term in Eq (17) obviously follows by again using this locally to get $t(\vec{r}) = c_k\{\rho(\vec{r})\}^{5/3}$ and hence the total kinetic energy $T = \int t(\vec{r})d\vec{r}$. $t(\vec{r})$ can also be obtained, transcending this result, from the March-Murray perturbation theory, the result being [12]

$$t - t_o = -V(\vec{r})\sum_{j=1}^{\infty}\left(\frac{j}{j+1}\right)\rho_j(\vec{r}) \tag{19}$$

where $\rho_j(\vec{r})$ is as in Eq. (15).

However, our focus below is to obtain the analogue of the Thomas-Fermi kinetic energy density $c_k\{\rho(\vec{r})\}^{5/3}$ for the exchange energy density $\varepsilon_x(\vec{r})$, related to the total exchange energy E_x by

$$E_x = \int \varepsilon_x(\vec{r})d\vec{r} \tag{20}$$

3.2 Exchange energy in terms of Dirac density matrix

Dirac [3] in his pioneering paper already wrote E_x in terms of his density matrix as

$$E_x = -\tfrac{1}{4}e^2\int \frac{|\gamma(\vec{r},\vec{r}')|^2}{|\vec{r}-\vec{r}'|}d\vec{r}d\vec{r}', \tag{21}$$

which is a basic formula for the present chapter.

To be brief, in setting the free electron form (11) of the Dirac matrix into Eq. (21) one finds

$$\varepsilon_x(\vec{r}) = -c_x\{\rho(\vec{r})\}^{1/3};\ c_x = \tfrac{3}{4}e^2\left(\tfrac{3}{\pi}\right)^{\frac{1}{3}}. \tag{22}$$

3.3. Chemical potential equation of Thomas-Fermi-Dirac theory: Dirac-Slater exchange potential

Inserting Eq. (22) into Eq. (20), adding the resulting exchange energy E_x to the Thomas-Fermi energy (18) and using again the variational principle (17) leads to the chemical potential μ in the Thomas-Fermi-Dirac theory as

$$\mu = \tfrac{5}{3}c_k\rho(\vec{r})^{2/3} + V_n(\vec{r}) + V_e(\vec{r}) - \tfrac{4}{3}c_x\rho(\vec{r})^{1/3}. \tag{23}$$

While Dirac [3] solved Eq. (23) as a quadratic equation for $\rho(\vec{r})^{1/3}$, Slater [11] made the very fruitful observation that Eq. (23) could be viewed as modifying the one-body potential $V_n(\vec{r}) + V_e(\vec{r})$ by the (non-classical) exchange potential $-\tfrac{4}{3}c_x\rho(\vec{r})^{1/3}$. This so called Dirac-Slater $\rho(\vec{r})^{1/3}$ exchange potential played an important role in reaching the present successful computational *state of the art* of modern density functional theory. Gombas [15] and others proposed the way to add correlation locally into the Thomas-Fermi-Dirac theory. The modern problem which remains to be solved is to construct the exchange-correlation potential $V_{xc}(\vec{r})$ without *local density* approximations. This problem will be tackled at some length below. But first let us consider the consequences of scaling the electron-electron interaction from e^2/r_{ij} to $\lambda e^2/r_{ij}$ on the Thomas-Fermi ground-state energy $E(Z,N)$ of positive atomic ions with large Z and $N \le Z$. The argument below follows the very recent work of Levy et al [16].

4. SCALING OF STRENGTH OF ELECTRON-ELECTRON INTERACTIONS

We follow the treatment of Levy et al. [16]. Consider the Hamiltonian H_λ^Z defined by

$$H_\lambda^Z = -\frac{\hbar^2}{2m}\sum_i^N \nabla_i^2 - Ze^2\sum_i^N \frac{1}{r_i} + \lambda e^2 \sum_{i<j}\frac{1}{r_{ij}}. \tag{24}$$

We shall consider the case $N \le Z$ and both N and Z will be taken to be large (see March [17]). Let us divide Eq. (24) by λ to obtain $H_\lambda^{Z\prime} = H_\lambda^Z/\lambda$, with an effective mass $m^* = m\lambda$ and effective nuclear charge $Z^*e = Ze/\lambda$.

4.1. Thomas-Fermi energy of positive ions

Let us apply the Thomas-Fermi approximation to this Hamiltonian $H_\lambda^{Z'}$ and calculate the ground-state energy $E'_\lambda(Z,N)$. Following the work of Grout et al. [18], this energy takes the form [16] (Levy et al. [16])

$$E'_\lambda(Z,N) = Z^{*7/3} f\left(N/Z^*\right)\left(e^2/a_o^*\right) \tag{25}$$

where $a_o^* = \hbar^2/m^*e^2$. The actual energy $E_\lambda(Z,N)$ is $E'_\lambda(Z,N)/\lambda$ and therefore

$$E_\lambda(Z,N) = \lambda^2 Z^{*7/3} f\left(N/Z^*\right)\left(e^2/a_o\right) \tag{26}$$

Some further rewriting yields the desired result (Levy et al. [16])

$$E_\lambda(Z,N) = \lambda^{-1/3} Z^{7/3} f\left(\lambda N/Z\right)\left(e^2/a_o\right) \tag{27}$$

$f\left(N/Z\right)$ can be obtained from the work of Grout et al. [18] for the range $0 < N/Z \leq 1$ and so $f\left(\lambda N/Z\right)$ is also known for $\lambda \leq 1$. It is given analytically near the neutral atom limit $N/Z = 1$ and also near $N/Z = 0$, and in numerical form in between. In particular one has $E_\lambda(Z,N) = F(q^*)N^{1/3}Z^2$ where the *degree of ionization* $q^* = 1 - \left(N/Z^*\right)$. From Table 1 of Grout et al [18], $F(0) = -0.768$, $F(0.5) = -0.929$ and $F(1) = -1.145$.

Following the approach of March and White [19] and the subsequent rigorous work of Tal and Levy [20], it is to be emphasized that contact can be made between such a formula and the important $1/Z$ expansion of atomic theory (Hylleraas [21]; Layzer [22]). As noted by Levy et al [16], it would be of interest to use an electronic structure ab initio program to examine how far the above Thomas-Fermi scaling carries through to the full many-electron case for fixed, and usually small, N and variable atomic number Z.

We next turn to the discussion of the electron-electron scaling, but with the external (i.e., nuclear) potential adjusted for each scale factor λ to preserve the same ground state density $\rho(\vec{r})$.

4.2 Electron-electron scaling with constant electron density

For this case we follow Levy et al [16] by introducing the Hamiltonian $\hat{H}_\lambda$ defined by

$$\hat{H}_\lambda = \hat{T} + \lambda\hat{V}_{ee} + \sum_i^N V_\lambda([\rho];\vec{r}_i) \tag{28}$$

As mentioned earlier, the ground-state density ρ is now fixed independent of λ and in atomic units

$$T = -\sum_i^N \tfrac{1}{2}\nabla_i^2; \tag{29a}$$

$$V_{ee} = \sum_{i<j}\frac{1}{r_{ij}}. \tag{29b}$$

4.2.1. Thomas-Fermi-Dirac ground-state energy

We now introduce the Thomas-Fermi-Dirac (TFD) ground-state (GS) energy corresponding to different λ and a definite density $\rho_o(\vec{r})$ say, as the minimum obtained from (Levy et al [16])

$$E^{TFD}_{GS,\lambda,\rho_o} = \min_\rho\left[\int V_\lambda([\rho_o];\vec{r})\rho(\vec{r})d\vec{r} + F^{TFD}_\lambda[\rho]\right]$$

$$= \int V_\lambda([\rho_o];\vec{r})\rho_o(\vec{r})d\vec{r} + F^{TFD}_\lambda[\rho_o] \tag{30}$$

Here $F^{TFD}_\lambda[\rho]$ takes the form

$$F^{TFD}_\lambda[\rho] = c_k\int \rho^{5/3}(\vec{r})d\vec{r} + \lambda J(\rho) - \lambda c_x\int \rho^{4/3}(\vec{r})d\vec{r}. \tag{31}$$

In Eq. (31), $J[\rho]$ is the coulomb energy

$$J[\rho] = \frac{1}{2}\int\frac{\rho(\vec{r}_1)\rho(\vec{r}_2)}{r_{12}}d\vec{r}_1 d\vec{r}_2. \tag{32}$$

Following Levy et al [16] the stationary principle yields

$$V_\lambda([\rho_o];\vec{r}) = -\frac{\delta F^{TFD}_\lambda[\rho]}{\delta\rho(\vec{r})} \tag{33}$$

evaluated at $\rho = \rho_o$. Then

$$V_\lambda([\rho_o];\vec{r}) = -\frac{5}{3}c_k\{\rho_o(\vec{r})\}^{2/3} - \lambda\int\frac{\rho_o(\vec{r}')}{|\vec{r}-\vec{r}'|}d\vec{r}' + \frac{4}{3}\lambda c_x\{\rho_o(\vec{r})\}^{1/3}. \tag{34}$$

One immediately notes that in the TFD limit the scaling of the one-body potential V_λ is such that V_λ is linear in the parameter λ. One can write the ground-state energy explicitly in the form

$$E_{GS,\lambda,\rho_o} = -\frac{2}{3}c_k\int \rho_o^{5/3}(\vec{r})d\vec{r} - \frac{\lambda}{2}\int \frac{\rho_o(\vec{r}_1)\rho_o(\vec{r}_2)}{r_{12}}d\vec{r}_1 d\vec{r}_2 + \frac{1}{3}\lambda c_x\int \rho_o^{4/3}(\vec{r})d\vec{r}. \tag{35}$$

As with V_λ, the ground-state energy $E^{TFD}_{GS,\lambda,\rho_o}$, corresponding to electron density $\rho_o(\vec{r})$, at a coupling constant λ, is linear in the scaling parameter λ.

4.2.2. Limiting results for full electron-electron interactions ($\lambda = 1$) and for electron-electron interactions "switched off" ($\lambda = 0$)

As Levy et al [16] point out, it is readily possible to construct two alternative forms of this result (35). The first of these relates the energy to its value $E^{TFD}_{GS,\lambda=0,\rho_o}$, when the electron-electron interactions are "switched off". One then finds

$$E^{TFD}_{GS,\lambda,\rho_o} = E^{TFD}_{GS,\lambda=0,\rho_o} - \frac{\lambda}{2}\int \frac{\rho_o(\vec{r}_1)\rho_o(\vec{r}_2)}{r_{12}}d\vec{r}_1 d\vec{r}_2 + \frac{1}{3}\lambda c_x\int \{\rho(\vec{r})\}^{4/3} d\vec{r}. \tag{36}$$

The second alternative "corrects" the full coupling strength energy corresponding to $\lambda = 1$:

$$E^{TFD}_{GS,\lambda,\rho_o} = E^{TFD}_{GS,\lambda=1,\rho_o} + \frac{(1-\lambda)}{2}\int \frac{\rho_o(\vec{r}_1)\rho_o(\vec{r}_2)}{r_{12}}d\vec{r}_1 d\vec{r}_2 + \frac{1}{3}(\lambda - 1)c_x\int \{\rho(\vec{r})\}^{4/3} d\vec{r}. \tag{37}$$

Thus, a rather complete analysis of the Thomas-Fermi-Dirac method proves possible [16].

5. ADIABATIC CONNECTION FORMULA

In modern density functional theory, the idea of an adiabatic connection to determine the exchange-correlation functional is of continuing importance. Early studies include those of Harris and Jones [23], Langreth and Perdew [24], Gunnarsson and Lundqvist [25], Levy [26], and Harris [27]. Following refs [24] and [25], and the book by Parr and Yang [28], the exchange-correlation functional $E_{xc}[\rho]$ may be expressed as (see also Levy et al [16]):

$$E_{xc}[\rho] = \int_0^1 d\lambda U^\lambda_{xc}[\rho] \tag{38}$$

where

$$U_{xc}^{\lambda}[\rho] = \langle \Psi_{\rho}^{\lambda} | \hat{V}_{ee} | \Psi_{\rho}^{\lambda} \rangle - J[\rho], \tag{39}$$

$J[\rho]$ being as usual the Coulomb energy. Here $\hat{V}_{ee}$ is the electron-electron repulsion operator while Ψ_{ρ}^{λ} is the GS wave function of a Hamiltonian $\hat{H}_{\lambda}$ which represents a system in which the electron-electron interaction is scaled as in section 4:

$$\hat{H}_{\lambda} = \hat{T} + \lambda \hat{V}_{ee} + \sum_{i} V_{\lambda}(\vec{r}_i). \tag{40}$$

It also minimizes for fixed ρ the constrained search expression Levy [29,30]: Levy et al [16]

$$\langle \Psi_{\rho}^{\lambda} | \hat{T} + \lambda \hat{V}_{ee} | \Psi_{\rho}^{\lambda} \rangle.$$

In this formula, $\rho(\vec{r})$ is to be understood to be the exact ground state density of $\hat{H}_{\lambda}$.

Becke [31,32] has utilized the above formalism in the context of a two point integration formula (see also Levy et al [16]). He approximates Eq. (38) as

$$E_{xc}[\rho] \approx \frac{1}{2}\left[U_{xc}^{0} + U_{xc}^{1}\right] \tag{41}$$

where U_{xc}^{0} is the exchange energy of the Slater-Kohn-Sham (SKS) determinant, while U_{xc}^{1} is the potential energy contribution to the exchange-correlation energy of the fully interacting system.

5.1. Two-point connection formula re-examined

Levy et al [16] noted further that the fully interacting potential energy U_{xc}^{1} has the form (Levy and Perdew [33]; Levy, [34]

$$U_{xc}^{1}[\rho] = 2E_{xc}[\rho] - \left.\frac{\partial E_{xc}[\rho_{\lambda}]}{\partial \lambda}\right|_{\lambda=1} \tag{42}$$

where

$$\rho_{\lambda}(\vec{r}) = \lambda^{3}\rho(\lambda\vec{r}) \tag{43}$$

Levy et al [16] generalized Eq. (42) to read

$$U_{xc}^{\lambda} = 2\lambda E_{xc}\left[\rho_{1/\lambda}\right] + \lambda^2 \frac{\partial E_{xc}\left[\rho_{1/\lambda}\right]}{\partial\lambda}. \tag{44}$$

Now in Eq. (41), one could employ the exact expression for U_{xc}^{0} and approximate U_{xc}^{1} by a model (mod) via Eq (42). The resulting approximation to E_{xc} will be written $\tilde{E}_{xc}$ and is given by

$$\tilde{E}_{xc} = \frac{1}{2}U_{xc}^{0}[\rho] + \frac{1}{2}U_{xc}^{mod.1}[\rho], \tag{45}$$

where

$$U_{xc}^{mod,1}[\rho] = 2E_{xc}^{mod}[\rho] - \left.\frac{\partial E_{xc}^{mod}\left[\rho_{\lambda}\right]}{\partial\lambda}\right|_{\lambda=1}. \tag{46}$$

Combining Eqs (45) and (46) gives [16]

$$\tilde{E}_{xc}[\rho] = \frac{1}{2}E_{x}[\rho] + \frac{1}{2}E_{x}^{mod}[\rho] + \frac{1}{2}E_{c}^{mod}[\rho] - \frac{1}{2}\left.\frac{\partial E_{c}^{mod}\left[\rho_{\lambda}\right]}{\partial\lambda}\right|_{\lambda=1} \tag{47}$$

where the requirement $E_{x}^{mod}\left[\rho_{\lambda}\right] = \lambda E_{x}^{mod}[\rho]$ has been imposed, which is the correct scaling requirement for exchange.

5.2. Relation between correlation energy $E_c[\rho]$ and density-functional definition of kinetic contribution $T_c[\rho]$

It is useful to summarize here the density-function definition of the kinetic contribution $T_c[\rho]$ to the correlation energy $E_c[\rho]$. Here

$$E_c[\rho] = \left\langle \Psi_{\rho}^{1}\middle|\hat{T} + \hat{V}_{ee}\middle|\Psi_{\rho}^{1}\right\rangle - \left\langle \Psi_{\rho}^{0}\middle|\hat{T} + \hat{V}_{ee}\middle|\Psi_{\rho}^{0}\right\rangle \tag{48}$$

and

$$T_c[\rho] = \left\langle \Psi_{\rho}^{1}\middle|\hat{T}\middle|\Psi_{\rho}^{1}\right\rangle - \left\langle \Psi_{\rho}^{0}\middle|\hat{T}\middle|\Psi_{\rho}^{0}\right\rangle. \tag{49}$$

It is expected [16], with these definitions, that $E_c[\rho] \approx -T_c[\rho]$. However, there is an exact relation connecting $E_c[\rho]$ and $T_c[\rho]$, namely [33]

$$E_c[\rho] = -T_c[\rho] + \left.\frac{\partial E_c[\rho_\lambda]}{\partial \lambda}\right|_{\lambda=1} . \tag{50}$$

Hence it follows that a necessary and sufficient condition for the two-point formula to be exact, when the exact U^λ_{xc} is used in Eq. (41), is

$$\left.\frac{\partial E_c[\rho_\lambda]}{\partial \lambda}\right|_{\lambda=1} = 0. \tag{51}$$

Levy et al [16] note, in support of the two-point formula, that the magnitude of $\left.\frac{\partial E_c[\rho_\lambda]}{\partial \lambda}\right|_{\lambda=1}$ has been found to be typically only about 10% of the magnitude of $E_c[\rho]$ in atoms.

Having discussed $E[\rho]$ at some length, we return to treat the generalization of the Thomas-Fermi force equation (5).

6. DIFFERENTIAL VIRIAL THEOREM FOR NON-INTERACTING ELECTRONS MOVING IN ONE-DIMENSION IN POTENTIAL ENERGY $V(x)$

March and Young [35] obtained a differential form of the virial theorem for non-interacting electrons moving in one-dimension in a potential energy $V(x)$ by starting from the equation of motion for the Dirac density matrix $\gamma(x,x')$. This is readily verified from the Schrödinger equation to be

$$\frac{\partial^2\gamma}{\partial x^2} - \frac{\partial^2\gamma}{\partial x'^2} = \frac{2m}{\hbar^2}\left[V(x) - V(x')\right]\gamma(x,x'). \tag{52}$$

Using sum and difference variables defined by

$$\xi = \frac{x+x'}{2}; \tag{53a}$$

and

$$\eta = \frac{x-x'}{2} \tag{53b}$$

leads to the form of Eq (52) as

$$\frac{\partial^2 \gamma(\xi,\eta)}{\partial\xi\partial\eta} = \frac{2m}{\hbar^2}\left[V(\xi+\eta) - V(\xi-\eta)\right]\gamma(\xi,\eta).$$

This form now prompts one to expand around the diagonal $\eta = 0$, when one obtains, with $t(x)$ as usual the kinetic energy "density"

$$\frac{\partial t(x)}{\partial x} = -\frac{1}{2}\rho(x)\frac{\partial V(x)}{\partial x} + \frac{1}{8}\frac{\partial^3 \rho(x)}{\partial x^3}. \tag{54}$$

As March and Young [35] emphasized, this is a differential form of the virial theorem. This is readily verified by multiplying both sides of Eq. (54) by x and integrating from $-\infty$ to $+\infty$. Then one recovers the usual (integral) form of the virial theorem, relating the total kinetic energy $T = \int_{-\infty}^{+\infty} t(x)dx$ to the virial of the forces.

While Eq. (54) is already an interesting result, as it can yield, by quadrature, the kinetic energy density $t(x)$) solely from the potential energy V(x) and the ground state electron density $\rho(x)$), the very recent study of Holas and March [36] motivates rewriting it as a force equation.

6.1. Exact Force Equation in One-dimension for Non-interacting Fermions

One merely must extract $-\dfrac{\partial V(x)}{\partial x}$ from Eq. (54) to find the force equation

$$F = -\frac{\partial V(x)}{\partial x} = \frac{2}{\rho(x)}\frac{\partial t(x)}{\partial x} - \frac{1}{4\rho(x)}\frac{\partial^3 \rho(x)}{\partial x^3}$$

$$\equiv F_K + F_L \tag{55}$$

Here, we have explicitly written the "external" force $-\dfrac{\partial V(x)}{\partial x}$ as a sum of a kinetic contribution F_K and a Laplacian contribution F_L, the latter notation anticipating the fact that, in three dimensions, the third derivative of the density ρ will generalize to $\dfrac{\partial}{\partial \bar{r}}(\nabla^2\rho)$, i.e. the gradient of the Laplacian of the density. Obviously, in three dimensions therefore, Eq. (55) becomes of vectorial character. Also, and equally important, for interacting electrons the two terms displayed explicitly on the RHS of Eq. (55) must be suplemented by a third term coming from the electron-electron interactions. This will have both a classical part already displayed in the Thomas-Fermi method in Eq. (5) and a non-classical part arising from the exchange and correlation hole that an electron moving through the molecule carries around with itself: already implicit in the discussion of $E_{xc}[\rho]$ in the previous section. But the above argument focusses already on the fact that one must give careful attention to the kinetic term in the force balance equation. To know this for non-interacting particles clearly requires knowledge of the off-diagonal idempotent Dirac matrix, or for interacting Fermions the fully correlated matrix satisfying $\gamma^2 < \gamma$. This inequality esentially imposes Pauli conditions on the correlated one-particle density matrix $\gamma(\vec{r},\vec{r}')$: for a definition from the many electron-wave function, see Appendix 1.

7. THREE-DIMENSIONAL DIFFERENTIAL VIRIAL THEOREM AND FORCE EQUATION INCLUDING FULL ELECTRON-ELECTRON INTERACTIONS

Holas and March [36] have generalized the argument of section 6 to three dimensions and have included fully electron-electron interactions. The differential virial theorem now requires not just the kinetic energy density $t(\vec{r})$ but the kinetic energy density tensor $t_{\alpha\beta}(\vec{r})$, which is directly related, in turn, to the fully correlated first-order density matrix. Then the external force $\vec{F} = -\frac{\partial V_n(\vec{r})}{\partial \vec{r}}$ generalizes from Eq. (55) to read

$$\vec{F} = \vec{F}_K + \vec{F}_L + \vec{F}_{ee}. \tag{56}$$

As in Eq. (55), $\vec{F}_L$ is known explicitly in terms of the gradient of the Laplacian of the density $\rho(\vec{r})$. But, as mentioned, $\vec{F}_K$ requires knowledge of the first order correlated density matrix (near its diagonal only) while $\vec{F}_{ee}$ is expressed in terms of the Coulomb interaction $\frac{e^2}{r_{ij}}$ and the pair function, which is the diagonal element of the second-order density matrix (see again Appendix 1).

Thus, to obtain the external force $\vec{F}$ in density functional form requires knowledge of $\gamma(\vec{r},\vec{r}')$ as a functional of its diagonal density. This problem was discussed earlier for the Dirac density matrix via plane wave perturbation theory. Furthermore, one needs to obtain $\vec{F}_{ee}$ in terms of $\rho(\vec{r})$ the pair function as a functional of ρ. These are, of course, difficult many-body problems for the future, though some progress will be referred to below.

7.1. External Potential as Line Integral

Holas and March [36] emphasize from Eq. (56) that the external potential $V_n(\vec{r})$ is a line integral on $\vec{F}$. Though Eq. (56) can not be handled exactly, for reasons discussed immediately above, if first- and second-order density matrices are approximated to the Slater-Kohn-Sham reference determinant, then one recovers the exchange-only potential as obtained from the work formalism of Harbola and Sahni [37]. Their formula is known to exhibit path dependence, but Levy and March [38] have utilized Eq. (56) of Holas and March [36] to demonstrate that there is a kinetic correction to be added to the exchange-only potential of Harbola and Sahni. This restores the path independence required of the exchange-only potential and Levy and March give arguments why the Harbola-Sahni term will generally dominate the kinetic correction. We shall discuss, having dealt with the exchange-only case, the correlation energy in terms of first-order density matrices later in the chapter. But first we will consider, in the spirit of the original electron density theory of Thomas-Fermi-Dirac, how the density $\rho(\vec{r})$ might be calculated directly, without invoking the semi-classical approximation of that theory.

8. DENSITY AMPLITUDE FOR INHOMOGENEOUS ELECTRON GAS

A long term aim of density functional theory has been to calculate the ground state electron density $\rho(\vec{r})$ of a many-electron molecule without recourse to wave functions. At the time of writing, the most promising route is to use the density amplitude $\{\rho(\vec{r})\}^{\frac{1}{2}}$, which satisfies a Schrödinger equation [39-42]:

$$\left\{-\frac{\hbar^2}{2m}\nabla^2 + v_{eff}(\vec{r})\right\}\rho^{\frac{1}{2}}(\vec{r}) = \varepsilon\rho^{\frac{1}{2}}(\vec{r}). \tag{57}$$

One must, as usual, impose normalization in an N-electron molecule:

$$\int d\vec{r}\,\rho(\vec{r}) = N$$

and also will need to choose that solution of Eq. (57) with the lowest eigenenergy. Of course, to exploit this attractive road to calculate the density amplitude, one has to have knowledge of the effective potential $v_{eff}(\vec{r})$ as a functional of the density. This potential is usually defined as the sum of a Pauli potential v_P and the one body potential of density functional theory- the Slater-Kohn-Sham potential v_{SKS}. The recent work of Holas and March [36] on the exchange-correlation contribution to $v_{SKS}(\vec{r})$ derives v_{xc} in terms of low-order density matrices. It remains, of course, a major problem for many-body theory to obtain the density matrices as functionals of $\rho(\vec{r})$. Although some properties of the Pauli potential v_P have been established, and a rather general procedure for its determination has been put forward [39] (the price to be paid being, however, the solution of N-1 differential equations) this quantity v_P is still far from being available for practical calculation.

Therefore, Holas and March [40] have demonstrated that $v_{eff}(\vec{r})$ can be characterized solely by the Hartree-Fock (HF) first-order density matrix (written as 1-DM below).

8.1. Equation of motion of Hartree-Fock density matrix

Following Holas and March [40] we shall below define the Hartree-Fock (HF) 1-DM via the HF equations and hence determine its equation of motion. The HF equations take the form

$$\{\hat{t}_1 + v_n(\vec{r}_1) + v_{es}(\vec{r}_1) + \hat{v}_x\}\Psi_j(\vec{x}_1) = \varepsilon_j \Psi_j(\vec{x}_1): \tag{58}$$

for $j = 1, 2, \ldots, N$. Here $\vec{x}_1$ is used for space plus spin coordinates $(\vec{r}_1, s_1)$, $\hat{t}_1 = -\frac{\hbar^2}{2m}\nabla_1^2$, $v_n(\vec{r})$ is the external potential, while v_{es} denotes the classical electrostatic potential

$$v_{es}(\vec{r}_1) = \int d\vec{r}_2\, u(\vec{r}_1 - \vec{r}_2)\rho_{HF}(\vec{r}_2) \tag{59}$$

with $u(\vec{r}) = \frac{e^2}{|\vec{r}|}$. The unique feature of the HF equations is the presence of $\hat{v}_x$, the HF exchange integral operator (a non-local potential).

The key tool used by Holas and March [40] is then

$$\gamma_{HF}(\vec{x}_1; \vec{x}_1') = \sum_{j=1}^{N} \Psi_j(\vec{x}_1)\Psi_j^*(\vec{x}_1') \tag{60}$$

and this leads to the HF electron density on the diagonal as

$$\rho_{HF}(\vec{r}) = \sum_s \gamma_{HF}(\vec{r}_s'; \vec{r}_s). \tag{61}$$

The N equations (58) have of course to be solved self-consistently, since both v_{es} and $\hat{v}_x$ depend on N normalized eigensolutions $\Psi_j(x)$.

Holas and March [40] then construct the equation of motion of the HF 1-DM. As a first step, one multiplies Eq. (58) by $\Psi_j^*(x_1')$ and utilizes Eqs. (59-62). Then one finds

$$\{\hat{t}_1 + v_n(\vec{r}_1) - \varepsilon_j\}\Psi_j(\vec{x}_1)\Psi_j^*(\vec{x}_1')$$

$$+\int dx_2\, u(\vec{r}_1 - \vec{r}_2)\left[\gamma_{HF}(\vec{x}_2; \vec{x}_2)\Psi_j(\vec{x}_1)\Psi_j^*(\vec{x}_1') - \gamma_{HF}(\vec{x}_1; \vec{x}_2)\Psi_j(\vec{x}_2)\Psi_j^*(\vec{x}_1')\right] = 0. \tag{62}$$

The next step (Holas and March [40]) requires taking the complex conjugate of the above equation and following that a relabelling of the variables according to $\vec{x}_1 \leftrightarrow \vec{x}_1'$, to find

$$\{\hat{t}_1 + v_n(\vec{r}_1) - \varepsilon_j\}\Psi_j(\vec{x}_1)\Psi_j^*(\vec{x}_1')$$

$$+\int dx_2\, u(\vec{r}_1' - \vec{r}_2)\left[\gamma_{HF}(\vec{x}_2;\vec{x}_2')\Psi_j(\vec{x}_1)\Psi_j^*(\vec{x}_1') - \gamma_{HF}(\vec{x}_2;\vec{x}_1')\Psi_j(\vec{x}_1)\Psi_j^*(\vec{x}_2)\right] = 0\,. \quad (63)$$

Finally, to get the equation of motion for γ_{HF}, one needs to substract Eq. (63) from Eq. (62) and then to sum over j up to N. The desired result is then

$$\{[\hat{t}_1 + v_n(\vec{r}_1)] - [\hat{t}_{1'} + v_n(\vec{r}_1')]\}\gamma_{HF}(\vec{x}_1;\vec{x}_1')$$

$$+2\int d\vec{x}_2\{u(\vec{r}_1 - \vec{r}_2) - u(\vec{r}_1' - \vec{r}_2)\}\gamma_{HF}^{(2)}(\vec{x}_1,\vec{x}_2;\vec{x}_1',\vec{x}_2) = 0 \quad (64)$$

In this result (64), $\gamma_{HF}^{(2)}$, the combination of the HF 1-DM known as the HF second-order density matrix (2-DM), has been introduced (see also Appendix 1)

$$\gamma_{HF}^{(2)}(\vec{x}_1,\vec{x}_2;\vec{x}_1',\vec{x}_2') = \frac{1}{2}\{\gamma_{HF}(\vec{x}_1;\vec{x}_1')\gamma_{HF}(\vec{x}_2;\vec{x}_2') - \gamma_{HF}(\vec{x}_1;\vec{x}_2')\gamma_{HF}(\vec{x}_2;\vec{x}_1')\} \quad (65)$$

Holas and March [40] stress that the equation of motion (65) for γ_{HF} is of exactly the same form as the equation for the exact interacting 1-DM obtained earlier by Dawson and March [41]. Of course, the exact $\gamma^{(2)}$ in their equation replaces the HF 2-DM ocurring in Eq. (64). Holas and March then stress that Eq. (64), when combined with Eq. (65) yields a relation between the external potential $v(\vec{r})$ and the HF 1-DM, namely

$$v_n(\vec{r}) = v_n(\vec{r}_1') - \frac{(\hat{t}_1 - \hat{t}_{1'})\gamma_{HF}(\vec{x}_1;\vec{x}_1') + 2\int dx_2\left[u(\vec{r}_1 - \vec{r}_2) - u(\vec{r}_1' - \vec{r}_2)\right]\gamma_{HF}^{(2)}(\vec{x}_1,\vec{x}_2;\vec{x}_1',\vec{x}_2)}{\gamma_{HF}(\vec{x}_1;\vec{x}_1')}. \quad (66)$$

The form of this equation indicates that the potential is determined to within an additive constant only. One may arbitrarily choose a reference point r_1' (e.g., at infinity), the value of the potential there (e.g., zero) and also a pair of spin indices s_1, s_1'. Then Eq. (66) permits the reconstruction of the external potential $v_n(\vec{r})$ from knowledge of the Hartree-Fock idempotent density matrix (1-DM).

8.2. Effective Potential in Density Amplitude Equation

In order to employ the above result to obtain v_{eff} in Eq. (57), Holas and March [40] reconstruct the formal derivation of this equation as follows. First the Euler equation

$$\mu = \frac{\delta F[\rho]}{\delta\rho(\vec{r})} + v_n(\vec{r}), \quad (67)$$

with μ as usual the chemical potential, follows from the variational principle (17). In Eq. (67), $F[\rho]$ is the universal (but unknown) functional formally proved to exist by Hohenberg and Kohn [4]. Eq. (67) shows, on one hand, that the external potential can be reconstructed, in principle, to within an additive constant, from a given ground-state density:

$$v_n(\vec{r}) = \mu - \frac{\delta F[\rho]}{\delta \rho(\vec{r})}. \tag{68}$$

On the other hand, should $F[\rho]$ eventually become established with sufficient accuracy, the ground state energy ρ of the inhomogeneous electron assembly in a molecule, characterized by external (nuclear) potential $v_n(\vec{r})$, could be found by solution of Eq. (67).

Holas and March [40] now transform Eq. (67) somewhat by adding and subtracting the functional derivative of the von Weizsacker kinetic energy (see March, [17]). This is defined as

$$T_W[\rho] = \int d\vec{r}\, t_W(\vec{r}) \tag{69}$$

where

$$t_W = \frac{\hbar^2}{2m}\left|\nabla \rho^{\frac{1}{2}}(\vec{r})\right|^2 = \frac{\hbar^2}{8m}\frac{|\nabla \rho(\vec{r})|^2}{\rho(\vec{r})}. \tag{70}$$

The functional derivative is readily obtained as

$$\frac{\delta T_W[\rho]}{\delta \rho(\vec{r})} = -\frac{\hbar^2}{2m}\frac{\nabla^2 \rho^{\frac{1}{2}}(\vec{r})}{\rho^{\frac{1}{2}}(\vec{r})} \tag{71}$$

and hence Eq. (67) can be recast into the form

$$-\frac{\hbar^2}{2m}\frac{\nabla^2 \rho^{\frac{1}{2}}(\vec{r})}{\rho^{\frac{1}{2}}(\vec{r})} + v_{eff}(\vec{r}) = \mu \tag{72}$$

which is identical to the Schrödinger Eq. (57) for the electron density amplitude. The effective potential is evidently

$$v_{eff}(\vec{r}) = \frac{\delta F[\rho]}{\delta \rho(\vec{r})} - \frac{\delta T_W}{\delta \rho(\vec{r})} \tag{73}$$

As Holas and March set out, the following chain of relations then afford the connection of the effective potential v_{eff} with the HF density matrix:

(i) From Eq. (68), it follows, as already mentioned above, that $v_n(\vec{r})$ is a functional of $\rho(\vec{r})$.

(ii) By solving the HF equations, the eigenfunctions and especially the combination $\gamma_{HF}(\vec{x}_1; x_1')$ in Eq. (60) are seen to be functionals of the external potential $v_n(\vec{r})$.

(iii) Therefore the HF first-order density matrix is a functional of the exact correlated ground-state density.

(iv) The effective potential entering the Schrödinger equation for the density amplitude can be expressed in terms of the HF first-order density matrix plus the known functional derivative of the von Weizsacker kinetic energy.

What remains is to develop approximate forms of the exact density in terms of the analogue of Moller-Plesset perturbation theory but posed in terms of HF density matrices as the zeroth order starting point. This study is currently being pursued in conjunction with Dr. A. Holas.

Having discussed a formulation in terms of the Hartree-Fock first-order density matrix, which focussed on the external potential, we shall turn to discuss very recent work by Savin [42], and by Levy and Görling [43], on the correlation energy functional in terms of exact, correlated density matrices. It has been known for a long time, from the work of Löwdin [44] and McWeeny [45], that the ground-state energy of a many-electron system is uniquely known in terms of the first-order correlated density matrix and the diagonal element of the second-order density matrix (compare the result of Holas and March [36] for the external potential). The price paid in the work of Savin and of Levy and Görling for the removal of the second-order density matrix of the interacting electronic assembly is that an integration over the scaling parameter λ of the electron-electron interaction is needed. This result will now be developed below.

9. CORRELATION ENERGY IN TERMS OF DENSITY MATRICES

As mentioned immediately above, Savin [42] has expressed the density-functional correlation energy in terms of first-order density matrices, a similar result having been obtained independently by Levy and Görling [43].

Savin writes the correlation energy functional $E_c[\rho]$ as

$$E_c[\rho] = \left\langle \Psi^{\lambda=1} \middle| \hat{T} + \hat{V}_{ee} \middle| \Psi^{\lambda=1} \right\rangle - \left\langle \Psi^{\lambda=0} \middle| \hat{T} + \hat{V}_{ee} \middle| \Psi^{\lambda=0} \right\rangle. \tag{74}$$

In this relation, Ψ^λ is the antisymmetric total electronic wave function yielding the electron density ρ and minimizing $\left(\hat{T} + \lambda \hat{V}_{ee}\right)$, $\hat{T}$ and $\hat{V}_{ee}$ being kinetic and electron-electron interaction operators respectively. Savin follows a proof given by Yasuhara [46] to show that the correlation energy density functional can be expressed solely in terms of first-order density matrices.

9.1. Constrained search procedure

Utilizing:

(i) the constrained search definition of the universal density functional for scaled electron-electron interaction, ie $F^{\lambda}[\rho]=\langle\Psi^{\lambda}|\hat{T}+\lambda\hat{V}_{ee}|\Psi^{\lambda}\rangle$, and

(ii) the Hellmann-Feynman theorem applied to F^{λ}, Savin derives the result

$$E_c[\rho]=-\int_0^1\frac{d\lambda}{\lambda^2}\left(\langle\Psi^{\lambda}|\hat{T}|\Psi^{\lambda}\rangle-\langle\Psi^{\lambda=0}|\hat{T}|\Psi^{\lambda=0}\rangle\right). \tag{75}$$

As the RHS of this equation contains only the one-electron operator $\hat{T}$, it can be expressed solely in terms of the first-order density matrix

$$\gamma_{\rho}^{\lambda}(\vec{r}_1;\vec{r}_1')=\sum_s\int d\vec{r}_2\ldots d\vec{r}_N\Psi^{\lambda*}(\vec{r}_1,\vec{r}_2,\ldots\vec{r}_N)\Psi^{\lambda}(\vec{r}_1',\vec{r}_2,\ldots\vec{r}_N) \tag{76}$$

the result being

$$E_c[\rho]=\frac{1}{2}\int_0^1 d\lambda\frac{1}{\lambda^2}\int d\vec{r}\,\nabla_{\vec{r}'}^2\left[\gamma_{\rho}^{\lambda}(\vec{r};\vec{r}')-\gamma_{\rho}^{\lambda=0}(\vec{r};\vec{r}')\right]\Big|_{\vec{r}'=\vec{r}}. \tag{77}$$

This formula avoids the so-called N-representability problem (see, for example the review by Coleman [47]) on the correlated second-order density matrix. The N-representability conditions for first-order density matrices, in contrast, can be written down (see, for example, Davidson [48]). Explicitly, $\gamma(\vec{r};\vec{r}')$ has the form

$$\gamma(\vec{r};\vec{r}')=\sum_i n_i\phi_i^*(\vec{r})\phi_i(\vec{r}') \tag{78}$$

where $\phi_i(\vec{r})$ are orthonormal natural orbitals and n_i, lying between 0 and 1, represent the Pauli conditions to be satisfied by an acceptable Fermion first-order matrix (see also Gilbert [49]).

9.2. Direct Determination of Second-order Density Matrix

In the above context of N-representability, the study of Nakatsuji and Yasuda [50] appeared as this Chapter was nearing completion. These authors have reported a direct determination of the second-order density matrix without recourse to wave functions. Their procedure is to decouple the third and fourth order density matrices into lower-order matrices, using the Green function as the main calculational tool. They have applied their method to Be, Ne, H_2O, H_3O+, NH_3, CH_4, BH_4^-, NH_4^+ and CH_3F. Their results for these systems compare favorably with full configuration interaction calculations. Nakatsuji and Yasuda [50] write that "the calculated second-order reduced density matrices (RDMs) almost satisfy the necessary conditions of the N representability ... and the first-order RDMs are exactly N representable".

These workers also establish contact with the earlier studies of Valdemoro [51] and co-workers [52-54] in this general area [55].

10. APPROXIMATE CONSTRUCTION OF IDEMPOTENT FIRST-ORDER DENSITY MATRIX FROM X-RAY DIFFRACTION EXPERIMENTS

There has been interesting progress on the use of X-ray diffraction data to build in to density functional and density matrix studies. Early work was that of Stoddart et al [56] who used the Bragg reflection data of Brown [57] on Be metal.

Later work by Massa et al [58] focussed on constructing an idempotent density matrix consistent with diffraction data (see also Karle et al [59]). Subsequent to the original study of Massa et al [58], Howard et al [60] have used a method proposed by McWeeny [61], to construct idempotent density matrices in molecular crystals, while the most recent work of which we are aware at the time of writing is that of Synder and Stevens [62] on the azide ion.

10.1. Model density matrices reproducing Hartree-Fock electron density

In the work of Holas et al [63], two different density matrices were set up, both of them satisfying the condition of idempotency, for the Be atom, which had the same (Hartree-Fock) density on the diagonal. One of these is derivable from a local potential $V(\vec{r})$, while the other is from the usual Fock potential, which is non-local or energy dependent as in Eq. (58). These two matrices obviously yield different kinetic energies if we calculate these in the usual way from the first-order density matrix near its diagonal.

What this example makes clear is that given the diagonal density $\rho(\vec{r})$ plus idempotency does not specify a unique first-order density matrix, even were the X-ray data able to specify the electron density perfectly.

11. SUMMARY AND FUTURE DIRECTIONS

In this Chapter, attention has been focused, in the main, on approaches that do not have recourse to wave functions. Rather these methods need to appeal to density matrix theory, as Dirac [3] already recognized in his pioneering work in which he introduced exchange into the Thomas-Fermi atom. For a given one-body potential $V(\vec{r})$, the Thomas-Fermi theory was transcended in 1961 in the study of March and Murray [7], who gave the general term in the perturbation expansion of the Dirac density matrix, satisfying idempotency, in powers of $V(\vec{r})$, using the plane wave density matrix for free particles in Eq. 11 as the unperturbed problem. They showed explicitely that, in the limit of slowly varying potentials $V(\vec{r})$, the perturbation expansion could be summed to infinity to yield the density-potential relation (4) of the Thomas-Fermi method. Stoddart and March [12], following the formal completion of the Thomas-Fermi method by Hohenberg and Kohn [4], showed that the perturbation theory for the density $\rho(\vec{r})$ in terms of $V(\vec{r})$ could be uniquely inverted term by term to yield $V(\vec{r})$ in terms of the density. In the present Chapter, a lot of attention has been focussed on the external potential (and force) as a functional of the electron density. The most fundamental result of this kind has been given by Holas and March [36]. In their work, the force, obtained as the (negative of) gradient of the external potential is extracted from the differential form of the virial theorem. This force divides into three parts: kinetic, Laplacian and electron-electron interaction (see Eq. (56)). Only the Laplacian part is as yet known explicitly in term of the electron density. The kinetic term requires knowledge of the correlated first-order density matrix $\gamma(\vec{r};\vec{r}')$ near to its diagonal $\vec{r}=\vec{r}'$. The electron-electron interaction contribution involves the diagonal element of the second-order density matrix.

Evidently, all this leads to the question as to the way to determine the first-order density matrix and the pair function as functionals of the electron density. Without correlation, that is at one-body potential level, this problem is, in principle, solved by the perturbation series for $\gamma(\vec{r};\vec{r}')$. Its diagonal gives $\rho(\vec{r})$ as a functional of the one body potential $V(\vec{r})$. But as already mentioned, this can be inverted term by term. Substitution of this results for $V(\vec{r})$ back into the perturbation expansion for the density matrix determines $\gamma(\vec{r};\vec{r}')$ in terms of $\rho(\vec{r})$ at this level of one-body potential. Alternatively, one can use Slater-Kohn-Sham approach and return to one-electron wave functions, but this is a step that, in the long term, density functional theory aims to bypass: i.e., one must continue to seek an "orbital-free" theory.

Recent progress in this general direction has been reviewed, the Schrödinger equation for the density amplitude in Eq. (57) being one focal point. If one has to approximate the one-body Slater-Kohn-Sham potential, then why not try to approximate $v_{eff}(\vec{r})$ in the density amplitude equation instead?

Finally, paralleling the work of Holas and March [36] on the external potential as a functional of low order density matrices, has been the study of Savin [42], and independently of Levy and Görling [43], on the correlation energy functional in terms of first-order density matrices. The second-order matrix has also be employed directly by Nakatsuji and Yasuda [50]. Each of these approaches has its (considerable) merits and all are certain to be worthy of fuller exploration in future work. What is already clear is that , in setting up a density functional theory, any approximations in the functional are likely to introduce some degree of violation of N representability. This must then mean that, applying the variation principle, a pillar of Thomas-Fermi-Dirac theory and methods that transcend this, there is a possibility of going below the exact ground-state energy. If therefore one is seeking rigor, one example of a functional going below the exact energy is enough to show that this functional violates N representability and that "correlation energy" has been gained by mixing in a Boson component, albeit small, in with a Fermion theory. While substantial advances have come, as is evident in practical results obtained from DFT, it is now the time to examine critically the foundations of what is fundamentally a variational theory. This Chapter has had this as one of its primary aims. It seems to the writer that Hartree-Fock theory still has great merit as starting point. As remarked above, Moller-Plesset perturbation theory needs developing in the framework of low-order density matrices, to yield the perturbation correction to the Hartree-Fock density from correlation in terms of the Hartree-Fock density matrix. This is an area which is now ripe for development.

ACKNOWLEDGEMENTS

Without the invaluable collaboration of my friends and colleagues, A. Holas, M. Levy and N. C. Handy, this Chapter could not possibly have taken shape, and I am most grateful to the three of them. The actual writing was greatly facilitated by a visit to the Department of Chemistry and the Theoretical Physics Institute, University of Alberta, Edmonton and thanks are due to menbers of both Departments for very stimulating discussion. I am most grateful to Professor G. R. Freeman for his generous hospitality during my stay in Edmonton. Finally I wish to acknowledge the award of an Emeritus Fellowship by the Leverhulme Trust, UK to support work on density matrices and density functionals.

APPENDIX 1. PROPERTIES OF DENSITY MATRICES

The presentation below follows closely that of Holas and March [64]. The N-th order density matrix (DM) generated from a normalized wave function Ψ of a N-electron system is defined as

$$\gamma_N(\vec{x}_1,\ldots,\vec{x}_N;\vec{x}_1{}',\ldots,\vec{x}_N{}')=\Psi(\vec{x}_1,\ldots,\vec{x}_N)\Psi^*(\vec{x}_1{}',\ldots,\vec{x}_N{}'), \tag{A1.1}$$

The p-th order reduced DM (for p < N) is given in terms of it as

$$\gamma_p(\vec{x}_1,\ldots,\vec{x}_p;\vec{x}_1{}',\ldots,\vec{x}_p{}')=\binom{N}{p}\int d\vec{x}_{p+1}\ldots d\vec{x}_N\gamma_N. \tag{A1.2}$$

where $\int d\vec{x}_i$ means integration $\int d^3r_i$ and summation over s_i together with the replacement of $\vec{x}_i{}'$ by $\vec{x}_i$ in the integrand. In many applications spinless DMs are sufficient

$$\rho_p(\vec{r}_1,\ldots,\vec{r}_p;\vec{r}_1{}',\ldots,\vec{r}_p{}')=\sum_{s_1,\ldots,s_p}\gamma_p(\vec{x}_1,\ldots,\vec{x}_p;\vec{x}_1{}',\ldots,\vec{x}_p{}')\Big|_{s_i'=s_i}. \tag{A1.3}$$

The diagonal elements of DMs are denoted

$$\xi_p(\vec{x}_1,\ldots,\vec{x}_p)=\gamma_p(\vec{x}_1,\ldots,\vec{x}_p;\vec{x}_1,\ldots,\vec{x}_p)>0, \tag{A1.4}$$

$$n_p(\vec{r}_1,\ldots,\vec{r}_p)=\rho_p(\vec{r}_1,\ldots,\vec{r}_p;\vec{r}_1,\ldots,\vec{r}_p)>0. \tag{A1.5}$$

The subscript '1' can be omitted. The basic quantity of DFT – the electron number density – is thus

$$n(\vec{r}) = \rho(\vec{r};\vec{r}) = \sum_s \xi(\vec{r}s) = \sum_s \gamma(\vec{r}s;\vec{r}s). \tag{A1.6}$$

From the definition (A1.2) of γ_p one finds

$$\int d\vec{x}_p \gamma_p(\vec{x}_1,...,\vec{x}_p;\vec{x}_1',...,\vec{x}_p') = \frac{N+1-p}{p}\gamma_{p-1}(\vec{x}_1,...,\vec{x}_{p-1};\vec{x}_1',...,\vec{x}_{p-1}'), \tag{A1.7}$$

since $\binom{N}{p}\binom{N}{p-1}^{-1} = \frac{N+1-p}{p}$.

For density matrices generated from a determinantal wave function Ψ^D (which occurs, e.g., in the HF method or the KS approach), the p-th order matrix can be obtained from the 1-st order one

$$\gamma^D(\vec{x};\vec{x}') = \gamma_1{}^D(\vec{x};\vec{x}') = \sum_{j=1}^{N} \phi_j(\vec{x})\phi_j^*(\vec{x}'), \tag{A1.8}$$

as the following determinant:

$$\gamma_p^D(\vec{x}_1,...,\vec{x}_p;\vec{x}_1',...,\vec{x}_p') = \frac{1}{p!}\begin{vmatrix} \gamma_1^D(\vec{x}_1;\vec{x}_1'), & ..., & \gamma_1^D(\vec{x}_1;\vec{x}_p') \\ \vdots & \ddots & \vdots \\ \gamma_1^D(\vec{x}_p;\vec{x}_1'), & ..., & \gamma_1^D(\vec{x}_p;\vec{x}_p') \end{vmatrix} \tag{A1.9}$$

The set $\{\phi_j(\vec{r}s), j=1,...,N\}$ of orbitals used for construction of Ψ^D can be separated into two subsets

$$\{\phi_{i\uparrow}(\vec{r})\alpha(s), i=1,...,N_\uparrow\} \tag{A1.10}$$

and

$$\{\phi_{i\downarrow}(\vec{r})\beta(s), i=1,...,N_\downarrow\} \tag{A1.11}$$

where

$$N_\uparrow + N_\downarrow = N. \tag{A1.12}$$

Then the corresponding 1-st order matrix is

$$\gamma_1^D(\vec{x};\vec{x}') = \rho_\uparrow^D(\vec{r};\vec{r}')\alpha(s)\alpha(s') + \rho_\downarrow^D(\vec{r};\vec{r}')\beta(s)\beta(s'), \tag{A1.13}$$

where

$$\rho_\sigma^D(\vec{r};\vec{r}') = \sum_{i=1}^{N_\sigma} \phi_{i\sigma}(\vec{r})\phi_{i\sigma}^*(\vec{r}'); \qquad \sigma = \uparrow,\downarrow \quad . \tag{A1.14}$$

The matrices $\rho_\uparrow^D(\vec{r};\vec{r}')$ and $\rho_\downarrow^D(\vec{r};\vec{r}')$ are idempotent with respect to indices $\vec{r},\vec{r}'$ while $\gamma_1^D(\vec{x};\vec{x}')$ is idempotent with respect to $\vec{x},\vec{x}'$. Eqs. (A1.13) and (A1.3) lead to an expression for the spinless 1-st order matrix and its diagonal elements follows, namely

$$\rho^D(\vec{r};\vec{r}') = \rho_\uparrow^D(\vec{r};\vec{r}') + \rho_\downarrow^D(\vec{r};\vec{r}'); \qquad n^D(\vec{r}) = n_\uparrow^D(\vec{r}) + n_\downarrow^D(\vec{r}). \tag{A1.15}$$

The diagonal element of the 2-nd order DM, calculated according to Eqs. (A1.9) and (A1.13) is known to be

$$n_2^D(\vec{r}_1;\vec{r}_2) = \frac{1}{2} n^D(\vec{r}_1) n^D(\vec{r}_2)\left\{1 + h_2^D(\vec{r}_1;\vec{r}_2)\right\}, \tag{A1.16}$$

where

$$h_2^D(\vec{r}_1;\vec{r}_2) = -\frac{\left\{\left|\rho_\uparrow^D(\vec{r}_1;\vec{r}_2)\right|^2 + \left|\rho_\downarrow^D(\vec{r}_1;\vec{r}_2)\right|^2\right\}}{n^D(\vec{r}_1) n^D(\vec{r}_2)} < 0. \tag{A1.17}$$

For spin-compensated systems (SCS), i.e, when

$$\rho_\uparrow^D = \rho_\downarrow^D = \frac{1}{2}\rho^D, \quad \text{(SCS)} \tag{A1.18}$$

Eq. (A1.17) simplifies to

$$h_2^D(\vec{r}_1;\vec{r}_2) = -\frac{1}{2}\frac{\left|\rho^D(\vec{r}_1;\vec{r}_2)\right|^2}{n^D(\vec{r}_1) n^D(\vec{r}_2)}, \qquad \text{(SCS)}. \tag{A1.19}$$

For DMs generated from a general wave function Ψ, relations analogous to Eq. (A1.16), namely

$$n_2(\vec{r}_1;\vec{r}_2) = \frac{1}{2} n(\vec{r}_1) n(\vec{r}_2)\left\{1 + h_2(\vec{r}_1;\vec{r}_2)\right\}, \tag{A1.20}$$

serve as definitions of $h_2(\vec{r}_1;\vec{r}_2)$– the pair correlation function. It should be noted that for the determinantal case correlations are between the same spins only [in Eq. (A1.17), $\left|\rho_\uparrow^D(\vec{r}_1;\vec{r}_2)\right|^2$ is just $\rho_\uparrow^D(\vec{r}_1;\vec{r}_2)\rho_\uparrow^D(\vec{r}_2;\vec{r}_1)$].

The following sum rule [stemming from Eq. A1.7] holds for h_2:

$$\int d^3r_2 h_2(\vec{r}_1;\vec{r}_2)n(\vec{r}_2) = -1; \qquad \text{for any } \vec{r}_1. \tag{A1.21}$$

$n_2(\vec{r}_1;\vec{r}_2)$, together with the electron-electron repulsion, determines the force contribution $\vec{F}_{ee}$ in the important result (56) of the main text (see also [36]).

APPENDIX II. EXCHANGE POTENTIAL FROM (PARTIAL) FUNCTIONAL DIFFERENTIATION IN INHOMOGENEOUS ELECTRON GAS AT JELLIUM-TYPE METAL SURFACE

Though this Chapter is aimed at treating molecules, it is of interest here to summarize progress that has been made recently in treating the inhomogeneous electron gas spilling out of a metal surface (for an early review, see Lang [65]). In particular, Solomatin et al [66] have derived a form of the exchange potential V_x for a semi-infinite jellium model via (partial) functional differentiation of the exchange energy functional

$$E_x = -\frac{1}{4}\int d\vec{r}d\vec{r}'\,|\gamma(\vec{r};\vec{r}')|^2/|\vec{r}-\vec{r}'| = \frac{1}{2}\int \rho(\vec{r})V_x^S(\vec{r})d\vec{r} \tag{A2.1}$$

where $V_x^S(\vec{r})$ is the Slater potential. This is generated by the Fermi hole $\rho_x(\vec{r};\vec{r}')$, ie

$$V_x^S(\vec{r}) = \int dr'\ \rho_x(\vec{r};\vec{r}')/|\vec{r}-\vec{r}'|, \tag{A2.2}$$

ρ_x being defined by $|\gamma(\vec{r};\vec{r}')|^2/2\rho(\vec{r})$. In DFT one needs

$$v_x(\vec{r}) = \delta E_x[\rho]/\delta\rho(\vec{r}). \tag{A2.3}$$

The work of Solomatin et al [66] is partial functional differentiation in the sense that, if we take the analogy of free electrons confined to move within a one-dimensional box of length L, the variations these workers consider are akin to those, in the jellium model, to changes in the electron density $\rho(\vec{r})$ induced by changing L to L+δL. However, the structure of the exchange potential they obtained, denoted below by $v_x^{(0)}(\vec{r})$, is worthy to note, in relation to the importance of density gradients in current usage in DFT. If we write

$$v_x(\vec{r}) = v_x^{(0)}(\vec{r}) + f(\vec{r}) \tag{A2.4}$$

then Solomatin et al calculate $v_x^{(0)}(\vec{r})$ as given below, whereas $f(\vec{r})$is shown to be short-ranged compared with $v_x^{(0)}(\vec{r})$ and to satisfy the sum rule

$$\int d\vec{r}\rho(\vec{r})\vec{r}.\nabla f(\vec{r}) = 0. \tag{A2.5}$$

The result for $v_x^{(0)}(\vec{r})$ is found to be

$$v_x^{(0)}(\vec{r}) = A(\vec{r})^{-1}\left\{\frac{2}{3}V_x^S(\vec{r}) + [6\rho(\vec{r})]^{-1}\vec{r}.\nabla[\rho(\vec{r})V_x^S(\vec{r})]\right\} \tag{A2.6}$$

where $A(\vec{r}) = 1 + \vec{r}.\nabla\rho(\vec{r})/3\rho(\vec{r})$. This result makes explicit the fact that this portion of the exchange potential $v_x^{(0)}(\vec{r})$ in Eq. (A2.6) is determined in closed form by $V_x^S(\vec{r})$ and $\rho(\vec{r})$ plus their first derivatives. Since the quantity $A(\vec{r})$ appears in the denominator of Eq. (A2.6), one could say that in $v_x^{(0)}(\vec{r})$ one has summed an infinite series in $\nabla\rho/\rho$ thereby. Of course, the part $f(\vec{r})$ is incompletely known, because one has performed only a partial functional differentiation of $E_x[\rho]$ in Eq (A2.1) may well contain higher derivatives than $\nabla\rho$, as well as the Slater potential itself, plus its higher derivatives.

It is relevant here to mention that in the context of atoms and molecules, Kleinman [67] has discussed the Slater potential itself as a partial functional derivative of the exchange energy functional $E_x[\rho]$; see also Holas and March [68].

REFERENCES

1. L. H. Thomas, Proc. Camb. Phil. Soc. **23**, 542 (1926).
2. E. Fermi, Zeit. fur Physik **48**, 73 (1928).
3. P. A. M. Dirac, Proc. Camb. Phil. Soc. **26**, 376 (1930).
4. P. Hohenberg and W. Kohn, Phys. Rev. **136,** B864 (1964).
5. N. H. March, Advances in Phys. **6**, 1 (1957).
6. N. H. March, *Self-consistent field in atoms* (Pergamon: Oxford, 1975).
7. N. H. March and A. M. Murray, Phys. Rev. **120**, 830 (1960).
8. N. H. March and A. M. Murray, Proc. Roy. Soc. **A261**, 119 (1961).
9. P. O. Löwdin Phys. Rev. **97** 1490, (1955).
10. R. Mc Weeny, *Methods of molecular quantum mechanics* (Academic Press, New York, 1989) p.119.
11. J. C. Slater, Phys. Rev. **81**, 385 (1951).
12. J. C. Stoddart and N. H. March, Proc. Roy. Soc. **A299**, 279 (1967).
13. N. H. March and W. H. Young, Proc. Phys. Soc. London **A 72**, 182 (1958).
14. N. H. March, Proc. Phys. Soc. **A68**, 726 (1955).
15. P. Gombas *Die statische theorie des atoms und ihre anwendungen* (Springer, Vienna 1949).
16. M. Levy, N. H. March and N. C. Handy, J. Chem. Phys. **104**, 1989 (1996)
17. N. H. March, *Electron density theory of atoms and molecules* (Academic Press, New York 1992).
18. P. J. Grout, Y. Tal and N. H. March, J. Chem. Phys. **79**,331 (1983).
19. N. H. March and R. J. White, J. Phys. **B5**,466 (1972).
20. Y. Tal and M. Levy, Phys. Rev. **A23**, 408 (1981).
21. E. A. Hylleraas, Z. Physik **65**, 209 (1930).

22. D. Layzer, Ann. of Phys. **8**, 271 (1959).
23. J. Harris and R. O. Jones, J. Phys. **F4**, 1170 (1974).
24. D. Langreth and J. P. Perdew, Phys. Rev. **B21**, 5469 (1980).
25. O. Gunnarsson and B. I. Lundqvist, Phys. Rev. **B13**, 4274 (1976).
26. M. Levy, in *Density Functional Theory*: Eds. J. Keller and J. L. Gasquez (Springer, New York 1983).
27. J. Harris, Phys. Rev. **A29**, 1648 (1984).
28. R. G. Parr and W. Yang, *Density functional theory of atoms and molecules* (University Press, Oxford 1989).
29. M. Levy, Proc. Nat. Acad. Sci. USA **76**, 6062 (1979).
30. M. Levy, Phys. Rev. **A26**, 1200 (1982).
31. A. D. Becke, J. Chem. Phys. **98**, 1372 (1993).
32. A. D. Becke, J. Chem. Phys. **88**, 1053 (1988).
33 M. Levy and J. P. Perdew, Phys. Rev. **A32**, 2010 (1985).
34. M. Levy, Phys. Rev. **A43**, 4637 (1991).
35. N. H. March and W. H. Young, Nuclear Physics **12**, 237 (1959).
36. A. Holas and N. H. March, Phys. Rev. **A51**, 2040 (1995).
37. M. K. Harbola and V. Sahni, Phys. Rev. Lett. **62**, 489 (1989).
38. M. Levy and N. H. March, 1996, in course of publication.
39. A. Holas and N. H. March, Phys. Rev. **A44**, 5521 (1991).
40. A. Holas and N. H. March, Phys. Chem. Liquids **31**, 15 (1996).
41. K. A. Dawson and N. H. March, J. Chem. Phys. **81**, 5850 (1984).
42. A. Savin, Phys. Rev. **A52**, R1805 (1995).
43. M. Levy and A. Görling, Phys. Rev. **A52**, R1808 (1995).
44. P. O. Löwdin, Phys. Rev. **97**, 1474 (1955).
45. R. McWeeny, *Methods of molecular quantum mechanics*, 2nd Edition (Academic Press, New York, 1989), p. 11.
46. H. Yasuhara, Lett. Nuovo Cim. **12**, 414, (1975).
47. A. J. Coleman, Rev. Mod. Phys. **35**, 668 (1963).
48. E. R. Davidson, *Reduced density matrices in quantum chemistry* (Academic Press, New York, 1976) p 51.
49. T. L. Gilbert, Phys. Rev. **B12**, 2111, (1975).
50. H. Nakatsuji and K. Yasuda, Phys. Rev. Lett. **76**, 1039 (1996).
51. C. Valdemoro, Phys. Rev. **A45**, 4462 (1992).
52. F. Colmenero, C. Perez del Valle and C. Valdemoro, Phys. Rev. **A47**, 971 (1993).
53. F. Colmenero and C. Valdemoro, Phys. Rev. **A47**, 979 (1993).
54. F. Colmenero and C. Valdemoro, Int. J. Quantum Chem. **51**, 369 (1994).
55. D. J. Klein and N. H. March, J. Mol. Structure (Theochem) 1996: to appear.
56. J. C. Stoddart, P. Stoney, N. H. March, and I. B. Ortenburger, Nuovo Cim. **23B**, 15 (1974).
57. P. J. Brown, Phil. Mag. **26**, 1377 (1972).
58. L. Massa, M. Goldberg, C. Frishberg, R. F. Boehme, and S. J. LaPlaca, Phys. Rev. Lett. **55**, 622 (1985).
59. J. Karle et. al. Sanibel Symposium, 1996.
60. S. T. Howard, J. P. Huke, P. R. Mallinson and C. S. Frampton, Phys. Rev. **B49**, 7124 (1994).
61. R. McWeeny, Rev. Mod. Phys. **32**, 335 (1960).

62. S. Synder and W. Stevens, Int. Cong. Theor. Chem. Physics, New Orleans, 1996.
63. A. Holas, N. H. March, Y. Takahashi and C. Zhang, Phys. Rev. **A48**, 2708 (1993).
64. A. Holas and N. H. March in *Topics in current chemistry*. Ed. R. Nalewajski (Springer-Verlag:Berlin) 1996: to appear.
65. N. D. Lang in *Theory of the inhomogeneous electron gas*. Eds. S. Lundquist and N. H. March (Plenum, New York 1985).
66. A. Solomatin, V. Sahni and N. H. March, Phys. Rev. **B49**, 16856 (1994).
67. L. Kleinman, Phys. Rev. **B49**, 14197 (1994).
68. A. Holas and N. H. March, to be published, (1996).

J.M. Seminario (Editor)
Recent Developments and Applications of Modern Density Functional Theory
Theoretical and Computational Chemistry, Vol. 4

An Introduction to High-precision Computational Methods for Simple Atomic and Molecular Systems

Frank C. Sanders

Department of Physics, Southern Illinois University at Carbondale
Carbondale, IL 62901, U.S.A.

The development of high-precision methods for calculating wave functions and energies of simple atoms and molecules is reviewed with emphasis on variational and variational perturbation methods. Most of the methods discussed here were developed in attempts to improve calculations on the helium atom and, in particular, its ground state. Hence, much of the discussion centers around computational methods for helium, including the development of more extensive and mathematically more sophisticated basis sets for helium-like systems and their extension to more complex systems. Beyond helium, simple atoms and molecules amenable to high-precision methods are discussed. Methods for excited states, including multiply-excited states, are also reviewed.

1. INTRODUCTION

In the implementation of density functional methods for atoms and molecules, accurate wave functions can be utilized to test and/or construct the energy density functional. Except for one-electron systems, however, these wave functions are necessarily approximate. For more complex systems, the most precise wave functions available are obtained by the use of variational methods. For the simplest, few-electron systems, such calculations are capable of producing energies and wave functions of very high accuracy, sufficient for the present requirements of density functional theory. In what follows, we review the development and present state of accurate calculations on simple atomic and molecular systems.

The history of the development of high-precision methods for calculating atomic energies and wave functions is essentially the history of calculations on helium; particularly, the ground state of helium. This history really begins, as do many other computational methods from the early days of quantum mechanics, with the work of Hylleraas[1]. A delightful account by Hylleraas, in the form of a personal memoir on this and other early work in atomic and molecular theory, can be found in Ref.[2]. A second milestone in this history occurs 20 years later with the work of unprecedented accuracy of Pekeris[3]. More recently, there has been a resurgence of interest in high accuracy calculations for many states of helium and other two-electron atoms. At the same time, there has been increasing effort given to accurate calculations on other simple atoms and molecules. Also of interest, but at a lower standard of accuracy, are studies of doubly-excited states of atoms, particularly the H^- ion and helium, as well as studies on multiply-excited states of other small atoms. As in the work of Pekeris, much of this development has been made possible by the availability of increased computing power. Nevertheless, "brute-force" application of established methods alone has not proven as successful as coupling increased computing power with the refinement and further development of existing methods. In this article, we review the methods that have been developed for calculations on such atomic systems, leading up to the most recent studies of extraordinarily high accuracy on helium. The presentation centers on the methods themselves. Hence, actual applications of particular methods to specific systems, although sometimes elaborated on here at some length, are of less interest to this discussion, and some significant calculations are here omitted despite their prominence. This, to some extent, is inevitable, given the considerable number of calculations carried out for systems such as helium. The "folklore" of computational methods in atomic physics is also extensive, and only the main threads are presented here. Similarly, we generally do not discuss what could be described as programming methods, except in so far as they are central to the computational methods discussed. Finally, we focus only on the calculation of non-relativistic energies and wave functions of few-electron systems with fixed nuclei; *i.e.* for atoms, the infinite-mass approximation, for molecules, the Born-Oppenheimer approximation. These approximations provide "benchmark" systems for purposes of comparison between different calculations. We also concentrate on those methods which can, in principle, achieve any desired accuracy in the solution of the Schrödinger equation.

The standard works on atomic structure contain only brief reviews of earlier calculations on atomic systems. Brief discussion of early calculations, containing some description of the methods utilized, can be found in Condon and Shortley[4], while Slater[5] contains some discussion of still more recent, accurate calculations up to the work of Pekeris[3]. For a definitive description of all this early work on helium, there is, of course, Bethe and Salpeter[6]. This work also covers much of the theory of one- and two-electron atoms, including relativistic and quantum electrodynamic effects, in a comprehensive manner.

In preparing a brief review such as this, the author can cover personal predilection in the cloak of necessity (for brevity and economy). In this instance, I have chosen to concentrate on variational methods, including the variational perturbation method, and have given little or no attention to other, very useful, non-variational methods. Not discussed or given only the briefest mention, are coupled cluster methods, many-body perturbation theory, diagrammatic perturbation theory, and finite-element and other numerical methods.

2. COMPUTATIONAL METHODS

For atomic systems, the non-relativistic Hamiltonian for an n-electron atom is (in atomic units)

$$H = \sum_{i}^{n} \left(-\frac{1}{2}\nabla_i^2 - \frac{Z}{r_i} + \sum_{j>i}^{n} \frac{1}{r_{ij}} \right). \tag{1}$$

where Z is the charge on the nucleus. The fundamental problem is to solve the Schrödinger equation

$$H\Psi_k = E_k\Psi_k, \tag{2}$$

for the energies (eigenvalues), E_k, and wave functions (eigenfunctions), $\Psi_k(\mathbf{r}_1, \ldots, \mathbf{r}_N)$ of the kth state of the system.

Approximate methods for the solution of the Schrödinger equation range from the purely analytical to the purely numerical. We shall discuss variational methods[7] primarily, with some discussion of perturbation methods and only brief remarks about other methods. This is simply because calculations of high accuracy on atomic systems have most often been carried out using variational methods.

Calculations capable of yielding extremely high accuracy are usually of such complexity that they can only be implemented for the simplest systems. One might call such methods "exact", not only because the most precise calculations of the genre are often so designated in the literature (usually as the last entry in a long table of figures), but also because such calculations have tended to stay just ahead of the most precise experimental measurements. Clearly, any such "exact" calculation must be placed in its historical context. One might have expected that, as computational capacity increased and as calculations of a certain degree of complexity or meeting a particular standard of accuracy were superseded by methods of yet greater complexity and accuracy, application of these earlier methods would have migrated to more complex systems. The fact that this has not occurred in any significant manner is indicative of the formidable barriers that exist in extending these methods to systems with only a few more electrons.

2.1. Ritz Variational Principle

The Ritz variational method, or, (as applied to Eq. 2), Schrödinger 's variational principle, has been the method of choice in calculating atomic energies and eigenfunctions. The variational functional $E_{\text{var}}(\tilde{\Psi})$ is defined as,

$$E_{\text{var}}(\tilde{\Psi}) = \frac{\langle \tilde{\Psi} | H | \tilde{\Psi} \rangle}{\langle \tilde{\Psi} | \tilde{\Psi} \rangle}. \tag{3}$$

with $\tilde{\Psi}(\mathbf{r}_1, ..., \mathbf{r}_N)$, an arbitrary function with the same dimensions as the eigenfunctions of H. Clearly, if $\tilde{\Psi}$ coincides exactly with a particular eigenfunction, $E_{\text{var}}(\tilde{\Psi})$ yields exactly the corresponding eigenvalue. Of course, in most cases, the exact eigenfunctions are not known. It is easy to see that if $\tilde{\Psi}$ differs from the exact eigenfunction by some infinitesimally small amount,

$$\tilde{\Psi} = \Psi + \epsilon\, \varphi, \tag{4}$$

then $E_{\text{var}}(\tilde{\Psi})$ differs from the corresponding eigenvalue by an amount of order ϵ^2[8].

The variational principle relies on the fact that, if the functional $E_{\text{var}}(\tilde{\Psi})$ is stationary for any arbitrary function $\tilde{\Psi}$, that function is a solution of the Schrödinger equation. In practical calculations, of course, the $\tilde{\Psi}$ actually utilized is *not* completely arbitrary, but is instead constrained by the limitations of trial wave functions and computing resources. Hence the stationary values of $E_{\text{var}}(\tilde{\Psi})$ will not correspond precisely to the solutions of the Schrödinger equation.

Almost universally one chooses what can be loosely described as a "basis" in which to expand $\tilde{\Psi}$,

$$\tilde{\Psi} = \sum_{i=1}^{N} c_i\, \phi_i(\alpha_1, \ldots, \alpha_k), \tag{5}$$

where the c_i and the α_j are variational parameters to be optimized. If the ϕ_i were complete in the Hilbert space, then one could, in principle, obtain the exact energies and wave functions of the system. One such complete basis is the set of eigenfunctions of H itself. Thus, we can write,

$$\tilde{\Psi} = \sum_{i=0} c_i \Psi_i, \tag{6}$$

where Ψ_0 is the ground state wave function. Then,

$$E_{\text{var}}(\tilde{\Psi}) = \frac{\sum_{i=0} |c_i|^2 E_i}{\sum_{i=0} |c_i|^2}, \tag{7}$$

which can be written as,

$$E_{\text{var}}(\tilde{\Psi}) = E_0 + \frac{\sum_{i=0} |c_i|^2 (E_i - E_0)}{\sum_{i=0} |c_i|^2} \geq E_0. \tag{8}$$

Hence, the variational principle must yield an upper bound to the ground state energy.

Clearly any appropriate trial wave function can be written as in Eq. 6, so that the "basis" used to describe the wave function need not be anything more than a simple collection of terms.

For excited states, this procedure can still be made to produce upper bounds. In the simplest case, one constructs a trial wave function that explicitly reflects some symmetry of the Hamiltonian different from that of the ground state. Then $\tilde{\Psi}$ will be orthogonal to Ψ_0 (and to all other wave functions with the same symmetry as Ψ_0). $\tilde{\Psi}$ can still be written as in Eq. 6, but Ψ_0 now represents the wave function for the state of lowest energy of that symmetry. This means that the variational method can be used to obtain rigorous upper bounds to the energies of all states of lowest energy of a particular spin multiplicity and/or angular momentum. This procedure can also be extended to even more highly-excited states by projecting out from the trial wave function all the exact wave functions of lower-lying states of the same symmetry. For the mth excited state of any system, this would require that the trial wave function satisfy

$$\left\langle \tilde{\Psi} | \Psi_j \right\rangle = 0, \qquad j = 0, \ldots, m-1. \tag{9}$$

As the wave functions for these lower-lying states are usually only known approximately, this procedure does not provide a practical method for accurate calculations on excited states. It is, however, possible to obtain rigorous upper bounds to any (finite) number of excited states via the Hylleraas-Undheim-MacDonald (HUM) theorem [9]. In HUM, a trial function of the form given by Eq. 5 is required. Optimization of the linear coefficients yields the system of equations,

$$\sum_{j=1}^{N} (H_{ij} - E S_{ij})\, c_j = 0, \quad i = 1, \ldots, N. \tag{10}$$

where

$$H_{ij} = \langle \phi_i | H | \phi_j \rangle \tag{11}$$

is the Hamiltonian matrix and

$$S_{ij} = \langle \phi_i | \phi_j \rangle. \tag{12}$$

is the overlap matrix. The HUM theorem states that the ordered eigenvalues which are roots of the secular determinant given by Eq. 10 are each upper bounds to the corresponding eigenvalues of the Schrödinger equation. Hence, a calculation for an excited state of a particular symmetry would begin by constructing a "basis" in which all terms reflect the symmetry of this state. The solution of the secular determinant then yields eigenvalues which are upper bounds to the energies of each corresponding state, while the eigenvectors contain the optimized linear coefficients for each state. These upper bounds can be further improved by optimization of any non-linear parameters (the α_j of Eq. 5) incorporated into the trial wave function. For a calculation to yield the best estimate of the energy of, say, the second excited state, the *third* eigenvalue of the secular determinant must be minimized, despite the fact that this procedure will generally yield a poorer energy for the ground and first excited states, and hence, presumably, a poorer approximation for the wave functions of these states. This procedure can, in principle, be extended to the last eigenvalue of the secular determinant, as this and all other eigenvalues are rigorous upper bounds. Typically, however, only a relatively few eigenvalues (in relation to the size of the basis) provide reasonably accurate estimates to the energies of the system.

The above discussion suggests that use of the variational method can provide eigenvalues for only a few excited states of a particular symmetry. In fact, it is possible to carry out variational calculations for excited states well beyond the limitations of the HUM theorem, although without an upper bound for the energy of the state. For isolated atomic systems, this problem commonly arises in the study of autoionizing states, where a very highly-excited, quasibound state lies above the ionization threshold of the atom. In this case, the autoionizing state is identified by the relative *stability* of a root of the secular determinant against variation in some parameter, typically a non-linear parameter of the trial wave function. Such a root is not completely stable and will undergo a sequence of "variational collapses" since a rigorous upper bound is not possible with a basis consisting of a finite number of terms. This type of variational procedure is discussed more fully below.

The conditions for optimizing the linear parameters of $\tilde{\Psi}$ yield a system of linear equations which are readily solved by standard numerical procedures, even for very large basis sets. The corresponding optimization conditions for the non-linear parameters, however, are not generally so simple to satisfy. In fact, there is no procedure that will locate the absolute minimum of this multi-dimensional energy surface. This task becomes increasingly difficult as the number of such non-linear parameters is increased. There is one "exception to the rule" however: one can introduce an overall scaling parameter into the trial function. This parameter, call it α_{sc}, is incorporated in the trial function via the transformation $r_i \rightarrow \alpha_{sc} r_i$. For such a parameter, the variational principle becomes

$$E_{\text{var}}(\tilde{\Psi}) = \alpha_{sc}^2 \langle T \rangle + \alpha_{sc} \langle V \rangle, \tag{13}$$

where we have assumed, with no loss of generality, that $\tilde{\Psi}$ is normalized. In the above expression, T is the kinetic energy operator, V is the potential energy operator (which includes only Coulomb potentials) and

$$\langle\Omega\rangle = \sum_{i,j=1}^{N} c_i^* c_j \; \langle\phi_i|\Omega|\phi_j\rangle. \tag{14}$$

The optimum α_{sc} is then

$$\alpha_{sc} = -\frac{\langle V\rangle}{2\langle T\rangle}. \tag{15}$$

If the linear coefficients, c_i, were independent of the choice of α_{sc}, the variational principle would then take the form,

$$E_{\rm var}(\tilde{\Psi}) = -\frac{\langle V\rangle^2}{4\langle T\rangle}. \tag{16}$$

Of course the optimized c_i are dependent on the value of the non-linear parameters. Nevertheless, starting with a reasonable choice of α_{sc}, it is possible to iteratively obtain a consistent set of variational parameters. The optimized trial wave function so obtained can be shown[1,10] to satisfy the quantum virial theorem,

$$2\langle T\rangle = -\langle V\rangle. \tag{17}$$

If multiple, non-linear parameters, α_i, (with *no* overall, scaling parameter) are included in the trial wave function, the fulfillment of the virial theorem is a necessary but not sufficient condition for the fulfillment of the variational principle[11]. In any case, it can be shown[12] that the set of points on the energy surface, $E(\alpha)$ for which the virial theorem is satisfied describe a path of steepest descent on the energy surface passing through the energy minimum:

$$\boldsymbol{\alpha} \cdot \boldsymbol{\nabla} E(\alpha) = 2\langle T\rangle + \langle V\rangle. \tag{18}$$

As noted above, the accuracy of the wave function obtained with the Ritz variational method is generally much lower than the corresponding accuracy in the energy. In fact, the accuracy of the wave function can vary widely over different portions of configuration space. This can have significant consequences for the calculation of expectation values of particular operators. The approximate expectation value is given by

$$\langle\Omega\rangle_{\rm var} = \left\langle\tilde{\Psi}\middle|\Omega\middle|\tilde{\Psi}\right\rangle. \tag{19}$$

Introducing the $\tilde{\Psi}$ of Eq. 4 into this expression yields an error for the expectation value of the same order as the error in the wave function.

A simple way of measuring local variations in the accuracy of the wave function over the entire configuration space is given by the local energy function,

$$E_{\rm loc}(\mathbf{r}_1 \ldots \mathbf{r}_n) = \frac{H\tilde{\Psi}(\mathbf{r}_1 \ldots \mathbf{r}_n)}{\tilde{\Psi}(\mathbf{r}_1 \ldots \mathbf{r}_n)}. \tag{20}$$

For the exact wave function, $E_{\rm loc}$ will be constant everywhere and equal to the exact energy. In actual practice, however, even for variational wave functions which produce accurate eigenvalues, there are considerable deviations in the local energy from the approximate eigenvalues, $E_{\rm var}(\tilde{\Psi})$, obtained with the same wave function. This is particularly so in regions of configuration space where the Hamiltonian is singular. For an approximate wave function, the local energy functional will itself be singular at these points, with $E_{\rm loc} \to -\infty$ when $r_i \to 0$, corresponding to an electron-nucleus collision, and $E_{\rm loc} \to +\infty$ for an electron-electron coalescence, $r_{ij} \to 0$. This behavior of the local energy is related to the Kato cusp conditions[13], and will be discussed more fully in the following sections.

The local energy is typically most nearly constant in a shell at "intermediate" distances from the nucleus; just those regions of configuration space where the integrand of the variational functional, $\tilde{\Psi}^* H \tilde{\Psi} d\tau$, is largest and where the variational functional would be most sensitive to the accuracy of $\tilde{\Psi}$. One of the best known examples of this characteristic of variational wave functions was noted by Chandrasekhar[14] in his calculation of the continuous absorption coefficient of the H^- ion with a 10-term Hylleraas wave function. It is well known that the matrix element for radiative transitions in the dipole approximation can be written in three equivalent (length, velocity, and acceleration) forms:

$$\frac{im}{\hbar}(E_j - E_k)\langle\Psi_j|\sum_{i=1}^{n}\mathbf{r_i}|\Psi_k\rangle = \langle\Psi_j|\sum_{i=1}^{n}\mathbf{p_i}|\Psi_k\rangle = -i\hbar(E_j - E_k)^{-1}\langle\Psi_j|\sum_{i=1}^{n}\frac{\mathbf{r_i}}{r_i^3}|\Psi_k\rangle. \tag{21}$$

If the wave functions for the initial and final states in the transition are exact, all three expressions give precisely the same result. For Chandrasekhar's variationally calculated wave function, however, this was not the case, and each formula gave different results. In general, for variational wave functions, the velocity form of the matrix element is the preferred expression, as its operator most closely matches the Hamiltonian operator. It has been suggested that a comparison of oscillator strengths calculated by these three expressions could serve as a measure of the accuracy of variational wave functions. However, there are many examples in which the agreement between different forms of the oscillator strength exceeds their actual accuracy.

In closing this section, we remark that variational principles for observables other than the energy can also be constructed[15]. The variational functional for an operator with very different behavior from that of the Hamiltonian will emphasize regions of configuration space different from those emphasized by the Hamiltonian. Such a functional might well produce wave functions more accurate than those produced by the Ritz principle, at least in some regions of configuration space.

2.2. Lower Bounds

A plethora of lower bound theorems exist, both for the energy and for other observables. It might appear surprising that they are not utilized more frequently. While it is true that the matrix elements encountered in these theorems are intrinsically more difficult to calculate (both analytically and numerically), the major impediment to their general use seems to be that the results so obtained seldom justify the effort required to produce them. Here, we will briefly discuss only lower bounds for the energy. The more familiar of these, the Temple[16] and Weinstein[17], and Stevenson-Crawford[18] bounds, are given below:

$$E_0 \geq E_{\text{Temple}}(\tilde{\Psi}) = \left\langle \tilde{\Psi} \middle| H \middle| \tilde{\Psi} \right\rangle - \frac{\left\langle \tilde{\Psi} \middle| H^2 \middle| \tilde{\Psi} \right\rangle - \left\langle \tilde{\Psi} \middle| H \middle| \tilde{\Psi} \right\rangle^2}{E_1 - \left\langle \tilde{\Psi} \middle| H \middle| \tilde{\Psi} \right\rangle}, \tag{22}$$

$$E_0 \geq E_{\text{Weinberg}}(\tilde{\Psi}) = \left\langle \tilde{\Psi} \middle| H \middle| \tilde{\Psi} \right\rangle - \left[\left\langle \tilde{\Psi} \middle| H^2 \middle| \tilde{\Psi} \right\rangle - \left\langle \tilde{\Psi} \middle| H \middle| \tilde{\Psi} \right\rangle^2 \right]^{\frac{1}{2}}, \tag{23}$$

$$E_0 \geq E_{\text{Stevenson}}(\tilde{\Psi}) = \bar{E}_{0,1} - \left[\left\langle \tilde{\Psi} \middle| H^2 \middle| \tilde{\Psi} \right\rangle - 2\bar{E}_{0,1} \left\langle \tilde{\Psi} \middle| H \middle| \tilde{\Psi} \right\rangle + \bar{E}_{0,1}^2 \right]^{\frac{1}{2}}, \tag{24}$$

where $\bar{E}_{0,1} \leq \frac{1}{2}(E_0 + E_1)$. In the above expressions we have assumed that $\tilde{\Psi}$ is normalized. Note that, typically, lower bound methods require knowledge of some property of the system, in this case of the first excited state, E_1, to some degree of accuracy. This limitation would appear to make it unlikely that these lower bounds could obtain the level of precision of which the Ritz principle is capable. In actuality, however, this is not the most significant problem in making use of these expressions. The simple fact is that the Ritz principle generally yields eigenvalues orders of magnitude more accurate than the corresponding lower bound. In practice, the Ritz upper bound obtained with a particular basis is usually so accurate in comparison to the lower bound obtained with the same basis, that the latter provides no useful information. Of course, in those cases where a *lower* bound is itself the object of interest, these theorems are indispensable.

Weinhold[19] has given a general inequality valid for any hermitian operator, F,

$$\left\langle \tilde{\Psi} \middle| F \middle| \Psi \right\rangle \lesseqgtr S \left\langle \tilde{\Psi} \middle| F \middle| \tilde{\Psi} \right\rangle \pm \left[\left\langle \tilde{\Psi} \middle| F^2 \middle| \tilde{\Psi} \right\rangle - (\left\langle \tilde{\Psi} \middle| F \middle| \tilde{\Psi} \right\rangle)^2 \right]^{\frac{1}{2}} (1 - S^2)^{\frac{1}{2}}, \tag{25}$$

In the above, $S = \left\langle \tilde{\Psi} | \Psi \right\rangle$ is the overlap of the approximate wave function with the exact wave function. Taking $F = H$, Weinhold obtains lower bounds for the energy which, using upper bound criteria on S, yield the three lower bounds to the energy given above[20].

3. PERTURBATION THEORY

Perturbation theory is a powerful method for the solution of the equations of motion in both classical and quantum mechanics. Its application in quantum mechanics is made simpler and more elegant by the mathematical properties of the hermitian operators that represent the observables of a system. In this section we discuss only Rayleigh-Schrödinger perturbation theory. A comprehensive survey of perturbation methods and their application to atomic and molecular systems has been made by Hirschfelder, Byers-Brown and Epstein[21]. In the present article, we present those aspects that are related to the calculation of the energies and wave functions for atomic systems.

For the Schrödinger wave equation, perturbation theory generates a family of inhomogeneous, second-order perturbation differential equations whose solution yields the perturbation expansions for both the wave function and energy of the system. In perturbation theory, the Hamiltonian of the system is first separated into an unperturbed part and a perturbation,

$$H = H_0 + \lambda H_1. \tag{26}$$

Here λ, the perturbation parameter, can be either a physical parameter of the system that controls the strength of the perturbation, or a dummy variable that is introduced simply to keep track of the order of various terms with regard to the perturbation. In any case, the perturbation parameter is treated as a completely arbitrary parameter. Leaving aside questions of convergence, it is assumed that the wave function and energy (and all other properties of the system) can be expressed as power series in the parameter λ:

$$\Psi = \sum_n \lambda^n \psi_n, \tag{27}$$

$$E = \sum_n \lambda^n \epsilon_n, \tag{28}$$

Note that, here, subscripts on the wave functions are used to denote the order of the perturbed wave function. The unspecified state of the system is understood to correspond to the state of the zero-order wave function. Introducing the perturbation series into the Schrödinger equation while recalling that λ is an arbitrary parameter, the perturbation differential equations are obtained by collecting terms of like powers in λ:

$$(H_0 - \epsilon_0)\psi_0 = 0, \tag{29}$$

$$(H_0 - \epsilon_0)\psi_1 + (H_1 - \epsilon_1)\psi_0 = 0, \tag{30}$$

$$(H_0 - \epsilon_0)\psi_2 + (H_1 - \epsilon_1)\psi_1 - \epsilon_2\psi_0 = 0, \tag{31}$$

and, in general,

$$(H_0 - \epsilon_0)\psi_n + (H_1 - \epsilon_1)\psi_{n-1} - \sum_{k=2}^{n} \epsilon_k \psi_{n-k} = 0, \qquad n \geq 2. \tag{32}$$

The first-order equation, Eq. 30, immediately yields the first-order correction to the energy on multiplying from the left by ψ_0^* and integrating:

$$\epsilon_1 = \langle \psi_0 | H_1 | \psi_0 \rangle. \tag{33}$$

ψ_1 can then be found by solving Eq. 30. The second-order energy coefficient is obtained from the second-order equation, Eq. 31, by the same procedure used to obtain the first-order energy,

$$\epsilon_2 = \langle \psi_1 | H_1 - \epsilon_1 | \psi_0 \rangle. \tag{34}$$

With a few more manipulations, a similar procedure on the third-order equation yields an expression for the third-order energy coefficient,

$$\epsilon_3 = \langle \psi_1 | H_1 - \epsilon_1 | \psi_1 \rangle - 2\epsilon_2 \langle \psi_1 | \psi_0 \rangle. \tag{35}$$

Thus the first-order wave function is sufficient to obtain the energy coefficients through third-order. Continuing this procedure[21], one obtains the ϵ_{2n} and ϵ_{2n+1} coefficients once the nth-order wave function is obtained. This $2n+1$-rule[22] could, of course, be anticipated from the variational principle: taking the truncated nth-order wave function to be the trial function yields an error in the energy of order $2n+2$.

Although this method yields upper bounds for the even-order energy coefficients, the truncated sum of these coefficients is not an upper bound to the energy. However, if the truncated sum of the perturbation wave functions is introduced into the Ritz variational method, we obtain a rigorous upper bound to the energy. The upper bound, $E_N \geq E$, obtained from $\tilde{\Psi}_N = \sum_{n=0}^{N} \lambda^n \psi_n$ can be expressed in terms of the energy coefficients[21],

$$E_0 = \epsilon_0 + \lambda\epsilon_1, \tag{36}$$

$$E_1 = \epsilon_0 + \lambda\epsilon_1 + \frac{\lambda^2\epsilon_2 + \lambda^3\epsilon_3}{D_1}, \tag{37}$$

$$E_n = \sum_{k=0}^{2n+1} \lambda^k \epsilon_k + \lambda^{2n+2}\frac{A_n}{D_n}, \qquad n \geq 2, \tag{38}$$

where

$$D_n = 1 + \lambda^{n+1} \sum_{k=1}^{n} \lambda^{k-1} \sum_{l=0}^{n-k} \langle \psi_{n-l} | \psi_{l+k} \rangle, \tag{39}$$

and

$$A_n = \langle \psi_n | \psi_n \rangle \sum_{k=2}^{n} \lambda^{k-2}\epsilon_k + 2\langle \psi_n | \psi_{n-1} \rangle \sum_{k=3}^{n} \lambda^{k-3}\epsilon_k +$$

$$\sum_{k=0}^{n-4}\sum_{l=0}^{k+2} \langle \psi_{n-l} | \psi_{n+l-2-k} \rangle \sum_{j=0}^{n-k-4} \lambda^j \epsilon_{k+4+j} + \sum_{k=1}^{n} \lambda^{k-1} \sum_{l=0}^{n-k} \langle \psi_{n-l} | \psi_{k+l} \rangle \sum_{j=0}^{n-1} \lambda^j \epsilon_{n+1+j}, \tag{40}$$

with slightly modified expressions required for A_2 and A_3[21]. Unfortunately, at low-order these expressions generally yield upper bounds which are much less accurate than the corresponding, truncated perturbation sum; so much so that they offer little useful information. Conversely, we see that this upper bound differs from the truncated perturbation sum by the addition of terms of order $2n + 2$. Hence, the truncated perturbation energy sum itself becomes approximately an upper bound as the order is increased.

3.1. Variational Perturbation Method

Perturbation theory has played an essential role in the development of atomic theory. For the most part, however, the use of perturbation methods has been limited to lowest-order calculations where a specific interaction acts as a small perturbation on an atomic or molecular system. Such low-order calculations have certainly proven extremely useful. In comparison to the variational method, however, there have been relatively few attempts in atomic and molecular physics to systematically apply higher-order perturbation theory as a computational tool. In part, this may be due to the complexity of the higher-order expressions in the conventional "textbook" approach[23]. In this approach, one expands the perturbation wave functions in the basis of the unperturbed Hamiltonian,

$$\psi_n = \sum_k c_n^{(k)} \psi_0^{(k)}, \tag{41}$$

where the superscripts here denote the states of the unperturbed problem. For atomic and molecular systems, the obvious choice of unperturbed Hamiltonian is often a sum of hydrogenic (or hydrogen-like) Hamiltonians. Unfortunately, the basis constructed from hydrogenic wave functions generally produces results which converge too slowly for high-precision calculations. In addition, this basis must include the continuum eigenfunctions to be complete. These complications present no insurmountable difficulties, but suggest that such calculations would involve very large expansions and the construction of some difficult integrals over the continuum functions. For an example of such a calculation the reader is referred to the work of Scherr[24] on the second-order energy of the ground state of helium. The point to note in this calculation is that only 40% of the second-order energy comes from bound states, a significant portion of the second-order energy (46%) comes from the $l = 0$ and $l = 1$ singly-ionized (bound-continuum) configurations and the one ($l = 0$) continuum-continuum contribution included in this calculation. Ultimately, about 14% of the second-order energy coefficient still remains to be calculated, and a good portion of that must come from the remaining continuum-continuum contributions.

An alternative to such an approach was suggested by Hylleraas[25] in another of his seminal papers. In this work, the perturbation differential equations are treated as the Euler-Lagrange equations of a variational problem. For the first-order equation, the variational functional is

$$J_1(\tilde{\psi}_1) = \left\langle \tilde{\psi}_1 \middle| H_0 - \epsilon_0 \middle| \tilde{\psi}_1 \right\rangle + 2\left\langle \tilde{\psi}_1 \middle| H_1 - \epsilon_1 \middle| \psi_0 \right\rangle. \tag{42}$$

Hylleraas also showed that this first-order functional yields an upper bound to the second-order energy coefficient for the ground state, $\epsilon_2 \leq J(\tilde{\psi}_1)$. Expanding $\tilde{\psi}_1$ in some basis,

$$\tilde{\psi}_1 = \sum_{i=1}^{N} c_i \, \phi_i(\alpha_1, \ldots, \alpha_k). \tag{43}$$

leads to a system of equations for the linear coefficients of the trial wave function,

$$\sum_{i=1}^{N} c_i G_{ij} + V_j = 0, \qquad j = 1 \ldots N. \tag{44}$$

In the above $G_{ij} = \langle \phi_i | H_0 - \epsilon_0 | \phi_j \rangle$ and $V_j = \langle \phi_j | H_1 - \epsilon_1 | \psi_0 \rangle$. The solution of this system of linear equations requires a diagonalization of the G_{ij} matrix and this can be accomplished by any of a variety of direct methods. Hence, these equations are much easier to solve than the secular determinant that appears in the variational method. If the ϕ_i are replaced by the unperturbed wave functions, the G_{ij} matrix will be diagonal and the usual expressions of conventional perturbation theory are recovered.

Variational perturbation calculations were first carried out by Hylleraas for the first-order wave function of the ground state of helium-like ions in 1930[25]. This paper is also interesting because Hylleraas gives every detail of the analytical and computational procedures followed in the calculation. An alternative procedure for solving the set of linear equations has been proposed by Dalgarno and Drake[26] and is contrasted with the original Hylleraas procedure by Sanders[27].

Variational functionals involving the higher-order wave functions can also be constructed which, if all lower-order wave functions are known exactly, yield upper bounds to the even-order energy coefficients[21]. These Hylleraas-Scherr-Knight (HSK) variational perturbation expressions can also be obtained directly from the Ritz variational principle[28], indicating clearly their connection with the usual variational method. An interesting problem arises in the calculation of the higher-order wave functions. The most efficient way to do this is to utilize the identical basis obtained in the first-order calculation for all the higher order wave functions[26]. This means that once the non-linear parameters in the basis are optimized so as to minimize ϵ_2, these parameters are held fixed for the remainder of the calculation. Hence, only a single diagonalization of the G_{ij} matrix has to be carried out, and the higher-order wave functions and energy coefficients can be calculated so efficiently that their computation time is negligible in comparison to the time required for the first-order calculation. Unfortunately, the optimum basis for the first-order wave function (as defined by the non-linear parameters) is not likely to be optimum for the higher-order wave functions. For the case of the ground state of helium, Knight and Scherr[29] have demonstrated that, as the order of the perturbation wave function increases, the wave function covers regions of configuration space at increasing distance from the nucleus. Hence, for a basis with a fixed number of terms, the optimum non-linear parameter (typically an exponential parameter) would tend to be quite different (smaller) from the first-order parameter. This means that the higher-order wave functions will become less accurate and the HSK method will no longer provide a meaningful upper bound for the even-order coefficients. The higher-order coefficients will then diverge rapidly beyond some order. Optimizing the non-linear parameters at each order of the calculation requires a re-diagonalization of G_{ij} but improves the stability of the higher-order energy coefficients considerably, permitting the calculation to be extended to much higher order. This increased stability can be seen by comparing Fig. 1 of Ref. [30] (fixed-basis) with Fig. 1 of Ref. [31] (order-by-order optimization). In both cases, exactly the same basis was utilized, but the fixed basis calculation has begun to diverge by eighth order while the optimized calculation yields energy coefficients through twenty-first order.

For excited states, upper bounds can be obtained with the variational perturbation method under constraints similar to those required in the Ritz variational principle. In this case, however, the conditions imposed on the trial wave functions involve only projections with the *zero*-order wave functions of all lower-lying states[31,32]. For the approximate nth-order wave function of the mth state, these are just the conditions that the exact nth-order wave function would have to satisfy:

$$\left\langle \tilde{\psi}_n^{(m)} | \psi_0^{(k)} \right\rangle = \left\langle \psi_n^{(m)} | \psi_0^{(k)} \right\rangle = -\frac{\left\langle \psi_0^{(m)} \middle| H_1 \middle| \psi_0^{(k)} \right\rangle}{\epsilon_0^{(m)} - \epsilon_0^{(k)}}, \qquad k = 1, \ldots, m-1. \tag{45}$$

The simplest way to satisfy these conditions is to explicitly include all zero-order wavefunctions for the lower-lying states in the trial wave function [32]. The variational perturbation method will then automatically construct a perturbation wave function which satisfies the upper bound conditions. If these conditions are not satisfied explicitly, upper bounds for the even-order energy coefficients will undergo the analogue of "variational collapse" that occurs in the usual variational method. In this case, however, there are no "avoided crossings" of multiple roots as there is no secular determinant and only a single optimization curve. Instead, this curve diverges at points along the optimization curve in a manner reminiscent of avoided crossings[33]. This can occur for even a low-lying excited state, regardless of the number of terms in the basis. If all the lower-lying, zero-order wave functions are not included explicitly in the basis, not only will the calculation not provide a rigorous upper bound to the even-order energy coefficients, but the optimization curve will have precisely the same number of divergences as the number of zero-order terms not included in the basis[34]. For very large bases, this might not present a serious loss of accuracy, as a sufficiently flexible basis will include the required zero-order functions implicitly[35], yielding an optimization curve in which the singular points are far from the minimum in the optimization curve, a minimum that is generally a very good approximation to an upper bound.

It is also possible to construct lower bounds to ϵ_2[21,36]. For example, the following lower bound is obtained from Temple's lower bound:

$$\epsilon_2 \geq J_1(\tilde{\psi}_1) - \frac{\left\langle \left[(H_0 - \epsilon_0)\tilde{\psi}_1 + H_1\psi_0 \right]^2 \right\rangle}{\epsilon_1^{(1)} - \epsilon_0}. \tag{46}$$

Since this expression only requires the first-order energy of the first excited state, $\epsilon_1^{(1)}$, it is considerably simpler to implement than the Temple upper bound itself.

3.2. Other Properties via Perturbation Theory

The perturbation expansion for the expectation value of any operator, Ω, is written as

$$\langle \Omega \rangle = \sum_n \lambda^n \langle \Omega \rangle_n, \tag{47}$$

where

$$\langle \Omega \rangle_n = \sum_{k=0}^{n} \langle \psi_{n-k} | \Omega | \psi_k \rangle. \tag{48}$$

The mathematical properties of the perturbation wave functions and matrix elements make it possible to derive many simplifying relations and alternative methods of computation for such matrix elements. One of the best known is the Dalgarno-Lewis interchange theorem[37,38]. This theorem has wide application in perturbation theory, but is perhaps best known as a means of obtaining exact, low-order results for certain classes of operators. The first-order expectation value of Ω is

$$\langle \Omega \rangle_1 = 2\langle \psi_0 | \Omega | \psi_1 \rangle. \tag{49}$$

Following Birss and Choi[39], we write

$$\psi_1 = RH_1\psi_0, \qquad R = \frac{1 - |\psi_0\rangle\langle\psi_0|}{\epsilon_0 - H_0}. \tag{50}$$

$\langle \Omega \rangle_1$ can then be written as,

$$2\langle \psi_0 | \Omega | RH_1\psi_0 \rangle = 2\langle R\Omega\psi_0 | H_1\psi_0 \rangle = 2\langle \omega_1 | H_1 | \psi_0 \rangle, \tag{51}$$

with $\omega_1 = R\Omega\psi_0$, the solution of a first-order, perturbation equation, but with Ω as the perturbation:

$$(H_0 - \epsilon_0)\omega_1 + (\Omega - \langle \psi_0 | \Omega | \psi_0 \rangle)\psi_0 = 0. \tag{52}$$

For one-body operators and some two-body operators, this equation can be solved exactly, making it possible to obtain the first-order expectation value exactly. This in turn makes it possible to study the accuracy of the first-order wave function, ψ_1, by comparing the approximate first-order expectation value of Eq. 49 with the exact value[29]. Similar expressions can be generated for second-order properties, but these will generally involve solving a differential equation that involves H_1 as well as Ω[21,40].

4. BASIS SETS

Central to advances in high-precision, variational calculations has been the development of basis sets appropriate for large-scale calculations. The limitations of contemporaneous computational "hardware" necessarily impose restrictions on the size of basis employed. Hence, the motivation for developing new bases for use in trial wave functions follows from the need to achieve higher levels of accuracy with roughly the same size basis. Using helium as a benchmark system, we will see in the sections that follow that the "typical" basis set has increased from roughly a half-dozen terms in the 1930's, with a numerical precision of no more than six (decimal) figures, to the present use of basis sets of over a thousand terms in calculations producing a precision of up to 12 significant figures. This change has also led to a difference in the way in which trial wave functions are constructed. The focus very early shifted from a careful selection of which particular terms to include in a trial function, a process often based on physical insight, to constructing entire bases based more on their mathematical characteristics.

In what follows, we divide basis sets into two classes: single-electron basis sets and correlated basis sets. We will discuss correlated bases in the context of calculations of the ground state of helium, a system in which most saw their earliest applications. In the next few sections we trace the development of single-electron basis sets, beginning with the prototypical basis set, the hydrogenic wave functions.

4.0.1. The hydrogen atom

Our intuitive understanding of the structure of atomic and molecular systems has developed out of our understanding of the hydrogen atom and its wave functions. Indeed, these wave functions contain the correct asymptotic behavior for very large nuclear charge (where the interelectron interaction can be neglected) for both atoms and molecules. The problem of hydrogen is related to the classical Kepler problem, is separable in a number of different coordinate systems[41], and its solutions are known. Full details of the relativistic as well as non-relativistic cases can be found in the classic article by Bethe and Salpeter[6]. Here we only sketch the essential features of the non-relativistic wave functions. The Hamiltonian for hydrogen is written (in atomic units) as

$$H = -\frac{1}{2}\nabla^2 - \frac{Z}{r}. \tag{53}$$

In spherical coordinates, the bound-state wave functions can be expressed in terms of Laguerre polynomials and spherical harmonics. Properly normalized, these wave functions are

$$\Psi_{n,l,m}(\mathbf{r}) = R_{n,l}(r)Y_l^m(\theta,\phi), \tag{54}$$

with

$$R_{n,l}(r) = \left(\frac{2Z}{n}\right)^{3/2} \left[\frac{(n-l-1)!}{2n[(n+l)!]^3}\right]^{1/2} \left(\frac{2Zr}{n}\right)^{l} L_{n+l}^{2l+1}(2Zr/n)\, \mathrm{e}^{-Zr/n}, \tag{55}$$

and with

$$Y_l^m(\theta,\phi) = \sqrt{\frac{(2l+1)(l-m)!}{4\pi(l+m)!}} P_l^m(\cos\theta) \mathrm{e}^{im\phi}, \tag{56}$$

the normalized spherical harmonics. The Laguerre polynomials, $L_{n+l}^{2l+1}(r)$, are polynomial in r of order $n-l-1$, while the entire radial wavefunction decays exponentially at a rate determined by Z/n.

The eigenvalue, in atomic units, is given by

$$E_n = -\frac{Z^2}{2n^2}. \tag{57}$$

The wave functions for continuum states ($E > 0$), can be obtained from the bound state functions by replacing n with $i\nu$, a continuous parameter. This gives an energy $\frac{1}{2}k^2 = \frac{1}{2}Z^2/\nu^2$ with wavenumber $k = Z/\nu$. Of course, the normalization of such continuum functions cannot be obtained by such a transformation[6].

The wave-function for an n-electron atom can be constructed from these hydrogenic orbitals. Such a basis is seldom utilized because of the slow convergence noted above. A single term (configuration) of these orbitals does form the zero-order wave function for the "bare-nucleus" unperturbed Hamiltonian in a perturbation expansion of the atomic wave function. Alternatives to hydrogenic orbitals exist that are both computationally more convenient and more flexible and, hence, better suited for use in variational wave functions.

4.1. Single-electron Basis

The wave function for an n-electron atom can be constructed from any set of one-electron orbitals by constructing an appropriately symmetrized sum over products (configurations) of n such orbitals. Since the Hamiltonian is symmetric under an interchange of particle labels, the wave function must itself be either symmetric or antisymmetric. The appropriate symmetry for electrons (spin $= \frac{1}{2}$) is, of course, the latter. In addition, the wave function must be consistent with the Pauli exclusion principle. In the conventional approach, an appropriate spin function is first constructed out of the two spin-$\frac{1}{2}$ states (component of spin angular momentum $m_s = \pm\frac{1}{2}$), conventionally denoted α and β. That is, an eigenfunction of S^2 and S_z, $\mathbf{S} = \sum_i^n \mathbf{S}_i$, is constructed for an n-particle system using the two spin states. The procedure calls for constructing an anti-symmetric space-spin wave function from an unsymmetrized product of one-electron orbitals and an appropriate spin function,

$$\tilde{\Psi} = \sum_{P_a} P_a(-)^{P_a} \left(\prod_i^n \chi_i(\mathbf{r}_i)\right) \Sigma(1,\ldots,n), \tag{58}$$

where the χ_i are one-electron orbitals and Σ is a spin eigenfunction. The P_a are the $n!$ permutations. For a two electron system, the appropriate spin functions are:

$$\Sigma(1,2) = \alpha(1)\beta(2) - \beta(1)\alpha(2), \tag{59}$$

with total spin $s = 0$ (singlet states), and

$$\Sigma(1,2) = \begin{cases} \alpha(1)\alpha(2), \\ \alpha(1)\beta(2) + \beta(1)\alpha(2), \\ \beta(1)\beta(2), \end{cases} \tag{60}$$

with $s = 1$ (triplet states). In the non-relativistic approximation, all triplet states are degenerate, so that any of the three spin functions will serve equally well for constructing the wave function. Introducing relativistic corrections, such as spin-orbit coupling, into the Hamiltonian breaks this degeneracy so that each triplet spin function yields a different energy. The complete two-electron trial wave function is then formed from this basis,

$$\phi(\mathbf{r}_1, \mathbf{r}_2) = [\chi_i(\mathbf{r}_1)\chi_j(\mathbf{r}_2) \pm \chi_j(\mathbf{r}_1)\chi_i(\mathbf{r}_2)]\ [\alpha(1)\beta(2) \mp \beta(1)\alpha(2)]\,, \tag{61}$$

with the upper sign corresponding to a singlet state. For three- (or more) electron systems, properly symmetrized space-spin functions can no longer be factored into this simple product form, but must instead be written as a sum over permutations of such functions.

For a spin-free Hamiltonian, such as our non-relativistic Hamiltonian, it is not necessary to introduce spin in order to obtain the appropriate symmetry for the wave function. Such a spin-free formulation[42] produces the same result as the conventional method by restricting the permutation symmetry of the wave function to symmetry states consistent with the Pauli exclusion principle. This procedure will serve as long as there are no spin-dependent (relativistic) terms in the Hamiltonian.

4.1.1. Hartree-Fock orbitals

If one introduces a *single*-term trial function constructed from a product of arbitrary one-electron orbitals into the Ritz variational functional and takes the variance with respect to each orbital, the coupled set of integro-differential equations generated are the Hartree-Fock equations. These equations yield the optimum (in a variational sense) single-configuration orbitals. Despite their apparent complexity, these equations are nevertheless no more than single-electron equations and reduce to one-dimensional equations in the radial coordinate. They can thus be solved very accurately by numerical or analytical means. HF orbitals in analytical form can be obtained by expressing the orbital in terms of another single-electron basis; for example, Slater-type orbitals.

As this is a single-configuration wave function, the energies obtained via the HF method are not very precise, being in error by typically a few percent in the energy. This discrepancy from the exact, non-relativistic energy is known as the correlation error. Orbitals obtained by this method can themselves be considered a one-electron basis, and utilized in constructing more elaborate, multi-configuration, trial wave functions.

4.1.2. Slater-type orbitals

A simple modification of the hydrogenic basis is the Slater-type orbital (STO),

$$\chi_{n,l,m}(\mathbf{r}) = r^{n+l}\, \mathrm{e}^{-\alpha r}\, Y_l^m(\theta, \phi). \tag{62}$$

STO's permit adjustment of the electron distribution by variation of the non-linear parameter, α. They also considerably simplify the analytical work by replacing the Laguerre polynomial with a single monomial of the same order. This also eliminates the problem of inappropriate, fixed nodes in the hydrogenic wave function. The single-electron STO's are coupled together to form terms with definite angular momentum, from which properly symmetrized products of STO's are constructed. For an n-electron atom or molecule, choosing an optimum set of n (or more) non-linear parameters represents a formidable problem for systems of more than a few electrons. Not surprisingly, this problem is often circumvented by relying on a predetermined set of orbital parameters.

4.1.3. Gaussian-type orbitals

Another form of one-electron orbital, constructed with Gaussian exponential functions, was proposed by Boys[43]. These orbitals lead to important simplifications in evaluating multi-center Coulomb integrals: such integrals over Gaussian-type orbitals (GTO's) can be expressed in terms of the error function. Hence, GTO's are extensively used in molecular calculations, either explicitly to construct the trial wave function, or implicitly, having been used to construct accurate fits to STO's.

4.1.4. Sturmian basis

Sturmian functions[44] can be thought of as modified hydrogenic wave functions. For hydrogenic wavefunctions, the energy appears implicitly in both the scaling parameters $(Z/n = \sqrt{-2E})$ and the indices of the Laguerre polynomials through the principal quantum number n. For Sturmian functions, this connection is relaxed so that the scaling parameter is no longer linked to the index of the Laguerre polynomial:

$$\chi_{n,l,m}(\mathbf{r}) = r^l L_n^{2l+1}(2\alpha r)\, e^{-\alpha r}\, Y_l^m(\theta,\phi). \tag{63}$$

This increased flexibility is a definite advantage in variational calculations. However, the main advantage of Sturmian functions over hydrogenic functions is that they form a complete, denumerable set over the same domain as the hydrogenic functions. Hence, this "discrete" basis is complete in and of itself, whereas the hydrogenic functions require both the discrete and continuum wave functions to be complete in the same domain.

The Sturmian functions can be considered solutions of an eigenvalue equation,

$$(-\frac{1}{2}\nabla^2 - E)\chi = \frac{\omega_n}{r}\chi, \tag{64}$$

where $E = -\frac{1}{2}\alpha^2$, is taken as a fixed parameter and the eigenvalues are the $\omega_n = n\alpha$. These functions are not orthogonal as are the hydrogenic functions, but they are orthogonal with respect to the weight function r^{-1},

$$\langle\chi_{n'l'm'}|r^{-1}|\chi_{nlm}\rangle = \delta_{n,n'}\,\delta_{l,l'}\,\delta_{m,m'}. \tag{65}$$

These functions have been utilized in helium atom calculations[45–47], for example, and in the multicenter integral problem[48].

4.2. Configuration Interaction

We will use the term configuration interaction (CI) to refer generically to calculations which only use one-electron functions in constructing the basis. The CI method takes single "configurations", properly symmetrized products of single-electron functions of the type discussed above, and combines them to form the CI basis. As an example, we can take the STO basis of Eq. 62 and construct a two-electron wave function for an S-state:

$$\tilde{\Psi} = \sum_i c_i(1 \pm P_{12})\ \chi_{n_i,l_i,m_i}(r_1)\ \chi_{n'_i,l_i,-m_i}(r_2). \tag{66}$$

In order to represent an S-state, orbitals with the same (n_i, n'_i, l_i) must be coupled together with the appropriate Clebsch-Gordan coefficients to form a state with zero total angular momentum. The sum over all terms with a particular value of l forms the lth partial wave of the wave function and each term in this partial wave has a metric (polynomial) order $\omega = n_i + n'_i + 2l_i$. From this and the addition theorem for the Legendre polynomials, it can be seen that all such CI functions can be represented by what would be the most general form of a CI expansion[5],

$$\tilde{\Psi} = \sum_{l=0} f_l(r_1, r_2)\ P_l(\cos\theta_{12}), \tag{67}$$

where

$$\cos\theta_{12} = \frac{\mathbf{r_1} \cdot \mathbf{r_2}}{r_1 r_2}. \tag{68}$$

For complex systems, using such basis sets provides considerable simplifications over the "correlated" basis sets discussed below, particularly in the analytical work. In a typical application, one adds terms to the basis up to a certain metric order, ω, within each set of terms constructed from spherical harmonics with a particular angular momentum, l.

The question of interest here is how large a trial wave function is required for a particular basis to produce a certain degree of convergence. The question of convergence of atomic and molecular basis sets is a difficult one. Based on empirical (computational) evidence or on more rigorous methods, certain concepts appear firmly established. For the basis sets utilized in these problems, the number of terms included in the trial wave function is not a good measure of the rate of convergence. Instead, the maximum metric order, ω_{max}, or the highest "partial wave", l_{max}, included in the basis set provides a much better index of convergence. Roughly speaking, convergence in ω can be associated with the description of radial, or "in-out", correlation of the electrons, while convergence in l is related to the description of angular correlation. It should be apparent that the number of terms required for a trial function to reach any particular maximum value of ω increases rapidly as ω increases. Hence, the number of terms required to reach a particular value of l in the basis while including all terms up to a maximum value of ω also increases rapidly, and the problem soon becomes intractable for any but the simplest systems. Finally, irrespective of the number of terms required for each partial wave, extensive investigation of the CI method has established the slow convergence of such calculations with l. An example of the application of the CI-method to helium-, lithium-, and beryllium-like atoms using a reasonably sized, "double-ζ" basis of STO's is found in Weiss[49]. In these calculations, Weiss used 35 configurations through $l = 5$ for helium, 45 configurations for lithium, and 55 configurations for beryllium. In commenting on these calculations, Weiss noted that this slow convergence could be attributed to the cusp-like behavior of the wave function[50] as $r_{12} \to 0$ and concluded that extending the calculations to include orbitals with higher l was not justified by the expected returns. The 45 configuration study of lithium indicated convergence strikingly similar to that of helium. Indeed, most of the slow convergence here was attributed by Weiss to the Li^+ core of the atom. Of course, by today's standards, Weiss' wave functions are relatively modest in size. Nevertheless, the slow convergence of the CI basis guarantees that matching the kind of precision possible with other methods will require enormous computing resources. Carroll, Silverstone and Metzger[51] estimate the minimum l required to reach a specified degree of precision for the ground state of helium. For precision at the level of a few parts in 10^5 they find that partial waves through $l = 11$ are required, with a total of over 100 terms. To achieve the level of precision of Pekeris' 1078-term calculation[3] (a few parts in 10^9) would require at least $l = 256$ with roughly 42,000 terms! For systems with more than three or four electrons, however, the CI approach has so far proved to be the only tenable method capable of producing accurate results. A particular variant of this method, the multi-configuration Hartree-Fock (MCHF) method, utilizing numerical Hartree-Fock orbitals, has been applied with considerable success by Fischer[52] and her co-workers.

An careful study of the convergence of typical basis sets and of the factors that influence the rate of convergence was carried out by Schwartz[53,54]. He found that the second-order energy coefficient of the ground state of helium converged as l^{-4} and guessed that the rate of convergence for singlet states would be similar while that for triplets would go as l^{-6}. He argued that this slow rate of convergence was due to the inability of the CI basis to deal with singularities in the wave function associated with coalescence of particles. The behavior of a wave function at such coalescence points is described by the so-called cusp conditions.

4.3. Cusp Conditions

The wave functions satisfying the Schrödinger equation are everywhere bounded. However, the Hamiltonian itself is singular at every two-particle coalescence. As a result, there exists a compensating divergence in the kinetic energy which will counteract the divergence in the potential energy. Hence, in order for the wave function to satisfy the Schrödinger equation and yet remain bounded, it will have to satisfy certain conditions at the singular points. These conditions take the form of a cusp in the wave function at a two-particle coalescence, giving rise to a discontinuity in the first derivative of the wave function. These cusp conditions were first obtained rigorously by Kato[13] and take the form,

$$\left\langle \frac{\partial \Psi}{\partial r}\Big|_{r=0} \right\rangle_{\mathrm{Ave}} = C_r \Psi|_{r=0} \tag{69}$$

where $C_r = -Z$ for $r = r_1, r_2$ and $C_r = \frac{1}{2}$ for $r = r_{12}$ and the left hand side is an average over a small sphere about $r = 0$. Pack and Byers-Brown[55] have derived similar conditions for molecular wave functions.

The partial wave expansion is able to reproduce these "cusps" in the wave function only very slowly, accounting for the inevitably slow convergence of all un-modified CI methods. From their CI calculations on the ground state, Carroll, Silverstone and Metzger[51] established the rate of convergence in l for this state empirically. They found that the incremental contribution of the energy for each partial wave behaved as $\Delta E_l \approx (l+\frac{1}{2})^{-4}$, which translates into $E_l - E \approx (l+1)^{-3}$. Klahn and Morgan[56] have examined the convergence in l of model problems while Hill[57] has been able to confirm this rate of convergence for the ground state of the helium atom rigorously.

In an analysis of the CI method, Bunge[58] reviews some of the characteristics of the method and discusses variants to CI that have proven more successful. A recent example of such a modified CI procedure with an enhanced rate of convergence is found in Goldman[59].

4.4. Correlated Basis

The slow convergence of the single-electron basis in a problem as simple as the ground state of helium led Hylleraas to consider a trial wave function in which the relative positions of all three particles could be described explicitly. For the case of an infinite-mass nucleus, the helium atom possesses six degrees of freedom. These could be taken to be r_1, r_2, θ_{12}, and the Euler angles, θ_1, ϕ_1, ϕ, where θ_1, ϕ_1 are the spherical polar coordinates of electron 1 and ϕ is an angle of rotation about r_1. It was realized by Hylleraas[1] that for the ground state (or for S-states in general) the wave function is independent of the three Euler angles that describe the orientation of the three-particle triangle, and the wave function can be described entirely by r_1, r_2, θ_{12},

$$\Psi = \psi(r_1, r_2, \theta_{12}) \pm \psi(r_2, r_1, \theta_{12}) \, . \tag{70}$$

For other angular momentum states, Breit[60] showed that the general form of a P-state wave function is

$$\Psi = \psi(r_1, r_2, \theta_{12}) \cos\theta_1 \pm \psi(r_2, r_1, \theta_{12}) \cos\theta_2 \tag{71}$$

while for D-states, the most general form would be

$$\begin{aligned} \Psi = {} & \psi(r_1, r_2, \theta_{12}) P_2(\cos\theta_1) \pm \psi(r_2, r_1, \theta_{12}) P_2(\cos\theta_2) \;\; + \\ & [\phi(r_1, r_2, \theta_{12}) \pm \phi(r_2, r_1, \theta_{12})] \left(\tfrac{3}{2}\cos\theta_1\cos\theta_2 - \tfrac{1}{2}\cos\theta_{12}\right). \end{aligned} \tag{72}$$

In fact, for D-states, either one of the two angular functions would suffice, but it is customary in variational calculations to include both terms, ensuring the greatest flexibility in the trial wave function. Expressions for wave functions of even higher angular momentum can be obtained in terms of vector spherical harmonics in the Euler angles of the three-particle triangle[61]. For finite nuclear mass helium atoms, the contribution of all particles to the total angular momentum must also be taken into account[62,63]. A similar decomposition of the three-electron wave function is possible[64].

All such functions, however, are still of CI form and thus suffer the same slow convergence. Given that

$$r_{12}^2 = r_1^2 + r_2^2 - 2r_1r_2\cos\theta_{12}, \tag{73}$$

these wave function could also be described in terms of r_1, r_2, r_{12}. However, a wave function of the form of Eq. 67 expressed in terms of r_{12} would contain only even powers of r_{12}. An obvious generalization would be to include odd powers of r_{12} in the wave function. This could be written as $(r_{12}^2)^l \times r_{12}$, where r_{12} is given by

$$r_{12} = \sum_{l=0}^{\infty} \left[\frac{r_<^{l+2}}{(2l+3)r_>^{l+1}} + \frac{r_<^l}{(2l-1)r_>^{l-1}} \right] P_l(\cos\theta_{12}). \tag{74}$$

Hence, odd powers of r_{12} will generate an expansion which, as in the case of r_{12}^{-1}, is different in the two half-spaces $r_1 < r_2, r_1 > r_2$. Inclusion of odd powers of r_{12} yields a rate of convergence for the basis far superior to that of Eq. 67.

5. TWO-ELECTRON ATOMS

The Hamiltonian for the non-relativistic, infinite mass, two-electron atom is

$$H = -\frac{1}{2}\left(\nabla_1^2 + \nabla_2^2\right) - \frac{Z}{r_1} - \frac{Z}{r_2} + \frac{1}{r_{12}}. \tag{75}$$

At the level of precision currently possible computationally and experimentally for this simple system, introduction of the finite-mass Rydberg and mass polarization corrections together with all relativistic and QED corrections are essential for a meaningful comparison between theory and experiment. Indeed, one of the prime motivations for calculations of such high precision for the non-relativistic energy is the opportunity to test higher-order QED corrections on this two-electron system. A discussion of all such corrections can be found in Bethe and Salpeter[6]. Calculation of these corrections is usually done by treating them as perturbations on the non-relativistic Hamiltonian. The lowest-order corrections are then just the expectation values of the appropriate relativistic operator on the non-relativistic variational wave function. For low-Z atoms, however, these corrections are so small that the non-relativistic energy must be calculated to roughly 10 to 12 significant figure accuracy in order that the concomitant error in the expectation value is reduced to the point that a comparison with experiment is meaningful. We shall see that, for two-electron atoms, this level of precision has been achieved.

5.1. The Ground State of Helium

The testing ground for new computational developments in atomic theory has traditionally been the ground state of helium. Perhaps the earliest application of the quantum variational method to the ground state of helium is due to Kellner[65] in 1927 using a four term wave function of the form of Eq. 66. Calculations by Slater[66] and Hylleraas[45] using the CI method followed in 1928

Results obtained for the ground state energy of helium by various calculations with correlated basis sets are tabulated in Table 1, where a general trend toward increasing accuracy and complexity is apparent.

5.1.1. Hylleraas coordinates

In his 1929 calculation for the ground state[1], Hylleraas constructed a three term wave function, later increased to six terms, utilizing a correlated basis set. Instead of using the coordinates (r_1, r_2, r_{12}), Hylleraas preferred the coordinates

$$s = r_1 + r_2, \quad t = r_2 - r_1, \quad u = r_{12}. \tag{76}$$

These coordinates have the domains $[0, \infty], [0, u], [0, s]$, respectively. For this work, Hylleraas chose a trial function of the form,

$$\tilde{\Psi} = \sum_{n,m,l} c_{nml}\, s^n t^m u^l\, e^{-s/2}. \tag{77}$$

Table 1
Energy of the Ground State of Helium (in a.u.)

	Energy	Method
Hylleraas[1]	-2.90324	6-term Hylleraas
Chandrasekhar and Herzberg[68]	-2.903715	18-term Hylleraas
Schwartz[83]	-2.90372437616	189-term half-integer exponents
Kinoshita[73]	-2.9037237	80-term Kinoshita
Pekeris[3]	-2.903724375	1078-term Pekeris
Scherr and Knight[28]	-2.90372433	100-term Kinoshita, Z^{-1}
Frankowski and Pekeris[87]	-2.9037243770326	246-term Fock
Thakkar and Smith[96]	-2.903724363	66-term exponentially correlated
Kono and Hattori[100]	-2.9037243766	186-term double-basis
Freund, *et al.*[88]	-2.9037243770340	230-term Fock
Baker, *et al.*[103]	-2.9037243770341184	400-term Fock, double-basis, Z^{-1}
Drake[104]	-2.903724377034119479	1262-term double-basis

Note that, for singlet states, only even powers of t are allowed in the wave function if it is to be a symmetric function, while for triplets, only odd powers are permitted. With only three terms, $(n, m, l) = (0,0,0), (0,0,1), (0,2,0)$, he was able to get an energy of -2.90244 compared to the exact value of -2.903724. "Chemical accuracy" would be achieved at the 1 kcal/mol level, or, in atomic units, at approximately 1.59×10^{-3} au. Hence the Hylleraas three-term wave function was already able to exceed this level of accuracy. With an additional three terms, $(1,0,0), (2,0,0), (0,0,2)$, he obtained an energy of -2.90324, an order of magnitude improvement in precision over what he had obtained earlier with a CI basis[45].

In 1955, improved experimental results[67] meant that, even with relativistic corrections, a discrepancy existed between experimental and theoretical values of the helium ionization energy. The source of this discrepancy was identified as the poor convergence of the variationally obtained, non-relativistic energy. By including 18 terms in a correlated basis, Chandrasekhar and Herzberg [68] were able to obtain a ground state energy of -2.903723. This gave an ionization potential only 0.1 cm^{-1} less than the experimental value.

5.1.2. Kinoshita variables

In 1957, Kinoshita[69] noted that the Chandrasekhar and Herzberg ionization energy was calculated using relativistic corrections obtained from less elaborate wave functions. At the same time, an upper bound calculation by Wilets and Cherry[70] with the Chandrasekhar-Herzberg wave function had produced a value 400 cm^{-1} lower than the upper bound obtained from the variational method. It was also known[71] that a Hylleraas series, i.e. a power series in Hylleraas variables (or powers of r_1, r_2, r_{12} for that matter), cannot satisfy the Schrödinger equation of helium in a formal sense, although the basis in Hylleraas coordinates, does form a complete set of functions in the domain $0 \leq |t| \leq u \leq s \leq \infty$. Seeking an improved wave function, Kinoshita proposed a more general basis for the trial wave function which included negative powers of the Hylleraas variables s and u. His basis would then use the new set of variables,

$$s, \quad p = \frac{u}{s}, \quad q = \frac{t}{u}, \tag{78}$$

with domains $[0, \infty], [0, 1], [-1, 1]$, respectively. Replacing s, t, u in Eq. 77 with these new variable, one obtains Kinoshita's wave function,

$$\tilde{\Psi} = \sum_{n,m,l} c_{nml}\, s^n p^m q^l\, e^{-s/2}. \tag{79}$$

To satisfy boundary conditions on the wave function, terms with $m = 0$, $l \neq 0$ are not permitted.

With a 39-term wave function of this form, Kinoshita obtained an upper bound to the ground state energy of -2.9037225 and a Temple lower bound of -2.9038737 . This lower bound is a significant improvement over that obtained with the Chandrasekhar-Herzberg wave function[70], being about 33 cm^{-1} below the upper bound. Kinoshita then obtained an extrapolated value of the ground state energy of -2.9037237. Adding relativistic, mass-polarization, and Lamb shift corrections[72] he calculated a value for the ionization potential of 198310.38 cm^{-1}compared to the then experimental value of 198310.5$\pm$1 cm^{-1}. A subsequent calculation[73], including a total of 80 terms in the trial wave function, yielded an energy of -2.9037237, a Stevenson lower bound of -2.9037467, and an ionization potential of 198310.77 cm^{-1}which could be compared to an improved experimental value of 198310.8$_2\pm$ 0.15 cm^{-1}[74]. Note that, although the error in the lower bound has been halved, it is still several orders of magnitude larger than the error in of the upper bound.

5.1.3. Perimetric coordinates

In 1958, Pekeris[75] published a result for the non-relativistic energy of the ground state of helium accurate to within 0.01 $cm^{-1}$and 0.40 cm^{-1}lower than the 39-term result of Kinoshita[69]. This work, based on a 203 term wave function, included all of the two-electron ions from $Z = 1$ to 10. In this work, Pekeris utilized perimetric coordinates[76],

$$u = \epsilon(r_2 + r_{12} - r_1), \qquad v = \epsilon(r_1 + r_{12} - r_2), \qquad w = 2\epsilon(r_1 + r_2 - r_{12}), \tag{80}$$

an independent set of coordinates all with common domains $[0, \infty]$. The trial function was constructed out of Laguerre polynomials,

$$\tilde{\Psi} = \sum_{n,m,l} c_{nml}\, L_n(u) L_m(v) L_l(w)\, \mathrm{e}^{-(u+v+w)/2}. \tag{81}$$

In the above, ϵ is an overall scale parameter like α_{sc}, but with ϵ set equal to $\sqrt{-E}$. Use of the Laguerre polynomials in the basis generated a 33-term recursion relation for the linear coefficients, c_{nml}, making it possible to carry out these calculations for very large trial functions. In subsequent calculation of the ground state, Pekeris[3] extended his trial function to 1078 terms, obtaining a result, -2.903724375, accurate to 2×10^{-6} a.u.

Examining the convergence of the eigenvalue with respect to the maximum metric order, ω_{max}, Pekeris obtained an extrapolated value of the non-relativistic energy of -2.903724376. This calculation remained the standard for helium-like systems for almost 10 years. With the inclusion of finite-mass, relativistic, and QED correction, it gave a value for the ionization energy of helium superior to the then current experimental value

5.1.4. The Fock expansion

In 1937, following Bartlett, Gibbons, and Dunn's[71] discovery that a Hylleraas series did not provide a formal solution to the Schrödinger equation, Bartlett[77] was able to show that a generalized power series which included powers of $\ln R$, $R = (r_1^2 + r_2^2)^{1/2}$ would provide a formal solution for the ground state. Fock[78] independently arrived at the same conclusions. In his work, Fock obtained an expansion of the ground state wave function about the origin in hyperspherical coordinates[79],

$$\Psi = \sum_{n}^{\infty} \sum_{m}^{[n/2]} C_{n,m}(\alpha, \cos\theta_{12})\, R^n\, (\ln R)^m, \tag{82}$$

with $\tan(\alpha/2) = r_2/r_1$ and $[n/2]$ the largest integer $\leq \frac{n}{2}$. Demkov and Ermolaev[80] have shown that a similar expansion can be applied to an n-electron system of arbitrary symmetry.

Fock developed a method to obtain the $C_{n,m}$ successively by means of expressions which rapidly become quite complicated as the order of the terms in the expansion increases. Fock was able to obtain a few of the lowest-order terms. Expressed in more usual coordinates, the expansion looks like,

$$\Psi = 1 - Z(r_1 + r_2) + \frac{1}{2} r_{12} - Z\, \frac{\pi - 2}{3\pi}\, r_1 r_2\, \cos\theta_{12}\, \ln R. \tag{83}$$

The second and third terms handle the singularities in the Hamiltonian that occur at two-body coalescence, and could have been anticipated from the Kato cusp conditions. The third term, which describes the electron-electron cusp, is absent from the CI basis but is explicitly included in a correlated basis and is thus responsible for the improved convergence of the Hylleraas method over the CI method. Along these lines, Hirschfelder[81] suggested a trial function of the form

$$\tilde{\Psi} = (1 + r_{12})\Psi_0 + \Phi, \tag{84}$$

which would yield a modified Schrödinger equation for Φ in which the singular potential energy terms have been eliminated. A version of this procedure can be constructed for excited states, care being taken to eliminate any nodes that exist in Ψ_0. Kutzelnigg[82] utilized this ansatz in a study of the ground state of two-electron ions via the Ritz variational method and the Z^{-1} expansion.

Including terms in a trial wave function like $R \ln R$, which handles the singularity at the three-particle coalescence, should further improve the convergence of the wave function. Schwartz[83] recognized the importance of logarithmic terms to the wave function, but found their inclusion in a trial wave function cumbersome. His study indicated that half-integer powers[84] of the variables would be a useful alternative, and so he constructed a trial wave function of the form,

$$\tilde{\Psi} = \sum_{n,m,l} c_{nml}\, s^{n/2} t^m u^l\, e^{-ks/2}. \tag{85}$$

Logarithmic terms were utilized by Hylleraas and Midtdal[85] in a small, 24-term wave function, while Ermolaev and Sochilin[86] carried out a 37-term calculation. Frankowski and Pekeris[87] extended this type of wave function to 246 terms, calculating energies for $Z = 1$ to 10 ions with energies superior to those obtained with Pekeris' 1078-term wave function. In fact, the 1078-term result was surpassed at the 101-term level by the Fock expansion, while their final result reduced the error in the energy by three orders of magnitude over that of the 1078-term calculation. Their trial function, with a single, overall scaling parameter, was of the form,

$$\tilde{\Psi} = \sum_{n,m,l,i,j} c_{nmlij}\; s^n t^m u^l\; R^i\; (\ln s)^j\; e^{-s/2}, \tag{86}$$

where $i = (0,1)$, m an even integer, $2j \leq n+m+l+i$, and negative powers of s, (which were essential to the accuracy of the calculation) appear only when $i = j = 0$, $n+m+l \geq 1$. More recently, Freund, Huxtable, and Morgan[88] have carried out a 230-term calculation utilizing a similar Fock basis, but with no factors of R^i, a more restrictive use of negative powers of s based on an examination of the low-order terms in the Fock expansion, and an emphasis on terms of low metric order in order to concentrate the trial function closer to the nucleus. With these modifications, they obtained a helium energy with an estimated error of a few parts in 10^{-13}, at least an order of magnitude smaller than the error in the Frankowski-Pekeris result.

In discussing the Fock expansion, Morgan[89] and Myers, *et al.*[90] have emphasized that satisfying the cusp conditions is of the greatest importance in constructing trial wave functions, of greater import than building in the correct asymptotic behavior for the wave function, and that requiring the correct behavior at the two-particle cusps takes precedence over the three-body coalescence.

We conclude this section by noting that all the terms of the Fock expansion required to correct the singular behavior at the origin are now known exactly[91], but incorporating them into a variational wave function has not yet been attempted[90].

5.1.5. Exponentially correlated functions

Correlation can also be introduced into the exponential functions of the basis. For example, a basis set can be constructed entirely of exponentially correlated functions of the form,

$$\phi_i = e^{-(\alpha_i r_1 + \beta_i r_2 + \gamma_i r_{12})}. \tag{87}$$

This form of correlated basis was first utilized by Hylleraas[92] with a single term of this form, and by Rosenthal[93], Somorjai and Power[94], and Winkler and Porter[95], all for helium-like atoms. The convergence of such a basis has proved to be extremely rapid. This is exemplified by the calculations on the helium isoelectronic sequence of Thakkar and Smith[96], who obtain an energy for the ground state of helium of -2.903724363 with only 66 such terms. In general such a basis would require the optimization of 3×66 non-linear parameters. However, Thakkar and Smith constrain their non-linear parameters, obtaining them from a set of relations involving only six parameters. A simple generalization of this basis would include a polynomial factor for each term in the basis.

A second form of exponentially correlated function was suggested by Boys[97] and by Singer[98]. This takes the form of a Gaussian correlated function,

$$\phi_i = e^{-(\alpha_i r_{1a}^2 + \beta_i r_{2b}^2 + \gamma_i r_{12}^2)}. \tag{88}$$

where $\mathbf{r}_{i\alpha} = \mathbf{r}_i - \alpha$ is the displacement of the ith electron coordinates from the center of the Gaussian at α. Like GTO's, these functions also lead to multi-center integrals expressible in terms of the error function. They can also be generalized by including an additional polynomial factor for each function. This could be a polynomial in (r_1, r_2) or (for a molecular system) Cartesian coordinates. Such functions have been referred to as exponentially correlated Gaussians (ECG's) or as Gaussian-type geminals (GTG's), and have been shown to be useful in both atomic and molecular problems. Morgan[89] has pointed out that, with sufficient flexibility in choosing the exponential parameters, ECG's can reproduce the cusps in the wave function very well, despite their formal inability to represent the two-particle coalescence.

5.1.6. "Double basis" method

"Double-basis" sets were introduced by Kono and Hattori[99,100] in 1984 in order to generate wave functions accurate enough to give reliable values for oscillator strengths of helium. Essentially, the doubled basis permits multiple length scales for the trial function so that it can more accurately represent the wave function over a wider region of configuration space. For the ground and excited S-states, the basis sets utilized in this work were of the form,

$$\phi_{nml}(\alpha, \beta) = (1 \pm P_{12})\, r_1^n r_2^m r_{12}^l\, \mathrm{e}^{-\alpha r_1 - \beta r_2}. \tag{89}$$

The trial wave function was then composed of two such basis sets with different non-linear parameters,

$$\tilde{\Psi} = \sum_{n,l,m} b_{nml}\, \phi_{nml}(\alpha, \beta) + \sum_{n,l,m} c_{nml}\, \phi_{nml}(\zeta, \zeta), \tag{90}$$

with the non-linear parameters α and β of the first basis fixed at $(-2E - Z^2)^{1/2}$ and Z, respectively, while ζ was variationally optimized. Kono and Hattori also placed restrictions on the kinds of terms that were to be included in the basis. The results for the ground state were exceptionally good; with an 196-term wave function, they achieved an accuracy of 4×10^{-10}[100].

The potential of this method was more fully realized in the work of Drake[101,102], who introduced two important features into the method: he included a single term in the wave function to represent the asymptotic behavior of the wave function in the absence of the electron-electron interaction, and he developed an efficient method to optimize all the non-linear parameters of the method. His trial wave functions are given by

$$\tilde{\Psi} = a_0 \Psi_0(1s, nl) + \sum_{n,l,m} b_{nml}\, \phi_{nml}(\alpha_1, \beta_1) + \sum_{n,l,m} c_{nml}\, \phi_{nml}(\alpha_2, \beta_2), \tag{91}$$

where $\Psi_0(1s, nl)$ is a screened hydrogenic wave function with nuclear charges Z and $Z-1$ for the $1s$ and nl electron, respectively. The addition of Ψ_0 alone accounts for a very large portion of the energy for the more highly excited states. The optimization of the non-linear parameters was carried out by calculating the derivative of the energy with respect to each of the non-linear variables;

$$\frac{\partial E}{\partial \alpha_k} = -2\left\langle \tilde{\Psi} \right| H - E \mid (1 \pm P_{12}) r_1 \Phi(\alpha_k)\rangle \tag{92}$$

Here $\Phi(\alpha_k)$ represents that part of $\tilde{\Psi}$ which depends explicitly on α_k. In the above, $\tilde{\Psi}$ is assumed to be normalized. The zero's of the first derivatives are then obtained numerically. The procedure, which is described in some detail in Ref. [102], depends on being able to locate the lowest root of the first derivatives of a complex energy surface defined by the four non-linear parameters. For the ground state, the dependence of the optimum values of the non-linear parameters on the size of the basis is quite simple, with the first pair of parameters staying roughly constant while the second pair increases linearly with $\omega_{\max}$ so that this part of the wave function maintains a maximum amplitude at roughly the same place in configuration space. Thus the two parts of $\tilde{\Psi}$ separate as ω increases, with the first part describing the long range behavior of the wave function while the second part is concentrated in the inner region of configuration space. This separation also helps to reduce complications due to possible linear dependence of the two basis sets.

For the ground state, a 660 term wave function containing terms up to metric order $\omega_{\max} = 13$, yielded an eigenvalue converged to 0.3×10^{-13} a.u., about the same accuracy as the 230-term, Fock-basis calculation of Freund *et al.*[88]. A similar calculation on the H^- ion agrees to about 14 significant figures with the results of Baker, *et al.*[103] based on the Z^{-1} expansion. A more recent calculation[104] for the helium S-states utilizing a wave function with $\omega_{\max} = 18$ and a total of 1262 terms yields a ground state energy in agreement with Baker, *et al.* to 15 significant figures!

A significant advantage of multiple basis sets over a single basis set lies in their ability to cover different regions of the configuration space through adjustment of the multiple, non-linear parameters. A typical two-electron, single-basis calculation would have only two such parameters to adjust. Hence, the only way for such a wave function to cover both large and small radial scales is by increasing the maximum metric order used in the basis. However, this leads to both computational and combinatorial problems; the largest such value of ω that can be utilized is effectively determined by the maximum precision available to the computer, and the total number of terms required for a complete set of a particular ω increases rapidly with ω. Consequently, the number of terms required for comparable convergence in a single basis calculation is much greater than with a multiple basis.

Given the extraordinary precision in the energy and, hence, to a lesser degree in the wave function, obtained with both Fock- and double-basis set calculations, mass-dependent, relativistic, and QED corrections to the non-relativistic energy can be calculated accurately and the results compared with experimental measurements. For the status of such relativistic and QED calculations on simple atomic systems as of 1985, see Ref.[105]. More recent work on this problem can be found in [106,107], where references to the corresponding experimental work can also be found.

5.2. The Z^{-1} Expansion

In his 1930 paper on the variational perturbation method[25], Hylleraas also introduced the Z^{-1} expansion. Defining a modified set of atomic units scaled on the nuclear charge, with units of length a_0/Z, a_0 the Bohr radius, and units of energy $2Z^2R_\infty hc$ with R_∞

the infinite-mass Rydberg constant, the Hamiltonian for an atom becomes

$$H = \sum_{i=1}^{n} \left(-\frac{1}{2}\nabla_i^2 - \frac{1}{r_i} + \frac{1}{Z}\sum_{j>i}^{n}\frac{1}{r_{ij}} \right). \tag{93}$$

Z^{-1} now appears as a natural perturbation parameter and the perturbation wave functions, energies, and all perturbation expansions of any observable, are independent of the nuclear charge. Hence, a single calculation suffices to calculate an entire isoelectronic sequence. Furthermore, scaling on Z indicates clearly how atomic and molecular properties scale with respect to the nuclear charges.

Having obtained the second-order energy coefficient, Hylleraas proceeded to fit computed energies of H^- and helium to obtain estimates of the third- and fourth-order coefficients, yielding the Z^{-1} expansion for the isoelectronic sequence through terms of order Z^{-2}. In 1956, Hylleraas and Midtdal[108] returned to this problem, obtaining an improved estimate to the second-order coefficient and the fourth and fifth-order energy coefficients from a second-order, variational perturbation wave function. This calculation was superseded in 1963 by the thirteenth-order calculation of Scherr and Knight[28]. This calculation was based on 100-term, Kinoshita-type wave functions. Similar calculations by Midtdal, Lyslo, and Aashamar[109] with 203 terms and Sanders and Scherr[31] with 100 terms extended the perturbation energy series to $21st$ and $25th$ order, respectively. In their work Scherr and Knight found that their energy coefficients converged in a roughly geometric sequence, with the ratio of consecutive coefficients approaching a limit of ≈ 0.78, giving a radius of convergence for the perturbation series, $\lambda^* \approx 1.25$. Hence the series would converge for the H^- ion. Kato[110] had much earlier shown that this perturbation series was, in fact, convergent and established a rough lower bound for the radius of convergence, $\lambda^* \approx 0.13$, which only showed the series converging as far as the O^{+6} ion. An improved estimate of the lower bound was obtained by Ahlrichs[111], $\lambda^* \approx 0.505$, showing that the series converges down to helium. In this work, Ahlrichs also studied the excited states of helium and the ground state of lithium, finding a larger radius of convergence in every case. This would imply a much slower rate of convergence for these series compared to the ground state of helium. Stillinger[112] attempted to characterize the nature of the nearest singularity in helium based on Midtdal's perturbation energy coefficients, obtaining $\lambda_* = 1.1184$. Other work on the radius of convergence by Brändas and Goscinski[113], was also based on Midtdal's coefficients.

Reinhardt[114] showed that if the radius of convergence were determined by a singularity on the positive, real Z-axis, then that singularity would have to occur at a value of Z where $E(Z)$ becomes degenerate with a threshold. This problem was taken up by Baker, *et al.*[103], who employed the HSK procedure with a Fock-type, double-basis set of 476 terms. Calculating the perturbation energy coefficient through $401st$ order(!), they obtained a radius of convergence for the Z^{-1} expansion of $\lambda^* = 1.09766$ and showed that the nearest singularity lies on the real Z-axis. They also used this same wave function in the Ritz variational method to find the value of λ_c where the energy of the ground state becomes degenerate with the first ionization threshold. This point could not be calculated directly of course, but their calculations set a lower bound to $\lambda_c = \lambda^* = 1.09766$, as predicted by Reinhardt. They conclude that the earlier discrepancies between λ^* and λ_c found by Stillinger and Brändas and Goscinski, were not so much due to errors in the

ϵ_n's of Midtdal, but to the simple fact that, at 21*st* order, the perturbation series is still very far from its true asymptotic behavior. These results of Baker, *et al.* were of such precision that they surpassed all other results for the isoelectronic sequence available at the time, including the case of the H^- ion. More recently, Ivanov[115] has made use of Baker, *et al.*'s ϵ_n's to construct the inverse function $\lambda(E)$ and obtain an improved estimate of $\lambda^* = 1.09766079$.

5.3. Excited States of Helium

In this section we survey some representative, high-precision calculations for the singly-excited states of helium and its isoelectronic sequence. Much of the work cited in the previous sections for the ground state of helium also reports calculations for various excited states of helium, or has been subsequently extended to such states. For example, Pekeris and co-workers carried out extensive studies of the excited S- and P-states of the helium isoelectronic sequence. There most comprehensive work[116] calculated the $1sns$, $1snp$ singlet and triplet states of the helium isoelectronic sequence for $n=2$ through 5 and $Z=2$ to 10. They also included calculations of the ^{1}S series through $n=15$ and the ^{3}S series through $n=17$ for the case of helium ($Z=2$). For the S states they were able to follow the original prescription of Pekeris[75] for the ground state. For the P-states, however, no simple recursion scheme can be constructed, although perimetric coordinates were utilized.

Applications of the HSK variational perturbation method by Sanders and Scherr[31] and by Midtdal and co-workers[117] for the $n=2$, singlet and triplet S and P states and by the latter for the $n=2$, 3 ^{3}S states[118] have been carried out. These calculations were all approximately 20*th* order, the former using 100-term trial functions, the latter between 162 and 235 terms, in "split-shell," correlated, basis sets.

Kono and Hattori[119] utilized their double basis set, 140-term wave functions to obtain energies generally superior to those of Pekeris and co-workers for the $1snl$, $n \leq 9$ S, P, and D singlet and triplet states of helium. They then extended these calculations for $3 \leq Z \leq 7$ and $n \leq 5$ for the singlet, triplet P and D states of the isoelectronic sequence[120]. Sims and Martin[121] also utilized the double basis set approach with 458-term expansions to obtain the $n=3$ - 8 singlet, triplet D and F states of helium.

The Fock-type basis was used by Frankowski[122] to obtain the energy of the 2^3S state of helium, while Baker, *et al.* utilized their Fock-type basis to obtain the Z^{-1} expansion coefficients through 26*th* order for this state.

Drake and Yan[123] carried out calculations for the $n \leq 10$, $1 \leq l \leq 7$ states of helium using their double basis set procedure with trial functions of metric order 17 and 18 and approximately 1000 terms. This number of terms would correspond to a metric order complete only through about $\omega = 11$, but Drake and Yan place restrictions on the terms selected for inclusion in their basis set. In so doing, they avoid instability due to "numerical" linear dependence in the bases, and are also able to tailor their wave function to the differing length scales present in the two basis sets. In a later article, Drake and Yan[104] improve on their earlier results for the $n \leq 10$ singlet, triplet S states. In this work they also compare their method of selecting terms to the alternative of simply including all terms up to a maximum metric order. The latter method is found to yield slightly improved results, but with a penalty in terms of the numerical stability of the calculation. This degraded stability limits the highest root of the secular determinant (and hence, the highest state) that can reliably be calculated.

5.4. Doubly-excited States

Doubly-excited states of atoms lie above the first ionization threshold and are degenerate with one or more single-electron ionization continua. They have been the subject of intense interest despite the difficulties, both theoretical and experimental, inherent in their study. They were first properly described over 30 years ago by Fano[124] who modeled them as bound states interacting with a single continuum. Such states would appear as resonances in the photoionization cross section or electron scattering cross section. They can also be described as quasibound, autoionizing states. For several years, computational results were more comprehensive than the corresponding experimental studies. More recently, however, experimental data of very high quality for a number of Rydberg series up to very high levels of excitation have been obtained[125]. This has challenged the theoretical calculation of energy levels and lifetimes. Where 4-5 significant figure accuracy for the energy and 2-3 significant figures for the width would have been adequate in the past, at least an additional two significant figures in both these properties would be considered state-of-the art today.

It was Fano who noted that, due to the high degree of degeneracy present in these states, they could not be described adequately by the usual designation of atomic configurations. Various classification schemes for doubly-excited states have been devised which more closely reflect the structure observed in the energy levels. These are generally based on proposed symmetries for the system of doubly-excited states and so yield new quantum numbers to designate the states. The most widely used classification scheme[126,127], introduces the quantum numbers K, T, and A which are related to angular(K, T) and radial (A) correlations of the system. A state with a configuration $Nlnl'\ ^{2s+1}L$ with $n \geq N$ could be designated as ${}_N[K,T]_n^A\ ^{2s+1}L$.

We should distinguish two classes of these states, depending on their parity. Doubly-excited states of unnatural parity, because of their symmetry, do not interact (in the non-relativistic limit) with any states of natural parity, including the singly-ionized continuum. In the non-relativistic approximation, these states can be treated as ordinary excited states. For example, the $2p^2\ ^3P^e$ state is the lowest state of this parity, and so can be calculated by ordinary variational methods[128].

Doubly-excited states of natural parity, however, present an unusual challenge for all computational methods discussed here. They interact strongly with any adjacent continuum and decay directly to it by autoionization, typically with lifetimes orders of magnitude shorter than for radiative decay. These states can be further separated into intrashell states, where both electrons are in the same shell ($n = N$) and the entire set form a "double Rydberg" series converging on the double ionization threshold, and intershell states ($n > N$), which for larger $n - N$ form a Rydberg series converging on the Nth ionization threshold. The intrashell states of He and $H^-$are particularly interesting, with the lower-lying members of a particular degenerate manifold most closely satisfying the group theoretical classification. These states exhibit highly correlated motion and, as such, are the prototypical example of electron correlation in atoms.

Because these states lie above an infinite number of discrete, singly-excited states and are embedded in one or multiple continua, the usual variational procedures for excited states will not serve. A variety of computational methods have been devised for treating these states. They include the stabilization method[129], Feshbach projection method[130,131], various L^2 discretization methods[132,133], saddle point method[134], complex scaling method[135], and hyperspherical close-coupling method[136]. These have all been implemented directly within the Ritz variational method. For some of these, equivalent methods have also been developed within the variational perturbation method[33,137]. Results for the 2s2p states of helium are tabulated in Table 2. In the following sections, we present a brief sketch of these methods, referring the reader to the more extensive articles noted above for a full discussion.

Table 2
Energies and Widths of the 2s2p States of Helium (in a.u.)

	Energy	Width$\times 10^{-3}$	Method
	2s2p ^{1}P		
Bhatia and Temkin[62]	-0.693157	1.330	Feshbach projection
Ho[152]	-0.6931355	1.366	Complex Scaling
Chung[144]	-0.693024	1.372	Saddle-point Method
Seminario and Sanders[33]	-0.69305	1.370	Feshbach projection, Z^{-1}
Manning and Sanders[137]	-0.6931349	1.373	Complex Scaling, Z^{-1}
Brage, *et al.*[160]	-0.69258	1.35	MCHF
	2s2p ^{3}P		
Bhatia and Temkin[62]	-0.760492	0.3270	Feshbach projection
Ho[152]	-0.7604925	0.2970	Complex Scaling
Seminario and Sanders[33]	-0.76044	0.300	Feshbach projection, Z^{-1}
Manning and Sanders[137]	-0.7604923	0.2988	Complex Scaling, Z^{-1}
Brage, *et al.*[160]	-0.76047	0.322	MCHF

5.4.1. Stabilization

The stabilization method[129] is essentially a straightforward variational method applied to a multiply-excited state. In the process of varying a non-linear parameter of the basis, the optimization curves for certain roots of the secular determinant may become stable over a range of values of the parameter. This stability against variational collapse does not persist long, and when a particular root collapses it typically stabilizes at the position of the next lowest stable root. Meanwhile, the stable root just above the original root drops down to replace it. In this way, a particular eigenvalue can be traced across a series of avoided crossings, identifying it as an eigenvalue of the Hamiltonian. The art in the use of this method lies in choosing the form of the trial wave function so that a stable root corresponding to the particular state of interest is generated. It is not always an insignificant task to identify the sought for stable root in a maze of avoided crossings, particularly when such a stable root is not altogether stable.

If this stabilization method is applied within the variational perturbation method, only a single optimization curve can exist. This opimization curve, however, will display a series of tangent-like singularities under the variation of any parameter of the problem[33,138].

An important variant of this method was developed by Hazi and Taylor[132], who noted that this stability should also appear as a function of the size of the basis. Thus a stable root can be traced as the size of the basis, N, is increased. Hazi and Taylor further showed that the degree of stability of the root as a function of N is proportional to the width of the corresponding state, thus extending the use of the stabilization method to the calculation of the energy and width of resonances.

5.4.2. Feshbach projection

The Feshbach projection method[130] divides the Hilbert space into two orthogonal subspaces with the help of projection operators, P and Q. These satisfy the relation, $P + Q = 1$, and the asymptotic conditions,

$$P\Psi \sim \Psi, \qquad Q\Psi \sim 0. \tag{94}$$

The Schrödinger equation can then be written

$$(E - H)(P\Psi + Q\Psi) = 0. \tag{95}$$

with $P\Psi$ being the open-channel, continuum-like part of the wave function and $Q\Psi$ being the closed-channel, boundlike part. Making use of the fact that $P^2 = P$, $Q^2 = Q$, and $PQ = QP = 0$, we can write[131],

$$\begin{aligned} (E - H_{PP})P\Psi &= H_{PQ}Q\Psi, \\ (E - H_{QQ})Q\Psi &= H_{QP}P\Psi, \end{aligned} \tag{96}$$

with $H_{PP} = PHP, H_{QQ} = QHQ$. Hence, we can write,

$$Q\Psi = \frac{1}{E - QHQ}QHP\Psi \tag{97}$$

One now calculates the eigenfunctions and eigenvalues of H_{QQ},

$$QHQ\Phi_n = \mathcal{E}_n\Phi_n, \qquad \Phi_n = Q\Phi_n, \tag{98}$$

and these eigenfunctions are utilized to construct $Q = \sum_n |\Phi_n\rangle\langle\Phi_n|$, so that

$$Q\Psi = \sum_n \frac{|\Phi_n\rangle\langle\Phi_n|H|P\Psi\rangle}{E - \mathcal{E}_n} \tag{99}$$

From this, a modified Schrödinger equation for $P\Psi$ can be obtained. Finally, expressions[131] for the width, Γ_n, and shift, Δ_n, of the resonance energy, $E = \mathcal{E}_n + \Delta_n$, can be obtained.

For the particular case of S-state resonances of two-electron atoms below the n=2 threshold, O'Malley and Geltman[139] took the projection operators to be,

$$P = P_1 + P_2 - P_1P_2 \qquad Q = 1 - P \qquad P_i = |1s(i)\rangle\langle 1s(i)| \tag{100}$$

where $1s$ is the hydrogenic, ground state orbital. With this operator, the eigenfunctions of H_{QQ}, which are square-integrable functions, can be calculated as in an ordinary variational calculation with one additional complication, the need to project out any $1s$ character from the basis set. How to do this for a Hylleraas-type basis is described in Ref. [140]. The optimization of any non-linear parameters in the basis is stable against variational collapse to lower-lying, singly-excited states since projection has made the trial function very nearly orthogonal to all lower-lying states, bound or continuum, at least in that region of configuration space where there is significant overlap between $Q\Psi$ and the wave functions for these singly-excited states. Nevertheless, there is a possibility for variational collapse to an energy lower than $\mathcal{E}_n$, since the singly-excited states are only approximately described as $1s$ in character.

A review of calculations based on this method can be found in Ref.[131]. Examples of high accuracy calculations for the lowest doubly-excited $P-$states of two-electron atoms can be found in Table 2. For autoionizing states of systems with more than two electrons, the projection operator is to be constructed from the singly-ionized (target) wave function, and these will no longer be known exactly. An example of such a calculation for three electron autoionizing states is Wakid, Bhatia, and Temkin[141] and a procedure for constructing such projection operators for three electron atoms is given in Ref. [142]

Applications of the Z^{-1} expansion to autoionizing states of atoms have been rather limited and most[33,138] have relied on use of the Feshbach projection method. Perturbation theory is perhaps computationally more naturally suited to the study of autoionizing states than approaches based on the variational method. The degeneracy between each doubly-excited, zero-order state and the singly-excited, zero-order continuum is one difficulty not present in a variational calculation that must be dealt with in perturbation theory. Application of the Feshbach projection method to the Z^{-1} expansion circumvents this problem by projecting out exactly these degenerate continuum functions. This results in a variational perturbation calculation for the closed-channel part of the resonance wave function similar to that for an ordinary bound-state problem. The condition for upper bounds in the variational perturbation method requires projection with respect to the exactly-known, zero-order wave functions of lower-lying states[31]. For two-electron atoms below the second ionization threshold, these are all precisely $1s$ in character. With this projection, the perturbation wave functions, $Q\psi_n$, are all independent of these conditions and will yield rigorous upper bounds to the even-order perturbation energy coefficients of $\mathcal{E}$ without the possibility of variational collapse[33]. For atoms with more than two electrons, the projection operators can still be constructed entirely from the exactly-known wave functions of the unperturbed Hamiltonian. Hence, rigorous upper bounds to the energy coefficients are also possible in this case. Calculating the width and shift of the autoionizing state, however, requires knowledge of the continuum-like part of the wave function that has been projected out.

5.4.3. Saddle-point method

In the saddle-point method[134], the projection operator of the Feshbach theory is replaced by a projection operator based on a variational representation of the "hole" function. For a two-electron atom below the $n = 2$ threshold, this would amount to replacing the hydrogenic $1s$ orbital with a single STO $1s$ orbital. The exponential parameter could then be varied so as to maximize the energy while the simultaneously minimizing the energy with respect to the other parameters of the trial wave function. This minimax procedure seems to give a better upper bound to the resonance energy than the straightforward Feshbach method[143].

Chung and co-workers have applied this method to several low-lying doubly-excited states of two-electron atoms. In particular, their calculation of the energy and width of the $2s2p\ ^1P$ state is very accurate for the size of the calculation[144]. This method has also been applied to three-[145] and four-electron[146] autoionizing states, while a variant of the method has been developed which combines the complex scaling method with the saddle-point method[147]. Chung and Davis[148] have applied the latter method to a great many lithium-like autoionizing states.

5.4.4. Complex scaling method

Of the methods developed to study such atomic states, the complex scaling method has been one of the most effective, yielding both resonance positions and widths of autoionizing states in a single calculation. Much of the earlier work applying this method to atomic systems has been reviewed by Ho[135], who also provides much information on the computational aspects of implementing this method. The complex scaling method involves a complex scaling of all the radial coordinates of the Hamiltonian for the system: $r \rightarrow e^{i\theta_r} r$. This scaling can be introduced into a typical variational program with only trivial modifications of the code. Hence, the method is extremely easy to implement on an established variational program. For the case of Coulomb potentials, the Hamiltonian becomes:

$$H(\theta) = e^{-2i\theta} T + e^{-i\theta} V. \tag{101}$$

The transformed Hamiltonian is non-hermitian with complex eigenvalues, W. The discrete, real eigenvalues of the rotated Hamiltonian are independent of θ and are the discrete, bound state spectrum of the system which, under this transformation, remain on the real axis. However, the continuous spectra will be rotated about their respective thresholds by -2θ into the lower energy half-plane. The resonances are the discrete, complex eigenvalues of the rotated Hamiltonian. For these states, the real part of W is the resonance position, E_r, while the imaginary part is the half-width, $\frac{1}{2}\Gamma$. As one increases θ and the associated continua sweep past a resonance, the wave-function for that resonance becomes square-integrable. With the wavefunction square-integrable, calculation of the eigenvalues can proceed in a manner similar to a standard variational calculation for a bound state. Difficulties in applying this method include the lack of any upper bound for the eigenvalue with associated difficulties in optimizing a basis set. Consequently, there is also no assurance that increasing the size of the basis set will improve the value of the energy, although, in practice, that is generally the case. Typically, optimization of the wave function is accomplished through the use of the complex stabilization method[135]. In this method, a stable point is sought in the "trajectory" of the eigenvalue in the complex-energy plane under the variation of the parameters of the method[149]. In Ho's implementation of the method, a preliminary selection of non-linear parameters is made through the usual (real) stabilization method. Once a stable, real root is located, complex scaling is introduced and stabilization is accomplished in the complex energy plane. There has been considerable discussion about the characteristic behavior of these variational trajectories,[150] although this discussion has focused primarily on their behavior in the vicinity of stable points. Less has been said about their global behavior over the entire range of the parameter space. This global behavior has recently been studied by Georgiev[151].

Complex stabilization calculations for a great many autoionizing states of two-electron atoms have been carried out by Ho using very large basis sets, up to and including states lying below the $n = 7$ threshold. For the lower-lying autoionizing states, most of this work has been done with correlated basis sets with up to 1000 terms[152,153], while the more highly-excited states have also been calculated with CI wave functions with a similar number of terms[154].

Adapting the complex scaling method to the Z^{-1} expansion results in few changes to the usual perturbation expressions. The rotated zero-order Hamiltonian, in charge-scaled atomic units, is now given by

$$H_0 = \sum_i \left[-\frac{1}{2} e^{-2i\theta} \Delta_i - e^{-i\theta} \frac{1}{r_i} \right], \tag{102}$$

while the perturbation becomes

$$\lambda H_1 = Z^{-1} \sum_{j>i} \frac{e^{-i\theta}}{r_{ij}}. \tag{103}$$

Under the complex rotation transformation, the resonant state has a complex energy, W. This is expanded in a power series in $1/Z$,

$$W = \sum_n \omega_n Z^{-n}. \tag{104}$$

The ω_n can themselves be expressed in terms of the expansion coefficients of the resonance position and width,

$$\omega_n = \epsilon_n - i \frac{\Gamma_n}{2} \tag{105}$$

Application of the complex rotation method removes the degeneracy with the continuum in zeroth order, resulting in a bound, zero-order wave function. This in turn means that the lowest-order contributions to the total energy are purely real, $\omega_0 = \epsilon_0$ and $\omega_1 = \epsilon_1$, and that Γ_2 is the leading contribution to the width[155]. In ordinary atomic units (obtained from the charge-scaled results by multiplying by Z^2), Γ_2 is the limiting value of Γ for high Z as well.

The first-order wave function may be found approximately by optimizing the complex analogue of the Hylleraas-type functional,

$$\omega_2 \approx J_1 = \left\langle \tilde{\psi}_1^\star \middle| (H_0 - \epsilon_0) \middle| \tilde{\psi}_1 \right\rangle + \left\langle \tilde{\psi}_1^\star \middle| (H_1 - \epsilon_1) | \psi_0(\theta) \right\rangle + \left\langle \psi_0^\star(\theta) \right| (H_1 - \epsilon_1) \left| \tilde{\psi}_1 \right\rangle, \tag{106}$$

where the complex conjugate of the ket is required by the bi-orthogonality of the wave functions in the complex scaling method. J_1 is stationary about the true ω_2 under variation of $\tilde{\psi}_1$. The variational perturbation expressions for the higher-order complex energy coefficients follow the usual $(2n+1)$ rule in which energy coefficients up to ω_{2n+1} are obtained from the approximate nth-order wave function.

Only a few applications of the Z^{-1} expansion within the complex rotation method have been carried out. Weinhold[156] studied the $2s^2\ ^1S$ state of two-electron atoms, but only through third-order in the energy while Moiseyev and Weinhold[157] analyzed the calculations of Ho[152] and Dalgarno and Drake[138] to extract the leading coefficients of the Z^{-1} expansion for several doubly-excited states of two-electron atoms. Manning and Sanders have studied a large number of doubly-excited P-states below the $n = 2$ threshold[158], but only to lowest-order in the energy and width. Manning and Sanders have also carried out a very accurate, higher-order study[137] of the $2s2p$ states. In this work, the approximate nth-order wave functions, $\tilde{\psi}_n$, were expanded in a correlated double-basis set with multiple non-linear parameters:

$$\tilde{\psi}_n = \sum_{i=1}^{N_1} a_i \, \phi_i(\alpha_1, \beta_1) + \sum_{i=1}^{N_2} b_i \, \phi_i(\alpha_2, \beta_2) \tag{107}$$

with each basis set chosen to be of the form

$$\phi_i(\alpha, \beta) = (1 \pm P_{12}) \, r_1^{n_i+l_i+1} r_2^{m_i+l_i} r_{12}^{c_i} \, e^{-\alpha r_1 - \beta r_2} \, P_{l_i}(\cos\theta_{12}) \cos\theta_1, \tag{108}$$

with $c_i = 0,\ 1$ and a total of 649 terms in the trial wave function. Minimizing the variance of the J_1 of Eq. 106 with respect to the linear coefficients, yields the usual system of linear equations of the Hylleraas variational perturbation method. The procedure followed in this work was to first stabilize ω_2 using the larger of the two basis sets. Keeping the first basis set fixed, the second basis set was then introduced and its non-linear parameters varied so as to yield well-behaved higher-order energy coefficients. This made it possible to extend the calculation through 19*th* order for the $2s2p\ {}^1P$ state and 15*th* order for the $2s2p\ {}^3P$ state. The behavior with increasing order for both the real and imaginary parts of these coefficients exhibits the damped, oscillatory behavior seen in many singly-excited states of two-electron atoms.[159]

6. THREE-ELECTRON ATOMS

One of the earliest calculations on the $1s^2 2s\ {}^2S$ ground state of lithium was that of Wilson[161] in 1933, with a one term, four parameter wave function. We have already noted the 45-term CI wave function of Weiss[49]. In three-electron systems, the slow convergence of a CI basis is exacerbated by the increasing number of terms required to reach a particular metric order, ω. The additional factor, r_3^p, in the CI basis means that convergence becomes an even greater problem. For a correlated basis, the number of terms required to reach a particular ω is further increased by the addition of two r_{ij}^p factors to the basis. In a two-electron CI calculation, 16 terms are required for $\omega = 6$ alone, including $l = 0, 1, 2, 3$ partial waves. Adding this number to the total number of terms needed for $\omega = 0$ through 5 yields a total of 50 terms to construct a wave function complete through $\omega_{\max} = 6$. The corresponding number of terms for a correlated basis would be 28 and 84. In comparison, a three-electron calculation will require 39 and 105 CI terms and 438 and 811 correlated terms to reach the same metric order. Clearly, this problem rapidly becomes unmanageable, and values of $\omega_{\max}$ comparable to those utilized in the largest two-electron calculations are presently not feasible.

More recent, three-electron, CI calculations include the MCHF calculation of Tong, Jönsson, and Fischer[162], the "full-core plus correlation" calculation of Chung[163] and the CI wave function of Jitrik and Bunge[164]. In his approach, Chung constructed a $1s^2$

Li^+ core wave function of 318 terms with $\omega \leq 15$ and $l \leq 6$. This core function was then held fixed and combined with 8 STO's to construct a $1s^2 2s$ function with no core-valence correlation. An additional CI basis of 1009 terms was then added for a combined "full-core plus correlation" wave function of 1017 terms. If, in this calculation, the core function had been left flexible, the total size of the wave function would have been 3553 terms. In their work, Jitrik and Bunge constructed a 3653-term, 60-configuration CI wave function. Their results were then extended to include corrections to the $1s^2$ and $1s2s$ correlation energies to arrive at an extrapolated non-relativistic energy for the ground state. This new estimate of the non-relativistic energy, while it too remains more negative than any calculated value for the energy, significantly reduces a long-standing discrepancy between calculated and estimated values of the lithium ground state energy. This discrepancy has had an interesting history, as it has been consistent enough to convince most of its reality and large enough to suggest that even correlated basis would be too slowly convergent for this problem. It now appears that all earlier estimates of the non-relativistic energy, whether based on experimental energies[165,166], on extrapolations of CI energies[167], or other methods[49] have arrived at similar but incorrect values for the energy.

For two-electron systems, a variety of basis sets based on modified r_1, r_2, r_{12} coordinates have been developed. For n-electron systems, such coordinates introduce additional complexity into an already difficult problem. In the specific case of perimetric coordinates, Wang[168] has shown that it is not possible to construct a system of independent coordinates analogous to perimetric coordinates for more than two particles. Hence, most correlated basis sets for three-electron systems are of the form,

$$\tilde{\Psi} = \mathcal{A} \sum_{i,j,k,l,m,n} c_{ijklmn}\ \phi_{ijklmn}(\mathbf{r}_1, \mathbf{r}_2, \mathbf{r}_3)\ \Sigma(1,2,3), \tag{109}$$

with $\mathcal{A}$ being the antisymmetrizing projection operator, Σ being a three-particle, doublet spin function, and

$$\phi_{ijklmn}(\mathbf{r}_1, \mathbf{r}_2, \mathbf{r}_3) = r_1^i r_2^j r_3^k\ r_{12}^l r_{13}^m r_{23}^n\ \mathrm{e}^{-(\alpha r_1 + \beta r_2 + \gamma r_3)} \tag{110}$$

Following the usual terminology, we shall call this a Hylleraas (HY) basis, to be distinguished from the Hylleraas coordinates of Eq. 76.

James and Coolidge[169] carried out the first calculation on three-electron systems with a small, correlated basis set in 1936. After disappointing results with several small, uncorrelated wave functions, they calculated the ground state energy of neutral lithium with a 17 term HY basis of the form given in Eq. 110. In order to undertake such a calculation, they first had to find a way to deal with the three-electron integrals that appear in these calculations.

6.1. Three-electron Integrals

The method James and Coolidge developed for three-electron integrals is similar to their method for two-electron integrals[170]. In this latter work, they computed two-electron integrals by means of V-integrals, essentially integrals over incomplete gamma functions,

$$V_{n,m}(\alpha, \beta) = \int_0^\infty dx x^n \mathrm{e}^{-\alpha x} \int_x^\infty dy y^m \mathrm{e}^{-\beta y}. \tag{111}$$

For three-electron systems, the required integral is,

$$I_{n,m,l,i,j,k}(\alpha,\beta,\gamma) = \int r_1^n r_2^m r_3^l r_{12}^i r_{13}^j r_{23}^k \, e^{-(\alpha r_1+\beta r_2+\gamma r_3)} \, d\mathbf{r_1} d\mathbf{r_2} d\mathbf{r_3} \tag{112}$$

The first step in reducing this integral is to expand the r_{ij}^k. This can be done, for example, by a formula obtained by Sack[171],

$$r_{ij}^k = \sum_{l=0}^{\infty} R_{kl}(r_i, r_j) P_l(\cos\theta_{ij}), \tag{113}$$

where the $P_l(\cos\theta_{ij})$ are Legendre polynomials. The R_{kl} are radial functions involving $r_<$, $r_>$, the smaller/greater of r_i, r_j. If k is even, the R_{kl} vanish for $p \geq k/2$ and the sum in Eq. 113 terminates, but, if k is odd, the sum is infinite. The Legendre polynomials are now expanded using the addition theorem for Legendre polynomials, and the angular integrations carried out. Orthogonality conditions between the resulting spherical harmonics reduce the three sums generated by the r_{ij} factors of Eq. 112 to a single sum. This remaining sum will be finite if no more than two of the (i,j,k) in this equation are odd, whereas if all are odd, an infinite sum results. The remaining integral over the r_i's is divided into six parts, each characterized by $0 < r_i < r_j < r_k < \infty$ and expressed as a sum over six W-integrals,

$$W_{n,m,l}(\alpha,\beta,\gamma) = \int_0^\infty dx x^n e^{-\alpha x} \int_x^\infty dy y^m e^{-\beta y} \int_y^\infty dz z^l e^{-\gamma z}. \tag{114}$$

For efficient calculation of these integrals, James and Coolidge developed a variety of recursion relations similar to those for the V-integrals. In their calculations, they restricted each term in their basis to no more than one r_{ij} factor so that the most general integral they encountered was of the form $I_{n,m,l,i,j,-1}(\alpha,\beta,\gamma)$, considerably simplifying the generation of the required integrals. Even with this restriction, of course, coupled three-electron integrals will still appear that can only be expressed as infinite sums over W-functions. Fortunately, these sums converge rapidly for reasonable values of the non-linear parameters. Some of James and Coolidge's recursion formulae are numerically unstable, however, and alternative methods to compute the W-integrals have been developed, as well as alternative computational schemes to generate the full three-electron integrals[172–177].

6.2. The Ground State of Lithium

Results of the calculations to be described below are found in Table 3, together with some CI results and empirical or extrapolated values of the non-relativistic energy.

Öhrn and Nordling[178] constructed a wave function for the ground state in 1966 by taking a six-term, HY-basis for the $1s^2$ core together with a $2s$ function, a core-geminal, valence-orbital prescription. A significant step forward in the lithium problem was made by Larsson[173], who took a 100-term HY-basis, utilizing the two degenerate doublet spin functions,

$$\tilde{\Psi} = \Psi_a(\alpha\beta\alpha - \beta\alpha\alpha) + \Psi_b(2\alpha\alpha\beta - \beta\alpha\alpha - \alpha\beta\alpha), \tag{115}$$

with 60 terms in Ψ_a and 40 in Ψ_b. These are of the form of Eq 110. This basis contained terms of metric order as high as 7 and included two terms with more than a single

Table 3
Energies of the Ground State of Lithium (in a.u.)

	E	Method
Wilson[161]	-7.4192	4-term CI
James and Coolidge[169]	-7.47607	17-term HY
A. W. Weiss[49]	-7.47710	45-term CI
Öhrn and Nordling[178]	-7.4760	12-term HY
Larsson[173]	-7.478025	100-term HY
Perkins[179]	-7.47793	30-term HY
Sims and Hagstrom[184]	-7.478023	150-term HY-CI
Ho[180]	-7.478031	92-term HY
Muszyńska, *et al.*[185]	-7.478044	139-term HY-CI
Pipin and Wòznicki[186]	-7.478044	170-term HY-CI
King and Shoup[181]	-7.478058	352-term HY
King[182]	-7.478059	602-term HY
Kleindienst and Beutner[187]	-7.47805824	310-term HY-CI
King and Bergsbaken[183]	-7.47805953	296-term HY
McKenzie and Drake[190]	-7.478060326	1134-term HY
Pipin and Bishop[188]	-7.4780601	1618-term HY-CI
Luchow and Kleindienst[189]	-7.47806025	976-term HY-CI
Yan and Drake[191]	-7.47806032156	1589-term HY
Tong, *et al.*[162]	-7.4780609	MCHF-extrapolated
Chung, *et al.*[163]	-7.4780597	CI-extrapolated
Jitrik and Bunge[164]	-7.477906662	3653-term CI
Estimates of ground state energy		
Scherr, Silverman, and Matsen[165]	-7.478073	semi-empirical
A. W. Weiss[49]	-7.47710	$Li^+ + 2s$
Bunge[166]	-7.478073	estimated
Jitrik and Bunge[164]	-7.47806241	CI extrapolated

r_{ij} factor in them. Although only one of these spin functions is necessary to obtain a complete ^2S wave function, both parts are included as the second spin function is known to be important in the calculation of the Fermi contact term. The non-linear parameters for the wavefunction were optimized on a small basis and from that point on taken to be taken to be $\alpha = \beta = 2.76, \gamma = 0.65$.

Similar HY trial functions, but with only a single spin function, were utilized by Perkins[179] (30-term HY) and Ho[180] (92-term HY). Perkins attempted to optimize his choice of 30 terms carefully, including two terms of metric order 6 but no terms with more than a single r_{ij} factor in it. The non-linear parameters were not optimized, but taken to be $\alpha = \beta = 2.80, \gamma = 0.65$. 19 of these 30 terms are also found in Larsson's basis set. Ho included in his basis all 60 of Larsson's terms, but utilized Perkins' values for the non-linear parameters. The 32 terms added to Larsson's basis set included one additional term with a pair of r_{ij} factors and terms up to $\omega = 10$. The effectiveness of these terms was tested by examining their contribution in groups, whereas Perkins had examined individual terms in selecting his 30 terms. All three of these applications of the HY basis required that higher order terms be selected carefully to improve the convergence of the energy within the limited number of terms utilized. Selecting terms to include in a wave function on the basis of their contribution to the energy is a difficult task. The contribution any one term makes to the energy depends on what other terms are present in the basis, so that simply replacing eliminated terms with new terms may mean that some eliminated terms might have assumed a greater importance in the new basis through their interaction with the new terms. Testing the effectiveness of a single term against all possible terms in a large basis for a system like lithium is clearly impractical, however, while simply including all terms up to a certain metric order is also not feasible except for very low values of $\omega_{\max}$.

In 1986, King and Shoup[181] extended the HY-basis to 352 terms. In this calculation, they included all 210 terms through metric order 4 and utilized both doublet spin functions. The remaining 142 terms included all terms included by Larsson and Perkins in their wave functions, plus additional terms which contributed to the convergence of the energy. Included in this wave function were a significant number of terms with multiple r_{ij} factors and terms up to metric order 11. Optimization of the non-linear parameters was carried out on small basis sets with only slight differences from the values of Larsson and Perkins. Hence, in the full basis set, the non-linear parameters of Larsson were used. The convergence of the energy, even with such a large basis, remained relatively slow once the wave function exceeded 30 terms, or so. King[182] extended this work to include a total of 602 terms, again including all 210 terms with $\omega \leq 4$. Although both spin functions were included in the calculation, it was noted that the second spin function contributed little to the energy or even to expectation values of spin-dependent operators, with the one exception being the fermi contact interaction. This calculation yielded the lowest variational energy yet for the ground state of lithium. The discrepancy with "empirical" estimates of the non-relativistic energy was noted. At this point, the differences between the various estimates of the non-relativistic energy were becoming significant in comparison to the discrepancy with the calculated energy and the need for a better estimate of the "exact" non-relativistic energy was apparent.

A different approach to the standard HY basis was followed in the work of Sims and Hagstrom[184]. Their 150-term Hylleraas-CI (HY-CI) basis consisted of products of 43 STO's multiplied by a single r_{ij} factor,

$$\phi_{klmn} = r_{ij}^{n}\, \varphi_k(\mathbf{r}_1)\varphi_l(\mathbf{r}_2)\varphi_m(\mathbf{r}_3) \tag{116}$$

with n ranging from 0 to 4. A single spin function was utilized, and all terms with $\omega \leq 2$ were included. Results were comparable to Larsson's 100-term results. Other examples of HY-CI calculations include Muszyńska, Papierowska, and Wòznicki[185] (139 terms), Pipin and Wòznicki[186] (170 terms), Kleindienst and Beutner[187] (310 terms), Pipin and D. M. Bishop[188](1618 terms), and Lüchow and Kleindienst[189] (976 terms).

The multiple basis set method, used so effectively for helium by Drake, was also applied to the lithium ground state by McKenzie and Drake[190]. With a 1134-term wave function which included terms with $\omega \leq 7$, they obtained a variational energy whose convergence clearly indicated that previous empirical and/or extrapolated estimates of the ground state energy had been too low. They estimated their result to be converged to within one part in 10^9. Yan and Drake[191] utilized the multiple basis set method with terms up to $\omega = 8$ in a 1589-term, single spin function wave function. Their wave function contained five basis sets, selected to describe different correlations among the electrons. The complete set of non-linear parameters were then optimized in a manner similar to Drake's method[102]. The enormous amount of computational effort embodied in this optimization of parameters was minimized by the use of their method[177] for rapidly evaluating three-electron integrals. Convergence of the energy in this calculation was studied as the metric order of the wave function was increased, permitting Yan and Drake to obtain an extrapolated value of the ground state energy of $-7.47806032310(31)$, estimated to be accurate up to a few parts in 10^{11}.

Much of the work cited above has also led to calculations of comparable accuracy on excited states of three-electron atoms. Of particular note are recent results by King[192] and Yan and Drake[191]. Yan and Drake have also calculated accurate oscillator strengths for transitions between some low-lying states[193]

6.3. Perturbation Methods

Perturbation methods for three-electron systems have not been developed to the point that they can produce the kind of precision obtained with the variational method. Although there are no intrinsic reasons why such a calculation could not provide the same degree of precision, it would need to be carried to high order in the perturbation. With the exception of relativistic, many-body perturbation calculations[194], only low-order calculations on lithium-like systems have been carried out. In Rayleigh-Schrödinger perturbation theory, the energy of the ground state of three-electron atoms has been obtained through third-order from calculations utilizing a Hartree-Fock zero-order Hamiltonian[195] as well as a bare-nucleus (hydrogenic) zero-order Hamiltonian[196]. There are also several second-order calculations for various excited states of lithium-like ions[197–199].

One very useful property of perturbation theory for multi-electron atoms, is that the complete first-order wavefunction of any atom can be constructed from the first-order pair functions of all two-electron configurations present in the zero-order wavefunction[200], while higher-order perturbation contributions similarly decompose into contributions from sub-configurations of the complete system[201]. The decomposition of the nth-order perturbation wave function into from two- to $n+1$-electron components also restricts the type of electron correlation that can occur in each component, with obvious computational advantages. This approach was exploited to obtain the first-order wave function and second-[197,199] and third-order[196] energy coefficients cited above, but only one calculation of the second-order wave functions seems to have been carried out using this method[202].

7. (SOMEWHAT MORE) COMPLEX ATOMS

For atoms or molecules more complex than lithium, the use of correlated basis sets of Hylleraas type rapidly becomes infeasible, both in terms of the complexity of the calculation and the amount of computational resources required. Hence, the CI method and its variants become more attractive as the number of electrons in the system increases.

The slow convergence in l for a CI basis is further complicated by the extremely large number of terms required to reach a given metric order, ω, for each partial wave. For a 10-electron problem like neon, a total of 55 terms are required to construct a CI wave function complete through $\omega \leq 2$ alone. Thus, in order to obtain accurate, non-relativistic energies for complex atoms with a CI basis, careful estimation of truncation errors is generally required. Introducing a correlated basis set further complicates this problem by introducing yet more combinations of exponents to complete a particular metric order. If only a single r_{ij} factor is permitted per term, 561 terms are required to complete $\omega = 2$ for neon, while a total of 1596 terms are required if this restriction is lifted. If, loosely speaking, one needs to reach equal values of ω or l to achieve roughly the same degree of convergence in n-electron systems as in two-electron systems, then convergence of the basis will present a major difficulty. Nevertheless, the significantly more rapid convergence of a correlated basis would make it an attractive alternative to CI if the magnitude of the calculation could be controlled. Extension to Hylleraas-type correlated basis sets for more complex systems can be accomplished in a fairly straightforward manner. Karl[203] has shown that n-electron integrals in which at most four linked r_{ij} factors appear result in integrals which can be evaluated in a manner similar to the methods of James and Coolidge's W-functions. However, computation time becomes a major issue in carrying out calculation of such integrals. Restricting the basis to only one r_{ij} factor per term means that no integrals more complex than these coupled, four-electron integrals can ever arise. Various schemes for calculating these integrals have been proposed, among which we cite some examples[204,205]. After reduction of the angular integrations, the radial integral that remains is divided into 24 subspaces, leading to sums over 24 X-integrals:

$$X_{n,l,m,p}(\alpha,\beta,\gamma,\delta) = \int_0^\infty dx x^n \mathrm{e}^{-\alpha x} \int_x^\infty dy y^m \mathrm{e}^{-\beta y} \int_y^\infty dz z^l \mathrm{e}^{-\gamma z} \int_z^\infty dv v^m \mathrm{e}^{-\delta v}. \tag{117}$$

This type of four-electron integral has been utilized in calculations of four-electron atoms with HY or HY-CI bases[206–209], but all such calculations have placed severe restrictions on the correlation factors in the basis set in order to avoid the most difficult of these integrals. Details of the evaluation of four electron integrals arising from such restricted basis sets are given by Sims and Hagstrom[210], while the more general, four-electron integrals with up to the maximum six r_{ij} factors are discussed in King[211].

Attempts at extending such calculations beyond beryllium have so far proved disappointing. As an example, we cite the HY-CI calculation of Clary and Handy[212] on the ground state of neon. They constructed an 83-term trial function using STO's while including correlation factors, r_{ij}^n, representing every pair of electrons, but restricted to one such factor per term. With this wave function, they were able to obtain only 73.5 % of the correlation energy of the ground state, where a similar calculation on beryllium had yielded 87 % of its correlation energy.

Explicitly correlated Gaussians (ECG's) have also been applied to four electron atoms. Komasa, *et al.*[213] used up to 1200 ECG's in their application of this method to the ground state of beryllium, yielding an energy of -14.6673550217, a significant improvement over any other variational calculation. The only other calculations of this size are all CI or MCHF calculations. Hence, the ECG basis shows greater promise of being extensible to complex atomic systems than other methods relying on correlated basis sets. Table 4 presents some recent results for the ground state of beryllium, including some CI and MCHF calculations for comparison.

Table 4
Energies of the Ground State of Beryllium (in a.u.)

	E	Method
Sims and Hagstrom[208]	-14.666547	107-term HY-CI
Clary and Handy[212]	-14.66587	40-term HY-CI
Komasa, *et al.*[213]	-14.6673550217	1200-term ECG
Chung, *et al.*[214]	-14.66704376	1016-term CI
Fischer[215]	-14.667113	3381-term MCHF

8. SIMPLE MOLECULES

The simplest molecule, H_2^+, is separable in confocal elliptic coordinates,

$$\lambda = (r_a + r_b)/R, \qquad \mu = (r_a - r_b)/R. \tag{118}$$

where r_a and r_b refer to radial coordinates centered on each proton and R is the distance between the protons. Being separable, this system can be solved to very high accuracy. For more complex systems, correlated, trial wave functions were introduced into calculations for molecular systems, although their use has been limited to the simplest systems. For the ground state of H_2, James and Coolidge[216] chose a correlated basis in confocal elliptic coordinates,

$$\phi_{nmlpq} = (1 + P_{12})\lambda_1^n \mu_1^m \lambda_2^l \mu_2^p r_{12}^q e^{-\alpha(\lambda_1+\lambda_2)}, \tag{119}$$

with $m + p$ even for this gerade state. With 13 terms in the basis set, they obtained an energy of -1.173465 at the equilibrium distance of $R = 1.4$, a value in error by one part in 10^4. Most subsequent calculations on molecules were not of the James and Coolidge type. However, in 1960, Kołos and Roothaan[217] utilized a James and Coolidge (JC) basis set with a total of 40 terms to obtain an improved energy of -1.17444. Following this, Kołos and Wolniewicz obtained energies of -1.1744746[218] and -1.17447498[219] with 80- and 100-term wave functions of this type. More recently, Kołos, *et al.*[220] have obtained an energy of -1.174475668 with a 249-term basis. The JC basis has the advantage of rapid convergence in comparison to other basis sets. However, it is not as easily generalized to multi-electron or polyatomic molecules because of the complexity of correlated integrals over elliptic coordinates. An example of the application of the James and Coolidge basis to four-electron, diatomic molecules can be found in Ref. [221], where calculations on He_2 and LiH are carried out.

Alternatives to the James and Coolidge prescription for molecular, correlated basis sets include the HY-CI method with either STO's, GTO's, or molecular orbitals (MO's) multiplied by a correlation factor. The MO's and STO's can themselves be constructed from GTO's to take advantage of the simplicity of multi-center integrals over GTO's. Exponentially correlated basis sets (ECG's) are also utilized, and for H_2, have surpassed the

JC-basis level of accuracy[222]. A discussion of these different forms of trial functions can be found in Rychlewski[223], who also compares recent calculations on simple molecules using such functions.

One particular way to generalize exponentially correlated Gaussians to more complex molecules is given in Ref. [224], where calculations on H_3 and LiH are carried out. For these ECG's, the large number of non-linear parameters to be optimized presents a problem as the method is applied to more complex systems. This can be circumvented by the use of a prescribed set of relations between the non-linear parameters. Such "even-tempered" basis are also utilized for simpler systems when the size of the ECG basis gets very large.

Table 5 presents results using a variety of these basis sets for the H_2, H_3^+, and H_3 molecules. For comparison purposes, a few CI calculations are also included in this table.

Table 5
Energies for H_2, H_3^+, and H_3, (in a.u.)

	Energy	Method
	H_2	
James and Coolidge[216]	-1.173465	13-term JC basis
Kołos and Roothaan[217]	-1.174440	40-term JC basis
Kołos and Wolniewicz[218]	-1.1744746	80-term JC basis
Kołos and Wolniewicz[219]	-1.17447498	100-term JC basis
Kołos, *et al.*[220]	-1.174475668	249-term JC basis
Jeziorski and Szalewicz[225]	-1.17442	101-term ECG basis
Urdaneta, *et al.*[226]	-1.174467	HY-CI (13s7p1d) GTO basis
Frye, *et al.*[227]	-1.17447467	HY-CI (15s7p2d1f) GTO basis
Alexander, *et al.*[228]	-1.17447485	900-term ECG basis
Komasa, *et al.*[229]	-1.174470	586-term ECG basis
Komasa, *et al.*[229]	-1.174104	504-term CI-GTO basis
Rychlewski, *et al.*[222]	-1.174475931	700-term ECG basis
Liu and Hagstrom[230]	-1.1743043	CI, elliptic orbitals
	H_3^+	
Salmon and Poshusta[231]	-1.34335	15-term ECG basis
Preiskorn and Woznicki[232]	-1.343422	192-term HY-CI basis
Urdaneta, *et al.*[226]	-1.343500	96-term HY-CI GTO basis
Frye, *et al.*[227]	-1.3438279	138-term HY-CI GTO basis
Alexander, *et al.*[228]	-1.3438220	700-term ECG basis
Meyer, *et al.*[233]	-1.34340	104-term CI-GTO basis
Wenzel, *et al.*[234]	-1.343814	4974-term CI-GTO basis
	H_3	
Frye, Preiskorn, and Clementi[236]	-1.636637	HY-CI (3s1p) GTO basis
Cencek and Rychlewski[224]	-1.658965	200-term ECG basis
Liu[235]	-1.658755	CI (5s3p3d2f) STO basis

An interesting discussion of HY-CI and other calculations on simple atomic and molecular systems is to be found in Clementi, *et al.*[237]. In particular, these authors give some examples of computational times (cpu and elapsed) for carrying out entire calculations as well as particular portions of a calculation. In discussing their calculations on H_3[236], they estimate the time to carry out a HY-CI calculation on this system on an IBM ES/3090-J/VF mainframe computer. They note that a single s-type orbital on each center of the H_3 molecule (six total orbitals, including correlated terms) requires almost one hour for the evaluation of the integrals, a (1s1p) basis (12 total orbitals) requires 191 hours, while a (3s1p) basis would require about 1000 hours to evaluate the integrals. They estimate that a calculation with a wave function capable of yielding 10 cm^{-1} accuracy would consume approximately 3500 years with their unoptimized code, while an improved code might reduce this to 10-20 years.

One possible way of limiting the complexity of a calculation at a particular level, regardless of the size of the system being considered, is through the use of perturbation theory. Depending on the choice of zero-order Hamiltonian, the maximum number of correlated electrons and the maximum number of nuclear centers that can appear in the calculation are determined by the perturbation order. If such a calculation can be expected to yield the desired accuracy when carried to a particular order, that order will determine the maximum complexity of the calculation. An example of such a perturbation theory[238–240], a variant of the Z^{-1} expansion method, requires calculation of $n+1$-electron single-center, n-electron two-center, ..., up to one-electron $n+1$-center contributions for the nth order perturbation coefficients. In addition, the method has the advantage of allowing results obtained for small systems to be transferred without modification to any larger system which contains the electron configuration of the smaller system as a sub-configuration. Effectively, this means that the individual components of the perturbation wave functions and energy coefficients are independent of the nuclear charges, the total number of electrons, and the overall electronic configuration of the system. This method has been applied[238] to the H_2, HeH^+, H_3^+ and HeH_2^{++} molecules through third-order in the energy, with promising results for such a low order calculation.

9. CONCLUSION

In this review we have seen how the development of better variational wave functions has resulted in the extraordinary accuracy now commonly achieved for the simplest atomic and molecular systems. At present, it is possible to achieve accuracy for three-electron systems superior to that once obtained for two-electron atoms by Pekeris. These advances have been accomplished, in part, by dealing explicitly with the most difficult points (singularities) of the wave function. A separate approach has led to an increased flexibility in the variational wave function, either through the use of multiple basis sets within a single trial function, or through the use of exponentially correlated functions with large numbers of non-linear parameters. In both these methods, very large basis sets have become common, reaching high metric order in the trial wave function. This can present computational difficulties, since the finite precision of a computer can mean that, in calculating integrals for large-metric terms or in the diagonalization of large matrices, the precision of the calculation can be compromised. These problems could be circumvented by multiple precision routines (software), permitting calculations to be carried out at arbitrary precision, or by increased word size (hardware), automatically increasing the precision of any calculation carried out on such a computer. The latter does not appear to be likely in the foreseeable future, while the former implies enormous increases in computing time. The history of these calculations suggests that, in the short term, the likely path to improved results for simple systems, and for extending these methods to more complex systems, will involve further evolution and refinement of basis sets for variational calculations. In the long term, however, basis set methods will need to rely on increasing computing power in order to deal with multi-electron systems more complex than those considered here, while for truly large systems, "hardware" solutions may ultimately not be possible.

REFERENCES

1. E. A. Hylleraas, Z. Physik 54 (1929) 347.
2. E. A. Hylleraas, Rev. Mod. Phys. 35 (1963) 421.
3. C. L. Pekeris, Phys. Rev. 115 (1959) 1216.
4. E. U. Condon and G. H. Shortley, "The Theory of Atomic Spectra," (Cambridge University Press, New York, 1935).
5. J. C. Slater, "Quantum Theory of Atomic Structure," (McGraw-Hill, New York, 1960)
6. H. Bethe and E. E. Salpeter, in "Handbuch der Physik," S. Flugge, ed., vol. XXXV, (Springer-Verlag, Berlin, 1957).
7. For a comprehensive review of variational methods for bound state and scattering problems, see M. A. Abdel-Raouf, Physics Repts. 84 (1982) 163.
8. C. Eckart, Phys. Rev. 36 (1930) 878.
9. E. A. Hylleraas and B. Undheim, Z. Physik 65 (1930) 759; J. K. L. McDonald, Phys. Rev. 43 (1933) 830.
10. V. Fock, Z. Physik 63 (1930) 855.
11. P. O. Löwdin, J. Mol. Spectry. 3 (1959) 46.
12. R. T. Brown, J. Chem. Phys. 48 (1968) 4698.
13. T. Kato, Commun. Pure Appl. Math. 10 (1957) 151.
14. S. Chandrasekhar, Ap. J. 102 (1945) 225.
15. See for example, F. Weinhold, J. Phys. A, 1 (1968) 305.
16. G. Temple, Proc. Lond. Math. Soc. 29 (1928) 257; Proc. Royal Soc. (London) A 119 (9128) 276.
17. D. H. Weinstein, Phys. Rev. 40 (1932) 797; 41 (1932) 839; Proc. Nat. Acad. Sci. (U.S.), 20 (1934) 529.
18. A. F. Stevenson, Phys. Rev. 53 (1938) 199; A. F. Stevenson and M. F. Crawford, *ibid.* 54 (1938) 375.
19. F. Weinhold, J. Phys. A 1 (1968) 305.
20. F. Weinhold, Intl. J. Quantum Chem. 3 (1969) 371.
21. J. O. Hirschfelder, W. Byers-Brown, and S. T. Epstein, Adv. Quantum Chem. 1 (1964) 352.
22. A. Dalgarno and A. L. Stewart, Proc. Royal Soc. (London) A 238 (1956) 269.
23. A. Messiah, "Quantum Mechanics", Vol. 2 (North Holland, Amsterdam, 1962) p.

686.
24. C. W. Scherr, J. Chem. Phys. 33 (1960) 317.
25. E. A. Hylleraas, Z. Physik 65 (1930) 209.
26. A. Dalgarno and G. W. F. Drake, Chem. Phys. Lett. 3 (1969) 349.
27. F. C. Sanders, Chem. Phys. Lett. 17 (1972) 291.
28. R. E. Knight and C. W. Scherr, Rev. Mod. Phys. 35 (1963) 431.
29. C. W. Scherr and R. E. Knight, J. Chem. Phys. 40 (1964) 3034.
30. C. W. Scherr, R. E. Knight, and F. C. Sanders, in "Perturbation Theory and its Applications in Quantum Mechanics," ed. C. H. Wilcox, (Wiley, New York, 1966) p. 101.
31. F. C. Sanders and C. W. Scherr, Phys. Rev. 181 (1969) 84.
32. O. Sinanoğlu, Phys. Rev. 122 (1961) 49.
33. J. M. Seminario and F. C. Sanders, Phys. Rev. A 42 (1990) 2562.
34. F. C. Sanders, unpublished
35. W. H. Miller, J. Chem. Phys. 45 (1966) 2198.
36. A. D. Stauffer, J. R. Mehaffey, and S. Woolf, J. Phys. B 3 (1970) 1; B. R. Burrows, *ibid.* 7 (1974) 881.
37. A. Dalgarno and J. T. Lewis, Proc. Royal Soc. (London) A 233 (1955) 70; A. Dalgarno and A. L. Stewart, *ibid.* A238 (1956) 276; A 240 (1957) 274; A 247 (1958) 245.
38. C. Schwartz, Ann. Phys. (N.Y.) 2 (1959) 170.
39. F. W. Birss and J. W. Choi, J. Chem. Phys. 60 (1974) 4090.
40. C. Schwartz, Ann. Phys. (N.Y.) 2 (1959) 156.
41. L. P. Eisenhart, Phys. Rev. 45 (1934) 427.
42. F. A. Matsen, Adv. in Quantum Chem. 1 (1964) 59.
43. S. F. Boys, Proc. Royal Soc. (London) A 200 (1950) 542.
44. M. Rotenberg, Ann. Phys. (NY) 19 (1962) 262.
45. E. A. Hylleraas, Z. Physik 48 (1928) 469.
46. H. Shull and P. O. Löwdin, J. Chem. Phys. 23 (1955) 1362 ; 30 (1959) 617.
47. J. P. Gazeau and A. Maquet, J. Chem. Phys. 73 (1980) 5147.
48. see, for example, E. J. Weniger and E. O. Steinborn, Phys. Rev. A 20 (1984) 2268.
49. A. W. Weiss, Phys. Rev. 122 (1961) 1826.
50. C. C. J. Roothaan and A. W. Weiss, Rev. Mod. Phys. 32(1960) 194.
51. D. P. Carroll, H. J. Silverstone, and R. M. Metzger, J. Chem. Phys. 71 (1979) 4142.
52. C. Froese Fischer, Comp. Phys. Rep. 3 (1986) 275.
53. C. Schwartz, Phys. Rev. 126 (1962) 1015.
54. C. Schwartz, Methods Comp. Phys. 2 (1963) 241.
55. R. T. Pack and W. Byers-Brown, J. Chem. Phys. 45 (1966) 556.
56. B. Klahn and J. D. Morgan III, J. Chem. Phys. 81 (1984) 410.
57. R. N. Hill, J. Chem. Phys. 83 (1985) 1173.
58. C. F. Bunge, Physica Scr. 21 (1980) 328.
59. S. P. Goldman, Phys. Rev. Lett. 73 (1994) 2547; Phys. Rev. A 52 (1995) 3718.
60. G. Breit, Phys. Rev. 35 (1930) 569.
61. A. K. Bhatia and A. Temkin, Rev. Mod. Phys. 36 (1964) 1050.
62. A. K. Bhatia and A. Temkin, Phys. Rev. 137 (1965) A1335.
63. B. T. Darling and R. M. Shoucri, Phys. Rev. A 12 (1975) 2264.

64. L. M. Delves, J. Chem. Phys. 46 (1967) 2563; T. Kalotas, M. H. Lloyd, and L. M. Delves, J. Chem. Phys. 48 (1968) 5494.
65. G. W. Kellner, Z. Physik 44 (1927) 91.
66. J. C. Slater, Phys. Rev. 31 (1928) 333.
67. G. Herzberg and Zbinden, quoted in Ref. [68].
68. S. Chandrasekhar and G. Herzberg, Phys. Rev. 98 (1955) 1050.
69. T. Kinoshita, Phys. Rev. 105 (1957) 1490.
70. L. Wilets and I. J. Cherry, Phys. Rev. 103 (1956) 112.
71. J. H. Bartlett, Gibbons, and Dunn, Phys. Rev. 47 (1935) 679.
72. P. K. Kabir and E. E. Salpeter, Phys. Rev. 108 (1957) 1256.
73. T. Kinoshita, Phys. Rev. 115 (1957) 366.
74. G. Herzberg, Proc. Royal Soc. (London) A 248 (1958) 309.
75. C. L. Pekeris, Phys. Rev. 112 (1958) 1649.
76. H. M. James and A. S. Coolidge, Phys. Rev. 51 (1937) 857.
77. J. H. Bartlett, Phys. Rev. 51 (1937) 661.
78. V. Fock, Izv. Akad. Nauk. SSSR, Ser. Fiz. 18 (1954) 161.
79. T. H. Gronwall, Phys. Rev. 51 (1937) 655.
80. Yu. N. Demkov and A. M. Ermolaev, Zh. Eksper. Teor. Fiz. 36 (1959) 896 [Sov. Phys. JEPT 36 (1959) 633].
81. J. O Hirschfelder, J. Chem. Phys. 39 (1963) 3145.
82. W. Kutzelnigg, Theor. Chim. Acta 68 (1985) 445.
83. C. Schwartz, Phys. Rev. 128 (1962) 1146.
84. see also H. M. Schwartz, Phys. Rev. 120 (1960) 483.
85. E. A. Hylleraas and J. Midtdal, Phys. Rev. 109 (1958) 1013.
86. A. M. Ermolaev and G. B. Sochilin, Dokl. Acad. Nauk. SSSR 155 (1964) 1050 [Sov. Phys. - Dokl. 9 (1964) 292].
87. K. Frankowski and C. L. Pekeris, Phys. Rev. 146 (1966) 46; 150 (1966) 366E.
88. D. E. Freund, B. D. Huxtable, and J. D. Morgan III, Phys. Rev. A 29 (1984) 980.
89. J. D. Morgan III in "Numerical Determination of the Electronic Structure of Atoms, Diatomic and Polyatomic Molecules", M. Defranchesi and J. Delhalle, eds. (Kulwer, Boston, 1989).
90. C. R. Myers, C. J. Umrigar, J. P. Sethna, and J. D. Morgan III, Phys. Rev. A 44 (1991) 5537.
91. J. E. Gottschalk, P. C. Abbott, and E. N. Maslen, J. Phys. A 20 (1987) 2077.
92. E. A. Hylleraas, Z. Physik 54 (1929) 347.
93. C. M. Rosenthal, Chem. Phys. Lett. 10 (1971) 381.
94. R. L. Somorjai and J. D. Power, Chem. Phys. Lett. 12 (1972) 502; J. D. Power and R. L. Somorjai, Phys. Rev. A 5 (1972) 2401; 6 (1972) 1996.
95. P. Winkler and R. N. Porter, J. Chem. Phys. 61 (1974) 2038.
96. A. J. Thakkar and V. H. Smith, Jr., Phys. Rev. A 15 (1977) 1; 15 (1977) 16.
97. S. F. Boys, Proc. Royal Soc. (London) A 258 (1960) 402.
98. K. Singer, Proc. Royal Soc. (London) A 258 (1960) 412.
99. A. Kono and S. Hattori, Phys. Rev. A 29 (1984) 1984; 30 (1984) 2093.
100. A. Kono and S. Hattori, Phys. Rev. A 34 (1986) 1727.
101. G. W. F. Drake, Phys. Rev. Lett. 59 (1987) 1549.

102. G. W. F. Drake, Nucl. Instrum. Methods B 31 (1988) 7.
103. J. D. Baker, D. E. Freund, R. N. Hill, and J. D. Morgan III, Phys. Rev. A 41 (1990) 1247.
104. G. W. F. Drake and Z.-C. Yan, Chem. Phys. Lett. 229 (1994) 486.
105. "Atomic Theory Workshop on Relativistic and QED Effects in Heavy Atoms" AIP Conference Proceedings No. 136, H. P. Kelley and Y.-K. Kim, eds., (American Institute of Physics, New York, 1985).
106. "Relativistic, Quantum Electrodynamic, and Weak Interaction Effects in Atoms" Proceedings of the Program held at the Institute of Theoretical Physics, Santa Barbara, CA, 1988, W. Johnson, P. Mohr, and J. Sucher, eds., AIP Conf. Proc. No. 189 (AIP, New York, 1988).
107. G. W. F. Drake, Nucl. Instrum. Methods 202 (1982) 273; G. W. F. Drake, I. P. Khriplovich, A. I. Milstein, and A. S. Yelkhovsky, Phys. Rev. A 48 (1993) R15.
108. E. A. Hylleraas and J. Midtdal, Phys. Rev. 103 (1956)
109. J. Midtdal, G. Lyslo, and K. Aashamar, Phys. Norv. 3 (1969) 163.
110. T. Kato, J. Faculty Sci. Univ. Tokyo, 6 (1951) 145, and "Perturbation Theory for Linear Operators," (Springer, Berlin, 1966)
111. R. Ahlrichs, Phys. Rev. A 5 (1972) 605.
112. F. H. Stillinger, Jr., J. Chem. Phys. 45 (1966) 3623.
113. Brändas and Goscinski, Intl. J. Quantum Chem. 4 (1970) 571; Intl. J. Quantum Chem. Symp. 6 (1972) 59.
114. W. P. Reinhardt, Phys. Rev. A 15 (197) 802.
115. I. A. Ivanov, Phys. Rev. A 51 (1995) 1080.
116. Y. Accad, C.L. Pekeris, and B. Schiff, Phys. Rev. A 4 (1971) 516.
117. K. Aashamar, G. Lyslo, and J. Midtdal, J. Chem. Phys. 52 (1970) 3324.
118. K. Aashamar, G. Lyslo, and J. Midtdal, J. Chem. Phys. 60 (1970) 3403; *ibid.* 61 (1974) 1345.
119. A. Kono and S. Hattori, Phys. Rev. A 29 (1984) 2981.
120. A. Kono and S. Hattori, Phys. Rev. A 30 (1984) 2093.
121. J. S. Sims and W. C. Martin, Phys. Rev. A 37 (1988) 2259.
122. K. Frankowski, Phys. Rev. 160 (1967) 1.
123. G. W. F. Drake and Z.-C. Yan, Phys. Rev. A 46 (1992) 2378.
124. U. Fano, Phys. Rev. 124 (1961) 1866.
125. M. Domke, C. Xue, A. Puschmann, T. Mandel, E. Hudson, D. A. Shirley, G. Kaindl, C. H. Greene, H. R. Sadeghpour, and H. Petersen, Phys. Rev. Lett. 66 (1991) 1306; M. Domke, K. Schulz, G. Remmers, A. Gutierrez, G. Kaindl, and D. Wintgen, Phys. Rev. A 51 (1995) R4309; M. Domke, K. Schulz, G. Remmers, G. Kaindl, and D. Wintgen, Phys. Rev. A 53 (1996) 1424.
126. D. R. Herrick and O. Sinanoğlu, Phys. Rev. A 11 (1975) 97.
127. C. D. Lin, Phys. Rev. A 29 (1984) 1019; Adv. At. Mol. Phys. 22 (1986) 77.
128. for a relativistic treatment of this state, see B. F. Davis and K. T. Chung, Phys. Rev. A 36 (1987) 1948; 37 (1988) 111.
129. E. Holøien, Proc. Phys. Soc. A 71 (1958) 337; 72 (1958) 141.
130. Feshbach, Ann. Phys. (N.Y.) 19 (1962) 287; Y. Hahn, T. F. O'Malley and L. Spruch, Phys. Rev. 128 (1962) 932.

131. A. Temkin and A. K. Bhatia, in "Autoionization", ed. A. Temkin (Mono, Baltimore, 1966), p. 1.
132. A. U. Hazi and H. S. Taylor Phys. Rev. A 1 (1970) 1109; V. A. Mandlestam, T. R. Ravuri, and H. S. Taylor, Phys. Rev. Lett. 70 (1992) 1932.
133. I. Sanchez and F. Martín, Phys. Rev. A 44 (1991) 7318; F. Martín, Phys. Rev. A 48 (1993) 331.
134. K. T. Chung, Phys. Rev. A 20 (1979) 1743.
135. Y. K. Ho, Phys. Rep. 99 (1983) 1.
136. J.-Z. Tang, S. Watanabe, M. Matsuzawa, and C. D. Lin, Phys. Rev. Lett. 69 (1992) 1633.
137. L. W. Manning and F. C. Sanders, Phys. Rev. A 44 (1991) 7206.
138. G. W. F. Drake and A. Dalgarno, Proc. Royal Soc. (London) A 320 (1971) 549.
139. T. F. O'Malley and S. Geltman, Phys. Rev. 137 (1965) A1344.
140. A. K. Bhatia, A. Temkin, and J. F. Perkins, Phys. Rev. 153 (1967) 177.
141. S. Wakid, A. K. Bhatia, and A. Temkin, Phys. Rev. A 22 (1980) 1395.
142. A. Temkin and A. K. Bhatia, Phys. Rev. A 31 (1985) 1259.
143. K. T. Chung and A. K. Bhatia, Phys. Rev. A 45 (1992) 1449.
144. K. T. Chung and B. F. Davis, Phys. Rev. A 26 (1982) 3278.
145. K. T. Chung, Phys. Rev. A 22 (1980) 1341; 23 (1981) 1079; 23 (1981) 2957; 24 (1981) 1350; B. F. Davis and K. T. Chung, *ibid.* 29 (1984) 682, 1871, and 2586; J.-J. Hsu, K. T. Chung, and K.-N. Huang, *ibid.* 44 (1991) 5485.
146. K. T. Chung, Phys. Rev. A 40 (1989) 4203; 42 (1990) 645.
147. K. T. Chung and B. F. Davis, Phys. Rev. A 26 (1982) 3278.
148. B. F. Davis and K. T. Chung, 29 (1984) 1878; 31 (1985) 3017; 41 (1990) 5844; K. T. Chung, *ibid.* 51 (1995) 844; K. T. Chung and B.-C. Gou, *ibid.* 52 (1995) 3669.
149. Other examples of such trajectories can be seen in Y. K. Ho, J. Phys. B 12 (1979) 387.
150. see, for example, N. Moiseyev, P. R. Certain, and F. Weinhold, Intl. J. Quantum Chem. 14 (1978) 727.
151. T. Georgiev, Dissertation, Southern Illinois University at Carbondale, 1996.
152. Y. K. Ho, Phys. Rev. A 23 (1981) 2137.
153. Y. K. Ho, J. Phys. B 12 (1979) 387; Phys. Rev. A 23 (1981) 2137; 44 (1991) 4154; Y. K. Ho and A. K. Bhatia, *ibid.* 44 (1991) 2895.
154. Y. K. Ho, Phys. Rev. A 41 (1990) 1492; 45 (1992) 148.
155. B. Simon, Ann. Math. 97 (1973) 247.
156. F. Weinhold, J. Phys. Chem. 83 (1979) 1517.
157. N. Moiseyev and F. Weinhold, Phys. Rev. A 20 (1978) 27.
158. L. W. Manning and F. C. Sanders, Phys. Rev. A 43 (1991) 272.
159. F. C. Sanders and R. E. Knight, Phys. Rev. A 27 (1983) 1279.
160. T. Brage, C. Froese Fischer, and N. Vaeck, J. Phys. B 26 (1993) 621.
161. E. B. Wilson, Jr., J. Chem. Phys. 1 (1933) 210.
162. M. Tong, P. Jönsson, and C. F. Fischer, Physica Scr. 48 (1993) 446.
163. K. T. Chung, Phys. Rev. A 44 (1991) 5421; K. T. Chung and X.-W. Zhu, Physica Scr. 48 (1993) 292.
164. O. Jitrik and C. F. Bunge, Phys. Rev. A 43 (1991) 5804.

165. C. W. Scherr, J. N. Silverman, and F. A. Matsen, Phys. Rev. 127 (1962) 830.
166. C. F. Bunge, Phys. Rev. A 16 (1977) 2496.
167. C. F. Bunge, Phys. Rev. A 19 (1979) 936.
168. P. S. C. Wang, Phys. Rev. 47 (1967) 2229.
169. H. M. James and A. S. Coolidge, Phys. Rev. 49 (1936) 688; 55 (1939) 873.
170. A. S. Coolidge and H. M. James , Phys. Rev. 49 (1936) 676.
171. R. A. Sack, J. Math. Phys. 5 (1964) 246.
172. Y. Öhrn and J. Nordling, J. Chem. Phys. 39 (1963) 1864.
173. S. Larsson, Phys. Rev. 169 (1968) 49.
174. J. F. Perkins, J. Chem. Phys. 48 (1968) 1985.
175. Y. K. Ho and B. A. P. Page, J. Comput. Phys. 17 (1975) 122.
176. D. M. Fromm and R. N. Hill, Phys. Rev. A 36 (1987) 1013.
177. G. W. F. Drake and Z.-C. Yan, Phys. Rev. 52 (1995) 3681.
178. Y. Öhrn and J. Nordling, Arkiv Fysik 31 (1966) 471.
179. J. F. Perkins, Phys. Rev. A 5 (1972) 514; 13 (1976) 915.
180. Y. K. Ho, Int. J. Quantum Chem. 20 (1981) 1077.
181. F. W. King and V. Shoup, Phys. Rev. A 33 (1986) 2940.
182. F. W. King, Phys. Rev. A 38 (1988) 6017; 40 (1989) 1735; 43 (1991) 3285.
183. F. W. King and M. P. Bergsbaken, J. Chem. Phys. 93 (1990) 2570.
184. J. S. Sims and S. A. Hagstrom, Phys. Rev. 11 (1975) 418.
185. J. Muszyńska, D. Papierowska, and W. Wòznicki, Chem. Phys. Lett. 76 (1980) 136.
186. J. Pipin and W. Wòznicki, Chem. Phys. Lett. 95 (1983) 392.
187. H. Kleindienst and S. Beutner, Chem. Phys. Lett. 164 (1989) 291.
188. J. Pipin and D. M. Bishop, Phys. Rev. A 45 (1992) 2736.
189. A. Lüchow and H. Kleindienst, Chem. Phys. Lett. 197 (1992) 105.
190. D. K. McKenzie and G. W. F. Drake, Phys. Rev. A 44 (1991) R6973 ; 48 (1993) 4809(E).
191. Z.-C. Yan and G. W. F. Drake, Phys. Rev. A 52 (1995) 3711.
192. F. W. King , Phys. Rev. A 52 (1995) 3711.
193. Z.-C. Yan and G. W. F. Drake, Phys. Rev. A 52 (1995) R4316.
194. W. R. Johnson, S. A. Blundell, and J. Sapirstein, Phys. Rev. A 37 (1988) 2764.
195. J. M. Schulman and W. S. Lee, Phys. Rev. A 5 (1972) 13; J. M. Schulman, W. S. Lee, S. Seung Hui, and J. I. Musher, Phys. Rev. A 6 (1972) 45.
196. Y. Y. Yang, F. C. Sanders, and R. E. Knight, Phys. Rev. A 15 (1977) 444.
197. C. D. H. Chisholm and A. Dalgarno, Proc. Royal Soc. (London) A 290 (1966) 264; C. D. H. Chisholm, A. Dalgarno, F. R. Innes, Phys. Rev. 167 (1968) 60; Z. J. Horak, N. M. Lewis, A. Dalgarno, and P. Blanchard, Phys. Rev. 185 (1969) 21.
198. U. I. Safronova and A. N. Ivanova, Opt. Spectrosk. 27 (1969) 193; U. I. Safronova and V. S. Senashenko, J. Phys. B 11 (1978) 2623.
199. R. E. Knight and F. C. Sanders, Phys. Rev. A 22 (1980) 1361.
200. O. Sinanoğlu, Phys. Rev. 122 (1961) 493.
201. F. C. Sanders, Phys. Rev. A 7 (1973) 1870.
202. Y. Y. Yung, Ph.D. Dissertation, Southern Illinois University at Carbondale, 1976.
203. J. H. Karl, J. Chem. Phys. 46 (1967) 4219.
204. J. F. Perkins, J. Chem. Phys. 50 (1969) 2819.

205. J. Grunninger, Intl. J. Quantum Chem. 4 (1971) 385.
206. L. Szasz and J. Byrne, Phys. Rev. 158 (1967) 34.
207. R. F. Getner and E. A. Burke, Phys. Rev. 176 (1968) 63.
208. J. S. Sims and S. A. Hagstrom, Phys. Rev. A 4 (1971) 908; Int. J. Quantum Chem. 9 (1975) 149.
209. J. F. Perkins, Phys. Rev. A 8 (1973) 700.
210. J. S. Sims and S. A. Hagstrom, J. Chem. Phys. 55 (1971) 4699.
211. F. W. King, J. Chem. Phys. 99 (1993) 3622.
212. D. C. Clary and N. C. Handy, Phys. Rev. A 14 (1976) 1607.
213. J. Komasa, W. Cencek, and J. Rychlewski, Phys. Rev. A 52 (1995) 4500.
214. K. T. Chung, X.-W. Zhu, and Z.-W. Wang, Phys. Rev. A 47 (1993) 1740.
215. C. Froese Fischer, J. Phys. B 26 (1993) 855.
216. James and Coolidge, J. Chem. Phys. 1 (1933) 825.
217. W. Kołos and C. C. J. Roothaan, Rev. Mod. Phys. 32 (1960) 219.
218. W. Kołos and L. Wolniewicz, J. Chem. Phys. 43 (1965) 2429.
219. W. Kołos and L. Wolniewicz, J. Chem. Phys. 49 (1968) 404.
220. W. Kołos, K. Szalewicz, and H. J. Monkhorst, J. Chem. Phys. 84 (1986) 3278.
221. D. C. Clary, Mol. Phys. 34 (1977) 793; D. C. Clary and N. C. Handy, Chem. Phys. Lett. 51 (1977) 483.
222. J. Rychlewski, W. Cencek, and J. Komasa, Chem. Phys. Lett. 229 (1994) 657.
223. J. Rychlewski, Intl. J. Quantum Chem. 49 (1994) 477.
224. W. Cencek and J. Rychlewski, J. Chem. Phys. 98 (1993) 1252.
225. B. Jeziorski and K. Szalewicz, Phys. Rev. A 19 (1979) 2360.
226. C. Urdaneta, A. Largo-Cabrerizo, J. Levin, G. C. Lie, and E. Clementi, J. Chem. Phys. 88 (1988) 2091.
227. D. Frye, G. C. Lie, and E. Clementi, J. Chem. Phys. 91 (1989) 2369.
228. S. A. Alexander, H. J. Monkhorst, R. Roeland, and K. Szalewicz, J. Chem. Phys. 93 (1990) 4230.
229. J. Komasa, W. Cencek, and J. Rychlewski, Phys. Rev. A 46 (1992) 2351.
230. J. W. Liu and S. Hagstrom, Phys. Rev. A 48 (1993) 166.
231. L. Salmon and R. D. Poshusta, J. Chem. Phys. 59 (1973) 3497.
232. A. Preiskorn and W. Woznicki, Mol. Phys. 52 (1984) 1291.
233. W. Meyer, P. Botschwina, and P. Burton, J. Chem. Phys. 84 (1986) 891.
234. K. B. Wenzel, quoted in Ref.[227].
235. B. Liu, J. Chem. Phys. 80 (1984) 581.
236. D. Frye, A. Preiskorn, and E. Clementi, J. Comput. Chem. 12 (1991) 560.
237. E. Clementi, G. Corongiu, D. Bhattacharya, B. Feuston, D. Frye, A. Preiskorn, A. Rizzo, and W. Xue, Chem. Rev. 91 (1991) 679.
238. D. A. Galvan, M. Abu-Jafar, and F. C. Sanders. J. Chem. Phys. 102 (1995) 4919.
239. C. D. H. Chisholm, Mol. Phys. 21 (1971) 769; C. D. H. Chisholm and K. B. Lodge, *ibid.* 21 (1971) 775.
240. H. E. Montgomery, B. L. Bruner, and R. E. Knight, J. Chem. Phys. 56 (1972) 1449.

J.M. Seminario (Editor)
Recent Developments and Applications of Modern Density Functional Theory
Theoretical and Computational Chemistry, Vol. 4

Density functional theory in the classical domain*

J.K. Percus

Courant Institute of Mathematical Sciences and Department of Physics, New York University, 251 Mercer St., New York, NY 10012

1. INTRODUCTION

1.1 Context

The proper description of physical matter depends upon both the circumstances under which the matter is to be found and the degree of resolution required for any use to which the description is to be put. The present discussion will be aimed at coherent many-particle matter in thermal equilibrium, amounting to a long-time photograph of the system, long enough that no change is met with on further extending the time frame. Many aspects of this discussion, which will take place in the domain of classical physics, transfer unchanged to the quantum domain, a major distinction being that we aim from the outset at systems of many particles. Very few aspects extend painlessly to non-equilibrium states or processes.

The study of strictly uniform bulk matter is of course a very old one, the realm of standard thermodynamics. For a single molecular species, all mechanical information in the fluid state is contained in the equation of state, pressure $P(n, \beta)$ as a function of particle density n and reciprocal temperature β. In the presence of a macroscopically varying conservative force field $F(r) = -\nabla u(r)$, the uniform temperature equilibrium state is found from the usual force balance on a unit volume to satisfy

$$-\nabla P(n(r), \beta) - n(r)\nabla u(r) = 0. \tag{1.1}$$

In terms of the thermodynamic quantity $\mu(n, \beta)$, the chemical potential, for which $dP = n\, d\mu$, (1.1) integrates at once to

$$\mu(n(r)) = \mu - u(r), \tag{1.2}$$

the constant μ being identified with the chemical potential of the whole system. Although it is the potential field $u(r)$ that causes the density non-uniformity $n(r)$, (1.2) expresses $u(r)$ in terms of $n(r)$. This *inverse* form is in a way more appropriate to a non-uniform state, because it readily allows for the possibility of large shifts of density with negligible applied force, as in multi-phase situations.

Finally, there is the energetic viewpoint. If $f(n, \beta)$ is the Helmholtz free energy per particle, then – hereafter dropping the argument β, which must be uniform in a state of thermal equilibrium –

$$\frac{\partial}{\partial n} n f(n) = \mu(n), \tag{1.3}$$

*Supported in part by grants from NSF and NASA

and so we can write (1.2) as

$$\mu(r) = \delta\overline{F}[n]/\delta n(r). \tag{1.4}$$

Here,

$$\mu(r) \equiv \mu - u(r) \tag{1.5}$$

is the *local* chemical potential, and

$$\overline{F}[n] = \int n(r) f(n(r)) dr \tag{1.6}$$

the *intrinsic* free energy. Alternatively, we have [SG93]

$$\delta/\delta n(r)\left(\overline{F}[n] + \int n(r)u(r)dr\right) = \mu, \tag{1.7}$$

in terms of the full Helmholtz free energy, including external field contribution, which is therefore a minimum at fixed $N = \int n(r)dr$.

The situations that we will focus on are those in which the locally constructed or *local thermodynamic* free energy $\overline{F}[n]$ of (1.6) is inadequate: container – bounded fluids, separated phase, crystalline spatial order. An effective treatment, the gradient-square correction that we will discuss in Sec. 2c, was given by van der Waals [W94] a century ago, and a formally similar correction for Fermion fluids by Weiszacker [W35] a few decades later. Systematic development of density functional theory, classical [SB62], [LP63b] and quantum [HK64], [KS65], did not hit the major journals until the 1960's, but since then has expanded with great rapidity [E92]. The intention here is to trace principal threads of this development, hewing closely to conceptual issues and approximation techniques, and let the reader fend for himself in the vast literature in which computations are carried out and comparisons made with physical and numerical experiments.

1.2 Quantities of interest

We will deal principally with fixed temperature pair interacting simple fluids in a grand ensemble, controlled therefore by local chemical potential $\mu(r)$ and interaction potential $\varphi(r, r')$. The traditional and directly applicable thermodynamic potential is then the grand potential $\Omega[\mu, \varphi]$ in terms of which density and pair density are given by

$$n(r) = -\delta\Omega/\delta\mu(r),\ n_2(r, r') = \frac{1}{2}\delta\Omega/\delta\varphi(r, r'), \tag{1.8}$$

and in fact the quantity

$$TS = -\Omega[\mu, \varphi] - \int n(r)\mu(r)dr + \frac{1}{2}\int\int n_2(r, r')\varphi(r, r')drdr' \tag{1.9}$$

is minimum with respect to variations in $\{\mu\}$ and $\{\varphi\}$ when $\{n\}$ and $\{n_2\}$ are given. Ω is a concave functional of its arguments ($\Omega[\alpha\mu_1 + \beta\mu_2, \varphi] \geq \alpha\Omega[\mu_1, \varphi] + \beta\Omega[\mu_2, \varphi]$ for $\alpha + \beta = 1$, $\alpha, \beta \geq 0$) which tells us that the compound continuous matrix $\begin{pmatrix} \delta^2\Omega/\delta\mu\delta\mu & \delta\Omega/\delta\mu\delta\varphi \\ \delta^2\Omega/\delta\varphi\delta\mu & \delta^2\Omega/\delta\varphi\delta\varphi \end{pmatrix}$ is negative definite.

The control variables can be replaced by Legendre transformation. Our major focus of interest will be the intrinsic free energy

$$\overline{F}[n, \varphi] = \Omega[\mu[n, \varphi], \varphi] + \int n(r)\mu(r, \{n, \varphi\})dr, \tag{1.10}$$

in terms of which

$$\mu(r) = \delta\overline{F}/\delta n(r), \quad n_2(r,r') = 2\delta\Omega/\delta\varphi(r,r'); \tag{1.11}$$

now $\overline{F}$ is a convex functional, and

$$\chi = -\overline{F}[n,\varphi] + \int \mu(r)n(r)dr + \frac{1}{2}\int\int n_2(r,r')\varphi(r,r')drdr' \tag{1.12}$$

(identified as enthalpy) is minimum with respect to $\{n\}$ and $\{\varphi\}$ at given $\{\mu\}$ and $\{n_2\}$. Lack of convexity of approximations to $\overline{F}$ is responsible for multiple solutions that have plagued many practitioners.

Little attention will be devoted to the enthalpy functional

$$\chi[\mu, n_2] = -\Omega[\mu, \varphi[\mu, n_2]] + \frac{1}{2}\int\int n_2(rr')\varphi(r,r',[n,n_2])drdr', \tag{1.13}$$

but the negentropy

$$-TS[n,n_2] = \Omega[\mu[n_1,n_2],\varphi[n_1,n_2]] + \int n(r)\mu(r(n_1,n_2))dr$$
$$-\frac{1}{2}\int\int n_2(r,r')\varphi(r,r',[n_1,n_2])drdr', \tag{1.14}$$

for which

$$\mu(r) = \delta(-TS)/\delta n(r), \ \frac{1}{2}\varphi(r,r') = \delta TS/\delta n_2(r,r') \tag{1.15}$$

will be of considerable use.

2. SOME RANDOM POINTS OF ENTRY

2.1 Weak correlation

We start our discussion somewhat pedantically with a *simple* (no internal degrees of freedom) non-uniform classical ideal gas in thermal equilibrium at reciprocal temperature β and represented as a grand ensemble of coordinate-component chemical potential μ (the momentum component is trivial and will hereafter be dropped). We can avoid deep dialogs on the meaning of β and μ by imagining that our system is in vanishingly weak interaction with an all-pervading foreign system with which it can exchange both energy and particles. But we do specify the external potential field $u(r)$, e.g. for the purpose of containment, from which we construct the local or relative chemical potential $\mu(r)$ of (1.5). It follows from standard kinetic theory or statistical mechanical arguments [H56] that the particle density is given by the Boltzmann factor

$$n(r) = \exp\beta\mu(r), \tag{2.1}$$

our *profile equation*, with of course the inverse profile

$$\mu(r) = \frac{1}{\beta}\ln n(r). \tag{2.2}$$

The grand potential

$$\Omega[\mu] = -\frac{1}{\beta}\int \exp\beta\mu(r)dr \tag{2.3}$$

serves as generating functional for the direct profile equation in the sense that

$$n(r) = -\delta\Omega/\delta\mu(r), \tag{2.4}$$

while the intrinsic Helmholtz free energy

$$\overline{F}[n] = \frac{1}{\beta}\int (n(r)\ln n(r) - n(r))dr \tag{2.5}$$

that in which the external field energy $\int n(r)u(r)dr$ has been subtracted out, likewise serves as generating functional for the inverse profile equation, (1.4). Ω and $\overline{F}$ are related here, and in general, by the Legendre transformation

$$\overline{F}[n] = \Omega[\mu] + \int n(r)\mu(r)dr, \tag{2.6}$$

and it follows that the linear response is given by

$$S(r,r') \equiv \delta n(r)/\delta\beta\mu(r') = \exp\beta\mu(r)\delta(r-r') = n(r)d(r-r'), \tag{2.7}$$

the inverse linear response by

$$C(r,r') \equiv \delta\beta\mu(r)/\delta n(r') = 1/n(r)\delta(r-r'). \tag{2.8}$$

C and S are always the kernels of inverse operators,

$$\int S(r,r')C(r'r'')dr' = \delta(r-r'') \tag{2.9}$$

but the decidedly local expressions for C and S in (2.7) are specific to the ideal gas.

The generating functionals $\overline{F}$ and Ω determine all of the measurable properties of the system in question. But this is difficult to get a feeling for in the case of non-interacting particles. So let us allow them to interact, still under the influence of an imposed external potential, $\overline{F}[n]$ and $\Omega[\mu]$ continuing to serve as generating functionals [P64], [HM86] in the sense of (1.4) and (2.4). Now write

$$\overline{F}[n] = \frac{1}{\beta}\int (n(r)\ln n(r) - n(r))dr + F^{ex}[n], \tag{2.10}$$

so that F^{ex} is the free energy excess over that of the ideal gas, and correspondingly

$$\mu(r) = \frac{1}{\beta}\ln n(r) + \mu^{ex}(r), \quad C(r,r') = \frac{1}{n(r)}\delta(r-r') - c^{(2)}(r,r') \tag{2.11}$$

with $\mu^{ex}(r) = \delta F^{ex}/\delta n(r)$, $c^{(2)}(r,r') = -\delta\mu^{ex}(r)/\delta n(r')$.

Among these excess quantities, $c^{(2)}(r,r')$ is termed the (Ornstein-Zernike) direct correlation function and plays a pivotal role in theoretical developments. Although we could similarly define $\Omega^{ex}[\mu]$ as the deviation from $-(1/\beta)\int \exp\beta\mu(r)dr$, this plays a lesser role; instead the "excess" quantities typically first appear at the pair level

$$S(r,r') = -\frac{1}{\beta}\delta^2\Omega/\delta\mu(r)\delta\mu(r') = n(r)\delta(r-r') + n(r)h(r,r')n(r'), \tag{2.12}$$

with $h(r,r')$ termed the correlation function. This is because in classical fluids, but alas not in the quantum case, linear response is directly related to density-density correlation; indeed, we will see that

$$n(r)h(r,r')n(r') = n_2(r,r') - n(r)n(r'), \tag{2.13}$$

where $n_2(r, r')$ is the density of particle pairs at (r, r'). An important consequence is that the ideal gas contribution to (2.11) and (2.12) exhausts the local δ-function singularities of these functions.

In the ideal gas, both correlation and direct correlation vanish, by definition. What happens when they are merely weak, as will be the case quite generally at low density? The most frequently used interaction model is that of pure pair interactions, a potential $\varphi(r_i, r_j)$ between particles at r_i and r_j. According to standard statistical mechanics, we then have

$$\Omega = -\frac{1}{\beta} \ln \Xi, \tag{2.14}$$

where the grand partition function Ξ takes the explicit form

$$\Xi[\mu] = \sum_N \frac{1}{N!} \int \cdots \int \exp \beta \sum_1^N \mu(r_j) \exp -\frac{1}{2}\beta \sum_{i \neq j=1}^N \varphi(r_i, r_j) dr^N. \tag{2.15}$$

Routine diagrammatic analysis [MM40], [S64] results in the formally explicit representation

$$-\beta\Omega[\mu] = \sum_1^\infty \frac{1}{s!} \sum_{\{A_s\}} \int \cdots \int \prod_1^s z(r_j) \prod_{(i,j)\in A_s} f(r_i, r_j) dr^s$$
$$z(r) \equiv \exp \beta\mu(r), \; f(r, r') \equiv e^{-\beta\varphi(r,r')} - 1, \tag{2.16}$$

where A_s signifies the set of links of a distinct connected graph on vertices $(r_1, \ldots, r_s)$. Consequently

$$n(r) = \sum \frac{1}{s!} \sum_{\{A_s(r)\}} \int \cdots \int z(r) \prod_1^s z(r_j) \prod_{(i,j)\in A_s(r)} f(r_i, r_j) dr^s \tag{2.17}$$

where $A_s(r)$ is a graph on $(r, r_1, \ldots, r_s)$. Elimination of the $z(r_j)$ from (2.16, 2.17) and (2.6) subsequently yields the expression

$$-\beta F^{ex}[n] = \sum_2^\infty \frac{1}{s!} \sum_{\{B_s\}} \int \cdots \int \prod_1^s n(r_j) \prod_{(i,j)\in B_s} f(r_i, r_j) dr^s \tag{2.18}$$

where B_s is a distinct graph on $\{r_1, \ldots, r_s\}$, connected and free of articulation points, i.e. vertices whose excision disconnects the graph. Eq. (2.8) starts out as

$$-\beta F^{ex}[n] = \frac{1}{2} \int\int n(r_1)n(r_2)f(r_1, r_2)dr^2$$
$$+\frac{1}{6} \int\int\int n(r_1)n(r_2)n(r_3)f(r_1, r_2)f(r_2, r_3)f(r_3, r_1)dr^3 + \cdots, \tag{2.19}$$

while the corresponding

$$-\beta\Omega[n] = \int z(r_1)dr + \frac{1}{2} \int\int z(r_1)z(r_2)f(r_1, r_2)dr^2$$
$$+\frac{1}{6} \int\int\int z(r_1)z(r_2)z(r_3)[f(r_1, r_2)f(r_2, r_3)f(r_3, r_1) \tag{2.20}$$
$$+3f(r_1, r_2)f(r_2, r_3)]dr^3 + \cdots$$

is less useful.

2.2 More formal background

The expressions (2.16), (2.18) are to be sure very explicit. Given the interaction potentials (and generalization to 3,4,... body interactions is immediate) and assuming convergence of the series, we have an in-principle procedure for finding the relevant thermodynamic potentials. But in-practice is a different story. High-dimensional integrals are quickly required, and convergence – in particular of (2.16) – for physically interesting systems not assured at all. For a given accuracy, one may be much better off by performing professional Monte Carlo simulations of a full many-particle system. But it is the non-uniformity that supplies the interest, as well as the difficulty. If we have reliable knowledge of the properties of the uniform extended fluid, and this is available from physical experiments, computer experiments, and sophisticated analytic approximations, than it is "only" the imposed non-uniformity that we have to deal. What properties do we need?

The non-uniform thermodynamic generating functionals encompass *all* of the (single-time) thermodynamic information on the system in question [P64]. This is clearly the case for the grand partition function $\Xi[\mu]$, for which one sees at once from (2.15) – and indeed for any interaction – that

$$z(r_1')\dots z(r_s')\delta^s\Xi[\mu]/\delta z(r_1')\dots\delta z(r_s')$$
$$=\sum\frac{1}{N!}\int\cdots\int N(N-1)\dots(N+1-s)\prod_1^N z(r_j)$$
$$\exp\frac{1}{2}\beta\sum_{i\neq j=1}^{N}\varphi(r_i,r_j)\prod_1^s\delta(r_j-r_j')dr^N=n_s(r_1',\dots,r_s')\Xi[\mu], \tag{2.21}$$

where n_s is the s-tuplet distribution function obtained by selecting s of the N particles in each term to fix at the prescribed locations. In terms of the generating functional

$$\mathcal{N}[\psi|\mu]=\sum 1/s!\int\cdots\int\prod_1^s\psi(r_j)n_s(r_1,\dots,n_s)dr^s, \tag{2.22}$$

this means that

$$\mathcal{N}[\psi|\mu]=\Xi\left[\mu+\frac{1}{\beta}\ln\psi\right]/\Xi[\mu]. \tag{2.23}$$

By dividing by $\prod_1^s z(r_j')$ and setting $z=0$ in (2.21), we draw the obvious but important conclusion that the full set of $\{n_s\}$ and hence $\Xi[\mu]$, implies all of the s-body Boltzmann factors. This would remain true in the event of many-body potentials, and so $\Xi[\mu]$ can in principle be inverted to obtain the full Hamiltonian (i.e. potential energy) of the particle system.

In the same way, the structure factor sequence arises from

$$S_p(r_1,\dots,r_p)=\delta^p-\beta\Omega[\mu]/\delta\beta\mu(r_1)\dots\int\beta\mu(r_p), \tag{2.24}$$

which means, according to (2.23), that

$$\mathcal{S}[\xi|\mu]\equiv\sum_1^\infty 1/p!\int\cdots\int\prod_1^p\xi(r_j)S_p(r_1,\dots,r_p)dr^p$$
$$=-\beta\Omega[\mu+\xi/\beta]+\beta\Omega[\mu]$$

$$= \ln \Xi[\mu + \xi/\beta] - \ln \Xi[\mu] \tag{2.25}$$
$$= \ln \mathcal{N}[e^{\xi}|\mu].$$

In particular,

$$S_1(r) = n(r),\ S_2(r,r') = n_2(r,r') - n(r)n(r') + n(r)\delta(r-r'), \dots \tag{2.26}$$

According to (2.15) and (2.25),

$$\mathcal{S}[\xi|\mu] = \ln\left\langle \exp \sum_j \xi(r_j)\right\rangle$$
$$= \ln\left\langle \exp \int \xi(r)\rho(r)dr\right\rangle \tag{2.27}$$

where $\rho(r) \equiv \sum_j \delta(r - r_j)$.

Thus the S_p are the cumulants of the microscopic density $\rho(r)$:

$$s_1(r) = \langle\rho(r)\rangle,\ s_2(r,r') = \langle\rho(r)\rho(r')\rangle - \langle\rho(r)\rangle\langle\rho(r')\rangle, \dots \tag{2.28}$$

The δ-functions that appear in the S_p represent "self-correlations", e.g.

$$\langle\rho(r)\rho(r')\rangle = \sum_{i,j}\langle\delta(r-r_i)\delta(r'-r_j)\rangle$$
$$= \sum_{i\neq j}\langle\delta(r-r_i)\delta(r'-r_j)\rangle + \delta(r-r')\sum_j\langle\delta(r-r_j)\rangle, \tag{2.29}$$

and would occur even for ideal gases. We will therefore frequently form instead the Ursell correlations, in which the δ-functions are missing:

$$W_2(r,r') = n_2(r,r') - n(r)n(r')$$
$$W_3(r,r',r'') = n_3(r,r',r'') - n_2(r,r')n(r'') - n_2(r',r'')n(r) \tag{2.30}$$
$$-n_2(r'',r)n(r') + 2n(r)n(r')n(r''), \dots$$

Furthermore, and far from least important, the free energy functional $\overline{F}[n]$ contains the same information as $\Omega[\mu]$, and generates the direct correlation sequence

$$C_p(r_1,\dots,r_p|n) = \delta^p\beta\overline{F}(n)/\delta n(r_1)\dots\delta n(r_p). \tag{2.31}$$

We have seen that, as operators, $C_2 = S_2^{-1}$, so that the full sequences $\{C_p\}$ and $\{S_p\}$ are related, e.g.

$$C_3(r_1r_2r_3) = \delta C_2(r_1r_2)/\delta n(r_3) = \int \delta\beta\mu(r_4)/\delta n(r_3)\delta S_2^{-1}(r_1r_2)/\delta\beta\mu(r_4)dr_4$$
$$= -\int S_2^{-1}(r_3r_4)S_2^{-1}(r_1r_5)S_3(r_4r_5r_6)S_2^{-1}(r_6r_2)dr_4dr_5dr_6.$$

The δ-function problem in the direct correlation function sequence is minor: we have seen that $C_2(r,r') = \delta(r-r')/n(r) - c^{(2)}(r,r')$, and the singularities in the higher C_p occur only from derivatives of the first term in C_2. We will define

$$C_p(r_1,\dots,r_p) = (-1)^{p-1}\prod_2^p \delta(r_j - r_1)/n(r_1)^{p-1} - c^{(p)}(r_1,\dots,r_p) \tag{2.32}$$

to indicate this explicitly.

2.3 Short-range correlations

Suppose then that we have reliable knowledge of the correlation structure of an interacting species in its bulk or uniform state. We will use the notation, e.g. $n_2(r_1, r_2; n)$ to indicate that n_2 is a function of the uniform density n and not a general functional of $\{n(r)\}$ as in $n_2(r_1, r_2|n)$. How we can extend the former to non-uniformity? In principle, this is trivial, because, (2.24) or (2.31), we have all the derivatives needed to carry out an expansion about uniformity. In practice, however, our firm knowledge stops at an early order in the distribution sequence. Nonetheless, we should not need more if the system is almost uniform in the sense that the Ursell correlations are negligible for particle separations on the scale of the density variation. This does not mean that the density $n(r)$ is nearly constant over the whole system; it can change enormously in the large but only slowly in the small. It therefore tells us that the obvious strategy of expanding $F[n]$ about some uniform density n_0 need not work. But if a local quantity such as $\mu(r|n)$ is expanded instead, we should presumably need information only up a "correlation length" away.

Consider $\mu(r|n)$ for non-uniform n, and let us expand about some uniform n_0 representing the environment of the point r. Hence

$$\begin{aligned}\beta\mu(r|n) &= \beta\mu(r|n_0) + \sum_1^\infty \frac{1}{s!}\int\cdots\int \frac{\delta^s\beta\mu(r)}{\delta n(r_1)\ldots\delta n(r_s)}\Big|_{n_0}\prod_1^s(n(r_j)-n_0)dr^s \\ &= \beta\mu(n_0) + \sum_1^\infty 1/s!\int\cdots\int C_{s+1}(rr_1\ldots r_s; n_0)\prod_1^s(n(r_j)-n_0)dr^s,\end{aligned} \tag{2.33}$$

in terms of the bulk chemical potential $\mu(n_0)$ and bulk direct correlations alone. n_0 can be chosen in various sophisticated ways, but simplest by far [LP63a] is to set it equal to the local density $n(r)$:

$$\beta\mu(r|n) = \beta\mu(n(r)) - \sum_1^\infty 1/s!\int - \int c^{(s+1)}(rr_1\ldots r_s; n(r))\prod_1^s(n(r_j)-n(r))dr^s. \tag{2.34}$$

On the spatial scale on which the factors $n(r_j) - n(r)$ differ significantly from zero, the direct correlation will be negligible, and so we can truncate the series (2.34) at an early stage. We will retain two terms (not one, for reasons of integrability that we will have to address later)

$$\begin{aligned}\beta\mu(r|n) &= \beta\mu(n(r)) - \int c^{(2)}(rr_1; n(r))(n(r_1)-n(r))dr_1 \\ &-\frac{1}{2}\int\int c^{(3)}(rr_1r_2; n(r))(n(r_1)-n(r))(n(r_2)-n(r))dr_1dr_2.\end{aligned} \tag{2.35}$$

The postulated show variation of $n(r)$ in the domain of $c^{(2)}, c^{(3)}$ then allows us to expand $n(r_{1,2})$ around $n(r)$ as well:

$$n(r') - n(r) = (r'-r)\cdot\nabla n(r) + \frac{1}{2}(r'-r)(r'-r)\colon\ \nabla\nabla n(r) + \cdots \tag{2.36}$$

so that, through second derivative order, (2.35) becomes

$$\begin{aligned}\beta\mu(r|n) &= \beta\mu(n(r)) - \int c^{(2)}(rr_1; n(r))[(r_1-r)\cdot\nabla n(r) + \frac{1}{2}(r_1-r)(r_1-r)\colon\ \nabla\nabla n(r)]dr_1 \\ &-\frac{1}{2}\int\int c^{(3)}(rr_1r_2; n(r))(r_1-r)(r_2-r)\colon\ \nabla n(r)\nabla n(r)dr_1dr_2.\end{aligned} \tag{2.37}$$

It is now necessary to process (2.37) a bit, using to advantage the specific properties of uniform systems (with the tacit assumption of translation and reflection invariant interaction energy). In particular, translation invariance tells us that $c^{(2)}(rr_1) = c^{(2)}(r_1 - r)$, $c^{(3)}(rr_1r_2) = c^{(3)}(r_1 - r_2, r_2 - r)$ depend only upon relative arguments. To start with, then, $\int c^{(2)}(rr_1)(r_1 - r)dr_1 = \int c^{(2)}(r')r'dr' = 0$ by reflection invariance. Further, using

$$(r_1 - r)(r_2 - r) + (r_2 - r)(r_1 - r)$$
$$= (r_1 - r)(r_1 - r) + (r_2 - r)(r_2 - r) - (r_1 - r_2)(r_1 - r_2), \tag{2.38}$$

the last term in (2.37) becomes $-\frac{1}{4}\int\int c^{(3)}(r_1-r_2, r_2-r)\cdot [(r_1-r)(r_1-r)+(r_2-r)(r_2-r)-(r_1-r_2)(r_1-r_2)]dr_1dr_2$. But $\int(r_1 - r_2)(r_1 - r_2)c^{(3)}(r_1 - r_2, r_2 - r)dr_2 = \int r_2r_2c^{(3)}(r_2, r_1 - r_2 - r)dr_2 = \int(r - r_2)(r - r_2)c^{(3)}(r_2 - r, r_1 - r_2)dr_2 = \int(r_2 - r)(r_2 - r)c^{(3)}(r_1 - r_2, r_2 - r)dr_2$. Thus, all three terms in the $c^{(3)}$ expression are equal, and it reduces to $-\frac{1}{4}\int\int c^{(3)}(r_1r_2r_3;n)(r_1 - r)(r_1 - r)dr_1dr_2 = -\frac{1}{4}\int\int(r_1 - r)(r_1 - r)\frac{\delta c^{(2)}(r_1r;n)}{\delta n(r_2)}|_{n(r)}dr_2$. Since a uniform distribution is characterized by $\partial n(r')/\partial n = 1$ at all r', we need $\int \delta c^{(2)}(r_1r|n)/\delta n(r_2)\partial n(r_2)/\partial n|_{n=n(r)}dr_2 = \frac{\partial}{\partial n}c^{(2)}(r_1r|n)|_{n=n(r)}$, and conclude that (2.37) simplifies to [P82a]

$$\beta\mu(r|n) = \beta\mu(n(r)) - \frac{1}{4}\int[2c^{(2)}(rr_1;n(r))(r_1 - r)(r_1 - r) \colon \nabla\nabla n(r)$$
$$-\frac{\partial}{\partial n(r)}c^{(2)}(rr_1;n(r))(r_1 - r)(r_1 - r) \colon \nabla n(r)\nabla n(r)]dr_1. \tag{2.39}$$

A more compact notation is desirable. In terms of the second moment tensor of $c^{(2)}$:

$$\Lambda(n) = \int c^{(2)}(r';n)r'r'dr', \tag{2.40}$$

we have

$$\beta\mu(r|n) = \beta\mu(n(r)) - \frac{1}{2}\Lambda(n(r)) \colon \nabla\nabla n(r) - \frac{1}{4}\Lambda'(n(r)) \colon \nabla n(r)\nabla n(r), \tag{2.41}$$

but this is recognized as the Euler-Lagrange variation of the intrinsic free energy [CH58], [BSD76]

$$\beta\overline{F}[n] = \int \beta f(n(r))n(r)dr + \frac{1}{4}\int \Lambda(n(r)) \colon \nabla n(r)\nabla n(r)dr$$
$$\text{where } \mu(n) = (nf(n))', \tag{2.42}$$

$f(n)$ being the free energy per particle in the uniform state. For very small $\nabla n(r)$, this is just $\int \beta f(n(r))n(r)dr$, the "local thermodynamics" integrated free energy per unit volume. For merely small $\nabla n(r)$, it is the modernized van der Waals corrected version of the same. Under two-phase separation conditions, $\nabla n(r)$ can be small outside of the interface, but not so within it. However, in practice, the correlation length is of the order of the repulsive interaction core for a typical molecular potential, while the interfacial thickness is of the order of the attractive tail. Thus the required condition remains valid, providing that one is excused in defining the uniform state in the phase transaction tie-line range.

2.4 Mean field format

The development of effective free energy density functionals starts universally with the choice of a suitable reference system or set of systems. In Sec. 2c, the reference systems were just the system under consideration, but restricted to the much more accessible case of total spatial uniformity. In a section to follow, a class of exactly solvable reference systems will be introduced. Here, we proceed from a very traditional viewpoint, dating at least from van der Waals, is which the intermolecular pair potential is divided into a repulsive core φ_0 and an attractive tail $-\varphi_1$. The reference system, in which φ_0 alone is acting, is to have a known, and uninteresting, phenomenology e.g. one with a very short correlation length, and certainly well above the critical point. One obvious advantage is that subtle questions, such as what uniform fluid should enter into (2.40) in the two-phase region, can be avoided.

For ease of manipulation, we represent the full interaction parametrically,

$$\varphi^\lambda(r-r') = \varphi_0(r-r') - \lambda\varphi_1(r-r'), \tag{2.43}$$

and correspondingly parametrize all of the thermodynamics. It is clear from (2.14) and (2.15) that

$$\partial\Omega^\lambda/\partial\lambda = -\left\langle \frac{1}{2}\sum_{i\neq j}\varphi_1(r_i-r_j)\right\rangle = -\frac{1}{2}\int\int \varphi_1(r-r')n_2^\lambda(r,r')drdr'. \tag{2.44}$$

From (2.6), we have $\delta\overline{F}|_n = \partial\Omega|_n + \int n\delta\mu|_n = \delta\Omega|_n - \int(\delta\Omega/\delta\mu)\delta\mu|_n = \delta\Omega|_\mu$, or

$$\delta\overline{F}|_n = \delta\Omega|_\mu, \tag{2.45}$$

we can write

$$\partial\overline{F}^\lambda[n]/\partial\lambda = -\frac{1}{2}\ \mathrm{Tr}\ \varphi_1 n_2^\lambda. \tag{2.46}$$

Here, φ_1 and n_2^λ are regarded as continuous matrices, so that the convenient trace operation becomes meaningful. Inserting (2.12) and (2.13), we then have

$$\partial\overline{F}^\lambda[n]/\partial\lambda = -\frac{1}{2}\ \mathrm{Tr}\ \varphi_1 S^\lambda - \frac{1}{2}\int\int n(r)\varphi_1(r-r')n(r')drdr' + \frac{1}{2}\varphi_1(0)\int n(r)dr. \tag{2.47}$$

Integrating from $\lambda = 0$, at which $\overline{F}^0[n] = \overline{F}_0[n]$ of the reference system, we obtain

$$\overline{F}^\lambda[n] = \overline{F}_0[n] + \frac{1}{2}\lambda\varphi_1(0)\int n(r)dr - \frac{\lambda}{2}\int\int n(r)\varphi_1(r-r')n(r')drdr'$$
$$-\frac{1}{2}\int_0^\lambda \mathrm{Tr}\ \varphi_1 S^{\lambda'}d\lambda'. \tag{2.48}$$

The corrections to $\overline{F}_0(n)$ in (2.48) consist first of a self-energy subtraction, which we will disregard either by imagining that $\varphi_1(0) = 0$, or by observing that this term simply amounts to an additive renormalization, $\mu \to \mu + \frac{1}{2}\lambda\varphi_1(0)$, of the system chemical potential. Next is the correlation-neglected or *mean field* interaction energy, and finally the correlation correction. Eq. (2.48) in fact constitutes a fully self-consistent set of conditions for the free energy functional. We need only note that $S^{\lambda'} = (C^{\lambda'})^{-1} = (\delta^2\beta\overline{F}^\lambda/\delta n\delta n)^{-1}$, in obvious notation. Thus [ZP90],

$$\overline{F}^\lambda[n] = \overline{F}^\lambda_{MF}[n] - \frac{1}{2\beta}\int_0^\lambda \mathrm{Tr}\ \varphi_1\left(\frac{\delta^2\overline{F}^{\lambda'}}{\delta n\delta n}\right)^{-1} d\lambda'$$

where $\overline{F}^\lambda_{MF}[n] = \overline{F}_0(n) - \frac{1}{2}\lambda\int\int n(r)\varphi_1(r-r')n(r')drdr'. \quad (2.49)$

A primitive but much used approximation [KP95] consists of dropping the correlation correction and representing $\overline{F}_0[n]$ by its local thermodynamic or vanishing relative correlation length approximation:

$$\overline{F}[n] = \int n(r) f_0(n(r)) dr - \frac{1}{2} \int \int n(r) \varphi_1(r - r') n(r') dr dr'. \tag{2.50}$$

But, in principle at least, iteration of (2.49) is not difficult. Lowest order would be $\overline{F}_{MF}$. For the next order, we would replace $\overline{F}^{\lambda'}$ in (2.49) by $\overline{F}^{\lambda'}_{MF}$. Then,

$$C^{\lambda'} = C_0 - \lambda' \beta \varphi_1, \tag{2.51}$$

so that $\int_0^\lambda \text{Tr } \varphi_1 (C^{\lambda'})^{-1} d\lambda' = -\frac{1}{\beta} \text{Tr}(\ln C^\lambda - \ln C_0)$, yielding

$$\overline{F}[n] = \overline{F}_{MF}[n] + \frac{1}{2\beta} \text{ Tr } \ln(I - \beta \varphi_1 S_0), \tag{2.52}$$

and so forth. Of course, "lowest order" is hardly unique; if we were to turn up the pair Boltzmann factor via $w^\lambda(r - r') = w_0(r - r') + \lambda w_1(r - r')$, where $w_1(r - r') = w_0(r - r')(e^{-\beta\varphi_1(r-r')} - 1)$, lowest order would replace (2.49)by

$$\overline{F}^{\lambda}_{MF}[n] = \overline{F}_0[n] + \frac{1}{2}(\lambda/\beta) \int \int n(r)(e^{-\beta\varphi_1(r-r')} - 1) n(r') dr' dr. \tag{2.49'}$$

3. THE INTEGRAL EQUATION ROUTE

3.1 Direct expansion

A popular path to the construction of approximate free energy density functionals involves making use of the uniform fluid information that can confidently be regarded as available. We have seen (2.37, 2.39) that even marginally available higher correlations can be used in special situations. In general, however, this is not the case, and so we will start this discussion by utilizing only the leading order of expansion about uniformity, generally in the form

$$a(r|n) = a(r; n_0) + \int \frac{\delta a(r|n)}{\delta b(r'|n)}\Big|_{n=n_0} (b(r'|n) - b(r'; n_0)) dr' \tag{3.1}$$

where n_0 is uniform. The art is that of choosing the fields a and b; we have already encountered the prototypical form

$$\beta\mu(r|n) = \beta\mu(n_0) + \int C(r, r'; n_0)(n(r') - n_0) dr'. \tag{3.2}$$

Eq. (3.2) certainly does come from the free energy functional

$$\beta\overline{F}[n] = \beta F(n_0) + \beta\mu(n_0) \int (n(r) - n_0) dr$$
$$+ \frac{1}{2} \int \int (n(r) - n_0) C(r, r'; n_0)(n(r') - n_0) dr dr', \tag{3.3}$$

whose validity is primarily restricted to weak deviations from uniformity.

An obvious improvement, following (2.34), is to choose n_0 as $n(r)$:

$$\beta\mu(r|n) = \beta\mu(n(r)) + \int C(r, r'; n(r))(n(r') - n(r)) dr', \tag{3.4}$$

but this does not arise from a free energy functional. That is, from $\mu(r) = \delta\overline{F}/\delta n(r)$, one has the *integrability* condition

$$\frac{\delta\mu(r)}{\delta n(r')} = \frac{\delta\mu(r')}{\delta n(r)}, \tag{3.5}$$

which is in fact sufficient for a generator $\overline{F}$ to exist (a vector whose curl vanishes must be a gradient). This is not satisfied by (3.4). However, we observe that if $C(\ ;n(r))$ depends only weakly on $n(r)$, then one cannot distinguish between $C(\ ;n(r))(n(r')-n(r))$ and $\int_{n(r)}^{n(r')} C(\ ;n)dn$, i.e. through first order. Thus, (3.4) can be replaced by

$$\beta\mu(r|n) = \beta\mu(n(r)) + \frac{1}{2}\int C(r,r';n(r))(n(r')-n(r))dr'$$
$$+\frac{1}{2}\int\int_{n(r)}^{n(r')} C(r,r';n)dndr', \tag{3.6}$$

which does indeed result from the free energy given by [P88]

$$\beta\overline{F}[n] = \int n(r)\beta f(n(r))d\rho$$
$$+\frac{1}{4}\int\int(n(r')-n(r))\int_{n(r)}^{n(r')} C(r,r';n)dndrdr'. \tag{3.7}$$

The singular part of $C(r,r') = \delta(r-r')/n(r) - c^{(2)}(r,r')$ does not contribute, and so we can replace (3.7) by

$$\beta\overline{F}[n] = \int n(r)\beta f(n(r))dr - \frac{1}{4}\int\int(n(r')-n(r))\int_{n(r)}^{n(r')} c^{(2)}(r,r';n)dndrdr'. \tag{3.8}$$

Eq. (3.8) has two defects, which must eventually be attended to. First, there remains a singularity to be examined. We can readily obtain the resulting $C(r,r''|n) = \delta\beta\mu(r|n)/\delta n(r'')$, which becomes

$$C(r,r''|n) = \left[\frac{1}{n(r)} - \frac{1}{2}\int c^{(2)\prime}(r,r';n(r))(n(r')-n(r))dr'\right]\delta(r-r'')$$
$$-\frac{1}{2}c^{(2)}(r,r'';n(r)) - \frac{1}{2}c^{(2)}(r,r'';n(r'')) \tag{3.9}$$

where $c^{(2)\prime}(n) = \partial c^{(2)}(n)/\partial n$. This does not quite have the required singularity $\delta(r-r'')/n(r)$, a common defect of mean field-type theories. And second, despite the reasonable resolution of the question of just which density should be inserted into $C(\ ;n)$ the result requires the microscopic densities $n(r)$, $n(r')$ to also appear as uniform densities. But a fluid with a hard core in its interaction has a maximum close-packed uniform density which can easily be exceeded by the point density, e.g. at a hard wall; this possibility is forbidden by (3.7).

A formulation such as (3.2) expresses the one-body profile relation in terms of a two-body quantity, which then must be estimated. But meanwhile the remaining terms in the series are neglected. The latter part of the approximation can in principle be avoided by use of exact "sum rule" relations between one and two-body quantities. Three somewhat differently motivated types have proved useful and will be elaborated upon in the sequel. First is the replacement of (3.2) by an expression with a formally exact remainder, and this proceeds as in (2.48). That is, we imagine a "path" $n^\lambda(r')$ between n_0 at $\lambda = 0$ and $n(r')$ at $\lambda = 1$, so that $\partial\beta\mu(r|n^\lambda)/\partial\lambda = \int \delta\beta\mu(r|n^\lambda)/\delta n^\lambda(r')\partial n^\lambda(r')/\partial\lambda dr' = \int C(r,r'|n^\lambda)\partial n^\lambda(r')/\partial\lambda dr'$, and conclude on integrating over λ that

$$\beta\mu(r|n) = \beta\mu(n_0) + \int\int_0^1 C(r,r'|n^\lambda)\partial n^\lambda(r')/\partial\lambda d\lambda dr'. \tag{3.10}$$

The path independence of (3.10) will of course not be maintained under approximations, and so the path should be carefully tuned to fit the application. But as a matter of fact, it is easy to show that for any path satisfying $n^\lambda(r) = n(r)$ for all λ, such as the simplest interpolation $n^\lambda(r) = (1-\lambda)n(r) + \lambda n(r')$, the local thermodynamic approximation

$$c^{(2)}(r,r'|n^\lambda) = \frac{1}{2}c^{(2)}(r,r';n^\lambda(r)) + \frac{1}{2}c^{(2)}(r,r';n^\lambda(r')) \tag{3.11}$$

leads precisely to (3.6).

A second version takes advantage of the structural characteristics of the system being studied. For example, we certainly have in mind a complete description, so that the external world has only a benign influence on the geometry of space. This means that the interaction, in accord with free space, is – as in (2.37) – translation invariant, rotation invariant, reflection invariant. Such invariance has immediate implications, perhaps more familiar in the form of the Ward-Takahashi identities [Z89], [CZB96] in the quantum domain. As prototypical example, translation invariance of the free energy density functional means that $\nabla_a \beta\overline{F}[n + a\cdot\nabla n]|_{a=0} = 0 = \nabla_a(\beta\overline{F}[n] + \int a\cdot\nabla n(r)\delta\beta\overline{F}[n]/\delta n(r)dr)|_{a=0} = \int \nabla n(r)\beta\mu(r|n)dr$. Thus

$$\int n(r')\nabla\beta\mu(r'|n)dr' = 0, \tag{3.12}$$

physically the vanishing in equilibrium of the total external force on the system. Since this is an identity, we can apply $\delta/\delta n(r)$, obtaining [W76], [LMB76]

$$\nabla\beta\mu(r|n) + \int n(r')\nabla' C(r,r'|n)dr' = 0 \tag{3.13}$$

$$\text{or } \nabla\beta\mu(r|n) = \int C(r,r'|n)\nabla n(r')dr',$$

to which one can apply approximations such as those used in treating (3.4). In fact, approximating $C(r,r'|n) = C(r,r';n_0)$ leads precisely to (3.2), while $C(r,r'|n) = \delta(r-r')/n(r) - c^{(2)}(r,r';n_0)$ results in mean field, e.g. (2.49), in which the *full* interaction is replaced by $-c^{(2)}(r,r';n_0)/\beta$. (See Sec. 4b.) Equally well, we can apply $\delta/\delta\beta\mu(r)$, so that

$$\nabla n(r) = \int S(r,r'|n)\nabla\beta\mu(r')dr'. \tag{3.14}$$

It is more customary to derive (3.13), as well as (3.14), directly from the observation that e.g. an infinitesimal translation of $n(r)$ will produce the same translation of $\mu(r)$: $\delta\mu(r|n) = \int (\delta\mu(r|n)/\delta n(r'))\delta n(r')dr'$ is satisfied by $\delta f(r) = \nabla f(r)$.

The simultaneous advantage and drawback of an approach such as (3.13) is the paucity of information actually used. But (3.12), (3.14), and their progeny are also reductions of a more detailed – and more familiar – profile sequence. This is the *YBG* hierarchy [F64], based upon the very general sum rule (an immediate consequence of $\int \nabla(Ae^{-\beta\phi})dr^N = 0$)

$$\langle\nabla A\rangle = \langle A\nabla\beta\phi\rangle \tag{3.15}$$

for any observable A, any coordinate gradient ∇, and total energy ϕ. Suppose that $\phi = \sum u(r_i) + \frac{1}{2}\sum'\varphi(r_i - r_j)$. Then if $\nabla = \sum_j \nabla_j$ and $A = 1$, we have

$$\int n(r)\nabla\beta u(r)dr = 0; \tag{3.16}$$

with $\nabla = \sum_j \nabla_j$ and $A = \rho(r) = \sum \delta(r - r_j)$,we find

$$\nabla n(r) + n(r)\nabla\beta u(r) + \int n_2(r,r')\nabla\beta\varphi(r-r')dr' = 0, \tag{3.17}$$

and higher microscopic densities produce higher members of the sequence, starting with

$$\nabla_r n_2(r,r') + n_2(r,r')(\nabla\beta u(r) + \nabla\beta\varphi(r-r')) + \int n_3(r,r',r'')\nabla\beta\varphi(r-r'')dr'' = 0 \tag{3.18}$$

In fact, they are all derivable from (3.17) either by successively applying the operation $\delta/\delta u$, or by the "bootstrap" operation that we will examine in Sec. 3b. Furthermore [P80], integration of (3.17) over r recovers (3.12), integration of (3.18) over r, coupled with (3.17), recovers (3.14), and so forth. We will return to (3.17) in Sec. 5a.

3.2 Classical integral equations and the mixed substrate format

The techniques mentioned in Sec. 3a received their major impetus from application to bulk matter. It is the use of the grand ensemble that makes the connection between conditional uniformity and non-uniformity so useful, the general idea being that another particle can be added to the system to probe its structure without changing the state of the system. The relationship is particularly simple in the classical domain, in which, spelling out the partition function as $\Xi[\mu,\varphi]$, we see that $\delta\Xi[\mu,\varphi]/\delta e^{\beta\mu(r)} = \sum \frac{1}{N-1!}\int\cdots\int \exp\beta\Big[\sum_2^N \mu(r_j) - \sum_2^N \varphi(r,r_j) - \frac{1}{2}\sum_2^N{}'\varphi(r_i,r_j)\Big]dr^{N-1}$, or simply

$$\frac{\delta\Xi[\mu,\varphi]}{\delta e^{\beta\mu(r)}} = \Xi[\mu-\varphi_r,\varphi] \text{ where } \varphi_r(r') \equiv \varphi(r,r'). \tag{3.19}$$

On iteration, then,

$$\frac{\delta^s\Xi[\mu,\varphi]}{\delta e^{\beta\mu(r_1)}\ldots\delta e^{\beta\mu(r_s)}} = e^{-\frac{\beta}{2}\sum_1^s{}'\varphi(r_i,r_j)}\,\Xi\left[\mu - \sum_1^s \varphi_{r_j},\varphi\right]. \tag{3.20}$$

Comparing with (2.21),

$$n_s(r_1,\ldots,r_s|\mu,\varphi) = e^{\beta\sum_1^s \mu(r_j)}\,e^{-\frac{1}{2}\beta\sum_1^s{}'\varphi(r_i,r_j)}\,\Xi\left[\mu - \sum_1^s \varphi_{r_j},\varphi\right]/\Xi[\mu,\varphi], \tag{3.21}$$

and as readily derived consequence

$$n_{s+t}(r_1,\ldots,r_s,r'_1,\ldots,r'_t|\mu,\varphi)/n_t(r'_1,\ldots,r'_t|\mu,\varphi) = n_s\left(r_1,\ldots,r_s \mid \mu - \sum_1^t \varphi_{r'_j},\varphi\right). \tag{3.22}$$

Eq. (3.22) is the "bootstrap" principle [P62] for multi-particle densities, stating the "obvious" fact that a classical system in thermal equilibrium does not know the difference between fixing particles at $r'_1,\ldots,r'_t$ and applying the external potential $v(r) = \sum_1^t \varphi(r,r'_j)$ that this fixing would produce.

The value of the observation (3.22) is that the added potential $\sum\varphi(r,r'_j)$ can be applied via a functional Taylor expansion, leaving us with a large set of integral equation relationships among the $\{n_s\}$. The method that generates most of the approximations that have been used is to truncate the Taylor expansion at low order, so that a closed set of relationships results. The linear Debye-Huckel approximation is prototypical: we take the first order expansion of $\beta\mu$ with respect to n,

$$\beta\mu(r|n) = \beta\mu(r;n_0) + \int C(r-r';n_0)(n(r')-n_0)dr',$$

starting with a uniform fluid at density n_0, and turn on a particle at the origin, so that $\mu(r|n) - \mu(r;n_0) = -\varphi(r)$, while $n(r') = n_2(r'-0;n_0)/n_0$. In other words, $-\beta\varphi(r) = \int C(r-r';n_0)n_0 h(r';n_0)$, so that on using the Ornstein-Zernike relation obtained from (2.9), (2.11), and (2.12)

$$h(r,r') - c^{(2)}(r,r') = \int c^{(2)}(r,r'')n(r'')h(r'',r')dr'', \tag{3.23}$$

we have simply

$$LDH\text{: } c^{(2)}(r) = -\beta\varphi(r). \tag{3.24}$$

Of course, since $\beta\mu = \ln n$ for a weakly coupled gas, it might make more sense to expand $\beta\mu$ in $\ln n$, with the additional advantage that choosing $\ln n$ as independent variable guarantees that n cannot go negative. Doing so, we now have $-\beta\varphi(r) = \int n_0 C(r,r';n_0)\ln g(r')dr'$, or $g(r) = \exp -\frac{1}{n_0}\int S(r-r';n_0)\beta\varphi(r')dr'$, so that

$$DH\text{: } g(r) = e^{-\beta\varphi(r)}e^{-n_0\int h(r-r';n_0)\beta\varphi(r')dr'}, \tag{3.25}$$

the non-linear Debye-Huckel equation. On the other hand, it makes no less sense to expand $e^{\beta\mu}$ with respect to n, replacing (3.24) by the simple

$$LDH'\text{: } c^{(2)}(r) = e^{-\beta\varphi(r)} - 1. \tag{3.26}$$

There is a non-trivial conceptual problem. If one is carrying out a series expansion only to low-order, there is the implicit assumption that the relationship is indeed one of low order, e.g. linear. This is not obviously the case for $\mu[n]$, leading to (3.24). But consider the combination $n(r)e^{\beta u(r)}$; it is slowly varying both for small and large u, and settles down to n when u vanishes. Perhaps this is more appropriate than μ itself. The "classical" integral equations make various choices for independent and dependent fields, and then follow the above argument by "turning on" a potential from $u_0 = 0$ to $u(r) = \varphi(r)$, while the density changes from n_0 to $n(r) = n_0 g(r;n_0)$. In fact, if (3.1) is rewritten as the general linear hypothesis

$$\int \frac{\delta b(r|n)}{\delta n(r')}\Big|_{n_0}(a(r'|n) - a(r';n_0))dr' = \int \frac{\delta a(r|n)}{\delta n(r')}\Big|_{n_0}(b(r'|n) - b(r';n_0))dr', \tag{3.27}$$

the distinction between independent and dependent fields is seen to be irrelevant. In the PY approximation, with identifications $a = ne^{\beta u}$, $b = n$, it follows precisely as did (3.24) that

$$PY\text{: } g(r)e^{\beta\varphi(r)} = g(r) - c^{(2)}(r), \tag{3.28}$$

whereas the $CHNC$ approximation, taking $a = \ln(ne^{\beta u})$ and $b = n$ results in

$$CHNC\text{: } \ln g(r) + \beta\varphi(r) = h(r) - c^{(2)}(r). \tag{3.29}$$

Finally, at this time, there is the YBG-type integral equation approximation. It can be cast in the same format by setting $a = \ln(ne^{\beta u})$, $b = n\nabla\beta u$, but this is quite artificial. The more usual derivation [F64] imposes the Kirkwood superposition approximation

$$n_3(r,r',r'') = n_0^3 g(r-r')g(r'-r'')g(r''-r) \tag{3.30}$$

to close (3.18), resulting therefore in

$$\nabla(\ln g(r) + \beta\varphi(r)) = -n_0 \int g(r-r')g(r')\nabla\beta\varphi(r-r')dr'. \tag{3.31}$$

A great deal of experience has been accumulated in the degree and domains of validity of these pair correlation integral equations in the bulk fluid context, but how do they relate to the non-uniform fluid problem? Probably the simplest such back reduction results from extension to a mixture of fluids. It is clear that very little in our discussion has depended crucially upon the nature of the parameters used to describe a particle – there could be an internal space in addition to location in physical space, and there certainly could be a discrete parameter as well, i.e. a species specification. If index α denotes species, the Ornstein-Zernike equation now reads, in obvious notation,

$$h_{\alpha\beta}(r) = c_{\alpha\beta}(r) + \sum_\gamma \int h_{\alpha\gamma}(r-r')n_\gamma c_{\gamma\beta}(r')dr' \tag{3.32}$$

(superscript 2 being dropped in $c^{(2)}_{\alpha\beta}$), while the LDH, DH, PY, $CHNC$, and YBG approximations read, respectively,

$$\begin{aligned}
&c_{\alpha\beta}(r) = -\beta\varphi_{\alpha\beta}(r)\\
&g_{\alpha\beta}(r)e^{\beta\varphi_{\alpha\beta}(r)} = \exp - \sum_\gamma \int h_{\alpha\gamma}(r-r')n_\gamma\beta\varphi_{\gamma\beta}(r')dr'\\
&g_{\alpha\beta}(r)e^{\beta\varphi_{\alpha\beta}(r)} = g_{\alpha\beta}(r) - c_{\alpha\beta}(r)\\
&\ln g_{\alpha\beta}(r) + \beta\varphi_{\alpha\beta}(r) = h_{\alpha\beta}(r) - c_{\alpha\beta}(r)\\
&\nabla(\ln g_{\alpha\beta}(r) + \beta\varphi_{\alpha\beta}(r) + \sum_\gamma \int g_{\alpha\gamma}(r-r')\beta\varphi_{\alpha\gamma}(r-r')n_\gamma g_{\gamma\beta}(r')dr'.
\end{aligned} \tag{3.33}$$

The trick then [P75], [P82a], [H92] is to regard an external potential $u(r)$ applied to a single species as the interaction potential of a single pseudoparticle fixed at the origin, i.e. a member of a field generating species of 0 density. If our approximations are to be believed under this wild extrapolation, we then have a two-fluid mixture with $n_0 = n_0$, $n_1 = 0$, $g_{00}(r) = g(r)$, $g_{01}(r) = n(r)/n_0, g_{11}(r) = 1$, $\varphi_{00}(r) = \varphi(r)$, $\varphi_{01}(r) = u(r)$, $\varphi_{11}(r)$ arbitrary. The Ornstein-Zernike equation (3.32) hence gives us, independent of any approximation,

$$c_{00}(r) = c(r), \quad c_{01}(r) = \frac{n(r)}{n_0} - 1 - \int c(r-r')(n(r') - n_0)dr'. \tag{3.34}$$

Consequently, the 00 components of (3.33) are just the density n_0 uniform fluid approximations for the presumably solved reference fluid, the 11 components specify the unneeded φ_{01}, the 01 components are the statements of the desired profile equations, which can be written as

$$\beta u(r) = 1 - \frac{n(r)}{n_0} + \int c(r-r')(n(r') - n_0)dr' \tag{3.35a}$$

$$\beta u(r) + \ln n(r)/n_0 = -n_0 \int h(r-r')\beta u(r')dr' \tag{3.35b}$$

$$\beta u(r) + \ln n(r)/n_0 = \ln\left(1 + \int c(r-r')(n(r') - n_0)dr'\right) \tag{3.35c}$$

$$\beta u(r) + \ln n(r)/n_0 = \int c(r-r')(n(r') - n_0)dr' \tag{3.35d}$$

$$\nabla(\beta u(r) + \ln n(r)) = -\int g(r-r')\nabla\beta\varphi(r-r')n(r')dr'. \tag{3.35e}$$

From the viewpoint of density functional formulations, (3.35a, 3.35d, 3.35e) are integrable (just compute $\delta u(r)/\delta n(r')$), but not (3.35b, 3.35c). The corresponding free energy changes take the form

$$\Delta\beta\overline{F}[n] = \frac{1}{2n_0}\int (n(r)-n_0)^2 dr - \frac{1}{2}\int\int (n(r)-n_0)c(r-r')(n(r')-n_0)drdr' \tag{3.36a}$$

$$\Delta\beta\overline{F}[n] = \int n(r)\ln(n(r)/n_0)dr - \frac{1}{2}\int\int (n(r)-n_0)c(r-r')(n(r')-n_0)drdr' \tag{3.36d}$$

$$\Delta\overline{F}[n] = \int n(r)\ln(n(r)/n_0)dr + \frac{1}{2}\int\int (n(r)-n_0)\beta\varphi_g(r-r')(n(r')-n_0)drdr' \tag{3.36e}$$

where $\varphi_g'(r) = g(r)\varphi'(r)$.

3.3 Singlet-pair construction

The traditional integral equation approximations for non-uniform pair-interacting fluids require a known particle density $n(r)$. However, except for the YBG type, which is anomalous in various ways, this enters only in the Ornstein-Zernike $c-h$ equation, and there is no built-in relation between $n(r)$ and the applied potential $u(r)$. This is unfortunate, because such equations do not labor under the necessity of using a particular uniform reference and corresponding distribution as input. If the equation could be recognized as arising from a variational principle at the 2-body level, then the missing pure one-body terms could be supplied at leisure and a more carefully tailored pair-singlet connection thereby created. It is understood that there are limitations as to how far one can go in combining the "φ-route" and "n-route" for the construction of $\overline{F}[n,\varphi]$. In fact, the argument of (2.49) shows that

$$\frac{\partial\overline{F}}{\delta\varphi(1,2)} = \frac{1}{2}\left(\frac{\delta^2\beta\overline{F}}{\delta n\delta n}\right)^{-1}(1,2) + \frac{1}{2}n(1)(n(2)-\delta(1,2)) \tag{3.37}$$

(r_j will be denoted by j when no confusion can arise) so that, starting from $\overline{F}[n,0] = \int n(1)(\ln n(1)-1)d1$, one would get the unique correct $\overline{F}$ if (3.37) is maintained. Approximations must introduce inconsistency at some level, unless one is simply dealing with some model Hamiltonian. [L88] We choose here to aim at satisfying

$$\delta\overline{F}/\delta\varphi(1,2) = \frac{1}{2}n_2(1,2) \tag{3.38}$$

and later will focus on the alternative $\delta^2\beta\overline{F}/\delta n(1)\delta n(2) = C(1,2|n)$.

The linear Debye-Huckel approximation derived in (3.24) is unchanged for non-uniform n_0:

$$c(1,2) = -\beta\varphi(1,2). \tag{3.39}$$

Since (3.38) holds at fixed $\{n\}$, integrability at the pair level requires

$$\delta n_2(1,2)/\delta\varphi(3,4) = \delta n_2(3,4)/\delta\varphi(1,2), \tag{3.40}$$

and this is satisfied by (3.38): $\delta n_2 = \delta S = \delta C^{-1} = -C^{-1}\delta C C^{-1}$, so $\delta n_2(1,2)/\delta\varphi(3,4) = S(1,3)S(2,4)$, which is symmetric. There are several ways to produce the pair part of $\overline{F}$ in such a fashion that (3.39) holds. Most directly, we use the Ornstein-Zernike equation in one of the equivalent matrix forms

$$(I - nc)(I + nh) = I,$$
$$c = h(I + nh)^{-1}, \quad h = c(I - nc)^{-1} \tag{3.41}$$

where n denotes the diagonal matrix $n(1,2) = n(1)\delta(1,2)$, to write $n_2(1,2) = n(1)n(2) - n(1)\delta(1,2) + (I + nh)n(1,2) = n(1)n(2) - n(1)\delta(1,2) + (I - nc)^{-1}n(1,2)$, or

$$n_2(1,2) = n(1)n(2) - n(1)\delta(1,2) + (I + n\beta\varphi)^{-1}n(1,2), \tag{3.42}$$

and then "integrate" with respect to φ to obtain

$$\overline{F}[n,\varphi] = \frac{1}{2}\int\int n(1)(n(2) - \delta(1,2))\varphi(1,2)d_1 d_2$$
$$+\frac{1}{2\beta} \text{ Tr } \ln(I + n\beta\varphi) + \Delta\overline{F}[n] \tag{3.43}$$

to within an additive singlet term. Since we know that

$$\overline{F}[n,0] = \frac{1}{\beta}\int n(1)(\ln n(1) - 1)d1, \tag{3.44}$$

the complete expression is

$$\beta\overline{F}[n,\varphi] = \int n(1)(\ln n(1) - 1)d1 + \frac{1}{2}\int\int n(1)(n(2) - \delta(1,2))\beta\varphi(1,2)d1d2$$
$$+\frac{1}{2} \text{ Tr } \ln(I + n\beta\varphi). \tag{3.45}$$

In deriving (3.43) from (3.42), we have used the cyclic invariance of the trace in the form

$$\delta \text{ Tr } f(A) = \text{ Tr}\left(\frac{\delta f(A)}{\delta A}\delta A\right). \tag{3.46}$$

There are a number of useful alternative forms, in which the interaction potential φ occurs more innocuously, and which have an expanded stationary character, particularly useful for computational purposes. For example, since the Ornstein-Zernike equation is equivalent to [O77]

$$ncn(1,2) = \frac{\delta}{\delta h(1,2)} \text{ Tr}[nh - \ln(I + nh)], \tag{3.47}$$

then (3.38) is equivalent to

$$\frac{\delta}{\delta h(1,2)} \text{ Tr}[n\beta\varphi ng + nh - \ln(I + nh)] = 0. \tag{3.48}$$

It follows that we can construct an extended $\overline{F}[n,\varphi,h]$ that generates $\mu = \delta\overline{F}/\delta n$, $\frac{1}{2}n_2 = \delta\overline{F}/\delta\varphi$, but is stationary with respect to h:

$$\beta\overline{F}[n,\varphi,h] = \int n(1)(\ln n(1) - 1)d1 + \frac{1}{2} \text{ Tr}[n\beta\varphi ng + nh - \ln(I + nh)]$$
$$\text{where } \delta\beta\overline{F}/\delta h(1,2) = 0. \tag{3.49}$$

The $CHNC$ approximation, $c + \beta\varphi + \ln g - h = 0$, is now only a step removed, since it becomes

$$\frac{\delta}{\delta h(1,2)}\Big\{\int\int n(1)\left[\beta\varphi(1,2)g(1,2) + g(1,2)\ln g(1,2) - h(1,2) - \frac{1}{2}h(1,2)^2\right]n(2)d1d2$$
$$+\text{Tr}(nh - \ln(I + nh))\Big\} = 0, \tag{3.50}$$

leading at once to

$$\beta F^{ex}[n,\varphi,h] = \frac{1}{2}\int\int n(1)[g(1,2)(\beta\varphi(1,2) + \ln g(1,2)) - h(1,2) - \frac{1}{2}h(1,2)^2)n(2)d1d2$$
$$+\frac{1}{2}\ \mathrm{Tr}(nh - ln(I+nh)) \tag{3.51}$$

and the associated profile equation

$$\beta\mu(1) = \ln n(1) + \left(gn\beta\varphi - \frac{1}{2}hnh\right)(1,1)$$
$$+\int(g(1,2)\ln g(1,2) - h(1,2))n(2)d2 + \frac{1}{2}hnh(I-nh)^{-1}(1,1). \tag{3.52}$$

Finding Tr $\ln(I+nh)$ can be a computational chore, so one might notice that

$$-\frac{1}{4}\int\int n(1)h(1,2)^2 n(1)d1d2 + \frac{1}{2}\ \mathrm{Tr}(nh - ln(I+nh)) = \frac{1}{2}\ \mathrm{Tr}(nh - \frac{1}{2}nhnh - \ln(I+nh)) =$$
$$-\frac{1}{6}\ \mathrm{Tr}\ nhnhnh + -,$$

and therefore replace (3.51) by

$$\beta F^{ex}[n,\varphi,h] = \frac{1}{2}\int\int n(1)[g(1,2)(\beta\varphi(1,2) + \ln g(1,2) - h(1,2))]n(2)d1d2$$
$$-\frac{1}{6}\int\int\int n(1)h(1,2)n(2)h(2,3)n(3)h(3,1)d1d2d3, \tag{3.53}$$

and the modified profile equation

$$\beta\mu(1) = \ln n(1) + \int g(1,2)[\beta\varphi(1,2) + \ln g(1,2) - h(1,2)]n(2)d2$$
$$-\frac{1}{2}\int\int h(1,2)n(2)h(2,3)n(3)h(3,1)d2d3. \tag{3.54}$$

These also arise from an entropy approximation method that we will describe later. They do not of course yield precisely the $CHNC$ approximation, since now

$$0 = \delta\beta\overline{F}/\delta h(1,2) = \frac{1}{2}n(1)n(2)[\beta\varphi(1,2) + \ln g(1,2) - \int h(1,3)n(3)h(3,2)d3] \tag{3.55}$$

implies that

$$\ln(ge^{\beta\varphi}) = hnh, \tag{3.56}$$

instead of $h - c$, and reduces (3.54) to

$$\beta\mu(1) = \ln n(1) + \int(hnh - h)(1,2)n(2)d2 - \frac{1}{2}hnhnh(1,1). \tag{3.57}$$

For the non-integrable PY approximation, something must give, and it is the consistency of the 2-body distributions, as attested by the relation $\delta\overline{F}/\delta\varphi(1,2) = \frac{1}{2}n_2(1,2)$ that is most dispensible (as we have discussed with respect to (3.37), $\delta\beta\overline{F}/\delta n\delta n = C$ is guaranteed to fail). Thus, we can couple (3.47) with the $P.Y.$ closure, $c = (1-e^{\beta\varphi})g$, presented as

$$n(1)c(1,2)n(2) = \frac{\delta}{\delta h(1,2)}\frac{1}{2}\int\int n(1)n(2)(1 - e^{\beta\varphi(1,2)})g(1,2)^2 d1d2, \tag{3.58}$$

to construct

$$\beta F^{ex}[n,\varphi,h] = -\frac{1}{2}\int\int n(1)g(1,2)^2\frac{f(1,2)}{1+f(1,2)}n(2)d1d2 + \frac{1}{2}\ \mathrm{Tr}(nh - \ln(I+nh)), \quad (3.59)$$

with profile equation, after inserting the closure

$$\beta\mu(1) = \ln n(1) - \int g(1,2)c(1,2)n(2)d2 + \frac{1}{2}(h-c)(1,1). \quad (3.60)$$

Equally well, c can be chosen as the auxiliary field. [AC72] $\delta\beta\overline{F}/\delta\varphi(1,2) = \frac{1}{2}\beta n_2(1,2)$ again requires the term

$$\int\int n_2(1,2)\beta\varphi(1,2)d1d2 = \int\int \beta\varphi(1,2)n(1)(n(2)-\delta(1,2))d1d2$$
$$+\mathrm{Tr}\ \beta\varphi(I-nc)^{-1}n, \quad (3.61)$$

but $\delta/\delta c(1,2)$ Tr $\beta\varphi(I-nc)^{-1}n = n\beta\varphi(I-nc)^{-1}n(1,2)$ must now be joined in the LDH case by

$$\frac{\delta}{\delta c(1,2)}\ \mathrm{Tr}[(I-nc)^{-1}nc + \ln(I-nc)] = nc(I-nc)^{-1}n(1,2). \quad (3.62)$$

Hence we have

$$\beta F^{ex}[n,\varphi,c] = \frac{1}{2}\int\int \beta\varphi(1,2)n(1)(n(2)-\delta(1,2))d1d2$$
$$+\frac{1}{2}\ \mathrm{Tr}[(\beta\varphi + c)(I-nc)^{-1}n + \ln(I-nc)]. \quad (3.63)$$

Once more, dropping the 2-body consistency allows for a $P.Y.$ transcription as well. From $h = c/(1-e^{\beta\varphi}) - 1$, we have

$$nhn(1,2) = \frac{\delta}{\delta c(1,2)}\int\int n(1)\left[\frac{1}{2}\frac{1+f(1,2)}{f(1,2)}c(1,2)^2 - c(1,2)\right]n(2)d1d2. \quad (3.64)$$

The free energy functional

$$\beta F^{ex}[n,\varphi,c] = \frac{1}{4}\int\int \frac{1+f(1,2)}{f(1,2)}c(1,2)^2 n(1)n(2)d1d2$$
$$-\frac{1}{2}\int\int n(1)c(1,2)n(2)d1d2 - \frac{1}{2}\ \mathrm{Tr}[(I-nc)^{-1}nc + \ln(I-nc)] \quad (3.65)$$

then follows from (3.62) and (3.64), resembling a form used by Baxter [B70] in a different context.

Both h and c can also be taken as auxiliary fields with respect to which $\overline{F}$ is stationary. This format may be established for example from the relation

$$\left(\frac{\delta}{\delta h(1,2)} + \frac{\delta}{\delta c(1,2)}\right)\ \mathrm{Tr}[\ln(I-nc) + nc + \frac{1}{2}nhnh]$$
$$= n(1)n(2)[h(1,2) - c(I-nc)^{-1}(1,2)], \quad (3.66)$$

whose vanishing is equivalent to Ornstein-Zernike. Hence any free energy expression

$$\beta F^{ex} = A[n,\varphi,h-c] + \frac{1}{2}\ \mathrm{Tr}\left[\ln(I-nc) + nc + \frac{1}{2}nhnh\right] \quad (3.67)$$

will impose Ornstein-Zernike via $(\delta/\delta c + \delta/\delta h)\beta\overline{F} = 0$. If A is given the local form

$$A[n,\varphi,h-c] = \frac{1}{2}\int\int n(1)\mathcal{A}(h(1,2)-c(1,2),\beta\varphi(1,2))n(2)d1d2, \quad (3.68)$$

then the remaining stationary condition $\delta\beta\overline{F}/\delta h = 0$ appears as

$$h + \mathcal{A}'_n(h - c, \beta\varphi) = 0. \tag{3.69}$$

And in addition, $\delta\beta F^{ex}/\delta\beta\varphi(1,2) = \frac{1}{2}n_2(1,2)$ is seen to imply that

$$1 + h = \mathcal{A}'_{\beta\varphi}(h - c, \beta\varphi). \tag{3.70}$$

Eqs. (3.69) and (3.7) together imply that $\mathcal{A} - \beta\varphi$ is a function of $h - c - \beta\varphi$ alone. This is certainly true for linear DH, $h = h - c - \beta\varphi$, so $\mathcal{A} = \beta\varphi - \frac{1}{2}(h - c - \beta\varphi)^2$, and for $CHNC$, $h = e^{h-c-\beta\varphi} - 1$, with $\mathcal{A} = \beta\varphi - e^{h-c-\beta\varphi} + (h - c - \beta\varphi) + 1$. In the latter case, we have

$$\beta F^{ex}[n, \varphi, c, h] = \frac{1}{2}\int\int n(1)(g(1,2) - c(1,2) - e^{h(1,2)-c(1,2)-\beta\varphi(1,2)})n(2)d1d2$$
$$+\frac{1}{2}\ \mathrm{Tr}\left[\ln(I - nc) + nc + \frac{1}{2}nhnh\right]. \tag{3.71}$$

Note that if the correct h and c are assumed known, then the profile equation $\beta\mu(1) = \ln n(1) + \delta\beta F^{ex}/\delta n(1)$ reduces to the simple

$$\beta\mu(1) = \ln n(1) - \int c(1,2)n(2)d2 + \frac{1}{2}hncnh(1,1); \tag{3.72}$$

the problem is then that of assuring the accuracy of h and c.

If we do not demand that n_2 information be consistently recovered from F^{ex}, then the condition (3.70) need not be imposed. Thus if the *P.Y.* closure $c = (1 - e^{\beta\varphi})g$ is written as

$$h = f + (1 + f)(h - c), \tag{3.73}$$

then $h = \frac{\partial}{\partial h}(f(h - c) + \frac{1}{2}(1 + f)(h - c)^2)$, so that we can choose

$$\beta F^{ex}[n, \varphi, c, h] = -\frac{1}{2}\int\int n(1)[f(1,2)(h(1,2) - c(1,2))$$
$$+\frac{1}{2}(1 + f(1,2))(h(1,2) - c(1,2))^2]n(2)d1d2 + \frac{1}{2}\ \mathrm{Tr}[\ln(I - nc) + nc + \frac{1}{2}nhnh], \tag{3.74}$$

with the profile equation

$$\beta\mu(1) = \ln n(1) - \frac{1}{4}\int(h - c)(1,2)(h + f)(1,2)n(2)d2 + \frac{1}{2}hncnh(1,1). \tag{3.75}$$

4. EXACTLY SOLVABLE MODELS

Our objective is to depict the equilibrium structure of classical many-particle systems as economically as possible, by means of caricatures that miss little in the way of essentials. The implicit assumption is that there are underlying themes against which the specifics fade into the background. An effective way of seeing these themes is by means of a spectrum of model systems which can be solved exactly, each system bringing out in relief a different qualitative aspect. In practice, the construction of solvable systems containing enough control parameters to be reasonably informative is very much of an art, and one is only very gradually picking up representative points of this spectrum. The logical structure of the following is hence very much a posteriori.

4.1 Lattice gases

Discrete models of space have the obvious advantage of converting statistical mechanics to combinatorics and the non-trivial advantage of cutting down drastically on the information that has to be supplied to characterize the system. We now imagine that physical space is represented by a set of discrete locations $\{x\}$, each of which can be regarded as the center of a mesh compartment that decomposes the spatial continuum. We leave undecided at the moment whether this is simply a model in which information is bunched into packages described by just a few parameters, or whether we really want to take the limit as the compartments shrink to zero. $\{x\}$ is not just a point set, since the geometry in the form of next neighbors establishes the interactions that characterizes a given system.

We start with an infinite one-dimensional integer lattice representing one-dimensional space. Each point x can either be empty, $\nu_x = 0$, or be occupied by a particle, $\nu_x = 1$, and so a particle configuration is specified by the set $\{\nu_x\}$, more convenient for our purposes than giving the locations of particles, and of course making a grand ensemble – in which that total number of particles $N = \sum_x \nu_x$ is not preordained – a very natural object. The simplest interaction energy is that between nest neighbor occupied sites, $J\nu_x\nu_{x+1}$; we will not bother making J inhomogeneous, but we will certainly require an inhomogeneous external potential u_x, coupled as $\mu_x = \mu - u_x$ with the chemical potential that controls the net occupation of the lattice. Hence the (grand) partition function for this system is

$$\Xi[\mu] = \sum_{\{\nu_x\}} \exp \sum_x \beta\mu_x\nu_x \exp -\beta J \sum_x \nu_x\nu_{x+1}. \tag{4.1}$$

This converges if $\mu_x \to -\infty$ fast enough as $|x| \to \infty$, since the ν_x as $|x| \to \infty$ will then be restricted to the value 0. A formal computation of $\Xi[\mu]$ is trivial. We define the 2×2 internal Boltzmann factor matrix

$$w_{\nu\nu'} = e^{-\beta J\nu\nu'} = \epsilon^{\nu\nu'} \text{ where } \epsilon = e^{-\beta J}, f = \epsilon - 1, \tag{4.2}$$

and the diagonal external Boltzmann factor matrix

$$z_{x,\nu\nu'} = e^{\beta\mu_x\nu}\delta_{\nu,\nu'} = (z_x)^\nu\delta_{\nu,\nu'} \tag{4.3}$$

where $z_x = e^{\beta\mu_x}$, so that

$$\Xi[\mu] = \left\langle 1 \left| \prod_x (wz_x) \right| 1 \right\rangle, \tag{4.4}$$

with 1 as the vector (1.1). But the 2×2 matrices do not commute, so that computation comes to a grinding halt.

Although $\Xi[n]$ – from which all distributions are available by μ_x differentiation – is not available, a change of representation works wonders. For this purpose, we first introduce the relative partition function in which the occupations at x and $x+1$ are fixed:

$$\xi_{\nu'\nu}^{x+\frac{1}{2}} = \left\langle 1 \left| \prod_{y \le x} (wz_y) \right| \nu \right\rangle w_{\nu\nu'} \left\langle \nu' \left| \prod_{y \ge x+1} (z_y(w) \right| 1 \right\rangle / \Xi. \tag{4.5}$$

The mean particle occupation $n_x = \langle \nu_x \rangle$ is clearly given by

$$n_x = \xi_{01}^{x+\frac{1}{2}} + \xi_{11}^{x+\frac{1}{2}}, \; n_{x+1} = \xi_{10}^{x+\frac{1}{2}} + \xi_{11}^{x+\frac{1}{2}}, \tag{4.6}$$

and we also have

$$\xi_{00}^{x+\frac{1}{2}} + \xi_{01}^{x+\frac{1}{2}} + \xi_{10}^{x+\frac{1}{2}} + \xi_{11}^{x+\frac{1}{2}} = 1, \quad \xi_{00}\xi_{11} - \epsilon\xi_{01}\xi_{10} = 0. \tag{4.7}$$

Solving, we find

$$\begin{aligned} \xi_{00}^{x+\frac{1}{2}} &= \frac{1}{2}\frac{\epsilon+f}{f} - \frac{1}{2}(n_x + n_{x+1}) + s^{1/2} \\ \xi_{10}^{x+\frac{1}{2}}, \xi_{01}^{x+\frac{1}{2}} &= -\frac{1}{2}\frac{1}{f} \pm \frac{1}{2}(n_{x+1} - n_x) - s^{1/2} \\ \xi_{11}^{x+\frac{1}{2}} &= \tfrac{1}{2}\tfrac{1}{f} + \tfrac{1}{2}(n_x + n_{x+1}) + s^{1/2} \end{aligned} \tag{4.8}$$

where $s = \dfrac{1}{4}\left(\dfrac{1}{f} + n_x - n_{x+1}\right)^2 - \dfrac{\epsilon}{f} n_x n_{x+1}$.

But we also see that if A^x fixes sites separated by two, omitting their connections

$$A^x_{\nu'\nu''} = \left\langle 1 \left| \prod_{y \le x-1} (w z_y) \right| \nu'' \right\rangle \left\langle \nu' \left| \prod_{y \ge x+1} (z_y w) \right| 1 \right\rangle \Big/ \Xi, \tag{4.9}$$

then $\xi_{\nu'\nu}^{x+\frac{1}{2}} = (A^x w y_x)_{\nu'\nu} w_{\nu'\nu'}$, $\quad \xi_{\nu\nu''}^{x-\frac{1}{2}} = (z_x\, w A^x)_{\nu\nu''}\, w_{\nu\nu''}$, or

$$z_x w (\xi/w)^{x+\frac{1}{2}} = (\xi/w)^{x-\frac{1}{2}} w z_x \tag{4.10}$$

(where $(z/w)_{\nu\nu'} \equiv z_{\nu\nu'}/w_{\nu\nu'}$). Taking the 01 element, we conclude that

$$z_x = [\xi_{01}^{x+\frac{1}{2}} + \xi_{11}^{x+\frac{1}{2}}/\epsilon]/[\xi_{00}^{x-\frac{1}{2}} + \epsilon\xi_{01}^{x-\frac{1}{2}}], \tag{4.11}$$

or using (4,6, 4.7)

$$z_x = \frac{\xi_{01}^{x+\frac{1}{2}}\,\xi_{10}^{x-\frac{1}{2}}}{\xi_{00}^{x+\frac{1}{2}}\,\xi_{00}^{x-\frac{1}{2}}}. \tag{4.12}$$

It follows that, in obvious notation,

$$\beta\mu_x = U_\epsilon(n_x, n_{x+1}) + U_\epsilon(n_x, n_{x-1}); \tag{4.13}$$

in other words, the profile μ_x is local in the field $\{n_x\}$, and explicit, if complicated. It also follows that we can write [P77]

$$\beta\overline{F} = \sum_x \varphi_\epsilon(n_x, n_{x+1}) \text{ where } U_\epsilon(n, n') = \partial\varphi_\epsilon(n, n')/\partial n, \tag{4.14}$$

φ_ϵ being a symmetric function. $\overline{F}$ is also *local* in the density field. In this case, then, the density functional format $\overline{F}[n]$ gives an explicit result and has a very simple structure.

But the one-dimensional infinite lattice with next neighbor interaction is a very restrictive physical vehicle. Can one deal with more complex physical structures? Looking back at (4.5), we see that the one-dimensionality entered in one crucial way: if we fix the state of one site, the system decomposes into independent subsystems. This is because – by definition – the lattice is simply connected, a special case of a Cayley tree (no loops) network, or its extreme version, a Bethe lattice (infinite tree, fixed coordination number). Indeed, the technique we have used goes through [S88], [P89a], with only minor changes, for a lattice gases on a tree or Bethe lattice. However, a further expanded viewpoint is helpful [P90], [ZP93]. Suppose we have a lattice gas on a Cayley tree, with both inhomogeneous external potential $\mu_x = \mu - u_x$ and inhomogeneous nearest neighbor interaction, J_{xy}. There is then a $1:1$ correspondence with the resulting densities $n_x = \langle \nu_x \rangle$ and next neighbor pair densities $n_{xy} = \langle \nu_x \nu_y \rangle$. What is it?

The most efficient way to proceed is by computing the entropy (in units of Boltzmann's constant) of the system,

$$S = -\sum_{\{\nu_x\}} \rho\{\nu_x\} \ln \rho\{\nu_x\}, \tag{4.15}$$

where ρ is the probability density on the full lattice

$$\rho\{\nu_x\} = W\{\nu_x\}/\Xi \text{ where } W\{\nu_x\} = \exp\beta\left(\sum_x \mu_x \nu_x - \frac{1}{2}\sum_{\langle x,y\rangle} J_{xy}\nu_x\nu_y\right); \tag{4.16}$$

Ξ is the grand partition function that serves to normalize ρ to a probability, and $\langle x,y\rangle$ signifies that x and y are next neighbors. If the control parameters $\{\mu_x, J_{xy}\}$ are varied, then it is seen that

$$\delta\Xi/\Xi = -\sum n_x \delta\mu_x + \frac{1}{2}\sum_{\langle x,y\rangle} n_{xy}\delta J_{xy}, \tag{4.17}$$

from which

$$\delta\rho\{\nu_x\} = \rho\{\nu_x\}\left(-\sum(\nu_x - n_x)\delta\mu_x + \frac{1}{2}\sum_{\langle x,y\rangle}(\nu_x\nu_y - n_{xy})\delta J_{xy}\right), \tag{4.18}$$

and consequently

$$\delta S = -\delta\langle \ln\rho\{\nu_x\}\rangle = \sum \mu_{x'}\delta\mu_x\langle(\nu_x - n_x)(\nu'_x - n_{x'})\rangle$$
$$-\frac{1}{2}\sum_{\langle x,y\rangle} J_{x'y'}\delta J_{xy}\langle(\nu_x\nu_y - n_{xy})(\nu_{x'}\nu_{y'} - n_{x'y'}\rangle, \tag{4.19}$$

readily identified as

$$\delta S = \sum \mu_x \delta n_x - \frac{1}{2}\sum_{\langle x,y\rangle} J_{xy}\delta n_{xy}. \tag{4.20}$$

It follows that if S is taken as a function of $\{n_x, n_{xy}\}$, then

$$\mu_x = \frac{\partial S}{\partial n_x}, \quad \frac{1}{2}J_{xy} = -\frac{\partial S}{\partial n_{xy}}. \tag{4.21}$$

It is then just one more step to solve for the n_{xy}, uniquely because of the known convexity of S, and carry out a Legendre transformation to the intrinsic Helmholtz free energy $\overline{F}\{n_x, J_{xy}\}$,

$$\beta\overline{F} = -S - \frac{1}{2}\sum_{\langle x,y\rangle} J_{xy}n_{xy}; \tag{4.22}$$

however, we will defer this operation.

Finding $S\{n_x, n_{xy}\}$ for a Cayley tree is not difficult. It is convenient to introduce the notation $n_\Lambda[\alpha]$ for the joint probability – or density – of the sites in any subset $\Lambda = \{x\}$ having the values $\{\alpha_x\}$, e.g.

$$n_x(1) = n_x, \qquad n_x(0) = 1 - n_x$$
$$n_{xy}(11) = n_{xy},\ n_{xy}(01) = n_y - n_{xy},\ n_{xy}(10) = n_x - n_{xy}, \tag{4.23}$$
$$n_{xy}(00) = 1 - n_x - n_y + n_{xy} \text{ for } \langle x, y\rangle.$$

Now if z and z' are two sites, there is only one path connecting them, and so the notation $y \in (z, z')$ makes perfect sense. Any such y splits the remaining lattice sites (not uniquely) into two disjoint sets A_y and B_y: $\{x\} = A_y + (y) + B_y$, on opposite sites of y in the sense that if $z \in A_y$, $z' \in B_y$, then necessarily $y \in (z, z')$. Hence, in obvious notation,

$$W\{\nu_x\} = e^{\beta\mu_y\nu_y} W_{\nu_y}\{\nu_z \mid z \in A_y\} W_{\nu_y}\{\nu_{z'} \mid z' \in B_y\}. \tag{4.24}$$

Consequently, if $z_i \in B_y$ for $i = 1, \ldots, s$, we have

$$\text{if } y \in (z, z_i) \text{ for } i = 1, \ldots, s, \text{ then}$$
$$n_{zz_1,\ldots,z_s}(\alpha_z, \alpha_{z_1}, \ldots, \alpha_{z_s} \mid \alpha_y) = n_z(\alpha_z|\alpha_y) n_{z_1\ldots z_s}(\alpha_{z_1} \ldots \alpha_{z_s}|\alpha_y). \tag{4.25}$$

Suppose then that Λ is a convex set of sites, i.e. contains every site on the paths between any two of its sites, and that $z \in \partial\Lambda$, the boundary of Λ, with $y \in \Lambda$ the unique nearest neighbor of z. It follows that

$$n_\Lambda[\alpha] = n_z(\alpha_z|\alpha_y) n_{\Lambda - z}[\alpha - \alpha_z]. \tag{4.26}$$

Using this in the form

$$n_\Lambda[\alpha] = \frac{n_{zy}(\alpha_z, \alpha_y)}{n_y(\alpha_y)} n_{\Lambda - z}[\alpha - \alpha_z], \tag{4.27}$$

we can nibble away at the boundary of any convex set, ultimately reducing it to a pair of sites. In particular, applying this to the full lattice $\Omega = \{x\}$, we arrive in this fashion at

$$\rho\{\alpha_x\} = \prod_{\langle x,y\rangle} n_{xy}(\alpha_x, \alpha_y) / \prod_x n_x(\alpha_x)^{q_x - 1} \tag{4.28}$$

where q_x is the coordination number (number of next neighbors) of x. Inserting into (4.15) now yields the desired expression, best written as

$$S = -\sum_{x,\alpha} n_x(\alpha) \ln n_x(\alpha) - \frac{1}{2} \sum_{\langle x,y\rangle \alpha,\alpha'} n_{xy}(\alpha, \alpha') \ln \frac{n_{xy}(\alpha\alpha')}{n_x(\alpha) n_y(\alpha')}, \tag{4.29}$$

a completely local form.

The lattice gas on a Cayley tree is not irrelevant to continuous systems. Let us imagine lattice points throughout physical space, with possible interaction links $\varphi(x, y)$ between any two points x, y. If we are restricted to a Cayley tree, only a small number of these can actually be realized, but if we let the mesh spacing go to zero, there will be an interaction between points arbitrarily close to a given pair of physical locations. Ignoring the local pathology of such a representation, we can proceed to compute the entropy. If the mesh volume is simply called dx, this entails the correspondence

$$n_x = n(x)dx, \quad n_{xy} = n_2(x, y)dxdy; \tag{4.30}$$

inserting into (4.29) and carrying out the limit $dx \to 0$, we obtain

$$S = -\int n(x)(\ln n(x) - 1)dx + \frac{1}{2} \int\int (n_2(x, y) - n(x)n(y))dxdy$$
$$-\frac{1}{2} \int\int n_2(x, y) \ln g(x, y) dxdy. \tag{4.31}$$

Now it is worthwhile to go over to the Helmholtz free energy. We have, unsurprisingly,

$$\frac{1}{2}\varphi(x,y) = -\delta S/\delta n_2(x,y) = -\frac{1}{2}\ln g(x,y), \tag{4.32}$$

from which

$$\begin{aligned}\beta\overline{F} &= -S - \frac{1}{2}\int\int \varphi(x,y) n_2(x,y)dxdy \\ &= \int n(x)(\ln n(x) - 1)dx - \frac{1}{2}\int\int n(x)f(x,y)n(y)dxdy,\end{aligned} \tag{4.33}$$

agreeing through second order in $\beta\varphi$ with the Debye-Huckel approximation (3.45).

A higher resolution local representation of the continuum can be achieved [P94] by first imagining a "tree" whose units are 3-vertex triangles, with at least virtual 3-body self-interaction, and mutual interaction on common edges. The entropy is then a function of 1,2, and 3-point densities, and in the continuum limit takes the symbolic form [R71], [BE89], [LWH93]

$$\begin{aligned}S &= -\int n(\ln n - 1) - \frac{1}{2}\int n_2 \ln\frac{n_2}{nn} + \frac{1}{2}\int (n_2 - nn) \\ &-\frac{1}{6}\int n_3 \ln\frac{n_3 nnn}{n_2 n_2 n_2} + \frac{1}{6}\int\left(n_3 - \sum\frac{n_2 n_2}{n} + \sum n_2 n - nnn\right).\end{aligned} \tag{4.34}$$

The condition $\frac{1}{6}\varphi_3(x,y,z) = 0 = -\delta S/\delta n_3(x,y,z)$ in the presence of pair interactions alone produces precisely the Kirkwood superposition approximation

$$n_3 = n_2 n_2 n_2/nnn, \tag{4.35}$$

thereby reducing (4.34) to

$$S = -\int n(\ln n - 1) - \frac{1}{2}\int ng\ln gn + \frac{1}{2}\int nhn + \frac{1}{6}\int nhnhnh, \tag{4.36}$$

which in turn implies βF^{ex} of (3.53).

4.2 A scattering of continuum models

What lattice models have taught us is that, with density as independent field, local interactions lend perhaps unexpected validity to locally described free energies, i.e. x and y being connected in the free energy only if they are directly connected by interaction. Let us see how much of this structure goes over rigorously to continuum fluids, and here we run up at once against the problem that so few of these systems can be solved.

One dimensional fluids are the most likely to be analytically tractable [P87], and among these the least structured interaction is undoubtedly that of hard cores of diameter a – i.e. a system of hard rods. In the absence of external fields, such systems map on to point particles in a reduced volume, but this tactic has not been employed successfully in the presence of other than gravitational forces. We therefore mimic the approach used in the one-dimensional lattice gas. Now, ordering the particles, so that the standard weight $1/N!$ drops out, the grand partition function takes the form

$$\begin{aligned}\Xi &= 1 + \sum_1^\infty \int e^{\beta\mu(x_1)}\theta(x_2 - x_1 - a)e^{\beta\mu(x_2)}\ldots\theta(x_N - x_{N-1} - a)e^{\beta\mu(x_N)}dx^N \\ &= 1 + \langle 1|z|1\rangle + \langle 1|zwz|1\rangle + \cdots \\ &= 1 + \langle 1|z(I - wz)^{-1}|1\rangle,\end{aligned} \tag{4.37}$$

where $z(x,y) = e^{\beta\mu(x)}\delta(x,y)$, $w(x,y) = \theta(y-x-a)$, $1(x) = 1$. Hence
$$n(x) = \delta\Xi/\delta\beta\mu(x)/\Xi$$
$$= \langle 1|(I-zw)^{-1}|x)z(x)\langle x|(I-wz)^{-1}|1\rangle/\Xi$$
$$\equiv \Xi_L(x)z(x)\Xi_R(x)/\Xi \tag{4.38}$$
where $(I-wz)\Xi_R = 1$, $\Xi_L(I-zw) = 1^T$.

Since $\partial w(x,y)/\partial x = -\delta(y-x-a)$, we have
$$\Xi_R'(x) = -z(x+a)\Xi_R(x+a),\ \Xi_L'(x) = z(x-a)\Xi_L(x-a), \tag{4.39}$$
or eliminating $z(x)$ via (4.38)
$$\Xi_R'(x) = -\frac{n(x+a)\Xi}{\Xi_L(x+a)},\ \Xi_L'(x) = \frac{n(x-a)\Xi}{\Xi_R(x-a)} \tag{4.40}$$
from which $n(x)/\Xi_L(x) = -(d/dx)(n(x-a)/\Xi_L'(x))$. Writing this as
$$n(x) = n(x-a) - \frac{d}{dx}\left(\frac{\Xi_L(x)}{\Xi_L'(x)}n(x-a)\right) \tag{4.41}$$
and fitting in conditions at ∞, this integrates to
$$\frac{\Xi_L'(x)}{\Xi_L(x)} = \frac{n(x-a)}{1-\int_{x-a}^{y} n(y)dy} \tag{4.42}$$
and hence to
$$\Xi = \Xi_L(\infty) = \exp\int_{\infty}^{\infty} \frac{n(x-\frac{a}{2})}{1-\int_{x-\frac{a}{2}}^{x+\frac{a}{2}} n(y)dy}dx. \tag{4.43}$$
The lack of manifest parity is cured by rewriting (4.43) as
$$\beta\Omega = -\ln\Xi = -\int_{-\infty}^{\infty} dx\frac{1}{2}\left(n\left(x+\frac{a}{2}\right)+n\left(x-\frac{a}{2}\right)\right)\Big/\left(1-\int_{x-\frac{a}{2}}^{x+\frac{a}{2}} n(y)dy\right). \tag{4.44}$$
From $\beta\Omega^{ex} = \beta\Omega + \int n(x)dx = \left(1-\int n(x)\frac{\delta}{\delta n(x)}dx\right)\overline{F}^{ex}$, we then find without difficulty [P76], [P82b]
$$\beta F^{ex}[n] = -\int n_\sigma(x)\ln(1-n_\tau(x)a)dx$$
$$\text{where } n_\sigma(x) = \frac{1}{2}\left(n\left(x+\frac{a}{2}\right)+n\left(x-\frac{a}{2}\right)\right),\ n_\tau(x) = \frac{1}{a}\int_{x-a/2}^{x+a/2} n(y)dy, \tag{4.45}$$
and correspondingly e.g.
$$\beta\mu(x) = \ln n(x)\Big/\left(1-\int_x^{x+a} n(y)dy\right) + \int_{x-a}^{x} \frac{n(z)}{1-\int_z^{z+a} n(y)dy}dx. \tag{4.46}$$

Proceeding to two-dimensional space, matters of course are much more difficult; only one special case has been solved in full [J81], [J92], and this, perhaps meaningfully, as a functional of potential, not density, and in a canonical ensemble. The case is that of Coulomb interactions as two dimensions. With the usual normalization, the interaction energy is
$$\Phi = -\frac{1}{2}v^2\sum{}'\ln((r_i-r_j)/R) \tag{4.47}$$

for particles of charge σ, with R a reference distance. It is convenient to make a temporary change of notation, defining the complete external Boltzmann factor

$$\zeta(r) = \exp \beta\mu(r) \tag{4.48}$$

where $\mu(r)$ will normally include a neutralizing background, i.e. a potential u_0 for which $\nabla^2 u_0(r) = 2\pi\sigma^2 n_0$, representing a background density n_0 of particles of charge $-\sigma$; then the system is referred to as a one-component plasma. It is convenient as well to parametrize physical space by complex locations and their conjugates

$$z_j = x_j + iy_j, \quad z_j^* = x_j - iy_j, \tag{4.49}$$

so that in particular

$$\Phi = -\frac{1}{4}\sigma^2 \sum{}' \ln[(z_i - z_j)(z_i^* - z_j^*)/R^2]. \tag{4.50}$$

The special case to be considered is that of $\Gamma = -\beta\sigma^2 = 2$, and we will take $R = 1$, an indirect normalization. Thus, the canonical partition function becomes

$$Z_N = \frac{1}{N!}\int \cdots \int \prod_{i<j}(z_i - z_j)(z_i^* - z_j^*) \prod_j \zeta(r_j) dr^N. \tag{4.51}$$

Z_N can now be "evaluated" by using the van der Monde determinant representation

$$\prod_{i<j}(a_i - a_j) = \text{Det } a_i^k|_{i,k=0,\ldots,N-1}, \tag{4.52}$$

which tells us that Z_N is just the normalization of the unnormalized Fermion wave function

$$\psi(r_1, \ldots, r_N) = (N!)^{-1/2} \text{ Det } \zeta(r_i)^{1/2} z_i^k). \tag{4.53}$$

This is simply the overlap determinant

$$Z_N = \text{ Det } \zeta_{jk}|_{0,\ldots,N-1} \text{ where } \zeta_{jk} = \int z^j \zeta(r) z^{*k} dr, \tag{4.54}$$

and the canonical free energy is correspondingly $F_N = -\frac{1}{\beta}\ln Z_N$, or

$$F_N = -\frac{1}{\beta} \text{ Tr } \ln \zeta^{(N)}, \tag{4.55}$$

superscript (N) indicates the order of the matrix. Now, expressions for the distributions follow at once [AL84], $n(r) = -\beta\zeta(r)\delta F_N/\delta\zeta(r)$, so

$$n(r)/\zeta(r) = \sum_{j,k} z^j z^{*k} (\zeta^{(N)})^{-1}_{kj}, \tag{4.56}$$

$[n_2(r, r') - n(r)n(r')]/n(r)n(r') = -\delta/\delta\zeta(r')n(r)/\zeta(r)$, or

$$\frac{n(r)}{\zeta(r)}\frac{n(r')}{\zeta(r')} h(r, r') = \sum z^j (\zeta^{(N)})^{-1}_{mj} z'^{*m} \cdot \sum z'^{\ell} (\zeta^{(N)})^{-1}_{k\ell} z^{*k}, \tag{4.57}$$

and so forth.

At higher dimensionality, very little is known as to exact solutions of systems with singlet and pair interactions. It is not terribly difficult – but perhaps overly artificial – to construct solvable models with carefully balanced many-body interactions [P86]. One need only write down a reasonable parametrized free energy functional, such as the familiar form

$$\beta\overline{F}[n] = \int n(r)(\ln n(r) - 1)dr - \frac{1}{2}\int\int n(r)c(r-r')n(r')drdr' \tag{4.58}$$

and ask what total interaction energy, if any, it arises from. The "if any" is because we can design a sequence of many-body Boltzmann factors giving rise to (4.58), only to find that some turn out to be negative! To check, the simplest procedure is to write down the grand potential $\beta\Omega = (1 - \int n(r)\frac{\delta}{\delta n(r)}dr)\beta\overline{F}$:

$$\beta\Omega[n] = -\int n(r)dr + \frac{1}{2}\int\int n(r)c(r-r')n(r')drdr', \tag{4.59}$$

then the profile equation $\beta\mu(r) = \ln z(r) = \delta\beta\overline{F}/\delta n(r)$ in the form

$$n(r) = z(r)\exp\int c(r-r')n(r')dr', \tag{4.60}$$

and eliminate $n(r)$ between the two, since

$$e^{-\beta\Omega} = \Xi = \sum\frac{1}{s!}\int\cdots\int\prod_1^s z(j)\exp -\beta V_s(1\ldots s)dr^s \tag{4.61}$$

is precisely the MacLaurin series generating function for the many-body Boltzmann factors. The analysis is best carried out graphically, and results in the sequence

$$e^{-\beta V_2(1,2)} = 1 + c(1,2) \tag{4.62a}$$

$$e^{-\beta V_s(1,2,3)} = e^{-\beta V_2(1,2)}e^{-\beta V_2(2,3)}e^{-\beta V_2(3,1)} \cdot\left(1 + \frac{c(1,2)}{1+c(1,2)} + \frac{c(2,3)}{1+c(2,3)} + \frac{c(3,1)}{1+c(3,1)}\right), \tag{4.62b}$$

giving rise to a sequence of sufficient conditions, $c(1,2) \geq -1$ from (4.62a), $c(1,2) \geq -\frac{1}{2}$ from (4.62b), etc. In fact, if c can become arbitrarily large and positive, the limit of this sequence shows that $c(1,2) \geq 0$ is both necessary and sufficient for (4.58) to be "realizable".

Oddly enough, construction of solvable models simplifies if one goes to arbitrarily high dimensionality, under suitable scaling [FRW85], [FP87]. The requirements are that interactions are strictly repulsive, that the interactions are isotropic in all but a finite number of dimensions, *and* that the density expansion of the free energy converges absolutely. First, since $|f(i,j)| \leq 1$, the absolute value contribution of any diagram in the expansion (2.18) is bounded from above by that of the same diagram with fewer occupied bonds, and hence by a maximal loop of bonds. Then remain only the second order contribution $-\frac{1}{2}\int\int n(r)f(r,r')n(r')drdr'$ to βF^{ex}, and 3 or more – site loops of the form $L_s = \int\cdots\int n(r_1)f(r_1,r_2)n(r_2)f(r_2,r_3)\ldots f(r_s,r_1)dr^s$. A simple fluid, each particle represented by a point in D-space, illustrates the argument:

$$|L_s| \leq n_{\max}^{s-1}\int\cdots\int n(r_s)|f(r_1-r_2)|\ldots|f(r_s-r_1)|dr^s =$$

$$n_{\max}^{s-1}\int n(r)dr\int|f(r_1-r_2)|\ldots|f(r_1)|dr^{s-1} = n_{\max}^{s-1}N\frac{1}{(2\pi)^D}\int|\tilde{f}(k)|^s dk,$$

where $\tilde{f}$ denotes Fourier transform. But if

$$v = \int|f(r)|dr \tag{4.63}$$

is the "interaction volume", a brief calculation shows that

$$\frac{1}{(2\pi)^D}\int |\tilde{f}(k)|^s dk/v^{s-1} \sim \frac{1}{\pi}e^{-\frac{1}{2}(s-2)D}. \tag{4.64}$$

Thus if $n(r) = \rho(r)/v$ at fixed $\rho(r)$, each loop term in $\overline{F}/N$ vanishes as $D \to \infty$. We conclude that we can take

$$\beta\overline{F}[\rho/v] = \int \frac{\rho(r)}{v}\ln\left(\frac{\rho(r)}{v}-1\right)dr - \frac{1}{2}\int\int\frac{\rho(r)}{v}\frac{\rho(r')}{v}f(r-r')drdr' \tag{4.65}$$

as the model limit, precisely the second virial truncation of (2.19). Eq. (4.65) is particularly useful for anisotropic molecular fluids, uniform in space but non-uniform in orientation. Then, the $D \to \infty$ limit allows very explicit computation, and shows e.g. a simple nematic transition for hard sphero-cylinders [CFP89].

4.3 Collective modes and expanded density space

Aside from the Coulomb fluid, with its very long range forces, the models we have examined thus far have been characterized by a blatantly local structure. The simplest case in which this motif is changed is very simple indeed. Let us return to the one-dimensional lattice gas, but suppose instead of infinite length, it is periodically bounded, of length N [PZ88]. The boundary connection means that the neat separation of (4.5) no longer obtains. However, it is still true that (4.10) holds, and it follows that

$$\text{Det}(\xi/w)^{x+\frac{1}{2}} = \text{Det}(\xi/w)^{x-\frac{1}{2}} = \cdots = K, \tag{4.66}$$

a constant independent of the site x. Thus, the second of (4.7) is now replaced by

$$\xi_{00}\xi_{11}/\epsilon - \xi_{01}\xi_{10} = K. \tag{4.67}$$

To see the consequences, without unenlightening algebraic complications, let us specialize to the case of hard core interaction $J = \infty$, or $\epsilon = 0$, in which case we had better use the combination

$$\bar{\xi}^{x+\frac{1}{2}} = (\xi/w)^{x+\frac{1}{2}}. \tag{4.68}$$

Solving as before, we readily find

$$\bar{\xi}_{00}^{x+\frac{1}{2}} = 1 - n_x - n_{x+1}, \quad \bar{\xi}_{01}^{x+\frac{1}{2}} = n_{x+1},$$
$$\bar{\xi}_{10}^{x+\frac{1}{2}} = n_x, \quad \bar{\xi}_{11}^{x+\frac{1}{2}} = (n_x n_{x+1} + K)/(1 - n_x - n_{x+1}), \tag{4.69}$$

and substituting into (4.10), as before, then

$$\beta\mu_x = \ln(K + n_x(1-n_x)) - \ln(1 - n_x - n_{x+1}) - \ln(1 - n_x - n_{x+1}), \tag{4.70}$$

There is certainly a free energy associated with (4.70). Writing $K = -\frac{1}{4} + C^2$ (since $K \geq -n_x n_{x+1}$ and $n_x + n_{x+1} \leq 1$) we can choose

$$\beta\overline{F}[n,C] = \sum_x \Big[\left(C - \frac{1}{2} + n_x\right)\ln\left(C - \frac{1}{2} + n_x\right) + \left(C + \frac{1}{2} - n_x\right)\ln\left(C + \frac{1}{2} - n_x\right)$$
$$-(1 - n_{x-1} - n_x)\ln(1 - n_{x-1} - n_x)\Big], \tag{4.71}$$

which if C were fixed would again be local in the density. But whatever C is, it must be some collective amplitude, cyclically invariant in the $\{n_x\}$. The situation is clarified if we quote a readily proved theorem, valid for a whole set of collective variables:

Theorem. If for the set $\{C_\alpha\}$, the profile equation $\mu_x = \partial\overline{\overline{F}}[n,C]/\partial n_x|_C$ is valid, then a pure collective contribution $\Delta[C]$ can be found such that on defining

$$\overline{F}[n,C] = \overline{\overline{F}}[n,C] + \Delta[C], \tag{4.72}$$

we have

$\mu_x = \partial\overline{F}[n,C]/\partial n_x|_C,\ 0 = 0\overline{F}[n,C]/\partial C_\alpha|_n,$ and $\overline{F}[n] = \overline{F}[n,C[n]]$.

Equivalently, on the expanded space $[n, C]$, one has the stationary (not necessarily minimum) principle at fixed $\{\mu_x\}$,

$$\delta\left[\overline{F}[n,C] - \sum_x \mu_x n_x\right] = 0, \tag{4.73}$$

the vanishing C variations being sufficient in principle to determine the expression $C_\alpha[n]$. For example, in the above nearest neighbor exclusion case, it is not hard to show (e.g. by restricting to uniformity) that

$$\beta\Delta[C] = \left(C + \frac{1}{2}\right)\ln\left(C + \frac{1}{2}\right) - \left(C - \frac{1}{2}\right)\ln\left(C - \frac{1}{2}\right). \tag{4.74}$$

For more complex networks, the use of suitable collective amplitudes, in particular one for each branch of sites, and one for each multi branch node, again allow the free energy structure to mirror the distinction between local behavior and that of successively higher levels of organization [PS94]. And a similar strategy has been employed in the continuum for describing simultaneously local fluid excitations and large scale surface waves. In these situations, it is clear that a multi-level descriptive vehicle is the most realistic one to use, and perhaps as a reward for virtue, that the resulting descriptions are particularly simple in form. But the use of densities other than the basic particle density to both clarify and better control the structure of the equilibrium state can be effective even without a blatant hierarchical spatial structure. A particularly illustrative example is that of one-dimensional fluids with nearest neighbor interactions. These can be fully analyzed [P89b] in the entropy format $S[n, n_2]$, but the necessity of working in pair space as well as singlet space makes this less appealing conceptually and far more difficult to handle computationally.

We already have the required machinery, in the context of the hard rod system in one dimension; it is then only necessary to replace the one-sided internal Boltzmann factor by

$$w(x,y) = e^{-\beta\varphi(x,y)}\theta(y-x) \tag{4.75}$$

for pair interaction φ. We process the basic

$$(I - wz)\Xi_R = 1, \quad \Xi_L(I - zw) = 1^T \tag{4.76}$$

of (4.38) by applying the operation w^{-1}, noting that $w^{-1} = \infty$ implies $w^{-1}1 = 0$:

$$z\Xi_R = w^{-1}\Xi_R, \quad \Xi_L z = \Xi_L w^{-1} \tag{4.77}$$

and eliminate z via (4.38):

$$n\Xi/\Xi_L - w^{-1}\Xi_R = 0, \quad n\Xi/\Xi_R - \Xi_L w^{-1} = 0. \tag{4.78}$$

Since we have lost information by applying w^{-1}, this has to be resupplied in the form of boundary conditions

$$\Xi_R(\infty) = \Xi_L(-\infty) = 1, \quad \Xi_R(-\infty) = \Xi_L(\infty) = \Xi. \tag{4.79}$$

Now if we introduce the free energy on a highly expanded space

$$\beta\overline{F}[n,\Xi_R,\Xi_L,\Xi] = \int n(x)(\ln n(x) - 1 - \ln \Xi_R(x) - \ln \Xi_L(x) + \ln \Xi)dx$$
$$+\frac{1}{2}\frac{1}{\Xi}\left(\int \Xi_L(x) w^{-1}\Xi_R(x)dx + \int \Xi_L w^{-1}(x)\Xi_R(x)dx\right)$$
$$-\frac{1}{2}\ln(\Xi_L(\infty)/\Xi_L(-\infty)) + \frac{1}{2}\ln(\Xi_R(\infty)/\Xi_R(-\infty)), \tag{4.80}$$

it is easy to see that $\delta\beta\overline{F}/\delta n(x) = \beta\mu(x)$, $\delta\beta\overline{F}/\delta\Xi_R(x) = \delta\beta\overline{F}/\delta\Xi_L(x) = \delta\beta\overline{F}/\delta\Xi = 0$, as well as $\delta\beta\overline{F}/\delta\Xi_L(\pm\infty) = \delta\beta\overline{F}/\delta\Xi_R(\pm\infty) = 0$ correctly reproduce all of our relationships (the fact that w^{-1} always has the singular component $\partial/\partial x$ is needed). But substantial simplification is possible; we set

$$\Xi_L(x) = e^{-\frac{1}{2}\beta v(x)}e^{-\beta\int_{-\infty}^{x}\omega(t)dt},$$
$$\Xi_R(x) = e^{-\frac{1}{2}\beta v(x)}e^{-\beta\int_{x}^{\infty}\omega(t)dt},$$
$$\Xi = e^{-\beta\int_{-\infty}^{\infty}\omega(t)dt}, \tag{4.81}$$

automatically satisfying boundary conditions, and obtain [P96a]

$$\beta\overline{F}[n,\omega,v] = \int n(x)(\ln n(x) - 1 + \beta v(x))dx + \beta\int \omega(x)dx$$
$$+\int\int e^{-\frac{\beta}{2}v(x)}e^{-\beta\int_y^x\omega(t)dt}w^{-1}(x,y)e^{-\frac{\beta}{2}v(y)}dxdy. \tag{4.82}$$

This is much more transparent: $\mu(x) = S\overline{F}/\delta n(x)$ identifies $v(x)$ as the local excess chemical potential, the complicated last term reduces to $N = \int n(x)dx$ when evaluated, and $\omega(x)$ is then identified as a specific grand potential.

4.4 Model interface

Another extreme in which explicit results are obtainable is that of the fluctuating interface between two fluids of constant density, say vapor at 0 density, incompressible liquid at density ρ. We can imagine a reference interfacial location $z = 0$, with gas to the left, liquid to the right, and the actual interface specified by

$$z = \xi(t), \tag{4.83}$$

t denoting the one or more transverse coordinates; ξ is here restricted to be single-valued. In the mesoscopic situation under consideration, the interfacial energy is simply that of surface tension

$$T = \sigma\int(1 + |\nabla_t\xi(t)|^2)^{1/2}dt, \tag{4.84}$$

while the external field – chemical potential energy is $U = \int\int_{z\geq\xi(t)}\mu(z,t)dzdt$, or

$$U = \int\int U(\xi(t),t)dt \text{ where } U(\xi,t) = -\int_{\xi}^{z_{\max}}\mu(z,t)dz. \tag{4.85}$$

If the density pattern were a fluctuational profile riding on the normal interface,

$$n(z,t) = n_0(z - \xi(t)) \tag{4.86}$$

where $n_0(\infty) = \rho$, $n_0(-\infty) = 0$, then because $\nabla n(z,t) = (\hat{z} - \nabla\xi(t))n_0'(z - \xi(t))$, or $|\nabla n_0(z,t)| = (1 + |\nabla\xi|^2)^{1/2}n_0'(z - \xi(t))$, we would just have [P84a]

$$T = \frac{\sigma}{\rho}\int|\nabla n(r)|dr. \tag{4.87}$$

Let us see how thermal fluctuations change this result.

For a shallow interfacial pattern, we can expand $(1+|\nabla_t\xi|^2)^{1/2} = 1+\frac{1}{2}|\nabla_t\xi|^2+\cdots$. Now restricting ourselves to two dimensions, so that t is one-dimensional, we have the energy per configuration

$$\phi[\xi] = \int \left(\frac{\sigma}{2}\dot{\xi}(t)^2 + \rho U(\xi(t),t)\right) dt, \tag{4.88}$$

where $\dot{f} \equiv \partial f/\partial t$, and, fixing the interface location at $t=-A$ and $+A$,

$$\Xi = \int_{\xi(-A)=z_-}^{\xi(A)=z_+} e^{-\beta\phi[\xi]} D\xi. \tag{4.89}$$

In the configuration ξ, we have

$$n_\xi(z,t) = \rho\theta(z-\xi(t)), \tag{4.90}$$

so that $n'_\xi(z,t) = \rho\delta(z-\xi(t))$, where $f' \equiv \partial f/\partial z$. Thus $n'(z,t) = \langle n'_\xi(z,t)\rangle = \langle \rho\delta(z-\xi(t)\rangle = \langle \delta U/\delta U(z,t)\rangle = -\frac{1}{\beta}\delta \ln \Xi/\delta U(z,t)$, or

$$\begin{aligned} n'(z,t) &= \rho \int_{\xi(-A)=z_-}^{\xi(t)=z} e^{-\beta\phi[\xi]} D\xi \int_{\xi(t)=z}^{\xi(A)=z_+} e^{-\beta\phi[\xi]} D\xi/\Xi \\ &\equiv \rho\Xi_-(z,t)\Xi_+(z,t)/\Xi. \end{aligned} \tag{4.91}$$

Now, $\Xi_-(x,t)$ and $\Xi_+(z,t)$ are recognized as path integral representations of solutions of the imaginary time Schrodinger, or Bloch equation, and we therefore have

$$\begin{aligned} &\pm\partial\Xi_\pm(z,t)/\partial t = \frac{1}{2\beta\sigma}\partial^2\Xi_\pm(z,t)/\partial z^2 - \beta\rho U(z,t)\Xi_\pm(z,t), \\ &\Xi_\pm(\pm A,z) = \delta(z-z_\pm). \end{aligned} \tag{4.92}$$

Some algebra allows us to eliminate Ξ_+ and Ξ_- between (4.91) and (4.92), and we have the exact

$$\frac{\rho}{\sigma}\mu = \frac{1}{2}\frac{\partial}{\partial z}\left(\frac{\dot{n}+n''/2\beta\sigma}{n'}\right)^2 - \left(\frac{\partial}{\partial t} - \frac{1}{2\beta\sigma}\frac{\partial^2}{\partial z^2}\right)\left(\frac{\dot{n}+n''/2\beta\sigma}{n'}\right), \tag{4.93}$$

establishing at once the free energy functional [P84b]

$$\overline{F}[n] = \frac{\sigma}{2\rho}\int\int\left(\frac{\dot{n}^2}{n'} + \frac{1}{4\beta^2\sigma^2}\frac{n''^2}{n'}\right) dzdt. \tag{4.94}$$

The first term is recognized as the energy emanating from T for weakly non-planar interface $|dz/dt| = |\dot{n}/n'| \ll 1$: $T = \sigma/\rho\int(n'^2+\dot{n}^2)^{1/2}dr = \sigma\rho\int(1+\frac{1}{2}\frac{\dot{n}^2}{n'^2}\ldots)n'dr$. Here the free energy is again local, the chemical potential involves only derivatives of the density, and $C = \delta\beta\mu/\delta n$ is a strictly differential operator, i.e. of zero range.

5. QUASI-UNIFORM FLUIDS

Now we want to make use of the manifold viewpoints presented above. Much of our discussion has concerned setting up solved reference systems for which the system under consideration can be regarded as a perturbation, and of course uniform references are those for which the most reliable information, theoretical, computational, and experimental, is available. There are two quite distinct categories for which a uniform reference seems suitable. First is that of a fluid bounded by a substrate, beyond which one can imagine that there is a vacuous fluid; a subclass might be that of two (or more) fluids bounding each other at an interface – and they could be two phases of the same fluid. The second class is that of systems that are macroscopically but not microscopically uniform, i.e. solids. One can perhaps visualize these as fluids with self-maintained non-uniformity, and the question is whether such an interpretation makes sense – are they in fact extensions of ordinary fluids?

5.1 Prototypes

Let us start with a one-dimensional situation in which – since the forces are of short range- no self-maintained non-uniformity is possible, but one can still test how effectively large microscopic inhomogeneities in the context of macroscopic uniformity can be handled. Suppose [RD85] then that we return to the one-dimensional hard rod fluid (4.45, 4.46) and apply an external potential of period $2\pi/k$

$$u(x) = u(1 + \cos kx). \tag{5.1}$$

If the core diameter a is a multiple of $2\pi/k$, the integrals in (4.46) are independent of x, and we have

$$\beta\mu(x) = \bar{n}a/(1 - \bar{n}a) - \ln(1 - \bar{n}a) + \ln n(x), \tag{5.2}$$

where $\bar{n}$ is the average density. Thus $\ln n(x)$ is trivial to find. If $ka/2\pi$ is not an integer, $\ln n(x)$ is less trivial, but one can for example express it as a Fourier series of fundamental wave number k and determine the coefficients numerically. How well do density functional approximation methods reproduce these results? We can restrict our attention to a unit volume. If $\bar{n} = \int n(x)dx$ is given, a uniform fluid of density $\bar{n}$ is a fine reference; expanding βF^{ex} about it and using $\int (n(x) - \bar{n})dx = 0$, one has to second order

$$\beta\overline{F} = \beta\bar{n}f_0(\bar{n}) + \int n(x)\ln(n(x)/\bar{n})dx$$
$$-\frac{1}{2}\int\int c(x - x', \bar{n})(n(x - \bar{n})(n(x') - \bar{n})dxdx' \tag{5.3}$$

with corresponding

$$\beta(\mu' - u(x)) = \ln n(x) - \int c(x - x'; \bar{n})(n(x') - \bar{n})dx' \tag{5.4}$$

($\mu' \neq \mu$ because of the implicit imposition of $\int n(x)dx = \bar{n}$). Even this unsophisticated approach does quite well. See also [MH89].

Now let us proceed to more realistic systems. If the transition to a solid phase is the objective, then one should have a decent solid phase prescription as an entree to the fluid-solid relationship. Simplest perhaps is the van der Waals mean-fluid approach in which the reference is a core-interaction fluid that is structureless on the scale of the attractive tail potential that invokes the condensation to a solid (and the core dominates distribution functions, too). We can go a bit further. Since $\delta\overline{F}/\delta\varphi(r,r') = \frac{1}{2}n_2(r,r')$, we have to first order in the tail potential $\Delta\varphi$,

$$\overline{F}(n) = F_0[n] + \frac{1}{2}\int\int n_{20}(rr'|n)\Delta\varphi(r,r')drdr' \tag{5.5}$$

The structureless reference assumption ensures that correlations can be neglected and that F_0 may be treated in local thermodynamic form

$$\overline{F}(n) = \int n(r) f_0(n(r)) dr + \frac{1}{2} \int\int n(r) n(r') \Delta\varphi(r, r') dr dr'. \tag{5.6}$$

Eq. (5.6) is then a model for both fluid and solid state. For fluid, $n(r)$ is uniform, while for a solid, a reasonable "perfect" crystal variational ansatz is to regard $n(r)$ as a sum of Gaussians [DTB94]. Phase diagrams obtained in this fashion have quite extensive phenomenology. Eq. (5.5) has also been used [WSB95] as is for wall-bounded fluids with the well-studied inhomogeneous hard sphere fluid as reference.

More typical use of reference fluids chooses the uniform state with full interaction as reference. In the presence of macroscopic inhomogeneity, a continuous set of reference is required, but for the present situation, the determination of the optimal reference density is one of the issues to be settled. Direct perturbation methods which expand in the local non-uniformity are certainly the first ones that come to mind. Several of these work by starting with exact relationships for the non-uniform fluid and then replace selected correlations by their reference values, thus easing practical solution. The identities used are of the force balance type. For example, the first order YBG equation (3.17) can be written as

$$\nabla\beta\mu(r) = \nabla n(r)/n(r) + \int g(r, r') n(r') \nabla\beta\varphi(r - r') dr'. \tag{5.7}$$

If $g(r, r'')$ is replaced by the reference $g_0(r - r')$ which, like $\varphi(r - r')$, is isotropic, then $g_0(r - r')\nabla\beta\varphi(r - r')$ is itself a gradient, say $\nabla\beta\varphi_g(r - r')$, and we can integrate at once (adjusting constants to reduce properly at n_0) to

$$\beta\mu(r) = \beta\mu_0 + \ln n(r)/n_0 + \int (n(r') - n_0)\beta\varphi_g(r - r') dr', \tag{5.8}$$

clearly derived from the free energy

$$\beta\overline{F}(n) = \int [n_0 - \beta p_0 + n(r)(\ln n(r)/n_0 + \beta\mu_0 - 1)] dr$$
$$+ \frac{1}{2} \int\int (n(r) - n_0)(n(r') - n_0)\beta\varphi_g(r - r') dr dr'. \tag{5.9}$$

A similar expression, in which the explicit internal interaction is not required, comes from the Ward, Takahashi, Wertheim, Lovett, Mou, Buff identity (3.13), which we write as

$$\nabla\beta\mu(r) = \nabla n(r)/n(r) - \int c(r, r') \nabla n(r') dr. \tag{5.10}$$

With $c(r, r')$ taken as $c_0(r - r')$, this integrates at once to

$$\beta\mu(r) = \beta\mu_0 + \ln n(r)/n_0 - \int c_0(r - r')(n(r') - n_0) dr', \tag{5.11}$$

now arising from

$$\beta\overline{F}[n] = \int [n_0 - \beta p_0 + n(r)(\ln n(r)/n_0 + \beta\mu_0 - 1)] dr$$
$$- \frac{1}{2} \int\int (n(r) - n_0)(n(r') - n_0) c_0(r - r') dr dr'. \tag{5.12}$$

And we may of course consider formal perturbation theory from the uniform state to the non-uniform. This too is not uniquely defined, since it depends upon what is expanded in terms of what. Even less sophisticated then (5.3) is $\overline{F}(n)$ expanded with respect to $n(r)$:

$$\overline{F}(n) = Nf(n_0) + \int \mu_0(n(r) - n_0)dr + \frac{1}{2}\int\int C_0(r-r')(n(r)-n_0)(n(r')-n_0)dr'dr, \quad (5.13)$$

and this in fact was used by Ramakrishnan and Yussouff [RY79] in the first modern treatment of the freezing transition. Equally well, one can expand with respect to $\ln n$,

$$\overline{F}[n] = Nf(n_0) + n_0\mu_0 \int \ln n(r)/n_0 dr$$
$$+\frac{1}{2}n_0^2 \int\int (\mu_0\delta(r-r') + C_0(r-r'))\ln n(r)/n_0 \ln n(r')/n_0 dr dr'. \quad (5.14)$$

Indirect expansions may run up against integrability problems. E.g. the $CHNC$ expansion of $\ln n(r) - \beta\mu(r)$ with respect to $n(r')$,

$$\ln n(r) - \beta\mu(r) = \ln n_0 - \beta\mu_0 \int C_0(r-r')(n(r')-n_0)dr' \quad (5.15)$$

leads to no problems – and to nothing new, because the quantity to be expanded is $f\delta\beta F^{ex}/\delta n(r)$, already an n-derivative. But the PY expansion of $n(r)e^{\beta u(r)}$ with respect to $n(r')$ can only be integrated by artificial means.

5.2 Effective liquid theory

Again with uniformity in the large as the general context, the question of precisely which uniform fluid is best to start with becomes important – it is not necessarily that of the mean density, since the grand ensemble allows us to change particle number without penalty. The effective uniform density to be used, $\rho[n]$, is of course a function of what one wants to use it for. Simplest would be for the thermodynamics, and the form

$$F^{ex}[n] = \bar{n}f^{ex}(\rho[n]) \quad (5.16)$$

is the one that has been used universally. Here again $\bar{n} = \int n(r)dr$ over a unit volume, and $f^{ex}(\rho)$ is the excess free energy per particle for a fluid of density ρ. Since (5.16) can be regarded as the definition of $\rho[n]$, no information is contained therein. Non-triviality is now imparted by adopting a model form for $\rho[n]$, with functional parameters determined, in a Galerkin approach, by demanding satisfaction of sufficiently many exact relationships. In particular, one can demand the reproduction of all available distribution information in any uniform state. In Sec. 3c, it was the relation $\delta\overline{F}/\delta\varphi(1,2) = 1/2n_2(1,2|n)$ that provided the principal restriction. But if the information available is that of the distributions rather than interaction potential, the alternative $\delta^2\beta\overline{F}/\delta n(1)\delta n(2) = C(1,2|n)$ must be used instead, and this is what is generally done. An important reason for modelling the *excess* free energy is that second derivatives of $\int(\rho\ln\rho - \rho)dr$ would otherwise irrevocably yield the wrong linear response singularity for any reasonable approximation to $\rho[n]$.

Now let us check successive derivatives of (5.16), i.e. its behavior in the vicinity of a uniform density n_0. To start with, we have of course from (5.16)

$$\rho[n_0] = n_0 \quad (5.17)$$

for any uniform n_0. Then, applying $\delta/\delta n(r)$ to (5.16),

$$\mu^{ex}(r|n) = f^{ex}(\rho[n]) + \bar{n}f^{ex\prime}(\rho[n])\frac{\delta\rho[n]}{\delta n(r)}, \quad (5.18)$$

so that

$$\mu^{ex}(n_0) = f^{ex}(n_0) + n_0 f^{ex\prime}(n_0)\frac{\delta\rho[n]}{\delta n(r)}\Big|_{n_0}, \quad (5.19)$$

but $n^{ex}(n) = \frac{d}{dn}(nf^{ex}(n))$, and hence

$$\delta\rho[n]/\delta n(r)|_{n_0} = 1. \tag{5.20}$$

Continuing, we apply $\delta/\delta n(r')$ to (5.18),

$$-\frac{1}{\beta}c(r,r'|n) = f^{ex'}(\rho[n])\frac{\delta\rho[n]}{\delta n(r')} + f^{ex'}(\rho[n])\frac{\delta\rho[n]}{\delta n(r)}$$

$$+Nf^{ex''}(\rho[n])\frac{\delta\rho[n]}{\delta n(r)}\frac{\delta\rho[n]}{\delta n(r')} + Nf^{ex'}(\rho[n])\frac{\delta^2\rho[n]}{\delta n(r)\delta n(r')}, \tag{5.21}$$

and set $n(r) = n_0$,

$$-\frac{1}{\beta}c(r-r';n_0) = n_0 f^{ex'}(n_0)\frac{\delta^2\rho[n]}{\delta n(r)\delta n(r')}\Big|_{n_0} + (n_0 f^{ex}(n_0)''. \tag{5.22}$$

Since $nf^{ex'}(n) = \mu^{ex}(n) - f^{ex}(n) = P^{ex}(n)/n$, and $(nf^{ex}(n))'' = \mu^{ex'}(n) = -\frac{1}{\beta}\int c(r)dr$, we have

$$\frac{\delta^2\rho[n]}{\delta n(r)\delta n(r')}\Big|_{n_0} = \frac{n_0}{\beta P^{ex}(n_0)}\left(\int c(r;n_0)dr - c(r-r';n_0)\right). \tag{5.23}$$

This is about all the information we can reliably make use of.

In order to take advantage of (5.17, 5.20, 5.23), we need a model for $\rho[n]$, which should be some sort of average density. The choice $\rho[n] = \bar{n}$ will certainly satisfy (5.17) and (5.20), but certainly not (5.23). The Denton-Ashcroft [DA89] form is (see also [LW93])

$$\rho = \frac{1}{n}\int\int n(r)n(r')W(r-r',\rho)drdr' \text{ where } \int W(r,\rho)dr = 1. \tag{5.24}$$

Since $\int W'(r,\rho)dr = 0$, $\int W''(r,\rho)dr = 0$, the prime always referring to density derivative, derivatives of (5.24) at uniform n_0 will pick up no W derivative. Thus $\delta\rho/\delta n(r)|_{n_0} = -(1/n_0^2)\int n_0^2 W + (1/n_0)2\int n_0 W = 1$ is automatic, but

$$\delta^2\rho/\delta n(r)\delta n(r')|_{n_0} = (2/n_0^3)\int n_0^2 W - (1/n_0^2)\int 2n_0 W + (1/n_0)W(r-r',n_0),$$

so that

$$W(r-r',\rho) = \frac{\rho^2}{\beta P^{ex}(\rho)}\left(\int c(r,\rho)dr - c(r-r',\rho)\right). \tag{5.25}$$

The form (5.24) depends only upon a single uniform density, perhaps inappropriate for the frozen state which is its major objective, in which very large local density variations appear. It has been joined by forms in which all densities up to that in question are required, but which do not change the validity of (5.17, 5.20, 5.23) in an arbitrary uniform state. Thus, Zeng and Oxtoby [ZO90] have used

$$W(r-r',\rho[n]) \to 2\int_0^1\int_0^\lambda\left[\int c(r;\rho[\lambda'n])dr - c(r-r';\rho[\lambda'n])\right]/\beta f^{ex'}(\rho[\lambda'n])d\lambda' d\lambda, \tag{5.26}$$

Lutsko and Baus [LB90] (GELA)

$$W(r-r',\rho[n]) \to \frac{-\rho[n]}{\beta f^{ex}(\rho[n])}\int_0^1\int_0^\lambda c(r-r';\rho[\lambda'n]d\lambda' d\lambda, \tag{5.27}$$

and a further approximation (MELA)

$$W(r-r',\rho[n]) \to \frac{-\rho[n]}{\beta f^{ex}(\rho[n])}\int_0^1\int_0^\lambda c(r-r';\lambda'\rho[n])d\lambda' d\lambda, \tag{5.28}$$

the last of which is again a function of $\rho[n]$ alone, but does not reduce consistently at uniformity. Forms (5.27) and (5.28) have the advantage that introduction of a scaled density $\rho[\alpha n]$, followed by a pair of α-differentiations, creates a specific α-trajectory for computation, e.g. [TC93] for GELA one has

$$\bar{n}\rho[\alpha n] = -\alpha\int\int n(r)n(r')\rho[\alpha n]/\beta f^{ex}(\rho[\alpha n])\int_0^1\int_0^\lambda c(r-r';\rho[\alpha\lambda' n])d\lambda d\lambda' dr dr',$$

or

$$\alpha\bar{n}\beta f^{ex}(\rho[\alpha n]) = -\int\int\int_0^\alpha\int_0^\lambda c(r-r';\rho[\lambda' n])n(r)n(r')d\lambda d\lambda' dr dr', \tag{5.29}$$

so that

$$\frac{\partial^2}{\partial\alpha^2}\alpha\rho f^{ex}(\rho[\alpha n]) = -\int\int c(r-r';\rho[\alpha n])n(r)n(r')dr dr', \tag{5.30}$$

suitable for iterative solution. GELA is remarkably effective at reproducing pure hard core freezing simulations, but less so for softer repulsive interactions [LK90], [TRP95].

5.3 Constrained fluids

A common situation in which the term "fluid" is more directly relevant is that in which it is indeed primarily uniform but constrained, e.g. by walls which rapidly reduce its density to the uniform value of zero. Although it is predominantly locally uniform, the system is certainly not globally uniform, and so it is not clear that expansion, perturbative or non-perturbative, about a uniform fluid is really appropriate. To test this, an extensive set of computations on wall – bounded fluids was carried out – with very reasonable results – by Sokolowski et al. [SF93]. They combined a primitive excess free energy for a repulsive core fluid with a simple mean field attractive contribution [NJF89], [TE84], [MTN87], a la van der Waals:

$$F^{ex}[n] = \int f(\tilde{n}(r))dr + \frac{1}{2}\int\int n(r)n(r')\varphi_A(r-r')dr dr' \tag{5.31}$$

where $\beta f(n) = \eta(4-3\eta)/(1-\eta)^2$, the Carnahan-Starling [CS69] approximation for a diameter d hard core reference fluid. Here $\eta = (\pi d^3/6)n$ and $\tilde{n}(r) = \int_{r'\leq d} n(r+r')dr'/(4\pi d^3/3)$.

Comparisons have more often been carried out with numerical simulations for basic geometries such as constraint by a flat wall. In a fashion which departs somewhat from historical accuracy, the array of approximations used for this purpose has as its common progenitor the exactly solvable 1-dimensional hard rod system of Sec. 4b, expressed in the form

$$\begin{aligned}
&\beta F^{ex}[n] = \int n_\sigma(x)\beta f^{ex}(n_\tau(x))dx \\
&\text{where } n_\sigma(x) = \int\sigma(x-x')n(x')dx', n_\tau(x) = \int\tau(x-x')n(x')dx', \\
&\sigma(x) = \frac{1}{2}\left(\delta\left(x-\frac{a}{2}\right)+\delta(x)+\frac{a}{2}\right), \tau(x) = \frac{1}{a}\theta\left(\frac{a}{2}-|x|\right), \\
&\beta f^{ex}(n) = -\ln(1-na).
\end{aligned} \tag{5.32}$$

Here, $f^{ex}(n)$ is the bulk excess free energy per particle, and $n_\sigma(x)$, $n_\tau(x)$ are linearly weighted densities, precisely the mean density over the two "surfaces" of a rod and the mean over its volume. Extrapolation to higher dimension and arbitrary interaction is to some extent obvious: one replaces $f^{ex}(n)$ by its bulk fluid value for the system at hand [P81]. But the choice of n_σ and n_τ is less obvious. In older work, $n_\sigma(r)$ was taken as $n(r)$ and $n_\tau(r)$ as the volume average over some reasonable volume – providing not unreasonable density profiles. In the special case of hard spheres, Robledo [RV81] used the surface average and volume average, together with a simple excluded volume free energy, with decent results. Fischer and Heinbuch [FH88] showed that these are related to an approximate closure for the first YBG equation.

A more rational approach, from the viewpoint of Sec. 5b, would be to determine n_σ and n_τ by insisting upon exact reproduction of $c(r-r')$ when the fluid is completely uniform, say at density n_0. From our generalization of (5.32) we have

$$c(r-r',n_0) = -B_2(n_0)\int \sigma(r-R)\tau(r'-R)dR$$
$$-n_0B_2'(n_0)\int \tau(r-R)\tau(r'-R)dR \tag{5.33}$$

where $B_2(n) = \beta P^{ex}(n)/n^2 = \beta f^{ex'}(n)$. Since this is not sufficient to determine both σ and τ, there are various options. One is to take σ as an infinitesimal scaling of τ:

$$\sigma(r) = -\frac{1}{D}r\cdot\nabla\tau(r) \tag{5.34}$$

which reduces properly for rods or spheres (D is the dimensionality of space) and τ is uniquely determined by (5.33). A second is to impose (5.33) at densities n_0 and n_1 whenever there are in fact two predominant densities, as in a phase-separated fluid, or even a wall-constrained fluid, bounded by zero density. Another is to at least reproduce the vicinity of n_0 by demanding that both $c(r;n_0)$ and $c'(r;n_0)$ are given consistently by the pair σ,τ; as a special case, one can even do this at $n_0 = 0$.

The convenience, and the conceptual disadvantage, of the above is that only two linearly averaged densities are used (and in fact this is demonstrably wrong [P82b] in the solved case of hard rods with "glue"). A whole set of averaged densities would certainly lead to a superior representation, but amassing sufficient restrictions for their determination is then a problem. A fairly explicit model representation is then called for. One way of proceeding is to generalize the uniform direct correlation to non-uniformity so as to allow for the construction of a free energy. For example, suppose that the w_α are a (normalized) basis for the full set of uniform direct correlation functions over all values of n_0:

$$c(r_1-r_2;n) = \sum_\alpha c_\alpha(n)w_\alpha(r_1-r_2); \tag{5.35}$$

the w_α are then to provide a graded sequence of coarse grainings of the density,

$$n_\alpha(r) = \int w_\alpha(r-r')n(r')dr'. \tag{5.36}$$

Corresponding to the partial direct correlations $c_\alpha(n)$, one can now construct partial *excess* free energies

$$\beta f_\alpha(n) = -\int_0^n\int_0^{n'} c_\alpha(n'')dn''dn'. \tag{5.37}$$

The non-uniform generalization of c chosen is then

$$c(r_1,r_2|n) = \sum_\alpha \frac{1}{2}[c_\alpha(n_\alpha(r_1) + c_\alpha(n_\alpha(r_2))]w_\alpha(r_1 - r_2)$$
$$-\frac{1}{2}\sum\int(n_\alpha(R) - n(R))c'_\alpha(n_\alpha(R))w_\alpha(r_1 - R)w_\alpha(r_2 - R)dR, \tag{5.38}$$

readily verified to arise from the free energy [GHD96]

$$F^{ex}[n] = \sum\int\left(f_\alpha(n_\alpha(r)) - \frac{1}{2}(n_\alpha(r) - n(r))f'_\alpha(n_\alpha(r))\right)dr. \tag{5.39}$$

Hard rods, with a basis of dimensionality two, are represented exactly in this fashion, while the *PY* hard sphere fluid, with only a 4-dimensional basis, is readily set up. Note that since the w_α are not unique – any linear transformation with constant coefficients can be applied – the explicit (5.39) is not unique either. Also note that w_α will typically have the range of the interaction, so that c of (5.38) extends to twice this range.

An older approach [P88] starts with the known positive definiteness of the complete direct correlation function to write

$$C(r_1 - r_2; n) = \int Q(r_1 - r, n)Q(r_2 - r, n)dr \tag{5.40}$$

and matches this to a free energy taken as a function of an arbitrary number of linearly weighted densities. The net effect is to produce a free energy of the form

$$\beta\overline{F} = \int\left[f(\nu(r)) + \int G(r - r', \nu(r))(n(r') - \nu(r))dr'\right]dr, \tag{5.41}$$

stationary with respect to the auxiliary density ν: $0 = \delta\beta\overline{F}/\delta\nu(r)$, or

$$f'(\nu(r)) + \int G'(r - r', \nu(r))(n(r') - \nu(r))dr' + \int G(r - r', \nu(r))dr' = 0, \tag{5.42}$$

and with a profile equation

$$\beta\mu(r') = \delta\beta\overline{F}/\delta n(r') = \int G(r - r', \nu(r))dr. \tag{5.43}$$

It follows that

$$C(r'r,''|n) = \delta\beta\mu(r')/\delta n(r'') = \int\frac{\delta\nu(r)}{\delta n(r'')}G'(r - r', \nu(r))dr. \tag{5.44}$$

But applying $\delta/\delta n(r'')$ to (5.42)

$$\frac{\delta\nu(r)}{\delta n(r'')}\left[f''(\nu(r)) + \int G''(r - r', \nu(r))(n(r') - \nu(r))dr' - 2\int G'(r - r', \nu(r))dr'\right]$$
$$+G'(r - r'', \nu(r)) = 0, \tag{5.45}$$

and so we conclude that

$$C(r', r''|n) = \int\left[2\int G'(r - r''', \nu(r))dr''' - \int G''(r - r''', \nu(r))(\nu(r) - n(r'''))dr'\right.$$
$$\left.-f''(\nu(r))\right]^{-1}G'(r - r', \nu(r))G'(r - r'', \nu(r))dr. \tag{5.46}$$

Specializing to the uniform case,

$$C(r' - r''; n) = \int G'(r - r', n)G'(r - r'', n)dr/\left(2\int G'(r|n)dr - f''(n)\right) \tag{5.47}$$

from which (with (5.43)), $G(r, n)$ can be found. The problem with this formalism, which stresses the convexity of $\overline{F}$ via the positive definiteness of C, is that the resulting $C(r, r'|n)$

does not necessarily have the required $\delta(r-r')/n(r)$ singularity. The obvious cure is to work with βF^{ex} instead, but then the positive definiteness of C is not guaranteed.

In (5.39), and in the genesis of (5.41), F^{ex} was written as the integral of a local function of linearly weighted densities. If there is reason to select only a finite number of these, the task of finding sufficient restrictions on the form of the function is clearly simplified. Since many practical implementations work with the mean field model

$$F^{ex}[n] = F^{ex}_{HS}[n] + \frac{1}{2}\int\int n(r)n(r')\Delta\varphi(r-r')drdr', \tag{5.48}$$

where HS refers to an optimal hard sphere system (optimized via one of number of possible schemes) it is then F^{ex}_{HS} that is really required. In the PY approximation for hard sphere bulk fluids, we have noted that in 3 dimensions, 3 additional densities suffice to linearly generate C. But this is not necessary. In general, with weighted densities defined as in (5.36), if

$$\beta F^{ex}[n] = \int \phi\{n_\alpha(n)\}dr, \tag{5.49}$$

then on differentiating twice,

$$-c(r,r'|n) = \sum_{\alpha\beta}\int w_\alpha(R-r)w_\beta(R-r')\phi''_{n_\alpha n_\beta}\{n_\gamma(R)\}dR, \tag{5.50}$$

and correct reduction to uniformity requires only that $c(r-r';n)$ be a linear combination of the $\{\int w_\alpha(R-r)w_\beta(R-r')dR\}$. The coefficients $\phi''_{n_\alpha n_\beta}\{n_\gamma\}|_{n_\gamma=n}$ then determine the function ϕ in a highly non-unique fashion, of which (5.39) is an example.

The $\{w_\alpha\}$ in the bilinear decomposition (5.50) are of course not unique either. One way to find them is to observe that the positive definiteness of C translates in Fourier space to the Baxter-Wertheim [W63], [B68] factorization of the PY bulk hard sphere fluid approximation,

$$\widetilde{C}_k(n) = \widetilde{Q}^*_k(n)\widetilde{Q}_k(n), \tag{5.51}$$

with Q_k a function only of the scalar wave number k. If $Q_k(n)$ were real, this would just be a rewriting of (5.40), in which $Q(r-r',n)$ depends only upon $|r-r'|$. Since it is not real, but we are dealing with real position space functions, we can write $\widetilde{Q}_k(n) = \tilde{A}_k(n)+i\widetilde{I}_k(n)$, where $\tilde{A}_k$ is even in k, $\widetilde{I}_k$ odd, and hence as

$$\widetilde{Q}_k(n) = \tilde{A}_k(n) + ik\widetilde{B}_k(n), \tag{5.52}$$

where both $\tilde{A}_k$ and $\widetilde{B}_k$ are even. Each function has the form $s(k)\cos ka/2 + t(k)\sin ka/2$ where s and t are polynomials of degree ≤ 3, and a is the core diameter. (5.52) then translates back in real space to the scalar-vector, and pure scalar, forms [P88]

$$C(r-r';n) = \int A(r-R;n)A(r'-R;n)dR + \int \nabla B(r-R;n)\cdot\nabla B(r'-R;n)dR$$
$$= \int A(r-R;n)A(r'-R';n)dR - \int B(r-R';n)\nabla^2 B(r'-R;n)dR. \tag{5.53}$$

We can hence choose as the $w_\alpha(r-R)$ a set that spans either the space of $\{A(r-R;n),\nabla B(r-R;n)\}$ as n varies [R88], [R89], or that of $\{A(r-R;n),\nabla^2 B(r-R;n)\}$. The pure scalar case is somewhat simpler [KR91] and gives rise to

$$w_3(r) = \frac{6}{\pi a^3}\theta\left(\frac{a}{2} - r\right), \quad w_2(r) = \frac{1}{\pi a^2}\delta\left(\frac{a}{2} - r\right),$$
$$w_1(r) = \frac{1}{4\pi a}\delta'\left(\frac{a}{2} - r\right), \quad w_0(r) = -\frac{1}{8\pi}\delta''\left(\frac{a}{2} - r\right) + \frac{1}{2\pi r}\delta'\left(\frac{a}{2} - r\right). \tag{5.54}$$

The function ϕ remains to be determined. Here the special nature of the uniform *PY* solution, now for a mixture $\{n^{(j)}(r)\}$, comes into play. One knows [L64] that the density dependence is only via the four combinations $\sum_j n^{(j)}a_j^\gamma$, $\gamma = 0, 1, 2, 3$, and now sets

$$n_\gamma(r) = \sum_j \int w_{j,\gamma}(r - r')a_j^\gamma n^{(j)}(r')dr', \tag{5.55}$$

quantities that scale under $r \to \lambda r$ just as the $n^{(j)}(r)$ and a_j do. In the uniform limit (since the $w_{j,\gamma}$ are normalized) the

$$n_\gamma = \sum_j n^{(j)}a_j^\gamma \tag{5.56}$$

are now independent, so that knowledge e.g. of the *PY* compressibilities $\partial n^{(i)}/\partial \beta p_j = \delta_{ij}/n^{(j)} - \int c_{ij}(r; \{n^{(j)}\})dr$ suffices to determine the function ϕ. One finds that

$$\phi\{n_\gamma\} = -n_0 \ln(1 - n_3) + \frac{n_1 n_2}{1 - n_3} + \frac{1}{24\pi}\frac{n_2^3}{(1 - n_3)^2}. \tag{5.57}$$

6. UNRESTRICTED INHOMOGENEITY

6.1 Weighted densities

We proceed to situations in which the assumption of a dominant density does not even lurk in the background. Then one must be able to encompass known microscopic information at "all" uniform densities simultaneously, e.g. $c(r; n)$, and a finite number of linearly weighted densities will not have enough parametric leeway – unless the r-space covered by $\{c(r; n)\}$ is of low dimensionality, which is certainly not true in general. This suggests a density-dependent weighted density, i.e. a non-linear relation, the simplest being [T85]

$$\nu(r) = \int w(r - r', \nu(r))n(r')dr \text{ where } \int w(r)dr = 1, \tag{6.1}$$

a generalization of the strategy (5.24) for non-uniformity, which historically preceded the latter. Eq. (5.32), or its Tarazona version with $n_\sigma \to n$, then has the form

$$F^{ex}[n] = \int n(r)f^{ex}(\nu(r))dr, \tag{6.2}$$

with separation of the ideal gas contribution as usual to guarantee the correct singularity when $\delta^2\beta\overline{F}/\delta n(r)\delta n(r')$ is computed. Imposition of the condition $\delta^2\beta F^{ex}(n)/\delta n(r)\delta n(r')|_{n_0} = -c(r - r'; n_0)$ then determines the weight function w by [CA85]

$$\tilde{c}_k(n) + 2\beta f^{ex'}(n)\tilde{w}_k(n) + n\frac{\partial}{\partial n}(\tilde{w}_k(n)^2\beta f^{ex'}(n)) = 0 \tag{6.3}$$

There is also an obvious local version of the self-consistent effective liquid theory (5.27), in which [KB92]

$$F^{ex}[n] = \int n(r)f^{ex}(\nu(r|n))dr \tag{6.4}$$

is coupled with

$$\beta f^{ex}(\nu(r|n)) = -\int\int_0^1 (1-\lambda)c(r-r';\nu(r|\lambda n))d\lambda n(r')dr'. \tag{6.5}$$

The enhanced non-locality of the relation between ν and n seems not to have dramatic practical advantages.

As we have stressed, the multiple density format has much to recommend it, from both conceptual and computational viewpoints. In other words, in the prototypical case of a single supplementary density $\nu(r)$, one would want the form of $\nu(r)$ to come automatically from the stationary property of $\overline{F}$ with respect to ν. One version of this approach has been discussed (5.41) in Sec. 5c. It appeared first in the continuum literature in the joint density-coarse grained density expansion of Meister and Kroll [MK85], and was put in a form satisfying bulk microscopic reproducibility by de Groot and van der Eerden [GE87]. See also [CR93]. The latter can be regarded as an extension of mean field van der Waals

$$F^{ex}[n] = \int n(r)f_0(n(r))dr + \frac{1}{2}\int\int n(r)n(r')\Delta\varphi(r-r')drdr' \tag{6.6}$$

to the expression

$$F^{ex}[n,\nu] = \int n(r)f_0(\nu(r))dr + \frac{1}{2}\int\int n(r)n(r')\Delta\varphi(r-r';\nu(r))drdr' \tag{6.7}$$

with the function $\Delta\varphi(r,\nu)$ being determined in the fashion of (6.3) by bulk microscopics. For a survey of applications to a hard sphere fluid at a hard wall, see [KL90].

In fact, although it is not necessary, even the explicit weighted density approach (6.2) with its model weighted density (6.1) can be cast in a very similar form, equivalent to replacing ν in (6.2) by its Newton-Raphson (or other) stationary representation, i.e. [P96a]

$$F^{ex}[n,\nu] = \int n(r)f\left[\frac{\int n(r')w^0(r-r',\nu(r))dr'}{1-\int n(r')w'(r-r',\nu(r))dr'}\right]dr$$

where $w^0(\nu) \equiv w(\nu) - \nu w'(\nu)$. (6.8)

It is easy enough to examine the general case of $\overline{F}[n,\nu]$. We then require

$$\begin{aligned}\beta\mu(r) &= \delta\beta\overline{F}[n,\nu]/\delta n(r)\\ 0 &= \delta\beta\overline{F}[n,\nu]/\delta\nu(r),\end{aligned} \tag{6.9}$$

and differentiating once more with respect to $n(r')$,

$$\begin{aligned}C(r,r'|n) &= \frac{\delta^2\beta\overline{F}[n,\nu]}{\delta n(r)\delta n(r')} + \int\frac{\delta^2\beta\overline{F}[n,\nu]}{\delta n(r)\delta\nu(R)}\frac{\delta\nu(R)}{\delta n(r')}dR\\ 0 &= \frac{\delta^2\beta\overline{F}[n,\nu]}{\delta\nu(r)\delta n(r')} + \int\frac{\delta^2\beta\overline{F}[n,\nu]}{\delta\nu(r)\delta\nu(R)}\frac{\delta\nu(R)}{\delta n(r')}dR.\end{aligned} \tag{6.10}$$

Elimination of $\delta\nu(R)/\delta n(r')$ yields the unsurprising

$$\begin{aligned}&C(r,r'|n) = \delta^2\beta\overline{F}[n,\nu]/\delta n(r)\delta n(r')\\ &-\int\frac{\delta^2\beta\overline{F}[n,\nu]}{\delta n(r)\delta\nu(R)}\left(\frac{\delta^2\beta\overline{F}}{\delta\nu\delta\nu}\right)^{-1}(R,R')\frac{\delta^2\beta\overline{F}[n,\nu]}{\delta\nu(R')\delta n(r')}dRdR',\end{aligned} \tag{6.11}$$

just a projected inverse. The formalism becomes more meaningful when special cases of $\overline{F}[n,\nu]$ are chosen, as above. A particularly simple form might separate out the ideal gas contribution and confine the remaining density dependence to first order:

$$\beta\overline{F}[n,\nu] = \int n(r)(\ln n(r) - 1)dr + \int h(\nu(r))dr + \int\int g(\nu(r), r - r')n(r')drdr', \quad (6.12)$$

so that

$$C(r, r'|n) = \frac{\delta(r - r')}{n(r)} - \int \frac{g'(\nu(R), R - r)g'(\nu(R), R - r')dR}{h''(\nu(R)) + \int g''(\nu(R), R - R')n(R')dR'}. \quad (6.13)$$

Restriction to uniform density thus requires

$$\begin{aligned} f^{ex}(n) &= h(n) + \tilde{g}_0(n) \\ \mu^{ex}(n) &= \tilde{g}_0(n) \\ \tilde{c}_k(n) &= \tilde{g}'_k(n)^2/(h''(n) + n\tilde{g}''_0(n)). \end{aligned} \quad (6.14)$$

6.2 Local association

The model of a homogeneous reference fluid is certainly less appropriate if temporary localized bound states occur frequently in major contributing configurations. Even for simple fluids where such local states will be dynamically short-lived, their neglect must miss some qualitative aspects in any approximation that is made. However, one can still take advantage of techniques that have been developed for fluids in which more permanent associations occur. A particularly convenient one for our purposes is that due to Wertheim [W84], using a format in which supplementary densities of association enter the description.

The situation we have in mind is once more that of a reference fluid that can be regarded as core-dominated, to which an attractive pair potential is appended, responsible for important qualitative aspects ranging from local to global coherence: $\varphi(i,j) = \varphi_0(i,j) + \Delta\varphi(i,j)$. Correspondingly, for the pair Boltzmann factor, $w(i,j) = w_0(i,j) + \Delta w(i,j)$, where

$$\Delta w(i,j) = w_0(i,j)(e^{-\beta\Delta\varphi(i,j)} - 1). \quad (6.15)$$

We will assume that the tail is literally of short range compared to the core upon which it is built (and the "tail" is even embedded in the core in some models), with the result that a given particle can interact simultaneously with at most t neighbors. Now the N-particle interaction Boltzmann factor can be written as

$$\prod_{\Lambda_N}(w(i,j) + \Delta w(i,j)) = \sum_{G' \in G_N}\left(\prod_{G_N - G'} w_0(i,j)\prod_{G'}\Delta w(i,j)\right) \quad (6.16)$$

where Λ_N denotes the full set of pairs (i,j), G_N the set of graphs on Λ_N. Under the above condition, we can restrict the summation to $G' \in G^t_N$, the graphs with at most $t\Delta w$'s at each vertex. Then, proceeding with the w_0-factors precisely as in the transcription from (2.15) to (2.16), we have

$$-\beta\Omega = \sum_N \sum_{G,G'} \left\{ \begin{array}{l} G \cup G' \in \{A_N\} \\ G \cdot G' = \emptyset, G' \in \{A^t_N\} \end{array} \right\} I^N_{GG'}/N! \quad (6.17)$$

where $I^N_{GG'} = \int \prod_G f_0(i,j) \prod_{G'} \Delta w(i,j) \prod_1^N z(j) dr^N$.

It follows that, in obvious notation,

$$n(r) = z(r)\sum_N \sum_{G,G'} \left\{ \begin{array}{l} G \cup G' \in \{A_N(r)\} \\ G \cdot G' = \emptyset, G' \in \{A^t_N(r)\} \end{array} \right\} I^N_{GG'}(r)/N!. \quad (6.18)$$

The restriction imposed by $\{A_N^t(r)\}$ requires a more careful dissection of the structure. Define the degree $d(j)$ of a vertex j as the number of Δw bonds attached to it, and then separate the r-rooted diagrams of (6.18) according to the number of Δw – bonds at the root:

$$n(r) = \sum_{s=0}^{t} \nu_s(r) \tag{6.19}$$

$$\nu_s(r) = z(r) \sum_N \sum_{G,G'} \left\{ \begin{array}{l} G \cup G' \in \{A_N(r)\}, d(r) = s \\ G \cdot G' = \emptyset, G' \in \{A_N^t(r)\} \end{array} \right\} I_{GG'}^N(r)/N!,$$

Further, separate out the connected branches hanging on vertex r by setting

$$\bar{\nu}_s(r) = \sum_{G,G'} \left\{ \begin{array}{l} G \cup G' \in \{A_N(r)\}, d(r) = s \\ G \cdot G' = \emptyset, G' \in \{A_N^t(r)\}, r \in \mathcal{A} \end{array} \right\} I_{GG'}^N(r)/N!, \tag{6.20}$$

where $\mathcal{A}$ is the class of articulation-free vertices, those whose excision results in a still-connected graph. It is clear that

$$\nu_0(r) = z(r) e^{\bar{\nu}_0(r)}$$

$$\nu_s(r)/\nu_0(r) = \sum_{\sum_{j=1}^{t} j a_j = s} \prod_{j=1}^{t} (\bar{\nu}_j(r)^{a_j}/a_j!). \tag{6.21}$$

Next, we focus on $\bar{\nu}_s(r)$, and in particular on the articulation-free hull of any of its component graphs, created by sequentially removing any subgraphs (but not the common vertex and weight) that is connected to r through an articulation point. The complete set of graphs can then be reconstructed by replacing $z(j)$ for any $j \neq r$ of the reduced set of hulls by a suitable $\nu_q(j)$. If $d(j) = d$, this means any $\nu_q(j)$ with $q \leq t - d$, leading to the new weight

$$n_{t-d}(j) = \sum_{q=0}^{t-d} \nu_q(j) \tag{6.22}$$

attached to j; note that $n_t(j) = n(j)$. We then have

$$\bar{\nu}_s(r) = \sum_N \sum_{G,G'} \left\{ \begin{array}{l} G \cup G' \in \{A_N(r)\}, G \cdot G' = \emptyset \\ G' \subset \{A_N^t(r)\}, d(r) = s, r \in \mathcal{A} \end{array} \right\} \tag{6.23}$$

$$\int \prod_G f_0(i,j) \prod_{G'} \Delta w(i,j) \prod_1^N n_{t-d(j)}(j) dr^N/N!,$$

recognized as implying that

$$\bar{\nu}_s(r) = \delta\Gamma\{n_q\}/\delta n_{t-s}(r)$$

$$\text{where } \Gamma\{n_q\} = \sum_N \sum_{G,G'} \left\{ \begin{array}{l} G \cup G' \in \{A_N\} \\ G \cdot G' = \emptyset, G' \in \{A_N^t\} \end{array} \right\} \tag{6.24}$$

$$\int \prod_G f_0(i,j) \prod_{G'} \Delta w(i,j) \prod_1^N n_{t-d(j)}(j) dr^N/N!.$$

Now let us gather together the loose ends. Although (6.21) is restricted to $s \leq t$, we can choose to *define* $\nu_s(r)$ for $s > t$ by the same relationship, and simultaneously define $n_s(r)$ for $s > t$ by (6.22). Doing so, we have

$$\sum_0^\infty \nu_s(r)\xi^s = z(r)\exp\sum_{q=0}^t \bar{\nu}_q(r)\xi^q, \tag{6.25}$$

so that

$$\sum_0^t \bar{\nu}_q(r)\xi^q = -\beta\mu(r) + \ln\left[(1-\xi)\sum_{q=0}^t n_q(r)\xi^q\right], \tag{6.26}$$

leading directly via (6.24) to

$$\beta\mu(r)\delta_{s,0} = \text{coef } \xi^s \ln\left[(1-\xi)\sum_{q=0}^t n_q(r)\xi^q\right] - \delta\Gamma/\delta n_{t-s}(r). \tag{6.27}$$

But $\delta\beta\overline{F} = \int \beta\mu(r)\delta\mu(r)dr = \sum_{s=0}^t \int \beta\mu(r)\delta_{s,0}\delta n_{t-s}(r)dr$. It then follows from (6.27) that $\delta\beta\overline{F} = \int \sum_{s=0}^t \delta n_{t-s}(r)$ coef $\xi^s \ln[(1-\xi)\sum_{q=0}^t n_q(r)\xi^q]dr - \sum_{s=0}^t \delta n_{t-s}(r)\delta\Gamma/\delta n_{t-s}(r)$, or simply

$$\delta\beta\overline{F} = \text{coef } \xi^t \int \sum_{s=0}^t \delta n_s(r)\xi^s \ln(1-\xi)\sum_{q=0}^t n_q(r)\xi^q dr - \delta\Gamma. \tag{6.28}$$

We conclude that, to within an additive constant,

$$\beta\overline{F}[n_0,\ldots,n_t] = \text{coef } \xi^t \int \sum_{s=0}^t n_s(r)\xi^s \ln\left[(1-\xi)\sum_{q=0}^t n_q(r)\xi^t\right] dr$$
$$-\int n(r)dr - \Gamma[n_0,\ldots,n_t]$$

where $n_t = n$. Furthermore, from (6.27) we have the set of profile equations

$$\mu(r) = \delta\overline{F}/\delta n(r)$$
$$0 = \delta\overline{F}/\delta n_s(r), \qquad 0 \le s \le t-1 \tag{6.30}$$

for the intrinsic free energy $\overline{F}[n_0, n_1, \ldots, n_{t-1}, n]$ on the extended space $(n_0, n_1, \ldots, n_t = n)$.

For a leading approximation, we can choose $t = 1$ in (6.29), as if at most one neighbor can be significantly bound, and correspondingly expand $\Gamma[n_0, n]$ around $\Gamma_0[n] = -\beta F_0^{ex}[n]$ to first order in Δw:

$$\Gamma[n_0, n] = \Gamma_0[n] + \frac{1}{2}\int\int n_0(1)n_0(2)y_0(1,2)\Delta w(1,2)d1d2$$

where $y_0(1,2) = g_0(1,2|n)e^{\beta\varphi_0(1,2)}$. (6.31)

Then

$$\beta\overline{F}[n_0, n] = \int (n(1)\ln n_0(1) - n_0(1))dr + \beta F_0^{ex}[n]$$
$$-\frac{1}{2}\int\int n_0(1)n_0(2)y_0(1,2)\Delta w(1,2)d1d2. \tag{6.32}$$

In this $TPT1$ approximation of Wertheim, the profile equations are then

$$\beta\mu(1) = \ln n_0(1) + \beta\mu_0^{ex}(1|n) - \frac{1}{2}\int\int n_0(2)n_0(3)\delta y_0(2,3)/\delta n(1)\Delta w(1,3)d2d3$$
$$0 = (n(1) - n_0(1))/n_0(1) - \int n_0(2)y_0(1,2)\Delta w(1,2)d2. \tag{6.33}$$

To get some feeling for the adequacy of (6.33), consider a special solvable system, that of 1-dimensional hard rods with "glue" on the boundary, i.e. with a zero-range attractive Boltzmann factor

$$\Delta w(1,2) = \gamma\delta(|x_1 - x_2| - a). \tag{6.34}$$

The exact $\overline{F}[n]$ has been found, and one can verify [KR92] that, here, (6.32) provides an exact representation of the free energy.

6.3 Two-phase coexistence

Physical phase separation in the gas-liquid coexistence region of a fluid certainly has two (or more) regions that are very close to uniform, but the interface itself is not, and is subject to interfacial fluctuations – capillary waves – on a scale much greater than that of the bulk correlation length. The standard way to deal with such restrictions [RJ81] is to decompose configuration space into subspaces each of which is sufficiently restricted that long range excitations are frozen out. In brief symbolic form, with the grand partition function as

$$\Xi[\mu] = \int e^{-\beta\phi[\rho]} e^{\beta\mu\cdot\rho} D\rho \tag{6.35}$$

where $\rho(r)$ denotes the microscopic density $\sum \delta(r - r_i)$ so that the appropriate measure $D\rho$ is highly singular; $\phi[\rho]$ is the total interaction energy. Then suppose constraining a set $\{\zeta[\rho]\}$ anchors relevant large scale fluctuations, leaving the system only locally correlated, and enforce the constraints to the values $\{\zeta\}$ to an adjustable extent by introducing the weight function $W\{\zeta[\rho] - \zeta\}$, normalized so that $\int W\{\zeta\} D\zeta = 1$. Hence [P96a]

$$\Xi[\mu] = \int \Xi[\mu|\zeta] D\zeta \tag{6.36}$$

$$\text{where } \Xi[\mu|\zeta] = \int W\{\zeta[\rho] - \zeta\} e^{-\beta\phi[\rho]} e^{\beta\mu\cdot\rho} D\rho,$$

and $\Xi[\mu|\zeta]$ can be evaluated by standard approximations.

Suppose we adopt the van der Waals pair interaction separation

$$\varphi(r - r') = \varphi_0(r - r') - \varphi_1(r - r'), \tag{6.37}$$

φ_0 a core, $-\varphi_1$ a (negative definite) attractive tail, and take the extreme example (a less extreme one is responsible for the original Ramakrishnan-Yussouff [RY79] solidification technique) of constraining the mean particle density at *every* point to be close to the field $\{\zeta(r)\}$ by setting

$$W\{\zeta[\rho] - \zeta\} = (\text{Det } \beta\varphi_1/2\pi)^{1/2} \exp -\frac{1}{2}\beta(\rho - \zeta)\cdot\varphi_1(\rho - \zeta). \tag{6.38}$$

Hence

$$\begin{aligned}\Xi[\mu|\zeta] &= \int e^{-\beta\phi_0[\rho]} e^{\beta(\mu - \frac{1}{2}\varphi_1(0) + \varphi_1\zeta)\cdot\rho} D\rho \\ &(\text{Det } \beta\varphi_1/2\pi)^{1/2} e^{-\frac{1}{2}\beta\zeta\cdot\varphi_1\zeta} \\ &= e^{-\frac{1}{2}\beta\zeta\cdot\varphi_1\zeta} (\text{Det } \beta\varphi_1/2\pi)^{1/2} \Xi_0\left[\mu - \frac{1}{2}\varphi_1(0) + \varphi_1\zeta\right],\end{aligned} \tag{6.39}$$

and

$$\Xi[\mu] = \langle \Xi_0[n' + \varphi_1\zeta] \rangle_{\mathcal{P}_\zeta} \tag{6.40}$$

where $\mathcal{P}_\zeta = e^{-\frac{1}{2}\beta\zeta\cdot\varphi_1\zeta}(\text{Det } \beta\varphi_1/2\pi)^{1/2}$;

$\mathcal{P}_\zeta$ is the probability distributions ζ space, and $\mu' = \mu - \frac{1}{2}\varphi_1(0)$. Now, we can even appropriate Ξ_0 in local thermodynamic form, so that [P91]

$$\Xi[\mu] = \int \exp\beta\left[\int p_0(\mu' + \varphi_1\zeta(r))dr - \frac{1}{2}\zeta\cdot\varphi_1\zeta\right](\text{Det } \beta\varphi_1/2\pi)^{1/2}D\zeta. \tag{6.41}$$

If there were no long range fluctuations, we could simply take the zeroth order of a steepest descent expansion in (6.41), meaning that only the maximum point in ζ-space contributes

$$\Omega[\mu] = \text{Min}_\zeta \frac{1}{2}\zeta\cdot\varphi_1\zeta - \int p_0(\mu' + \varphi_1\zeta(r))dr$$
$$\text{at } n_0(r;\mu' + \varphi_1\zeta(r)) = \zeta(r), \tag{6.42}$$

and this is precisely the usual mean field result, with the usual mean field interfacial structure. Higher terms in the expansion indeed pick up the capillary wave fluctuations [BLS65], but steepest descent in function space is more than a bit tricky, and in particular the free energy functional $\overline{F}[n]$ corresponding to (6.4)) has not been determined (but see Sec. 4d).

If the long range capillary waves are quenched, either by a weak external field that anchors the interface, or simply by limiting the system to a finite volume, then many approximations for $\overline{F}[n]$ become relevant. Suppose that such an $\overline{F}[n]$ has been obtained. What kind of information pertinent to the interfacial structure should one extract? Probably most basic is the local pressure tensor $P_{\alpha\beta}(r)$ of the non-uniform system, from which all required mechanical properties can be found. The defining equation for $P_{\alpha\beta}(r)$ is that

$$\sum_\alpha \nabla_\alpha P_{\alpha\beta}(r) = n(r)\nabla_\beta\mu(r) \tag{6.43}$$

and in a mesoscopic dynamical context, $\nabla\mu$ can include inertial forces as well. For a locally uniform fluid, one has $P_{\alpha\beta}(r) = p(r)\delta_{\alpha\beta}$, where $p(r)$ is the local pressure, and the relation between thermal and mechanical properties is then expressed by

$$\Omega = -\int p(r)dr$$
$$= -\frac{1}{3}\int \text{Tr } P(r)dr. \tag{6.44}$$

In the pressure of microscopic inhomogeneity, this is no longer true, although the mechanical and thermal properties are related e.g. by [PPG95]

$$\lambda\frac{\partial}{\partial\lambda}\Omega[\mu_\lambda] - \int \text{Tr } P(r|\mu_\lambda)dr \tag{6.45}$$

where $\mu_\lambda(r) = \mu(r/\lambda)$. Assuming that the free energy is known in the local form

$$F^{ex}[n] = \int n(r)f^{ex}(r|n)dr \tag{6.46}$$

(for example, $f^{ex}(r|n) = \int_0^1 \mu^{ex}(r|\lambda n)d\lambda$), actually finding $P(r)$ as a symmetric density functional is not trivial, but one can readily demonstrate the explicit realization [P96b]

$$P_{\alpha\beta}(r) = n(r)(\mu(r) - f(r|n))\delta_{\alpha\beta} + P'_{\alpha\beta}(r)$$

$$\text{where } P'_{\alpha\beta}(r) = \int dr' \int_0^1 d\lambda \frac{\delta f(r-(1-\lambda)r';n)}{\delta n(r+\lambda r')} r'_\alpha \nabla_\beta n(r+\lambda r') \tag{6.47}$$

$$+\sum_\nu \nabla_\nu \int dr' \int d\lambda \frac{\delta f(r-(1-\lambda)r';n)}{\delta n(r+\lambda r')} r'_\beta (r'_\alpha \nabla_\nu n(r+\lambda r') - r'_\nu \nabla_\alpha n(r+\lambda r')).$$

The first term or fact integrates to $\int n(r)\mu(r) - \overline{F}[n] = -\Omega[n]$, while the second is the inhomogeneity correction, nonvanishing only in the interfacial region.

Surface tension for various geometries can be computed from (6.47). But in fact the complete "interface Hamiltonian", the free energy change due to existence and form of the interface, can be found directly. One needs to define the location of the interface and be reasonably explicit as to the form of $\overline{F}[n]$. A typical identification of the interface points Λ_s is that of mean phase density: $n(r_s) = \frac{1}{2}(\rho_0 + \rho_1)$, while free energy functionals ranging from (2.42) to more sophisticated versions give very similar results as to the way in which area, mean curvature, Gaussian curvature,... contribute [?], [?], [?]. When the scale of variation of the interface is very small, e.g. small droplets or high wave number ripples, these expansions become inadequate, but recalculation for the specific circumstances in mind is not difficult.

REFERENCES

[AC72] H.C. Anderson and D. Chandler, J. Chem. Phys. 57 (1972) 1918.

[AL84] A. Alastuey and J.L. Lebowitz, J. Phys. (France) 45 (1984) 1859.

[B68] R.J. Baxter, J. Chem. Phys. 49 (1968) 2770.

[B70] R.J. Baxter, J. Chem. Phys. 52 (1970) 4559.

[BE89] A. Baranyi and D.J. Evans, Phys. Rev. A40 (1989) 3817.

[BLS65] Buff, Lovett, and Stillinger, Phys. Rev. Lett. 15 (1965) 621.

[BSD76] Bongiorno, Scriven, and Davis, J. Coll. Interface Sci. 57 (1976) 472.

[CA85] W.A. Curtin and N.W. Ashcroft, Phys. Rev. A32 (1985) 2909.

[CFP89] Carmesin, Frisch, and Percus, Phys. Rev. B40 (1989) 9416.

[CH58] J.W. Cahn and J.E. Hilliard, J. Chem. Phys. 28 (1958) 258.

[CR93] M. Calleja and G. Rickayzen, Mol. Phys (1993).

[CS69] N.F. Carnahan and K.E. Starling, J. Chem. Phys. 51 (1969) 635.

[CZB96] di Caprio, Zhang, Badiali, Phys. Rev. E 53 (1996) 2320.

[DA89] A.R. Denton and N.W. Ashcroft, Phys. Rev. A39 (1989) 4701.

[DN91] S. Dietrich and M. Napiorkowski, Physica A177 (1991) 437.

[DTB94] Daanoun, Tejero, and Baus, Phys. Rev. E50 (1994) 2913.

[E92] R. Evans, in "Inhomogeneous Fluids", ed. D. Henderson, Dekker (1992).

[F64] I.Z. Fisher, "Statistical Theory of Liquids", U. Chicago (1964).

[FH88] J. Fischer and U. Heinbuch, J. Chem. Phys. 88 (1988) 1909.

[FP87] H.L. Frisch and J.K. Percus, Phys. Rev. A35 (1987) 4696.

[FRW85] Frisch, Rivier, and Wyler, Phys. Rev. Lett. 54 (1985) 2061.

[GE87] R. de Groot and J.P. van der Eerden, Phys. Rev. A36 (1987) 4356.

[GHD96] Gotzelmann, Haase, and Dietrich, Phys. Rev. E53 (1996) 3456.

[H56] T.L. Hill, "Statistical Mechanics", McGraw-Hill (1956).

[H92] D. Henderson, in "Inhomogeneous Fluids", ed. D. Henderson, Dekker, (1992).

[HK64] P. Hohenberg and W. Kohn, Phys. Rev. 136 (1964) B864.

[HM86] J.P. Hansen and I.R. McDonald, "Theory of Simple Liquids", Academic (1986).

[J81] B. Jancovici, Phys. Rev. Lett. 46 (1981) 386.

[J92] B. Jancovici, in "Inhomogeneous Fluids", ed. D. Henderson, Dekker (1992).

[KB92] A. Kyrlidis and R.A. Brown, Phys. Rev. A45 (1992) 5654.

[KL90] D.M. Kroll and B.B. Laird, Phys. Rev. A42 (1990) 4803.

[KP95] MK-H. Kiessling and J.K. Percus, J. Stat. Phys. 78 (1985) 1337.

[KR91] E. Kierlik and M.C. Rosinberg, Phys. Rev. A44 (1991) 5025.

[KR92] E. Kierlik and M.C Rosinberg, J. Stat. Phys. 68 (1992) 1037.

[KS65] W. Kohn and L.J. Sham, Phys. Rev. 140 (1965) A1133.

[L64] J.L. Lebowitz, Phys. Rev. 133 (1964) A895.

[L88] R. Lovett, J. Chem. Phys 88 (1988) 7739.

[LB90] J.F. Lutsko and M. Baus, Phys. Rev. Lett. 64 (1990) 761; Phys. Rev. A41 (1990) 6647.

[LK90] B.B. Laird and D.M. Kroll, Phys. Rev. A42 (1990) 4810.

[LMB76] Lovett, Mou, and Buff, J. Chem. Phys. 65 (1976) 570.

[LP63a] J.L. Lebowitz and J.K. Percus, J. Math. Phys. 4 (1963) 248.

[LP63b] J.L. Lebowitz and J.K. Percus, J. Math. Phys. 4 (1963) 1495.

[LW93] R. Leidl and H. Wagner, J. Chem. Phys. 98 (1993) 4142.

[LWH93] Laird, Wang, and Haymet, Phys. Rev. E47 (1993) 2491.

[MH89] J.D. McCoy and A.D.J. Haymet, Int. J. Thermophys. 10 (1989) 87.

[MK85] T.F. Meister and D.M. Kroll, Phys. Rev. A31 (1985) 4055.

[MM40] J.E. Mayer and M.G. Mayer, "Statistical Mechanics", Wiley (1940).

[MTN87] Mederos, Tarazona, and Navascues, Phys. Rev. B35 (1987) 3376.

[ND95] M. Napiorkowski and S. Dietrich, Z. Phys. B97 (1995) 511.

[NJF89] Nordholm, Johnson, and Freasier, Aust. J. Chem. 33 (1980) 2139.

[O77] W. Olivares, J. Stat. Phys (1977).

[P62] J.K. Percus, Phys. Rev. Lett. 8 (1962) 462.

[P64] J.K. Percus, in "Equilibrium Theory of Classical Fluids"', ed. H.L. Frisch and J.L. Lebowitz, Benjamin (1964).

[P75] J.K. Percus, "Non-uniform Fluids", 3eme Cycle de Physique en Suisse Romande (1975).

[P76] J.K. Percus, J. Stat. Phys. 15 (1976) 505.

[P77] J.K. Percus, J. Stat. Phys. 16 (1977) 299.

[P80] J.K. Percus, J. Stat. Phys. 23 (1980) 657.

[P81] J.K. Percus, J. Chem. Phys. 75 (1981) 1316.

[P82a] J.K. Percus, in "The Liquid State of Matter", ed. E.W. Montroll and J.L. Lebowitz, North-Holland (1982).

[P82b] J.K. Percus, J. Stat. Phys. 28 (1982) 67.

[P84a] J.K. Percus, J. Phys. Chem. 88 (1984) 6499.

[P84b] J.K. Percus, J. Chem. Phys. 81 (1984) 452.

[P86] J.K. Percus, J. Stat. Phys. 42 (1986) 921.

[P87] J.K. Percus, in "Simple Models of Equilibrium and Non-Equilibrium Phenomena", ed. J.L. Lebowitz, North-Holland (1987).

[P88] J.K. Percus, J. Stat. Phys. 52 (1988) 1157.

[P89a] J.K. Percus, J. Stat. Phys. 55 (1989) 1263.

[P89b] J.K. Percus, J. Phys. Cond. Matt. 1 (1989) 2911.

[P90] J.K. Percus, J. Stat. Phys. 60 (1990) 221.

[P91] J.K. Percus, Physica A172 (1991) 1.

[P94] J.K. Percus, Accts. Chem. Res. 27 (1994) 224. item[P96a] J.K. Percus, in "Chemical Applications of Density Functional Theory", ed. Laird, Ross, and Ziegler, ACS Symposium Series No. 629.

[P96b] J.K. Percus, J. Math. Phys. 37 (1996), 1259.

[PPG95] Percus, Pozhar, Gubbins, Phys. Rev. E51 (1995) 261.

[PS94] J.K. Percus and L. Samaj, J. Stat. Phys. 77 (1994) 421.

[PZ88] J.K. Percus and M.Q. Zhang, Phys. Rev. B38, (1988) 11737.

[R71] H.J. Raveche, J. Chem. Phys. 55 (1971) 2242.

[R88] Y. Rosenfeld, J. Chem. Phys. 89 (1988) 4272.

[R89] Y. Rosenfeld, Phys. Rev. Lett. 63 (1989) 980.

[RD85] E. Roman and W. Dietrich, Phys. Rev. A32 (1985) 3276.

[RJ81] J. Rudnick and D. Jasnow, Phys. Rev. B24 (1981) 2760.

[RV81] A. Robledo and C. Varea, J. Stat. Phys. 26 (1981) 573.

[RVR91] Romero-Rochin, Varea, and Robledo, Phys. Rev. A44 (1991) 8417.

[RY79] T.V. Ramakrishman and M. Yussouff, Phys. Rev. B19 (1979) 2775.

[S64] G. Stell, in "Classical Fluids", ed. H.L. Frisch and J.L. Lebowitz, Benjamin (1964).

[S88] L. Samaj, Physica A153 (1988) 517.

[SB62] F.H. Stillinger and F.P. Buff, J. Chem. Phys. 37 (1962) 1.

[SF93] S. Sokolowski and J. Fischer, J. Chem. Soc. Faraday Trans. 89 (1993) 789.

[SG93] G.S. Simmons and C. Garrod, J. Math. Phys. 14 (1973) 1075.

[T85] P. Tarazona, Phys. Rev. A31 (1985).

[TC93] C.F. Tejero and J.A. Cuesta, Phys. Rev. E47 (1993) 490.

[TE84] P. Tarazona and R. Evans, Mol. Phys. 52 (1984) 84.

[TRP95] Tejero, Ripoil, Perez, Phys. Rev. E52 (1995) 3632.

[W35] C.V. von Weiszacker, Z. Phys. 95 (1935) 43.

[W63] M.S. Wertheim, Phys. Rev. Lett. 10 (1963) 321.

[W76] M.S. Wertheim, J. Chem. Phys. 65 (1976) 2377.

[W84] M.S. Wertheim, J. Stat. Phys. 35 (1984) 19, 35.

[W94] J.B. van der Waals, Z. Phys. Chem. 13 (1894) 657.

[WSB95] Weeks, Selinger, and Broughton, Phys. Rev. Lett. 75 (1995) 2694.

[Z89] J. Zinn-Justin, "Quantum Field Theory and Critical Phenomena", Oxford (1989).

[ZO90] X.C. Zeng and D.W. Oxtoby, Phys. Rev. A41 (1990) 7094.

[ZP90] M.Q. Zhang and J.K. Percus, J. Chem. Phys. 92 (1990) 6779.

[ZP93] G. Zhang and J.K. Percus, J. Stat. Phys. 70 (1993) 1365.

PART II:

FUNCTIONALS AND THEIR PROBLEMS

J.M. Seminario (Editor)
Recent Developments and Applications of Modern Density Functional Theory
Theoretical and Computational Chemistry, Vol. 4

Density functional theory, the exchange hole, and the molecular bond

Matthias Ernzerhof [a], Kieron Burke [b], and John P. Perdew [a]

[a]Department of Physics and Quantum Theory Group,
Tulane University, New Orleans, LA 70118

[b] Department of Chemistry,
Rutgers University, Camden, NJ 08102

The Perdew-Wang 91 (PW91) generalized gradient approximation (GGA) derives from a model for the exchange-correlation hole. The basic concepts of the exchange-correlation hole are explained, and the relation to wavefunction methods such as Hartree-Fock is emphasized. To verify the construction of the PW91 functional, we compare the PW91 model for the exchange hole with the exact exchange hole. We study the change of the exchange hole upon atomization of the hydrogen molecule, and find that the PW91 model gives an accurate description of this change. For multiply-bonded molecules like N_2, PW91 provides too much exchange interaction between electrons in neighboring bonds. The largest PW91 errors occur not in the atom but in the molecule; we discuss how errors in the exchange and correlation energies tend to cancel. The fundamentals of the coupling-constant integration are explained, and the coupling-constant dependence of the atomization energy is investigated. Finally, we discuss hybrids which admix a fraction of the exact exchange energy with density functional approximations.

1. Introduction

Density functional theory [1–3] has become a valuable tool for chemistry, one which complements and augments the traditional quantum chemical methods which implicitly or explicitly construct interacting wavefunctions. The workhorse of the wavefunction methods is the Hartree-Fock (HF) approximation. The inherent limitations of Hartree-Fock have been studied over the years, and much is known about the reliability of this and other wavefunction methods when applied to specific systems [4]. Furthermore, the theory of the HF method is also well understood, and the performance of this and other wavefunction methods can be predicted from an understanding of the basic elements of the theory [4].

Density functional methods have become popular in chemistry only recently, so the formal framework of this theory and consequences for the performance of the approximate density functionals are less well understood. In the present article we therefore discuss the basic elements of density functional theory, and illustrate the formalism with calculations on small representative systems.

Kohn and Sham [3] showed that the exact ground-state energy E and spin-densities $\rho_\sigma(\mathbf{r})$ for N electrons in the presence of an external potential $v(\mathbf{r})$ can be found in principle by solving a selfconsistent one-electron Schrödinger equation:

$$\left[-\frac{1}{2}\nabla^2 + v^{\sigma}_{eff}(\mathbf{r})\right] \phi_i(\mathbf{r},\sigma) = \epsilon_i \ \phi_i(\mathbf{r},\sigma), \tag{1}$$

$$\rho_\sigma(\mathbf{r}) = \sum_i^{occ} |\phi_i(\mathbf{r},\sigma)|^2, \tag{2}$$

where the sum runs over all occupied Kohn-Sham orbitals ϕ_i and $\sigma = \uparrow$ or $\downarrow$. Thus Kohn-Sham density functional theory resembles Hartree-Fock theory, with two changes: the Hartree-Fock exchange energy E_x^{HF} is replaced by the density functional $E_{xc}[\rho_\uparrow, \rho_\downarrow]$ for the exchange-correlation energy, and the nonlocal Hartree-Fock exchange potential $v_x^\sigma(\mathbf{r},\mathbf{r}')$ is replaced by the local exchange correlation potential $v_{xc}^\sigma(\mathbf{r})$ as a component of the selfconsistent effective potential $v_{eff}^\sigma(\mathbf{r})$. This local potential $v_{xc}^\sigma(\mathbf{r})$ is the functional derivative $\delta E_{xc}/\delta\rho_\sigma(\mathbf{r})$, and so depends upon $\rho_\uparrow$ and $\rho_\downarrow$.

Typically 50% or more of the energy needed to atomize a molecule is exchange-correlation energy. Because the exact density functional $E_{xc}[\rho_\uparrow, \rho_\downarrow]$ is unknown, it must be approximated. The local spin density approximation (LSD) [3]

$$E_{xc}^{LSD}[\rho_\uparrow, \rho_\downarrow] = \int d^3r \ \rho(\mathbf{r})\epsilon_{xc}(\rho_\uparrow(\mathbf{r}), \rho_\downarrow(\mathbf{r})), \tag{3}$$

where $\rho = \rho_\uparrow + \rho_\downarrow$, utilizes the exchange-correlation energy per electron of a uniform electron gas, $\epsilon_{xc}(\rho_\uparrow, \rho_\downarrow)$, which is accurately known [5]. (In the uniform-gas or jellium model, the electron density is neutralized by a rigid infinite uniform background of positive charge.) Clearly this approximation is valid in the limit of spin densities $\rho_\sigma(\mathbf{r})$ that vary slowly over space. LSD has a long and mostly triumphant history in solid state physics, where it routinely predicts correct crystal structures, lattice constants, bulk moduli, and vibrational frequencies. But LSD has been much less popular in quantum chemistry, where it seriously overestimates the atomization energies of molecules.

The first systematic correction to LSD is the second-order gradient expansion approximation (GEA) [6,7]:

$$E_{xc}^{GEA}[\rho_\uparrow, \rho_\downarrow] = E_{xc}^{LSD}[\rho_\uparrow, \rho_\downarrow] + \sum_{\sigma,\sigma'} \int d^3r \ C_{xc}^{\sigma,\sigma'}(\rho_\uparrow, \rho_\downarrow)\nabla\rho_\sigma \cdot \nabla\rho_{\sigma'}/\rho_\sigma^{2/3}\rho_{\sigma'}^{2/3}. \tag{4}$$

But, for real atoms, molecules, and solids, GEA is *less* accurate [6] than LSD. This fact shows that the electron densities of real systems are *not* close to the slowly-varying limit. The true reasons for the success of LSD in solid state calculations were identified by Gunnarsson and Lundqvist [8], who showed that LSD preserves certain properties of the exact exchange-correlation hole which constrain the energy to a reasonable value. Later Langreth and Perdew [9] showed that GEA does *not* preserve these properties.

The GEA form of Eq. 4 is semi-local, as is the generalized gradient approximation (GGA)

$$E_{xc}^{GGA}[\rho_\uparrow, \rho_\downarrow] = \int d^3r \ \rho \ \epsilon_{xc}(\rho_\uparrow, \rho_\downarrow, \nabla\rho_\uparrow, \nabla\rho_\downarrow), \tag{5}$$

which approximately sums up an infinite series in powers of $\nabla\rho_\sigma$. It is the success of the GGA's that has attracted the current intense interest in density functionals within the quantum chemistry community. Some GGA's have been constructed semi-empirically [6,10–15], with parameters fitted to experiment, while others have been constructed *ab initio* [16–19], by starting from the gradient expansion and then restoring those properties of the exchange-correlation hole that are responsible for the success of the LSD in solids.

All of the *ab initio* GGA's, and some but not all of the semi-empirical ones, are "locally-based" or exact for the uniform electron gas, i.e.,

$$\epsilon_{xc}(\rho_\uparrow, \rho_\downarrow, \nabla\rho_\uparrow = 0, \nabla\rho_\downarrow = 0) = \epsilon_{xc}(\rho_\uparrow, \rho_\downarrow). \tag{6}$$

Eq. 6 is a powerful constraint which keeps a locally-based GGA from going astray, a constraint that is particularly (but not only) relevant to solid state applications.

While the Hartree-Fock approximation underbinds atoms in molecules, and LSD overbinds them, GGA's achieve a useful accuracy [18,20–22]. This accuracy can be further boosted by constructing hybrids [14,23–28] that mix GGA with a fraction of the exact or Hartree-Fock exchange. While these methods do not compete in accuracy with Configuration Interaction and Coupled-Cluster techniques [29] applied to small systems, they do bring medium- and large-size molecules within the reach of quantitative theory.

Among the gradient-corrected density functionals, the generalized gradient approximation of Perdew, Wang, and Burke [16–19] offers the possibility for a detailed analysis, since it is based on a model for the exchange-correlation hole (or equivalently a model for the pair density). This makes it possible to compare exchange-correlation holes predicted by the GGA approximation with exchange-correlation holes obtained from accurate wavefunction calculations for small systems.

In density functional theory, as opposed to wavefunction theory, there are few practical reasons to distinguish the exchange limit, and in fact there are many arguments not to do so. The exchange limit is usually not well approximated by local and semi-local density functionals, in contrast to the full interacting case. Somewhat paradoxically, density functional approximations to the exchange limit are often a good approximation to experimental results, while an exact description of exchange is not. In particular, atomization energies calculated within the local density approximation for exchange are in fair agreement with experimental results [26]. The conventional wisdom is that we have little reason to put much effort into the construction of functionals which reproduce the exact exchange energy; it is more promising to improve functionals for exchange and correlation together. On the other hand, progress in density functional theory is achieved by analyzing density functionals to understand how they work and to isolate features of functionals which can be further improved. In this sense, it is of theoretical interest to analyze the exchange limit of density functionals. In this work we therefore focus first on exchange. We hope that this approach will make the article more accessible to the reader familiar with traditional quantum chemistry methods, in which the Hartree-Fock approximation often serves as a well-defined starting point.

The Hartree-Fock exchange energy is

$$E_x = -\frac{1}{2}\sum_\sigma \int d^3r \int d^3r' |\sum_i^{occ} \phi_i^*(\mathbf{r}, \sigma)\phi_i(\mathbf{r}', \sigma)|^2/|\mathbf{r}' - \mathbf{r}|, \tag{7}$$

where the ϕ_i are Hartree-Fock orbitals, while the exact exchange energy of Kohn-Sham theory is given by the same expression evaluated using Kohn-Sham orbitals. Clearly the exchange energy is the sum of two contributions, one from the spin-up electrons and the other from spin-down. It contains a self-interaction correction to the Hartree approximation, plus a further reduction of the Coulomb energy V_{ee} due to the Pauli principle, which keeps parallel-spin electrons apart. The exact Kohn-Sham density functional for exchange has two very simple scaling properties:

$$E_x[\rho_\uparrow, \rho_\downarrow] = \frac{1}{2}E_x[2\rho_\uparrow] + \frac{1}{2}E_x[2\rho_\downarrow] \tag{8}$$

where $E_x[\rho]$ is the exchange energy of a spin-unpolarized system [30], and

$$E_x[\rho_{\uparrow\lambda}, \rho_{\downarrow\lambda}] = \lambda E_x[\rho_\uparrow, \rho_\downarrow] \tag{9}$$

where $\rho_{\sigma\lambda}(\mathbf{r}) = \lambda^3 \rho_\sigma(\lambda\mathbf{r})$ is a uniformly-scaled density [31]. As a consequence of Eqs. 8 and 9, the GGA for exchange assumes the simple form

$$E_x^{GGA}[\rho] = \int d^3r\ \rho\epsilon_x(\rho)F_x(s), \tag{10}$$

where $\epsilon_x(\rho) = -3k_f/4\pi,\ \ k_f = (3\pi^2\rho)^{1/3}$ is the local Fermi wavevector, and $s = |\nabla\rho|/2k_f\rho$ is a dimensionless density gradient. Different GGA's for exchange correspond to different choices for the function $F_x(s)$.

Starting from the exchange-correlation energy functional, the exchange component can be picked out by uniform scaling to the high-density limit, i.e., [32]

$$E_x[\rho_\uparrow, \rho_\downarrow] = \lim_{\lambda\to\infty} \lambda^{-1} E_{xc}[\rho_{\uparrow\lambda}, \rho_{\downarrow\lambda}] \tag{11}$$

for a density whose ground-state wavefunction is non-degenerate. However, the valence electrons of a molecule are far from the high-density limit, so the contribution of Coulomb correlation to the atomization energy is comparable to that of exchange. Coulomb correlation (the tendency of any two electrons to stay apart because of the repulsive force between them) further lowers the electron-electron repulsion energy V_{ee}, and further stabilizes the molecule relative to the separated atoms.

If correlation is neglected, the only difference between Hartree-Fock and Kohn-Sham theory is the appearance of a *local* exchange potential $v_x^\sigma(\mathbf{r})$ in the latter theory. Recently a number of methods have been presented for the construction of this potential from orbitals, either exactly [33,34] or to a good approximation [35,36].

2. The ground-state energy and its exchange component

The ground-state energy of a Coulomb system can be expressed in terms of the one- and two-particle density matrices γ and Γ,

$$\begin{aligned}\gamma(\mathbf{r},\sigma,\mathbf{r}',\sigma') &= N\int d\sigma_2\ldots d\sigma_N, d^3r_2\ldots d^3r_N \\ &\times\ \psi^*(\mathbf{r},\sigma,\mathbf{r}_2,\sigma_2,\ldots,\mathbf{r}_N,\sigma_N)\psi(\mathbf{r}',\sigma',\mathbf{r}_2,\sigma_2,\ldots,\mathbf{r}_N,\sigma_N),\end{aligned} \tag{12}$$

$$\begin{aligned}\Gamma(\mathbf{r}_1,\sigma_1,\mathbf{r}_2,\sigma_2;\mathbf{r}_1',\sigma_1',\mathbf{r}_2',\sigma_2') &= N(N-1)\int d\sigma_3\ldots d\sigma_N, d^3r_3\ldots d^3r_N \\ &\times\ \psi^*(\mathbf{r}_1,\sigma_1,\mathbf{r}_2,\sigma_2,\mathbf{r}_3,\sigma_3,\ldots,\mathbf{r}_N,\sigma_N) \\ &\times\ \psi(\mathbf{r}_1',\sigma_1',\mathbf{r}_2',\sigma_2',\mathbf{r}_3,\sigma_3,\ldots,\mathbf{r}_N,\sigma_N),\end{aligned} \tag{13}$$

where $\psi(\mathbf{r}_1, \sigma_1, \mathbf{r}_2, \sigma_2, \ldots, \mathbf{r}_N, \sigma_N)$ denotes the ground-state wavefunction of the N-electron system of interest and $\mathbf{r}_i$ and σ_i are the spatial- and spin coordinates, respectively, of the electrons. Now the ground-state expectation value of the Hamiltonian is

$$\begin{aligned} E &= T + v + V_{ee} \\ &= -\frac{1}{2}\int d^3r d\sigma \; \Delta_{\mathbf{r}'}\gamma(\mathbf{r},\sigma,\mathbf{r}',\sigma)|_{\mathbf{r}=\mathbf{r}'} + \int d^3r d\sigma \; \gamma(\mathbf{r},\sigma,\mathbf{r},\sigma)v(\mathbf{r}) \\ &+ \frac{1}{2}\int d^3\mathbf{r}_1 d^3\mathbf{r}_2 d^3\mathbf{r}_1' d^3\mathbf{r}_2' d\sigma_1 d\sigma_2 d\sigma_1' d\sigma_2' \; \Gamma(\mathbf{r}_1,\sigma_1,\mathbf{r}_2,\sigma_2;\mathbf{r}_1',\sigma_1',\mathbf{r}_2',\sigma_2') \\ &\times \frac{\delta_{\sigma_1,\sigma_1'}\delta_{\sigma_2,\sigma_2'}\delta_{\mathbf{r}_1,\mathbf{r}_1'}\delta_{\mathbf{r}_2,\mathbf{r}_2'}}{|\mathbf{r}_1 - \mathbf{r}_2|}. \end{aligned} \tag{14}$$

$v(\mathbf{r})$ denotes the external local potential. The diagonal part of γ, which appears in the term $\int d^3r d\sigma \; \gamma(\mathbf{r},\sigma,\mathbf{r},\sigma)v(\mathbf{r})$ in the expression of Eq. 14 for the energy, is equal to the spin density

$$\rho_\sigma(\mathbf{r}) = \gamma(\mathbf{r},\sigma,\mathbf{r},\sigma). \tag{15}$$

Wavefunction methods such as Configuration Interaction techniques, Coupled-Cluster, or Møller-Plesset perturbation theory are usually implemented by introducing a one-particle basis of orbitals. As a consequence, no advantage can be taken of the fact that the electron-electron repulsion operator is a bi-local operator. I.e., the basis-set representation of the Coulomb operator leads to four index quantities, whereas the real-space representation leads to a matrix with two indices, namely $\mathbf{r}_1$ and $\mathbf{r}_2$. In actual calculations the spin-averaged diagonal part of Γ, the so called pair density $P(\mathbf{r}_1, \mathbf{r}_2)$

$$P(\mathbf{r}_1,\mathbf{r}_2) = \int d\sigma_1 d\sigma_2 \; \Gamma(\mathbf{r}_1,\sigma_1,\mathbf{r}_2,\sigma_2;\mathbf{r}_1,\sigma_1,\mathbf{r}_2,\sigma_2), \tag{16}$$

needs to be approximated[1]. Again, wavefunction methods lead to a representation of $P(\mathbf{r}_1, \mathbf{r}_2)$ in terms of four index quantities. Recent developments in density functional theory have shown that the relevant component of $P(\mathbf{r}_1, \mathbf{r}_2)$ can be accurately approximated in real space, since a number of important conditions are known, which must be satisfied by any real-space representation of $P(\mathbf{r}_1, \mathbf{r}_2)$. What is meant by the relevant part of $P(\mathbf{r}_1, \mathbf{r}_2)$ will be explained now. We note that the electron-electron interaction energy V_{ee} is given in terms of the pair density $P(\mathbf{r}_1, \mathbf{r}_2)$ by

$$V_{ee} = \frac{1}{2}\int d^3r_1 d^3r_2 \; \frac{P(\mathbf{r}_1,\mathbf{r}_2)}{|\mathbf{r}_1 - \mathbf{r}_2|}. \tag{17}$$

Variable substitution $\mathbf{r}_2 = \mathbf{r}_1 + \mathbf{u}$ leads to

$$\begin{aligned} V_{ee} &= \frac{1}{2}\int d^3u \; \frac{\int d^3r_1 \; P(\mathbf{r}_1,\mathbf{r}_1+\mathbf{u})}{u} \\ &= \frac{1}{2}\int_0^\infty 4\pi u^2 \; du \; \frac{\int \frac{d\Omega_{\mathbf{u}}}{4\pi} \int d^3r_1 \; P(\mathbf{r}_1,\mathbf{r}_1+\mathbf{u})}{u}. \end{aligned} \tag{18}$$

[1]The one-matrix γ also needs to be approximated. We will come back to this point in chapter 4.

This shows that only the system- and spherically-averaged pair density $P(u)$,

$$P(u) = \int \frac{d\Omega_{\mathbf{u}}}{4\pi} \int d^3r_1 \; P(\mathbf{r}_1, \mathbf{r}_1 + \mathbf{u}), \tag{19}$$

is needed to calculate V_{ee}.

2.1. Exchange energy

2.1.1. The exchange hole

In this section we introduce basic quantities in density functional theory such as the exchange hole and its system- and spherical-average. The definition of these quantities can readily be generalized to include electron correlation effects (see the following section). To keep the discussion as simple as possible, we focus on exchange. The exchange limit is probably the weakest point of all density functionals, i.e., we do not expect local or semi-local density functionals to work so well for exchange. Coulomb correlation (section 3), which deepens and narrows the exchange-correlation hole, is needed to improve the performance of these approximations.

The reader familiar with the details of density functional theory will know that the Hartree-Fock exchange energy is in general not the same as the exchange energy obtained from the exact functional for exchange, $E_x[\rho]$ with $\rho = \rho^{HF}$, because of the small difference between Hartree-Fock and Kohn-Sham orbitals. Likewise the pair density of the exchange-only Kohn-Sham determinant corresponding to ρ^{HF} will not be equal to the pair density of the Hartree-Fock determinant. In practice, however, the difference between $E_x[\rho^{HF}]$ and the Hartree-Fock exchange energy is often negligible [37], and for the sake of simplicity we do not distinguish between the Hartree-Fock exchange energy and the density functional exchange energy evaluated on the Hartree-Fock density, $E_x[\rho^{HF}]$. Furthermore, we do not distinguish between the Hartree-Fock pair density and the pair density of the Kohn-Sham determinant which yields ρ^{HF}.

In the Hartree-Fock approximation the two-particle density matrix Γ is that of a Slater determinant, which is given in terms of the one-matrix γ^{HF} by

$$\begin{aligned}\Gamma^{HF}(\mathbf{r}_1, \sigma_1, \mathbf{r}_2, \sigma_2; \mathbf{r}_1', \sigma_1', \mathbf{r}_2', \sigma_2') &= \gamma^{HF}(\mathbf{r}_1, \sigma_1, \mathbf{r}_1', \sigma_1')\gamma^{HF}(\mathbf{r}_2, \sigma_2, \mathbf{r}_2', \sigma_2') \\ &- \gamma^{HF}(\mathbf{r}_1, \sigma_1, \mathbf{r}_2', \sigma_2')\gamma^{HF}(\mathbf{r}_2, \sigma_2, \mathbf{r}_1', \sigma_1'),\end{aligned} \tag{20}$$

with

$$\gamma^{HF}(\mathbf{r}, \sigma, \mathbf{r}', \sigma') = \sum_i^{occ} \varphi_i^*(\mathbf{r}, \sigma)\varphi_i(\mathbf{r}', \sigma')\delta_{\sigma,\sigma'}, \tag{21}$$

where the sum runs over the occupied Hartree-Fock spin-orbitals $\varphi_i(\mathbf{r}, \sigma)$. The Hartree-Fock pair density becomes

$$P^{HF}(\mathbf{r}_1, \mathbf{r}_2) = \rho^{HF}(\mathbf{r}_1)\rho^{HF}(\mathbf{r}_2) - \sum_\sigma \gamma^{HF}(\mathbf{r}_1, \sigma, \mathbf{r}_2, \sigma)\gamma^{HF}(\mathbf{r}_2, \sigma, \mathbf{r}_1, \sigma), \tag{22}$$

where $\rho(\mathbf{r}) = \sum_\sigma \rho_\sigma(\mathbf{r})$. The first term on the right hand side is the uncorrelated part of the pair density and the second term describes the exchange contribution. Dividing the pair density by the electron density at the point $\mathbf{r}_1$, we obtain a quantity which describes the probability density for finding an electron at a point $\mathbf{r}_2$ given an electron at point $\mathbf{r}_1$:

$$\frac{P^{HF}(\mathbf{r}_1,\mathbf{r}_2)}{\rho^{HF}(\mathbf{r}_1)} = \rho^{HF}(\mathbf{r}_2) - \frac{1}{\rho^{HF}(\mathbf{r}_1)}\sum_\sigma \gamma^{HF}(\mathbf{r}_1,\sigma,\mathbf{r}_2,\sigma)\gamma^{HF}(\mathbf{r}_2,\sigma,\mathbf{r}_1,\sigma). \tag{23}$$

The second term on the right hand side accounts for a reduction in the probability density due to non-classical effects. To familiarize ourselves with this expression, we think of a two-electron system, with one doubly occupied spatial orbital. In this case the expression of Eq. 23 can be readily simplified to give

$$\frac{P^{HF}(\mathbf{r}_1,\mathbf{r}_2)}{\rho^{HF}(\mathbf{r}_1)} = \rho^{HF}(\mathbf{r}_2) - \frac{1}{2}\rho^{HF}(\mathbf{r}_2). \tag{24}$$

This shows that, if we assume that one electron is at position $\mathbf{r}_1$, then the probability density for finding a second electron at point $\mathbf{r}_2$ is independent of $\mathbf{r}_1$ and equal to half the electron density at $\mathbf{r}_2$. Both electrons are in the same orbital, and if one electron is assumed to be at position $\mathbf{r}_1$ it cannot be anywhere else in the system, so that the probability density is simply reduced by $\rho^{HF}(\mathbf{r}_2)/2$. In this simple example the hole does not follow the position of the electron labelled 1, i.e., the hole is independent of $\mathbf{r}_1$.

If more than one orbital is occupied, $P(\mathbf{r}_1,\mathbf{r}_2)/\rho(\mathbf{r}_1)$ in general becomes a function of $\mathbf{r}_1$ and $\mathbf{r}_2$. In cases where the two electrons have the same spin orientation and occupy different orbitals (which are assumed to be real functions), the hole becomes

$$\begin{aligned}\frac{P^{HF}(\mathbf{r}_1,\mathbf{r}_2)}{\rho^{HF}(\mathbf{r}_1)} = \rho^{HF}(\mathbf{r}_2) \quad &- \quad \frac{1}{\rho^{HF}(\mathbf{r}_1)}\left[\varphi_1^2(\mathbf{r}_1,\sigma)\varphi_1^2(\mathbf{r}_2,\sigma) + \varphi_2^2(\mathbf{r}_1,\sigma)\varphi_2^2(\mathbf{r}_2,\sigma)\right.\\ &+ \quad \left.2\varphi_1(\mathbf{r}_1,\sigma)\varphi_2(\mathbf{r}_1,\sigma)\varphi_1(\mathbf{r}_2,\sigma)\varphi_2(\mathbf{r}_2,\sigma)\right].\end{aligned} \tag{25}$$

The non-classical term of the probability distribution takes a more complicated form than in the previous example. The second term on the right hand side of Eq. 25 involves the probability density that an electron in the orbital φ_1 is at point $\mathbf{r}_1$ times the probability density that the second electron in the orbital φ_1 is at point $\mathbf{r}_2$. The third term on the right hand side involves a similar product except that φ_1 has been replaced by φ_2. The cross term in the above equation becomes important only in regions of space where there is a significant overlap between the orbitals φ_1 and φ_2. Everywhere else either the second or third term on the right hand side will dominate, so that the hole in the density will be localized in one of the orbitals. In regions of space where the cross terms are important, we obtain a hybrid hole [38], i.e., a hole which is localized in a hybrid orbital. An example of such a hybrid hole is shown in Fig. 9 of Ref. [39], where cuts through the exchange hole in the Ne atom are plotted. The reference electron is localized in the L shell and the exchange hole is a hybrid hole in the $2s$ and $2p$ orbitals. $P(\mathbf{r}_1,\mathbf{r}_2)/\rho(\mathbf{r}_1)$ is now a function of $\mathbf{r}_1$ and $\mathbf{r}_2$.

At this point it is natural to define the Hartree-Fock exchange hole $\rho_x^{HF}(\mathbf{r}_1,\mathbf{r}_2)$:

$$\rho_x^{HF}(\mathbf{r}_1,\mathbf{r}_2) \quad = \quad \frac{P^{HF}(\mathbf{r}_1,\mathbf{r}_2)}{\rho^{HF}(\mathbf{r}_1)} - \rho^{HF}(\mathbf{r}_2)$$

$$= -\frac{1}{\rho^{HF}(\mathbf{r}_1)} \sum_\sigma \gamma^{HF}(\mathbf{r}_1, \sigma, \mathbf{r}_2, \sigma)\gamma^{HF}(\mathbf{r}_2, \sigma, \mathbf{r}_1, \sigma). \tag{26}$$

This quantity describes the reduction in the electron density at point $\mathbf{r}_2$ due to the presence of an electron at $\mathbf{r}_1$. We expect that

$$\int d^3\mathbf{r}_2\ \rho_x^{HF}(\mathbf{r}_1, \mathbf{r}_2) = -1, \tag{27}$$

since the integrated reduction of the electron density should be one electron, to account for the fact that one electron is at $\mathbf{r}_1$. By making use of the idempotency of the one-particle density matrix of a Slater determinant, i.e.,

$$\begin{aligned} \int d^3r_2 d\sigma'\ \gamma(\mathbf{r}_1, \sigma, \mathbf{r}_2, \sigma')\gamma(\mathbf{r}_2, \sigma', \mathbf{r}_1, \sigma) &= \gamma(\mathbf{r}_1, \sigma, \mathbf{r}_1, \sigma) \\ &= \rho_\sigma(\mathbf{r}_1), \end{aligned} \tag{28}$$

we see that this is indeed the case. Furthermore, it is easy to read from the definition of Eq. 26 that

$$\rho_x^{HF}(\mathbf{r}_1, \mathbf{r}_2) \leq 0. \tag{29}$$

An important property of the exchange hole (of a single determinant) is that the on-top value (i.e., $\mathbf{r}_1 = \mathbf{r}_2$) of the exchange hole is given exactly in terms of the spin density [40]. From Eq. 26 we obtain

$$\rho_x^{HF}(\mathbf{r}_1, \mathbf{r}_1) = -\frac{1}{\rho^{HF}(\mathbf{r}_1)} \sum_\sigma (\rho_\sigma^{HF}(\mathbf{r}_1))^2. \tag{30}$$

2.1.2. Local density approximation to the exchange hole

A simple approximation to the exact exchange hole should fulfill the properties of the exact exchange hole derived above. The correct on-top value and the non-positivity and normalization conditions ensure a qualitative agreement of approximate and exact exchange holes.

Before we introduce a model for the exchange hole, we derive a simplification which states that the exchange hole of a spin-polarized system can be expressed in terms of the exchange hole of a spin-unpolarized system [19]. From Eq. 26 we see that the exchange hole is the sum of a hole in the α- and another in the β-spin density:

$$\rho_x(\mathbf{r}_1, \mathbf{r}_1 + \mathbf{u}) = \rho_{x\uparrow}(\mathbf{r}_1, \mathbf{r}_1 + \mathbf{u}) + \rho_{x\downarrow}(\mathbf{r}_1, \mathbf{r}_1 + \mathbf{u}), \tag{31}$$

where $\rho_{x\uparrow}(\mathbf{r}_1, \mathbf{r}_1 + \mathbf{u}) = -\frac{1}{\rho(\mathbf{r}_1)}|\gamma(\mathbf{r}_1, \uparrow, \mathbf{r}_2, \uparrow)|^2$. The hole in the up-spin density $\rho_\uparrow(\mathbf{r})$ is essentially the exchange hole in a spin-unpolarized system with a density $\rho(\mathbf{r}) = 2\rho_\uparrow(\mathbf{r})$. An analogous result holds for the hole in the down-spin density, so that we have

$$\rho_x(\mathbf{r}_1, \mathbf{r}_1 + \mathbf{u}) = c_1\rho_x([2\rho_\uparrow], \mathbf{r}_1, \mathbf{r}_1 + \mathbf{u}) + c_2\rho_x([2\rho_\downarrow], \mathbf{r}_1, \mathbf{r}_1 + \mathbf{u}). \tag{32}$$

The coefficients c_1 and c_2 are determined by Eq. 30 to be

$$c_1 = \rho_\uparrow(\mathbf{r}_1)/\rho(\mathbf{r}_1), \quad c_2 = \rho_\downarrow(\mathbf{r}_1)/\rho(\mathbf{r}_1). \tag{33}$$

The only system-dependent property of the above-listed properties of the exchange hole is the on-top value of the hole, which in turn can be expressed through the spin density. It is therefore reasonable to model the exchange hole in terms of the local spin density $\rho_\sigma(\mathbf{r})$. The exchange hole $\rho_x^{LSD}(\rho_\uparrow(\mathbf{r}_1), \rho_\downarrow(\mathbf{r}_1), \mathbf{r}_1, \mathbf{r}_1 + \mathbf{u})$ of a homogeneous electron gas of spin density ρ_σ fulfills all the conditions given above, including the condition on the on-top value of the hole. Furthermore, the exchange hole of the homogeneous electron gas is analytically known. For a spin-unpolarized system we have

$$\rho_x^{LSD}(\rho/2, \rho/2, \mathbf{r}_1, \mathbf{r}_1 + \mathbf{u}) = \frac{-3k_f}{2\pi^2 u^2}\left(\frac{\sin(uk_f)}{(uk_f)^2} - \frac{\cos(uk_f)}{uk_f}\right)^2, \qquad k_f = [3\pi^2\rho(\mathbf{r})]^{1/3}. \tag{34}$$

This model for the exchange hole yields a corresponding model for the pair density:

$$P^{LSD}(\mathbf{r}_1, \mathbf{r}_1 + \mathbf{u}) = \rho(\mathbf{r}_1)\rho(\mathbf{r}_1 + \mathbf{u}) + \rho(\mathbf{r}_1)\rho_x^{LSD}(\mathbf{r}_1, \mathbf{r}_1 + \mathbf{u}). \tag{35}$$

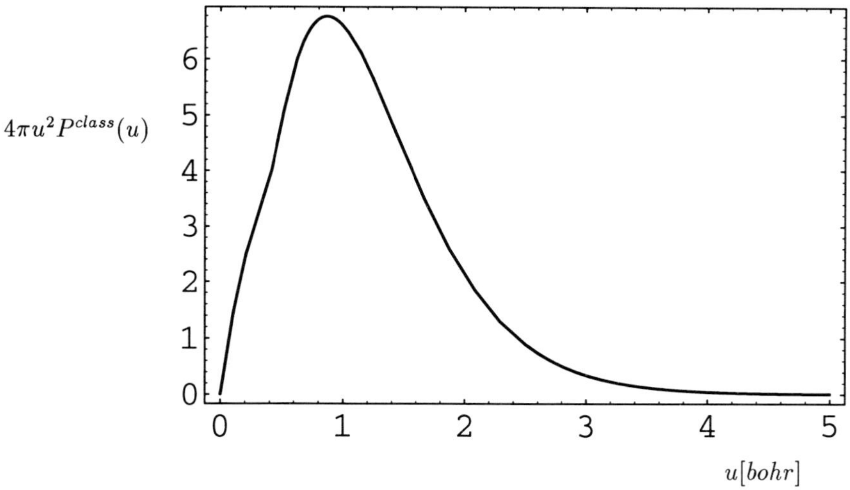

Figure 1. Uncorrelated contribution $P^{class}(u)$ to the system-and spherically-averaged pair density of the Ne atom.

We do not expect this approximation to describe the details of $P(\mathbf{r}_1, \mathbf{r}_1 + \mathbf{u})$ as a function of both variables correctly, but in order to estimate the electron-electron interaction we only need the system- and spherically-averaged pair density $P(u)$,

$$\begin{aligned} P(u) &= \int \frac{d\Omega_{\mathbf{u}}}{4\pi} \int d^3r_1\, \rho(\mathbf{r}_1)\rho(\mathbf{r}_1 + \mathbf{u}) \\ &+ \int \frac{d\Omega_{\mathbf{u}}}{4\pi} \int d^3r_1\, \rho(\mathbf{r}_1)\rho_x(\mathbf{r}_1, \mathbf{r}_1 + \mathbf{u}) \\ &= P^{class}(u) + N\langle\rho_x\rangle(u), \end{aligned} \tag{36}$$

in which many fine details are washed away. N in this equation stands for the particle number. The first term on the right hand side is the classical or uncorrelated part of the pair density, which is straightforward to calculate as a functional of the electron density. The second or non-classical term, which is only approximately known as a functional of the density, will be the quantity we are concerned with. Fig. 1 shows the radial uncorrelated part of the pair density, i.e., $4\pi u^2 P^{class}(u)$. This function gives the classical probability density to find an electron at a distance u from another electron. The shell structure of the Ne atom is still visible in this quantity. The shoulder at about $0.2a_0$ is due to the electrons in the K shell.

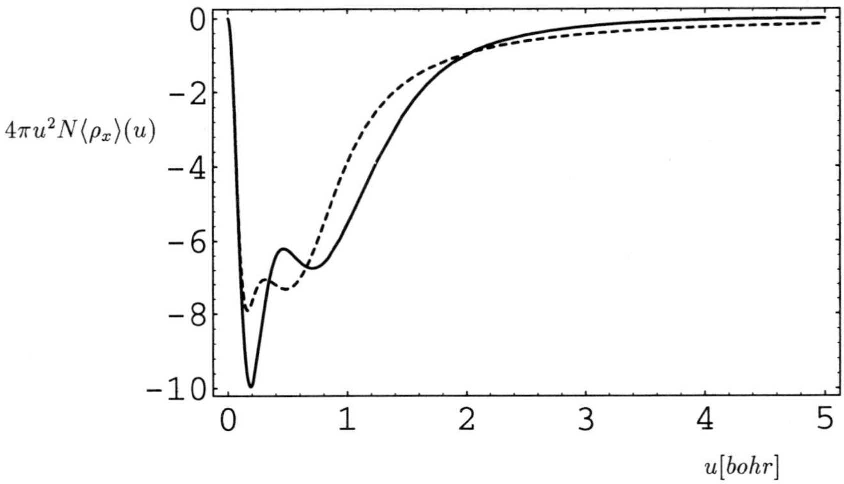

Figure 2. Comparison of the exact (solid line) and the LSD approximation (dashed line) to the system- and spherically-averaged exchange hole of the Ne atom.

In Fig. 2 the quantities $4\pi u^2 N\langle\rho_x^{HF}\rangle(u)$ and $4\pi u^2 N\langle\rho_x^{LSD}\rangle(u)$ are compared. $4\pi u^2 N\langle\rho_x^{HF}\rangle(u)$ clearly shows the shell structure of the Ne atom, and the LSD approximation is in semiquantitative agreement with the Hartree-Fock result.

For intermediate u values on the length scale of the K shell (i.e., $u \in [0.05, 0.15]$), the LSD exchange hole is not negative enough and the shell oscillation in $\langle\rho_x^{HF}(u)\rangle$ is only qualitatively reproduced. The same is true for the hole at intermediate u-values on the scale of the L shell. It is interesting to note that the shell oscillations are not averaged out in the spherical- and system-average. The explanation for this feature of the radial exchange hole is that, for small enough u, $\mathbf{r}$ and $\mathbf{r}+\mathbf{u}$ are in the $1s$ shell for all values of $\mathbf{r}$ inside the $1s$ shell, which leads to large negative contributions to the system-averaged hole. For $u > 0.1a_0$, where $0.1a_0$ is roughly half the radius of the $1s$ shell, $\mathbf{r}$ and $\mathbf{r}+\mathbf{u}$ fall in different shells for a significant fraction of $r-$space, even if $\mathbf{r}$ is located in the $1s$ shell. Since the hole is basically localized in a certain shell, the contribution to the system-averaged hole suddenly drops for $u > 0.1a_0$.

To understand the LSD underestimation of the depth of the hole at intermediate u values, we need to look at the local hole $\rho_x(\mathbf{r}, \mathbf{r}+\mathbf{u})$. In an atom the minimum value of $\rho_x^{HF}(\dot{\mathbf{r}}, \mathbf{r}+\mathbf{u})$ as a function of $\mathbf{u}$ is shifted towards the nucleus (where the density is higher), whereas $\rho_x^{LSD}(\mathbf{r}, \mathbf{r}+\mathbf{u})$ has its minimum at $u = 0$. This different behavior of $\rho_x^{LSD}(\mathbf{r}, \mathbf{r}+\mathbf{u})$ and $\rho_x^{HF}(\mathbf{r}, \mathbf{r}+\mathbf{u})$ becomes particularly obvious if we look at a closed-shell system with two electrons, where $\rho_x^{HF}(\mathbf{r}, \mathbf{r}+\mathbf{u}) = -\rho(\mathbf{r}+\mathbf{u})/2$. This observation explains why $\langle\rho_x^{LSD}\rangle(u)$ is too shallow compared to the exact exchange hole at intermediate u-values, which in turn results in an underestimation of the magnitude of the exchange energy. The LSD exchange hole always 'follows' the electron at point $\mathbf{r}$, in the sense that the minimum of $\rho_x^{LSD}(\mathbf{r}, \mathbf{r}+\mathbf{u})$ is at $u = 0$. In the tail region the LSD and Hartree-Fock holes cross over and the LSD hole shows a slowly decaying tail. The LSD exchange hole has a power law decay as $u \to \infty$, which is unrealistic for systems with exponentially decaying densities. Thus, the LSD approximation predicts a hole in the electron density where there is no electron density. This long-range tail is tied via the normalization constraint on the exchange hole with the LSD hole being too shallow for intermediate u-values. (Note however that the large-u Friedel oscillations of the uniform-gas exchange hole, which have no counterpart in atoms and molecules, are properly washed away by the system- and spherical averaging.)

2.1.3. Exchange energy in terms of the exchange hole

Usually we are more interested in the energetics of the system than in the pair density and the hole. Therefore we decompose the electron-electron interaction energy into the contribution coming from the uncorrelated part of the pair density and the contribution coming from the hole around an electron. In the exchange-only approximation the

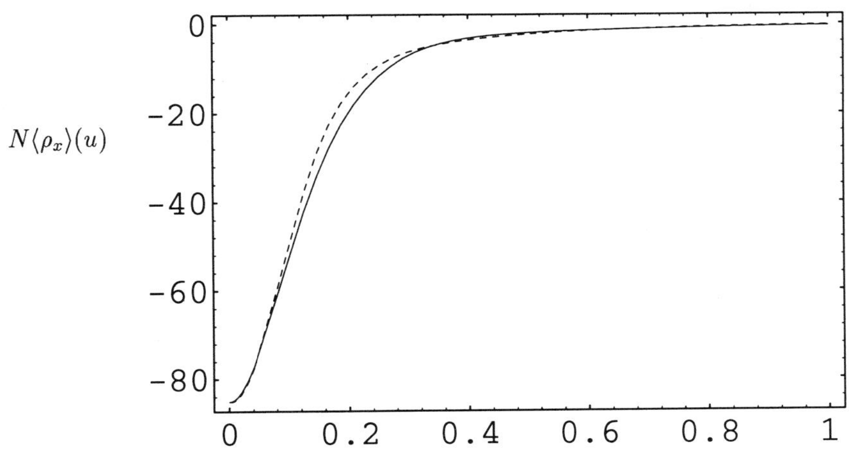

Figure 3. Exact (solid line) and LSD approximation (dashed line) to the system- and spherically-averaged exchange hole of the Ne atom. Unlike the correlation holes, these exchange holes have no electron-electron cusp at $u = 0$.

electron-electron interaction energy can be written as

$$\begin{aligned} V_{ee} &= 2\pi \int_0^\infty du\ uP(u) \\ &= 2\pi \int_0^\infty du\ uP^{class}(u) + 2\pi N \int_0^\infty du\ u\langle\rho_x\rangle(u). \end{aligned} \tag{37}$$

The second term on the right hand side gives the exchange energy, i.e.,

$$E_x = 2\pi N \int_0^\infty du\ u\langle\rho_x\rangle(u). \tag{38}$$

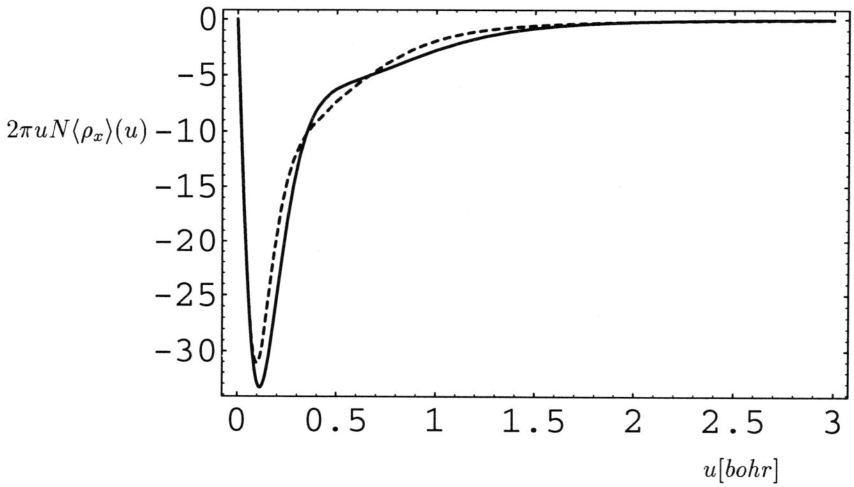

Figure 4. Exact (solid line) and LSD approximation (dashed line) to the weighted exchange hole or energy density $2\pi uN\langle\rho_x\rangle(u)$ in the Ne atom.

Since the second moment $4\pi \int_0^\infty du\ u^2\langle\rho_x\rangle(u) = -1$ is fixed by Eq. 27, the first moment or exchange energy of Eq. 38 becomes more negative when the hole becomes more localized about $u = 0$. Fig. 4 shows the integrand of the Hartree-Fock exchange energy, i.e., $2\pi uN\langle\rho_x^{HF}\rangle(u)$, and compares it with the corresponding LSD approximation $2\pi uN\langle\rho_x^{LSD}\rangle(u)$. $2\pi uN\langle\rho_x\rangle(u)$ is a more interesting quantity to study than the hole itself (see Fig. 3), because the hole is highly dominated by contributions from the core electrons at short u-values and this contribution gets suppressed by the factor u. We will refer to $2\pi uN\langle\rho_x\rangle(u)$ as the weighted hole. The above-discussed failure of the LSD hole to describe the intermediate region of the hole in the K and in the L shell leads to an underestimation of the magnitude of the exchange energy. The Hartree-Fock exchange energy of the Ne atom is -12.108 hartree, whereas the LSD exchange energy is -11.033 hartree.

2.1.4. The LSD approximation for the exchange contribution to atomization energies

The effect that the LSD exchange hole follows the reference electron contributes to the differential error in the exchange energy between a molecule and the separated atoms.

Table 1
Exchange contribution ΔE_x to the atomization energy, evaluated exactly and within the LSD approximation on the LSD densities. $\Delta E_x = E_x(atoms) - E_x(molecule)$. The mean absolute errors (m.a.e.) are also reported.

System	ΔE_{xc} exact	E_x molecule	E_x atoms	ΔE_x exact	E_x^{LSD} molecule	E_x^{LSD} atoms	ΔE_x^{LSD}
H_2	.090	-.648	-.597	.051	-.559	-.513	.045
CH_4	.468	-6.536	-6.215	.320	-5.864	-5.476	.388
H_2O	.262	-8.876	-8.736	.140	-8.075	-7.845	.231
HF	.160	-10.343	-10.253	.090	-9.438	-9.286	.152
m.a.e.	-	-	-	-	.617	.670	.057
CO	.213	-13.229	-13.159	.070	-11.998	-11.780	.217
N_2	.137	-13.035	-13.088	-.053	-11.824	-11.715	.109
NO	.106	-14.616	-14.682	-.066	-13.300	-13.189	.112
O_2	.099	-16.224	-16.277	-.053	-14.805	-14.663	.142
m.a.e.	-	-	-	-	1.294	1.465	.170

The minimum in the exact exchange hole in a homonuclear diatomic molecule is to a lesser extent displaced towards one or the other nuclei than is that of the exact exchange hole in the separated atoms. In the midbond region, the two nuclei compete for the electron density and the hole becomes more isotropic compared to the hole of an atom. The LSD approximation neglects the displacement of the hole completely and therefore introduces a differential error in the exchange energy. In general as shown in Table 1 the exchange energy of the molecule is better described within the LSD approximation than the exchange energy of the separated atoms. The data compiled in this table have been obtained with the CADPAC program [41]. Self-consistent local spin density calculations have been carried out, and the exact exchange results have been obtained by evaluating the exchange energy of the Kohn-Sham determinant. Details of the calculations can be found in the appendix.

Another way of explaining the better performance of the LSD approximation for molecules as opposed to the separated atoms goes as follows. The density gradients in the molecule tend to be smaller than the density gradients in the separated atoms, so that an approximation which neglects the gradient (and all higher derivatives) of the density works better for molecules. LSD fails to make the exchange energy negative enough. This failure, which is more pronounced in the atom, leads to an LSD overestimation of the binding energy at the exchange-only level. Table 1 shows that the LSD approximation to ΔE_x does not compare well with the exact ΔE_x, especially for the multiply-bonded molecules collected in the bottom half of the table. However, ΔE_x^{LSD} is a much better approximation to the exact exchange-correlation energy difference ΔE_{xc}. ΔE_{xc} is obtained by subtracting $\Delta(E^{LSD} - E_{xc}^{LSD})$ from the experimental atomization energies.

2.2. Improving the model for the exchange hole by including the gradient of the density

2.2.1. The gradient expansion of the exchange hole

To improve the accuracy of the LSD exchange hole, we seek a model for the exchange hole which is able to account for its anisotropy around the position of the reference electron. The above-described local spin density approximation to the one-particle density matrix emerges naturally as the zeroth-order approximation from the semiclassical expansion [2] of the one-particle density matrix. In this semiclassical expansion, the one-particle density matrix is expanded in powers of $\hbar$ and subsequently the external potential in this expansion is eliminated in favor of the electron density and its derivatives. The resulting gradient expansion approximation $\gamma^{GEA}(\mathbf{r}_1, \sigma, \mathbf{r}_1+\mathbf{u}, \sigma)$ is then used to construct the GEA hole. The resulting expression for ρ_x^{GEA} is further simplified by making use of the fact that, in the system-average, Laplacian terms can be transformed to gradient terms, i.e., $\int d^3r f(\rho(\mathbf{r}))\Delta\rho(\mathbf{r}) = -\int d^3r f'(\rho(\mathbf{r}))(\nabla\rho(\mathbf{r}))^2$. The important consequence of this modification of the GEA exchange hole is that the modified hole is no longer an approximation to the local hole but only to the system-averaged hole. Therefore the system-averaged exchange hole introduced in section 2 is the quantity which gets approximated in the present formalism. Note, however, that the integration by parts leaves the terms of zeroth and first order in the density gradient unchanged. The resulting hole for a spin-unpolarized system is [2,16]

$$\begin{aligned}\rho_x^{GEA}(\mathbf{r}, \mathbf{r}+\mathbf{u}) &= -\frac{1}{2}\rho(\mathbf{r}) \\ &\times \left[J(z) + \tilde{L}(z)\, \hat{\mathbf{u}}\cdot\mathbf{s}(\mathbf{r}) - \tilde{M}(z)\,(\hat{\mathbf{u}}\cdot\mathbf{s}(\mathbf{r}))^2 - \tilde{N}(z)s^2(\mathbf{r})\right],\end{aligned} \tag{39}$$

where

$$\mathbf{s}(\mathbf{r}) = \frac{\nabla n(\mathbf{r})}{2k_f(\mathbf{r})n(\mathbf{r})}, \tag{40}$$

and $J, \tilde{L}, \tilde{M}$, and $\tilde{N}$ are given by

$$\begin{aligned}J &= 72\frac{[4+z^2-(4-z^2)\cos(z)-4z\sin(z)]}{z^6}, \\ \tilde{L} &= 12\frac{[2-2\cos(z)-z\sin(z)]}{z^3}, \\ \tilde{M} &= \frac{1}{3}\frac{[\sin(z)-z\cos(z)]}{z}, \\ \tilde{N} &= \frac{[8-(8-4z^2)\cos(z)-(8z-z^3)\sin(z)]}{z^4},\end{aligned} \tag{41}$$

with $z = 2k_f(\mathbf{r})u$. The zeroth-order term in $\mathbf{s}(\mathbf{r})$ gives the exchange hole in the LSD approximation, and the first-order term in $\mathbf{s}(\mathbf{r})$ ensures that the minimum in the hole is shifted towards the high-density side of the reference electron, i.e., towards $\nabla\rho(\mathbf{r})$. To first order in u we have

$$\rho_x^{GEA}(\mathbf{r},\mathbf{r}+\mathbf{u}) = -\frac{1}{2}\rho(\mathbf{r}+\mathbf{u}), \tag{42}$$

which is the correct behavior of the exchange hole. Note that this first-order term does not contribute to the system-averaged hole for systems where the density gradient $\nabla\rho(\mathbf{r})$ at each point in the system can be mapped onto its negative by a symmetry transformation, such as the H_2 molecule or any spherical atom.

The quadratic term in $\mathbf{u}$ does not cause a displacement of the minimum of the hole but does modify the angle dependence of the hole.

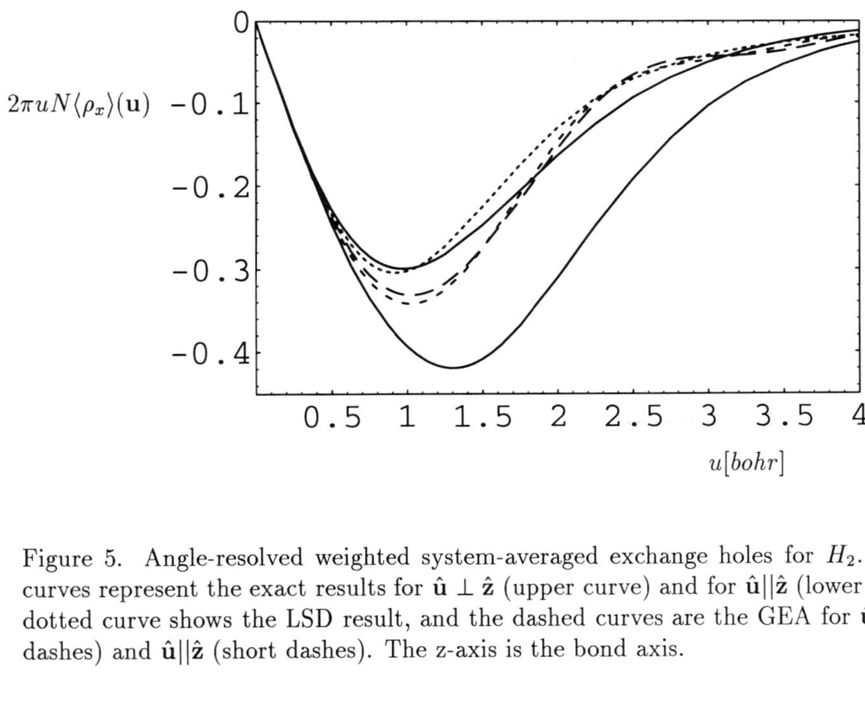

Figure 5. Angle-resolved weighted system-averaged exchange holes for H_2. The solid curves represent the exact results for $\hat{\mathbf{u}} \perp \hat{\mathbf{z}}$ (upper curve) and for $\hat{\mathbf{u}}||\hat{\mathbf{z}}$ (lower curve), the dotted curve shows the LSD result, and the dashed curves are the GEA for $\hat{\mathbf{u}} \perp \hat{\mathbf{z}}$ (long dashes) and $\hat{\mathbf{u}}||\hat{\mathbf{z}}$ (short dashes). The z-axis is the bond axis.

To investigate the GEA hole and its dependence on the orientation of $\mathbf{u}$, we define the system-averaged hole $\langle\rho_x\rangle(\mathbf{u})$ as the system- (but not spherical) average of the exchange hole:

$$\langle\rho_x\rangle(\mathbf{u}) = \frac{1}{N}\int d^3r\rho(\mathbf{r})\rho_x(\mathbf{r},\mathbf{r}+\mathbf{u}). \tag{43}$$

The system of interest is now the H_2 molecule at the experimental equilibrium bond length of $1.401a_0$. In Fig. 5 a comparison is made of the GEA and the exact holes weighted by

$2\pi uN$. A fuller account of both the GEA and GGA system-averaged hole calculations appears in Ref. [42]. The weighted system-averaged hole in the LSD approximation, which is independent of $\hat{\mathbf{u}}$, is also shown. The orientation of $\hat{\mathbf{u}}$ is chosen such that $\hat{\mathbf{u}}||\hat{\mathbf{z}}$ and $\hat{\mathbf{u}} \perp \hat{\mathbf{z}}$, where the z-axis is the bond axis. As already mentioned, the linear term in $\mathbf{s}$ does not contribute to the system average of the exchange hole for this system. If we compare the GEA curves along the different directions, we observe a small splitting which can be traced back to the term quadratic in $\mathbf{u} \cdot \mathbf{s}$ in Eq. 39. The deepening of the weighted GEA hole compared to the weighted LSD hole must be attributed to the term quadratic in $\mathbf{s}$. Note that $\langle \rho_x \rangle(\mathbf{u})$ is independent of $\hat{\mathbf{u}}$ for spherical atoms, since there is no distinguished direction in such a system. The GEA hole reproduces part of the angle dependence of the

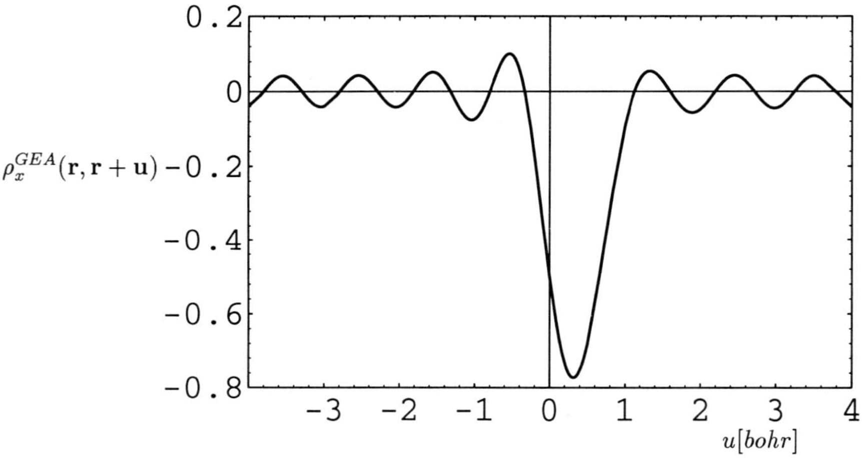

Figure 6. The GEA hole for $s = 1$ and $\rho = 1$, plotted along the direction of $\mathbf{s}$, i.e., $\hat{\mathbf{u}}||\hat{\mathbf{s}}$.

exact hole. The hole along the bond axis is slightly deeper than the hole perpendicular to the bond axis for intermediate u values, i.e., for $u < 1.5a_0$. For $u > 1.5a_0$ the GEA holes oscillate around each other, as a consequence of unphysical oscillations in the GEA hole as u becomes large. Also for $u > 1.5a_0$, the GEA curves are no longer bracketed by the exact angle-resolved holes, which are the solid curves in Fig. 5. The exact angular variation of the weighted exchange hole is much bigger than predicted by the GEA expansion.

To illustrate the unrealistic large-u behavior of the GEA to the exchange hole, we plot the hole for $\rho = 1$, and $s = 1/2$ along the direction of $\mathbf{s}$ in Fig 6. The GEA hole is shifted toward the high-density side, as expected. Unphysical features of the GEA hole, visible in Fig. 6 but not in the system-averaged holes of Fig. 5, are that it is not always negative and that it shows a long-range undamped oscillation which leads to a violation of the normalization condition. These unphysical features of the GEA hole arise because each successive term in the gradient expansion, while improving the hole at small u, worsens it at large u [43]. They are responsible for an order-of-limits dependence [7,44] of the gradient coefficient C_x in Eq. 4.

2.2.2. The exchange hole in the generalized gradient approximation

In view of the findings in the previous section, we construct a GGA by modifying the large-u behavior of the GEA hole in such a way that the modified hole does not become positive and that it is normalized, i.e., we restore the exact conditions of Eqs. 27 and 29. This is achieved by the so-called real-space cutoff procedure [16–19]. The GEA hole as a function of $\mathbf{u}$ is set to zero if it becomes positive, and it is also set to zero for $u > u_c$, where the radial cutoff u_c is chosen such that the normalization condition for the exchange hole is satisfied. Furthermore, the GEA exchange hole is modified so that the long-range

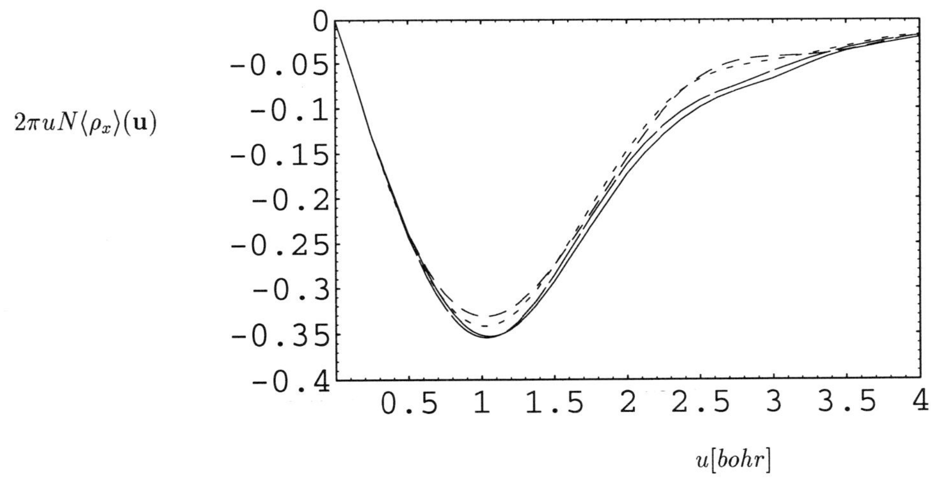

Figure 7. Comparison of the angle-dependent quantity $2\pi u N\langle\rho_x\rangle(\mathbf{u})$ for H_2 in GEA and GGA for $\hat{\mathbf{u}}||\hat{\mathbf{z}}$ (shortest and longest dashes, respectively) and $\hat{\mathbf{u}} \perp \hat{\mathbf{z}}$ (intermediate dashed and solid line, respectively).

oscillations are damped by a factor such as

$$\frac{1}{[1+(0.13z)^3]}, \tag{44}$$

which multiplies the terms in the GEA hole which depend on s. In formulas,

$$\rho_x^{GGA}(\mathbf{r},\mathbf{r}+\mathbf{u}) = \Theta(u_c(\mathbf{r})-u)\,\Theta(-\tilde{\rho}_x^{GEA}(\mathbf{r},\mathbf{r}+\mathbf{u}))\,\tilde{\rho}_x^{GEA}(\mathbf{r},\mathbf{r}+\mathbf{u}), \tag{45}$$

where the damped GEA hole $\tilde{\rho}_x^{GEA}(\mathbf{r}, \mathbf{r}+\mathbf{u})$ is given by

$$\begin{aligned}\tilde{\rho}_x^{GEA}(\mathbf{r}, \mathbf{r}+\mathbf{u}) &= -\frac{1}{2}\rho(\mathbf{r}) \times \Bigg[J(z) \\ &+ \frac{1}{[1+(0.13z)^3]} \left\{ \tilde{L}(z)\ \hat{\mathbf{u}} \cdot \mathbf{s}(\mathbf{r}) - \tilde{M}(z)\,(\hat{\mathbf{u}} \cdot \mathbf{s}(\mathbf{r}))^2 - \tilde{N}(z) s^2(\mathbf{r}) \right\} \Bigg] .\end{aligned} \tag{46}$$

(Depictions of the GGA exchange hole, without damping or system averaging, have

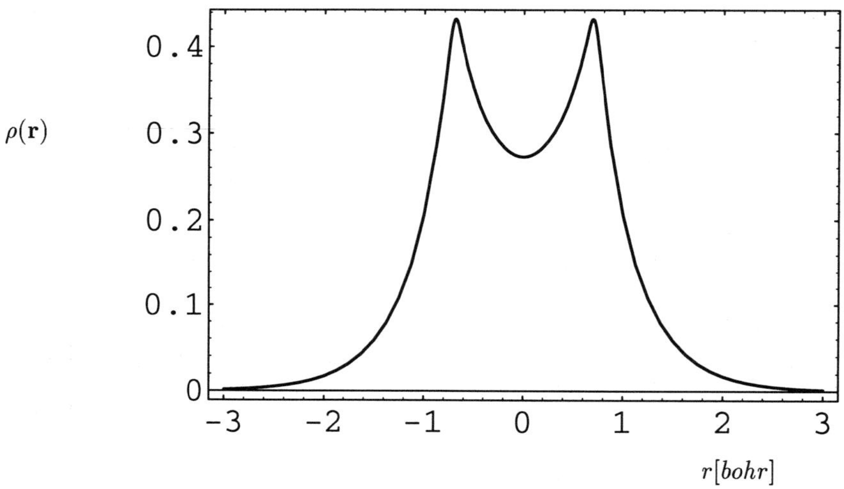

Figure 8. The electron density in H_2 along the bond axis.

been given by Slamet and Sahni [45].) With this GGA model for the exchange hole, we evaluate the energy integral of Eq. 38, yielding a numerically-defined function $F_x(s)$ for Eq. 10, to which we fit an analytic form [16,17,19]. This and other choices for $F_x(s)$ are plotted in Refs. [46] and [47].

In Fig. 7 we compare GGA to GEA for the angle-dependent weighted hole. For small values of u we again find the correct ordering of the curves parallel and perpendicular to the bond axis. The cutoff procedure correctly removes the long-range oscillations in the GEA hole and leads to a deeper hole at $u \approx 2.5$.

To address the question why the angle dependence in the GEA and in the GGA expansion is underestimated, we plot the electron density of the H_2 molecule along the z-axis in Fig. 8. For this unpolarized two-electron system, the exact exchange hole is simply given by $-\rho(\mathbf{r}_1 + \mathbf{u})/2$. We place the reference electron at some point between the left nucleus and the bond center. A large portion of the exact hole is localized at the right nucleus, making the system-averaged hole $\langle\rho_x\rangle(\mathbf{u})$ much deeper for $\mathbf{u}$ parallel to the bond axis

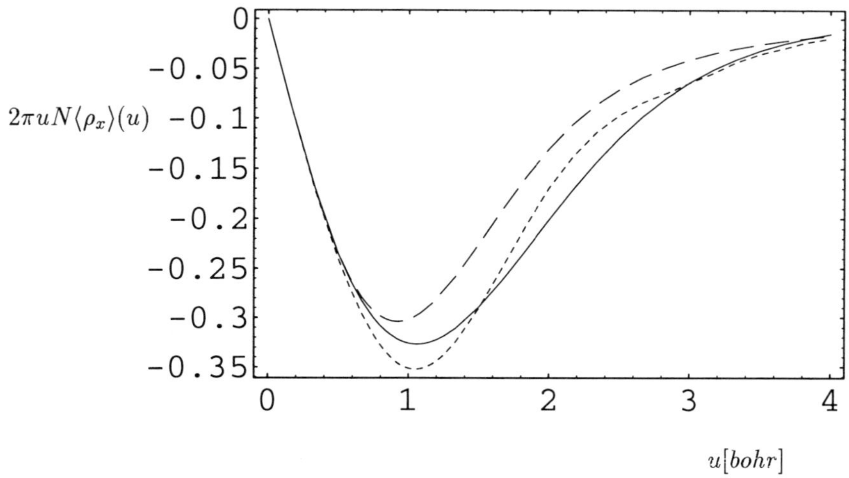

Figure 9. System- and spherically-averaged exchange holes for H_2 weighted by $2\pi uN$ in LSD (long dashes) and PW91 (short dashes). The solid curve represents the exact result.

than for **u** perpendicular to this axis. GEA and GGA, which use only information about the density and the gradient of the density, do not recognize and so miss this effect. GGA for exchange already contains correlation effects in the sense that it leads to an exchange hole which is more localized at the nearer nucleus. The GGA exchange error which shows up in the molecule is not a limitation of one particular GGA, but of the restricted ansatz for any GGA, which uses only the density and the gradient of the density. The exchange hole in this system is not really a semi-local quantity, and the degree of non-semi-locality is bigger in the H_2 molecule than in the separated atoms.

To see what the consequences for the exchange energy of the H_2 molecule are, we plot the weighted system- and spherically-averaged exchange holes in LSD and GGA and compare them with the exact result in Fig. 9. GGA gives a much better description of the hole than does LSD. The GGA hole is more realistic than the LSD hole for intermediate u values, and the unrealistic long-range decay of the LSD hole is suppressed.

2.2.3. Change in the spherically- and system-averaged hole upon breaking of the bond

To investigate the impact of the various approximations on the binding energy of the H_2 molecule, we introduce a differential exchange hole $\Delta\langle\rho_x\rangle(u)$, which describes the change in the exchange hole upon atomization of the molecule:

$$-\Delta\langle\rho_x\rangle(u) = \langle\rho_x\rangle_{H_2}(u) - \langle\rho_x\rangle_{2H}(u). \tag{47}$$

In Fig. 10 we compare the weighted differential spherically- and system-averaged holes in LSD and GGA with the exact result. The electron density becomes more compact upon bond formation, so that the hole becomes deeper for small u-values. Furthermore,

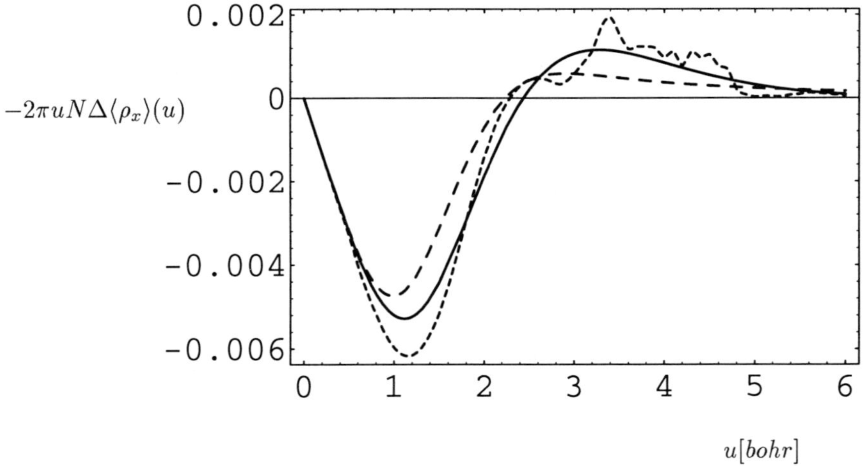

Figure 10. Differential spherically-and system-averaged holes upon atomization of H_2. The solid line represents the exact result, the long dashes show LSD, and the short dashed line is the GGA curve.

the density falls off faster in the molecule than in the separated atoms, so that the corresponding exchange holes show the same behavior. GGA clearly reproduces the exact curve much better than LSD. As in Fig.9 we again find that the weighted hole in the GGA oscillates around the exact curve, so that integration gives a value of the exchange energy difference of 0.0359 hartree in close agreement with the exact value of 0.0336 hartree. The exchange energy difference at the LSD level is 0.0294 hartree. The non-smooth features of the GGA curve for $u \in [3.5, 5]$ are due in part to numerical problems in the integration scheme which result from the normalization cutoffs.

We find that GGA underestimates the angle dependence of the system-averaged hole. However, it gives a good approximation to the system- and spherically averaged hole.

2.2.4. The PW91 approximation for the exchange contribution to atomization energies

The above analysis of the GGA hole suggests that the exchange energy of the separated atoms may be better approximated within a semi-local density functional than is the exchange energy of the molecule. Table 2 shows that this is indeed the case for the multiply-bonded molecules in the bottom half of the table: The PW91 GGA overestimates the magnitude of the exchange energy of the molecule by about 0.7%, while the corresponding error for the separated atoms is about 0.0%. The exact exchange hole in the molecule must be somewhat more delocalized than the PW91 exchange hole. Thus the general trend is reversed compared to LSD, which works better for the molecule than for the separated atoms. The total exchange energies as well as the differential exchange energies are significantly improved within GGA.

Table 2
Exchange contribution ΔE_x to the atomization energy, evaluated exactly and in the PW91 approximation of Refs. [17–19], on the LSD densities. The mean absolute errors (m.a.e.) are also reported.

System	E_x molecule	E_x atoms	ΔE_x exact	E_x^{PW91} molecule	E_x^{PW91} atoms	ΔE_x^{PW91}	ΔE_x^{LSD}
H_2	-.648	-.597	.051	-.641	-.589	.052	.045
CH_4	-6.536	-6.215	.320	-6.521	-6.185	.336	.388
H_2O	-8.876	-8.736	.140	-8.919	-8.732	.187	.231
HF	-10.343	-10.253	.090	-10.395	-10.271	.124	.152
m.a.e.	-	-	-	.029	.015	.025	.057
CO	-13.229	-13.159	.070	-13.309	-13.151	.158	.217
N_2	-13.035	-13.088	-.053	-13.123	-13.058	.065	.109
NO	-14.616	-14.682	-.066	-14.729	-14.672	.057	.112
O_2	-16.224	-16.277	-.053	-16.361	-16.287	.074	.142
m.a.e.	-	-	-	.104	.014	.114	.170

3. Exchange-correlation energy at full coupling strength

As already mentioned, the definition of quantities such as the exchange hole can readily be generalized if electron correlation is considered in addition to exchange effects. However, the approximations for the exchange-correlation holes become much more involved than those for the exchange holes. The exchange-correlation hole at full coupling strength[2] is defined in terms of the exact pair density of the interacting system by

$$\rho_{xc,\lambda=1}(\mathbf{r}_1,\mathbf{r}_2) = \frac{P(\mathbf{r}_1,\mathbf{r}_2)}{\rho(\mathbf{r}_1)} - \rho(\mathbf{r}_2). \tag{48}$$

The potential energy of exchange and correlation is then

$$E_{xc,\lambda=1} = \frac{1}{2}\int d^3r\ \rho(\mathbf{r})\int d^3u\ \rho_{xc,\lambda=1}(\mathbf{r},\mathbf{r}+\mathbf{u})/u. \tag{49}$$

To successfully model the exchange-correlation hole, we need to know relevant properties of this hole to be included in the model. For example, it is easily shown that the sum rule which holds for the exchange hole is also a constraint on the exchange-correlation hole, i.e.,

$$\int d^3u\ \rho_{xc,\lambda=1}(\mathbf{r},\mathbf{r}+\mathbf{u}) = -1. \tag{50}$$

In the preceding section we saw that the on-top value of the exchange hole can be simply expressed in terms of the spin density. For a correlated system there is no such exact relation between the on-top value of the hole and the local spin density [48]. However,

[2]For the definition of exchange-correlation holes at other values of the coupling constant λ, we refer to the following section.

it has been demonstrated that the on-top value of the exchange-correlation hole as a function of the local density shows a nearly universal behavior [49], and thus to a good approximation can be transferred from the uniform electron gas to real systems. Furthermore the behavior of the exchange-correlation hole $\rho_{xc,\lambda=1}(\mathbf{r},\mathbf{r}+\mathbf{u})$ for small values of u is determined by the on-top value of the hole through the cusp condition

$$\frac{\partial}{\partial u}\int\frac{d\Omega_{\mathbf{u}}}{4\pi}\rho_{xc,\lambda}(\mathbf{r},\mathbf{r}+\mathbf{u})|_{u=0}=\lambda\left[\rho_{xc,\lambda}(\mathbf{r},\mathbf{r})+\rho(\mathbf{r})\right]. \tag{51}$$

The exchange hole ($\lambda=0$) of course has no cusp. The cusp also vanishes in the strongly-interacting limit ($\lambda\rightarrow\infty$), where $\rho_{xc,\lambda}(\mathbf{r},\mathbf{r}+\mathbf{u})\rightarrow-\rho(\mathbf{r})$; the singularity of the Coulomb interaction λ/u as $u\rightarrow 0$ is not by itself sufficient to achieve the latter condition at finite λ. The PW91 model for the exchange-correlation hole respects all constraints discussed above and gives a realistic description of the system-averaged exchange-correlation hole.

For additional pictures of the exchange hole ρ_x and the exchange-correlation hole $\rho_{xc,\lambda=1}$, see Refs. [38,42]

4. Coupling-constant integration

In the preceding sections we discussed density functional models for the exchange hole and the exchange-correlation hole at full coupling strength. However, to evaluate the total energy $E=T+V_{ee}+v$ of an interacting system we also need to approximate the kinetic energy contribution. To this purpose the kinetic energy is written as the kinetic energy T_s of a non-interacting system which has the same density as the interacting system, plus a correction term T_c which accounts for the kinetic energy due to electron correlation. In this section we explain how this non-interacting system is defined and how the corresponding T_c is calculated [50,51,8]. We define a modified Hamiltonian

$$H_\lambda=T+\lambda V_{ee}+v_\lambda, \tag{52}$$

where the electron-electron repulsion operator is multiplied by the coupling constant λ. The $\lambda-$dependent potential v_λ, which is identical to the Coulomb potential of the nuclei at $\lambda=1$, is added to keep the density fixed and equal to the physical spin density for all values of λ. For $\lambda=0$ we obtain the Kohn-Sham non-interacting Hamiltonian, and for $\lambda=1$ we obtain the physical Hamiltonian. Obviously we have

$$\begin{aligned}E&=\langle\psi_{\lambda=1}|H_{\lambda=1}|\psi_{\lambda=1}\rangle\\&=\langle\psi_{\lambda=0}|H_{\lambda=0}|\psi_{\lambda=0}\rangle+\int_0^1 d\lambda\frac{d\langle\psi_\lambda|H_\lambda|\psi_\lambda\rangle}{d\lambda},\end{aligned} \tag{53}$$

where ψ_λ is the ground state of H_λ. From the Hellmann-Feynman theorem it follows that

$$\frac{d\langle\psi_\lambda|H_\lambda|\psi_\lambda\rangle}{d\lambda}=\frac{1}{2}\int d^3r d^3r'\,\frac{P_\lambda(\mathbf{r},\mathbf{r}')}{|\mathbf{r}-\mathbf{r}'|}+\int d^3r\rho(\mathbf{r})\frac{d}{d\lambda}v_\lambda. \tag{54}$$

$P_\lambda(\mathbf{r},\mathbf{r}')$ is the pair density calculated from ψ_λ. Furthermore we have

$$\langle\psi_{\lambda=0}|H_{\lambda=0}|\psi_{\lambda=0}\rangle=T_s+\int d^3r\ \rho(\mathbf{r})v_{\lambda=0}(\mathbf{r}). \tag{55}$$

T_s is the kinetic energy of a non-interacting ($\lambda = 0$) ground-state wavefunction which yields the exact density $\rho(\mathbf{r})$ of the interacting system. From Eqs. 53, 54, and 55, we find

$$E = T_s + \int d^3r \; \rho(\mathbf{r})v(\mathbf{r}) + \frac{1}{2}\int_0^1 d\lambda \int d^3rd^3r' \; \frac{P_\lambda(\mathbf{r},\mathbf{r}')}{|\mathbf{r}-\mathbf{r}'|}. \tag{56}$$

The λ -dependent pair density is split up into a λ-independent classical term and the λ -dependent exchange-correlation hole $\rho_{xc,\lambda}(\mathbf{r},\mathbf{r}')$:

$$\frac{1}{2}\int d^3rd^3r' \frac{P_\lambda(\mathbf{r},\mathbf{r}')}{|\mathbf{r}-\mathbf{r}'|} = \frac{1}{2}\int d^3rd^3r' \; \frac{\rho(\mathbf{r})\rho(\mathbf{r}')}{|\mathbf{r}-\mathbf{r}'|} + \frac{1}{2}\int d^3rd^3r' \; \frac{\rho(\mathbf{r})\rho_{xc,\lambda}(\mathbf{r},\mathbf{r}')}{|\mathbf{r}-\mathbf{r}'|}. \tag{57}$$

Thus we obtain [50,51,8]

$$E = T_s + \int d^3r \; \rho(\mathbf{r})v(\mathbf{r}) + \frac{1}{2}\int d^3rd^3r' \; \frac{\rho(\mathbf{r})\rho(\mathbf{r}')}{|\mathbf{r}-\mathbf{r}'|} + \int_0^1 d\lambda \; E_{xc,\lambda}, \tag{58}$$

where

$$E_{xc,\lambda} = \frac{1}{2}\int d^3rd^3r' \; \frac{\rho(\mathbf{r})\rho_{xc,\lambda}(\mathbf{r},\mathbf{r}')}{|\mathbf{r}-\mathbf{r}'|}. \tag{59}$$

The important result of Eq. 58 shows that the kinetic energy contribution T_c to the correlation energy can be deduced from the λ-dependence of the pair density, and so made to "look like" a potential energy. At the $\lambda = 0$ end of the coupling-constant integration $\rho_{xc,\lambda}(\mathbf{r},\mathbf{r}')$ reduces to the exchange hole, and at $\lambda = 1$ to the exchange-correlation hole at full coupling strength. The exchange hole is more shallow and less centered around its electron than is the exchange-correlation hole at $\lambda = 1$. When the interaction is turned on, the hole around an electron becomes deeper and the sum rule ensures that the hole becomes more short-ranged. For the H_2 molecule we saw that the exchange hole extends over the entire molecule. For non-zero λ-values, the "left-right" correlation of the electrons ensures that the exact exchange-correlation hole becomes more localized around its electron. (When one electron is to the left of the bond midpoint, the other is very probably to the right.)

The exchange-correlation energy E_{xc} of density functional theory is now defined by

$$E_{xc} = \int_0^1 d\lambda \; E_{xc,\lambda}, \tag{60}$$

and its kinetic energy contribution is

$$T_c = E_{xc} - E_{xc,\lambda=1} = E_c - E_{c,\lambda=1}. \tag{61}$$

The Euler equation for the expression of Eq. 58 is

$$\mu_\sigma = \frac{\delta E}{\delta\rho_\sigma(\mathbf{r})} = \frac{\delta T_s}{\delta\rho_\sigma(\mathbf{r})} + v(\mathbf{r}) + \int d^3r' \frac{\rho(\mathbf{r}')}{|\mathbf{r}-\mathbf{r}'|} + \frac{\delta E_{xc}}{\delta\rho_\sigma(\mathbf{r})}. \tag{62}$$

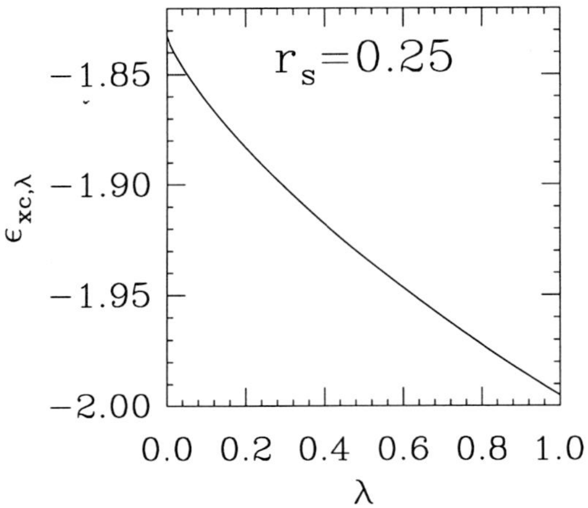

Figure 11. The λ-dependent exchange-correlation energy per particle of a uniform electron gas at a high density, $r_s = 0.25$ *bohr*, typical of core electrons. (Energy in hartrees).

This equation can be interpreted as the Euler equation of an independent-particle (Kohn-Sham) problem with local potential

$$v^{\sigma}_{eff}(\mathbf{r}) = v(\mathbf{r}) + \int d^3r' \frac{\rho(\mathbf{r}')}{|\mathbf{r}-\mathbf{r}'|} + \frac{\delta E_{xc}}{\delta \rho_\sigma(\mathbf{r})}, \tag{63}$$

as in Eq. 1.

Usually E_{xc} is approximated directly, starting from an electron gas of uniform or slowly-varying density. The λ-dependence of a given approximation to the integrated quantity E_{xc} can be extracted from $E_{xc}[\rho]$ via the scaling relation[31,52]

$$E_{xc,\lambda}[\rho] = \frac{d}{d\lambda}\Big(\lambda^2 E_{xc}[\rho(\mathbf{r}/\lambda)/\lambda^3]\Big). \tag{64}$$

For the homogeneous electron gas, $\epsilon_{xc,\lambda} = E_{xc,\lambda}/N$ is known accurately as a function of the electron density. Although $\epsilon_{xc,\lambda}$ for the uniform gas is not an analytic function of λ (due to the presence of terms like $\lambda \ln \lambda$), its coupling-constant dependence is otherwise similar to that of a finite system. In Figures 11 and 12 the coupling-constant dependence of $\epsilon_{xc,\lambda} = E_{xc,\lambda}/N$ for the homogeneous electron gas, as parametrized by Perdew and Wang[5], is shown. We consider two electron densities characterized by the Seitz radius $r_s = (3/4\pi\rho)^{1/3}$. The first, $r_s = 0.25$, is a typical density of core electrons in an atom or molecule. For high densities, the kinetic energy dominates the exchange-correlation energy. In finite systems such as atoms or molecules, we expect that $E_{xc,\lambda}[\rho]$ becomes

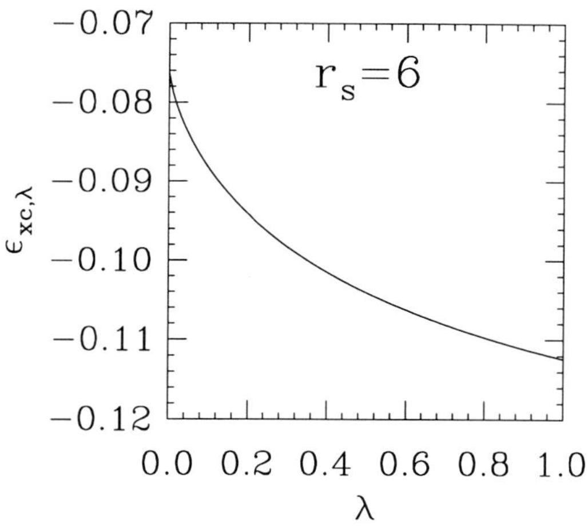

Figure 12. Same as Fig. 11, but at a low density, $r_s = 6$ *bohr*, typical of valence electrons.

linear in λ when the electron-electron interaction can be treated as a small perturbation, i.e., in the high-density limit. The correlation energy E_c can then be accounted for by second-order perturbation theory [53]. The curve in Fig. 11 is nearly linear, so the behavior of the homogeneous electron gas at high density is similar to that of a high-density finite system. For low densities, correlation effects become more important both in the homogeneous electron gas and in finite systems. In finite systems we then need higher-order perturbation theory to account for the correlation energy; second-order perturbation theory by itself overestimates the correlation energy, which saturates as $\lambda \to \infty$. Fig. 12 shows $\epsilon_{xc,\lambda}$ for a homogeneous electron gas with $r_s = 6$, which is typical of valence electrons in a low-density metal. The curve is strongly bent upward, a behavior expected also for a finite system of low density.

4.1. Coupling-constant averaged exchange-correlation energies

In the preceding section, we saw how to describe a physical system in terms of a fictitious non-interacting system and a correction term E_{xc} which accounts for the exchange-correlation energy of the electrons. The crucial point in this formulation of the many-electron problem is the approximation to be employed for E_{xc}. In Section 2 we saw how the PW91 approximation to E_x is constructed, so it remains to describe the approximation to $E_c = E_{xc} - E_x$. A detailed analysis of the PW91 construction of the correlation hole $\rho_c(\mathbf{r}, \mathbf{r}+\mathbf{u}) = \rho_{xc}(\mathbf{r}, \mathbf{r}+\mathbf{u}) - \rho_x(\mathbf{r}, \mathbf{r}+\mathbf{u})$ will be given in a forthcoming publication [42], so here we only state the main steps of this construction. As in the construction of the exchange hole, the correlation hole of the homogeneous electron gas of spin densities $\rho_\uparrow(\mathbf{r})$ and $\rho_\downarrow(\mathbf{r})$ defines the zeroth-order approximation to the correlation hole at $\mathbf{r}$ of an inhomogeneous system with spin densities $\rho_\uparrow(\mathbf{r})$ and $\rho_\downarrow(\mathbf{r})$. The correlation hole of the homogeneous electron gas is not exactly known, but an accurate parametrization has been constructed [54]. To this hole a gradient correction is added. The resulting GEA correlation hole violates the normalization condition

$$\int d^3u \; \rho_c(\mathbf{r}, \mathbf{r} + \mathbf{u}) = 0 \tag{65}$$

on the correlation hole. This deficiency is rectified by a cutoff procedure completely analogous to the cutoff procedure which restores the normalization condition on the GEA expansion of the exchange hole. (Note however that the correlation hole does not have the constraint to be negative everywhere.) The numerical correlation energy functional defined by this procedure is then parametrized [17,19]. Again, the LSD on-top exchange correlation hole, while not exact [48], is highly accurate, and so helps explain the reliability of LSD exchange-correlation energies [49].

Table 3
Atomization energies (D_e) in the LSD and PW91 approximation. Unrestricted Hartree-Fock (UHF) and hybrid (as explained in section 5) results are also shown. LSD and PW91 energies are evaluated on LSD densities. The mean absolute errors (m.a.e.) are given separately for the singly- and multiply-bonded systems.

System	ΔE^{exact}	ΔE^{UHF}	ΔE^{LSD}	ΔE^{PW91}	ΔE^{hybrid}
H_2	.174	.136	.180	.168	.167
CH_4	.668	.523	.736	.672	.668
H_2O	.370	.248	.424	.376	.363
HF	.224	.154	.258	.228	.220
m.a.e.	-	.095	.040	.005	.005
CO	.413	.277	.476	.429	.406
N_2	.364	.183	.426	.386	.359
NO	.244	.084	.316	.273	.242
O_2	.192	.052	.279	.229	.198
m.a.e.	-	.154	.071	.026	.005

Local and semi-local approximations to $E_{xc,\lambda}$ are expected to work better at the upper end of the coupling-constant integration than at the lower end. Table 3 shows a remarkable improvement in the density functional approximations to E_{xc} over the approximations to E_x. As can be seen from Table 3, atomization energies are noticeably better in the PW91 approximation than in the LSD approximation. In general, we find that overestimation of the atomization energies at the exchange level carries over to the coupling-constant averaged quantities (but to a less severe extent), so that atomization energies of multiply-bonded systems are less accurately reproduced than atomization energies of singly-bonded systems.

With the coupling-constant integration in mind, we now propose an explanation of the results for the atomization energies. The atomization of a singly-bonded system will be interpreted as a process involving an effective two-electron system with paired spins. The two-electron system is ideally realized in the H_2 molecule, and approximately in systems such as CH_4 where the binding electron pairs are separated from each other. For these systems we observe a smaller PW91 error at the exchange level than in the multiply-bonded systems, where the binding electron pairs interact strongly. In an effective two-electron process, the change in the exchange energy reduces to a change in the self-interaction energy, which seems to be well reproduced by the PW91 functional for exchange. In cases where two or more bonds are close to each other, we no longer have an effective two-electron system because there are several electrons with the same spin orientation in the same region of space. The LSD and to a lesser extent the PW91 approximation significantly overestimate the magnitude of the exchange energy in these cases, and this in turn leads to an overestimation of the coupling-constant averaged atomization energies. Presumably the density functionals overestimate the magnitude of the interelectronic exchange [26] here because they do not account for subtle effects of orbital nodality [55]. These functionals may also overestimate the magnitude of the self-exchange energy for the delocalized π orbitals.

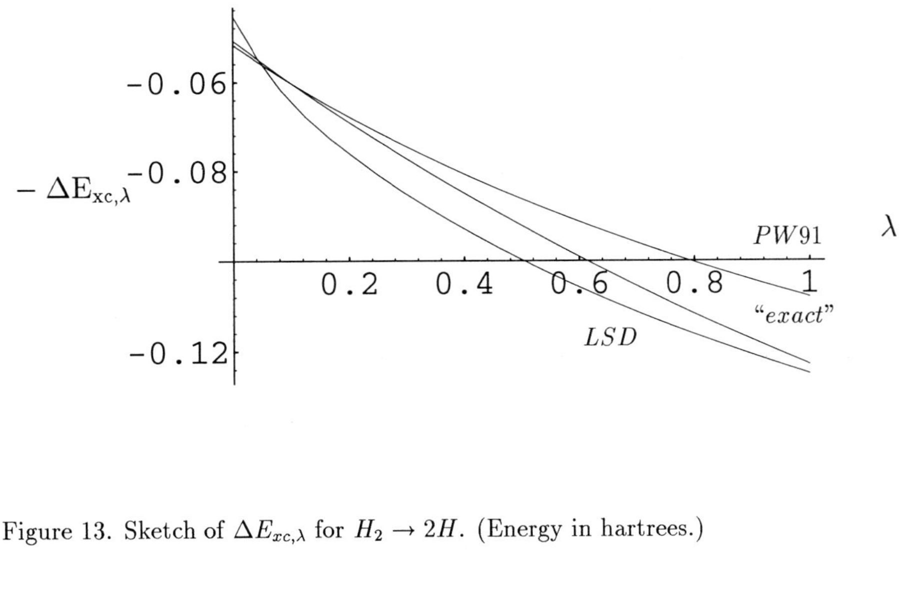

Figure 13. Sketch of $\Delta E_{xc,\lambda}$ for $H_2 \rightarrow 2H$. (Energy in hartrees.)

To understand the correlation contribution to the atomization energy, we consider the H_2 molecule where we have two anti-parallel-spin electrons in the molecular bond. To a limited extent the H_2 molecule already shows static or 'left-right' correlation, which becomes more pronounced if we stretch the bond length, and less pronounced if we shorten it. Static correlation causes a more rapid drop of $-\Delta E_{xc,\lambda}$ for small λ, followed by a flattening of the curve, i.e., a stronger concavity. Local and semi-local approximations do not accurately account for this static correlation, but are more appropriate for the description of dynamic correlation. In the presence of static correlation, since $-\Delta E_{c,\lambda}$

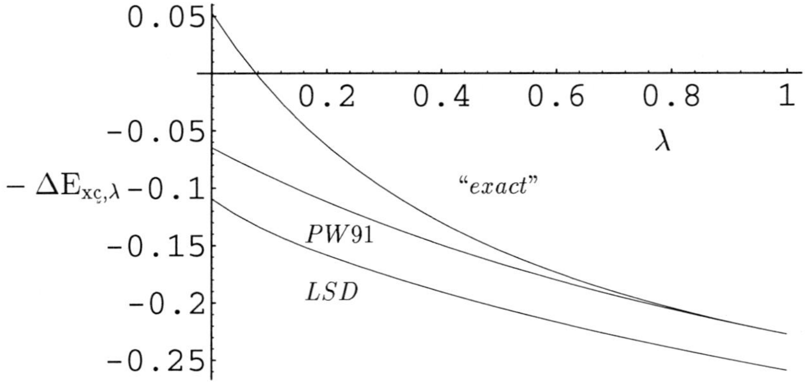

Figure 14. Sketch of $\Delta E_{xc,\lambda}$ for $N_2 \rightarrow 2N$. (Energy in hartrees.)

for the approximate functionals is less concave as a function of λ than the exact curve, such functionals are likely to underestimate the magnitude of the correlation energy in the molecule: The exact slope of $-\Delta E_{xc,\lambda}$ at $\lambda = 0$ is more negative than the approximate one. In cases where the exchange contribution to the atomization energy is accurately reproduced (as in H_2 at the equilibrium bond length), the total atomization energy is then underestimated.

For multiply-bonded systems like N_2, the LSD and PW91 overestimation of the ΔE_x contribution to the atomization energy is only partly compensated by underestimation of ΔE_c. Cases intermediate between the H_2 molecule and the N_2 molecule are for instance CH_4 and H_2O, where the described errors in the differential exchange-correlation effects approximately cancel, i.e., the overestimation of ΔE_x is compensated by an underestimation of ΔE_c.

To illustrate these arguments, we sketch the coupling-constant dependence of $\Delta E_{xc,\lambda}$ for H_2 and N_2. $\Delta E_{xc,\lambda}^{LSD}$ and $\Delta E_{xc,\lambda}^{PW91}$ in Figs. 13 and 14 are calculated using Eq. 64, while the "exact" curve for H_2 is a sketch which reproduces the correct exchange limit, the experimental value for ΔE_{xc} and the correct value for $\Delta E_{xc,\lambda=1}$ [56]. The "exact" curve for N_2 is an interpolation between the exact $\Delta E_{xc,\lambda=0}$ and $\Delta E_{xc,\lambda=1}^{PW91}$, which yields the experimental ΔE_{xc}.

For the H_2 molecule the PW91 approximation gives good results for ΔE_x. Due to the missing weak static correlation, $-\Delta E_{xc,\lambda=1}$ is insufficiently negative in the PW91 approximation, so that ΔE_{xc} is slightly underestimated. The LSD approximation overestimates the dynamic correlation contribution to ΔE_{xc}.

The N_2 molecule shows much stronger static correlation (expected to be most pronounced within each π bond) which is mimicked by the large PW91 exchange energy error, so that $\Delta E_{xc,\lambda=1}^{PW91}$ is very accurate. Because the PW91 curve in Fig. 14 is too flat, it predicts an atomization energy for the N_2 molecule which is too large. Strong static correlation in the molecule causes the exact $-\Delta E_{xc,\lambda}$ to drop rapidly from its $\lambda = 0$ value towards its $\lambda = 1$ value. The separated atoms do not have much static correlation, so $E_{xc,\lambda}$ can be accurately obtained from second-order perturbation theory, which implies that $E_{xc,\lambda}^{atoms}$ is close to a straight line. The LSD curve overestimates $\Delta E_{xc,\lambda}$ at $\lambda = 1$, but its error is smaller there than at the $\lambda = 0$ end.

5. Hybrids of Hartree-Fock with density functional theory

The Hartree-Fock approximation is exact in the very limit ($\lambda = 0$ or exchange-only) where local or semi-local density functionals are often least accurate. This realization led Becke to propose the approximation [14,23]

$$E_{xc}^{hybrid} = E_{xc}^{GGA} + a_0(E_x - E_x^{GGA}), \tag{66}$$

where E_x is the exact exchange energy (the Fock integral of Eq. 7, evaluated with LSD or GGA orbitals), and $a_o \approx 0.28$ is an empirical parameter. The "hybrid" of Eq.66 is also known as the "adiabatic connection" or "exact exchange mixing" formula.

Eq. 66 is a way to refine the coupling constant integral of Eq. 60 through replacement of GGA by exact information about the lower limit $\lambda = 0$, as in our construction of the "exact" curve of Fig. 14. We have recently argued [28] that this construction leads to $a_0 \approx 1/n$, where n is the lowest order of density functional perturbation theory that yields accurate atomization energies, whence $n = 4$ for typical molecules described by a fourth-order Møller-Plesset (MP4)-like expansion. At this level, $\Delta E_{xc,\lambda}$ is a cubic in λ. Table 3 shows that this nonempirical $n = 4$ hybrid brings the PW91 atomization energies close to the goal of chemical accuracy.

Hybrids more sophisticated than Eq. 66 can be constructed by using additional information about $\Delta E_{xc,\lambda}$, such as the GGA value at $\lambda = 1$ [27] or the exact derivative $d\Delta E_{xc,\lambda}/d\lambda$ at $\lambda = 0$ from second-order density functional perturbation theory [28,57]. Although further improvements in accuracy may result, this accuracy is ultimately limited by the residual GGA errors that persist far from the $\lambda \to 0$ limit.

6. Conclusions

Both exchange and correlation contribute heavily to the atomization energy of a molecule. While exchange is treated exactly in Hartree-Fock theory, even a simple local-spin density (LSD) approximation to exchange and correlation does better than Hartree-Fock alone. Further improvements in atomization energies are achieved by adding gradient corrections (GGA's or semi-local approximations), and by hybridizing 25% (but not 100 %!) of the exact exchange energy with GGA.

Local and semi-local density functionals work, not only because they are exact for a uniform or slowly-varying density, but more directly because they respect exact constraints on the exchange-correlation hole (Eqs. 27,29,65). For LSD and for the PW91 GGA, we have extracted the model exchange holes for comparison with the exact one (Figs. 9 and 10). After (but only after) system- and spherical averaging, we find that these holes are realistic. The inhomogeneity or gradient corrections to LSD correctly produce a hole which is deeper close to the electron and shallower far away, and thus an exchange energy that is more negative than that of LSD.

The exchange-correlation energy is a coupling-constant integral (Eq. 60), and local or semi-local approximations typically work better at the fully-interacting or $\lambda = 1$ end, where the hole is deepest and most short-ranged, than at the $\lambda = 0$ (exchange) end. GGA exchange is more accurate than LSD exchange, but the pattern of error is different: While LSD is more accurate for the molecule than for the separated atoms, GGA is more accurate for the atoms. Moreover, we can distinguish between singly-bonded molecules, for which GGA exchange is quite accurate, and multiply-bonded molecules, for which GGA makes the exchange energy too negative. We interpret this to mean that GGA describes "self-exchange" accurately (at least for the σ bonds) but exaggerates the "interelectronic exchange" between parallel-spin electrons in neighboring bonds.

On the other hand, the GGA for the correlation energy misses some of the static or "left-right" correlation between two anti-parallel spin electrons in a bond, and so does not make the correlation energy of the molecule negative enough. The two GGA errors (in exchange and in correlation) tend to cancel, but the cancellation is not uniform over all molecules. In H_2, the exchange error is negligible while the correlation error is not, so GGA underestimates the atomization energy. In N_2, the exchange error is dominant, so GGA overestimates the atomization energy. In CH_4, the two errors very neatly cancel.

Because the exchange errors of GGA tend to be the most severe ones, quite accurate atomization energies can be found by mixing a fraction of the exact exchange energy with GGA. Moreover, this fraction can be estimated non-empirically.

Acknowledgments

This work was supported in part by the Deutsche Forschungsgemeinschaft, and in part by the National Science Foundation under Grant DMR95-21353.

Appendix: Technical details of the calculations

The Kohn-Sham calculations are performed with a modified version of the CADPAC program [41]. The electron densities are obtained from unrestricted Kohn-Sham calculations in the LSD approximation, and the various functionals have been evaluated on these densities. Nonspherical densities and Kohn-Sham potentials have been used for open-shell atoms [60]. The experimental geometries employed in our work are listed in Refs. [61]. The D_e values are obtained from the experimental atomization energies and the zero point energies given in Refs. [62]. The gaussian basis sets used are of triple zeta quality with p- and d- type polarization functions for the hydrogen and d- and f-type polarization functions for the second-row elements.

The results for the exchange holes are based on Hartree-Fock calculations performed with the COLUMBUS program system [58,59]. Gaussian basis sets close to the Hartree-Fock limit are used, and the resulting two-particle density matrix in the basis set representation is transformed to the real-space representaion of the exchange hole.

REFERENCES

1. R. G. Parr and W. Yang, *Density-Functional Theory of Atoms and Molecules* (Oxford University Press, Oxford, 1989).

2. R. M. Dreizler and E. K. U. Gross, *Density Functional Theory* (Springer Verlag, Berlin, 1990).
3. W. Kohn and L. J. Sham, Phys. Rev. **140**, A1133 (1965).
4. A. Szabo and N. S. Ostlund, *Modern Quantum Chemistry* (Macmillan Publishing, New York, 1982).
5. J. P. Perdew and Y. Wang, Phys. Rev. B **45**, 13244 (1992).
6. S.-K. Ma and K. A. Brueckner, Phys. Rev. **165**, 18 (1968).
7. L. J. Sham, in *Computational Methods in Band Theory*, edited by P. Marcus, J. F. Janak, and A. R. Williams (Plenum, New York, 1971).
8. O. Gunnarsson and B. I. Lundqvist, Phys. Rev. B **13**, 4274 (1976).
9. D. C. Langreth and J. P. Perdew, Phys. Rev. B **21**, 5469 (1980).
10. D. C. Langreth and M. J. Mehl, Phys. Rev. B **28**, 1809 (1983).
11. A. D. Becke, J. Chem. Phys. **84**, 4524 (1986).
12. A. D. Becke, Phys. Rev. A **38**, 3098 (1988).
13. C. Lee, W. Yang, and R. G. Parr, Phys. Rev. B **37**, 785 (1988).
14. A. D. Becke, J. Chem. Phys. **104**, 1040 (1996).
15. J. P. Perdew, Phys. Rev. B **33**, 8822 (1986).
16. J. P. Perdew and Y. Wang, Phys. Rev. B **33**, 8800 (1986).
17. J. P. Perdew, in *Electronic Structure of Solids 91*, edited by P. Ziesche and H. Eschrig (Akademie Verlag, Berlin, 1991).
18. J. P. Perdew, J. A. Chevary, S. H. Vosko, K. A. Jackson, M. R. Pederson, D.J. Singh, and C. Fiolhais, Phys. Rev. B **46**, 6671 (1992); **48**, 4978 (1993) (E).
19. J. P. Perdew, K. Burke, and Y. Wang, submitted to Physical Review B.
20. A. D. Becke, J. Chem. Phys. **97**, 9173 (1992).
21. B. G. Johnson, P. M. W. Gill, and J. A. Pople, J. Chem. Phys. **98**, 5612 (1993).
22. J. M. Seminario, Chem. Phys. Lett. **206**, 547 (1993).
23. A. D. Becke, J. Chem. Phys. **98**, 1372 (1993).
24. A. D. Becke, J. Chem. Phys. **98**, 5648 (1993).
25. A. Görling and M. Levy, unpublished.
26. M. Ernzerhof, K. Burke, and J. P. Perdew, submitted to the International Journal of Quantum Chemistry Symposium.
27. K. Burke, M. Ernzerhof, and J. P. Perdew, submitted to Chemical Physics Letters.
28. J. P. Perdew, K. Burke, and M. Ernzerhof, submitted to the Journal of Chemical Physics.
29. *Ab Initio Methods in Quantum Chemistry I and II*, *Advances in Physical Chemistry*, edited by K. Lawley (Wiley, New York, 1987).
30. G. L. Oliver and J. P. Perdew, Phys. Rev. A **20**, 397 (1979).
31. M. Levy and J. P. Perdew, Phys. Rev. A **32**, 2010 (1985).
32. M. Levy, Phys. Rev. A **43**, 4637 (1991).
33. J. D. Talman and W. F. Shadwick, Phys. Rev. A **14**, 36 (1976).
34. V. Sahni, J. Gruenebaum, and J. P. Perdew, Phys. Rev. B **26**, 4371 (1982).
35. M. K. Harbola and V. Sahni, Phys. Rev. Lett. **62**, 489 (1989).
36. J. B. Krieger, Y. Li, and G. J. Iafrate, Phys. Rev. A **45**, 101 (1992).
37. A. Görling and M. Ernzerhof, Phys. Rev. A **51**, 4501 (1995).
38. M. A. Buijse and E. J. Baerends, in *Density Functional Theory of Molecules, Clusters,*

and Solids, edited by D. E. Ellis (Kluwer Academic, Dordrecht, 1995).
39. M. Ernzerhof, J. P. Perdew, and K. Burke, in *Density Functional Theory*, edited by R. Nalewajski (Springer-Verlag, Berlin, 1996), to appear.
40. T. Ziegler, A. Rauk, and E. J. Baerends, Theoret. Chim. Acta **43**, 261 (1977).
41. CADPAC6: The Cambridge Analytical Derivatives Package Issue 6.0 Cambridge (1995) A suite for quantum chemistry programs developed by, R. D. Amos, with contributions from I.L. Alberts, J.S. Andrews, S.M. Cowell, N.C. Handy, D. Jayatilaka, P.J. Knowles, R. Kobayashi, N. Noga, K.E. Laidig, P.E. Maslan, C.W. Murray, J.E. Rice, J. Sanz, E.D. Simandrias, A.J. Stone, M.-D. Su, and D.J. Tozer.
42. K. Burke, M. Ernzerhof, and J. P. Perdew, in preparation.
43. Y. Wang, J. P. Perdew, J. A. Chevary, L. D. MacDonald, and S. H. Vosko, Phys. Rev. A **41**, 78 (1990).
44. L. Kleinman and S. Lee, Phys. Rev. B **37**, 4634 (1988).
45. M. Slamet and V. Sahni, Phys. Rev. B **44**, 10921 (1991).
46. J. P. Perdew, K. Burke, and M. Ernzerhof, submitted to Physical Review Letters.
47. J. P. Perdew and K. Burke, Int. J. Quantum Chem. Symposium **57**, 309 (1996).
48. K. Burke, J. P. Perdew, and D. C. Langreth, Phys. Rev. Lett. **73**, 1283 (1994).
49. K. Burke, J. P. Perdew, and M. Ernzerhof, in preparation.
50. D. C. Langreth and J. P. Perdew, Solid State Comm. **17**, 1425 (1975).
51. D. C. Langreth and J. P. Perdew, Phys. Rev. B **15**, 2884 (1977).
52. M. Levy, N.H. March, and N.C. Handy, J. Chem. Phys. **104**, 1989 (1996).
53. A. Görling and M. Levy, Phys. Rev. A **50**, 196 (1994).
54. J. P. Perdew and Y. Wang, Phys. Rev. B **46**, 12947 (1992).
55. O. Gunnarsson and R. O. Jones, Phys. Rev. B **31**, 7588 (1985).
56. O. Gritsenko, R. van Leeuwen, and E. J. Baerends, preprint.
57. M. Ernzerhof, J. P. Perdew, and K. Burke, in preparation.
58. R. Shepard, I. Shavitt, R. M. Pitzer, D. C. Comeau, M. Pepper, H. Lischkau, P. G. Szalay, R. Ahlrichs, F. B. Brown, and J.-G. Zhoa, Int. J. Quantum Chem. **142**, 22 (1988).
59. R. Shepard, H. Lischka, P. G. Szalay, T. Kovar, and M. Ernzerhof, J. Chem. Phys. **96**, 2085 (1992).
60. F.W. Kutzler and G.S. Painter, Phys. Rev. Lett. **59**, 1285 (1987).
61. D.J. DeFrees, B.A. Levi, S.K. Pollack, W.J. Hehre, J.S. Binkley, and J.A. Pople, J. Am. Chem. Soc. **101**, 4085 (1979). Geometry of *NO*: K.P. Huber and G. Herzberg, *Molecular Spectra and Molecular Structure IV: Constants of Diatomic Molecules*, Van Nostrand Reinhold, New York, 1979.
62. J.A. Pople, M. Head-Gordon, D.J. Fox, K. Raghavachari, L.A. Curtiss, J. Chem. Phys. **90**, 5622 (1989); L.A. Curtiss, C. Jones, G.W. Trucks, K. Raghavachari, and J.A. Pople, J. Chem. Phys. **93**, 2537 (1990).

J.M. Seminario (Editor)
Recent Developments and Applications of Modern Density Functional Theory
Theoretical and Computational Chemistry, Vol. 4

Nonlocal Weighted Density Approximation to exchange, correlation and kinetic energy in Density Functional Theory

J.A. Alonso and N.A. Cordero

Departamento de Física Teórica. Universidad de Valladolid.
E-47011 Valladolid, Spain

The local density approximation (LDA) to exchange and correlation has been very successful in the applications of density functional theory to atomic and condensed systems, but it has well-known limitations. The nonlocal weighted density approximation (WDA) corrects some of those defects. The WDA is reviewed here in a critical way. Applications to bulk solids, surfaces, clusters and atoms are presented and discussed. A key ingredient in the WDA is the pair-correlation function, which controls the shape of the Fermi-Coulomb hole. The form of this function for an inhomogeneous system is in general unknown, and the approximations so far developed lack generality, although they appear to be adequate for restricted classes of systems. A proper modeling of the Fermi-Coulomb hole for the system of interest is then crucial for the success of the WDA. Possible improvements and applications are also discussed.

1. INTRODUCTION

The Local Density Approximation (LDA) [1] to exchange and correlation has enjoyed an enormous success in the applications of the Density Functional Theory (DFT) [1,2] to atoms, molecules, clusters and condensed matter [3–6]. However, the LDA has well-known limitations, and several new approaches have been proposed as an improvement over the LDA. In spite of early doubts about the utility of the density gradient expansions, the so called Generalized Gradient Approximations (GGA), developed by Langreth [7], Perdew [8], Becke [9] and others, have demonstrated in many recent applications to atoms and molecules that a substantial improvement can be obtained compared to the LDA results. One of the problems that the gradient approximations still have is the imperfect cancellation of self-interaction effects between the Hartree and exchange-correlation terms. This is corrected in the Self-Interaction Corrected (SIC) method [10]. In this method the spurious self-interaction of one electron with itself is explicitly subtracted from both the classical Coulomb and exchange energies. A formal drawback of the SIC approach is that it leads to an orbital-dependent effective potential and then departs from the original spirit of the Kohn-Sham equations, in which the effective potential is state-independent.

Another approach where self-interaction is properly handled is the Weighted Density Approximation (WDA), proposed some time ago by Alonso and Girifalco [11] and by Gunnarsson *et al.* [12]. This is a fully nonlocal description of exchange and correlation, based on an explicit modeling of the Fermi-Coulomb hole around an electron in an inhomogeneous system. An accurate modeling of this hole is a basic requirement for the WDA. In this review we concentrate on the WDA, reviewing in a critical way the applications to bulk solids, surfaces, atomic clusters and atoms. The LDA already gives fairly good results for the ground state properties of all those systems, so the improvements that the WDA, or any other nonlocal description of exchange and correlation, should achieve are rather subtle. Only for some particular properties the nonlocal approximations could lead to qualitative changes with respect to the LDA. One of those examples concerns atomic or cluster anions, whose outermost electron is often unbound in the LDA. Some subtle effects in metallic clusters also appear to require the introduction of nonlocal effects.

The applications of DFT aiming at chemical accuracy use the Kohn-Sham version of the formalism [1]. But, in principle, if an accurate enough kinetic energy functional expressed in terms of the electron density existed, such a functional would lead to a competitive order-N method. The difficulty is, of course, finding such an accurate functional. The WDA allows to derive a kinetic energy functional that has been tested for atoms, producing good total energies. One of the characteristics of this functional is that it produces an incipient shell structure in atoms, so it could be useful for future applications.

2. EXCHANGE AND CORRELATION

2.1. Exchange-correlation energy

The crucial simplification of DFT is the relationship between the interacting system and a fictitious non-interacting system with the same electronic density $n(\mathbf{r})$. This relationship can be studied by considering the electron-electron interaction (Hartree atomic units will be used throughout this review unless explicitly stated otherwise)

$$\lambda V_{ee} = \frac{\lambda}{|\mathbf{r}_1 - \mathbf{r}_2|} \tag{1}$$

and varying the coupling λ from 0 (non interacting system) to 1 (physical system). The corresponding hamiltonian

$$H_\lambda = -\tfrac{1}{2}\nabla^2 + V_{ext} + V_\lambda + \lambda V_{ee} \tag{2}$$

contains a piece V_λ designed such that the ground state of H_λ has density $n(\mathbf{r})$ for all λ [13]. In our case V_{ext} is the nuclear potential.

It is now convenient to introduce the pair-correlation function $g_{xc}(\mathbf{r}_1, \mathbf{r}_2)$. In the system with interaction $\frac{\lambda}{|\mathbf{r}_1 - \mathbf{r}_2|}$, this is defined as

$$g_{xc}^{\lambda}(\mathbf{r}_1, \mathbf{r}_2) = \frac{2\Gamma_\lambda(\mathbf{r}_1, \mathbf{r}_2)}{n(\mathbf{r}_1)n(\mathbf{r}_2)} \tag{3}$$

where $\Gamma_\lambda(\mathbf{r}_1, \mathbf{r}_2)$ is the two-particle density matrix (we neglect spin polarization in this presentation, although the extension is immediate). The exchange-correlation energy can be written

$$E_{xc}[n] = \tfrac{1}{2} \iint \frac{n(\mathbf{r}_1)\left\{n(\mathbf{r}_2)\left[g_{xc}(\mathbf{r}_1, \mathbf{r}_2) - 1\right]\right\}}{|\mathbf{r}_1 - \mathbf{r}_2|} d^3r_1\, d^3r_2 \;, \tag{4}$$

where

$$g_{xc}(\mathbf{r}_1, \mathbf{r}_2) = \int_0^1 g_{xc}^{\lambda}(\mathbf{r}_1, \mathbf{r}_2)\, d\lambda \ . \tag{5}$$

The function $g_{xc}(\mathbf{r}_1, \mathbf{r}_2)$ is symmetric

$$g_{xc}(\mathbf{r}_1, \mathbf{r}_2) = g_{xc}(\mathbf{r}_2, \mathbf{r}_1) \ , \tag{6}$$

and it tends to 1 when $|\mathbf{r}_1 - \mathbf{r}_2| \to \infty$. The exchange-correlation energy of eq. (4) can be interpreted as the electrostatic interaction between two charge density distributions: the first one, $n(\mathbf{r}_1)$, is the electronic density of the system and the other,

$$n_{xc}(\mathbf{r}_1, \mathbf{r}_2) = n(\mathbf{r}_2)\left[g_{xc}(\mathbf{r}_1, \mathbf{r}_2) - 1\right] \ , \tag{7}$$

represents the exchange-correlation hole around an electron at $\mathbf{r}_1$ (Fermi hole if we neglect correlation). This hole, which accounts for the instantaneous decrease in the electron density $n(\mathbf{r}_2)$ due to the presence of a reference electron at $\mathbf{r}_1$, satisfies the sum-rule

$$\int n(\mathbf{r}_2)\left[g_{xc}(\mathbf{r}_1, \mathbf{r}_2) - 1\right] d^3 r_2 = -1 \ . \tag{8}$$

Evidently, the sign of the hole charge distribution is opposite to that of the electron charge density.

The Local Density Approximation (LDA) arises by setting $n(\mathbf{r}_2) = n(\mathbf{r}_1)$ in eq. (7) and approximating the pair-correlation function $g_{xc}(\mathbf{r}_1, \mathbf{r}_2)$ by that $g_{xc}^{\text{hom}}(|\mathbf{r}_1 - \mathbf{r}_2|; n(\mathbf{r}_1))$ for a homogeneous (hom) electron gas with "constant" density $n \equiv n(\mathbf{r}_1)$. Evidently, the LDA satisfies the sum rule (8), and this is the main reason for its practical success. We can say even more. If we define the spherical average of $n_{xc}(\mathbf{r}_1, \mathbf{r}_2)$ at a distance R from $\mathbf{r}_1$

$$n_{xc}^{\text{av}}(\mathbf{r}_1, R) = \int n_{xc}(\mathbf{r}_1, \mathbf{r}_1 + \mathbf{R}) \frac{d\Omega_R}{4\pi} \tag{9}$$

it is then easy to show that E_{xc} can be written as

$$E_{xc}[n] = \tfrac{1}{2} \int n(\mathbf{r}_1) \int \frac{1}{R} n_{xc}^{\text{av}}(\mathbf{r}_1, R)\, dR\, d^3 r_1 \ . \tag{10}$$

This means that the exact exchange-correlation energy depends only on the spherical average of the exchange-correlation hole. Calculations by Gunnarsson and coworkers [12] have shown that this spherical average is reproduced rather well by the LDA; this fact helps to explain the success of the LDA.

If we now restrict our attention to the exchange-only limit (Fermi hole), the form of $g_x^{\text{hom}}(|\mathbf{r}_1 - \mathbf{r}_2|; n)$ is known exactly

$$g_x^{\text{hom}}(r_{12}; n) = 1 - \frac{9}{2}\left[\frac{\sin(r_{12}k_F) - r_{12}k_F \cos(r_{12}k_F)}{(r_{12}k_F)^3}\right]^2 \tag{11}$$

where $k_F = (3\pi^2 n)^{1/3}$ is the Fermi momentum and $r_{12} = |\mathbf{r}_1 - \mathbf{r}_2|$. In this case one arrives at the well known result

$$E_x^{\text{LDA}}[n] = -\tfrac{3}{4}\left(\tfrac{3}{\pi}\right)^{1/3} \int n(\mathbf{r}_1)^{4/3}\, d^3 r_1 \ . \tag{12}$$

2.2. Exchange-correlation potential

Before introducing the WDA it is useful to derive the expression for the exchange-correlation potential in DFT. This is given by

$$V_{xc}(\mathbf{r}_1) = \frac{\delta E_{xc}[n]}{\delta n(\mathbf{r}_1)} , \tag{13}$$

that is, by the functional derivative of the exact exchange-correlation energy of eq. (4). The result can be expressed as a sum of three terms:

$$V_{xc}(\mathbf{r}_1) = V_1(\mathbf{r}_1) + V_2(\mathbf{r}_1) + V_3(\mathbf{r}_1) \tag{14}$$

where

$$V_1(\mathbf{r}_1) = \tfrac{1}{2}\int \frac{n(\mathbf{r}_1)}{|\mathbf{r}_1 - \mathbf{r}_2|}\,[g_{xc}(\mathbf{r}_1,\mathbf{r}_2) - 1]\; d^3r_2 \tag{15}$$

$$V_2(\mathbf{r}_1) = \tfrac{1}{2}\int \frac{n(\mathbf{r}_2)}{|\mathbf{r}_1 - \mathbf{r}_2|}\,[g_{xc}(\mathbf{r}_2,\mathbf{r}_1) - 1]\; d^3r_2 \tag{16}$$

$$V_3(\mathbf{r}_1) = \tfrac{1}{2}\iint \frac{n(\mathbf{r}_2)n(\mathbf{r}_3)}{|\mathbf{r}_2 - \mathbf{r}_3|}\,\frac{\delta g_{xc}(\mathbf{r}_2,\mathbf{r}_3)}{\delta n(\mathbf{r}_1)}\; d^3r_2\, d^3r_3 . \tag{17}$$

The symmetry of the pair-correlation function (see eq. (6)) leads to

$$V_{xc}(\mathbf{r}_1) = 2V_1(\mathbf{r}_1) + V_3(\mathbf{r}_1) . \tag{18}$$

Baerends and coworkers [14,15] have performed a detailed analysis of the two terms in eq. (18) for the case of neutral atoms. The first term, $2V_1(\mathbf{r}_1)$, is equal to the potential of the exchange-correlation hole and represents the screening of the full Hartree potential by exchange-correlation effects. For this reason it was called the screening potential, $V_{scr}(\mathbf{r}_1)$. Since the exchange-correlation hole density integrates to (minus) one electron (see eq. (8)), this screening potential has a long-range coulombic behavior

$$V_{scr}(\mathbf{r}_1) \sim -\frac{1}{r_1} \qquad (r_1 \rightarrow \infty) . \tag{19}$$

The screening potential varies smoothly with r_1 and reflects only weakly the electronic shell structure of atoms. The other term, $V_3(\mathbf{r}_1)$, which depends on the functional derivative of $g_{xc}(\mathbf{r}_1, \mathbf{r}_2)$, was called the screening-response potential, $V_{resp}(\mathbf{r}_1)$. This potential is short-ranged and displays a step structure: it is constant within regions corresponding to the different electronic shells and changes abruptly at the shell boundaries. The long range behavior of $V_{xc}(\mathbf{r}_1)$ is thus controlled by the screening part $V_{scr}(\mathbf{r}_1)$, that is, it is given by eq. (19). This asymptotic behavior is difficult to keep in practical implementations of the exchange-correlation functional. For instance, in the most usual form, the LDA, V_{xc} decays exponentially, that is, much faster than the exact form of eq. (19). This is because, if we restrict to the exchange-only case, we can see from eq. (12) that

$$V_x^{\mathrm{LDA}}(\mathbf{r}_1) = -\left(\tfrac{3}{\pi}\right)^{1/3} n(\mathbf{r}_1)^{1/3} , \tag{20}$$

and it is well known that the tail of the electron density decays exponentially.

3. THE WEIGHTED DENSITY APPROXIMATION (WDA)

3.1. Formalism of the Weighted Density Approximation

The idea of the Weighted Density Approximation (WDA) is to preserve $n(\mathbf{r}_2)$ in eq. (7) and to use for $g_{xc}(\mathbf{r}_1, \mathbf{r}_2)$ a representation appropriate to an "inhomogeneous" system [11,12]. This is the crucial point in the application of the WDA, and a representation of $g_{xc}(\mathbf{r}_1, \mathbf{r}_2)$ which is accurate enough for all different situations (atoms, molecules, surfaces, solids, valence electrons, core electrons, ...) is not yet available. Different workers have proposed functional forms for $g_{xc}(\mathbf{r}_1, \mathbf{r}_2)$ which are valid for particular classes of problems, so care must be taken in the applications of the WDA. The first proposal for the pair-correlation function [11,12] relied again on our knowledge of the homogeneous electron gas: $g_{xc}(\mathbf{r}_1, \mathbf{r}_2)$ was modeled by its form for a homogeneous electron gas with an effective "weighted" density $\tilde{n}(\mathbf{r}_1)$, that is

$$g_{xc}^{\mathrm{WDA}}(\mathbf{r}_1, \mathbf{r}_2) = g_{xc}^{\mathrm{hom}}(r_{12}, \tilde{n}(\mathbf{r}_1)) \ . \tag{21}$$

This weighted density has a nonlocal dependence on the actual density $n(\mathbf{r}_1)$ over a region around $\mathbf{r}_1$ with the dimensions of the exchange-correlation hole. The weighted density is obtained by requiring the exact normalization of the hole-charge

$$\int n(\mathbf{r}_2) \left[g_{xc}^{\mathrm{hom}}(r_{12}; \tilde{n}(\mathbf{r}_1)) - 1 \right] d^3 r_2 = -1 \ . \tag{22}$$

The weighted density $\tilde{n}(\mathbf{r}_1)$ is different from the local density $n(\mathbf{r}_1)$. Substituting g_{xc}^{WDA} in eq. (4) we then obtain the exchange-correlation energy

$$E_{xc}^{\mathrm{WDA}}[n] = \tfrac{1}{2} \iint \frac{n(\mathbf{r}_1) \left\{ n(\mathbf{r}_2) \left[g_{xc}^{\mathrm{hom}}(r_{12}; \tilde{n}(\mathbf{r}_1)) - 1 \right] \right\}}{|\mathbf{r}_1 - \mathbf{r}_2|} d^3 r_1 \, d^3 r_2 \ . \tag{23}$$

The WDA reproduces the exact Hartree-Fock exchange energy of the Helium atom (or any two-electron ion) in its ground state. To see this, consider the exchange-only version of the sum rule in eq. (22), and use for $g_x^{\mathrm{hom}}(r_{12}; \tilde{n}(\mathbf{r}_1))$ the functional form given in eq. (11). The condition imposed by the sum rule requires $\tilde{n}(\mathbf{r}_1) = 0$ for all values of $\mathbf{r}_1$, and this leads to $g_x^{\mathrm{WDA}}(\mathbf{r}_1, \mathbf{r}_2)) = \frac{1}{2}$. We then find from eq. (23) that the exchange energy cancels one half of the classical electron-electron energy of the He atom, which corresponds to the spurious self-interaction of each electron with itself. Similar arguments can be used to show that the WDA also gives the exact E_{xc} for the hydrogen atom and for one-electron ions (in this case E_{xc} cancels the spurious self-interaction of the electron).

The application of the WDA needs as input the form of $g_{xc}^{\mathrm{hom}}(r_{12}; n)$ as a function of the parameter n. A particularly useful analytical representation of $g_{xc}^{\mathrm{hom}}(r_{12}; n)$ has been given by Chacón and Tarazona [16]. In fact, these authors have noticed that the WDA provides a direct link between the pair-correlation function and the response function of the homogeneous electron liquid. They have used this relation as a closure for the fluctuation-dissipation equations to obtain in a closed form all the bulk properties of the homogeneous system [16,17]. One of the outcomes of their work is, as mentioned above, a result for the pair-correlation function which is in good agreement with the variational Fermi-hypernetted-chain (FHN) theory of Lantto [18]. Also, the correlation energy agrees well with the results of Singwi *et al.* [19], FHN theory [18] and Montecarlo simulations

[20]. The advantage of the pair-correlation function of Chacón and Tarazona is that this is given in analytical form. Another realistic parameterization has been recently presented by Perdew and Wang [21].

The exchange-correlation potential, which is an important component of the effective potential in the self-consistent Kohn-Sham equations, becomes

$$V_{xc}^{\mathrm{WDA}}(\mathbf{r}_1) \equiv \frac{\delta E_{xc}^{\mathrm{WDA}}[n]}{\delta n(\mathbf{r}_1)} = V_1(\mathbf{r}_1) + V_2(\mathbf{r}_1) + V_3(\mathbf{r}_1) \tag{24}$$

with

$$V_1(\mathbf{r}_1) = \tfrac{1}{2}\int \frac{n(\mathbf{r}_2)}{|\mathbf{r}_1 - \mathbf{r}_2|}\left[g_{xc}^{\mathrm{hom}}(r_{12};\tilde{n}(\mathbf{r}_1)) - 1\right] d^3r_2 \tag{25}$$

$$V_2(\mathbf{r}_1) = \tfrac{1}{2}\int \frac{n(\mathbf{r}_2)}{|\mathbf{r}_1 - \mathbf{r}_2|}\left[g_{xc}^{\mathrm{hom}}(r_{12};\tilde{n}(\mathbf{r}_2)) - 1\right] d^3r_2 \tag{26}$$

$$V_3(\mathbf{r}_1) = \tfrac{1}{2}\iint \frac{n(\mathbf{r}_2)n(\mathbf{r}_3)}{|\mathbf{r}_2 - \mathbf{r}_3|}\left[\frac{\delta g_{xc}^{\mathrm{hom}}(r_{23};n)}{\delta n}\right]_{n=\tilde{n}(\mathbf{r}_2)} \frac{\delta \tilde{n}(\mathbf{r}_2)}{\delta n(\mathbf{r}_1)}\, d^3r_2\, d^3r_3 \,. \tag{27}$$

The WDA improves the asymptotic behavior of $V_{xc}(\mathbf{r}_1)$ for a finite system. This behavior is controlled by $V_{scr}(\mathbf{r}_1) = V_1(\mathbf{r}_1) + V_2(\mathbf{r}_1)$. However, $g_{xc}^{\mathrm{WDA}}(\mathbf{r}_1, \mathbf{r}_2)) = g_{xc}^{\mathrm{hom}}(r_{12}; \tilde{n}(\mathbf{r}_1))$ is nonsymmetric with respect to the interchange of $\mathbf{r}_1$ and $\mathbf{r}_2$, and this leads to the inequality $V_1(\mathbf{r}_1) \neq V_1(\mathbf{r}_2)$. The asymptotic behavior becomes in this case dominated by V_1, as we can observe in Figure 1, that is

$$V_{scr}^{\mathrm{WDA}}(\mathbf{r}_1) \sim -\frac{1}{2r_1} \qquad (r_1 \to \infty) \,. \tag{28}$$

This represents an improvement over the LDA, which is reflected in properties depending on the behavior of the density tail in atoms, molecules and clusters. However, the exact $V_{scr}(\mathbf{r}_1)$ is still more attractive.

The functional form chosen for $g_{xc}^{\mathrm{WDA}}(\mathbf{r}_1, \mathbf{r}_2)$ in eq. (21) is not the only possible election. Other ones have also been tested, which are described later.

3.2. Spin-polarized WDA

If the spin-dependence of the exchange-correlation energy is considered in the LDA, one arrives at the LSDA (Local Spin-Density Approximation). In a similar way, the spin dependence can be introduced into the WDA in a straightforward way if we restrict ourselves to the exchange-only case. Then the exchange energy functional can be written

$$E_x^{\mathrm{WDA}}[n_\uparrow, n_\downarrow] = \tfrac{1}{2}\sum_\sigma \iint \frac{n_\sigma(\mathbf{r}_1)\left\{n_\sigma(\mathbf{r}_2)\left[g_{\sigma\sigma}^{\mathrm{hom}}(r_{12};\tilde{n}_\sigma(\mathbf{r}_1)) - 1\right]\right\}}{|\mathbf{r}_1 - \mathbf{r}_2|}\, d^3r_1\, d^3r_2 \,. \tag{29}$$

The summation is over the two spin directions $(\uparrow, \downarrow)$ and $g_{\sigma\sigma}^{\mathrm{hom}}(r_{12}; n_\sigma)$ is now the pair-correlation function between electrons with parallel spins in a homogeneous, fully polarized, electron gas with spin-density n_σ at the Hartree-Fock level, that is when Coulomb correlations are neglected:

$$g_{\sigma\sigma}^{\mathrm{hom}}(r_{12}; n_\sigma) = 1 - 9\left[\frac{\sin(r_{12}k_{F\sigma}) - r_{12}k_{F\sigma}\cos(r_{12}k_{F\sigma})}{(r_{12}k_{F\sigma})^3}\right]^2 , \tag{30}$$

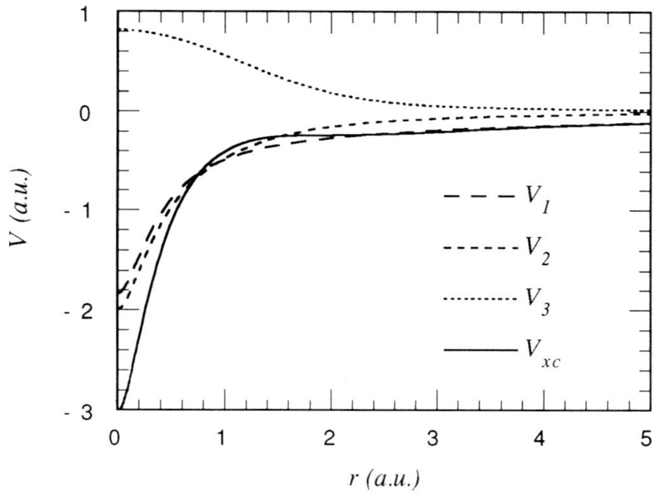

Figure 1. Components V_1, V_2 and V_3 of the exchange-correlation potential in the WDA approximation for the case of the Be atom. Their sum V_{xc} is also given by the solid line.

where $k_{F\sigma}$ is the Fermi momentum $k_{F\sigma} = (6\pi^2 n_\sigma)^{1/3}$. From eq. (29) we can observe that the exchange hole $n_\sigma(\mathbf{r}_2)\left[g_{\sigma\sigma}^{\rm hom}(r_{12}; \tilde{n}_\sigma(\mathbf{r}_1)) - 1\right]$ around an electron with spin σ at $\mathbf{r}_1$ is built from contributions by electrons with the same spin. In other words, electrons with antiparallel spins are not subjected to Pauli correlations. The weighted density $\tilde{n}_\sigma(\mathbf{r}_1)$ in eq. (29) is fixed again by the normalization condition of the exchange hole charge

$$\int n_\sigma(\mathbf{r}_2)\left[g_{\sigma\sigma}^{\rm hom}(r_{12}; \tilde{n}_\sigma(\mathbf{r}_1)) - 1\right] d^3 r_2 = -1 \; . \tag{31}$$

3.3. Shell partitioning in atoms

Early in the development of the WDA, Gunnarsson, Jonson and Lunqvist pointed out the distinction between intrashell and intershell exchange-correlation interactions for the case of atoms [12]. The replacement of the true pair-correlation function $g_{xc}(\mathbf{r}_1, \mathbf{r}_2)$ by $g_{xc}^{\rm WDA}(\mathbf{r}_1, \mathbf{r}_2)$ of eq. (21) involves two basic assumptions. The first one is that the only forces that keep electrons apart are due to exchange and correlation. The second one is that the effects of these forces are the same as in a homogeneous electron liquid with the appropriate "weighted" density. Other important mechanism that keeps electrons apart, namely external potentials that break the translational symmetry of the homogeneous liquid and localize the electrons in particular regions, are not well modeled by $g_{xc}^{hom}(r_{12}; \tilde{n}(\mathbf{r}_1))$. An example is the formation of electronic shells in atoms, which is due to the dominant role of the nuclear potential in the inner region of the atom. The electrons within a given shell are spatially and energetically well separated from those of other shells, and may be viewed as forming a separate electron liquid. The intrashell effects should therefore be reasonably well described by the WDA ansatz for $g_{xc}(\mathbf{r}_1, \mathbf{r}_2)$. However the intershell effects involve two different subsystems and the WDA weighting of the density may exaggerate intershell correlations. Gunnarsson and coworkers proposed that, rather than using the total density, the weighting of the density should be performed only within each shell, and that the LDA should be used for intershell effects.

The corresponding energy functional is

$$E_{xc}[n] = \frac{1}{2}\sum_i \iint n_i(\mathbf{r}_1)\frac{1}{r_{12}}\Big\{ n_i(\mathbf{r}_2)\left[g_{xc}^{\text{hom}}(r_{12};\tilde{n}_i(\mathbf{r}_1)) - 1\right] + \sum_{j\neq i} n_j(\mathbf{r}_1)\left[g_{xc}^{\text{hom}}(r_{12};n(\mathbf{r}_1)) - 1\right]\Big\}\, d^3r_1\, d^3r_2\ , \tag{32}$$

where the sums are over shells: K(1s), L(2s,2p), M(3s,3p),.... The weighted density $\tilde{n}_i(\mathbf{r}_1)$ should now be chosen so that the charge-conservation sum rule is satisfied

$$\int \Big\{ n_i(\mathbf{r}_2)\left[g_{xc}^{\text{hom}}(r_{12};\tilde{n}_i(\mathbf{r}_1)) - 1\right] + \sum_{j\neq i} n_j(\mathbf{r}_1)\left[g_{xc}^{\text{hom}}(r_{12};n(\mathbf{r}_1)) - 1\right]\Big\}\, d^3r_2 = -1\ . \tag{33}$$

This form of $E_{xc}[n]$ cuts off most of the unphysical intershell contributions. As a first illustration of the WDA, we present in Table 1 results for the exchange energy of a few spherically symmetric atoms calculated using the WDA functional for exchange-only (that is neglecting Coulomb correlation) with and without shell partitioning. The electron densities were, in all cases, constructed from the Hartree-Fock wave functions of Clementi [22]. The results are compared with those of the LDA, and the spin-polarized formalism was used for both LDA and WDA. There is a striking improvement over the LDA for light atoms. For heavy atoms, the WDA improves little with respect to the LDA. This is ascribed to an overestimate of intershell contributions. Applying a shell partitioning scheme to the WDA (this is the column labeled WDA-shell) leads to a substantial improvement. Although the densities used for this test are not the self-consistent ones corresponding to V_x^{WDA}, the use of consistent densities does preserve the conclusions.

Table 1
Exchange energy, in Rydbergs, of some spherically symmetric atoms using the LDA, WDA and WDA with shell partitioning, compared to the Hartree-Fock result. Numbers in parentheses are errors in percent. All the calculations used spin-polarized functionals [12].

	LDA	WDA	WDA-shell	Exact
He	-1.768 (13.8)	-2.052 (0.0)	-2.052 (0.0)	-2.052
Li	-3.08 (13.5)	-3.57 (0.3)	-3.54 (0.6)	-3.56
Be	-4.63 (13.1)	-5.38 (0.9)	-5.31 (0.4)	-5.33
N	-11.79 (10.6)	-13.68 (3.7)	-13.37 (1.4)	-13.19
Ne	-22.07 (8.9)	-25.59 (5.7)	-24.87 (2.7)	-24.22
Mg	-29.22 (8.7)	-33.98 (6.2)	-32.72 (2.3)	-31.99

3.4. The model of Gunnarsson and Jones for the pair-correlation function

In the WDA functional of eq. (23), the pair-correlation function $g_{xc}^{\text{hom}}(r_{12}; n)$ for a homogeneous system is used, though it is not guaranteed that this is a good approximation for the corresponding quantity in an inhomogeneous system. For this reason, Gunnarsson and Jones [23] have proposed an alternative form of for $g_{xc}^{\text{WDA}}(\mathbf{r}_1, \mathbf{r}_2)$ which attempts to describe not only the limit of a homogeneous electron gas but also the metallic surface. The proposed form is:

$$g_{xc}^{\text{GJ}}(r_{12}; n) = 1 - C(n) \exp\left[-\left(\frac{\lambda(n)}{r_{12}}\right)^5\right] , \tag{34}$$

where $C(n)$ and $\lambda(n)$ are parameters (which depend on the density n of the homogeneous gas) to be determined. g_{xc}^{GJ} behaves as $-r_{12}^{-5}$ for large distances r_{12}, which is needed to obtain the correct image potential form for $V_{xc}(r_1)$ outside a surface. This point will be discussed in more detail in Section 5. On the other hand, the dependence of λ and C on n is fixed by requiring the fulfillment of the two following conditions "for the homogeneous gas" of density n: the first one is the sum rule for the hole

$$n \int \left[g_{xc}^{\text{GJ}}(r_{12}; n) - 1\right] d^3r_{12} = -1 , \tag{35}$$

and the second one is the requirement of a correct exchange-correlation energy

$$\tfrac{1}{2} \int \frac{g_{xc}^{\text{GJ}}(r_{12}; n) - 1}{r_{12}} d^3r_{12} = \epsilon_{xc}^{\text{hom}}(n) . \tag{36}$$

Here $\epsilon_{xc}^{\text{hom}}(n)$ represents the exchange-correlation "energy per particle" obtained from the best theories of the interacting homogeneous electron gas [20]. Of course, when this model, g_{xc}^{GJ}, is used in the context of the WDA, the exchange-correlation hole density becomes $n_{xc}^{\text{WDA}}(\mathbf{r}_1, \mathbf{r}_2) = n(\mathbf{r}_2)\left[g_{xc}^{\text{GJ}}(r_{12}; \tilde{n}(\mathbf{r}_1)) - 1\right]$ and the weighted density $\tilde{n}(\mathbf{r}_1)$ is fixed at each point $\mathbf{r}_1$ in the usual way, that is, by the normalization sum rule

$$\int n(\mathbf{r}_2)\left[g_{xc}^{\text{GJ}}(r_{12}; \tilde{n}(\mathbf{r}_1)) - 1\right] d^3r_2 = -1 . \tag{37}$$

An indication that g_{xc}^{GJ} can be useful for inhomogeneous systems is provided in Figure 2, which compares the errors of the total energy (measured as a fraction of E_{xc}) for some selected atoms, obtained by the (spin dependent) LDA and the WDA. The LSDA errors are typically 5–8%, while the WDA (using g_{xc}^{GJ}) reduces the errors to values smaller than 1%. However, other properties of free atoms, like the first ionization potential, or energy differences between configurations $3d^{n-1}4s^1$ and $3d^{n-2}4s^2$ in elements of the 3d series, do not show improvement with respect to the LDA.

4. BULK SOLIDS

Several applications of the WDA have been reported for bulk metals and semiconductors [24–34]. Borstel and coworkers [24] have performed band structure calculations for Rhodium. The unoccupied bands are shifted to higher energies with respect to the Fermi energy when compared to band calculations based on the LDA. This shift improves

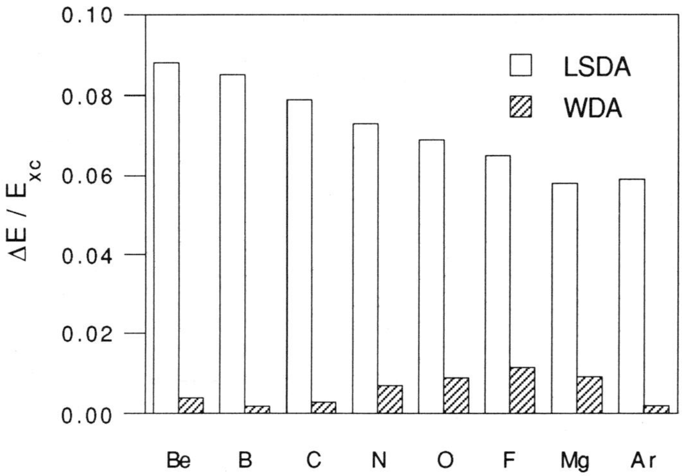

Figure 2. Error in the total energy of a free atom as a fraction of the exchange-correlation energy in the local spin-density and weighted density approximations [23].

the agreement with the results of angle resolved photoemission experiments. For copper, Przybylski and Borstel obtained an improved description of the *d* band at the Fermi surface, although for vanadium the LDA errors were not corrected [28,29]. Wang [27] performed spin-polarized band calculations for Ni. The exchange splitting (0.4 eV) was much closer to the experimental measurement (0.3 eV) than the LDA prediction (0.6 eV). Manghi and coworkers [26,32] have studied the band structure of Si, Ge and GaAs. They found no significant changes in the valence band states. Although some small increase was reported in the band gaps, the gaps remained much smaller than the measured ones. Hybertsen and Louie [30,31] found a small increase in the lattice parameter of Si and Ge, which constitutes an improvement over the LDA.

Singh [34] has recently pointed out that, with the exception of the work of Hybertsen and Louie [30,31], which reports ground-state properties of Si and Ge, the other studies focussed on band structures and related properties. However, this is not the best testing ground, since the (unknown) exact DFT functional may give an equally poor description of excited states. For this reason, Singh has performed calculations for the ground state properties (lattice constant and bulk modulus) of a variety of materials including transition metals. Singh's work is, in our view, the most consistent test of the WDA for bulk solids. Singh performed calculations using two different models for $g_{xc}^{\mathrm{WDA}}(\mathbf{r}_1, \mathbf{r}_2)$. One of them is the usual one of eq. (21), based on the realistic parameterization of the pair-correlation function in a homogeneous electron gas given by Perdew and Wang (PW) [21]. The second model for g_{xc}^{WDA} in the ansatz proposed by Gunnarsson and Jones [23,35], that we have described in Section 3.4. One of the key features in Singh's work is the use of a core-valence separation. Based on the physical ideas leading to the shell partitioning of exchange-correlation interactions, Singh separated the valence and core components of the electron density and treated the exchange-correlation interactions between valence and core electrons by the LDA. Since the core electrons are localized and do not contribute to the bonding, exchange and correlation between the core electrons was also treated by the LDA. In summary, the exchange-correlation functional was approximated in the following way

$$E_{xc}[n] = E_{xc}^{\mathrm{LDA}}[n] + \left\{E_{xc}^{\mathrm{WDA}}[n_v] - E_{xc}^{\mathrm{LDA}}[n_v]\right\} \tag{38}$$

where $n_v(\mathbf{r})$ is the valence electron density and $n(\mathbf{r}) = n_c(\mathbf{r}) + n_v(\mathbf{r})$ is the total density (sum of core and valence contributions). In practice, matters were made even simpler. In the case of C, Si and GaAs, pseudopotentials were used, so n_c is not explicitly present. For the transition metals V and Mo, upper core states (3s and 3p for V, 4s and 4p for Mo) were explicitly included and their effect was treated in the LDA according to eq. (38). The calculations were performed with a plane-wave electronic structure code [36] and state-of-the-art pseudopotentials were employed [37]. The results for the lattice parameters and bulk moduli of diamond, Si, GaAs, Mo and V are reproduced in Table 2. The first four of these materials are already well described by the LDA. Consequently the purpose of the calculations was to determine if the WDA gives sufficiently small corrections so that it does not degrade the good LDA results. For C and Si this is the case for both WDA calculations. Even more, for these two materials the WDA results are significantly closer to experiment than the LDA ones. GaAs is also well described in the LDA, and it was reported that some degradation of its ground-state properties occurs when GGA functionals are used [38]. The WDA with the PW form of g_{xc}^{hom} yields results with small errors comparable to the LDA, but in the opposite direction, while the GJ parameterization shows a degradation in the results relative to the LDA. Singh's opinion, that we fully subscribe, is that this is an indication that the electron gas parameterization may be preferable for solids. Transition metals form an interesting class of materials because of the coexistence of d and sp states in the valence region. Mo is very well described in the LDA, but V is poorly described. For Mo, the WDA yields small corrections to the LDA that improve the agreement with experiment. Finally, for V, the WDA gives large corrections which significantly improve the agreement with experiment. As a global trend from Table 2, the WDA increases the lattice parameter compared to the LDA. This correction is in the right direction, because the underestimation of interatomic distances is a well known error of the LDA [35]. This improvement in lattice constants is, in part, responsible for a similar improvement of the bulk moduli, which is overestimated in the LDA for all the materials in Table 2, and decreases towards values much closer to experiment in the WDA calculations.

The work of Singh provides a nice example of a "proper" application of the WDA. We stress that the most crucial point is the model taken for $g^{\mathrm{WDA}}(\mathbf{r}_1, \mathbf{r}_2)$ in inhomogeneous systems. The valence electron cloud in an infinite solid is one of the few systems where one can be fairly confident of having a reasonable approximation for the exchange-correlation hole around an electron placed at $\mathbf{r}_1$. This is $n_{xc}^{\mathrm{WDA}}(\mathbf{r}_1, \mathbf{r}_2) = n(\mathbf{r}_2)\left[g_{xc}^{hom}(r_{12}; \tilde{n}(\mathbf{r}_1)) - 1\right]$. For this reason, we encourage the application of the WDA to study of the ground state properties of bulk solids. Other systems, namely those in which the electron density decays to zero, like in surfaces, provide a much harder test to the WDA, because less is known about the form of $g_{xc}(\mathbf{r}_1, \mathbf{r}_2)$ at or near a surface. However, some progress has also

Table 2
Lattice parameter (a) in a.u. and bulk modulus (B) in GPa of C, Si, GaAs, Mo and V using the LDA and the WDA, compared to experimental values. WDA-GJ and WDA-PW denote the WDA with the ansatz for g^{WDA} given by Gunnarsson-Jones [23] and by the electron gas parameterization of Perdew and Wang [21] respectively.

	LDA		WDA-GJ		WDA-PW		Expt.	
	a	B	a	B	a	B	a	B
C	1.87	483	1.89	404	1.88	423	1.89	442
Si	2.84	111	2.87	99	2.86	105	2.87	99
GaAs	2.97	85	3.04	62	3.01	68	2.99	75
Mo	1.65	296	1.66	253	1.66	262	1.67	265
V	1.55	215	1.59	145	1.58	149	1.60	157

been made for systems with a surface, which is reviewed in Section 5.

The GGA has also been applied recently to the ground state properties of bulk solids: metals and semiconductors [38,39]. The conclusion of these works is that the GGA results for cohesive properties (lattice constant, bulk modulus and binding energy) are not consistently better and tend to overcorrect the LDA results. A direct comparison between the GGA and the WDA for solids is, nevertheless difficult, because this would require to compare calculations performed with the same computational scheme, since we are looking in general at rather small changes with respect to the LDA.

Table 2 also serves to illustrate that, for many ground state properties, the LDA gives in general good results, so the requirements imposed to descriptions of exchange and correlation beyond the LDA are rather tough: those improved descriptions should agree with the LDA in cases when the performance of the LDA is good, and should correct the LDA errors when these are not so small.

5. SURFACE PROPERTIES

The LDA has been used to evaluate many properties of surfaces with remarkable success [40]. However, some problems of interest like tunneling [41] and the formation of surface image states [42] are beyond the capability of the LDA. This fact has motivated the search for a correct description of the image potential outside the metal surface.

An electron far outside a metal surface is attracted towards the metal by the classical image potential [43]

$$V(z) = -\frac{1}{4(z - z_{\rm im})} , \qquad (39)$$

where the z-axis is normal to the surface, the bulk metal is at $z < 0$, and $z_{\rm im}$ is the position of the image plane. $z_{\rm im}$ can be considered as the actual termination of the crystal. It also represents the reference plane position for the van der Waals force between an atom and the surface, and its knowledge is crucial in order to obtain a reliable estimate of this interaction [44]. In principle, the image potential should be recovered by the DFT as the limit of $V_{xc}(\mathbf{r})$ when $z \to \infty$. However, this is a tough test for any approximate model for $E_{xc}[n]$ because it requires a very accurate treatment of the electron-electron correlations in a strongly inhomogeneous system. The LDA fails to reproduce the image potential behavior. In the LDA, the exchange-correlation hole is attached to the moving electron, and follows it as the electron departs from the metal. Consequently, $V_{xc}^{\rm LDA}(z)$ decays exponentially far outside the surface. Instead, the exchange-correlation hole should remain attached to the surface when the electron escapes from the metal. This is correctly predicted by the WDA. The WDA hole contains the prefactor $n(\mathbf{r}_2)$ multiplying the pair-correlation function (see, for instance, eq. (23)). Furthermore, the pair-correlation function itself contains the weighted density $\tilde{n}(\mathbf{r}_1)$. This indicates that the hole is digged on the real density, and that it remains attached to regions with a nonneglegible electron density, in this case to the surface of the system.

Gunnarsson and coworkers [12] had noticed that the models for $g_{xc}^{\rm WDA}(\mathbf{r}_1, \mathbf{r}_2)$ that decay faster than any inverse power of r_{12} for large r_{12} give the correct power law for the image potential, but with a coefficient $\frac{1}{2}$ instead of the correct $\frac{1}{4}$ of eq. (39). However, the value of this coefficient changes if the model for $g_{xc}^{\rm WDA}(\mathbf{r}_1, \mathbf{r}_2)$ has a power law decay for large r_{12}. In this case it may be shown [17] that for large z:

$$V_{xc}^{\rm WDA}(z) = -\frac{\nu - 3}{2(\nu - 1)z} , \qquad (40)$$

where ν is the decay power of $g_{xc}(r_{12})$, and the correct coefficient is obtained for $\nu = 5$. On this basis, Gunnarsson and Jones [23] introduced the ansatz given in eq. (34) above, for which $\nu = 5$.

The WDA with $g_{xc}^{\rm GJ}(r_{12}, n)$ has been used to evaluate the properties of metal surfaces in the jellium model [45–48], including the position of the image plane, and Manghi [49] has studied a realistic GaAs (100) surface using a pseudopotential scheme. The use of $g_{xc}^{\rm GJ}(r_{12}, n)$ still needs as an important input the exchange-correlation "energy" of the homogeneous electron gas (see eq. (36)) in order to fix $C(n)$ and $\lambda(n)$. Manghi has discovered that the introduction of nonlocal effects has a small influence on many surface properties (compared to the LDA) when the same form of the exchange-correlation energy functional for the homogeneous gas is used in the construction of the local and nonlocal V_{xc} potentials. This is because in this case the local and nonlocal constructions give rise to a similar surface barrier. The surface barrier is the "total" potential difference between bulk and vacuum. In the jellium model of a metallic surface, the average potential deep inside the bulk is the same for the LDA and the WDA, so the surface barrier potential is also nearly the same. When the metal is treated in a more realistic way (that is, beyond jellium) the average LDA and WDA potentials deep inside the bulk will be different, and the surface barriers will also differ. What Manghi has shown is that this difference is small, or more precisely, the differences in many surface properties are expected to be of the same order as the small differences found in bulk properties (of course, $V_{xc}^{\rm LDA}(z)$ and $V_{xc}^{\rm WDA}(z)$ are not similar in the outermost region, where $V_{xc}^{\rm LDA}$ decays much faster; properties depending on the asymptotic behavior of V_{xc} are sensitive to nonlocal effects).

As a consequence, the occupied energy levels and the surface electron density change little. The surface electron density determines the corrugation in the diffractive scattering of low energy He atoms by the surface. In the effective medium theory [50] the repulsive potential acting upon a He atom is proportional to the charge density

$$V^{\rm He}(\mathbf{r}) = c\, n(\mathbf{r}) , \qquad (41)$$

and values of the empirical constant c between 200 and 750 have been used in the literature. The corrugation function is defined as the vertical position z for a given (x, y) at which the He atom with energy E^{He} is scattered back. z is calculated from the equation

$$V^{\mathrm{He}}(x, y, z) - E^{\mathrm{He}} = 0 \ . \tag{42}$$

Manghi expressed the corrugation $z(x, y)$ of the surface in terms of the first Fourier components

$$z(x, y) = D_{10} \exp \left[\frac{2\pi i x}{L_x}\right] + D_{01} \left[\frac{2\pi i y}{L_y}\right] , \tag{43}$$

where L_x and L_y are the surface-cell dimensions and D_{10}, D_{01} are the corrugation parameters. The corrugation parameters obtained by Manghi for $c = 750$ are presented in Table 3. LDA and WDA corrugations using the same expression for the exchange-correlation energy of the homogeneous gas are practically the same (in particular, Manghi used the Ceperley-Alder data [20]). Nevertheless, the same Table shows the results obtained by an LDA calculation in which a Slater form of V_{xc} was used:

$$V_{xc}^{\mathrm{Slater}}(\mathbf{r}) = -\alpha \frac{3}{2} \left(\frac{3}{\pi}\right)^{1/3} [n(\mathbf{r})]^{1/3} , \tag{44}$$

with $\alpha = 1$. The conclusion is that the difference between the two LDA calculations is much larger than the difference between LDA and WDA, and that $\epsilon_{xc}^{\mathrm{hom}}(n)$ used either in the LDA or the WDA theories is the crucial input. Consistent with this discussion, Manghi also obtained small differences in the energy positions of surface states between LDA and WDA calculations.

Table 3
Corrugation parameters D_{10} and D_{01} for the (100) surface of GaAs [49].

	LDA Ceperley-Alder		WDA Ceperley-Alder		LDA Slater	
$E^{\mathrm{He}} = 0.045$ eV	0.413	0.158	0.416	0.154	0.481	0.166
$E^{\mathrm{He}} = 0.063$ eV	0.439	0.161	0.439	0.159	0.505	0.177

Chacón and Tarazona [17] have arrived at similar conclusions concerning the surface energy σ. These authors have performed a variational calculation of σ for a finite barrier model of the surface. The surface energies obtained by the LDA and the WDA using the same input for $\epsilon_{xc}^{\mathrm{hom}}(n)$ are very similar (see Table II of that article). They have also analyzed the influence of the form of $g_{xc}^{\mathrm{WDA}}(r_{12})$ on the surface energy. Using an infinite barrier model for the surface, they conclude that a form for $g_{xc}^{\mathrm{WDA}}(r_{12})$ taken from an accurate theory for the homogeneous electron gas, that is $g_{xc}^{\mathrm{hom}}(r_{12}; \tilde{n}(\mathbf{r}_1))$, leads to better surface energies than g_{xc}^{GJ}.

The motivation for introducing g_{xc}^{GJ} of eq. (34) was to obtain the correct coefficient $\frac{1}{4}$ in the asymptotic potential outside a metallic surface. However, Chacón and Tarazona have argued that this is obtained in an artificial way that leads to unphysical results for the position $z_{\rm im}$ of the image plane. When g_{xc} has a power law decay for large r_{12}, the center of gravity of the hole does not remain in the surface and is driven towards the bulk by the slow decay of $g_{xc}(r_{12})$, so that the surface structure becomes irrelevant. In other words, the WDA would give the same image plane position for any electronic profile (for the same bulk). In summary, although the WDA is qualitatively better than the LDA in the description of the potential outside a metal surface, it is still not able to give full account of the image potential. Chacón and Tarazona ascribe this fact to the use of a spherical $g_{xc}(\mathbf{r}_1, \mathbf{r}_2)$, that is, depending on r_{12}, while the actual g_{xc} for an electron outside the metal should flatten in the surface. They propose that g_{xc} should be taken to decay exponentially and to accept the behavior $V_{xc}^{\rm WDA} \to -\frac{1}{2z}$ as an intrinsic limitation. Nevertheless, they expect that the WDA potential may still be realistic in the intermediate region between 0 and 15 a.u., which is the most relevant for electron-surface and atom-surface interactions.

A lot of work remains to be done in order to explore the proper form of $g_{xc}(\mathbf{r}_1, \mathbf{r}_2)$ in a surface. Also, a systematic calculation of the ground state properties of a surface, comparing LDA and WDA, similar to what Singh [34] did for bulk solids is still to be done. One of the problems is, no doubt, that the ground state properties of a surface are not experimentally known with the same level of precision as the corresponding bulk properties.

6. AN ALTERNATIVE TO IMPROVE THE ASYMPTOTIC BEHAVIOR OF V_{xc}

The exact pair correlation function is symmetric, that is $g_{xc}(\mathbf{r}_1, \mathbf{r}_2) = g_{xc}(\mathbf{r}_2, \mathbf{r}_1)$. This is a very strong requirement for approximate models. In fact, the models used in the WDA do not fulfill this property of symmetry. The consequence is that $V_1(\mathbf{r}_1)$ and $V_1(\mathbf{r}_2)$ given by eqs. (25) and (26) respectively are not equal, and this results in the wrong constant, $\frac{1}{2}$ instead of 1, in the asymptotic behavior of $V_{xc}(\mathbf{r})$ in the tail of an atom. An *ad hoc* way of forcing the correct constant is by applying the WDA directly to the potential V_{xc} of eq. (14), that is

$$V_{xc}^{\rm WDAS}(\mathbf{r}) = 2V_1^{\rm WDA}(\mathbf{r}) + V_3^{\rm WDA}(\mathbf{r}) \ , \tag{45}$$

where $V_1^{\rm WDA}$ and $V_3^{\rm WDA}$ are the potentials given by eqs. (25) and (27) respectively. The symbol WDAS stands for "Symmetrized" WDA. This prescription was introduced by Przybylski and Borstel [28,29], and used in band structure calculations for Cu and V. More recently, the symmetrization has been used in different applications to atomic clusters, which will be reviewed in Section 7. The improvement of the asymptotic decay of the potential becomes reflected in an improvement in the description of those properties that are sensitive to the tail of the electron density. However, the prescription of eq. (45) has a drawback: $V_{xc}^{\rm WDAS}(\mathbf{r})$ is not the functional derivative of $E_{xc}^{\rm WDA}$ of eq. (23). This point will be further discussed later in this review.

7. GROUND STATE PROPERTIES OF METALLIC CLUSTERS WITHIN THE SPHERICAL JELLIUM MODEL

The WDA has been used in the study of the electronic properties of metallic clusters, although only within the context of simple structural models. The success of the jellium model in the description of the electronic properties of simple metal clusters is well established [51]. The spherical jellium model explains the magic numbers observed in the abundance mass spectra (or in the ionization potential) as due to the closing of electronic shells, which confers the cluster a special stability. In the spherical jellium model, the density of the positive ionic background is assumed to be of the form

$$n_+(\mathbf{r}) = \begin{cases} \overline{n_+}\,, & r \leq R \\ 0\,, & r > R \end{cases} \tag{46}$$

that is, constant inside an spherical volume of radius R and zero outside. The details of the abundance population spectra (or the ionization potential) in size regions corresponding to open electronic shells are also explained by allowing for nonspherical deformations of the jellium background [52,53]. In the present context, the interest of finite clusters arises from the fact that the electron density changes rapidly at the surface. Clusters, like the surface of a semi-infinite metal, provide a stringent test for theories of exchange and correlation in systems with a highly inhomogeneous density. Besides, because of their finite size, clusters are convenient systems to test the influence of the asymptotic behavior or $V_{xc}(\mathbf{r})$ on some properties. On the other hand, a difference between finite clusters and the surface of a semi-infinite metal is that a truly bulk behavior in the interior of the cluster is only achieved for large cluster sizes. Spherical jellium calculations by Ekardt show that density oscillations induced by the surface step develop in the inner region of sodium clusters, Na_N, for sizes as large as $N = 200$ [54]. One can also view the density oscillations as a consequence of the discreteness of the electronic levels in a potential well of finite radius.

Table 4 gives the one-electron energy eigenvalues of the occupied orbitals of Na_{20}, obtained by solving the Kohn-Sham equations in the spherical jellium model [55]. In fact, all the calculations reported in Sections 7, 8 and 9 use the spherical jellium model. Strictly speaking, DFT eigenvalues other than that for the highest occupied orbital (HOMO) have no precise physical meaning, although it is customary to use the calculated band structure of solids in the discussion of their optical properties. Instead, the concept of quasi-particles provides the adequate framework to discuss excitations [56]. The quasi-particle energies calculated by Saito *et al.* [57] using the so called GW approximation [56] (and the jellium model) are also given in Table 4. The same authors have also calculated LDA eigenvalues. We can observe that the WDAS eigenvalues are much closer to the quasi-particle energies than the LDA ones (the pair-correlation function used in the WDA calculations of this Section is $g_{xc}^{\mathrm{hom}}(r_{12}, n)$ of Chacón and Tarazona [16]). This is particularly interesting for the HOMO level. In this case, the WDAS eigenvalue and the quasi-particle energy closely bracket the experimental ionization potential, 3.76 eV.

A result valid in DFT is that

$$-\varepsilon_{\mathrm{HOMO}} = \mathrm{IP}\ , \tag{47}$$

Table 4
DFT one-electron energy eigenvalues (in eV) of Na_{20} compared to quasi-particle energies. All the calculations use the spherical jellium model. The experimental ionization potential is 3.76 eV.

	LDA[57]	WDAS[55]	Quasi-particle[57]
ε_{1s}	-5.1	-5.84	-5.8 ± 0.4
ε_{1p}	-4.4	-5.14	-5.2 ± 0.4
ε_{1d}	-3.5	-4.23	-4.4 ± 0.1
ε_{2s}	-2.9	-3.72	-3.8 ± 0.03

where IP is the first ionization potential calculated subtracting the total energies of the cation and the neutral atom (or cluster). If one employs an approximate version of DFT, as usual, eq. (47) is not fulfilled exactly, but is still useful. On the other hand, the closeness between the two quantities can give a hint of the goodness of the approximation used for exchange and correlation. Table 5 gives both $-\varepsilon_{\mathrm{HOMO}}$ and IP for Na_N and K_N with $N = 8, 18, 20$ [55]. In addition, we have included quasi-particle energies [57] and the measured ionization potentials [51,58]. The nonlocal HOMO eigenvalues are closer to quasi-particle energies, and also to the measured IPs, than the LDA eigenvalues, confirming the conclusion derived earlier for Na_{20}. Before discussing the calculated IPs, we have to clarify how total energies have been computed within the WDAS. In this case $E_{xc}[n]$ was calculated from the usual WDA energy functional of eq. (23), inserting the density obtained by solving the Kohn-Sham equations for the V_{xc}^{WDAS} potential, that is,

$$E_{xc}[n] = E_{xc}^{\mathrm{WDA}}[n^{\mathrm{WDAS}}] \,. \tag{48}$$

In spite of the lack of selfconsistency between E_{xc}^{WDA} and V_{xc}^{WDAS}, the errors in the calculated IPs with respect to the experimental values decrease by about 0.2 eV compared to the LDA. This suggests that the *ad hoc* combination of V_{xc}^{WDAS} and E_{xc}^{WDA} can be useful.

Table 5
Highest occupied orbital eigenvalues ($\varepsilon_{\mathrm{HOMO}}$) and ionization potential (IP) of Na and K clusters [55] compared to quasi-particle energies and to experimental ionization potentials. All quantities are in eV.

	LDA		WDAS		Quasipart.[57]	Exp.[51,58]
	$-\varepsilon_{\mathrm{HOMO}}$	IP	$-\varepsilon_{\mathrm{HOMO}}$	IP		
Na_8	3.37	4.88	4.04	4.69	4.59 ± 0.05	4.22
Na_{18}	3.24	4.41	4.02	4.29	4.20 ± 0.08	3.83
Na_{20}	2.87	4.03	3.72	3.97	3.84 ± 0.03	3.76
K_8	2.96	4.21	3.50	4.01	3.95 ± 0.05	3.35
K_{18}	2.85	3.83	3.47	3.68	3.63 ± 0.09	3.09
K_{20}	2.52	3.50	3.24	3.32	3.29 ± 0.05	3.02

Figure 3 shows that the shift of the occupied WDAS eigenvalues to larger binding energies also occurs for the unoccupied levels [59]. The number of unoccupied bound states increases and this will have important consequences in the treatment of collective excitations (see Sections 8 and 9). In the same Figure we can see that, in the nonsymmetrized version of the WDA, the change of eigenvalues with respect to the LDA is very small for occupied states, although it is large for some unoccupied states.

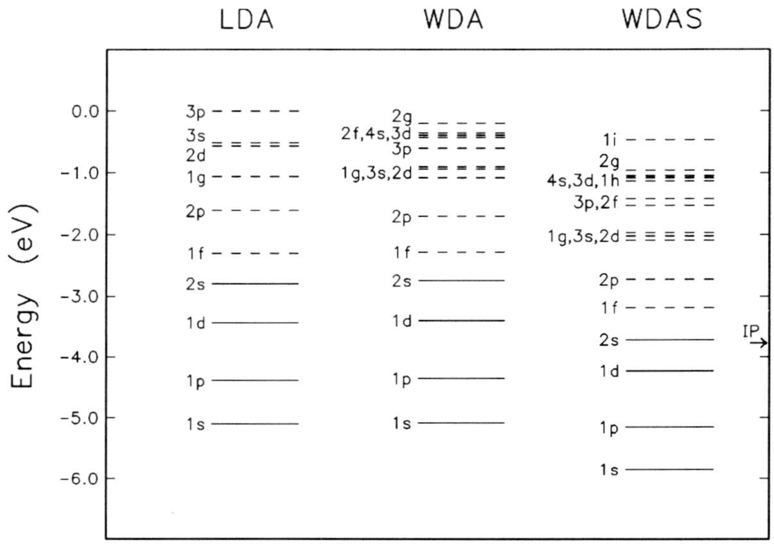

Figure 3. Single-particle energy eigenvalues of Na_{20} within the spherical jellium model. Occupied and unoccupied levels are represented by continuous and dashed lines respectively. The arrow on the right-hand side gives the experimental ionization potential (with the opposite sign).

The use of nonlocal effects leads to qualitative differences in the calculations of the electron affinity. This magnitude is defined

$$\mathrm{EA}(\mathrm{X}_N) = E(\mathrm{X}_N) - E(\mathrm{X}_N^-) , \tag{49}$$

where $E(\mathrm{X}_N)$ and $E(\mathrm{X}_N^-)$ are the total energies of the neutral cluster X_N and the anionic one X_N^- (cluster with an extra electron added). Spherical jellium calculations show that many small Na_N^- and Ag_N^- clusters with an outer electronic shell less than half-filled are unstable against detachment of the extra electron within the LDA, that is, this extra electron is not bound [60]. Electron affinities of aluminum clusters are compared in Figure 4 with the experimental values of Taylor *et al.* [61], obtained by ultraviolet photoelectron spectroscopy of mass selected cluster anions. The most salient feature revealed by the experiments is that Al_6, Al_{13}, Al_{19} and Al_{23} have especially high affinities compared to their neighbors. The WDAS calculation [62] reproduces this feature (notice the pronounced drops of the affinity after those clusters). The large affinities at those particular sizes and the associated drops are due to shell-closing effects. The number of electrons (excluding core electrons) in the corresponding cluster anions is $N_e = 3N + 1 = 19, 40, 58, 70$ respectively. Electronic shells in Al_N clusters are filled in the sequence: 1s, 1p, 1d, 2s, 1f, 2p,1g, 2d, 3s, 1h, ... The shell closing numbers are $N_e = 2, 8, 18, 20, 34, 40, 58, 68, 70, 92, \ldots$. Then the large affinities of Al_6, Al_{13}, Al_{19} and Al_{23} are linked to the shell closings at N_e =18-20, 40, 58 and 70 respectively. The calculation overestimates the magnitude of the oscillations of the affinity. This is a well known deficiency of the jellium model that is only corrected by introducing the granularity of the ions.

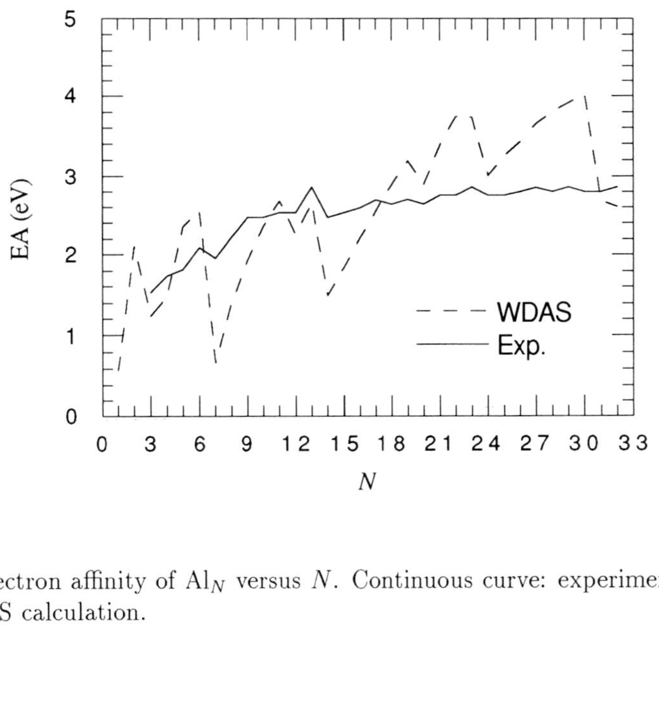

Figure 4. Electron affinity of Al_N versus N. Continuous curve: experiment [61]. Dashed curve: WDAS calculation.

Although the *ad hoc* symmetrization of V_{xc}^{WDA} appears to be useful for the study of properties critically depending of the asymptotic behavior of V_{xc}, we stress that it is not a fully satisfactory alternative, since the variational consistency between E_{xc}^{WDA} and V_{xc}^{WDA} is lost. A more adequate approach should be to symmetrize $g_{xc}(\mathbf{r}_1, \mathbf{r}_2)$, although developing a symmetric and "accurate" $g_{xc}(\mathbf{r}_1, \mathbf{r}_2)$ seems to be a difficult task.

8. ELECTRONIC RESPONSE OF METAL CLUSTERS TO A STATIC ELECTRIC FIELD

The net force acting on an electrically neutral system in the presence of a static electric field $\mathcal{E}(\mathbf{r})$ can be expanded as

$$\mathbf{F} = \mathcal{E}(\mathbf{r}_0) \int \Delta n(\mathbf{r})\, d^3r + [\nabla \cdot \mathcal{E}](\mathbf{r}_0) \int \Delta n(\mathbf{r})(\mathbf{r} - \mathbf{r}_0)\, d^3r + \cdots \,, \tag{50}$$

where $\Delta n(\mathbf{r}) = n(\mathbf{r}) - n_+(\mathbf{r})$, and $n(\mathbf{r})$ and $n_+(\mathbf{r})$ are the electronic and ionic densities. The first term does not contribute for a neutral system and the second one is the product of the field divergence $[\nabla \cdot \mathcal{E}](\mathbf{r}_0)$ times the induced dipole moment $\mathbf{p}$. If the deformation of the ionic density due to the applied field is small, then the force on the system is, to first order,

$$\mathbf{F} \approx \mathbf{p}[\nabla \cdot \mathcal{E}](\mathbf{r}_0) \,, \tag{51}$$

where $\mathbf{p}$ can be written as the product $\mathbf{p} = \alpha \mathcal{E}(\mathbf{r}_0)$ of the polarizability α and the applied field. Knight and coworkers [63] have determined the polarizability of neutral alkali clusters by measuring the deviation of a mass selected cluster beam that travels through a region where an inhomogeneous electric field is applied. The polarizabilities of Al_N clusters have been determined by the same method [64].

Within DFT, the polarizability α of a cluster can be calculated using linear response theory [65]. If a static electric field characterized by a multipole potential $\delta V_L = \mathcal{E}_0 r^L Y_L^0(\theta)$, where $\mathcal{E}_0$ is a small number and $Y_L^0(\theta)$ is a spherical harmonic, is applied to the system, this develops an induced moment $\mathbf{p}_L$ of magnitude $p_L = \alpha_L \mathcal{E}_0$ in response to the field. The first order response of the system is characterized by a small change in the one-electron wave functions $\phi_i(\mathbf{r}) \rightarrow \phi_i(\mathbf{r}) + \delta\phi_i(\mathbf{r})$. The corresponding change in the electron density can be written

$$n(\mathbf{r}) = \sum_{i=1}^{\mathrm{occ}} |\phi_i(\mathbf{r})|^2 \rightarrow n(\mathbf{r}) + \delta n(\mathbf{r}) \,, \tag{52}$$

where $\delta n(\mathbf{r})$ is given by $\delta n(\mathbf{r}) = 2\,\mathrm{Re}\,[\sum_{i=1}^{\mathrm{occ}} \phi_i^*(\mathbf{r})\delta\phi_i(\mathbf{r})]$. Using first order perturbation theory and the Kohn-Sham equations one obtains a set of equations for the changes $\delta\phi_i(\mathbf{r})$ in the wave functions

$$\left[\tfrac{1}{2}\nabla^2 + V_{\mathrm{eff}}(\mathbf{r}) - \varepsilon_i\right] \delta\phi_i(\mathbf{r}) = \delta V_{\mathrm{eff}}(\mathbf{r})\phi_i(\mathbf{r}) \,, \tag{53}$$

where

$$\delta V_{\mathrm{eff}}(\mathbf{r}) = \mathcal{E}_0 r^L Y_L^0(\theta) + \int \frac{\delta n(\mathbf{r}')}{|\mathbf{r} - \mathbf{r}'|}\, d^3r' + \int \frac{\delta V_{xc}(\mathbf{r})}{\delta n(\mathbf{r}')}\, \delta n(\mathbf{r}')\, d^3r' \,, \tag{54}$$

is the change in the self-consistent potential associated with the change in electron density due to the external field.

The calculation of the polarizability α_L proceeds by first solving the Kohn-Sham equations for the field-free cluster, and thus obtaining ϕ_i and ε_i. Then, after solving self-consistently equations (53) and (54) to get the set of functions $\delta\phi_i(\mathbf{r})$, the density change $\delta n(\mathbf{r})$ is evaluated, and the L-order polarizability is obtained

$$\alpha_L = \frac{1}{\mathcal{E}_0} \int r^L\, Y_L^0(\theta)\, \delta n(\mathbf{r})\, d^3r \,. \tag{55}$$

In the particular case of $L = 1$ (dipole polarizability)

$$\alpha_{L=1} = \frac{1}{\mathcal{E}_0} \int z\, \delta n(\mathbf{r})\, d^3r \; . \tag{56}$$

The dipole polarizabilities of some Na clusters with closed electronic shells are given in Table 6. The calculations used the spherical jellium model and the LDA [66]. The results are given in units of the classical polarizability R^3 ($R = r_s N^{1/3}, r_s = 4$ a.u.). The enhancement of α over its classical value is directly proportional to the amount of charge that extends beyond the positive background in the field-free system. The agreement with experiment is reasonable, although the theory systematically underestimates the polarizabilities.

Table 6
Static dipole polarizabilities of Na_N clusters in units of the classical polarizability $\alpha_{\text{classical}} = R^3$, of a metallic sphere of radius $R = r_s N^{1/3}, r_s = 4$ a.u.

N	LDA[66]	WDAS[66]	SIC[67]	Exp.[63]
8	1.41	1.81	1.70	1.72 ± 0.03
20	1.34	1.63	1.61	1.58 ± 0.04
40	1.30	1.53	1.51	1.56 ± 0.04

The effect of nonlocal descriptions of exchange and correlation has been studied in [66]. The slower decay of V_{xc}^{WDAS} with respect to V_{xc}^{LDA} in the outer part of the cluster gives a more extended density tail and also a higher number of unoccupied bound states in the discrete one-particle spectrum (see Figure 3). These effects give a higher static polarizability, improving the agreement with experiment. Self-interaction corrections (SIC), applied by Pacheco and Ekardt [67], also improve the results.

9. DYNAMICAL RESPONSE

The linear response theory can also be developed for the case when the external electric field, characterized by the potential $V_{\text{ext}}(\mathbf{r};\omega) = \mathcal{E}_0\, r^L\, Y_L^0\, e^{i\omega t}$, is time-dependent. This leads to the time-dependent DFT [68].

The external field induces, in this case, a time-dependent perturbation of the electron density, $\delta n(\mathbf{r}, t)$, with Fourier components $\delta n(\mathbf{r}, \omega)$. The key quantity to calculate the response of the system in the linear regime is the dynamical susceptibility $\chi(\mathbf{r}, \mathbf{r}'; \omega)$, which relates the individual components $\delta n(\mathbf{r}, \omega)$ of the induced density to those of the external potential

$$\delta n(\mathbf{r}, \omega) = \int \chi(\mathbf{r}, \mathbf{r}'; \omega) V_{\text{ext}}(\mathbf{r}'; \omega)\, d^3r' \; . \tag{57}$$

Again, the main interest is on a dipole field, that is, $V_{\text{ext}}(\mathbf{r};\omega) = \mathcal{E}_0\, z\, e^{i\omega t}$. The dynamic polarizability, $\alpha(\omega)$, which is the ratio of the induced dipole moment and the applied field, becomes in this case ($L = 1$):

$$\alpha(\omega) = \frac{1}{\mathcal{E}_0} \int z\, \delta n(\mathbf{r};\omega)\, d^3r \; . \tag{58}$$

The dynamical polarizability evidently reduces to the static one of eq. (56) in the case $\omega = 0$.

Let us now return to the calculation of $\chi(\mathbf{r}, \mathbf{r}'; \omega)$. In its Kohn-Sham formulation, DFT is a theory of independent particles moving in an effective self-consistent field. Thus eq. (57) can be rewritten

$$\delta n(\mathbf{r}, \omega) = \int \chi_0(\mathbf{r}, \mathbf{r}'; \omega)\, \delta V_{\text{eff}}(\mathbf{r}'; \omega)\, d^3r' \ . \tag{59}$$

where χ_0 is the independent-particle (or noninteracting) susceptibility and $\delta V_{\text{eff}}(\mathbf{r}'; \omega)$ is the self-consistent perturbing potential (compare with eq. (54))

$$\delta V_{\text{eff}}(\mathbf{r}'; \omega) = V_{\text{ext}}(\mathbf{r}', \omega) + \int \frac{\delta n(\mathbf{r}''; \omega)}{|\mathbf{r}' - \mathbf{r}''|}\, d^3r'' + \int \frac{\delta^2 E_{xc}}{\delta n(\mathbf{r}')\delta n(\mathbf{r}'')}\, \delta n(\mathbf{r}''; \omega)\, d^3r'' \ , \tag{60}$$

where the "local field correction" $K_{xc}(\mathbf{r}', \mathbf{r}''; \omega)$

$$K_{xc}(\mathbf{r}', \mathbf{r}''; \omega) = \frac{\delta^2 E_{xc}[n]}{\delta n(\mathbf{r}'; \omega)\delta n(\mathbf{r}''; \omega)} \tag{61}$$

has been approximated in eq. (60) as usual by its frequency-independent version

$$K_{xc}(\mathbf{r}', \mathbf{r}'') = \frac{\delta^2 E_{xc}[n]}{\delta n(\mathbf{r}')\delta n(\mathbf{r}'')} \ . \tag{62}$$

A convenient way of writing δV_{eff} is

$$\delta V_{\text{eff}}(\mathbf{r}'; \omega) = V_{\text{ext}}(\mathbf{r}'; \omega) + \int K(\mathbf{r}', \mathbf{r}'')\, \delta n(\mathbf{r}''; \omega)\, d^3r'' \ , \tag{63}$$

where we have introduced the nonlocal kernel (or residual particle-hole interaction)

$$K(\mathbf{r}', \mathbf{r}'') = \frac{1}{|\mathbf{r}' - \mathbf{r}''|} + \frac{\delta^2 E_{xc}[n]}{\delta n(\mathbf{r}')\delta n(\mathbf{r}'')} \ . \tag{64}$$

Inserting the expression (63) for δV_{eff} in eq. (59) and using the form (57) for $\delta n(\mathbf{r}; \omega)$ in both sides of the resulting equation, the following Dyson-type equation is obtained for the interacting dynamic susceptibility [68,69]

$$\chi(\mathbf{r}, \mathbf{r}'; \omega) = \chi_0(\mathbf{r}, \mathbf{r}'; \omega) + \int \chi_0(\mathbf{r}, \mathbf{r}_1; \omega)\, K(\mathbf{r}_1, \mathbf{r}_2)\, \chi_0(\mathbf{r}_2, \mathbf{r}'; \omega)\, d^3r_1\, d^3r_2 \ , \tag{65}$$

which has to be solved iteratively.

Finally, the non-interacting susceptibility, which is needed to solve this equation, has the form

$$\chi_0(\mathbf{r}, \mathbf{r}'; \omega) = \sum_{i=1}^{\text{occ}} \left[\phi_i^*(\mathbf{r})\phi_i(\mathbf{r}')G(\mathbf{r}, \mathbf{r}'; \varepsilon_i + \hbar\omega) + \phi_i(\mathbf{r})\phi_i^*(\mathbf{r}')G^*(\mathbf{r}, \mathbf{r}'; \varepsilon_i - \hbar\omega)\right] \ , \tag{66}$$

where ϕ_i are the occupied single-particle orbitals of the ground state Kohn-Sham Hamiltonian for the field-free system, ε_i are the corresponding single-particle energies and

$G(\mathbf{r},\mathbf{r}';\varepsilon_i \pm \hbar\omega)$ are the retarded Green functions associated to the effective Kohn-Sham potential

$$\left[E + \tfrac{1}{2}\nabla^2 - V_{\text{eff}}(\mathbf{r})\right] G(\mathbf{r},\mathbf{r}';E) = \delta(\mathbf{r}-\mathbf{r}') \ . \tag{67}$$

If the LDA is used in this formalism, one arrives at the TDLDA (time-dependent LDA)[68]. In this case, the local field correction becomes a function of the local density

$$K_{xc}^{\text{LDA}}(\mathbf{r}',\mathbf{r}'') = \frac{dV_{xc}^{\text{LDA}}(\mathbf{r}')}{dn(\mathbf{r}')}\,\delta(\mathbf{r}'-\mathbf{r}'') \ , \tag{68}$$

and the LDA kernel reads

$$K^{\text{LDA}}(\mathbf{r}',\mathbf{r}'') = \frac{1}{|\mathbf{r}'-\mathbf{r}''|} + \frac{dV_{xc}^{\text{LDA}}(\mathbf{r}')}{dn(\mathbf{r}')}\,\delta(\mathbf{r}'-\mathbf{r}'') \ . \tag{69}$$

In order to connect with measurable properties, Fermi's golden rule is used to obtain the photoabsorption cross section

$$\sigma(\omega) = \frac{4\pi\omega}{c}\,\mathrm{Im}\,\alpha(\omega) \ , \tag{70}$$

where c is the velocity of light and $\mathrm{Im}\,\alpha(\omega)$ is the imaginary part of $\alpha(\omega)$.

Figure 5 gives the result of a TDLDA calculation of $\sigma(\omega)$ for Na_{20} within the spherical jellium model [59]. The lines are not δ functions because a complex photon energy $\hbar\omega + i\beta$ with $\beta = 0.013$ eV was used for numerical convenience. The dotted line represents the results obtained for the independent-particle dynamical susceptibility, that consists in using χ_0 instead of χ in eq. (57), and the continuous line the results for the interacting susceptibility χ. The peaks in the non-interacting spectrum correspond to single particle-hole excitations. When electron-electron interactions are switched on (that is, using the fully interacting χ) some particle-hole transitions are shifted and others lose their individual identities to merge into a collective resonance.

Before comparing to experiment the results of the theoretical calculations, let us present the way the experimentalists measure the photoabsorption spectrum of metallic clusters. The clusters in a molecular beam fragment upon absorption of light of the appropriate frequency, and this effect induces a deviation of the cluster away from the initial direction of motion of the beam [70]. The ratio between the number of clusters (of a given size) arriving at the detector with and without light excitation gives the value of the absorption cross section. The process involved in the fragmentation is the excitation of a collective mode (surface plasmon) at an energy that for sodium clusters is about 3 eV. Since this energy is higher than the binding energy of an atom in the aggregate (which is below 1.1 eV) the excited cluster decays by evaporating single atoms. Using a statistical model, the time required to evaporate an atom if we assume that the energy of the collective mode is converted into vibrations, turns out to be very small compared to the time of flight of the molecular beam in the spectrometer. Consequently one can assume that the photoabsorption and photoevaporation cross sections are equal.

The integral of $\sigma(\omega)$ gives rise to the dipole sum rule

$$\int_0^\infty \sigma(\omega)\,d\omega = \frac{4\pi^2}{\hbar c}\,m_1 \ , \tag{71}$$

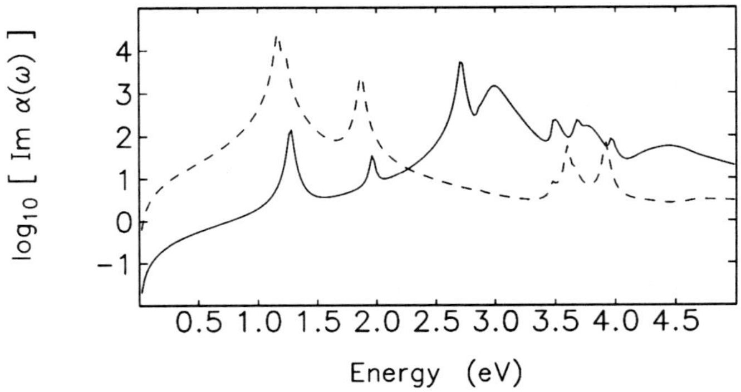

Figure 5. Imaginary part of the dynamical polarizability (per electron) of Na_{20}. The calculations employ the LDA and the spherical jellium model. Dotted and continuous lines correspond to the use of the non-interacting (χ_0) and fully interacting (χ) dynamical susceptibilities respectively. Notice the logarithmic scale used for $\mathrm{Im}\,\alpha(\omega)$.

with $m_1 = \frac{e^2\hbar^2 Z}{2m}$. Z is the number of electrons taking part in the collective motion and m is the electron mass. Consequently, the experimental determination of $\sigma(\omega)$ helps to identify the collective nature of a resonance.

The classical theory of the dynamic polarizability of a metallic sphere of radius R developed by Mie [71] predicts a single dipole resonance at a frequency given by

$$\omega_{\mathrm{Mie}} = \sqrt{\frac{Z\hbar^3 c^2}{mR^3}} \, . \tag{72}$$

If R is identified with the cluster radius of the spherical jellium model, then a value for ω_{Mie} is obtained equal to one third of the bulk plasma frequency ω_{pl}.

The TDLDA applied to the jellium model follows the Mie result, but only in a qualitatively way: the dipole absorption cross sections of spherical alkali clusters normally exhibit a dominant peak which exhausts some 70-90% of the dipole sum rule and is red-shifted by 10-20% with respect to Mie formula. The centroid of the strength distribution tends towards the Mie resonance in the limit of a macroscopic metal sphere. Its red shift in clusters is a quantum mechanical finite size effect closely related to the spill-out of the electronic cloud beyond the jellium edge. Some 10-25% of the dipole strength is typically found at higher energies and can be interpreted as a reminiscence of a strongly fragmented volume plasmon.

Sometimes the dominant peak is fragmented in two or more lines. When this occurs in "spherical" clusters it can be attributed to an interference of specific particle-hole excitations with the predominant collective mode. This fragmentation can be compared to Landau damping in the solid.

As compared to experiment, the LDA spherical jellium calculations yield an insufficient red shift of the Mie resonance [72]. This fact can be appreciated in Figure 6, where the photoabsorption cross section of Na_{20} has been represented [59]. The calculated resonance lines have been folded with a Lorentzian of width 0.2 eV to simulate finite temperature effects. The experimental points [73] have been scaled by a constant factor so that the maximum of this curve has a magnitude similar to the calculated ones. The main LDA peak occurs at an energy 0.3 eV larger than the experimental one. Although shifted to larger energies, the form of the LDA spectrum of Na_{20} is rather similar to the experimental one. However, the physics behind the fragmentation of the plasmon line is not well described by the LDA. This fragmentation occurs in the LDA calculation because of the proximity of a single particle-hole transition (2s→3p) to the plasmon line. The energy of this transition is 2.8 eV (see left panel of Figure 3). It occurs that the 3p level is practically degenerate with the vacuum level, and the fragmentation line is broadened by the proximity of one-electron transitions from the 2s states to scattering states in the continuum. In contrast, the experimental ionization threshold is 3.76 eV (see Table 5).

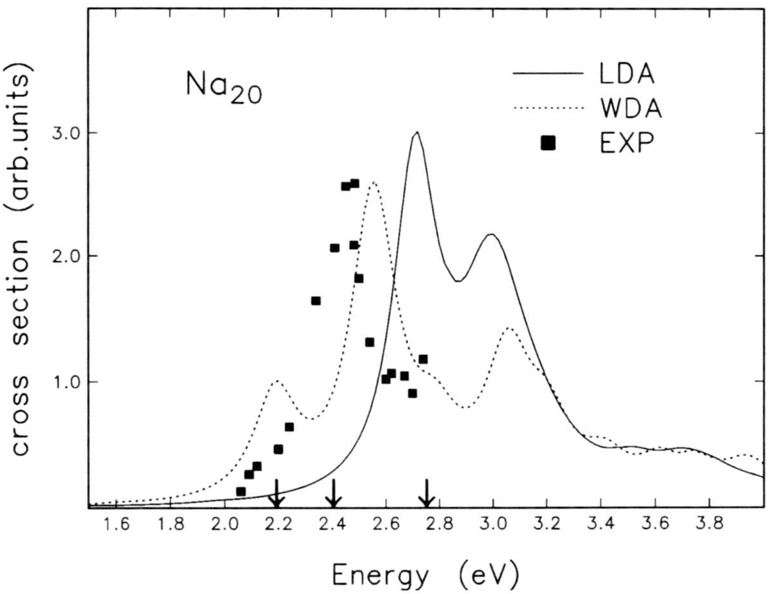

Figure 6. LDA, WDA and experimental [73] photoabsorption cross sections (per electron) of Na_{20} [59]. Calculations were made within the spherical jellium model. The three arrows correspond to the position of the peaks observed in recent experiments [75].

The WDA calculation [59] shifts the plasmon line to lower energies (2.56 eV), improving the agreement with experiment. The shift with respect to experiment is only 0.1 eV and further improvement would require the explicit consideration of the geometry, transcending the spherical jellium model. The improved description of the asymptotic behavior of V_{xc} in the nonlocal calculation leads to an accurate ionization threshold (at 3.72 eV), which does not interact with the plasmon because of the large separation between the two features. Instead, a small shoulder develops at ≈2.7-2.8 eV, and a pronounced feature appears on the low energy side (≈2.2 eV) of the resonance, both features being due to the interaction of the plasmon with one particle-hole transitions. Although only the two high energy features were clearly identified in early photoabsorption experiments [73,74], three resonances at ≈2.19, 2.41 and 2.76 eV have been observed in recent experiments by Thalweiser *et al.* [75] which support the spectrum predicted by the WDA. Calculations of the dynamic polarizabilities using SIC corrections also improve the results with respect to the LDA [67,76].

It may be useful to clarify some technical details about the WDA calculations in ref. [59]. The exact $K_{xc}(\mathbf{r}_1, \mathbf{r}_2)$ is symmetric under the interchange of $\mathbf{r}_1$ and $\mathbf{r}_2$. $K_{xc}^{\mathrm{WDA}}(\mathbf{r}_1, \mathbf{r}_2)$ fulfills this property, so its functional form (as a functional of the density) was used in the calculation of the interacting susceptibility χ. On the other hand, the ground state wave functions, energy eigenvalues, electron density, ... required in the evaluation of the Green functions, noninteracting χ_0, K_{xc}^{WDA} and the interacting χ, were obtained with the symmetrized V_{xc}^{WDAS} potential. Another point is that the form used for $g_{xc}^{\mathrm{WDA}}(\mathbf{r}_1, \mathbf{r}_2)$ was that for the homogeneous electron gas, that is $g_{xc}^{\mathrm{hom}}(r_{12}; \tilde{n}(\mathbf{r}_1))$.

It was pointed out above that one of the consequences of the improved asymptotic behavior of V_{xc}^{WDAS} is that the binding energy of the HOMO becomes very close to the ionization potential, in contrast to the LDA. For this reason, interesting effects can be expected whenever the plasmon excitation occurs near the ionization threshold. This should be the case for negatively charged clusters of large enough size. For small negative clusters the calculations indicate that the collective resonance frequency lies in the region of transitions to the continuum of states, where Landau damping produces a broadening of the resonance. As the cluster size increases, the plasmon frequently approaches the region of discrete states because its value changes at a lower rate than the increase of the ionization threshold. Then, when the plasmon frequency becomes close to the ionization threshold, fragmentation of the plasmon resonance is likely to occur.

These expectations have been confirmed by calculations for negative K_N^- and Na_N^- clusters [59,77]. We show in Figure 7 the calculated photoabsorption cross section for Na_{197}^-. Notice the fragmentation of the plasmon in the WDA, which is absent in the LDA calculation. The reason is the proximity between $-\varepsilon_{\mathrm{HOMO}}$ (indicated by the arrow in the figure) and the plasmon resonance for the WDA case. On the other hand, the HOMO level is far below the plasmon energy in the LDA. It would be interesting to perform experiments for large cluster anions to test this prediction.

10. EFFECT OF STRUCTURAL CORRECTIONS BEYOND THE JELLIUM MODEL

Although the jellium model had an enormous success in explaining qualitatively or semiquantitatively many static and dynamical properties of simple metal clusters, consideration of the granularity of the cluster is needed to explain in full detail the electronic properties. The measured photoabsorption cross section of Na_{21}^+ shows a surface plasmon resonance in the region 2.65-2.74 eV (more precisely, at 2.65 eV in [78], 2.68 eV in [79], and 2.74 eV in [80]). Although Na_{21}^+ is isoelectronic with Na_{20}, the plasmon resonance

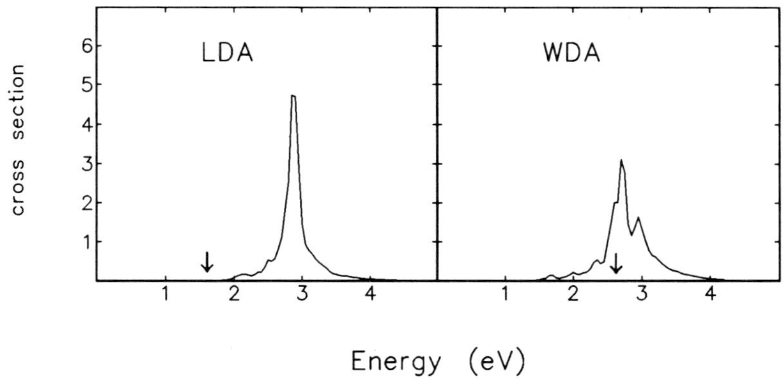

Figure 7. Calculated photoabsorption cross section (per electron) of Na_{197}^-. Arrows give the binding energy of the HOMO level [77].

occurs at slightly higher energy in Na_{21}^+ due to the stronger confining potential. Figure 8 gives the results of a spherical jellium calculation of the photoabsorption cross section of this cluster. The LDA value (2.95 eV), as usual, overestimates the energy of the plasmon (notice that the value of the plasmon energy, averaged over the three experimental results is 2.69 eV). By introducing nonlocal corrections through the WDA, a plasmon resonance is found at 2.63 eV, but also a well separated fragmentation peak appears at 2.95 eV. This fragmentation has not been observed in the experiments (only a shoulder on the blue side of the plasmon peak has been observed by Borgreen *et al.* [79] and by Reiners *et al.* [80]). The discrepancy can be removed by introducing geometrical corrections to the jellium model. The introduction of the full geometry is a formidable task that, so far, has only been implemented in conjunction with the LDA. A simpler way which achieves a significant improvement over the spherical jellium model is the SAPS (spherically averaged pseudopotential) model [81]. This model replaces the ion cores by local pseudopotentials (a pseudopotential often used in this context is the empty-core model potential developed by Ashcroft [82]). The main approximation of the SAPS model is the replacement of the complicated tree-dimensional potential of the ionic skeleton

$$V_{\text{ionic}}(\mathbf{r}) = \sum_j v_{\text{ec}}(|\mathbf{r} - \mathbf{R}_j|) \ , \tag{73}$$

written here as the superposition of empty-core (ec) pseudopotentials centered on the individual ion sites, by its spherical average about the cluster center:

$$V_{\text{ionic}}(\mathbf{r}) \rightarrow V^{\text{SAPS}}(r) \ . \tag{74}$$

In spite of the SAPS approximation for the total ionic potential (and thus for the effective potential), the geometry of the cluster is fully taken into account in the computation of the ion-ion interaction, which provides an important contribution to the total energy of

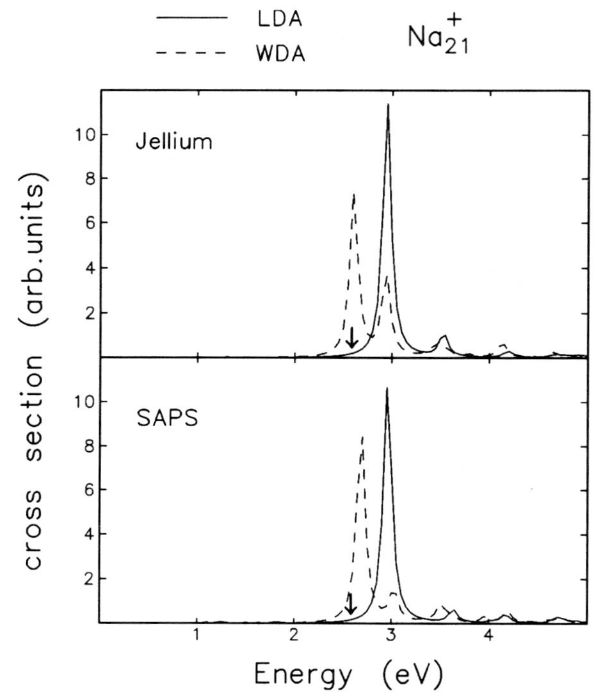

Figure 8. Calculated photoabsorption cross section (per electron) of Na_{21}^{+}. The arrows indicate the position of the plasmon resonance averaged over three experimental values [78–80].

the cluster. In this model clusters develop a structure formed by concentric atomic layers, and the overall shape is rather spherical as a consequence of the SAPS approximation.

The resulting geometry for Na_{21}^{+} (obtained by minimization of the cluster energy, using the LDA) can be described as an inner core of three atoms, surrounded by a rather spherical shell composed by 18 atoms. In the bottom panel of Figure 8 we have plotted the LDA and WDA cross sections, both obtained with the LDA geometry. The LDA plasmon is found at 2.94 eV, which once more gives an overestimation of the experimental energy. Finally, the WDA gives the plasmon at 2.70 eV, but this time the strong fragmentation peak found in the WDA-jellium calculation has almost disappeared, and only small features remain at 3.0 and 3.5 eV, that correlate with the shoulder observed by Borgreen *et al.* [79] and by Reiners *et al.* [80]. The photoabsorption spectrum of Na_{41}^{+} has also been calculated using the SAPS model and the WDA [83]. The position and the shape of the plasmon resonance is also in good agreement with experiment.

In summary, consideration of both effects, geometrical corrections and nonlocal corrections to exchange and correlation, leads to agreement with experiment in the calculation of the photoabsorption spectrum. Such a good agreement is not achieved for Na_{21}^+ or Na_{41}^+ if only one of those corrections is taken into account.

11. ATOMIC SYSTEMS

11.1. Ground state energies

Most applications discussed in previous Sections concern the valence electrons in condensed systems. In those cases the electron density is not very inhomogeneous except at the surface, and exchange and correlation effects are equally important. Now we consider free atoms or ions, which provide examples of a system with an extremely inhomogeneous density: this is a piece-wise exponentially decaying function, reflecting the electronic shell structure. In atoms it is customary to separate exchange and correlation effects by writing

$$E_{xc} = E_x + E_c \ , \tag{75}$$

and exchange dominates over correlation. Let us consider one example. The errors in the exchange energy from LDA calculations for He, Ne and Ar are 0.18, 1.18 and 2.43 a.u. respectively [84]. On the other hand, the empirical correlation energies of those atoms are -0.042, -0.381 and -0.732 a.u. respectively [85]. The conclusion is that the LDA exchange errors are larger than the absolute values of the correlation energy (in practice the LDA errors in the exchange and correlation energies tend to compensate each other in part). This example indicates the need to derive nonlocal approximations for E_x in atoms. Again the candidates are the GGA, SIC and WDA.

The WDA for exchange-only, based on $g_x^{\text{hom}},(r_{12};\tilde{n}(\mathbf{r}_1))$ whose functional form is known exactly (see eq. (30)), has been applied to atoms [84]. Somehow surprisingly, neither the exchange nor the total energies provide a significant improvement with respect to the LDA. The reason for the poor performance is that free atoms have a highly inhomogeneous density and the homogeneous gas form used for g_x^{WDA} is not adequate. Lorentzian and Gaussian forms have been explored by Fritsche and coworkers [86–89] and exponential forms by Pedroza [90]. Taking into account that $g_x - 1$ can be well represented near the exchange hole center by a Gaussian function and at large interelectronic distances by a Slater function, Gritsenko *et al.* [84] proposed an exponential form which in its spin-polarized version [91] reads

$$g_{\sigma\sigma}^{\text{exp}}(r_{12};r_x^\sigma(\mathbf{r}_1)) = 1 - \exp\left[-\left(\frac{r_{12}}{r_x^\sigma(\mathbf{r}_1)}\right)^{\frac{3}{2}}\right] \ , \tag{76}$$

where the exponent $\frac{3}{2}$ is intermediate between those of Slater and Gaussian functions. The radius $r_x^\sigma(\mathbf{r}_1)$ of the exchange hole around an electron with spin σ placed at $\mathbf{r}_1$ is, of course, fixed by the sum rule

$$\int n_\sigma(\mathbf{r}_2)\,[g_{\sigma\sigma}^{\text{exp}}(r_{12};r_x^\sigma(\mathbf{r}_1)) - 1]\; d^3r_2 = -1 \ . \tag{77}$$

This model fulfills the following rigorous conditions at $r_{12} = 0$

$$g_{\sigma\sigma}^{\text{exp}}(r_{12}=0;r_x^\sigma(\mathbf{r}_1)) = 0 \ ; \quad \left.\frac{dg_{\sigma\sigma}^{\text{exp}}(r_{12};r_x^\sigma(\mathbf{r}_1))}{dr_{12}}\right|_{r_{12}=0} = 0 \ , \tag{78}$$

and is free from empirical parameters. The WDA exchange energy associated to $g^{\exp}$ becomes

$$E_x^{\mathrm{WDA}}[n_\uparrow, n_\downarrow] = \frac{1}{2}\sum_\sigma \iint \frac{n_\sigma(\mathbf{r}_1)\left\{n_\sigma(\mathbf{r}_2)\left[g_{\sigma\sigma}^{\exp}(r_{12}; r_x^\sigma(\mathbf{r}_1)) - 1\right]\right\}}{|\mathbf{r}_1 - \mathbf{r}_2|} d^3r_1\, d^3r_2\ . \tag{79}$$

$g_{\sigma\sigma}^{\exp}$ is more localized than $g_{\sigma\sigma}^{\mathrm{hom}}$ at short interelectronic distances, but it decays more slowly and does not exhibit the spurious Bessel-like oscillations of the latter. Evidently $g_x^{\exp}$ does not reproduce exactly the exchange energy per electron in the homogeneous limit, although the error is small (≈2.5%). Unless explicitly stated otherwise, all the WDA calculations presented in this Section were performed with the spin-polarized functional of eq. (79).

Table 7 shows the exchange energies from selfconsistent calculations for noble gas atoms. The average error of the WDA decreases from 7% to 0.4% when $g_{\sigma\sigma}^{\exp}$ is used instead of $g_{\sigma\sigma}^{\mathrm{hom}}$. The results improve over the LDA (average error 8.3%) although there is no improvement over Becke's GGA (average error 0.08%); one has to remember that Becke's functional contains one parameter (fitted, in fact, to reproduce the exchange energies of the inert gases) while the WDA is parameter free. The improvement over the LDA is bigger for total energies due to cancellation of errors among the different components of the total energy.

Table 7
Selfconsistent exchange energies of noble gas atoms (with the opposite sign and in a.u.) using LDA, WDA (with homogeneous $g_{\sigma\sigma}^{\mathrm{hom}}$ and exponential $g_{\sigma\sigma}^{\exp}$ functions), and Becke's GGA exchange-only functionals. Hartree-Fock results from [92,93] are given as a reference.

Atom	LDA	WDA		GGA	HF [92]
		$g_{\sigma\sigma}^{\mathrm{hom}}$	$g_{\sigma\sigma}^{\exp}$		
He	0.85	1.03	1.03	1.03	1.03
Ne	10.93	12.90	12.03	12.09	12.11
Ar	27.76	32.50	30.25	30.13	30.19
Kr	88.45	101.81	93.66	93.79	93.89
Xe	170.38	194.63	177.97	179.00	179.19
Rn	372.66	423.20	384.03	387.38	387.5 *

* from ref. [93]

In order to compare with empirical nonrelativistic total energies it is necessary to include Coulomb correlation effects. All the WDA calculations including correlation presented in this Section have been performed with an Optimized Local Approximation (OLA) for the correlation energy functional [91]. This functional is obtained by approximating the correlation hole around an electron at $\mathbf{r}_1$ by that in a homogeneous polarized electron gas with spin densities $(n_\uparrow(\mathbf{r}_1), n_\downarrow(\mathbf{r}_1))$. Using this approximation, the correlation energy can be written:

$$E_c = \sum_\sigma \int n_\sigma(\mathbf{r}_1)\varepsilon_c^\sigma\left(n_\uparrow(\mathbf{r}_1), n_\downarrow(\mathbf{r}_1)\right) d^3r_1\ , \tag{80}$$

where $\varepsilon_c^\sigma(n_\uparrow(\mathbf{r}_1), n_\downarrow(\mathbf{r}_1))$ is the spin-polarized Coulomb correlation energy density:

$$\varepsilon_c^\sigma(n_\uparrow(\mathbf{r}_1), n_\downarrow(\mathbf{r}_1)) = \tfrac{1}{2}\sum_{\sigma'} n_{\sigma'}(\mathbf{r}_1)\int \frac{1}{r_{12}}\int_0^1 \left[g_c^{\sigma\sigma'}(r_{12}; n_\uparrow(\mathbf{r}_1), n_\downarrow(\mathbf{r}_1); \lambda) - 1\right] d\lambda\, d^3r_2\,. \quad (81)$$

The function $g_c^{\sigma\sigma'}(r_{12}; n_\uparrow(\mathbf{r}_1), n_\downarrow(\mathbf{r}_1); \lambda)$ is the pair correlation function for electrons with spins σ and σ' in a polarized homogeneous electron gas with spin densities $(n_\uparrow, n_\downarrow) = (n_\uparrow(\mathbf{r}_1), n_\downarrow(\mathbf{r}_1))$. Those pair correlation functions $g_c^{\sigma\sigma'}$ are approximated in the following way (see ref. [91] for details)

$$g_c^{\sigma\sigma'}(r_{12}; n_\uparrow(\mathbf{r}_1), n_\downarrow(\mathbf{r}_1); \lambda) = 1 + \frac{e^{-k^2 r_{12}^2}[-1 + a^{\sigma\sigma'}(\mathbf{r}_1; \lambda)(1 + \lambda r_{12})]}{1 + b^{\sigma\sigma'} n_{\sigma'}(\mathbf{r}_1) r_{12}^2}\,, \quad (82)$$

$$g_c^{\sigma\sigma}(r_{12}; n_\uparrow(\mathbf{r}_1), n_\downarrow(\mathbf{r}_1); \lambda) = 1 + \frac{g_x^{\sigma\sigma}(r_{12}; n_\uparrow(\mathbf{r}_1), n_\downarrow(\mathbf{r}_1)) e^{-k^2 r_{12}^2}[-1 + a^{\sigma\sigma}(\mathbf{r}_1; \lambda)(1 + \frac{\lambda}{2} r_{12})]}{1 + b^{\sigma\sigma} n_\sigma(\mathbf{r}_1) r_{12}^2}\,. \quad (83)$$

The model functions $a^{\sigma\sigma'}(\mathbf{r}_1; \lambda)$ and $b^{\sigma\sigma'}(k)$, whose explicit expressions are given in ref. [91], are built up to be consistent with the fulfillment of some local and integral properties of g_c. On the other hand $g_x^{\sigma\sigma}(r_{12}; n_\uparrow(\mathbf{r}_1), n_\downarrow(\mathbf{r}_1))$ is the Gaussian approximation of Berkowitz [4] to the exchange-only pair correlation function in a homogeneous electron gas with density equal to the local density at $\mathbf{r}_1$. The meaning of $g_x^{\sigma\sigma}$ in (83) is to suppress the short-range Coulomb correlation between parallel spin electrons, because this short-range correlation is already accounted for in the exchange part. The only parameter in the model is k, which was fixed in [91] by a global fit to the empirical correlation energies of small closed shell neutral atoms. Effects beyond the usual LDA enter the OLA functional through this empirical fitting of E_c.

In Table 8 the total non-relativistic ground-state energies calculated self-consistently using the WDA exchange functional and the OLA correlation functional, are compared with the empirical data [85,94], available for atoms of the first two rows of the Periodic Table, and with the results of Perdew and Zunger [10], who also used an explicit separation of exchange and correlation energies. These authors calculated total energies by means of the SIC method. This method also uses a non-local exchange functional free of empirical parameters and a local correlation spin-density functional. The WDA+OLA method provides a definite improvement for the total energies for all atoms with the only exceptions of Be and B. In all cases the SIC method overestimates total energies (in absolute value), while the errors of WDA+OLA reflect the errors of the WDA exchange functional.

A detailed analysis of the energies for the He, Be and Ne isoelectronic series has been performed in [95]. At the exchange-only level WDA results are better than GGA ones for the He series both for exchange and total energies (due to the exact cancellation of self-interaction for the WDA functional in this case), but slightly worse for the Be and Ne series. When correlation is taken into account, WDA+OLA total energies for the He and Be series are comparable to energies from Perdew's GGA91 functional [8] but worse than energies obtained from Becke's exchange plus Wilson-Levy (B-WL) [96] or Lee-Yang-Parr (B-LYP) [97] correlation functionals. For the Ne series, WDA+OLA results are better due to a cancellation between errors of exchange and correlation energies.

Table 8
Total energies (with the opposite sign and in atomic units) obtained by using WDA for exchange and OLA for correlation, compared to experiment [85,94] and to the self-interaction-corrected (SIC) results of Perdew and Zunger [10]. The relative errors of the WDA+OLA results are also given.

Atom	Exp.	WDA+OLA	Error %	SIC
He	2.904	2.909	0.17	2.918
Li	7.478	7.473	-0.07	7.504
Be	14.667	14.605	-0.42	14.692
B	24.653	24.583	-0.28	24.695
C	37.844	37.779	-0.17	—
N	54.587	54.530	-0.10	54.716
O	75.015	75.011	-0.00	—
F	99.725	99.695	-0.03	99.983
Ne	128.928	128.914	-0.01	129.268
Na	162.245	162.270	0.01	162.655
Mg	200.043	200.108	0.03	200.517
Al	242.336	242.425	0.04	242.867
Si	289.374	289.452	0.04	—
P	341.240	341.356	0.03	341.902
S	398.139	398.231	0.03	—
Cl	460.196	460.286	0.03	—
Ar	527.550	527.680	0.02	528.389

11.2. Interconfigurational energies

It has been pointed out by a number of workers [35,98,99], that the successful calculation of interconfigurational energies (ICE) between the two valence electron configurations $d^{n-1}s$ and $d^{n-2}s^2$ of transition metal atoms:

$$\Delta_{sd} = E[d^{n-1}s] - E[d^{n-2}s^2] \tag{84}$$

is a key point for the accurate description of the metallic bonding in transition metals and clusters and thus for the calculation of various bonding characteristics, such as the cohesive energies, heats of formation of alloys, and dissociation energies of clusters and molecules. However, a successful description of the interconfigurational energies has not yet been achieved by the existent DFT schemes [100,101], and it poses a critical test for novel descriptions of exchange and correlation.

The ICE in (84) corresponds to the transition of an electron from the diffuse s-orbital, where it is spin-paired with another s electron, to the localized d-orbital where, depending on the electron configuration of the actual atom, it can be spin-paired or not. Owing to this, ICE is a result of the subtle interplay between the exchange and Coulomb correlation effects at various regions in the atom. To calculate accurately this quantity, one should have a method with a highly balanced description of these effects. Relativistic contributions to ICE are also important [100,101].

ICEs calculated for atoms of the first two transition metal rows by using the sping polarized WDA+OLA scheme, within the scalar-relativistic [102] approximation, are presented in Table 9 and in Fig. 9 [91]. In the scalar-relativistic approximation the relativistic effects are taken into account by solving linear relativistic equations for the radial parts of the atomic orbitals, with the angular parts being fixed in their standard non-relativistic form. The spin configurations used correspond, in all cases, to the maximum total spin. For example, in the case of Vanadium, the ICE corresponds to the $4s^2(\uparrow\downarrow)\ 3d^3(\uparrow\uparrow\uparrow) \rightarrow 4s^1(\uparrow)\ 3d^4(\uparrow\uparrow\uparrow\uparrow)$ transition.

Table 9
s → d interconfigurational energies (in eV) calculated with the scalar relativistic scheme for atoms of the first and second transition-metal rows. Experimental results [103] are given as a reference.

Atom	LSDA*	LSDX	WDA+OLA	Exp.
Ca	1.86	2.57	2.29	2.52
Sc	0.73	1.49	1.44	1.43
Ti	-0.07	1.18	0.60	0.81
V	-0.80	0.87	-0.16	0.26
Cr	-1.92	-0.38	-0.93	-0.96
Mn	1.31	3.07	1.82	2.12
Fe	0.25	1.78	1.01	0.86
Co	-0.35	1.65	0.35	0.43
Ni	-0.91	1.51	-0.43	0.03
Cu	-2.06	0.04	-1.22	-1.39
Sr	1.9	—	2.31	2.39
Y	0.9	—	1.63	1.41
Zr	-0.2	—	0.92	0.60
Nb	-1.2	—	0.03	-0.24
Mo	-2.3	—	-0.93	-1.36
Tc	-0.1	—	0.49	0.46
Ru	-0.9	—	-0.41	-0.82
Rh	-2.0	—	-0.98	-1.71
Pd	-3.0	—	-1.90	-2.34
Ag	-4.0	—	-2.91	-3.73

*First transition metal series from ref. [101]; second transition metal series from ref. [100].

The ICEs are compared in this table with the experimental data [103] and also with the ICEs calculated within the standard LSDA for exchange and correlation and by the LSDX method [104,105]. The latter combines the Hartree-Fock description of exchange with the standard LSDA description of correlation. For the first transition-metal row the LSDA and LSDX calculations of refs. [101,104] used the correlation functional interpolated by Perdew and Zunger [10] on the basis of the Ceperley-Alder energies [20]. The relativistic LSDA and LSDX calculations were performed in [101] with the relativistic kinetic energy functional obtained from the modified Dirac equation. The LSDA results for the second transition-metal row are from ref. [100]; the calculations were performed with the correlation functional of von Barth and Hedin [106]. To include relativistic contributions, the fully relativistic Dirac equation was solved in [100] for the core levels of those atoms, while the scalar-relativistic approximation [102] was used for the valence levels.

In the case of the LSDX there is a substantial overestimation of the ICEs for most atoms of the first transition row. In contrast, the scalar-relativistic variant of the WDA+OLA nicely approximates the experimental ICEs (see Figure 9). For these atoms it yields a much better agreement with experiment than the LSDA and LSDX. The only exception is Ca, for which the LSDX estimate is the best one. The WDA+OLA error is only 0.2 eV and for Sc, Cr and Co the ICEs are calculated with chemical accuracy, i.e. the corresponding errors are less than 0.1 eV. Only for V and Ni the errors exceed 0.4 eV.

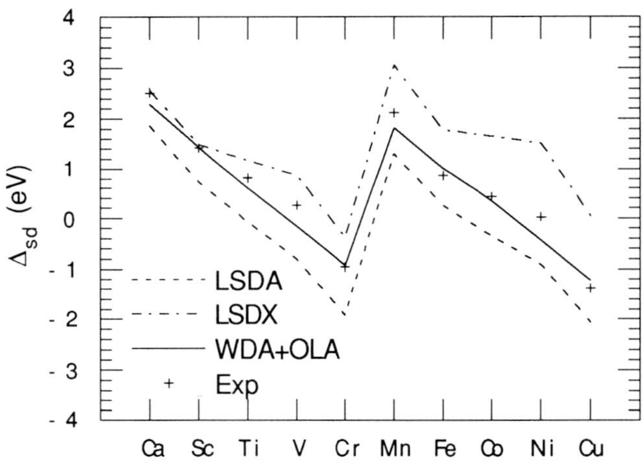

Figure 9. Scalar relativistic calculations of the s → d interconfigurational energies for the first series of transition metal atoms. The experimental values are also included.

It is interesting to compare the WDA+OLA results with *ab initio* calculations available for the 3d atoms. In Table 10 the ICEs for Ti, Cr, Fe and Ni [91] are compared to those from various *ab initio* approaches to the electron correlation problem [107]. In the third column of this table we report results involving single and double configuration interactions from the Hartree-Fock wave function (HF-SDCI). The fourth column corresponds to multireference CI calculations (MRCI), with the multiconfigurational self-consistent field (MCSCF) reference wave functions providing angular correlation of 4s electrons and radial correlation of 3d electrons at the reference level. The fifth column gives results from the generalized Möller-Plesset (MP) second-order perturbation theory using the MCSCF reference wave functions. Finally MP values with relativistic corrections [107,108] are presented in the seventh column. All *ab initio* schemes in the table give ICEs which are close to each other for a given atom. At the non-relativistic level they yield better agreement with experiment (as compared to WDA+OLA) for Ti and Ni, while for Fe the WDA+OLA estimate is better and for Cr all approaches are of similar quality. However, the relativistic corrections worsen the *ab initio* MP results. At the relativistic level the WDA+OLA method yields better results than MP for the four atoms. The errors in the *ab initio* estimates have been attributed to the neglect of higher excitations of the reference wave function and to the neglect or very restricted account of the Coulomb correlation of core electrons, while the WDA+OLA accounts for the global correlation effect via the model pair-correlation function.

Table 10
Comparison of s $\rightarrow$ d interconfigurational energies (in eV) calculated with various *ab initio* methods [107] and with the WDA+OLA scheme at the non-relativistic and relativistic (rel.) levels. Experimental results [101,103] are given as a reference; experimental values estimated in [107] are given in parentheses. See text for the meaning of symbols of the *ab initio* calculations.

Atom	WDA+OLA	HF-SDCI	MRCI	MP	WDA+OLA (rel.)	MP (rel.)	Exp.
Ti	0.46	1.03	1.06	1.04	0.60	1.14	0.81
Cr	-1.12	-0.99	-0.93	-1.04	-0.93	-0.83	-0.96 (-1.00)
Fe	0.82	1.28	—	1.32	1.01	1.58	0.86 (0.88)
Ni	-0.87	0.44	—	0.34	-0.43	0.70	0.03 (-0.03)

For second transition row atoms, WDA+OLA scalar relativistic calculations overestimate the ICEs [91]. Although the method reproduces well the data for Sr, Y and Tc, the errors are larger than those for the first transition row. Nevertheless, the errors are smaller than the LSDA errors, with the exception of the three heavy atoms Ru, Rh and Ag. The evident reasons for the overestimation of the ICEs are the approximations introduced in the exchange-correlation energy functionals and the use of the scalar-relativistic scheme. Bearing in mind the good results obtained for the first transition-row atoms, one may argue that the scalar-relativistic approximation is quite enough for these atoms. But for the second transition row, where the relativistic effects play a more important role, a more accurate relativistic treatment appears to be needed.

It is also useful to compare further the efficiency of the exchange-only WDA and Becke's GGA functionals. For this purpose we present in Table 11 $\Delta_{\rm sd}$ for Sc, Ti, V and Cr calculated at the exchange-only level. Hartree-Fock results are also shown as reference. The WDA values are much closer to HF, and the GGA consistently overestimates the stability of the $d^{n-1}s$ configuration. Notice also that, combined with the OLA correlation functional and the relativistic approach, the WDA itself overestimates the $d^{n-1}s$ stability for Ti and V, but yields ICEs in close agreement with the experiment for Sc and Cr. One can expect, therefore, that correcting the GGA functional by correlation and relativistic effects should still lead to a substantial overestimation of the $d^{n-1}s$ stability, in spite of the remarkable accuracy of the total ground-state and exchange energies noticed earlier.

Table 11
Δ_{sd} values (in eV) calculated for several atoms of the first transition metal row with the exchange-only WDA and Becke's GGA [109] functionals. The energy difference between the corresponding lowest Hartree-Fock states [94] is used as a reference.

Atom	WDA	GGA	HF
Sc	1.27	0.38	1.00
Ti	0.39	-0.63	0.54
V	-0.44	-1.58	0.13
Cr	-1.24	-2.48	-1.24

Strictly speaking, Kohn-Sham theory does not justify its use for excited states, although this is done in practice in a routinary way. As an alternative, Shirley and Martin [110] have performed quasiparticle calculations of the s $\rightarrow$ d promotion energies in the 3d series using the GW approximation of Hedin [56]. Without going into a detailed comparison, we just point out that the errors of this calculation are larger, specially in the second half of the series, than the WDA errors.

To sum up, the present density functional scheme with the WDA exchange and OLA correlation provides an unusual example of a successful DFT calculation of the s $\rightarrow$ d ICEs, especially for the most frequently discussed case of atoms of the first transition-metal row. This method appears to produce a balanced description of exchange and correlation effects both in the region where the d-states are localized, and in the outer region occupied by the s-electrons. One may hope that the scheme will be capable of describing the metallic binding in transition metals, clusters and complexes, because the $d^{n-1}s$ configuration is one of the dominating valence configurations for transition metal atoms in those condensed phases, while $d^{n-2}s^2$ is the ground-state configuration for the majority of the free transition metal atoms.

11.3. Ionization potentials

The WDA exchange functional of eq. (79) has been used in [95] to calculate atomic ionization potentials $\mathrm{IP(X)}=E(\mathrm{X}^+)-E(\mathrm{X})$. We present in Table 12 results for the first ionization potential of the atoms Li to Ar from exchange-only LDA, WDA and GGA calculations compared to Hartree-Fock. Both GGA and WDA improve over LDA. The WDA tends to overestimate the ionization potential with respect to HF and the GGA results are better. Nevertheless one has to remember that the WDA functional is parameter free, while GGA contains one fitted parameter.

Inclusion of correlation improves the WDA performance. The IPs obtained using the WDA+OLA [95] scheme are compared in Table 13 to experimental values [111], to local

Table 12
Ionization potentials calculated using several spin-polarized exchange-only approximations: Local Density Approximation (LDA), Weighted Density Approximation (WDA) and Becke's generalized gradient approximation (GGA). Hartree-Fock results from Clementi and Roetti [94] are given for comparison. All results are in eV.

Atom	LDA	WDA	GGA	HF
Li	5.4	4.62	5.36	5.34
Be		7.16	8.07	8.04
B		8.11	7.83	7.93
C		11.27	11.06	10.78
N		14.46	14.30	13.96
O	16.5	13.04	12.28	11.88
F		17.22	16.65	15.72
Ne	22.5	21.33	20.89	19.84
Na		6.16	5.09	4.90
Mg		8.04	6.75	6.61
Al		5.87	5.35	5.51
Si		7.83	7.64	7.66
P		9.90	9.90	10.04
S		10.06	9.03	9.03
Cl		12.44	11.93	11.79
Ar	16.1	14.93	14.73	14.78

spin density (LSD) results [112] and to the HF calculations of ref. [113] in which the correlation contribution was evaluated perturbatively using Perdew's GGA91 [8]. WDA+OLA gives IPs higher than the experimental values with the exception of Li, Be, Si, P, Cl and Ar. This is mainly due to the errors in the exchange contribution to the IPs. However, the IPs are slightly better than those from non-self-consistent HF+GGA91 calculations, given in Table 13 for the first row elements, with the exception of Li and Be

A full perspective of the trends in the IPs can be seen in Figure 10, where we have plotted the WDA results, with and without local correlation. The trends are well reproduced in both cases, and the effect of correlation is, in general, small, increasing the IPs and usually improving the agreement with experiment.

11.4. Electron affinities

Kendall *et al.* [114] have pointed out that the calculation of accurate electron affinities EA=$E(\mathrm{X})-E(\mathrm{X}^{+})$ remains one of the most difficult problems for computational quantum chemistry in general, and DFT in particular. The reasons are the wrong asymptotic behavior of the common exchange-correlation potentials and the fact that the outermost electron of the anion is loosely bound and samples the region far from the nucleus. This wrong asymptotics is due to an insufficient cancellation of electron self-interaction and often leads, as a consequence, to the extra electron of the anion being unbound. Lagowski and Vosko [115,116] have presented a detailed analysis of the ionization potentials and electron affinities in DFT. In particular, exchange and correlation contributions to electron

Table 13
Ionization potentials (in eV) calculated using different exchange-correlation density functionals: local spin density (LSD) [112], spin polarized weighted density approximation plus local correlation (WDA+OLA) [95] and HF exchange with Perdew's GGA91 correlation [113]. Experimental values are taken from ref.[111].

Atom	LSD	WDA+OLA	HF+GGA91	Exp.
Li	5.7	4.67	5.63	5.39
Be	9.1	7.59	9.07	9.32
B	8.8	8.42	8.55	8.30
C	12.1	11.61	11.75	11.26
N	15.3	14.84	15.00	14.53
O	14.2	13.73	13.75	13.61
F	18.4	17.88	17.95	17.42
Ne	22.6	22.00		21.56
Na	5.6	6.26		5.14
Mg		8.38		7.64
Al		6.09		5.98
Si		8.06		8.15
P		10.15		10.48
S		10.59		10.36
Cl		12.94		13.01
Ar	16.2	15.44		15.76

affinities are discussed separately in these articles. One method for binding this electron in the LDA is to enclose the system in a box, so that the highest-lying orbitals remain localized [117]. Another way is to use the Hartree-Fock method (in which self-interaction is automatically canceled) and apply a local or gradient corrected correlation functional in a perturbative way to obtain the correlation energy [92,118]. Using this procedure, the proper sign of the electron affinity (EA) is obtained [119]. EAs have also been quite successfully calculated using the self-interaction corrected (SIC) method [120–122] but this method leads to an orbital-dependent effective potential departing from DFT spirit.

A way of correcting for self-interaction effects within DFT is to use a non-local theory, like the weighted density approximation. As we saw in Section 3.1, the WDA improves the asymptotic behavior of V_{xc}, which becomes $-\frac{1}{2r}$ for neutral atoms, but this improvement is not enough to describe well negative ions. An alternative that has been explored is Slater's transition state method [123]. In this method the energy difference $E(\mathrm{X})-E(\mathrm{X}^+)$ is approximated by the energy eigenvalue (with the opposite sign) of the lowest unoccupied orbital of the neutral atom obtained in a selfconsistent calculation in which the occupation of that orbital is fixed at $\frac{1}{2}$ (instead of 1 for the anion). The results of WDA+OLA and LSDA calculations using this method [124] are compared in Table 14. There is a slight overall improvement over LSDA but with several exceptions, notably those of alkalis. Nevertheless, these results are not definitive since Slater's transition state method is a first order approximation and allows to give only qualitative EAs.

Another alternative is to symmetrize the WDA potential following the ideas of Przy-

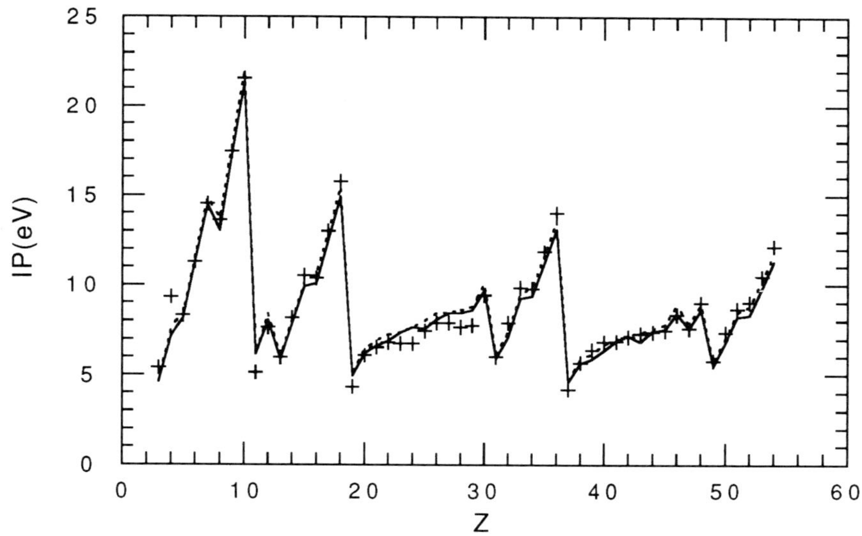

Figure 10. Atomic ionization potentials calculated using WDA exchange with (dashed line) and without (solid line) OLA correlation compared to experiment [111] (crosses).

bylski and Borstel [28,29,126] presented in Section 6. We show in Table 15 the results of spin-polarized exchange-only calculations for neutral atoms and negative ions using the WDAS method; more precisely, the Przybylski-Borstel prescription of Eq. (45) in its spin-polarized version was used for the exchange potential, but exchange energies were obtained with the original WDA functional (79). The HOMO eigenvalue of the anion is given in the second column, and the electron affinity, calculated subtracting the energy of the anion from that of the neutral atom, in the third column. The ionization potential theorem [129] guarantees the exact agreement between $-\varepsilon_{\mathrm{HOMO}}$ and EA, but only for the exact exchange-correlation functional. The agreement is only rough in the WDAS calculation. Although the experimental trends are well reproduced by the calculated affinities (column WDAS), these are, except for F, smaller than the experimental ones (given in the last column of the Table). This is due to the lack of Coulomb correlation in the calculation. In the fourth column of the Table we report the results obtained by Guo and Whitehead [120] using the generalized-exchange local-spin-density functional theory with self-interaction correction (LSD-GX-SIC). Comparing these results with the WDAS column we observe that the WDAS affinities are larger, and closer to experiment, except for Ga, Ge and Br.

By symmetrizing V_x^{WDA} one regains the correct asymptotics but loses variational consistency between the exchange-correlation energy and potential. In order to recover this con-

Table 14
Electron Affinities (in eV) estimated using Slater's transition state method within spin-polarized DFT. Experimental values from [125] are given for comparison.

Atom	WDA+OLA	LSDA	Exp.
Li	0.05	0.62	0.62
B	0.68	0.72	0.28
C	1.77	1.78	1.26
O	1.88	2.10	1.46
F	3.94	4.11	3.40
Na	0.84	0.63	0.55
Al	0.59	0.32	0.46
Si	1.42	1.59	1.39
P	0.94	0.97	0.75
S	2.15	2.42	2.08
Cl	3.51	3.85	3.62
K	0.67	0.58	0.50
Br	3.18	3.52	3.36
I	2.86	3.17	3.06
At	2.65	2.53	2.8

sistency one can use a formula derived by Levy and Perdew [130], based on the Hellmann-Feynman theorem that expresses the exchange energy in terms of the exchange potential and the density:

$$E_x^{\rm LP}[n] = -\int n(\mathbf{r})\,\mathbf{r}\cdot\nabla V_x(\mathbf{r})\,d^3r\ . \qquad (85)$$

We present in Table 16 electron affinities calculated with the WDAS method (that is, using Eq. (45) for $V_x(\mathbf{r})$ and the WDA exchange energy functional including Coulomb correlation (OLA). The results are given in the column labeled WDAS. Another set of results is shown in which the exchange energies were calculated from the Levy-Perdew relation of Eq. (85). These are given under the heading WDAS-LP. Since the exchange-correlation potential is the same for both methods, the single-particle eigenvalues, and in particular $\varepsilon_{\rm HOMO}$ for the anion, are also the same. This eigenvalue is given in column 2.

Inclusion of correlation leads to larger electron affinities (compare the WDAS columns in Tables 15 and 16), bringing the results in closer agreement with experiment, except for F, Na and P. The increase of EA induced by correlation is well understood. Correlation acts mainly between electrons with antiparallel spins, and the number of antiparallel pairs increases in the anion [131]. The comparison between $-\varepsilon_{\rm HOMO}$(anion) and the EAs obtained by subtracting energies shows again only a rough correlation. Comparing the third and fourth columns of Table 16 we observe that the WDAS-LP affinities are less accurate. This suggests to us that, although the *ad hoc* symmetrization of $V_x^{\rm WDA}$ introduces the correct asymptotic behavior in $V_x^{\rm WDAS}(\mathbf{r})$, which is crucial for loosely bound states, the substitution of $V_2^{\rm WDA}$ by $V_1^{\rm WDA}$ may have a drastic effect in the internal part of the atom, leading through eq. (85) to an exchange energy functional not necessarily better than the original WDA functional. Some confirmation for this comes from a comparison between

Table 15
Results of WDAS calculations [127] of the electron affinity compared to experiment [125] and to calculations by Guo and Whitehead [120] using a generalized-exchange local-spin-density functional with self-interaction correction (LSD-GX-SIC). All the calculations neglect Coulomb correlation. The HOMO eigenvalue of the WDA calculation for the anion is also given. All the results are in eV.

Atom	$-\varepsilon_{\mathrm{HOMO}}^{\mathrm{WDAS}}$	WDAS	LSD-GX-SIC	Exp.
F	2.68	3.72	2.37	3.4
Na	1.07	0.48	-0.01	0.55
Mg	0.16	-0.30		< 0
Al	0.58	0.38	0.22	0.46
Si	1.18	1.20	1.09	1.39
P	1.49	0.59		0.74
S	2.26	1.84	1.26	2.08
Cl	3.20	3.17	2.87	3.61
K	0.90	0.34	0.03	0.5
Ca	0.21	-0.22		$< 0^*$
Ga	0.70	0.15	0.20	0.3
Ge	1.36	0.95	1.08	1.2
As	1.55	0.52		0.8
See	2.36	1.58	1.35	2.02
Br	3.27	2.72	2.78	3.36
Rb	0.85	0.26	0.05	0.49
Sr	0.21	-0.25		$< 0^*$

*Recent experiments give positive electron affinities for these elements of about a few hundredths of eV (See [128] and references therein).

the exchange energies of neutral atoms calculated with these two functionals and exact Hartree-Fock energies. We found that the WDAS-LP exchange energies are less accurate. In summary, in spite of the lack of variational consistency, a good compromise seems to be achieved by using the V_x^{WDAS} potential, which incorporates the correct asymptotic behavior, combined with the original WDA exchange energy functional.

In order to compare WDAS results with other recent calculations, we present in the fifth column of Table 16 the EAs calculated by Guo and Whitehead [120] using the generalized exchange local-spin-density-functional theory with self-interaction correction. The correlation contribution to the energy was obtained as a perturbation, using the LDA of Vosko-Wilk-Nusair [132]. The results of Guo and Whitehead are slightly better than WDAS results, reversing the trend found when correlation was not included.

A comparison with other recent self-consistent calculations is provided in Fig. 11, where WDA+OLA affinities for some atoms with $Z < 18$ are plotted together with the results of Grabo and Gross [133]. These authors used the optimized effective potential (OEP) method [134,135] with an approximation due to Krieger *et al.* [136]. The OEP method gives the variationally best local exchange potential. Grabo and Gross included correlation using an energy functional developed by Colle and Salvetti [137], which contains density

Table 16
Results of WDAS+OLA calculations of the electron affinity [127] compared to experiment [125] and to calculations by Guo and Whitehead [120] using a generalized-exchange local-spin-density functional with self-interaction correction (LSDA-GX-SIC). A second set of results is given (WDAS-LP) in which the exchange energies are computed from Eq. (85) of the text. All the calculations in this Table include Coulomb correlation at the LDA level (see text). The HOMO eigenvalue in the WDAS calculation for the anion is also given in column 2. All the results are in eV.

Atom	$-\varepsilon_{\mathrm{HOMO}}^{\mathrm{WDAS}}$	WDAS	WDAS-LP	LSDA-GX-SIC	Exp.
C	0.88	1.74	1.52	1.19	1.27
N	0.32	0.33	1.80		-0.07
O	1.50	2.14	2.98	1.11	1.46
F	2.96	4.08	4.38	3.13	3.4
Na	1.18	0.67	3.57	0.56	0.55
Mg	0.18	-0.27	1.73		< 0
Al	0.65	0.43	2.26	0.31	0.46
Si	1.31	1.24	2.85	1.21	1.39
P	1.69	0.98	4.96		0.74
S	2.49	2.15	5.35	2.02	2.08
Cl	3.45	3.46	6.00	3.53	3.61
K	0.99	0.47	3.18	0.52	0.5
Ca	0.24	-0.18	1.84		$< 0^*$
Ga	0.79	0.16	3.50	0.28	0.3
Ge	1.51	0.96	4.26	1.19	1.2
As	1.76	0.87	6.05		0.8
Se	2.59	1.89	6.67	2.08	2.02
Br	3.52	2.99	7.38	3.41	3.36
Rb	0.94	0.38	3.27	0.52	0.49
Sr	0.23	-0.22	1.93		$< 0^*$

*Recent experiments give positive electron affinities for these elements of about a few hundredths of eV (See [128] and references therein).

gradients. First of all, the figure shows the already discussed fact that the use of Eq. (85) for the exchange energies worsens the results of the WDAS calculation. Let us now focus on the comparison between WDAS and OEP. If we compare EAs obtained as energy differences, we see that both methods reproduce well the experimental trend (notice that the elements with $Z = 7, 10, 12$ are missing from the plot), and that the WDAS affinities are closer to experiment. With respect to $-\varepsilon_{\mathrm{HOMO}}$, we see that the experimental trends are reproduced in both cases.

11.5. Molecules

The same reasoning that lead Gunnarsson and coworkers [12] to separate intershell and intrashell effects in atoms (see Section 3.3) indicates that it is advisable to separate valence from core electrons to study molecules, since only the former ones play a role in

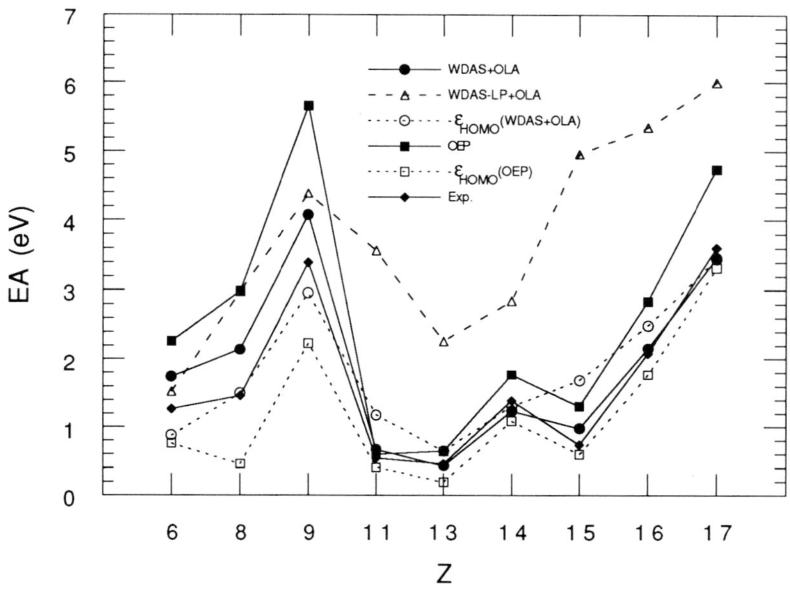

Figure 11. Experimental [125] and calculated electron affinities (in eV). All calculations include correlation (see text). An estimation of the affinity given by the negative of the eigenvalue of the highest occupied orbital of the anion is also given. Results of similar calculations [133] using the optimized effective potential (OEP) method are given for comparison.

molecular bonding. There is evidently one case in which this is not necessary: the H_2 molecule, because in this dimer the two electrons are valence electrons. We present in this Section some preliminary results on the dissociation energy (D_e) at the experimental bond length (1.4 a.u.) of this molecule.

At the exchange-only level, the WDAS functional gives a binding energy of 3.65 eV while Becke's GGA [109] functional gives 3.70 eV. The WDAS result is slightly closer to the unrestricted HF value of 3.62 eV [138] than the GGA one. Table 17 shows that inclusion of Coulomb correlation brings the computed binding energies in better agreement with the experimental value. We present in this Table results for the dissociation energy obtained with several exchange-correlation functionals: Slater's X_α, LDA, Becke's GGA [109] (B) and WDAS exchange-only functionals, combined with Vosko-Wilk-Nusair [132] (VWN) and Lee-Yang-Parr [97] (LYP) correlation functionals. WDAS results are similar to GGA and better than LDA values. Nevertheless, it is necessary to perform further calculations to test the WDA in molecules, not only for binding energies but also for the rest of the spectroscopic constants. Work is now under progress in this direction.

Table 17
Dissociation energy at experimental bond length (D_e) of the H_2 molecule computed using several exchange-correlation functionals.

Functional	X_α	LDA	B-VWN	B-LYP	WDAS-LYP	Exp.
D_e (eV)	3.58	4.90	5.08	4.73	4.69	4.75

12. KINETIC ENERGY

A WDA approximation has also been developed for the kinetic energy of an inhomogeneous electron system [11]. Let us begin with the usual expression for the kinetic energy of a many-electron system

$$T = -\tfrac{1}{2}\int \left[\nabla^2_{\mathbf{r}}\gamma(\mathbf{r}',\mathbf{r})\right]_{\mathbf{r}=\mathbf{r}'} d^3r \ , \tag{86}$$

where $\gamma(\mathbf{r},\mathbf{r}')$ is the one-particle density matrix (its diagonal part is the electron density $n(\mathbf{r})$). We now restrict to Hartree-Fock theory, that is, we neglect Coulomb correlation. Then, the one-particle and two-particle density matrices are related through the following equation [139]

$$\gamma(\mathbf{r}_1,\mathbf{r}_2)\gamma(\mathbf{r}_2,\mathbf{r}_1) = 2n(\mathbf{r}_1)n(\mathbf{r}_2) - 4\Gamma(\mathbf{r}_1,\mathbf{r}_2) \ . \tag{87}$$

On the other hand, from eq. (3) we have a relation between the density, the two-particle density matrix $\Gamma(\mathbf{r}_1,\mathbf{r}_2)$ and the pair-correlation function

$$g_x(\mathbf{r}_1,\mathbf{r}_2)n(\mathbf{r}_1)n(\mathbf{r}_2) = 2\Gamma(\mathbf{r}_1,\mathbf{r}_2) \ . \tag{88}$$

Combining eqs. (87) and (88) we obtain

$$\gamma(\mathbf{r}_1,\mathbf{r}_2)\gamma(\mathbf{r}_2,\mathbf{r}_1) = 2n(\mathbf{r}_1)n(\mathbf{r}_2)\left[1 - g_x(\mathbf{r}_1,\mathbf{r}_2)\right] \ . \tag{89}$$

We now assume the symmetry of $\gamma(\mathbf{r}_1,\mathbf{r}_2)$, that is

$$\gamma(\mathbf{r}_1,\mathbf{r}_2) = \gamma(\mathbf{r}_2,\mathbf{r}_1) \ . \tag{90}$$

that, although not exactly fulfilled in general, is correct in some limiting cases, like for closed shell atoms or for the homogeneous electron gas. Using this symmetry condition, eq. (89) becomes

$$\gamma(\mathbf{r}_1,\mathbf{r}_2) = \left\{2n(\mathbf{r}_1)n(\mathbf{r}_2)\left[1 - g_x(\mathbf{r}_1,\mathbf{r}_2)\right]\right\}^{\frac{1}{2}} \ . \tag{91}$$

Defining the kinetic energy density $t(\mathbf{r}_1)$ by

$$T = \int t(\mathbf{r}_1)\, d^3r_1 \ , \tag{92}$$

we then obtain $t(\mathbf{r}_1)$ from the Laplacian of (91)

$$\begin{aligned} t(\mathbf{r}_1) = & - \frac{1}{2}n(\mathbf{r}_1)\left[\nabla^2_{\mathbf{r}_1}A(\mathbf{r}_1,\mathbf{r}_2)\right]_{\mathbf{r}_1=\mathbf{r}_2} + \frac{1}{8}\frac{(\nabla n(\mathbf{r}_1))^2}{n(\mathbf{r}_1)} \\ & - \frac{1}{4}\nabla^2 n(\mathbf{r}_1) - \frac{1}{2}\left[\nabla_{\mathbf{r}_1}A(\mathbf{r}_1,\mathbf{r}_2)\right]_{\mathbf{r}_1=\mathbf{r}_2}\nabla n(\mathbf{r}_1) \ , \end{aligned} \tag{93}$$

where

$$A(\mathbf{r}_1, \mathbf{r}_2) = \{2\,[1 - g_x(\mathbf{r}_1, \mathbf{r}_2)]\}^{\frac{1}{2}} \ . \tag{94}$$

In arriving at eq. (93), the relation $A(\mathbf{r}_1, \mathbf{r}_1) = 1$, valid in Hartree-Fock, has been used. Relation (93) is formally exact for cases when the symmetry relation (90) is fulfilled, and it displays the close relationship between kinetic and exchange energies through the pair-correlation function. Thus, approximations to the exchange energy formulated in terms of an approximate $g_x(\mathbf{r}_1, \mathbf{r}_2)$ can be translated into approximations to the kinetic energy.

The simplest approximation to $t(\mathbf{r}_1)$ is obtained by applying a local density approximation to $A(\mathbf{r}_1, \mathbf{r}_2)$, that is

$$A^{\mathrm{LDA}}(\mathbf{r}_1, \mathbf{r}_2) = \left\{2\left[1 - g_x^{\mathrm{hom}}(r_{12}, n(\mathbf{r}_1))\right]\right\}^{\frac{1}{2}} \ . \tag{95}$$

In this case, the kinetic energy density becomes

$$t(\mathbf{r}_1) = c_k n(\mathbf{r}_1)^{\frac{5}{3}} + \frac{1}{8}\frac{(\nabla n(\mathbf{r}_1))^2}{n(\mathbf{r}_1)} - \frac{1}{4}\nabla^2 n(\mathbf{r}_1) \ , \tag{96}$$

where the first term is the Thomas-Fermi approximation ($c_k = \frac{3}{10}(3\pi^2)^{2/3}$), and the second one is the quantum correction of Weizsäcker [140]. The last term integrates to zero in cases of physical interest.

A nonlocal approximation to $t(\mathbf{r}_1)$ can now be derived if we use in (94) the WDA form for $g_x(\mathbf{r}_1, \mathbf{r}_2)$

$$g_x^{\mathrm{WDA}}(\mathbf{r}_1, \mathbf{r}_2) = g_x^{\mathrm{hom}}(r_{12}, \tilde{n}(\mathbf{r}_1)) \ , \tag{97}$$

where the weighted density is determined in the usual way from eq. (22). The following functional is then obtained

$$T^{\mathrm{WDA}}[n] = \int \left[c_k n(\mathbf{r})\tilde{n}(\mathbf{r})^{\frac{2}{3}} + \frac{1}{8}\frac{(\nabla n(\mathbf{r}))^2}{n(\mathbf{r})} - \frac{1}{4}\nabla^2 n(\mathbf{r}_1)\right] d^3r \ . \tag{98}$$

Notice that the improvement in $g_x(\mathbf{r}_1, \mathbf{r}_2)$ only affects the first term in eq. (96).

It may be useful to recall that the Weizsäcker term, without any other contribution, gives the exact kinetic energy of the hydrogen atom (or any one-electron ion) and of any two-electron system at the Hartree-Fock level. Also, the Weizsäcker kinetic energy density is exact in the tail region of atoms, molecules and clusters [141]. This kinetic energy functional was first tested nonselfconsistently using Hartree-Fock atomic densities, and promising results were obtained [11]. However, a proper test of energy density functionals is only done through a selfconsistent calculation, and this is treated in the next Section.

13. UNIFIED WDA METHOD FOR KINETIC AND EXCHANGE ENERGIES

By combining $T^{\mathrm{WDA}}[n]$ and $E^{\mathrm{WDA}}[n]$ we arrive at the following approximation for the total energy of a many-electron system in an external potential $V_{\mathrm{ext}}(\mathbf{r}_1)$ (neglecting Coulomb correlation) [142]

$$E[n] = T^{\mathrm{WDA}}[n] + E_x^{\mathrm{WDA}}[n] + \frac{1}{2}\iint \frac{n(\mathbf{r}_1)n(\mathbf{r}_2)}{|\mathbf{r}_1 - \mathbf{r}_2|}\, d^3r_1\, d^3r_2 + \int n(\mathbf{r}_1)V_{\mathrm{ext}}(\mathbf{r}_1)\, d^3r_1 \ . \tag{99}$$

The last two terms are the classical electron-electron interaction and the interaction of the electrons with the external potential respectively. Minimization of the total energy, keeping the density normalized to the correct number of electrons, gives the Euler equation

$$\mu = V_T(\mathbf{r}) + \frac{1}{8}\left[\frac{\nabla n(\mathbf{r})}{n(\mathbf{r})}\right]^2 - \frac{1}{4}\frac{\nabla^2 n(\mathbf{r})}{n(\mathbf{r})} + c_k\left[n(\mathbf{r})^{\frac{2}{3}} + \frac{2}{3}\int n(\mathbf{r}')\tilde{n}(\mathbf{r}')^{-\frac{1}{3}}\frac{\delta\tilde{n}(\mathbf{r}')}{\delta n(\mathbf{r})}d^3r'\right] , \quad (100)$$

where $V_T(\mathbf{r})$ is the total potential:

$$V_T(\mathbf{r}) = V_{\text{ext}}(\mathbf{r}) + V_H(\mathbf{r}) + V_x^{\text{WDA}}(\mathbf{r}) , \quad (101)$$

sum of the external potential, the Hartree potential of the electronic cloud and the exchange potential.

Deb and Gosh [143] were the first authors who studied equation (100). Following them, we define the function

$$b(\mathbf{r}) = \frac{3}{5}\left[\frac{\tilde{n}(\mathbf{r})}{n(\mathbf{r})}\right]^{\frac{2}{3}} + \frac{2}{5}n(\mathbf{r})^{-\frac{2}{3}}\int n(\mathbf{r}')\tilde{n}(\mathbf{r}')^{-\frac{1}{3}}\frac{\delta\tilde{n}(\mathbf{r}')}{\delta n(\mathbf{r})}d^3r' . \quad (102)$$

Using this expression, the Euler equation (100) takes the form

$$\mu = V_T(\mathbf{r}) + \frac{1}{8}\left[\frac{\nabla n(\mathbf{r})}{n(\mathbf{r})}\right]^2 - \frac{1}{4}\frac{\nabla^2 n(\mathbf{r})}{n(\mathbf{r})} + \frac{5}{3}c_k\, b(\mathbf{r})\, n(\mathbf{r})^{\frac{2}{3}} , \quad (103)$$

which closely resembles the form of the Euler equation for the Thomas-Fermi-Weizsäcker functional. Deb and Gosh solved this equation for inert gas atoms using two approximations: first, $b(\mathbf{r})$ was evaluated from Hartree-Fock densities [94], and second, the exchange potential $V_x(\mathbf{r})$ was approximated by the LDA. Instead, Glossman *et al.* have solved eq. (103) selfconsistently. By defining $\Psi(\mathbf{r}) = n(\mathbf{r})^{\frac{1}{2}}$, and setting

$$V_{\text{eff}}(\mathbf{r}) = V_T(\mathbf{r}) + \frac{5}{3}c_k\, b(\mathbf{r})\, n(\mathbf{r})^{\frac{2}{3}} , \quad (104)$$

eq. (103) can be viewed as a Schrödinger-like equation

$$\left[-\tfrac{1}{2}\nabla^2 + V_{\text{eff}}(\mathbf{r})\right]\Psi(\mathbf{r}) = \mu\Psi(\mathbf{r}) , \quad (105)$$

which can be solved by conventional methods for one "fictitious" s-type orbital normalized to Z electrons for a neutral atom (or $Z-1$ for a singly positively charged ion; Z indicates the atomic number) [142,144].

14. APPLICATION OF THE UNIFIED WDA METHOD TO ATOMS

The Euler equation (105) has been first solved for the inert gas atoms [144], and the calculated total energies are reported in Table 18, together with the Hartree-Fock energies [94]. The energy of He is reproduced exactly. This is understandable from the arguments presented in Section 3.1 above for two-electron atoms. In addition, this result provides a check of the numerical code used. Also for the other inert gases the total energies are very accurate (the average relative error is only 0.4%). The Table also reports the energies that have been calculated with two popular energy functionals. Both functionals use the LDA for exchange. For the kinetic energy, one of the functionals (TFDK) includes the local Thomas-Fermi term and the second-order gradient correction deduced by Kirzhnits [145] in a series expansion of the kinetic energy in the gradients of the electron density

$$T[n] = T_0[n] + T_2[n] = \int c_k n(\mathbf{r})^{\frac{5}{3}}\, d^3r + \frac{1}{72}\int \frac{(\nabla n(\mathbf{r}))^2}{n(\mathbf{r})}\, d^3r\ , \tag{106}$$

and the other (TFDW) is analogous to eq. (106) but with the original coefficient ($\frac{1}{8}$) of Weizsäcker [140] in the second term (instead of $\frac{1}{72}$). The difference in the value of this coefficient is now understood: Kirzhnits value ($\frac{1}{72}$) is the correct one when $T[n]$ is expressed as a series expansion in ∇n for slowly varying densities. The corrections to eq. (106) are terms of higher order in ∇n. On the other hand, the Thomas-Fermi-Weizsäcker functional is the starting point for other functionals in which the Weizsäcker term is a fundamental component and improvements arise by modifying the Thomas-Fermi term [11]. One of those improved functionals is T^{WDA} of eq. (98). The results of Table 18 indicate that E^{TFDK} overestimates the binding of the inert gas atoms, whereas E^{TFDW} underestimates it. The errors of both functionals are large in comparison to the errors of E^{WDA}. The Table also contains results obtained by Engel and Dreizler [146], with a functional including up to fourth-order gradient corrections ($T_0+T_2+T_4$) in the kinetic energy (in this calculation exchange was treated by the LDA). Although the inclusion of the fourth-order term (T_4) reduces the errors of the TFDK to about one half of their original value, the WDA functional is still substantially more accurate. The success of the WDA functional extends to other atoms (or ions) with filled and half-filled shells: results for some alkali and alkaline-earth atoms and singly-charged ions are given in Table 19.

Table 18
Total ground state energy of inert gas atoms (in a.u.) for several density functionals: WDA, TFDK, TFDW and $T_0+T_2+T_4$ (see text). Hartree-Fock (HF) energies [94] are given as a reference.

Atom	HF	TFDK	TFDW	WDA	$T_0+T_2+T_4$
He	-2.86	-3.22	-1.48	-2.86	-2.96
Ne	-128.55	-139.88	-85.73	-130.01	-133.74
Ar	-526.82	-561.80	-374.15	-526.83	-542.12
Kr	-2752.05	-2896.93	-2095.43	-2754.79	-2819.12
Xe	-7232.13	-7495.22	-5695.87	-7261.80	-7383.55
Rn	-21866.8		-17823.32	-22099.06	-22256.3

The electron density provides a stringent test about the quality of approximate kinetic energy functionals. Table 20 gives the electron density at the nucleus, $n(0)$, for the same functionals compared in Table 18. The functional TFDK drastically overestimates $n(0)$, missing even the order of magnitude. TFDW brings the values into the correct order of magnitude, although it underestimates $n(0)$. To understand this improvement we should notice that a kinetic energy functional containing the full Weizsäcker term

Table 19
Total ground state energies (in a.u.) of atoms and ions of the first and second rows of the Periodic Table, compared to Hartree-Fock energies [94].

Atom	HF	WDA	Ion	HF	WDA
Li	-7.43	-7.40	Li^+	-7.24	-7.24
Be	-14.57	-14.68	Be^+	-14.28	-14.21
Na	-161.86	-163.10	Na^+	-161.68	-162.82
Mg	-199.62	-200.60	Mg^+	-199.37	-200.33
K	-599.17	-599.26	K^+	-599.02	-599.01
Ca	-676.76	-677.24			

"only", reproduces the cusp condition at the position of the nucleus (of atomic number Z) [147], that is, the condition $\frac{n'(0)}{2Zn(0)} = 1$. The WDA functional fulfills this cusp condition reasonably well [148]. The expectation value $\langle r^{-1}\rangle = \int \frac{n(\mathbf{r})}{r}\, d^3r$ is given in Table 21. The good agreement between the WDA and HF results is again due to the good description of the electron density near the nucleus.

Table 20
Electron densities at the site of the nucleus (in a.u.) for noble gases ($\frac{n(0)}{Z^3}$).

Atom	HF	TFDK	TFDW	WDA
He	0.450	6.296	0.118	0.450
Ne	0.620	6.437	0.169	0.556
Ar	0.658	6.438	0.187	0.577
Kr	0.691	6.449	0.204	0.596
Xe	0.706	6.470	0.212	0.603
Rn	0.720		0.220	0.605

Table 21
$\langle r^{-1}\rangle$ expectation values (in a.u.).

Atom	HF	TFDK	TFDW	WDA
He	3.37	3.65	1.75	3.38
Ne	31.11	33.00	21.38	30.84
Ar	69.72	73.79	51.72	69.60
Kr	183.85	190.26	143.76	182.56
Xe	317.88	328.44	259.18	321.30
Rn		604.39	506.22	613.28

The shape of the WDA electron densities is also worth to be discussed [148]. The radial densities ($4\pi r^2 n(r)$) of the inert gas atoms have been plotted in Fig. 12. An analysis of this and similar plots for all the atoms in the main groups of the Periodic Table reveals the existence of two local maxima at radial positions close to those of the first two maxima of the HF density. This is true for all but the lightest atoms, where a shoulder appears instead of a well developed second maximum. The evolution of the relative height of the two maxima is also in qualitative agreement with the HF trend. A direct comparison of the krypton densities obtained by several methods is presented in Figure 13. The TFDK density shows only one maximum. The radial positions of the two maxima of the WDA density agree well with the positions of the first two HF maxima.

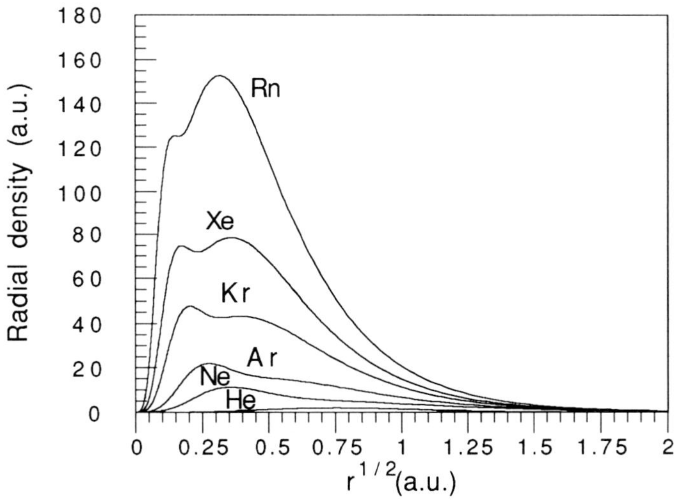

Figure 12. Radial density ($4\pi r^2 n(r)$) of noble gas atoms obtained by the WDA.

In order to interpret the two maxima of the WDA density we first recall that the Weizsäcker term alone, that is, the second term on the r.h.s of eq. (96), without any other contribution gives the exact kinetic energy of a two-electron system at the exchange-only (or HF) level. In addition, the value of the weighted density $\tilde{n}(\mathbf{r})$ for the He atom, obtained from the sum rule (8) equals zero at all points $\mathbf{r}$. This effect, discussed in Section 3.1 above, is the basis of the result $E_x^{\mathrm{WDA}}(\mathrm{He}) = E_x^{\mathrm{HF}}(\mathrm{He})$. But we now focus on another consequence of the result $\tilde{n}(\mathbf{r}) = 0$ for He. The term in the kinetic energy containing $\tilde{n}(\mathbf{r})^{\frac{2}{3}}$ is in this case zero, and $T^{\mathrm{WDA}}[n]$ reduces to the pure Weizsäcker term. For an atom with more than two electrons we can still expect $\tilde{n}(\mathbf{r}) \ll n(\mathbf{r})$ in the region of 1s core shell, so the Weizsäcker term will dominate the kinetic energy density in that region. This argument suggests that the Weizsäcker term in the WDA functional describes the kinetic energy of the 1s core shell, and that this term, plus the nonlocal term $\int c_k n(\mathbf{r})\tilde{n}(\mathbf{r})^{\frac{2}{3}}\, d^3r$ accounts for the kinetic energy of the remaining shells. According to this interpretation,

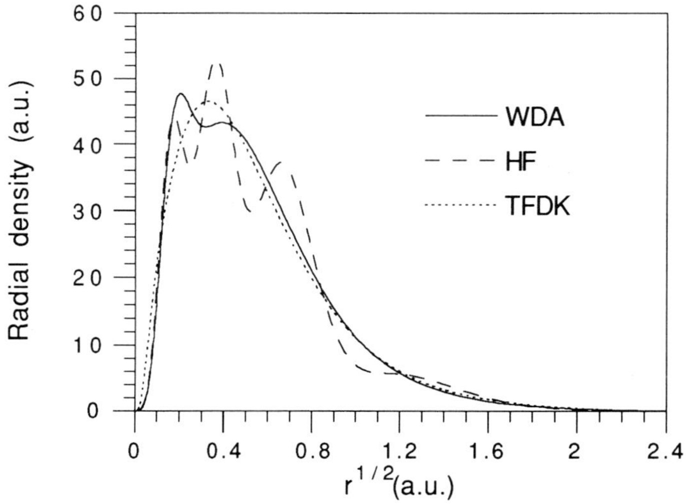

Figure 13. Radial density of Krypton, obtained by the WDA, gradient expansion (T_0+T_2) and HF methods.

the first oscillation in the radial densities plotted in Fig 12 is to be associated with the 1s shell.

To substantiate this interpretation we represent in Figure 14 the position of the maximum of the radial density of the 1s shell obtained from HF calculations versus the position of the first maximum of the WDA radial density. The atoms included in the plot correspond to the main groups of the Periodic Table, that is, the data set excludes transition metal elements. The points of this plot are fitted very well by a linear relation

$$r_{\max}^{\rm HF}(1s) = r_{\max}^{\rm WDA}(1s) - 0.0201 \ , \tag{107}$$

that is, the overall shift is only 0.0201 a.u..

15. FURTHER APPLICATIONS AND POSSIBLE IMPROVEMENTS

In this review we have presented a number of applications of the WDA to bulk solids, surfaces, clusters and atoms. A Kohn-Sham formulation of DFT is, of course, required in any attempt at achieving chemical accuracy but a formulation based on the electron density may help to obtain trends and qualitative answers.

Some workers have performed DFT molecular dynamics (MD) simulations for liquid and solid metals [149,150], surfaces [151] and clusters [152–154] employing kinetic energy functionals which are extensions of the Thomas-Fermi theory including gradient corrections. We expect that the WDA functional for kinetic and exchange energies presented in Section 13, combined with a description of the core electrons by local pseudopotentials, could be useful for performing MD simulations in similar materials. The time-consuming

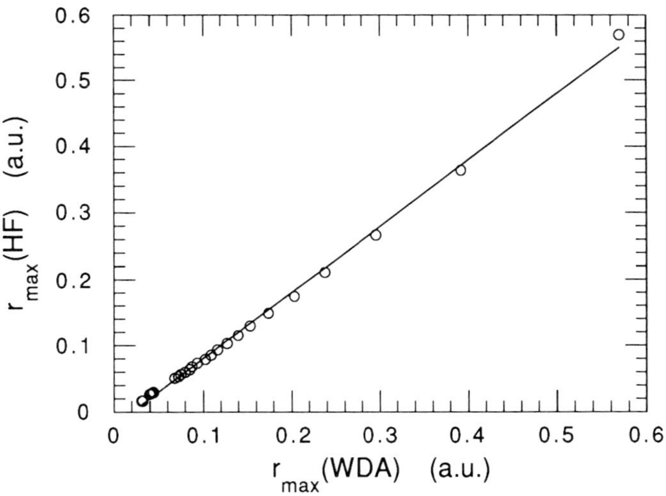

Figure 14. Position of the maximum of the radial density of the 1s shell obtained in a HF calculation, versus the position of the first maximum of the WDA radial density. The data set includes atoms from the main groups of the Periodic Table.

step in those simulations would be the calculation of the weighted density $\tilde{n}(\mathbf{r})$, so devising practical ways to approximate this quantity with a modest computational effort could be of great potential interest.

Turning now to the Kohn-Sham formulation, a problem that appears to be beyond the capabilities of LDA or GGA functionals is that of van der Waals forces. The potential energy curves for rare-gas pairs have been calculated by Pérez-Jordá and Becke [155]. The LDA severely overestimates the dissociation energies and the GGA functionals give a purely repulsive interatomic potential. Similar conclusions have been reached by Nardelli [156] in a study of the interaction of He with a semiconductor surface (InSb). van der Waals forces arise from the nonlocality of the correlation energy and this suggests the application of WDA for exchange and correlation.

The most difficult task in looking for improvements of the present versions of the WDA is that of devising accurate forms of the pair-correlation function $g_{xc}(\mathbf{r}_1, \mathbf{r}_2)$ in systems with an inhomogeneous density. Most problems (although not all of them) of interest in condensed matter physics, involving bonding between atoms, only concern the valence electrons. For those systems, a core-valence separation is recommended. In this case, modeling $g_{xc}(\mathbf{r}_1, \mathbf{r}_2)$ for the valence electron cloud appears to be feasible with the present approximations. The results of Singh for metals and semiconductors [34] substantiate this recommendation.

A big step forward would be achieved if one comes out with an "accurate" symmetrized form for $g_{xc}(\mathbf{r}_1, \mathbf{r}_2)$, fulfilling eq. (6). However, this is expected to be a difficult task.

ACKNOWLEDGMENTS

This work has been supported by DGICYT (Grant PB92-0645). One of us (N.A.C.) is a DGICYT Doctoral Fellow. We acknowledge the help of A. Rubio and D. Glossman in the preparation of the figures.

REFERENCES

1. W. Kohn and L.J. Sham, Phys. Rev. **140**, A1133 (1965).
2. P. Hohenberg and W. Kohn, Phys. Rev. **136**, B864 (1964).
3. N.H. March and S. Lundqvist (eds.), *Theory of the Inhomogeneous Electron Gas*, Plenum Press, New York, 1983.
4. R.G. Parr and W. Yang, *Density Functional Theory of Atoms and Molecules*, Oxford University Press, New York, 1989.
5. R.M. Dreizler and E.K.U. Gross, *Density Functional Theory*, Springer Verlag, Berlin, 1990.
6. E.K.U. Gross and R.M. Dreizler (eds.), *Density Functional Theory*, NATO ASI Series B, Vol. 337, Plenum Press, New York, 1995.
7. D.C. Langreth and M.J. Mehl, Phys. Rev. B **28**, 1809 (1983).
8. J.P. Perdew, Physica B **172**, 1 (1991).
9. A.D. Becke, J. Chem. Phys. **84**, 4524 (1986).
10. J.P. Perdew and A. Zunger, Phys. Rev. B**23**, 5048 (1981).
11. J.A. Alonso and L.A. Girifalco, Solid State Commun. **24**, 135 (1977); Phys. Rev. B **17**, 3735 (1978).
12. O. Gunnarsson, M. Jonson, and B. Lundqvist, Solid State Commun. **24**, 765 (1977); Phys. Rev. B**20**, 3136 (1979).
13. J. Harris and R.O. Jones, J. Phys. F**4**, 1170 (1974).
14. O.V. Gritsenko, R. van Leeuwen, and E.J. Baerends, J. Chem. Phys. **101**, 8955 (1994).
15. R. van Leeuwen, O.V. Gritsenko, and E.J. Baerends, Z. Phys. D **33**, 229 (1995).
16. E. Chacón and P. Tarazona, Phys. Rev. B **37**, 4013 (1988).
17. E. Chacón and P. Tarazona, Phys. Rev. B **37**, 4020 (1988).
18. L. Lantto, Phys. Rev. B **22**, 1380 (1980).
19. K.S. Singwi, M.P. Tosi, R.H. Land, and A. Sjölander, Phys. Rev. **176**, 589 (1968).
20. D.M. Ceperley and B.J. Alder, Phys. Rev. Lett. **45**, 566 (1980).
21. J.P. Perdew and Y. Wang, Phys. Rev. B **46**, 12947 (1992).
22. E. Clementi, IBM Jour. Res. Develop. **9**, 2 (1965).
23. O. Gunnarsson and R.O. Jones, Phys. Scripta **21**, 394 (1980).
24. G. Borstel, M. Newman, and W. Braun, Phys. Rev. B **23**, 3113 (1981).
25. G.P. Kerker, Phys. Rev. B **24**, 3468 (1981).
26. F. Manghi, G. Riegler, C.M. Bertoni, C. Calandra, and G.B. Bachelet, Phys. Rev. B **28**, 6157 (1983).
27. C.S. Wang, J. Magn. Mag. Mater. **31-34**, 95 (1983).
28. H. Przybylski and G. Borstel, Solid State Commun. **49**, 317 (1984).
29. H. Przybylski and G. Borstel, Solid State Commun. **52**, 713 (1984).

30. M.S. Hybertsen and S.G. Louie, Solid State Commun. **51**, 451 (1984).
31. M.S. Hybertsen and S.G. Louie, Phys. Rev. B **30**, 5777 (1984).
32. F. Manghi, G. Riegler, C.M. Bertoni, and G. Bachelet, Phys. Rev. B **31**, 3680 (1985).
33. P. Kruger, G. Wolfgarten, and J. Pollman, Solid State Commun. **53**, 885 (1985).
34. D.J. Singh, Phys. Rev. B **48**, 14099 (1993).
35. R.O. Jones and O. Gunnarsson, Rev. Mod. Phys. **61**, 689 (1989).
36. D.J. Singh, H. Krakauer, C. Haas, and A.Y. Liu, Phys. Rev. B **46**, 13065 (1992).
37. N. Troullier and J.L. Martins, Phys. Rev. B **43**, 1993 (1991).
38. A. Garcia, C. Elsasser, J. Zhu, S.G. Louie, and M.L. Cohen, Phys. Rev. B **46**, 9829 (1992).
39. Y.M. Juan and E. Kaxiras, Phys. Rev. B **48**, 14944 (1993).
40. A.R. Williams and U. von Barth, in ref. [3], p. 189.
41. G. Binning, N. Garcia, H. Rohrer, J.M. Soler, and F. Flores, Phys. Rev. B **30**, 4816 (1984).
42. N. Garcia, B. Reihl, K.H. Frank, and A.R. Williams, Phys. Rev. Lett. **54**, 591 (1985).
43. J. Bardeen, Phys. Rev. **38**, 727 (1940).
44. E. Zaremba and W. Kohn, Phys. Rev. B **13**, 2270 (1976).
45. S. Ossicini, C.M. Bertoni, and P. Gies, Europhys. Letters **1**, 661 (1986).
46. S. Ossicini, C.M. Bertoni, and P. Gies, Surf. Sci. **178**, 244 (1986).
47. S. Ossicini, F. Finocchi, and C.M. Bertoni, Surf. Sci. **189/190**, 776 (1987).
48. S. Ossicini and C.M. Bertoni, J. Vac. Sci. Technol. A **5**, 727 (1987).
49. F. Manghi, Phys. Rev. B **33**, 2554 (1986).
50. N. Esbjerg and J.K. Norskov, Phys. Rev. Lett. **45**, 807 (1980).
51. W.A. de Heer, W.D. Knight, M. Chou, and M.L. Cohen, Solid State Phys. Vol. 40, eds. H. Ehrenreich, F. Seitz and D. Turnbull (Academic Press, New York, 1987) p. 93.
52. W.Ekardt and Z. Penzar, Phys. Rev. B **38**, 4273 (1988); *Ibid.* **43** 1322 (1991).
53. M. Koskinen, P.O. Lipas, and M. Manninen, Europhys. Lett. **30**, 519 (1995).
54. W. Ekardt, Phys. Rev. B **29**, 1558 (1989).
55. L.C. Balbás, J.A. Alonso, and A. Rubio, Europhys. Letters **14**, 323 (1991).
56. L. Hedin and S. Lundqvist, Solid State Phys. Vol. 23, eds. F. Seitz and D. Turnbull (Academic Press, New York, 1969) p. 1.
57. S. Saito, S.B. Zhang, S.G. Louie, and M.L. Cohen, Phys. Rev. B **40**, 3643 (1989).
58. M.M. Kappes, M. Schär, U. Röthlisberger, C. Yeretzian, and E. Schumacher, Chem. Phys. Lett. **143**, 25 (1988).
59. A. Rubio, L.C. Balbás, and J.A. Alonso, Phys. Rev. B **46**, 4891 (1992).
60. L.C. Balbás, A. Rubio, J.A. Alonso, and G. Borstel, Chem. Phys. **120**, 239 (1988).
61. K.J. Taylor, C.L. Pettiete, M.J. Craycraft, O. Chesnovsky, and R.E. Smalley, Chem. Phys. Lett. **152**, 347 (1988).
62. A. Rubio, L.C. Balbás, and J.A. Alonso, Physica B **168**, 32 (1991).
63. W.D. Knight, K. Clemenger, W.A. de Heer, and W. Saunders, Phys. Rev. B **31**, 2539 (1985).
64. W.A. de Heer, P. Milani, and A. Châtelain, Phys. Rev. Lett. **63**, 2834 (1989).
65. D.E. Beck, Phys. Rev. B **30**, 6935 (1989).
66. À. Rubio, L.C. Balbás Ll. Serra, and M. Barranco, Phys. Rev. B **42**, 10950 (1990).
67. J.M. Pacheco and W. Ekardt, Ann. Phys. (Leipzig) **1**, 255 (1992).

68. M.J. Stott and E. Zaremba, Phys. Rev. A **21**, 12 (1980); A. Zangwill and P. Soven, Phys. Rev. A **21**, 1561 (1980).
69. M Petersilka, U.J. Grossman, and E.K.U. Gross, Phys. Rev. Lett. **76**, 1212 (1996).
70. W.A. de Heer, K. Selby, V. Kresin, J. Masui, M. Vollmer, A. Châtelain, and W.D. Knight, Phys. Rev. Lett. **59**, 1805 (1987).
71. G. Mie, Ann. Phys. (Leipzig) [Folge 4] **25**, 377 (1908).
72. M. Brack, Rev. Mod. Phys. **65**, 677 (1993).
73. K. Selby, V. Kresin, J. Masui, M. Vollmer, W.A. de Heer, A. Scheidemann, and W.D. Knight, Phys. Rev. B **43**, 4565 (1991).
74. C.R.C. Wang, S. Pollack, D. Cameron, and M.M. Kappes, J. Chem. Phys. **93**, 3789 (1990); S. Pollack, C.R.C. Wang, and M.M. Kappes, J. Chem. Phys. **94**, 2496 (1991).
75. R. Thalweiser, C. Rothenfusser, V. Weiss, and G. Gerber, work quoted in V. Bonacic-Koutecky, P. Fantucci, C. Fuchs, C. Gatti, J. Pittner, and S. Polezzo, Chem. Phys. Lett. **213**, 522 (1993).
76. S. Saito, G.F. Bertsch, and D. Tomanek, Phys. Rev. B **43**, 6804 (1991).
77. J.A. Alonso, A. Rubio and L.C. Balbás, Philos. Mag. B **69**, 1037 (1994).
78. C. Bréchignac, Ph. Cahuzac, F. Carlier, M. de Frutos, and J. Leygnier, Chem. Phys. Lett. **189**, 28 (1992).
79. J. Borgreen, P. Chowdhury, N. Kebaïli, L. Lundsberg-Nielsen, K. Lützenkirchen, M.B. Nielsen, J. Pedersen, and M.D. Rasmussen, Phys. Rev. B **48**, 17507 (1993).
80. Th. Reiners, W. Orlik, Ch. Ellert, M. Schmidt, and H. Haberland, Chem. Phys. Lett. **215**, 357 (1993).
81. M.P. Iñiguez, M.J. López, J.A. Alonso, and J.M. Soler, Z. Phys. D **11**, 163 (1989).
82. N.W. Ashcroft, Phys. Letters **23**, 48 (1966).
83. J.A. Alonso, L.C. Balbás, and A. Rubio, Int. J. Quantum Chem. **56**, 499 (1995).
84. O.V. Gritsenko, A. Rubio, L.C. Balbás, and J.A. Alonso, Chem. Phys. Lett. **205**, 348 (1993).
85. A. Veillard and E. Clementi, J. Chem. Phys. **49**, 2415 (1968).
86. L. Fritsche, J. Noffke, and H. Gollisch, J. Phys. B **17**, 1637 (1984).
87. L. Fritsche and H. Gollisch, *Local density approximations in quantum chemistry and solid state physics*, eds. J.P. Dahl and J. Avery, Plenum Press, New York (1984) p. 245.
88. L. Fritsche, Phys. Rev. B **33**, 3976 (1986).
89. J. Cordes and L. Fritsche, Z. Phys. D **13**, 345 (1989).
90. A.C. Pedroza, Phys. Rev. A **33**, 804 (1986)
91. O.V. Gritsenko, N.A. Cordero, A. Rubio, L.C. Balbás, and J.A. Alonso, Phys. Rev. A **48**, 4197 (1993).
92. J.P. Perdew, J.A. Chevary, S.H. Vosko, K.A. Jackson, M.R. Pederson, D.J. Singh, and C. Fiolhais, Phys. Rev. B **46**, 6671 (1992).
93. C. Froese-Fischer, *The Hartree-Fock method for atoms*, Wiley, New York, 1977; Computer Phys. Commun. **14**, 145 (1978).
94. E. Clementi and C. Roetti, At. Data Nucl. Data Tables **14**, 3 (1974); ib. **14**, 177 (1974).
95. N.A. Cordero, O.V. Gritsenko, A. Rubio, L.C. Balbás, and J.A. Alonso, Int. J. Quant. Chem. **52**, 993 (1994).

96. L.C. Wilson and M. Levy, Phys. Rev. B **41**, 12930 (1990).
97. C. Lee, W. Yang, and R.G. Parr, Phys. Rev. B **37**, 785 (1988).
98. J. Harris and R.O. Jones, J. Chem. Phys. **70**, 830 (1979).
99. O. Gunnarsson and R.O. Jones, Phys. Rev. B **31**, 7588 (1985).
100. Y.J. Wang and J.W. Davenport, Phys. Rev. A **41**, 4690 (1990).
101. F.R. Vukajlovic, E.L. Shirley, and R.M. Martin, Phys. Rev. B **43**, 3994 (1991).
102. D.D. Koelling and B.N. Harmon, J. Phys. C **10**, 3107 (1977).
103. C.E. Moore, Atomic Energy Levels, Natl. Bur. Stand. (U.S.) Spec. Publ. No. 476 (U.S. GPO, Washington, D.C.,1971).
104. S. Baroni, J. Chem. Phys. **80**, 5703 (1984).
105. S. Baroni and E. Tuncel, J. Chem. Phys. **79**, 6140 (1983).
106. U. von Barth and L. Hedin, J. Phys. C **5**, 1629 (1972).
107. R.B. Murphy and R.P. Messmer, J. Chem. Phys. **97**, 4974 (1992).
108. R.L. Martin and P.J. Hay, J. Chem. Phys. **75**, 4539 (1981).
109. A.D. Becke, Phys. Rev. A **38**, 3098 (1988).
110. E.L. Shirley and R.L. Martin, Phys. Rev B **47**, 15404 (1993).
111. *Handbook of Chemistry and Physics*, 58th ed. (C.R.C. Press, Boca Raton 1977).
112. O. Gunnarsson and B.I. Lundqvist, Phys. Rev. B **13**, 4274 (1976).
113. R. Merkle, A. Savin, and H. Preuss, Chem. Phys. Lett. **194**, 32 (1992).
114. R.A. Kendall, T.H. Dunning, and R.J. Harrison, J. Chem. Phys. **96**, 6796 (1992).
115. S.H. Vosko and J.B. Lagowski, in *Density Matrices and Density Functionals*, Ed. by R. Erdahl and V.H. Smith (Reidel, Dordrecht, 1987), p. 391.
116. J.B. Lagowski and S.H. Vosko, J. Phys. B: At. Mol. Opt. Phys. **21**, 203 (1988).
117. H.B. Shore, J.H. Rose, and E. Zaremba, Phys. Rev. B **15**, 2858 (1977).
118. L.A. Cole and J.P. Perdew, Phys. Rev. B **25**, 1265 (1982).
119. E. San Fabián, F. Moscardó, and J.M. Pérez Jordá, Int. J. Quantum Chem. **52**, 1027 (1994).
120. Y. Guo and M.A. Whitehead, Phys. Rev. A **38**, 3166 (1988); Phys. Rev. A **39**, 28 (1989).
121. J.G. Harrison, J. Chem. Phys. **86**, 2849 (1987).
122. P. Cortona, G. Böbel, and F.G. Fumi, J. Physique **50**, 2647 (1989).
123. J.C. Slater, Phys. Rev. **81**, 385 (1951).
124. N.A. Cordero, J.A. Alonso, L.C. Balbás, and K.D. Sen, Anales de Física (in press; 1996).
125. H. Hotop and W.C. Lineberger, J. Phys. Chem. Ref. Data B **14**, 731 (1985).
126. L.C. Balbás G. Borstel, and J.A. Alonso, Phys. Lett. A **144**, 326 (1986).
127. N.A. Cordero and J.A. Alonso, Int. J. Quant. Chem. (in press; 1996).
128. D. Sundholm, J. Phys. B: At. Mol. Opt. Phys. **28**, L393 (1995).
129. J.P. Perdew, R.G. Parr, M. Levy, and J.L. Balduz, Phys. Rev. Lett. **49**, 1691 (1982).
130. M. Levy and J.P. Perdew, Phys. Rev. A **32**, 2010 (1985).
131. J.A. Alonso and N.A. Cordero, J. Phys. B: At. Mol. Opt. Phys. (in press; 1996).
132. S.H. Vosko, L. Wilk, and M. Nusair, Can. J. Phys. **58**, 1200 (1980).
133. T. Grabo and E.K.U. Gross, Chem. Phys. Lett. **240**, 141 (1995).
134. R.T. Sharp and G.K. Horton, Phys. Rev. **90**, 317 (1953).
135. J.D. Talman and W.F. Shadwick, Phys. Rev. A **14**, 36 (1976).

136. J.B. Krieger, Y. Li and G.J. Iafrate, Phys. Rev. A **45**, 101 (1992); *Ibid.* **46**, 5453 (1992).
137. R. Colle and D. Salvetti, Theor. Chim. Acta **37**, 329 (1975); *Ibid.* **53**, 55 (1979).
138. J.M. Pérez-Jordá, E. San-Fabián, and F. Moscardó Phys. Rev. A **45**, 4407 (1992).
139. R. McWeeny and B. Sutcliffe, *Methods of Molecular Quantum Chemistry*, Academic, London, 1969.
140. C.F. Weizsäcker, Z Phys. **96**, 431 (1935).
141. Y. Tal and R.F.W. Bader, Int. J. Quantum Chem.: Quantum Chem. Symp. **12**, 153 (1978).
142. M.D. Glossman, A. Rubio, L.C. Balbás, and J.A. Alonso, Int. J. Quantum Chem.: Quantum Chem. Symp. **26**, 347 (1992).
143. B.M. Deb and S.K. Gosh, Int. J. Quantum Chem. **23**, 1 (1983).
144. M.D. Glossman, A. Rubio, L.C. Balbás, and J.A. Alonso, Phys. Rev. A **47**, 1804 (1993).
145. D.E. Kirzhnits, Sov. Phys. J.E.T.P. **5**, 64 (1957).
146. E. Engel and R.M. Dreizler, J. Phys. B **22**, 1901 (1989).
147. R.F.W. Bader and P.M. Bedall, J. Chem. Phys. **56**, 3320 (1972).
148. M.D. Glossman, L.C. Balbás, and J.A. Alonso, Chem. Phys. **196**, 455 (1995).
149. M. Pearson, E. Smargiassi, and P.A. Madden, J. Phys.: Condens. Matter **6**, 10773 (1994).
150. E. Smargiassi and P.A. Madden, Phys. Rev. B **51**, 117 (1995); *Ibid.* **51**, 129 (1995).
151. N. Govind, J. Wang, and H. Guo, Phys. Rev. B **50**, 11175 (1994).
152. N. Govind, J.L. Mozos, and H. Guo, Phys. Rev. B **51**, 7101 (1995).
153. D. Nehete, V. Shah, and D.G. Kanhere, Phys. Rev. B **53**, 2126 (1996).
154. M.J. Stott (private communication).
155. J.M. Pérez-Jordá and A.D. Becke, Chem. Phys. Lett. **233**, 134 (1995).
156. M.B. Nardelli, Solid State Commun. **97**, 215 (1996).

J.M. Seminario (Editor)
Recent Developments and Applications of Modern Density Functional Theory
Theoretical and Computational Chemistry, Vol. 4

Generalized gradient approximations to density functional theory: comparison with exact results

Claudia Filippi[a], Xavier Gonze[b] and C. J. Umrigar[c,a]

[a]Laboratory of Atomic and Solid State Physics, Cornell University, Ithaca, NY 14853

[b]Unité P.C.P.M., Université Catholique de Louvain, B-1348 Louvain-la-Neuve, Belgium

[c]Cornell Theory Center, Cornell University, Ithaca, NY 14853

In order to assess the accuracy of commonly used approximate exchange-correlation density functionals, we present a comparison of accurate exchange and correlation potentials, exchange energy densities and energy components with the corresponding approximate quantities. Four systems are used as illustrative examples: the model system of two electrons in a harmonic potential and the He, Be and Ne atoms. A new ingredient in the paper is the separation of the exchange-correlation potential into exchange and correlation according to the density functional theory definition.

1. INTRODUCTION

Within density functional theory (DFT), the ground state energy of an interacting system of electrons in an external potential can be written as a functional of the ground state electronic density [1]. When compared to conventional quantum chemistry methods, this approach is particularly appealing since it does not rely on the knowledge of the complete N-electron wave function but only of the electronic density. Unfortunately, although the theory is in principle exact, the energy functional contains an unknown quantity, called the exchange-correlation energy, $E_{\rm xc}[\rho]$, that must be approximated in any practical implementation of the method. Several approximate exchange-correlation functionals have been proposed in the literature, the most commonly used ones being the local density approximation (LDA) and the generalized gradient approximation (GGA).

The local density approximation [2] is the simplest and most widely used exchange-correlation functional:

$$E_{\rm xc}^{\rm LDA}[\rho] = \int \rho(\mathbf{r})\, \epsilon_{\rm xc}^{\rm LDA}(\rho(\mathbf{r}))\, d\mathbf{r}, \tag{1}$$

where $\epsilon_{\rm xc}^{\rm LDA}(\rho)$ is the exchange-correlation energy per particle of a homogeneous electron gas of density ρ. $E_{\rm xc}^{\rm LDA}[\rho]$ is by definition the exact functional for a homogeneous electron gas and has been shown to give also a qualitatively good description of the ground state properties of a variety of highly inhomogeneous systems [3]. However, LDA does not always provide sufficiently accurate results. For example, it almost always overestimates the binding energy and underestimates the bond-length of weakly bound molecules and solids [3]. LDA fails also to predict the ground state structure of iron, although the error is quantitatively small [4].

In an attempt to improve upon LDA, a dependence of the exchange-correlation energy on the derivatives of the electronic density can be introduced. A simple and systematic extension of LDA is the gradient expansion approximation (GEA) [2],

$$E_{\mathrm{xc}}^{\mathrm{GEA}}[\rho] = \int \rho(\mathbf{r})\epsilon_{\mathrm{xc}}^{\mathrm{LDA}}(\rho(\mathbf{r}))\,\mathrm{d}\mathbf{r} + \int B_{\mathrm{xc}}(\rho(\mathbf{r}))|\nabla\rho(\mathbf{r})|^2\,\mathrm{d}\mathbf{r} + \cdots, \qquad (2)$$

which is asymptotically valid for densities that vary slowly over space. For finite systems, the terms in the expansion of the exchange energy of order greater or equal than four diverge while the exchange potential diverges already at second order. By using a constant B_{xc} determined variationally for each atom, Herman *et al.* showed that the gradient expansion up to second order yields improved total energies of atoms [5]. An improvement in the exchange energies is also obtained when the exact expansion coefficient for exchange, B_{x}, is used in an exchange-only calculation [6]. However, the inclusion of correlation in the second order expansion gives energies that are less accurate than LDA [7,8].

The analysis of this failure has lead some authors [9,10] to introduce a generalization of the gradient expansion, whose generic functional form (here restricted to second-order derivative) is

$$E_{\mathrm{xc}}^{\mathrm{GGA}}[\rho] = \int \rho(\mathbf{r})\,\epsilon_{\mathrm{xc}}^{\mathrm{GGA}}(\rho(\mathbf{r}), |\nabla\rho(\mathbf{r})|, \nabla^2\rho(\mathbf{r}))\,\mathrm{d}\mathbf{r}. \qquad (3)$$

Many different GGA's have been proposed since then [9,11–18] but none of them is clearly better than the others. Other classes of approximate functionals have also been tested, including schemes that remove the self-interaction of electrons [19,20], the weighted density approximation [21,22] and the average density approximation [21] that attempt to model the exchange-correlation hole. However, generalized gradient approximations have recently received particular attention both in the physics and chemistry communities since they do not add appreciable computational complexity to the LDA scheme.

The GGA functionals seem to give a good description of several finite systems: they significantly improve total energies of atoms [23], as well as binding energies [23–26] and vibrational frequencies [24] of first and second-row molecules. They also give better estimates of bond-lengths and binding energies of weakly bound systems such as IIA and IIB homonuclear dimers [27] and hydrogen bonded systems [28], which are greatly overbound in LDA. Lattice constants, bulk moduli and cohesive energies of simple metals [23] and $3d$ transition metals [29] are also improved. For iron, the correct ferromagnetic bcc ground state is predicted [29]. On the other hand, the lattice constants and the bulk moduli of the semiconductors Si, Ge and GaAs are less accurately predicted by the GGA's than LDA, the effect being particularly large for the bulk moduli [30]. Calculations for $4d$ and $5d$ metals indicate that, while LDA tends to slightly underestimate lattice constants and overestimate bulk moduli, GGA's often overcorrect, sometimes leading to results in worse agreement with experiment [31]. The tendency of the GGA's to yield smaller binding energies and larger bond-lengths than LDA is explained by the fact that the GGA's favor inhomogeneity in the density. Since isolated atoms are more inhomogeneous than molecules or solids, the energy of the atoms is lowered more, resulting in smaller binding energies. Similarly, for solids and weakly bound molecules, larger bond-lengths lead to increased inhomogeneity and are therefore favored by the GGA's.

Studies that compare the predicted structural and energetic properties of various systems with the corresponding experimental quantities add very little to the understanding of the reasons for the successes or the failures of the approximate functionals. In order to gain more insight into the limitations of the approximate functionals, many researchers [32–47] have compared accurate density functional properties, calculated by more sophisticated and more computationally demanding methods, with the corresponding density functional properties calculated from the approximate functionals. In this review, we limit ourselves to some of our earlier work and discuss four systems, for which it is possible to obtain exact or very accurate wave functions, as illustrative examples: a model system of two electrons in a harmonic potential and the He, Be and Ne atoms. While, we mostly concentrate on the description of our previous work, there is a new ingredient in this paper: the separation of the exchange-correlation potential of many-electron systems into separate exchange and correlation components according to the DFT definition.

The outline of the rest of the paper is as follows. In Sec. 2, we briefly introduce density functional theory and its Kohn-Sham formulation. In Sec. 3, we list some properties of the exact exchange-correlation functional and we determine which properties are satisfied or violated by LDA and various GGA's. The short and long distance asymptotic behavior of the exchange-correlation potential and local exchange energy per electron is also discussed. In Sec. 4, we derive the formulae used to determine the exact density functional quantities and, in Sec. 5, we describe the systems we study and the corresponding densities. The exact exchange and correlation potentials, local exchange energies per electron and components of the total energy of these systems are compared with the corresponding approximate LDA and GGA quantities in Secs. 6, 7 and 8, respectively. In Sec. 9, we summarize our conclusions and discuss some prospects for better approximate functionals. In Appendix A, we list the approximate functionals used in the comparison.

2. THEORETICAL BACKGROUND

Density functional theory provides an expression for the ground state energy of a system of interacting electrons in an external potential as a functional of the ground state electronic density [1]. Let us assume for simplicity that the spin polarization of the system of interest is identically zero. In the Kohn-Sham formulation of density functional theory [2], the ground state density is written in terms of single-particle orbitals obeying the equations in atomic units ($\hbar = e = m = 1$):

$$\left\{-\frac{1}{2}\nabla^2 + v_{\text{ext}}(\mathbf{r}) + \int \frac{\rho(\mathbf{r}')}{|\mathbf{r}-\mathbf{r}'|}d\mathbf{r}' + v_{\text{xc}}([\rho];\mathbf{r})\right\}\psi_i = \epsilon_i\psi_i, \quad (4)$$

where

$$\rho(\mathbf{r}) = \sum_{i=1}^{N} |\psi_i(\mathbf{r})|^2 . \tag{5}$$

The electronic density is constructed by summing over the N lowest energy orbitals where N is the number of electrons. $v_{\text{ext}}(\mathbf{r})$ is the external potential. The exchange-correlation potential $v_{\text{xc}}([\rho];\mathbf{r})$ is the functional derivative of the exchange-correlation energy $E_{\text{xc}}[\rho]$ that enters in the expression for the total energy of the system:

$$E = -\frac{1}{2}\sum_{i=1}^{N}\int \psi_i \nabla^2 \psi_i \, d\mathbf{r} + \int \rho(\mathbf{r})\, v_{\text{ext}}(\mathbf{r})\, d\mathbf{r} + \frac{1}{2}\int\int \frac{\rho(\mathbf{r})\rho(\mathbf{r}')}{|\mathbf{r}-\mathbf{r}'|} d\mathbf{r}\, d\mathbf{r}' + E_{\text{xc}}[\rho] . \tag{6}$$

The exchange-correlation functional is written as the sum of two separate contributions for exchange and correlation,

$$E_{\text{xc}}[\rho] = E_{\text{x}}[\rho] + E_{\text{c}}[\rho] . \tag{7}$$

The definition of the exchange energy is in terms of the non-interacting wave function Φ_0, the Slater determinant constructed from the Kohn-Sham orbitals, as

$$E_{\text{x}}[\rho] = \langle \Phi_0 | V_{\text{ee}} | \Phi_0 \rangle - \frac{1}{2}\int\int \frac{\rho(\mathbf{r})\rho(\mathbf{r}')}{|\mathbf{r}-\mathbf{r}'|} d\mathbf{r}\, d\mathbf{r}', \tag{8}$$

where V_{ee} is the electron-electron interaction. This definition differs from the conventional quantum chemistry definition of E_{x} as the exchange energy in a Hartree-Fock calculation, given by the same expression as in Eq. 8 but with the Kohn-Sham determinant replaced by the Hartree-Fock determinant. The separation of the exchange-correlation energy functional into the separate exchange and correlation components yields a corresponding splitting of the exchange-correlation potential into $v_{\text{x}}([\rho];\mathbf{r})$ and $v_{\text{c}}([\rho];\mathbf{r})$. In this formulation, the essential unknown quantity is the exchange-correlation energy $E_{\text{xc}}[\rho]$. If the functional form of $E_{\text{xc}}[\rho]$, and consequently the exchange-correlation potential, were available, we could solve the N-electron problem by finding the solution of a set of single-particle equations. However, as mentioned in Sec. 1, the exact functional form of $E_{\text{xc}}[\rho]$ is not known and it is necessary to make approximations for this term.

3. PROPERTIES OF THE EXACT DENSITY FUNCTIONAL

In order to understand if it is possible to construct an approximate functional that does not suffer from the weaknesses of the existing GGA's, it is useful to study the known properties of the exact density functional and which of these are violated by the commonly used approximate functionals. There are actually a surprisingly large number of known properties [48–53] and some of these are listed in Table 1. The GGA functionals compared in the table can be found in Appendix A.

The first group of properties in Table 1 are sum rules that are satisfied by the exchange and correlation holes. The exchange-correlation energy functionals can be interpreted as

Table 1
Some properties of the exact exchange-correlation functional. Comparison with approximate functionals. Taken from Ref. [45] and modified.

	Property	E_{xc}^{LDA} [15]	E_{xc}^{LM} [9]	E_{xc}^{PW91} [14]	E_{x}^{B88} [11]	E_{x}^{ECMV} [18]	E_{c}^{WL} [16]	E_{c}^{LYP} [17]
1	$\rho_x(\mathbf{r},\mathbf{r}') \leq 0$	Y	—	Y	—	—	—	—
2	$\int \rho_x(\mathbf{r},\mathbf{r}')\, d\mathbf{r}' = -1$	Y	—	Y	—	—	—	—
3	$\int \rho_c(\mathbf{r},\mathbf{r}')\, d\mathbf{r}' = 0$	Y	—	Y	—	—	—	—
4	$E_x[\rho] < 0$	Y	Y	Y	Y	Y	—	—
5	$E_c[\rho] \leq 0$	Y	N	N	—	—	N	N
6	$E_x[\rho], E_{xc}[\rho] \geq -c\int \rho^{4/3} d\mathbf{r}$ [a]	Y	N	Y	N	N	—	—
7	$E_x[\rho_\lambda] = \lambda E_x[\rho]$	Y	Y	Y	Y	Y	—	—
8	$E_c[\rho_\lambda] < \lambda E_c[\rho],\ \lambda < 1$ [b]	Y	N	Y	—	—	N	N
9	$\lim_{\lambda\to\infty} E_c[\rho_\lambda] > -\infty$	N	Y [c]	Y [c]	—	—	Y	Y
10	$\lim_{\lambda\to 0} \frac{1}{\lambda} E_c[\rho_\lambda] > -\infty$	Y	N	Y	—	—	Y	Y
11	$\lim_{\lambda\to\infty} E_x[\rho_\lambda^x] > -\infty$	N	N	Y	N	N	—	—
12	$\lim_{\lambda\to 0} E_x[\rho_\lambda^x] > -\infty$	Y	N	Y	Y	Y	—	—
13	$\lim_{\lambda\to\infty} \frac{1}{\lambda} E_x[\rho_\lambda^{xy}] > -\infty$	Y	N	Y	Y	Y	—	—
14	$\lim_{\lambda\to 0} \frac{1}{\lambda} E_x[\rho_{\lambda\lambda}^{xy}] > -\infty$	N	N	Y	N	N	—	—
15	$\lim_{\lambda\to\infty} \lambda E_c[\rho_\lambda^x] > -\infty$	N	Y [c]	Y	—	—	N	N
16	$\lim_{\lambda\to 0} \frac{1}{\lambda} E_c[\rho_\lambda^x] = 0$	N	N	Y	—	—	N	N
17	$\lim_{\lambda\to\infty} E_c[\rho_{\lambda\lambda}^{xy}] = 0$	N	N	Y	—	—	N	N
18	$\lim_{\lambda\to 0} \frac{1}{\lambda^2} E_c[\rho_{\lambda\lambda}^{xy}] > -\infty$	N	Y [c]	Y	—	—	Y [c]	N
19	$\epsilon_x(r) \to -1/2r,\ r\to\infty$	N	N	N	YN [d]	N	—	—
20	$v_x(r) \to -1/r,\ r\to\infty$	N	N	N	N	N	—	—
21	$v_x(r), v_c(r)$ finite, $r\to 0$	Y	N	N	N	N	N	N
22	LDA limit for constant ρ	Y	N	Y	Y	Y	N	N
23	GEA limit for slowly varying ρ	—	N	Y	N	N	N	N

[a] $1.44 < c < 1.68$

[b] Note that $E_c[\rho_\lambda] < \lambda E_c[\rho],\ \lambda < 1$ is equivalent to $E_c[\rho_\lambda] > \lambda E_c[\rho],\ \lambda > 1$.

[c] But it diverges to $+\infty$. The PW91 GGA can be modified to satisfy this relation [49].

[d] "Y" for exponential $\rho(\mathbf{r})$ but "N" in general, *e.g.* $\epsilon_x^{B88}(r) \to -1/r$ for a Gaussian density.

the energy arising form the interaction of an electron at $\mathbf{r}$ and its exchange-correlation hole at $\mathbf{r}'$,

$$E_{\mathrm{xc}}[\rho] = \frac{1}{2}\iint \frac{\rho(\mathbf{r})\,\rho_{\mathrm{xc}}(\mathbf{r},\mathbf{r}')}{|\mathbf{r}-\mathbf{r}'|}\,\mathrm{d}\mathbf{r}\,\mathrm{d}\mathbf{r}'. \tag{9}$$

The exchange-correlation hole is then separated into exchange and correlation, $\rho_{\mathrm{x}}(\mathbf{r},\mathbf{r}')$ and $\rho_{\mathrm{c}}(\mathbf{r},\mathbf{r}')$ respectively, where the exchange contribution comes from the non-interacting system according to Eq. 8. The exchange-correlation hole can be expressed as an integral over the coupling constant of the density-density correlation function of the interacting system [48].

The second group of properties provide bounds on the functionals. Condition 6 is known as the Lieb-Oxford bound and a tighter version of it is given in Ref. [49]:

$$\lim_{\lambda\to 0}\frac{1}{\lambda}E_{\mathrm{xc}}[\rho_\lambda] = \inf_{\Psi\to\rho}\langle\Psi|V_{\mathrm{ee}}|\Psi\rangle - \frac{1}{2}\iint\frac{\rho(\mathbf{r})\rho(\mathbf{r}')}{|\mathbf{r}-\mathbf{r}'|}\mathrm{d}\mathbf{r}\,\mathrm{d}\mathbf{r}' \geq -c\int\rho(\mathbf{r})^{4/3}\,\mathrm{d}\mathbf{r}, \tag{10}$$

where the density ρ_λ is obtained by uniformly scaling the density ρ in all three spatial directions:

$$\rho_\lambda(\mathbf{r}) = \lambda^3\rho(\lambda\mathbf{r}). \tag{11}$$

This scaled density integrates to the same number of electrons as the unscaled one.

The scaling of the exchange and correlation functionals when the charge density is scaled uniformly [50,51] is described by the third group of properties. It now becomes evident why it is useful to separate the exchange component from the entire functional according to Eq. 8. For exchange, an exact relation (condition 7) exists under uniform density scaling that determines how derivatives of the density combine with the density in an exchange GGA functional:

$$E_{\mathrm{x}}^{\mathrm{GGA}}[\rho] = \int\rho(\mathbf{r})^{4/3}\,F(|\nabla\rho(\mathbf{r})|/\rho(\mathbf{r})^{4/3}, \nabla^2\rho(\mathbf{r})/\rho(\mathbf{r})^{5/3},\ldots)\,\mathrm{d}\mathbf{r}. \tag{12}$$

The fourth set of properties consists of the relations under non-uniform scaling of the density in one or two of the spatial directions:

$$\rho_\lambda^{\mathrm{x}}(\mathbf{r}) = \lambda\,\rho(\lambda x, y, z), \quad \rho_{\lambda\lambda}^{\mathrm{xy}}(\mathbf{r}) = \lambda^2\,\rho(\lambda x, \lambda y, z). \tag{13}$$

The non-uniform scaling relations for the correlation energy are here given in the tightest form and are derived under the assumption of the existence of a Taylor series in λ or $1/\lambda$ [52]. A summary of the scaling relations and their derivation is provided in Ref. [53].

The fifth group of properties describes the long-distance asymptotic behavior (for finite systems) of the exchange potential and the local exchange energy per electron (see Sec. 3) and the short-distance behavior of the exchange and correlation potential.

The last two conditions are that the functionals must reduce to the LDA functional in the limit of a homogeneous density and to the correct second-order expansion for a slowly varying density.

We also mention an additional relation not included in the table, the convexity constraint [49], which is given by

$$\left.\frac{\partial^2 B\left[\rho+\epsilon\Delta\rho\right]}{\partial\epsilon^2}\right|_{\epsilon=0}+\iint\frac{\Delta\rho(\mathbf{r})\Delta\rho(\mathbf{r}')}{|\mathbf{r}-\mathbf{r}'|}\mathrm{d}\mathbf{r}\,\mathrm{d}\mathbf{r}'\geq 0, \tag{14}$$

for arbitrary $\Delta\rho$ integrating to zero, where

$$B\left[\rho\right]=\lim_{\lambda\to 0}\frac{1}{\lambda}E_{\mathrm{xc}}\left[\rho_\lambda\right]. \tag{15}$$

This constraint is very stringent: it is violated by the LDA, by the Perdew-Wang '91 GGA and, likely, by any other GGA functional [49].

From Table 1, it appears that most of the functionals violate the asymptotic behavior of the exchange-correlation potential and local exchange energy per electron. Within GGA, the exchange-correlation potential is given by the functional derivative of $E_{\mathrm{xc}}^{\mathrm{GGA}}\left[\rho\right]$ (Eq. 3):

$$v_{\mathrm{xc}}\left(\left[\rho\right];\mathbf{r}\right)=\left[\frac{\partial\, e_{\mathrm{xc}}}{\partial\,\rho}-\nabla\cdot\left(\frac{\partial\, e_{\mathrm{xc}}}{\partial\,\nabla\rho}\right)+\nabla^2\left(\frac{\partial\, e_{\mathrm{xc}}}{\partial\,\nabla^2\rho}\right)\right]_{\rho(\mathbf{r}),\nabla\rho(\mathbf{r}),\nabla^2\rho(\mathbf{r})}, \tag{16}$$

where e_{xc} is the exchange-correlation energy density, $e_{\mathrm{xc}}=\rho\,\epsilon_{\mathrm{xc}}$. As pointed out in our earlier papers [54,55], any GGA that has terms containing $\nabla\rho$ but no higher derivatives of ρ must yield an exchange-correlation potential that diverges at nuclei. If $\rho\sim\mathrm{constant}+r^s$ near the origin, then $|\nabla\rho|\sim r^{s-1}$. If the leading behavior of e_{xc} in the gradient, near an extremum of the density, is $|\nabla\rho|^m$, then

$$\frac{\partial\, e_{\mathrm{xc}}}{\partial\,\nabla\rho}\sim|\nabla\rho|^{m-1}\sim r^{(s-1)(m-1)}, \tag{17}$$

and

$$\nabla\cdot\left(\frac{\partial\, e_{\mathrm{xc}}}{\partial\,\nabla\rho}\right)\sim r^{(s-1)(m-1)-1}. \tag{18}$$

Therefore, the exchange-correlation potential diverges at the origin if

$$(s-1)(m-1)-1<0, \tag{19}$$

where s and m are positive. Thus, the exchange-correlation potential diverges at extrema of the density for all values of s if $m\leq 1$, as is the case for the Wilson-Levy potential [16], and for all values of m if $s\leq 1$. At nuclei, $s=1$ so that the exchange-correlation potential always diverges.

As far as the large distance asymptotics are concerned, any GGA that includes no higher than first derivatives of the density cannot simultaneously satisfy both the correct $-1/r$ behavior of the exchange-correlation potential and the correct $-1/2r$ behavior of the local exchange-correlation energy per electron [18].

It is however possible, as mentioned in Refs. [45,55], to include the Laplacian of the density in an appropriate way [56] and construct a GGA that has the triple advantage that it satisfies both of the long-distance asymptotic conditions and also does not suffer from a spurious divergence at the nuclei. There are an infinite number of possible variations in the functional form that preserve the desired short and long range asymptotic behaviors. We are presently searching for the most physically reasonable functional expression and for a universal set of parameter values in such a functional. In the same line of research, Jemmer and Knowles have recently published a simple expression for an exchange functional depending on the Laplacian of the density and reproducing the correct long and short distance asymptotics of the exchange-correlation potential and local exchange-correlation energy per electron [57]. However, they conclude that such a functional is not a suitable choice as a general purpose density functional. The importance of Laplacian terms has also been pointed out by Engel and Vosko [58] who show that inclusion of Laplacian terms in a fourth order Taylor expansion of the exchange-correlation functional results in improved exchange potentials for atoms and jellium spheres. However, their expression does not obey any of the above three asymptotic conditions.

4. DETERMINATION OF ACCURATE DFT QUANTITIES

For atomic systems, it is possible to determine accurate exchange-only quantities by using the solution of the optimized effective potential (OEP) method which represents the exact solution in an exchange-only DFT scheme [20]. However, these quantities are determined not for the exact density but for the OEP density which corresponds to the self-consistent solution in an exchange-only approach. The performance of exchange-only approximate functionals and the corresponding exchange potentials can therefore be routinely checked for atomic systems by using the OEP method. Approximate quantities in exchange-only GGA, evaluated either for the OEP density or the self-consistent density, can be compared with the corresponding OEP quantities.

Obtaining reliable exchange-correlation potentials and energies is instead a more difficult task. One has to generate an accurate density and then compute an exchange-correlation potential that yields the desired density as a solution of the Kohn-Sham equations (Eq. 4 and 5). In this context, researchers have used charge densities, of varying degrees of accuracy, calculated by quantum chemistry methods for atoms [32–41] and molecules [42–44], as well as Quantum Monte Carlo methods for atoms [45,47] and for a model semiconductor [46]. The subsequent inverse problem, namely the search of the corresponding exchange-correlation potential, has been performed using a variety of different techniques. For example, in the special case of the singlet ground state of a two-electron system, the exchange-correlation potential can be obtained simply from Eq. 4 while, for systems with more than two electrons, v_{xc} can be determined by expanding it in a complete set of basis functions and varying the expansion coefficients such that Eqs. 4 and 5 yield the accurate density[34,47]. Whatever the technique used, an accurate density is a key ingredient for the determination of the exchange-correlation potential, because small errors in the density are greatly magnified by the inversion procedures used to obtain the potential.

By knowing the exchange-correlation potential, we are able to calculate the Kohn-Sham orbitals and, if the total energy of the system can be estimated, the exchange-correlation energy is obtained by inversion of the expression for the total energy (Eq. 6):

$$E_{\mathrm{xc}}[\rho] = E - \frac{1}{2}\sum_{i=1}^{N}\int \psi_i \nabla^2 \psi_i \, d\mathbf{r} - \frac{1}{2}\iint \frac{\rho(\mathbf{r})\rho(\mathbf{r}')}{|\mathbf{r}-\mathbf{r}'|} d\mathbf{r}\, d\mathbf{r}' - \int \rho(\mathbf{r})\, v_{\mathrm{ext}}(\mathbf{r}) d\mathbf{r}. \quad (20)$$

Since the Kohn-sham orbitals are known, the exchange energy (Eq. 8) can be calculated as

$$E_{\mathrm{x}}[\rho] = -\frac{1}{2}\sum_{i=1}^{N}\sum_{j=1}^{N} \delta_{m_{s_i}, m_{s_j}} \iint \frac{\psi_i^*(\mathbf{r})\psi_j^*(\mathbf{r}')\psi_j(\mathbf{r})\psi_i(\mathbf{r}')}{|\mathbf{r}-\mathbf{r}'|} d\mathbf{r}\, d\mathbf{r}', \quad (21)$$

where the δ-function is over the spin quantum numbers of the i-th and j-th spin-orbitals. The correlation energy is finally obtained as the difference of the exchange-correlation energy (Eq. 20) and the exchange energy (Eq. 21):

$$E_{\mathrm{c}}[\rho] = E_{\mathrm{xc}}[\rho] - E_{\mathrm{x}}[\rho]. \quad (22)$$

The functional derivative of the exchange energy (Eq. 21) with respect to the density yields the exchange potential. By knowing the exchange-correlation and the exchange potentials, we obtain the correlation potential as the difference

$$v_{\mathrm{c}}([\rho];\mathbf{r}) = v_{\mathrm{xc}}([\rho];\mathbf{r}) - v_{\mathrm{x}}([\rho];\mathbf{r}). \quad (23)$$

We show in Sub-sec. 4.1 that it is easy to calculate the exchange potential for a system of two electron in a singlet state. The procedure is instead much more involved for a many-electron system, as we outline in Sub-sec. 4.2.

As already introduced in Sec. 1, the exchange-correlation functional can be written as an integral over the exchange-correlation energy density, $e_{\mathrm{xc}}(\mathbf{r})$, or the local exchange-correlation energy per electron, $\epsilon_{\mathrm{xc}}(\mathbf{r})$:

$$E_{\mathrm{xc}}[\rho] = \int e_{\mathrm{xc}}(\mathbf{r})\, d\mathbf{r} = \int \rho(\mathbf{r})\, \epsilon_{\mathrm{xc}}(\mathbf{r})\, d\mathbf{r}. \quad (24)$$

The separation of E_{xc} into exchange and correlation (Eq. 7) corresponds to an equivalent decomposition of ϵ_{xc} into ϵ_{x} and ϵ_{c}. The definition of ϵ_{xc} and its components is however not unique since two ϵ_{xc}'s whose difference is a function f with $\int \rho(\mathbf{r}) f(\mathbf{r})\, d\mathbf{r} = 0$ yield the same exchange-correlation energy functional.

We observe that a natural definition of the local exchange energy per electron follows from Eq. 21:

$$\epsilon_{\mathrm{x}}(\mathbf{r}) = -\frac{1}{2}\frac{1}{\rho(\mathbf{r})}\sum_{i=1}^{N}\sum_{j=1}^{N} \delta_{m_{s_i}, m_{s_j}} \psi_i^*(\mathbf{r})\psi_j(\mathbf{r}) \int \frac{\psi_j^*(\mathbf{r}')\psi_i(\mathbf{r}')}{|\mathbf{r}-\mathbf{r}'|} d\mathbf{r}'. \quad (25)$$

This definition results in an ϵ_{x} that at large distances goes as $-1/2r$ [59]. A procedure for constructing ϵ_{c} has been recently proposed using first and second order density matrices from correlated wave functions and applied to He and H_2 [60].

In the remaining sections, the accurate exchange-correlation energies and potentials, the separate exchange and correlation components and the local exchange energy per electron are used to test the accuracy of the approximate exchange-correlation energy functionals listed in Appendix A.

4.1. Two-electron systems

For two electrons of opposite spin, there is a single spatial Kohn-Sham orbital ψ and is simply related to the electronic density (Eq. 5) as

$$\psi(\mathbf{r}) = \left[\frac{\rho(\mathbf{r})}{2}\right]^{1/2}. \tag{26}$$

Knowing the exact density, the exchange-correlation potential can be obtained by inversion of the Kohn-Sham equation (Eq. 4),

$$v_{\mathrm{xc}}([\rho];\mathbf{r}) = \epsilon_{\mathrm{KS}} + \frac{1}{2}\frac{\nabla^2\psi}{\psi} - v_{\mathrm{ext}}(\mathbf{r}) - \int \frac{\rho(\mathbf{r}')}{|\mathbf{r}-\mathbf{r}'|}\mathrm{d}\mathbf{r}'. \tag{27}$$

The eigenvalue ϵ_{KS} is equal to minus the ionization energy if we impose that the exchange-correlation potential goes to zero at infinity. Therefore, the exchange-correlation potential is completely determined. From Eq. 21, it follows that the exchange energy is given by

$$E_{\mathrm{x}}[\rho] = -\frac{1}{4}\int\int \frac{\rho(\mathbf{r})\rho(\mathbf{r}')}{|\mathbf{r}-\mathbf{r}'|}\mathrm{d}\mathbf{r}\,\mathrm{d}\mathbf{r}', \tag{28}$$

and the functional derivative of this expression with respect to the density yields the exact exchange potential:

$$v_{\mathrm{x}}([\rho];\mathbf{r}) = -\frac{1}{2}\int \frac{\rho(\mathbf{r}')}{|\mathbf{r}-\mathbf{r}'|}\mathrm{d}\mathbf{r}'. \tag{29}$$

Observe that $v_{\mathrm{x}}([\rho];\mathbf{r})$ simply follows from the condition that it cancels the self-interaction term in the Hartree potential.

For a system of two interacting electrons in a singlet state, Eq. 25 for the local exchange energy per electron reduces to

$$\epsilon_{\mathrm{x}}(\mathrm{r}) = -\frac{1}{4}\int \frac{\rho(\mathbf{r}')}{|\mathbf{r}-\mathbf{r}'|}\mathrm{d}\mathbf{r}'. \tag{30}$$

The same expression could have also been obtained from Eq. 28.

4.2. Many-electron systems

While the separation of the exchange-correlation potential into exchange plus correlation is quite simple in the two-electron systems, this decomposition according to the DFT definition has never been obtained for many-electron systems. In previous work [34,35,38,45,41], the exchange potential was defined as the difference of the effective Kohn-Sham potential yielding the Hartree-Fock density and the sum of the Hartree and the external potentials. The correlation potential was then obtained as the difference of the exchange-correlation potential corresponding to the exact density and the above potential. Note that this separation does not correspond to the DFT definition of Eq. 23: it involves two densities, the exact and the Hartre-Fock densities, while the potential used for exchange is only approximately equal to the exchange potential corresponding to the Hartree-Fock density (although very close to it) since it is not the functional derivative with respect to the density of the exchange energy evaluated for the orbitals obtained from the effective potential yielding the Hartree-Fock density. Here, we obtain the correct separation of the exchange and correlation components, according to their DFT definition, for the exact density. The difference between the exchange potential we obtain and the approximate "exchange" potential described above is almost not detectable on the scale of the exchange potential. On the other hand, on the more expanded scale of the correlation potential, the difference is clearly visible, although the shapes are very similar.

We follow Görling and Levy [61] in showing how to separate the exchange-correlation potential into exchange and correlation. We consider a spin unpolarized system. If we assume that the density ρ is non-interacting v-representable, it can be expressed as in Eq. 5 in terms of single-particle orbitals $\{\psi_i\}$ of the Kohn-Sham potential $v_s(\mathbf{r})$,

$$v_s(\mathbf{r}) = v_{ext}(\mathbf{r}) + \int \frac{\rho(\mathbf{r}')}{|\mathbf{r}-\mathbf{r}'|} d\mathbf{r}' + v_{xc}([\rho];\mathbf{r}). \tag{31}$$

We evaluate the functional derivative of the exchange energy functional with respect to the Kohn-Sham potential as

$$\frac{\delta E_x[\rho]}{\delta v_s(\mathbf{r})} = \int \frac{\delta E_x[\rho]}{\delta \rho(\mathbf{r}')} \frac{\delta \rho(\mathbf{r}')}{\delta v_s(\mathbf{r})} d\mathbf{r}' = \int v_x([\rho];\mathbf{r}') \sum_{i=1}^{N} \left(\psi_i^*(\mathbf{r}') \frac{\delta \psi_i(\mathbf{r}')}{\delta v_s(\mathbf{r})} + \frac{\delta \psi_i^*(\mathbf{r}')}{\delta v_s(\mathbf{r})} \psi_i(\mathbf{r}') \right) d\mathbf{r}'. \tag{32}$$

On the other hand, since the exchange functional can be written as a function of the orbitals (Eq. 21), we also have

$$\frac{\delta E_x[\rho]}{\delta v_s(\mathbf{r})} = \sum_{i=1}^{N} \int \left(\frac{\delta E_x[\rho]}{\delta \psi_i(\mathbf{r}')} \frac{\delta \psi_i(\mathbf{r}')}{\delta v_s(\mathbf{r})} + \frac{\delta E_x[\rho]}{\delta \psi_i^*(\mathbf{r}')} \frac{\delta \psi_i^*(\mathbf{r}')}{\delta v_s(\mathbf{r})} \right) d\mathbf{r}'. \tag{33}$$

If we combine Eqs. 32 and 33, we obtain the integral equation

$$\int v_x([\rho];\mathbf{r}') \mathcal{K}(\mathbf{r}',\mathbf{r}) d\mathbf{r}' = \mathcal{Q}(\mathbf{r}), \tag{34}$$

where the kernel $\mathcal{K}(\mathbf{r}',\mathbf{r})$ and the right hand side $\mathcal{Q}(\mathbf{r})$ depend on the orbital $\{\psi_i\}$ and their functional derivative with respect to the potential $v_s(\mathbf{r})$. This integral equation is equivalent to the one solved in the OEP method with the KS orbitals replaced by the OEP orbitals [20]. The functional derivatives of the orbitals $\delta\psi_i(\mathbf{r})/\delta v_s(\mathbf{r}')$ can be expressed in terms of the Green's function $G_i(\mathbf{r},\mathbf{r}')$ as

$$\frac{\delta \psi_i(\mathbf{r})}{\delta v_s(\mathbf{r}')} = -G_i(\mathbf{r},\mathbf{r}')\psi_i(\mathbf{r}'), \tag{35}$$

where $G_i(\mathbf{r},\mathbf{r}')$ satisfies the differential equation

$$\left(-\frac{1}{2}\nabla^2 + v_s(\mathbf{r}) - \epsilon_i\right) G_i(\mathbf{r},\mathbf{r}') = \delta(\mathbf{r}-\mathbf{r}') - \psi_i(\mathbf{r})\psi_i^*(\mathbf{r}'). \tag{36}$$

By knowing the exchange-correlation potential, the KS orbitals and eigenvalues, we can compute the Green's functions $\{G_i\}$ and, consequently, the kernel $\mathcal{K}$ and the function $\mathcal{Q}$. If we express the exchange potential as a linear combination of basis functions, Eq. 34 can be rewritten as a non-homogeneous set of linear equations for the coefficients of the expansion of the potential in the basis set. More details on the separation procedure can be found in Ref. [62].

5. SYSTEMS STUDIED AND CORRESPONDING DENSITIES

Experimental and computed energies are often compared in order to assess the accuracy of approximate density functionals. However, this procedure is not very reliable since energies are integrated quantities and are therefore subject to cancellation of errors. In order to better understand the performance of approximate exchange-correlation functionals, we will examine not only total energies but also the components of the energy, the self-consistent charge densities, the exchange-correlation potentials and the local exchange energies per electron. We will compare these quantities, obtained from approximate schemes, with the corresponding accurate quantities derived from (1) the exact solution of a model system of two interacting electrons in a harmonic potential (we call this system "Harmonium"), (2) a nearly exact wave function for the He atom and (3) accurate quantum Monte Carlo calculations for the Be and Ne atoms.

5.1. Harmonium

A simple two-electron system where the exchange-correlation potential and energy is calculable exactly consists of two interacting electrons in a harmonic potential. The model can be solved analytically for a discrete and infinite set of values of the spring constant k [63] and the degree of correlation within the system can be varied by simply tuning the value of k.

In Fig. 1, we show the density in a low correlation case, corresponding to a spring constant $k = 0.25$ a.u., and in a high correlation case with a spring constant $k \approx 3.6\times10^{-6}$ a.u. In the low correlation case, the density has a maximum at the origin while, in the high correlation case, the density has a local minimum at the origin and an absolute maximum at a finite distance from the origin [54]. Although the high correlation density is too low to be physically relevant for electronic structure calculations ($\langle r_s\rangle = 57.9$ a.u.), the system is interesting because it is strongly correlated and its density differs qualitatively from an atomic density due to the presence of a maximum at a finite distance from the origin. Consequently, in the present paper, we will only discuss the high correlation limit. For an analysis of the system with $k = 0.25$ a.u., see Ref. [54].

5.2. He atom

It is necessary to employ very accurate wave functions in order to observe the true short-distance and long-distance behavior to $\mathcal{O}(1/r)$ of the exchange-correlation potential.

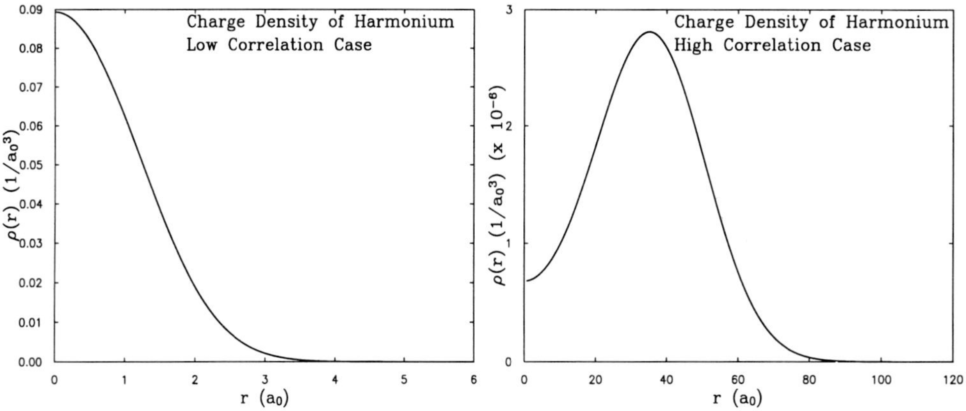

Figure 1. Ground state electronic densities: the spring constants are $k = 0.25$ a.u. (left, low correlation case) and $k \approx 3.6 \times 10^{-6}$ a.u. (right, high correlation case). Taken from Ref. [54] and modified.

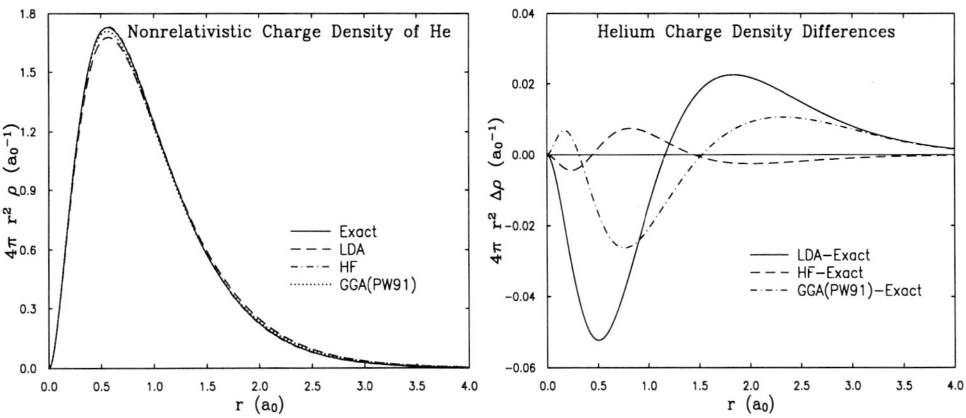

Figure 2. Comparison of the LDA, GGA (PW91), HF and exact densities of He (left). Errors in the LDA, GGA, and HF densities of He (right). Taken from Ref. [55] and modified.

For the He iso-electronic series, it is possible to achieve this accuracy. The form of the wave function is described in Refs. [55,64]. The energy for He corresponding to this wave function was estimated to be about 1 part in 10^{16}. With this nearly exact wave function, the exchange-correlation potential is resolved with unprecedented accuracy: it is possible to observe [55] not only the $-1/r$ behavior of the exchange potential but also the asymptotic behavior of the correlation potential, $-9/(4Z^4r^4)$, predicted by Almbladh and von Barth [65].

In Fig. 2, we compare the density obtained with our accurate wave function with the self-consistent densities obtained from LDA, the Perdew-Wang '91 GGA[14] and Hartree-Fock (HF). The LDA density is less peaked than the true density while the HF density is very close but slightly more peaked than the true density. We also show the error in the LDA, GGA, and HF densities. The GGA density is more accurate than the LDA density but less accurate than the HF one. Note that the error in the HF density is considerably smaller than the LDA and GGA errors; however, as shown below, this is true only in the core region for heavier systems.

5.3. Be atom

For many-electron systems, an accurate determination of the charge densities is obtained by combining the results from variational Monte Carlo (VMC) and diffusion Monte Carlo (DMC) methods [66]. In VMC, the square of a trial wave function is sampled and the expectation values for the trial wave function are calculated as averages over the sampled configurations. An advantage of this technique versus other conventional quantum chemistry methods is that the many-dimensional integrals can be performed for any given form of the trial wave function. Therefore, one is freed from the constraint of having to express the trial wave function in some restricted form such as a linear combination of determinants of single-particle orbitals. The starting point of the DMC method is also a good trial wave function from which DMC projects out an improved estimate of the true wave function.

To calculate the charge density, the integral of the square of the true wave function over all but one of the electrons has to be estimated. VMC samples the square of the trial wave function. DMC samples the product of the trial wave function and the true wave function. If the error in the wave function is of $\mathcal{O}(\epsilon)$, the errors in the densities obtained from VMC and DMC are of the same order. A density with an error of $\mathcal{O}(\epsilon^2)$ (that we will refer to as the quantum Monte Carlo (QMC) density) is obtained by taking twice the DMC density and subtracting the VMC density[67]. The QMC density is obtained in the form of an histogram and then fitted to a sum of products of monomials, exponentials and an appropriate asymptotic function [68].

In Fig. 3, we show the densities (left) obtained from LDA, the Perdew-Wang '91 GGA, HF and QMC for Be and the errors in these densities (right), with the QMC density as the reference. The first thing to notice is that all densities are very similar. The LDA and HF errors are of comparable magnitude in the valence region whereas, as in the case of He (Fig. 2), the HF error is smaller than the LDA error in the region near the nucleus. A hand-waving explanation of this behavior is that, in the core region, exchange dominates correlation and HF, by definition, is exact for exchange only. The Perdew-

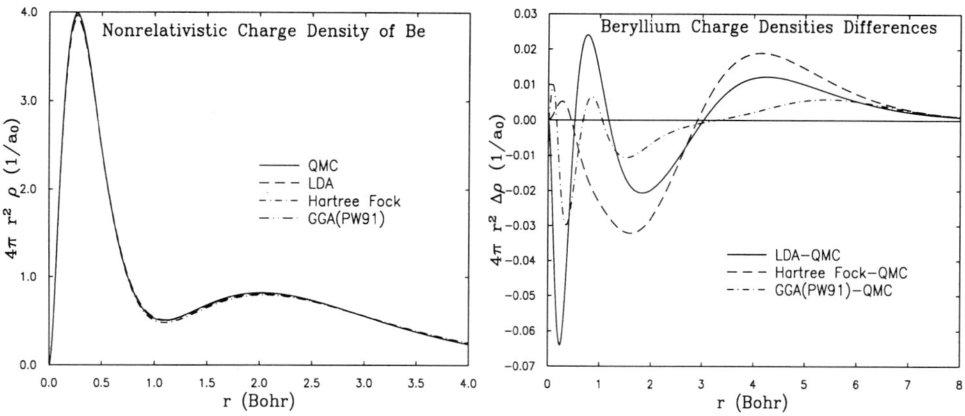

Figure 3. Comparison of the LDA, GGA (PW91), HF and "exact" (QMC) densities of Be (left). Errors in the LDA, GGA, and HF densities of Be (right). Taken from Ref. [47].

Wang '91 density is somewhat more accurate than the LDA density. The GGA potential has a spurious negative divergence at the nucleus that is partially responsible for the improvement in the density. The LDA density is too low at the nucleus and the negative divergence of the GGA potential results in increased charge density there, a feature we have observed for Ne as well.

6. A CLOSER LOOK AT EXISTING APPROXIMATIONS: v_{XC}, v_X AND v_C

The accurate exchange-correlation potentials and their separation in exchange and correlation components were obtained for the four systems described in this paper, using the techniques mentioned in Sec. 4. The approximate exchange-correlation potentials are evaluated numerically as functional derivatives of the approximate exchange-correlation energy functionals (Eq. 16) and compared with the corresponding accurate quantities.

6.1. Harmonium

In Fig. 4, we plot the exact exchange potential and several approximate exchange potentials evaluated for the exact density of the model system. The approximate potentials differ from the exact potential over all the spatial range. At large r, all the potentials do not reproduce the correct $-1/r$ asymptotic behavior: they go exponentially to zero with the exception of the Becke functional which behaves as $(-\text{constant}/r^2)$ [18] and of the Langreth-Mehl functional which diverges. At the origin, the exact exchange potential has a minimum because it is simply proportional to the Hartree potential through Eq. 29 while the approximate exchange potentials exhibit a maximum (they do not diverge since, from Eq. 19, $s = 2$ and $m > 1$). The Perdew-Wang '91 exchange potential is very close to the Becke '88 potential except that it displays additional oscillations at the extrema

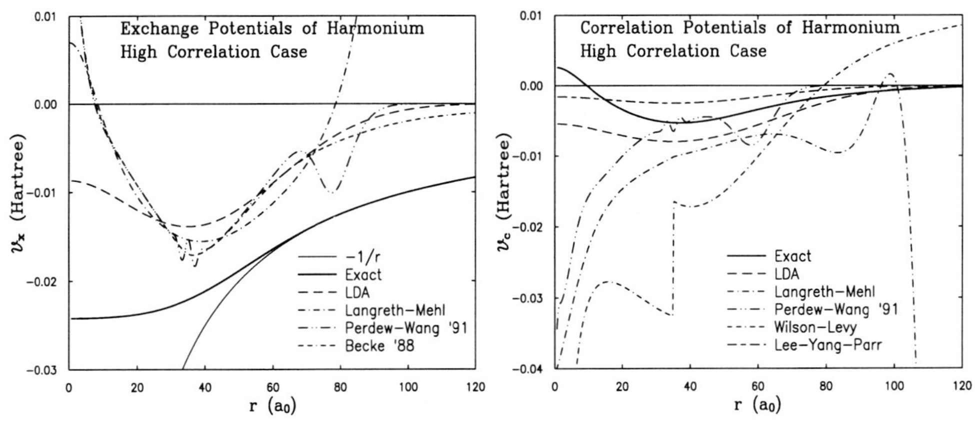

Figure 4. Exact and approximate exchange (left) and correlation (right) potentials for the spring constant $k \approx 3.6 \times 10^{-6}$ a.u. Taken from Ref. [54] and modified.

of the densities and in the tail of the density. In addition, as mentioned earlier, the long-distance asymptotic behaviors are different. As pointed out by Engel and Vosko [69], the oscillations of the Perdew-Wang '91 potential at the quadratic extrema of the density are due to the fact that the effective coefficient of the $|\nabla\rho|^2$ term increases very rapidly from the known value at $\xi = 0$, specified by the second-order gradient expansion, to a 2.1 times larger value at only $\xi = 0.04$, where $\xi = [\nabla\rho/(2k_F\rho)]^2$ and k_F is the Fermi wave vector. Observe that the Becke '88 potential does not have spurious oscillations at the extremum of the density nor in the tail of density.

In Fig. 4, we also show the exact correlation potential and several approximate correlation potentials evaluated for the exact density. At large r, the approximate potentials go to zero exponentially with the exception of the Wilson-Levy potential which goes to a positive constant and the Langreth-Mehl functional which diverges. As predicted by Eq. 19, the Wilson-Levy potential diverges at the origin: the energy density, $e_{xc}(\rho, |\nabla\rho|)$, depends on $|\nabla\rho|$ as $\mathcal{O}(|\nabla\rho|))$ for small values of the gradient of the density (Eq. 19, $m = 1$). All the other functionals assume a finite value at the origin (Eq. 19, $s = 2$ and $m > 1$). The Wilson-Levy functional has a discontinuity at the maximum of the electronic density: the functional derivative of the exchange-correlation energy contains the sign of the radial derivative of the density and is discontinuous if the energy density depends on the absolute value of gradient of the density as $\mathcal{O}(|\nabla\rho|))$, for small values of $|\nabla\rho|$. As in the case of exchange the Perdew-Wang '91 correlation potential shows an oscillatory behavior near the maximum at a finite distance from the origin and also in the tail of the potential.

6.2. He atom

For atomic systems, the exchange potential is the dominant part of the exchange-correlation potential. It shares with it most of its characteristics: the asymptotic behavior, the intershell bump (see Sub-sec. 6.3) and the finite short-range behavior.

In Fig. 5 we compare the exchange potential obtained from the LDA and the various GGA's evaluated for the exact charge density with the exact potential obtained from Eq. 29. The approximate potentials differ significantly from the exact one. At large r, all the potentials do not reproduce the correct $-1/r$ asymptotic behavior: they go exponentially to zero with the exception of the Becke '88 functional which behaves as $(-\text{constant}/r^2)$ [18] and of the Langreth-Mehl functional which diverges. The Perdew-Wang '91 functional has a spurious minimum between 2.5 and 3.5 a_0.

At the origin, the exact exchange potential has a quadratic minimum because it is simply proportional to minus the Hartree potential. In contrast all the proposed GGA exchange potentials diverge at the nucleus while the LDA exchange potential has a finite value and slope there. Hence, very close to the nucleus, the various GGA potentials are an even poorer approximation to the true potential than LDA but the more negative values of the GGA potentials at short and intermediate distances is a step in the right direction. As shown in Eq. 19, any approximate functional that has terms containing the gradient but no higher derivatives of ρ must yield an exchange-correlation potential that diverges at nuclei since, at nuclei, $s = 1$.

In Fig. 5, we also compare approximate correlation potentials of He to the exact one. In the case of an n-electron atom or ion with orbitally non-degenerate n and $n-1$ electron ground states, the correlation potential goes as $-\alpha/2r^4$ at large distances, where α is the dipole-polarizability of the $n-1$ electron system[65]. In the case of the He iso-electronic series, the dipole-polarizability of the residual 1-electron system is $\alpha = 9/(2Z^4)$. Due to the rapid fall-off of $1/r^4$, this asymptotic behavior is not discernible in Fig. 5 but is evident in a plot of $r^4 v_c$ [55]. At large distances, the approximate potentials go to zero exponentially with the exception of the Wilson-Levy potential which goes to a positive constant and the Langreth-Mehl potential which diverges. All the GGA correlation potentials diverge at the origin as follows from Eq. 19. Of the GGA's proposed so far, only the Lee-Yang-Parr GGA contains Laplacian terms but not in the form necessary to eliminate the divergence at the nucleus.

6.3. Be and Ne atoms

In Fig. 6, we show the accurate exchange potentials of Be and Ne and compare them with the approximate exchange potentials from the local density approximation and several generalized gradient approximations, all of them being evaluated for the corresponding accurate QMC density.

At short and large distances, all GGA's are similar and qualitatively wrong and the same analysis applies as in the case of the He atom. The Perdew-Wang '91 exchange potential has a spurious dip at about 6 a.u. (outside the plotted range) for Be and at about 3.3 a.u. for Ne. This dip is not present in either Perdew-Wang '86 or the Becke '88 potentials. The exact exchange potential has a peak in the intershell region (around 1 a.u. for Be and around 0.3 a.u. for Ne). There, the various GGA potentials exhibit some differences

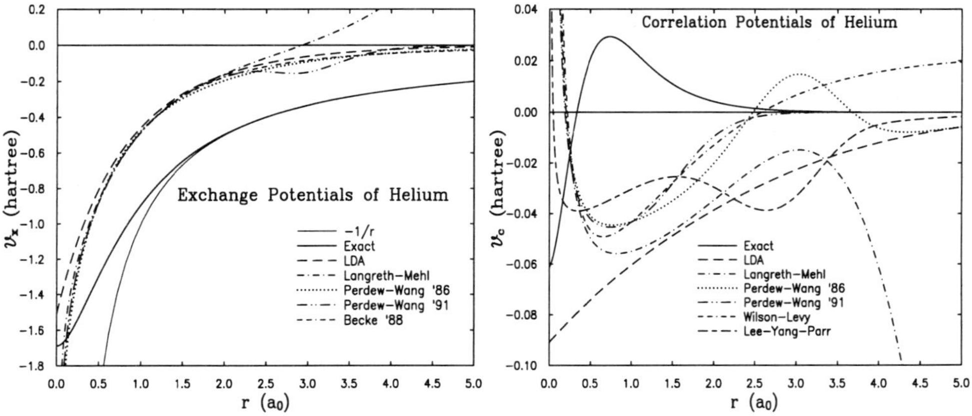

Figure 5. Exact and approximate exchange (left) and correlation (right) potentials for He. Taken from Ref. [55] and modified.

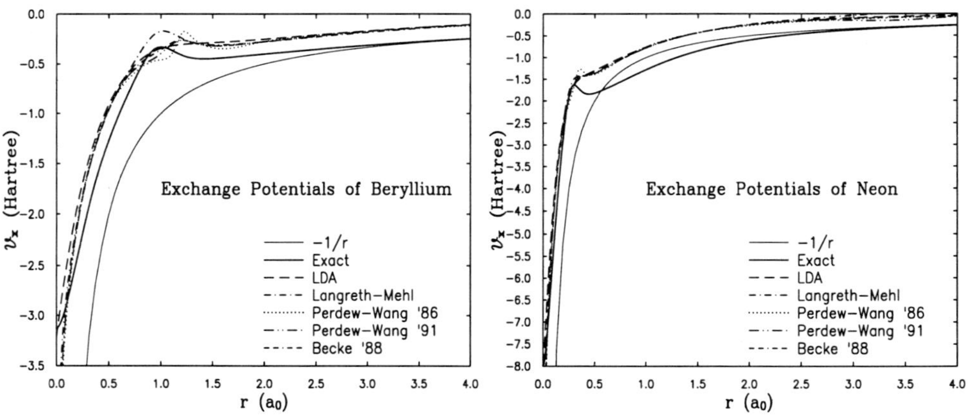

Figure 6. Exchange potentials of Be (left) and Ne (right) from QMC (labeled exact - see text), LDA and various GGA's. Taken from Ref. [47].

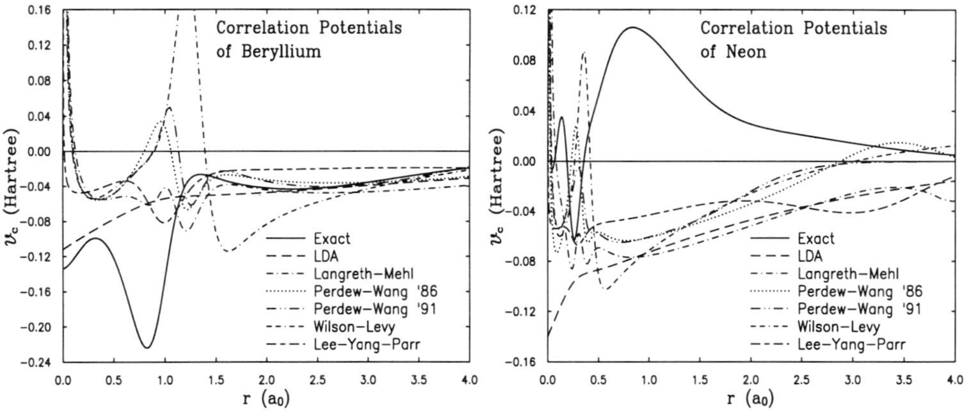

Figure 7. Correlation potentials of Be (left) and Ne (right) from QMC (labeled exact - see text), LDA and various GGA's. The "exact" Ne potential has significant uncertainty for $r < 0.4\ a_0$. Taken from Ref. [47].

but they all have a peak at approximately the right position. The GGA potentials are more accurate than the LDA potential at distances smaller than the intershell radius, except at very short distances where they have a spurious divergence. To summarize, the exchange potentials from the GGA's show some tendency for improvement upon LDA in the intershell region and at shorter distances but they have a spurious divergence at the nucleus and do not improve the long-distance behavior.

In Fig. 7, we show the correlation potentials for Be and Ne from QMC, LDA and the different GGA's. The LDA correlation potential is everywhere negative, monotonically increasing with the radius (since the density is monotonically decreasing) and rather smooth. Except for the part of the potential beyond 1.5 a.u. for Be, it bears no resemblance to the exact correlation potential. The GGA correlation potentials exhibit more structure but none of these does better than LDA. Nevertheless, as we will see in Section 8 the GGA's yield more accurate correlation energies than does LDA. The true correlation potential is everywhere negative for Be but is predominantly positive for Ne.

7. LOCAL EXCHANGE ENERGY PER ELECTRON: $\epsilon_{\rm x}$

Using the expression in Eq. 25 for the local exchange energy per electron, $\epsilon_{\rm x}$, we present accurate $\epsilon_{\rm x}$'s for the two-electron systems and the Be atom. Since $\epsilon_{\rm x}$ is not uniquely defined (see Sec. 4), the differences of approximate $\epsilon_{\rm x}$'s from each other and from the "true" one are not necessarily indicative of a failure of the approximate functionals.

In the left panel of Fig. 8, we plot the exact and approximate local exchange energies per electron for Harmonium. At large r, the approximate exchange functionals go to zero exponentially with the exception of the Langreth-Mehl functional which diverges and of

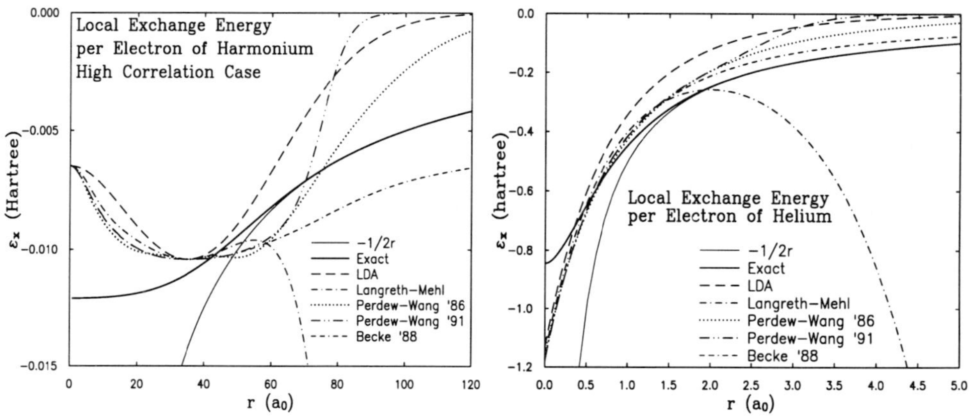

Figure 8. Exact and approximate local exchange energy per electron for Harmonium ($k \approx 3.6 \times 10^{-6}$ a.u.) (left) and He (right), evaluated for the respective exact densities. Taken from Refs. [54] and [55] and modified.

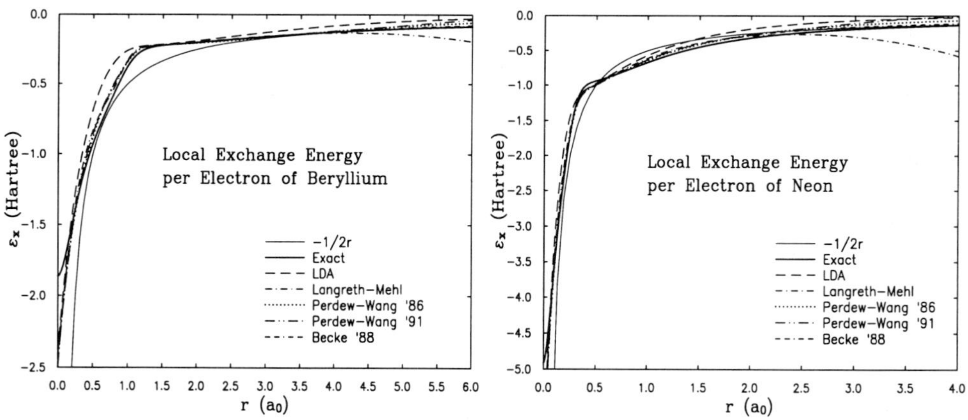

Figure 9. Exact and approximate local exchange energy per electron for Be (left) and Ne (right), evaluated for the respective exact densities.

the Becke functional which has the wrong asymptotic behavior $-1/r$ for Harmonium although it has the correct asymptotic behavior $-1/2r$ for atoms. The Becke functional was constructed to have the correct asymptotic behavior for exponentially decaying density but it fails for Gaussian densities (see Table 1). At the extrema of the density, all the functionals go to the LDA value.

In the right panel of Fig. 8 and in Fig. 9, we compare the exact and approximate local exchange energies per electron of the He, Be and Ne atoms. At large r, the approximate ϵ_x go to zero exponentially with the exception of the Langreth-Mehl functional, which diverges, and the Becke functional which has the correct asymptotic behavior $-1/2r$. However, as also noted by Engel *et al.* [18], the Becke ϵ_x achieves its asymptotic behavior at much larger distances than does the true ϵ_x. This is because the correction to the leading $-1/2r$ term is exponentially small for the true functional but $\mathcal{O}(1/r^2)$ for the Becke functional. The Becke exchange energy per electron is however closer to the correct one over all the range than any other approximate ϵ_x. At intermediate distances, the various GGA's represent an improvement upon LDA, yielding a deeper ϵ_x than LDA, but, at very short distances, they are significantly lower than the true one.

8. COMPARISON OF THE COMPONENTS OF THE TOTAL ENERGY

For the model system of two electrons in a harmonic potential, the He iso-electronic series and the Be atom, we present a comparison of the accurate values of E_x, E_c and E_{xc} with the ones corresponding to approximate functionals evaluated for the accurate densities. In order to assess the change that results from employing the self-consistent rather than the exact density, we also perform, for Be, self-consistent calculations with LDA and the Perdew-Wang '91 GGA and compare the different components of the total energy (Eq. 6) with the accurate values.

For the model system of two interacting electrons in a harmonic potential, we compare in Table 2 the values of the exact exchange and correlation energies with those obtained from the approximate functionals evaluated for the exact electronic density. The GGA functionals, except for the Langreth-Mehl functional, improves considerably upon LDA. The major corrections are achieved for the exchange contribution. The failure of the Langreth-Mehl functional is a result of the diverging energy per electron at large r. Most of the approximate functionals are constructed to give accurate values of E_{xc} for atoms and it is encouraging that they also improve E_{xc} for this rather different model system. However, we can see in Tables 3 and 4 that the improvement upon LDA is much larger for atomic systems: for instance, the percentage error in the LDA exchange energy evaluated for the electronic density of He is 14% and in the LDA correlation energy, -167%, while the percentage errors for the other approximate functionals we list in the appendix range from -1 to 1% for exchange and from -19 to 1% for correlation. If we consider the sum of the exchange energy and the correlation energy, LDA yields a good result due to a cancellation of errors and the GGA's do not improve upon LDA.

In Tables 3 and 4, we compare the approximate E_x, E_c and E_{xc}, evaluated for the accurate densities of the He iso-electronic series and Be respectively, with the corresponding accurate quantities. As is well known, LDA yields values of E_x that are too small in

Table 2
Model system with spring constant $k \approx 3.6 \times 10^{-6}$ a.u. Exchange-correlation energy for the approximate functionals listed in Appendix A. The functionals are evaluated for the exact density. The exact results are also reported. The numbers in parentheses are the percentage errors. The total energy of the system is 0.0228 a.u. Energies are in Hartree atomic units. Taken from Ref. [54] and modified.

Functional	E_x	E_c	E_{xc}
Exact KS	−0.0195	−0.0064	−0.0259
LDA (Perdew-Wang)	−0.0174(11%)	−0.0108(−69%)	−0.0282(−9%)
Langreth-Mehl	−0.0217(−11%)	−0.0119(−86%)	−0.0336(−30%)
Perdew-Wang '86	−0.0198(−2%)	−0.0096(−50%)	−0.0294(−14%)
Perdew-Wang '91	−0.0195(0.2%)	−0.0084(−31%)	−0.0279(−8%)
Becke '88	−0.0201(−3%)	—	—
Wilson-Levy	—	−0.0145(−127%)	—
Lee-Yang-Parr	—	−0.0032(50%)	—

absolute magnitude and values of E_c that are too large in absolute magnitude, resulting in a cancellation of errors for E_{xc}. All the GGA's yield improved exchange, correlation and exchange-correlation energies compared to the LDA.

For the He iso-electronic series, the true correlation energies are nearly constant, whereas for the Be iso-electronic series there is a linear component in the nuclear charge Z. It is difficult for any GGA functional to mimic the correct Z dependence of both iso-electronic series.

In Table 5, we show various quantities, computed by LDA and the Perdew-Wang '91 GGA, and the corresponding accurate values for Be. LDA and GGA values were obtained through a self-consistent procedure and are therefore evaluated for the LDA and the GGA density respectively. The change in E_x and E_c due to self-consistency are not negligible. The values of E_x and E_c in Table 5 differ from those in Table 4 by a larger amount for LDA than for the Perdew-Wang '91 GGA reflecting our earlier observation that the GGA yields somewhat more accurate self-consistent densities than does LDA. The last two columns show the errors in LDA and GGA. The LDA components of the total energy are all smaller in absolute magnitude than the corresponding accurate values. Both the total energy and its components have much smaller errors in GGA than in LDA. In Fig. 3, we observed that, near the nucleus, the LDA density is too small while GGA gives an increased charge density due to the lower value at short distances and the negative divergence at the nucleus of the GGA exchange-correlation potential. This increase in the GGA density results in an increase in the magnitude of all the components of the total energy. Consequently, the GGA yields improvements not only in E_{xc} but also in other components of the total energy, despite, or may be one should say because of, the spurious divergence of the exchange-correlation at the nucleus. On the other hand, quantities related to the single-particle eigenvalues (shown in the last two rows) are not improved in GGA relative to LDA. In particular, the highest lying eigenvalue which, for the true density functional, should equal the ionization energy, is not much improved.

Table 3: Comparison of exchange and correlation energies of the He iso-electronic series for LDA and the various GGA's with the corresponding exact values. The approximate functionals are evaluated for the exact (not the self-consistent) densities. The numbers in parentheses are the percentage errors. All numbers are accurate to all digits presented with the exception of the "exact" results for Hg^{+78} which probably have an error in the last digit. Taken from Ref. [55] and modified.

	Ion	Exact	LDA [15]	LM [9]	PW86 [12,13]	PW91 [14]	B88 [11]	WL [16]	LYP [17]
	H^-	−0.380898	−0.337(12%)	−0.387(−2%)	−0.400(−5%)	−0.393(−3%)	−0.395(−4%)	—	—
	He	−1.024568	−0.883(14%)	−1.014(1%)	−1.033(−1%)	−1.016(1%)	−1.025(−0%)	—	—
E_x	Be^{+2}	−2.276628	−1.957(14%)	−2.246(1%)	−2.279(−0%)	−2.245(1%)	−2.265(1%)	—	—
	Ne^{+8}	−6.027484	−5.173(14%)	−5.940(1%)	−6.015(0%)	−5.927(2%)	−5.982(1%)	—	—
	Hg^{+78}	−49.777931	−42.699(14%)	−49.036(1%)	−49.605(0%)	−48.889(2%)	−49.348(1%)	—	—
	H^-	−0.041995	−0.072(−71%)	−0.049(−16%)	−0.038(10%)	−0.032(24%)	—	−0.032(25%)	−0.030(29%)
	He	−0.042107	−0.112(−167%)	−0.050(−19%)	−0.044(−4%)	−0.046(−9%)	—	−0.042(1%)	−0.044(−4%)
E_c	Be^{+2}	−0.044274	−0.150(−240%)	−0.025(43%)	−0.049(−12%)	−0.053(−21%)	—	−0.045(−2%)	−0.049(−11%)
	Ne^{+8}	−0.045694	−0.202(−342%)	0.082(279%)	−0.084(−83%)	−0.062(−35%)	—	−0.047(−3%)	−0.050(−10%)
	Hg^{+78}	−0.046538	−0.326(−600%)	1.833(4038%)	−0.299(−543%)	−0.081(−73%)	—	−0.048(−3%)	−0.051(−9%)
	H^-	−0.422893	−0.408(3%)	−0.435(−3%)	−0.438(−4%)	−0.425(−0%)	—	—	—
	He	−1.066675	−0.996(7%)	−1.064(0%)	−1.077(−1%)	−1.062(0%)	—	—	—
E_{xc}	Be^{+2}	−2.320902	−2.107(9%)	−2.271(2%)	−2.328(−0%)	−2.298(1%)	—	—	—
	Ne^{+8}	−6.073178	−5.375(11%)	−5.858(4%)	−6.099(−0%)	−5.989(1%)	—	—	—
	Hg^{+78}	−49.824470	−43.025(14%)	−47.203(5%)	−49.904(−0%)	−48.970(2%)	—	—	—

Table 4
Comparison of exchange and correlation energies of Be for LDA and the various GGA's with the corresponding accurate values. The approximate functionals are evaluated for the exact (not the self-consistent) densities. The numbers in parentheses are the percentage errors. Energies are in Hartree atomic units. Taken from Ref. [47].

Functional	E_x	E_c	E_{xc}
Accurate	−2.67398	−0.09615	−2.77013
LDA (Perdew-Wang)	−2.32125(13%)	−0.22515(−134%)	−2.54641(8%)
Langreth-Mehl	−2.60576(3%)	−0.10012(−4%)	−2.70588(2%)
Perdew-Wang '86	−2.69050(−1%)	−0.09425(2%)	−2.78475(−1%)
Perdew-Wang '91	−2.65434(1%)	−0.09496(1%)	−2.74929(1%)
Becke '88	−2.66694(0%)	—	—
Wilson-Levy	—	−0.09572(0%)	—
Lee-Yang-Parr	—	−0.09546(1%)	—

Table 5
Comparison of LDA, Perdew-Wang '91 GGA and accurate values for various components of the Kohn-Sham total energy, the density-weighted integral of v_{xc} and single-particle eigenvalues of Be. The approximate functionals are evaluated for the respective self-consistent densities. The accurate values probably have errors in the last digit quoted. The accurate value of the total energy is from Ref. [70]. Energies are in Hartree atomic units. Taken from Ref. [47].

Property	LDA	PW91	Accurate	Δ_{LDA}	Δ_{PW91}
$E = T_s + E_{en} + E_H + E_{xc}$	−14.44647	−14.64794	−14.66736	0.22089	0.01942
T_s	14.30879	14.57461	14.5942	−0.2854	−0.0196
E_{en}	−33.35610	−33.65670	−33.7099	0.3538	0.0532
E_H	7.11487	7.17261	7.2185	−0.1036	−0.0459
E_{xc}	−2.51403	−2.73845	−2.77013	0.25610	0.03168
E_x	−2.29033	−2.64428	−2.67398	0.38365	0.02970
E_c	−0.22370	−0.09417	−0.09615	−0.12754	0.00198
$E_{en} + E_H$	−26.24123	−26.48410	−26.4914	0.2502	0.0073
$\int d\mathbf{r}\, \rho(\mathbf{r})\, v_{xc}(\mathbf{r})$	−3.30614	−3.50058	−4.4595	1.1533	0.9589
ϵ_{1s}	−3.85609	−3.91154	−4.2265	0.3704	0.3149
ϵ_{2s}	−0.20577	−0.20719	−0.3426	0.1368	0.1354

This is to be expected since the long-distance behavior of exchange-correlation potential, which strongly influences the highest eigenvalue, is not appreciably improved. All the observations made here for the Be atom are also true for the Ne atom (see Refs. [45] and [47]).

9. PROSPECTS

Although the exchange and correlation energies calculated from the various GGA's (with the exception of the Langreth-Mehl GGA for the model system and for the correlation energies in the He iso-electronic series) are considerably more accurate than those from LDA, the situation is not as clear for the exchange and correlation potentials. Since E_x and E_c are integrated quantities, it is possible for GGA's to improve upon them without making comparable improvements to the exchange and correlation potentials which contain more detailed information.

The various GGA exchange functionals have certain features in common. By favoring regions with large density gradients, they correctly lead to more negative total energies and reduced binding energies relative to LDA. They yield a lower potential than LDA at short and intermediate distances from nuclei and approximately reproduce the intershell structure – two steps in the right direction – but they also introduce a spurious divergence at the nucleus and fail to improve the long-range behavior. Whereas the various GGA exchange functionals have some common features, the GGA correlation functionals are very different from each other and none of them yields a correlation potential resembling the true one. Correlation is more subtle than exchange as manifested in several ways. 1) The exchange potential is everywhere negative, whereas the correlation potential can be either negative or positive in different regions of space. 2) For He iso-electronic series, the correlation potential varies less over space for heavy ions than for light atoms [55]. This is opposite to the behavior in LDA. 3) The correlation potential has qualitatively the same behavior for all members of the helium iso-electronic series [55] (negative at short distances and long distance, positive at intermediate distances, see Fig. 5) but it has the opposite qualitative behavior for the closely related model system of two electrons in a harmonic potential (Fig. 4).

As discussed in Sec. 2, any GGA that includes no higher than first derivatives of the density cannot simultaneously satisfy both the correct $-1/r$ behavior of the exchange-correlation potential and the correct $-1/2r$ behavior of exchange energy per electron at large distances. Moreover, any such GGA must have a spurious divergence in the exchange-correlation potential at nuclei. We mentioned that, by including the Laplacian of the density in an appropriate way, it is possible to construct a GGA that has the correct long-range asymptotics of the exchange potential and energy per electron and the proper short-range behavior of the exchange-correlation potential. Nevertheless, it may turn out that even GGA functionals containing Laplacian terms are not flexible enough to achieve a major improvement and it may be necessary to consider more general functionals. We discuss here some possibilities.

Since a data-base on accurate exchange-correlation potentials has being built in recent years and none of the existing GGA functionals passes the test of the comparison with the true potentials, more effort is now being devoted to developing schemes that focus on reproducing the main features of the true exchange-correlation potential. As mentioned above, the potential controls the quality of the self-consistent density and is the key ingredient in evaluating energy gradients [71],

$$\frac{\partial E_{xc}[\rho]}{\partial X} = \int v_{xc}([\rho];\mathbf{r}) \frac{\partial \rho(\mathbf{r})}{\partial X} \mathrm{d}\mathbf{r}, \tag{37}$$

whose accurate estimation is for instance necessary in determining potential energy surfaces.

Instead of starting from an exchange-correlation energy functional and then deriving the corresponding potential, some authors start with an expression for the potential which has the correct asymptotic behavior [39,73]. While it is easy to construct a potential with the correct asymptotics, one encounters the problem of how to obtain the exchange-correlation energy functional that corresponds to this potential. In fact, the potential may not be the functional derivative of an energy expression. If the approximate exchange potential satisfies the uniform scaling property $v_x([\rho_\lambda];\mathbf{r}) = \lambda v_x([\rho];\lambda\mathbf{r})$, a possibility is to use the Levy-Perdew relation to assign a corresponding exchange energy [50]:

$$E_x = \int v_x([\rho];\mathbf{r})\,(3\rho(\mathbf{r}) + \mathbf{r}\cdot\nabla\rho(\mathbf{r}))\,\mathrm{d}\mathbf{r}. \tag{38}$$

The relation is derived by considering a line integral of the potential along the density path ρ_λ (Eq. 11) and is exact only for a potential that is a functional derivative and has the correct scaling property:

$$E_x[\rho] = \int_0^1 \mathrm{d}\lambda \int \mathrm{d}\mathbf{r}\; v_x([\rho_\lambda];\mathbf{r}) \frac{\mathrm{d}\rho_\lambda(\mathbf{r})}{\mathrm{d}\lambda} = \int v_x([\rho];\mathbf{r})\,(3\rho(\mathbf{r}) + \mathbf{r}\cdot\nabla\rho(\mathbf{r}))\,\mathrm{d}\mathbf{r}. \tag{39}$$

The first equality always holds for a potential that is a functional derivative while the second one is valid only if the potential has the correct behavior under uniform scaling. Van Leeuwen and Baerends [71] discuss why Eq. 38 is a sensible choice for a potential that is not a functional derivative but has the proper scaling property and is constructed to mimic the true potential at the given density. Unfortunately, such a relation does not exist in the case of exchange and correlation since the scaling properties of the correlation potential are not homogeneous [49]. Assigning an exchange-correlation energy to a given potential becomes more complicated because we need to know the potential not only for the density of interest, as in the case of exchange-only, but along some path in density space [71].

Another attempt to generate a DFT scheme yielding a better description of the exchange-correlation potential has as a starting point the OEP method. The OEP method provides the exact solution for the problem of exchange-only DFT but, as shown by Norman and Koelling [72], it can be generalized to treat exchange and correlation if the correlation energy is also expressed as a functional of the Kohn-Sham orbitals. The resulting problem is technically difficult to solve but Krieger, Li and Iafrate proposed an approximation that greatly simplify the approach [74]. Although the resulting exchange-correlation potential is no longer equivalent to the functional derivative with respect to the density of the original exchange-correlation energy, it closely corresponds to the exact generalized OEP exchange-correlation potential if the expression $\delta E_{\mathrm{xc}}\left[\{\psi_i\}\right]/\delta\psi_i(\mathbf{r})/\psi_i^*(\mathbf{r})$ goes to zero at large distances. Since the true exchange energy is known as a functional of the Kohn-Sham orbitals (Eq. 21) and can therefore be employed, the approximate exchange potential, evaluated according to the procedure of Ref. [74], has the correct short and long-distance asymptotic behavior although the intershell bump is underestimated. An improved approximation that yields a more accurate intershell bump is also discussed in Ref. [74]. This approach is of practical interest since it is easier to construct self-interaction free, exchange-correlation functionals that depend on the orbitals instead of the density. Grabo and Gross [75] recently implemented this scheme with the approximation of Krieger, Li and Iafrate and the orbital-dependent correlation functional of Colle and Salvetti [76].

A different approximate scheme that cures the long-distance asymptotics of the exchange-correlation potential is obtained by combining a given approximate exchange-correlation functional with a self-interaction correction (SIC) procedure [19]. A GGA depending on the Laplacian of the density could be easily constructed so that the exchange-correlation potential does not have a spurious divergence at nuclei and could then be implemented in a SIC scheme to yield a potential with also the correct long-range asymptotic behavior.

Acknowledgments

We thank John Morgan and Jonathan Baker for making available their program for calculating very accurate wave functions for the He iso-electronic series, Manfred Taut for providing the program for calculating the densities of Harmonium and John Perdew the routine for calculating the Perdew-Wang 91 functional. We benefited from several useful discussions with Mel Levy and Eberhard Gross. The calculations were performed on the IBM SP2 computer at the Cornell Theory Center. This work is supported by the Office of Naval Research and NATO (grant number CRG 940594). X.G. acknowledges financial support from FNRS-Belgium and the European Union (Human Capital and Mobility Program contract CHRX-CT940462).

A. APPROXIMATE FUNCTIONAL FORMS OF $E_{\mathrm{xc}}[\rho]$

The second order generalized gradient approximation exchange-correlation energy is written as

$$E_{\mathrm{xc}}^{\mathrm{GGA}}[\rho] = \int d\mathbf{r}\, e_{\mathrm{xc}}(\rho(\mathbf{r}), |\nabla\rho(\mathbf{r})|, \nabla^2\rho(\mathbf{r})). \tag{40}$$

We define the following density dependent variables:

$$k_{\mathrm{F}} = \left(3\pi^2\rho\right)^{1/3}, \quad k_s = \left(\frac{4}{\pi}k_{\mathrm{F}}\right)^{1/2}, \quad s = \frac{|\nabla\rho|}{2\,k_{\mathrm{F}}\,\rho}, \quad t = \frac{|\nabla\rho|}{2k_s\,\rho}, \quad r_s = \left(\frac{3}{4\pi\rho}\right)^{1/3}.$$

We present the spin unpolarized version of the approximate functionals, $\zeta = (\rho_\uparrow - \rho_\downarrow)/\rho = 0$. All the parameters that appear in the following functionals are in atomic units.

LDA exchange functional

$$e_{\mathrm{x}}^{\mathrm{LDA}} = A_x \rho^{4/3}\,, \tag{41}$$

where $A_{\mathrm{x}} = -(3/4)\,(3/\pi)^{1/3}$.

LDA correlation functional (Perdew-Wang [15])

$$e_{\mathrm{c}}^{\mathrm{LDA}} = -2a\rho\,(1+\alpha_1 r_s)\log\left[1+\frac{1}{2a(\beta_1 r_s^{1/2}+\beta_2 r_s+\beta_3 r_s^{3/2}+\beta_4 r_s^2)}\right], \tag{42}$$

where $a = 0.0310907$, $\alpha_1 = 0.21370$, $\beta_1 = 7.5957$, $\beta_2 = 3.5876$, $\beta_3 = 1.6382$ and $\beta_4 = 0.49294$.

RPA correlation functional [77]

$$\begin{aligned} e_{\mathrm{c}} &= -\rho\, c^{\mathrm{P}} f(r_s/r^{\mathrm{P}}), \qquad (43)\\ f(z) &= (1+z^3)\ln\left(1+\frac{1}{z}\right)+\frac{z}{2}-z^2-\frac{1}{3}, \end{aligned}$$

where $c^{\mathrm{P}} = 0.0252$ and $r^{\mathrm{P}} = 30$.

Langreth-Mehl exchange-correlation functional [9]

$$e_{\mathrm{x}} = e_{\mathrm{x}}^{\mathrm{LDA}} - a\frac{|\nabla\rho|^2}{\rho^{4/3}}\left(\frac{7}{9}+18\,f^2\right), \tag{44}$$

$$e_{\mathrm{c}} = e_{\mathrm{c}}^{\mathrm{RPA}}(\rho) + a\,\frac{|\nabla\rho|^2}{\rho^{4/3}}\left(2e^{-F}+18f^2\right), \tag{45}$$

where $F = b\,|\nabla\rho|\,/\rho^{7/6}$, $b = (9\pi)^{1/6} f$, $a = \pi/(16(3\pi^2)^{4/3})$ and $f = 0.15$.

Perdew-Wang '86 exchange functional [12]

$$e_{\mathrm{x}} = e_{\mathrm{x}}^{\mathrm{LDA}}(\rho)\left(1+0.0864\,\frac{s^2}{m}+b\,s^4+c\,s^6\right)^m, \tag{46}$$

where $m = 1/15$, $b = 14$ and $c = 0.2$.

Perdew-Wang '86 correlation functional [13]

$$e_{\mathrm{c}} = e_{\mathrm{c}}^{\mathrm{LDA}}(\rho) + e^{-\Phi} C_{\mathrm{c}}(\rho) \frac{|\nabla \rho|^2}{\rho^{4/3}}, \tag{47}$$

where

$$\begin{aligned} \Phi &= 1.745\, \tilde{f}\, \frac{C_{\mathrm{c}}(\rho = \infty)}{C_{\mathrm{c}}(\rho)} \frac{|\nabla \rho|}{\rho^{7/6}}, \\ C_{\mathrm{c}}(\rho) &= C_1 + \frac{C_2 + C_3 r_s + C_4 r_s^2}{1 + C_5 r_s + C_6 r_s^2 + C_7 r_s^3} \end{aligned}$$

and $\tilde{f} = 0.11$, $C_1 = 0.001667$, $C_2 = 0.002568$, $C_3 = 0.023266$, $C_4 = 7.389 \cdot 10^{-6}$, $C_5 = 8.723$, $C_6 = 0.472$, $C_7 = 7.389 \cdot 10^{-2}$.

Perdew-Wang '91 exchange functional [14]

$$e_{\mathrm{x}} = e_{\mathrm{x}}^{\mathrm{LDA}}(\rho) \left[\frac{1 + a_1 s \sinh^{-1}(a_2 s) + \left(a_3 + a_4 e^{-100 s^2}\right) s^2}{1 + a_1 s \sinh^{-1}(a_2 s) + a_5 s^4} \right], \tag{48}$$

where $a_1 = 0.19645$, $a_2 = 7.7956$, $a_3 = 0.2743$, $a_4 = -0.1508$ and $a_5 = 0.004$.

Perdew-Wang '91 correlation functional [14]

$$e_{\mathrm{c}} = \left[e_{\mathrm{c}}^{LDA}(\rho) + \rho\, H(\rho, s, t) \right], \tag{49}$$

where

$$\begin{aligned} H &= \frac{\beta^2}{2\alpha} \log \left[1 + \frac{2\alpha}{\beta} \frac{t^2 + A\, t^4}{1 + A\, t^2 + A^2 t^4} \right] + C_{\mathrm{c0}} \left[C_{\mathrm{c}}(\rho) - C_{\mathrm{c1}} \right] t^2 e^{-100 s^2}, \\ A &= \frac{2\alpha}{\beta} \left[e^{-2\alpha \epsilon_{\mathrm{c}}^{\mathrm{LDA}}(\rho)/\beta^2} - 1 \right]^{-1} \end{aligned}$$

and $\alpha = 0.09$, $\beta = 0.0667263212$, $C_{\mathrm{c0}} = 15.7559$, $C_{\mathrm{c1}} = 0.003521$. The function $C_{\mathrm{c}}(\rho)$ is the same as for the Perdew-Wang '86 correlation functional. $\epsilon_{\mathrm{c}}^{\mathrm{LDA}}(\rho) = e_{\mathrm{c}}^{LDA}(\rho)/\rho$.

Becke '88 exchange functional [11]

$$e_{\mathrm{x}} = e_{\mathrm{x}}^{\mathrm{LDA}}(\rho) \left[1 - \frac{\beta}{2^{1/3} A_{\mathrm{x}}} \frac{x^2}{1 + 6\beta\, x \sinh^{-1}(x)} \right], \tag{50}$$

where $x = 2\,(6\pi^2)^{1/3}\, s = 2^{1/3}\, |\nabla \rho| / \rho^{4/3}$, $A_{\mathrm{x}} = -(3/4)\,(3/\pi)^{1/3}$ and $\beta = 0.0042$.

Engel-Chevary-Macdonald-Vosko exchange functional [18]

$$e_{\rm x} = e_{\rm x}^{\rm LDA}(\rho)\frac{1+a_1\xi+a_2\xi^2}{1+b_1\xi+b_2\xi^2}, \tag{51}$$

where $\xi = s^2$ and $a_1 = 27.8428$, $a_2 = 11.7683$, $b_1 = 27.5026$ and $b_2 = 5.7728$.

Wilson-Levy correlation functional [16]

$$e_{\rm c} = \frac{a\,\rho + b\,|\nabla\rho|\,/\rho^{1/3}}{c + d\,|\nabla\rho|\,/(\rho/2)^{4/3} + r_s}, \tag{52}$$

where $a = -0.74860$, $b = 0.06001$, $c = 3.60073$ and $d = 0.90000$.

Closed shell Lee-Yang-Parr correlation functional [17]

$$e_{\rm c} = -a\frac{1}{1+d\,\rho^{-1/3}}\left\{\rho + b\,\rho^{-2/3}\left[C_F\,\rho^{5/3} - 2t_W + \frac{1}{9}\left(t_W + \frac{1}{2}\nabla^2\rho\right)\right]e^{-c\,\rho^{-1/3}}\right\}, \tag{53}$$

where

$$t_W = \frac{1}{8}\left(\frac{|\nabla\rho|^2}{\rho} - \nabla^2\rho\right)$$

and $C_F = 3/10\,(3\pi^2)^{2/3}$, $a = 0.04918$, $b = 0.132$, $c = 0.2533$ and $d = 0.349$.

REFERENCES

1. P. Hohenberg and W. Kohn, Phys. Rev. **136**, B864 (1964).
2. W. Kohn and L. J. Sham, Phys. Rev. **140**, A1133 (1965).
3. R. O. Jones and O. Gunnarsson, Rev. Mod. Phys. **61**, 689 (1989).
4. C. S. Wang, B. M. Klein and H. Krakauer, Phys. Rev. Lett. **54**, 1852 (1985).
5. F. Herman, J. P. van Dyke and I. B. Ortenburger, Phys. Rev. Lett. **22**, 807 (1969); F. Herman, I. B. Ortenburger and J. P. Van Dyke, Int. J. Quantum Chem. Symp. **3**, 827 (1970).
6. A. E. DePristo and J. D. Kress, J. Chem. Phys. **86**, 1425 (1987).
7. S. -K. Ma and K. A. Brueckner, Phys. Rev. **165**, 19 (1968).
8. J. P. Perdew, D. C. Langreth and V. Sahni, Phys. Rev. Lett. **38**, 1030 (1977).
9. D. C. Langreth and M. J. Mehl, Phys. Rev. B **28**, 1809 (1983).
10. J. P. Perdew, Phys. Rev. Lett. **55**, 1665 (1985).
11. A. D. Becke, Phys. Rev. A **33**, 3098 (1988).
12. J. P. Perdew and Y. Wang, Phys. Rev. B **33**, 8800 (1986).
13. J. P. Perdew, Phys. Rev. B **33**, 8822 (1986); erratum *ibid.* **34**, 7406 (1986).

14. J. P. Perdew in *Electronic structure of solids '91*, edited by P. Ziesche and H. Eschrig (Akademie Verlag, Berlin, 1991); J. P. Perdew, K. Burke and Y. Wang, submitted to Phys. Rev. B.
15. J. P. Perdew and Y. Wang, Phys. Rev. B **45**, 13244 (1992).
16. L. C. Wilson and M. Levy, Phys. Rev. B **41**, 12930 (1990).
17. C. Lee, W. Yang and R. G. Parr, Phys. Rev. B **37**, 785 (1988).
18. E. Engel, J. A. Chevary, L. D. Macdonald and S. H. Vosko, Z. Phys. D **23**, 7 (1992).
19. J. P. Perdew and A. Zunger, Phys. Rev. B **23**, 5048 (1981); H. Stoll, C. M. E. Pavlidou and H. Preuss, Theor. Chim. Acta **49**, 143 (1978); S. H. Vosko and L. Wilk, J. Phys. B **16**, 3687 (1983).
20. R. T. Sharp and G. K. Horton, Phys. Rev. **90**, 317 (1953); J. D. Talman, and W. F. Shadwick, Phys. Rev. A **14**, 36 (1976).
21. O. Gunnarsson, M. Jonson and B. I. Lundqvist, Phys. Rev. B **20**, 3136 (1979).
22. O. Gunnarsson and R. O. Jones, Phys. Scr. **21**, 394 (1980).
23. J. P. Perdew, J. A. Chevary, S. H. Vosko, K. A. Jackson, M. R. Pederson, D. J. Singh and C. Fiolhhais, Phys. Rev. B **46**, 6671 (1992).
24. F. W. Kutzler and G. S. Painter, Phys. Rev. B **45**, 3236 (1992).
25. A. D. Becke, J. Chem. Phys. **96**, 2155 (1992); *ibid.* **97**, 9173 (1992).
26. B. G. Johnson, P. M. W. Gill and J. A. Pople, J. Chem. Phys. **98**, 5612 (1993).
27. G. Ortiz and P. Ballone, Phys. Rev. B **43**, 6376 (1991).
28. K. Laasonen, F. Csajka and M. Parrinello, Chem. Phys. Lett. **194**, 172 (1992); C. Lee, D. Vanderbildt, K. Laasonen, R. Car and M. Parrinello, Phys. Rev. B **47**, 4863 (1993).
29. P. Bagno, O. Jepsen and O. Gunnarsson, Phys. Rev. B **40**, 1997 (1989).
30. C. Filippi, D. J. Singh and C. J. Umrigar, Phys. Rev. B **50**, 19947 (1994), and references therein.
31. A. Khein, D. J. Singh and C. J. Umrigar, Phys. Rev. B **51**, 4105 (1995), and references therein.
32. D. W. Smith, S. Jagannathan, and G. S. Handler, Int. J. Quantum Chem. Symp. **13**, 103 (1979).
33. U. von Barth in *Electronic Structure of Complex Systems*, edited by P. Phariseau and W. M. Temmerman, NATO, ASI Series B, vol. 113 (1984).
34. C. -O. Almbladh and A. C. Pedroza, Phys. Rev. B **29**, 2322 (1984); A. C. Pedroza, Phys. Rev. A **33**, 804 (1986).
35. F. Aryasetiawan and M. J. Stott, Phys. Rev. B **34**, 4401 (1986); *ibid.* **38**, 2974 (1988).
36. A. Nagy and N. H. March, Phys. Rev. A **39**, 5512 (1989); *ibid.* **40**, 554 (1989).
37. E. R. Davidson, Int. J. Quant. Chem. **37**, 811 (1990).
38. J. Chen, R. O. Esquivel, and M. J. Stott, Philos. Mag. B **69**, 1001 (1994).
39. R. van Leeuwen and E. J. Baerends, Phys. Rev. A **49**, 2421 (1994).
40. R. C. Morrison and Q. Zhao, Phys. Rev. A **51**, 1980 (1995).
41. E. V. Ludeña, J. Maldonado, R. López-Boada, T. Koga, and E. S. Kryachko, J. Chem. Phys. **102**, 318 (1995).
42. M. A. Buijse, E. J. Baerends and J. G. Snijders, Phys. Rev. A **40**, 4190 (1989).
43. O. V. Gritsenko, R. van Leeuwen and E. J. Baerends, Phys. Rev. A **52**, 1870 (1995).
44. V. E. Ingamells and N. C. Handy, Chem. Phys. Lett. **248**, 373 (1996).

45. C. J. Umrigar and X. Gonze, in *High Performance Computing and its Application to the Physical Sciences*, proceedings of the Mardi Gras '93 Conference, edited by D. A. Browne *et al.*, (World Scientific, Singapore, 1993).
46. W. Knorr and R. W. Godby, Phys. Rev. Lett. **68**, 639 (1992).
47. C. J. Umrigar and X. Gonze, unpublished (1996).
48. O. Gunnarsson and B. I. Lundqvist, Phys. Rev. B **13**, 4274 (1976)
49. M. Levy and J. P. Perdew, Phys. Rev. B **48**, 11638 (1993).
50. M. Levy and J. P. Perdew, Phys. Rev. A **32**, 2010 (1985).
51. M. Levy, Phys. Rev. A **43**, 4637 (1991).
52. A. Görling and M. Levy, Phys. Rev. A **45**, 1509 (1992).
53. M. Levy and J. P. Perdew, Int. J. Quant. Chem. **49**, 539 (1993), and references therein.
54. C. Filippi, C. J. Umrigar and M. Taut, J. Chem. Phys. **100**, 1290 (1994).
55. C. J. Umrigar and X. Gonze, Phys. Rev. A **50**, 3827 (1994).
56. C. J. Umrigar, unpublished notes (1993).
57. P. Jemmer and P. J. Knowles, Phys. Rev. A **51**, 3571 (1995).
58. E. Engel and S. H. Vosko, Phys. Rev. B **50**, 10498 (1994).
59. N. H. March, Phys. Rev. A **36**, 5077 (1987).
60. P. Süle, O. V. Gritsenko, A. Nagy and E. J. Baerends, J. Chem. Phys. **103**, 10085 (1995).
61. A. Görling and M. Levy, Int. J. Quant. Chem, Symp. **29**, 93 (1995); Phys. Rev. A **50**, 196 (1994).
62. C. Filippi, C. J. Umrigar and X. Gonze, submitted to Phys. Rev. A.
63. M. Taut, Phys. Rev. A **48**, 3561 (1993).
64. The wave function used is a minor modification of that in D. E. Freund, B. D. Huxtable and J. D. Morgan, Phys. Rev. A **29**, 980 (1984).
65. C. -O. Almbladh and U. von Barth, Phys. Rev. B **31**, 3231 (1985).
66. C. J. Umrigar, P. Nightingale and K. J. Runge, J. Chem. Phys. **99**, 2865 (1993).
67. D. M. Ceperley and M. H. Kalos in *Monte Carlo Methods in Statistical Physics* edited by K. Binder, Topics Current Phys., Vol.7 (Springer, Berlin, Heidelberg, 1979).
68. S. H. Patil, J. Phys. B **23**, 1 (1990).
69. E. Engel and S. H. Vosko, Phys. Rev. B **47**, 13164 (1993).
70. E. R. Davidson, S. A. Hagstrom, S. J. Chakravorty, V. M. Umar and C. F. Fischer, Phys. Rev. A **44**, 7071 (1991); S. J. Chakravorty, S. R. Gwaltney, E. R. Davidson, F. A. Parpia and C. F. Fischer, Phys. Rev. A **47**, 3649 (1993).
71. R. van Leeuwen and E. J. Baerends, Phys. Rev. A **51**, 170 (1995).
72. M. R. Norman and D. D. Koelling, Phys. Rev. B **30**, 5530 (1984).
73. A. Lembarki, F. Rogemond and H. Chermette, Phys. Rev. A **52**, 3704 (1995).
74. J. B. Krieger, Y. Li and G. J. Iafrate, Phys. Rev. A **46**, 5453 (1992).
75. T. Grabo and E. K. U. Gross, Chem. Phys. Lett. **240**, 141 (1995).
76. R. Colle and D. Salvetti, Theoret. Chim. Acta **37**, 329 (1975); *ibid.* **53**, 55 (1979).
77. U. von Barth and L. Hedin, J. Phys. C **5**, 1629 (1972).

J.M. Seminario (Editor)
Recent Developments and Applications of Modern Density Functional Theory
Theoretical and Computational Chemistry, Vol. 4

On degeneracy, near-degeneracy and density functional theory

A. Savin[a] *

[a]Laboratoire de Chimie Théorique (CNRS)
Université P. et M. Curie
4,place Jussieu, 75252 Paris, France

While density functional theory can deal in theory with degeneracy problems, several ill-behaved cases arise at different levels of approximation. A few remedies to these problems are presented (ensembles, multi-determinant wavefunctions), which are extensions of the widely used Kohn-Sham formalism.

1. INTRODUCTION

In an elegant proof, Hohenberg and Kohn laid down the theoretical foundation of density functional theory[1]. They did not treat explicitly degeneracy: 'We shall in all that follows assume for simplicity that we are dealing with situations in which the ground state is nondegenerate'. While it was clear that the theorem was also valid for degenerate states, the publications discussing aspects of degeneracy started only to show up around 1980. Some of the questions related to the implementation of density functional theory still have no answer.

Density functional theory is mainly a ground state theory, where degeneracy is less frequent, but still present, e.g., for most atoms. In density functional practice, one has the predicament presenting the choice between artificially:

i) lifting the degeneracy (breaking symmetry) by minimizing the energy, or

ii) imposing symmetry constraints.

The first approach appears to be physically wrong, but both approaches may produce fictitious results when near-degeneracy is present:

i) One can always imagine a physical perturbation which lowers the energy of a state different from the one with the lowest energy obtained by breaking the symmetry. When the physical perturbation is small enough, the energy lowering is smaller than that falsely induced by the density functional approximation which lifts the

*Many of the ideas presented benefited from the stimulative atmosphere created at the Density Functional Workshop at the Institute for Theoretical Physics at Santa Barbara animated by W. Kohn's numerous ideas and suggestions. I would like to thank also A. Becke, D. Ceperley, F. Colonna, C. Daul, H.-J. Flad, E.K.U. Gross, M. Levy, R. Merkle, B. Miehlich, J. Perdew, H. Stoll, C. Stückl for the valuable discussions. Computing time was made available by IDRIS (Orsay, France). This work was supported in part by the National Science Foundation under Grant No. PHY89-04035.

degeneracy. This situation is shown schematically in Figure 1, where it is assumed (in addition) that the approximate density functional exactly follows one of the states (which of course is not true in general).

ii) Imposing the symmetry is impossible when a perturbation destroys symmetry. In the limit of a vanishing perturbation the result obtained is that of the symmetry broken solution in the case of no perturbation (and thus different from the one obtained by artificially imposing symmetry).

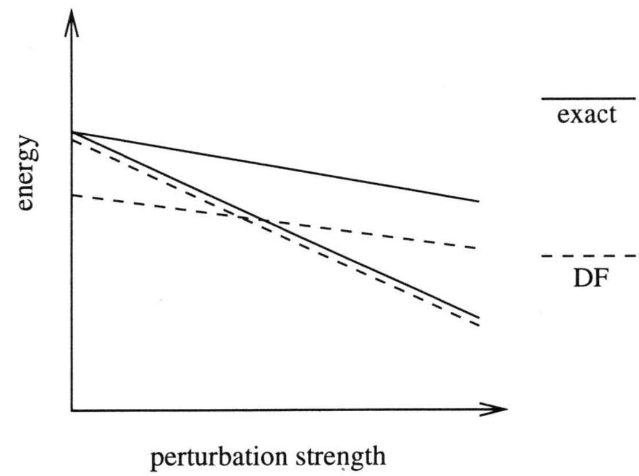

Figure 1. Change of the two lowest energy levels with a perturbing potential (schematical). Exact behavior shown with full lines. An approximate density functional (dashed lines) does not reproduce the degeneracy present in the absence of the perturbation.

Size-consistency can be viewed as a special case of degeneracy. Consider two isolated atomic or molecular systems, *left* and *right*, characterized by their ground state energies (E_{left} and E_{right}), densities and spin-densities. Let them now form a composite system without interaction between them. In order to prepare the interaction of the fragments, the global wavefunction presents the correct global symmetry properties. Size-consistency requires that the energy of the composite system, $E_{left+right}$ equals the sum of the fragment energies. Furthermore, the densities in the composite system should be obtainable from that of the fragment ensembles. Examples are given in Section 3.

When $E_{left+right} \neq E_{left} + E_{right}$, it is unclear how to define the dissociation energy of the system (in interaction) as the dissociated limit is unspecified. When $E_{left+right} > E_{left} + E_{right}$, the breaking of the symmetry of the composite system lowers the energy and brings $E_{left+right}$ equal to $E_{left} + E_{right}$. Giving up symmetry constraints on the composite system is, however, no solution when $E_{left+right} < E_{left} + E_{right}$.

The interaction of the left and right fragment can be viewed as a perturbation, the inverse distance being the perturbation strength parameter. Thus the problems of near-degeneracy can affect the potential curves, for example producing fake ground states for certain domains of the interatomic separation.

In the following, a few fundamental notions of density functional theory will be recalled. (The interested reader can find a more detailed description in books like Ref. [2–4]). The second part exemplifies the nature of problems which appear due to degeneracy. Only the simplest cases (atoms and homonuclear diatomic molecules) are shown below. Polyatomic molecules can present more complex problems of degeneracy (see, e.g., [5]). Finally, some remedies are presented, without raising the claim of having solved the degeneracy problem. A profound analysis of the subject treated in this article can be found in Reference [6].

2. A SHORT REVIEW OF DENSITY FUNCTIONAL THEORY

2.1. The Hohenberg-Kohn theorem

Instead of obtaining the ground state energy, E_0, by minimizing a functional of the (antisymmetric) wavefunction Ψ , the expectation value of the Hamiltonian, $\hat{H}$:

$$E_0 = \min_{\Psi} E[\Psi] = \min_{\Psi} < \Psi \mid \hat{H} \mid \Psi > \tag{1}$$

a functional of the density ρ, $E[\rho]$, can be minimized. The existence of such a functional was proven by Hohenberg and Kohn[1]. It is most easily understood in the context of the constrained search formulation[7–9] by performing the minimization in Equation 1 in two steps, first over all wavefunctions yielding a given density ($\Psi \rightarrow \rho$) and then over all densities:

$$E_0 = \min_{\rho} \min_{\Psi \rightarrow \rho} E[\Psi]$$

and defining the energy density functional

$$E[\rho] = \min_{\Psi \rightarrow \rho} E[\Psi] \tag{2}$$

The equation which shows that the density functional $E[\rho]$, is always greater or equal to the ground state energy:

$$E_0 = \min_{\rho} E[\rho] \leq E[\rho] \tag{3}$$

E_0 is attained by $E[\rho]$ only by using densities which can be produced from ground state wavefunctions: Take any ground state wavefunction Ψ_0 yielding the ground state energy $E[\Psi_0] = E_0$ and a ground state density, ρ_0. Using Ψ_0 as the minimizing one in Equation 2 (at $\rho = \rho_0$) shows that E_0 can be attained for $E[\rho_0]$. A density which is not a ground state density is produced by a wavefunction Ψ which is not a ground state density: $E[\Psi] > E_0$. This holds also for the minimizing one in Equation 2. To conclude, the variational principle can be used to optimize $E[\rho]$ to get the ground state energy, E_0 and a density which comes from a ground state wavefunction, ρ_0. For the discussion below it is important to remember that $E[\rho]$ depends on the Hamiltonian $\hat{H}$ (as does $E[\Psi]$).

The disadvantage of this definition of the energy density functional given in Equation 2 is that densities obtained from ground state ensembles:

$$\bar{\rho} = \sum w_i \rho_i \tag{4}$$

may not be obtainable from any wavefunction. Here the densities ρ_i are obtained from a set of degenerate wavefunctions and

$$\sum_i w_i = 1$$
$$0 \leq w_i \leq 1$$

An example is given, e.g., in Ref. [8], section IV.C. As a result, $E[\rho]$ of Equation 2 may be higher than E_0 even if the density comes from a ground state. A few comments are now due.

1. Why not restrict density functional theory to densities obtainable from ground state wavefunctions? It turns out that dissociation produces fragments with ensemble densities (specific examples are shown in Section 3.7). In a molecule some parts can be practically isolated from the rest, producing similar effects. Consider Ψ given by the antisymmetrized product of wavefunctions on the *left* and *right* fragments. Then

$$\begin{aligned} E[\rho] &= \min_{\Psi \to \rho} < \Psi|\hat{H}|\Psi > \\ &\to \min_{\Psi_{left} \to \rho_{left}} < \Psi_{left}|\hat{H}|\Psi_{left} > + \min_{\Psi_{right} \to \rho_{right}} < \Psi_{right}|\hat{H}|\Psi_{right} > \end{aligned} \tag{5}$$

   If, in accordance with the global symmetry of the system, ρ_{left} and ρ_{right} are forced to be ensemble fragment densities, the minimizing Ψ_{left} and Ψ_{right} yield $< \hat{H} >$ larger than the value for this expectation value obtained after the search over densities not constrained to be symmetrical.

2. In practice real orbitals are widely used. Constructing Ψ in Equation 2 from them may, however, restrict its domain of applicability. Let us consider a system without electron-electron interaction with a single p-orbital occupied. A linear combination of real p-orbitals

$$(p_x \cos\alpha + p_y \sin\alpha)^2 = p_x^2 \cos^2\alpha + p_y^2 \sin^2\alpha + 2p_x p_y \cos\alpha \sin\alpha$$

   will never be equal to that of the equi-ensemble of p_x and p_y:

$$\frac{1}{2}(p_x^2 + p_y^2)$$

   With complex orbitals, however,

$$|p_x c_x + p_y c_y|^2 = p_x^2|c_x|^2 + p_y^2|c_y|^2 + p_x p_y(c_x^* c_y + c_y^* c_x)$$

   can be made equal to that of the equi-ensemble, e.g., by choosing $c_x = ic_y = 1/\sqrt{2}$.

Let us now consider the equi-ensemble of the p-set that yields the spherical density

$$\frac{1}{3}(p_x^2 + p_y^2 + p_z^2)$$

No linear combination (even with complex orbitals) will satisfy the three conditions:

$$c_x^* c_y + c_y^* c_x = c_y^* c_z + c_z^* c_y = c_z^* c_x + c_x^* c_z = 0$$

If we consider, however, spinors (or equivalently replace spin-orbitals by one-particle functions which are not eigenvalues of $\hat{S}_z$) a spherically symmetric density can be obtained (the hydrogen-like eigenfunctions with $j = 1/2$ yield spherically symmetric densities, see, e.g., Ref. [10]).

3. Density functional theory can be extended to ensembles (see, e.g., [8,11–16]). In our case, one simply has to consider functionals of the density matrix

$$\Gamma = \sum_i w_i |\Psi_i >< \Psi_i|$$

which yields $\bar{\rho}$ (Equation 4). In order to get the variational principle for the density functional, a correct functional to be minimized has to be chosen and the procedure shown for pure states can be followed. For the ground state, E_0 is obtained by minimizing $\sum w_i < \Psi_i|\hat{H}|\Psi_i >= Tr\hat{H}\Gamma$,

$$E_0 = \min_{\Gamma} Tr\hat{H}\Gamma = \min_{\bar{\rho}} \min_{\Gamma \to \bar{\rho}} Tr\hat{H}\Gamma = \min_{\bar{\rho}} \bar{E}[\bar{\rho}] \tag{6}$$

with

$$\bar{E}[\bar{\rho}] = \min_{\Gamma \to \bar{\rho}} Tr\hat{H}\Gamma$$

With this definition, E_0 can be reached for ensemble ground state densities by $\bar{E}[\bar{\rho}]$.

Hohenberg and Kohn made the very important observation that $E[\rho]$ (or $\bar{E}[\bar{\rho}]$) can be split into two parts, one specific for the system under consideration, and one which is universal (independent of the specific electronic system, or $\hat{H}$). The first part is given by:

$$< \Psi|\hat{V}_{ne}|\Psi >= \int \rho(\mathbf{r}) v_{ne}(\mathbf{r}) d^3r \tag{7}$$

where

$$\hat{V}_{ne} = v_{ne}(\mathbf{r}_1) + v_{ne}(\mathbf{r}_2) + \ldots v_{ne}(\mathbf{r}_N)$$

is a local external potential, such as the one describing the interaction between nuclei and electrons. The remaining part is

$$\begin{aligned} F[\rho] &= E[\rho] - \int \rho(\mathbf{r}) v_{ne}(\mathbf{r}) d^3r \\ &= \min_{\Psi \to \rho} < \Psi|\hat{T} + \hat{V}_{ee}|\Psi > \end{aligned} \tag{8}$$

because $\int v_{ne}\rho$ does not change during the search over the wavefunctions constrained to yield the same density:

$$\begin{aligned}\min_{\Psi\to\rho} E[\Psi] &= \min_{\Psi\to\rho}\{<\Psi|\hat{H}-\hat{V}_{ne}|\Psi>+<\Psi|V_{ne}|\Psi>\} \\ &= \min_{\Psi\to\rho}\{<\Psi|\hat{T}+\hat{V}_{ee}|\Psi>\}+\int\rho(\mathbf{r})v_{ne}(\mathbf{r})d^3r \end{aligned} \tag{9}$$

In order to see that $F[\rho]$ is universal (i.e., does not depend on the external potential, but solely on the density) one has to notice that besides ρ, only the operators $\hat{T}$ and $\hat{V}_{ee}$ are needed in order to obtain $F[\rho]$ in Equation 8. But, as

$$\int\rho(\mathbf{r})d^3r = N \tag{10}$$

both $\hat{T}$ and $\hat{V}_{ee}$ can be constructed by the knowledge of ρ. The proof for an ensemble follows the same lines as the proof for the pure state shown above.

Using Equation 8 to obtain $F[\rho]$ in practice would be even more difficult than minimizing $E[\Psi]$ (Equation 1). It is possible to find successful approximations for $F[\rho]$. The usual approach is to make an ansatz for it, compute its value for one or several selected systems (the uniform electron gas is a favorite choice) in order to fix the degrees of freedom present in the ansatz, and then use the functional for all systems, relying on the property of universality.

Finding an ansatz can be greatly simplified by cutting of parts which can be simply described as functionals of the density. For example, the classical Coulomb interaction of charge clouds, the Coulomb energy, is given by:

$$U[\rho] = \frac{1}{2}\int\frac{\rho(\mathbf{r}_1)\rho(\mathbf{r}_2)}{r_{12}}d^3r_1d^3r_2 \tag{11}$$

As the Coulomb interaction between electrons and nuclei is treated exactly, and largely compensates that between the electrons, it is important to treat properly $U[\rho]$, as given by Equation 11. On the other hand, a subtle compensation of $U[\rho]$ with other parts of $F[\rho]$ may appear. The typical example is the case of the one-electron system, when no electron-electron interaction is present, and $U[\rho]$ should should be exactly compensated by other terms of the density functional (there should be no electron self-interaction). For ensembles the problem is even more serious as using $\bar{\rho}$ instead of ρ in Equation 11 will produce Coulomb interaction between members of the ensemble:

$$U[\bar{\rho}] = \sum_{i,j} w_i w_j \int\frac{\rho_i(\mathbf{r}_1)\rho_j(\mathbf{r}_2)}{r_{12}}d^3r_1d^3r_2 \tag{12}$$

instead of

$$\sum_i w_i \int\frac{\rho_i(\mathbf{r}_1)\rho_i(\mathbf{r}_2)}{r_{12}}d^3r_1d^3r_2$$

demanding other terms of the density functional to guarantee the compensation.

A commonly made ansatz to approximate a density functional has the form

$$\int f(\rho(\mathbf{r}))d^3r \tag{13}$$

(the local density approximation), or

$$\int f(\rho(\mathbf{r}, |\nabla\rho(\mathbf{r}|)d^3r \tag{14}$$

(gradient corrected). It is important to realize that different degenerate densities may not yield the same value for such functionals. (In fact, none of the approximate density functionals now in use are able to solve this problem.) At dissociation, ρ_{left} and ρ_{right} do not overlap and

$$\int f(\rho_{left} + \rho_{right}) = \int f(\rho_{left}) + \int f(\rho_{right}) \tag{15}$$

When degeneracy is present in the fragments, the density of the dissociated products may be different from that obtained by energy minimization on each of the individual fragments and this produces problems (see examples in Section 3.7).

2.2. The Kohn-Sham formalism.

The most useful density functionals used nowadays rely on a construction for a kinetic energy density functional proposed by Kohn and Sham which explicitly makes use of the Pauli exclusion principle. It is [18,19]:

$$T_s[\rho] = \min_{\Phi\rightarrow\rho} < \Phi|\hat{T}|\Phi > \tag{16}$$

Here Φ is an antisymmetric wavefunction. While the prescription for obtaining $T_s[\rho]$ may seem as complicated as that for determining $F[\rho]$ (Equation 8), a considerable simplification comes from the fact that the wavefunction Φ can be chosen to be a Slater determinant. In order to understand this, it should be noticed that $\hat{T}$ is a one-body operator and that the restriction on Φ to yield a given density can be insured by the Lagrangian multiplier technique. As there is a restriction for every point in space $\mathbf{r}$, one has for each point in space a Lagrangian multiplier, which can be assimilated to a potential, $v_{KS}(\mathbf{r})$, the Kohn-Sham potential. Furthermore, it has been proven that every N-particle density can be constructed by using a single determinant [20,21]. With this, one deduces that the minimizing Φ can be chosen to be a Slater determinant (Φ_{KS}, the Kohn-Sham determinant): Φ_{KS} is an eigenvector of the Kohn-Sham Hamiltonian, $\hat{H}_{KS} = \sum_i^N \hat{h}_{KS}(\mathbf{r}_i)$ where $\hat{h}_{KS}(\mathbf{r}) = -1/2\nabla^2 + v_{KS}(\mathbf{r})$, in atomic units. The (Kohn-Sham) orbitals φ_i used in constructing Φ_{KS} are the eigenvectors corresponding to the lowest eigenvalues of $\hat{h}_{KS}$. The Kohn-Sham equations

$$\hat{h}_{KS}\varphi_i = \epsilon_i\varphi_i$$

also appear by minimization of the energy with respect to the orbitals (ϵ_i appears from the normalization constraint on φ). The energy is calculated according to:

$$\begin{aligned} E &= \min_{\rho}\{T_s[\rho] + \int v_{ne}\rho + U[\rho] + E_{xc}[\rho]\} \\ &= \min_{\Phi}\{< \Phi|\hat{T}|\Phi > + \tilde{F}[\rho(\Phi)]\} \end{aligned} \tag{17}$$

where the functional

$$F[\tilde{\rho(\Phi)}] = \int v_{ne}\rho(\Phi) + U[\rho(\Phi)] + E_{xc}[\rho(\Phi)]$$

and the density ρ is obtained from Φ.

The remaining term in $F[\rho]$ is called the exchange-correlation functional and is given by:

$$E_{xc}[\rho] = F[\rho] - T_s[\rho] - U[\rho] \tag{18}$$

In most practical calculations, only $E_{xc}[\rho]$ is approximated (cf. Equations 13,14).

As the Kohn-Sham procedure uses a Slater determinant, it is also possible to obtain an exchange energy with it:

$$E_x[\rho] = <\Phi_{KS}|\hat{V}_{ee}|\Phi_{KS}> - U[\rho] \tag{19}$$

defining a correlation energy

$$E_c[\rho] = E_{xc}[\rho] - E_x[\rho] \tag{20}$$

as the only term to be approximated (cf. Equations 13,14).

It should be remembered, that Φ_{KS} is (in general) different from the Hartree-Fock determinant (even for the Hartree-Fock, or the exact density). Thus the exchange energy, and the correlation energy are not the same as in the usual definitions. (Numerical comparisons have been made, and also exact inequalities are known, see, e.g., Ref. [22–24].)

It is more important, however, to notice that the separation into exchange and correlation requires the selection of one Kohn-Sham determinant. In the case of degeneracy, this choice is not unique, meaning that $<\Phi_{KS}|\hat{V}_{ee}|\Phi_{KS}>$, E_x and E_c may not be well-defined (an example is given in Section 3.5).

2.3. Spin-density functionals

In an extension of density functional theory, ground state spin-densities can be produced in addition to ground state energies and densities. Spin densities appear naturally (like spin) in density functional theory by using the relativistic formalism [25–27]. Another possibility is to consider external potentials (like magnetic fields) which couple to the spin-density [28]. Spin densities are also used when the magnetic field is absent, with the argument that an infinitesimal magnetic field can be introduced for stabilization [29]. Note, however, that a magnetic field will split the multiplet , while usual spin-density functionals yield the same value for the state M_S and $-M_S$, as spin-density functionals are constructed to depend on ζ^2,

$$\zeta = \frac{\sigma(\mathbf{r})}{\rho(\mathbf{r})} \tag{21}$$

Here

$$\begin{aligned} \rho(\mathbf{r}) &= \rho_\uparrow + \rho_\downarrow \\ \sigma(\mathbf{r}) &= \rho_\uparrow - \rho_\downarrow \end{aligned} \tag{22}$$

where $\rho_\uparrow, \rho_\downarrow$ are the partial densities for spin-up and spin-down, respectively, and σ is the spin-density, and ζ is called spin-polarization.

One property of the spin-density is that it changes within a set of degenerate states having the same S, but having different M_S:

$$\sigma(\mathbf{r}) = \frac{M_S}{S}\sigma_S(\mathbf{r}) \qquad (S \neq 0) \tag{23}$$

where σ_S is invariant with M_S (see, e.g., [30]). Therefore, density functionals depending on σ will change with $|M_S|$ within a degenerate set. It is, however, widely acknowledged that using density functionals improve the results.

The ensemble spin-density may be radically different from that obtained from pure states. For example, an equi-ensemble of open-shell atoms or molecules has zero spin-density. Thus, a functional depending on the spin-density takes different values for the ground state ensemble and the pure states. Ensemble spin-densities appear (like ensemble densities) at dissociation.

3. EXAMPLES

3.1. Degeneracy in atoms

A simple case of degeneracy appears for atoms where p-orbitals are occupied. Suppose that the p subshell is not closed. Occupying a real, say p_x, orbital will generate a density with a different shape than that generated by using complex orbitals (say $p_x + ip_y$) and different again from that obtained from an ensemble, e.g., with equal weights. Such contributions to the densities are shown in Figure 2.

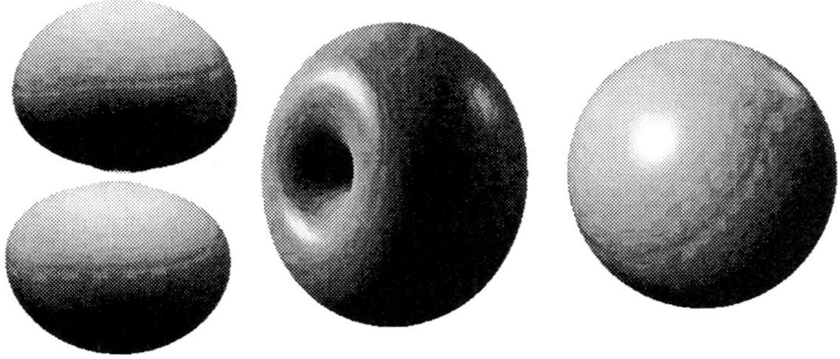

Figure 2. Shapes of different densities obtained from p orbitals; shown as isodensity surfaces; from left to right: a) p_x^2, b) $|p_x + ip_y|^2$, c) $1/3(p_x^2 + p_y^2 + p_z^2)$.

The approximate density functional has the difficult task to generate the same energy from the three different densities which are produced by using the different density contributions. Let us consider for example a functional of the form $\int \rho^{4/3}$. In a region where the p orbitals dominate, $\rho \approx x^2 f(r)$ for a density generated by a p_x orbital, and $\rho \approx (1/3) r^2 f(r)$ for the equi-ensemble density. The integral over this region of space is in the first case

$$\int [x^2 f]^{4/3} d^3 r = \frac{12\pi}{11} \int [r^2 f]^{4/3} r^2 dr$$

while in the second case:

$$\int [\frac{1}{3} r^2 f]^{4/3} d^3 r = \frac{4\pi}{3^{4/3}} \int [r^2 f]^{4/3} r^2 dr$$

The problem is even more pronounced for the transition elements, as not even all real d-type orbitals produce equivalent densities, while p_x produces a density which is equivalent to that of p_y or p_z. Numerical examples can be found, for example, in Ref. [31].

Suppose that an atomic program generates radial orbitals and computes the atomic energy from the spherically symmetric ensemble density. The use of this program for producing a table of atomic energies for the calculation of atomization energies is questionable: In many cases the (spherically symmetric, ensemble) density generated by the atomic program differs from the (non-spherical) density obtained from the molecular code; thus the energies obtained in the two calculations will be different when using approximate density functionals (see Section 3.7).

3.2. Near-degeneracy in the Be series

In the limit of infinite nuclear charge ($Z \to \infty$) the atomic energy levels of the atom become hydrogen-like. This can be most easily seen by scaling the radial coordinate with Z, the nuclear charge. The scaled Hamiltonian equation gets a prefactor of Z^2 in its one-electron part, while the electron-electron interaction is multiplied by Z, and can be treated as a perturbation for $Z \to \infty$ [32]. The zeroth-order energy is thus proportional to Z^2. As the Hamiltonian is mono-electronic at zeroth order, no correlation energy appears at this order. When electron-electron interaction is treated perturbationally correlation effects will appear: the first term is proportional to Z when degeneracy is present at zeroth order, and otherwise proportional to Z^0(= constant) [33]. For example, due to the $2s - 2p$ degeneracy in the non-interacting limit, the correlation energy is decreasing linearly with Z for $N = 4$ (the Be series). On the other hand, for $N = 2$ (the He series) the correlation energy tends to a constant, since there is no degeneracy present at zeroth order for $1s^2$.

It was pointed out that the local density approximation gives in both cases a dependence proportional to $\log Z$ for $Z \to \infty$. [34] A density functional which changes little with Z is more or less easily obtained (see, e.g., [35]), but it does not show the linear behavior in the Be series. This is readily obtained in a configuration interaction calculation (just by adding a supplementary configuration and thus eliminating the near-degeneracy problem), but coupling the configuration interaction calculation and the density functional calculation is not a trivial task. For example, the method of Colle and Salvetti [36] which is able to couple configuration interaction with density functionals gives in general very good results for the correlation energy, but does not show a correct Z-dependence in the Be series, even after combining a correlation energy density functional with the configuration interaction calculation which correctly describes the infinite Z limit [37].

3.3. Dissociation of ionic molecules: H_2^+

The simplest molecule is the hydrogen molecular ion. Due to $D_{\infty h}$ symmetry, the densities are equal on both atoms, for all interatomic separations. This holds also in the dissociation limit, when it becomes that of an equi-ensemble of H and H^+.

In usual approximations (Equations 13,14) the energy of the molecule at infinite internuclear separation and that of an hydrogen atom (plus a proton) are different. If the energy of the symmetric density would have been higher than that of the unsymmetric one, one might have expected to obtain the size-consistent result by symmetry breaking. It turns out, however, that the molecular solution, with the charge symmetrically distributed on the two centers, has lower energy. Thus the variational solution yields the density with the correct symmetry, but is not size-consistent.

In order to understand this error, let us distribute the total density among the left atom and right atom with a weight λ:

$$\rho_{total} = \lambda\rho_{left} + (1-\lambda)\rho_{right}$$

($0 \leq \lambda \leq 1$). For $\lambda = 1/2$ we have a symmetric density, for $\lambda = 1$, the electron is on the right atom. The Coulomb repulsion is at infinite separation

$$U = \lambda^2 \frac{1}{2}\int \frac{\rho_{left}(\mathbf{r}_1)\rho_{left}(\mathbf{r}_2)}{r_{12}} + (1-\lambda)^2 \frac{1}{2}\int \frac{\rho_{right}(\mathbf{r}_1)\rho_{right}(\mathbf{r}_2)}{r_{12}} = [\lambda^2 + (1-\lambda)^2]U_{at}$$

where

$$U_{at} = \frac{1}{2}\int \frac{\rho_{left}(\mathbf{r}_1)\rho_{left}(\mathbf{r}_2)}{r_{12}} = \frac{1}{2}\int \frac{\rho_{right}(\mathbf{r}_1)\rho_{right}(\mathbf{r}_2)}{r_{12}}$$

U will show a minimum at $\lambda = 1/2$. The exchange-correlation functional must also have a symmetric behavior when the left atom and the right atom are exchanged ($\lambda \to 1-\lambda$). For example, a functional proportional to $\int \rho^{4/3}$ will yield a prefactor $\lambda^{4/3} + (1-\lambda)^{4/3}$. The Coulomb and exchange-correlation energy exactly compensate when no approximation is made. If the Coulomb part dominates (as it happens in the common approximations) a minimum is obtained at $\lambda = 1/2$.

As a consequence of the too low energy at infinite internuclear separation, the energy is also too low at large (but finite) separations. One can understand this by comparing a Morse curve (fitted to reproduce the correct dissociation energy, equilibrium distance and force constant of the H_2 molecule leading at infinity to the size-consistent solution) and the curve obtained from the interaction of two fragments with net charge $+1/2$ having at infinity a lower energy than the sum of the fragment energies (within the approximation, see Figure 3). A variational solution will choose the lower lying curve.

Numerical results were produced (not only for the H_2^+, but also for other X_2^+ molecules, where the same effect distorts the potential curves) in Reference [38].

In a less severe form, the problem is also present in neutral molecules. While diatomic molecules should dissociate into neutral atoms, density functionals may produce slightly charged fragments, a problem already noticed by Slater in the context of the X_α method [39].

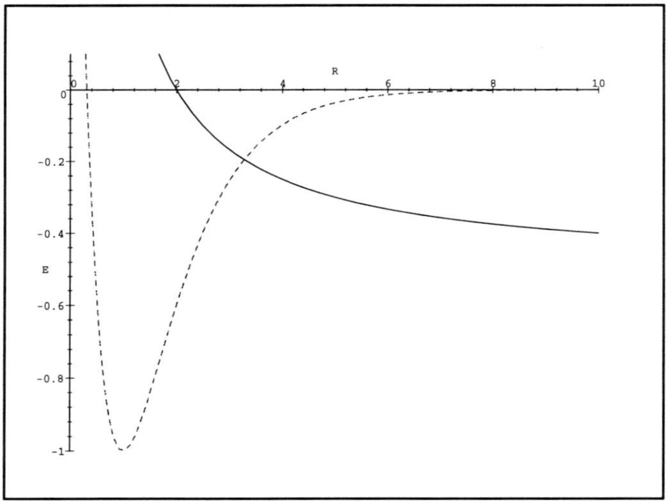

Figure 3. Morse curve for an X_2^+ molecule (dashed curve), and the repulsion of two $X^{+1/2}$ fragments(full curve), their approximate energy being lower than that of the sum of the energies of X and X^+.

3.4. Spin density in H_2

The often quoted example of the hydrogen molecule raises the question of spin-density. As the distance between the nuclei in the hydrogen molecule in its ground state tends to infinity, the energy should reach twice that of an isolated hydrogen atom. Consider first the hydrogen molecule in its singlet ground state. Its spin density at any internuclear separation is, of course, zero. Consider now an isolated hydrogen atom: it has non-zero spin-density. A functional which explores the spin-density around each atom will find zero spin-density in the dissociated hydrogen molecule, but a non-zero spin-density in the isolated atom. Thus, an energy functional that depends on the spin-density gives different values for the isolated atom and each of the atoms of the dissociated hydrogen molecule.

If we accept a wrong spin-density for H_2 (i.e., a spin-density only as a computational artifact) we can get the correct energy. Please notice that even when the spin-density is wrong, the eigenvalue of the $\hat{S}_z$-operator, M_S, is in the case of H_2 correctly zero.

The fake spin-density is not only present for the hydrogen molecule; it can always occur when a bond (described by a doubly occupied orbital) is broken. It will appear for Li_2, N_2, F_2, etc.

3.5. The arbitrariness in the correlation energy definition: The H_2 molecule at dissociation

It was mentioned that the correlation energy functional may be ill-defined when degeneracy is present for the Kohn-Sham problem: Several Φ_i yield the same $< \hat{T} >$ for all Φ_i), but $< \hat{V}_{ee} >$ may be different. In the H_2 molecule, for example, at infinite internuclear distance, both doubly occupying the σ_g or the σ_u orbitals

$$\sigma_g \sim \chi_{left} + \chi_{right}$$

$$\sigma_u \sim \chi_{left} - \chi_{right}$$

will lead to the same Kohn-Sham energy, and the same density,

$$\rho = \chi^2_{left} + \chi^2_{right}$$

(χ_{left} and χ_{right} are the hydrogen wavefunctions on the left and right atom, respectively). Furthermore, both Slater determinants (Φ_g and Φ_u, built from σ_g and σ_u, respectively) correctly yield zero spin-density. The linear combination of Φ_g and Φ_u will also yield the same density, spin-density and kinetic energy. The correlation energy may be different, however, depending on the mixing of Φ_g and Φ_u. Let us write down

$$\Phi = \cos\theta \cdot \Phi_g + \sin\theta \cdot \Phi_u$$

Although $< \hat{T} >$ will be unchanged $< \hat{V}_{ee} >$ may change:

$$< \Phi|\hat{V}_{ee}|\Phi >= \cos^2\theta < \Phi_g|\hat{V}_{ee}|\Phi_g > + \sin^2\theta < \Phi_u|\hat{V}_{ee}|\Phi_u > +2\sin\theta\cos\theta < \Phi_u|\hat{V}_{ee}|\Phi_g >$$

As for infinite separation

$$< \Phi_g|\hat{V}_{ee}|\Phi_g >=< \Phi_u|\hat{V}_{ee}|\Phi_u >=< \Phi_u|\hat{V}_{ee}|\Phi_g >$$

the density functional definition of the correlation energy will change with θ according to $[\cos\theta + \sin\theta]^2$.

3.6. N dependence and He_2

It may seem that no problem can appear for the He_2 molecule, as no size-consistency problem within the Hartree-Fock model which is formally close to the Kohn-Sham formalism. In this case there is, however, a typical size-consistency error which is worth being mentioned.

In the proof of the Hohenberg and Kohn theorem, a fundamental point is that the electron number, N, is a simple functional of the density, ρ (see Equation 10). The exact density functional depends explicitly on N and one might seem free to use N in the construction of approximate density functionals. Indeed, such functionals have been proposed and improve the results by including an explicit dependence on the electron number. N-dependent functionals were constructed for example to correct the self-interaction present in most density functionals (see, e.g., the approach of Fermi and Amaldi [40] in the context of the Thomas-Fermi model). More recent papers show that a significant improvement of the correlation energy of atoms and ions can be obtained this way[41] [2].

[2] $E_c[\rho]$, Equation 20, should be zero in a one-electron system.

Using functionals explicitly depending on N leeds to size-consistency problems. Let us consider two He atoms at infinite separation. In this case $N = 4$ while for each isolated atom N=2. Let us use the densities of the isolated atoms to compute the energy of the composite system:

$$E[\rho_{left} + \rho_{right}] = E[\rho_{left}] + E[\rho_{right}]$$

In our composite system, $N = 4$. The two terms on the r.h.s. are equal, but if $E[\rho]$ explicitly depends on N each term is different from that obtained for the isolated atom, where $N = 2$.

To be more specific, consider a correction making $U[\rho]$, Equation 11, disappear for $N = 1$ [40]:

$$\tilde{U}[\rho] = U[\rho] - NU[\rho/N]$$

The Coulomb energy, U, of the composite system is equal to that of the fragments:

$$\begin{aligned} U[\rho_{left} + \rho_{right}] &= \frac{1}{2}\int \frac{[\rho_{left}(\mathbf{r}_1) + \rho_{right}(\mathbf{r}_1)][\rho_{left}(\mathbf{r}_2) + \rho_{right}(\mathbf{r}_2)]}{r_{12}} \\ &= \frac{1}{2}\int \frac{\rho_{left}(\mathbf{r}_1)\rho_{left}(\mathbf{r}_2)}{r_{12}} + \frac{1}{2}\int \frac{\rho_{right}(\mathbf{r}_1)\rho_{right}(\mathbf{r}_2)}{r_{12}} \end{aligned}$$

due to the infinite separation. On the other hand,

$$NU[\rho/N] = \frac{1}{N}U[\rho]$$

is lower for $N = 4$ than for N=2, so that size-consistency is violated.

The analogy between He_2 and Be_2 or Ne_2 is evident. Note, however, that this type of problem is characteristic for dissociation.

3.7. Atomic ensembles from dissociation: B_2, C_2

The following examples show how ensemble densities (and not spin-densities, like in the case of H_2) are produced at infinite separation.

B_2 in the $^3\Sigma_g^-$ state is mainly described by the orbital occupancy $\ldots\pi^2$. The correct density of the molecule is for all interatomic separations cylindrically symmetric.

The Kohn-Sham methods generates the density from Slater determinants (see previous section). The cylindrically symmetric density can be easily described in the molecule by using a Slater determinant, where the two π orbitals are mainly given by the linear combination of the p-orbitals centered on the atoms:

$$\begin{aligned} \pi_x &\approx \frac{1}{\sqrt{2}}(p_{x,left} + p_{x,right}) \\ \pi_y &\approx \frac{1}{\sqrt{2}}(p_{y,left} + p_{y,right}) \end{aligned}$$

which contributes to the density with

$$\frac{1}{2}[p^2_{x,left}(\mathbf{r}) + p^2_{x,right}(\mathbf{r}) + 2p_{x,left}(\mathbf{r})p_{x,right}(\mathbf{r}) + p^2_{y,left}(\mathbf{r}) + p^2_{y,right}(\mathbf{r}) + 2p_{y,left}(\mathbf{r})p_{y,right}(\mathbf{r})]$$

At infinite separation the $left-right$ mixed terms disappear, and this yields an ensemble density $p_x^2 + p_y^2$ on each atom. As mentioned above, this ensemble density cannot be produced by real orbitals. Thus, a program which works with real orbitals and uses the exact density functional defined according to Equation 8 by using only real functions will yield a too high energy, as there is no real ground state wavefunction which yields this density.

It was also mentioned that with complex orbitals this problem will not appear. We are still left with the problem (mentioned in the atomic case) of different densities which might yield the different energy. Thus this cylindrically symmetric density may yield not the lowest energy (which might come from the density produced by real orbitals) or that produced from a spherically symmetric density (which one might be tempted to use in atomic calculations); cf. Figure 2.

The solution usually adopted in a situation like that presented for the B_2 molecule is to choose the lowest energy solution, which happens to break symmetry (see Figure 4). Thus, the choice between the correct density and the size-consistent energy is normally made in favor of the latter.

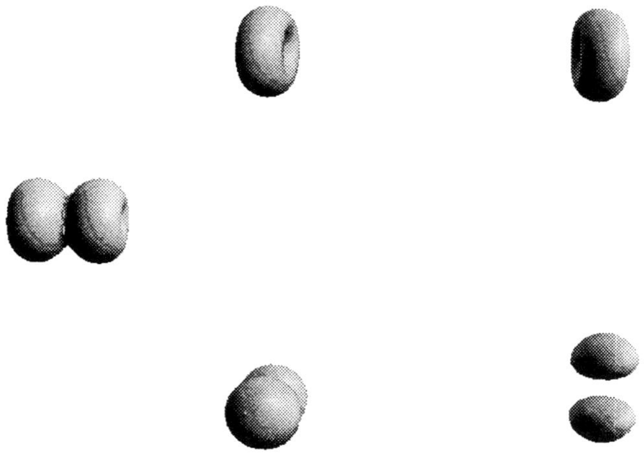

Figure 4. Symmetry breaking in B_2: isodensity surfaces. The $\pi_1^2 + \pi_{-1}^2 = \pi_x^2 + \pi_y^2$ density (left) becomes at dissociation equal to that obtained from complex orbitals $|p_x \pm i p_y|^2$ or of the ensemble $p_{x,left}^2 + p_{y,right}^2$ (top); the symmetry broken solution (bottom) may have a lower energy.

The C_2 molecule presents a closely related case. It has been a frequently studied molecule, due to the proximity of two states near the equilibrium distance (see, e.g., [42, 43]). It attains its lowest energy for the ${}^1\Sigma_g^+$ state (at equilibrium internuclear separation, R_e). The ${}^3\Pi_u$ state ($\ldots\sigma\pi$ occupancy) has its minimum for R_e' slightly larger than R_e and probably has there an energy which lies below that of the ${}^1\Sigma_g^+$ state [44] (see Figure 5). As one considers in density functional theory the state with the lowest energy, the symmetry of the ground state changes at a certain R, lying between R_e and R'_e. We will now suppose that the $^3\Pi_u$ state has of all states the lowest energy from this R to the dissociation limit.

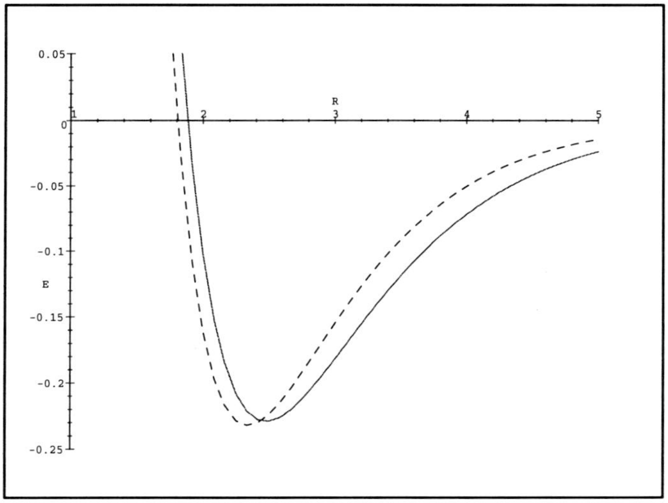

Figure 5. Morse curves of the $^1\Sigma_g^+$ and the $^3\Pi_u$ state in the C_2 molecule obtained from experimental data (dashed curve: $^1\Sigma_g^+$, full curve: $^3\Pi_u$).

At dissociation, the molecular occupancy $\ldots\pi^3\sigma$ will be transformed into $3/4p_x^2 + 3/4p_y^2 + 1/2p_z^2$ on each of the atoms. This density cannot be obtained from an atomic ground-state wavefunction. Then Equation 8 yields a value for the energy in an atomic calculation for the correct ensemble density which is higher than the lowest which can be obtained (for a non-symmetric density). We are now faced with the problem discussed after Equation 5: The atomic calculations yield $E[\rho_{left}]+E[\rho_{right}]$ larger than the ground-state energy.

Let us mention in passing another problem appearing in our example. The exact ground state density functional should jump from the $^1\Sigma_g^+$ to the $^3\Pi_u$ potential curve at the crossing point. One may ask whether it would not be useful to have a theory which follows a given potential curve [43].

3.8. Spin-densities and M_S in O_2

As for the hydrogen molecule, a spin-density problem is present for the oxygen molecule ($^3\Sigma_g^-$ state, $\ldots\sigma^2\pi^2$ occupancy) [45]. While for the hydrogen molecule it was possible to keep the eigenvalue of $\hat{S}_z$ (M_S) invariant (equal to the correct value, 0), it will be shortly shown that this does not happen for the oxygen molecule. We thus have an even more difficult case of size-consistency violation.

A comparison of the energies of states obtained by flipping the spins in the degenerate open shell orbitals shows that Hund's rule is respected: the parallel orientation of spins is preferred. This is induced in approximate density functionals mainly by the exchange energy, which is lower for parallel spins. Thus, an energy functional which depends on the spin-density will have different values for states with different M_S. There are, however, states with different M_S which are degenerate, and usual density functionals are not able to reproduce that, inducing an artificial lifting of degeneracy.

Let us suppose that both the oxygen molecule (at equilibrium distance) and the oxygen atom both have their lowest energy for $|M_S| = 1$.

This means that the lowest energy of the system formed by two oxygen atoms is attained if the global $|M_S|$ is $|1+1| = |-1-1| = 2$ (or $|1-1| = 0$, by breaking left-right symmetry). This M_S cannot be realized for the O_2 molecule in its triplet ground state. [3] In order to remove this contradiction, one can either keep $M_S = 1$ for the molecule, but give up following the ground-state potential curve, or unphysically change the M_S at some point of the potential curve, where a fake 'phase transition' occurs.

This type of problem is less uncommon as it may appear (e.g., Si_2, two transition metal atoms hold at some distance by their environment, etc.).

4. REMEDIES

As there is no question about the existence of the exact density functional even for degenerate states, the objective is to find the level(s) of approximation where the problems mention in the previous section can be avoided.

In some cases, the prescriptions avoiding the problems are trivial. For example, one is on the safe side when not using expressions which explicitly depend on the electron number N. The usual approximations of the exchange-correlation density functional (like the local density or the generalized gradient approximation) are well-behaved in this sense. The problems of not obtaining the correct spin-density (or not even the correct density) if one is not willing to violate the size-consistency requirement, need a more detailed discussion.

4.1. Ensembles

The examples above showed that avoiding the correct treatment of ensembles seems unavoidable: The dissociation energy of H_2^+ produces $H^{+1/2}$ fragments (with densities given by an equi-ensemble of H^+ and H), C_2 yields densities, H_2 and O_2 spin-densities which belong to atomic ensembles. A first step (which correctly describes all isolated one-electron systems, like H_2^+) is to introduce a self-interaction correction. An explicit N dependence can be avoided by introducing an orbital dependence [19]. For example, the Coulomb energy and the exchange correlation energy can be self-interaction corrected by:

$$\begin{array}{lcl} U[\rho] & \rightarrow & U[\rho] - \sum_i U[|\varphi_i|^2] \\ E_{xc}[\rho_\uparrow, \rho_\downarrow] & \rightarrow & E_{xc}[\rho_\uparrow, \rho_\downarrow] - \sum_i E_{xc}[[|\varphi_i|^2, 0] \end{array}$$

[3]No $M_S = 2$ state can be generated with two singly occupied open shell orbitals. This state can be obtained by promoting one electron from the $2p\sigma$ bonding to the anti-bonding orbital.

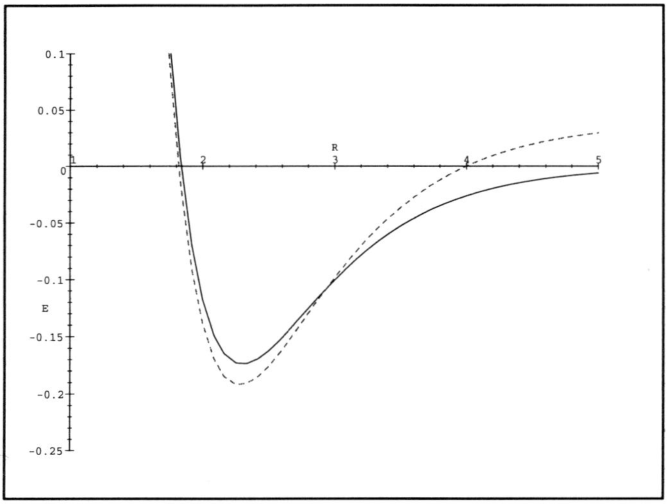

Figure 6. In a molecule having like O_2, ${}^3\Sigma_g^-$, spin-density functionals yield different energies for different M_S: The energy may be lower for $M_S = 1$ (dashed curve) than for $M_S = 0$ (full curve) at the equilibrium distance, but the opposite may happen at large interatomic separation.

where φ_i are the orbitals of the Kohn-Sham determinant. In the original Kohn-Sham scheme, invariance is guaranteed for the energy at unitary transformation among the orbitals. This is not the case for the equations above. Thus, a supplementary clarification is needed in these schemes in order to explain the significance of the orbitals. Although not exact for X_2^+ systems different from H_2^+, this method is expected to significantly reduce the error in the potential curves of these molecules.

Let us now come back to the general ensemble problem. As the widespread philosophy in density functional calculations is to break symmetry, let us see if it can be reconciled with the physical requirement of obtaining a correct (symmetric) density. In many cases the answer is positive. The solution is to write down the energy as in Equation 6, where each of the ρ_i is one of the equivalent densities obtained by symmetry breaking.

$$
\begin{aligned}
E[\rho_1, \rho_2, \ldots] &= \sum_i w_i E[\rho_i] \\
\rho &= \sum_i w_i \rho_i
\end{aligned}
\qquad (24)
$$

By a suitable choice of the weights w_i very often a correct density (and by extension spin-density) can be obtained.

This procedure has some similitude with that of using fractional occupation, as the ensemble density can be in general obtained only by using the latter. The difference between the two is in the calculation of the energy: in the usual treatment spurious terms both in the Coulomb and exchange-correlation terms (in analogy to Equation 12) may appear, as the total density, and not the individual terms, is used for computing the functional. [4]

Another parallel can be drawn to self-interaction corrections. As in the latter, one does not expect invariance of the energy functional of the weights and individual symmetry-broken densities when choosing different weights or densities in common-type approximations.

In order to clarify the success and failures of this procedure let us consider some examples. Let us take again the B_2 molecule. Usual Kohn-Sham calculations yield Slater determinants:

$$\Phi_1 = |\ldots p_{x,left} p_{y,right}|$$
$$\Phi_2 = |\ldots p_{y,left} p_{x,right}|$$

which produce unsymmetric, but equivalent densities:

$$\rho_1 = \ldots + p_{x,left}^2 + p_{y,right}^2$$
$$\rho_2 = \ldots + p_{y,left}^2 + p_{x,right}^2$$

and equal energies $E[\rho_1] = E[\rho_2]$. $\bar{E}$ (Equation 6) will be independent of the choice of the weights w_i, but not the density, which is symmetric only for $w_1 = w_2 = 1/2$.

The dissociated C_2 molecule can be discussed in a similar manner.

Let us see how this method works for the spin-density. In the case of the H_2 molecule we can consider the exact spin-density as an ensemble of Kohn-Sham spin-densities, use

$$\Phi_1 = |\chi_{left} \bar{\chi}_{right}|$$
$$\Phi_2 = |\bar{\chi}_{left} \chi_{right}| \tag{25}$$

which yields

$$E[\rho_{\uparrow,left}, \rho_{\downarrow,right}] = E[\rho_{\uparrow,right}, \rho_{\downarrow,left}]$$

and fix the weights to $w_1 = w_2 = 1/2$ in order to get

$$\bar{\rho}_\uparrow - \bar{\rho}_\downarrow = \frac{1}{2}[(\rho_{\uparrow,left} - \rho_{\downarrow,right}) + (\rho_{\uparrow,right} - \rho_{\downarrow,left})] = 0$$

The situation is more complex in the O_2 molecule. For our purpose it is sufficient to consider the contribution of the 'singly occupied' orbitals to the density and spin-density, In the molecular calculation their origin will be the two *pi*-orbitals while for an atom they will be produced by two *p*-orbitals. At dissociation limit, on each of the atoms this density will be:

$$\rho_{s.o.} = \frac{1}{2}(p_x^2 + p_y^2) + p_z^2$$

[4]Fractional occupation appears in Kohn-Sham theory when the highest occupied level of $\hat{h}_{KS}$ is degenerate. [19,46]

(z is the molecular axis), while the spin-density on an atom in the molecule is

$$\sigma_{s.o.} = \frac{1}{2}(p_x^2 + p_y^2)$$

This is different from the isolated atom case, where $\rho_{s.o.} = \sigma_{s.o.} = 1/2(p_x^2 + p_y^2)$. In Section 3.7, we considered the case when a M_S-dependent functional stabilizes the state where $|M_S| = 1$ on each of the atoms. It follows that in such a case energy minimization produces densities and spin-densities which cannot have $\rho_{s.o.} \neq \sigma_{s.o.}$ as it is in the correct description. Using an ensemble of the energy-minimizing solutions will thus never yield both the correct density and correct spin-density.

Density functionals which wrongly favor a subset of the degenerate densities can make Equation 24 useless when the density of the ensemble cannot be generated by the densities of the artificially lowered states.

The idea of using broken-symmetry ensembles is not so successful in the case of the H_2^+ molecule. As the symmetric solution has the lower energy, symmetry breaking cannot be induced by the variational principle.

It is interesting to notice, however, that using

$$w_1 E_c[\rho_{\uparrow,left}, \rho_{\downarrow,right}] + w_2 E_c[\rho_{\uparrow,right}, \rho_{\downarrow,left}]$$

will be zero for any weights w_1, w_2 in the case of the H_2 molecule. Thus no correlation energy problem is present any more in this case. (The origin lies in $< \Phi_i|\hat{V}_{ee}|\Phi_j >= 0$, when the symmetry broken Slater determinants of Equation 25 are used.)

We thus see that the philosophy of breaking symmetry and restore the correct density and spin-density by constructing ensembles (Equation 24) could solve only part of the problems due to degeneracy.

4.2. Getting around the spin-density problem

The source of the O_2 problem seems to be the lack of flexibility in the choice of the wavefunction in the Kohn-Sham scheme. In order to get a proper $\hat{S}^2$ eigenfunction several determinants are needed at dissociation. $\hat{S}^2$ can be always used to project out the proper component out of the Kohn-Sham determinant. In the constrained search, this means: 'Find the antisymmetric wavefunction, eigenfunction of $\hat{S}^2$, which yields the density ρ, and minimizes ...'. This does not affect the universality of functionals, as one assumes that the external potential is spin-independent. (The restriction is just on the form of the wavefunction, like antisymmetry.)

What is the density functional for multi-determinant wavefunctions? As the spin-projection does not alter the uniform gas Kohn-Sham determinants, it may be argued that $E_{xc}[\rho]$ can be simply used in its usual (local density approximation) form.

As long as the density functional depends on M_S the O_2 problem would not be solved. In order to get rid of this dependence, Ziegler et al. [47]. argued that the usual approximations should only be used for single determinants. They use the density functional expression for the state with maximal M_S, which is described by single Slater determinant, and then require that the states with different M_S should have the same energy. A generalization of this prescription has been given later[48]. In this procedure all diagonal elements of the Hamiltonian matrix in the given basis of Slater determinants are computed as in Equation 17. The non-diagonal elements need integrals of the type:

$$< ij|ji >= \int\int \phi_i(\mathbf{r}_1)\phi_j(\mathbf{r}_2)\frac{1}{r_{12}}\phi_j(\mathbf{r}_1)\phi_i(\mathbf{r}_2)d^3r_1d^3r_2$$

By observing that these integrals can be obtained from expectation values of the Hamiltonian over different Slater determinants:

$$< ij|ji >= \frac{1}{2}[E(|\ldots i\bar{j}\ldots|) + E(|\ldots \bar{i}j\ldots|) - E(|\ldots ij\ldots|) - E(|\ldots \bar{i}\bar{j}\ldots|)]$$

and that only the exchange part differs in the various $E(|\ldots|)$

$$< ij|ji >= \frac{1}{2}[E_x(|\ldots i\bar{j}\ldots|) + E_x(|\ldots \bar{i}j\ldots|) - E_x(|\ldots ij\ldots|) - E_x(|\ldots \bar{i}\bar{j}\ldots|)] \quad (26)$$

But $E_x(|\ldots|)$ can be computed by using the exchange approximation for the density functional. (This scheme was also used with E_{xc} replacing E_x, see, e.g., [50]). The success obtained is considerable, although the reason of using a ground state density functional for the diagonal elements of the Hamiltonian of the excited state determinants is not clear.

An alternative to this scheme would be to give up the use spin-densities. But how can the quality of spin-density functionals be kept? A substitute for the spin-density would be certainly needed. A possibility which keeps the advantage of getting better energies, and avoiding the problems of spin-density was first applied for correlation energies, and then for exchange energies too. Its key observation is that there is a one-to-one correspondence between the pair $(\rho, \rho_\uparrow - \rho_\downarrow)$ and the pair (ρ, P_2), where P_2 is the on-top pair density, all quantities being obtained from the same Slater determinant [37,51,52,45]. (The on-top pair density in $\mathbf{r}$ is obtained by integrating the absolute value of the wavefunction squared over all spatial coordinates, except that of electrons 1 and 2, by summing over all the spins, and by setting the positions of electrons 1 and 2 equal to $\mathbf{r}$; for normalization it is multiplied by $N(N-1)$.) For a Slater determinant, the on-top pair density can be calculated (see, e.g., [49]). A simple relationship to ρ and σ (or $\rho_\uparrow$ and $\rho_\downarrow$, see Equation 22) exists:

$$\begin{aligned} P_2(\mathbf{r},\mathbf{r}) &= \rho(\mathbf{r})^2 - \rho_\uparrow^2(\mathbf{r}) - \rho_\downarrow^2(\mathbf{r}) \\ &= \rho_\uparrow(\mathbf{r})\rho_\downarrow(\mathbf{r}) \end{aligned}$$

Solving this equation (together with $\rho = \rho_\uparrow + \rho_\downarrow$) for $\rho_\uparrow - \rho_\downarrow$ yields:

$$\rho_\uparrow - \rho_\downarrow = \sqrt{\rho^2 - 2P_2}$$

One can thus consider instead of $E[\rho, \rho_\uparrow - \rho_\downarrow]$ the equivalent $E[\rho, P_2]$ as long as one deals with a single Slater determinant. In general, one can view P_2 as the better variable to refine $E[\rho]$ than $\rho_\uparrow - \rho_\downarrow$.

Ordinary (one-determinant) spin-density functional calculations can be re-interpreted as not attempting to obtain the spin-densities, but producing besides ρ also P_2 [45]. This is supported by the observation that P_2 seems to be transferable from the uniform electron gas.

As P_2 is unchanged when going from a state with one value of M_S to another degenerate state characterized by another M_S, so will be the energy [54]. In general, in order to get the correct wavefunction, more than one determinant may be necessary. Explicit formulas exist, however, to produce first-order and on-top second-order density matrix by projection [55–57].

A reasonable condition to impose to $E[\rho, P_2]$ is the recovery of usual density functionals when only one Slater determinant is used. If more than one determinant are present (by spin-projection), there is more than one prescription which satisfies the previous condition, when the number of determinants becomes equal to one. It can be simply decided to use the dependence on P_2 as if it would have been obtained from a single determinant as the spin-projection does not alter the uniform electron gas exchange-correlation energy. The main problem here resides in the fact that for a single determinant $0 \leq P_2 \leq \rho^2$, while for many-determinant wavefunctions only $0 \leq P_2$ is guaranteed. (An example where $P_2 > \rho^2$ is given in Reference [52].) In order to find an approximation for $E[\rho, P_2]$ in this case, the single-determinant uniform electron gas calculations are not well suited, as this system (polarized or not) does not present $P_2 > \rho^2$. $E[\rho, P_2]$ can be approximated for $P_2 > \rho^2$ by using $P_2 \approx \rho^2$, by extrapolating, or by determining the density functional by new calculations. While the latter would be certainly the best way, it has yet to be done.

Let us consider our specific examples, and see how $E[\rho, P_2]$ works. Let us first analyze the hydrogen atoms and their ensemble. P_2 is zero in both cases. Let us next consider the problem of the hydrogen molecule (that the spin-density is not correct at the separated atom limit). Looking at the spin-density in density functional theory only as an intermediate quantity needed to generate P_2 no more claim is made that the spin-density can be correctly obtained. Finally, let us consider the O_2 problem. As the dependence of the energy on M_S has been removed, no more problem is present.

It should be remembered that problems related to multi-determinant wavefunction also appear due to spatial symmetry. The arguments given at the beginning of this section concerning $\hat{S}^2$ cannot be repeated as the spatial symmetry is determined by the external potential and the density functional could not be universal. [5] It seems thus desirable to have a mechanism that automatically selects the density related to the many-determinant wavefunction.

4.3. Two-body operators in density functional theory

Many-determinant wavefunctions can be introduced through two-body operators. These were hidden when producing the on-top pair density (S^2 is a two-body operator), and will now be used to pin down the definition of the correlation energy when degeneracy is present. (Remember that choosing different linear combinations of the degenerate Kohn-Sham determinants may leed to different expectation values of the electron-electron interaction, $< \hat{V}_{ee} >$.) One may, however, fix the definition according to convenience, e.g., by minimizing $|E_c[\rho]|$, the part of the correlation energy described by the density functional. From the constrained search perspective this means choosing the the Kohn-Sham wavefunction by minimizing $< \hat{V}_{ee} >$ in the space of the degenerate determinants ($< \hat{T} >$

[5]Using different functionals for different symmetries is possible in principle[29,58], but does not allow avoiding the problem of insuring invariance when passing from a symmetry group to one of its subgroups, e.g. by an infinitesimal perturbation.

does not change during this search. [6] This definition is equivalent with the minimization of $< \hat{T} + \hat{V}_{ee} >$ in that space as well as finding the minimum of $< \hat{T} + \lambda\hat{V}_{ee} >$ for arbitrary λ. The special case $\lambda \to 0$ (with no supplementary restriction on the space of wavefunctions) is of special interest, due to the analogy to the usual definition when no degeneracy is present.

Using $\hat{V}_{ee}$ to couple the determinants is also useful for the dissociation problems (like that of the $\ldots\pi^2$ occupancy), as it will select the symmetric solution for any finite distance, yielding a symmetric density.

As two-electron integrals are required in order to obtain $< \hat{V}_{ee} >$ one may want to avoid the supplementary effort. The prescription given in Equation 26 (or approximations resulting from it) might be useful, as it avoids the explicit computation of the integrals.

If one considers degeneracy as a symmetry property, one may also replace $\hat{V}_{ee}$ by another two-body operator, $\hat{W}$, which may have more convenient properties for finding approximations or from the computational point of view. For example, one can expect that short-range interaction can be more easily transferred from the uniform electron gas than the long-range ones.

In the remaining part of this section it will be shown that such a choice is possible, and that it allows an automatic selection of Slater determinants.

In the Kohn-Sham scheme one minimizes the energy as a functional of the a determinant, the optimal one being the Kohn-Sham determinant (see Equation 17). Let us consider its extension where one minimizes the energy expression

$$E[\Phi] = < \Phi|\hat{T} + \hat{W}|\Phi > + \tilde{F}_W[\rho(\Phi)])]$$

where the wavefunction Φ is now not restricted to a single determinant [54,59]. The density functional $\tilde{F}_W$ depends on the choice of the operator $\hat{W}$. Please notice that for $\hat{W} = 0$ one recovers the Kohn-Sham method. While choosing the operator $\hat{W}$ as a local potential does not produce any change with respect to the Kohn-Sham scheme, a local two-body operator induces wavefunctions Φ which are linear combinations of Slater determinants. For example, using $\hat{W} = \hat{V}_{ee}$, one would get for $\tilde{F}_W = \int \rho v_{ne}$ (just the interaction with the external potential) and thus the standard configuration interaction method. For a given nonzero $\hat{W}$ no restriction is present for the selection of the wavefunction. Of course, there are more convenient choices for $\hat{W}$. For example, one can use $\hat{W}$ to depend on the electron-electron distances r_{ij}, and an arbitrary parameter μ

$$\hat{W} = \sum_{i<j} w(\mathbf{r}_i, \mathbf{r}_j)$$

$$w(\mathbf{r}_i, \mathbf{r}_j) = \frac{erf(\mu r_{ij})}{r_{ij}}$$

In Figure 7 $w(r)$ is shown together with $1/r$; please notice that $w(r)$ is finite for $r = 0$, and $w \to 1/r$ for $r \to \infty$. The parameter μ can be used to switch between a Kohn-Sham calculation ($\mu = 0$) and a configuration interaction calculation ($\mu \to \infty$). The advantage (over configuration interaction calculations) is that with such an operator the

[6] The density may change in the general case when making a linear combination of degenerate Kohn-Sham determinants.

short-range part of the electron-electron interaction is described by a density functional. (The transferability of approximate density functionals can be expected to be much better for short-range than for long-range interaction of the electrons.) Please notice that due to the disappearance of the divergence of the electron-electron interaction potential for $r_{12} \to 0$ no cusp will appear in Φ and thus the expansion in terms of Slater determinants can be expected to be shorter than in usual configuration interaction calculations. The new two-electron integrals can be computed quite easily with the w operator, as shown in Appendix.

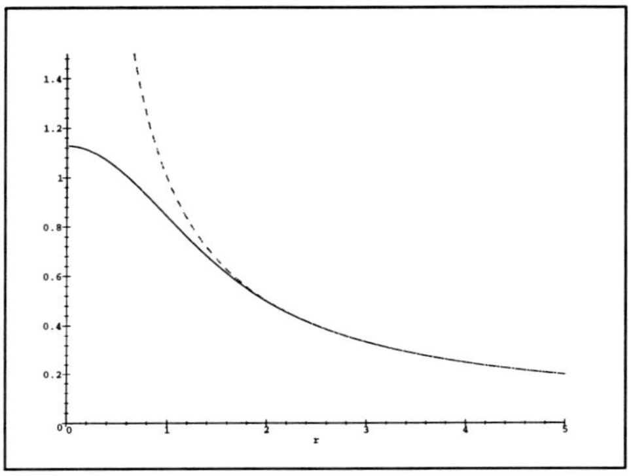

Figure 7. The long-range potential $w(r)$ (for $\mu = 1$, full curve) and $1/r$ (dashed curve).

The energy depends on the choice of the arbitrary parameter μ and is calculated according to:

$$E_\mu = \min_\Phi < \Phi|\hat{T} + \hat{W}(\mu)|\Phi > + U_\mu[\rho] + E_{xc,\mu}[\rho]$$

where

$$U_\mu[\rho] = \frac{1}{2}\int \rho(\mathbf{r}_1)\rho(\mathbf{r}_2)[1 - w(r_{12})]d^3r_1 d^3r_2$$

and the short-range exchange-correlation functional is used in the local density approximation

$$E_{xc,\mu} \approx \int \rho(\mathbf{r})\epsilon_{xc,\mu}(\rho(\mathbf{r}))d^3r$$

and $\epsilon_{xc,\mu}$ is obtained, as usual, from uniform electron gas calculations. The exchange energy can be computed analytically in the homogeneous electron gas but the correlation energy must be computed numerically for the same system. Details are given in the Appendix.

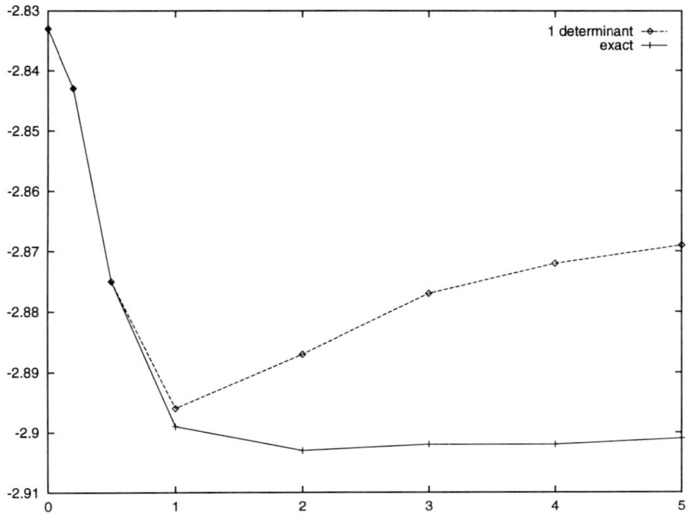

Figure 8. Energy of the He atom for a short-range density functional (in the local density approximation) as a function of the parameter μ controlling the separation between the wavefunction and density functional treatment ($\mu = 0$: Kohn-Sham local density approximation, $\mu \to \infty$: full wavefunction calculation). The upper curve shows the result obtained with a single determinant, the lower one with a many-determinant trial function.

Figure 8 shows the total energy computed for the He atom as a function of μ, by using a many-determinant wavefunction and a single-determinant one. Figure 9 shows the same curves for the Be atom in a pseudopotential [60] calculation.

Let us first analyze the behavior of the energy with respect to μ when a good many-body basis is used. Increasing μ from zero (which corresponds to the Kohn-Sham calculation in the usual local-density approximation) produces a rapid change in the energy, which stabilizes after a threshold value, yielding for the energy the exact value (corresponding to $\mu \to \infty$). A detailed analysis shows that this is not due to the disappearance of the density functional contribution to the exchange-correlation energy, but to the validity of the transfer of $E_{xc,\mu}$ from the uniform electron gas. (See Reference [59] for more details.)

If we now restrict the wavefunction to a single determinant the value for $\mu \to \infty$ will be the Hartree-Fock value, and thus certainly in error. In the He series a good compromise can be found shortly before the threshold value of μ just discussed. In the Be series good results are not obtained without adding a second configuration (p^2), what is not surprising when considering the near-degeneracy in the Be series.

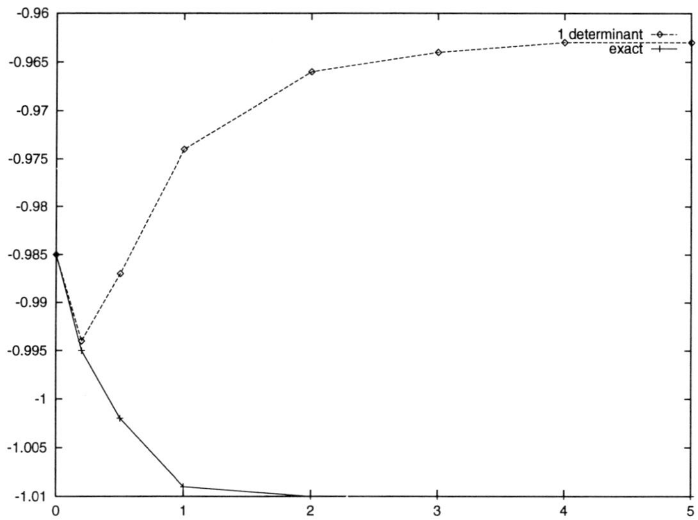

Figure 9. Total energy of the Be atom for a short-range density functional in the local density approximation (in the local density approximation, with a pseudopotential) as a function of the parameter μ controlling the separation between the wavefunction and density functional treatment ($\mu = 0$: Kohn-Sham local density approximation, $\mu \to \infty$: full wavefunction calculation). The lower curve shows the result obtained with a many-determinant wavefunction which is very close to that obtained from a two-configuration trial function, which is very close to the the curve just above it is obtained with two configurations; the upper curve is obtained from a single-determinant calculation.

5. CONCLUSION

Degeneracy can produce problems in the application of density functional theory. Due to errors inherent in the class of approximations the essential quantities which are the object of density functional calculations (the ground state energy and density) may be wrong. Including the dependence on spin-density even accentuates these problems. There are possibilities to avoid at least some of these problems, but not without additional effort:

- the broken symmetry ensemble expression
- many-determinant wavefunctions within the degenerate space
- two-body operators

No general solution was given for constructing density functionals which correctly treat:

- the explicit N-dependence

- different densities which yield the same energy.

Up to now, it seems that a combination of corrections to the presently widely used schemes is needed in order to avoid the problems raised by degeneracy.

6. APPENDICES

6.1. Two-electron integrals: $erfc(\mu r_{12})/r_{12}$

In this appendix it will be shown that integrals like:

$$(AB|w|CD) >= \int\int d^3r_1 d^3r_2 \chi_A(\mathbf{r}_1)\chi_B(\mathbf{r}_1)w(r_{12})\chi_C(\mathbf{r}_2)\chi_D(\mathbf{r}_2)$$

where $w(r) = erf(\mu r)/r$, and χ are Gaussian type orbitals can be easily computed, following standard procedures (see, e.g., [61]. We will consider below the integral $(AB|w'|CD)$ where $w'(r_{12}) = 1/r_{12} - w(r) = erfc(\mu r_{12})/r_{12}$, as $(AB|w(r_{12})|CD)$ is just $(AB|1/r_{12}|CD) - (AB|w'(r_{12})|CD)$. By using the Fourier transforms of χ and w'

$$\tilde{w}'(\mathbf{k}) = \int w'(\mathbf{r}) exp(-i\mathbf{k}\cdot\mathbf{r}) d^3r$$

we get for

$$\chi_A(\mathbf{r}) = e^{-\alpha_A|\mathbf{r}-\mathbf{R_A}|^2}$$
$$\chi_B(\mathbf{r}) = e^{-\alpha_B|\mathbf{r}-\mathbf{R_B}|^2}$$
$$\chi_C(\mathbf{r}) = e^{-\alpha_C|\mathbf{r}-\mathbf{R_C}|^2}$$
$$\chi_D(\mathbf{r}) = e^{-\alpha_D|\mathbf{r}-\mathbf{R_D}|^2}$$

$$(AB|w'|CD) = K_{AB}K_{CD}(2\pi)^{-3}\left(\frac{\pi^2}{\alpha_P\alpha_Q}\right)^{3/2}\frac{4\pi}{|\mathbf{R}_P\mathbf{R}_Q|}\int_0^\infty k\tilde{w}'(k)e^{-\frac{\alpha_P+\alpha_Q}{4\alpha_P\alpha_Q}k^2}\sin(k|\mathbf{R_P}-\mathbf{R_Q}|)dk$$

with

$$\begin{aligned}
\alpha_P &= \alpha_A + \alpha_B \\
\alpha_Q &= \alpha_C + \alpha_D \\
R_P &= \frac{\alpha_A\mathbf{R}_A + \alpha_B\mathbf{R}_B}{\alpha_A + \alpha_B} \\
R_Q &= \frac{\alpha_C\mathbf{R}_C + \alpha_D\mathbf{R}_D}{\alpha_C + \alpha_D} \\
K_{AB} &= e^{-\frac{\alpha_A\alpha_B}{\alpha_A+\alpha_B}|\mathbf{R_A}-\mathbf{R_B}|^2} \\
K_{CD} &= e^{-\frac{\alpha_C\alpha_D}{\alpha_C+\alpha_D}|\mathbf{R_C}-\mathbf{R_D}|^2}
\end{aligned}$$

By using

$$\tilde{w}'(\mathbf{k}) = \frac{4\pi}{k^2}e^{-\frac{k^2}{4\mu}}$$

one obtains for the integral on the r.h.s.:

$$4\pi\int_0^\infty exp(-ak^2)\frac{\sin(kx)}{k}dk$$

with

$$a = \frac{\alpha_P + \alpha_Q}{4\alpha_P\alpha_Q} + \frac{1}{4\mu}$$
$$x = |\mathbf{R_P} - \mathbf{R_Q}|$$

which in turn integrates to

$$2\pi^2 erf(\frac{|\mathbf{R_P} - \mathbf{R_Q}|}{\sqrt{1/\alpha_P + 1/\alpha_Q + 1/\mu}})$$

with the final result

$$(AB|w'|CD) =$$
$$e^{-\frac{\alpha_A\alpha_B}{\alpha_A+\alpha_B}} e^{-\frac{\alpha_C\alpha_D}{\alpha_C+\alpha_D}} \frac{\pi^3}{[(\alpha_A + \alpha_B)(\alpha_C + \alpha_D)]^{3/2}} \frac{1}{|\mathbf{R_P} - \mathbf{R_Q}|} erf(\frac{|\mathbf{R_P} - \mathbf{R_Q}|}{\sqrt{1/\alpha_P + 1/\alpha_Q + 1/\mu}})$$

This is just the two-electron integral over spherical Gaussians, up to a change of the argument of the error function:

$$1/\alpha_P + 1/\alpha_Q \rightarrow 1/\alpha_P + 1/\alpha_Q + 1/\mu$$

It is natural to recover the usual two-electron integral for $\mu \rightarrow \infty$ as in this limit $w' \rightarrow 1/r$.

Integrals over Gaussians with higher angular momentum can be obtained, for example, by calculating the derivative of the above integral with respect to the components of the vectors **R**.

6.2. Density functionals for the short-range interaction $w = erf(r_{12})/r_{12}$

The expression for the exchange energy per particle for the uniform electron gas for the short-range interaction $w = erf(r_{12})/r_{12}$ is given by:

$$\epsilon_{x,\mu}(\rho) = -(\frac{24\rho}{\pi})^{1/3}\{\frac{3}{8} - a[\sqrt{\pi}erf(\frac{1}{2a}) + (2a - 4a^3)exp(-\frac{1}{4a^2}) - 3a + 4a^3]\}$$

where $a = \mu/(2k_F)$, $k_F = (3\pi^2\rho)^{1/3}$. The correlation energies for various μ were obtained in a coupled-cluster calculation of the type described in Reference [62]. The ratio between the correlation energy for μ and that for $\mu = 0$ is given in the Table. The correlation energy is obtained by multiplying the local density approximation value given by the usual expression[63] with this ratio.

Table 1
Ratio between the short-range and total correlation energy of the uniform electron gas obtained from a coupled cluster calculation; $r_s = [3/(4\pi\rho)]^{1/3}$, μ in atomic units.

	r_s								
μ	0.2	0.5	1.0	2.0	3.0	4.0	5.0	6.0	10.
0.2	0.987	0.963	0.919	0.822	0.723	0.629	0.548	0.483	0.323
0.5	0.944	0.858	0.722	0.494	0.354	0.271	0.216	0.179	0.099
1.0	0.856	0.679	0.453	0.238	0.153	0.107	0.080	0.062	0.031
2.0	0.689	0.412	0.209	0.090	0.053	0.034	0.024	0.017	0.010
3.0	0.552	0.264	0.119	0.046	0.026	0.016	0.012	0.008	0.005
4.0	0.443	0.183	0.076	0.028	0.015	0.009	0.007	0.004	0.003
5.0	0.358	0.134	0.053	0.018	0.011	0.006	0.005	0.002	0.003
6.0	0.294	0.103	0.039	0.013	0.007	0.005	0.003	0.002	0.003
7.0	0.245	0.081	0.030	0.010	0.005	0.003	0.002	0.000	0.000
8.0	0.207	0.066	0.023	0.008	0.004	0.002	0.002	0.000	0.000
9.0	0.178	0.054	0.019	0.006	0.004	0.002	0.002	0.000	0.000
10.0	0.154	0.046	0.016	0.004	0.003	0.002	0.002	0.000	0.000
11.0	0.135	0.038	0.013	0.003	0.003	0.002	0.002	0.000	0.000
12.0	0.119	0.033	0.011	0.003	0.003	0.002	0.002	0.000	0.000
13.0	0.106	0.029	0.010	0.002	0.001	0.000	0.000	0.000	0.000
14.0	0.095	0.025	0.008	0.002	0.001	0.000	0.000	0.000	0.000
15.0	0.085	0.023	0.008	0.002	0.001	0.000	0.000	0.000	0.000
16.0	0.077	0.020	0.007	0.001	0.001	0.000	0.000	0.000	0.000
17.0	0.070	0.018	0.006	0.001	0.001	0.000	0.000	0.000	0.000
18.0	0.064	0.016	0.005	0.001	0.001	0.000	0.000	0.000	0.000
19.0	0.059	0.015	0.005	0.001	0.001	0.000	0.000	0.000	0.000
20.0	0.054	0.014	0.004	0.001	0.001	0.000	0.000	0.000	0.000
21.0	0.050	0.012	0.004	0.001	0.001	0.000	0.000	0.000	0.000
22.0	0.046	0.011	0.003	0.001	0.001	0.000	0.000	0.000	0.000
23.0	0.043	0.010	0.003	0.001	0.001	0.000	0.000	0.000	0.000
24.0	0.040	0.010	0.003	0.001	0.001	0.000	0.000	0.000	0.000

REFERENCES

1. P. Hohenberg, W. Kohn, Phys. Rev. **B136**, 864 (1964).
2. R. Dreizler, E.K.U. Gross, *Density Functional Theory*, Springer-Verlag, Berlin, 1990.
3. E.S. Kryachko, E.V. Ludeña, *Energy Density Functional Theory of Many-Electron Systems*, Kluwer, Academic Publishers, Dordrecht , 1990.
4. R.G. Parr, W. Yang, 'Density-Functional Theory of Atoms and Molecules', Oxford University Press, New York, 1989.
5. R.F. Frey, E.R. Davidson in: Th. H. Dunning, Jr. (ed.), *Advances in Molecular Structure Theory*, JAI Press Inc., Greenwich, 1990, p.213.
6. J.P. Perdew, in: R.M. Dreizler, J. da Providencia (eds.), *Density Functionals Methods in Physics*, Plenum Press, New York, 1985, p.265.
7. M. Levy, Proc. Natl. Acad. Sci. (USA) **76**, 6062 (1979).
8. E.H. Lieb, Int. J. Quantum Chem, **24**, 243 (1983); E.H. Lieb, in: R.M. Dreizler, J. da Providencia (eds.), *Density Functionals Methods in Physics*, Plenum Press, New York, 1985.
9. J.K. Percus, Int. J. Quantum Chem., **13**, 89 (1978).
10. R.E. Moss, *Advanced Molecular Quantum Mechanics*, Chapman and Hall, London, 1976, p. 216.
11. N.D. Mermin, Phys. Rev. **137**, A1441 (1965).
12. S.M. Valone, J. Chem. Phys. **73**, 4653 (1980).
13. M. Levy, Phys. Rev. **A 26**, 1200 (1982).
14. Englisch H., Englisch R., Physica, **121 A**, 253 (1983).
15. E.K.U. Gross, L.N. Oliveira, W. Kohn, Phys. Rev. **A 37**, 2805 (1988).
16. A.K. Rajagopal, F.A. Buot, Phys. Rev. **A51**, 1170 (1995).
17. J.A. Pople, D.L. Beveridge, *Approximate Molecular Orbital Theory*, McGraw Hill Book Co., New York, 1970.
18. W. Kohn and L. Sham, Phys. Rev. **A140**, 1133 (1965)
19. J.P. Perdew, A. Zunger, Phys. Rev. **B23**, 5048 (1981).
20. T.L. Gilbert, Phys. Rev. **B 12**, 2111 (1975).
21. J.E. Harriman, Phys. Rev. **A 24**, 680 (1981).
22. A. Savin, H. Stoll, H. Preuss, Theor. Chim. Acta **70**, 407 (1986).
23. V. Sahni, M. Levy, Phys. Rev. B**33**, 3869 (1986).
24. E.K.U. Gross, M. Petersilka, T. Grabo, in: T. Ziegler (ed.), 'Density Functional Methods in Chemistry', ACS Symposium Series, 1996.
25. U. von Barth, L. Hedin, J. Phys. **C 5**, 1629 (1972).
26. M.M. Pant, A.K. Rajagopal, Solid. St. Comm. **10**, 1157 (1972)
27. A.K. Rajagopal, J. Callaway, Phys. Rev. **B 7**, 1912 (1973).
28. J.C. Stoddart, N. H. March, Ann. Phys. New York **64**, 174 (1971).
29. O. Gunnarsson, B.I. Lundqvist, Phys. Rev. **B13**, 4724 (1976).
30. R. McWeeny, *Methods of Molecular Quantum Mechanics*, Academic Press, London, 1002, p.148.
31. F.W. Kutzler, G.S. Painter, Phys. Rev. **B 43**, 6865, 1991.
32. D. Layzer, Annals of Phys. **8**, 271 (1959).
33. J. Linderberg, H. Shull, J. Mol. Spectroscopy, **5**, 1 (1960).

34. J.P. Perdew, E. R. McMullen, A. Zunger, Phys. Rev. A**23**, 2785 (1981).
35. H. Stoll, E. Golka, H. Preuss, Theor. Chim. Acta **55**, 29 (1980).
36. R. Colle and O. Salvetti, Theor. Chim. Acta **79**, 1404 (1979).
37. B. Miehlich, Diplomarbeit, Universität Stuttgart, 1989
38. R. Merkle, A. Savin, H. Preuss, J. Chem. Phys. **97**, 9216 (1992).
39. J.C. Slater, *The Self-Consitent Field for Molecules and Solids*, vol. 4, McGraw Hill Book Co., New York, 1974.
40. E. Fermi, E. Amaldi, Accad. Ital. Roma **6**, 119 (1934).
41. S.H. Vosko, L. Wilk, J. Phys. B: At. Mol. Phys. **16**, 3687 (1983).
42. B.I. Dunlap, J.W.D. Connolly, J.R. Sabin, J. Chem. Phys. **71**, 4993 (1979).
43. J. Harris, Int. J. Quantum Chem. **S 13**, 189 (1979).
44. K.P. Huber, G. Herzberg, *Molecular Spectra and Molecular Structure*, Vol. 4, Van Norstrand, New York 1979.
45. J.P. Perdew, A. Savin, K. Burke, Phys. Rev. A **51**, 4531 (1995).
46. J.F. Janak, Phys. Rev. **B18**, 7165 (1978).
47. T. Ziegler, A. Rauk, E.J. Baerends, Theor. Chim. Acta **43**, 261 (1977).
48. T. Ziegler, in: J.P. Dahl, J. Avery, (eds.) *Local density approximations in quantum chemistry and solid state physics,* Plenum Press, New York, 1984.
49. P.-O. Löwdin, Phys. Rev. **97**, 1474 (1955).
50. U. von Barth, Phys. Rev. **A20**, 1693 (1979).
51. F. Moscardo and E. San-Fabian, Phys. Rev. A **44**, 1549 (1991).
52. A.D. Becke, A. Savin and H. Stoll, Theor. Chim. Acta, Theor. Chim. Acta, **91**, 147 (1995).
53. J.E. Harriman, in: *Density Matrices and Density Functionals*, R. Erdahl and V.H. Smith, Jr., eds., Reidel, Dordrecht, 1987, p.359.
54. A. Savin, in D.P. Chong (ed.), *Recent Advances in Density Functional Methods*, World Scientific, Singapore, 1996.
55. J.E. Harriman, J. Chem. Phys. **40**, 2827 (1964).
56. A. Hardisson and J.E. Harriman, J. Chem. Phys. **46**, 3639 (1967).
57. K.M. Sando and J.E. Harriman, J. Chem. Phys. **47**, 180(1967).
58. A. Görling, Phys. Rev. **A 47**, 2783 (1993).
59. A. Savin and H.-J. Flad, Int. J. Quantum Chem. **56**, 327 (1995).
60. P. Fuentealba, L. von Szentpaly, H. Preuss, H. Stoll, J. Phys. B, **18**, 1287 (1985).
61. A. Szabo, N.S. Ostlund, *Modern Quantum Chemistry*, McMillan Publishing Co, New York, 1982, appendix A.
62. D.L. Freeman, Phys. Rev. **B 15**, 5512 (1977).
63. S.H. Vosko, L. Wilk and M. Nusair, Can. J. Phys. **58**, 1200 (1980).

J.M. Seminario (Editor)
Recent Developments and Applications of Modern Density Functional Theory
Theoretical and Computational Chemistry, Vol. 4

A Simple Method of Removing Spin Contamination From Unrestricted Kohn-Sham Density Functional Calculations

Alexander A. Ovchinnikov,[a]* Charles F. Bender,[b] Jan K. Łabanowski[b] †

[a]Institute of Chemical Physics, Russian Academy of Sciences, 4 Kosygin St., 117334 Moscow, Russia

[b]Ohio Supercomputer Center, 1224 Kinnear Rd., Columbus, OH 43212, USA

A simple method of removing spin contamination from Kohn-Sham orbitals in the Unrestricted Kohn-Sham DFT calculations is presented. The method corrects energies of mixed spin solutions, allowing for meaningful comparisons of energy of different multiplets. Computational details of the current implementation are included. Preliminary results for molecules: C_2, CH_2, O_2, O_2^-, and small oxygen clusters: $(O_2)_2$, $(O_2)_2^-$, and $(O_2)_6$ provide examples of practical application of the method. It has been demonstrated that this method can be useful in describing not only isolated molecules, but also provide solid basis for computations of the magnetic properties of larger clusters, solids, and other extended systems. The method is also used to detect the kinematic spin alignment in the anion of O_2 dimer.

1. INTRODUCTION

Unrestricted Kohn-Sham DFT (UKS) calculations often result in solutions that are not pure spin states with respect to the desired multiplicity. This situation is especially acute for radical and polyradical species. Comparisons of energy splits between different multiplets are quite unreliable if solutions are heavily contaminated with undesired spin states. Moreover, the properties (e.g., geometry) calculated from spin contaminated solutions are often artificial. We have recently derived a method that removes contaminating contributions from higher spin states[1] present in the unrestricted Kohn-Sham results. Obviously, such a method can be only approximate because Kohn-Sham orbitals do not correspond to a real system of interacting electrons but rather to a fictitious system of noninteracting electrons of the same charge density. How to calculate the spin contamination accurately for the real system is still not known, though some important research in this area was recently reported[2]. In this chapter, we focus on the practical issues concerning computer implementation of our approach.

*Supported by the International Science and Technology Center in Moscow (Grant No. ISTC 015-94).
†Supported by the grant of computer time from the Ohio Supercomputer Center.

2. COMPUTATIONAL APPROACH

The derivation of the method was described in detail in our previous paper[1] on this topic. In essence, the unrestricted (Hartree-Fock, or Kohn-Sham) wave function, $\Psi_{Unr.,S}$, calculated for the total spin S can be, in principle, contaminated with components corresponding to higher spin states:

$$\Psi_{Unr.,S} = a_S(S)\Psi_S + a_S(S+1)\Psi_{S+1} + \cdots + a_S(S_{max})\Psi_{S_{max}} \tag{1}$$

where the wave functions Ψ_S, Ψ_{S+1}, ... $\Psi_{S_{max}}$ correspond to pure spin states with spins: S, $S+1$, ... S_{max}, respectively. It is convenient to introduce spin contamination as:

$$\mathcal{A}_S(S') = a_S^2(S') \tag{2}$$

The $\mathcal{A}_S(S')$ provides a measure of contamination of the unrestricted wave function $\Psi_{Unr.,S}$ with the wave function $\Psi_{S'}$, which describes the pure spin state[3] S'. To calculate spin contaminations present in $\Psi_{Unr.,S}$ we introduce the hermitian operator:

$$\hat{B}_L = \left(\hat{S}^-\right)^L \left(\hat{S}^+\right)^L \tag{3}$$

where $L = 0, 1, 2, \cdots$, and $\hat{S}^-$, and $\hat{S}^+$ are the well known step-up and step-down operators[3]:

$$\begin{aligned} \hat{S}^+ &= \hat{S}^x + i\hat{S}^y \\ \hat{S}^- &= \hat{S}^x - i\hat{S}^y \end{aligned} \tag{4}$$

The $\hat{S}^x$ and $\hat{S}^y$ are the x and y components of the operator of the total spin of the system. The expectation values of $\hat{B}_L$ for $L = 0, 1, \cdots S_{max} - S$ can be found by using either the left- or the right-hand side of eq. 1. Such a procedure leads to a system of equations for spin contaminations $\mathcal{A}_S(S')$. To calculate the expectation value:

$$<\hat{B}_L> = \left\langle \Psi_{Unr.,S} \middle| \hat{B}_L \middle| \Psi_{Unr.,S} \right\rangle = \left\langle \Psi_{Unr.,S} \middle| \left(\hat{S}^-\right)^L \left(\hat{S}^+\right)^L \middle| \Psi_{Unr.,S} \right\rangle \tag{5}$$

we will use the second quantization approach. In this approach, the $\hat{S}^x$ and $\hat{S}^y$ can be represented as:

$$\begin{aligned} \hat{S}^+ &= \int \Psi_\alpha^\dagger(x)\Psi_\beta(x)dx \\ \hat{S}^- &= \int \Psi_\beta^\dagger(x)\Psi_\alpha(x)dx \end{aligned} \tag{6}$$

and by expressing creation (†) and annihilation operators $\Psi_\sigma(x)$ via a set of orbitals, we arrive at:

$$\begin{aligned} \hat{S}^+ &= \sum_{i=1}^{N_\alpha}\sum_{j=1}^{N_\beta} a_{i\alpha}^\dagger a_{j\beta} S_{ij}^{\alpha\beta} \\ \hat{S}^- &= \sum_{i=1}^{N_\beta}\sum_{j=1}^{N_\alpha} a_{i\beta}^\dagger a_{j\alpha} S_{ji}^{\alpha\beta} \end{aligned} \tag{7}$$

[3]The $\mathcal{A}_S(S)$ represents the contents of the parent spin state in the wave function and for the uncontaminated solution $\mathcal{A}_S(S) = 1$.

where the $a_{i\alpha}$ and $a_{i\beta}$ are the creation (†) and annihilation operators for electrons α and β on given molecular orbitals, respectively, and the overlap integral between ith α and jth β molecular orbital, $S_{ij}^{\alpha\beta}$ is:

$$S_{ij}^{\alpha\beta} = \left\langle \phi_i^\alpha | \phi_j^\beta \right\rangle = \sum_{k=1}^{N} \sum_{l=1}^{N} C_{ki}^\alpha C_{lj}^\beta S_{kl} \tag{8}$$

Generally, $S_{ij}^{\alpha\beta}$ is not a unit matrix because the α and β spin orbitals need not be orthogonal to each other within the framework of the UKS approach. In the above equation, the C_{ki}^σ are the KS orbital coefficients, i.e., coefficients that expand spin orbitals σ (i.e, α or β) into the linear combination of basis functions $\phi_i^\sigma = \sum_k C_{ki}^\sigma \chi_k$, and $S_{kl} = \langle \chi_k | \chi_l \rangle$ is the overlap integral between kth and lth basis function.

After cumbersome derivation using the permutation group theory[9], one arrives at the final formula for the expectation value of the $\hat{B}_L$:

$$< \hat{B}_L >= B_L = (L!)^2 \sum_{\{k_i\}} \frac{1}{k_1!} \left(\frac{v_1}{1} \right)^{k_1} \frac{1}{k_2!} \left(\frac{v_2}{2} \right)^{k_2} \cdots \frac{1}{k_L!} \left(\frac{v_L}{L} \right)^{k_L} (-1)^{L-k_1-k_2-\cdots k_L} \tag{9}$$

where the summation is carried over all sets of $\{k_i\} = \{k_1, k_2, \ldots k_L\}$, obeying the condition: $k_1 + 2k_2 + 3k_3 + \cdots Lk_L = L$, for nonnegative k_i's. The $v_m = Tr(V^m)$, and the V matrix is defined as[4]:

$$V(j, j') = \delta_{j,j'} - \sum_k S_{jk}^{\alpha\beta} S_{j'k}^{\alpha\beta} \tag{10}$$

The expectation value of $< \hat{B}_L >$ for the linear combination of pure spin wave functions on the right side of eq. 1 can be calculated from the Eckart-Wigner relations:

$$\begin{aligned} \hat{S}^+ \Psi(S; M) &= \sqrt{S(S+1) - M(M+1)} \Psi(S; M+1) \\ \hat{S}^- \Psi(S; M) &= \sqrt{S(S+1) - M(M-1)} \Psi(S; M-1) \end{aligned} \tag{11}$$

where $\Psi(S; M)$ is the function having spin S and its z-projection equal to M; $S \geq M$. Note also that:

$$\hat{S}^+ \Psi(S; S) = 0 \tag{12}$$

By applying eq. (11) L times one arrives at the expression:

$$\left\langle \Psi(S; M) \left| \hat{B}_L \right| \Psi(S; M) \right\rangle = \prod_{i=0}^{L-1} [S(S+1) - (M+i)(M+i+1)] \tag{13}$$

It is convenient to introduce the following symbol:

$$T_K(P, S) = \prod_{i=0}^{P-1} [S(S+1) - (K+i)(K+i+1)] \tag{14}$$

[4]For actual calculations it is convenient to use eigenvalues of V, so the computation of traces of the mth power of V is reduced to the summation of mth powers of eigenvalues.

The $T_K(0,S) \equiv 1$ by definition. It can be easily verified that for $S-K<P$ the product above contains a zero factor, i.e., the whole product is equal to zero.

Combining equations for $L=0,1,\cdots S_{max}-S$ leads to a triangular system of equations (i.e., these equations can be solved by a simple back substitution):

$$\begin{array}{rl} \mathcal{A}_S(S) + \mathcal{A}_S(S+1) + \mathcal{A}_S(S+2) + \cdots \mathcal{A}_S(S+L) & = 1 \\ T_S(1,S+1)\mathcal{A}_S(S+1) + T_S(1,S+2)\mathcal{A}_S(S+2) + \cdots T_S(1,S+L)\mathcal{A}_S(S+L) & = B_1 \\ T_S(2,S+2)\mathcal{A}_S(S+2) + \cdots T_S(2,S+L)\mathcal{A}_S(S+L) & = B_2 \\ \ddots \quad \vdots & \\ T_S(L,S+L)\mathcal{A}_S(S+L) & = B_L \end{array} \tag{15}$$

Once the contaminations $\mathcal{A}_S(S')$ are known, the energy of the the pure spin state S can be derived from the expectation value of the hamiltonian for the wave function on both sides of eq. 1:

$$E_{corr}(S) = \frac{E_{uncorr}(S) - \mathcal{A}_S(S+1)E_{corr}(S+1) - \mathcal{A}_S(S+2)E_{corr}(S+2) - \cdots}{1 - \mathcal{A}_S(S+1) - \mathcal{A}_S(S+2) - \mathcal{A}_S(S+3) - \cdots} \tag{16}$$

where $E_{corr}(S')$ is the energy corresponding to the pure spin state S', and $E_{uncorr}(S)$ is the energy corresponding to the mixed spin state described by $\Psi_{Unr.,S}$.

This equation requires that all pure spin state energies for contaminating spin states are known. In practice, it may be solved by starting from the highest spin state that comes pure from unrestricted calculations[5] and successively estimating the corrected energies for the spin states with the lower S.

Another approach for deriving pure spin energies would be to run the calculations for each multiplicity $M=2S+1$ for $S=S_{min}, S_{min}+1,\ldots,S_{max}$ and then solve a triangular system of equations via back substitution:

$$E_{uncorr}(S_i) = \sum_{S_j=S_i}^{S_{max}} \mathcal{A}_{S_i}(S_j)E_{corr}(S_j); \qquad S_{min} \leq S_i \leq S_{max} \tag{17}$$

in which $\mathcal{A}_{S_{max}}(S_{max})=1$ and therefore $E_{uncorr}(S_{max} = E_{corr}(S_{max})$. In principle, there may be several different mixed spin solutions for a given S_i or several similar solutions differing slightly due to limited accuracy of calculations. In this case, the system of equations (17) may have more equations than $S_{max}-S_{min}+1$ and needs to be solved as a multiple linear regression "through origin" problem where the unknown energies $E_{corr}(S_j)$'s are required to fit best the available data by minimizing the sum

$$\sigma^2 = \sum_{j=1}^{N_{equations}} \left(E^{UKS}_{uncorr}(S_j) - E^{Regression}_{uncorr}(S_j)\right)^2 \tag{18}$$

where σ^2 is the sum of squared deviations, $E^{UKS}_{uncorr}(S_j)$ is the mixed spin state energy resulting from UKS DFT calculations for the spin state S_j, and the $E^{Regression}_{uncorr}(S_j)$ denotes the corresponding value of energy fitted by regression. However, it is possible (and likely)

[5]While theoretically such a state corresponds to a system in which all electrons are unpaired, in reality unpairing electrons from inner shells is not practical in chemical situations, and such spin states do not have to be taken into account in most situations.

that different UKS solutions for the given spin S_j may contain some contamination coming from the excited spin states. One may detect such UKS solutions by careful analysis of KS orbitals and/or by using statistical techniques: principal component analysis or partial least squares. However, we found the cross-validation quite useful. Cross-validation is a statistical procedure in which multiple regression is performed on the data table from which one or more rows have been removed. The $E_{uncorr}^{Regression}(S_k)$ for the removed kth row is then predicted from the calculated regression coefficients and compared with the original value of $E_{uncorr}^{UKS}(S_k)$. If substantial discrepancy exists, it may suggest that the kth state contains an admixture of some excited spin state(s), and therefore, one (or more) of the $E_{corr}(S_i)$ in eq. 17 corresponds to some excited spin state, and such UKS result(s) should be removed from the fitting set.

3. COMPUTER IMPLEMENTATION

The extensions to calculate pure spin states were implemented within *deMon* program[4], whose source was kindly provided to us. This program uses contracted gaussians as basis sets and Vosko-Wilk-Nusair[5] local spin-density functional which can be augmented with a variety of gradient corrections. In examples presented in this chapter we used Perdew correlation corrections[6] and and Becke exchange correction[7]. We used TZP quality atomic basis sets[8] optimized specifically for the DFT use with a split of (7111/411/1*) for C and O, and (41/1*) for H. As to the auxiliary basis sets consisting of uncontracted gaussians, the same quality sets (though having different exponents) were used to fit the charge density and the exchange-correlation potential. For the C and O atoms they included 4 uncontracted s-type gaussians and 4 sets of cartesian gaussians consisting of s, p, and d-type primitives. For the H atom, the 3 s-type gaussians, and one set of spd cartesian gaussians were used. In most calculations presented here we used Pulay DIIS SCF acceleration scheme for the SCF and the level shift method[15], with a 3 to 5 eV gap to stabilize SCF convergence. Depending on the oscillatory tendency of SCF for a given case, we were also damping, to a various degree, the updates from the new iteration by choosing the value of $DMIX$ between 0.005 and 0.2.

Beside programming in the calculations described in the previous section, we also had to modify the way in which singlets are treated in *deMon*. The original code assumes the restricted procedure for singlets, i.e., the same KS orbitals describe α and β electrons. Majority of stable classical molecules in their equilibrium geometry are adequately described by the restricted solution, and even when run in the unrestricted mode, they naturally converge to it. One needs to realize, however, that for typical SCF initialization for singlets, the Fock matrices for α and β are identical, and the SCF will produce the same KS orbitals for electrons of opposite spin even if the unrestricted solution corresponds to a lower energy. To allow different KS orbitals for α and β electrons one needs to start with different exchange-correlation potentials for α, and β electrons at the SCF initialization stage.

The unrestricted solution, may yield lower energy solutions for some radical species and clusters. Similarly, deformed molecules far from typical geometry may converge to the unrestricted solution. The unrestricted solution may also have the advantage over restricted one, because the restricted SCF calculations will not converge for systems with degenerate HOMO energy levels, while lifting the orbital degeneracy as result of the unrestricted mode, may allow successful completion of the SCF procedure.

Other modifications allowed us to read only selected information from the restart files, which is useful if the restart file from a run for a different multiplicity, charge and/or geometry is used to seed the calculations for the current run. For seeding the KS SCF procedure in *deMon* one needs essentially only the coefficients of the fit to the exchange-correlation potential, and to a lesser degree, the fit to the density. The restart file is written to the disk after SCF has converged. We provided also a keyword for saving the restart information after each SCF iteration. We find this option quite useful for the difficult to converge cases, which are usually interrupted before they complete, when one probes for the best convergence conditions. In such a case the interrupted run is not totally lost, but one can recover from the "iteration dump" file.

We also provided an option to request the occupation numbers for the KS orbitals to be fixed at the desired values during the SCF calculations. It is sometimes useful in achieving the SCF convergence, especially in the case of singlets with a close gap between HOMO and LUMO. While such solutions may have HOMO higher in energy than LUMO, due to a known problem of self-interaction in DFT, they are still useful to analyze the problem and asses if SCF convergence in restricted mode is possible at all.

Other modifications allowed reading atom coordinates from a separate file rather than the regular input file. This option is used during SIMPLEX geometry optimization of pure spin states. The cartesian coordinates corresponding to a SIMPLEX vertex are saved in the file CARTCOOR by the top level routine. Then, for each multiplicity run, the same molecular geometry is read in. Finally, after calculations are completed for the lowest multiplicity, the pure spin state energy is calculated from eq 16. The new SIMPLEX vertex is chosen, coordinates corresponding to this vertex saved in CARTCOOR, and the new cycle starts.

Beside these new keywords which we added to the dictionary of the *deMon* input file, we also provide an option for calculating spin contaminations for the range of highest contaminating states, rather than calculate them for the single selected highest multiplet. It is convenient, in this respect, that at initial stages of calculations one may not be sure as to the highest relevant multiplet for which the UKS energies are needed. Since the computational effort of calculating spin contaminations is small compared to solving SCF KS equations, it allows the user to defer this decision when there is enough data, or to try several high cut-off multiplicities in the quest for the consistency of results.

To perform geometry optimization of the pure spin state, we implemented a procedure that starts each cycle of geometry optimization from the highest uncontaminated spin state and computes successive pure spin states energies until it reaches the state for which the geometry is optimized. Because the run for each multiplicity is controlled via its own input file, SCF convergence options can be tuned for the particular multiplicity. Moreover, the restart information for each multiplicity is saved as a separate file, which ensures that the crossing to another state will not occur (provided the geometry is changed in small steps).

The basic code for calculation of a pure spin state energy for multiplicity `MINMUL` is:

```
      KK = 0
      DO 2000 I = MAXMUL,MINMUL,-2
        KK = KK + 1
C -- open input file for a given multiplicity
        OPEN(UNIT=5, FILE=INFILS(KK), STATUS='OLD',ERR=9010)
C -- Read and process input
        CALL DEINIT
        CLOSE(UNIT=5)
C -- Read in basis sets and restart information
        CALL DEINPT
C -- Run SCF
        CALL DEDRVE
C -- Calculate overlap betweem KS MOs
        CALL SABCMP(0)
C -- Calculate spin contaminations
        CALL CONTCL
C -- Calculate pure energy from previous energies and contaminations
        CALL PURIFE
2000  CONTINUE
```

The last routine, PURIFE, calculates pure spin state energy for current multiplicity and saves it in the array. Its main code is:

```
      X = ENERGY
      Y = 1.0D0
C ---   number of higher states
      L = (MAXMUL - IMULT)/2
C ---   pointer to pure energies for higher states
      K = (IMULT - MINMUL)/2 + 1
C ---   pointer to contaminations
      J = 1
C --- Subtract contamination energies
      DO 100 I = 1, L
        K = K + 1
        J = J + 1
        X = X - PURENG(K)*SPCNTM(J)
        Y = Y - SPCNTM(J)
 100  CONTINUE
      K = (IMULT - MINMUL)/2 + 1
      IF(ABS(Y) .GT. 0.0000001)THEN
        X = X/Y
        PURENG(K) = X
      ELSE
        WRITE(6,*)'Error....'
        STOP
      ENDIF
```

The current implementation is far from optimal, but, at least at this stage, we reuse the two-electron integrals because they do not depend upon the multiplicity of the system but upon the basis set and geometry. More stages of the calculations can be reused for different multiplicities at the same geometry (e.g., integration grids, overlap integrals, etc.), but at this moment these improvements are not implemented.

We do not have energy gradients at this time, and therefore we use SIMPLEX optimization[10] if needed. Obviously, this is a very inefficient procedure of geometry optimization and can be applied only to small molecules, or molecules with high symmetry[6]. Because our original version of *deMon* optimizes geometry in cartesian coordinates, we had to add some code that allowed optimization of selected internal coordinates so only nonredundant variables are being optimized. To ensure that the starting simplex is convex we used the following code to generate a regular `NOPT+1`-hedron in `NOPT`-dimensional space centered around the origin of coordinate system:

```
         DO 200 I = 1, NOPT
           X = 1.0*SQRT(0.5/(I*(I+1)))
           DO 150 J = 1, I
             P(I,J) = X
150        CONTINUE
           P(I,I+1) = -I*X
200      CONTINUE
```

where `NOPT` is the number of optimized variables and columns of `P` represent sets of coordinates for simplex vertices. The initial regular simplex was then scaled by multiplying each coordinate by its sensitivity (0.01 Å for bond lengths, 1.0° for valence angles, and 2.0° for torsional angles). Then, the simplex figure was translated to the starting geometry (i.e., corresponding coordinates of the starting conformation were added to the appropriate coordinates of initial simplex vertices). The program also allowed us to read in the desired coordinates simplex vertices, if the defaults were inappropriate.

The computation of $S_{ij}^{\alpha\beta}$ in eq. 8 was done as two subsequent matrix multiplications, to avoid the $O(N^4)$ scaling. The sets of $\{k_i\}$ in eq. 9 were calculated only once at the initialization phase of the program using the following code:

```
C *** MAXSTS - largest number of sets array ITABLE can hold
C *** MAXL   - longest set that can be stored in ITABLE
C *** MXL    - largest value of L in current calculations

      DO 20 M1 = 1, MAXSTS
        DO 10 M2= 1, MAXL
           ITABL(M1,M2)=0
 10     CONTINUE
 20   CONTINUE
```

[6]One needs to remember, however, about the Jahn-Teller effect, which for lower multiplicities often produces distorted structures in the ground state.

```
C  initialize for l=1
      ITABL(1,1) = 1
      NLSC = 1
      ISTART(1) = 1
      IEND(1) = 1
      DO 1000 IL = 2, MXL
        ISTART(IL) = NLSC+1
C  There is always a set  0 0 .... 0 1
        NNLSC = 1
        ITABL(NLSC+NNLSC,IL) = 1
C  Now check all previous sets
        DO 700 IL1 = 1, IL/2
          IL2 = IL - IL1
          DO 600 I1 = ISTART(IL1), IEND(IL1)
            DO 500 I2 = ISTART(IL2), IEND(IL2)
              NNLSC = NNLSC + 1
              ICUR = NLSC + NNLSC
              DO 300 K1 = 1, IL
                ITABL(ICUR,K1) = ITABL(I1,K1) + ITABL(I2,K1)
  300         CONTINUE
C         Check if the set was found before
              DO 400 K1 = NLSC+1, ICUR-1
                DO 350 K2 = 1, IL
                  IF(ITABL(K1, K2) .NE. ITABL(ICUR, K2))GOTO 400
  350           CONTINUE
                NNLSC = NNLSC-1
                DO 360 K2 = 1, IL
                  ITABL(ICUR, K2) = 0
  360           CONTINUE
                GOTO 500
  400         CONTINUE
  500       CONTINUE
  600     CONTINUE
  700   CONTINUE
        NLSC = NLSC + NNLSC
        IEND(IL) = NLSC
 1000 CONTINUE
```

The sets were accumulated as rows of table ITABL, and tables ISTART(L) and IEND(L) kept pointers to the first and last row of ITABLE for the given L. Obviously, the above algorithm was not efficient but was quite simple. The efficient algorithm would have to be based on the list of prime dividers of L, but because this step does not consume much time, this simplistic algorithm was quite adequate.

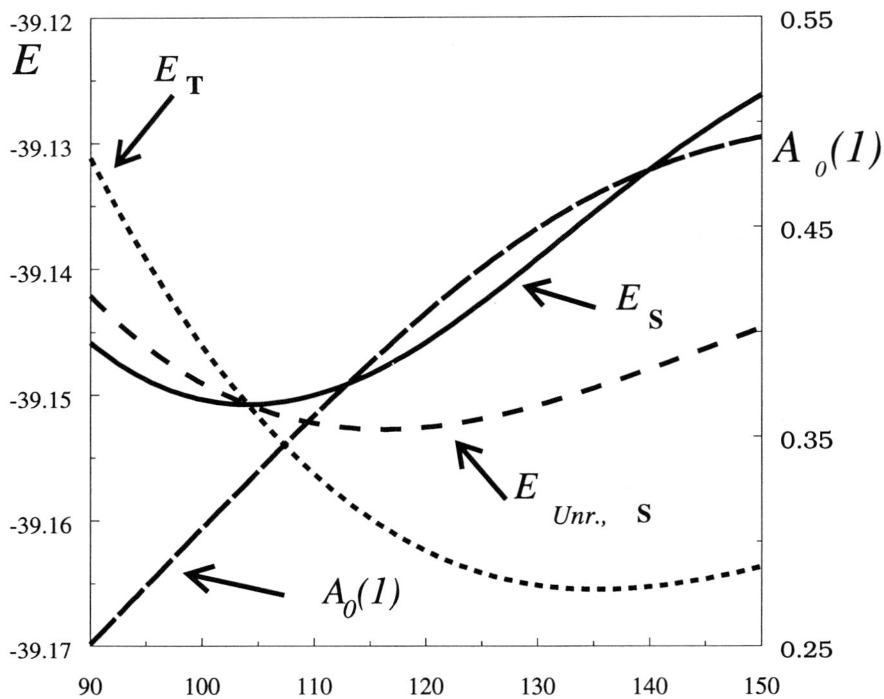

Figure 1. The energy (hartrees) of CH_2 singlet and triplet states versus H–C–H angle: $E_{\mathbf{S}}$ – energy of a pure singlet, $E_{\mathbf{T}}$ – pure triplet, and $E_{Unr.,\mathbf{S}}$ – triplet contaminated singlet for which the spin contamination is $\mathcal{A}_0(1)$.

4. METHYLENE RADICAL CH_2

Methylene radical is often viewed as an elementary building block of hydrocarbon based biradicals and polyradicals. For this reason it is intensively studied, though only higher level correlated *ab initio* methods[11] are able to produce consistent results which agree with experiment[12]. For this molecule, the triplet is a ground state, and it is located 0.391 eV below the singlet[12]. If one tries to use DFT to estimate the singlet-triplet splitting, Δ_{ST}, from restricted singlet calculations and the unrestricted triplet calculations, the calculated gap is much too high (Δ_{ST} = 0.716eV). The restricted singlet calculations produce the following CH_2 geometry: d_{C-H} = 1.125Å, and θ_{H-C-H} = 100.9°, and the total energy is −39.139327 hartree. The unrestricted triplet calculations result in d_{C-H} = 1.087Å, θ_{H-C-H} = 136.1°, the total energy of −39.165647 hartree, and solution is only slightly spin contaminated.

If one starts the unrestricted calculations for a singlet of CH_2 without seeding the SCF, even assuming that the program allows for unrestricted calculations for singlets, the outcome will be a restricted solution, because the KS equivalents of Fock matrices for α and β electrons are identical. To arrive at the unrestricted solution in which α and β electrons are described by different sets of KS orbitals, one needs to perturb the initial potentials seen by the electrons of different spin[13]. One way of doing this, is to start with the artificially perturbed exchange-correlation potential, e.g. by setting the first few coefficients for the exchange-correlation fit to, say, −100 for α electrons and +100 for β electrons. It is the way in which we first obtained the mixed spin solution as a result of the formally singlet calculations.

More elegant method is to seed the unrestricted singlet calculations with the solutions for higher multiplicities for which the exchange-correlation potential naturally differs for α and β electrons. For the current example, we performed single point calculations for CH_2 with d_{C-H} = 1.1Å, and θ_{H-C-H} = 134° for the triplet. We then used the triplet fit coefficients for the exchange-correlation potential and the charge density, to seed the calculations for quintet. Subsequently, we used these quintet results to seed the septet calculations. Seeding the singlet calculations with a triplet or quintet did not produce the mixed spin solution and the SCF converged to the restricted singlet with the total energy, $E_{RKS}(0) = -39.1201422$ hartrees. The triplet energy, corresponding to a practically pure spin state, $<\hat{S}^2>= 2.00543$, was $E(1) = -39.1654372$ hartrees. The energies and expectation values of square of spin for quintet and septet were: $E(2) = -38.7434901$ hartrees, $<\hat{S}^2>_2= 6.00108$, and $E(3) = -38.1168874$ hartrees, $<\hat{S}^2>_3= 12.00001$, respectively. However, using the septet coefficients for the exchange-correlation potential fit resulted in the mixed spin solution for this CH_2 geometry with $E(1) = -39.1497636$ hartree, $<\hat{S}^2>_0= 0.92377$, and the following contaminations with spin states up to $M = 7$: $\mathcal{A}_0(1) = 0.4609592$, $\mathcal{A}_0(2) = 0.0003084$, and $\mathcal{A}_0(3) = 0.0000000$.

Once the mixed spin state for singlet is produced by the appropriate SCF seeding, it is possible to change geometry gradually and use the SCF results from the previous geometry to seed the SCF for the next geometry to study the relationship between different states. We performed such calculations for CH_2 by changing its valence angle and the results are shown in Figure 1. Analysis of the HOMO KS orbitals for α and β electrons, for the mixed spin solution for singlet, reveals that they do not overlap to the great extent

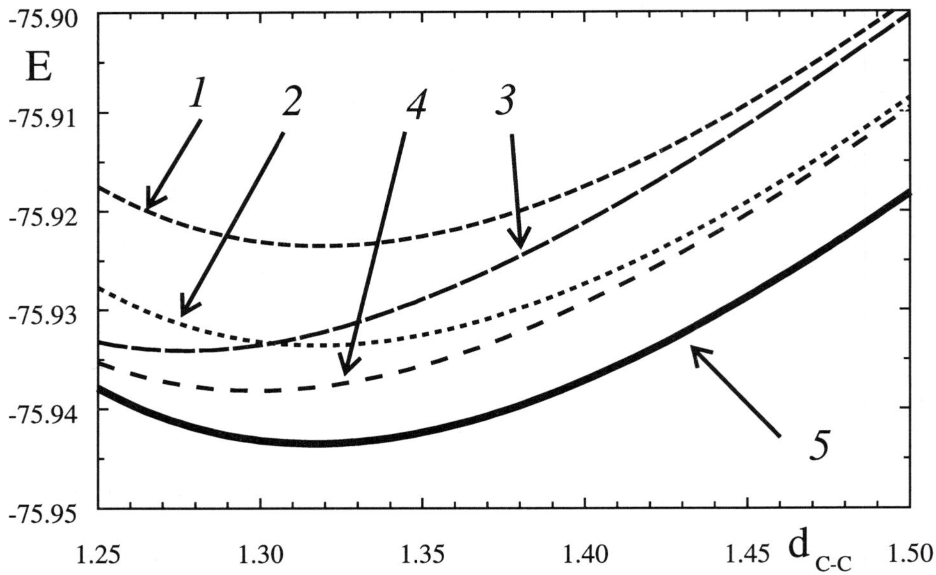

Figure 2. Energies (in hartrees) of two singlet states and a triplet for dicarbon as a function of C–C distance (in Å). **1**) contaminated singlet with higher energy, **2**) pure singlet derived from **1** by subtracting triplet **5**, $d_{opt} = 1.318$Å, **3**) lower energy contaminated singlet, **4**) pure singlet energies for **3**, $d_{opt} = 1.278$Å, **5**) triplet energy.

(for the geometry given above, the $S_{4,4}^{\alpha\beta} = 0.28076$) and the corresponding spin densities for the spin up, and spin down electrons are located below and above the carbon atom, on the opposite sides of the plane containing the molecule, rather than on the hydrogen atoms as sometimes implied.

We also performed the SIMPLEX optimization of geometry for the pure singlet and triplet starting from the above geometry. In this optimization we were correcting triplet for the very small quintet contamination ($\mathcal{A}_1(2) = 0.0014$ for the starting geometry), and singlet was corrected for its substantial triplet, and negligible ($\mathcal{A}_0(2) < 0.0004$) quintet, contaminations. The results were very similar to our previous grid search[1], however, in this case the C–H bond lengths were also allowed to change[7]. The resulting geometries and energies were: $d_{C-H} = 1.120$Å, $\theta_{H-C-H} = 103.4°$, $E_{\mathrm{S}} = -39.1510465$ hartree for the pure singlet, and $d_{C-H} = 1.087$Å, $\theta_{H-C-H} = 136.4°$, $E_{\mathrm{T}} = -39.1662438$ hartree for the pure triplet. The resulting singlet-triplet splitting, $\Delta_{ST} = 0.4135$ eV is in very good agreement with the experiment.

[7]The C–H bond lengths are not, however, particularly well reproduced by the currently used density functionals.

5. SINGLET-TRIPLET SPLITTING IN C_2 MOLECULE

The dicarbon molecule is of great interest to quantum chemists not only because it plays an important role in some chemical reaction mechanisms and is present in the interstellar matter, but also because it is challenging from the theoretical point of view. Herzberg[14] in his classical volume provides a schematic diagram of states for C_2. The KS procedure for this molecule is very sensitive to the starting point of calculations. In *deMon*, the initial SCF run starts with a null exchange-correlation potential, and when the first charge density (calculated from solving KS equations with a null exchange-correlation potential) is available, then the calculated exchange-correlation potential is fitted on a grid to a set of uncontracted gaussian functions centered on atoms. By initializing coefficients of this expansion to desired nonzero values, one may explore different possible solutions of the KS SCF procedure. Obviously, the KS SCF solution depends not only quantitatively, but also qualitatively upon the geometry, and one may obtain different ordering of KS orbitals depending on the geometry.

By using different seeds for the SCF run, we obtained several singlet solutions, two of which are shown in Figure 2. Both of them were about 50% contaminated with the triplet ($< \hat{S}^2 > \approx 1.0$). The calculations were performed with a level shift[15] of 5 eV, and the Pulay acceleration procedure (DIIS) was used to speed up the SCF convergence.

These results are still preliminary, and no attempt was made to assign the symmetry to these solutions. Moreover, there are also restricted solutions for the C_2 molecule. The C_2 example is a clear illustration that, for molecules with a close gap between HOMO and LUMO, one needs to be very careful in selecting the starting SCF conditions. Moreover, for smaller C–C distances (in the vicinity of 1.20Å) the higher energy singlet spontaneously reverts to the lower energy one. Therefore it is important not to start geometry optimization with a large initial SIMPLEX, and allow it to adjusts its size and shape along the directions of the least resistance. Before one starts the optimization, region for the geometry optimization should be at least roughly checked for the possibility that the starting solution may spontaneously revert to another state.

6. OXYGEN MOLECULE O_2

The oxygen molecule is a textbook example of the triplet ground state resulting from the two electrons occupying, in accordance with the Hund's rule, doubly degenerate antibonding π_g^* orbitals.

To explore possible mixed spin solutions of UKS calculations for O_2 molecule we started from the fresh triplet run for $d_{O-O} = 1.2$Å. By seeding the new calculations with the results of previous runs we could identify several solutions which differed in energy and spin contamination. These solutions are collected in Table 1. The first entry in this table corresponds to the UKS calculation which converged to a practically restricted singlet solution. The HOMO orbital energies of this solution are located above LUMO energies due to self-interaction effect which is characteristic to the KS DFT procedure. Another interesting finding is that triplet solutions in this study are identical in spite of the different seeds used at SCF initialization.

Only singlet of O_2 is substantially contaminated with higher spin states. On the other hand, all other solutions, beside high energy singlet, contain the trace of higher spin states.

Table 1
UKS energies (in hartrees, in descending order) for different multiplets of O_2 molecule ($d_{O-O} = 1.2$Å) with corresponding expectation values of the square of spin, and the seeding path which lead to a particular solution.

M	UKS Energy	$<\hat{S}^2>$	last.last-1...next.first
1	−150.333193829	0.00000	1.9.5
1	−150.377757027	1.00164	1.7.3
1	−150.377757092	1.00164	1.7.5
1	−150.377764393	1.00190	1.5
1	−150.377764422	1.00189	1.5.3
1	−150.377764467	1.00189	1.3
1	−150.377764486	1.00190	1.7.1.3
1	−150.377764519	1.00189	1.9.5.1.3
3	−150.392725796	2.00404	3
3	−150.392725797	2.00404	3.1.3
3	−150.392725799	2.00404	3.5.3
3	−150.392725811	2.00404	3.7.3
3	−150.392725815	2.00404	3.5
3	−150.392725815	2.00404	3.7.5
3	−150.392725818	2.00404	3.9.5
3	−150.392725833	2.00405	3.9.5.1.3
3	−150.392725840	2.00405	3.7.1.3
5	−149.845344794	6.00762	5
5	−149.872310382	6.00365	5.9.5
5	−149.872310389	6.00365	5.7.5
5	−149.872312520	6.00366	5.7.1.3
5	−149.872312660	6.00366	5.3.7.1.3
5	−149.872312671	6.00366	5.7.5.3
5	−149.872313559	6.00366	5.3
5	−149.872317492	6.00369	5.9.5.1.3
5	−149.872318636	6.00370	5.1.3
7	−149.166819579	12.00531	7
7	−149.166821004	12.00532	7.1.3
7	−149.171980218	12.00462	7.5
7	−149.190701969	12.00221	7.3
7	−149.190701985	12.00221	7.5.3
9	−148.442264713	20.00279	9
9	−148.442264718	20.00279	9.1.3
9	−148.442264721	20.00279	9.3.1.3
9	−148.442264733	20.00279	9.7.3
9	−148.442264780	20.00279	9.5.1.3
9	−148.443561213	20.00182	9.7.1.3
9	−148.443564196	20.00182	9.7.5
9	−148.443564201	20.00182	9.5
11	−147.446951726	30.00144	11

Table 2
Lowest energy UKS solutions for the range of multiplicities, and their spin contaminations for the O_2 molecule.

$S_{max}=5$	$\mathcal{A}_S(0)$	$\mathcal{A}_S(1)$	$\mathcal{A}_S(2)$	$\mathcal{A}_S(3)$	$\mathcal{A}_S(4)$	$\mathcal{A}_S(5)$
11						1.0
9.5					.999817931	.000182069
7.5.3				.999723344	.000276627	.000000029
5.1.3			.999382777	.000617078	.000000145	.000000000
3.7.1.3		.998989338	.001010274	.000000387	.000000000	.000000000
1.9.5.1.3	.499684794	.499999929	.000315212	.000000066	.000000000	.000000000

$S_{max}=4$	$\mathcal{A}_S(0)$	$\mathcal{A}_S(1)$	$\mathcal{A}_S(2)$	$\mathcal{A}_S(3)$	$\mathcal{A}_S(4)$
9.5					1.0
7.5.3				.999723307	.000276693
5.1.3			.999382777	.000617078	.000000145
3.7.1.3		.998989338	.001010274	.000000387	.000000000
1.9.5.1.3	.499684794	.499999929	.000315212	.000000066	.000000000

$S_{max}=3$	$\mathcal{A}_S(0)$	$\mathcal{A}_S(1)$	$\mathcal{A}_S(2)$	$\mathcal{A}_S(3)$
7.5.3				1.0
5.1.3			.999382584	.000617416
3.7.1.3		.998989338	.001010274	.000000388
1.9.5.1.3	.499684794	.499999929	.000315211	.000000066

$S_{max}=2$	$\mathcal{A}_S(0)$	$\mathcal{A}_S(1)$	$\mathcal{A}_S(2)$
5.1.3			1.0
3.7.1.3		.998988757	.001011243
1.9.5.1.3	.499685122	.499999338	.000315540

$S_{max}=1$	$\mathcal{A}_S(0)$	$\mathcal{A}_S(1)$
3.7.1.3		1.0
1.9.5.1.3	.499054043	.500945957

Table 3
Pure spin state energies (in hartrees) calculated for O_2 molecule for a range of contaminating spin states.

S	E_{uncor}	$S_{max}=5$	$S_{max}=4$	$S_{max}=3$	$S_{max}=2$	$S_{max}=1$
$E(5)$	−147.446951	−147.446951				
$E(4)$	−148.443564	−148.443745	−148.443564			
$E(3)$	−149.190701	−149.190908	−149.190908	−149.190701		
$E(2)$	−149.872318	−149.872739	−149.872739	−149.872739	−149.872318	
$E(1)$	−150.392725	−150.393252	−150.393252	−150.393252	−150.393252	−150.392725
$E(0)$	−150.377764	−150.362585	−150.362585	−150.362585	−150.362585	−150.362746

To check how these minute contaminations affect the calculated pure spin energies, we performed the calculations, using eq 16, of all pure spin state energies up to multiplicity $M = 11$. To calculate pure spin state energies, the lowest energy solutions for each multiplicity were taken. Their spin contaminations and seeding history are shown in Table 2 for multiplicities between 1 and 11. Within each range of contaminating spin states, the solution corresponding to the highest multiplicity is assumed pure ($\mathcal{A}_{S_{max}}(S_{max}) = 1$).

The resulting pure spin energies are presented in Table 3. As can be seen from this table, the energies in this case do not change significantly with the number of states included. In the case of a small spin contamination of the highest spin state of interest it is quite adequate to account only for one higher multiplicity above it. While the geometry of O_2 in this case is not optimal (the bond length is too short), nevertheless, the calculated $\Delta_{ST} = 0.835$eV compares favorably with the experimental value of 0.97eV for singlet-triplet splitting.

We also performed geometry optimization of the O_2 molecule using quintet as the highest contaminating state. The lowest energy of pure singlet of O_2 was -150.364572 hartrees and occured at $d_{O-O} = 1.2405$Å. The pure triplet energy was -150.395024 hartrees and corresponded to the bond length, $d_{O-O} = 1.2368$Å. The calculated singlet triplet splitting is therefore $\Delta_{ST} = 0.828641$ eV, i.e., in good agreement with experiment.

Some preliminary calculations were also performed for the oxygen molecule anion, O_2^-. We tried several different seeding for possible multiplets, including results for neutral O_2 molecule, but we only got single solution for each multiplicity studied, namely, doublet, quartet, and sextet. These solutions were only slightly contaminated. Geometry optimization for doublet state of O_2^- resulted in $d_{O-O} = 1.3842$Å and the total pure spin state energy for doublet was -150.407497 hartree. The quartet geometry was $d_{O-O} = 1.8484$Å and corresponded to the pure spin state energy of -150.328405 hartree. The experimental vibrational frequencies for O_2 and O_2^- are 1580 cm^{-1}[16,17] and 1089 cm^{-1}[18], respectively; which corresponds to the zero point energy change of $\Delta ZPE = 0.702$ kcal/mol for the reaction $O_2 + e^- \rightarrow O_2^-$. The electronic energy difference between ground states of oxygen and its anion is 7.826 kcal/mol and hence, the calculated electron affinity of oxygen molecule is 8.528 kcal/mol which compares favorably with the experimental value of about 10.1 kcal/mol[19].

This agreement with experimental results provides quite an encouraging confirmation that the method of removing spin contamination presented in this chapter has a potential of providing reliable results in the the area which is notoriously difficult to study by other available methods. Moreover, the approach is straightforward computationally, and opens new possibilities for further analysis of the energetics and magnetic properties of interesting and important systems.

7. OXYGEN DIMER, $(O_2)_2$

Oxygen dimer receives much attention in studies of van der Waals potential[20] between oxygen molecules as well as possible paths to formation and dissociation of O_4 molecule, a potential high energy density material[21–23]. It is also an intriguing species to study the spin alignment phenomenon. To find the lowest pure spin states for multiplicities $M =$

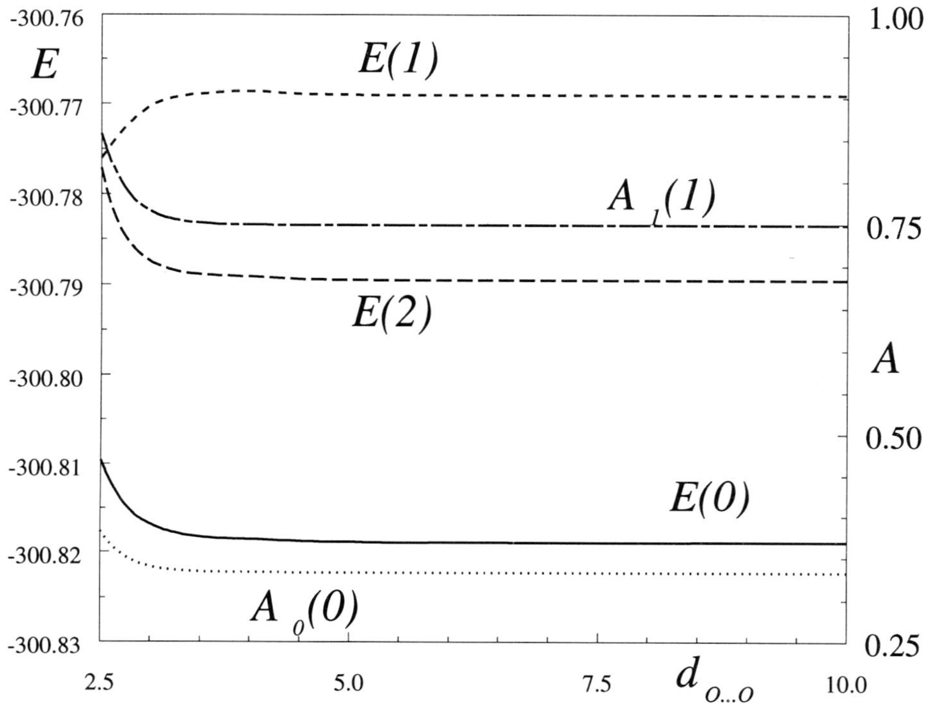

Figure 3. Energies (in hartrees) for the three lowest multiplicities of oxygen dimer (O–O bond length kept at 1.234Å) as a function of distance between oxygen molecules in a rectangular arrangement. The parent spin contents, $\mathcal{A}_S(S)$ are also shown for singlet and triplet.

$1, 3, \ldots 11$ we started with a symmetric, parallel dimer with equal bond lengths, $d_{O-O} = 1.2$Å and the distance between oxygen molecules of $d_{O_2 \cdots O_2} = 3.2$Å. After checking results of possible seedings, we selected the lowest energy solutions for each of the multiplets studied: 1.5.3.1.3.5, 3.9.1.3.5, 5.3.1.3.5, 7.1.5, 9.7.1.5, and 11.9.7.1.5. These served as seeds for future calculations for other geometries. The spin contamination was calculated for states up to $M = 15$, however, after examining calculated contaminations it was clear that only multiplicities up to $M = 9$ are needed for deriving the consistent pure spin state energies for the multiplets of interest, i.e, 1, 3, and 5.

As the next step, we used the lowest energy multiplets to seed the first point ($d_{O_2 \cdots O_2} = 3.2$Å) in a series of calculations in which we used realistic bond length $d_{O-O} = 1.234$Å and varied the distance, $d_{O_2 \cdots O_2}$, between oxygen molecules, in the rectangular arrangement between 1.2 and 5.0Å, by starting from the initial 3.2Å distance and going outwards in small steps. The results of these calculations are shown in Figure 3.

There are no evident van der Waals minima in Figure 3 and we did not perform calculations on the grid dense enough to consider energy differences of the order of 0.1 kcal/mol, however, shallow minima may be present for triplet and quintet at the distance of about 5Å.

These results allow us, however, to study magnetic interactions of essentially isolated species using the DFT energies corrected for spin contamination. The oxygen molecule has a ground state with a spin of 1. One form of crystal oxygen, found at low temperatures (β phase[24]), consists of oxygen molecules located at a distance of 3.272Å with O–O bonds perpendicular to the plane which passes through the midpoints of O–O bonds. The magnetic properties of the crystal can be better understood by studying interactions between the pair of oxygen molecules separated by the distance characteristic for this phase.

Table 4 shows the original spin contaminated energies for the neutral oxygen dimer, $(O_2 \cdots O_2)$, and its anion, $(O_2 \cdots O_2)^-$, as well as, spin contaminations, and the pure spin state energies derived from these data by using eq 16. The singlet is a ground state of the neutral dimer which agrees with the experimental finding that at very low-temperatures, the β phase oxygen crystal exhibits antiferromagnetic ordering of spins. The results in Table 4 can be used to construct model spin hamiltonian which will produce the same energy levels as those given in the table. Since the spin of a single molecule of O_2 in its ground state is 1, it is possible to use spin-1 operators to construct the model hamiltonian. The most general form of the rotationally invariant expression, describing interaction of two spin centers is given by the following formula:

$$H = A(S_1, S_2) + B(S_1, S_2)^2 + C \qquad (19)$$

where A, B, and C are energy parameters which depend strongly upon the distance between spin centers and their relative geometry, and (S_1, S_2) is a scalar product of spins on the centers.

There are 3 possible multiplets (the higher multiplets are so high in energy that they cannot be observed) for this system, namely singlet, triplet, and quintet. Their respective energies are: E_0, E_1, and E_2. Their energies are can be therefore expressed as:

$$E_0 = -2A + 4B + C \qquad (20)$$

Table 4
Pure spin state energies (hartrees) for the neutral and negatively charged (-1) dimer of oxygen molecules (d_{O-O} = 1.234Å) located at a distance of $d_{O_2 \cdots O_2}$ = 3.272Å. The molecules are parallel to each other and coincide with the shorter sides of the rectangle.

S	E_{corr}	E_{uncorr}	Spin contaminations	
		Results for $(O_2 \cdots O_2)$		
4	-299.8373209	-299.8373209		
3	-300.3135485	-300.3130265	$\mathcal{A}_3(3) = 0.99890395482$	$\mathcal{A}_3(4) = 0.00109604518$
2	-300.7885779	-300.7878715	$\mathcal{A}_2(2) = 0.99851415148$	$\mathcal{A}_2(3) = 0.00148464191$
			$\mathcal{A}_2(4) = 0.00000120662$	
1	-300.7692155	-300.7737624	$\mathcal{A}_1(1) = 0.75486452430$	$\mathcal{A}_1(2) = 0.24471572805$
			$\mathcal{A}_1(3) = 0.00041944832$	$\mathcal{A}_2(4) = 0.00000029933$
0	-300.8179242	-300.7885333	$\mathcal{A}_0(0) = 0.33521678459$	$\mathcal{A}_0(1) = 0.49955693784$
			$\mathcal{A}_0(2) = 0.16478597175$	$\mathcal{A}_0(3) = 0.00043982985$
			$\mathcal{A}_0(4) = 0.00000047597$	
		Results for anion $(O_2 \cdots O_2)^{-1}$		
$\frac{9}{2}$	-299.5694003	-299.5694003		
$\frac{7}{2}$	-300.1477423	-300.1469779	$\mathcal{A}_{\frac{7}{2}}(\frac{7}{2}) = 0.99867829250$	$\mathcal{A}_{\frac{7}{2}}(\frac{9}{2}) = 0.00132170750$
$\frac{5}{2}$	-300.6110692	-300.6104352	$\mathcal{A}_{\frac{5}{2}}(\frac{5}{2}) = 0.99863297311$	$\mathcal{A}_{\frac{5}{2}}(\frac{7}{2}) = 0.00136598603$
			$\mathcal{A}_{\frac{5}{2}}(\frac{0}{2}) = 0.00000104086$	
$\frac{3}{2}$	-300.8397967	-300.8394783	$\mathcal{A}_{\frac{3}{2}}(\frac{3}{2}) = 0.99860964690$	$\mathcal{A}_{\frac{3}{2}}(\frac{5}{2}) = 0.00138937665$
			$\mathcal{A}_{\frac{3}{2}}(\frac{7}{2}) = 0.00000097605$	$\mathcal{A}_{\frac{3}{2}}(\frac{9}{2}) = 0.00000000040$
$\frac{1}{2}$	-300.8257500	-300.8301413	$\mathcal{A}_{\frac{1}{2}}(\frac{1}{2}) = 0.67991598398$	$\mathcal{A}_{\frac{1}{2}}(\frac{3}{2}) = 0.31962608582$
			$\mathcal{A}_{\frac{1}{2}}(\frac{5}{2}) = 0.00045767332$	$\mathcal{A}_{\frac{1}{2}}(\frac{7}{2}) = 0.00000025680$
			$\mathcal{A}_{\frac{1}{2}}(\frac{9}{2}) = 0.00000000008$	

$$E_1 = -A + B + C \tag{21}$$
$$E_2 = A + B + C \tag{22}$$

Using pure state energies from Table 4 one arrives at the following values of these constants:

$$A = -0.0096811832 \text{ hartree}$$
$$B = -0.0194632942 \text{ hartree}$$
$$C = -300.75943344 \text{ hartree}$$

The parameters just derived can be used to construct the spin hamiltonian for the oxygen crystal.

We will also show how this type of calculation can be used for the quantitative study of the kinematic mechanism of ferromagnetic ordering in the paramagnetic medium. There are many different mechanisms of the ferromagnetic spin alignment which occur in solids and molecular systems. Most popular is the conventional Heisenberg exchange interaction for the case of negative exchange integral. However, there are many cases when the direct exchange interaction between two paramagnetic species has the antiferromagnetic character. In our case it is clearly illustrated by the results presented in Table 4. It has been shown recently[25] that an electron or the hole which propagates through paramagnetic medium can cause the spin alignment in the medium. It is highly probable that such a phenomenon will be experimentally confirmed in the low temperature oxygen crystals.

One can use the simple model of oxygen dimer to show that in general such a phenomenon can occur. In this case, we have two identical paramagnetic centers A, with a nonzero spin, $S = 1$, in the ground state, and they are located at a relatively large distance from each other. For the neutral oxygen dimer the interaction between paramagnetic centers is antiferromagnetic and results in a zero total spin. The situation should change dramatically if one adds a negative charge to this system. In this case, the energy of the $(A \cdots A)^-$ system as a function of total spin, S_0, is given by the following expression:

$$E(S_0) = \pm t \frac{S_0 + \frac{1}{2}}{S_{max} + \frac{1}{2}} \tag{23}$$

where t is the hopping integral which depends exponentially on the distance R between the two centers, and S_{max} is the highest possible spin of the combined system. The $S_{max} = 2S + \frac{1}{2}$ if the ground state of A^- has spin $S + \frac{1}{2}$, and $S_{max} = 2S - \frac{1}{2}$ if the ground state of A^- has spin $S - \frac{1}{2}$. The latter case corresponds to our current example, namely, the spin of O_2^- in the ground state which equal to $\frac{1}{2}$.

For the ground state of the combined system, one has to choose the minus sign in eq 4 because in our case $t > 0$. For the energy $E(S_0)$ to be the lowest possible (it is a negative number), it needs to have the largest possible value, i.e., the S_0 has to be equal to S_{max} which corresponds to the complete spin alignment of the anion of oxygen dimer, $(A \cdots A)^-$ The physical picture behind this alignment mechanism is that it helps electron "hopping".

It is clear from eq 23 that full alignment allows hopping energy to achieve its initial value $-t$. This mechanism of spin alignment (which we call kinematic, as it is connected only with the motion of electron in the surrounding medium, and depends solely on the hoping/resonance integral) is valid regardless of the source of extra electrons and the nature of particular paramagnetic center. However, in reality, there are many other factors which can affect the magnetic interactions and electronic structure, and need to be taken into account. The *ab initio* and DFT calculations can be used to provide the understanding of essential factors involved in these interactions. In our case, Table 4 provides energies for multiplets of the dimer anion together with their spin contaminations. The results of DFT calculations agree with this simple theory, and indeed, the spin of the oxygen dimer anion in its ground state is $\frac{3}{2}$. These calculations provide us not only with a relative ordering of multiplets, but also the energetic scale of the interactions. Moreover, the parameters derived from these model systems can be used to predict the spin in more complicated system, e.g., large clusters or oxygen solid phases doped with electron donor.

8. MULTIPLET ENERGIES IN $(O_2)_6$ CLUSTER

Oxygen clusters have a potential of possessing a ground state with a high spin. In this case we performed some pilot calculations on a cluster of six oxygen molecules that were placed at the corners of a regular hexagon whose geometry corresponds to the single plane of β form of solid oxygen[24] with a central O_2 removed. In this structure, the side of the hexagon is equal to 3.272Å and we used 1.2Å for the distance between oxygen atoms in the O_2 molecule. The O_2 molecules are perpendicular to the plane of the hexagon.

The highest practical multiplicity for such a system is $M = 13$ and corresponds to a ferromagnetic arrangement of all oxygen molecules in their ground, triplet state. Only the highest multiplicity solution, $M = 13$, converges easily at the SCF level. The SCF oscillations are characteristic of all lower multiplicities due to the close proximity of HOMO and LUMO orbitals. To reach convergence we seeded the calculations for lower multiplicities with already converged results obtained for other multiplicities. Table 5 lists the energies of various solutions obtained for this cluster together with the spin contaminations coming from higher spin states. The spin contaminations and UKS energies were calculated for the range of $M = 1, 3, \ldots 17$.

For each solution the seeding method is given. For example, 5.11.7.13 means that the quintet ($M = 5$) was seeded with the results of 11.7.13 calculations. This particular 11.7.13 solution was obtained by seeding the $M = 11$ calculations with the results of the 7.13 run. The 7.13 corresponds to a septet ($M = 7$) calculations seeded with the results of initial, unseeded run for $M = 13$. It should be stressed that not all seedings lead to SCF convergence, and failures are not listed in Table 5. In these calculations we used 0.0001 for density convergence, 0.00001 a.u. for energy convergence criterion, the level shift option with 5 eV gap, Pulay acceleration method, and $0.01 \leq DMIX \leq 0.05$[8]. for most calculations

As can be seen from Table 5, there are several possible solutions for many of the multiplicities studied. There may be more of them, though we are quite confident that we were

[8] $DMIX$ gives a proportion of the new density and exchange-correlation potential, which is mixed with values from the previous iteration of SCF procedure.

Table 5

UKS solutions for different multiplicities of $(O_2)_6$ cluster, expectation values of $<\hat{S}^2>$ and spin contaminations from multiplicities up to $M = 17$.

1.3.5.7.9.13	01=0.144166435	03=0.323324383	05=0.297487994
$E(01) = -902.323736120$	07=0.165103674	09=0.057285286	11=0.011565902
$<\hat{S}^2>= 5.95037$	13=0.001060798	15=0.000005514	17=0.000000014
1.5.13	01=0.145255730	03=0.324936161	05=0.297452384
$E(01) = -902.325188380$	07=0.163835143	09=0.056277682	11=0.011222911
$<\hat{S}^2>= 5.90576$	13=0.001014696	15=0.000005280	17=0.000000013
1.5.7.9.13	01=0.145431854	03=0.325195499	05=0.297444602
$E(01) = -902.325404786$	07=0.163630163	09=0.056116573	11=0.011168601
$<\hat{S}^2>= 5.89862$	13=0.001007458	15=0.000005238	17=0.000000013
1.7.9.13	01=0.145431342	03=0.325194515	05=0.297444195
$E(01) = -902.325421857$	07=0.163630654	09=0.056117357	11=0.011169072
$<\hat{S}^2>= 5.89866$	13=0.001007597	15=0.000005254	17=0.000000013
1.9.13	01=0.145432933	03=0.325196849	05=0.297444108
$E(01) = -902.325423578$	07=0.163628798	09=0.056115913	11=0.011168594
$<\hat{S}^2>= 5.89859$	13=0.001007537	15=0.000005255	17=0.000000013
1.7.13	01=0.143994270	03=0.323054335	05=0.297464823
$E(01) = -902.353814987$	07=0.165296073	09=0.057465222	11=0.011641772
$<\hat{S}^2>= 5.95861$	13=0.001076662	15=0.000006823	17=0.000000022
1.9.11.7.9.13	01=0.144772519	03=0.324208078	05=0.297442932
$E(01) = -902.355171869$	07=0.164389567	09=0.056742173	11=0.011394703
$<\hat{S}^2>= 5.92663$	13=0.001043343	15=0.000006664	17=0.000000021
1.3.11.7.13	01=0.144771054	03=0.324205904	05=0.297442968
$E(01) = -902.355171977$	07=0.164391270	09=0.056743537	11=0.011395173
$<\hat{S}^2>= 5.92669$	13=0.001043408	15=0.000006664	17=0.000000021
1.11.7.13	01=0.144771038	03=0.324205882	05=0.297442970
$E(01) = -902.355172019$	07=0.164391288	09=0.056743550	11=0.011395177
$<\hat{S}^2>= 5.92669$	13=0.001043409	15=0.000006664	17=0.000000021
3.1.11.7.13	01=0.000000000	03=0.378312708	05=0.347678485
$E(03) = -902.339305520$	07=0.192635213	09=0.066697991	11=0.013437109
$<\hat{S}^2>= 6.94351$	13=0.001231496	15=0.000006979	17=0.000000019
3.7.13	01=0.000000000	03=0.378789140	05=0.347760025
$E(03) = -902.339436729$	07=0.192386193	09=0.066477370	11=0.013359429
$<\hat{S}^2>= 6.93477$	13=0.001220894	15=0.000006931	17=0.000000019
3.1.7.13	01=0.000000000	03=0.379104435	05=0.347815872
$E(03) = -902.339582737$	07=0.192222630	09=0.066330458	11=0.013306581
$<\hat{S}^2>= 6.92892$	13=0.001213216	15=0.000006789	17=0.000000019
3.5.7.9.13	01=0.000000000	03=0.379110708	05=0.347815242
$E(03) = -902.339595527$	07=0.192218324	09=0.066328423	11=0.013306800
$<\hat{S}^2>= 6.92886$	13=0.001213612	15=0.000006872	17=0.000000019

3.7.5.7.9.13	01=0.000000000	03=0.379097777	05=0.347812815
$E(03) = -902.339612198$	07=0.192224966	09=0.066334527	11=0.013309055
$<\hat{S}^2>= 6.92911$	13=0.001213958	15=0.000006882	17=0.000000019
3.9.13	01=0.000000000	03=0.379102976	05=0.347813710
$E(03) = -902.339612348$	07=0.192222252	09=0.066332117	11=0.013308204
$<\hat{S}^2>= 6.92901$	13=0.001213841	15=0.000006881	17=0.000000019
3.1.5.7.9.13	01=0.000000000	03=0.379194836	05=0.347830242
$E(03) = -902.339626288$	07=0.192174749	09=0.066289177	11=0.013292621
$<\hat{S}^2>= 6.92730$	13=0.001211527	15=0.000006828	17=0.000000019
3.11.7.13	01=0.000000000	03=0.379581525	05=0.347898343
$E(03) = -902.340063502$	07=0.191973056	09=0.066108721	11=0.013228675
$<\hat{S}^2>= 6.92017$	13=0.001202837	15=0.000006825	17=0.000000019
3.9.11.7.9.13	01=0.000000000	03=0.379588178	05=0.347899522
$E(03) = -902.340063860$	07=0.191969603	09=0.066105619	11=0.013227561
$<\hat{S}^2>= 6.92004$	13=0.001202676	15=0.000006822	17=0.000000019
5.13	01=0.000000000	03=0.000000000	05=0.563460760
$E(05) = -902.294876520$	07=0.308693752	09=0.105235889	11=0.020760898
$<\hat{S}^2>= 9.89037$	13=0.001842012	15=0.000006676	17=0.000000012
5.1.11.7.13	01=0.000000000	03=0.000000000	05=0.558865218
$E(05) = -902.323529989$	07=0.310004856	09=0.107472421	11=0.021667917
$<\hat{S}^2>= 9.95643$	13=0.001980264	15=0.000009303	17=0.000000022
5.7.13	01=0.000000000	03=0.000000000	05=0.560658229
$E(05) = -902.324248433$	07=0.309484003	09=0.106597839	11=0.021319937
$<\hat{S}^2>= 9.93092$	13=0.001930739	15=0.000009231	17=0.000000022
5.9.11.7.9.13	01=0.000000000	03=0.000000000	05=0.562089050
$E(05) = -902.324936422$	07=0.309069387	09=0.105901060	11=0.021041478
$<\hat{S}^2>= 9.91052$	13=0.001890128	15=0.000008876	17=0.000000021
5.11.7.13	01=0.000000000	03=0.000000000	05=0.562087723
$E(05) = -902.324936455$	07=0.309070407	09=0.105901713	11=0.021041352
$<\hat{S}^2>= 9.91052$	13=0.001889947	15=0.000008837	17=0.000000020
5.3.11.7.13	01=0.000000000	03=0.000000000	05=0.558784494
$E(05) = -902.353809077$	07=0.309991236	09=0.107511955	11=0.021705921
$<\hat{S}^2>= 9.95845$	13=0.001994878	15=0.000011483	17=0.000000033
5.7.9.13	01=0.000000000	03=0.000000000	05=0.558748198
$E(05) = -902.353821377$	07=0.309996801	09=0.107530055	11=0.021715915
$<\hat{S}^2>= 9.95907$	13=0.001997298	15=0.000011699	17=0.000000034
5.1.5.7.9.13	01=0.000000000	03=0.000000000	05=0.558811401
$E(05) = -902.353823034$	07=0.309982691	09=0.107499149	11=0.021701096
$<\hat{S}^2>= 9.95808$	13=0.001994171	15=0.000011458	17=0.000000033
5.1.7.13	01=0.000000000	03=0.000000000	05=0.558823125
$E(05) = -902.353824799$	07=0.309978332	09=0.107493448	11=0.021699394
$<\hat{S}^2>= 9.95793$	13=0.001994157	15=0.000011511	17=0.000000033

5.9.13	01=0.000000000	03=0.000000000	05=0.558850071
$E(05) = -902.353824968$	07=0.309971595	09=0.107480309	11=0.021693506
$<\hat{S}^2>= 9.95752$	13=0.001993042	15=0.000011444	17=0.000000033
7.13	01=0.000000000	03=0.000000000	05=0.000000000
$E(07) = -902.309127287$	07=0.706082603	09=0.241692553	11=0.047927929
$<\hat{S}^2>= 14.92538$	13=0.004280699	15=0.000016186	17=0.000000031
7.5.7.9.13	01=0.000000000	03=0.000000000	05=0.000000000
$E(07) = -902.338617991$	07=0.703774112	09=0.243018455	11=0.048755528
$<\hat{S}^2>= 14.95560$	13=0.004430817	15=0.000021037	17=0.000000051
7.1.5.7.9.13	01=0.000000000	03=0.000000000	05=0.000000000
$E(07) = -902.338694846$	07=0.703940572	09=0.242920291	11=0.048696787
$<\hat{S}^2>= 14.95347$	13=0.004421299	15=0.000021000	17=0.000000051
7.9.13	01=0.000000000	03=0.000000000	05=0.000000000
$E(07) = -902.338703003$	07=0.703965749	09=0.242905613	11=0.048687830
$<\hat{S}^2>= 14.95315$	13=0.004419781	15=0.000020977	17=0.000000051
9.5.13	01=0.000000000	03=0.000000000	05=0.000000000
$E(09) = -902.323490344$	07=0.000000000	09=0.822519929	11=0.162913592
$<\hat{S}^2>= 21.95032$	13=0.014514913	15=0.000051469	17=0.000000097
9.1.5.7.9.13	01=0.000000000	03=0.000000000	05=0.000000000
$E(09) = -902.323546272$	07=0.000000000	09=0.822674664	11=0.162783877
$<\hat{S}^2>= 21.94847$	13=0.014490039	15=0.000051323	17=0.000000097
9.13	01=0.000000000	03=0.000000000	05=0.000000000
$E(09) = -902.323594561$	07=0.000000000	09=0.822836745	11=0.162651884
$<\hat{S}^2>= 21.94649$	13=0.014460293	15=0.000050982	17=0.000000096
9.1.11.7.13	01=0.000000000	03=0.000000000	05=0.000000000
$E(09) = -902.352451438$	07=0.000000000	09=0.819235297	11=0.165552169
$<\hat{S}^2>= 21.99111$	13=0.015147203	15=0.000065180	17=0.000000151
9.3.11.7.13	01=0.000000000	03=0.000000000	05=0.000000000
$E(09) = -902.352451498$	07=0.000000000	09=0.819235791	11=0.165551772
$<\hat{S}^2>= 21.99111$	13=0.015147108	15=0.000065178	17=0.000000151
9.11.7.9.13	01=0.000000000	03=0.000000000	05=0.000000000
$E(09) = -902.352459248$	07=0.000000000	09=0.819200143	11=0.165580188
$<\hat{S}^2>= 21.99155$	13=0.015154158	15=0.000065360	17=0.000000151
9.1.7.13	01=0.000000000	03=0.000000000	05=0.000000000
$E(09) = -902.352459320$	07=0.000000000	09=0.819201467	11=0.165579117
$<\hat{S}^2>= 21.99154$	13=0.015153908	15=0.000065358	17=0.000000151
11.1.7.13	01=0.000000000	03=0.000000000	05=0.000000000
$E(11) = -902.337185675$	07=0.000000000	09=0.000000000	11=0.917794420
$<\hat{S}^2>= 30.98974$	13=0.081972256	15=0.000232933	17=0.000000392
11.7.9.13	01=0.000000000	03=0.000000000	05=0.000000000
$E(11) = -902.337245801$	07=0.000000000	09=0.000000000	11=0.917914443
$<\hat{S}^2>= 30.98832$	13=0.081850805	15=0.000234355	17=0.000000397

11.3.5.7.9.13	01=0.000000000	03=0.000000000	05=0.000000000
$E(11) = -902.337323070$	07=0.000000000	09=0.000000000	11=0.918090376
$<\hat{S}^2>= 30.98621$	13=0.081675051	15=0.000234175	17=0.000000398
11.1.5.7.9.13	01=0.000000000	03=0.000000000	05=0.000000000
$E(11) = -902.337330248$	07=0.000000000	09=0.000000000	11=0.918109855
$<\hat{S}^2>= 30.98597$	13=0.081655712	15=0.000234036	17=0.000000397
11.13	01=0.000000000	03=0.000000000	05=0.000000000
$E(11) = -902.337353066$	07=0.000000000	09=0.000000000	11=0.918200707
$<\hat{S}^2>= 30.98486$	13=0.081566188	15=0.000232712	17=0.000000393
11.7.13	01=0.000000000	03=0.000000000	05=0.000000000
$E(11) = -902.337354538$	07=0.000000000	09=0.000000000	11=0.918214465
$<\hat{S}^2>= 30.98469$	13=0.081552615	15=0.000232527	17=0.000000392
11.3.11.7.13	01=0.000000000	03=0.000000000	05=0.000000000
$E(11) = -902.337354693$	07=0.000000000	09=0.000000000	11=0.918207190
$<\hat{S}^2>= 30.98478$	13=0.081559691	15=0.000232726	17=0.000000393
13	01=0.000000000	03=0.000000000	05=0.000000000
$E(13) = -902.351095867$	07=0.000000000	09=0.000000000	11=0.000000000
$<\hat{S}^2>= 42.02409$	13=0.998281841	15=0.001715859	17=0.000002299
13.5.3.11.7.13	01=0.000000000	03=0.000000000	05=0.000000000
$E(13) = -902.351096015$	07=0.000000000	09=0.000000000	11=0.000000000
$<\hat{S}^2>= 42.02409$	13=0.998281881	15=0.001715819	17=0.000002299
13.9.11.7.9.13	01=0.000000000	03=0.000000000	05=0.000000000
$E(13) = -902.351096086$	07=0.000000000	09=0.000000000	11=0.000000000
$<\hat{S}^2>= 42.02409$	13=0.998281863	15=0.001715837	17=0.000002299
13.9.13	01=0.000000000	03=0.000000000	05=0.000000000
$E(13) = -902.351096233$	07=0.000000000	09=0.000000000	11=0.000000000
$<\hat{S}^2>= 42.02409$	13=0.998281918	15=0.001715782	17=0.000002299
13.11.7.9.13	01=0.000000000	03=0.000000000	05=0.000000000
$E(13) = -902.351096279$	07=0.000000000	09=0.000000000	11=0.000000000
$<\hat{S}^2>= 42.02409$	13=0.998281888	15=0.001715813	17=0.000002299
13.7.9.13	01=0.000000000	03=0.000000000	05=0.000000000
$E(13) = -902.351096428$	07=0.000000000	09=0.000000000	11=0.000000000
$<\hat{S}^2>= 42.02409$	13=0.998281859	15=0.001715842	17=0.000002299
13.7.13	01=0.000000000	03=0.000000000	05=0.000000000
$E(13) = -902.351096643$	07=0.000000000	09=0.000000000	11=0.000000000
$<\hat{S}^2>= 42.02409$	13=0.998281884	15=0.001715817	17=0.000002299
13.5.13	01=0.000000000	03=0.000000000	05=0.000000000
$E(13) = -902.351096653$	07=0.000000000	09=0.000000000	11=0.000000000
$<\hat{S}^2>= 42.02409$	13=0.998281892	15=0.001715809	17=0.000002299
15	01=0.000000000	03=0.000000000	05=0.000000000
$E(15) = -901.834836033$	07=0.000000000	09=0.000000000	11=0.000000000
$<\hat{S}^2>= 56.02583$	13=0.000000000	15=0.998385354	17=0.001614646
17	01=0.000000000	03=0.000000000	05=0.000000000
$E(17) = -901.318015060$	07=0.000000000	09=0.000000000	11=0.000000000
$<\hat{S}^2>= 72.02791$	13=0.000000000	15=0.000000000	17=1.000000000

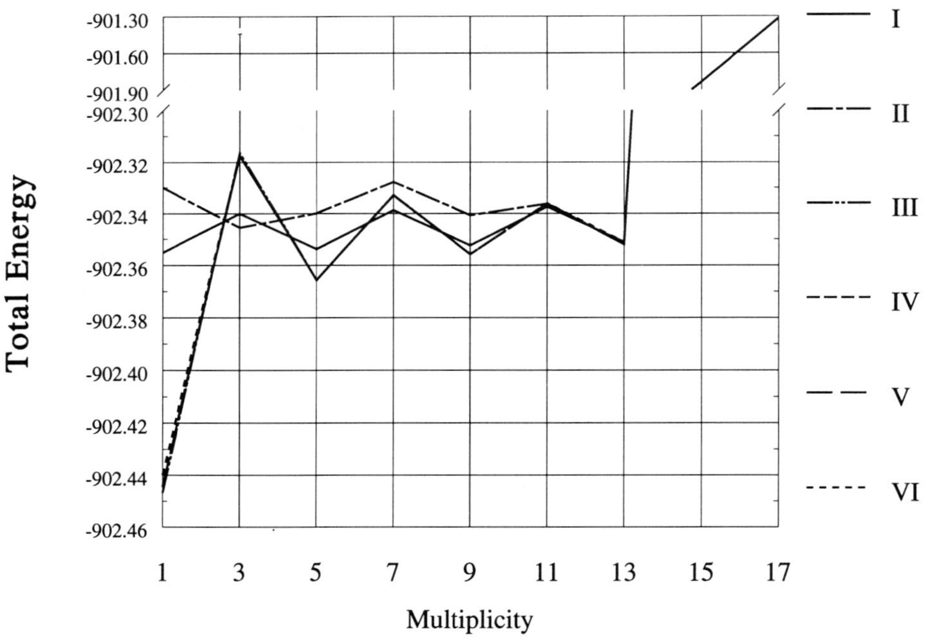

Figure 4. Energies for $(O_2)_6$ calculated in a number of ways: **I**) lowest contaminated energies for each multiplicity; **II**) multiple regression results for all solutions (high and low energy); **III**) multiple regression results for lower energy solutions with energies and spin contaminations up to multiplicity $M = 17$; **IV**) energies derived by using eq 16 and taking the lowest energy within each multiplicity up to $M = 17$; **V**) like in **IV** but $M = 15$ was the highest multiplicity for which contaminations and energies were calculated; **VI**) similar to **V** but the highest multiplicity included was 13. Note that plots **III**, **IV**, **V**, and **IV** are practically on top of each other within the scale of this figure, and one needs to examine entries in Table 6 to grasp the differences between them.

Table 6
Pure spin state energies (in hartrees) of $(O_2)_6$ cluster calculated for different maximum multiplicities (M).

M	$M = 13$ (eq 16)	$M = 15$ (eq 16)	$M = 17$ (eq 16)	$M = 17$ (regression)
1	−902.43982909850	−902.44441505828	−902.44436334192	−902.4468406100
3	−902.31741178930	−902.31750893682	−902.31750803962	−902.3164090989
5	−902.36572848950	−902.36574836191	−902.36574817044	−902.3657679127
7	−902.33290290266	−902.33290994159	−902.33290969640	−902.3328652550
9	−902.35578026666	−902.35579964754	−902.35579935848	−902.3558047254
11	−902.33612613317	−902.33618314893	−902.33618285333	−902.3361286004
13	−902.35109665329	−902.35198653277	−902.35198492510	−902.3519849951
15		−901.83483600763	−901.83567177532	−901.8356718324
17			−901.31801506000	−901.3180151000

able to locate the lowest energy ones. Because spin contaminations for all solutions for the given multiplicity are similar, we assume that the higher energy solutions incorporate excited spin states. Examining Table 5 reveals that lower multiplicities tend to be more contaminated than the higher ones. Singlet is a drastic example of how strong the contamination can get. The results of unrestricted calculations for the singlet contain only about 14% of pure singlet, more than twice this number of triplet and quintet, and more than 16% of septet. Triplet is less contaminated, but still, the amount of triplet in the UKS solution is only about 35%. Obviously, any properties derived directly from strongly contaminated KS wave function and density, will be far from the ones calculated with the pure singlet solution. On the other hand, it is most unlikely that the SCF convergence in the restricted mode can be achieved for the singlet of this cluster.

There is also an annoying feature, which is quite visible for larger size molecules. Our hope was that for the oxygen cluster $(O_2)_n$, the highest spin state which needs to be addressed, corresponds to the multiplicity $M = 2n + 1$. And in fact, the contributions from $M = 15$ are small, however, not negligible. There is some sort of a "spin trailing effect", and it is not clear to us, if it is related to the accuracy of the method[9] or is it a real phenomenon. This effect is also more acute for larger molecules due to a simple fact that their total energies in absolute numbers are larger, and even small fractions of the large total energy value produce differences which are significant on the chemical scale. For those molecules it seems unavoidable to include multiplets that involve electrons located below the valence shell.

On the other hand, crossing to the higher multiplets, involving inner shell electrons, may open a Pandora's box of a new manifold of spins which are heavily intermixed together. Luckily it did not happen in this case, though it may be an issue with some other molecules. For example: $\mathcal{A}_{M=11}(M = 15) = 0.000232726$, $\mathcal{A}_{M=13}(M = 15) =$

[9]The exchange-correlation potential used during SCF iterations is an approximated gaussian based expansion, fitted on a lesser density grid, in contrast to the accurate, but computationally demanding, numerically calculated quantity which is used to calculate energy after the SCF converged.

0.001715809. Since the multiplets above $M = 13$ correspond to unpairing electrons from inner shells, they are much higher in energy, e.g., $E(15) - E(13) = 0.516260610$ and $E(15) - E(13) = 0.502518660$ hartrees (i.e., over 300 kcal/mol), and therefore their contributions to the pure spin state energies substantially bias higher multiplets, and as a second order effect, also the lower multiplets which mix substantially with the higher multiplets within the "valence electrons spin system".

The energies calculated in a variety of ways for $(O_2)_6$ are shown in Figure 4 and Table 6. It is obvious that contaminated UKS energies (**I**) cannot be used directly for studying the multiplet energy splits, especially for $M = 1$ and $M = 3$, because, from other considerations, it is most likely that the singlet should have the lowest, and triplet the highest energy among the spin states of this cluster of oxygen molecules in their ground state.

The "black box" regression on all states calculated (**II**) is similarly inadequate, and the standard deviations (not given here) for the calculated pure spin state energies, as well as, the sum of squared deviations between energies predicted from regression and the actual UKS calculated values, are quite high, suggesting that the fit is of poor quality. On the other hand, the regression **III** which includes only lowest energy UKS solutions for each multiplicity up to $M = 17$, yields values of $E_{corr}(S_i)$ which are statistically very relevant. For all practical purposes, the results of direct solution of eq 16 using the lowest energy solution for each multiplet up to $M = 17$ (**IV**) are identical with energies **III** and within the scale of Figure 4 cannot be distinguished from the former. The analogous results of direct solution of eq 16 (**V**), as well as, the regression (not shown here) for multiplicities up to $M = 15$ are practically identical to the $M = 17$ case. The differences between corresponding pure energy values for **III**, **IV**, and **V** are at most 2 kcal/mol, i.e., probably much lower than the accuracy of the method in this case. Pure spin state energies derived with the assumption that $M = 13$ is the highest contaminating multiplet, are not, however, satisfactory. The fact that small, but not negligible, contamination $\mathcal{A}_{13}(15)$ is now "redistributed" among other spin states (since all contaminations have to sum up to unity), affects pure singlet energy quite substantially, namely, for the case of the contaminated singlet with the lowest energy: 01=0.14773623375, 03=0.31728911605, 05=0.30435532355, 07=0.15999743222, 09=0.05862306649, 11=0.01086685194, 13=0.00113197600, and the pure singlet energy comes as $E_{corr}(0) = -902.43982909850$ hartrees, which is noticeably different than the value of $E_{corr}(0) = -902.44436334192$ hartrees when one uses $M = 17$ as the highest multiplicity when solving eq 16. It is therefore a good practice to calculate a few higher spin multiplets and check at which level the energies corrected for spin contamination become stable. In this particular case, the satisfactory result can be obtained by including just the lowest multiplet involving electrons from inner shells.

9. CONCLUSIONS

Calculations for different spin states of the same molecule/radical are becoming increasingly important in studies of magnetism and reaction mechanisms. While there are, to our knowledge, some recently introduced software packages which allow for correcting spin contamination in a similar fashion as it is done in the Restricted Open Shell Hartree-Fock approach[26,27], the method presented above is more general and allows for the additional insight concerning the sources of spin contamination. Moreover, the calculations of this type may be used to derive parameters which are then used to study larger systems (e.g., clusters, or solids) which cannot be approached directly with first principle methods.

The method is still at its development stage but has a potential of providing much more reliable results than those obtained from the spin uncorrected UKS solution. Our preliminary results for simple organic radicals (to be presented elsewhere) show that this method provides reliable energy differences between energies of different multiplets. At this moment, we do not yet have gradients for this method, and they are not necessarily simple, because the spin contaminations depend strongly upon the molecular geometry. Also, the efficient code is needed, since several stages of UKS DFT calculations are unnecessarily repeated for all multiplets in the current scheme. It may be achieved by rearranging flowchart of the calculations and/or storing needed results which are common across all multiplets on a disk file, and retrieving them, rather than recomputing, for all but the first multiplet at a given geometry.

REFERENCES

1. A.A. Ovchinnikov and J.K. Labanowski, *Phys. Rev.* A., **56**, 3946, (1996).
2. J. Wang, A.D. Becke, V.H. Smith Jr., *J. Chem. Phys.* **102**, 3477 (1995).
3. R. Paunz, *Spin Eigenfunctions: Construction and Use* (Plenum, New York, 1979).
4. D.R. Salahub, R. Fournier, P. Młynarski, I. Papai, A. St-Amant, and J. Ushio, in *Density Functional Methods in Chemistry*, edited by J. Labanowski and J. Andzelm (Springer-Verlag, New York, 1991) p. 77.
5. S.H. Vosko, L. Wilk, and M. Nusair, Can. J. Phys. **58**, 1200 (1980).
6. J.P. Perdew, Phys. Rev. B **33**, 8822 (1986); *erratum:* **38**, 7406 (1986).
7. A.D. Becke, Phys. Rev. A **38**, 3098 (1988).
8. N. Godbout, D.R. Salahub, J. Andzelm, and E. Wimmer, Can. J. Chem. **70**, 560 (1992).
9. H. Weyl, *The classical groups, their invariants and representations* (Princeton University Press, Princeton, 1946).
10. R. O'Neill, Appl. Statistics, **20**, 338 (1971), including remarks to the original algorithm: **23**, 240 (1974); **25**, 97 (1976); and **27**, 380 (1978).
11. D.C. Comeau, I. Shavitt, P. Jensen, P.R. Bunker, J. Chem. Phys. **90**, 6491 (1989).
12. P.R. Bunker, P. Jensen, W.P. Kraemer, R. Beardsworth, J. Chem. Phys. **85**, 3724 (1986).
13. A. St-Amant, *private communication.*
14. G. Herzberg, *Molecular Spectra and Molecular Structure. I. Spectra of Diatomic Molecules*, (Van Nostrand, New York, 1959) p. 454.
15. R. Carbo, J.A. Hernandez, and F. Janz, *Chem. Phys. Lett.* **47**, 581, 1977.
16. M. Bones, G. Schilz, Phys. Rev. A, **2**, 2182 (1970).
17. F. Linder, H. Schmidt, Z. Naturforsch. A, **26**, 1617 (1971).
18. G. Parlant, F. Fiquet-Fayard, J. Phys. B, **9**, 1617 (1976).
19. R.H. Celotta, R. Bennet, J. Hall, M. Siegel, J. Levine, Phys. Rev. A, **6**, 631 (1972).
20. B. Bussery, P.E.S. Wormer, J. Chem. Phys. **99**, 1230 (1993).
21. H. Helm, C.W. Walter, J. Chem. Phys. **98**, 5444 (1993).
22. E.T. Seidl, H.F. Shaefer III, J. Chem. Phys. **96**, 1176 (1992).

23. A.C. Pavão, G.M. Seabra, C.A. Taft, J. Molec. Struct. (Theochem) **335**, 59 (1995).
24. A.P.J. Jansen, A. van der Avoird, *J. Chem. Phys.* **86**, 3583 (1987).
25. A.A. Ovchinnikov, Phys. Lett. A, **164**, 218 (1992).
26. A. Szabo, N.S. Ostlund, *Modern Quantum Chemistry. Introduction to Advanced Electronic Structure Theory,* (McGraw-Hill, New York, 1989)
27. M. Dupuis, S.A. Maluendes, in: *Modern Techniques in Computational Chemistry* edited by E. Clementi (ESCOM, Leiden, 1991) p. 469.

PART III:

APPROACHES AND METHODS

J.M. Seminario (Editor)
Recent Developments and Applications of Modern Density Functional Theory
Theoretical and Computational Chemistry, Vol. 4

Time-Dependent Density Functional Response Theory of Molecular Systems: Theory, Computational Methods, and Functionals

Mark E. Casida[a]

[a]Département de chimie, L'Université de Montréal, C.P. 6128, succursale centre-ville, Montréal, Québec H3C 3J7, Canada

This article reviews recent developments in time-dependent density-functional response theory (TD-DFRT) from a molecular perspective. The formal foundations, molecular formulation, and molecular algorithm are summarized. The remainder of the article is devoted to an examination of the features of exchange-correlation functionals that are most important for the calculation of molecular electronic spectra and response properties. Thus attention is given to the long-range behavior of the exchange-correlation potential and to the TD-DFRT ionization threshold. First the strengths and limitations of the time-dependent local density approximation (TDLDA) are examined, followed by a review of some of the methods which have been used to go beyond the TDLDA. The Sham–Schlüter and optimized effective potential methods for the calculation of the exact exchange-correlation potential, v_{xc} are reviewed, and some new results are presented for the derivative $\delta v_{xc}(\mathbf{r})/\delta\rho(\mathbf{r}') = \delta^2 E_{xc}/\delta\rho(\mathbf{r})\delta\rho(\mathbf{r}')$ which plays key rôle in TD-DFRT within the adiabatic approximation. Finally heuristic connections are made between Hartree–Fock, exchange-only TD-DFRT, and exchange-only ΔSCF-based density-functional theory for the calculation of excitation energies.

1. Introduction

The recent popularity of density-functional theory (DFT) for molecular applications attests to the success of the classic Hohenberg–Kohn [1], Kohn–Sham [2] formalism when combined with modern density functionals. Any property which is a function only of the ground state energy and/or charge density can in principle be obtained exactly in this formalism, in the limit of the exact exchange-correlation functional. Of course no practical form of the exact functional is known. However, modern approximate functionals allow quite good results to be obtained for many properties. This includes a wide variety of properties of chemical interest, including equilibrium geometries, reaction energies, dipole moments and static dipole polarizabilities. Nevertheless, the Hohenberg–Kohn–Sham theory is an incomplete theory in so far as it is not in general applicable to excited states or to problems involving non-local potentials or time-dependent fields, thus excluding the calculation of optical response properties, electronic spectra, and the treatment of excited states and photochemistry, to mention only a few problems of current chemical interest.

Fortunately, thanks to the efforts of many workers [3–12], the Hohenberg–Kohn–Sham theory has been generalized to treat time-dependent systems. Theorems have now been established for time-dependent DFT which parallel those that form the foundation of classic time-independent DFT. These are reviewed in Sec. 2 of the present chapter. Further details may be found in the review articles of Gross and Kohn [12], Gross, Ullrich, and Gossmann [13], and Gross, Dobson, and Petersilka [14], which are particularly notable for their treatment of formal aspects of time-dependent DFT.

The first applications of time-dependent density-functional response theory (TD-DFRT) [15] actually predate much of the formal development, and relied on analogies with the time-dependent Hartree–Fock (TDHF) method, rather than any rigorous formal justification. Of course there was no problem of justifying calculations of static response properties [16,17] since these are within the domain of applicability of time-independent DFT. Early TD-DFRT calculations, both static and dynamic, [16,17,15] were motivated partly by algorithmic considerations. It was shown that, compared to Hartree–Fock (HF), the simple multiplicative nature of the DFT exchange-correlation potential allows a particularly simple numerical solution of the response theory equations for systems with spherical symmetry such as closed-shell atoms. Two algorithms were of particular importance, namely the so-called Green function method of Stott and Zaremba [16] and the modified Sternheimer method of Mahan [17]. Neither involves summations over virtual orbitals and calculations are reduced to relatively simple numerical integrations in a single radial coordinate. These methods later became the workhorses for calculations of metal cluster response properties within the jellium sphere and spherically averaged pseudopotential (SAPS) models. (See for example Refs. [18] and [19].) In comparison, TDHF calculations for systems with spherical symmetry are limited to much smaller clusters [19].

A good deal has been learned from calculations on atoms and spherical models of clusters about the strengths and limitations of TD-DFRT. (See Ref. [20] for a review of the atomic literature.) Many of the conclusions drawn from this experience have also been found to apply to molecules. But of course the nonspherical nature and chemical bonding of molecules make them rather different from atoms, both from a physical and a computational standpoint.

Although the first applications of TD-DFRT to molecules [21,22] used single-center expansions in order to borrow atomic algorithms, the well-known difficulties with using single-center expansions to describe multicenter systems make this approach impractical for general molecular applications. The first truly molecular formulation of TD-DFRT was given much more recently [23,24]. Although the derivation necessarily differs from that usually given for TDHF, the resulting equations are analogous to those of molecular TDHF. This approach is now being pursued in several groups [23,25,24,26–29]. The method is reviewed in Sec. 3 of the present chapter. See the previous paper [23] for a detailed presentation. The mutiplicative nature of the DFT exchange-correlation potential makes the basic equations somewhat simpler than the analogous TDHF equations, resulting in certain computational advantages. Auxiliary function techniques permit a more complete exploitation of the potential computational advantages of a purely mutiplicative exchange-correlation term. An auxiliary function algorithm for molecular TD-DFRT is given in Sec. 4, improving on the one given previously [23]. These techniques open the way to calculating a wide variety of properties previously either inaccessible or difficult to obtain using molecular DFT codes, including dynamic polarizabilities [25,26] for optical materials, van der Waals coefficients [25] for describing long range intermolecular forces, and excitation spectra [26,29] for spectroscopic and photochemical applications.

It should be noted that the TD-DFRT method for calculating spectra offers an automatic and formally rigorous wave function-free approach for the simultaneous calculation of an entire electronic excitation spectrum. In practice, the method produces both spin and spatial multiplet splittings associated with molecular spin and spatial symmetries, as well as coupling between zero-order states which would not be anticipated on the basis of symmetry alone. This is illustrated in Sec. 5.1. In contrast, the ΔSCF-based procedure [30,31] traditionally used in DFT for treating excited states lacks a solid formal foundation, requires assumptions to be made about the form of excited state wave functions, and is much less automatic in practice. These difficulties with the method are particularly evident when symmetry alone is insufficient for predicting configuration mixing [32,23].

Due to the sum-over-states relation between the excitation spectrum and the polarizability, TD-DFRT calculation of these two properties yields complementary information and thus provides a more detailed tool for analyzing the functionals used than would the calculation of either property alone. Thus Sec. 5 examines functionals from the point of view of calculating both polarizabilities and spectra. This begins (Sec. 5.1) with an illustration of the strengths and weaknesses of the traditional workhorse functional of TD-DFRT, namely the time-dependent local-density approximation (TDLDA). This is followed by an analysis of the long-range behavior and related properties pertinent to TD-DFRT calcultions that the exact exchange-correlation potential should have (Sec. 5.2), and a survey of some of the ways that have been devised to try to improve practical approximate functionals with regard to these criteria (Sec. 5.3). Most, though not all, of these approximate corrections are drawn from the TD-DFRT literature for atoms and spherical clusters, but should also prove useful for molecules if properly reformulated to meet the needs of molecular calculations. An even deeper examination of properties of the exact exchange potential is given in Sec. 5.4 where equations for the exact exchange potential and its derivative are derived and then solved within the reasonably accurate Krieger–Li–Iafrate (KLI) approximation. This is used in Sec. 6 to help provide a physical interpretation of heuristic relationships between HF and exchange-only DFT (both TD-DFRT and the traditional ΔSCF-based method), clarifying the relation between how coulomb and exchange effects enter into the calculation of excitation spectra in these methods.

The treatment of functionals in this paper is within the adiabatic approximation, with an emphasis on examining those aspects of adiabatic exchange-correlation functionals that are most in need of improvement from the point of view of TD-DFRT calculations. A few comments on the import of the adiabatic approximation itself are given in the chapter conclusion.

2. Formal foundations

The classic Hohenberg–Kohn–Sham formulation of density-functional theory is restricted both to local potentials and to the time-independent case. Thus the treatment of time-dependent systems or of magnetic fields requires a generalization of the basic DFT formalism. These two problems are related insofar as the presence of a time-dependent electric field necessarily implies the existence of a time-dependent magnetic field. However, this latter field can often be neglected, as a good first approximation, in many applications, such as in the treatment of electronic excitation spectra. Also the magnetic field does not enter into the calculation of purely electrical response properties, such as the polarizability $\boldsymbol{\alpha} = [\partial\vec{\mu}(\vec{\mathcal{E}},\vec{B})/\partial\vec{\mathcal{E}}]_{\vec{\mathcal{E}}=\vec{B}=\vec{0}}$. Thus it is reasonable to treat the generalization of DFT to handle time-dependent systems (with local potentials) separately from that to handle (static) magnetic fields, and this has been the course of the historical development. The need to separate the external fields from the internal currents while simultaneously maintaining gauge invariance makes the rigorous introduction of magnetic fields into DFT somewhat intricate [33,14]. Nevertheless, a manifestly gauge invariant theory has been developed for the time-independent case [33]. The development of time-dependent DFT has been pursued by a number of authors [3–12,34,35], providing time-dependent analogs of the Hohenberg–Kohn theorems and Kohn–Sham equation. The combination of these two formal generalizations to handle time-dependent electromagnetic properties has been tackled by Ghosh and Dhara [36] and we will touch on this briefly at the end of this section. However, the primary emphasis here will be on the formal foundations of a time-dependent DFT involving only local external potentials, notably electrical perturbations. The brief synopsis of the essential elements of the time-dependent theory given here is based primarily on the work of Runge and Gross [9] and Gross and Kohn [12]. The presentation is essentially the same as that given in [23], but is reproduced here for the sake of completeness. A more in depth review of the formal time-dependent theory may be found in the recent review of Gross, Dobson, and Petersilka [14].

Consider an N_e-electron system described by the Schrödinger equation (in hartree atomic units),

$$\hat{H}(t)\Psi(t) = i\frac{\partial}{\partial t}\Psi(t)\,, \tag{1}$$

with Hamiltonian

$$\hat{H}(t) = \hat{T} + U + V(t) \tag{2}$$

where

$$\hat{T} = -\frac{1}{2}\sum_{i=1}^{N_e}\nabla_i^2 \tag{3}$$

is the kinetic energy,

$$U = \sum_{i<j}\frac{1}{|\mathbf{r}_i - \mathbf{r}_j|} \tag{4}$$

is the electron repulsion, and

$$V(t) = \sum_{i=1}^{N_e} v(\mathbf{r}_i, t) \tag{5}$$

is the external potential, assumed constant for $t < t_0$. Thus we may think of a system interacting with a time-dependent field switched on at time t_0. The first problem is to obtain (time-dependent) expectation values as functionals of the time-dependent charge density. To this end, Runge and Gross [9] showed that the time-dependent charge density, ρ, determines the wave function up to a time-dependent phase factor,

$$\Psi(t) = e^{-i\phi(t)}\Psi[\rho, \Psi_0](t)\,, \tag{6}$$

where Ψ_0 denotes the initial condition $\Psi(t_0)$.[1] More specifically, they showed that the external potential, $v(\mathbf{r}, t)$, is determined by the charge density, ρ, up to a spatially constant time-dependent function $c(t)$, provided that ρ arises from a system with initial state Ψ_0 and that the external potential (*i*) can be represented as the electric potential due to a normalizable charge distribution, and (*ii*) has a time dependence which can be expressed in a Taylor's series about $t = t_0$. Equation (6) follows as a consequence. The proof of the Runge-Gross theorem proceeds by first showing that the external potential is determined by the current density, and then showing that the current density can be eliminated in favor of the charge density alone, provided a suitable boundary condition is satisfied. Although Xu and Rajagopal [37] pointed out a difficulty with the elimination of the current density in Runge and Gross's original formulation of this theorem [9], the clarification of condition (*i*) above by Gross and Kohn [12] and Dhara and Ghosh [38] has resolved this problem. Gross and Kohn [12] also note that the conditions of the Runge–Gross theorem are not strictly satisfied in time-dependent response theory, because of the difficulty of expanding $v(\mathbf{r}, t)$ about $t_0 = -\infty$, as should be done when the perturbation is turned on adiabatically starting at $t = -\infty$. In addition, the restriction to an electric potential due to a normalizable charge distribution excludes the *exact* representation of a uniform field. However, these remarks only concern idealizations involved in the theory, not the physical systems it is intended to model. Thus they do not constitute limitations on the applicability of a response formulation of time-dependent DFT to physical fields (e.g. due to a large but finite charged plate) introduced at some finite time in the distant past.

If it is further assumed, as is usual in response theory, that the initial state Ψ_0 is a stationary ground state, then Ψ_0 is completely determined by the initial charge density, $\rho_0 = \rho(t_0)$, according to the first Hohenberg–Kohn theorem [1], provided Ψ_0 is also nondegenerate. (Of course, any initial degeneracy can usually be lifted by a small perturbation, so the degeneracy question is not a major concern.) Hence Ψ_0 can be eliminated from Eq. (6) so that Ψ is then determined up to a phase factor by ρ alone. For an operator $\hat{O}(t)$ which may be a function of time but which contains no derivative or integral operators on t, this phase factor cancels out on taking the expectation value. Thus expectation values are functionals only of ρ,

$$\langle \Psi(t)|\hat{O}(t)|\Psi(t)\rangle = O[\rho](t)\,, \tag{7}$$

analogous to the time-independent case.

The role of the second Hohenberg–Kohn theorem, in the time-independent case, is filled, in the time-dependent theory, by a variational principle involving the action,

$$\mathcal{A} = \int_{t_0}^{t_1} \langle \Psi(t)|i\frac{\partial}{\partial t} - \hat{H}(t)|\Psi(t)\rangle\, dt\,. \tag{8}$$

[1] In the present notation, functionals are distinguished by the use of square brackets while the independent variables of a function appear in parentheses. Thus $\Psi[\rho, \Psi_0](t)$ indicates that Ψ is both a functional of the functions $\rho(\mathbf{r}, t)$ and $\Psi_0(\mathbf{r})$, and a function of t.

The true time-dependent density is the one which makes the action stationary,

$$0 = \frac{\delta \mathcal{A}}{\delta \rho(\mathbf{r},t)} = \int_{t_0}^{t_1} \langle \frac{\delta \Psi(t')}{\delta \rho(\mathbf{r},t)} | i\frac{\partial}{\partial t'} - \hat{H}(t') | \Psi(t') \rangle \, dt' + \text{c.c.} \,. \tag{9}$$

The expectation value in Eq. (8) is not of the form required for the phase factor to cancel out as in Eq. (7). Nevertheless, it is easy to see that in this case the effect of the phase factor is simply to contribute an additive constant,

$$\mathcal{A} = \int_{t_0}^{t_1} \langle \Psi[\rho](t) | i\frac{\partial}{\partial t} - \hat{H}(t) | \Psi[\rho](t) \rangle \, dt + \phi(t_1) - \phi(t_0) = \mathcal{A}[\rho] + \text{const} \,. \tag{10}$$

Thus the time-dependent density determines the action, up to an additive constant. Of course, when the variational condition (9) is used, the additive constant is immaterial, so this provides the analog of the second Hohenberg–Kohn theorem, for the time-dependent theory.

The action functional can be rewritten as

$$\mathcal{A}[\rho] = B[\rho] - \int_{t_0}^{t_1} \int v(\mathbf{r},t)\rho(\mathbf{r},t) \, d\mathbf{r}dt \,, \tag{11}$$

where the functional B is independent of the external potential v. A time-dependent Kohn–Sham equation can then be derived in a manner analogous to that for the time-independent case by *assuming* the existence of a potential $v_{\text{eff}}(\mathbf{r},t)$, for an independent particle system, whose orbitals $\psi_i(\mathbf{r},t)$ yield the same charge density $\rho(\mathbf{r},t)$ as for the interacting system,

$$\rho(\mathbf{r},t) = \sum_i f_i |\psi_i(\mathbf{r},t)|^2 \,, \tag{12}$$

where the f_i are orbital occupation numbers. The question of whether such a potential exists is known as the "time-dependent v-representability problem." As in the time-independent case, this problem is not entirely resolved, although $v_{\text{eff}}(\mathbf{r},t)$ can be shown to exist [11] for adiabatically introduced perturbations whose frequency does not exceed the HOMO-LUMO gap of the unperturbed system. Mearns and Kohn [11] give examples which suggest that $v_{\text{eff}}(\mathbf{r},t)$ will also exist for higher frequency perturbations as well, except at isolated frequencies. Assuming $v_{\text{eff}}(\mathbf{r},t)$ does exist, then the universal functional B can be written as

$$\begin{aligned} B[\rho] &= \sum_i f_i \int_{t_0}^{t_1} \langle \psi_i(t) | i\frac{\partial}{\partial t} - \frac{1}{2}\nabla^2 | \psi_i(t) \rangle \, dt \\ &\quad - \frac{1}{2} \int_{t_0}^{t_1} \iint \frac{\rho(\mathbf{r}_1,t)\rho(\mathbf{r}_2,t)}{|\mathbf{r}_1 - \mathbf{r}_2|} \, d\mathbf{r}_1 d\mathbf{r}_2 dt - \mathcal{A}_{\text{xc}}[\rho] \end{aligned} \tag{13}$$

thereby defining the exchange-correlation action functional $\mathcal{A}_{\text{xc}}$ which plays a role analogous to the exchange-correlation energy functional in the time-independent theory. Minimizing the action (11) subject to the condition (12), results in the time-dependent Kohn–Sham equation

$$\left[-\frac{1}{2}\nabla^2 + v_{\text{eff}}(\mathbf{r},t)\right] \psi_i(\mathbf{r},t) = i\frac{\partial}{\partial t}\psi_i(\mathbf{r},t) \,, \tag{14}$$

where

$$v_{\text{eff}}(\mathbf{r},t) = v(\mathbf{r},t) + \int \frac{\rho(\mathbf{r},t)}{|\mathbf{r}-\mathbf{r}'|}\,d\mathbf{r}' + v_{\text{xc}}(\mathbf{r},t) \tag{15}$$

and

$$v_{\text{xc}}(\mathbf{r},t) = \frac{\delta \mathcal{A}_{\text{xc}}[\rho]}{\delta\rho(\mathbf{r},t)}\,. \tag{16}$$

Although the functional $\mathcal{A}_{\text{xc}}$ is unknown, in the limit of an external potential which varies slowly in time it must reduce to

$$\mathcal{A}_{\text{xc}} = \int_{t_0}^{t_1} E_{\text{xc}}[\rho_t]\,dt\,, \tag{17}$$

where E_{xc} is the exchange-correlation functional of time-independent Kohn–Sham theory, and ρ_t denotes ρ evaluated at the time t. This is known as the "adiabatic approximation." Notice that, whereas $\mathcal{A}_{\text{xc}}$ is a functional of a function ρ over both time and space, E_{xc} is a functional of a function ρ_t over only space (since t is fixed).

The adiabatic approximation is a local approximation in time. Like the familiar local density approximation from time-independent DFT, which is a local approximation in space, the adiabatic approximation works well beyond its domain of rigorous justification. For this reason, and due to its relative simplicity, the adiabatic approximation has come to be the work horse of time-dependent DFT. In the adiabatic approximation, the first derivative of $\mathcal{A}_{\text{xc}}$,

$$v_{\text{xc}}[\rho](\mathbf{r},t) = \frac{\delta \mathcal{A}_{\text{xc}}[\rho]}{\delta\rho(\mathbf{r},t)} \cong \frac{\delta E_{\text{xc}}[\rho_t]}{\delta\rho_t(\mathbf{r})} = v_{\text{xc}}[\rho_t](\mathbf{r})\,, \tag{18}$$

is just the familiar exchange-correlation potential of time-independent DFT evaluated with the density at a particular time. (Note the difference in notation between the right and left hand sides.) Since $v_{\text{xc}}(t)$ is then determined solely by the density $\rho(t)$ at the same time, the adiabatic approximation neglects all retardation effects and assumes an instantaneous reaction of the self-consistent field to temporal changes in ρ. This is particularly obvious from the second derivative of $\mathcal{A}_{\text{xc}}$ which gives the response of the exchange-correlation potential to a change in the charge density,

$$\frac{\delta v_{\text{xc}}[\rho](\mathbf{r},t)}{\delta\rho(\mathbf{r}',t')} \cong \delta(t-t')\frac{\delta v_{\text{xc}}[\rho_t](\mathbf{r})}{\delta\rho_t(\mathbf{r}')}\,. \tag{19}$$

While the adiabatic approximation is a physically reasonable first approximation, and is useful in practical applications, it is also important to develop functionals which go beyond this approximation. Such functionals are discussed by Gross *et al.* [14]. However, the focus in the present chapter will be on a treatment within the adiabatic approximation.

The extension of TD-DFT to handle time-dependent magnetic fields by way of CDFT has been treated by Ghosh and Dhara [36]. In the presence of a magnetic field, Eq. (14) becomes

$$\left\{\frac{1}{2}\left|-i\boldsymbol{\nabla} + \frac{1}{c}\boldsymbol{A}_{\text{eff}}(t)\right|^2 + v_{\text{eff}}(\mathbf{r},t)\right\}\psi_i(\mathbf{r},t) = i\frac{\partial}{\partial t}\psi_i(\mathbf{r},t)\,, \tag{20}$$

where

$$\frac{1}{c}\boldsymbol{A}_{\text{eff}}(\mathbf{r},t) = \frac{1}{c}\boldsymbol{A}(\mathbf{r},t) + \int \frac{\mathbf{j}(\mathbf{r}',t)}{|\mathbf{r}-\mathbf{r}'|}d\mathbf{r}' + \frac{\delta \mathcal{A}_{\text{xc}}[\rho,\mathbf{j}]}{\delta \mathbf{j}(\mathbf{r},t)} \tag{21}$$

and

$$v_{\text{eff}}(\mathbf{r},t) = v(\mathbf{r},t) + \int \frac{\rho(\mathbf{r}',t)}{|\mathbf{r}-\mathbf{r}'|}d\mathbf{r}' + \frac{\delta \mathcal{A}_{\text{xc}}[\rho,\mathbf{j}]}{\delta \rho(\mathbf{r},t)} + \frac{1}{2c^2}\left[A^2(\mathbf{r},t) - A^2_{\text{eff}}(\mathbf{r},t)\right] . \tag{22}$$

Here, $\boldsymbol{A}$ is the external vector potential and the exchange-correlation action-functional $\mathcal{A}_{\text{xc}}$ depends on both the charge density ρ and the current density

$$\mathbf{j}(\mathbf{r},t) = -\frac{i}{2}\sum_i f_i[\psi_i^*(\mathbf{r},t)\boldsymbol{\nabla}\psi_i(\mathbf{r},t) - \psi_i(\mathbf{r},t)\boldsymbol{\nabla}\psi_i^*(\mathbf{r},t)] + \frac{1}{c}\rho(\mathbf{r},t)\boldsymbol{A}_{\text{eff}}(\mathbf{r},t) . \tag{23}$$

Note that, although the variation of the action functional is with respect to the total current density (which is gauge invariant), the fact that the external vector potential is fixed means that the diamagnetic component of the current density is constant, so the variation reduces to one with respect to the (non gauge invariant) paramagnetic component, $\mathbf{j}_p$, only. Thus, although the Ghosh and Dhara theory would be gauge invariant in the limit of the exact functional, it is not *manifestly* gauge invariant. This makes the development of approximate functionals more difficult. Vignale and Rasolt [39,40,33] have shown that E_{xc} is a functional of the gauge invariant combination

$$\boldsymbol{\nu}(\mathbf{r}) = \boldsymbol{\nabla} \times [\mathbf{j}_p(\mathbf{r})/\rho(\mathbf{r})] . \tag{24}$$

This permitted them to develop a manifestly gauge invariant formulation for CDFT, in the static case. However, as was noted by Gross *et al.* [14], a similar development is still needed for the time-dependent case.

Proceeding from the Ghosh and Dhara formulation, we note that Eq. (20) can also be written in the form

$$\left[-\frac{1}{2}\nabla^2 + \hat{u}_{\text{eff}}(\mathbf{r},t)\right]\psi_i(\mathbf{r},t) = i\frac{\partial}{\partial t}\psi_i(\mathbf{r},t) , \tag{25}$$

where

$$\begin{aligned} \hat{u}_{\text{eff}}[\rho,\mathbf{j}](\mathbf{r},t) = v(\mathbf{r},t) \quad &+ \quad \int \frac{\rho(\mathbf{r}',t)}{|\mathbf{r}-\mathbf{r}'|}d\mathbf{r}' + \frac{\delta \mathcal{A}_{\text{xc}}[\rho,\mathbf{j}]}{\delta \rho(\mathbf{r},t)} \\ &- \quad \frac{i}{2c}\left[\boldsymbol{\nabla}, \boldsymbol{A}_{\text{eff}}[\rho,\mathbf{j}](\mathbf{r},t)\right]_+ + \frac{1}{2c^2}A^2(\mathbf{r},t) . \end{aligned} \tag{26}$$

The orbital equation (25), and the fact that both the charge and current densities are trivial functions of the density matrix, means that the TD-DFRT formalism of Ref. [23] can be generalized to include time-dependent current-density functional response theory (TD-CDFRT). Naturally, this is computationally more demanding. In particular, terms involving derivatives with respect to the current density and derivatives of the current density with respect to the density matrix have to be introduced, and simplifications due to the multiplicative nature of the TD-DFRT effective potential are lost in TD-CDFRT. Nevertheless, the effective potential, $\hat{u}_{\text{eff}}$, divides into external and self-consistent field parts, a coupling matrix can be defined, and the remaining development is very similar to that given in Sec. 4 of Ref. [23]. In the adiabatic approximation, the TD-CDFRT coupling matrix is frequency independent, so the static result of Colwell and Handy [41] could be taken over directly. A practical implementation of such a scheme has been given by Colwell, Handy, and Lee [28]. Note that they follow Vignale and Rasolt and use the combination (24). Given the close interrelationship between time-dependent electric and magnetic fields, it will be a more challenging problem how to combine current-density functionals and time-dependent functionals beyond the adiabatic approximation without overcounting. Since the present paper is not concerned with magnetic fields, the details of a time-dependent current-density functional theory will not be further pursued here.

3. Time-dependent density-functional response theory

The formal development of TD-DFT, reviewed in the last section, opened the way for a sound treatment of time-dependent properties within DFT. Since, just as in the time-independent case, the densities of the orbitals obtained from the time-dependent Kohn–Sham equation sum to the true (time-dependent) charge density, any property depending only on the time-dependent charge density can be obtained exactly, in the limit of the exact functional. In particular, we will be interested in properties involving the linear response of the charge density to an applied field. Not only does this allow the treatment of response properties such as dynamic polarizabilities, but, since excitation energies and oscillator strengths can be obtained as the poles and residues of the dynamic polarizability, it makes possible a formally well-founded treatment of excited states in DFT.

A formulation of time-dependent density-functional response theory (TD-DFRT) suitable for calculations of molecular properties has been given in some detail in a previous paper [23]. The essentials of the method are reviewed here, assuming the adiabatic approximation. The interested reader is refered to the earlier article for a more detailed derivation, including a treatment that is not restricted to the adiabatic approximation. The emphasis is on the calculation of dynamic dipole polarizabilities and excitation spectra within the dipole approximation, but the method can certainly be applied to the calculation of other types of response properties.

The starting point is the time-dependent Kohn–Sham equation (14), in spin-unrestricted form,

$$\left\{-\frac{1}{2}\nabla^2 + v(\mathbf{r},t) + \int \frac{\rho(\mathbf{r}',t)}{|\mathbf{r}-\mathbf{r}'|}\,d\mathbf{r}' + v_{\text{xc}}^{\sigma}[\rho_\uparrow,\rho_\downarrow](\mathbf{r},t)\right\}\psi_{i\sigma}(\mathbf{r},t) = i\frac{\partial}{\partial t}\psi_{i\sigma}(\mathbf{r},t)\,. \tag{27}$$

Here, the charge densities for each type of spin are given by

$$\rho_\sigma(\mathbf{r},t) = \sum_i f_{i\sigma}|\psi_{i\sigma}(\mathbf{r},t)|^2\,, \tag{28}$$

and the total charge density is

$$\rho(\mathbf{r},t) = \rho_\uparrow(\mathbf{r},t) + \rho_\downarrow(\mathbf{r},t)\,, \tag{29}$$

where the $f_{i\sigma}$ are (possibly fractional) spin-orbital occupation numbers. In the adiabatic approximation,

$$v_{xc}^{\sigma}[\rho_t^{\uparrow},\rho_t^{\downarrow}](\mathbf{r}) = \frac{\delta E_{xc}[\rho_t^{\uparrow},\rho_t^{\downarrow}]}{\delta \rho_t^{\sigma}(\mathbf{r})}, \tag{30}$$

where E_{xc} is the exchange-correlation energy functional from the time-independent theory.

Consider a system initially in its ground electronic state. The linear response to a perturbation $w(t)$ turned on slowly at some time in the distant past is just given by

$$\delta\rho_\sigma(\mathbf{r},\omega) = \sum_{ij} \psi_{i\sigma}(\mathbf{r})\delta P_{ij\sigma}(\omega)\psi_{j\sigma}^*(\mathbf{r}), \tag{31}$$

where $\delta P_{ij\sigma}(\omega)$ is the linear response of the Kohn–Sham density matrix in the basis of the unperturbed molecular orbitals. It is obtained by taking advantage of the quasi-independent particle nature of the Kohn–Sham equation. The fact that the particles are not truly independent is reflected in the perturbation introduced into the Kohn–Sham Hamiltonian by turning on an applied field $w(t)$ being, not $w(t)$, but

$$w^{\text{eff}}(t) = w(t) + \delta v^{\text{SCF}}(\mathbf{r},t). \tag{32}$$

Here the self-consistent field, v^{SCF}, is the last two terms of the orbital Hamiltonian in Eq. (27). The response of the Kohn–Sham density matrix to the perturbation w^{eff} is then given by the generalized susceptibility for a system of independent particles,

$$\frac{\partial P_{ij\sigma}}{\partial w_{i'j'\sigma'}^{\text{eff}}} = \delta_{i,i'}\delta_{j,j'}\delta_{\sigma,\sigma'}\frac{f_{j\sigma}-f_{i\sigma}}{\omega-(\epsilon_{i\sigma}-\epsilon_{j\sigma})}. \tag{33}$$

Thus

$$\delta P_{ij\sigma}(\omega) = \frac{f_{j\sigma}-f_{i\sigma}}{\omega-(\epsilon_{i\sigma}-\epsilon_{j\sigma})}\left[w_{ij\sigma}(\omega) + \sum_{kl\tau} K_{ij\sigma,kl\tau}\delta P_{kl\tau}(\omega)\right], \tag{34}$$

where the coupling matrix $\boldsymbol{K}$ describes the linear response of the self-consistent field to changes in the charge density,

$$\delta v_{ij\sigma}^{\text{SCF}}(\omega) = \sum_{kl\tau} K_{ij\sigma,kl\tau}\delta P_{kl\tau}(\omega), \tag{35}$$

$$\begin{aligned} K_{ij\sigma,kl\tau} &= \frac{\partial v_{ij\sigma}^{\text{SCF}}}{\partial P_{kl\tau}} \\ &= \int\!\!\int \psi_{i\sigma}^*(\mathbf{r})\psi_{j\sigma}(\mathbf{r})\frac{1}{|\mathbf{r}-\mathbf{r}'|}\psi_{k\tau}(\mathbf{r}')\psi_{l\tau}^*(\mathbf{r}')\,d\mathbf{r}d\mathbf{r}' \\ &\quad + \int\!\!\int \psi_{i\sigma}^*(\mathbf{r})\psi_{j\sigma}(\mathbf{r})\frac{\delta^2 E_{xc}[\rho_\uparrow,\rho_\downarrow]}{\delta\rho_\sigma(\mathbf{r})\delta\rho_\tau(\mathbf{r}')}\psi_{k\tau}(\mathbf{r}')\psi_{l\tau}^*(\mathbf{r}')\,d\mathbf{r}d\mathbf{r}'. \end{aligned} \tag{36}$$

Note that the functional derivative is evaluated for the unperturbed charge densities. This is where the adiabatic approximation [Eq. (30)] manifests itself: Only E_{xc} (rather than $\mathcal{A}_{xc}$) enters into the equation, and $\boldsymbol{K}$ is time- (frequency-)independent.

The dynamic polarizability is now easily obtained from $\delta \boldsymbol{P}$. The induced dipole moment due to the electric perturbation

$$w(\mathbf{r},t) = z\mathcal{E}_z(t) \tag{37}$$

is just

$$\delta\mu_x(\omega) = -\sum_{ij\sigma} x_{ij\sigma}\delta P_{ij\sigma}(\omega)\,, \tag{38}$$

where $x_{ji\sigma}$ is the matrix element of $\hat{x}$ in the basis of the unperturbed molecular orbitals. Thus the dynamic polarizability is

$$\alpha_{xz}(\omega) = -2 \sum_{ij\sigma}^{f_{i\sigma}-f_{j\sigma}>0} x_{ji\sigma}\left(\mathrm{Re}\,\delta P_{ij\sigma}\right)(\omega)\,/\,\mathcal{E}_z(\omega)\,, \tag{39}$$

where $(\mathrm{Re}\,\delta P_{ij\sigma})(\omega)$ denotes the Fourier transform of the real part of $\delta P_{ij\sigma}(t)$. Note that only the real part of the response of the density matrix is needed for present purposes. Solving Eq. (34) for $\delta \boldsymbol{P}$ and separating the real and imaginary parts then allows the polarizability to be expressed as

$$\alpha_{xz}(\omega) = 2\vec{x}^{\dagger}\boldsymbol{\mathcal{S}}^{-1/2}\left\{\boldsymbol{\Omega} - \omega^2\mathbf{1}\right\}^{-1}\boldsymbol{\mathcal{S}}^{-1/2}\vec{z}\,, \tag{40}$$

where

$$\mathcal{S}_{ij\sigma,kl\tau} = \frac{\delta_{\sigma,\tau}\delta_{i,k}\delta_{j,l}}{(f_{k\tau}-f_{l\tau})(\epsilon_{l\tau}-\epsilon_{k\tau})} > 0 \tag{41}$$

and

$$\Omega_{ij\sigma,kl\tau} = \delta_{\sigma,\tau}\delta_{i,k}\delta_{j,l}(\epsilon_{l\tau}-\epsilon_{k\tau})^2 + 2\sqrt{(f_{i\sigma}-f_{j\sigma})(\epsilon_{j\sigma}-\epsilon_{i\sigma})}K_{ij\sigma,kl\tau}\sqrt{(f_{k\tau}-f_{l\tau})(\epsilon_{l\tau}-\epsilon_{k\tau})}\,. \tag{42}$$

Equation (40) has been deliberately cast in the form of the exact sum-over-states (SOS) formula for the mean dynamic polarizability,

$$\bar{\alpha}(\omega) = \frac{1}{3}\mathrm{tr}\,\boldsymbol{\alpha}(\omega) = \sum_I \frac{f_I}{\omega_I^2 - \omega^2}\,, \tag{43}$$

where the poles

$$\omega_I = E_I - E_0 \tag{44}$$

are vertical excitation energies and the residues

$$f_I = \frac{2}{3}(E_I - E_0)\left(|\langle\Psi_0|\hat{x}|\Psi_I\rangle|^2 + |\langle\Psi_0|\hat{y}|\Psi_I\rangle|^2 + |\langle\Psi_0|\hat{z}|\Psi_I\rangle|^2\right) \tag{45}$$

are the corresponding oscillator strengths. Identifying the poles and residues of the dynamic polarizability in the form (40) shows that excitation energies and oscillator strengths can be obtained from TD-DFRT by solving

$$\boldsymbol{\Omega}\vec{F}_I = \omega_I^2\vec{F}_I \tag{46}$$

and calculating

$$f_I = \frac{2}{3}\left(|\vec{x}^\dagger \mathcal{S}^{-1/2}\vec{F}_I|^2 + |\vec{y}^\dagger \mathcal{S}^{-1/2}\vec{F}_I|^2 + |\vec{z}^\dagger \mathcal{S}^{-1/2}\vec{F}_I|^2\right) / |\vec{F}_I|^2 . \tag{47}$$

It is not difficult to show that the TRK sum rule, $\sum_I f_I = N_e$ in the limit of a complete basis set, holds here [23,26].

Note that the TD-DFRT equations have been derived within a strictly density-functional formalism and no reference has been made to a wave function. However, for purposes of assigning excitations after they have been calculated, it is convenient, if not strictly necessary, to make some *ansatz* concerning the wave function. The choice

$$\Psi_I = \sum_{ij\sigma}^{f_{i\sigma}-f_{j\sigma}>0} \sqrt{\frac{\epsilon_{j\sigma} - \epsilon_{i\sigma}}{\omega_I}} F^I_{ij\sigma} \hat{a}^\dagger_{j\sigma} \hat{a}_{i\sigma} \Phi + \cdots \tag{48}$$

for the excited state wave function, where Φ is the single determinant of Kohn–Sham (ground state) occupied orbitals, seems to be a reasonable approach for the qualitative work of assigning state labels, at least for simple closed-shell systems. Notice that a multideterminantal description of the density response and the transition densities has arisen naturally, with coefficients determined from the eigenvectors in Eq. (46), independent of the *ansatz* (48).

The TD-DFRT equations are quite similar to those of TDHF. In fact, TDHF can be derived in the same way as has been done here for TD-DFRT, the only difference being that the exchange-correlation potential, v_{xc}, of DFT is replaced by the (nonlocal) HF exchange operator, $\hat{\Sigma}_x$, which results in the second term in the coupling matrix (36) being different. This difference between TDHF and TD-DFRT does have some important consequences. In TD-DFRT, the simple diagonal form (41) for $\mathcal{S}$, and thus the simple relation (42) between $\boldsymbol{\Omega}$ and $\boldsymbol{K}$, result from the local nature of the DFT exchange-correlation potential. In TDHF however, the corresponding matrix $\mathcal{S}$ is not diagonal and $\boldsymbol{\Omega}$ involves $\mathcal{S}^{-1/2}$, which is not only computationally more demanding but also sometimes presents more serious difficulties, since $\mathcal{S}$ is not necessarily positive definite in TDHF. (See Ref. [42] pp. 144-148. Note that their matrix $-\boldsymbol{S}(\boldsymbol{A}_{11} - \boldsymbol{B}_{11})^{-1}\boldsymbol{S}$ is analogous to our matrix $\mathcal{S}$.) Further computational advantages can be obtained through the use of auxiliary function techniques, to which TD-DFRT lends itself readily, thanks again to the locality of v_{xc}. This will be discussed further in the next section. Finally, as noted above, the TD-DFRT method becomes exact in the limit of the exact exchange-correlation functional, $\mathcal{A}_{xc}$, whereas TDHF is inherently approximate. Of course in practice approximate functionals must be used, whether for $\mathcal{A}_{xc}$, or in making the adiabatic approximation and then approximating the E_{xc} of time-independent DFT, and these approximate functionals are decisive for the quality of the results. The important problem of practical functionals for use in TD-DFRT calculations will be discussed in Section 5.

4. Computational method

One of the attractions of using TD-DFRT for calculations of molecular response properties and spectra is that, even while including some correlation effects, calculations can be done in a manner that is less computationally demanding than with other *ab initio* methods. The least computationally demanding, and least accurate, of the traditional *ab initio* methods for calculating dynamic polarizabilities and excitation spectra are TDHF and the Tamm–Dancoff approximation (TDA, also known as CI singles). These methods, even when carefully implemented (see e.g. Ref. [43]), scale formally as $\mathcal{O}(N_A^4)$ with the number of atoms. In contrast, TD-DFRT offers the possibility of a formally $\mathcal{O}(N_A^3)$ method for these properties. (Of course formal scalings are only a rough guide for comparison purposes. Techniques such as integral cutoffs can be used to yield improved scalings in practice.) Van Gisbergen, Snijders, and Baerends report having implemented a TD-DFRT method with $\mathcal{O}(N_A^3)$ scaling, for calculation of dynamic polarizabilities and van der Waals coefficients [25], though no algorithmic details were given. The present section describes an $\mathcal{O}(N_A^3)$ TD-DFRT method for calculating excitation spectra and dynamic polarizabilities. This is accomplished by extending the auxiliary function techniques commonly used in DFT SCF programs [44,45] to the treatment of the TD-DFRT coupling matrix. Thus the section begins with a summary of the use of auxiliary function expansions in the SCF step.

4.1. SCF calculations

The introduction of auxiliary function techniques into Gaussian-type orbital (GTO)-based DFT SCF programs by Dunlap, Connolly, and Sabin [44] and Sambe and Felton [45], provided a computationally more efficient way to handle exchange-correlation integrals and allowed the costly four-center electron-repulsion integrals used in evaluating the coulomb interaction to be reduced to three-center integrals. The method described here is that used in the program deMon-KS. The discussion is general for any atomic orbital basis $\{\chi_\mu\}$, presumed real.

Two sets of auxiliary basis functions are introduced, to handle the coulomb and exchange-correlation integrals, respectively. Consider first the coulomb integrals,

$$\langle \chi_\mu | v_{\text{coul}} | \chi_\nu \rangle = [\chi_\mu \chi_\nu | \rho] \,, \tag{49}$$

where the square brackets denote the inner product

$$[f|g] = \int\int f^*(\mathbf{r}) \frac{1}{|\mathbf{r}-\mathbf{r}'|} g(\mathbf{r}') \, d\mathbf{r} d\mathbf{r}' \,. \tag{50}$$

By expanding the charge density in Eq. (49) in an atom-centered GTO auxiliary basis set,

$$\rho(\mathbf{r}) \cong \tilde{\rho}(\mathbf{r}) = \sum_I g_I^{\text{cd}}(\mathbf{r}) a_I \,, \tag{51}$$

the need to evaluate four-center integrals is eliminated and the number of coulomb integrals is reduced from N^4 to MN^2 where N is the number of orbital and M is the number of auxiliary basis functions. The tilde serves as a reminder that, in practice, the auxiliary basis is incomplete and this expansion is thus an approximation. The expansion coefficients a_I are obtained from a least squares fit, i.e. by minimizing $[\rho - \tilde{\rho}|\rho - \tilde{\rho}]$. Normally the minimization is done subject to the constraint

$$\int \tilde{\rho}(\mathbf{r}) \, d\mathbf{r} = n \,, \tag{52}$$

where n is the number of electrons. This leads to

$$a_I = \sum_J \left(S^{\rm cd}\right)^{-1}_{I,J} \left([g_J^{\rm cd}|\rho] - \lambda \int g_J^{\rm cd}(\mathbf{r})\, d\mathbf{r} \right) , \tag{53}$$

where

$$S^{\rm cd}_{I,J} = [g_I^{\rm cd}|g_J^{\rm cd}] \tag{54}$$

is the charge density overlap matrix, and the Lagrange multiplier

$$\lambda = \frac{\sum_{I,J} \int g_I^{\rm cd}(\mathbf{r})\, d\mathbf{r} \left(S^{\rm cd}\right)^{-1}_{I,J} [g_J^{\rm cd}|\rho] - n}{\sum_{I,J} \int g_I^{\rm cd}(\mathbf{r})\, d\mathbf{r}\, (S^{\rm cd})^{-1}_{I,J} \int g_J^{\rm cd}(\mathbf{r}')\, d\mathbf{r}'} . \tag{55}$$

In the absence of the constraint, $\tilde{\rho}$ is simply the projection of ρ onto the space spanned by the auxiliary basis set and may integrate to a total charge slightly different from n. However, in the limit of a complete charge density auxiliary basis set, $\lambda = 0$ and minor differences in fitting philosophy become irrelevant.

Now consider the exchange-correlation integrals. The exchange-correlation potential,

$$v_{\rm xc}^\sigma(\mathbf{r}) = \frac{\delta E_{\rm xc}[\rho_\uparrow, \rho_\downarrow]}{\delta \rho_\sigma(\mathbf{r})} , \tag{56}$$

is expanded in a second atom-centered GTO auxiliary basis set

$$v_{\rm xc}^\sigma(\mathbf{r}) \cong \tilde{v}_{\rm xc}^\sigma(\mathbf{r}) = \sum_I g_I^{\rm xc}(\mathbf{r}) b_I^\sigma , \tag{57}$$

where again the tilde serves to distinguish between $v_{\rm xc}^\sigma(\mathbf{r})$ and its expansion in an auxiliary basis which is, in practice, incomplete. The expansion coefficients are again obtained by a least squares fit, minimizing $\{v_{\rm xc}^\sigma - \tilde{v}_{\rm xc}^\sigma | v_{\rm xc}^\sigma - \tilde{v}_{\rm xc}^\sigma\}$, where the inner product

$$\{f|g\} = \sum_i w_i f^*(\mathbf{r}_i) g(\mathbf{r}_i) \tag{58}$$

is defined by numerical quadrature, with weights w_i, over a molecular grid. This gives

$$b_I^\sigma = \sum_J (S^{\rm xc})^{-1}_{I,J} \{g_J^{\rm xc}|v_{\rm xc}^\sigma\} , \tag{59}$$

where the exchange-correlation overlap matrix

$$S^{\rm xc}_{I,J} = \{g_I^{\rm xc}|g_J^{\rm xc}\} . \tag{60}$$

Thus the exchange-correlation integrals become

$$\langle \chi_\mu | \tilde{v}_{\rm xc}^\sigma | \chi_\nu \rangle = \sum_I \langle \chi_\mu | g_I^{\rm xc} | \chi_\nu \rangle b_I^\sigma = \sum_{I,J} \langle \chi_\mu | g_I^{\rm xc} | \chi_\nu \rangle \, (S^{\rm xc})^{-1}_{I,J} \, \{g_J^{\rm xc}|v_{\rm xc}^\sigma\} . \tag{61}$$

4.2. Post-SCF calculations

The coupling matrix (36) is the sum of a coulomb and an exchange-correlation part,

$$K_{\mu\nu\sigma,\mu'\nu'\sigma'} = K^{\text{coul}}_{\mu\nu\sigma,\mu'\nu'\sigma'} + K^{\text{xc}}_{\mu\nu\sigma,\mu'\nu'\sigma'} = \frac{\partial v^{\text{coul}}_{\mu\nu\sigma}}{\partial P_{\mu'\nu'\sigma'}} + \frac{\partial v^{\text{xc}}_{\mu\nu\sigma}}{\partial P_{\mu'\nu'\sigma'}} . \tag{62}$$

This would normally involve four-center integrals, but these can be avoided by extending the use of the auxiliary functions already introduced in the SCF step to the calculation of the coupling matrix. Explicit differentiation of the auxiliary function expressions for the matrix elements of the coulomb and exchange-correlation potentials yield expressions, involving only three-center integrals, for the coulomb and exchange-correlation parts of the coupling matrix. The result, however, is only approximately hermitian. Although the coupling matrix of Eq. (36) is real and symmetric, a slight loss of hermitivity arises from the auxiliary function expression involving the use of different inner products (denoted by the two types of brackets $\langle\ \rangle$ and $\{\ \}$), in the case of the exchange correlation part, and from the Lagrange multiplier in the case of the coulomb part. Of course this is not a problem in principle since, in the limit of complete auxiliary basis sets and exact numerical integration, there would be no loss of hermitivity. Nevertheless, it is inconvenient in practice, particularly when calculating excitation spectra, since it results in the occasional mixture of nearly degenerate states. Thus a slightly different approach is prefered.

For the coulomb part of the coupling matrix, hermitivity is easily restored by dropping the constraint (52) in determining the auxiliary function expansion coefficients. We have not observed any significant difference in the mean polarizability calculated with and without this constraint, with the auxiliary basis sets we normally use [26]. In the absence of the constraint, the second term in Eq. (53) vanishes, and the coulomb matrix elements are given by

$$v^{\text{coul}}_{\mu\nu\sigma} = \sum_{I,J} [\chi_\mu \chi^*_\nu | g^{\text{cd}}_I] \left(S^{\text{cd}}\right)^{-1}_{I,J} \sum_{\mu',\nu',\sigma'} [g^{\text{cd}}_J | \chi_{\mu'} \chi^*_{\nu'}] P_{\mu'\nu'\sigma'} . \tag{63}$$

Straightforward differentiation then gives

$$K^{\text{coul}}_{\mu\nu\sigma,\mu'\nu'\sigma'} = \sum_{I,J} [\chi_\mu \chi_\nu | g^{\text{cd}}_I] A_{I,J} [g^{\text{cd}}_J | \chi_{\mu'} \chi_{\nu'}] \tag{64}$$

where

$$\boldsymbol{A} = \left(\boldsymbol{S}^{\text{cd}}\right)^{-1} . \tag{65}$$

For the exchange-correlation part of the coupling matrix, rather than differentiating the matrix elements of the exchange-correlation potential (61), the same auxilary basis that was used for the exchange-correlation potential can be used again to expand the derivative of the exchange-correlation potential,

$$\frac{\delta v^\sigma_{\text{xc}}(\mathbf{r})}{\delta \rho_\tau(\mathbf{r}')} \cong \sum_{I,J} g^{\text{xc}}_I(\mathbf{r}) B^{\sigma,\tau}_{I,J}\, g^{\text{xc}}_J(\mathbf{r}') , \tag{66}$$

where the $B^{\sigma,\tau}_{I,J}$ are again determined from a least squares fit over the grid [using the metric implied by the one defined in Eq. (58)]. Thus

$$B^{\sigma,\tau}_{I,J} = \sum_{K,L} (S^{\text{xc}})^{-1}_{I,K} \left\{ g^{\text{xc}}_K \middle| \frac{\delta v^\sigma_{\text{xc}}}{\delta \rho_\tau} \middle| g^{\text{xc}}_L \right\} (S^{\text{xc}})^{-1}_{L,J} \tag{67}$$

where

$$\left\{g_K^{\rm xc}|\frac{\delta v_{\rm xc}^{\sigma}}{\delta\rho_\tau}|g_L^{\rm xc}\right\} = \sum_{i,j} w_i g_K^{\rm xc}(\mathbf{r}_i)\frac{\delta v_{\rm xc}^{\sigma}(\mathbf{r}_i)}{\delta\rho_\tau(\mathbf{r}_j)} w_j g_L^{\rm xc}(\mathbf{r}_j)\,. \tag{68}$$

Just as was the case in the SCF step, this becomes exact in the limit of a complete auxiliary basis set, provided the grid is chosen so that $\boldsymbol{S}^{\rm xc}$ remains invertible. The exchange-correlation part of the coupling matrix can then be written in a form quite analogous to the coulomb part,

$$K^{\rm xc}_{\mu\nu\sigma,\mu'\nu'\sigma'} = \sum_{I,J}\langle\chi_\mu\chi_\nu^*|g_I^{\rm xc}\rangle B_{I,J}^{\sigma,\sigma'}\langle g_J^{\rm xc}|\chi_{\mu'}\chi_{\nu'}\rangle\,. \tag{69}$$

The construction of $\boldsymbol{B}$ can be simplified for local and gradient-corrected functionals, since in this case

$$\frac{\delta v_{\rm xc}^{\sigma}(\mathbf{r})}{\delta\rho_\tau(\mathbf{r}')} = \delta(\mathbf{r}-\mathbf{r}')\frac{\delta v_{\rm xc}^{\sigma}(\mathbf{r})}{\delta\rho_\tau(\mathbf{r})}\,, \tag{70}$$

so that the double sum in Eq. (68) can be reduced to the single sum

$$\left\{g_K^{\rm xc}|\frac{\delta v_{\rm xc}^{\sigma}}{\delta\rho_\tau}|g_L^{\rm xc}\right\} = \sum_{i} w_i g_K^{\rm xc}(\mathbf{r}_i)\frac{\delta v_{\rm xc}^{\sigma}(\mathbf{r}_i)}{\delta\rho_\tau(\mathbf{r}_i)} g_L^{\rm xc}(\mathbf{r}_i) \tag{71}$$

by using the approximate resolution of the identity

$$\delta(\mathbf{r}-\mathbf{r}') = \sum_{I,J} g_I^{\rm xc}(\mathbf{r})\,(S^{\rm xc})^{-1}_{I,J}\,g_J^{\rm xc}(\mathbf{r}')\,. \tag{72}$$

This too becomes exact in the limit of a complete auxiliary basis. Note that there are functionals for which Eq. (70) does not hold, and the double sum (68) must be evaluated in these cases. The Slater coupling matrix of Sec. 5.4 is one example.

The auxiliary function method given here for handling the coupling matrix is most powerful when used in conjunction with semidirect algorithms where the coupling matrix is never actually constructed. Semidirect algorithms are used in TDHF and in other *ab initio* mehtods, both to calculate dynamic polarizabilities and to calculate excitation spectra. For example, in the present case, Davidson's algorithm [46,47] can be used to solve for the first several excitation energies of the eigenvalue problem (46) without ever constructing $\boldsymbol{\Omega}$. All that is required is the ability to evaluate the action of $\boldsymbol{\Omega}$ on a vector, which is trivial except for the contribution of the coupling matrix. However, the action of the coupling matrix on a vector, $\vec{F}$, can be written as the sum of

$$\begin{aligned}\left(\boldsymbol{K}^{\rm coul}\vec{F}\right)_{ij\sigma} &= \sum_\mu c_{\mu,i\sigma}\sum_\nu c_{\nu,j\sigma}\sum_I[\chi_\mu\chi_\nu|g_I^{\rm cd}]\sum_J A_{I,J} \\ &\quad\times\sum_{\mu',\nu'}[g_J^{\rm cd}|\chi_{\mu'}\chi_{\nu'}]\sum_{i'\sigma'}c_{\mu',i'\sigma'}\sum_{j'\sigma'}c_{\nu',j'\sigma'}F_{i'j'\sigma'}\end{aligned} \tag{73}$$

and

$$\begin{aligned}\left(\boldsymbol{K}^{\rm xc}\vec{F}\right)_{ij\sigma} &= \sum_\mu c_{\mu,i\sigma}\sum_\nu c_{\nu,j\sigma}\sum_I\langle\chi_\mu\chi_\nu|g_I^{\rm xc}\rangle\sum_J B_{I,J}^{\sigma,\sigma'} \\ &\quad\times\sum_{\mu',\nu'}\langle g_J^{\rm xc}|\chi_{\mu'}\chi_{\nu'}\rangle\sum_{i'\sigma'}c_{\mu',i'\sigma'}\sum_{j'\sigma'}c_{\nu',j'\sigma'}F_{i'j'\sigma'}\end{aligned} \tag{74}$$

using the expansion of the molecular orbitals in the atomic orbital basis set,

$$\psi_{i\sigma}(\mathbf{r}) = \sum_{\mu} \chi_{\mu}(\mathbf{r}) c_{\mu,i\sigma} \,. \tag{75}$$

By using the method of partial summation as indicated by the arrangement of summation signs in Eqs. (73) and (74), the action of the coupling matrix on a vector can be evaluated rapidly in $\mathcal{O}(N^2M)$ steps with only moderate storage requirements, creating a method which scales formally as the cube of the number of atoms.

5. Functionals

In presenting a molecular formulation and computational method, half of the problem of developing a practical TD-DFRT method for molecules has been addressed. The other half of the problem concerns the exchange-correlation functional, $\mathcal{A}_{\rm xc}$, for which no practical exact expression is known. As indicated earlier, the present work focuses on what can be done within the limits of the adiabatic approximation [Eq. (30)]. This, in effect, serves to focus attention on those aspects of conventional exchange-correlation functionals, $E_{\rm xc}$, where improvement is most needed for calculating response properties.

The present section begins with an illustration of the sort of results that can be obtained in the time-dependent local density approximation (TDLDA). The most important difficulties with results using the LDA functional in TD-DFRT calculations are highlighted. These are seen to be due to the long range behavior of the LDA exchange-correlation potential and the intimately related problem of the TDLDA ionization threshold. A cursory survey is then given of some of the practical approaches which have been taken to alleviate these problems. The section concludes with a detailed treatment of analytic models for the exact exchange potential and its derivative, $\delta v_{\rm x}/\delta\rho$, including a new result for the derivative in the Krieger–Li–Iafrate approximation.

5.1. Illustration of the TDLDA

The present subsection summarizes results for the excitation energies and dynamic polarizability of N_2 in order to illustrate both some general features of the TD-DFRT method and the performance of the TDLDA and the contributions which compose it. We have observed roughly similar behavior for several other small molecules [48]. A more detailed account of the results for N_2 is given in Ref. [26]. Note that the computational method used there differs slightly from that described in the preceding section.

An important feature of TD-DFRT is that it provides a natural way to treat excitations involving excited state multiplets, within DFT. This is an important part of the more general multiplet problem in DFT. The simplest case of an excited-state multiplet is mixing of single excitations out of a doubly occupied orbital in a closed-shell molecule to form a singlet and three degenerate triplet states. A more complicated example is illustrated in Fig. 1. Here a spatial multiplet arises from the $\pi \rightarrow \pi^*$ excitations in N_2, due to the coupling between the x- and y-components of the π orbitals. This results in three distinct many-electron symmetries, which yield eight states (including two pairs of states degenerate by symmetry) when spin is taken into account. In wave function parlance, a linear combination of (at least) four determinants is required for the description of each of these states. The treatment of such a multiplet via the *ad hoc* ΔSCF-based DFT approach would require a detailed group-theoretical analysis. In addition, configurational mixing other than that dictated by symmetry requirements would have to be handled *a posteriori*, guided by comparison with experiment or *ab initio* results, in the ΔSCF-based approach. This latter case includes the commonly encountered situation of mixing of nearly degenerate states arising from a nearly symmetric ("distorted") structure. In contrast, multiplet splittings and assignments of spin multiplicity are obtained naturally in TD-DFRT, without reference to a wave function (just as in any *ab initio* Green function method.) In Fig. 1, the solid lines lead to states which are primarily (but not entirely) symmetry-dictated combinations of $1\pi_u \rightarrow 1\pi_g$ excitations. The dotted lines indicate that a substantial mixing occurs between the $1\pi_u \rightarrow 1\pi_g$ and $2\sigma_u \rightarrow 4\sigma_g$ transitions to form two ${}^1\Sigma_u^+$ states, neither of which can be attributed predominantly to either the $1\pi_u \rightarrow 1\pi_g$ or $2\sigma_u \rightarrow 4\sigma_g$ transitions alone.

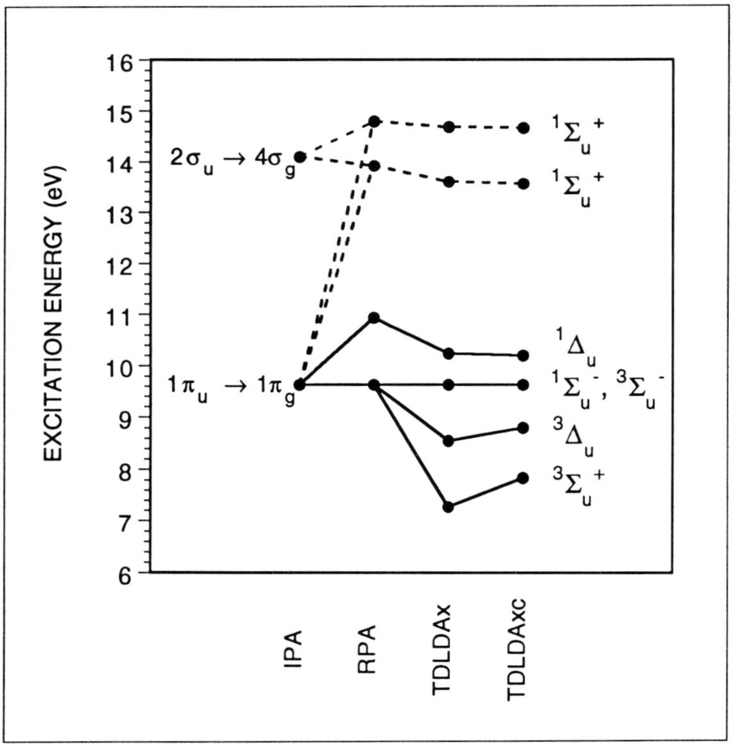

Figure 1. Correlation between coupled and uncoupled states for the $\pi \rightarrow \pi^*$ transitions in nitrogen. Functionals X/LDA for X = IPA, RPA, TDLDAx, and TDLDAxc (see text). Reprinted from Ref. [26].

Table 1
Comparison of TDLDAxc vertical excitation energies for the first eight excited states of N_2 with experiment and with other *ab initio* results. $R_{NN} = 2.074$ bohr. Table reprinted from Ref. [26].

State	Excitation	TDLDAxc[a]	Excitation Energy (eV) Expt[b]	MRCCSD[c]	SOPPA[d]	MCTDHF[f]	TDHF[g]	TDA[h]
			Singlet → Singlet Transitions					
$w^1\Delta_u$	$1\pi_u \rightarrow 1\pi_g$	10.22	10.27	10.54	10.51	10.76	8.75	9.09
$a'^1\Sigma_u^-$	$1\pi_u \rightarrow 1\pi_g$	9.66	9.92	10.09	10.02	10.40	7.94	8.51
$a^1\Pi_g$	$3\sigma_g \rightarrow 1\pi_g$	9.10	9.31	9.28	9.32	9.60	9.76	9.60
			Singlet → Triplet Transitions					
$C^3\Pi_u$	$2\sigma_u \rightarrow 1\pi_g$	10.36	11.19	11.19	11.05	11.43	11.26	11.85[i]
$B'^3\Sigma_u^-$	$1\pi_u \rightarrow 1\pi_g$	9.66	9.67	9.87	9.96	10.07	7.94	8.51
$W^3\Delta_u$	$1\pi_u \rightarrow 1\pi_g$	8.83	8.88	8.93	8.93	8.86	5.80	7.35
$B^3\Pi_g$	$3\sigma_g \rightarrow 1\pi_g$	7.60	8.04	8.05	7.87	8.17	7.62	7.94
$A^3\Sigma_u^+$	$1\pi_u \rightarrow 1\pi_g$	7.88	7.75	7.56	7.91[e]	7.64	3.47	6.25
			Average Error for All Eight States					
		0.25		0.11	0.14	0.27	1.69	1.02

[a] 106CGTO basis.
[b] Computed in Ref. [50] from the spectroscopic constants of Huber and Herzberg [51].
[c] Multireference coupled cluster singles and doubles, BK90 basis set (52 CGTO), Ref. [52].
[d] Second-order polarization propagator approximation, 58 CGTO basis set, $R_{NN} = 2.068$ bohr, Ref. [50].
[e] 36 CGTO basis set, may differ significantly from the 58 CGTO result.
[f] Multiconfigurational time-dependent Hartree–Fock, [11s8p4d1f] basis set, Ref. [53].
[g] Time-dependent Hartree–Fock, [11s8p4d1f] basis set, Ref. [53].
[h] Tamm–Dancoff approximation, DZP (30 CGTO) basis, Ref. [54].
[i] 66 CGTO basis, $R_{NN} = 2.067$ bohr, Ref. [55]; the four other values available from this calculation differ from the ones shown here by ≤ 0.07 eV.

Another interesting aspect of Fig. 1 is that it shows the effect of successive contributions to the coupling matrix on the splitting of the $\pi \rightarrow \pi^*$ excitations. It will be convenient to describe TD-DFRT calculations by denoting the functional as X/Y where X is the level of approximation (functional) used in the coupling matrix and Y is the functional used in the SCF step. For the moment, consider Y = LDA, and four levels of approximation of the coupling matrix: X = IPA, RPA, TDLDAx, or TDLDAxc, which correspond to complete neglect of the coupling matrix, inclusion of the coulomb term in $\boldsymbol{K}$, inclusion of the coulomb and LDA exchange terms in $\boldsymbol{K}$, and inclusion of coulomb and LDA exchange and correlation terms in $\boldsymbol{K}$, respectively. Thus, for example, in the IPA/LDA the coupling matrix is zero so the excitation energies are just simple differences of the LDA orbital energies obtained in the SCF step. For the singlet → singlet transitions, the largest correction to the IPA excitation energies comes from the coupling of the coulomb term (RPA/LDA). Although this already gives reasonable excitation energies, it does significantly over-estimate them. The response of the exchange term (TDLDAx/LDA) reduces the excitation energies, and the coupling of the correlation has little effect. In contrast, the large effect of the coulomb coupling cancels out in the singlet → triplet transitions, so no correction to the IPA/LDA is seen until the response of the exchange is included, and the response of the correlation term is more important in this case. Petersilka and Gross have confirmed the importance of correlation for singlet → triplet energies in their TD-DFRT study of atomic excitation energies [49].

As illustrated in Table 1, the TDLDA (i.e. TDLDAxc/LDA in the notation defined above, also known as the adiabatic local density approximation) does pretty well for the low-lying excitation energies, in comparison with experiment and with other *ab initio* methods. We have not encountered any particular difficulties with triplet excitation energies (at least as long as the correlation functional is included) in contrast to the well-known shortcomings of TDHF for these transitions. The TDLDA also out-performs the TDA, even though the latter is more computationally demanding. Tests on several other molecules have also found the TDLDA to do remarkably well for low-lying excitation energies [48,29]. On the other hand, the TDLDA does not do as well as the more expensive correlated *ab initio* methods. Thus the TDLDA looks likely to provide a cost-effective approach to the calculation of low-lying excitation energies of large molecules. At the same time, it is hoped that improvements in the functional will eventually allow even better results to be obtained, without substantially escalating the cost.

The effect of the several contributions to the coupling matrix on the mean dynamic polarizability parallels that for the singlet excitations, as would be expected from their interrelation through the Cauchy expansion

$$\bar{\alpha}(\omega) = \sum_{l=1}^{\infty} S_{-2l}\,\omega^{2(l-1)}\,, \tag{76}$$

where the coefficients

$$S_{-2l} = \sum_{I} \omega_I^{-2l} f_I \tag{77}$$

are obtained from the SOS formula [Eq. (11)]. This is illustrated in Table 2. Note that the TRK sum, S_0, is independent of the level of approximation of the coupling matrix, (though it does of course depend on the basis set). While decontracting the Sadlej basis gives a dramatic improvement in the TRK sum, it has relatively little effect on the other Cauchy moments. While the TDLDA does pretty well for the mean static polarizability, S_{-2}, it typically (but not always) slightly overestimates this quantity. This overestimate is increasingly magnified in the successively higher Cauchy moments, and unlike the case for the low-lying excitation energies, the TDLDA clearly does not do better than TDHF for the higher Cauchy moments. Similar results concerning the dynamic polarizability have been observed for several other molecules [48,27]. This difficulty arises from limitations of the LDA, which are critical for the higher excitations. This is the subject of the following subsections.

5.2. Location of the TD-DFRT ionization threshold

The character of the higher excitation energies obtained in TD-DFRT depends, of course, on the location of the TD-DFRT ionization threshold. Thus the present subsection starts by formally identifying where ionization occurs in the TD-DFRT excitation manifold. This was first done by Zangwill and Soven [15]. The present treatment is an elaboration on that of Mahan and Subbaswamy (Ref. [20] pp. 97-98). The result holds for essentially any exchange-correlation functional. This is followed by a discussion, based on the work of Almbladh and von Barth [60] and of Russier [61], of the relation between the ionization potential and the HOMO energy in the limit of the exact functional. The

Table 2
Comparison of TD-DFRT Cauchy coefficients (in a.u.) with those from other *ab initio* methods and with experiment. $R_{\text{NN}} = 2.074$ bohr. Adapted from Ref. [26].

Method	Basis Set	S_0	S_{-2}^a	S_{-4}	S_{-6}
	TD-DFRT Calculations				
IPA/LDA	Sadlej	10.44	21.56	115.1	769.2
RPA/LDA	Sadlej	10.44	10.95	28.42	97.79
TDLDAx/LDA	Sadlej	10.44	12.00	33.99	125.9
TDLDAxc/LDA	Sadlej	10.44	12.11	34.83	131.3
	DSadlej[k]	13.99	12.19	34.00	128.2
	Experiment[b]				
		14[c]	11.74	30.11	101.8
	Other theory				
TDHF[d]	[11s8p4d1f]	13.96	11.56	27.60	100.51
TDHF[e]	[7s5p3d]		11.44	27.53	83.56
TDHF[f]	DSadlej[k]	14.01	11.54	27.39	83.84
MCTDHF[g]	[11s8p4d1f]		11.06	25.61	68.53
MCTDHF[h]	[8s6p4d1f]			28.92	94.51
SOPPA[i]	DSadlej[k]		11.42	29.24	98.6

[a] Mean static polarizability.
[b] Experimentally derived values of Ref. [56].
[c] Exact value equals number of electrons.
[d] Time-dependent Hartree–Fock calculation of Ref. [53].
[e] Time-dependent Hartree–Fock calculation of Ref. [57].
[f] Time-dependent Hartree–Fock calculation of Ref. [58].
[g] Multiconfigurational time-dependent Hartree–Fock calculation of Ref. [53]. Six orbital active space.
[i] Multiconfigurational time-dependent Hartree–Fock calculation of Ref. [59]. Twelve orbital active space.
[j] Second-order polarization propagator approximation calculation of Ref. [58].
[k] Decontracted Sadlej basis.

discussion of exact functionals serves to illuminate improvements needed in approximate functionals. Note that the treatment is for systems whose spatial extent is finite.

Consider adding a time-dependent perturbation $\hat{w}(\mathbf{r},t)$ to a system initially in a bound, stationary state, satisfying the time-independent Kohn–Sham equation

$$\hat{F}_\sigma \psi_{i\sigma}(\mathbf{r}) = \epsilon_{i\sigma}\psi_{i\sigma}(\mathbf{r})\,. \tag{78}$$

Writing the perturbed orbitals as

$$\psi_{i\sigma}(\mathbf{r},t) = e^{-i\epsilon_{i\sigma}t}\left[\psi_{i\sigma}(\mathbf{r}) + \delta\psi_{i\sigma}(\mathbf{r},t) + \delta^2\psi_{i\sigma}(\mathbf{r},t) + \mathcal{O}(w^3)\right]\,, \tag{79}$$

the charge density is

$$\begin{aligned}\rho(\mathbf{r},t) \;=\; & \sum_{i\sigma} f_{i\sigma}\left[|\psi_{i\sigma}(\mathbf{r})|^2 + \psi_{i\sigma}(\mathbf{r})\delta\psi^*_{i\sigma}(\mathbf{r},t) + \delta\psi_{i\sigma}(\mathbf{r},t)\psi^*_{i\sigma}(\mathbf{r})\right. \\ & \left. + \psi_{i\sigma}(\mathbf{r})\delta^2\psi^*_{i\sigma}(\mathbf{r},t) + \delta^2\psi_{i\sigma}(\mathbf{r},t)\psi^*_{i\sigma}(\mathbf{r}) + |\delta\psi_{i\sigma}(\mathbf{r},t)|^2 + \mathcal{O}(w^3)\right]\,. \end{aligned} \tag{80}$$

Increasing the perturbing frequency, ω, from zero will first lead to a simple perturbation of the density, then to excitations, and finally to excitations to continuum (i.e. ionized) states. We will use the asymptotic behavior of the charge density as a criterion to identify the ionization threshold: Since ionized electrons obey free-particle boundary conditions, the charge density will become oscillatory, instead of exponentially decaying, at very large r, when ionization has occured. Since the unperturbed (bound) orbitals are exponentially decaying, it is evident from Eq. (80) that this change in the behavior of the charge density occurs when the first-order response of some orbital, $\delta\psi_{i\sigma}(\mathbf{r},t)$, becomes oscillatory asymptotically. In order to avoid working with singular equations, the system should be imagined as being in a large but finite box with ω chosen off resonance so that $\delta\rho$ and hence δv_{SCF} are finite and are in fact infinitessimal when w is infinitessimal.

Substituting Eq. (79) for the perturbed orbitals into the time-dependent Kohn–Sham equation (27) leads to the first-order perturbation equation,

$$\left(i\frac{\partial}{\partial t} + \epsilon_{i\sigma} - \hat{F}_\sigma\right)\delta\psi_{i\sigma}(\mathbf{r},t) = \left[\hat{w}(\mathbf{r},t) + \delta v_{\mathrm{SCF}}(\mathbf{r},t)\right]\psi_{i\sigma}(\mathbf{r})\,. \tag{81}$$

Fourier transforming with respect to t gives

$$\left(\omega + \epsilon_{i\sigma} - \hat{F}_\sigma\right)\delta\psi_{i\sigma}(\mathbf{r},\omega) = \left[\hat{w}(\mathbf{r},\omega) + \delta v_{\mathrm{SCF}}(\mathbf{r},\omega)\right]\psi_{i\sigma}(\mathbf{r}) \tag{82}$$

(since $\delta\psi_{i\sigma}(t=\pm\infty)=0$ in adiabatic perturbation theory.) The description of light as a classical electromagnetic field is a good example of $\hat{w}(\mathbf{r},t)$ to have in mind. Suppose for simplicity that the perturbation is monochromatic with frequency ω (energy $\hbar\omega$),

$$\hat{w}(\mathbf{r},t) = \hat{w}(\mathbf{r})\cos\omega t\,, \tag{83}$$

but infinitesimal amplitude, $\hat{w}(\mathbf{r})$, and that the phases of the unperturbed orbitals are chosen so that the unperturbed orbitals are real. Then it is easy to show using Eq. (82) that $\delta\psi_{i\sigma}$ is also monochromatic with the same frequency,

$$\delta\psi_{i\sigma}(\mathbf{r},t) = \delta\psi^{\mathrm{R}}_{i\sigma}(\mathbf{r})\cos\omega t + i\delta\psi^{\mathrm{I}}_{i\sigma}(\mathbf{r})\sin\omega t\,, \tag{84}$$

for some real functions $\delta\psi_{i\sigma}^{\rm R}$ and $\delta\psi_{i\sigma}^{\rm I}$. Note that $\delta v_{\rm SCF}$ has the same time-dependence as

$$\delta\rho(\mathbf{r},t) = 2\sum_{i\sigma} \psi_{i\sigma}(\mathbf{r})\delta\psi_{i\sigma}^{\rm R}(\mathbf{r}) \cos\omega t \,. \tag{85}$$

In order to determine the (spatial) asymptotic behavior of $\delta\psi_{i\sigma}$, consider the asymptotic form of the first-order perturbation equation (82). As long as $[v_{\rm xc}^{\sigma}(\mathbf{r}) - v_{\rm xc}^{\sigma,\infty}]$ for the unperturbed system [and the RHS of Eq. (82)] fall off at least as fast as $1/r$, then the asymptotic form of Eq. (82) is

$$\left(\omega + \epsilon_{i\sigma} + \frac{1}{2r}\frac{\partial^2}{\partial r^2}r - v_{\rm xc}^{\sigma,\infty} + \mathcal{O}(1/r)\right)\delta\psi_{i\sigma}(\mathbf{r},\omega) = 0 \,, \tag{86}$$

where the constant $v_{\rm xc}^{\sigma,\infty}$ is the asymptotic value of the exchange-correlation potential for the unperturbed system. (More will be said about $v_{\rm xc}^{\sigma,\infty}$ at the end of this subsection.) The asymptotic equation (86) has the solution,

$$\delta\psi_{i\sigma}(\mathbf{r},\pm\omega) \rightarrow \delta\Upsilon_{i\sigma}^{(\pm\omega)}(\theta,\phi)\exp\left[-\sqrt{2(v_{\rm xc}^{\sigma,\infty} - \epsilon_{i\sigma} \mp \omega)}\; r\right] \,, \tag{87}$$

where $\delta\Upsilon_{i\sigma}^{(\pm\omega)}(\theta,\phi)$ accounts for wave function normalization and angular effects. This is oscillatory when

$$v_{\rm xc}^{\sigma,\infty} - \epsilon_{i\sigma} \pm \omega \leq 0 \,. \tag{88}$$

Thus the TD-DFRT ionization threshold is

$$v_{\rm xc}^{\sigma,\infty} - \epsilon_{{\rm HOMO},\sigma} \equiv -\omega_{{\rm HOMO},\sigma}^{\rm TD-DFRT} \,. \tag{89}$$

This important theorem was first enunciated in the seminal work of Zangwill and Soven [15]. Note that this relation is independent of the choice of functional (within the adiabatic approximation), as long as $v_{\rm xc}^{\sigma}$ does not approach its asymptotic value more slowly than $1/r$.

This raises the question of the relation between the ionization potential and the HOMO energy in the case of the exact exchange-correlation potential, which will now be considered in some detail. These are properties of the unperturbed system and will be treated entirely within the context of time-independent Kohn–Sham theory. The treatment is basically that of Almbladh and von Barth [60] as revisited by Russier [61]. Again, the argument hinges on the asymptotic behavior. Since correlation is usually short ranged,[2] the long range behavior of the exact $v_{\rm xc}^{\sigma}$ will be the same as for the HF exchange, namely

$$\begin{aligned} v_{\rm xc}^{\sigma}(\mathbf{r}) &\stackrel{r\to\infty}{\longrightarrow} -\int \frac{\rho_{{\rm HOMO},\sigma}(\mathbf{r}')}{|\mathbf{r}-\mathbf{r}'|}\,d\mathbf{r}' + v_{\rm xc}^{\sigma,\infty} \\ &\stackrel{r\to\infty}{\longrightarrow} -\frac{f_{{\rm HOMO},\sigma}}{r} + v_{\rm xc}^{\sigma,\infty} \,, \end{aligned} \tag{90}$$

where

$$\rho_{i\sigma}(\mathbf{r}) = f_{i\sigma}|\psi_{i\sigma}(\mathbf{r})|^2 \tag{91}$$

[2]Superconductors are an obvious exception where correlation lengths assume macroscopic proportions. They are also far from the molecular applications we have in mind.

and of course v_{xc}^{σ} depends only on the charge density for spin σ. Note that, being the most diffuse occupied orbital, only the HOMO is important at very large r. Also, in this region, v_{xc}^{σ} acts primarily to correct the self-interaction error for the HOMO in the coulomb term.

The asymptotic form of the static Kohn–Sham equation (78) is thus

$$\left(-\frac{1}{2r}\frac{\partial^2}{\partial r^2}r - \frac{Q + f_{\text{HOMO},\sigma}}{r} + v_{\text{xc}}^{\sigma,\infty} + \mathcal{O}(1/r^2)\right)\psi_{i\sigma}(\mathbf{r}) = \epsilon_{i\sigma}\psi_{i\sigma}(\mathbf{r}), \tag{92}$$

where Q is the net charge on the system. Neglecting terms of $\mathcal{O}(1/r)$ and solving yields

$$\psi_{i\sigma}(\mathbf{r}) \overset{r\to\infty}{\longrightarrow} \Upsilon_{i\sigma}(\theta,\phi)\exp\left(-\sqrt{2(v_{\text{xc}}^{\sigma,\infty} - \epsilon_{i\sigma})}\,r\right). \tag{93}$$

Thus the asymptotic behavior of the charge density is

$$\rho_{\sigma}(\mathbf{r}) \overset{r\to\infty}{\longrightarrow} |\Upsilon_{\text{HOMO},\sigma}(\theta,\phi)|^2 \exp\left(-2\sqrt{2(v_{\text{xc}}^{\sigma,\infty} - \epsilon_{\text{HOMO},\sigma})}\,r\right). \tag{94}$$

Now, the asymptotic behavior of the exact charge density, obtained from Dyson's quasi-particle equation, has the same form but with $(v_{\text{xc}}^{\sigma,\infty} - \epsilon_{\text{HOMO},\sigma})$ replaced by the true ionization potential, $-\omega_{\text{HOMO},\sigma}$ [60]. Since the Kohn–Sham exchange-correlation potential is defined by the requirement that it yield the true charge density, it follows that the exact ionization potential,

$$-\omega_{\text{HOMO},\sigma} = v_{\text{xc}}^{\sigma,\infty} - \epsilon_{\text{HOMO},\sigma}, \tag{95}$$

in the limit of the exact exchange-correlation functional. Note that the RHS is identical to the TD-DFRT ionization threshold [Eq. (89)]. Thus, in the limit of the exact functional, the TD-DFRT ionization threshold is at the true ionization potential. Of course, when approximate functionals are used in TD-DFRT, the observed ionization threshold will depend on the properties of the particular functional, according to Eq. (89). This will be discussed further in Sec. 5.3.

Up until this point, no explanation for the introduction of the constant $v_{\text{xc}}^{\sigma,\infty}$ has been given. The key to this mystery is a theorem due to Russier [61] which shows that the constant, $v_{\text{xc}}^{\sigma,\infty}(N_e)$, depends upon the number of electrons, N_e, and is completely determined within the fractional occupation number formalism. His theorem,

$$v_{\text{xc}}^{\sigma,\infty}(N_e) = \int_0^1 \left[\epsilon_{\text{HOMO},\sigma}(N_e) - \epsilon_{\text{HOMO},\sigma}(N_e - f_{\text{HOMO},\sigma})\right] df_{\text{HOMO},\sigma}, \tag{96}$$

follows trivially from Janak's theorem [62],

$$\begin{aligned}\omega_{\text{HOMO},\sigma} &= \int_0^1 \frac{\partial E}{\partial f_{\text{HOMO},\sigma}}\, df_{\text{HOMO},\sigma} \\ &= \int_0^1 \epsilon_{\text{HOMO},\sigma}(N_e - f_{\text{HOMO},\sigma})\, df_{\text{HOMO},\sigma},\end{aligned} \tag{97}$$

and Eq. (95), for the exact exchange-correlation potential. However the ease of derivation does not make Russier's theorem any less significant in that it expresses in a particularly clear form that the choice of "energy zero" for $v_{\text{xc}}^{\sigma,\infty}$ is intimately related to the choice of how to treat fractional occupation numbers in DFT. It is clear from Eq. (96) that $v_{\text{xc}}^{\sigma,\infty}(N_e)$

will not normally be zero for the exact functional, unless $\epsilon_{\text{HOMO},\sigma}$ were independent of occupation number, and will typically vary with N_e.

Alternatively, the conclusion that the value of $v_{\text{xc}}^{\sigma,\infty}$ is tied to the treatment of fractional particle number would also follow from the assumption that the functional E_{xc} contains a term, $F(N_e)$, dependent only on particle number, N_e. This amounts to a choice of the energy zero in E_{xc}, but one which may depend on particle number. This term could, for example, be zero at integer values of N_e, though Russier's theorem suggests $F(N_e) = \int_0^{N_e} v_{\text{xc}}^{\sigma,\infty}(M)\,dM$ in the usual fractional occupation number formalism. Whether or not $F(N_e)$ is zero for integer N_e, v_{xc}^{σ} will then contain a constant term, $F'(N_e)$, which depends on the rate of variation of E_{xc} with fractional particle number.

Note that this resembles the infamous derivative discontinuity in DFT [63,64] in that there is a constant shift of the exchange-correlation potential for whose theoretical treatment the treatment of fractional particle number also plays a decisive rôle. However, no supposition regarding derivative discontinuities (or their existence or non-existence) has been used here. (See Ref. [65] pp. 19-24, 145-148 for a discussion of derivative discontinuities and their possible importance for the calculation of bandgaps using DFT.)

5.3. Survey of approximate functionals for TD-DFRT

TD-DFRT calculations are very sensitive to the large r behavior of the exchange-correlation potential. This is evident on physical grounds for polarizabilities since the most polarizable part of the charge density is at large r. It is also clear from the previous subsection that the asymptotic behavior of the exchange-correlation potential determines the placement of the ionization threshold in TD-DFRT. This, in turn, is important for the higher excitations and Cauchy moments, including the static polarizability. According to Eq. (90), the exact exchange-correlation potential should decay asymptotically as $-1/r$. In the LDA, the exchange-correlation potential falls off much faster, roughly as $\sim \rho^{1/3}$, which is to say exponentially [Eq. (94)]. One result is that v_{xc}^{σ} is less attractive than it should be so that $-\epsilon_{\text{HOMO},\sigma}$ is significantly smaller than the ionization potential. Consequently the TDLDA ionization threshold is too low. Another consequence is that the charge density tends to be too diffuse, and the electrons are too weakly bound, leading to polarizabilities which may be too large in comparison with experiment. The sum-over-states expression for the polarizability (43), which is the basis for the calculation of TD-DFRT excitation energies, provides a rigorous link between the underestimation of the ionization threshold and overestimation of the polarizability. It should be noted, however, that the polarizability is *not* always overestimated in the LDA. In fact, the LDA polarizability appears to be an underestimate for the alkaline-earth atoms as well as for Zn and Cd [16] and for the sodium atoms and dimer [66]. Such observations are compatible with the TD-DFRT ionization threshold being too low, once the contribution of oscillator strengths to the TD-DFRT SOS expression for the polarizability is taken into account. Thus one would expect that the substantial errors in the high-lying excitations in the TDLDA would be resolved, and the dynamic polarizability, which is already fairly good in the TDLDA, be improved, by rectifying the large r behavior of v_{xc}. Unfortunately, other functionals commonly used in molecular DFT calculations do not correct the asymptotic behavior of v_{xc}. Thus they lead to results similar to those obtained in the LDA, as illustrated for water in the first five lines of Table 3, and documented more

Table 3
Static properties of water calculated with various functionals (Sadlej basis set), from Ref. [68]. Values in parentheses taken from Ref. [27].

Functional	H_2O $-\epsilon_{\mathrm{HOMO}}$ (eV)	μ (D)	$\bar{\alpha}$ (bohr^3)	
LDA[a]	7.33	1.87	10.56	(10.53)
B88x[b]+P86c[c]	7.29	1.82	10.44	
TDLDA/(B88x+P86c)[g]				(10.28)
PW86x[d]+P86c	7.39	1.81	10.70	
B88x+LYP88[e]				(10.63)
LB94[f]	13.05	1.97	8.15	
TDLDA/LB94[g]			9.19	(9.20)
Experimental values				
	12.61[h]	1.85[j]	9.64[l]	
Hartree–Fock results				
	13.84[i]	1.98[k]	8.51[k]	

[a] Local density approximation using the parameterization of Vosko, Wilk, and Nusair [79].
[b] Becke's 1988 gradient-correction for exchange [80].
[c] Perdew's 1986 gradient-correction for correlation [81].
[d] Perdew and Wang's 1986 gradient-correction for exchange [82].
[e] Lee Yang and Perdew's 1988 gradient correction for correlation [83].
[f] van Leeuwen and Baerends exchange-correlation functional [84].
[g] TDLDA/Y means that the functional Y was used in the SCF step and the coupling matrix in the post-SCF calculation was evaluated in the TDLDA.
[h] First ionization potential [85].
[i] Ref. [86]. [j] Ref. [87]. [k] Ref. [88]. [l] Ref. [56].

extensively elsewhere [67,27,68]. Hybrid schemes[69,70] which mix local and gradient-corrected density functionals with Hartree-Fock exchange may be expected to improve the asymptotic behavior, although the asymptotic behavior will go as $-a_0/r$ where a_0 is some constant other than one (it is 0.5 in Becke's original half-and-half functional[69]) and, of course, computing the Hartree-Fock exchange is computationally more demanding than is a purely local density approximation or gradient-corrected functional method alone. On the other hand, errors in the plasmon frequency and in the polarizability calculated using the LDA have helped to fuel a re-examination of functionals for calculating response properties in the jellium sphere, and other spherical, models [71–73,19,74–78,18]. Note, however, that there may be other reasons for the observed differences between experimental polarizabilities and those calculated in these models (see e.g. Ref. [66]). The present subsection contains a brief summary of a few of the methods that have been devised to improve the large r behavior of v_{xc}, with an emphasis on applicability to molecular calculations.

The scissors operator approach offers a simple *ad hoc* first correction to the TDLDA. For a given molecule, the $i\sigma$ ionization potential, $-\omega_{i\sigma}$, is observed to differ from the negative of the corresponding LDA orbital energy by roughly an orbital-independent constant shift,

$$-\omega_{i\sigma} = -\epsilon_{i\sigma} + \Delta \,. \tag{98}$$

This shift, whose resemblence to the $v_{\text{xc}}^{\sigma,\infty}$ of the previous section is striking, is about 5 eV in small main group molecules [68]. Levine and Alan have suggested that TDLDA calculations be corrected by increasing by Δ (whose value is obtained empirically) the energy difference between occupied and unoccupied orbitals and then using these shifted energies together with the unmodified orbitals in a post-SCF TDLDA calculation [89]. They used this scissor operator approach to improve the frequency-dependent dielectric functions of silicon and germanium. Since then, Zhong, Levine and Wilkins have used this approach in the calculation of dynamic polarizabilities for atoms [90], and Stener, Decleva, and Lisini did something similar in their TD-DFRT calculation of the partial photoionization cross-sections in noble gases using the LB94 functional [91] when they replaced the occupied orbital energies with the corresponding negative ionization potentials. Note however that scissors-operator type approaches do not change the *shape* of v_{xc}^{σ}, which is the fundamental problem. This is why Stener, Decleva, and Lisini also used the LB94 functional discussed below. Also note that the scissors operator approach is expected to lead to a further decrease of those polarizabilities that are already underestimated in the LDA.

A second approach is based on the observation that the fundamental action of v_{xc}^{σ} in the asymptotic region is to correct the self-interaction error in the coulomb term. The presence of a self-interaction error is most evident for one-electron systems where the coulomb and exchange-correlation energies should exactly cancel each other. Typically this is not the case for approximate exchange-correlation functionals and so offers an obvious target for efforts aimed at improving functionals. Several forms of self-interaction corrections (SICs) have been proposed, beginning with the very old SIC of Fermi and Amaldi [92],

$$E_{\text{SIC}}[\rho] = -\frac{1}{N_e}\left(\frac{1}{2}\int\int \frac{\rho(\mathbf{r})\rho(\mathbf{r}')}{|\mathbf{r}-\mathbf{r}'|}\, d\mathbf{r}d\mathbf{r}' + E_{\text{xc}}\right), \tag{99}$$

which has at least the advantage of avoiding some of the computational problems of more sophisticated SICs. However the most popular SIC is that of Perdew and Zunger [93] who suggested adding the SIC,

$$E_{\text{SIC}}[\{\rho_{i\sigma}\}] = -\sum_{i\sigma}\left(E_{\text{xc}}[\rho_{i\sigma}] + \frac{1}{2}\int\int \frac{\rho_{i\sigma}(\mathbf{r})\rho_{i\sigma}(\mathbf{r}')}{|\mathbf{r}-\mathbf{r}'|}\, d\mathbf{r}d\mathbf{r}'\right), \tag{100}$$

to the DFT energy expression. The associated exchange-correlation potential SIC entering into the Kohn–Sham equation for orbital $\psi_{i\sigma}$ is then orbital-dependent,

$$v_{\text{SIC}}^{i\sigma}(\mathbf{r}) = -\left(v_{\text{xc}}^{\sigma}[\rho_{i\sigma}](\mathbf{r}) + \int \frac{\rho_{i\sigma}(\mathbf{r}')}{|\mathbf{r}-\mathbf{r}'|}\, d\mathbf{r}'\right). \tag{101}$$

Straightforward implementation of the Perdew–Zunger approach increases the computational cost considerably since one must now work with N_e different orbital Hamiltonians and the total energy is no longer invariant under a unitary transformation of the occupied orbitals. Thus many minor modifications of the Perdew–Zunger idea may be found in the literature (e.g. orthogonalization of the orbitals[72] or use of the HOMO SIC for all the orbitals [73].) Despite these difficulties, SICs have been useful in TD-DFRT calculations for atoms [94] and the jellium sphere model of clusters [71–73]. Since the Perdew–Zunger SIC leads to singularities in the derivative of the LDA exchange potential,

$$\frac{\delta v_{\mathbf{x}}^{\sigma}[\rho_{i\sigma}](\mathbf{r})}{\delta \rho_{i\sigma}(\mathbf{r})} = \text{const} \times \rho_{i\sigma}^{-1/3}(\mathbf{r}) , \tag{102}$$

at the positions of the nodes of the orbitals Mahan and Subbaswamy ([20] p. 88) recommended that the SIC be used only during the SCF step, i.e. that a TDLDA/SIC-LDA calculation be performed. A simple Fermi–Amaldi-type SIC to the coupling matrix,

$$\boldsymbol{K}_{\text{SIC}} = -\frac{1}{N_e}\boldsymbol{K} , \tag{103}$$

has also been used [73]. Interestingly, SICs can lead to either increased [71] or decreased [95] polarizabilities, depending on the nature of the system (and possibly on the precise formulation of the SIC). Thus, although the Perdew–Zunger SIC generally leads to more negative orbital energies and hence to more tightly bound electrons (Ref. [20] p. 72), when the same SIC is applied to the jellium sphere model of sodium clusters at the exchange-only level, the SIC is actually found to decrease the cluster volume for small clusters but to increase the cluster volume for large clusters [19].

A third approach to improving TD-DFRT calculations is to design orbital-independent exchange-correlation potentials which fall off asymptotically as $-1/r$. One such approach is the modified weighted density approximation. The original weighted density approximation (WDA) of Alonso and Girifalco [96] and Gunnarson, Jonson, and Lundqvist [97] is based on the observation that the exact exchange-correlation energy can be written as

$$E_{\text{xc}} = \frac{1}{2}\int\int \frac{\rho(\mathbf{r})\rho(\mathbf{r}')G(\mathbf{r},\mathbf{r}')}{|\mathbf{r}-\mathbf{r}'|}\, d\mathbf{r}d\mathbf{r}' , \tag{104}$$

where $G(\mathbf{r},\mathbf{r}')$ is the so-called exchange-correlation hole. It satisfies the normalization condition

$$\int \rho(\mathbf{r})G(\mathbf{r},\mathbf{r}')\, d\mathbf{r} = -1 . \tag{105}$$

The WDA is obtained from the *ansatz*,

$$G(\mathbf{r},\mathbf{r}') = G_{\text{HEG}}(|\mathbf{r}-\mathbf{r}'|; \tilde{\rho}(\mathbf{r})) , \tag{106}$$

where G_{HEG} is the exchange-correlation hole for a homogeneous electron gas of density $\tilde{\rho}$. The LDA is obtained by taking $\tilde{\rho}$ to be the density ρ. The WDA goes a step beyond the LDA by obtaining $\tilde{\rho}$ as the solution of the integral equation obtained by substituting Eq. (106) into (105). This equation can be solved for systems with spherical symmetry [74–77,18]. The exchange-correlation potential corresponding to Eq. (104) is given by

$$v_{\text{xc}}(\mathbf{r}) = \int \frac{\rho(\mathbf{r}')G(\mathbf{r},\mathbf{r}')}{|\mathbf{r}-\mathbf{r}'|}\, d\mathbf{r}' + \frac{1}{2}\int\int \frac{\rho(\mathbf{r}')\rho(\mathbf{r}'')}{|\mathbf{r}'-\mathbf{r}''|}\frac{\delta G(\mathbf{r}',\mathbf{r}'')}{\delta \rho(\mathbf{r})}\, d\mathbf{r}'d\mathbf{r}'' . \tag{107}$$

Substituting the WDA *ansatz* (106) into Eq. (107) results in a potential which falls off asymptotically as $-1/r$ [98,99]. This is refered to as the modified WDA because it is not the same as first making the WDA *ansatz* for E_{xc} and then obtaining the exchange-correlation potential as the functional derivative, which does not yield $-1/r$ asymptotic behavior for v_{xc}. Interestingly, the WDA coupling matrix is not diagonal in the position representation [74]. The modified WDA has been used to good effect in calculations of the optical properties of alkali metal clusters using the jellium sphere and spherically averaged pseudopotential models [78,18]. Unfortunately, it is too elaborate to be of much practical interest for molecular calculations.

There have been several recent efforts [84,100,101] to combine computationally simple forms having $-1/r$ asymptotic behavior with the LDA.

Thus, motivated by the asymptotic form of the HF exchange energy density, Santamaria [101] has proposed adding a correction

$$\epsilon_{\text{x3}}^{\sigma} = -\frac{1}{2\rho_\sigma(\mathbf{r})N_a^\sigma}\int \frac{\rho_a^\sigma(\mathbf{r})\rho_a^\sigma(\mathbf{r}')}{|\mathbf{r}-\mathbf{r}'|}\,d\mathbf{r}', \tag{108}$$

to the spin σ exchange energy density, where

$$\rho_a^\sigma = \rho_\sigma\left[1-\left(1+\gamma\rho_\sigma^{-1/3}\right)\exp\left[-\gamma\rho_\sigma^{-1/3}\right]\right], \tag{109}$$

$$N_a^\sigma = \int \rho_a^\sigma(\mathbf{r})\,d\mathbf{r}, \tag{110}$$

and γ is a fitting parameter. He has demonstrated that the resulting exchange potential has $-1/r$ asymptotic behavior.

In another approach, Baerends and coworkers [84] have proposed that the term,

$$\Delta v_{\text{xc}}^\sigma = -\beta\rho_\sigma^{1/3}\frac{x_\sigma^2}{1+3\beta x_\sigma \sinh^{-1}(x_\sigma)}, \tag{111}$$

nearly the same as that originally used by Becke as a correction to the exchange energy-density [80], be added to the LDA exchange-correlation potential. Here

$$x_\sigma = \frac{|\vec{\nabla}\rho_\sigma|}{\rho_\sigma^{4/3}} \tag{112}$$

and the parameter β was obtained by a least squares fit to an accurate exchange-correlation potential for the beryllium atom. We refer to this as the LB94 functional. A similar approach has been pursued by Lembarki, Rogemond, and Chermette [100]. The LB94 functional has been used to calculate photoionization cross-sections in noble gas atoms in a mixed LB94 and scissors operator type of approach [91]. Figure 2 shows our TD-DFRT calculations of the excitation energies of N_2 comparing the TDLDA (i.e. TDLDA/LDA) and TDLDA/LB94 functionals [68]. Although, as pointed out in Subsec. 5.1, the lower excitation energies from the TDLDA are quite good, there is a clear collapse of the TDLDA spectrum above the TD-DFRT LDA ionization threshold of $-\epsilon_{\text{HOMO}}^{\text{LDA}} = 10.38$ eV. The LB94 functional substantially improves $-\epsilon_{\text{HOMO}}$ in comparison to experimental ionization potentials of both atoms [84] and molecules [68]. Since the TD-DFRT LB94 ionization threshold is at $-\epsilon_{\mathrm{HOMO}}^{\mathrm{LB94}} = 15.90$ eV, the higher excitation energies in Figure 2 are dramatically improved. Unfortunately, this is accompanied by a deterioration in the quality of the lower excitation energies in going from the TDLDA/LDA to the TDLDA/LB94, indicating a probable deterioration of v_{xc}^{σ} in the valence region of space. This is consistent with the deterioration of the LB94 relative to LDA dipole moment seen for H_2O in Table 3. In addition, as illustrated in Table 3, although the LB94 functional often improves the polarizability when used only in the SCF step (i.e. TDLDA/LB94) [27,68], even this mixed calculation often overcorrects the LDA, and purely LB94 calculations often overcorrect the polarizability to such an extent that the agreement with experiment is worse than for the LDA [68]. Thus, while the LB94 functional represents a valuable step in the right direction, further work is still needed to develop improved functionals with the correct asymptotic behavior of v_{xc}.

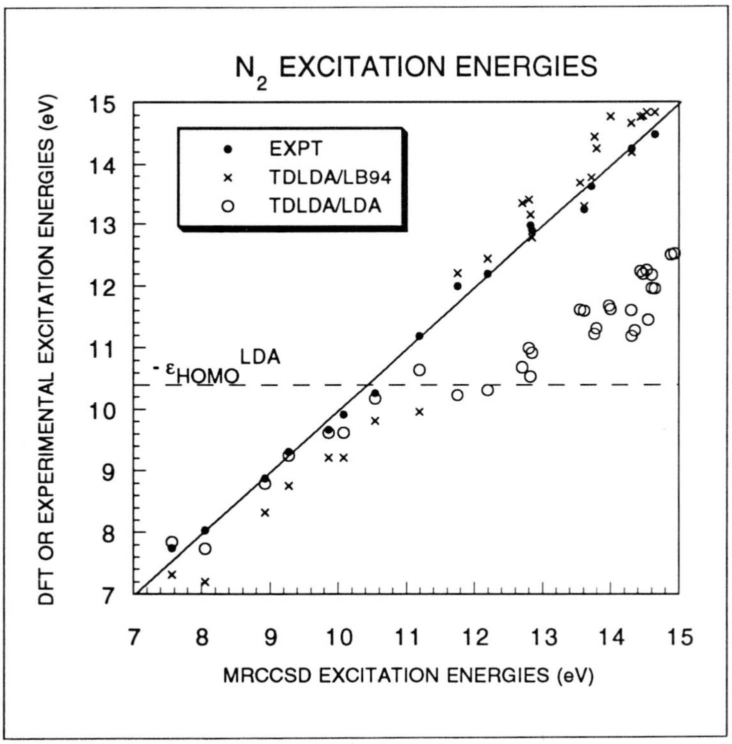

Figure 2. From Ref. [68]. Comparison of the first several TDLDA/LDA and TDLDA/LB94 excitation energies for N_2 with those from the multireference coupled cluster singles and doubles calculation of Ref. [52].

Thus far, none of the proposals for improving the exchange-correlation potential, except possibly the scissors operator approach, take into account Russier's theorem that the potential does not always go to zero at infinity. Future attempts to graft the correct asymptotic form onto existant exchange-correlation potentials should take care neither to disturb too much the form of v_{xc}^{σ} in the valence region of space where the density is known to be reasonably good, nor to arbitrarily force the potential to go to zero at infinity when the charge density goes to zero.

5.4. Exact exchange-only coupling matrix evaluated in the Krieger–Li–Iafrate approximation

Analyses of the exact exchange-correlation potential have long played a valuable rôle in the development and refinement of practical exchange-correlation functionals. This has been due both to analytic conditions and numerical examples resulting from exact equations for v_{xc}, and to the insight gained from approximate model potentials derived as approximate analytic solutions to the exact v_{xc} equations. Although correlated equations have been given [63,102], most of the work deals with exchange potentials and encompasses an extensive literature refining on Slater's initial concept [103] of v_x as a local approximation to the nonlocal exchange operator, $\hat{\Sigma}_x$, in Hartree–Fock. As already noted, the asymptotic behavior of v_{xc}, which is so crucial for the high-lying excitations in TD-DFRT, and affects the dynamic polarizability as well, is primarily a question of the exchange potential since correlation is normally short ranged. Thus the present subsection will be restricted to the exchange-only case. While much has been learned about v_x itself, very little is known about $\delta v_x/\delta\rho$, which enters into the coupling matrix in TD-DFRT, the only treatment being that of Petersilka, Gossman, and Gross [104] who give $\delta v_x/\delta\rho$ corresponding to Slater's approximation [103] for v_x. Yet attempts to improve practical functionals do not necessarily improve their derivatives, as has been emphasized by van Gisbergen *et al.* [27].

The present subsection begins with a review of the Sham–Schlüter and optimized effective potential (OEP) equations for v_x and of their relation to the historic approximate model of Slater and the more recent, more accurate model of Krieger, Li, and Iafrate (KLI) [105,106]. A treatment of $\delta v_x/\delta\rho$, within the context of the adiabatic approximation, is then given, by taking the derivative of the Sham–Schlüter and OEP equations for v_x. The derivative is evaluated first in the Slater approximation, giving the same result as obtained by Gross and coworkers [104], and the treatment is then extended to the KLI approximation.

An abbreviated spin-orbital notation, $i \equiv (\mathbf{r}_i, \sigma_i)$, will be used in this subsection, in order to keep the equations as compact as possible. In this notation, the Kohn–Sham equation is,

$$\left(-\frac{1}{2}\nabla^2 + v_{\text{eff}}(1)\right)\psi_p^{\text{KS}}(1) = \epsilon_p^{\text{KS}}\psi_p^{\text{KS}}(1)\,, \tag{113}$$

where, restricting to the exchange-only case,

$$v_{\text{eff}}(1) = v(1) + \int \frac{\rho^{\text{HF}}(2)}{r_{12}}\,d2 + v_x(1)\,. \tag{114}$$

Note that the equivalence of the Hartree–Fock and Kohn–Sham densities,

$$0 = \rho^{\text{HF}}(1) - \rho^{\text{KS}}(1)\,, \tag{115}$$

has been used. This follows from the equality of the Kohn–Sham and exact charge densities, which is a defining characteristic of the Kohn–Sham method [2], and the fact that the HF density is the exact density in an exchange-only formalism. It will be convenient to write the Hartree–Fock equation as

$$\left(-\frac{1}{2}\nabla^2 + v_{\text{eff}}(1) + \hat{M}_{\text{x}}\right)\psi_p^{\text{HF}}(1) = \epsilon_p^{\text{HF}}\psi_p^{\text{HF}}(1)\,, \tag{116}$$

where

$$\hat{M}_{\text{x}} = \hat{\Sigma}_{\text{x}}^{\text{HF}} - v_{\text{x}}\,, \tag{117}$$

and the kernal[3] of the exchange operator,

$$\Sigma_{\text{x}}^{\text{HF}}(1,2) = -\frac{\gamma^{\text{HF}}(1,2)}{r_{12}}\,, \tag{118}$$

may be expressed in terms of the density matrix

$$\gamma^{\text{HF}}(1,2) = \sum_p f_p \psi_p^{\text{HF}}(1)\psi_p^{\text{HF}*}(2)\,. \tag{119}$$

The essential idea of the Sham–Schlüter [63] approach to obtaining the exact v_{x} is elegantly simple: Since the Kohn–Sham density is exact, the change in the density resulting from the "perturbation" $\hat{M}_{\text{x}}$ must be zero. Sham and Schlüter actually treated the more general correlated case, using not $\hat{M}_{\text{x}}$ but $\hat{M}_{\text{xc}} = \hat{\Sigma}_{\text{xc}} - v_{\text{xc}}$, where $\hat{\Sigma}_{\text{xc}}$ is the full self-energy, and used many-body Green function techniques to obtain an equation for v_{xc} [63]. A derivation will now be given for the exchange-only case, without the use of Green functions. First note that the change in the density matrix due to $\hat{M}_{\text{x}}$ is

$$\begin{aligned}\Delta\gamma(1,2) &\equiv \gamma^{\text{HF}}(1,2) - \gamma^{\text{KS}}(1,2)\\ &= \sum_p f_p\left(\Delta\psi_p(1)\psi_p^{\text{KS}*}(2) + \psi_p^{\text{HF}}(1)\Delta\psi_p^*(2)\right)\,,\end{aligned} \tag{120}$$

where

$$\Delta\psi_p(1) = \psi_p^{\text{HF}}(1) - \psi_p^{\text{KS}}(1)\,. \tag{121}$$

The orbital difference can be expressed in two different ways,

$$\Delta\psi_p(1) = \sum_q \psi_q^{\text{KS}}(1)\left(\frac{\langle\psi_q^{\text{KS}}|\hat{M}_{\text{x}}|\psi_p^{\text{HF}}\rangle}{\epsilon_p^{\text{HF}} - \epsilon_q^{\text{KS}}} - \delta_{pq}\right)\,, \tag{122}$$

and

$$\Delta\psi_p(2) = \sum_q \psi_q^{\text{HF}}(2)\left(\frac{\langle\psi_q^{\text{HF}}|\hat{M}_{\text{x}}|\psi_p^{\text{KS}}\rangle}{\epsilon_p^{\text{KS}} - \epsilon_q^{\text{HF}}} + \delta_{pq}\right)\,, \tag{123}$$

as is readily verified, taking into account that

$$\langle\psi_q^{\text{KS}}|\hat{M}_{\text{x}}|\psi_p^{\text{HF}}\rangle = (\epsilon_p^{\text{HF}} - \epsilon_q^{\text{KS}})\langle\psi_q^{\text{KS}}|\psi_p^{\text{HF}}\rangle \tag{124}$$

[3] The kernal $\Omega(1,2)$ of an operator $\hat{\Omega}$ is defined by $\hat{\Omega}f(1) = \int \Omega(1,2)f(2)\,d2$ for all functions f.

and

$$\hat{1} = \sum_q |\psi_q^{\text{KS}}\rangle\langle\psi_q^{\text{KS}}| = \sum_p |\psi_p^{\text{HF}}\rangle\langle\psi_p^{\text{HF}}| . \tag{125}$$

Replacing $\Delta\psi_p(1)$ in Eq. (120) with Eq. (122) and $\Delta\psi_p(2)$ in Eq. (120) with Eq. (123) gives,

$$\Delta\gamma(1,2) = \sum_{p,q} \frac{f_p - f_q}{\epsilon_p^{\text{KS}} - \epsilon_q^{\text{HF}}} \psi_q^{\text{HF}}(1)\psi_p^{\text{KS}*}(2)\langle\psi_p^{\text{HF}}|\hat{M}_{\text{x}}|\psi_q^{\text{KS}}\rangle , \tag{126}$$

or

$$\Delta\gamma(1,1') = \int\int X(1,1';2,2')M_{\text{x}}(2,2')\, d2d2' , \tag{127}$$

where

$$X(1,1';2,2') = \sum_{p,q} \frac{f_p - f_q}{\epsilon_p^{\text{KS}} - \epsilon_q^{\text{HF}}} \psi_q^{\text{HF}}(1)\psi_p^{\text{KS}*}(1')\psi_p^{\text{HF}*}(2)\psi_q^{\text{KS}}(2') . \tag{128}$$

Note that X is analogous to the generalized susceptibility, but gives the total change rather than the linear response. Now, using the condition that the change in the density be zero ($\Delta\gamma(1,1) = 0$) gives the Sham–Schlüter equation,

$$0 = \int\int X(1,1;2,2')\left(\Sigma_{\text{x}}^{\text{HF}}(2,2') - \delta(2-2')v_{\text{x}}(2)\right)\, d2d2' . \tag{129}$$

This allows v_{x} to be determined up to an arbitrary additive constant,

$$v_{\text{x}}(1) = v_{\text{x}}'(1) + C . \tag{130}$$

The Sham–Schlüter equation (129) is generally solved in the linear response approximation. That is, $\hat{M}_{\text{x}}$ is treated as a perturbation and the linear (rather than the total) response of the density is set equal to zero. In addition, $\hat{\Sigma}_{\text{x}}$ is constructed from the KS, instead of the HF orbitals. This results in the equation,

$$0 = \int\int \chi(1,1;2,2')\left(\Sigma_{\text{x}}(2,2') - \delta(2-2')v_{\text{x}}^{\text{OEP}}(2)\right)\, d2d2' , \tag{131}$$

where the superscripts HF and KS have been dropped since only the KS quantities enter here, and

$$\chi(1,1';2,2') = \sum_{p,q} \frac{f_p - f_q}{\epsilon_p - \epsilon_q} \psi_q(1)\psi_p^*(1')\psi_p^*(2)\psi_q(2') \tag{132}$$

is the generalized susceptibility of linear response theory, i.e.

$$\chi(1,1';2,2') = \frac{\delta\gamma(1,1')}{\delta w(2,2')} \tag{133}$$

describes the linear response of the density matrix to an arbitrary perturbation, $\hat{w}$. Note that it is the independent-particle susceptibility which appears here. This is as expected on physical grounds since the HF and KS equations are quasi-independent particle equations (cf. Sec. 3). The superscript OEP (for "optimized effective potential") has been added to distinguish this v_x from the exact one obtained by solving Eq. (129). Equation (131) was first given by Sharp and Horton [107] as the solution to the problem of finding the (local) potential ($v_x^{\rm OEP}$) whose orbitals minimize the Hartree–Fock energy expression. This solution was subsequently rediscovered by Talman and Shadwick [108], and Talman and coworkers went on to make extensive applications of Eq. (131) and refinements thereupon [109–115]. Correlation effects in the exact Sham–Schlüter equation for v_{xc} [correlated version of Eq. (129)] complicate its numerical solution. Thus Sham and coworkers habitually solve the correlated Sham–Schlüter equation in the linear response approximation [116] [identical to Eq. (131) in the exchange-only case.] Casida has shown [102] that the linear response form of the correlated Sham–Schlüter equation may be obtained in a variational manner, thus generalizing the optimized effective potential approach to the correlated case. Ullrich, Gossmann, and Gross have given a time-dependent generalization of the optimized effective potential[117]. Given an orbital-dependent action expression, (from, for example, TDHF) they find the (time-dependent) potential v_{xc} whose orbitals make the action stationary.

Although the analogous equation is inconvenient in the correlated case, Eq. (129) for the exact v_x could be solved numerically in the exchange-only case. However, the linear response equation (131), derived via the OEP approach, predates the exact Sham–Schlüter equation, and has been the form generally used in studies of v_x. Numerical solutions of Eq. (131) have been useful for determining the properties of the exact Kohn–Sham orbitals and orbital energies, and testing approximate functionals. But further approximation is useful, both in order to use the resulting potential for computational purposes and to gain additional insight regarding the potential. It is natural to do this by approximating χ.

An appropriate approximation to use for this purpose is an average energy approximation suggested by Sharp and Horton [107]. Rewriting Eq. (132) (for integer occupation numbers) as

$$\chi(1,1';2,2') = \sum_i^{\rm occ}\sum_p^{p\neq i} \frac{1}{\epsilon_i-\epsilon_p}\left(\psi_p(1)\psi_i^*(1')\psi_i^*(2)\psi_p(2') + \psi_i(1)\psi_p^*(1')\psi_p^*(2)\psi_i(2')\right)\,, \tag{134}$$

the approximation suggested by Sharp and Horton consists of replacing the $\epsilon_p - \epsilon_i$ with an average energy, $\Delta\epsilon$. Using the completeness relation,

$$\delta(1-2) = \sum_p \psi_p(1)\psi_p^*(2)\,, \tag{135}$$

then gives

$$\chi^{\rm KLI}(1,1';2,2') = \frac{2}{\Delta\epsilon}\left(\sum_i^{\rm occ}\gamma_i(1,2)\gamma_i(2',1') - \frac{\gamma(1,2)\delta(1'-2') + \delta(1-2)\gamma(2',1')}{2}\right)\,, \tag{136}$$

where $\gamma_i(1,2) = \psi_i(1)\psi_i^*(2)$ is the density matrix for orbital i. An additional approximation is just to drop the first term, to obtain

$$\chi^{\rm S}(1,1';2,2') = -\frac{1}{\Delta\epsilon}\left(\gamma(1,2)\delta(1'-2') + \delta(1-2)\gamma(2',1')\right)\,. \tag{137}$$

The "KS" orbitals used in constructing χ and other quantities are to be understood to be those appropriate to the approximation under consideration. Approximating χ in Eq. (131) by $\chi^{\rm S}$ and solving for $v_{\rm x}$ gives Slater's potential [103] (hence the S in $\chi^{\rm S}$),

$$\rho(1)v_{\rm x}^{\rm S}(1) = -\int \frac{|\gamma(1,2)|^2}{r_{12}}\, d2\,, \tag{138}$$

as was first pointed out by Sharp and Horton [107].

Krieger, Li, and Iafrate (hence KLI) [105] improved on this by instead approximating χ in Eq. (131) with $\chi^{\rm KLI}$. This gives

$$\rho(1)v_{\rm x}^{\rm KLI}(1) = -\int \frac{|\gamma(1,2)|^2}{r_{12}}\, d2 - \sum_i^{\rm occ} \rho_i(1)\langle\psi_i|\hat{\Sigma}_{\rm x} - v_{\rm x}^{\rm KLI}|\psi_i\rangle\,, \tag{139}$$

where $\rho_i(1) = |\psi_i(1)|^2$ is just the density for orbital i. Krieger, Li, and Iafrate have extensively investigated this approximation (see Ref. [106] for a review). They find that, unlike $v_{\rm x}^{\rm S}$, $v_{\rm x}^{\rm KLI}$ is an excellent approximation to $v_{\rm x}^{\rm OEP}$. The KLI potential is also easier to compute than is $v_{\rm x}^{\rm OEP}$, and it satisfies the exact asymptotic condition (90), making it an interesting candidate for applied work. Gross, Grabo, and Petersilka [49,118,119] have recently combined $v_{\rm x}^{\rm KLI}$ with correlation functionals for use in calculations on atoms. Note however that KLI calculations require a computational effort comparable to HF since $\hat{\Sigma}_{\rm x}$ must still be constructed. The KLI potential has also motivated an approximate functional recently proposed by Gritsenko *et al.* [120].

One of the most interesting aspects of the KLI potential is that it is readily interpreted as an average of orbital-dependent potentials,

$$v_{\rm x}^{\rm KLI}(1) \cong \frac{\sum_i \rho_i(1)v_{\rm x}^i(1)}{\rho(1)}\,. \tag{140}$$

To see this, define the potential for orbital i such that

$$\left(-\frac{1}{2}\nabla^2 + v(1) + \int \frac{\rho^{\rm HF}(2)}{r_{12}}\, d2 + v_{\rm x}^i(1)\right)\psi_i^{\rm HF}(1) = \epsilon_i\psi_i^{\rm HF}(1)\,. \tag{141}$$

Then it is easy to show that

$$\begin{aligned}\rho_i^{\rm HF}(1)v_{\rm x}^i(1) &= \frac{1}{2}\left(\int \Sigma_{\rm x}^{\rm HF}(1,2)\gamma_i^{\rm HF}(2,1)\, d2 + \int \gamma_i^{\rm HF}(1,2)\Sigma_{\rm x}^{\rm HF}(2,1)\, d2\right)\\ &- \rho_i^{\rm HF}\langle\psi_i^{\rm HF}|\hat{\Sigma}_{\rm x}^{\rm HF} - v_{\rm x}^i(1)|\psi_i^{\rm HF}\rangle\,.\end{aligned} \tag{142}$$

Summing over i and dividing by $\rho(1)$ gives the average potential (140). If $v_{\rm x}^i$ is then approximated by the average potential [Eq. (140)] and the HF orbitals are approximated by the resulting Kohn–Sham orbitals (dropping the superscript HF), the result is equation (139) for the KLI potential. Note that the difference between the KLI potential [Eq. (139)] and Slater's potential [Eq. (138)] is essentially just the average of the orbital energy differences $(\epsilon_i^{\rm HF} - \epsilon_i^{\rm KS})$. If the $v_{\rm x}^i$ were truely independent of i, then the HF and (exchange-only) KS orbitals would have been the same, and no approximation would have been involved in Eq. (140) or (139). The approximation of HF orbitals by KS orbitals is interesting from this point of view. We have termed the correlated analogue of this approximation the target Kohn–Sham approximation (TKSA) and found it to be useful for modeling electron momentum spectroscopy triple-differential cross-sections [121]. Refs. [102] and [121] include generalizations of the KLI approximation to the correlated case and discuss their interpretation.

We now turn to the much less studied problem of finding an equation for the derivative $\delta v_x/\delta\rho$, which plays a central rôle in TD-DFRT, and of gaining some insight into this more complicated form by evaluating the derivative in the Slater and in the KLI approximations. Gross and coworkers have generalized the OEP approach to give the exchange potential resulting from the exchange action-functional of (non-adiabatic) time-dependent DFT [117] and have used this to obtain an expression for $\delta v_x/\delta\rho$ in the Slater approximation [104]. This generalized OEP approach requires an orbital-dependent action expression from traditional *ab initio* theory, and allows the investigation of potentials beyond the adiabatic approximation. The present chapter is concerned only with functionals within the adiabatic approximation, thus a different approach is taken here. In the adiabatic approximation, the exchange-correlation functional is the same as that of the time-independent theory. Since the coupling matrix is evaluated for the unperturbed orbitals (which are solutions of the time-independent Kohn–Sham equation) (c.f. Sec. 3), this means that only $\delta v_x/\delta\rho$ of the time-independent theory is needed in the adiabatic approximation. Thus equations for $\delta v_x/\delta\rho$ are here derived from the Sham–Schlüter equation for the v_x of time-independent Kohn–Sham theory. Being based on the Sham–Schlüter formalism makes this approach trivial to extend to the correlated problem. Solving for the derivative in the Slater approximation gives the same expression found by Gross and coworkers in this approximation. We then go beyond previous work to obtain an equation for $\delta v_x/\delta\rho$ in the KLI approximation, and show it to be conceptually similar in structure to the now familiar KLI equation for v_x.

Since the Sham–Schlüter equation (129) holds for all (static) external potentials v, it must hold for small variations δv. Taking the functional derivative of the Sham–Schlüter equation with respect to the external potential gives

$$\begin{aligned} 0 &= \int\int \frac{\delta X(1,1;2,2')}{\delta v(4)} \left(\Sigma_x^{\mathrm{HF}}(2,2') - \delta(2-2')v_x(2)\right) d2d2' \\ &+ \int\int\int\int X(1,1;2,2') \left(\frac{\delta\Sigma_x^{\mathrm{HF}}(2,2')}{\delta\gamma^{\mathrm{HF}}(3,3')} - \delta(2-2')\frac{\delta v_x(2)}{\delta\rho(3)}\delta(3-3')\right) \\ &\times \chi^{\mathrm{HF}}(3,3';4,4)\, d2d2'd3d3' , \end{aligned} \tag{143}$$

where Eq. (133) for the generalized susceptibility has been used. This determines $\delta v_x/\delta\rho$ up to an additive function of the form $f(1)+f(2)+C$:

$$\frac{\delta v_x(1)}{\delta\rho(2)} = \left(\frac{\delta v_x(1)}{\delta\rho(2)}\right)' + f(1) + f(2) + C\,. \tag{144}$$

In the linear response approximation, Eq. (143) becomes

$$\begin{aligned} 0 &= \int\int \frac{\delta\chi(1,1;2,2')}{\delta v_{\mathrm{eff}}(4)} \left(\Sigma_x(2,2') - \delta(2-2')v_x^{\mathrm{OEP}}(2)\right) d2d2' \\ &+ \int\int\int\int \chi(1,1;2,2') \left(\frac{\delta\Sigma_x(2,2')}{\delta\gamma(3,3')} - \delta(2-2')\frac{\delta v_x^{\mathrm{OEP}}(2)}{\delta\rho(3)}\delta(3-3')\right) \end{aligned}$$

$$\times \quad \chi(3,3';4,4)\, d2d2'd3d3' \,, \tag{145}$$

where the chain rule,

$$\frac{\delta F}{\delta v(4)} = \int \frac{\delta F}{\delta v_{\text{eff}}(3)} \frac{\delta v_{\text{eff}}(3)}{\delta v(4)}\, d4 \,, \tag{146}$$

has been used and $\delta v_{\text{eff}}/\delta v$ has been cancelled[4]. Petersilka, Gossmann, and Gross [104] have derived a nonadiabatic analogue of Eq. (145) based on the time-dependent generalization of the OEP equation (131) given by Ullrich, Gossmann, and Gross [117].

In order to obtain a consistent approximate solution of Eq. (145), consistent approximations must be used for $\delta\chi/\delta v_{\text{eff}}$, χ, and v_{x}. Starting with the Slater approximation for χ [Eq. (137)] and taking the derivative gives

$$\frac{\delta \chi^{\text{S}}(1,1';2,2')}{\delta w(3,3')} = \frac{1}{(\Delta\epsilon)^2} \left(\chi(1,2;3,3')\delta(1'-2') + \delta(1-2)\chi(2',1';3,3')\right) . \tag{147}$$

For consistency, the result should be evaluated in the Slater approximation,

$$\left(\frac{\delta \chi^{\text{S}}(1,1';2,2')}{\delta w(3,3')}\right)^{\text{S}} = \frac{1}{(\Delta\epsilon)^2} \left(\chi^{\text{S}}(1,2;3,3')\delta(1'-2') + \delta(1-2)\chi^{\text{S}}(2',1';3,3')\right) . \tag{148}$$

Solving Eq. (145) for $\delta v_{\text{x}}/\delta\rho$ in the Slater approximation (i.e. using v_{x}^{S}, χ^{S}, and $\left(\delta\chi^{\text{S}}/\delta v_{\text{eff}}\right)^{\text{S}}$) leads to

$$\rho(1)\left(\frac{\delta v_{\text{x}}^{\text{S}}(1)}{\delta\rho(4)}\right)^{\text{S}} \rho(4) = -\frac{|\gamma(1,4)|^2}{r_{14}} + \delta(1-4)\int \left(\Sigma_{\text{x}}(1,2) - \delta(1-2)v_{\text{x}}^{\text{S}}(1)\right)\gamma(2,1)\, d2 \,. \tag{149}$$

The second term on the RHS vanishes due to the use of the Slater approximation for the potential [Eq. (138)]. Thus,

$$\rho(1)\left(\frac{\delta v_{\text{x}}^{\text{S}}(1)}{\delta\rho(4)}\right)^{\text{S}} \rho(4) = -\frac{|\gamma(1,4)|^2}{r_{14}} . \tag{150}$$

This result is the same as that obtained in the Slater approximation by Petersilka, Gossmann, and Gross [104] via their time-dependent (non-adiabatic) OEP approach. They have used this formula for the derivative in conjunction with an SCF calculation in the KLI approximation (i.e. Slater/KLI, in the notation of Sec. 5.1) to calculate excitation energies for atoms [49,104].

Note that Eq. (150) can also be derived in a very simple manner from Eq. (138) by direct functional differentiation with respect to γ, using

$$\begin{aligned} \frac{\delta v_{\text{x}}^{\text{S}}(1)}{\delta\gamma(2,2')} &= \int \frac{\delta v_{\text{x}}^{\text{S}}(1)}{\delta\rho(3)} \frac{\delta\rho(3)}{\delta\gamma(2,2')}\, d2d2' \\ &= \frac{\delta v_{\text{x}}^{\text{S}}(1)}{\delta\rho(2)} \delta(2-2') . \end{aligned} \tag{151}$$

[4]The invertibility $\delta v_{\text{eff}}/\delta v$ follows from the first Hohenberg–Kohn theorem (so that the mappings $v \rightarrow \rho$ and $v_{\text{eff}} \rightarrow \rho$ are 1-1, up to the usual additive constant) and from v-representibility (so that $v \rightarrow \rho$ and $v_{\text{eff}} \rightarrow \rho$ are onto.) Under these conditions, the mapping $v_{\text{eff}} \rightarrow v$ (defined only up to an additive constant) is a 1-1 correspondance and the derivative is invertible.

This gives

$$\rho(1)\frac{\delta v_{\rm x}^{\rm S}(1)}{\delta\rho(2)}\delta(2-2') = -\delta(1-2')\frac{\gamma(1,2)}{r_{12}} + \delta(1-2)\left(\Sigma_{\rm x}(2',1) - \delta(2-2')v_{\rm x}^{\rm S}(1)\right) . \qquad (152)$$

Multiplication by $\gamma(2,2')$ followed by integration over $2'$ then gives Eq. (149) [with $\left(\delta v_{\rm x}^{\rm S}/\delta\rho\right)^{\rm S}$ replaced by $\delta v_{\rm x}^{\rm S}/\delta\rho$].

Since the KLI potential has been found to be a much better approximation to $v_{\rm x}$ than the Slater potential is, it is of interest to consider the derivative $\delta v_{\rm x}/\delta\rho$ in the KLI approximation. The derivative can be evaluated at the KLI level using $\chi^{\rm KLI}$ [Eq. (136)] and its derivative,

$$\begin{aligned}\left(\frac{\delta\chi^{\rm KLI}(1,1';2,2')}{\delta w(3,3')}\right)^{\rm KLI} &= \frac{4}{(\Delta\epsilon)^2}\left[\sum_i^{\rm occ}\left(\chi_i^{\rm KLI}(1,2;3,3')\gamma_i(2',1') + \gamma_i(1,2)\chi_i^{\rm KLI}(2',1';3,3')\right)\right. \\ &- \left.\frac{\chi^{\rm KLI}(1,2;3,3')\delta(1'-2') + \delta(1-2)\chi^{\rm KLI}(2',1';3,3')}{2}\right] , \qquad (153)\end{aligned}$$

where

$$\chi_i^{\rm KLI}(1,1';2,2') = \frac{2}{\Delta\epsilon}\left(\gamma_i(1,2)\gamma_i(2',1') - \frac{\gamma_i(1,2)\delta(1'-2') + \delta(1-2)\gamma_i(2',1')}{2}\right) . \qquad (154)$$

Inserting $\chi^{\rm KLI}$, $\left(\delta\chi^{\rm KLI}/\delta w\right)^{\rm KLI}$, and $v_{\rm x}^{\rm KLI}$ into Eq. (145) leads, after lengthy algebra, to[5]

$$\begin{aligned}\rho(1)\left(\frac{\delta v_{\rm x}(1)}{\delta\rho(4)}\right)^{\rm KLI}\rho(4) &= -\frac{|\gamma(1,4)|^2}{r_{14}} \\ &- \left\{\frac{1}{2}\sum_i^{\rm occ}\rho_i(1)\left[\int\gamma_i(4,2)\Sigma_{\rm x}(2,4)\,d2 + \int\Sigma_{\rm x}(4,2)\gamma_i(2,4)\,d2\right]\right. \\ &- \left.\rho(4)\sum_i^{\rm occ}\rho_i(1)\int\rho_i(2)\left(\frac{\delta v_{\rm x}(2)}{\delta\rho(4)}\right)^{\rm KLI}d2\right\} + (1\leftrightarrow 4) \\ &- \sum_{i,j}^{\rm occ}\rho_i(1)\rho_j(4)\int\!\!\int\frac{\gamma_i(2,3)\gamma_j(3,2)}{r_{23}}\,d2d3 \\ &- \sum_{i,j}^{\rm occ}\rho_i(1)\rho_j(4)\int\!\!\int\rho_i(2)\left(\frac{\delta v_{\rm x}(2)}{\delta\rho(3)}\right)^{\rm KLI}\rho_j(3)\,d2d3 , \qquad (155)\end{aligned}$$

plus two other terms,

$$\begin{aligned}T_1 &= \delta(1-4)\left[\int\left(\Sigma_{\rm x}(1,2) - v_{\rm x}^{\rm KLI}(1)\delta(1-2)\right)\gamma(2,1)\,d2\right. \\ &- \left.\sum_i^{\rm occ}\rho_i(1)\langle\psi_i|\hat{\Sigma}_{\rm x} - v_{\rm x}^{\rm KLI}|\psi_i\rangle\right] , \qquad (156)\end{aligned}$$

[5]The notation $(1\leftrightarrow 4)$ means "same as the previous term but with 1 and 4 interchanged."

and

$$\begin{aligned}
T_2 &= 2\sum_i^{\text{occ}} |\gamma_i(1,4)|^2 \langle \psi_i | \hat{\Sigma}_{\text{x}} - v_{\text{x}}^{\text{KLI}} | \psi_i \rangle \\
&- \frac{1}{2}\sum_i^{\text{occ}} \gamma_i(1,4) \Big[\int \left(\Sigma_{\text{x}}(4,2) - \delta(4-2) v_{\text{x}}^{\text{KLI}}(2) \right) \gamma_i(2,1)\, d2 \\
&+ \int \gamma_i(4,2) \left(\Sigma_{\text{x}}(2,1) - \delta(1-2) v_{\text{x}}^{\text{KLI}}(1) \right) d2 \Big] \\
&+ (1 \leftrightarrow 4)\,.
\end{aligned} \tag{157}$$

The term T_1 is zero for the KLI potential [Eq. (139)] and the term T_2 is expected to be small (it is zero in the TKSA). Given the nature of the average-energy approximation on which the KLI method is based, it seems reasonable to neglect T_2.

Note that the first term of the KLI approximation for the derivative is just the Slater approximation for the derivative, as was the case for the potential. Although Eq. (155) may at first seem rather complicated, it has a rather simple interpretation, similar to that given for $v_{\text{x}}^{\text{KLI}}$, as the average over an orbital-dependent quantity,

$$\left(\frac{\delta v_{\text{x}}(1)}{\delta \rho(4)} \right)^{\text{KLI}} \cong \sum_{i,j} \rho_i(1) \left(\frac{\delta v_{\text{x}}(1)}{\delta \rho(4)} \right)_{ij} \rho_j(4) / \rho(1)\rho(4)\,. \tag{158}$$

To see this, suppose that Eq. (144) holds at the orbital level,

$$\left(\frac{\delta v_{\text{x}}(1)}{\delta \rho(4)} \right)_{ij} = F_{ij}(1,4) + f_i(1) + f_j(4) + C_{ij}\,, \tag{159}$$

where f_i and f_j have the same functional form except for the interchange of i and j. We may assume without loss of generality that

$$\int \rho_i(1) f_i(1)\, d1 = \int f_j(4) \rho_j(4)\, d4 = 0\,, \tag{160}$$

since $f_i(1)$ can always be redefined as, $f_i(1) = f_i^{\text{old}}(1) - \int f_i^{\text{old}}(2)\rho_i(2)\, d2$, etc. to make Eq. (160) true. Then

$$\int\!\!\int \rho_i(1) \left(\frac{\delta v_{\text{x}}(1)}{\delta \rho(4)} \right)_{ij} \rho_j(4)\, d1 d4 = \int\!\!\int \rho_i(1) F_{ij}(1,4) \rho_j(4)\, d1 d4 + C_{ij}\,, \tag{161}$$

$$\int \rho_i(1) \left(\frac{\delta v_{\text{x}}(1)}{\delta \rho(4)} \right)_{ij} d1 = \int \rho_i(1) F_{ij}(1,4)\, d1 + f_j(4) + C_{ij}\,, \tag{162}$$

and

$$\int \left(\frac{\delta v_{\text{x}}(1)}{\delta \rho(4)} \right)_{ij} \rho_j(4)\, d4 = \int F_{ij}(1,4) \rho_j(4)\, d4 + f_i(1) + C_{ij}\,. \tag{163}$$

Using these three equations to eliminate $f_i(1)$, $f_j(2)$, and C_{ij} in Eq. (159) gives

$$\begin{aligned}\left(\frac{\delta v_{\mathrm{x}}(1)}{\delta\rho(4)}\right)_{ij} &= F_{ij}(1,4) \\ &+ \int \rho_i(2)\left[\left(\frac{\delta v_{\mathrm{x}}(2)}{\delta\rho(4)}\right)_{ij} - F_{ij}(2,4)\right] d2 + \left(\begin{array}{c} i\leftrightarrow j \\ 1\leftrightarrow 4\end{array}\right) \\ &- \int\!\!\int \rho_i(2)\left[\left(\frac{\delta v_{\mathrm{x}}(2)}{\delta\rho(3)}\right)_{ij} - F_{ij}(2,3)\right] \rho_j(3)\, d2d3 . \end{aligned} \tag{164}$$

Multiplying by $\rho_i(1)\rho_j(4)/\rho(1)\rho(4)$ and summing over i and j gives the orbital-averaged derivative (158). Approximating $(\delta v_{\mathrm{x}}/\delta\rho)_{ij}$ by the orbital average (158) then results in

$$\begin{aligned}\rho(1)\left(\frac{\delta v_{\mathrm{x}}(1)}{\delta\rho(4)}\right)^{\mathrm{KLI}}\rho(4) &= \sum_{i,j}^{\mathrm{occ}} \rho_i(1)F_{ij}(1,4)\rho_j(4) \\ &- \left[\sum_i^{\mathrm{occ}} \rho_i(1)\int \rho_i(2)\left(\sum_j^{\mathrm{occ}} F_{ij}(2,4)\rho_j(4)\right) d2\right. \\ &- \left.\rho(4)\sum_i^{\mathrm{occ}}\rho_i(1)\int\rho_i(2)\left(\frac{\delta v_{\mathrm{x}}(2)}{\delta\rho(4)}\right)^{\mathrm{KLI}} d2\right] + (1\leftrightarrow 4) \\ &- \sum_{i,j}^{\mathrm{occ}}\rho_i(1)\rho_j(4)\int\!\!\int\rho_i(2)F_{ij}(2,3)\rho_j(3)\, d2d3 \\ &- \sum_{i,j}^{\mathrm{occ}}\rho_i(1)\rho_j(4)\int\!\!\int\rho_i(2)\left(\frac{\delta v_{\mathrm{x}}(2)}{\delta\rho(3)}\right)^{\mathrm{KLI}}\rho_j(3)\, d2d3 . \end{aligned} \tag{165}$$

This is identical to Eq. (155) if F_{ij} is chosen to be an orbital analogue of the Slater form,

$$F_{ij}(1,4) = -\frac{\gamma_i(1,4)\gamma_j(4,1) + \gamma_j(1,4)\gamma_i(4,1)}{2\rho_i(1)r_{14}\rho_j(4)} . \tag{166}$$

Thus the seemingly complicated form of Eq. (155) for the derivative of v_{x} in the KLI approximation can be understood as an average over the relatively simple orbital expression (166), yielding the Slater approximation for the derivative, plus averages over orbital-dependent boundary functions f_i, f_j, and C_{ij}. This is, of course, analogous to the KLI potential itself which is the average over an orbital-dependent potential, giving Slater's v_{x} plus an average over orbital-dependent energy shifts.

Another interesting observation about the expressions for the derivative $\delta v_{\mathrm{x}}/\delta\rho$ in the Slater and KLI models is that, unlike the case for local or gradient-corrected functionals, $\delta v_{\mathrm{x}}/\delta\rho$ is not diagonal in the position representation. This suggests an unexplored direction for improving functionals for the calculation of response properties.

In addition to its direct value, for conditions for improved functionals or for computation, the present result for the derivative of the KLI potential is interesting from the point of view of physical interpretation and relationships with HF. In HF (or correlated *ab initio* methods) the exchange operator (or self-energy) is constructed in terms of the orbitals, whereas in DFT the exchange (or exchange-correlation) potential is an orbital-independent potential which is a functional of the total charge density. This tends to obscure the details of the relationship between DFT and traditional *ab initio* methods. Thus, given that the KLI potential is a reasonably faithful model of the exact v_x, the interpretation of it and its derivative in terms of relatively simple orbital expressions provides a means for obtaining a more detailed physical picture of the relationship between (static or time-dependent) DFT and HF. This is pursued in the next section.

6. Heuristic relationships between DFT and Hartree–Fock

According to the initial ideas of Slater [103], subsequently developed in the optimized effective potential approach of Sharp and Horton [107], and Talman and Shadwick [108], and the exact v_{xc} equation of Sham and Schlüter [63], the Kohn–Sham v_x is an approximation to the Hartree–Fock $\hat{\Sigma}_x$, but one whose orbitals yield the same density and properties derived therefrom. Thus one would expect to find some approximate relations between the orbital energies and integrals entering into Hartree–Fock and exchange-only DFT excitation energy and oscillator strength expressions, at least within the limits of some additional assumptions. This is done here by considering a simplified model involving transitions between a single doubly occupied spatial orbital, ψ_i, and a single unoccupied spatial orbital, ψ_a, and assuming that HF, exchange-only TD-DFRT, and ΔSCF-based exchange-only DFT should give the same excitation energies and oscillator strengths. It is also assumed that differences between HF and DFT orbitals are small enough to be neglected, in a first approximation. The evidence that this is often the case for occupied orbitals has been reviewed in Ref. [121]. However, this is a shakier assumption for unoccupied orbitals.

In a previous paper [23], a comparison of TDHF, TD-DFRT, and ΔSCF-based expressons for excitaton energies and oscillator strengths was made for H_2 in a minimal basis set. The expressions of Ref. [23] are easily adapted (Table 4) to the more general two-state model and will form the basis for the present discussion, which focusses on bringing out approximate physical relationships between HF and DFT, in light of what has been learned in the previous section regarding $\delta v_x/\delta\rho$. Although the formulae in Table 4 are more general, the discussion in this section will be restricted to the exchange-only case.

First, a few preliminary remarks about the formulae in Table 4 are in order. The response theory (TDHF and TD-DFRT) excitation energy expressions were derived exactly and then linearized to facilitate comparison with the corresponding ΔSCF-based expressions. In contrast, the oscillator strength expressions are the exact (i.e. unmodified) response theory expressions for the two-level model. The ΔSCF-based expressions were derived within a simple frozen-orbital model that corrects for singlet-triplet spin symmetry, but neglects orbital relaxation completely. Liberal use of low-order Taylor's expansions of E_{xc} and of v_{xc}^{σ} was made in deriving the ΔSCF-based DFT expressions. The ΔSCF-based oscillator strength expressions follow from the corresponding ΔSCF-based singlet excitation energy and the assumed form of the wave functions. Additional details may be found in Ref. [23].

It is natural to expect that TDHF and (exchange-only) TD-DFRT should give the same results for excitation energies and oscillator strengths. Identifying the TDHF and

Table 4
Approximate excitation energy and oscillator strength expressions for excitations from a single doubly occupied spatial orbital ψ_i to a single unoccupied spatial orbital ψ_a. See text.

	Excitation energies
	Triplet
TD-DFRT	$\epsilon_a^{\mathrm{KS}} - \epsilon_i^{\mathrm{KS}} + \int\int \psi_i(\mathbf{r})\psi_a(\mathbf{r}) \left[\frac{\delta v_{\mathrm{xc}}^{\uparrow}(\mathbf{r})}{\delta\rho_{\uparrow}(\mathbf{r}')} - \frac{\delta v_{\mathrm{xc}}^{\uparrow}(\mathbf{r})}{\delta\rho_{\downarrow}(\mathbf{r}')}\right] \psi_i(\mathbf{r}')\psi_a(\mathbf{r}')\, d\mathbf{r}d\mathbf{r}'$
TDHF and ΔSCF HF	$\epsilon_a^{\mathrm{HF}} - \epsilon_i^{\mathrm{HF}} - [\psi_i\psi_i\|\psi_a\psi_a]$
ΔSCF-based DFT	$\omega_a - \omega_i - \int\int \|\psi_i(\mathbf{r})\|^2 \frac{\delta v_{\mathrm{xc}}^{\uparrow}(\mathbf{r})}{\delta\rho_{\downarrow}(\mathbf{r}')} \|\psi_a(\mathbf{r}')\|^2\, d\mathbf{r}d\mathbf{r}' - [\psi_i\psi_i\|\psi_a\psi_a]$
	Singlet
TD-DFRT	$\epsilon_a^{\mathrm{KS}} - \epsilon_i^{\mathrm{KS}} + 2[\psi_i\psi_a\|\psi_i\psi_a] + \int\int \psi_i(\mathbf{r})\psi_a(\mathbf{r}) \left[\frac{\delta v_{\mathrm{xc}}^{\uparrow}(\mathbf{r})}{\delta\rho_{\uparrow}(\mathbf{r}')} + \frac{\delta v_{\mathrm{xc}}^{\uparrow}(\mathbf{r})}{\delta\rho_{\downarrow}(\mathbf{r}')}\right] \psi_i(\mathbf{r}')\psi_a(\mathbf{r}')\, d\mathbf{r}d\mathbf{r}'$
TDHF and ΔSCF HF	$\epsilon_a^{\mathrm{HF}} - \epsilon_i^{\mathrm{HF}} + 2[\psi_i\psi_a\|\psi_i\psi_a] - [\psi_i\psi_i\|\psi_a\psi_a]$
ΔSCF-based DFT	$\omega_a - \omega_i - \int\int \|\psi_i(\mathbf{r})\|^2 \left[2\frac{\delta v_{\mathrm{xc}}^{\uparrow}(\mathbf{r})}{\delta\rho_{\uparrow}(\mathbf{r}')} - \frac{\delta v_{\mathrm{xc}}^{\uparrow}(\mathbf{r})}{\delta\rho_{\downarrow}(\mathbf{r}')}\right] \|\psi_a(\mathbf{r}')\|^2\, d\mathbf{r}d\mathbf{r}' - [\psi_i\psi_i\|\psi_a\psi_a]$
	Oscillator strengths
	$f_S = \frac{4}{3} D \left(\|\langle\psi_i\|\hat{x}\|\psi_a\rangle\|^2 + \|\langle\psi_i\|\hat{y}\|\psi_a\rangle\|^2 + \|\langle\psi_i\|\hat{z}\|\psi_a\rangle\|^2\right)$
	D
TD-DFRT	$\epsilon_a^{\mathrm{KS}} - \epsilon_i^{\mathrm{KS}}$
TDHF	$\epsilon_a^{\mathrm{HF}} - \epsilon_i^{\mathrm{HF}} + [\psi_i\psi_a\|\psi_i\psi_a] - [\psi_i\psi_i\|\psi_a\psi_a]$
ΔSCF HF	$\epsilon_a^{\mathrm{HF}} - \epsilon_i^{\mathrm{HF}} + 2[\psi_i\psi_a\|\psi_i\psi_a] - [\psi_i\psi_i\|\psi_a\psi_a]$
ΔSCF-based DFT	$\omega_a - \omega_i - \int\int \|\psi_i(\mathbf{r})\|^2 \left[2\frac{\delta v_{\mathrm{xc}}^{\uparrow}(\mathbf{r})}{\delta\rho_{\uparrow}(\mathbf{r}')} - \frac{\delta v_{\mathrm{xc}}^{\uparrow}(\mathbf{r})}{\delta\rho_{\downarrow}(\mathbf{r}')}\right] \|\psi_a(\mathbf{r}')\|^2\, d\mathbf{r}d\mathbf{r}' - [\psi_i\psi_i\|\psi_a\psi_a]$
	ΔSCF DFT quasiparticle energies
Ionization	$\omega_i = \epsilon_i^{\mathrm{KS}} - \frac{1}{2}[\psi_i\psi_i\|\psi_i\psi_i] - \frac{1}{2}\int\int \|\psi_i(\mathbf{r})\|^2 \frac{\delta v_{\mathrm{xc}}^{\uparrow}(\mathbf{r})}{\delta\rho_{\uparrow}(\mathbf{r}')} \|\psi_i(\mathbf{r}')\|^2\, d\mathbf{r}d\mathbf{r}'$
Electron attachment	$\omega_a = \epsilon_a^{\mathrm{KS}} + \frac{1}{2}[\psi_a\psi_a\|\psi_a\psi_a] + \frac{1}{2}\int\int \|\psi_a(\mathbf{r})\|^2 \frac{\delta v_{\mathrm{xc}}^{\uparrow}(\mathbf{r})}{\delta\rho_{\uparrow}(\mathbf{r}')} \|\psi_a(\mathbf{r}')\|^2\, d\mathbf{r}d\mathbf{r}'$

TD-DFRT oscillator strength expressions in Table 4 leads to

$$\epsilon_a^{\mathrm{KS}} - \epsilon_i^{\mathrm{KS}} \cong \left(\epsilon_a^{\mathrm{HF}} - [\psi_i\psi_i|\psi_a\psi_a] + [\psi_i\psi_a|\psi_i\psi_a]\right) - \epsilon_i^{\mathrm{HF}} . \tag{167}$$

This is easily interpreted physically, and underlines what is perhaps the most important difference between HF and (exchange-only) DFT quantities entering into the calculation of excitation spectra. That is, while the potential due to the other electrons experienced by the occupied and unoccupied orbitals is the same in DFT (since v_{x} is orbital independent), that experienced by the unoccupied orbital in HF is for N_e electrons while that for the occupied orbitals is for $N_e - 1$ electrons. If the energy zero is chosen so that $v_{\mathrm{x}}^{\infty} = 0$, then ϵ_i^{KS} is the true (exchange-only) ionization potential (Sec. 5.2), so Koopman's theorem gives $\epsilon_i^{\mathrm{KS}} \cong \epsilon_i^{\mathrm{HF}}$. Thus the difference is entirely in the behavior of the unoccupied orbital energies. Subtracting a coulomb integral from, and adding an exchange integral to, the Hartree–Fock unoccupied orbital energy acts to compensate for the difference in self-consistent field for occupied and unoccupied orbitals. Of course, this difference between HF and DFT as regards the potential felt by the occupied vs. unoccupied orbitals is why one should not take too seriously the supposition that the differences between the unoccupied HF and DFT orbitals are small.

In comparing the TD-DFRT and TDHF excitation energy expressions, it should be kept in mind that $\delta v_{\mathrm{x}}^{\uparrow}/\delta\rho_{\downarrow} = 0$ in an exchange-only theory. Using Eq. (167) for the relation between the KS and HF orbital energy differences, and identifying the corresponding TD-DFRT and TDHF excitation energies, leads to the relation

$$\iint \psi_i(\mathbf{r})\psi_a(\mathbf{r})\frac{\delta v_{\mathrm{x}}^{\uparrow}(\mathbf{r})}{\delta\rho_{\uparrow}(\mathbf{r}')}\psi_i(\mathbf{r}')\psi_a(\mathbf{r}')\, d\mathbf{r}d\mathbf{r}' \cong -[\psi_i\psi_a|\psi_i\psi_a] . \tag{168}$$

This can readily be understood in terms of the orbital-average interpretation of $\delta v_{\mathrm{x}}/\delta\rho$ in the KLI approximation, given in the previous section. Using the orbital-dependent derivative [Eq. (159)], with the form (166) for F_{ij}, gives (for real spin orbitals, in the notation of Sec. 5.4)

$$\begin{aligned}\iint \psi_i(1)\psi_p(1)\left(\frac{\delta v_{\mathrm{x}}(1)}{\delta\rho(2)}\right)_{ij}\psi_j(2)\psi_q(2)\, d1d2 &= \iint \psi_i(1)\psi_p(1)F_{ij}(1,2)\psi_j(2)\psi_q(2)\, d1d2 \\ &= -[\psi_i\psi_q|\psi_j\psi_p] ,\end{aligned} \tag{169}$$

if the orbitals are chosen to be orthogonal ($\langle\psi_i|\psi_p\rangle = \langle\psi_j|\psi_q\rangle = 0$). If, as was done in Sec. 5.4, $(\delta v_{\mathrm{x}}/\delta\rho)_{ij}$ is approximated by the orbital-average, $(\delta v_{\mathrm{x}}/\delta\rho)^{KLI}$, Eq. (169) becomes the same as (168).

Finally, in comparing the ΔSCF HF and ΔSCF-based DFT expressions, it must be kept in mind that $\omega_i \cong \epsilon_i^{\mathrm{HF}}$ and $\omega_a \cong \epsilon_a^{\mathrm{HF}}$ in the exchange-only case. Identification of the corresponding excitation energies then leads to

$$\iint |\psi_i(\mathbf{r})|^2\frac{\delta v_{\mathrm{x}}^{\uparrow}(\mathbf{r})}{\delta\rho_{\uparrow}(\mathbf{r}')}|\psi_a(\mathbf{r}')|^2\, d\mathbf{r}d\mathbf{r}' \cong -[\psi_i\psi_a|\psi_i\psi_a] , \tag{170}$$

which has an interesting consistency with Eq. (169).

It is interesting to note that, if the energy zero is such that $v_x^\infty = 0$, and we assume $\epsilon_i^{KS} \cong \omega_i$ (which holds exactly for the HOMO in this case), the second last line in Table 4 suggests

$$\int\int |\psi_i(\mathbf{r})|^2 \frac{\delta v_x^\uparrow(\mathbf{r})}{\delta \rho_\uparrow(\mathbf{r}')} |\psi_i(\mathbf{r}')|^2 \, d\mathbf{r}d\mathbf{r}' \cong -[\psi_i\psi_i|\psi_i\psi_i] \, . \tag{171}$$

However, note that this assumption is only reasonable for the occupied orbitals. Since $\epsilon_a^{KS} \neq \omega_a$ it would be a mistake to try to draw an analogous conclusion for the integrals involving unoccupied orbitals, based on the last line in Table 4.

The approximate relations between DFT and HF given in this section should not be taken too literally, if only because the difference between the HF and DFT unoccupied orbitals is expected to be nonnegligible. Nevertheless, I hope that this short heuristic discussion helps to shed further light on the sometimes mysterious relationships between DFT and traditional *ab initio* theory, and may help to harmonize the TD-DFRT and ΔSCF-based DFT approaches to treating electronic excitations.

7. Some concluding remarks

In the area of response properties and spectra, as in other areas of quantum chemistry, there is a need both for rigorous, and sometimes heroically difficult, calculations on small molecules (or on large, but often simplified, model molecules) and for accurate but simple methods to be able to treat practical problems on a reasonable time scale. Density-functional theory has a special rôle to play here — that of being simultaneously computationally simple and yet formally rigorous. This asset of DFT is no longer restricted to the classic, time-independent Hohenberg–Kohn–Sham theory, but is now also true of time-dependent density-functional response theory, thus opening the way to many interesting possibilities in the area of dynamic response properties and spectra. The essential formal concepts and numerical methods of time-dependent density-functional response theory have been described in this article. No doubt, they will continue to evolve. But the theory and its implementation have reached the point where one can and should focus attention on the problem of exchange-correlation functionals for response theory calculations. Thus the latter half of the chapter has been devoted to this problem.

The focus has been on functionals within the adiabatic approximation. As pointed out in a previous paper [26], simple counting arguments in a finite basis set show that the adiabatic approximation can only describe excitations corresponding to linear combinations of singly-excited determinants. But that does not mean that the adiabatic approximation is restricted to describing electron correlation at the configuration interaction singles (CIS) or, more precisely, at the TDHF level. Instead one should think of the adiabatic approximation as being something like a TDHF (or CIS) calculation but with additional correlation effects not included at the TDHF (or CIS) level. One way to understand how this can be would be to begin with an exact formal equation for the electron polarization propagator and explictly derive the correction terms to TDHF, but we can just as well keep to the configuration interaction (CI) formalism which is conceptually simpler and adequate for understanding the basic idea. So, imagine repartitionning the full CI equation,

$$\begin{bmatrix} \boldsymbol{H}_{P,P} & \boldsymbol{H}_{P,Q} \\ \boldsymbol{H}_{Q,P} & \boldsymbol{H}_{Q,Q} \end{bmatrix} \begin{pmatrix} \vec{C}_P \\ \vec{C}_Q \end{pmatrix} = \omega \begin{pmatrix} \vec{C}_P \\ \vec{C}_Q \end{pmatrix}, \tag{172}$$

as

$$\left[\boldsymbol{H}_{P,P} + \boldsymbol{H}_{P,Q}\left(\omega\mathbf{1} - \boldsymbol{H}_{Q,Q}\right)^{-1}\boldsymbol{H}_{Q,P}\right]\vec{C}_P = \omega\vec{C}_P, \tag{173}$$

where $\boldsymbol{H}$ is defined with respect to the ground state energy as zero of energy, P indicates the principle space of single excitations and Q is the secondary space of other excitations. ($\boldsymbol{H}_{P,P}$ is the CIS matrix.) The second term on the LHS may be thought of as our coupling matrix. Notice that it has a complicated pole structure with poles roughly at all the two-electron, three-electron, etc. excitation energies. Such a coupling matrix seems likely to defy the ingenuity of even the best designers of density functionals for at least a few years to come. However, we can expand it about $\omega = 0$ to obtain

$$\left[\boldsymbol{H}_{P,P} - \boldsymbol{H}_{P,Q}\left(\boldsymbol{H}_{Q,Q}^{-1} + \boldsymbol{H}_{Q,Q}^{-2}\omega + \cdots\right)\boldsymbol{H}_{Q,P}\right]\vec{C}_P = \omega\vec{C}_P. \tag{174}$$

This is what is most amenable to modelling with an appropriate $\mathcal{A}_{\text{xc}}$ because it is a smooth function. The first term in parentheses is ω independent. It is the CI equivalent of the adiabatic approximation. Notice how it includes higher-order correlation effects not present in configuration interaction singles. The next terms in parentheses represent nonadiabatic effects and they may also be modeled, with a smooth function, at least for small ω. It is precisely nonadiabatic time-dependent functionals of this type which are discussed in Ref. [14].

Although the leading (adiabatic) term is formally a low-frequency approximation, its useful range of applicability evidently extends considerably beyond this limiting regime in practice, not unlike the case for the local density approximation. Investigation of the limits of the domain of practical applicability of, or corrections to, the adiabatic approximation presupposes that the adiabatic exchange-correlaton functional being used is quite accurate as regards the properties that are important for TD-DFRT calculations. As we have seen, this is not yet the case for the functionals used in molecular calculations at the present time.

The simplest adiabatic exchange-correlation functional is the time-dependent local density approximation (TDLDA). We have seen that the TDLDA works rather well for polarizabilities, though they tend to be a bit larger than the experimental values, and that the TDLDA works especially well for low-lying excitation energies but encounters difficulties for high-lying excitation energies. These problems can ultimately be traced back to difficulties with the asymptotic behavior of the exchange-correlation potential, v_{xc}. The consequences of this behavior for response properties have been analyzed here in detail, with particular attention to the location of the TD-DFRT ionization threshold. This analysis, together with considerations on models for the exact exchange-correlation potential, leads to the identification of three interrelated difficulties with the LDA, that are of particular significance for TD-DFRT calculations. Namely: (1)The exchange part of the potential falls off too rapidly at large r, leading to an effective potential which is too shallow. (2) The absolute magnitude of the orbital energies relative to the value of the potential at infinity is too small compared to the ionization potentials. (3) The exchange part of the coupling matrix, $K^{\sigma,\tau}(\mathbf{r},\mathbf{r}')$, is diagonal — i.e. proportional to $\delta(\mathbf{r}-\mathbf{r}')$ — when it should not be diagonal. These problems are not corrected by the gradient-corrected functionals usually used in molecular calculations. Several approaches to alleviating these problems, including scissors operators, self-interaction corrections, the weighted-density approximation, and specially designed gradient-corrected functionals have been reviewed. In view of what has already been learned from these approaches, and the rapidly increasing understanding of functionals and of the properties of the exact exchange-correlation functional, the outlook is optimistic for further improvements upon functionals for adiabatic time-dependent density-functional response theory and its molecular applications.

Acknowledgement

I am grateful to Dennis Salahub for his generous support of this work, including financial support through grants from the Natural Sciences and Engineering Research Council of Canada, the Canadian Centre of Excellence in Molecular and Interfacial Dynamics, and the Fonds pour la formation des chercheurs et l'aide à la recherche of Québec. I would especially like to thank Christine Jamorski for her work on the programming of deMon-DynaRho. Thanks to Emil Proynov and Kim Casida for reading parts of this manuscript. World Scientific Publishing is acknowledged for permission to reprint part of Sec. 2.

REFERENCES

1. P. Hohenberg and W. Kohn, Phys. Rev. **136**, B864 (1964).
2. W. Kohn and L.J. Sham, Phys. Rev. **140**, A1133 (1965).
3. A.K. Rajagopal, Adv. Chem. Phys. **41**, 59 (1980).
4. B.M. Deb and S.K. Ghosh, J. Chem. Phys. **77**, 342 (1982).
5. S.K. Ghosh and B.M. Deb, Chem. Phys. **71**, 295 (1982).
6. L.J. Bartolotti, Phys. Rev. A **24**, 1661 (1981).
7. L.J. Bartolotti, Phys. Rev. A **26**, 2243 (1982).
8. S. Chakravarty, M.B. Fogel, and W. Kohn, Phys. Rev. Lett. **43**, 775 (1979).
9. E. Runge and E.K.U. Gross, Phys. Rev. Lett. **52**, 997 (1984).
10. E.K.U. Gross and W. Kohn, Phys. Rev. Lett. **55**, 2850 (1985); Erratum, *ibid* **57**, 923 (1985).
11. D. Mearns and W. Kohn, Phys. Rev. A **35**, 4796 (1987).
12. E.K.U. Gross and W. Kohn, Adv. Quant. Chem. **21**, 255 (1990).
13. E.K.U. Gross, C.A. Ullrich, and U.J. Gossmann, in *Density Functional Theory*, E.K.U. Gross and R.M. Dreizler, Eds. (NATO ASI series, Pelnum, 1994), p. 149.
14. E.K.U. Gross, J.F. Dobson, and M. Petersilka, to appear in *Density Functional Theory*, R.F. Nalewajski, Ed. (Springer: New York, 1996).
15. A. Zangwill and P. Soven, Phys. Rev. A **21**, 1561 (1980).
16. M.J. Stott and E. Zaremba, Phys. Rev. A **21**, 12 (1980); Erratum, *ibid* **22**, 2293 (1980).
17. G.D. Mahan, Phys. Rev. A **22**, 1780 (1980).
18. J.A. Alonso, L.C. Balbás, and A. Rubio, Int. J. Quant. Chem. **56**, 499 (1995).
19. M. Madjet, C. Guet, and W.R. Johnson, Phys. Rev. A **51**, 1327 (1995).

20. G.D. Mahan and K.R. Subbaswamy *Local Density Theory of Polarizability* (Plenum Press: New York, New York, 1990).
21. Z.H. Levine and P. Soven, Phys. Rev. Lett. **50**, 2074 (1983).
22. Z.H. Levine, and P. Soven, Phys. Rev. A **29**, 625 (1984).
23. M.E. Casida, in *Recent Advances in Density Functional Methods, Part I*, D.P. Chong, Ed. (World Scientific: Singapore, 1995), p. 155.
24. M.E. Casida, C. Jamorski, F. Bohr, J. Guan, and D.R. Salahub, in *Nonlinear Optical Materials: Theory and Modeling*, S.P. Karna and A.T. Yeates, Eds. (ACS Press: Washington, D.C., 1996), (Proceedings of ACS Symposium, Washington, D.C., 1994) p. 145.
25. S.J.A. van Gisbergen, J.G. Snijders, and E.J. Baerends, J. Chem. Phys. **103**, 9347 (1995).
26. C. Jamorski, M.E. Casida, and D.R. Salahub, J. Chem. Phys. **104**, 5134 (1996).
27. S.J.A. van Gisbergen, V.P. Osinga, O.V. Gritsenko, R. van Leeuwen, J.G. Snijders, and E.J. Baerends, *unpublished.*
28. S.M. Collwell, N.C. Handy, and A.M. Lee, Phys. Rev. A **53**, 1316 (1996).
29. R. Bauernschmitt and R. Ahlrichs, Chem. Phys. Lett. (in press).
30. T. Ziegler, A. Rauk, and E.J. Baerends, Theor. Chim. Acta **43**, 877 (1977).
31. C. Daul, Int. J. Quant. Chem. **52**, 867 (1994).
32. G. Gardet, F. Rogemond, and H. Chermette, Theor. Chim. Acta **91**, 91 (1995).
33. G. Vignale, M. Rasolt, and D.J.W. Geldard, Adv. Quant. Chem. **21**, 235 (1990).
34. S.C. Ying, Nuovo Cimento **B23**, 270 (1974).
35. V. Peuckert, J. Phys. C **11**, 4945 (1978).
36. S.K. Ghosh and A.K. Dhara, Phys. Rev. A **38**, 1149 (1988).
37. B.-X. Xu and Rajagopal, Phys. Rev. A **31**, 2682 (1985).
38. A.K. Dhara and S.K. Ghosh, Phys. Rev. A **35**, 442 (1987).
39. G. Vignale and M. Rasolt, Phys. Rev. Lett. **59**, 2360 (1987).
40. G. Vignale and M. Rasolt, Phys. Rev. B **37**, 10685 (1988).
41. S.M. Colwell and N.C. Handy, Chem. Phys. Lett. **217**, 271 (1994).
42. P. Jørgensen and J. Simons, *Second Quantization-based Methods in Quantum Chemistry* (Academic, New York, 1981).
43. J.B. Foresman, M. Head-Gordon, J.A. Pople, and M.J. Frisch, J. Phys. Chem. **96**, 135 (1992).
44. B.I. Dunlap, J.W.D. Connolly, and J.R. Sabin, J. Chem. Phys. **71**, 3396 (1971).
45. H. Sambe and R.H. Felton, J. Chem. Phys. **62**, 1122 (1975).
46. E.R. Davidson, J. Comput. Phys. **17**, 87 (1975).
47. C.W. Murray, S.C. Racine, and E.R. Davidson, J. Comput. Phys. **103**, 382 (1992).
48. C. Jamorski, M.E. Casida, and D.R. Salahub, *unpublished.*
49. M. Petersilka and E.K.U. Gross, Int. J. Quant. Chem. Symp. (in press).
50. J. Oddershede, N.E. Grüner, and G.H.F. Diercksen, Chem. Phys. **97**, 303 (1985).
51. G. Huber and G. Herzberg, *Spectra of Diatomic Molecules* (Van Nostrand: Princeton, 1979).
52. S.B. Ben-Shlomo and U. Kaldor, J. Chem. Phys. **92**, 3680 (1990).
53. M. Jaszunski, A. Rizzo, and D.L. Yeager, Chem. Phys. **136**, 385 (1989).
54. S. Pal, M. Rittby, R.J. Bartlett, D. Sinha, and D. Mukherjee, J. Chem. Phys. **88**,

4357 (1988).
55. J. Odershedde, *Adv. Chem. Phys.* **69**, 201 (1987).
56. G.D. Zeiss and W.J. Meath, Molec. Phys. **33**, 1155 (1977).
57. M.A. Spackman, J. Chem. Phys. **94**, 1288 (1991).
58. M.J. Packer, S.P.A. Sauer, and J. Oddershede, J. Chem. Phys. **100**, 8969 (1994).
59. V. Carravetta, Y. Luo, and H. Ågren, Chem. Phys. **174**, 141 (1993).
60. C.O. Almbladh and U. von Barth, Phys. Rev. B **31**, 3231 (1985).
61. V. Russier, Phys. Rev. B **45**, 8894 (1992).
62. J.F. Janak, Phys. Rev. B **18**, 7165 (1978).
63. L.J. Sham and M. Schlüter, Phys. Rev. Lett. **51**, 1888 (1983).
64. J.P. Perdew and M. Levy, Phys. Rev. Lett. **51**, 1884 (1983).
65. R.M. Dreizler and E.K.U. Gross, *Density Functional Theory* (Springer–Verlag, New York, 1990).
66. J. Guan, M. Casida, A. Köster, and D.R. Salahub, Phys. Rev. B **52**, 2184 (1995).
67. S.A.C. McDowell, R.D. Amos, and N.C. Handy, Chem. Phys. Lett. **235**, 1 (1995).
68. M. Castro, M.E. Casida, C. Jamorski, K.C. Casida, and D.R. Salahub, *unpublished.*
69. A.D. Becke, J. Chem. Phys. **98**, 1372 (1993).
70. A.D. Becke, J. Chem. Phys. **98**, 5648 (1993).
71. P. Stampfli and K.H. Bennemann, Phys. Rev. A **39**, 1007 (1989).
72. J.M. Pachecho and W. Eckart, Z. Phys. D **24**, 65 (1992).
73. S. Saito, G.F. Bersch, and D. Tománek, Phys. Rev. B **43**, 6804 (1991).
74. A. Rubio, L.C. Balbás, and J.A. Alonso, Phys. Rev. B **46**, 4891 (1992).
75. L.C. Balbás, J.A. Alonso, and A. Rubio, Europhys. Lett. **14**, 323 (1991).
76. L.C. Balbás, A. Rubio, J. Alonso, and G. Borstel, J. Chem. Phys. **86**, 799 (1989).
77. A. Rubio, L.C. Balbás, and A. Vega, Z. Phys. D **12**, 209 (1989).
78. A. Rubio, L.C. Balbás, Ll. Serra, and M. Barranco, Phys. Rev. B **42**, 10950 (1990).
79. S.H. Vosko, L. Wilk, and M. Nusair, Can. J. Phys. **58**, 1200 (1980).
80. A.D. Becke, Phys. Rev. A **38**, 3098 (1988).
81. J.P. Perdew, Phys. Rev. B **33**, 8822 (1986).
82. J.P. Perdew and Y. Wang, Phys. Rev. B **33**, 8800 (1986).
83. C. Lee, W. Yang, and R.G. Parr, Phys. Rev. B **37**, 785 (1988).
84. R. van Leeuwen and E.J. Baerends, Phys. Rev. A **49**, 2421 (1994).
85. C.R. Brundle and D.W. Turner, Proc. R. Soc. London, Ser. A **307**, 27 (1968).
86. M.E. Casida and D.P. Chong, Phys. Rev. A **40** 4837 (1989).
87. S.A. Clough, Y. Beers, G.P. Klein, and L.S. Rothman, J. Chem. Phys. **59**, 2254 (1973).
88. J. Guan, P. Duffy, J.T. Carter, D.P. Chong, K.C. Casida, M.E. Casida, and M. Wrinn, J. Chem. Phys. **98**, 4753 (1993).
89. Z.H. Levine and D.C. Allan, Phys. Rev. Lett. **63**, 1719 (1989).
90. H. Zhong, Z.H. Levine, and J.W. Wilkins, Phys. Rev. A **43**, 4629 (1991).
91. M. Stener, P. Decleva, and A. Lisini, J. Phys. B **28**, 4973 (1995).
92. E. Fermi and E. Amaldi, Mem. Acc. Ital. **6**, 117 (1934).
93. J.P. Perdew and A. Zunger, Phys. Rev. B **23**, 5048 (1981).
94. G. Senatore and K.R. Subbaswamy, Phys. Rev. A **34**, 3619 (1986).
95. I. Moullet and J.L. Martins, J. Chem. Phys. **92**, 527 (1990).

96. J.A. Alonso and L.A. Girifalco, Phys. Rev. B **17**, 3735 (1978).
97. O. Gunnarsson, M. Jonson, and B.I. Lundqvist, Phys. Rev. B **20**, 3136 (1979).
98. H. Przybilski and G. Borstil, Solid State Commun. **49**, 317 (1984).
99. H. Przybilski and G. Borstil, Solid State Commun. **52**, 713 (1984).
100. A. Lembarki, F. Rogemond, and H. Chermette, Phys. Rev. A **52**, 3704 (1995).
101. R. Santamaria, Int. J. Quant. Chem. (in press).
102. M.E. Casida, Phys. Rev. A **51**, 2005 (1995).
103. J.C. Slater, Phys. Rev. **81**, 385 (1951).
104. M. Petersilka, U.J. Gossmann, and E.K.U. Gross, Phys. Rev. Lett. **76**, 1212 (1996).
105. J.B. Krieger, Y. Li, and G.J. Iafrate, Phys. Lett. A **146**, 256 (1990).
106. J.B. Krieger, Y. Li, and G.J. Iafrate, in *Density Functional Theory*, E.K.U. Gross and R.M. Dreizler, Eds. (Plenum Press, New York, 1995), p. 191.
107. R.T. Sharp and G.K. Horton, Phys. Rev. **90**, 317 (1953).
108. J.D. Talman and W.F. Shadwick, Phys. Rev. A **14**, 36 (1976).
109. M.M. Pant and J.D. Talman, Phys. Rev. A **17**, 1819 (1978).
110. M.M. Pant and J.D. Talman, Phys. Lett. **68A**, 154 (1978).
111. K. Aashamar, T.M. Luke, and J.D. Talman, At. Data and Nuclear Data Tables **22**, 443 (1978).
112. K. Aashamar, T.M. Luke, and J.D. Talman, J. Phys. B **12**, 3455 (1979).
113. K. Aashamar, T.M. Luke, and J.D. Talman, Phys. Rev. A **19**, 6 (1979).
114. K. Aashamar, T.M. Luke, and J.D. Talman, J. Phys. B **14**, 803 (1981).
115. J.D. Talman, Computer Physics Comm. **54**, 85 (1989).
116. R.W. Godby, M. Schlüter, and L.J. Sham, Phys. Rev. B **37**, 10159 (1988).
117. C.A. Ullrich, U.J. Gossman, and E.K.U. Gross, Phys. Rev. Lett. **74**, 892 (1995).
118. T. Grabo and E.K.U. Gross, Chem. Phys. Lett. **240**, 141 (1995).
119. E.K.U. Gross, M. Petersilka, and T. Grabo, in *Density Functional Methods in Chemistry*, T. Ziegler, Ed. (American Chemical Society, Washington D.C., 1996).
120. O. Gritsenko, R. van Leeuwen, E. van Lenthe, and E.J. Baerends, Phys. Rev. A **51**, 1944 (1995).
121. P. Duffy, D.P. Chong, M.E. Casida, and D.R. Salahub, Phys. Rev. A **50**, 4707 (1994).

J.M. Seminario (Editor)
Recent Developments and Applications of Modern Density Functional Theory
Theoretical and Computational Chemistry, Vol. 4

Advances in methodologies for linear-scaling density functional calculations

Benny G. Johnson[a], Christopher A. White[b], Qiming Zhang[c], Bin Chen[c], Richard L. Graham[c], Peter M. W. Gill[d] and Martin Head-Gordon[b]

[a]Q-Chem, Inc., 317 Whipple St., Pittsburgh, PA 15218, USA
E-mail: johnson@q-chem.com
[b]Department of Chemistry, University of California, Berkeley, CA 94720, USA
[c]Cray Research, Inc., 655 Lone Oak Dr., Eagan, MN 55121, USA
[d]Department of Chemistry, Massey University, Palmerston North, New Zealand

1. INTRODUCTION

In recent years we have witnessed an explosion of interest in the Density Functional Theory [1-6] (DFT) as a promising alternative to conventional *ab initio* methods in quantum chemistry. Modern DFT is founded upon the Hohenberg-Kohn Theorem [1] which states that the exact ground-state energy is a unique functional of the exact one-electron charge density. Though this important theorem only demonstrates the existence of such an exact functional and does not give its form, much progress has been made in developing useful approximate functionals [7-13] for practical work. In a number of systematic validation studies [14-20], DFT has been shown to be successful in predicting various molecular properties, often giving results of a quality comparable to or better than second-order Møller-Plesset (MP2) perturbation theory. One must be cautious not to apply DFT indiscriminately, as there are also examples in the literature of dramatic failures of DFT [21-23], but nonetheless, the overall results so far are most encouraging and the future of DFT methods in computational quantum chemistry appears bright. The cost of a DFT calculation is on the same order as that of Hartree-Fock (HF), but with a description of electron correlation included, and is substantially less expensive than traditional correlation techniques; hence DFT may be applied to a much larger range of molecular systems.

Given the attractiveness of density functional methods, there is considerable interest in advancing the computational methodologies available for applying DFT to chemical problems. The Kohn-Sham (KS) formulation [2] of DFT, which is the most well-suited for practical computation, is closely analogous to HF theory in that a set of molecular orbitals is derived from an effective one-electron potential via a self-consistent procedure. In the previous volume in this series, we gave an account [23] of the ongoing work in our research group in developing an efficient Gaussian-orbital LCAO KS program, including the KS SCF procedure and analytic derivatives of the KS energy. The motivation of this effort is to embody a consistent approach to DFT calculations to facilitate, in the first instance, the systematic validation of DFT methods for chemical studies, and in the second instance, the practical application of the useful methods to actual research problems, especially those involving very large molecules.

The focus of the previous installment was on the formulation of a KS DFT methodology which is more comparable to conventional *ab initio* methods, and on computational aspects of efficient implementation of the exchange-correlation terms. Here, we describe the progress of our continued work in this area, in the development of linear-cost methods which enable DFT studies of much larger molecules than is currently possible with available computational methodologies. Both the theory and the development work, carried out in the Q-Chem quantum chemistry program [24], will be reviewed, and results from practical calculations will be presented to assess the efficiency and efficacy of the approach for large calculations. The presentation is at a level of detail which should be useful to those interested in the developmental aspects of DFT; however, it is not intended to be a complete exposition, and for further details the reader is directed to the various original publications referenced at the appropriate points. A particularly interesting aspect, discussed in the later sections, is the implementation of our method on parallel computers and the potential that these platforms hold for expanding the applicability of DFT.

2. COMPUTATIONAL KOHN-SHAM DENSITY FUNCTIONAL THEORY

The first step in endeavoring to devise an improved-cost DFT approach is to examine the components of a practical DFT calculation to see how the costs go. Therefore, we briefly discuss the relevant equations of KS theory.

In the KS formalism, the system is connected with a reference state consisting of a set of *non-interacting* one-particle orthonormal orbitals yielding the same density as that of the actual system. Introducing the usual expansion of these orbitals ψ_i in a finite basis set ϕ_μ of atomic orbital (AO) basis functions,

$$\psi_i = \sum_{\mu=1}^{N} c_{\mu i}\phi_\mu \tag{1}$$

where the $c_{\mu i}$ are the molecular orbital (MO) coefficients and N is the number of basis functions, the charge density is given by

$$\rho = \sum_{i=1}^{n} |\psi_i|^2 = \sum_{\mu\nu} P_{\mu\nu}\phi_\mu\phi_\nu \tag{2}$$

where n is the number of electrons. The second equality expresses the density in the AO basis, through the density matrix

$$P_{\mu\nu} = \sum_{i=1}^{n} c_{i\mu} c_{i\nu} \tag{3}$$

(where we have used real orbitals for convenience).

Then, the total electronic energy is given by

$$E = \sum_{\mu\nu} P_{\mu\nu} H_{\mu\nu} + \frac{1}{2} \sum_{\mu\nu\lambda\sigma} P_{\mu\nu} P_{\lambda\sigma} (\mu\nu \mid \lambda\sigma) + E_{XC} \tag{4}$$

The first term is the "one-electron" energy (kinetic energy, electron-nucleus Coulomb energy, interaction with applied fields) and the $H_{\mu\nu}$ are the matrix elements of the one-electron Hamiltonian. The second term is the electron-electron Coulomb energy, and a standard notation is used for the two-electron repulsion integrals. The final term is the exchange-correlation (XC) energy

$$E_{XC} = \int f[\rho(\mathbf{r})]\, d\mathbf{r} \tag{5}$$

given as an integral over all space of the density functional f.

Variational optimization of the energy with respect to the density leads to the equations

$$FC = SC\varepsilon \tag{6}$$

where C is the matrix of MO coefficients, S is the AO overlap matrix, ε is the diagonal matrix of orbital energies, and F is KS potential matrix, whose elements are

$$F_{\mu\nu} = \frac{\partial E}{\partial P_{\mu\nu}} \tag{7}$$

These equations are then solved iteratively via a self-consistent procedure. It is interesting to note that Eq. (6) has exactly the same form as the familiar Roothaan-Hall equations [25, 26] of HF theory, except that the exact HF exchange potential has been replaced by an approximate DFT XC potential. The matrix F is the analogue of the Fock matrix in HF theory, and so for convenience we shall refer to it as the Fock matrix from now on.

From Eqs. (4), (5) and (7), the Fock matrix has the form

$$F_{\mu\nu} = H_{\mu\nu} + \sum_{\lambda\sigma} P_{\lambda\sigma} (\mu\nu \mid \lambda\sigma) + \int \phi_\mu \phi_\nu \frac{\partial f}{\partial \rho}\, d\mathbf{r} \tag{8}$$

We now examine the one-electron, Coulomb and XC contributions in turn. When considering large molecules, the computational expense is usefully analyzed in terms of its scaling with the size of the system. Formally, the size of H is $O(N^2)$, quadratic in the size of the orbital basis and hence in the size of the molecule. However, if the overlap of two basis functions ϕ_μ and ϕ_ν is negligibly small, their matrix element can be considered equal to zero to within an acceptable error tolerance and does not need to be evaluated. This is merely an example of "integral cutoff" strategy, employed with great success in molecular quantum mechanical calculations for many years. As the molecule grows larger, eventually the number of functions ϕ_ν overlapping with a given ϕ_μ becomes independent of N and tends to a constant, as only functions within a certain distance of the center of ϕ_μ will overlap. Therefore, asymptotically there are only O(N) *significant* elements of H that must be calculated. The nuclear attraction matrix, which is the most costly component of H, has a contribution to each element from each nucleus in the system, and so with cutoffs the cost of H for large molecules is $O(NN_A)$, where N_A is the number of atoms, i.e. quadratic in system size.

The Coulomb contribution to F, usually termed the J matrix, formally scales as $O(N^4)$ due to the number of two-electron integrals appearing in Eq, (8). This is the same source of the much-ballyhooed $O(N^4)$ often claimed as the cost of HF. While this is correct for small calculations, as soon as the system is large enough to contain basis functions which do not overlap (which is not very large) the cost scaling of the integrals decreases, asymptotically approaching $O(N^2)$. By the same reasoning as for H above, using cutoffs in large molecules J contains only O(N) significant elements, evaluated at a cost of $O(N^2)$.

Whereas the integrals in H and J are evaluated analytically, those in the XC matrix must be done with numerical quadrature due to their complexity. This requires a grid for representing the density and the functional whose total number of points is directly proportional to the molecular size. Thus, in a very simplistic implementation the cost of the XC step would go as the cube of the molecule size, as it would seem that every matrix element must be calculated at every grid point. However, it is possible to do much better than this, and in our earlier work [23, 27, 28] we have developed a method for calculating the XC contribution to F in only O(N) computation. While for small systems the numerical XC work can often dominate, for the analysis here it is sufficient to note that soon it is overtaken by the Coulomb cost as the system becomes larger, as we have observed in practice [27] and as would be expected from the respective scaling behaviors.

The only other significant step in the SCF procedure is the solution of Eq. (6), accomplished by matrix diagonalization or an equivalent approach. This scales as $O(N^3)$ formally as well as in practice (without considering the possibility of taking advantage of the asymptotic sparsity of the KS Fock matrix). But, for systems at the limit of the capabilities of commonly available programs (a few hundred atoms), the diagonalization problem is not yet dominant nor generally even significant.

Therefore, the Coulomb potential is the computational bottleneck for large molecules. There has been dramatic progress in recent years in improved algorithms for integral evaluation, which has been reviewed elsewhere [29], but all these developments are still based on evaluating the electron interactions in a pairwise fashion, and hence can scale no better than $O(N^2)$. Likewise, advances in faster computer hardware, though progressing at a remarkable pace, cannot be depended upon solely to provide a satisfactory resolution. If our goal is to be able to treat systems which are, say, an order of magnitude larger than is currently feasible, clearly then it is a breakthrough in the *scaling behavior* of the Coulomb contribution is required; it must be brought below quadratic. The remainder of this chapter is devoted to addressing this problem.

Before moving on to this most exciting topic, however, a couple of other relevant points should be mentioned. To complete the total molecular energy evaluation, purely classical nuclear terms such as the nuclear repulsion energy must be calculated in addition to the electronic terms; this scales quadratically in the number of atoms. This is completely insignificant computationally by comparison, though, and thus warrants no further comment beyond pointing out that these terms are also Coulombic in nature and thus any method for reducing the scaling of the electronic problem may also be applied to this simpler version of problem. This also applies to the nuclear attraction matrix in the one-electron Hamiltonian.

It is also worth stating that we have chosen not to incorporate any of the various fitting techniques (see, for example, Ref. [30]) that are common in several DFT programs. These arose in the early days of practical DFT calculations on molecules out of necessity to extend the scope of applicability of DFT, and generally involve the introduction of a second, auxiliary basis set and projection of the density and/or the XC potential onto this basis in order to speed up the J and XC evaluation. There is now a myriad of variations on the fitting theme employed though, and this is a major source of disparity among DFT programs. We previously discussed [23] the need for consistency among DFT programs in that the methods implemented should be well-defined and the results reproducible from one program to another, as is the case e.g., with HF programs. This has not yet been achieved with DFT as it has with conventional *ab initio* programs.

Density fitting does provide an efficiency advantage for the Coulomb contribution, but introduces extra considerations which must be weighed. For example, the fitted densities do not automatically conserve the total charge in the system. Though it is straightforward to make them do so by explicit constraint, in general the multipole moments of the true charge distribution are not preserved. This is a specific example related to the largely unexplored issue of the effect of fitting basis incompleteness. With the advent of more sophisticated algorithms for the XC energy and potential, it has been established that XC fitting, unlike density fitting, no longer offers any meaningful computational advantage, but does substantially complicate the theory of analytic derivatives. Komornicki and Fitzgerald [31] have given a useful comparison of the energy and derivative equations involved when fitting is carried out or not. Finally, fitting can indeed make the calculation faster, but it does not change the asymptotic scaling of the cost, and so in itself is not a solution to the problem posed for large molecules.

These remarks are not intended to denigrate the fitting approaches unduly, but it should be kept in mind that they amount to additional *approximations* that are made for the sake of increased computational speed. They could be viewed alternate "theoretical model chemistries" [32], which could then proceed to be validated and their accuracies and limitations characterized through systematic studies, but it does not seem that this has yet been done satisfactorily. We would like to accomplish our goal without resorting to further approximation if at all possible.

3. THE FAST MULTIPOLE METHOD

To gain insight into attacking the electronic Coulomb problem, we now digress into discussion of a related classical problem which is conceptually simpler than the one we set out to solve, but critically fundamental nonetheless.

The evaluation of Coulomb interactions in large-scale ensembles of classical particles is a central task in the numerical treatment of many physical processes. Traditionally, a quadratic amount of work has been required to evaluate all pairwise interactions between the particles, unless some truncation method or other sort of approximation was used. Consequently, very large-scale problems have appeared prohibitively expensive. However, in 1987, while studying the evaluation of the potential energy of a large system of point particles, Greengard [33] developed a powerful algorithm, the Fast Multipole Method (FMM), whose cost increases only *linearly* with the number of particles.

Greengard's breakthrough method built upon the earlier "tree code" methods of Appel [34] and Barnes and Hut [35], which solved the classical Coulomb problem in O(N log N) work. We do not need to be painstakingly detailed about the FMM here, but we do wish present a heuristic explanation of the FMM which should give the reader a basic understanding of the approach and convince him of its O(N) scaling. For the rigorous derivation of the FMM Greengard's original papers [33, 36, 37] can be consulted. The algorithm consists of several steps, briefly described as follows.

1. The particles are placed within a 1x1x1 "computational box" with all coordinates in the interval [0, 1]. To fit all particles within the box requires coordinate scaling for most problems; however this is necessary for the numerical stability of the algorithm when using high-order multipoles.

   The computational box is then divided in half along each Cartesian axis to form "children" of the parent box. This process may be continued recursively, yielding a tree structure. At the n-th level of the tree, there are 8^n boxes in three dimensions (with the original box being level 0). The total depth of the tree is chosen so that number of particles in each of the lowest level boxes is approximately constant and independent of the total number of particles.

2. The particles are radix sorted by their coordinates into the individual boxes on the lowest level of the tree. Multipole expansions of the charges on the particles within each lowest-level box are then constructed about the center of the box. The degree of the expansion is determined by the accuracy desired in the calculation, through well-defined error bounds given by Greengard.

   The next phase of the calculation is termed the "upward pass", in which the multipole expansion of each box is translated to the center of its parent box, using mathematical "translation operators". The translated moments from all children of a particular box are summed and stored in the parent box. This procedure is likewise continued up the tree until level 2. At this point we have a multipole expansion for each box representing of all particles contained in its children.

3. The next step, or the "downward pass", is the most critical aspect of the FMM. At this point we could evaluate the interaction of the particles within a given box with the multipole expansions within each well-separated box (i.e., the boxes far enough away that their multipole expansions are valid within the given box). Instead, however, a mathematical lemma due to Greengard is used to convert the multipole expansion of a well-separated box into a *local* Taylor-like expansion of this potential about the center of the current box. Such local expansions could be constructed for all well-separated boxes and summed, yielding an expansion of the far-field potential from all well-separated boxes.

   This procedure need not be carried out explicitly. By beginning at the top of the tree and traversing downward, it is only necessary to perform the conversion for boxes which are well-separated from a given box B, but for which their *parent* is *not* well-separated from B's parent. This is because the information contained in well-separated boxes not translated to B has already been translated to an ancestor of B. This neglected information is transferred from parent to child by using another simple translation operator to translate each parent's Taylor expansion to the center of each of its children, which is then summed with the child's Taylor expansion. This procedure is continued until we reach the lowest level of the tree, whereupon each box will contain a Taylor expansion representing the potential of *all* particles contained in well-separated boxes.

   The above procedure avoids the handling of every particle at every level in the tree, and is the distinguishing feature making the FMM an O(N) algorithm. The computational cost of the external to local conversion operation is $O(L^4)$, where L is the degree of the expansions, making this the most expensive step in the FMM.

4. For each box on the lowest level of the tree, the local Taylor expansion is evaluated at each particle within the box to obtain the far-field potential at these points. This represents the effects of all particles which are well-separated from the current box.

5. The final step is to evaluate the interaction of each particle with the particles within its own box, and within neighboring boxes (i.e., the non-well-separated particles). The multipole/local expansions are not valid here, and these interactions are therefore calculated directly as in an $O(N^2)$ approach. However, there are only O(1) particles which are treated directly with each particle, and so the number of direct interactions has been reduced from $O(N^2)$ to O(N). This near-field potential is added to the far-field potential calculated above to obtain the total potential at each particle.

Greengard's method has been profitably applied to numerous problems in classical mechanics [37-42], from studying the evolution of galaxies to simulating the behavior of biomolecules, plasmas and fluids using molecular dynamics. Heretofore, though, it has not been possible to apply it to the quantum mechanical realm. The next section shows how to unlock its power in DFT calculations as well.

4. GENERALIZATION TO CONTINUOUS CHARGE DISTRIBUTIONS: THE CONTINUOUS FAST MULTIPOLE METHOD

The FMM in its original formulation is not appropriate for electronic structure calculations as it deals only with discrete (point-source) charge distributions. A generalization is necessary in order to be able to treat the interactions of continuous charge distributions (most commonly Gaussian basis functions) used to represent the electrons. Here we describe our work in extending the classical FMM to accomplish this goal, resulting in the Continuous Fast Multipole Method (CFMM). Again, as with the FMM we present this at an overview level which should give a conceptual understanding of the approach, but for more details the reader is directed to our original papers [43, 44].

4.1. Well-separatedness

The first step requires the concept of well-separatedness to be revisited. The fundamental difference in the quantum mechanical case is that the charge distributions have spatial extent.

The multipole and local expansions have well-defined regions of validity. The reason space is partitioned into boxes is to establish which particles are well-separated, and can hence be treated via multipole expansions. The definition for well-separated (WS) boxes requires all particles contained within the first box must be well-separated from all points in the second box. The multipole expansion about the center of a given box converges for points outside a sphere circumscribed around the box (i.e. a sphere touching each corner of the box).

Since this sphere overlaps with nearest-neighbor boxes, the "minimal" definition for well-separated boxes requires that they be separated by one box (i.e. next-nearest neighbors). It is also possible to use a WS definition of greater than one. This reduces the number of interactions evaluated explicitly through multipole expansions; however, the expansions converge more quickly and the error bound is improved. In his book [33], Greengard proposes a WS definition of two boxes for the classical FMM, and we have found this to be the most appropriate in the CFMM as well.

One important additional concern in developing a CFMM is to ensure that overlapping continuous distributions are not allowed to interact through the use of the multipole expansions. More specifically, one must ensure the rigorous error bound of the FMM remains intact for the continuous case. To do so, we introduce the definition of the extent of a distribution. The extent of a distribution is taken as the number of boxes required for the distribution to decay from its maximum to a negligible value. Basically, if a distribution leaves the box where it is centered and overlaps with a neighboring box, it has an extent of 1. For example, to ensure the validity of an error bound corresponding to a WS definition of 2, the centers of two distributions having extent 1 must be separated by at least 4 boxes. This is illustrated with a one-dimensional case in Figure 1.

Figure 1
Two continuous distributions which are well-separated

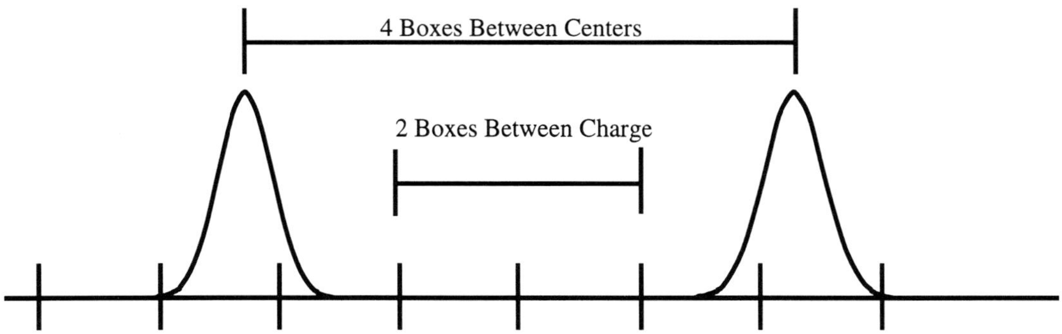

In general, two distributions of extent X may be treated via multipoles if their centers are separated by at least $2X + WS_{box}$ boxes. Therefore, we shall take this as the WS definition of a distribution. In the classical (point-charge) case, $X = 0$ and so WS_{dist} reduces to WS_{box}, as it should.

4.2. Placement of distributions in the tree structure

To ensure that a maximal number of distributions are treated through multipole expansions, the tree structure used for holding the point-particles in the FMM has been elaborated. At the lowest level, the distributions in each box are classified and grouped according to extent, from 0 to 2^d-1, where d is the depth of the tree. Each such collection has a different WS definition. Those be treated directly with *all* distributions, are placed in a special separate category. These distributions effectively have a WS definition of infinity.

The above classification scheme is a sufficient for dealing with distributions consisting of a single, or "primitive", Gaussian function. In calculations employing "contracted" basis functions, which are linear combinations of concentric primitives, an additional consideration arises. The charge distributions are constructed out of products of pairs of basis functions, and, as is well known, the product of two contracted Gaussians is a sum of primitive Gaussians distributed along the line joining the two centers. The difficulty occurs when this line crosses a boundary between adjacent boxes, as the set of distributions which are well-separated from the primitives whose centers are in the first box is different from the set of distributions well-separated from the primitives in the second box. To deal with this situation efficiently, the contracted distribution is split into multiple distributions according to the location of the centers of the primitives involved. Each such distribution is assigned an index keeping track of which original distribution the split distributions came from, as these all contribute to the same elements of the Coulomb matrix.

The same approach may be likewise employed to separate primitives in the same box which belong to the same distribution but have different extents, as a large variation in extent can lead to inefficiency if the compact primitive is constrained to the same WS definition as the diffuse primitive.

4.3. Upward pass

The upward pass of the CFMM begins straightforwardly as in the original FMM. At the lowest level, multipole moments of the distributions in each box are formed about the center of the box. The moments of the distributions in each WS category are maintained separately. (Of course, moments need not be constructed for WS = ∞.) To contrast with the FMM, the moments in the CFMM correspond to multipole integrals instead of classical formulae.

The next step is to shift the moments to the parent boxes at the next level up the tree, where the multipole moments are likewise separated by WS definition. This is slightly more elaborate than in the classical FMM. First, at each level a set of moments should be shifted up to the next level only if the corresponding distributions can potentially be well-separated from some other distribution in the system *at the next level*. For example, with WS_{box} = 2, on level 4 we need not shift the moments of the distributions with extent 12 (WS = 26) to level 3, as these distributions have extent 6 (WS = 14) at this level, and hence there can be no other distributions more than 2 boxes away from these distributions at level 3.

Second, it is noted that if one uses only even WS definitions for the distributions, then it is possible for each upward translation to reduce the WS definition of the moments by 2. This allows the information contained in collections with different WS definitions to be combined on higher levels, efficiently sharing information between levels. For example, the WS 6 multipole expansions at level 4 become WS 4 expansions when shifted to the parent boxes at level 3. The WS 2 and WS 4 expansions at levels 4 and 3 are combined into the WS 2 expansions at levels 3 and 2, respectively.

4.4. Downward pass

Here the multipole expansions are converted to local expansions in well-separated boxes. The difficulty lies in the proper treatment of the interaction of distributions having *different* WS definitions. The way the WS definitions have been set up makes this relatively straightforward, however, as it is evident that the interaction of boxes B_1 having WS definition WS_1 and B_2 having WS definition WS_2 may be treated via multipoles if and only if they are separated by at least $(WS_1+WS_2)/2$ boxes. That is, the appropriate effective WS definition to use is the *average* of the WS definitions of the two boxes. A conceptual loop structure for this procedure is as follows.

```
For each level of the tree:
    For each WS definition WS1 at the current level:
        For each WS definition WS2 at the current level:
            For each WS1 box B1 at the current level :
                For each WS2 box B2 at the current level, such that the parents of
                B1 and B2 are not well-separated:
                    If distance between B1 and B2 ≥ (WS1+WS2)/2
                        Perform external to local conversion between B1 and B2.
                Next B2
            Next B1
        Next WS2
    Next WS1
Next level
```

The local expansions must now be translated from parent to child. This is accomplished in a manner strictly analogous to the diagonal upward shifting of multipole moments, except performed in reverse.

4.5. Evaluation of long-range interactions

This step is conceptually the same as in the classical version. For each distribution, at the lowest level the local expansion in the distribution's box represents the potential of all well-separated distributions. This local expansion is evaluated for each distribution by computing the appropriate multipole-type integrals, to yield the far-field Coulomb potential for the distribution. This is similar to the first step in the CFMM where the moments are formed.

4.6. Evaluation of short-range interactions

It remains to evaluate set of short-range interactions, which is the complement of the set of long-range interactions discussed above, explicitly. The interactions which must be evaluated explicitly are easily determined from the classification of distributions by position and extent at the lowest level of the tree.

The algorithm for the short-range interactions requires the evaluation of a reduced set of two-electron repulsion integrals, and so the loop structure is similar to that used in the $O(N^2)$ case. This loop structure will be examined in more detail in Section 6 where parallelism is discussed. For now it suffices to say that the number of explicit integrals required is reduced to O(N).

4.7. Error analysis

A very important feature of the FMM which distinguishes it from many other methods for large-scale calculations is that it is *not* an approximation. The FMM has well-defined error bounds associated with all steps, such that the error can be rigorously controlled to within an acceptable tolerance by choosing the degree of the multipole expansions appropriately. Therefore, the same result is produced as would be obtained by an explicit calculation (even though for the largest systems, explicit calculation would not be feasible). This carries over to the CFMM as well, and as we discussed earlier is a very desirable characteristic in developing a method for very large DFT calculations.

For example, in his book [33], Greengard gives the following error bound for the potential using an Lth-degree multipole expansion:

$$|\Phi_L - \Phi_{Exact}| \leq \frac{Q}{R}\left(\frac{1}{2}\right)^{L+1} \tag{9}$$

Here Q is the sum of the magnitudes of the charges whose potential is being expanded, and R is the radius of a sphere enclosing the charges (essentially the distance from the center of the box to one of its corners). That is, the dependence on the error ε with expansion degree is

$$\varepsilon \propto \left(\frac{1}{2}\right)^{L+1} \tag{10}$$

In the continuous case, we have taken care to ensure that distributions treated via multipole expansions are well-separated, and hence their interactions could just as well be evaluated by classical expressions (i.e. neglecting penetration effects) to within acceptable error. Therefore, though it is possible to develop a CFMM error bound which takes into account the effect of using a range of WS definitions, etc., in the first instance one can use the classical error bound as a error bound for the CFMM case.

Using Greengard's bound, one arrives at the conclusion that a multipole expansion of degree roughly L = 33 is necessary to achieve an error $\varepsilon \leq 10^{-10}$ in the far-field potentials. This is quite a high degree of expansion compared with those used in classical applications, and is not encouraging given the steep cost of the FMM with L. However, the above bound represents the worse case possible and for the majority of the interactions treated the bound is very loose.

To obtain a practical measure of the error trends that can be expected in actual calculations, we have examined the empirical behavior of the error. A detailed characterization of the error can found in Ref. [44]. From representative accuracies we have observed, it is seen that for a typical accuracy of 10^{-10} required for *ab initio* calculations, an expansion degree of 25 is appropriate. In the direct SCF scheme this high-level accuracy is necessary only in the latter cycles as the wavefunction nears convergence; in early cycles far from convergence a much looser error tolerance may be used (e.g. 10^{-6}), and there expansions of much lower degree, say 10-15, would suffice. This lower accuracy is also often used for single-point energy evaluations, i.e. when only the energy is desired and a highly accurate wavefunction is not needed. For tight SCF convergence (necessary, for example, when computing energy derivatives), the degree can be gradually increased to the final value as the convergence progresses.

5. PERFORMANCE OF THE CFMM

Of course, the most important aspect of the CFMM is how it performs in practice, to see whether it is actually useful in facilitating much larger DFT calculations. We have implemented the CFMM within the Q-Chem [24] program and applied it to the calculation of the Coulomb energy and potential matrix in DFT. Before presenting the results, a few comments about the expected performance in the quantum mechanical case.

Greengard states that the "crossover point" (the point at which the FMM becomes faster than explicit pairwise calculation) for a well-implemented (classical) FMM code is on the order of 1000 particles. Obviously if this carried over literally to the CFMM it would be useless, since 1000 atoms is already well beyond the limits of current programs. Fortunately, in the quantum case the situation is much more favorable because the charge distributions (the analogue of the point charges in the classical case) are generated by taking products of *pairs* of orbital basis functions, and so their number grows large even for relatively small molecules. Also, one would expect the CFMM to become useful much earlier than the FMM because the explicit pairwise interactions which are avoided by collectivization in the CFMM (two-electron integrals) are much more expensive than their counterparts in the FMM.

Figure 2 gives a direct comparison of conventional integral methods with the CFMM for evaluating the J matrix. The test calculations were performed on a series of graphitic sheets using the 3-21G basis set (9 basis functions per carbon atom). A mid-level workstation (DEC Alpha 3000 Model 600) was used. This figure is from our earlier work [44], where we presented calculations on one-, two- and three-dimensional model systems for completeness; the results shown here are representative.

The four curves in Figure 2 show timings for four different methods for calculating the *same* result, the J matrix for the graphite system. The "Conventional" timings are for the usual approach of calculating all significant two-electron integrals, here done with the HGP-PRISM method [29], scaling as $O(N^2)$ for large molecules. The "J Engine" curve is for a new method we have developed [45] which is an improved way to evaluate the integrals in the special

Figure 2
CPU times for J matrix formation by various methods

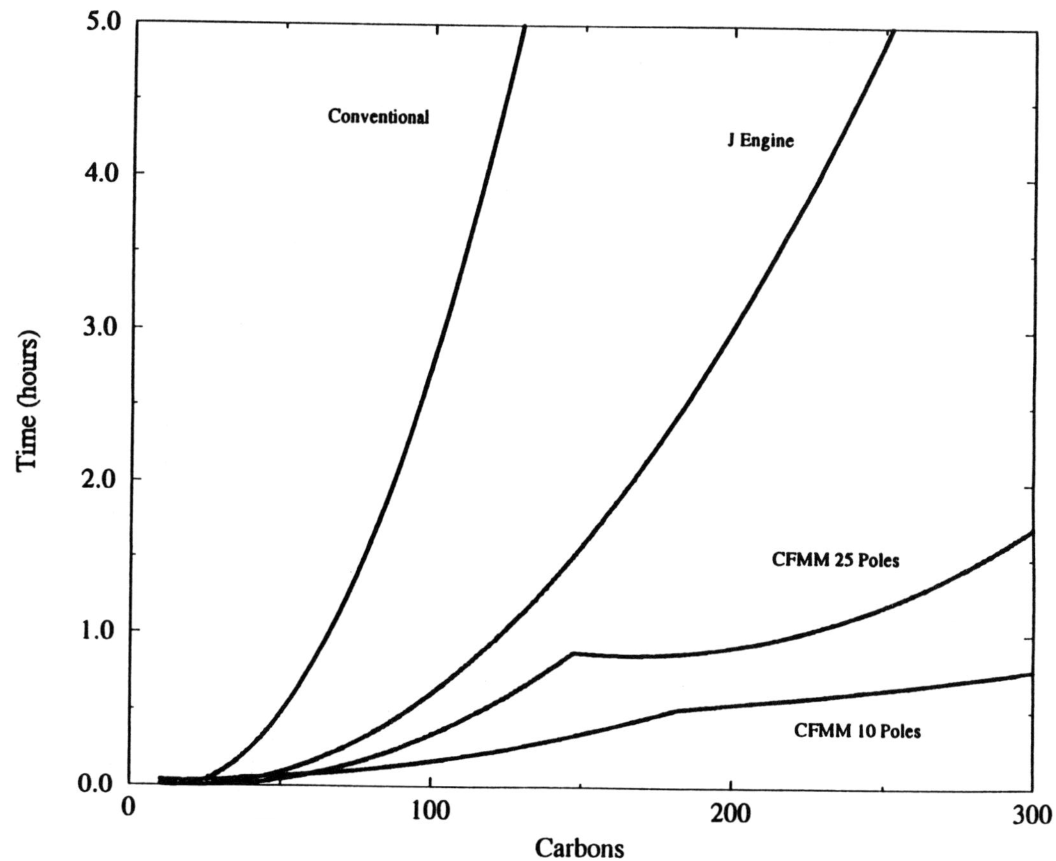

case when only the Coulomb (and not exchange terms) are required. While more efficient than the conventional method, on its own the J engine still scales asymptotically as $O(N^2)$ (it may be used in conjunction with the CFMM for the short-range contribution, however). The bottom two curves are for the CFMM using 10-th degree and 25-th degree expansions, corresponding to "early" and "late" SCF cycle accuracies, respectively. The cusp in the CFMM curves corresponds to the transition from a 3-level to a 4-level tree.

The timings speak for themselves. Taking a time limit of 5 hours for a J matrix formation on this particular workstation, the conventional method runs out of steam at about 125 carbon atoms. The J engine improves the situation considerably, allowing 250 carbons to be calculated in the same amount of time, but its scaling behavior at this point is the same as the conventional method and it is seen that the J engine curve is becoming fairly steep. With the CFMM, it is a different story altogether. While the usual approach takes 5 hours to calculate a J matrix for approximately 125 carbons (1125 basis functions), the CFMM can handle 300 carbons (2700 basis functions) in only an hour and a half (45 minutes for initial accuracy)! We did not carry out the conventional calculation for the 300 atom system, but it is evident from extrapolating the curves that the performance advantage of the CFMM at this point would be around an order of magnitude. Furthermore, both CFMM curves are much flatter than either of their explicit-interaction counterparts, and it is clear that, due to the O(N) scaling, the CFMM methods will reach the arbitrary 5-hour time limit for a system that is qualitatively much larger than the 125 or so atoms possible with the conventional method. We have estimated the coefficient of linear scaling from our results, finding it to be approximately 10 MFLOPs (millions of floating-point operations) [44] per basis function for the graphite sheets. Put another way, this corresponds to less than a second per basis function on most modern workstations.

It is difficult to discern directly from Figure 2, but the crossover point for these graphite sheets occurs at around 40 carbons, which is well within the realm of applicability of current programs. This is very important because it means the CFMM affords a performance advantage for a great many of the systems *already* studied by DFT, in addition to extending that range of accessible systems considerably. Even for 3-dimensional (diamond) structures, which are compact and completely space-filling and thus represent the biggest challenge for a multipole-based method like the CFMM, the crossover point occurs around 65 carbons [44].

Before leaving the topic of CFMM performance, we note that other workers have proposed and often also implemented other sub-quadratic methods for the quantum chemical Coulomb problem [46-49]. These are designed to have an asymptotic scaling behavior of O(N) or O(N log N).

Of these other methods, the work of Scuseria *et al.* [48] is the most directly comparable to our own. Their implementation within the Gaussian [50] series of programs, which they have termed the Gaussian Very Fast Multipole Method (GvFMM), is a variant upon our early work [43] consisting of the CFMM plus the "very fast" (vFMM) variation on the far-field multipole shifting proposed by

Petersen *et al.* [51] as an improvement in the classical context. In the case of the GvFMM however, the vFMM modification is only of marginal value, as it affects only the far-field CPU time, while the GvFMM is completely dominated by near-field work. As a comparison of the efficiency of our implementation and theirs, we note that Scuseria *et al.* [48] quote a CPU timing of 9 hours on an IBM RS/6000 370 workstation for a single Coulomb matrix evaluation for a $C_{384}H_{48}$ graphitic sheet with the 3-21G basis set (3552 basis functions). A five-level tree was used, with an integral neglect threshold of 10^{-6} atomic units (initial SCF cycle accuracy). Approximately 95% of their timing is due to the near-field contribution. For the same calculation under exactly the same conditions, our implementation of the CFMM requires only 51 minutes (0.86 hours) to evaluate the Coulomb matrix, with the near-field timing comprising only 63% of the total, and thus has a full factor of 10 advantage in speed over the GvFMM for this large test case.

6. THE CFMM ON PARALLEL COMPUTERS

The CFMM is an example of reducing the intrinsic scaling behavior of the abstract quantum mechanical method, and as such represents a definite advance even before the details of its efficient computer implementation are addressed. At the other end of the development spectrum are the complementary opportunities afforded by recent advances in the world of high-performance computing, with perhaps the most notable of these being the advent of parallel and massively parallel computer architectures. This new class of computing platforms offers enormous potential for scientific applications, much greater than is able to be achieved by increasing the speeds of the individual processors alone. However, the writing of programs which can take maximum advantage of such architectures is, at least presently, quite a difficult problem.

There have now been parallel implementations of one variety or another of nearly all prevalent electronic structure methods; one survey of the current state of affairs can be found in Ref. [52]. For all the revolutionary promises made at the onset, however, experiences with parallelizing pre-existing quantum chemistry codes have generally not lived up to expectations, and it has been realized that one can expect acceptable performance only on a few processors at best when starting with a "dusty deck". It is becoming generally accepted that in order to harness the power of parallel processing fully, then it is necessary to re-design and re-implement the software "from the ground up" with parallelism considered and included from the outset.

The FMM is an algorithm which contains a high degree of parallelism, and moreover, several groups have already succeeded in developing parallel versions [42, 53, 54] of the classical FMM. A comparable parallelization of the CFMM does not pose significant additional theoretical difficulties, and we have implemented a parallel version of the CFMM within Q-Chem [24].

Before describing the parallelization of the various CFMM phases, there are some unique features of our implementation which greatly facilitate efficient parallelism. One of the most important of these is the full exploitation of sparsity in the density and Coulomb matrices, allowing the use of *replicated* data for large systems. In this way, the only communication between processors that need occur comes only at the beginning (broadcast density matrix) and at the end (global sum of J matrix) of the algorithm, since no processor need ever consult another processor for any particular data needed for its portion of the computation. This is a dramatic simplification which greatly enhances parallel efficiency.

When sparsity is not taken into account, as is the case with most programs, replicated data puts a severe limit on the maximum problem size that can be handled with a fixed amount of memory per node, and larger problems need a distributed data approach, which generally comes with a large increase in communication cost. However, the full exploitation of sparsity allows the replicated data model to extend to much larger systems than would otherwise be possible, even in a modest amount of memory. For example, a system with 2000 basis functions requires 16 MB of memory to store the full (lower triangular) density matrix, but by storing only the significant elements this is reduced to less than 1 MB, which is much more manageable.

6.1. Parallelization of long-range CFMM

As we have already discussed, the long-range part of the CFMM is similar in many respects to the classical FMM, and hence so is its parallelization. One consideration is that for the CFMM the tree depths required are typically smaller than for the FMM, due to the greater rate of increase of total distributions. This permits the replicated data approach to be used for the shallower tree depths (3 and 4 levels) that are optimal for molecules with a few hundred atoms. Thus for these systems a distributed-data approach via a spatial decomposition is avoided. With fully replicated data, the only major issue is load balance. At this stage we have completed an implementation which uses a dynamic load-balance strategy for the external-to-local conversion (the dominant component of the long-range step) distributing the shifts between boxes over the processors. A simpler static load balance is currently used for all other phases. This yields reasonable initial parallel performance, which will be improved by upgrading all phases of the long-range step to use dynamic load balance. Details of our initial parallel long-range approach may be found elsewhere [55].

6.2. Parallelization of short-range CFMM

The short-range CFMM step is considerably different than its classical counterpart, and thus warrants more attention here. We begin with our parallel implementation of the PRISM algorithm [29] for two-electron integrals in the context of "conventional" direct SCF. PRISM required some reformulation at a fundamental level before implementation in Q-Chem [24] in order to make it suitable for parallelism on a large number of processors, as in a previous implementation of an earlier version of PRISM parallelism was not considered. However, upon making critical modifications we have now observed parallel scaling with greater than 90% efficiency on 128 processors for a calculation on more than 3900 basis functions [56]. Therefore, this is a reasonable starting point for parallel CFMM.

Figure 3 is a depiction of the near-field CFMM algorithm, in which is evident the classification by integral angular momentum (L) and contraction (K) which is the fundamental strength of the PRISM approach. For the details of PRISM, the reader is directed to Ref. [29]. We would like to preserve this efficient structure as much as possible, and so for the CFMM, loops over the spatial boxes were inserted after the classification of the integrals into L and K types. With the use of replicated data, this necessitates that the ket shell-pairs be gathered in each integral batch to interact with the bra shell-pairs; however, this is not a significant step in practice.

Figure 3
The near-field CFMM loop structure

```
Loop over bra L and K types
    Loop over ket L and K types
        Select optimal path for generating integrals of current type
        Loop over (non-empty) lowest-level boxes
            Gather ket shell pairs in current and neighboring boxes to be treated
                explicitly with bra shell-pairs in current box
            Compute batches of integrals of current type and spatial position
            Use batches of integrals and scatter back short-range Coulomb matrix
                contributions
        Next box
    Next KKet, LKet
Next KBra, LBra
```

One critical difference between the short-range CFMM and usual quadratic integrals evaluation is that shell-pairs are paired only those in neighboring boxes to obtain the integral batches. This is significant since it results in batch sizes which do not grow with the size of the molecule, and thus the available parallelism in our previous approach [56] is limited compared to the $O(N^2)$ case. There is the opportunity to switch the parallelism to distributing the boxes over the processors, which is perhaps the approach naturally suggested by the FMM. This has the advantage of being coarser-grained than splitting integrals loops, but has the large drawback that the load-balancing problem is much more severe than in our earlier work [56] since each box contains a different number of integrals to be computed and the advantage of constructing integrals batches of exactly the same cost which can be handed out to the processors is partially lost.

Fortunately, this can easily be smoothed out in our implementation, since PRISM [29] makes reliable estimates of the computational cost of the integrals ahead of time, and so it is relatively straightforward to arrange for the most expensive integrals batches to come first. This combined with a dynamic load-balancing approach ensures that the amount of work assigned to each processor is kept relatively even. Our parallel short-range implementation is also discussed in more detail in another publication [57].

7. PERFORMANCE OF THE PARALLEL CFMM

The exploitation of parallelism in the CFMM only amplifies the performance advantage on a single processor, making for a powerful combination for calculations on large molecules. Figure 4 shows the parallel speedup curves for calculations on three representative graphite sheets, C_{50}, C_{100} and C_{150}, with the largest system having 1350 basis functions. These calculations were carried out on a Cray T3D machine.

The speedup curves stay close to the "ideal" linear speedup through 16 processors, and by 32 processors the parallel efficiency is starting to diminish for these particular molecules. For the smaller molecules this is partly because the available parallelism is becoming exhausted. Also contributing is the current non-optimal load-balancing algorithm in part of the long-range phase; this is easily rectified. However, the most important point of this graph is to note that as the system becomes larger, the parallel efficiency is increased. C_{50}, C_{100} and C_{150} get progressively closer to ideal speedup line at 32 processors, indicating that as the system grows, more and more processors may be effectively employed on the problem, as it should be.

Finally, Table 1 presents some representative timings of calculations of the CTCGAG molecule, a six base-pair oligonucleotide with 378 atoms. With the 3-21G basis set this system contains 2476 basis functions. While the timings on various numbers of processors are impressive, perhaps the bottom line from this table is the 64-processor result in which the J matrix is evaluated in *less than one minute* of CPU time. Clearly this demonstrates that the CFMM combined with parallel platforms greatly enhances the investigatory capabilities at the disposal of the researcher. This problem is essentially out of reach of conventional methods on a workstation for all practical purposes, while with the new approach such calculations can obviously be made quite routine.

Table 1
Parallel CFMM timings for CTCGAG molecule on Cray T3D

NPES	Near-Field (s)	Far-Field (s)	Total (s)
8	213.7	147.8	361.5
16	109.2	79.8	189.1
64	31.0	28.4	59.4

Figure 4
Parallel CFMM speedups on Cray T3D

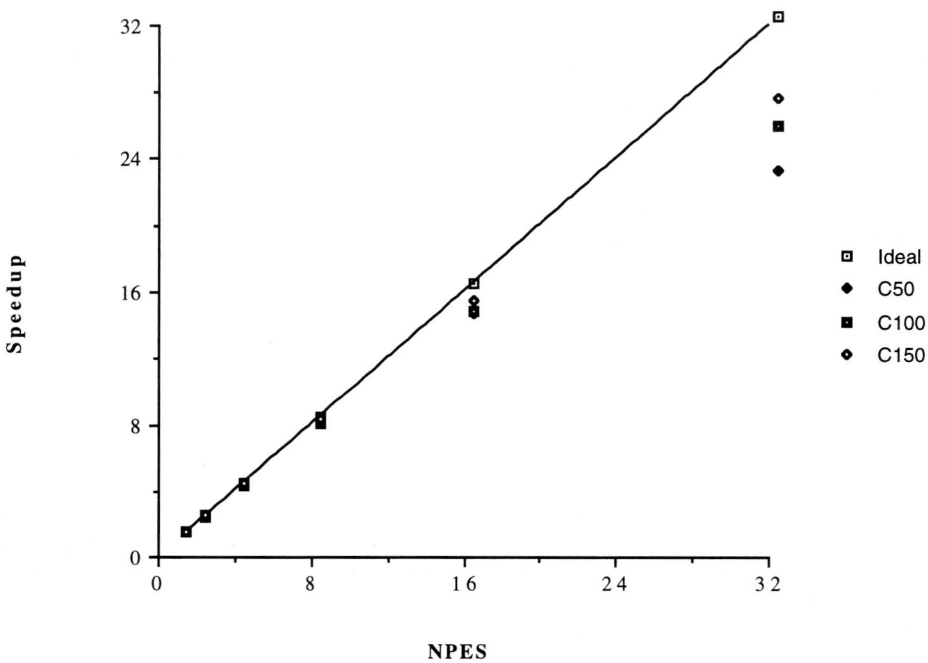

8. CLOSING REMARKS

It is clear that DFT is establishing a useful and significant niche in modern computational quantum chemistry, such that investment in development of algorithms for very large molecules is prudent. However, it is also clear that in order to extend the capabilities of DFT significantly, a radical departure from existing methods is called for.

We have presented an account of our work in developing and implementing the Continuous Fast Multipole Method, a new computational methodology enabling the rate-liming step in large-scale DFT calculations to be carried out in only linear work in system size. The necessary advance is a generalization of Greengard's algorithm for classical discrete particles to the case of continuous charge distributions required to treat the electrons. In our ongoing work on the CFMM within Q-Chem [24] we are continuing to optimize and refine our algorithm for both serial and parallel computers. In particular, the next stage of development for the parallel version is an algorithm which requires only a constant amount of memory per processor with respect to system size, thus enabling scalability on a much larger number of processors than with our initial parallel version.

By making the Coulomb problem linear, the frontier of molecular size limitation for DFT has been pushed back considerably, at which point the diagonalization step actually becomes the dominant one at the new end of the spectrum of tractable chemical problems (as is already the case in semi-empirical quantum mechanical approaches). Efficient diagonalization of large matrices is a topic of much active research, as the results are applicable in virtually all areas of computational science; however, we feel that the most promise in our continued development efforts for large-molecule methods lies in so-called "diagonalization-free" methods, and the formulation of such methods which scale linearly (instead of cubically) in system size is one focus of our present research efforts.

ACKNOWLEDGMENT

The authors are grateful for funding and support for this work provided by the following sources: the U. S. Department of Energy (SBIR Grant DE-FG05-93ER81643), Cray Research, Inc., the Ben Franklin Partnership of Western Pennsylvania (Challenge Grant 95W.CC00152R-1) and the Pittsburgh Supercomputing Center.

REFERENCES

1. P. Hohenberg and W. Kohn, Phys. Rev. B 136, 864 (1964).
2. W. Kohn and L. J. Sham, Phys. Rev. A 140, 1133 (1965).

3. R. G. Parr and W. Yang, *Density-Functional Theory of Atoms and Molecules* (Oxford Univ. Press, New York, 1989).
4. R. O. Jones and O. Gunnarsson, Rev. Mod. Phys. 61, 689 (1989).
5. J. K. Labanowski and J. W. Andzelm, Eds., *Density Functional Methods in Chemistry* (Springer-Verlag, New York, 1991).
6. T. Ziegler, Chem. Rev. 91, 651 (1991).
7. P. A. M. Dirac, Proc. Cam. Phil. Soc. 26, 376 (1930).
8. J. C. Slater, *The Self-Consistent Field for Molecules and Solids,* Quantum Theory of Molecules and Solids (McGraw-Hill, New York, 1974), vol. 4.
9. S. H. Vosko, L. Wilk and M. Nusair, Can. J. Phys. 58, 1200 (1980).
10. J. P. Perdew, Phys. Rev. B 33, 8822 (1986).
11. A. D. Becke, Phys. Rev. A 38, 3098 (1988).
12. C. Lee, W. Yang and R. G. Parr, Phys. Rev. B 37, 785 (1988).
13. J. P. Perdew, in *Proceedings of the 21st Annual International Symposium "Electronic Structure of Solids"* (Nova Science, Commack, NY, 1991).
14. A. D. Becke, J. Chem. Phys. 96, 2155 (1992).
15. A. D. Becke, J. Chem. Phys. 97, 9173 (1992).
16. A. D. Becke, J. Chem. Phys. 98, 5648 (1993).
17. J. Andzelm and E. Wimmer, J. Chem. Phys. 96, 1280 (1992).
18. C. W. Murray, G. J. Laming, N. C. Handy and R. D. Amos, Chem. Phys. Lett. 199, 551 (1992).
19. B. G. Johnson, P. M. W. Gill and J. A. Pople, J. Chem. Phys. 98, 5612 (1993).
20. J. M. Seminario, Chem. Phys. Lett. 206, 547 (1993).
21. K. Raghavachari, D. L. Strout, G. K. Odom, G. E. Scuseria, J. A. Pople, B. G. Johnson and P. M. W. Gill, Chem. Phys. Lett. 214, 357 (1993).
22. B. G. Johnson, C. A. Gonzalez, P. M. W. Gill and J. A. Pople, Chem. Phys. Lett. 221, 100 (1994).
23. B. G. Johnson, in *Modern Density Functional Theory: A Tool for Chemistry* J. M. Seminario and P. Politzer, Eds. (Elsevier Science B.V., Amsterdam, 1995), vol. 2, p. 169.
24. B. G. Johnson, P. M. W. Gill, M. Head-Gordon, C. A. White, D. R. Maurice, R. D. Adamson, T. R. Adams, M. Oumi and N. Ishikawa, Q-Chem, Q-Chem, Inc., Pittsburgh, PA (1996).
25. C. C. J. Roothaan, Rev. Mod. Phys. 23, 69 (1951).
26. G. G. Hall, Proc. R. Soc. London, Ser. A 205, 541 (1951).
27. B. G. Johnson, Ph.D. thesis, Carnegie Mellon University (1993).
28. B. G. Johnson, unpublished work.
29. P. M. W. Gill, Advan. Quantum Chem. 25, 141 (1994).
30. B. I. Dunlap, J. W. D. Connolly and J. R. Sabin, J. Chem. Phys. 71, 3396 (1979).
31. A. Komornicki and G. Fitzgerald, J. Chem. Phys. 98, 1398 (1993).
32. W. J. Hehre, L. Radom, P. v. R. Schleyer and J. A. Pople, *Ab Initio Molecular Orbital Theory* (Wiley, New York, 1986).
33. L. Greengard, *The Rapid Evaluation of Potential Fields in Particle Systems* (MIT Press, Cambridge, 1987).
34. A. W. Appel, SIAM J. Sci. Stat. Comput. 6, 85 (1985).
35. J. Barnes and P. Hut, Nature 324, 446 (1986).

36. L. Greengard and V. Rokhlin, J. Comput. Phys. 73, 325 (1987).
37. L. Greengard, Comput. Phys. 4, 142 (1990).
38. J. Ambrosiano, L. Greengard and V. Rokhlin, in *Third International Conference on Supercomputing Proceedings*, L. P. Kartashev and S. I. Kartashev, Eds. (Supercomputing '88, Boston, 1988), vol. 1, p. 334.
39. L. Greengard, SIAM J. Sci. Stat. Comput. 11, 603 (1990).
40. C. R. Anderson, SIAM J. Sci. Stat. Comput. 13, 923 (1992).
41. N. Engheta, W. D. Murphy, V. Rokhlin and M. S. Vassilou, IEEE Trans. Antennas Propag. 40, 634 (1992).
42. J. A. Board, J. W. Causey, J. F. Leathrum, A. Windemuth and K. Schulten, Chem. Phys. Lett. 198, 89 (1992).
43. C. A. White, B. G. Johnson, P. M. W. Gill and M. Head-Gordon, Chem. Phys. Lett. 230, 8 (1994).
44. C. A. White, B. G. Johnson, P. M. W. Gill and M. Head-Gordon, Chem. Phys. Lett., in press.
45. C. A. White and M. Head-Gordon, J. Chem. Phys. 104, 2620 (1996).
46. R. Kutteh, E. Apra and J. Nichols, Chem. Phys. Lett. 238, 173 (1995).
47. M. Challacombe, E. Schwegler and J. Almlöf, in *Computational Chemistry: Review of Current Trends*, J. Leszszynski, Ed. (Sinapore Publishing, in press).
48. M. C. Strain, G. E. Scuseria and M. J. Frisch, Science, in press.
49. J. P. Dombroski, S. W. Taylor and P. M. W. Gill, J. Phys. Chem. 100, 6272 (1996).
50. M. J. Frisch, G. W. Trucks, M. Head-Gordon, P. M. W. Gill, M. W. Wong, J. B. Foresman, B. G. Johnson, H. B. Schlegel, M. A. Robb, E. S. Replogle, R. Gomperts, J. L. Andres, K. Raghavachari, J. S. Binkley, C. Gonzalez, R. L. Martin, D. J. Fox, D. J. Defrees, J. Baker, J. J. P. Stewart and J. A. Pople, GAUSSIAN 92, Gaussian, Inc., Pittsburgh, PA (1992).
51. H. G. Petersen, D. Soelvason, J. W. Perram and E. R. Smith, J. Chem. Phys. 101, 8870 (1994).
52. T. G. Mattson, Ed., *Parallel Computing in Computational Chemistry* (American Chemical Society, Washington, DC, 1995).
53. L. Greengard and W. D. Kropp, Computers Math. Applic. 20, 63 (1990).
54. E. Zyszkowski, Sci. Comput. Autom. 8, 21 (1992).
55. Q. Zhang, B. G. Johnson and R. L. Graham, Chem. Phys. Lett., submitted.
56. B. G. Johnson, J. W. Kyle, T. R. Adams and P. M. W. Gill, Chem. Phys. Lett., submitted.
57. B. G. Johnson, B. Chen and R. L. Graham, Chem. Phys. Lett., submitted.

J.M. Seminario (Editor)
Recent Developments and Applications of Modern Density Functional Theory
Theoretical and Computational Chemistry, Vol. 4

A divide-and-conquer implementation of the linear combination of gaussian-type orbitals density functional (LCGTO-DF) method

Alain St-Amant, Sor Koon Goh, and Roger T. Gallant

Department of Chemistry, University of Ottawa,
10 Marie Curie, Ottawa, Ontario, K1N 6N5, Canada

A divide-and-conquer implementation of the linear combination of gaussian-type orbitals density functional formalism has been developed within the DeFT software package. Use of both Yang's density and density matrix formulations of divide-and-conquer density functional theory are incorporated in this scheme. The latter is used for the construction of the density and the subsequent fit of that density which in turn eases the computational burden associated with the assembly of subsystem Kohn-Sham matrices. The former is used to fit the exchange-correlation terms. This software package should achieve linear scaling for very large systems and the algorithms developed are ideally suited for parallel architectures. The conventional linear combination of gaussian-type orbitals density functional method, already implemented in DeFT, is outlined. The same is done for both the density and density matrix formulations of divide-and-conquer density functional theory. The synthesis of these concepts within DeFT is presented. Test calculations on the glycine heptapeptide establish the errors to be expected given a certain criterion for assigning buffer atoms in each of our divide-and-conquer procedures. The concept of extended buffer space, where lesser quality bases are placed on faraway buffer atoms, is also discussed.

1. INTRODUCTION

The linear combination of gaussian-type orbitals density functional (LCGTO-DF) approach originally proposed by Sambe and Felton [1], and subsequently refined by Dunlap, Connolly, and Sabin [2], is implemented in such programs as DGauss [3], deMon [4], and DeFT [5]. The LCGTO-DF method has been successfully applied in a wide range of applications [3, 6, 7]. A key aspect of the LCGTO-DF approach is its use of fits to the electronic density and the exchange and correlation (XC) potential. These fits dramatically reduce the computational burden associated with the evaluation of the Kohn-Sham (KS) matrix elements [2, 3]. However, for large systems, these fitting procedures do not particularly scale well and quickly become the most time-consuming part of the LCGTO-DF calculation. Consequently, gaussian density functional approaches that do not make use of fits have emerged and have become very popular and extremely viable options [8]. Unfortunately, these calculations can no longer be carried out faster than their Hartree-Fock (HF) counterparts as both the DF and HF approaches face the same bottleneck: the four-center two-electron integrals arising from electron/electron repulsion. The computational advantages traditionally held by DF methods *vis à vis* HF methods are lost. Even if this were to now always remain the case, we must never forget the tremendous CPU savings afforded by current DF methods over conventional correlated post-HF *ab initio* methods [9, 10].

Recently, Yang has proposed a pair of divide-and-conquer (DAC) approaches to DF theory [11, 12]. In a DAC approach, a large molecule is divided into a set of subsystems. For each subsystem, subsystem KS equations are solved within a truncated orbital basis localized about the subsystem. The results arising from these subsystem calculations are then pieced together to construct the electronic density, $\rho(\mathbf{r})$, of the complete molecule. For molecules that are sufficiently large, the amount of CPU time spent on any one subsystem will eventually plateau at a given value. At this point, the cost of the entire calculation will depend only on the number of subsystems. As the number of subsystems is linearly proportional to the size of the molecule being studied, linear scaling in overall CPU time is achieved. Also, the DAC approach is ideally suited for parallel architectures as the vast majority of the CPU time is devoted to the independent solution of the KS equations in each subsystem. Consequently, this type of DAC approach has garnered a great deal of attention by those seeking to tackle very large molecules [13].

The LCGTO-DF method's fitting procedures present a new challenge to the DAC philosophy. There is no point in developing software to perform a DAC construction of $\rho(\mathbf{r})$ within an LCGTO-DF program if a DAC approach cannot be adopted for the fits as well. Yang's work is done within a software package [14] that is similar to the more familiar DMol program [15]. No fits are performed within such programs. In this report, we will show how each of Yang's DAC formulations, the original density formulation [11] and the subsequent density matrix formulation [12], can be used to tackle the fitting procedures. The density formulation can be modified to tackle the fit of the XC potential while the same can be done with the density matrix formulation when dealing with the fit of $\rho(\mathbf{r})$. The density matrix formulation has also been used to solve the subsystem KS equations and reconstruct $\rho(\mathbf{r})$ over the entire molecule.

In this paper, we will first outline the conventional LCGTO-DF formalism as it is programmed within our DeFT code. We will also briefly compare our techniques with those used in other popular DF programs so as to highlight the strengths and weaknesses of the LCGTO-DF methodology. We will then take a look at the two DAC schemes: the density formulation and the density matrix formulation. The concept of buffer atoms will be discussed. We will then take an in-depth look at our DAC implementation of the LCGTO-DF formalism within our conventional LCGTO-DF software package, DeFT. We will also introduce our concept of extended buffer space. Some of the individual components of this work can be taken and implemented within other programs to make them more efficient. For example, our DAC approach to fitting $\rho(\mathbf{r})$ can potentially be of great use to those developing LCGTO-HF codes. Test calculations on a small peptide will give a feel for the errors introduced by each one of our DAC procedures.

2. THE LCGTO-DF METHOD

The Kohn-Sham equations [16, 17] are the cornerstone of all modern DF applications. Within the KS approach, the electronic density, $\rho(\mathbf{r})$, of an N-electron system is expressed as the sum of the square moduli of doubly-occupied KS orbitals $\{\psi_i(\mathbf{r})\}$ (for simplicity, we will deal with closed shell systems — the method can be readily generalized to open shell systems),

$$\rho(\mathbf{r}) = 2\sum_{i}^{N/2} |\psi_i(\mathbf{r})|^2 . \tag{1}$$

Also within the KS approach, Hohenberg and Kohn's unknown, exact, total energy functional [18], $E[\rho(\mathbf{r})]$, is split into three terms,

$$E[\rho(\mathbf{r})] = T[\rho(\mathbf{r})] + U[\rho(\mathbf{r})] + E_{xc}[\rho(\mathbf{r})], \tag{2}$$

The first term, $T[\rho(\mathbf{r})]$,

$$T[\rho(\mathbf{r})] = 2\sum_{i}^{N/2} \int \psi_i(\mathbf{r})\frac{-\nabla^2}{2}\psi_i(\mathbf{r})\mathbf{dr}, \tag{3}$$

is the kinetic energy of a set of non-interacting electrons occupying the KS orbitals $\{\psi_i(\mathbf{r})\}$. The difference between the true kinetic energy of the electrons in the system and this definition for $T[\rho(\mathbf{r})]$ is swallowed up, as we shall see, by the third term. Unfortunately, any functional for $T[\rho(\mathbf{r})]$ that does not make use of KS orbitals and makes use of only $\rho(\mathbf{r})$ is not sufficiently accurate to be of any practical use in chemistry [17]. KS orbitals are thus required to accurately handle the kinetic energy of the electrons. The second term, $U[\rho(\mathbf{r})]$,

$$U[\rho(\mathbf{r})] = -\sum_{A}^{nuclei} \int \frac{\rho(\mathbf{r})Z_A}{|\mathbf{r}-\mathbf{R_A}|}\mathbf{dr} + \frac{1}{2}\int\int \frac{\rho(\mathbf{r})\rho(\mathbf{r}')}{|\mathbf{r}-\mathbf{r}'|}\mathbf{dr}\,\mathbf{dr}', \tag{4}$$

represents the classical Coulomb interactions arising from the repulsion between the electrons and the attraction between the electrons and the set of nuclei with charges $\{Z_A\}$ positioned at $\{\mathbf{R_A}\}$. The final term, $E_{xc}[\rho(\mathbf{r})]$, is referred to as the exchange and correlation energy functional and contains any contributions to the total energy not handled by either of the first two terms. The precise nature of $E_{xc}[\rho(\mathbf{r})]$ remains unknown and the accuracy of any DF calculation employing the KS approach is limited by the quality of the approximation made concerning its nature. Consequently, a great deal of work has been, and continues to be, invested in the construction of better XC functionals [19, 20]. In this work, we will limit ourselves to the simple local spin density approximation (LSDA) [21]. Our findings should not be affected once we begin to use more sophisticated XC functionals.

The KS equations are in turn derived by invoking Hohenberg and Kohn's [18] proof that the true ground state electronic density minimizes the total energy functional, i.e.,

$$\frac{\delta E[\rho(\mathbf{r})]}{\delta\rho(\mathbf{r})} = 0, \tag{5}$$

subject to the constraint that the density remains properly normalized, i.e.,

$$\int \rho(\mathbf{r})\mathbf{dr} = N. \tag{6}$$

This yields the KS equations [16, 17],

$$\left\{\frac{-\nabla^2}{2} - \sum_{A}^{nuclei}\frac{Z_A}{|\mathbf{r}-\mathbf{R_A}|} + \int\frac{\rho(\mathbf{r}')}{|\mathbf{r}-\mathbf{r}'|}\mathbf{dr}' + v_{xc}(\mathbf{r})\right\}\psi_i(\mathbf{r}) = \varepsilon_i\psi_i(\mathbf{r}), \tag{7}$$

where $\{\varepsilon_i\}$ is the set of KS eigenvalues and $v_{xc}(\mathbf{r})$ is the XC potential and is the functional derivative of the XC energy with respect to the density,

$$v_{xc}(\mathbf{r}) = \frac{\delta E_{xc}[\rho(\mathbf{r})]}{\delta \rho(\mathbf{r})}. \tag{8}$$

Though the KS equations appear to be very similar to the HF equations, and, upon visual inspection, the KS orbitals closely resemble the HF orbitals, we must remember that the KS and HF equations have different origins. Therefore, KS orbitals are not HF orbitals, and the corresponding eigenvalues, $\{\varepsilon_i\}$, do not carry the same Koopmans' theorem significance that their HF counterparts carry [9].

2.1. Solving the Kohn-Sham equations

In the LCGTO-DF formalism, the KS orbitals are expanded within a basis of K atom-centered contracted gaussians [9, 10, 22], $\{\chi_\nu(\mathbf{r})\}$,

$$\psi_i(\mathbf{r}) = \sum_{\nu}^{K} C_{\nu i}\chi_\nu(\mathbf{r}) \tag{9}$$

where $\{C_{\nu i}\}$ is the set of expansion coefficients for the ith KS orbital. The KS equations can be recast in matrix form in exactly the same way that the HF equations have been by Roothaan [9, 10]. If we insert this expansion for $\psi_i(\mathbf{r})$ into equation (7), multiply through by an arbitrary basis function $\chi_\mu(\mathbf{r})$, and integrate, we obtain the following:

$$\mathbf{HC} = \mathbf{SC}\varepsilon. \tag{10}$$

The elements of the KS matrix, $\mathbf{H}$, are given by

$$H_{\mu\nu} = \int \chi_\mu(\mathbf{r})\left\{\frac{-\nabla^2}{2} - \sum_{A}^{nuclei}\frac{Z_A}{|\mathbf{r}-\mathbf{R_A}|} + \int \frac{\rho(\mathbf{r}')}{|\mathbf{r}-\mathbf{r}'|}\mathbf{dr}' + v_{xc}(\mathbf{r})\right\}\chi_\nu(\mathbf{r})\mathbf{dr}. \tag{11}$$

The elements of the overlap matrix, $\mathbf{S}$, are given by

$$S_{\mu\nu} = \int \chi_\mu(\mathbf{r})\chi_\nu(\mathbf{r})\mathbf{dr}. \tag{12}$$

The ith column of the coefficient matrix, $\mathbf{C}$, contains the expansion coefficients of the KS orbital $\psi_i(\mathbf{r})$. The associated eigenvalue is the ith element of the eigenvalue vector, ε. Once the KS equations have been solved, the electronic density can be conveniently expressed in terms of the orbital basis functions,

$$\rho(\mathbf{r}) = \sum_{\mu}^{K}\sum_{\nu}^{K} P_{\mu\nu}\chi_\mu(\mathbf{r})\chi_\nu(\mathbf{r}), \tag{13}$$

where the elements of the density matrix, $\mathbf{P}$, are given by

$$P_{\mu\nu} = 2\sum_{i}^{N/2} C_{\mu i}C_{\nu i}. \tag{14}$$

The construction of the KS matrix, $\mathbf{H}$, dominates the CPU time in any DF calculation. The kinetic energy and nuclear attraction contributions to $\mathbf{H}$ are comparatively very easy to evaluate [23]. As in LCGTO-HF applications, Coulomb repulsion and exchange pose the greatest challenge [23]. Fortunately, by using contracted gaussians, LCGTO-DF programs can make use of the wealth of experience gained in the efficient evaluation of two-electron integrals by those developing LCGTO-HF programs [23, 24]. In the upcoming two sections, we shall show how fitting $\rho(\mathbf{r})$ and $v_{xc}(\mathbf{r})$ within auxiliary bases further allows us to ease the computational burden associated with these terms.

2.2. Fitting the electronic density

The evaluation of the Coulomb repulsion component of the KS matrix elements requires us to evaluate the same four-center two-electron integrals

$$\int \chi_\mu(\mathbf{r}) \left\{ \int \frac{\rho(\mathbf{r}')}{|\mathbf{r}-\mathbf{r}'|} \mathbf{dr}' \right\} \chi_\nu(\mathbf{r}) \mathbf{dr} = \int \chi_\mu(\mathbf{r}) \left\{ \sum_\sigma^K \sum_\lambda^K P_{\sigma\lambda} \int \frac{\chi_\sigma(\mathbf{r}')\chi_\lambda(\mathbf{r}')}{|\mathbf{r}-\mathbf{r}'|} \mathbf{dr}' \right\} \chi_\nu(\mathbf{r}) \mathbf{dr}$$
$$= \sum_\sigma^K \sum_\lambda^K P_{\sigma\lambda} \int\int \frac{\chi_\mu(\mathbf{r})\chi_\nu(\mathbf{r})\chi_\sigma(\mathbf{r}')\chi_\lambda(\mathbf{r}')}{|\mathbf{r}-\mathbf{r}'|} \mathbf{dr}\,\mathbf{dr}' \qquad (15)$$

that are the bottleneck of conventional HF methods [9], making them scale formally as K^4. The computational burden is eased within the LCGTO-DF approach by employing a fitted density, $\tilde{\rho}(\mathbf{r})$ [2]. The density is fit within an auxiliary basis of L atom-centered uncontracted gaussians [22], $\{\chi_k'(\mathbf{r})\}$,

$$\tilde{\rho}(\mathbf{r}) = \sum_k^L c_k \chi_k'(\mathbf{r}). \qquad (16)$$

Since the density is expressed as the sum of the square moduli of occupied KS orbitals, the exponents in the density fitting auxiliary basis typically span values that are roughly double those found in the orbital basis [22]. The number of auxiliary basis functions associated with an atom is roughly two times greater than that found in the corresponding orbital basis set [22]. What is important is that the number of auxiliary basis functions is of the same order as the number of orbital basis functions.

Sambe and Felton originally proposed to least squares fit the density directly [1], minimizing the discrepancies between the true density and the fitted density over all of space. Dunlap *et al.* dramatically improved the LCGTO-DF formalism by stating that the optimal set of $\rho(\mathbf{r})$ fitting coefficients, $\{c_k\}$, is obtained by minimizing the error in the Coulomb repulsion energy [2],

$$\int\int \frac{[\rho(\mathbf{r}) - \tilde{\rho}(\mathbf{r})][\rho(\mathbf{r}') - \tilde{\rho}(\mathbf{r}')]}{|\mathbf{r}-\mathbf{r}'|} \mathbf{dr}\,\mathbf{dr}' \qquad (17)$$

subject to the constraint that the fitted density is properly normalized to the total number of electrons in the system,

$$\int \tilde{\rho}(\mathbf{r})\mathbf{dr} = N. \qquad (18)$$

This yields the following solution for the $\rho(\mathbf{r})$ fit coefficient vector, $\mathbf{c}$,

$$\mathbf{c} = \mathbf{S}'^{-1}(\mathbf{t} + \lambda\mathbf{n}), \qquad (19)$$

where the elements of $\mathbf{S}'$ are given by

$$S_{kl}' = \int\int \frac{\chi_k'(\mathbf{r})\chi_l'(\mathbf{r}')}{|\mathbf{r}-\mathbf{r}'|} \mathbf{dr}\,\mathbf{dr}', \qquad (20)$$

the elements of $\mathbf{t}$ are given by

$$t_k = \int\int \frac{\chi_k'(\mathbf{r})\rho(\mathbf{r}')}{|\mathbf{r}-\mathbf{r}'|}\mathbf{dr}\,\mathbf{dr}' = \sum_\mu^K\sum_\nu^K P_{\mu\nu}\int\int \frac{\chi_k'(\mathbf{r})\chi_\mu(\mathbf{r}')\chi_\nu(\mathbf{r}')}{|\mathbf{r}-\mathbf{r}'|}\mathbf{dr}\,\mathbf{dr}', \tag{21}$$

and the elements of $\mathbf{n}$ are given by

$$n_k = \int \chi_k'(\mathbf{r})\mathbf{dr}. \tag{22}$$

The Lagrange multiplier, λ, is given by

$$\lambda = N - \frac{\mathbf{nS}'^{-1}\mathbf{t}}{\mathbf{nS}'^{-1}\mathbf{n}}. \tag{23}$$

The time-consuming part of this fitting procedure is the evaluation of the three-center two-electron integrals required for the construction of $\mathbf{t}$. The CPU time associated with this step scales as $\mathrm{K}^2\mathrm{L}$. Since L is of the same order as K, we can say that the scaling goes as K^3.

Once we have fitted $\rho(\mathbf{r})$, $\tilde{\rho}(\mathbf{r})$ is used to ease the computational burden associated with the assembly of the Coulomb repulsion component of the KS matrix,

$$\begin{aligned}\int \chi_\mu(\mathbf{r})\left\{\int \frac{\rho(\mathbf{r}')}{|\mathbf{r}-\mathbf{r}'|}\mathbf{dr}'\right\}\chi_\nu(\mathbf{r})\mathbf{dr} &\approx \int \chi_\mu(\mathbf{r})\left\{\int \frac{\tilde{\rho}(\mathbf{r}')}{|\mathbf{r}-\mathbf{r}'|}\mathbf{dr}'\right\}\chi_\nu(\mathbf{r})\mathbf{dr} \\ &\approx \sum_k^L c_k \int\int \frac{\chi_\mu(\mathbf{r})\chi_\nu(\mathbf{r})\chi_k'(\mathbf{r}')}{|\mathbf{r}-\mathbf{r}'|}\mathbf{dr}\,\mathbf{dr}'.\end{aligned} \tag{24}$$

This step will now also scale as $\mathrm{K}^2\mathrm{L}$ (or K^3) rather than as K^4, as we had at the onset. Therefore, for small and intermediate size systems, the use of $\tilde{\rho}(\mathbf{r})$ will speed up a calculation by replacing a single K^4 operation by a pair of K^3 operations.

However, for large systems, efficient cutoff schemes reduce the evaluation of the two-electron integrals to a procedure that scales as K^2 in practise [25], whether or not one chooses to fit or to not fit $\rho(\mathbf{r})$. At this point, fitting is not advantageous as we have to go through the integral evaluations *twice* per SCF cycle: once for the construction of $\mathbf{t}$ and once for the construction of the KS matrix. We are also faced with the inversion of $\mathbf{S}'$ which formally scales as L^3. Therefore, for large systems, the LCGTO-DF method, as originally proposed by Dunlap, Connolly, and Sabin [2], is not particularly well-suited for calculations on very large systems. We will address this issue in our divide-and-conquer implementation of the LCGTO-DF method.

2.3. Fitting the exchange-correlation terms

The density fitting procedure just outlined would not be of particularly great benefit if our DF approach required us to then evaluate the same exchange contribution as found in HF calculations,

$$-\frac{1}{2}\sum_\sigma^K\sum_\lambda^K P_{\sigma\lambda}\int\int \frac{\chi_\mu(\mathbf{r})\chi_\sigma(\mathbf{r})\chi_\nu(\mathbf{r}')\chi_\lambda(\mathbf{r}')}{|\mathbf{r}-\mathbf{r}'|}\mathbf{dr}\,\mathbf{dr}'. \tag{25}$$

The above expression cannot be simplified by a fitting procedure as was the Coulomb repulsion contribution since the basis functions χ_σ and χ_λ are now associated with the positions of two different electrons after the exchange operation. One is therefore forced to evaluate the four-center two-electron integrals for exchange and would then therefore be best advised to forego fitting $\rho(\mathbf{r})$ and use these integrals to evaluate the exact contribution to the KS matrix arising from electron/electron repulsion. There would still potentially be a benefit to fitting $\rho(\mathbf{r})$ in these circumstances since the integrals contributing to exchange vanish at a far greater rate than do those contributing to Coulomb repulsion as system size is increased. With the advent of XC potentials that incorporate "exact" exchange from HF theory [20], the benefits of using $\tilde{\rho}(\mathbf{r})$ to accelerate the evaluation of the Coulomb repulsion contribution to the KS matrix should be investigated.

Fortunately, in a wide range of applications, the XC energy is expressed as [19, 21]

$$E_{xc}\left[\rho(\mathbf{r})\right] = \int \varepsilon_{xc}(\mathbf{r})\rho(\mathbf{r})d\mathbf{r} \tag{26}$$

where $\varepsilon_{xc}(\mathbf{r})$ is the XC energy density. Within the LSDA, $\varepsilon_{xc}(\mathbf{r})$ is a function of the local spin up and spin down densities, $\rho^{\alpha}(\mathbf{r})$ and $\rho^{\beta}(\mathbf{r})$ [21]. Gradient-corrected XC functionals express $\varepsilon_{xc}(\mathbf{r})$ as a function of the local values of $\rho^{\alpha}(\mathbf{r})$ and $\rho^{\beta}(\mathbf{r})$ as well as the local values of their gradients, $\nabla\rho^{\alpha}(\mathbf{r})$ and $\nabla\rho^{\beta}(\mathbf{r})$ [19]. Since $\varepsilon_{xc}(\mathbf{r})$, and in turn, $v_{xc}(\mathbf{r})$, are expressed as functions of only local properties of the spin densities, the evaluation of $E_{xc}\left[\rho(\mathbf{r})\right]$ and the XC contributions to the KS matrix,

$$\int \chi_{\nu}(\mathbf{r})v_{xc}(\mathbf{r})\chi_{\nu}(\mathbf{r})d\mathbf{r}, \tag{27}$$

are in principle easy to evaluate. Unfortunately, the complex nature of the XC energy functionals precludes the direct evaluation of equations (26) and (27).

We must therefore resort to the use of a grid. The one specifically employed in DeFT follows Becke's algorithm for generating grids [26]. The grid is atom-centered and uses more points near the nuclei where $\rho(\mathbf{r})$ is greatest and variations in $\rho(\mathbf{r})$ are the most dramatic. DeFT's default grid has 32 radial shells per atom and 26 angular points per radial shell. As we shall see later on, the angular grid is augmented when evaluating total energies and energy gradients at the end of an SCF cycle.

Now that a grid has been laid down, the XC terms could be integrated numerically:

$$\int \varepsilon_{xc}(\mathbf{r})\rho(\mathbf{r})d\mathbf{r} \approx \sum_{I}^{grid} \varepsilon_{xc}(\mathbf{R_I})\rho(\mathbf{R_I})W_I \tag{28}$$

$$\int \chi_{\mu}(\mathbf{r})v_{xc}(\mathbf{r})\chi_{\nu}(\mathbf{r})d\mathbf{r} \approx \sum_{I}^{grid} \chi_{\mu}(\mathbf{R_I})v_{xc}(\mathbf{R_I})\chi_{\nu}(\mathbf{R_I})W_I \tag{29}$$

where $\{\mathbf{R_I}\}$ and $\{W_I\}$ are the coordinates and associated quadruture weights of the grid points. Some programs choose to do this, albeit on a much finer grid [27]. In accord with the original LCGTO-DF formalism of Dunlap *et al.* [2], DeFT takes a different approach and first fits both $\varepsilon_{xc}(\mathbf{r})$ and $v_{xc}(\mathbf{r})$, minimizing

$$\sum_{I}^{grid} \left[\varepsilon_{xc}(\mathbf{R_I}) - \tilde{\varepsilon}_{xc}(\mathbf{R_I})\right]^2 W_I \tag{30}$$

and

$$\sum_{I}^{grid} \left[v_{xc}(\mathbf{R_I}) - \tilde{v}_{xc}(\mathbf{R_I})\right]^2 W_I \tag{31}$$

over the set of grid points. The XC terms are fit by a common second auxiliary basis of M atom-centered uncontracted gaussians, $\{\chi_l''(\mathbf{r})\}$:

$$\tilde{\varepsilon}_{xc}(\mathbf{r}) = \sum_l^M e_l \chi_l''(\mathbf{r}) \tag{32}$$

$$\tilde{v}_{xc}(\mathbf{r}) = \sum_l^M b_l \chi_l''(\mathbf{r}). \tag{33}$$

Typically, M is equal to L, the number of basis functions in the $\rho(\mathbf{r})$ fitting basis. Since the XC terms are roughly proportional to $\rho^{1/3}(\mathbf{r})$, the values of the exponents in the XC fitting basis are simply set equal to one third those found in the associated $\rho(\mathbf{r})$ fitting basis [22].

Unlike the fit of $\rho(\mathbf{r})$, no constraints are imposed on these fits. This simplifies the equations for the fit coefficient vectors to

$$\mathbf{e} = \mathbf{S}''^{-1}\mathbf{t}^{\mathbf{e}} \tag{34}$$

and

$$\mathbf{b} = \mathbf{S}''^{-1}\mathbf{t}^{\mathbf{b}} \tag{35}$$

where elements of the matrix $\mathbf{S}''$ are given by

$$S_{kl}'' = \sum_I^{grid} \chi_k''(\mathbf{R_I})\chi_l''(\mathbf{R_I})W_I, \tag{36}$$

and the elements of $\mathbf{t}^{\mathbf{e}}$ and $\mathbf{t}^{\mathbf{b}}$ are given by

$$t_l^e = \sum_I^{grid} \chi_l''(\mathbf{R_I})\varepsilon_{xc}(\mathbf{R_I})W_I \tag{37}$$

and

$$t_l^b = \sum_I^{grid} \chi_l''(\mathbf{R_I})v_{xc}(\mathbf{R_I})W_I. \tag{38}$$

Once the XC fit coefficients are known, $\tilde{\varepsilon}_{xc}(\mathbf{r})$ and $\tilde{v}_{xc}(\mathbf{r})$ can be synthesized via equations (32) and (33) and the XC terms of equations (26) and (27) can be approximated as

$$\int \varepsilon_{xc}(\mathbf{r})\rho(\mathbf{r})d\mathbf{r} \approx \int \tilde{\varepsilon}_{xc}(\mathbf{r})\rho(\mathbf{r})d\mathbf{r} = \sum_\mu^K \sum_\mu^K P_{\mu\nu} \sum_l^M e_l \int \chi_l''(\mathbf{r})\chi_\mu(\mathbf{r})\chi_\nu(\mathbf{r})d\mathbf{r} \tag{39}$$

and

$$\int \chi_\nu(\mathbf{r})v_{xc}(\mathbf{r})\chi_\nu(\mathbf{r})d\mathbf{r} \approx \int \chi_\nu(\mathbf{r})\tilde{v}_{xc}(\mathbf{r})\chi_\nu(\mathbf{r})d\mathbf{r} = \sum_l^M b_l \int \chi_l''(\mathbf{r})\chi_\mu(\mathbf{r})\chi_\nu(\mathbf{r})d\mathbf{r}. \tag{40}$$

These involve the evaluation of rather simple three-center overlap integrals. This step also formally scales as K^3 (more precisely, K^2M), but is significantly less demanding than the evaluation of the three-center two-electron integrals of the preceding section. By far, the most time-consuming part of this fitting procedure is not integral evaluation, but rather the synthesis of the density (and its gradient in the case of gradient-corrected functionals) over the set of grid points. This scales as K^2N_g where N_g is the total number of grid points. Schemes that can reduce the number of grid points [27] can therefore dramatically improve the efficiency of LCGTO-DF programs.

One may ask why we even bother fitting $\varepsilon_{xc}(\mathbf{r})$ and $v_{xc}(\mathbf{r})$ and choose not to just simply numerically integrate the above terms as in equations (28) and (29). Given the same number of points, direct numerical integration is more efficient. However, we must remember that the use of a grid introduces noise in the total energy, i.e., the total energy of the molecule is no longer rotationally invariant [28]. Calculations are less susceptible to fluctuations in the total energy arising from grid noise when we are only using the grid for fitting purposes. Therefore, the fitting procedure allows us to achieve greater stability in the total energy given the same number of grid points, or conversely, it allows us to achieve a desired level of precision with a smaller number of grid points.

2.4. Evaluating the total energy

The traditional expression for the LCGTO-DF total energy is given by [2]

$$E = \sum_{\mu}^{K}\sum_{\nu}^{K} P_{\mu\nu} \int \chi_\mu(\mathbf{r}) \left\{ \frac{-\nabla^2}{2} - \sum_{A}^{nuclei} \frac{Z_A}{|\mathbf{r}-\mathbf{R_A}|} \right\} \chi_\nu(\mathbf{r}) d\mathbf{r}$$
$$+ \sum_{\mu}^{K}\sum_{\nu}^{K} P_{\mu\nu} \sum_{k} c_k \int\int \frac{\chi_\mu(\mathbf{r})\chi_\nu(\mathbf{r})\chi_k^{'}(\mathbf{r}^{'})}{|\mathbf{r}-\mathbf{r}^{'}|} d\mathbf{r}\, d\mathbf{r}^{'} - \frac{1}{2}\sum_{k}^{L}\sum_{l}^{L} c_k c_l \int \frac{\chi_k^{'}(\mathbf{r})\chi_l^{'}(\mathbf{r}^{'})}{|\mathbf{r}-\mathbf{r}^{'}|} d\mathbf{r}\, d\mathbf{r}^{'}$$
$$+ \sum_{\mu}^{K}\sum_{\nu}^{K} P_{\mu\nu} \sum_{m}^{M} e_m \int \chi_\mu(\mathbf{r})\chi_\nu(\mathbf{r})\chi_m^{''}(\mathbf{r}) d\mathbf{r} + \sum_{A}^{nuclei}\sum_{B<A} \frac{Z_A Z_B}{|\mathbf{R_A}-\mathbf{R_B}|}. \quad (41)$$

The Coulomb repulsion energy is expressed as

$$\sum_{\mu}^{K}\sum_{\nu}^{K} P_{\mu\nu} \sum_{k} c_k \int\int \frac{\chi_\mu(\mathbf{r})\chi_\nu(\mathbf{r})\chi_k^{'}(\mathbf{r}^{'})}{|\mathbf{r}-\mathbf{r}^{'}|} d\mathbf{r}\, d\mathbf{r}^{'} - \frac{1}{2}\sum_{k}^{L}\sum_{l}^{L} c_k c_l \int \frac{\chi_k^{'}(\mathbf{r})\chi_l^{'}(\mathbf{r}^{'})}{|\mathbf{r}-\mathbf{r}^{'}|} d\mathbf{r}\, d\mathbf{r}^{'}, \quad (42)$$

rather than just

$$\frac{1}{2}\sum_{\mu}^{K}\sum_{\nu}^{K} P_{\mu\nu} \sum_{k}^{L} c_k \int\int \frac{\chi_\mu(\mathbf{r})\chi_\nu(\mathbf{r})\chi_k^{'}(\mathbf{r}^{'})}{|\mathbf{r}-\mathbf{r}^{'}|} d\mathbf{r}\, d\mathbf{r}^{'}, \quad (43)$$

since the former will make the total energy variational with respect to the fit of $\rho(\mathbf{r})$, provided it is carried out in the fashion described earlier [2]. However, in this situation, the total energy obtained with a perfect fit to $\rho(\mathbf{r})$ would serve as an *upper* bound since we are doing the best possible job on a repulsive component of the total energy.

In our particular LCGTO-DF implementation, DeFT [5], the XC term is modified. Rather than using the fit to the XC energy density and performing the analytical integration, we perform a numerical integration,

$$\sum_{\mu}^{K}\sum_{\nu}^{K} P_{\mu\nu} \sum_{I}^{grid} \chi_\mu(\mathbf{R_I})\chi_\nu(\mathbf{R_I})\epsilon_{xc}(\mathbf{R_I}) W_I. \quad (44)$$

However, this numerical integration is carried out over an augmented set of grid points. Rather than having approximately 800 points per atom, as we do throughout the course of the SCF where we are only performing fits, a grid with well over 2500 points per atom is used. The same number of radial points, 32, remain. However, the number of angular grid points is augmented to either one of 50, 110, or 194 per radial shell [4]. The number used depends on the location of the radial shell. More points are used near the atom's covalent radius as this is where grid points from different atoms also carry appreciable quadruture weights [26] and we are most susceptible to fluctuations in quadruture weights which in turn lead to fluctuations in the total energy. Fewer angular points are needed near the nuclei where spherical symmetry exists and very far from the nuclei where the total density is very small and thus a very small contribution is made to the above sum.

Tests have shown that this numerical integration reduces the noise in the total energy when small molecules are rotated in space [4]. This procedure does increase the amount of CPU time required. However, it is done only once per SCF procedure. Equation (41), with the fit, is used to evaluate an approximate total energy throughout the course of the SCF procedure in order to monitor whether or not we have converged. Once self-consistency is achieved, equation (44) is used to evaluate a more precise $E_{xc}[\rho(\mathbf{r})]$. Though numerical integration on the augmented grid would be ideally carried out for each iteration in the SCF procedure, for $v_{xc}(\mathbf{r})$ as well as for $\varepsilon_{xc}(\mathbf{r})$, the cost is, in our opinion, too high. This combination of fitting procedures on a reduced grid and numerical integration on an augmented grid strikes a reasonable compromise between computational efficiency and precision.

2.5. Evaluating the energy gradient

Within DeFT, the following expression for the LCGTO-DF energy gradient is used (we have gone to the bra-ket notation for conciseness),

$$\begin{aligned} E^{(\lambda)} = \sum_{\mu}\sum_{\nu} P_{\mu\nu} \left(\left\{ \left\langle \chi_\mu \left| \frac{-\nabla^2}{2} - \sum_A \frac{Z_A}{|\mathbf{r}-\mathbf{R_A}|} \right| \chi_\nu \right\rangle \right\}^{(\lambda)} + \sum_k c_k \left[\chi_\mu\chi_\nu \left| \frac{1}{r_{12}} \right| \chi_k^{'} \right]^{(\lambda)} \right) \\ - \sum_k\sum_l c_k c_l \left[\chi_k^{'} \left| \frac{1}{r_{12}} \right| \chi_l^{'} \right]^{(\lambda)} - \sum_{\mu\nu} W_{\mu\nu} \langle \chi_\mu | \chi_\nu \rangle^{(\lambda)} + \left(\sum_A^{nuclei} \sum_{B<A} \frac{Z_A Z_B}{|\mathbf{R_A}-\mathbf{R_B}|} \right)^{(\lambda)} \\ + \sum_\mu\sum_\nu P_{\mu\nu} \sum_I^{grid} \left\{ \chi_\mu^{(\lambda)}(\mathbf{R_I})\, \chi_\nu(\mathbf{R_I})\, v_{xc}(\mathbf{R_I}) + \chi_\mu(\mathbf{R_I})\, \chi_\nu^{(\lambda)}(\mathbf{R_I})\, v_{xc}(\mathbf{R_I}) \right\} W_I, \end{aligned} \tag{45}$$

where elements of the energy-weighted density matrix, $\mathbf{W}$, are given by

$$W_{\mu\nu} = 2\sum_i^{\frac{N}{2}} \varepsilon_i C_{\mu i} C_{\nu i}. \tag{46}$$

The (λ) superscript denotes differentiation with respect to the nuclear coordinate λ. The above expression for the energy gradient is a modification of the original expression derived by Fournier, Andzelm, and Salahub [29]. As with the total energy, the XC term, originally treated by analytical integration with fitted quantities, is now handled by numerical integration over an augmented set of grid points. It is important to note that this gradient is not exact. However, the differences in bonds, bond angles, and torsions between the geometry at which the total energy is minimized and the geometry at which the gradient vanishes (if the gradients were exact, the two would be the same) are invariably smaller than the errors inherent in the LCGTO-DF formalism with present day XC functionals [4]. This error in the gradient could be removed by going to much larger XC auxiliary fitting bases. However, the extra cost is not warranted. Also, these errors in the gradient do not seem to introduce discontinuities in the gradient that would make analyzing potential energy surfaces very difficult.

DeFT does not have analytical second derivatives. They have however been implemented within another LCGTO-DF program, DGauss [30]. Vibrational analyses within DeFT are carried out by constructing the second derivative matrix through numerical differentiation of the gradient [31].

2.6. Other approaches

The vast majority of DF chemistry applications make use of a linear combination of atomic orbitals (LCAO) approach. Basis functions other than contracted gaussians are commonly used in these approaches. Alternate methods for constructing the KS matrix elements are also taken, each exploiting the nature of the basis functions being used. These approaches will be outlined in this section. In the following discussion, two increasingly viable approaches will be overlooked. The first, plane wave approaches, have been extremely popular in the physics community. However, the ability of plane waves to accurately describe the electronic structure of a simple main group molecule within a reasonably sized basis has always been a source of concern among computational chemists. With the recent advances in the plane wave methodology [32], this problem has been addressed and prelinary results on simple molecules are promising [33]. Plane wave bases also have the advantage of being ideally suited for performing *ab initio* molecular dynamics [35] via the Car-Parrinello approach [36] since these delocalized basis functions do not require the so-called Pulay forces [34] since the Hellman-Feynman theorem [34] can be used to evaluate energy gradients. The second scheme that is emerging is Becke's fully numerical approach which is being implemented in the program NUMOL [37]. A basis-set-free program such as NUMOL is ideal for testing out the accuracy of new gradient-corrected XC energy functionals because, provided we are using a sufficiently accurate grid, all errors lie in the approximate XC energy functional.

Three types of basis functions are used in LCAO-DF applications: contracted gaussian basis functions, Slater basis functions, and numerical basis functions. The latter is not to be confused with Becke's fully numerical approach. Numerical basis functions are obtained from atomic calculations and are subsequently used as basis functions in molecular calculations in exactly the same fashion in which we would use gaussians or Slaters. Contracted gaussians are used in such programs as DGauss [3], deMon [4], DeFT [5], and GAUSSIAN [8]. Slaters are used in AMOL [38]. Finally, DMol [15] makes use of the numerical basis functions. This wealth of possibilities for viable types of basis functions is in sharp contrast to conventional *ab initio* HF-based codes where contracted gaussian basis sets clearly dominate [9]. This most likely arises from the four-center two-electron integrals in the exchange component of the Fock matrix element, $F_{\mu\nu}$,

$$-\frac{1}{2}\sum_{\sigma}^{K}\sum_{\lambda}^{K}P_{\sigma\lambda}\int\int\frac{\chi_\mu(\mathbf{r})\chi_\sigma(\mathbf{r})\chi_\nu(\mathbf{r}')\chi_\lambda(\mathbf{r}')}{|\mathbf{r}-\mathbf{r}'|}\mathbf{dr}\,\mathbf{dr}'. \tag{47}$$

In methods based on the HF approximation, nothing can be done to simplify this exchange contribution to the Fock matrix. It must be explicitly evaluated. We therefore need to adopt an approach where the efficiency of the evaluation of the four-center two-electron integrals is of primary concern. For this reason, contracted gaussian basis functions are clearly the basis set of choice [10]. Many efficient algorithms [23, 24] have been developed to exploit several of the unique properties of gaussian functions that make integral evaluation with gaussians far more efficient than with any other type of basis function.

However, in DF methods, we have seen how easily the local nature of the LSDA and gradient-corrected XC functionals can be exploited to greatly simplify the exchange (and correlation) contribution to the KS matrix elements. It has been reduced to either a straightforward numerical integration, equations (28) and (29), or a sum of three-center overlap integrals, equations (39) and (40). Having eliminated the need for two-electron integrals for exchange, the cost associated with the evaluation of the Coulomb repulsion contribution to the KS matrix elements,

$$H_{\mu\nu}^{ee} = \int \chi_\mu(\mathbf{r}) \left\{ \int \frac{\rho(\mathbf{r}')}{|\mathbf{r}-\mathbf{r}'|} d\mathbf{r}' \right\} \chi_\nu(\mathbf{r}) d\mathbf{r} \tag{48}$$

can be greatly reduced with the use of a fitted density, as previously outlined. Other approaches may also be adopted to render this part of the calculation more efficient. Clearly, eliminating the need for the explicit evaluation of four-center two-electron integrals will give us more leeway in the matter of choosing the type of basis functions to be used in a particular LCAO-DF program. As we shall see, the major differences between the various LCAO-DF codes arise from the particular approach taken by each to simplify the evaluation of $H_{\mu\nu}^{ee}$. Subsequently, the approach taken will determine which type of basis function is most appropriate for that program.

In GAUSSIAN, Coulomb repulsion is handled exactly by explicitly evaluating the four-center two-electron contributions to $H_{\mu\nu}^{ee}$,

$$H_{\mu\nu}^{ee} = \sum_{\sigma}^{K} \sum_{\lambda}^{K} P_{\sigma\lambda} \int \int \frac{\chi_\mu(\mathbf{r})\chi_\nu(\mathbf{r})\chi_\sigma(\mathbf{r}')\chi_\lambda(\mathbf{r}')}{|\mathbf{r}-\mathbf{r}'|} d\mathbf{r}\, d\mathbf{r}'. \tag{49}$$

As previously outlined, LCGTO-DF program such as DGauss, deMon, and DeFT make use of fitted densities and $H_{\mu\nu}^{ee}$ is reduced to a sum of three-center two-electron integrals, equation (24). This cuts the formal scaling of the method down from K^4 to K^3. The fitting procedure used has the distinct advantage of being variational, but unfortunately, as has already been shown, it does not particularly scale well. Despite reducing the number of two-electron integrals to be evaluated, they do still remain the most time-consuming part of a calculation. Efficient evaluation of these integrals remains crucial and should remain the major consideration in choosing a type of basis function. Contracted gaussians are therefore used when such a fitting procedure is adopted.

The AMOL code also makes use of a fitted density. However, rather than minimizing the error in the Coulomb repulsion energy, equation (17), its fit is obtained by minimizing the error in the density, as originally proposed for the LCGTO-DF approach by Sambe and Felton [1],

$$\int (\rho(\mathbf{r}) - \tilde{\rho}(\mathbf{r}))^2 d\mathbf{r}, \tag{50}$$

subject to the constraint that $\tilde{\rho}(\mathbf{r})$ is appropriately normalized to the total number of electrons in the system. The fit is further simplified by splitting the total density into a sum of pairwise atomic contributions,

$$\begin{aligned}\rho(\mathbf{r}) = \sum_{\mu}^{K} \sum_{\nu}^{K} P_{\mu\nu}\chi_\mu(\mathbf{r})\chi_\nu(\mathbf{r}) &= \sum_{A}^{nuclei} \sum_{B}^{nuclei} \rho^{AB}(\mathbf{r}) \\ &= \sum_{A}^{nuclei} \sum_{B}^{nuclei} \left[\sum_{\mu\in A}^{K} \sum_{\nu\in B}^{K} P_{\mu\nu}\chi_\mu(\mathbf{r})\chi_\nu(\mathbf{r}) \right],\end{aligned} \tag{51}$$

where the last two summations run over only those orbital basis functions centered on atoms A and B, respectively (A and B may be the same atom). These pairwise contributions are fit with only the auxiliary basis functions centered on either one of the two atoms. Therefore, this fitting procedure scales particularly well. Regardless of the size of the molecule, the fitting procedure is carried out as a series of fits over each pair of atoms. This procedure therefore has a quadratic scaling, as opposed to the cubic scaling associated with the fitting procedure in the conventional LCGTO-DF codes. However, AMOL's fitting proacedure has the disadvantage of not being variational. Once the density has been fit, $H_{\mu\nu}^{ee}$ could be evaluated numerically, via equation (24). However, it is not. With the fitted density in hand, an approximate Coulomb potential, $\tilde{v}_c(\mathbf{r})$, can be constructed,

$$\tilde{v}_c(\mathbf{r}) = \int \frac{\tilde{\rho}(\mathbf{r}')}{|\mathbf{r}-\mathbf{r}'|}\mathbf{dr}' = \sum_k^L c_k \int \frac{\chi'(\mathbf{r}')}{|\mathbf{r}-\mathbf{r}'|}\mathbf{dr}'. \tag{52}$$

Note that the fitted density is again expressed as in equation (16). In this manner, $\tilde{v}_c$ is evaluated over a set of grid points. The exact v_c could be evaluated from the exact density. However, its computation is considerably more complex [38] since the exact density is an expansion over two centers, equation (51), rather than one, as with the fitted density, equation (16). Afterwards, the approximate contribution to $H_{\mu\nu}^{ee}$ is evaluated by numerical integration over the complete set of grid points,

$$H_{\mu\nu}^{ee} = \int \chi_\mu(\mathbf{r})v_c(\mathbf{r})\chi_\nu(\mathbf{r})\mathbf{dr} \approx \sum_I^{points} \chi_\mu(\mathbf{R_I})\tilde{v}_c(\mathbf{R_I})\chi_\nu(\mathbf{R_I})W_I. \tag{53}$$

We see that this approach is ideally suited for Slater basis functions. Gaussians are no longer required: at no point in this procedure do we ever have to evaluate a two-electron integral. The choice of basis functions now only hinges on the question of whether or not they accurately depict the atomic orbitals. However, the fitting procedures and the synthesis of the Coulomb potential do require us to use analytic basis functions. Slaters thus emerge as the ideal choice.

DMol is similar to AMOL in that equation (53) is used to numerically integrate $H_{\mu\nu}^{ee}$. However, DMol does not make use of a fitted density. Rather than obtaining $v_c(\mathbf{r})$ via equation (52), it is obtained by solving Poisson's equation,

$$\nabla^2 v_c(\mathbf{r}) = -4\pi\rho(\mathbf{r}). \tag{54}$$

In practise, Poisson's equation must be solved on a grid [15, 39]. This is done by decomposing the problem into a series of single-center problems, solving Poisson's equation for each center, and then repiecing these single-center solutions to get the final $v_c(\mathbf{r})$ over the complete set of grid points. Equation (53) is then used to evaluate the contributions to the KS matrix arising from Coulomb repulsion. Note that the entire procedure was carried out on a grid. There is very little point in using analytic basis functions. Therefore, numerical basis functions obtained from atomic calculations are used as basis functions since no other basis set could possibly be as compact. Since atomic results with such bases are now "exact", within the limitations of the XC functional being used, basis set superposition error [40] is much less of an issue.

Therefore, we have seen that a great deal of options open themselves once exchange is simplified with the use of LSDA and gradient-corrected XC functionals. However, with the popularity of the recently introduced hybrid functionals that incorporate certain aspects of "exact exchange" [20], evaluation of the four-center two-electron exchange integrals, equation (47), may become a necessary feature of any popular DF program. If this indeed becomes the case, then the programs already using contracted gaussian basis functions will clearly have an important advantage, particularly those, such as GAUSSIAN, that already evaluate these integrals to construct the exact contributions to the KS elements arising from electron/electron repulsion.

3. DIVIDE-AND-CONQUER APPROACHES

In a DAC approach, the system under study is divided into a set of subsystems. We will be using peptides in our test calculations. These molecules could be divided as finely as one atom per subsystem, or as coarsely as, for example, one amino acid residue per subsystem. As we shall see, the goal of the DAC approaches is to achieve linear scaling. The CPU time will rise linearly with the number of subsystems since the CPU time associated with any one subsystem calculation should plateau at a certain value and become independent of the overall size of the system under study. Therefore, we'd like to avoid having an excessive number of susbystems. On the other hand, even though it is independent of the overall system size, the cost of the individual subsystem calculations will, as in conventional DF calculations, scale cubicly with subsystem size. This discourages having too few subsystems as they will necessarily have to be quite large. A balance between these two considerations will have to be struck. This should improve with time as we accumulate more experience with DAC calculations.

Two separate DAC schemes have been proposed by Yang: the density formulation [11] and the density matrix formulation [12]. Both will be outlined in detail as both are used in our DAC LCGTO-DF program. Both formulations decompose the total electronic density into a sum of subsystem contributions,

$$\rho(\mathbf{r}) = \sum_{\alpha}^{subsystems} \rho^{\alpha}(\mathbf{r}). \tag{55}$$

This decomposition of $\rho(\mathbf{r})$ is not a source of error. As we shall see, errors arise from the manner in which we construct these subsystem densities. The two DAC formulations take very different approaches to constructing the set of subsystem densities, $\{\rho^{\alpha}(\mathbf{r})\}$.

3.1. Density formulation

In the density formulation of DAC [11], a subsystem density is given by

$$\rho^{\alpha}(\mathbf{r}) = p^{\alpha}(\mathbf{r})\rho(\mathbf{r}) \tag{56}$$

where $p^{\alpha}(\mathbf{r})$ is a function that obeys the constraint

$$\sum_{\alpha}^{subsystems} p^{\alpha}(\mathbf{r}) = 1 \tag{57}$$

at each point in space. We can now take our expression for $\rho(\mathbf{r})$ in equation (1) and transform it into the equivalent expression,

$$\rho(\mathbf{r}) = 2\sum_{i}^{K} \eta(\varepsilon_F - \varepsilon_i)\, |\psi_i(\mathbf{r})|^2 . \tag{58}$$

In the above definition, ε_F is the Fermi energy and η is a Heavyside functions that returns a value of 1 if $\varepsilon_i \leq \varepsilon_F$, and 0 if $\varepsilon_i > \varepsilon_F$. The sum in the new expression now runs over all

K KS molecular orbitals, both occupied and virtual, that can be generated within a basis of K atomic orbitals. However, the Heavyside function assures that these virtual orbitals do not contribute to the total density. Clearly, equations (1) and (58) are identical.

Equations (56) and (58) can now be combined to give a new expression for a subsystem density,

$$\rho^{\alpha}(\mathbf{r}) = 2p^{\alpha}(\mathbf{r})\sum_{i}^{K}\eta(\varepsilon_F - \varepsilon_i)\,|\psi_i(\mathbf{r})|^2\,, \tag{59}$$

and $\rho(\mathbf{r})$ within this DAC approach is now given by

$$\rho(\mathbf{r}) = 2\sum_{\alpha}^{subsystems} p^{\alpha}(\mathbf{r})\sum_{i}^{K}\eta(\varepsilon_F - \varepsilon_i)\,|\psi_i(\mathbf{r})|^2\,. \tag{60}$$

The partition function $p^{\alpha}(\mathbf{r})$ is constructed in such a fashion that it vanishes rather rapidly as we move away from the atoms of subsystem α. This therefore localizes the subsystem's contribution to the total density, $\rho^{\alpha}(\mathbf{r})$, about the atoms of subsystem α. This can be done by defining $p^{\alpha}(\mathbf{r})$ as [11]

$$p^{\alpha}(\mathbf{r}) = \frac{\sum_{A\in\alpha}^{atoms}\left[\rho_{atom}^{A}(|\mathbf{r}-\mathbf{R_A}|)\right]^2}{\sum_{B}^{atoms}\left[\rho_{atom}^{B}(|\mathbf{r}-\mathbf{R_B}|)\right]^2} \tag{61}$$

where the sum in the denominator runs over all the atoms in the system while the sum in the numerator runs over only those atoms in subsystem α, and $\rho_{atom}^{A}(|\mathbf{r}-\mathbf{R_A}|)$ is the density obtained from a calculation on the isolated atom A, with spherical symmetry being imposed. The value of $p^{\alpha}(\mathbf{r})$ will vary from a value very close to, but no greater than, one in the vicinity of a nucleus of an atom in subsystem α to a value very close to, but no less than, zero in regions far removed from any atom in subsystem α. A power greater than two in equation (61) will better localize the subsystem density. However, this would lead to greater noise in the total energy arising from steps involving numerical quadruture.

Equations (60) and (61) will return the true $\rho(\mathbf{r})$ since the former was simply obtained by manipulating the original KS expression for $\rho(\mathbf{r})$, equation (1), and the latter satisfies the constraint we imposed in equation (57). It is at this point that we begin making approximations. Since $p^{\alpha}(\mathbf{r})$ localizes the contributions to $\rho^{\alpha}(\mathbf{r})$ about the subsystem atoms, the nature of the KS orbitals, $\{\psi_i(\mathbf{r})\}$, need only be known in the vicinity of the atoms of subsystem α when constructing $\rho^{\alpha}(\mathbf{r})$. Therefore, rather than solving equation (10) and getting the system's true KS orbitals, the KS equations are solved only within the orbital bases of the subsystem atoms,

$$\mathbf{H}^{\alpha}\mathbf{C}^{\alpha} = \mathbf{S}^{\alpha}\mathbf{C}^{\alpha}\varepsilon^{\alpha} \tag{62}$$

where each matrix and vector is as defined in equation (10), save for the fact that they span only the basis functions of the atoms of subsystem α. It is important to realize that we are not making any approximations as to the value of any single matrix element in $\mathbf{H}^{\alpha}$ or $\mathbf{S}^{\alpha}$. The values of these subsystem matrix elements are equal to the corresponding matrix elements in the full matrices, $\mathbf{H}$ and $\mathbf{S}$. The subsystem KS orbitals, $\{\psi_i^{\alpha}(\mathbf{r})\}$, can then be constructed with the elements of $\mathbf{C}^{\alpha}$,

$$\psi_i^{\alpha}(\mathbf{r}) = \sum_{\mu\epsilon\alpha}^{K} C_{\mu i}^{\alpha}\chi_{\mu}(\mathbf{r}), \tag{63}$$

where the summation now runs over only those basis functions, $\{\chi_\mu(\mathbf{r})\}$, centered on an atom in subsystem α.

In each subsystem, equation(62) is solved and the corresponding subsystem KS orbitals, $\{\psi_i^\alpha(\mathbf{r})\}$, are constructed. The total density can then be approximately reconstructed by substituting subsystem KS orbitals into equation (60), giving

$$\rho(\mathbf{r}) \approx 2 \sum_{\alpha}^{subsystems} p^\alpha(\mathbf{r}) \sum_{i}^{K^\alpha} \eta(\varepsilon_F - \varepsilon_i^\alpha) \left|\psi_i^\alpha(\mathbf{r})\right|^2 . \tag{64}$$

We no longer have the exact $\rho(\mathbf{r})$ since we are not using the true KS orbitals. The sum over the subsystem KS orbitals runs up to K^α, the total number of orbital basis functions centered on atoms in subsystem α.

Equation (64) has a Fermi energy, ε_F, common to all subsystems. It is adjusted so that the approximate $\rho(\mathbf{r})$ obtained within the DAC scheme is appropriately normalized to the total number of electrons in the system. In practise, a Heavyside function will not allow an SCF procedure to converge [11]. The DAC expression for $\rho(\mathbf{r})$ is further approximated by replacing the discontinuous Heavyside function by a steep, but continuous, Fermi function [11],

$$\rho(\mathbf{r}) \approx 2 \sum_{\alpha}^{subsystems} p^\alpha(\mathbf{r}) \sum_{i}^{K^\alpha} \frac{1}{1 + e^{-\beta(\varepsilon_F - \varepsilon_i^\alpha)}} \left|\psi_i^\alpha(\mathbf{r})\right|^2 . \tag{65}$$

Provided the adjustable parameter β is set such that the Fermi function is sufficiently steep, its specific value has little effect on the total energy. ε_F is now set such that the new expression for the DAC $\rho(\mathbf{r})$ is properly normalized,

$$2 \sum_{\alpha}^{subsystems} \sum_{i}^{K^\alpha} \frac{1}{1 + e^{-\beta(\varepsilon_F - \varepsilon_i^\alpha)}} \int p^\alpha(\mathbf{r}) \left|\psi_i^\alpha(\mathbf{r})\right|^2 d\mathbf{r} = N. \tag{66}$$

The common ε_F permits a fraction of the electronic density to be displaced from one subsystem to another. No restrictions on the electronic density are imposed by our arbitrary partition of the molecule into subsystems. The only approximations being made are the replacement of the Heavyside function by the Fermi function and the replacement of the true KS orbitals by the subsystem KS orbitals obtained via equation (62).

3.2. Density matrix formulation

A second DAC approach, based this time on the density matrix, has been proposed [12]. The total electronic density can be expressed, as in equation (13), in terms of the density matrix, $\mathbf{P}$, and the orbital basis functions, $\{\chi_\mu(\mathbf{r})\}$. In the density matrix DAC formulation, $\rho(\mathbf{r})$ is again expressed as a sum of subsystem densities, as in equation (55). However, a subsystem density is now defined as

$$\rho^\alpha(\mathbf{r}) = \sum_{\mu}^{K} \sum_{\nu}^{K} p_{\mu\nu}^\alpha P_{\mu\nu} \chi_\mu(\mathbf{r}) \chi_\nu(\mathbf{r}) \tag{67}$$

where $p_{\mu\nu}^\alpha$ is set equal to 1 if both $\chi_\mu(\mathbf{r})$ and $\chi_\nu(\mathbf{r})$ are centered on atoms in subsystem α, $\frac{1}{2}$ if one of the two is, and 0 if neither of the two is. Note that the partition functions, $\left\{p_{\mu\nu}^\alpha\right\}$, obey a constraint,

$$\sum_{\alpha}^{subsystems} p_{\mu\nu}^\alpha = 1, \tag{68}$$

similar to that in equation (57) for the DAC density formulation. However, rather than having it held at each point in space, it is held for each pair of basis functions.

At this point in the density matrix formulation of DAC, approximations have yet to be made. Combining equations (14) and (67), and again using a sum over all the orbitals, both occupied and virtual, the following expression for $\rho^\alpha(\mathbf{r})$ is obtained,

$$\rho^\alpha(\mathbf{r}) = 2\sum_\mu^K \sum_\nu^K p_{\mu\nu}^\alpha \sum_i^K \eta(\varepsilon_F - \varepsilon_i) C_{\mu i} C_{\nu i} \chi_\mu(\mathbf{r}) \chi_\nu(\mathbf{r}). \tag{69}$$

At this point, the same approximations are made as in the density formulation. The true KS orbital coefficients, $\{C_{\mu i}\}$, are replaced by solutions to the subsystem problem, equation (62), $\left\{C_{\mu i}^\alpha\right\}$. The Heavyside function is again replaced by a Fermi function. This yields

$$\rho^\alpha(\mathbf{r}) = 2\sum_\mu^K \sum_\nu^K p_{\mu\nu}^\alpha \sum_i^{K^\alpha} \frac{1}{1 + e^{-\beta(\varepsilon_F - \varepsilon_i^\alpha)}} C_{\mu i}^\alpha C_{\nu i}^\alpha \chi_\mu(\mathbf{r}) \chi_\nu(\mathbf{r}), \tag{70}$$

and ε_F is again adjusted such that the sum of the subsystem densities is appropriately normalized to the total number of electrons in the system,

$$\sum_\alpha^{subsystems} 2\sum_\mu^K \sum_\nu^K p_{\mu\nu}^\alpha \sum_i^{K^\alpha} \frac{1}{1 + e^{-\beta(\varepsilon_F - \varepsilon_i^\alpha)}} C_{\mu i}^\alpha C_{\nu i}^\alpha \int \chi_\mu(\mathbf{r}) \chi_\nu(\mathbf{r}) d\mathbf{r} = N. \tag{71}$$

The density matrix formulation is far easier to implement within a conventional program as we need only modify the construction of the density matrix, $\mathbf{P}$, which is already in place. In fact, this scheme can be implemented in any quantum mechanical program [12] with relatively very little effort. On the other hand, the density formulation requires us to construct the new partition functions $p^\alpha(\mathbf{r})$. The nature of these partition functions then force us to resort to numerical integration of equation (66) when normalizing the density. The density formulation does however have the advantage of better localizing the subsystem density contributions [12]: the divide-and-conquer approximations are more valid within the density formulation. However, as we shall see, the use of buffer atoms will allow us to achieve whatever level of accuracy we wish to have in our calculations, provided we are willing to pay the computational cost.

3.3. Buffer atoms

At this point, linear scaling in CPU time is possible within both DAC procedures. Rather than constructing the entire KS matrix, $\mathbf{H}$, we need only construct the subsystem counterparts, $\{\mathbf{H}^\alpha\}$. Their construction remains the most time-consuming part of the calculation. The goal is to have each procedure constructing a particular $\mathbf{H}^\alpha$ becoming independent of the total size of the system. However, formally, the CPU time spent on any one subsystem calculation will scale linearly with the total size of the system since the Coulomb interactions with faraway atoms still need to be taken into account. Fortunately, multipole expansions may be used to handle long-range electrostatics [41]. If this is done, the CPU time spent on a specific subsystem should almost become independent of the total system size. At this point, the CPU time will then only increase linearly with the number of subsystems, or equivalently, with the total size of the system. Implementation of the continuous fast multipole method [41] will become part of the final DAC version of DeFT.

Unfortunately, the DAC methods, as presented, impose approximations that are far too severe. As stated, replacing the Heavyside function by a steep Fermi function is not a problem. In our tests where the true KS orbitals are still used (as opposed to the subsystem sets of orbitals) but the Heavyside function is replaced by the Fermi function, the error in the final total energy is of the order of one microhartree. The real errors in the DAC approaches arise from the fact that subsystem KS orbitals that span only the basis functions of the subsystem atoms are not good enough representations of the true KS orbitals in the vicinity of the subsystem. The orbital bases used in the subsystem calculations must somehow be expanded. This is done by introducing buffer atoms [11]. Basis functions on the buffer atoms supplement those on the subsystem atoms. Equation (62) is now solved within the orbital bases of the subsystem and buffer atoms, and K^{α} is now the total number of basis functions centered on the subsystem's atoms and its associated buffer atoms. As the number of buffer atoms increases, the subsystem KS orbitals begin to more closely resemble the true KS orbitals in the vicinity of the subsystem. Of course, in the limit of an infinitely large buffer space, every atom in the system will be a part of each subsystem calculation and each set of subsystem KS orbitals will be equal to the true set of KS orbitals. At this point, each DAC approach would yield the exact $\rho(\mathbf{r})$.

It is important to note that the addition of buffer atoms in no way affects the subsystem partition functions, i.e., the definitions of $\{p^{\alpha}(\mathbf{r})\}$ in the density formulation and $\left\{p^{\alpha}_{\mu\nu}\right\}$ in the density matrix formulation remain as before. A subsystem density is still localized about the subsystem atoms and the subsystem KS orbitals constructed within the orbital bases of the subsystem's atoms and buffer atoms only go towards the construction of the total density in regions where the value of the subsystem's partition function is appreciable. An atom which is a buffer atom to one subsystem in turn has those subsystem's atoms in its own subsystem's buffer. Buffer atoms can reasonably be assigned following two criteria. An atom may be made a buffer atom provided it is within a certain distance of a subsystem atom. Buffer atoms could also be assigned based on the number of bonds separating them from subsystem atoms. In the test cases published to date [11, 12, 13], extended one and two dimensional systems were studied. Therefore, in these instances, our two buffer space criteria essentially amount to the same.

4. THE DIVIDE-AND-CONQUER LCGTO-DF METHOD

As we outline our DAC implementation of the LCGTO-DF formalism, we will focus on one specific test molecule, the extended conformation of the glycine heptapeptide, illustrated in Figure 1. The heptapeptide is split into eight subsystems: a terminal HCO group, a terminal NH_2 group, and six $NHCH_2CO$ groups. Previous implementations of DAC methods [11, 12] have clearly established that the DAC approximations improve, i.e., buffer space need not be as extensive, when minimal basis sets are used. However, to have a truly realistic and representative test of our LCGTO-DF DAC protocols, we use the same bases that we would use in a typical application [7]. Therefore, in our tests, we use a triple-ζ plus polarization orbital basis on C, N, and O atoms [22] and a double-ζ plus polarization basis on H atoms [22]. For each of the auxiliary fitting bases, C, N, and O atoms have four lone s functions and three sets of s, p, and d functions constrained to have the same exponent [22]. For H atoms, each auxiliary basis has three lone s functions and one such constrained-exponent s, p, and d set [22]. Our test molecule therefore has a total of 618 orbital basis functions, 1191 auxiliary $\rho(\mathbf{r})$ fitting basis functions, and 1191 auxiliary XC fitting basis functions. When performing calculations on systems of biochemical interest, gradient-corrected XC functionals must be used to achieve any meaningful results [7, 42]. However, for testing purposes only, calculations were performed within the LSDA so as to save computer time. Since gradient-corrected XC functionals

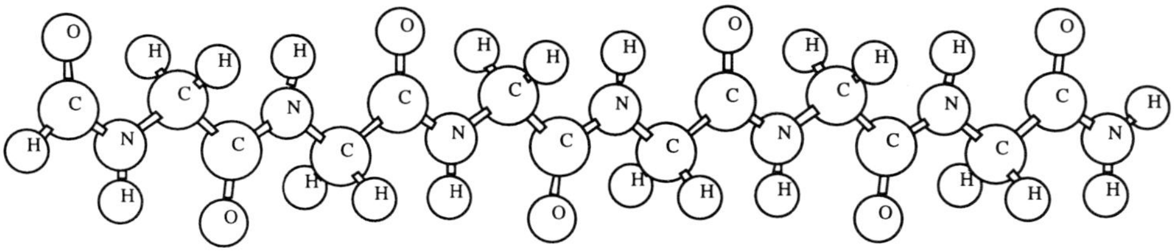

Figure 1. Glycine heptapeptide.

still only rely on local properties of $\rho(\mathbf{r})$ and the LSDA electronic distribution is little affected when gradient corrections are added [43], they should not prove to be any greater a challenge for DAC procedures.

In our tests, buffer atoms are included based on a distance criterion rather than upon the number of bonds separating them from subsystem atoms. For an extended system such as our test case heptapeptide, both schemes are more or less equivalent. However, we feel that the former approach is more appropriate in general. Therefore, if any atom is within a certain cutoff distance of any one atom in a subsystem, that atom becomes a buffer atom to that subsystem. Buffer space cutoffs ranging from 2.0 Å to 14.0 Å are investigated. The possibility of having no buffer atoms whatsoever is also tested. In our heptapeptide test case, a 2.0 Å cutoff is equivalent to a buffer space containing only atoms directly bound to any one of the subsystem atoms. Meanwhile, a 14.0 Å cutoff sees the central $NHCH_2CO$ groups' buffer spaces encompassing the entire molecule. At this point, there is clearly no possible advantage to adopting a DAC scheme since these central subsystems' calculations will be as time-consuming as the conventional calculation on the entire molecule. However, it is important to note that even with a 14.0 Å cutoff, the exact results will not be recovered. This is because the subsystems near each end of the heptapeptide still see their buffer spaces covering only a fraction of the molecule.

4.1. Fitting the electronic density

Divide-and-conquer approaches can only be effective within the LCGTO-DF formalism if the fitting procedures can be done in a DAC fashion as well. Fortunately, this is the case. One of the key features to the LCGTO-DF formalism is its analytical and variational fit of the electronic density. Since we wish to retain the analytical apect of this fitting procedure, the density matrix formulation will be used in the DAC fit of $\rho(\mathbf{r})$. Rather than fitting $\rho(\mathbf{r})$ directly, we carry out a series of calculations, fitting each of the subsystem densities, $\rho^{\alpha}(\mathbf{r})$, given by equation (67). The density is fit within an auxiliary basis limited to those $\rho(\mathbf{r})$ fitting functions centered on the atoms and buffer atoms of the subsystem in question. The only approximation that we are making is that these subsystem fits can be carried out effectively within these truncated auxiliary bases. In this DAC procedure, we need not invoke the idea of subsystem KS orbitals or the replacement of the Heavyside function by the Fermi function. We are merely borrowing the concept of partitioning the density with the aid of the partition elements, $\left\{p^{\alpha}_{\mu\nu}\right\}$, that take on the values of 1, $\frac{1}{2}$, and 0, depending on whether or not the basis functions $\chi_{\mu}(\mathbf{r})$ and $\chi_{\nu}(\mathbf{r})$ are centered on subsystem atoms. Provided we appropriately normalize each subsystem fit, as outlined below, then the exact LCGTO-DF results will be obtained if the buffer space cutoff is set so high that truncation of the auxiliary basis does not occur.

The approach taken to fitting $\rho^\alpha(\mathbf{r})$ follows the conventional fit to $\rho(\mathbf{r})$ as closely as possible. Rather than minimizing the error in the Coulomb repulsion in the entire molecule, equation (17), we minimize the error in the Coulomb energy arising from the interaction of the subsystem density with itself,

$$\int\int \frac{[\rho^\alpha(\mathbf{r}) - \tilde{\rho}^\alpha(\mathbf{r})][\rho^\alpha(\mathbf{r}') - \tilde{\rho}^\alpha(\mathbf{r}')]}{|\mathbf{r}-\mathbf{r}'|}\mathbf{dr}\,\mathbf{dr}' \tag{72}$$

A new constraint must also be imposed on the fits of $\rho^\alpha(\mathbf{r})$. The fit in the conventional LCGTO-DF calculation was normalized to the total number of electrons in the molecule. Clearly, this cannot be done for any one subsystem fit. A series of unconstrained fits could be carried out and normalization could occur once they have been all completed. Fortunately, we can exploit the fact that the DAC density matrix formulation splits the total electronic density into a sum of subsystem contributions in the same manner by which a Mulliken population analysis splits electronic density into a sum of atomic contributions [9, 12]. The following constraint is imposed on each subsystem fit:

$$\int \tilde{\rho}^\alpha(\mathbf{r})\mathbf{dr} = \sum_{A\epsilon\alpha} Q_A \tag{73}$$

where Q_A is the Mulliken population of subsystem atom A [9]. This yields the following solution for the set of $\rho^\alpha(\mathbf{r})$ fit coefficients, $\{c_k^\alpha\}$,

$$\mathbf{c}^\alpha = \mathbf{S}_\alpha'^{-1}\,(\mathbf{t}^\alpha + \lambda^\alpha\mathbf{n}^\alpha) \tag{74}$$

where $\mathbf{S}_\alpha'$, $\mathbf{t}^\alpha$, and $\mathbf{n}^\alpha$ are defined in exactly the same fashion as equations (20) to (22). However, the α superscript has been added to remind us that these vectors and matrices now span only those $\rho(\mathbf{r})$ auxiliary fitting functions centered on atoms or buffer atoms of the subsystem whose density is being fitted. The Lagrange multiplier, λ^α, is now given by

$$\lambda^\alpha = \sum_{A\epsilon\alpha} Q_A - \frac{\mathbf{n}^\alpha(\mathbf{S}^\alpha)^{-1}\mathbf{t}^\alpha}{\mathbf{n}^\alpha(\mathbf{S}^\alpha)^{-1}\mathbf{n}^\alpha}, \tag{75}$$

with the sum of Mulliken populations on subsystem atoms now replacing the total number of electrons in equation (23).

The need to invert the overlap matrix for the $\rho(\mathbf{r})$ fitting basis of the entire molecule is clearly averted. We need only do this for the subsystem and buffer atoms in each subsystem calculation and the total amount of time spent on this aspect of the calculation will simply scale linearly with the number of subsystems, and hence linearly with total system size. The construction of the subsystem $\mathbf{t}$ vectors, $\mathbf{t}^\alpha$, will be the time-consuming part of the calculation and it is this step of the procedure that we will need to focus upon to achieve true linear scaling. The elements of $\mathbf{t}^\alpha$ are given by

$$t_i^\alpha = \int\int \frac{\chi_i'(\mathbf{r})\rho^\alpha(\mathbf{r}')}{|\mathbf{r}-\mathbf{r}'|}\mathbf{dr}\,\mathbf{dr}' = \sum_\mu^N\sum_\nu^N p_{\mu\nu}^\alpha P_{\mu\nu}\int\int \frac{\chi_i'(\mathbf{r})\chi_\mu(\mathbf{r}')\chi_\nu(\mathbf{r}')}{|\mathbf{r}-\mathbf{r}'|}\mathbf{dr}\,\mathbf{dr}'. \tag{76}$$

Recalling that $p_{\mu\nu}^\alpha$ vanishes unless at least one of the two orbital basis functions is centered on a subsystem atom, the above expression can be reduced to

$$t_i^\alpha = \sum_{\mu\epsilon\alpha}\sum_\nu^N p_{\mu\nu}^\alpha P_{\mu\nu}\int\int \frac{\chi_i'(\mathbf{r})\chi_\mu(\mathbf{r}')\chi_\nu(\mathbf{r}')}{|\mathbf{r}-\mathbf{r}'|}\mathbf{dr}\,\mathbf{dr}' \tag{77}$$

where $p^{\alpha}_{\mu\nu}$ will equal 1 if $\chi_{\nu}(\mathbf{r})$ is also centered on a subsystem atom, and $\frac{1}{2}$ if it is not. The work associated with the construction of $\mathbf{t}^{\alpha}$ will eventually plateau and be independent of the total size of the system under study as we go to bigger and bigger systems. The elements of $\mathbf{t}^{\alpha}$ span only the subsystem and buffer atoms. Their number is therefore set once the buffer space cutoff is defined. The first summation in equation (77) runs over only the orbital basis functions of the actual subsystem atoms. This first summation is therefore not only independent of the total size of the system: it is also independent of the buffer space cutoff with which we choose to work. The second summation might however lead us to believe that we'll see linear scaling in each subsystem fit as the sum over basis functions $\{\chi_{\nu}(\mathbf{r})\}$ still runs over the entire extent of the molecule. However, only those functions with appreciable overlap with basis functions centered on subsystem atoms will contribute appreciably to $\mathbf{t}^{\alpha}$. Therefore, at some point, increasing the total size of the system will have absolutely no effect on the number of orbital basis functions that need to be accounted for in the second summation. This is the point at which construction of $\mathbf{t}^{\alpha}$ is independent of the total system size and the whole fitting procedure will scale linearly with the number of subsystems,

$$\tilde{\rho}(\mathbf{r}) = \sum_{\alpha}^{subsystems} \tilde{\rho}^{\alpha}(\mathbf{r}). \tag{78}$$

Within DeFT, the subsystem density fit coefficient vectors, $\mathbf{c}^{\alpha}$, are added up together to yield the DAC approximation to the conventional LCGTO-DF density fit coefficient vector, $\mathbf{c}$. Once the approximate DAC $\mathbf{c}$ vector is constructed, the calculation may proceed as any other conventional LCGTO-DF calculation would proceed.

The results obtained using a DAC fit of $\rho(\mathbf{r})$ in calculations on the glycine heptapeptide are presented in Table 1. All other aspects of the calculation were carried out as in a conventional LCGTO-DF calculation. When no buffer atoms are used in the DAC fit, an error of just over 10.0 kcal mol^{-1} in the total energy is seen. A 2.0 Å buffer cutoff (the equivalent of having only atoms directly bound to a subsystem atom added to the buffer space) cuts the error dramatically, down to 0.46 kcal mol^{-1}. In some instances, this may be an acceptable error, but for most chemistry applications, this is too inaccurate. The error may be cut down by another order of magnitude, to just under 0.05 kcal mol^{-1}, by taking the buffer space cutoff out to 3.0 Å. This is most likely an acceptable error in the vast majority of chemistry applications. The second column of Table 1 lists the total number of atoms, subsystem and buffer, associated with each fit of a subsystem density, progressing from left to right in Figure 1. With a 3.0 Å cutoff, there are at most 17 atoms associated with any one subsystem fit. This is well under the total number of atoms in the heptapeptide, 48. Of course, if we were to extend the polypeptide, our largest subsystem calculation would remain fixed at 17 atoms and the advantages of the DAC approach would dramatically increase. To bring the error down under 0.01 kcal mol^{-1}, the buffer space must be taken out to 5.0 Å. At this point, the central subsystems' density fits each involve a relatively small number of atoms, 23. To bring down the error by another full order of magnitude, down to 0.001 kcal mol^{-1}, a 9.0 Å buffer cutoff must be employed. For this particular molecule, a 9.0 Å cutoff means some subsystem fits encompass nearly the entire molecule. There is little advantage to using the DAC fitting procedure at this point for the heptapeptide as the central subsystems' buffers involve 39 of the 48 atoms. Of course, there would be a CPU savings to be had if we were to go to a larger molecule. By the time the central subsystems' buffer spaces span the entire heptapeptide, the error is cut down to 0.0001 kcal mol^{-1}. We recall that we do not reproduce the exact, conventional, LCGTO-DF result yet since the other subsystems' buffer spaces still span only a fraction of the heptapeptide.

Table 1
Effect of buffer space on the error in total energy introduced by the DAC fit of $\rho(\mathbf{r})$

buffer space cutoff (Å)	# atoms (subsystem + buffer)	total energy (hartrees)	error (kcal mol^{-1})
true[a]		-1407.1545851	
no buffer space	3, 7, 7, 7, 7, 7, 7, 3	-1407.1709566	10.2742
2.0	4, 9, 9, 9, 9, 9, 9, 4	-1407.1553139	0.4574
3.0	8, 15, 17, 17, 17, 17, 15, 8	-1407.1546644	0.0498
4.0	9, 16, 20, 20, 20, 20, 17, 10	-1407.1546199	0.0218
5.0	11, 18, 23, 23, 23, 23, 18, 11	-1407.1545932	0.0051
6.0	12, 19, 26, 25, 25, 26, 19, 12	-1407.1545920	0.0043
7.0	15, 22, 29, 31, 32, 29, 22, 15	-1407.1545892	0.0026
8.0	17, 24, 31, 35, 35, 31, 24, 17	-1407.1545877	0.0016
9.0	19, 26, 33, 39, 39, 33, 26, 19	-1407.1545856	0.0003
10.0	22, 29, 36, 43, 43, 36, 29, 22	-1407.1545856	0.0003
11.0	22, 29, 36, 43, 43, 36, 29, 22	-1407.1545856	0.0003
12.0	24, 31, 38, 45, 45, 38, 31, 24	-1407.1545854	0.0002
13.0	26, 33, 40, 47, 47, 40, 33, 26	-1407.1545854	0.0002
14.0	29, 36, 43, 48, 48, 43, 36, 29	-1407.1545853	0.0001

[a]Energy from the conventional LCGTO-DF calculation on the glycine heptapeptide.

In every instance, the total energy of the heptapeptide is too low with a DAC fit to $\rho(\mathbf{r})$. This is consistent with the earlier statement that the ideal fit to $\rho(\mathbf{r})$ will yield a total energy that serves as an upper bound to LCGTO-DF calculations with imperfect fits. The sum of the subsystem fits, equation (78), can never be quite as good as the fit to $\rho(\mathbf{r})$ that one obtains from the conventional procedure. Since we can rely on a systematic underestimation of the total energy to arise, we can expect a systematic cancellation of errors to occur when we are looking at relative energies. These errors should be considerably lower than those reported here for *total* energies. Unfortunately, we will see that the other DAC procedures do not have a variational aspect, and therefore, the advantages associated with the variational nature of the DAC fit to $\rho(\mathbf{r})$ is lost within our overall DAC approach. However, the DAC fit to $\rho(\mathbf{r})$ may be used, on its own, in any *ab initio* program. The DAC procedure just outlined is as applicable to HF calculations as it is to DF calculations. Therefore, this procedure may also be used to speed up conventional *ab initio* calculations in LCGTO-HF programs. In these instances, the variational nature of the fit will be an important consideration.

4.2. Fitting the exchange-correlation terms

Fitting the XC terms by a DAC procedure presents a new problem. Since the XC terms are not linear, one cannot construct subsystem densities, $\{\rho^{\alpha}(\mathbf{r})\}$, and then simply evaluate the corresponding subsystem XC potentials, $\{v_{xc}^{\alpha}(\mathbf{r})\}$. Rather, the total density, $\rho(\mathbf{r})$, must be constructed and the XC potential, $v_{xc}(\mathbf{r})$, must be evaluated. At this point, one may choose to decompose $v_{xc}(\mathbf{r})$ into a sum of subsystem contributions,

$$v_{xc}(\mathbf{r}) = \sum_{\alpha}^{subsystems} v_{xc}^{\alpha}(\mathbf{r}). \tag{79}$$

The total density must therefore first be synthesized over the entire set of grid points,

$$\rho(\mathbf{R_I}) = \sum_{\mu}^{K}\sum_{\nu}^{K} P_{\mu\nu}\chi_{\mu}(\mathbf{R_I})\chi_{\nu}(\mathbf{R_I}). \tag{80}$$

Fortunately, when we go to very large systems, this procedure will scale linearly. Contributions to $\rho(\mathbf{R_I})$ will only come when both $\chi_{\mu}(\mathbf{R_I})$ and $\chi_{\nu}(\mathbf{R_I})$ have an appreciable value at $\mathbf{R_I}$. As system size increases, there will come a point where the addition of new atoms will have no effect on the number of orbital basis functions making an appreciable contribution at grid points far removed from these new atoms. This would also apply to the derivatives of $\rho(\mathbf{r})$ when gradient-corrected XC functionals are employed.

Once $\rho(\mathbf{R_I})$ is known at each grid point, $v_{xc}(\mathbf{R_I})$ may be evaluated. The XC potential at each point may now be partitioned into a sum of subsystem contributions,

$$v_{xc}(\mathbf{R_I}) = \sum_{\alpha}^{subsystems} v_{xc}^{\alpha}(\mathbf{R_I}). \tag{81}$$

This is done with the aid of the same partition functions used in the density formulation of DAC, $\{p^{\alpha}(\mathbf{r})\}$, and which are defined in equation (61),

$$v_{xc}^{\alpha}(\mathbf{R_I}) = p^{\alpha}(\mathbf{R_I})v_{xc}(\mathbf{R_I}). \tag{82}$$

A partition function localizes the corresponding subsystem contribution, $v_{xc}^{\alpha}(\mathbf{r})$, to the total XC potential, $v_{xc}(\mathbf{r})$, about the atoms of the subsystem. A fitting procedure for $v_{xc}^{\alpha}(\mathbf{r})$ should therefore only require auxiliary XC fitting functions from subsystem atoms and nearby buffer atoms. The same can be done for the XC energy density, $\varepsilon_{xc}(\mathbf{r})$.

Subsystem fits are indeed carried out within our modified version of DeFT. Equations (34) to (38) are only slightly modified. Now, rather than spanning the entire auxiliary basis, the matrices and vectors span only those fitting functions centered on subsystem or buffer atoms. Once the ensemble of subsystem coefficient vectors, $\{\mathbf{b}^{\alpha}\}$ and $\{\mathbf{e}^{\alpha}\}$, have been determined, DeFT sums them together to get the DAC approximations to the conventional LCGTO-DF $\mathbf{b}$ and $\mathbf{e}$ vectors. These approximate DAC fits to the XC terms can then be used to evaluate the XC contributions to the total energy, equation (39), and to the KS matrix elements, equation (40), as in a conventional LCGTO-DF calculation.

The results obtained with DAC fits to both $\rho(\mathbf{r})$ and $v_{xc}(\mathbf{r})$ are presented in Table 2. A common buffer space cutoff is used for both fits. When no buffer atoms are used, a 4.6 kcal mol^{-1} error arises. This seems to be better than when we only fit $\rho(\mathbf{r})$ by DAC means. However, the total energy with both DAC fits overestimates the total energy. When only $\rho(\mathbf{r})$ was fit by a DAC approach, the total energy was *underestimated* by over 10 kcal mol^{-1}. The DAC fit to $v_{xc}(\mathbf{r})$ therefore has a 15 kcal mol^{-1} error. That the XC DAC fitting procedure generates a greater error than its density counterpart is borne out by the subsequent calculations at longer buffer space cutoffs. A 2.0 Å buffer sees the total energy overestimated by 2.0 kcal mol^{-1}. With only the $\rho(\mathbf{r})$ DAC fit, a 0.45 kcal mol^{-1} error existed. The buffer space cutoff must be taken out to 6.0 Å to have the error in the energy brought under 0.05 kcal mol^{-1}. This was achieved at a mere 3.0 Å with only the DAC fit to $\rho(\mathbf{r})$. At 7.0 Å and beyond, the error is consistently under 0.005 kcal mol^{-1}

Table 2
Effect of buffer space on the error in total energy introduced by the DAC fit of $v_{xc}(\mathbf{r})$

buffer space cutoff (Å)	# atoms (subsystem + buffer)	total energy (hartrees)	error (kcal mol^{-1})
true[a]		-1407.1545851	
no buffer space	3, 7, 7, 7, 7, 7, 7, 3	-1407.1472272	4.6175
2.0	4, 9, 9, 9, 9, 9, 9, 4	-1407.1513734	2.0155
3.0	8, 15, 17, 17, 17, 17, 15, 8	-1407.1542530	0.2084
4.0	9, 16, 20, 20, 20, 20, 17, 10	-1407.1544430	0.0892
5.0	11, 18, 23, 23, 23, 23, 18, 11	-1407.1544780	0.0672
6.0	12, 19, 26, 25, 25, 26, 19, 12	-1407.1545177	0.0423
7.0	15, 22, 29, 31, 32, 29, 22, 15	-1407.1545830	0.0013
8.0	17, 24, 31, 35, 35, 31, 24, 17	-1407.1545796	0.0034
9.0	19, 26, 33, 39, 39, 33, 26, 19	-1407.1545822	0.0018
10.0	22, 29, 36, 43, 43, 36, 29, 22	-1407.1545855	0.0003
11.0	22, 29, 36, 43, 43, 36, 29, 22	-1407.1545855	0.0003
12.0	24, 31, 38, 45, 45, 38, 31, 24	-1407.1545854	0.0002
13.0	26, 33, 40, 47, 47, 40, 33, 26	-1407.1545861	0.0006
14.0	29, 36, 43, 48, 48, 43, 36, 29	-1407.1545858	0.0004

[a]Energy from the conventional LCGTO-DF calculation on the glycine heptapeptide.

with both DAC fitting procedures. This type of accuracy should be sufficient for the vast majority of chemical applications. At 10.0 Å and beyond, the error is consistently well under 0.001 kcal mol^{-1}.

When one recalls that the total XC energy is much smaller than the Coulomb energy, the relative errors in the DAC fits to $v_{xc}(\mathbf{r})$ are much larger than those in the corresponding DAC fits to $\rho(\mathbf{r})$. This may in large part be due to the fact that the XC fitting functions, $\left\{\chi_l''(\mathbf{r})\right\}$ are much more diffuse than the density fitting functions, $\left\{\chi_k'(\mathbf{r})\right\}$. The XC fitting functions have exponents with values one third those of the density fitting functions. The effect of fitting functions on faraway atoms is thus much more important in the XC case. Tests should be carried out to see if it is advantageous to augment the size of the auxiliary XC bases with tight gaussians so that the effect of XC fitting functions from faraway atoms is lessenned. This would increase the total number of XC functions in the system, but it could lead to a decrease in the average number of XC functions in a subsystem calculation since the buffer space cutoff would be cut down. We must remember that it is the latter consideration that controls the overall CPU time in a DAC calculation.

We also note that there is not a systematic decrease in the errors found in Table 2 as we go down the columns and go to large buffer space cutoffs. This is understandable since the fits to the XC terms are not variational. This also explains why the DAC fits to $v_{xc}(\mathbf{r})$ cause the total energy to be sometimes overestimated, and sometimes underestimated. Once DAC fits to $v_{xc}(\mathbf{r})$ are introduced, we can no longer rely on a systematic cancellation

Table 3
Effect of buffer space on the error in total energy introduced by the DAC construction of the electronic density

buffer space cutoff (Å)	# atoms (subsystem + buffer)	total energy (hartrees)	error (kcal mol^{-1})
true[a]		-1407.1545851	
3.0	8, 15, 17, 17, 17, 17, 15, 8	-1407.0689441	53.7
4.0	9, 16, 20, 20, 20, 20, 17, 10	-1407.0326819	76.5
5.0	11, 18, 23, 23, 23, 23, 18, 11	-1407.0736704	50.8
6.0	12, 19, 26, 25, 25, 26, 19, 12	-1407.0930417	38.6
7.0	15, 22, 29, 31, 32, 29, 22, 15	-1407.0998947	34.3
8.0	17, 24, 31, 35, 35, 31, 24, 17	-1407.1242829	19.0
9.0	19, 26, 33, 39, 39, 33, 26, 19	-1407.1365557	11.3
10.0	22, 29, 36, 43, 43, 36, 29, 22	-1407.1486799	3.7
11.0	22, 29, 36, 43, 43, 36, 29, 22	-1407.1486799	3.7
12.0	24, 31, 38, 45, 45, 38, 31, 24	-1407.1499527	2.9
13.0	26, 33, 40, 47, 47, 40, 33, 26	-1407.1522842	1.4
14.0	29, 36, 43, 48, 48, 43, 36, 29	-1407.1535599	0.6

[a]Energy from the conventional LCGTO-DF calculation on the glycine heptapeptide.

of errors to occur when looking at relative energies, as was the case when we only had a DAC fit to $\rho(\mathbf{r})$.

4.3. Constructing the density

No modification to the basic density matrix formulation of DAC, outlined in a previous section, need be made within our LCGTO-DF DAC program to solve the set of subsystem KS equations and construct the DAC $\rho(\mathbf{r})$ of the system. The density matrix formulation is chosen over the density formulation as the latter requires a numerical integration of the integral in equation (66),

$$\int p^{\alpha}(\mathbf{r})\,|\psi_i^{\alpha}(\mathbf{r})|^2\,d\mathbf{r} \approx \sum_I^{grid} p^{\alpha}(\mathbf{R_I})\,|\psi_i^{\alpha}(\mathbf{R_I})|^2\,W_I, \tag{83}$$

since $p^{\alpha}(\mathbf{r})$'s nature precludes an analytical integration. The density matrix formulation does not require any such procedure. In fact, the density matrix formulation requires surprisingly few modifications to be made in DeFT. This should hold true for any other quantum mechanical program.

The results of LCGTO-DF DAC calculations on the glycine heptapeptide are presented in Table 3. In these calculations, DAC procedures are used to construct the density and to fit both the density and the XC terms. The buffer space cutoffs listed in the first column apply to each of the DAC procedures. Immediately, we can see that the DAC construction of $\rho(\mathbf{r})$ introduces by far the greatest errors. The fitting procedures unique to the LCGTO-DF formalism introduce very small errors in comparison. The success of

DAC implementations within the LCGTO-DF framework, like any other DF framework, will depend on the ability of DAC techniques to accurately reconstruct the electronic density of the molecule with a sum of subsystem contributions.

Even when we go to our largest buffer space cutoff, 14.0 Å, where the errors in the fitting procedures are of the order of 0.0005 kcal mol^{-1}, the error is still quite large at 0.6 kcal mol^{-1}. This is consistent with the results of Yang and Lee [12]. For their test molecule, the glycine pentapeptide, they report an error of 1.02×10^{-5} hartrees per atom with a double-ζ plus polarization basis when buffer spaces include up to seventh nearest neighbors. For the subsystem at the center of the pentapeptide, buffer space therefore runs over the entire molecule. Their error corresponds to a total error of about 0.2 kcal mol^{-1}. This is better than our results. However, Yang and Lee also report [12] that the error in the same DAC calculation with a double-ζ (no polarization) basis is an order of magnitude smaller. A simple minimal basis, or single-ζ basis, gives an error that is yet another order of magnitude smaller. In our calculations, heavy atoms have triple-ζ plus polarization bases and we are dealing with a larger system, albeit we are using a cutoff of 14 Å which includes much more than only seventh nearest neighbors. A larger error is not totally unexpected.

A 0.6 kcal mol^{-1} error is disappointing. It is too large to be acceptable. There is no assurance that there will be a systematic improvement in results as the buffer space is increased. For example, the error with a 3.0 Å buffer space cutoff is better than the result with a 4.0 Å buffer space cutoff. Errors in the fits cannot be used to explain this discrepancy as the errors associated with the fits have been shown to be comparatively quite small. However, we do notice that once the results become *somewhat* reasonable, the true LCGTO-DF results are approached from above. From a 4.0 Å cutoff to a 14.0 Å cutoff, there is a systematic improvement in the results. Hopefully, one will be able to rely on such a systematic improvement to occur and that when taking relative energies, the resulting errors will be much smaller than 0.6 kcal mol^{-1}.

It should be obvious now that a common buffer space cutoff for all three DAC procedures is not efficient. The two DAC fitting procedures require much smaller cutoffs than the DAC construction of $\rho(\mathbf{r})$, given the same error tolerance. The density fitting procedure could most likely take on an even smaller buffer space cutoff than the XC fitting procedure. Something still needs to be done with the DAC construction of $\rho(\mathbf{r})$ as a cutoff of 14.0 Å (or even greater) will most likely make the LCGTO-DF DAC scheme less feasible than conventional LCGTO-DF calculations for anything other than extremely large systems. One potential way out of this dilemna is the use of an extended buffer space. It will not alter the spatial extent of buffer space, but it will dramatically reduce the number of basis functions enclosed within buffer space.

4.4. Extended buffer atoms

In previous DAC implementations, an "all or nothing" approach has been taken with regards to buffer atoms. An atom has its full basis added to a subsystem calculation if it is within the buffer space cutoff. Otherwise, none of its basis functions are used. Rather than taking this approach, multiple cutoff values could be implemented for the buffer space of any one DAC procedure. For example, in the DAC solution of the KS equations within a double-ζ, or split valence, plus polarization (DZP) basis, nearby buffer atoms would add their DZP basis sets to the subsystem problem. Buffer atoms that were further away would only add double-ζ (DZ) basis sets, and those atoms even further away would add only single-ζ (SZ), or minimal, basis sets. In essence, each and every atom in this molecule would carry three levels of basis set sophistication and all three would be used in the course of the DAC calculation. The basis used for any one subsystem calculation

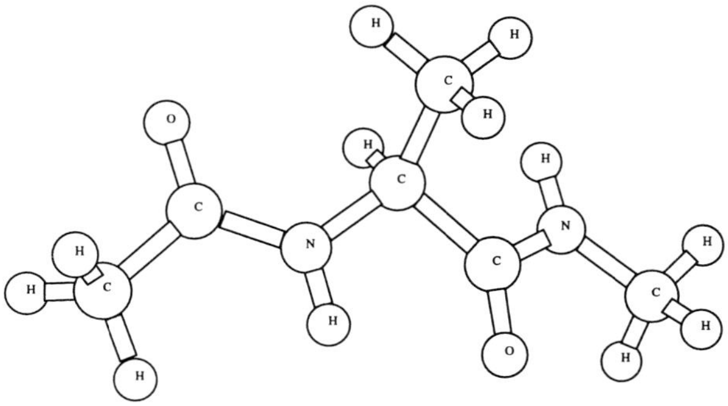

Figure 2. Alanine dipeptide.

would depend on the atom's proximity to the particular subsystem under study. The nearer it is, the more sophisticated is the chosen basis set. Of course, if the subsystem is very far away, the atom is entirely excluded from this subsystem's buffer. Despite the various levels of basis set sophistication, we must remember that the goal of the DAC calculation with these various levels of buffer space is to reproduce the results one would obtain with a conventional LCGTO-DF calculation with the *most sophisticated* basis set in our basis set hierarchy.

We call this region of buffer space with low-quality bases the extended buffer space. Though the concept may be used for any of the DAC procedures, extended buffer space will be most useful for the DAC construction of $\rho(\mathbf{r})$ since we have seen that it is this DAC procedure that requires, by far, the largest buffer space. Also, it is the easiest DAC procedure in which to incorporate this type of approach since we have extensive experience with increasing and decreasing levels of orbital basis set sophistication. The extended buffer space allows us to expand the spatial extent of buffer space while keeping the number of basis functions in check. Since it is the latter point that ultimately controls the cost of a subsystem calculation in any LCAO approach, extended buffer space may become a powerful time-saver in any practical DAC implementation.

Extended buffer space has been tested out within an LCGTO-DF implementation fully based on the density formulation of DAC [44] (this procedure is however not as efficient as the one outlined in this paper, where we incorporate aspects of the density matrix formulation). The alanine dipeptide, shown in Figure 2, was the test system. Test calculations were carried out on four of the dipeptide's conformations within the LSDA with DZP basis sets. The dipeptide was divided into three subsystems for testing purposes: two terminal methyl groups, and the remaining interior. The central group had both methyl groups in its buffer, i.e., its subsystem calculation encompassed the entire molecule. The terminal methyl groups each lacked only the other in their respective subsystem calculations. When extended buffer space was introduced, each terminal methyl group was placed in the other's extended buffer. In each of the four conformers, no atom in one terminal methyl group was within 5.3 Å of an atom in the opposing terminal methyl group.

The results of test calculations with extended orbital buffer spaces are presented in Table

Table 4
Effect of an extended orbital buffer on the DAC construction of the electronic density

orbital basis[a]	extended buffer basis[a]	$rms\Delta(\Delta E^b)$ (kcal mol^{-1})
DZP	none	0.34
DZP	SZ	0.02
DZP	DZ	0.00

[a]DZP = double-ζ + polarization; DZ = double-ζ; SZ = single-ζ.
[b]Root mean square error in relative conformational energies of the alanine dipeptide.

4. The auxiliary bases were not truncated in any way so all the errors arise from the DAC solution of the subsystem KS equations. The third column is listed as $rms\Delta(\Delta E)$ as it is the root mean square error in the six unique relative energy values between any two conformers. Without an extended buffer, an average error of 0.34 kcal mol^{-1} is observed. As this is of the same order as the inherent error of gradient-corrected XC functionals for this type of problem [7], this error is disconcerting. However, the mere inclusion of a SZ extended buffer cuts the error from 0.34 kcal mol^{-1} all the way down to 0.02 kcal mol^{-1}. This type of precision should suffice in the vast majority of cases, provided the potential energy surfaces remain smooth. If the quality of the extended buffer is raised to DZ, the average error falls below 0.005 kcal mol^{-1}.

A SZ basis for a methyl group consists of 8 basis functions. Meanwhile, a DZP basis for this same methyl group has 30 basis functions. The extended buffer concept therefore could allow us to have, at exactly the same cost, roughly three times more atoms in the buffer and extended buffer spaces, combined, than we previously had in our "all or nothing" buffer space. Our test cases show that we can go to lesser basis sets at as little as 5.0 Å away from the subsystem without introducing a significant error. No tests have yet been carried out at even smaller distances. On the other hand, no tests to date have either clearly shown at what point extended buffer space may be terminated. Our tests with the heptapeptide show that extended buffer space must be taken out to well over 14.0 Å: our tests with regular buffer atoms up to 14.0 Å provided disappointing results. Since extended buffer atoms do not have as high a quality basis set as regular buffer atoms, we therefore know that, given the same error tolerance, extended buffer space must be taken out to a further distance than regular buffer space would have to be taken out. However, the number of basis functions will be dramatically reduced.

Extended buffer space will be implemented within the DAC approach outlined in this paper. Tests will also be run to establish the optimal strategy for setting up an extended buffer. The optimal number of levels of basis set sophistication in the extended buffer space will have to be discovered through trial and error. As we accumulate this experience with extended buffer space, we will also establish the distances at which we may safely drop from one level of basis set sophistication to a lesser one. As with any one of our DAC procedures, we would most likely wish to err on the side of caution rather than automating the DAC procedures with aggressive cutoffs that may be too small and occasionally lead to significant errors. The possibility of leaving matrix elements between a pair of faraway buffer atoms within a subsystem calculation set to zero will also be investigated [12]. Since the CPU time spent diagonalizing subsystem KS matrices is negligeable compared to the time required to construct them, this last option has tremendous possibilities.

5. CONCLUSION

A divide-and-conquer implementation of the LCGTO-DF formalism has been outlined. In all of our work to date, we have been interested in developing a feel for the errors introduced by our DAC procedures. No great effort has yet been made in optimizing the code for parallel computer architectures or in totally rewriting whole sections of the conventional LCGTO-DF DeFT code to take full advantage of the DAC scheme. This work will be done in the near future. Only at that time can we begin to provide meaningful insights into the computational savings afforded by the DAC version of DeFT versus the conventional version of DeFT. The key question will be to see how big a system must become before the DAC version is more efficient than the conventional version. We do not foresee this being any lower than in the range of 50 to 100 atoms. Potentially, it may be much greater. Of course, this will also depend on how compact is the system under study.

We have not yet discussed energy gradients within our DAC LCGTO-DF scheme. We do not see ourselves deriving and programming equations for the derivatives of the DAC LCGTO-DF total energies. Following Yang's lead [45], we will use a DAC approach to the evaluation of the conventional LCGTO-DF gradient. Discrepancies in the conventional and DAC LCGTO-DF potential energy surfaces should not be taken as the sign of a need to develop an exact expression for the DAC LCGTO-DF gradient. Rather, it tells us that we need to expand buffer space to bring the DAC results back in line with the conventional results. Implementing the DAC evaluation of the LCGTO-DF gradient will not require many changes in the conventional code. Testing the errors introduced in equilibrium bond lengths, bond angles, and dihedrals will however be time-consuming as we will necessarily be working with large systems and geometry optimizations will require many lengthy steps.

Though this work is meant to be part of a complete LCGTO-DF package, individual components of this work may be used in conventional codes. DAC approaches to fitting $\rho(\mathbf{r})$, $\varepsilon_{xc}(\mathbf{r})$, and $v_{xc}(\mathbf{r})$ may be used to speed up LCGTO-DF calculations that choose not to use a DAC approach to constructing the electronic density of the system. Our DAC approach to fitting $\rho(\mathbf{r})$ may in fact be used in any longstanding LCGTO-HF code to ease the burden of assembling the Coulomb repulsion component of the Fock matrix. Even if algorithms such as fast multipole methods are in place to handle long-range electrostatics, a fitted $\rho(\mathbf{r})$ may immensely reduce the computational burden associated with evaluating electrostatic contributions to the Fock matrix arising from regions that are not sufficiently well separated to be treated by multipole expansions [41]. For very large systems, the linear scaling of our DAC $\rho(\mathbf{r})$ fitting procedure will nicely complement these fast multipole methods which scale linearly as well. The overall scaling will be the same, but with the fitted density, we can expect the pre-exponential factor to be much smaller. Also, our extended buffer space concept may be used to improve any DAC program, regardless of whether contracted gaussian, Slater, or numerical basis functions are used. This concept may also be used to make our own DAC fitting procedures more efficient by reducing the quality of fitting bases placed on buffer atoms that are relatively far from the particular subsystem under study.

ACKNOWLEDGEMENT

We wish to thank the Natural Sciences and Engineering Research Council of Canada and the University of Ottawa for financial support.

REFERENCES

1. H. Sambe and R. H. Felton, *J. Chem. Phys.* **62**, 1122 (1975).
2. B. I. Dunlap, J. W. D. Connolly, and J. R. Sabin, *J. Chem. Phys.* **71**, 3396 (1979).
3. J. Andzelm and E. Wimmer, *J. Chem. Phys.* **96**, 1280 (1992).
4. A. St-Amant, Ph.D. thesis (Université de Montréal, 1992).
5. A. St-Amant, the complete source code and documentation for the DeFT program package may be downloaded from http://www.chem.uottawa.ca/DeFT.html.
6. D. Salahub, R. Fournier, P. Mlynarski, I. Papai, A. St-Amant, and J. Ushio, in *Density Functional Methods in Chemistry*, J. K. Labanowski and J. W. Andzelm, Eds. (Springer-Verlag, New York, 1991).
7. A. St-Amant, W. D. Cornell, T. A. Halgren, and P. A. Kollman, *J. Comput. Chem.* **16**, 1483 (1995).
8. J. A. Pople, P. M. W. Gill, and B. G. Johnson, *Chem. Phys. Lett.* **199**, 557 (1992).
9. W. J. Hehre, L. Radom, P. v. R. Schleyer, and J. A. Pople, *Ab Initio Molecular Orbital Theory* (Wiley, New York, 1986).
10. A. Szabo and N. S. Ostlund, *Modern Quantum Chemistry: Introduction to Advanced Electronic Structure Theory*, First Edition, Revised (McGraw-Hill, New York, 1989).
11. W. Yang, *J. Mol. Struct. (THEOCHEM)* **225**, 461 (1992).
12. W. Yang and T. S. Lee, *J. Chem. Phys.* **103**, 5674 (1995).
13. J. P. Lu and W. Yang, *Phys. Rev. B* **49**, 11421 (1994).
14. W. Yang, *Phys. Rev. A* **44**, 7823 (1991).
15. B. Delley, *J. Chem. Phys.* **92**, 508 (1990).
16. W. Kohn and L. J. Sham, *Phys. Rev. A* **140**, 1133 (1965).
17. R. G. Parr and W. Yang, *Density-Functional Theory of Atoms and Molecules* (Oxford University Press, New York, 1989).
18. P. Hohenberg and W. Kohn, *Phys. Rev. B* **136**, 864 (1964).
19. J. P. Perdew, J. A. Chevary, S. H. Vosko, K. A. Jackson, M. R. Pederson, D. J. Singh, and C. Fiolhais, *Phys. Rev. A* **46**, 6671 (1992).
20. A. D. Becke, *J. Chem. Phys.* **98**, 5648 (1993).
21. S. H. Vosko, L. Wilk, and M. Nusair, *Can. J. Phys.* **58**, 1200 (1980).
22. N. Godbout, D. R. Salahub, J. Andzelm, and E. Wimmer, *Can. J. Chem.* **70**, 560 (1992).
23. S. Obara and A. Saika, *J. Chem. Phys.* **84**, 3963 (1986).
24. M. Head-Gordon and J. A. Pople, *J. Chem. Phys.* **89**, 5777 (1988).
25. M. Haser and R. Ahlrichs, *J. Comput. Chem.* **10**, 104 (1989).
26. A. D. Becke, *J. Chem. Phys.* **88**, 2547 (1988).
27. P. M. W. Gill, B. G. Johnson, and J. A. Pople, *Chem. Phys. Lett.* **209**, 506 (1993).
28. R. S. Jones, J. W. Mintmire, and B. I. Dunlap, *Int. J. Quantum Chem.* **S22**, 77 (1988).
29. R. Fournier, J. Andzelm, and D. R. Salahub, *J. Chem. Phys.* **90**, 6371 (1989).
30. A. Komornicki and G. Fitzgerald, *J. Chem. Phys.* **98**, 1398 (1993).
31. I. Papai, A. St-Amant, J. Ushio, and D. R. Salahub, *Int. J. Quantum Chem.* **S24**, 29 (1990).
32. K. Laasonen, R. Car, C. Lee, and D. Vanderbilt, *Phys. Rev. B* **43**, 6796 (1991).

33. A. M. Rappe, J. D. Joannopoulos, and P. A. Bash, *J. Am. Chem. Soc.* **114**, 6466 (1992).
34. P. Pulay, in *Applications of Electronic Structure Theory*, H. F. Schaefer III, Ed. (Plenum, New York, 1977).
35. K. Laasonen, M. Sprik, M. Parrinello, and R. Car, *J. Chem. Phys.* **99**, 9080 (1993).
36. R. Car and M. Parrinello, *Phys. Rev. Lett.* **55**, 2471 (1985).
37. R. M. Dickson and A. D. Becke, *J. Chem. Phys.* **99**, 3898 (1993).
38. E. J. Baerends, D. E. Ellis, and P. Ros, *Chem. Phys.* **2**, 41 (1973).
39. A. D. Becke and R. M. Dickson, *J. Chem. Phys.* **89**, 2993 (1988).
40. D. Feller and E. R. Davidson, in *Reviews in Computational Chemistry*, **Volume 1**, K. B. Lipkowitz and D. B. Boyd, Eds. (VCH Publishers, New York, 1990).
41. C. A. White, B. G. Johnson, P. M. W. Gill, and M. Head-Gordon, *Chem. Phys. Lett.* **230**, 8 (1994).
42. A. St-Amant, in *Reviews in Computational Chemistry*, **Volume 7**, K. B. Lipkowitz and D. B. Boyd, Eds. (VCH Publishers, New York, 1995).
43. L. Y. Fan and T. Ziegler, *J. Chem. Phys.* **94**, 6057 (1991).
44. A. St-Amant, *ACS Symp. Series*, B. Laird, R. Ross, and T. Ziegler, Eds., *in press*.
45. Q. Zhao and W. Yang, *J. Chem. Phys.* **102**, 9598 (1995).

J.M. Seminario (Editor)
Recent Developments and Applications of Modern Density Functional Theory
Theoretical and Computational Chemistry, Vol. 4

The Douglas-Kroll-Hess Approach to Relativistic Density Functional Theory: Methodological Aspects and Applications to Metal Complexes and Clusters

N. Rösch,[a] S. Krüger,[a] M. Mayer[a], and V. A. Nasluzov[b]

[a]Lehrstuhl für Theoretische Chemie, Technische Universität München, D-85747 Garching, Germany

[b]Permanent address: Institute of Chemistry of Natural Organic Materials, K. Marx Str. 42, 660049 Krasnoyarsk, Russian Federation

Basic theoretical aspects of relativistic density functional theory as well as methodological details and key applications of the relativistic variant of the linear combination of Gaussian-type orbitals density functional (RLCGTO-DF) method are presented. Starting from the quantum electrodynamical Hamiltonian, an overview over relativistic density functional theory is given. The adaption of the second-order Douglas-Kroll-Hess scheme for solving the Dirac-Kohn-Sham equations results in a computationally efficient and numerically stable two-component Kohn-Sham formalism which is well suited for molecular applications, in particular due to the availability of analytical energy gradients. Finally, various applications of the RLCGTO-DF method to metal clusters, to metal-ligand interactions, and to metal complex reactions are reviewed.

1. INTRODUCTION

The investigation of relativistic effects in quantum chemistry deals with the differences that are introduced if one replaces the nonrelativistic Schrödinger equation by the relativistic Dirac equation or, as it is often formulated, with the consequences of reducing the speed of light from an infinite to its finite value. Although one reasonably expects relativistic effects to become relevant for sufficiently heavy atoms, such phenomena were for a long time considered to be of peripheral nature. However, during the last decade relativistic effects in chemistry have been identified for elements starting even from a nuclear charge as low as $Z = 30$ [1]. Nevertheless, relativistic effects on the valence electrons are in general expected to be important only for higher values of the nuclear charge. In chemistry, many of these effects are connected to the relativistic stabilization of the valence s and p orbitals and from a concomitant expansion of the valence d and f orbitals [1]. The energetic lowering of the s and p levels is due to two effects. On the one hand, the s and p orbitals extend into the vicinity of the nucleus which leads to an increased speed of the particles, hence of their effective mass, and thus to a reduction of the effective orbital radii. The second relativistic effect is an indirect one introduced by the orthogonality condition of the atomic orbitals. The contraction of the s and p orbitals results in an improved screening of the effective nuclear attraction experienced by the orbitals of higher angular momentum.

A detailed consideration of relativistic effects shows that many well-known properties of materials can only be explained within relativistic quantum mechanics [1]. A very prominent example is the unique position of gold among the elements. If one considers the relativistic contraction of the valence 6s orbital, one discovers a maximum for gold (Z = 79) [1,2]. This is reflected in the valence configuration $5d^{10}6s^1$ where the 6s and 5d levels are very close in energy and therefore enable hybridization. Also the spectacular color of gold metal compared to its lighter homologue silver has been connected to relativistic effects [1]. On the other hand, gold has a rich chemistry which has been intensely explored in recent years [3]; the bond distances of these compounds clearly reflect the relativistic contraction of the atomic radius of Au. Another prominent example is platinum for which nonrelativistic calculations predict the ground state configuration $5d^{10}6s^0$, in analogy to its lighter homologue palladium. Relativistic effects stabilize the 6s level and lead to the ground state $5d^9 6s^1$ [4]. As for many other third-row transition elements, relativistic effects contribute to the particular reactivity of platinum and thus cause its ability to act as a catalyst.

Relativistic effects may only be discovered by comparison of nonrelativistic with relativistic calculations. This brings up the question of how to derive an appropriate many-electron description at the relativistic level. In a naive attempt one might simply supplement the one-particle Dirac Hamiltonian by the electron-electron interaction which leads to the so-called Dirac-Coulomb Hamiltonian. As will be explained later on, this Hamiltonian has no bound state solutions due to the fact that positive and negative energy solutions can interact [5]. It turns out that an appropriate relativistic treatment requires reference to quantum electrodynamics (QED) [6]. However, it is possible to derive a Hamiltonian with bound states by restricting the space of solutions to wave functions of positive energy (including the electron rest mass), a concept known as the no-pair approach [7]. Once the Hamiltonian is obtained one may proceed to solve the wave equation, for instance by the Hartree-Fock (HF) method and related formalisms (ab initio theories) which aim at explicitly constructing a wave function.

Density functional (DF) theory provides an alternative approach to the many-electron problem which by now constitutes a rather widely used "first principles" method in chemistry [8], especially since the generalized gradient approximation (GGA) for the exchange-correlation energy [9–12] has been developed and since analytic energy gradients have been introduced to assist in the geometry optimization (see for example [13] and references therein). For many chemical problems, DF methods furnish a sufficiently accurate, yet computationally efficient description of molecular structure and energetics.

An adaption of the DF approach to relativistic systems requires the use of a Hamiltonian which is bounded from below, a condition fulfilled by the QED Hamiltonian and presumably also by the no-pair Hamiltonian [6,14]. Essentially all relativistic DF implementations have been implicitly based on the QED Hamiltonian. Within the DF approach the energy of a system is minimized with respect to the electron density [15]. This variational procedure is usually carried out with the help of the Kohn-Sham (KS) approach which leads to a one-electron problem governed by an effective potential [15,16]. In many cases, these KS equations are solved algebraically by expanding the KS orbitals in a finite basis set. The relativistic analogue of the KS Hamiltonian, the Dirac-Kohn-Sham Hamiltonian, is not bounded from below and thus subject to the so-called "variational collapse" [17–19] which implies a wrong "Schrödinger" limit of the Hamiltonian [19]. These problems may be circumvented by choosing kinetically balanced basis sets for the large and small components of the Dirac spinors [17,20]. However, it is also possible to project out the solutions of negative energy, in analogy to the no-pair approach for many-electron systems [7]. This decoupling of large and small components may be performed by the Foldy-Wouthuysen transformation (FW) [19]; in this way, the four-component Dirac problem is reduced to a two-component problem that has only solutions of positive energy and is thus bounded from below [21]. In general, this transformation cannot be carried out exactly. The attempt to approximate the FW transformation by an expansion in p^2/c^2 leads to numerical instabilities for the Coulomb potential [22], but the resulting Hamiltonian is nevertheless suitable for a perturbational treatment [23]. If one assumes "frozen" atomic cores and requires the valence orbitals to be orthogonal to the core orbitals one obtains a variationally stable Hamiltonian, the quasi-relativistic Pauli Hamiltonian [22]. A very promising alternative – the one favored in this presentation – relies on the Douglas-Kroll transformation (DK) for decoupling large and small components which leads to a numerically stable Hamiltonian and eliminates the problems associated with the existence of negative energy solutions [24–26]. This scheme utilizes an expansion up to the second order in powers of the effective potential [24]. It has first been employed for wave function based ab initio methods in combination with a no-pair Hamiltonian [25,27]. The accuracy and numercial efficiency of this formalism derives from a suggestion [27,28] of how to conveniently evaluate all "relativistic" matrix elements in momentum space. We have employed this Douglas-Kroll-Hess (DKH) approach for solving the Dirac-Kohn-Sham problem in the scalar relativistic variant of the linear combination of Gaussian-type orbitals density functional (LCGTO-DF) method [29,30]. Recently, another relativistic DF method, the variationally stable ZORA scheme, has been introduced [31,32]. It is based on a regular potential-dependent expansion [33,34] and exhibits an accuracy comparable to that of the DKH scheme. For many chemical problems the possibility of computing analytical energy gradients is highly advantageous. With the help of these forces it becomes possible to locate stationary points of multidimensional potential energy surfaces in an efficient way and thus to explicitly address questions of chemical reactivity. Corresponding extensions of the Pauli [35] and DKH LCGTO-DF [36] schemes have been implemented.

This article is structured as follows. In chapter 2 the foundations of relativistic DF theory are reviewed. Especially the choice of an appropriate Hamiltonian, suitable for deriving the Hohenberg-Kohn theorem, is discussed. Chapter 3 elaborates how the Dirac-Kohn-Sham equations can be solved algebraically with the help of the Douglas-Kroll transformation, avoiding a "variational collapse". In chapter 4 details of the second-order DKH scheme used in the RLCGTO-DF program are expounded and some examples are presented to illustrate the accuracy of the method. Finally, chapter 5 is devoted to presenting a collection of applications of the RLCGTO-DF method to various areas of physics and chemistry of heavy elements. These examples are intended to demonstrate the reliability and the computational efficiency which may be achieved in molecular applications of the relativistic variant of the LCGTO-DF method.

2. RELATIVISTIC DENSITY FUNCTIONAL THEORY

2.1. Relativistic Hamiltonians and the Hohenberg-Kohn theorem

Nonrelativistic density functional theory is based on the Hohenberg-Kohn theorem which aims at establishing a one-to-one map between the external potential and the ground state density [15]. The corresponding argument rests on the fact that the Schrödinger Hamiltonian is bounded from below, such that an N-electron ground state may be determined by minimizing the energy expectation value of normalized square-integrable N-electron wave functions [15]. In addition, this theorem serves as a basis for the Kohn-Sham approach to DF theory which provides an efficient scheme for determining the ground state density [15,16]. Attempts at constructing a relativistic analogue to the nonrelativistic Hohenberg-Kohn-Sham formalism have been undertaken early on in the development of DF theory [37–39]. However, it is not straightforward how to choose a relativistic Hamiltonian which is bounded from below. The one-electron Dirac-Hamiltonian, for example, has a continuum of negative energy eigenstates and therefore is not bounded from below. Of course, when dealing with chemical questions, one is not interested in these states of negative energy. The problem is thus to construct a Hamiltonian which avoids the difficulties associated with the states of negative energy. For the general development of relativistic density DF theory we refer the reader to a recent excellent review [6]. In the following we will discuss several alternatives for a relativistic Hamiltonian in some detail in order to set the stage for the Douglas-Kroll-Hess approach to relativistic DF theory.

2.1.1. The QED Hamiltonian

The proper choice for a relativistic Hamiltonian is the covariant Hamiltonian of quantum electrodynamics (QED) which in second quantization reads [14]:

$$\begin{aligned} H^{QED} &= \int d^3x\, \psi^\dagger(x)\left[c\vec{\alpha}\vec{p} + (\beta-1)c^2\right]\psi(x) + \frac{1}{2}\int d^3x \int d^3x' \frac{\rho(x)\rho(x')}{|\vec{x}-\vec{x}'|} \\ &+ \frac{1}{2}\int d^3x \int d^3x' \frac{\vec{j}(x)\vec{j}(x')}{|\vec{x}-\vec{x}'|} + \int d^3x \rho(x) v_{ext}(\vec{x}) + H_{em}\,. \end{aligned} \tag{1}$$

Here $\psi(x)$ is the four-component second-quantized Dirac spinor operator and $\rho(x)$ and $\vec{j}(x)$ are the second-quantized charge and current density operators, respectively, which constitute the four-current density operator:

$$j^\mu(x) = \psi^\dagger(x)\beta\gamma^\mu\psi(x)\,, \quad j^\mu(x) = \left(\rho(x), \vec{j}(x)\right)\,. \tag{2}$$

The 4×4 Dirac matrices α_i, $i = 1,2,3$, β and γ_μ, $\mu = 0,1,2,3$ where $\gamma_\mu = (\beta, \beta\vec{\alpha})$, are taken in the standard fashion [40]. The first term in eq. (1) is the relativistic kinetic energy operator where the rest mass of the system is subtracted. The second term is the Coulomb interaction operator, the third term is the interaction operator of the currents with their own magnetic fields, the fourth term is the interaction operator of the system with an external field v_{ext}, where we have excluded an external vector potential [14]. Finally, H_{em} is the energy of the radiation field. The speed of light c is 137.04 in atomic units [1].

The Hamiltonian H^{QED} is not yet suitable for establishing the Hohenberg-Kohn theorem since the evaluation of its expectation value for some arbitrary ψ leads to divergencies. First, this is due to the so-called UV divergencies which arise from infinite integrals connected to the selfinteraction of electrons and the interaction of the electrons with photons and with the external field v_{ext} [6]. These divergencies may be dealt with by renormalizing all fields and physical constants [6]. Note that the renormalization is independent of the external potential v_{ext} [6]. With this procedure we arrive at the renormalized Hamiltonian H_R^{QED} which is formally identical to H^{QED}, but with all fields and constants replaced by their renormalized counterparts. In order to keep the energy expectation value $\langle\psi|H_R^{QED}|\psi\rangle$ finite the renormalization procedure requires additional counter terms $\delta E[v_{ext}]$ for the energy and $\delta j^\mu[v_{ext}]$ for the current density. These counter terms depend on the external potential v_{ext}, but not on the wave function ψ [6]. This fact is crucial for the proof of the relativistic Hohenberg-Kohn theorem as will be seen later on.

The Hamiltonian derived up to this point leads to finite expectation values as long as wave functions for a finite number N of electrons are considered. Yet, the operator is not bounded from below since the electrons may still occupy states of the negative energy continuum. One way to deal with this problem proceeds by considering the negative continuum as completely occupied [40]. From physical intuition one might expect that the "effective" energy expectation values are then bounded from below because the lower energy states of the filled negative continuum are not accessible. However, no rigorous proof is known so far. Filling the negative continuum introduces a further divergence, namely an infinite negative energy expectation value which may be removed either by subtraction of the infinite vacuum expectation value $\langle 0|H_R^{QED}|0\rangle$ or by normal ordering of the operators [6,14]. If one assumes boundedness from below [6], one can determine the ground state according to the Ritz variational principle by minimizing the energy functional

$$E_R[\psi] = \langle\psi|H_R^{QED}|\psi\rangle - \langle 0|H_R^{QED}|0\rangle + \delta E[v_{ext}] \tag{3}$$

for arbitrary normalized wave functions ψ.

Starting from this energy expectation value the relativistic Hohenberg-Kohn theorem can then be derived. In a first step where the minimum principle is not yet needed one establishes a one-to-one mapping between the external potential v_{ext} and the ground state wave function ψ_0 [6]. However, in the second step, the proof of the one-to-one mapping between the ground state density ρ and v_{ext} relies on the minimum property of the energy expression (3). Similar to the nonrelativistic case the proof proceeds by reductio ad absurdum [6,37]. One assumes that two different external potentials v_{ext} and v'_{ext} lead to the same ground state density ρ.

The minimum property then implies

$$E_R = E_R[\psi_0] < E_R[\psi_0'] = E_R' + \int d^3x \rho_R(v_{ext} - v_{ext}') + \delta E[v_{ext}] - \delta E[v_{ext}'] \\ - \int d^3x \left(\delta\rho[v_{ext}]v_{ext} - \delta\rho[v_{ext}']v_{ext}'\right), \tag{4}$$

where E_R, ψ_0 and E_R', ψ_0' are the ground state energies and ground state wave functions of v_{ext} and v_{ext}', respectively. Interchanging the roles of v_{ext} and v_{ext}' and adding the resulting equations leads to the well-known contradiction [6,15]

$$E_R + E_R' < E_R + E_R' , \tag{5}$$

establishing the relativistic Hohenberg-Kohn theorem.

2.1.2. The four-component no-pair Hamiltonian

An alternative way for obtaining a relativistic Hamiltonian which is bounded from below, at least for a certain range of nuclear charges, starts from the Dirac-Coulomb (DC) Hamiltonian

$$H_{DC}^{(4)} = \sum_i^N h_{D,ext}^{(4)}(i) + \sum_{i<j}^N \frac{1}{r_{ij}} , \tag{6}$$

where

$$h_{D,ext}^{(4)}(i) = c\vec{\alpha}\vec{p} + \beta c^2 + v_{ext} \tag{7}$$

is the relativistic one-particle Hamiltonian for an electron in the external potential v_{ext}. The electron-electron interaction appears in its nonrelativistic form.

This approach and similar ones which are not derived from QED suffer from the continuum dissolution ("Brown-Ravenhall disease"): the electron-electron interaction scatters bound states of $h_{D,ext}^{(4)}$ into a continuum of states where one or more electrons occupy states of negative energy [5].

Thus, the DC Hamiltonian neither has well defined bound states nor is it bounded from below. Both problems vanish if one restricts the wave functions to the space spanned by the products of one-electron states with positive energy eigenvalue of the Hamiltonian $h_{D,ext}^{(4)}$ (or of some less involved one-electron Hamiltonian like the free-particle Hamiltonian). This strategy leads to the so-called no-pair Hamiltonians; it has first been explored in the framework of wave function based ab initio theory [7].

The resulting Hamiltonian is [7,41]

$$H_{+,ext}^{(4)} = \mathcal{L}_+ \left(\sum_i^N h_{D,ext}^{(4)}(i) + \sum_{i<j}^N \frac{1}{r_{ij}} \right) \mathcal{L}_+ , \tag{8}$$

where

$$\mathcal{L}_+ = \mathcal{L}_+(1)\mathcal{L}_+(2)\cdots\mathcal{L}_+(\mathrm{N}) \tag{9}$$

with

$$\mathcal{L}_+(i) = \frac{1}{2}\Big(1 + h^{(4)}_{D,ext}(i)/|h^{(4)}_{D,ext}(i)|\Big)\,. \tag{10}$$

The projector $\mathcal{L}_+$ onto the space spanned by the of positive energy solutions of $h^{(4)}_{D,ext}$ ensures that positive and negative energy states do not mix. Nevertheless, the operator $H^{(4)}_{+,ext}$ is not yet suitable for deriving the Hohenberg-Kohn theorem because all wave functions which belong to the space of negative energy solutions of $h^{(4)}_{D,ext}$ have an energy eigenvalue of zero. For potentials with Coulomb singularities the bound state solutions of $H^{(4)}_{DC}$ have energy eigenvalues in the interval $\left[0, Nc^2\right]$. Therefore $H^{(4)}_{+,ext}$ has an infinitely degenerate artificial "ground state" of energy zero. This has been considered as a barrier for deriving the Hohenberg-Kohn theorem [42]. However, this deficiency may easily be removed by subtracting the electron rest mass c^2 from $h^{(4)}_{D,ext}$ before the projection onto the space of electronic solutions is carried out. In this way, the energy eigenvalues of $H^{(4)}_{+,ext}$ which correspond to bound electronic solutions are shifted to the interval $\left[-Nc^2, 0\right]$. After this modification the fundamental inequality

$$\langle\psi_0|H^{(4)}_{+,ext}(v_{ext})|\psi_0\rangle < \langle\psi|H^{(4)}_{+,ext}(v_{ext})|\psi\rangle \tag{11}$$

may be derived for a nondegenerate ground state ψ_0 and for any normalized wave function ψ. As pointed out previously [42], one has

$$\begin{aligned}\langle\psi_0|H^{(4)}_{+,ext}(v_{ext})|\psi_0\rangle &< \frac{\langle\mathcal{L}_+(v_{ext})\psi|H^{(4)}_{+,ext}(v_{ext})|\mathcal{L}_+(v_{ext})\psi\rangle}{\langle\mathcal{L}_+(v_{ext})\psi|\mathcal{L}_+(v_{ext})\psi\rangle} \\ &= \frac{\langle\psi|H^{(4)}_{+,ext}(v_{ext})|\psi\rangle}{\langle\mathcal{L}_+(v_{ext})\psi|\mathcal{L}_+(v_{ext})\psi\rangle}\,.\end{aligned} \tag{12}$$

Thus, for bound states where

$$\langle\psi|H^{(4)}_{+,ext}(v_{ext})|\psi\rangle < 0, \tag{13}$$

one obtains

$$\langle\psi_0|H^{(4)}_{+,ext}(v_{ext})|\psi_0\rangle < \frac{\langle\psi|H^{(4)}_{+,ext}(v_{ext})|\psi\rangle}{\langle\mathcal{L}_+(v_{ext})\psi|\mathcal{L}_+(v_{ext})\psi\rangle} < \langle\psi|H^{(4)}_{+,ext}(v_{ext})|\psi\rangle < 0\,, \tag{14}$$

since

$$0 < \langle\mathcal{L}_+(v_{ext})\psi|\mathcal{L}_+(v_{ext})\psi\rangle \ \leq 1\,. \tag{15}$$

For wave functions ψ with a positive energy expectation value the inequality (11) holds trivially. Then the Hohenberg-Kohn theorem may be derived along the familiar lines [15].

2.1.3. The two-component no-pair Hamiltonian

The wave functions ψ considered so far are four-component Dirac spinors. When dealing with electronic problems, it is often desirable to work just with the two "electronic", the so-called large components. This may be achieved with help of the Foldy-Wouthuysen transformation $U_{FW}(i)$ [21] which decouples large and small components. The (one-particle) Foldy-Wouthuysen transformation U_{FW} is defined as the unitary transformation that solves the equation

$$B_+(i) = U_{FW}(i)\mathcal{L}_+(i)U^\dagger_{FW}(i)\,, \tag{16}$$

where

$$B_+(i) = \frac{1}{2}(1+\beta)\,. \tag{17}$$

The four-component Hamiltonian $H^{(4)}_{+,ext}$ can be reduced to a two-component Hamiltonian $H^{(2)}_{+,ext}$:

$$\begin{aligned} H^{(2)}_{+,ext} &= U_{FW} H^{(4)}_{+,ext} U^\dagger_{FW} \\ &= \sum_i^N B_+(i) U_{FW}(i) h^{(4)}_{D,ext} U^\dagger_{FW}(i) B_+(i) \\ &+ \sum_{i<j}^N B_+(i)B_+(j)U_{FW}(i)U_{FW}(j)\frac{1}{r_{ij}}U^\dagger_{FW}(j)U^\dagger_{FW}(i)B_+(j)B_+(i)\,. \end{aligned} \tag{18}$$

Note that the one-electron operators have to be transformed once, while the two-electron operators must be transformed twice in order to yield a two-component Hamiltonian in the space of electronic solutions of $h^{(4)}_{D,ext}$. So far no formal proof exists that the two-component no-pair Hamiltonian is bounded as required for a derivation of the Hohenberg-Kohn theorem. Nevertheless experience from many atomic and molecular calculations [25,28,43,44] shows that it is numerically well behaved.

2.2. The relativistic Kohn-Sham equations in the "no-sea" approximation

We now turn to a discussion of the relativistic analogue of the Kohn-Sham (KS) approach to DF theory [16]. In the following, we will restrict ourselves to a treatment of the QED Hamiltonian which is the physically most appropriate Hamiltonian. The one-to-one mapping between the ground state density and the external potential established by the Hohenberg-Kohn theorem implies that one may determine the ground state of the system by minimizing the density functional $E_R[\rho]$. In order to achieve this, one considers a reference system of noninteracting electrons. In the relativistic framework, one additionally assumes the so-called "no-sea" approximation where all vacuum contributions are neglected [6].

Similar to the nonrelativistic case, the renormalized density functional $E_R[\rho]$ is partitioned as follows [6,14]:

$$E_R[\rho] = T_S[\rho] + E_H[\rho] + E_{xc}[\rho] + E_{ext}[\rho] \;, \tag{19}$$

where

$$T_S[\rho] = \sum_i^N \int d^3x \varphi_i^\dagger \left(c\vec{\alpha}\vec{p} + (\beta - 1)c^2\right)\varphi_i \tag{20}$$

is the kinetic energy of the noninteracting reference system of density

$$\rho = \sum_i^N |\varphi_i|^2 \,. \tag{21}$$

The Hartree energy

$$E_H = \frac{1}{2}\int d^3x \int d^3x' \frac{J_\mu(\vec{x})J^\mu(\vec{x}')}{|\vec{x} - \vec{x}'|} \tag{22}$$

is given in terms of the four-current density [38,39]

$$J^\mu = \sum_i^N \varphi_i^\dagger \beta\gamma^\mu \varphi_i \;, \tag{23}$$

where the summation is carried out over all N occupied one-particle states φ_i. The Hartree energy may be further partioned [6,45] into the contribution of the longitudinal Coulomb interaction

$$E_{Coul} = \frac{1}{2}\int d^3x \int d^3x' \frac{\rho(\vec{x})\rho(\vec{x}')}{|\vec{x} - \vec{x}'|} \tag{24}$$

and the contribution of the transversal magnetic interaction

$$E_{magn} = \frac{1}{2}\int d^3x \int d^3x' \frac{\vec{J}(\vec{x})\vec{J}(\vec{x}')}{|\vec{x} - \vec{x}'|} \;. \tag{25}$$

The exchange-correlation functional $E_{xc}[\rho]$ contains exchange and correlation contributions corresponding to the longitudinal as well as to the transversal interactions [6,45]. In addition, it contains all retardation terms which are due to effects of the finite velocity of light [6]; they are usually included in the transversal exchange part of the correlation potential. Finally,

$$E_{ext} = \int d^3x v_{ext}(\vec{x})\rho(\vec{x}) \tag{26}$$

represents the energy due to the interaction with the external potential. We shall simplify the following discussion by assuming the currents to vanish. This is approximately the case if most of the electrons reside in closed shells, as in all systems considered here, since the current contributions of closed shells vanish. However, the fact that E_{magn} vanishes does not imply that the corresponding exchange-correlation energy contribution vanishes as well. The minimization with respect to the density ρ may be carried out by varying the one-electron orbitals φ_i. This procedure leads to the relativistic Dirac-Kohn-Sham (DKS) equations [38,39]:

$$h^{(4)}_{DKS}\varphi_i = \epsilon_i\varphi_i \ . \tag{27}$$

The one-particle Dirac-type Hamiltonian

$$h^{(4)}_{DKS} = \left(c\vec{\alpha}\vec{p} + (\beta - 1)c^2 + v_{eff}\right) \tag{28}$$

contains the effective one-particle potential

$$v_{eff}(\vec{x}) = v_{ext}(\vec{x}) + \int d^3x' \frac{\rho(\vec{x}')}{|\vec{x} - \vec{x}'|} + \frac{\delta E_{xc}[\rho]}{\delta\rho(\vec{x})} \ . \tag{29}$$

Since we neglected the vacuum contributions, only electronic solutions, either bound or in the corresponding continuum, have to be considered in the DKS procedure. Therefore, it is sufficient to deal with the two-component DKS Hamiltonian

$$h^{(2)}_{+,DKS} = U_{FW}\mathcal{L}_+ h^{(4)}_{DKS}\mathcal{L}_+ U^\dagger_{FW} = B_+ U_{FW} h^{(4)}_{DKS} U^\dagger_{FW} B_+ \ . \tag{30}$$

Either form of the DKS equations, with four- or two-component solutions, has to be solved selfconsistently. Because the exact form of the exchange-correlation energy functional $E_{xc}[\rho]$ is unknown, approximations have to be applied. Just as in the nonrelativistic case, the appropriate choice of this approximation is crucial. In the next section we will discuss this in more detail.

Before proceeding, we briefly discuss the no-pair alternatives for deriving relativistic KS equations. Of course, both the four-component and the two-component no-pair Hamiltonians mentioned in the previous section may be employed [46,47]. The KS equations of the four-component no-pair Hamiltonian are essentially equivalent to those just derived for the QED Hamiltonian. However, in this case the exchange-correlation functional only contains longitudinal contributions and thus exhibits the form familiar from the nonrelativistic formalism.

On the other hand, the energy functional based on the two-component no-pair Hamiltonian $H^{(2)}_{+,ext}$ will lead to a KS Hamiltonian $h^{(2)}_{+,FW}$ that differs somewhat from the KS Hamiltonian $h^{(2)}_{+,DKS}$.

Two pathways for estabilishing a two-component relativistic KS Hamiltonian are compared in Figure 1. In order to obtain $h^{(2)}_{+,DKS}$ one first constructs a density functional $E^{(4)}[\rho]$ employing the QED Hamiltonian (or the DC Hamiltonian) and then derives four-component DKS equations. The two-component form $h^{(2)}_{+,DKS}$ may then be obtained by applying a one-particle Foldy-Wouthuysen transformation. On the other hand, to arrive at $h^{(2)}_{+,FW}$ one first

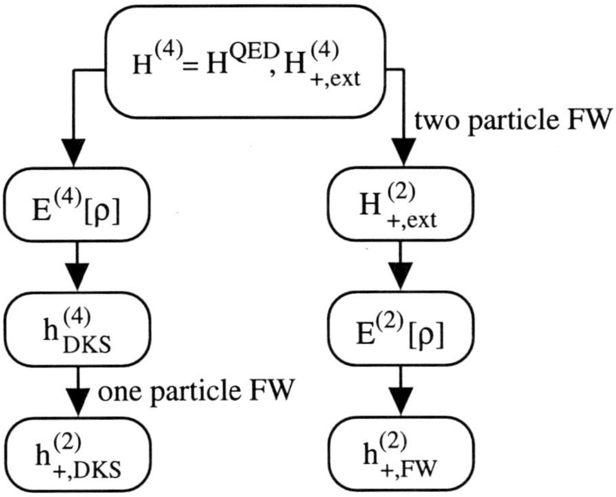

Figure 1. Alternative pathways for deriving a two-component relativistic Kohn-Sham formalism. For details see text.

applies a two-particle Foldy-Wouthuysen transformation to the two-electron part of $H^{(4)}_{+,ext}$ in order to derive the two-component many-particle Hamiltonian $H^{(2)}_{+,ext}$. In the next step, this Hamiltonian is used to construct the corresponding energy functional $E^{(2)}[\rho]$ which finally leads to the KS Hamiltonian $h^{(2)}_{+,FW}$. A crucial difference between the two alternatives lies in the form of the Foldy-Wouthuysen transformation [46,47]: a one-particle transformation suffices in the first case, but a two-particle transformation is required for the electron-electron interaction in the second case. Therefore, the latter pathway leads to a different relativistic Coulomb potential and, concomitantly, to a different exchange-correlation potential in $h^{(2)}_{+,FW}$. Since relativistic corrections to the exchange-correlation functional are in general determined along the first pathway (as discussed in more detail in the following section), the first variant of deriving the relativistic KS equations is preferable.

2.3. Relativistic exchange-correlation functionals

Relativistic effects enter the relativistic density functional $E_R[\rho]$ (eq. 19) in two ways. By far the most important change results from replacing the nonrelativistic kinetic energy by its relativistic counterpart, i. e. by introducing relativistic kinematics. The exchange-correlation energy term also contains relativistic effects due to the transversal interaction and to retardation contributions. The other terms in eq. (19), the Hartree energy and the energy of the interaction with the external potential, take the same form as in the nonrelativistic case as long as no currents are considered.

Within the generalized gradient approximation (GGA) [9,12,48–50] the exchange-correlation energy is written as

$$E_{xc}[\rho] = \int d^3x \, e_{xc}(\rho, \nabla\rho) \,. \tag{31}$$

Here, e_{xc} is the exchange-correlation energy density which in GGA fashion is assumed to depend only on the electron density ρ and on its gradient $\nabla\rho$. This form of the exchange-correlation energy leads to the potential

$$v_{xc}(\vec{x}) = \frac{\delta E_{xc}[\rho]}{\delta\rho(\vec{x})} = \left(\frac{\partial}{\partial\rho} - \nabla\frac{\partial}{\partial\nabla\rho}\right) e_{xc}(\rho, \nabla\rho) \,. \tag{32}$$

When dealing with the relativistic GGA (RGGA) exchange energy functional, it is convenient to split the exchange energy density into the following four contributions:

$$e_x^{RGGA} = e_x^{LDA} + e_x^{\Delta GGA} + e_x^{\Delta RLDA} + e_x^{\Delta RGGA} \,. \tag{33}$$

Here

$$e_x^{LDA} = -\frac{3c}{4\pi}\beta \tag{34}$$

with

$$\beta = \left(3\pi^2\rho\right)^{1/3}/c \tag{35}$$

is the nonrelativistic exchange energy density of the local density approximation (LDA). The nonrelativistic GGA correction to LDA $e_x^{\Delta GGA}$ may be written as [9,12]

$$e_x^{\Delta GGA} = g(\zeta) e_x^{LDA} \tag{36}$$

with

$$\zeta = \left(|\nabla\rho|^2/\left(2\left(3\pi^2\rho\right)^{1/3}\rho\right)\right)^2. \tag{37}$$

The relativistic correction to the LDA $e_x^{\Delta RLDA}$ is available in analytical form from calculations on the relativistic homogeneous electron gas [39,51]:

$$e_x^{\Delta RLDA} = e_x^{LDA}\left(\Phi_0^L(\beta) + \Phi_0^T(\beta)\right) \,. \tag{38}$$

Here, $\phi_0^L(\beta)$ and $\phi_0^T(\beta)$ denote the longitudinal and transversal relativistic LDA correction factors [39,51]. The decomposition of the RLDA energy density function $\varepsilon_x = e_x/\rho$ into the LDA contribution and the longitudinal as well as the transversal relativistic ΔLDA corrrection terms is shown in Figure 2. For large values of the density ρ (or β) the transversal relativistic term soon dominates the exchange energy. The relativistic corrections are noticeable only in the immediate vicinity of the nuclei; as an example, the density values of various core levels of the gold atom are indicated in Figure 2.

Finally, the relativistic correction to GGA $e_x^{\Delta RGGA}$ may be written as

$$e_x^{\Delta RGGA} = g(\zeta) e_x^{LDA}\left(\Phi_2^L(\beta) + \Phi_2^T(\beta)\right) , \tag{39}$$

where $\phi_2^L(\beta)$ and $\phi_2^T(\beta)$ are the longitudinal and transversal relativistic GGA correction factors which have been determined [52] for the exchange GGA of Becke [9].

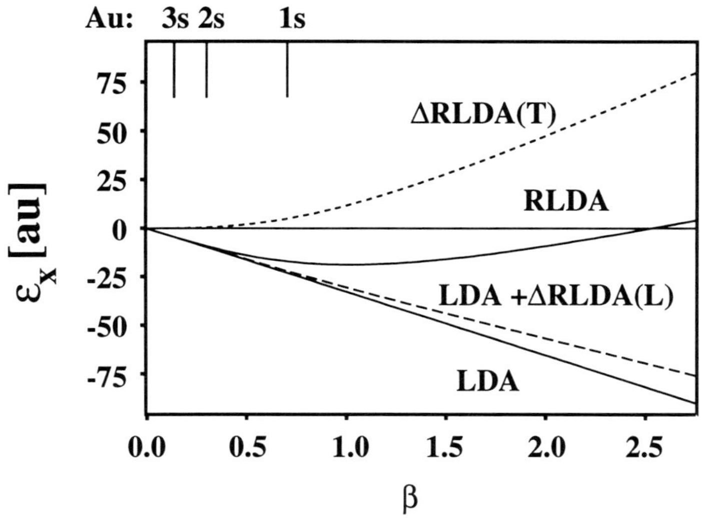

Figure 2. Decomposition of the RLDA exchange kernel $\varepsilon_x = e_x/\rho$ into its LDA, longitudinal ΔRLDA(L) and transversal ΔRLDA(T) relativistic contributions. The density values at the radial expectation values of some s core orbitals of the Au atom are also indicated.

Relativistic calculations on finite systems like atoms and molecules where the exchange energy is described by the ΔRLDA only,

$$e_x^{RLDA} = e_x^{LDA} + e_x^{\Delta RLDA} , \tag{40}$$

overestimate the transversal contributions [45]. This is due to the fact that, at variance with the situation in the relativistic homogeneous electron gas from which the RLD approximation is derived, magnetic long-range interactions are less important in finite systems [45]. This deficiency of RLDA for the exchange is to some extent corrected by the RGGA correction [52]. In particular, this latter approximation reproduces the balance between the transversal and longitudinal relativistic exchange correction to LDA [52].

The correlation contribution to the energy density may be partitioned in analogy to eq. (33):

$$e_c^{RGGA} = e_c^{LDA} + e_c^{\Delta GGA} + e_c^{\Delta RLDA} + e_c^{\Delta RGGA} . \tag{41}$$

It is common to employ the nonrelativistic LDA in the parametrization proposed by Vosko et al. [53]. In the following, various forms of the nonrelativistic GGA correlation contribution will be compared [10,11,50,54]. So far, only the relativistic LDA correlation correction has been derived [55], but to our knowledge none is available for the generalized gradient correlation term. However, it should be noted that already the relativistic local density correlation energy is rather small compared to other exchange-correlation contributions.

Table 1. Incremental changes on observables of various diatomic molecules due to applying the RLDA, the GGA, and the RGGA corrections in the exchange-correlation energy functional: bond length d, harmonic vibrational frequency ω_e, and binding energy D_e.

	Δd (Å)			$\Delta\omega_e$ (cm^{-1})			ΔD_e (eV)		
	ΔRLDA[a]	ΔGGA[b]	ΔRGGA[c]	ΔRLDA[a]	ΔGGA[b]	ΔRGGA[c]	ΔRLDA[a]	ΔGGA[b]	ΔRGGA[c]
AuH	0.004	0.010	-0.001	-18	-59	7	-0.044	-0.450	0.016
AuCl	0.006	0.050	0	-3	-35	1	-0.015	-0.612	0.002
Au_2	0.007	0.059	-0.002	-2	-15	0	-0.037	-0.629	0.011
Ag_2	0.004	0.074	-0.002	-1	-25	1	-0.012	-0.560	0.003
Cu_2	0.002	0.058	-0.001	-2	-22	0	-0.014	-0.580	0.005

[a] Relativistic correction to LDA [39,55].
[b] Gradient correction to LDA by Becke and Perdew [9,10].
[c] Relativistic correction to ΔGGA [52].

It has been shown that relativistic corrections to the exchange-correlation energy should be taken into account in order to properly reproduce total atomic energies and atomic core level spectra, although the main relativistic effects are due to relativistic kinematics [45]. To illustrate the impact of relativistic corrections to the exchange-correlation energy functional on molecular observables, their influence on the bond length, the vibrational frequency, and the binding energy of the diatomics Au_2, AuH, AuCl, Ag_2 and Cu_2 is compared in Table 1. For gold compounds, all effects of the various corrections to the LDA on the bond length, the vibrational frequency and the binding energy are small in comparison to those of relativistic kinematics. For instance, the relativistic contraction of the Au_2 bond distance amounts to 0.20 Å at the LDA level [30]. For the examples shown in Table 1, the corrections due to ΔGGA are about one order of magnitude larger than those due to the relativistic corrections of the exchange-correlation functional [56]. Furthermore, it is remarkable that the ΔRGGA correction has almost no effect on molecular observables although it plays an important role in correcting the balance of transversal and longitudinal corrections in atoms [45]. Apparently, relativistic corrections to exchange-correlation potentials only affect the high density regime, but their overall influence is not sufficient to significantly influence the electron density [45]. This implies that these relativistic corrections have only a rather small direct effect on the valence region, hence they hardly affect molecular bonding [56].

3. SOLUTIONS OF THE DIRAC-KOHN-SHAM EQUATIONS

In the previous chapter the four-component DKS equations were derived from QED:

$$h^{(4)}_{DKS}\varphi_i = \epsilon_i\varphi_i \; , \tag{42}$$

with

$$h^{(4)}_{DKS} = h^{(4)}_{D} + v_{eff} \tag{43}$$

as sum of the free-electron Hamiltonian

$$h^{(4)}_{D} = c\vec{\alpha}\vec{p} + (\beta - 1)c^2 \ , \tag{44}$$

and the effective one-electron potential v_{eff}. These equations exhibit the form of a one-particle Dirac equation with the external potential v_{eff}. Hence there exists a continuum of eigensolutions to the DKS equations with negative energy, i. e. the spectrum of $h^{(4)}_{DKS}$ is not bounded from below. Therefore any attempt to directly solve the DKS equations by means of a finite basis set representation would be confronted with serious problems [19]. In contrast to the Schrödinger equation where the variational principle guarantees all energy solutions obtained from a finite basis approach to be upper bounds for the corresponding exact one-particle energies, no such relations exist for the eigensolutions of the DKS operator, not even for those of positive energy [19,20]. Instead, approximate algebraic solutions of the DKS equations tend to be spurious and difficult to relate to the exact solution; no convergence with increasing flexibility of the basis sets is observed [19]. Furthermore, in the nonrelativistic limit, i. e. for an infinite speed of light, the eigensolutions do not approach those of the corresponding Schrödinger equation. In order to avoid these problems which generally are referred to as "variational collapse" of the four-component DKS solutions, one can construct a Hamiltonian which is bounded from below and which reduces to the correct Schrödinger operator in the nonrelativistic limit. Alternatively one may use the Hamiltonian $h^{(4)}_{DKS}$, but take precautions against the "variational collapse". For example, one can employ so-called kinetically balanced basis sets, which take the relationship between large and small components of the Dirac spinors into account and thus explicitly ensure the correct "Schrödinger limit" [19,20]. With this special choice of the basis set, often used in ab initio methods [57–59], the Dirac Hamiltonian is found to be variationally stable for solutions of positive energy [19,20]. Another strategy for a numerically stable procedure utilizes basis sets generated from numerical atomic calculations [60].

In principle, the required transformation from the four-component DKS operator $h^{(4)}_{DKS}$ to the appropriate two-component KS Hamiltonian $h^{(2)}_{+,DKS}$ which exhibits a positive eigenvalue spectrum only, may be achieved by the Foldy-Wouthuysen transformation

$$h^{(2)}_{+,DKS} = U_{FW}\mathcal{L}_{+}h^{(4)}_{DKS}\mathcal{L}_{+}U^{\dagger}_{FW} = B_{+}U_{FW}h^{(4)}_{DKS}U^{\dagger}_{FW}B_{+} \ . \tag{45}$$

However, only approximations to the transformation matrix U_{FW} may be constructed. In the following, we will discuss four strategies for accomplishing this approximate transformation.

3.1. The Hamiltonian with free-particle projectors

The simplest way of deriving a variationally stable Hamiltonian is by restricting the solutions of the DKS equations to the space of the free-electron states with positive energy.

This corresponds to employing the following two-component Hamiltonian:

$$h^{(2)}_{+,FE} = U_0 L_+ h^{(4)}_{DKS} L_+ U_0^{\dagger}\,, \tag{46}$$

The projector onto the free-electron states of positive energy,

$$L_+ = \frac{1}{2}\Big(1 + h^{(4)}_D / E_p\Big)\,, \tag{47}$$

and the free-particle Foldy-Wouthuysen transformation

$$U_0 = A_p(1 + \beta R_p)\,, \tag{48}$$

are related by:

$$B_+ = U_0 L_+ U_0^{\dagger}\,. \tag{49}$$

Furthermore, E_p is the relativistic energy of a free particle with momentum $\vec{p}$,

$$E_p = c\big(p^2 + c^2\big)^{1/2}, \tag{50}$$

and the operator R_p is defined as

$$R_p = K_p(\vec{\sigma}\cdot\vec{p})\,, \tag{51}$$

where $\vec{\sigma}$ designates the vector of the Pauli spin matrices. The kinematic factors A_p and K_p are given by

$$A_p = \big[(c^2 + E_p)/2E_p\big]^{1/2} \tag{52}$$

and

$$K_p = c/\big(E_p + c^2\big). \tag{53}$$

If one works out $h^{(2)}_{+,FE}$ in detail one finally arrives at

$$h^{(2)}_{+,FE} = E_p + E_1 \tag{54}$$

with

$$E_1 = A_p v_{eff} A_p + A_p R_p v_{eff} R_p A_p\,. \tag{55}$$

The operator E_1 is even in the sense that it does not couple the large and the small components of Dirac spinors. The corresponding odd contribution

$$O_1 = \beta A_p\big(R_p v_{eff} - v_{eff} R_p\big) A_p \tag{56}$$

to the KS Hamiltonian

$$h^{(4)}_{FE} = U_0 h^{(4)}_{DKS} U_0^{+} = \beta E_p + E_1 + O_1 \tag{57}$$

vanishes in eq. (46) since it contains only off-diagonal terms between the large and small components which are removed by the projection operators L_+. It was found [28,43] that $h^{(2)}_{+,FE}$ is bounded from below for all elements of chemical interest. However, the eigenvalues of this operator are somewhat lower than those of the four-component Hamiltonian $h^{(4)}_{DKS}$. This is not surprising because the eigenfunctions of positive energy to $h^{(4)}_D$ are, in general, contaminated by eigenfunctions of negative energy to $h^{(4)}_{DKS}$ [34].

3.2. The Pauli and the Zeroth Order Regular Approximated (ZORA) Hamiltonians

An often used approach to constructing a variational stable two-component Hamiltonian is to employ a Talyor series of some suitable parameter to expand the Foldy-Wouthuysen transformed, and thus decoupled Hamiltonian,

$$h_{FW}^{(4)} = U_{FW} h_{DKS}^{(4)} U_{FW}^{\dagger} \, . \tag{58}$$

The expansion is then truncated after a certain order n of the expansion parameter. The usual expansions have the property that the resulting Hamiltonian is even in the sense described in the previous section, except for an odd contribution of order $n+1$. Foldy and Wouthuysen proposed a Taylor series in $(p/c)^2$ which to first order yields [21]

$$h_{FW}^{(4)} = \beta(p^2/2) + v - \beta p^4/8c^2 - p^2 v/8c^2 - i\vec{\sigma}(\vec{p}v \times \vec{p})/4c^2 + O(p^4/c^4) \, . \tag{59}$$

The first two terms of this expansion constitute the Schrödinger Hamiltonian which also furnishes the corresponding nonrelativistic limit for $(p/c)^2 \to 0$. The remaining terms describe relativistic effects. The third term $-\beta p^4/8c^2$ constitutes the well-known correction to the kinetic energy, while the fourth term, the so called Darwin term $-p^2 v/8c^2$, is a correction to the potential. For a Coulomb potential the Darwin term becomes a delta function at the location of the point nucleus. Finally, the fifth term $-i\vec{\sigma}(\vec{p}v \times \vec{p})/4c^2$ describes the spin-orbit interaction which for a Coulomb potential becomes proportional to r^{-3}.

The Schrödinger Hamiltonian together with these three correction terms constitutes the Pauli Hamiltonian. However, the expansion in $(p/c)^2$ is not justified for an attractive Coulomb potential where the electron momentum may become very large in the vicinity of the nuclei [32]. As a consequence, the Pauli Hamiltonian is not bounded from below and thus cannot be used variationally [22]. Nevertheles, it has been successfully employed within a perturbational treatment of relativistic effects [22].

Another procedure to arrive at a Hamiltonian which exhibits the correct "Schrödinger" limit of the DKS operator is based on a Taylor expansion of the operator that arises during the elimination of the small components from the DKS equations (eq. 42) [61]:

$$\bar{X} = \frac{1}{2}\left(1 + (E - v_{eff})/2c^2\right)^{-1} . \tag{60}$$

The expansion parameter $(E - v_{eff})/2c^2$ relates the kinetic energy to twice the electron rest mass $2c^2$ [32]. This expansion is again not valid in the region close to the nucleus where the kinetic energy may dominate the energy eigenvalue E [32]. Lenthe et al. [32] suggest an alternative "regular" expansion in terms of the parameter

$$E/(2c^2 - v_{eff}) \tag{61}$$

which is always small. Therefore, a well behaved Hamiltonian may be expected which in the zeroth order regular approximation (ZORA),

$$H^{ZORA} = v_{eff} + \vec{\sigma}\vec{p}\frac{c^2}{2c^2 - v_{eff}}\vec{\sigma}\vec{p} \, , \tag{62}$$

reproduces all important relativistic effects including the spin-orbit interaction [32,62]. This Hamiltonian is bounded from below for $Z < c$ [32,62] and approaches the relativistic free-particle Hamiltonian for weak potentials. Thus, orbitals of higher angular momentum which "avoid" the singularity of the nuclear Coulomb potential are actually better described by the Pauli Hamiltonian [62]. This deficiency of H^{ZORA} is corrected if the first-order term in $E/\left(2c^2 - v_{eff}\right)$ is taken into account. However, the resulting FORA Hamiltonian (first order regular approximation) is unbounded from below and may thus only be applied perturbationally, like the Pauli Hamiltonian [32,42].

3.3. The Douglas-Kroll transformation

All expansions proposed so far lead to divergent terms, starting from some order on. Another expansion strategy for decoupling large and small components of the Dirac spinors which avoids singularities has been suggested by Douglas and Kroll (DK) [24]. Thereby, one expands $h_{FW}^{(4)}$ in powers the effective potential v_{eff}. This is formally equivalent to an expansion in terms of the Coulomb coupling parameter e^2 [61]. The expansion procedure starts with the free-electron projected Hamiltonian $h_{FE}^{(4)}$ (eq. 57) and applies a unitary transformation U_1 in order to eliminate the odd term O_1 which is of first order in the potential v_{eff}. The resulting Hamiltonian contains an odd term of second order in v_{eff}. The DK ansatz [24]

$$U_1 = \sqrt{1 + W_1^2} + W_1 = W_1 - \frac{1}{2}W_1^2 + \mathcal{O}(W_1^4)\,, \tag{63}$$

where W_1 is an antihermitian operator leads to

$$\begin{aligned} h_{DK,1}^{(4)} = U_1 h_{FE}^{(4)} U_1^\dagger &= \beta E_p + E_1 + O_1 - [\beta E_p, W_1]_- \\ &+ \frac{1}{2}\beta\left(\left[E_p, W_1^2\right]_+ + W_1 E_p W_1\right) \\ &+ [W_1, E_1]_- + [W_1, O_1]_- + O\left(W_1^3\right). \end{aligned} \tag{64}$$

As usual,

$$[\,X,\,Y]_\pm = XY \pm YX \tag{65}$$

denote the commutator and anticommutator of two operators X and Y. To eliminate O_1, the odd first-order contribution in v_{eff}, one now requires

$$[\beta E_p, W_1]_- = O_1 \tag{66}$$

which after transformation into momentum space reads

$$W_{1,p,p'} = \frac{\beta O_{1,p,p'}}{E_p + E_{p'}} = A_p(R_p - R_{p'})A_{p'}\frac{v_{eff,p,p'}}{E_p + E_{p'}}\,. \tag{67}$$

Here, $v_{eff;p,p'}$ is the Fourier transform of the effective potential. The operator W_1 is of first order in v_{eff}/E_p and thus the first-order DK Hamiltonian $h_{DK,1}^{(4)}$ (eq. 64) may be recast as

$$h_{DK,1}^{(4)} = \beta E_p + E_1 + E_2 + O_2 + \mathcal{O}\left(v_{eff}^3\right), \tag{68}$$

with

$$E_2 = -\frac{1}{2}\beta\Big(\big[E_p, W_1^2\big]_+ + 2W_1 E_p W_1\Big) \tag{69}$$

and

$$O_2 = [W_1, E_1]_- \,. \tag{70}$$

The operators E_2 and O_2 are even and odd operators, respectively, of second order in v_{eff}/E_p. Therefore, the first-order DK transformed DKS operator $h^{(4)}_{DK,1}$ is an even operator up to first order in the effective potential. The transformation scheme may be extended to arrive at Hamiltonians $h^{(4)}_{DK,n}$ which are even up to order n:

$$\begin{aligned} h^{(4)}_{DK,n} &= U_{n-1} h^{(4)}_{DK,n-1} U^{\dagger}_{n-1} = \\ &= \beta E_p + E_1 + E_2 + \cdots + E_n + E_{n+1} + O_{n+1} + \mathcal{O}\Big(\big(v_{eff}/E_p\big)^{n+2}\Big). \end{aligned} \tag{71}$$

The unitary transformation

$$U_n = \sqrt{1 + W_n^2} + W_n \tag{72}$$

with

$$W_{n,p,p'} = \frac{\beta O_{n,p,p'}}{E_p + E_{p'}} \tag{73}$$

removes the odd term O_n of order $\big(v_{eff}/E_p\big)^n$ from $h^{(4)}_{DK,n-1}$. The two-component Hamiltonian $h^{(2)}_{DK,n}$ for the positive energy solutions is obtained by considering the large components only,

$$h^{(2)}_{DK,n} = B_+ h^{(4)}_{DK,n} B_+ = E_p + E_1 + \cdots + E_n \tag{74}$$

where terms of order $n+1$ and higher are neglected. If the DK expansion is justified then one may argue that the Hamiltonian $h^{(2)}_{DK,n}$ is bounded from below and thus variationally stable. However, no formal proof of convergence is yet available. Nevertheless calculations with the second-order DK Hamiltonian $h^{(2)}_{DK,2}$ yield numerically stable results [25,30]. Therefore it seems plausible to assume that $h^{(2)}_{DK,2}$ is bounded from below, at least for the elements of interest in chemistry.

4. DENSITY FUNCTIONAL IMPLEMENTATION OF THE DOUGLAS-KROLL-HESS SCHEME

In this chapter we describe how the Douglas-Kroll based relativistic DF scheme is implemented within the LCGTO-DF method [63]. We start with a brief overview of the nonrelativistic LCGTO-DF method.

4.1. The LCGTO-DF method

After introducing a finite basis set χ_k of (real) atom-centered Gaussian-type functions to represent the one-particle KS orbitals

$$\varphi_i = \sum_k \chi_k c_{ki} \tag{75}$$

the KS equations may be cast into matrix form:

$$(F - \varepsilon_i S)c_i = 0\,. \tag{76}$$

F is the corresponding matrix representation of the KS Hamiltonian and S is the overlap matrix of the GTO basis. In principle the density representation

$$\rho(\vec{x}) = \sum_{kl} P_{kl}\chi_k(\vec{x})\chi_l(\vec{x}) \tag{77}$$

has to be used when F is evaluated. Here,

$$P_{kl} = \sum_i^{occ} n_i c_{ki} c_{li} \tag{78}$$

is the density matrix and n_i are the occupation numbers of the molecular orbitals φ_i. The computational effort required for evaluating the electronic Coulomb interaction may be reduced by introducing an auxiliary basis $p(\vec{x})$ for representing the electron density [64]

$$\rho(\vec{x}) \approx \overline{\rho}(\vec{x}) = \sum_p b_p p(\vec{x})\,. \tag{79}$$

It is important to note that the coefficients b_p are determined via a variational principle [63,64]. Then, in the spin unrestricted version of the method, the KS Hamiltonian is

$$F^\sigma = T + N + \sum_p b_p \int d^3x' \frac{p(\vec{x}')}{|\vec{x} - \vec{x}'|} + F^{xc,\sigma}. \tag{80}$$

Here, the matrices T and N represent the kinetic energy and the nuclear potential, respectively. The elements of the exchange-correlation energy matrix may be written as [65,66]

$$F_{kl}^{xc,\sigma} = \int d^3x \left\{ \frac{\partial e_{xc}}{\partial \rho_\sigma}\chi_k\chi_l + \left[2\frac{\partial e_{xc}}{\partial \gamma_{\sigma\sigma}}\nabla\rho_\sigma + \frac{\partial e_{xc}}{\partial \gamma_{\sigma\sigma'}}\nabla\rho_{\sigma'} \right] \cdot \nabla(\chi_k\chi_l) \right\}. \tag{81}$$

The exchange-correlation energy density e_{xc} (eq. 31) is a function of the electron density ρ and of the electron density gradient invariants $\gamma_{\sigma\sigma'} = \nabla\rho_\sigma \cdot \nabla\rho_{\sigma'}$ where $\sigma, \sigma' \in \{\alpha, \beta\}$. After integration by parts this expression can be rearranged to a form where the exchange-correlation potential acts on the basis functions. The resulting integrals have to be evaluated by a numerical quadrature scheme [67,68]. Alternatively, a further auxiliary basis may be introduced for fitting the exchange-correlation energy density [64,69,70].

4.2. An efficient Douglas-Kroll Kohn-Sham Hamiltonian for molecular calculations

Next we discuss the implementation of the second-order Douglas-Kroll Kohn-Sham Hamiltonian $h^{(2)}_{DK;2}$ in the LCGTO-DF method. In order to obtain a numerically efficient procedure one has to introduce further approximations. First the DK decoupling transformation is restricted to terms connected to the nuclear potential. In a second step the spin-orbit interaction will be neglected, resulting in a scalar relativistic variant of the method. We will illustrate the accuracy of these approximations by various example calculations.

The second-order DK KS Hamiltonian reads

$$h^{(2)}_{DK;2} = E_p + A_p v_{eff} A_p + A_p R_p v_{eff} R_p A_p \\ - \frac{1}{2}\left(E_p W^2_{eff} + W^2_{eff} E_p + 2 W_{eff} E_p W_{eff}\right). \tag{82}$$

The evaluation of these terms is quite involved since v_{eff} and therefore $W_{eff} = W_1$ change during the selfconsistency process. Therefore, a large part of the Hamiltonian must be recalculated from cycle to cycle. A substantial saving is achieved if the DK transformation is applied only to the singular part of the potential, i. e. to the nuclear potential v_{nuc}, and if only the nuclear potential is taken into account when constructing the DK transformation. This approximation leads to

$$h^{(2)}_{DKnuc;2} = E_p + A_p v_{nuc} A_p + A_p R_p v_{nuc} R_p A_p + v_{ee} \\ - \frac{1}{2}\left(E_p W^2_{nuc} + W^2_{nuc} E_p + 2 W_{nuc} E_p W_{nuc}\right), \tag{83}$$

where v_{ee} represents the untransformed electron-electron interaction, comprising both the classical Coulomb as well as the exchange-correlation contributions. One formally obtains the relativistic KS Hamiltoninan $h^{(2)}_{DKnuc;2}$ if one replaces the kinetic energy in the nonrelativistic KS Hamiltonian by the relativistic kinetic energy

$${}^{r}T = E_p - c^2 \tag{84}$$

and the nuclear potential by the effective nuclear potential

$$v^{eff}_{nuc}(\vec{x}) = A_p[v_{nuc} + R_p v_{nuc} R_p] A_p \\ - \frac{1}{2}\left(E_p W^2_{nuc} + W^2_{nuc} E_p + 2 W_{nuc} E_p W_{nuc}\right) =: {}^{r}N. \tag{85}$$

In this approximation, only the integral part of the LCGTO-DF program had to be modified.

We still have a two-component formalism which includes the spin-orbit interaction. This becomes obvious if one expands the terms in eq. (83) with the help of the Pauli relation for the spin matrices

$$(\vec{\sigma}\vec{p})v(\vec{\sigma}\vec{p}) = \vec{p}v \cdot \vec{p} + i\vec{\sigma}(\vec{p}v \times \vec{p}), \tag{86}$$

for example the third term reads

$$A_p R_p v R_p A_p = A_p K_p (\vec{p} \cdot v\vec{p}) K_p A_p + A_p K_p i\vec{\sigma}(\vec{p}v \times \vec{p}) K_p A_p\,. \tag{87}$$

All terms which couple the two spin components depend on $(\vec{p}v \times \vec{p})$. In the following we shall restrict the discussion to the scalar relativistic formalism which is obtained by neglecting these coupling terms [25]. However, even in such a scalar relativistic version some of these rather costly terms of the type $(\vec{p}v \times \vec{p})$ have to be evaluated as they occur in scalar products of the second-order DK constributions. The approximation where such vector product terms are dropped alltogether would thus be very convenient. We shall denote these consecutive scalar relativistic approximations to the second-order DK transformed KS Hamiltonians by $h^{(2)}_{DKnuc;2}$ (transformation restricted to the nuclear potential) and by $\bar{h}^{(2)}_{DKnuc;2}$ (all vector product terms dropped in addition).

First, we shall illustrate the accuracy of the nuclear potential approximation for a DF calculation on the cerium atom in the configuration $6s^2 5d^1 4f^1$. Table 2 provides the total energies and the one-electron energies of the KS orbitals. Later on, we shall demonstrate that dropping all vector products is a very good approximation when calculating molecular observables (Table 3).

In the first two columns of Table 2 the nonrelativistic and scalar relativistic results obtained by a numerical procedure [29] are compared. The next two columns of Table 2 show results for the scalar relativistic variants of the full second-order DK Hamiltonian $h^{(2)}_{DK;2}$ and of the DK Hamiltonian where the DK transformation is restricted to the nuclear potential, $h^{(2)}_{DKnuc;2}$. In the last column the results are displayed for the scalar relativistic KS Hamiltonian with free-particle projectors

$$h^{(2)}_{FEnuc} = E_p + A_p v_{nuc} A_p + A_p R_p v_{nuc} R_p A_p \tag{88}$$

which is obtained by a first-order DK decoupling procedure (eq. 54), the transformation again restricted to terms connected to the nuclear potential.

Inspection of Table 2 reveals that the method based on free-particle projected Hamiltonian $h^{(2)}_{FEnuc}$ overestimates relativistic effects [27,43,71]. The decrease of the total energy is overestimated by about 25%. The relativistic shifts of the one-electron energies, to lower values for s orbitals and to higher energies for d orbitals, are overestimated by about 15% and 5%, respectively. The downward shift of the p orbitals is underestimated by about 10%. For the 1s orbital the relativistic shift is calculated even about 40% too large. Relativistic shifts of s orbitals are not well reproduced due to admixtures of negative energy eigenstates of the free-particle projected Hamiltonian $h^{(2)}_{FEnuc}$. Apparently, the screening of the p levels due to the s orbitals is overestimated, reducing the relativistic downward shift.

The Hamiltonians based on second-order DK transformation provide an obvious improvement of the results: all s orbitals now lie above the values obtained in the numerical procedure. This indicates that the second-order DK transformation reduces the admixture of negative energy solutions which is typical for the free-particle projected first-order Hamiltonian [34]. For s levels the largest error in the relativistic shifts is about 5% (for the 1s orbital). Due to the improved representation of the "inner" orbitals, the "outer" shells are now also much better described. With the exception of the 5p and 5d orbitals which exhibit only very small shifts, the errors in the relativistic shifts of the one-electron energies are smaller than about 2 %.

Table 2. Calculated orbital energies (ϵ, in eV) and total energies E_{tot} (in atomic units) of cerium in the configuration $6s^2 5d^1 4f^1$ [29].

	Nonrelativistic	Relativistic			
Orbital	Numerical	Numerical	$h^{(2)}_{DK;2}$	$h^{(2)}_{DKnuc;2}$	$h^{(2)}_{FEnuc}$
1s	-38286.20	-40219.13	-40186.16	-40124.29	-41010.20
2s	-5956.84	-6441.93	-6438.79	-6431.44	-6522.20
3s	-1281.39	-1387.22	-1386.56	-1385.25	-1403.06
4s	-256.50	-279.90	-279.78	-279.53	-283.51
5s	-36.54	-39.99	-40.01	-39.98	-40.61
6s	-2.99	-3.22	-3.23	-3.22	-2.26
2p	-5637.49	-5771.19	-5773.68	-5768.52	-5769.40
3p	-1142.79	-1174.05	-1174.55	-1173.76	-1173.52
4p	-204.75	-210.26	-210.38	-210.26	-210.07
5p	-22.38	-22.54	-22.61	-22.60	-22.51
3d	-883.62	-871.13	-871.10	-870.88	-869.51
4d	-113.76	-110.05	-110.07	-110.08	-109.7
5d	-3.08	-2.44	-2.45	-2.46	-2.39
4f	-8.58	-5.05	-5.09	-5.14	-4.92
E_{tot}	-8567.30	-8858.14	-8855.29	-8851.72	-8929.79

The agreement is even better for the relativistic shift of the total energy. The errors of corresponding relativistic shifts are 2.2% and 1.0% for methods based on $h^{(2)}_{DKnuc;2}$ and $h^{(2)}_{DK;2}$, respectively. Both total energy values lie above the numerical result. Thus the results of both second-order DK KS Hamiltonians $h^{(2)}_{DKnuc;2}$ and $h^{(2)}_{DK;2}$ are very similar. Nevertheless, the more sophisticated Hamiltonian $h^{(2)}_{DK;2}$ leads to an improved description of the one-electron levels, in particular for the s orbitals. Since the approximate KS Hamiltonian $h^{(2)}_{DKnuc;2}$ reproduces the overall relativistic changes to better than 96% for one-electron energies and to better than 98% for the total energy, the strategy used in many calculations [25,72] to neglect the relativistic corrections which are due to the electron-electron interaction is justified, especially if highly accurate results are not required.

We now discuss the implications of the relativistic KS Hamiltonians on the structure of molecules. As an example we investigate the gold diatomics Au_2, AuCl and AuH [30]. In Table 3 the spectroscopic constants and dipole moments for the KS Hamiltonians $h^{(2)}_{FEnuc}$,

Table 3. Bond length d_e, vibrational frequency ω_e, binding energy D_e, and dipole moment μ of some gold diatomics as calculated by the scalar relativistic LCGTO-LDF method in different approximations: nonrelativistic (nrel.) Hamiltonian as well as relativistic Hamiltonians $h^{(2)}_{FEnuc}$, $h^{(2)}_{DKnuc;2}$, and $\bar{h}^{(2)}_{DKnuc;2}$ (see text).

	Method	d_e (Å)	ω_e (cm^{-1})	D_e (eV)	μ (D)
AuH	nrel.	1.653	1749	2.59	2.577
	$h^{(2)}_{FEnuc}$	1.491	2722	4.00	0.503
	$\bar{h}^{(2)}_{DKnuc;2}$	1.505	2621	3.76	1.160
	$h^{(2)}_{DKnuc;2}$	1.505	2619	3.75	1.155
	exp.[a]	1.524	2305	3.22	–
AuCl	nrel.	2.403	291	3.56	5.341
	$h^{(2)}_{FEnuc}$	2.181	382	3.64	1.264
	$\bar{h}^{(2)}_{DKnuc;2}$	2.206	362	3.57	2.768
	$h^{(2)}_{DKnuc;2}$	2.207	362	3.56	2.772
	exp.[a]	–	383	3.52	–
Au_2	nrel.	2.697	135	2.02	–
	$h^{(2)}_{FEnuc}$	2.429	206	3.10	–
	$\bar{h}^{(2)}_{DKnuc;2}$	2.465	195	2.89	–
	$h^{(2)}_{DKnuc;2}$	2.466	195	2.88	–
	exp.[a]	2.472	191	2.30	–

[a] Experimental data from ref. 73.

$h^{(2)}_{DKnuc;2}$, and the modified Hamiltonian $\bar{h}^{(2)}_{DKnuc;2}$ are compared with each other and to experiment.

Even the approximate KS Hamiltonian $h^{(2)}_{FEnuc}$ reproduces the spectroscopic constants quite satisfactorily. Nevertheless, the quality of the results obtained with the second-order DK Hamiltonians is obvious. Clearly, omission of the costly matrix elements of the operator $(\vec{p}v \times \vec{p})$ leads only to negligible differences. These examples provide support for employing the KS Hamiltonian $\bar{h}^{(2)}_{DKnuc;2}$ in molecular applications. Therefore, this KS Hamiltonian will be used exclusively in the following discussions and in all applications reported in chaper 5.

4.3. Evaluation of the relativistic matrix elements in momentum space – The Douglas-Kroll-Hess scheme

The DK transformation introduces new operators like R_p, A_p, K_p, and W which are naturally defined in momentum space. These operators enter into the KS Hamiltonian $h^{(2)}_{DKnuc;2}$ and it is thus convenient to evaluate its matrix elements and those of related relativistic operators, like the relativistic kinetic energy ${}^{r}T$ and the nuclear potential ${}^{r}N$, in the momentum space.

Since we do not want to complicate the computation by introducing the plane wave basis of the momentum space we follow a strategy originally proposed by Hess et al. [27,28]. Thereby, the space spanned by the eigenfunctions of the nonrelativistic kinetic energy matrix T is interpreted as an approximation to the momentum space. Note that this technique requires a basis set which is very flexible; in particular, this condition implies that the basis functions of the "momentum space" have to be obtained from the uncontracted orbital basis functions [25,30].

As an example, we describe how the matrix elements of the relativistic kinetic energy ${}^{r}T$ are calculated. First the orbital basis is orthonormalized, e. g. by the canonical orthogonalization

$$Q = Us^{-1/2}. \tag{89}$$

We denote diagonal matrices by lower case letters. The unitary (orthogonal) matrix U diagonalizes the overlap matrix S

$$s = U^{\dagger}SU. \tag{90}$$

The corresponding basis sets $\mathcal{Q}$ and $\mathcal{U}$ lead to the following two matrix representations of the kinetic energy:

$$T_{\mathcal{U}} = U^{\dagger}TU, \quad T_{\mathcal{Q}} = s^{-1/2}T_{\mathcal{U}}s^{-1/2}. \tag{91}$$

The matrix which diagonalizes $T_{\mathcal{Q}}$ defines the transformation V to momentum space:

$$t_p = V^{\dagger}T_{\mathcal{Q}}V. \tag{92}$$

Now we construct the matrix of the relativistic kinetic energy ${}^{r}T$ by evaluating $\kappa(p^2) = E_p - c^2$ for the diagonal elements of the matrix t_p, followed by the transformations back to coordinate space:

$${}^{r}T_{\mathcal{Q}} = V\kappa(2t_p)\,V^{\dagger}, \quad {}^{r}T_{\mathcal{U}} = s^{1/2}\,{}^{r}T_{\mathcal{Q}}s^{1/2}, \quad {}^{r}T = U\,{}^{r}T_{\mathcal{U}}U^{\dagger}. \tag{93}$$

The matrices of the effective (Darwin-type corrected) nuclear potential ${}^{r}N$ are calculated along similar lines; however, the procedure is much more involved.

4.4. Relativistic energy gradients

In this section we present the basic concepts that allow the evaluation of analytic energy gradients for DK Hamiltonians. These gradients are essential for a geometry optimization of molecular systems exhibiting many geometrical degrees of freedom. Crucial for the evaluation of gradients of the relativistic total energy in the DK approach to the DKS problem is the observation that in the approximation used so far in all molecular applications the nonrelativistic operators of the kinetic energy and of the nuclear potential are substituted by their relativistic equivalents, as described in section 4.2. Therefore, most of the computational procedure may follow the strategies employed for obtaining the analytic gradients in the framework of the nonrelativistic LCGTO-DF method; a more detailed discussion of energy gradients for DF methods may be found elsewhere [63,65,66,74,75].

The derivative of the nonrelativistic total energy E_{tot} with respect to a displacement of nucleus a is given by [36,65,66]

$$\begin{aligned}(E_{tot})^a := \nabla_a E_{tot} = E^a_{nuc} &+ \sum_{kl} P_{kl} T^a_{kl} + \sum_{kl} P_{kl} N^a_{kl} + \sum_p b_p \int d^3x d^3x' p^a(\vec{x}')(\rho - \overline{\rho})(\vec{x}) \\ &+ \sum_{kl} P_{kl} \sum_p b_p \int d^3x d^3x' (\chi_k(\vec{x}')\chi_l(\vec{x}'))^a (\rho - \overline{\rho})(\vec{x}) \\ &- \sum_{kl} \overline{P}_{kl} S^a_{kl} + \int d^3x g_a^{xc,\alpha}(\vec{x}) + \int d^3x g_a^{xc,\beta}(\vec{x}) \,.\end{aligned} \tag{94}$$

Here, $\overline{P}$ is the one-electron eigenvalue weighted density matrix:

$$\overline{P}_{kl} = \sum_i^{occ} n_i \varepsilon_i c_{ki} c_{li}. \tag{95}$$

A gradient with respect to the position of nucleus *a* is denoted by an upper index. Furthermore, the shorthand notation

$$g_a^{xc,\sigma} = -2 \sum_k{}^a \sum_l P^\sigma_{kl} \left\{ \frac{\partial e_{xc}}{\partial \rho_\sigma} \chi_l \nabla \chi_k + X_{kl} \left(2 \frac{\partial e_{xc}}{\partial \gamma_{\sigma\sigma}} \nabla \rho_\sigma + \frac{\partial e_{xc}}{\partial \gamma_{\sigma\sigma'}} \nabla \rho_{\sigma'} \right) \right\} \tag{96}$$

for the gradient of the exchange-correlation energy kernel is used (see eq. 81) where the upper index *a* of the sum denotes a restriction of the summation over *k* to basis functions centered on atom *a*. The quantity X_{kl} is defined as

$$X_{kl} = \chi_l \nabla \left(\nabla^\dagger \chi_k \right) + (\nabla \chi_k) \left(\nabla^\dagger \chi_l \right). \tag{97}$$

Consistent analytical gradients, i. e. derivatives of the true energy hypersurface, are only obtained if the approximation errors that arise when the analytical integration is carried out via a numerical quadrature scheme [13,66,76] or via fitting of the exchange-correlation energy density [69,70] are taken into account.

As indicated above, one obtains the analytical gradients of the relativistic total energy in the scalar relativistic variant of the LCGTO-DF method if one substitutes the kinetic energy and the nuclear potental matrix elements in eq. (94) by their relativistic counterparts. We will illustrate the procedure by evaluating the nuclear displacement derivative of the relativistic kinetic energy ${}^{r}T$ [36]. Thereby, one has to take into account the matrix technique connecting quantities in the coordinate and in the approximate momentum space (eqs. 91, 92 and 93).

To obtain the derivative ${}^{r}T^{a}$ with respect to nucleus a we observe that

$$T_{\mathcal{U}}^{a} = U^{\dagger}T^{a}U - [P_{\mathcal{U}}, T_{\mathcal{U}}]_{-} \,. \tag{98}$$

The real antisymmetric matrix

$$P_{\mathcal{U}} = U^{\dagger}U^{a} = -U^{\dagger a}U = -P_{U}^{\dagger} \tag{99}$$

is determined by differentiating eq. (90),

$$s^{a} = U^{\dagger}S^{a}U - [P_{\mathcal{U}}, s]_{-} \,. \tag{100}$$

Then the matrix elements of $P_{\mathcal{U}}$ may be read off from off-diagonal elements:

$$0 = \left(U^{\dagger}S^{a}U\right)_{kl} - (P_{\mathcal{U}})_{kl}\,(s_{k} - s_{l}). \tag{101}$$

For the second transformation in eq. (91) one has

$$T_{\mathcal{Q}}^{a} = s^{-1/2}T_{\mathcal{U}}^{a}s^{-1/2} + \left[-\tfrac{1}{2}s^{-1}s^{a}, T_{\mathcal{Q}}\right]_{+} \tag{102}$$

and the third step yields in analogy to eq. (100):

$$t_{p}^{a} = V^{\dagger}T_{\mathcal{Q}}^{a}V - [P_{\mathcal{V}}, t_{p}]_{-} \,. \tag{103}$$

Using this equation, the matrix

$$P_{\mathcal{V}} = V^{\dagger}V^{a} = -V^{\dagger a}V = P_{\mathcal{V}}^{\dagger} \tag{104}$$

is constructed elementwise, similarly to $P_{\mathcal{U}}$. To continue the evaluation of ${}^{r}T^{a}$, one starts a series of backward transformations:

$$ {}^{r}T_{\mathcal{Q}}^{a} = V\kappa'(2t_{p})\left(2t_{p}^{a}\right)V^{\dagger} + \left[VP_{\mathcal{V}}V^{\dagger}, {}^{r}T_{\mathcal{Q}}\right]_{-}, \tag{105}$$

$$ {}^{r}T_{\mathcal{U}}^{a} = s^{1/2}\,{}^{r}T_{\mathcal{Q}}^{a}s^{1/2} + \left[\tfrac{1}{2}s^{-1}s^{a}, {}^{r}T_{\mathcal{U}}\right]_{+}, \tag{106}$$

and

$$ {}^{r}T^{a} = U\,{}^{r}T_{\mathcal{U}}^{a}U^{\dagger} + \left[UP_{\mathcal{U}}U^{\dagger}, {}^{r}T\right]_{-} \,. \tag{107}$$

The displacement derivatives of the various kinematic factors and of the relativistically corrected nuclear potential ${}^{r}N$ are calculated in similar fashion.

Table 4. Comparison of various observables of gold containing diatomic molecules: bond lengths d_e, harmonic vibrational frequencies ω_e, and the binding energies D_e as calculated by DF methods using the generalized gradient approximation (GGA) and by post-HF methods.

Molecule	Method	d_e (Å)	ω_e (cm^{-1})	D_e (eV)
Au_2	DKH BLYP[a]	2.549	172	2.22
	Pauli BP[b]	2.556	175	2.19
	ZORA BP[c]	2.517	174	2.26
	DKH BP[d]	2.515	180	2.29
	DKH PW[e]	2.513	180	2.54
	ARPR-CEPA[f]	2.54	170	1.86
	RPP-QCI(T)[g]	2.54	178	21.3
	exp.[h]	2.472	191	2.31
AuH	DKH BLYP[a]	1.547	2230	3.22
	ZORA BP[c]	1.539	2290	3.33
	DKH BP[d]	1.539	2270	3.33
	DKH PW[e]	1.538	2279	3.31
	DKH CI[i]	1.52	2381	3.34
	ARPR-CEPA[f]	1.51	2254	3.12
	DKH CC[j]	1.525	2288	2.92
	exp.[h]	1.524	2305	3.21
AuCl	DKH BLYP[a]	2.260	343	2.85
	DKH BP[d]	2.230	366	3.06
	DKH PW[e]	2.227	368	3.24
	ARPR-CEPA[f]	2.28	383	2.52
	exp.		383[h]	2.82[k], 3.54[h]

[a] Becke-Lee-Yang-Parr GGA DF calculations with the DKH scalar relativistic (SR) Kohn-Sham Hamiltonian, [36]. [b] Pauli SR DF method [31].
[c] Zero order regular approximation SR DF method [31].
[d] Becke-Perdew GGA DF calculations with the DKH SR Hamiltonian, [36].
[e] Perdew-Wang GGA DF calculations with the DKH SR Hamiltonian, [36].
[f] Spin-orbit averaged relativistic pseudopotential method with the coupled electron pair approximation [77].
[g] Relativistic pseudopotential-based quadratic configuration interaction method with single and double substitutions as well as some triple contributions [78]. [h] Ref. [73].
[i] SR HF configuration interaction method with a spin-free no-pair Hamiltonian [79].
[j] Ref. [80]. [k] SR HF coupled–cluster method with a spin-free no-pair Hamiltonian [81].

4.5. Results of benchmark calculations on diatomics in the GGA to DF theory

We conclude the methodological sections of this review with a comparison of the second-order DKH scheme to other DF and HF methods. We will consider the gold diatomics Au_2, AuCl and AuH which, due to the strong influence of relativistic effects on their electronic and geometric structure, are well suited as benchmark systems [1,31,60,77,78,82–84]. The DKH LCGTO-DF calculations were performed at the GGA level, utilizing the following exchange-correlation functionals [36]: Becke-Perdew (BP) [9,10,85], Perdew-Wang (PW) [12] (scheme IIa), and Becke-Lee-Yang-Parr (BLYP) [9,86].

The results are shown in Table 4. The LCGTO-DF results in combination with all three GGA variants are very similar. The bond distances do not exibit a strong dependence on the GGA variant. For Au_2 and AuCl the binding energies for the different GGA approaches vary by about 5%, where the trend $D_e(\mathrm{BLYP}) < D_e(\mathrm{BP}) < D_e(\mathrm{PW})$ can be observed in full agreement with a larger set of benchmark results [36]. For AuH the binding energy is essentially the same for all the three GG approximations.

For Au_2 the present DKH scheme is compared to the alternative relativistic treatments based on the ZORA [31,32] and on the Pauli KS Hamiltonian [22]. Recall that in order to obtain a variationally stable procedure the Pauli Hamiltonian has been applied in combination with "frozen" atomic cores and only interacting nonrelativistic valence orbitals are considered [22]. Apparently, the variational DKH and ZORA approaches are very close to each other and both agree better with experiment than the Pauli scheme. The comparison of the DKH results with the post-HF methods also shows very good agreement.

5. APPLICATIONS

In the following sections some applications of the RLCGTO-DF method to various fields of the chemistry of heavy elements are discussed. For most of the applications presented the need of a relativistic approach is obvious, for example for gold compounds. For transition metal elements of the second row relativistic effects are normally small, but often not negligible, especially if strong bonds are examined as will be shown for the Mo-N triple bond. In many cases special attention will be paid to the effects of relativity by comparison to the corresponding nonrelativistic calculations.

5.1. Metal Clusters

5.1.1. Gold clusters

Clusters represent an intermediate state of matter which smoothly bridges the more familiar regimes of molecules and solids [87,88]. While clusters consisting of a few atoms exhibit properties which depend on the details of the geometrical and electronic structure, a more uniform behavior has to be expected for particles comprising hundreds of atoms. The question of how the properties of clusters evolve from molecular to bulklike behavior is of considerable current interest [89–91]. It has been studied for gold by means of the scalar relativistic

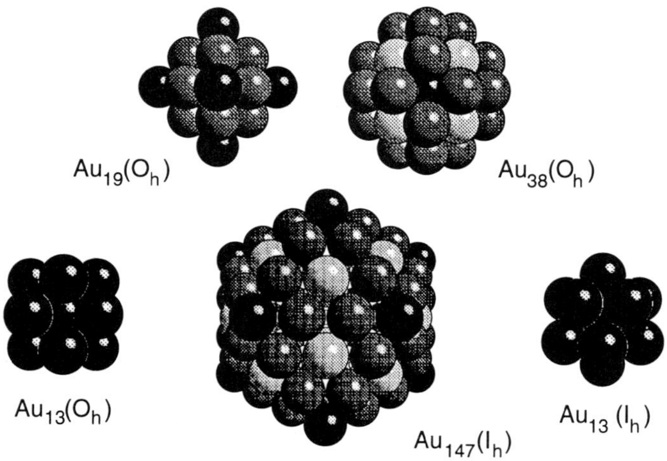

Figure 3. Examples of highly symmetric gold clusters (point group labels are given in parentheses). Atoms are shaded according to their position: inner and corner atoms black, edge atoms dark gray, face atoms light gray.

LCGTO-DF approach. The geometric and electronic structure of a series of highly symmetric gold clusters has recently been examined [92]. Six clusters of octahedral (O_h) symmetry (Au_n, n = 6, 13, 19, 38, 44, 55), representing sections of the fcc lattice of bulk gold, and three clusters of icosahedral (I_h) symmetry (Au_n, n = 13, 55, 147) have been chosen; examples for the different structures are shown in Figure 3. The largest cluster of this series, Au_{147}, contains 11613 electrons which have been described by 10584 contracted Gaussian-type basis functions. A relativistic all-electron first-principles calculation of this size should be among the largest quantum chemical calculations performed up to now at this level of accuracy. The feasibility of such a calculation clearly exemplifies the effectiveness of the RLCGTO-DF method.

The cluster series examined spans a range of particle diameters between 0.7 to 1.7 nm and comprises species of one to three atomic shells (Fig. 3). For example, the cluster Au_{147} consists of a single central atom, a first shell of 12 atoms, a second shell of 42 and a surface layer of 92 atoms. For all clusters the average Au-Au distance has been determined by optimizing the totally symmetric breathing mode. In case of the icosahedral clusters, the ideal geometry has been preserved where the inter-shell bond distance is by 5% shorter than the intra-shell distance [93]. As expected, the interatomic distances increase with growing cluster size (Fig. 4). The values optimized at the LDA level of theory lie between the bond distance of the diatomic molecule Au_2 (2.47 Å [73]) and the experimental Au-Au distance of bulk gold (2.88 Å [94]). The average Au-Au distance correlates almost linearly with the average coordination number of the cluster atoms (Fig. 4). An extrapolation to the bulk coordination number of 12 leads to a bond length of 2.89 Å if the local density approach

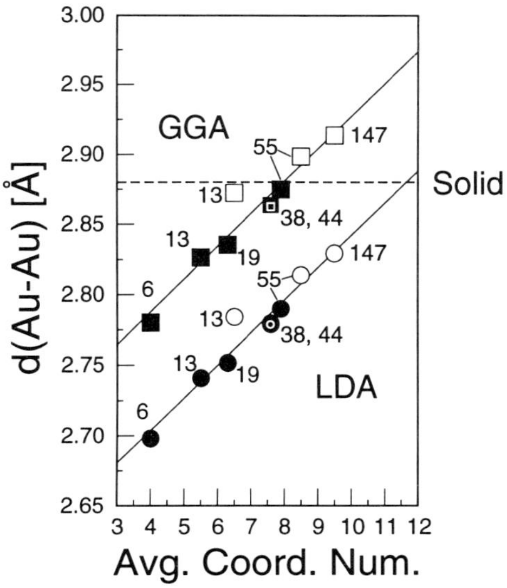

Figure 4. Optimized bond lengths as a function of the average coordination for the cluster series Au_n, n = 6 – 147. Results obtained with a local density (LDA, circles) as well as with a gradient corrected (GGA, squares) exchange-correlation functional are shown. Filled symbols represent clusters of O_h symmetry, open symbols clusters of I_h symmetry. The symbols for the clusters Au_{38} (O_h) and Au_{44} (O_h) are superimposed.

(LDA) is employed, in very good agreement with the experimental value. The generalized gradient form of the exchange-correlation potential, used here in the parametrization according to Becke and Perdew [9,10], results in longer bond distances, in line with lower binding energies compared to LDA (see below). Since the same linear trend is observed as a function of the average coordination, the extrapolated bulk Au-Au distance of 2.97 Å exceeds the experimental value.

While geometries and other spectroscopic parameters are accurately represented by the LDA approach, binding energies are often overestimated [8]. More reliable energies are calculated using gradient corrected exchange-correlation functionals (GGA) [9,10]. The calculated binding energy per atom E_b/n increases with growing cluster size n over the whole gold cluster series (Fig. 5). In LDA, the bulk cohesive energy per atom (3.80 eV [94]) is already reached for Au_{147} while the corresponding GGA binding energy values are considerably lower. Gold clusters are mainly bound due to the delocalized 6s electrons with some 5d contributions. Thus the increasing values of E_b/n may be rationalized by a decreasing surface to volume ratio S/V in the clusters. Indeed the corresponding linear relationship between E_b/n and S/V which is proportional to $n^{-1/3}$

$$E_b/n = E_{coh} + n^{-1/3} \Delta E_{surf} \tag{108}$$

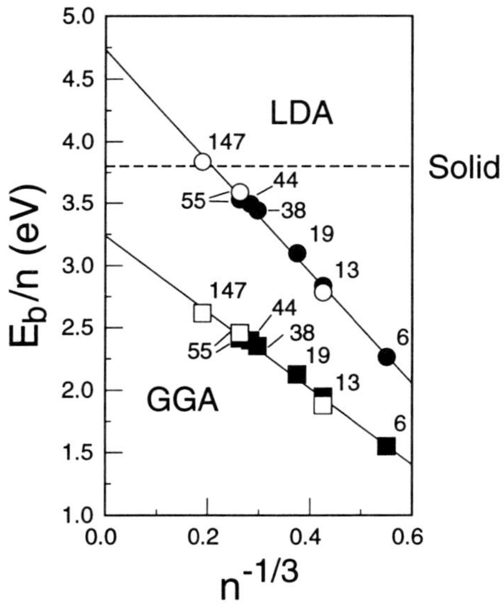

Figure 5. Binding energies per atom E_b/n as a function of $n^{-1/3}$ for the cluster series Au_n, n = 6 – 147. Results obtained with a local density (LDA, circles) as well as a gradient corrected (GGA, squares) exchange-correlation functional are shown. Filled symbols represent clusters of O_h symmetry, open symbols clusters of I_h symmetry.

is very well reproduced by the calculations (Fig. 5) [92]. Extrapolation of the binding energies toward the bulk ($n^{-1/3} = 0$) yields a value of 4.73 eV for LDA, overestimating the experimental cohesive energy by about 1 eV. The more accurate GGA data lead to an extrapolated cohesive energy of 3.24 eV which underestimates the experimental result by about 0.5 eV. If one compares clusters of the same size, but of different symmetries (Au_{13}, Au_{55}, see Fig. 5) one finds only small binding energy differences of less than 0.1 eV per atom. The cluster Au_{13} is slightly more stable in icosahedral symmetry, while for Au_{55} the octahedral geometry is preferred. Thus, the present calculations for gold clusters do not yield an energetic preference for either symmetry in the examined range of particle size.

The comparison of the results of accurate all-electron calculations with simple models for metallic particles, like the spherical droplet model [95], provides some insight into the question of metallic behavior of clusters with growing size. According to the droplet model, the successive ionization potentials IP_n and electron affinities EA_n of a metal particle of radius R are determined by the work function W_∞ of the bulk metal and the classical Coulomb energy needed to add or remove n electrons:

$$
\begin{aligned}
IP_n(R) &= W_\infty + \alpha\frac{e}{R} + (n-1)\frac{e}{R} \\
EA_n(R) &= W_\infty - \beta\frac{e}{R} - (n-1)\frac{e}{R}
\end{aligned}
\tag{109}
$$

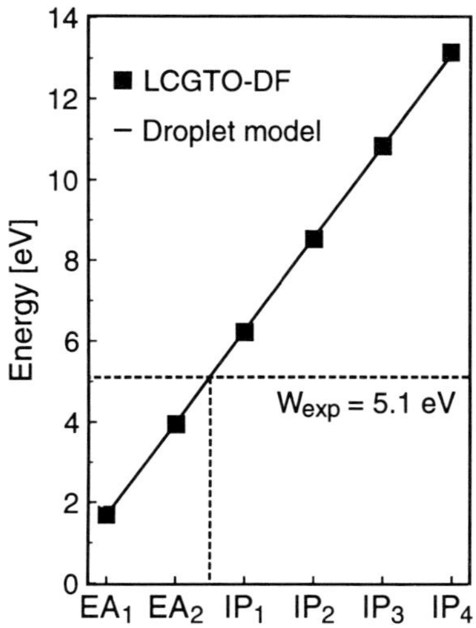

Figure 6. Successive vertical ionization potentials and electron affinities of $Au_{55}(I_h)$ in comparison to the droplet model of a classical metal particle of the same size (radius 6.29 Å).

The coefficients α and β are constrained to $\alpha + \beta = 1$. While the theoretical determination of these coefficients is still a subject of discussion [95–99], the values of α = 3/8 [100] and α = 1/2 [101,102] have been found to compare favorably with experimental data for alkali and noble metal clusters, respectively.

In Figure 6 the calculated successive ionization potentials and electron affinities of the icosahedral cluster Au_{55} are compared to the droplet model [103]. As model parameters the bulk work function of gold, W_∞ = 5.1 eV [104], and the calculated effective radius of Au_{55}, R = 6.29 Å, have been chosen. The perfect agreement of the RLCGTO-DF data to the droplet model may be to some extent fortuitous since it depends on the exact values of α and R. On the other hand, the linearity of the calculated results shows that Au_{55} acts as a perfect electron reservoir due to its delocalized s-derived level manifold. In this regard, the cluster Au_{55} exhibits the characteristic behavior of a metallic particle.

The effect of relativity on the one-electron level spectra of gold clusters may also be illustrated for $Au_{55}(I_h)$ [103,105]. As gold is a monovalent metal ($5d^{10}6s^1$), some similarity to the spectra of alkaline metal clusters which are well described by the jellium model [106,107] may be expected. On the other hand, the d orbitals of gold will also be present in the cluster valence spectra and may contribute to the bonding [108,109]. Recall that relativity leads to a radial contraction and energetic stabilization of s orbitals and to a destabilization of d orbitals [1]. In the case of gold, this also improves the possibility for 5d-6s hybridization and thus may, at least to some extent, lead to a modification of a pure s metal description. In Figure 7, the Kohn-Sham level spectrum of a spherical jellium model for Au_{55}, filled by 55 valence s-type electrons, is compared to the nonrelativistic and the relativistic valence spectra of all-electron LCGTO-DF calculations of Au_{55}. The classical ordering of the jellium

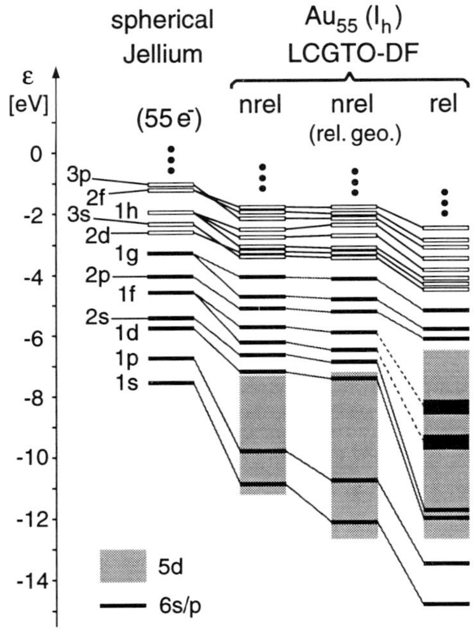

Figure 7. Relativistic effects on the electronic structure of $Au_{55}(I_h)$ and comparison with the spherical jellium model. Compared are, from left to right, the Kohn-Sham one-electron valence spectra of the nonrelativistic jellium model, the nonrelativistic all-electron cluster calculation at the equilibrium geometry, the nonrelativistic cluster calculation at the relativistic geometry, and the scalar relativistic calculation.

levels (1s < 1p < 1d < 2s < 1f < 2p < 1g < 2d < ...) is obtained, with the 1g level as the cluster HOMO. Comparison with the cluster spectrum from a nonrelativistic LCGTO-DF calculation reveals that the ordering of occupied s-type levels is preserved although some of the levels are split due to lower symmetry (I_h). A clear overall stabilization of the orbital energies is observed, in particular for the two lowest 6s-derived orbitals at the low-energy end of the d manifold which undergo some s-d hybridization. If the relativistic geometry with its shorter average Au-Au distances (nonrel. $\bar{d}_{AuAu}$ = 2.99 Å, rel. $\bar{d}_{AuAu}$ = 2.81 Å) is used in a nonrelativistic calculation, a broadening of the d manifold is observed as the main effect (Fig. 7). This is a consequence of the stronger d-d overlap. The level spectrum of the scalar relativistic calculation exhibits an even larger spreading of the d manifold and a significantly stronger mixing of s and d orbitals. The relativistic stabilization leads even to s-type orbitals with energies well below the d manifold. Nevertheless, the jellium-like ordering of 6s-derived levels is essentially preserved. Thus, except for a considerable s-d hybridization due to relativistic kinematics, a significant amount of preserved s-metal characteristics may be deduced from the one-electron level spectrum of Au_{55}.

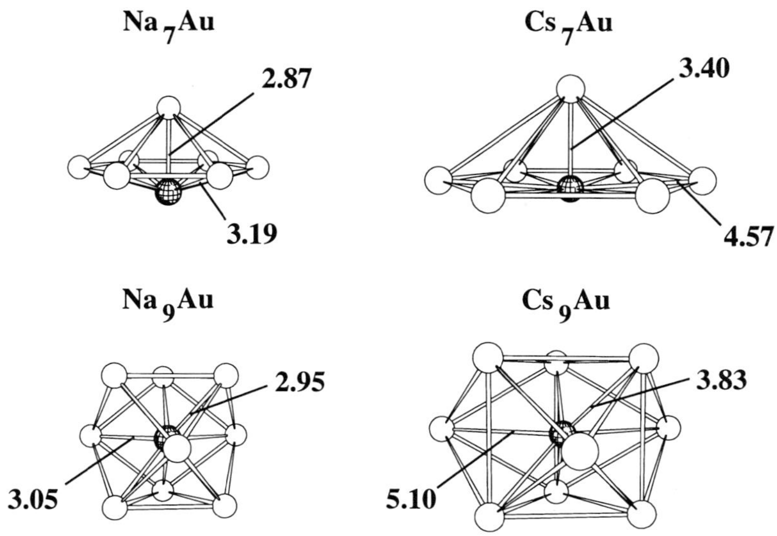

Figure 8. Optimized geometries for the clusters Na_7Au and Cs_7Au in C_{6v} symmetry as well as for Na_9Au and Cs_9Au in D_{3h} symmetry. Alkali-gold distances in Å.

5.1.2. Alkali-gold clusters

Relativistic DF calculations have been employed to investigate the electronic structure of mixed alkali-gold clusters Na_xAu and Cs_xAu and to clarify some puzzling findings of molecular beam experiments [110]. Such clusters are of special interest because they may be considered as finite-size building blocks of the stoichiometric 1:1 alkali-gold intermetallic phases which exhibit quite different behavior [111–113]. For example, NaAu displays metal properties while CsAu is a semiconductor [114].

In the mass spectra of Na_xAu clusters high abundances have been found for Na_8Au and Na_9Au and no signs of larger clusters have been observed. For the Cs_xAu clusters a rather different mass spectrum was found featuring peaks for the small clusters CsAu and Cs_2Au and a relative maximum for $x = 8$ and 9. Additionally, small amounts of even larger clusters have been observed. The ionization potentials of the sodium-gold clusters display the well-known even-odd oscillations with clusters of even nuclearity exhibiting the larger values. On the other hand, the ionization potentials of cesium-gold clusters show even-odd oscillations of reversed amplitudes. Besides a rationalization of this unexpected observation, another goal of the theoretical investigation [110] was to identify possible differences of the electronic and geometric structures of the clusters that may reflect the different behavior of the corresponding bulk intermetallic phases.

The geometries of various clusters were optimized at the GGA level [9,11] using the scalar relativistic variant of the all-electron RLCGTO-DF method. The energy minimization was performed for each degree of freedom permitted by the imposed symmetry constraint. The following clusters have been considered: Na_7Au and Cs_7Au in D_{5h} and C_{6v} symmetry,

Table 5. Calculated atomization energies ΔE_{at}, calculated and experimental ionization potentials IP, and relative abundances of the two cluster series Na_xAu and Cs_xAu.

Cluster		ΔE_{at} (eV)	Rel. ab.[a] (%)	calc. IP (eV)	exp. IP (eV)
Na_7Au	C_{6v}	6.73	22	4.28	4.13
Na_7Au	D_{5h}	7.21		4.02	
Na_8Au	D_{4d}	8.22	55	3.70	3.75
Na_9Au	D_{3h}	9.15	100	3.87	3.97
Cs_7Au	C_{6v}	4.94	33	3.24	3.09
Cs_7Au	D_{5h}	5.27		3.17	
Cs_8Au	D_{4d}	5.89	100	3.22	3.16
Cs_9Au	D_{3h}	6.49	93	3.09	2.91

[a] Relative abundances, normalized to the maximum of the corresponding mass range.

Na_8Au and Cs_8Au in D_{4d} symmetry and Na_9Au and Cs_9Au in D_{3h} symmetry. The symmetry constraints were imposed for computational convenience since the minima on the potential energy surfaces are rather shallow. It may be expected that the electronic structure of these clusters is not much influenced by moderate geometry changes. The results of the geometry optimization can be summarized as follows: the Na–Au bond lengths are rather constant in the series (ranging from 2.81 to 3.19 Å) while the Cs-Au bond distances spread over quite a wide range (from 3.40 to 5.10 Å). This finding indicates that in Na_xAu clusters all heteroatomic bonds are essentially equivalent, while in Cs_xAu large differences occur among the various Cs-Au bonds. These structural results support the hypothesis suggested by the experimental high abundances of CsAu and Cs_2Au in the molecular beam that specific CsAu subunits are present within the larger Cs_xAu clusters. A direct comparison of the optimized geometries of Na_7Au and Cs_7Au in C_{6v} symmetry demonstrates the existence of such subunits in the case of the cesium-gold clusters (Fig. 8): while in Na_7Au all the Na-Au bond lengths are rather similar, in Cs_7Au the cesium atom on the six-fold axis is considerably closer to the gold center than the other Cs atoms, resulting in a Cs-Au subunit. In Fig. 8 the optimized geometries of Na_9Au and Cs_9Au (D_{3h} symmetry) are reported as well, since their direct comparison is also informative. While the sodium cluster shows close similarity between the two different types of Na-Au bonds, the cesium cluster exhibits two types of coordination shells: the first one with a closer Cs-Au distance of 3.83 Å consists of six "quasi-axial" cesium atoms, while the three "equatorial" cesium atoms capping the three rectangular faces of the trigonal prism formed by the first Cs shell constitute a second shell of neighbors. The presence of only one coordination sphere in Na_xAu clusters in contrast to multiple coordination shells in Cs_xAu correlates with the experimental mass spectra in the range of higher masses.

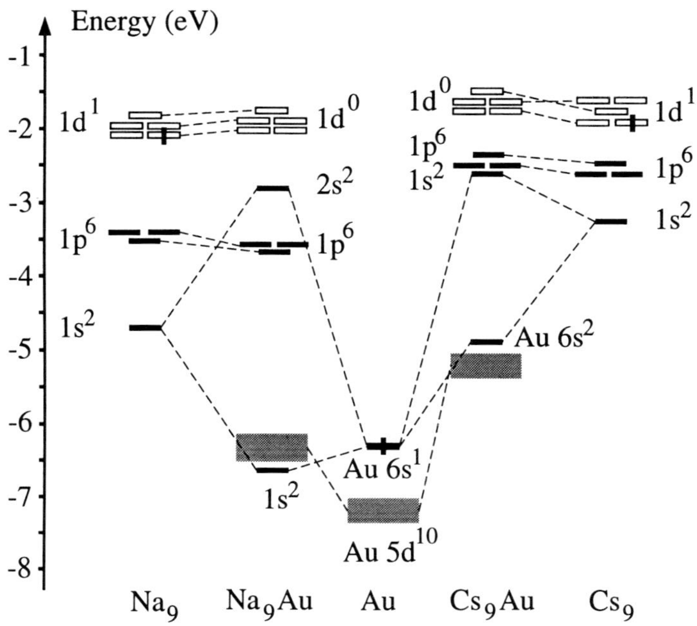

Figure 9. Kohn-Sham orbital correlation diagram for assembling the clusters Na_9Au and Cs_9Au (D_{3h}) from their fragments Au, Na_9, and Cs_9 (D_{3h}). Where meaningful, the levels are labeled according to the jellium model; the manifold of Au 5d orbitals is represented by shadowed grey boxes. Filled and open levels indicate doubly occupied and empty orbitals, respectively; for open shells, the unpaired electron is explicitly given as a vertical bar.

In Table 5, the calculated cluster atomization energies and the ionization potentials are compared to experimental abundances and ionization potentials. The atomization energies of the sodium clusters increase with cluster size, in accordance with the abundances which indicate that the thermodynamic stabilities follow the same trend. The atomization energies of the cesium clusters also increase with size, but in a less pronounced fashion than found for the sodium series. This is at variance with experiment which seems to indicate that Cs_8Au is the most stable species in the Cs_xAu series. However, the abundance difference to Cs_9Au is small enough that this discrepancy may also be due to experimental inaccuracies.

The calculated ionization potentials of the clusters Na_xAu are in good agreement with experiment (deviating at most by 0.1 eV) and they exhibit the same odd-even effect as the experiment. For Cs_xAu the rather surprising reversed odd-even effect is successfully reproduced by the theoretical results, with a maximum deviation between calculated and measured values of less than 0.2 eV. The behavior of the ionization potentials may profitably be discussed with the help of an orbital correlation diagram based on the KS energies, as reported in Figure 9 for Na_9Au and Cs_9Au. Whenever possible, the levels have been classified according to the jellium model. The classification of the Au 6s derived orbital differs for the two clusters: for Na_9Au it is assigned as the jellium 1s level but as an atomic level, not involved in the jellium, for Cs_9Au. Of course, this is a simplification which nevertheless reflects the dominant character of this orbital. Thus, while the bonding in Na_9Au may be described as polar covalent, in Cs_9Au it is essentially ionic. Moreover, it is worth noting that for Na_9Au the jellium shell structure is evident. The gap of 0.8 eV between the levels 1p and 2s rationalizes the even-odd behavior of the ionization potentials. On the other hand, for Cs_9Au no shell effect is present at all since the levels 1p and 1s are almost degenerate, thus explaining the failure of the jellium model for the cesium-gold clusters. Another indication for the ionic bonding of the clusters Cs_xAu is the destabilization of Au 5d orbitals as a consequence of the negative charge located on the gold atom. It has been verified that this comparison, discussed here in detail for Na_9Au and Cs_9Au, remains also valid for the other clusters in the series.

In summary, the calculations on the clusters Na_xAu and Cs_xAu are in good accordance with experiment. They allowed for a rather accurate investigation of the electronic structure of these mixed metal systems. Alkali-gold clusters of Na and Cs were found to exhibit qualitatively different electronic and structural properties, reflecting the different behavior of the corresponding bulk phases already at the cluster level.

5.1.3. The sodium-lead cluster Na_6Pb

A number of divalent metal/alkali systems known to be non-miscible in the bulk phase, have been shown to readily aggregate in the gas phase to form finite-size mixed-metal clusters over an extended size range [115]. Interestingly, particular mixed clusters, like Na_6Eu and $Na_{18}Yb$, can dominate the mass spectral intensity distribution by more than one order of magnitude thus reflecting a rather large relative stability [115]. Under certain experimental conditions a cluster consisting of one lead and six sodium atoms is the only product of the expansion of a mixed sodium/lead vapor [116]. In the absence of detailed experimental and theoretical information on the cluster Na_6Pb its geometric and electronic structure as well as its striking stability could not be rationalized [116]. Therefore a thorough theoretical investigation of Na_6Pb and a comparison with Na_6Mg has been performed by means of the RLCGTO-DF approach [117].

Lead is one of the elements whose chemistry is believed to be strongly influenced by relativistic effects [1]. For example, depending on whether relativity is taken into account or not, the atomic ground state of lead is s^2p^2 or p^4. In order to quantify the assumed relativistic stabilization of Na_6Pb, relativistic and nonrelativistic calculations on the electronic and geometric structure of this compound have been carried out [117].

The results of the geometry optimizations under various symmetry constraints are summarized in Figure 10. The corresponding atomization energies are given in Table 6. For the cluster Na_6Mg, a preference for open structures with the Mg atom located outside the Na_6 core is found in accordance with MRD-CI results [118]. The energy difference between the most stable C_{5v} structure and the cluster with octahedral symmetry is 0.55 eV. In contrast, for the lead compound, a closed structure is slightly preferred; the atomization energy varies only by about 0.1 eV over the whole series of symmetries examined. Inclusion of relativistic effects leads to a small stabilization of only 0.1-0.2 eV and a rather uniform contraction of the

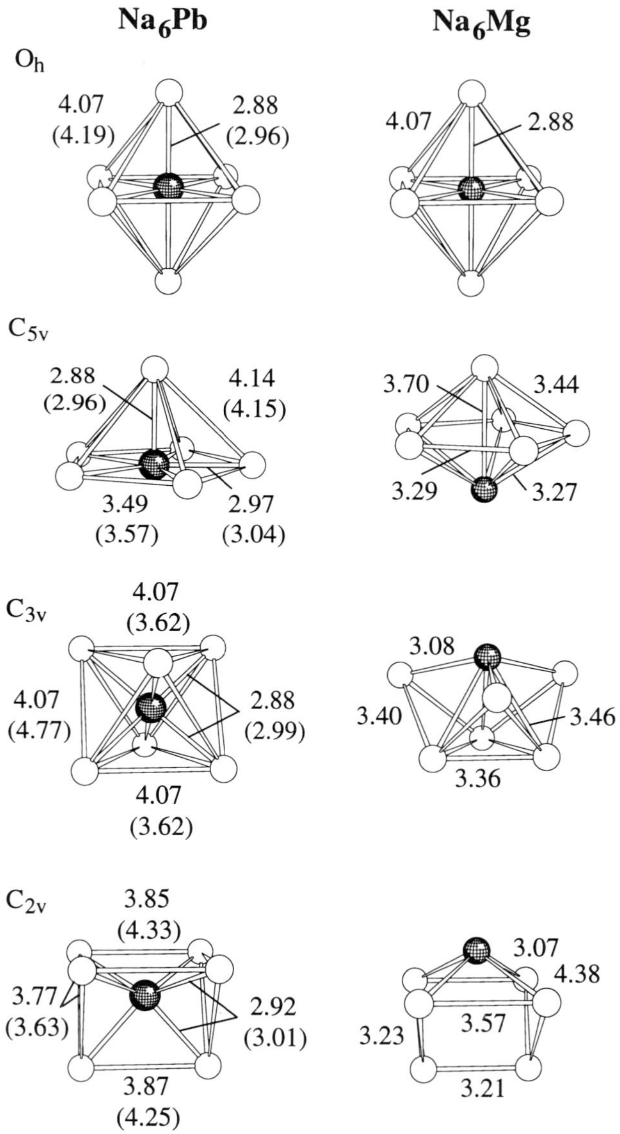

Figure 10. Sketches of some cluster geometries of Na_6Pb (on the left) and Na_6Mg (on the right) optimized under various symmetry restrictions. The Pb and Mg atoms are indicated by shadowed spheres. Bond lengths are given in Å. For the Na_6Pb cluster the results of scalar relativistic geometry optimizations are displayed with the corresponding nonrelativistic data listed in parentheses. Only nonrelativistic geometries are shown for the cluster Na_6Mg.

Table 6. Atomization energies (in eV) for the clusters Na_6Pb and Na_6Mg computed at the relativistic (rel.) and nonrelativistic (nrel.) levels of theory.

Cluster symmetry	Na_6Pb rel.	Na_6Pb nrel.	Na_6Mg nrel.	
O_h	6.11	5.91	2.99	(3.70)[a]
D_{3d}	6.11	5.96	2.99	
D_{3h}	6.13	5.99	3.09	
C_{5v}	6.05	5.94	3.54	(4.08)[a]
C_{3v}	6.11	5.96	3.48	(3.93)[a]
C_{2v}	6.14	5.99	3.30	

[a] MRD-CI results from ref. [118].

Na-Pb distances by 0.1 Å (Fig. 10). Interestingly, in the nonrelativistic case, there is a larger dispersion of Na-Na bond lengths. The smaller Pb radius in the relativistic calculation allows for a uniform distribution of the Na atoms around the central atom, whereas in the nonrelativistic case, the Na-Na interaction is optimized by the formation of two different Na-Na bond lengths (e. g. see the C_{3v} geometry, Fig. 10). The various structures of the cluster Na_6Pb exhibit atomization energies that are 2.5-3.0 eV higher than those of their Mg analogues.

To quantify different contributions to the binding energy the interaction of the Na_6 cage with the Mg and Pb atom has been analyzed with the help of the constrained space orbital variation (CSOV) procedure [119] for the O_h geometry [117]. Two factors govern the enhanced stability of Na_6Pb compared to Na_6Mg: the larger $Na_6 \rightarrow Pb$ charge transfer and the higher polarizability of Pb. Surprisingly, no significant difference has been found between the relativistic and nonrelativistic results for Na_6Pb. Thus, the particular stability of Na_6Pb has to be ascribed to the relatively high electronegativity of Pb and to the optimal number of six Na atoms forming a closed coordination sphere around the Pb atom.

5.2. Metal-Ligand Interactions

5.2.1. Mononuclear gold phosphine complexes

Triphenylphosphine ligands (PPh_3) are often employed to stabilize metal clusters; a large number of polynuclear gold clusters, synthesized in recent years [3], illustrate this effect. In most of these gold clusters the unit $Au(I)PPh_3$ may be considered as the basic building block and thus a quantitative description of this metal-ligand interaction is of fundamental interest for the understanding the electronic structure of these gold clusters. In theoretical studies PPh_3 ligands would considerably increase the computational effort, therefore model systems comprising only the bare cluster cores [120] or containing simple phosphines PH_3

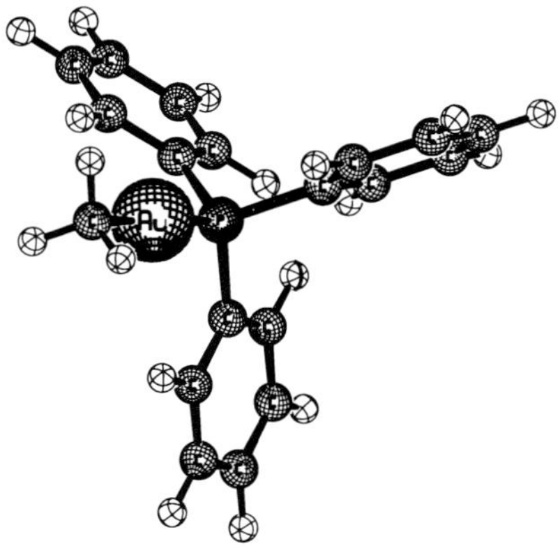

Figure 11. Methyl(triphenylphosphine)gold $MeAuPPh_3$.

[108,109,121] have been treated instead. In the subsequent section we will show that the inclusion of phosphine ligands is mandatory for achieving a qualitatively correct description of the electronic structure of element-centered gold cluster compounds. Here, we discuss the influence of phosphine substituents R on the electronic structure by comparing the molecules $MeAuPR_3$ (R = H, Me, Ph; Me = methyl, Ph = phenyl) (Fig. 11).

The results of a scalar relativistic local density functional study [127] are summarized in Table 7. While the optimized geometry of $MeAuPMe_3$ agrees well with experiment, only the Au-P distance is accurately reproduced in $MeAuPPh_3$, casting some doubt on the experimental Au-C distance [127]. The computational results demonstrate that, as far as structural data are concerned, PH_3 is an acceptable model of the larger phosphine PPh_3. On the other hand, PMe_3 exhibits a slightly shorter bond to the Au-P atom. Other properties of PMe_3 and PPh_3, like the phosphine binding energy, the dipole moment, and the first ionization potential, are calculated quite similar while the simple phosphine ligand shows clear deviations: the binding energy as well as the dipole moment are significantly lower than for the larger phosphines. The methyl binding is quite similar in all three compounds. Thus, PMe_3 should be employed as a ligand if an accurate model of the energetics and the charge distribution of triphenylphosphine gold compounds is desired. However, geometry optimizations may economically be carried out with the help of the simple phosphine ligand PH_3 (Table 7).

The increasing dipole moment of $MeAuPR_3$ along the ligand series R = H, Me, Ph points towards an increasing electron donation ability of the ligands. The gold-phosphine binding energy increases concomitantly (Table 7). This effect becomes even more pronounced if one compares the species $AuPR_3$ and $AuPR_3^+$. While for the Au(0) systems $AuPR_3$ phosphine binding energies are calculated comparable in size and trend to $MeAuPR_3$ (R = H: 124 kJ/mol, Me: 176 kJ/mol, Ph: 184 kJ/mol), a considerable strengthening has to be noted for the Au(I) moieties $AuPR_3^+$ (R = H: 401 kJ/mol, Me: 553 kJ/mol, Ph: 580 kJ/mol) [127]. Even if one takes into account that the binding energy values are most likely overestimated due to the local density approximation used in these calculations, the results clearly demonstrate the

Table 7. Comparison of calculated and experimental properties of $MeAuPR_3$ (R = H, Me, Ph): bond lengths d(Au-C) and d(Au-P), phosphine binding energy $E_b(PR_3)$ ($MeAuPR_3 \rightarrow MeAu + PR_3$), methyl binding energy $E_b(Me)$ ($MeAuPR_3 \rightarrow Me + AuPR_3$), dipole moment μ and first ionization potential IP.

	d(Au-C) (Å)	d(Au-P) (Å)	$E_b(Me)$ (kJ/mol)	$E_b(PR_3)$ (kJ/mol)	μ^a (D)	IP (eV)
Theory						
MeAu	1.96	–	–	–	-0.14	9.96
$MeAuPH_3$	2.00	2.29	432	182	3.84	9.10
$MeAuPMe_3$	2.03	2.25	434	236	5.57	8.34
$MeAuPPh_3$	2.02	2.28	436	246	5.80	8.13
Experiment						
$MeAuPMe_3$	2.03[b]	2.28[b]			5.4-5.6[c]	8.24[d]
$MeAuPPh_3$	2.12[e]	2.27[e]			5.6[f]	

[a] The positive dipole orientation corresponds to $(Me)^{\delta-}(AuPR_3)^{\delta+}$. [b] Ref. [122]. [c] Ref. [123]. [d] Ref. [124]. [e] Ref. [125]. [f] Ref. [126].

ability of phosphine ligands to stabilize gold cluster compounds, especially in the case of monovalent gold species.

5.2.2. Element-centered gold clusters

Main group element-centered gold clusters [128–131] (Fig. 12) are a class of compounds where stabilization effects of phosphine ligands are essential. The octahedral species $[(LAu)_6X_m]^{m+}$ (X_1 = B, X_2 = C, X_3 = N), where m designates the cluster charge and the type of central atom, constitute an interesting series of complexes. The carbon-centered cluster of this series, predicted by simple molecular orbital arguments [132], has been synthesized [129]. However, the existence of the nitrogen-centered isoelectronic analogue [133] is discussed controversially [134] while attempts to synthesize the boron-centered isoelectronic complex have so far not been successful.

To clarify the special role of the carbon-centered cluster RLCGTO-DF calculations [135] using the LDA energy functional have been carried out on a larger class of gold clusters comprising four to six gold phosphine units around one of the elements B, C, or N: $[(LAu)_nX_m]^{(n+m-6)+}$ (L = PH_3; n = 4, 5, 6; X_1 = B, X_2 = C, X_3 = N) (Fig. 12). The charge on these isoelectronic closed shell cluster cations ranges from 0 to +3.

In Table 8 the effect of relativity on the geometry of nonligated empty and element-centered gold octahedra is shown. In the nonrelativistic case the empty gold cluster expands

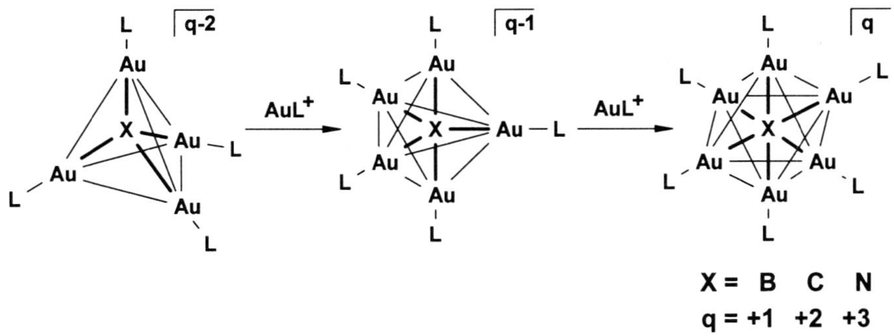

Figure 12. Main group element-centered gold clusters.

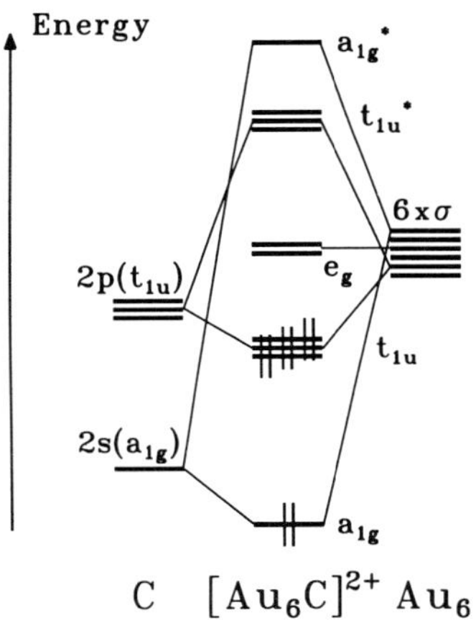

Figure 13. Schematic orbital interaction diagram for the carbon-centered octahedral cluster $[Au_6C]^{2+}$. Only orbitals contributing to the bonding of the central C atom are shown.

with growing charge due to the Coulomb repulsion. The relativistic strengthening of the metal-metal bonds leads to a considerable bond contraction which is well known also for other compounds [1]. Even the variation of the size of the octahedron with the cluster charge is reduced. Thus, the Coulomb repulsion is nearly compensated by the relativistic increase of the bond strength. In all cases the gold cage grows by roughly 0.15 Å after introduction of a central atom whether relativistic effects are accounted for or not (Table 8). Again the variation of the cluster size with the cluster charge is large in case of a nonrelativistic treatment. These results deviate considerably from those of a Hartree-Fock quasirelativistic pseudopotential calculation [120] which yielded longer distances for the element-centered naked clusters (d(Au-X) = 2.20 Å for B, 2.24 Å for C and 2.35 Å for N) similar to the nonrelativistic data of Table 8. The major difference between both methods is the missing treatment of

Table 8. Comparison of calculated (nrel. = nonrelativistic, rel. = relativistic) distances d(Au-X) (in Å) of the gold atoms from the cluster center for the empty clusters $[Au_6]^{m+}$ (m = 1, 2, 3) and for the element-centered clusters $[Au_6X_m]^{m+}$ (m = 1, 2, 3).

Cluster	nrel.	rel.	Cluster	nrel.	rel.
$[Au_6]^{1+}$	2.06	1.91	$[Au_6B]^{1+}$	2.20	2.05
$[Au_6]^{2+}$	2.08	1.91	$[Au_6C]^{2+}$	2.23	2.04
$[Au_6]^{3+}$	2.12	1.92	$[Au_6N]^{3+}$	2.30	2.07

Table 9. Comparison of calculated (nrel. = nonrelativistic, rel. = relativistic) and experimental bond distances (in Å) for the clusters $[LAu_6X_m]^{m+}$, (X_1 = B, X_2 = C, X_3 = N). L refers to the phosphine ligands, PH_3 in the calculations and PPh_3 in experiment.

	d(Au-X)			d(Au-P)		
Cluster	nrel.	rel.	exp.	nrel.	rel.	exp.
$[(LAu)_6B]^{1+}$	2.21	2.11	–	2.46	2.27	–
$[(LAu)_6C]^{2+}$	2.19	2.08	2.12[a]	2.44	2.26	2.27[a]
$[(LAu)_6N]^{3+}$	2.19	2.10	2.13[b]	2.44	224	–

[a] Ref. [137]. [b] Ref. [133].

correlation in the Hartree-Fock method. This finding underlines that both electron correlation and relativistic effects have to be taken into account to achieve an adequate description of the electronic structure of gold compounds.

The addition of phosphine ligands to the naked clusters leads to a slight elongation of the Au-X distance, see Table 9. The relativistic results are in good agreement with experimental findings for PPh_3 ligated clusters, although only simple phosphines had been employed as model ligands. The experimental Au-P bond length of 2.27 Å, which is quite typical for gold-phosphorus bonds [136], is accurately reproduced (see section 5.2.1). For these bonds, the necessity of a relativistic treatment is particularly obvious (Table 9). The only structural sign for a special stability of the carbon-centered cluster is the slightly shorter Au-X bond compared to its isoelectronic neighbors. The nonrelativistic results for $[(PH_3Au)_6X_m]^{m+}$ show no increase of the cluster size with growing charge m. This effect had been ascribed to an opening of the Au d^{10} shell by the phosphine ligands [135,109] allowing for a d contribution to the Au-Au bond. Thus the ligands act in a similar way as the enhanced s-d hybridization caused by relativistic effects.

Table 10. Total binding energies[a] (in kJ/mol) of various empty and element centered octahedral model gold clusters with and without phosphine ligands. The maximum of the binding energy along each row of varying charge *m* is underlined.

Charge m =	0	+1	+2	+3	+4
Empty clusters					
$[q_6]^{m+}$ [b]	<u>0</u>	202	808	1818	3232
$[Au_6]^{m+}$	–1289	<u>–1504</u>	–1266	–593	523
$[(H_3PAu)_6]^{m+}$	–2083	–2588	<u>–2766</u>	–2631	–2143
$[(Me_3PAu)_6]^{m+}$	–2159	–2793	–3121	<u>–3176</u>	–2928
Element-centered clusters		X=B	X=C	X=N	
$[Au_6X_m]^{m+}$		<u>–2294</u>	–1932	–805	
$[(H_3PAu)_6X_m]^{m+}$		–3491	<u>–3715</u>	–3287	
$[(Me_3PAu)_6X_m]^{m+}$		–3693	<u>–4084</u>	–3848	

[a] Total binding energy with respect to the constituents X_m, $m \times Au^+$, $(6–m) \times Au$, and $6 \times PR_3$ (R = H, Me). [b] Array of point charges (charge $q = +m/6$) in the geometry of $[Au_6]^{2+}$ (d_{qq} = 2.70 Å).

Several contributions to the cluster binding have to be separated to rationalize the bonding in element-centered gold cluster cations: there is an electrostatic repulsion due to the overall cluster charge, a tangential intermetal bonding in the gold octahedron, a radial bonding to the central atom, and the gold-ligand interaction. For empty Au_6 octahedral cations a considerable stabilization due to the Au-Au bonds is calculated offset by the Coulomb repulsion which has been quantified by a point charge model (Table 10). The most stable cation calculated is $[Au_6]^+$, but even for a charge of +3 the intermetal bonding exceeds the Coulomb repulsion. Addition of phosphine ligands which act as electron donors stabilizes the Au_6 cations. In line with the discussion in section 5.2.1, the amount of stabilization grows as the effective oxidation state of the gold atoms increases with the cluster charge. This shifts the maximum of stability to higher cation charges. As is shown in Table 10, for PH_3 ligands the species with a charge of +2 is most stable, while for the more effective donor PMe_3 the maximum of stability is even shifted to a cluster charge of +3.

Addition of a central atom to the non-ligated gold cage stabilizes the cluster; this contribution decreases from B to N (cf. $[Au_6]^{m+}$ to $[Au_6X_m]^{m+}$ in Table 10). This trend counteracts the ligand effect. Therefore, the maximum of stability for the element-centered ligated clusters is shifted back to lower charges (Table 10). In summary, for the phosphine ligated clusters a maximum of stability is calculated for the carbon-centered species as a result of the interplay of Coulomb repulsion, stabilization due to phosphine ligands and different bond strength of the central atoms. Nevertheless, the differences to the B and N centered clusters are small.

The extension of the above analysis of energetics to the smaller trigonal bipyramidal $[(LAu)_5X_m]^{(m-1)+}$ and tetrahedral $[(LAu)_4X_m]^{(m-2)+}$ (Fig. 12) element-centered gold clusters has shown that the same concepts as discussed here for the octahedral clusters determine their relative stability [135]. For the five-coordinate cluster series the carbon-centered species has been calculated to be most stable, while the tetrahedral gold cage more favorably incorporates a nitrogen atom. Also these results are in agreement with experimental evidence [128,138,139].

5.2.3. Other gold cluster compounds

Interesting effects of phosphine ligands on the structure of gold compounds have also been observed for trigold oxonium cluster cations $[(LAu)_3O]^+$ (L = tertiary phosphine) [140–144]. The trigold oxonium monomer consists of a gold triangle capped with an oxygen atom, forming a trigonal pyramid. Phosphine ligands are attached to the Au atoms such that almost linear O-Au-P moieties are formed (Fig. 14). Depending on the phosphine substituents, these cluster compounds form crystals composed of isolated moieties or of dimers, the latter occurring in two different forms. For bulky phosphine ligands like tri(isopropyl)phosphine P^iPr_3 [143] or tri(ortho-tolyl)phosphine $P(o\text{-}tol)_3$ [141] trigold oxonium monomers are found in the crystal. For smaller ligands like triphenylphosphine PPh_3 [140,141] and methyl-diphenylphosphine PPh_2Me [141] the trigold oxonium clusters dimerize via parallel edges of the gold triangles (Fig. 14a), forming a rectangular contact (rectangular dimer). On the other hand, the trimethylphosphine ligated $[Au_3O]^+$ clusters form a new dimer structure with crossed edges of the two gold triangles (Fig. 14b, tetrahedral dimer) [142]. In the latter structure, four Au-Au contacts exist between the two Au_3 subunits. This propensity to form larger aggregates with intermolecular Au-Au contacts which are comparable to Au-Au bonds in clusters is known for quite a large number of gold compounds [3,145]. The term "aurophilic attraction" [3] has been coined for the underlying interaction of formally closed shell Au(I) atoms which occurs both in intra- and intermolecular form and is one of the sources of the surprisingly rich chemistry of gold.

In a recent study of the factors which determine the dimer geometries the structure of the trigold oxonium monomer has been optimized with the help of the RLCGTO-DF method employing a LDA energy functional [146]. The "naked" monomer $[Au_3O]^+$ as well as the phosphine ligated model cluster $[(PH_3Au)_3O]^+$ have been considered. The only marked difference between experimental and theoretical geometries are the Au-Au distances which are calculated too short (2.78 Å for the naked and 2.88 Å for the ligated cluster) compared to experiment. For monomers 3.09 Å (L = P^iPr_3 [143]) and 3.20 Å (L = $P(o\text{-}tol)_3$ [141]) have been found; in the dimers the intramolecular Au-Au distances range from 3.05 to 3.31 Å [140–142]. These widely scattering experimental values provide a strong indication for a very soft Au-O-Au potential. The difference between calculated and measured Au-Au distances may very well be due to the fact that in the model cluster no effect of the crystalline surrounding has been taken into account.

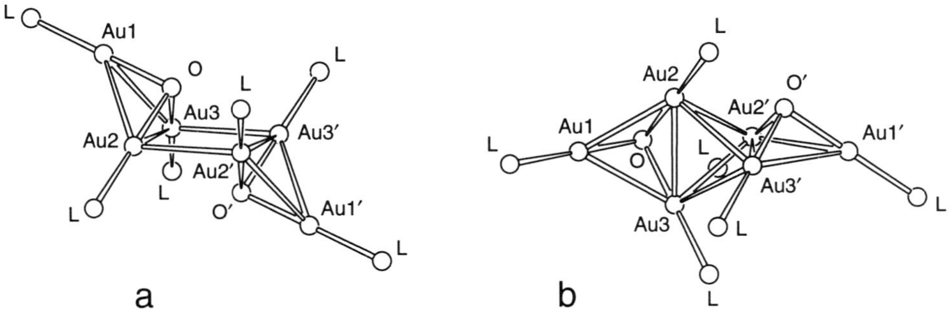

Figure 14. Dimer structures of trigold oxonium cations: a) rectangular dimer, b) tetrahedral dimer. Phosphine ligands are abbreviated by L.

Using the optimized monomer geometries as a starting point, both dimer structures, rectangular and tetrahedral, have been compared by optimizing the intermonomer distance [146]. While the tetrahedral geometry was found to be slightly favored (by 26 kJ/mol) for the non-ligated cluster dimer $[Au_3O]_2^{2+}$, a surprisingly strong preference for the rectangular dimerization (by 311 kJ/mol) was calculated for the ligated species $[(PH_3Au)_3O]_2^{2+}$. A strong influence of the crystal field on this preference could be ruled out since the same qualitative results were achieved by applying different point charge models of the crystalline surrounding. The experimental finding of a tetrahedral dimer structure for the sterically more demanding PMe_3 ligated trigold oxonium $[(PMe_3Au)_3O]^+$ seems to be in contradiction to this result because the rectangular structure is better suited for avoiding the inter-monomer steric ligand-ligand repulsion (cf. Fig. 12). A calculation on the total energy of the two isolated ligand shells $(PH_3)_6$ revealed that a large part of the energetic difference (185 kJ/mol) between the two optimized dimer structures is due to ligand repulsion in the tetrahedral dimer. Using the experimental geometry this value for the steric repulsion yielded even a preference for the tetrahedral structure by 43 kJ/mol. This difference may be rationalized by the longer experimental Au-Au distances (3.31 Å in $[(PMe_3Au)_3O]_2^{2+}$ [142]) which leads to an opening of the Au-O-Au angle and thus to a decrease of the direct ligand-ligand interaction. With the help of force field calculations [146] it was shown that also for PMe_3 ligands there is no significant difference in steric repulsion between the two dimer structures if the experimental geometry is used. However, for the PPh_3 ligands the rectangular dimer structure is clearly preferred for steric reasons, in agreement with experiment [140,141].

As a result of these investigations, the suggestive picture which relates the appearance of more compact crystal structures to a decreasing ligand size has to be refined [146]. The tetrahedral dimer structure is favored for the bare $[Au_3O]^+$ cations due to the larger number of Au-Au contacts (Fig. 14). For small phosphine ligands like PH_3 or PMe_3 the weak inter-monomer steric interaction is comparable for both dimer structures. Thus crystal packing effects will be decisive for the dimer geometry. In case of sterically more demanding phosphines, ligand-ligand repulsion becomes important, resulting in a preference of the rectangular dimerization. For very bulky ligands like P^iPr_3 dimerization is no longer possible.

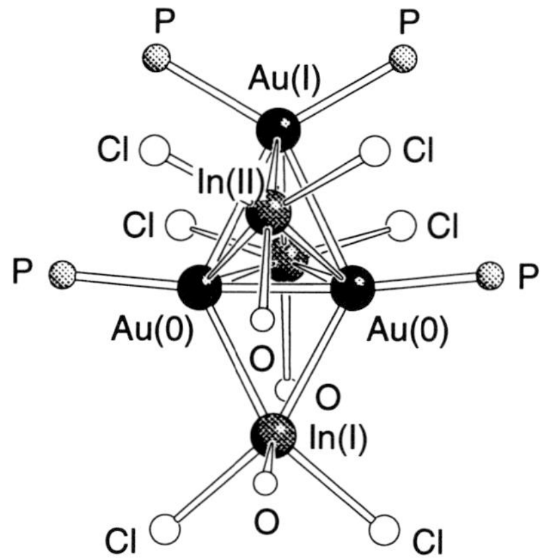

Figure 15. Cluster $(PH_3)_4Au_3(InCl_2H_2O)_3$ modeling the experimental Au-In complex $(dppe)_2Au_3(InCl_2thf)_3$ [147]. Hydrogen atoms are omitted for clarity. Au and In atoms are labeled by their tentative oxidation states.

So far, the chemistry of mixed-metal gold clusters [148] is less developed than other parts of gold chemistry [145]. As an example for this emerging field, the recently synthesized indium-gold compound $(dppe)_2Au_3(InCl_2thf)_3$ (dppe = 1,2-bis(diphenylphospino)ethane, thf = tetrahydrofuran) [147] has been studied by the RLCGTO-DF approach [149] using simplified models (Fig. 15). A striking feature of this complex is the fact that two rather different Au-Au bond lengths have been found (2.56 Å, 2.94 Å) in the triangular gold core. As shown in Figure 15, the gold triangle is capped above and below by two $InCl_2$(thf) groups (indium groups); a third indium group bridges the short Au-Au bond. Since six chlorine atoms are present in the cluster, a formal charge of +6 has been ascribed to the Au_3In_3 core [147]. The short Au-Au distance as well as the linear structure of the P-Au-Au-P moiety suggest oxidation state 0 for these gold atoms. Oxidation state +1 has been assigned to the third gold atom whereas the indium atoms have been tentatively classified as μ_3-In(II) and μ_2-In(I) (Fig. 15). The aim of the theoretical study was to provide support for this interpretation of the cluster electronic structure as well as to shed some light on structural and energetic aspects of this interesting mixed-metal compound.

The charge distribution and the ligand binding have been examined for the model compound $(PH_3)_4Au_3(InCl_2H_2O)_3$ [149] employing the experimental geometry [147]. For the gold atoms a distinct difference of the Mulliken charges has been calculated for the two types of Au atoms: −0.03 au for Au(0) and +0.19 au for Au(I). The latter is comparable to the charge of gold in AuCl (+0.29 au), thus confirming the experimental assignment. For the indium atoms, on the other hand, no significant differences have been found, μ_2-In +0.42 au and μ_3-In +0.44 au, at variance with the experimental proposition [147].

The phosphine ligands are bound to the isolated Au_3 triangle with a mean bond energy of 120 kJ/mol. The electron withdrawing groups $InCl_2H_2O$ strengthen the phosphine-gold bonds (179 kJ/mol). This ligand synergy effect is similar to the situation in element-centered

Table 11. Optimized Au-Au distances (in Å) and their difference Δ for the model compound $(PH_3)Au_3I_3$ and for its subunits.

Cluster	Au(0)-Au(0)	Au(0)-Au(I)	Δ
Au_3	2.52	2.63	0.11
$Au_3(PH_3)_4$	2.53	2.79	0.26
$Au_3(PH_3)_4(\mu_2\text{-I})$	2.56	2.75	0.19
$Au_3(PH_3)_4(\mu_3\text{-I})_2$	2.57	2.90	0.33
$Au_3(PH_3)_4I_3$	2.57	2.90	0.33
exp.[a]	2.56	2.94	0.37

[a] Ref. [147].

gold clusters (section 5.2.2) where the cluster stabilization due to the phosphine ligands was found to increase with positive cluster charge. At the same time, the binding of the In groups is strengthened by the electron donating phosphine ligands: the averaged binding energy of $InCl_2H_2O$ to Au_3 amounts to 226 kJ/mol, but is enhanced to 304 kJ/mol for the phosphine ligated cluster. Thus, the bonds of both types of ligands to the Au_3 core are mutually stabilized.

The evolution of the particular geometry of the Au_3 moiety in the gold-indium cluster has been modeled by $(PH_3)_4Au_3I_3$. The iodine atom has approximately the same size as the In atom and, using a formal electron count, also the same number of valence electrons as In in $InCl_2H_2O$. The geometry of the cluster and its fragments have been optimized with the RLCGTO-DF method at the BLYP level of theory (Table 11) [149]. The neutral gold triangle is slightly Jahn-Teller distorted, Δ = 0.11 Å. The phosphine ligands enhance this distortion by 0.15 Å since two phosphines are coordinated to Au(I). Addition of the bridging iodine atom slightly elongates the Au(0)-Au(0) bond and shortens the Au(I)-Au(0) bond. The second strong elongation of the latter bond is caused by the two μ_3-I atoms, establishing the final difference Δ = 0.33 Å, in good agreement with experiment (Table 11). As a result the considerably different Au-Au bond distances in the Au-In cluster can be rationalized as follows: The small Jahn-Teller distortion in the Au_3 triangle is considerably enhanced by a weakening of the Au-Au bonds due to the coordination of electron withdrawing In groups. The In groups act more strongly on the weak Au(0)-Au(I) bonds and leave the relatively strong Au(0)-Au(0) bond unaffected.

5.2.4. Monocarbonyls of Ni, Pd, and Pt

The CO molecule is a very common ligand in organometallic complexes of middle to late transition metals [150] and also in transition metal cluster compounds [89]. Moreover, CO as an adsorbate has been studied extensively in surface science where it is the standard probe molecule for the properties of surface sites [151,152]. The classical model of metal-carbonyl bonding (5σ donation, $2\pi^*$ backdonation) [153,154] has been adapted for surfaces [155] and later on considerably refined [119]. Recently, the relativistic effect on the metal-carbonyl bond has been systematically studied for the monocarbonyls of Ni, Pd, and Pt employing the relativistic RLCGTO-DF method, both at the LDA and GGA level of theory [4,156]. These monocarbonyls may serve as model systems which allow a fairly accurate treatment and a detailed analysis.

The calculated spectroscopic constants of MCO (M = Ni, Pd, Pt) shown in Table 12 demonstrate the increasing influence of relativity down the column of the periodic table. While relativistic effects are negligible for NiCO, distinct changes can be observed for PdCO and strong effects are visible for PtCO, especially for the M-C distance and the corresponding vibrational frequency as well as for the carbonyl binding energy. On the other hand, the bonding to a heavy element seems to exert only a minor indirect relativistic effect on the C-O bond (PtCO, Table 12). Except for the bond distances, the nonrelativistic properties of PdCO and PtCO are very similar. Clearly, for the Pt complex relativistic treatment is essential.

For the M-C bond length the nonrelativistic calculations yield increasing distances in the order Ni < Pd < Pt as might be expected due to the mass increase (Table 12). Due to relativity this trend is changed and the Pt-CO bond is found to be considerably shorter than the Pd-CO distance. The trends for the carbonyl binding energy and for the M-C vibrational frequency are in line with this finding: both indicate a strengthening of the Pt-CO bond due to relativistic effects. These effects are the result of the relativistic contraction of the metal s orbital and of the subsequent destabilization of d orbitals [1]. At the nonrelativistic level, both Pd and Pt exhibit a closed shell d^{10} ground state configuration. PdCO exhibits the weakest metal-carbonyl bond in the triad since the closed shell $^1S(d^{10})$ atomic ground state is also calculated for Pd at the relativistic level (Table 12). In case of Pt, relativity changes the ground state to $^3D(d^9s^1)$ leading to the strongest bond and to the highest M-C frequency in the group examined. Similar results with respect to geometry and bond strength have been obtained at the GGA level for the monocarbonyls and for the monoxides of Ni, Pd, and Pt [4]. Shorter metal ligand bonds for third row compared to second row transition metal elements are quite common [1] and have also been found for other triads, e. g. for the hexacarbonyls of Cr, Mo, and W [157].

The carbonyl C-O frequency is often used experimentally as an indicator for the properties of the carbonyl bonding site. Yet, the results presented in Table 12 show that the vibrational frequency may be fairly unaffected for different metals. This lack of sensitivity may be rationalized as an interplay of two effects [158,159]: the bonding of CO to a metal center weakens the C-O bond due to π backdonation, resulting in a red-shift of the C-O frequency. A short M-C bond length leads to a counteracting blue-shift via an enhanced Pauli repulsion between the 5σ orbital of CO and the metal orbitals ("wall effect"). For Pd the relativistic increase in bond strength and the shortening of the M-C bond result in a cancellation of both effects on the C-O frequency (Table 12). In PtCO, relativistic effects cause a sizeable shortening of the M-C bond, hence a strong increase of the π backdonation, but the counteracting increase of the Pauli repulsion leads only to an overall small blue-shift of the C-O frequency compared to the nonrelativistic value. Thus, the C-O frequency

Table 12. Comparison of nonrelativistic (nrel.) and relativistic (rel.) spectroscopic constants for the monocarbonyls MCO (M = Ni, Pd, Pt) from RLCGTO-LDA calculations: equilibrium distances d, vibrational frequencies ω and binding energies E_b of the carbonyl ligand.

		d(M-C) (Å)	d(C-O) (Å)	ω(M-C) (cm^{-1})	ω(C-O) (cm^{-1})	E_b (kJ/mol)
NiCO	nrel.	1.66	1.16	641	2056	295
	rel.	1.64	1.16	668	2064	302
PdCO	nrel.	1.87	1.15	468	2057	223
	rel.	1.81	1.15	523	2058	286
PtCO	nrel.	1.93	1.15	458	2051	221
	rel.	1.74	1.16	680	2085	414

remains essentially unaffected by relativisitc effects although the Pt-CO bond is considerably strengthened (Table 12).

A detailed analysis of the various mechanisms contributing to the metal-carbonyl bonding in MCO has been carried out [156] with the help of a CSOV analysis [119]. Most of the Pauli repulsion between the metal M and the CO fragment is compensated by a polarization and rehybridization of the metal atom. By far the largest contribution to the M-CO binding stems from the π backdonation [119]; for all three monocarbonyls it is at least three times larger than the σ donation.

5.2.5. Carbonyl cluster compounds

After discussing the metal carbonyl bond for the model compounds MCO, two types of carbonylated mixed-metal clusters will be treated as examples of the metal carbonyl interaction in chemical systems.

Since nickel is a ferromagnetic element, gas phase Ni clusters are highly magnetic, as has been shown by molecular beam experiments [160] and confirmed by theoretical calculations [90,91,161–163]. The magnetization vanishes for low nuclearity clusters which are coordinated with CO ligands [91,161–164], but survives in the interior of larger metal species [90,161–163]. A similar phenomenon has been observed for Ni surfaces when CO molecules are adsorbed [165]. The quenching of the magnetism due to CO coordination has been rationalized by the repulsive interaction of the CO 5σ lone pair with the Ni 4s electrons which leads to a change of the effective Ni configuration from d^9s^1 to $d^{10}s^0$ and thus to a pairing of the Ni valence electrons [91,161].

A recent magnetization measurement for the cluster compound $[Ni_{38}Pt_6(CO)_{48}H]^{5-}$ (Fig. 16) found a magnetic moment of only 0.03 μ_B per atom. To explain the very low magnetic moment of a cluster of this size a relativistic LCGTO-DF study has been performed

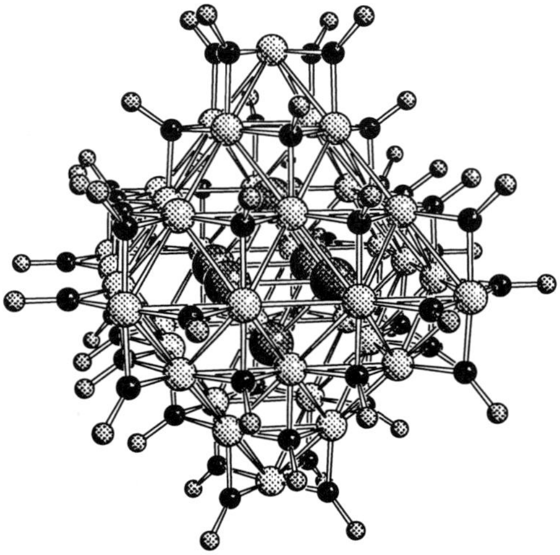

Figure 16. Structure of the carbonylated mixed-metal cluster $Ni_{38}Pt_6CO_{48}$.

Table 13. Averaged number of unpaired electrons n_s per atom for surface (*s*) and core (*c*) atoms in free and carbonylated Ni and Ni-Pt clusters.

	n_s/atom			
Cluster	Ni (*s*)	Ni (*c*)	Pt(*s*)	Pt(*c*)
$Ni(s)_{38}Ni(c)_6$	0.75	0.72		
$Ni(s)_{38}Pt(c)_6$	0.76			0.22
$[Ni(s)_{38}Ni(c)_6(CO)_{48}]^{6-}$	0.00	0.59		
$[Ni(s)_{38}Pt(c)_6(CO)_{48}]^{6-}$	0.00			0.00
$[Ni(s)_{32}Pt(s)_6Ni(c)_6(CO)_{48}]^{6-}$	0.01	0.58	0.01	

[164]. The octahedral cluster $Ni_{38}Pt_6(CO)_{48}$ consists of a Pt_6 core, surrounded by a shell of 38 Ni atoms, to which the CO ligands are attached (Fig. 16). The free metal cluster Ni_{44}, where the Pt atoms have been replaced by Ni, carries a high magnetic moment, even exceeding the bulk value of 0.6 μ_B per atom [166] (Table 13). The high magnetization of the Ni atoms persists if the cluster core contains Pt and although metallic Pt is non-magnetic, a small magnetic moment is induced in the Pt core of the cluster. As the data in Table 13 show, the addition of CO ligands quenches the magnetism on the surface of the Ni_{44} cluster, but the number of unpaired electrons in the cluster core (atoms Ni(*c*), Table 13) is only reduced. For the Pt-centered cluster the magnetic quenching due to the CO ligand shell is complete,

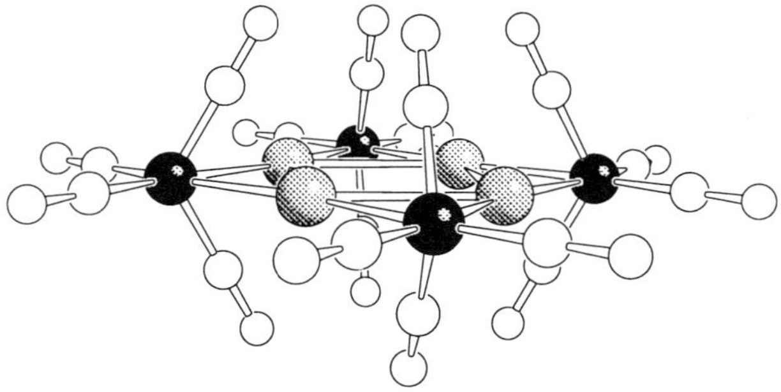

Figure 17. Mixed-metal clusters $[M_4\{Fe(CO_4)\}_4]^{4-}$ (M = Cu, Ag, Au). Noble metal atoms are shaded, iron atoms black, and CO ligands white.

in agreement with experiment [164], since there are no unpaired electrons on Ni atoms left to induce a magnetic moment on Pt. The very small magnetic moment measured may be explained by a small number of cluster isomers with Pt atoms positioned at the cluster surface instead of the interior. If the six Pt atoms of the cluster $Ni_{38}Pt_6(CO)_{48}$ are situated at the cluster corners, the central Ni atoms exhibit the same number of unpaired electrons as a pure Ni cluster (Table 13).

In the last years, a group of mixed-metal clusters containing group 11 elements stabilized by $Fe(CO)_4$ units have been synthesized [167–170]. The smaller members, e. g. $[Cu_3\{Fe(CO)_4\}_3]^{3-}$ [168], $[M_4\{Fe(CO)_4\}_4]^{4-}$ (M = Ag, Au) [170,171], and $[M_5\{Fe(CO)_4\}_4]^{3-}$ (M = Cu, Ag) [168,170], exhibit a planar geometry (Fig. 17) while larger members of this group, with 6 and 13 noble metal atoms, show an idealized tetrahedral or octahedral structure [167,169]. The discussion of the bonding in these compounds focused on the existence of direct M-M bonding and on the question why some species could not be synthesized for one or several members of the triad Cu, Ag, and Au. As a contribution to this discussion, a LCGTO-DF study of the clusters $[M_4\{Fe(CO)_4\}_4]^{4-}$ (M = Cu, Ag, Au) was undertaken [117] employing a relativistic approach for the Ag and Au containing species. The geometries were optimized at the LDA level, but the energetics was determined using the GGA suggested by Becke and Perdew [9,10].

The clusters $[M_4\{Fe(CO)_4\}_4]^{4-}$ consist of a square of noble metal atoms where the sides M-M are bridged by $Fe(CO)_4$ units (Fig. 17). For each $Fe(CO)_4$ moiety, two CO ligands are oriented equatorially (in the M_4 plane) and the other two axial carbonyls are tilted toward the cluster core. The optimization of the M_4Fe_4 cluster core yielded M-M and M-Fe bond lengths which follow the characteristic trend of group 11 elements [172] (see also the preceding section). The shortest bonds are found for Cu while for Ag bond lengths comparable to Au or even longer have been calculated (Fig. 18). As already discussed above, this finding is rationalized by the relativistic stabilization of the metal s orbital and the subsequent destabilization of d orbitals [1]. The CO ligands induce a considerable elongation

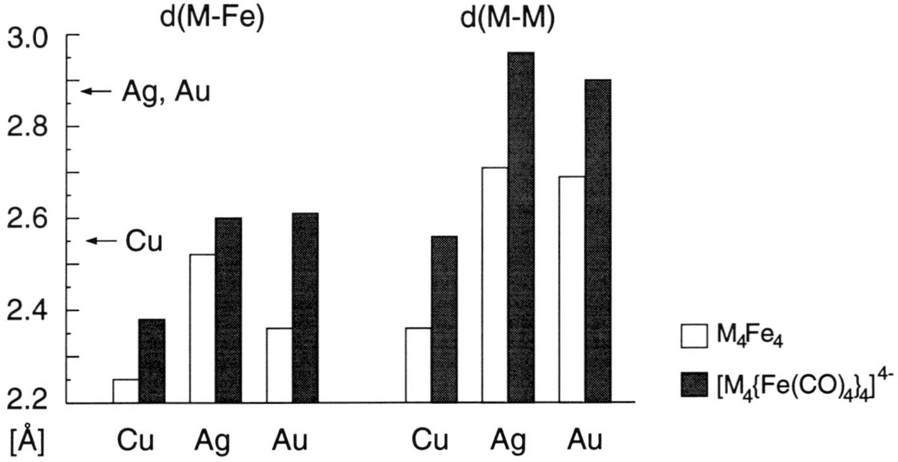

Figure 18. M-M and Fe-M bond lengths of M_4Fe_4 and $[M_4\{Fe(CO)_4\}_4]^{4-}$ (M = Cu, Ag, Au). The corresponding M-M bulk distances are indicated by arrows.

of the metal-metal bonds (Fig. 18) for the following reason. The s derived metal orbitals contributing to the M-M bond lie close to the Fermi edge for the ligand free clusters. Due to the interaction with the CO ligands these levels are pushed up and the M-M bonds are weakened. The particularly long M-M distance in the Ag complex reflects the larger energy separation of the atomic valence s and d orbitals in comparison to Cu and Au, essentially preventing Ag s-d hybridization. As a consequence, the metal-metal binding capability of silver is lower; this is reflected in the atomization energies of the M_4 subunit (Ag: 385 kJ/mol, Cu: 500 kJ/mol, Au: 520 kJ/mol).

An explanation of the planar structure of the clusters $[M_4\{Fe(CO)_4\}_4]^{4-}$ (Fig. 17) may be derived from the charge distribution. There is no charge separation in the bare clusters M_4Fe_4. However, the interaction of M_4 with the $Fe(CO)_4$ units leads to a charge transfer from M to Fe via the destabilization of the s-derived bonding orbitals of M_4, leaving the noble metal atoms in an effective +1 oxidation state. The Mulliken charges of the noble metals amount to 0.22, 0.35, and 0.44 au for Au, Cu, and Ag, respectively. As in other complexes of these coinage metals, linear coordination is favored in the +1 oxidation state, rationalizing the approximately quadratic planar structure of the M_4Fe_4 unit. The deviation from an ideal square, which shows up in Fe-M-Fe angles of about 160°, indicates a rather weak direct metal-metal interaction; the calculations show that this direct interaction is again weakest for Ag [117].

The importance of s-d hybridization is also reflected in the relative stabilities of the clusters. Comparable atomization energies have been calculated for Cu_4Fe_4 and Au_4Fe_4 (about 1525 kJ/mol) while Ag_4Fe_4 is less stable by 370 kJ/mol. A strong contribution to the stability of the carbonylated clusters is due to the binding of the carbonyl ligands. Furthermore, there is a small destabilization due to the addition of 4 electrons to form the anionic clusters for Cu (50

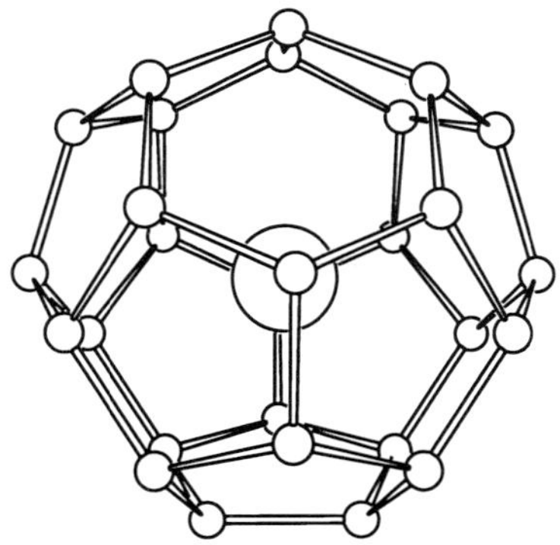

Figure 19. The endohedral carbon cluster compound Ce@C_{28}.

kJ/mol) and Au (200 kJ/mol), but an additional stabilization for the Ag cluster (135 kJ/mol). Overall these contributions lead to comparable fragmentation energies for the carbonylated clusters (Cu: 4225 kJ/mol, Ag: 4285 kJ/mol, Au: 4160 kJ/mol), with the Ag species being slightly more stable than its neighbors. The latter finding includes a relativistic increase of the fragmentation energy $[Ag_4\{Fe(CO)_4\}_4]^{4-} \rightarrow 4\ Ag + 4\ Fe + 16\ CO + 4\ e$ by 175 kJ/mol. Thus, if the reason for the failure of the synthesis of $[Cu_4\{Fe(CO)_4\}_4]^{4-}$ is of thermodynamic nature, it must be looked for in the relative stability of the reactants and not in the lower stability of the product [117].

5.3. Endohedral carbon cluster compounds

Ever since the discovery of fullerenes [173,174] the chemistry of these carbon cage molecules has attracted considerable interest [175,176] which was founded in the drive to prepare novel materials, but which undoubtedly was also connected to the appealing geometric structure of these compounds. By means of mass spectrometry one of the smallest of these cages, C_{28}, was found to form endohedral complexes with tetravalent metals, Ti@C_{28}, Zr@C_{28}, Hf@C_{28}, and U@C_{28}, with increasing abundances from Ti to U [177]. The notation M@C_{28} indicates a metal atom situated inside the carbon cage (Fig. 19). RLCGTO-DF calculations have been performed on Ti@C_{28} [178] and on Ce@C_{28} [179,180] to compare their relative stabilities and to analyze the bonding of the metal inside the carbon cage. Ce@C_{28} was taken as a model for U@C_{28} since the organometallic chemistry of the lanthanide cerium shows some similarities to the actinide uranium.

The electronic structure of fullerenes may qualitatively be understood by viewing the carbon cage as a spherical shell, held together by sp^2 hybridized σ orbitals. Furthermore, each carbon atom contributes one electron in a p orbital to form π orbitals delocalized over the cage whose nodal structure may approximately be classified according to the spherical harmonics. As usual, the delocalized π orbitals are filled in the order s, p, d, f, etc. For C_{28}

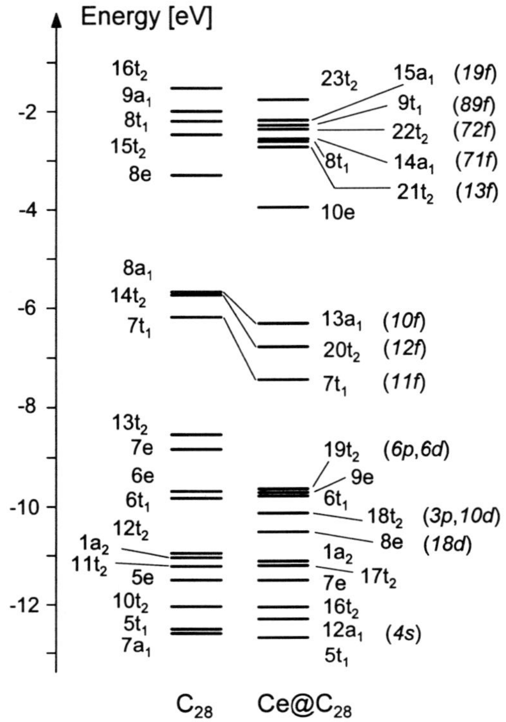

Figure 20. One-electron energy level diagram for C_{28} and Ce@C_{28}. The LUMOs are 8e and 10e, respectively. C_{28} has a quintet ground state (the levels $14t_2$ and $8t_1$ are half filled), Ce@C_{28} has a closed shell. The Mulliken contributions of Ce (in %) to each level of Ce@C_{28} are given in parentheses. For C_{28}, the spin restricted spectrum is shown for clarity.

this model leads to a tetravalent open shell electronic structure since four electrons are missing to complete the f shell. Indeed the open f shell is manifest as an energetically well separated group of levels in the one-electron spectrum of C_{28} (Fig. 20) and the corresponding ground state is a quintet [180]. The proposed tetrahedral symmetry of C_{28} and σ-π mixing leads to a splitting of the shell structure of the simple spherical model, thus the lower π-derived shells are not separated into groups according to increasing angular momentum.

As the main difference between Ti@C_{28} and Ce@C_{28} it has been found that Ti is too small to efficiently bind to all sides of the carbon cage from the central position [178]. Rather, its equilibrium position is displaced by 0.52 Å from the center of C_{28} (diameter about 4.5 Å) along a C_3 axis and is stabilized by 198 kJ/mol relative to the central position. The binding occurs predominantly between the d orbitals of Ti and the valence π orbitals of C_{28} with a high binding energy of 1083 kJ/mol (in LDA). In line with the experimental evidence, a higher binding energy (1322 kJ/mol) is calculated for Ce in C_{28}, and the Ce atome is found to prefer the central position in the cage. Due to the interaction with the Ce atom, the open f-like shell of C_{28} is completely filled and stabilized (Fig. 20). Thus, a formal oxidation state

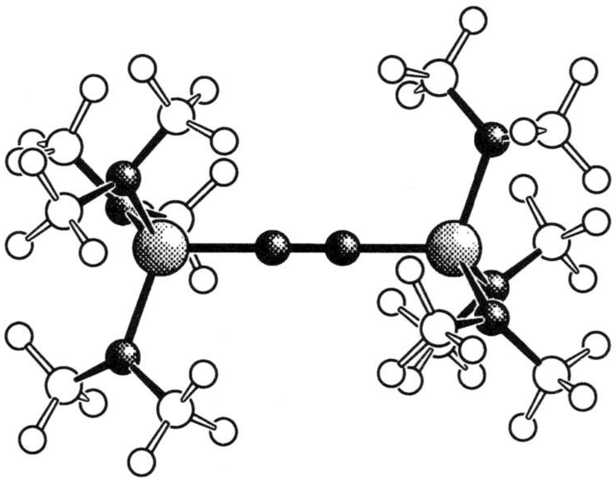

Figure 21. Model of the intermediate complex $(NMe_2)_3Mo\text{-}N\text{-}N\text{-}Mo(NMe_2)_3$ of the N_2 cleavage reaction catalyzed by $[(ArRN)_3Mo]$.

of +4 has to be ascribed to Ce. It is worth noting that the valence shell of the lanthanide Ce (as well as the actinide U) comprises f orbitals, just like the frontier group of the C_{28} cage. Thus, one may expect some extra stabilization for these f elements. Indeed, clear indications of f-covalency in the metal-carbon bonding can be seen in the Mulliken populations (Fig. 20). These f covalency contributions to the frontier orbitals are comparable to those found in the uranium sandwich compound uranocene, $U(C_8H_8)_2$ [181]. Also the π levels of the d shell are lowered by the interaction with Ce and contribute to the strong metal-carbon bonding. Compared to Ti, not only the size of Ce is better suited for simultaneous bonding to all sides of the carbon cage, but also the valence electronic structure of the f element provides orbitals of all angular momentum values to match the π orbitals of the cage C_{28}.

5.4. Metal complex reactions and catalysis

5.4.1. Nitrogen activation by molybdenum amide complexes

It is well documented in the literature [1] and demonstrated in the present article that a relativistic description is crucial to adequately reproduce properties of compounds and materials containing atoms of the third transition row or heavier ones. The presence of a second-row transition metal atom in a chemical species is usually accompanied by significantly smaller relativistic effects (see section 5.2.4). However, these effects are often still noticeable. For instance, a relativistic increase of 16-32 kJ/mol (or 20-30%) has recently been computed for various Pd-ligand bonds which show only moderately large binding energies of 65-130 kJ/mol [182]. It is interesting to search for 4d-metal systems which feature stronger relativistic effects. Suitable candidates are likely to be found among complexes with very strong metal-ligand bonds as the relativistic effects are expected to approximately correlate with the binding energy. In this section we briefly report a RLCGTO-DF study [183] which was aimed at elucidating how the extremely strong triple bond $Mo{\equiv}N$ is formed, stressing the unusually

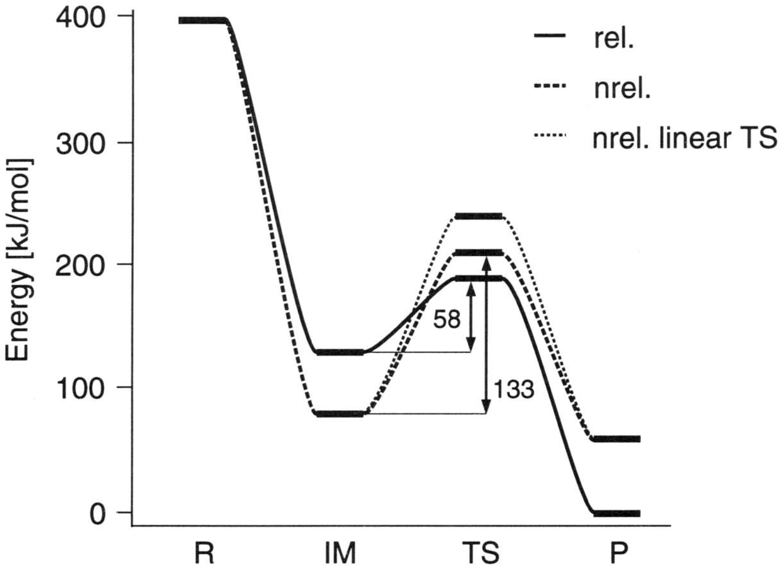

Figure 22. Sketches of energy profiles for the reaction $2[(Me_2N)_3Mo] + N_2 \rightarrow 2[(Me_2N)_3Mo{\equiv}N]$. The relative energetic positions for the reactants (R), the intermediate state (IS), the transition state (TS) and the products (P) are shown as obtained from nonrelativistic (nrel.) and relativistic (rel.) LCGTO-DF calculations. The energy for the transition state under a linear geometry constraint (nrel. linear TS) is also shown.

strong relativistic effects. The latter issue had not been addressed in a similar DF investigation [184] which employed a relativistic effective core potential approach.

The recent discovery of N_2 scission under mild conditions by the three-coordinate Mo(III) amide complex $[(ArRN)_3Mo]$ [185] (R = $C(CD_3)_2Me$, Ar = 3,5-$C_6H_3Me_2$ and Me = CH_3, methyl), has benefited from the existence of stable complexes $[L_3Mo{\equiv}N]$ which exhibit a strong metal-nitrogen triple bond. This novel direct way [186–188] to cleave one of the strongest known bonds, N≡N, at the expense of forming two Mo≡N bonds involves a dramatic spin quenching from a quartet state of the two reactants $[L_3Mo]$ and N_2 to the closed shell products.

The Mo-N distance obtained in RLCGTO-DF calculations on the reactant models $[(H_2N)_3Mo]$, **1**, and $[(Me_2N)_3Mo]$, **2**, 2.00–2.02 Å, is close to the mean value of 1.97 Å measured for complexes with more bulky ligands [189]. It is instructive to view the reactant as isolobal to a nitrogen atom. The calculated electronic configuration of **1** and **2**, a^1e^2, exhibits three unpaired electrons, in line with the experimental findings. The strong triple bond in the diamagnetic complexes $[L_3Mo{\equiv}N]$ is manifested by a short bond distance (measured 1.63 Å [190]; calculated 1.67 Å for the model complexes **1'** with L = NH_2 and **2'** with L = NMe_2) as well as by a rather high vibrational frequency ω(Mo≡N) of 1070 cm^{-1}, calculated for **2'** (experiment: 1042 cm^{-1} [185,189]). No experimental estimates seem

Table 14. Characteristic energy changes (in kJ/mol, negative values indicate an exothermic process) of the nitrogen cleavage reaction $2[L_3Mo] + N_2 \rightarrow 2[L_3Mo{\equiv}N]$ calculated by nonrelativistic (nrel.) and by relativistic[a] (rel.) DF methods: conversion of the reactants (R) into the intermediate (IM) $[L_3Mo\text{-}N\text{-}N\text{-}MoL_3]$, $\Delta E(\mathrm{R}\rightarrow\mathrm{IM})$; conversion of IM into the products (P), $\Delta E(\mathrm{IM}\rightarrow\mathrm{P})$; barrier toward the corresponding transition state[b] (TS), $\Delta E(\mathrm{IM}\rightarrow\mathrm{TS})$.

Ligand L	NH_2		NMe_2	
	nrel.	rel.	nrel.	rel.
$\Delta E(\mathrm{R}\rightarrow\mathrm{IM})$	-299	-295	-316	-267
$\Delta E(\mathrm{IM}\rightarrow\mathrm{P})$	-17	-95	-32	-128
$\Delta E(\mathrm{IM}\rightarrow\mathrm{TS})$	134	75	133	58

[a] Computed for the structures optimized at the nonrelativistic level of theory.
[b] The TS energy and geometry estimates are based on a stepwise variation of the N-N distance.

to be available for the corresponding bond energy $E_b(Mo{\equiv}N)$; the nonrelativistic values are 643 kJ/mol for **1'** and 657 kJ/mol for **2'**. Relativistic effects increase these energies to 678 and 681 kJ/mol, respectively. In total the formation of two Mo≡N bonds provides sufficient energy to induce the cleavage of N_2: $2E_b([L_3Mo{\equiv}N]) \approx 1360$ kJ mol^{-1} (calc.) $> E_b(N{\equiv}N) = 950$ kJ mol^{-1} (calc.).

The N_2 cleavage has been suggested [185] to proceed via a stable paramagnetic intermediate $[L_3Mo\text{-}N\text{-}N\text{-}MoL_3]$. Calculations for **1'-1'** and **2'-2'** indeed resulted in a linear intermediate complex (Fig. 21). It exhibits D_{3d} symmetry and has an open-shell electron configuration, e^2. This intermediate is stabilized by about 300 kJ/mol compared to the reactant system $2[L_3Mo] + N_2$ (Table 14, Fig. 22). The structures of the intermediates are similar to those of known analogous complexes $[L_nMo\text{-}N\text{-}N\text{-}MoL_n]$ [191,192]. The corresponding Mo-N bond length is calculated to 1.77 for **1'-1'** and 1.80 Å for **2'-2'**, whereas experimental distances range from 1.82 to 1.91 Å [191,192]. Even better agreement is achieved for the N-N distance which is calculated to 1.20–1.21 Å for all model dimers; the measured values fall into the range of 1.20–1.24 Å [191,192].

For N-N distances from about 1.1 Å (free N_2, reactants) to about 1.3 Å relativistic effects on the potential energy curves of the dimers are of subordinate importance only (Table 14). A further stretching of the N-N bond leads to the transition state complex $[L_3Mo\text{-}N\text{-}N\text{-}MoL_3]$ where relativistic effects become crucial (Fig. 22). Relativistic effects also seem to facilitate an earlier cross-over from the open-shell to the closed shell electronic structure, around d(N-N) = 1.4 Å, and they increase the exothermicity of the conversion of the intermediate into the products by 80–100 kJ/mol (Table 14).

A quite unusual zig-zag structure of the Mo-N-N-Mo unit with Mo-N-N angles of 137° for **1'-1'** and 135° for **2'-2'** has been calculated in the vicinity of the transition state (closed-shell

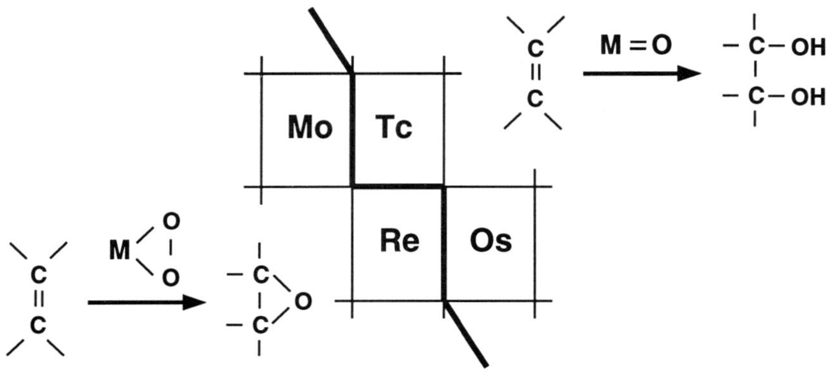

Figure 23. Schematic representation of the borderline within the periodic table of elements that separates the two regimes of oxo compounds catalyzing epoxidation (left side) or cis-dihydroxylation (right side).

configuration, d(N-N) = ~1.5 Å). This structure suggests a bonding scheme of the type Mo=N-N=Mo, with sp^2 hybridization and one lone-pair on each nitrogen. Interestingly, the calculated Mo-Mo distance remains almost unchanged, 4.74–4.81 Å, when the N-N bond stretches from the equilibrium of the intermediate to the transition state. Along the reaction path, the N-N moiety simultaneously stretches and turns around its center by an angle of up to ~30°. The implied breaking of the effective D_{3d} symmetry entails a lowering of the activation barrier by ~60 kJ/mol for **2'-2'** (Table 14). This effect together with that of the amide substituents (cf. ΔE(IM→TS) for **1'-1'** and **2'-2'**, Table 14) and the relativistic stabilization of the product species brings the calculated height of the reaction barrier in reasonable agreement with the measured kinetic parameters [185]. The nonrelativistic height of the activation barrier is about a factor of two larger, 133 kJ/mol (Fig. 22), which corresponds to a kinetically forbidden reaction in clear contradiction to experiment.

5.4.2. Oxygen transfer by metal complexes

Methyltrioxorhenium CH_3ReO_3, its derivatives and other oxo complexes of early transition metals are powerful catalysts for olefin epoxidation [193–195]. Several hypotheses on the mechanism have been advanced that differ in particulars, but agree on the two basic steps: First, the metal oxo complex is activated by an oxidant to the peroxo species which then reacts further with an olefin under epoxidation [196–204]. The lighter homologue CH_3TcO_3, as well as OsO_4 and other oxo complexes of the late transition metals, reveal a totally different oxidation behavior by mediating cis-dihydroxylation [193,205,206]. In this case it is assumed that the olefin reacts first with an M=O double bond to an intermediate ester-type structure that is then cleaved by the oxidant [207–210]. The borderline between the two regimes of different reactivity passes diagonally between the element pairs V/Cr, Mo/Tc, and Re/Os as indicated in Figure 23.

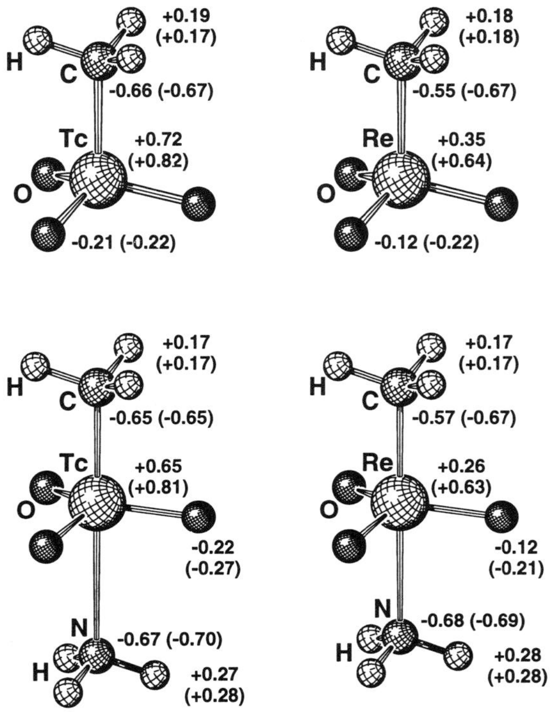

Figure 24. Structures and partial charges (in au) from a Mulliken analysis for the compounds CH_3MO_3 (M = Tc, Re) and of their NH_3 adducts. The charge values from nonrelativistic calculations are given in parentheses.

Despite the fundamental difference in reactivity towards oxidant and olefin, there are strong experimental indications that the two compounds CH_3TcO_3 and CH_3ReO_3 are isostructural [205]. Steric effects should be of equal magnitude for both molecules and therefore these compounds constitute an ideal model system to work out the electronic factors relevant for their different reactivity. In order to provide a concise quantitative picture of the influence which relativistic effects exert on the electronic structure of these complexes, both scalar relativistic and nonrelativistic density functional calculations employing the Becke-Perdew GGA [9,10] have recently been carried out for the compounds CH_3TcO_3 and CH_3ReO_3 [211]. Furthermore, NH_3 was chosen as a model for the hard Lewis base H_2O_2 which is required to activate CH_3ReO_3, but not in case of the corresponding Tc compound for olefin oxidation (Fig. 24).

On relativistic treatment excellent agreement with experiment is obtained for geometrical and vibrational data of CH_3MO_3 (M = Tc, Re) and of their NH_3 base adducts (Table 15); the Re-C bond is markedly elongated if relativistic effects are not accounted for. A significantly stabilized base adduct is calculated only for CH_3ReO_3 (E_b = 46 kJ/mol) with a Re-N distance of 2.51 Å (exp. 2.40–2.50 Å) if relativistic effects are taken into account. The calculations

Table 15. Observable quantities of the base adducts of CH_3MO_3 (M = Tc, Re) with NH_3. Comparison of nonrelativistic (nrel.) and scalar relativistic (rel.) GGA results for bond lengths d, binding energies E_b, vibrational frequencies ω, and force constants k.

	$CH_3ReO_3{\cdot}NH_3$		$CH_3TcO_3{\cdot}NH_3$	
	rel.	nrel.	rel.	nrel.
d(M-C) (Å)	2.10	2.14	2.11	2.12
d(M-O) (Å)	1.73	1.76	1.73	1.74
d(M-N) (Å)	2.51	2.57	2.58	2.60
E_b (kJ/mol)	46	28	26	25
ω(M-N) (cm^{-1})	302	254	219	220
k(M-N) (mdyne/Å)	0.79	0.60	0.37	0.43

for the Re complex at the nonrelativistic level yield results that resemble closely the ones obtained for the Tc complex at both levels of theory. Longer M-N distances (by about 0.1 Å) and lower stretching frequencies for the M-N bond are found in line with the fairly small values of the association energy (Table 15). According to an analysis of Mulliken populations and of core and valence level shifts, charge transfer from NH_3 occurs preferentially to the Re center in CH_3ReO_3 and to the oxygen atoms in CH_3TcO_3.

From the charge distribution in CH_3MO_3 and in the corresponding NH_3 adducts it is obvious that Re has in general a larger effective electronegativity within the molecular framework than Tc. Especially if treated relativistically, CH_3ReO_3 becomes the strongest and most reactive Lewis acidic metal center among the models compared (Fig. 24). The hardness, as measured by the HOMO-LUMO splitting and by the M-O displacement derivatives of the charges as well as of the dipole moment [212,213], increases in the order $CH_3TcO_3 < CH_3ReO_3$ (Fig. 25). This finding rationalizes the fact that CH_3TcO_3 (and OsO_4) reacts preferentially with softer bases, whereas CH_3ReO_3 prefers to react with hard bases. As the strongest and most reactive Lewis acidic metal center, Re withdraws electronic charge not only from the oxygen centers of the ReO_3 moiety, but also from the methyl ligand and from the donor base. The experimentally observed reduction of the catalytic activity upon adduct formation between CH_3ReO_3 and a donor base can be explained by the synergistic effect of three factors: enhanced steric crowding at the metal center and reduction of both the Lewis acid strength and the hardness due to charge transfer from the NH_3 fragment (Fig. 25).

Viewing together all these results one is able to explain the observed chemoselectivities in olefin oxidation by reference to the HSAB principle (hard and soft acids and bases [214,215]) stated above: CH_3ReO_3 exhibits a propensity to react with a harder Lewis base such as H_2O_2 to the well-known catalytically active mono- and bis-peroxo complexes [216]. Instead of oxidizing olefins directly it has even been observed to react with epoxides and further to glycolate complexes if no H_2O_2 is present [195,217]. From the present findings one

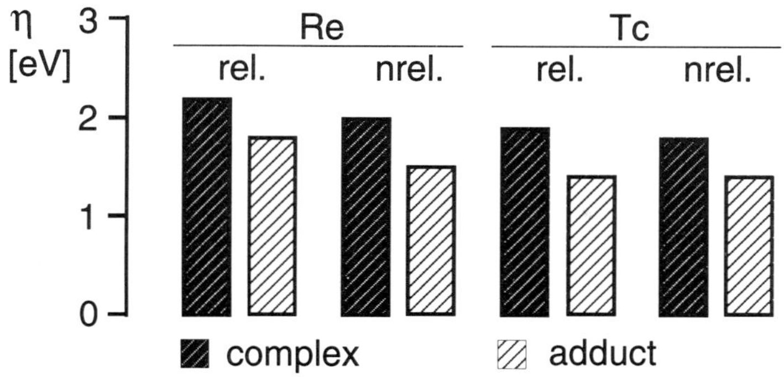

Figure 25. Hardness η from the HOMO – LUMO splitting (in eV) of the compounds CH_3MO_3 (M = Tc, Re) and of their NH_3 adducts.

expects the same reactivity pattern for other transition metal complexes on the left hand side of the above mentioned borderline between V/Cr, Mo/Tc, and Re/Os, at least as far as electronegativity and hardness are comparable to the corresponding values for Re in the relativistic case. Upon nonrelativistic treatment, CH_3ReO_3 exhibits a higher polarizability comparable to that of the MO_3 moiety calculated for CH_3TcO_3 and OsO_4. Hence it is plausible that the preferred reaction partners of late transition metal oxo complexes are soft Lewis bases, e. g. as provided by the π bond of an olefin. In the special case of the element pair Tc/Re we are able to attribute the crucial difference in Lewis acidity strength and in the polarizability to relativistic effects. However, the same reactivity trends should hold for other transition metal oxo compounds along the borderline as well.

6. OUTLOOK

Starting with a brief review of relativistic density functional theory we have presented pertinent ideas for realizing a numerically stable relativistic Kohn-Sham procedure. The accuracy and efficiency of the RLCGTO-DF method, based on these ideas, has been demonstrated by benchmark results as well as by a series of applications ranging from metal clusters to metal complexes and their reactions. Presently, we are supplementing the RLCGTO-DF program by a treatment of the spin-orbit interaction. In the Douglas-Kroll-Hess scheme this is straightforward although it requires the introduction of a two-component complex-valued description for the orbital wave functions. This extension of the method will open new areas of application, albeit at a larger computational effort.

The present survey of some recent applications demonstrates the wide applicability of molecular density functional methods, their practicability and accuracy. Density functional methods provide today a very promising electronic structure approach, especially for systems with several heavy transition metal atoms. In the future, we expect to see an increasing number of applications based on these and related techniques in the field of catalysis where third-row transition metal compounds are often involved.

ACKNOWLEDGMENT

The authors are indepted to Dr. O. D. Häberlen, Prof. W. A. Herrmann, Prof. G. Pacchioni, Dr. S. Ph. Ruzankin, and Prof. H. Schmidbaur for valuable contributions to various joint projects. We thank Dipl.-Chem. K. Albert and Dipl.-Chem. S. Köstlmeier for their assistance during the preparation of this manuscript. We have also profited much from the contributions made by numerous gifted members of the Munich group. The investigations described in this review have been made possible by grants from the Deutsche Forschungsgemeinschaft, from the Bayerische Forschungsverbund Katalyse FORKAT, and from the Fonds der Chemischen Industrie.

REFERENCES

1. P. Pyykkö, Chem. Rev. **88**, 563 (1988).
2. J. P. Desclaux, At. Data Nucl. Data Tables **12**, 311 (1973).
3. H. Schmidbaur, Gold Bull. **23**, 11 (1990).
4. S.-C. Chung, S. Krüger, G. Pacchioni, and N. Rösch, J. Chem. Phys. **102**, 3695 (1995).
5. G. E. Brown and D. G. Ravenhall, Proc. R. Soc. London Ser. A **208**, 552 (1951).
6. E. Engel, H. Müller, C. Speicher, and R. M. Dreizler, in *Density Functional Theory*, edited by E. K. U. Gross and R. M. Dreizler, Physics Vol. 337 of *NATO ASI Series B* (Plenum Press, New York, 1995), p. 65.
7. J. Sucher, Phys. Rev. A **22**, 348 (1980).
8. T. Ziegler, Chem. Rev. **91**, 651 (1991).
9. A. D. Becke, Phys. Rev. A **38**, 3098 (1988).
10. J. P. Perdew, Phys. Rev. B **33**, 8822 (1986), erratum ibid. **34** (1986) 7406.
11. C. Lee, W. Yang, and R. G. Parr, Phys. Rev. B **37**, 785 (1988).
12. J. P. Perdew, J. A. Chevary, S. H. Vosko, K. A. Jackson, M. R. Pederson, D. J. Singh, and C. Fiolhais, Phys. Rev. B **46**, 6671 (1992).
13. N. C. Handy, D. J. Tozer, G. J. Laming, C. W. Murray, and R. D. Amos, Isr. J. Chem. **33**, 331 (1993).
14. M. V. Ramana and A. K. Rajagopal, Adv. Chem. Phys. **54**, 231 (1983).
15. R. G. Parr and W. Yang, *Density Functional Theory for Atoms and Molecules* (Oxford University Press, New York, 1989).
16. W. Kohn and L. J. Sham, Phys. Rev. A **140**, 1133 (1965).
17. W. H. E. Schwarz and H. Wallmeier, Mol. Phys. **46**, 1045 (1982).
18. W. H. E. Schwarz and E. Wechsel-Trakowski, Chem. Phys. Lett. **85**, 94 (1982).
19. W. Kutzelnigg, Int. J. Quantum Chem. **25**, 107 (1984).
20. R. E. Stanton and S. Havriliak, J. Phys. Chem. **81**, 1910 (1984).
21. L. L. Foldy and S. A. Wouthuysen, Phys. Rev. **78**, 29 (1950).
22. T. Ziegler, V. Tschinke, E. J. Baerends, J. G. Snijder, and W. Ravenek, J. Phys. Chem. **93**, 3050 (1989).
23. J. G. Snijders, E. J. Baerends, and P. Ros, Mol. Phys. **38**, 1909 (1979).
24. M. Douglas and N. M. Kroll, Ann. Phys. **82**, 89 (1974).

25. B. A. Hess, Phys. Rev. A **33**, 3472 (1986).
26. B. A. Hess and P. Chandra, Phys. Scr. **36**, 412 (1987).
27. B. A. Hess, Phys. Rev. A **32**, 756 (1985).
28. B. A. Hess, R. J. Buenker, and P. Chandra, Int. J. Quantum Chem. **29**, 737 (1986).
29. P. Knappe and N. Rösch, J. Chem. Phys. **92**, 1153 (1990).
30. O. D. Häberlen and N. Rösch, Chem. Phys. Lett. **199**, 491 (1992).
31. E. van Lenthe, E. J. Baerends, and J. G. Snijders, J. Chem. Phys. **101**, 9783 (1994).
32. E. van Lenthe, E. J. Baerends, and J. G. Snijders, J. Chem. Phys. **99**, 4597 (1993).
33. C. Chang, M. Pélissier, and P. Durand, Physica Scr. **34**, 394 (1986).
34. J. L. Heully, I. Lindgren, E. Lindroth, S. Lundqvist, and A. M. Mårtensson-Pendrill, J. Phys. B **19**, 2799 (1986).
35. G. Schreckenbach, T. Ziegler, and J. Li, Int. J. Quantum Chem. **56**, 477 (1995).
36. V. A. Nasluzov and N. Rösch, to be published.
37. A. K. Rajagopal and J. Callaway, Phys. Rev. B **7**, 1912 (1973).
38. A. K. Rajagopal, J. Phys. C **11**, L943 (1978).
39. A. H. MacDonald and S. H. Vosko, J. Phys. Chem. **12**, 2977 (1979).
40. J. D. Bjorken and S. D. Drell, *Relativistic Quantum Mechanics*, International Series in Pure and Applied Physics (McGraw-Hill, New York, 1964).
41. J. Sucher, Int. J. Quantum Chem. **25**, 3 (1984).
42. E. van Lenthe, PhD thesis, The Free University of Amsterdam.
43. G. Hardekopf and J. Sucher, Phys. Rev. A **30**, 703 (1984).
44. M. M. Gleichmann and B. A. Hess, J. Chem. Phys. **101**, 9691 (1994).
45. E. Engel, S. Keller, A. F. Bonetti, H. Müller, and R. M. Dreizler, Phys. Rev. A **52**, 2750 (1995).
46. N. Rösch and O. D. Häberlen, J. Chem. Phys. **96**, 6322 (1992).
47. R. Samzow, B. A. Hess, and G. Jansen, J. Chem. Phys. **96**, 6320 (1992).
48. D. C. Langreth and M. J. Mehl, Phys. Rev. B **28**, 1809 (1983).
49. J. P. Perdew and Y. Wang, Phys. Rev. B **33**, 8800 (1986).
50. J. P. Perdew, in *Electronic Structure of Solids '91*, edited by P. Ziesche and H. Eschrig (Akademie Verlag Berlin, 1991).
51. D. E. Ellis, J. Phys. B **10**, 1 (1977).
52. E. Engel, S. Keller, and R. M. Dreizler, Phys. Rev. A **53**, 1367 (1996).
53. S. H. Vosko, L. Wilk, and M. Nusair, Can. J. Phys. **58**, 1200 (1980).
54. J. P. Perdew and Y. Wang, Phys. Rev. B **45**, 13244 (1992).
55. M. V. Ramana and A. K. Rajagopal, Phys. Rev. A **24**, 1689 (1981).
56. M. Mayer, O. D. Häberlen, and N. Rösch, to be published.
57. P. J. C. Aerts and W. C. Nieuwpoort, Int. J. Quantum Chem. Symp. **19**, 267 (1986).
58. K. G. Dyall, P. R. Taylor, K. Faegri, and H. Partridge, J. Chem. Phys. **95**, 2583 (1993).
59. K. G. Dyall, J. Chem. Phys. **100**, 2118 (1993).
60. T. Bastug, D. Heinmann, W. D. Sepp, D. Kolb, and B. Fricke, Chem. Phys. Lett. **211**, 119 (1993).
61. W. Kutzelnigg, Z. Phys. D **15**, 27 (1990).

62. R. van Leeuwen, E. van Lenthe, E. J. Baerends, and J. G. Snijders, J. Chem. Phys. **101**, 1272 (1994).
63. B. I. Dunlap and N. Rösch, Adv. Quantum Chem. **21**, 317 (1990).
64. B. I. Dunlap, J. W. D. Connolly, and J. R. Sabin, J. Chem. Phys. **71**, 3396 (1979).
65. J. A. Pople, P. M. W. Gill, and B. G. Johnson, Chem. Phys. Lett. **199**, 557 (1992).
66. B. G. Johnson, P. M. W. Gill, and J. A. Pople, J. Chem. Phys. **98**, 5612 (1993).
67. A. D. Becke, J. Chem. Phys **88**, 2547 (1988).
68. P. M. W. Gill, B. G. Johnson, and J. A. Pople, Chem. Phys. Lett. **209**, 506 (1993).
69. B. I. Dunlap, J. Andzelm, and J. W. Mintmire, Phys. Rev. A **42**, 6354 (1990).
70. B. I. Dunlap and J. Andzelm, Phys. Rev. A **45**, 81 (1992).
71. J. Almlöf, K. Fægri, and H. H. Grelland, Chem. Phys. Lett. **114**, 53 (1985).
72. I. P. Grant, In *Methods in Computational Chemistry*, S. Wilson, Ed., vol. 2 (Plenum Press, New York, 1988), p. 1.
73. K. P. Huber and G. Herzberg, *Molecular Spectra and Molecular Structure*, vol. 4 (Van Nostrand, New York, 1979).
74. B. I. Dunlap and N. Rösch, J. Chim. Phys. Physico-Chim. Biol. **86**, 671 (1989).
75. R. Fournier, J. Andzelm, and D. R. Salahub, J. Chem. Phys. **90**, 6371 (1989).
76. B. Delley and D. Ellis, J. Chem. Phys. **76**, 1949 (1982).
77. P. Schwerdtfeger, M. Dolg, W. H. E. Schwarz, G. A. Bowmaker, and P. D. W. Boyd, J. Chem. Phys. **91**, 1762 (1989).
78. P. Schwerdtfeger, Chem. Phys. Lett. **183**, 457 (1991).
79. G. Jansen and B. A. Hess, Z. Physik D **13**, 363 (1989).
80. T. L. Allen, J. Chem. Phys. **26**, 1644 (1957).
81. U. Kaldor and B. A. Hess, Chem. Phys. Lett. **230**, 1 (1994).
82. P. J. Hay, W. R. Wadt, L. R. Kahn, and F. W. Bobrowicz, J. Chem. Phys. **69**, 984 (1978).
83. Y. S. Lee, W. C. Ermler, K. S. Pitzer, and A. D. McLean, J. Chem. Phys **70**, 288 (1979).
84. Y. S. Lee, W. C. Ermler, K. S. Pitzer, and A. D. McLean, J. Chem. Phys. **70**, 293 (1979).
85. J. P. Perdew, Phys. Rev. B **34**, 7046 (1986).
86. B. Miehlich, A. Savin, H. Stoll, and H. Preuss, Chem. Phys. Lett. **157**, 200 (1989).
87. G. Schmid, Ed., *Clusters and Colloids* (Verlag Chemie, Weinheim, 1994).
88. G. Benedek, T. P. Martin, and G. Pacchioni, Eds., *Elemental and Molecular Clusters* (Springer, Berlin, 1988).
89. N. Rösch and G. Pacchioni, In *Clusters and Colloids*, G. Schmid, Ed. (Verlag Chemie, Weinheim, 1994), p. 5.
90. G. Pacchioni, S.-C. Chung, S. Krüger, and N. Rösch, Chem. Phys. **184**, 125 (1994).
91. G. Pacchioni and N. Rösch, Acc. Chem. Res. **28**, 390 (1995).
92. O. D. Häberlen, M. Stener, S.-C. Chung and N. Rösch, in preparation.
93. R. Williams, *The Geometrical Foundation of Natural Structure* (Dover Publications, New York, 1979).
94. R. C. Weast, Ed., *CRC Handbook of Chemistry and Physics*, 64 ed. (CRC Press, Inc., Florida, 1983).
95. D. M. Wood, Phys. Rev. Lett. **46**, 749 (1981).

96. M. P. J. van Staveren, H. B. Brom, L. J. de Jongh, and Y. Ishii, Phys. Rev. B **35**, 7749 (1987).
97. J. P. Perdew, Phys. Rev. B **37**, 6175 (1988).
98. G. Makov, A. Nitzan, and L. E. Brus, J. Chem. Phys. **88**, 5076 (1988).
99. G. Makov and A. Nitzan, J. Chem. Phys. **85**, 9024 (1991).
100. M. M. Kappes, M. Schär, U. Röthlisberger, C. Yeretzian, and E. Schumacher, Chem. Phys. Lett. **143**, 251 (1988).
101. C. Jackschath, I. Rabin, and W. Schulze, Z. Phys. D **22**, 517 (1992).
102. I. Rabin and W. Schulze, Chem. Phys. Lett. **201**, 265 (1993).
103. O. D. Häberlen, S.-C. Chung, and N. Rösch, Ber. Bunsenges. Phys. Chem. **98**, 882 (1994).
104. C. Jackschath, I. Rabin, and W. Schulze, Ber. Bunsenges. Phys. Chem. **96**, 1200 (1992).
105. O. D. Häberlen, S.-C. Chung, and N. Rösch, Int. J. Quantum Chem. Quant. Chem. Symp. **28**, 595 (1994).
106. W. Ekardt, Phys. Rev. B **29**, 1558 (1984).
107. M. Brack, Rev. Mod. Phys. **65**, 677 (1993).
108. N. Rösch, A. Görling, D. E. Ellis, and H. Schmidbaur, Angew. Chem. Int. Ed. Engl. **28**, 1357 (1989).
109. A. Görling, N. Rösch, D. E. Ellis, and H. Schmidbaur, Inorg. Chem. **30**, 3986 (1991).
110. U. Heiz, A. Vayloyan, E. Schumacher, C. Yeretzian, M. Stener, P. Gisdakis, and N. Rösch, J. Chem. Phys. in print.
111. A. H. Sommer, Nature **152**, 215 (1943).
112. W. E. Spicer, A. H. Sommer, and J. G. White, Phys. Rev. **115**, 57 (1959).
113. F. Hensel, Adv. Phys. **28**, 555 (1979).
114. W. E. Spicer, Phys. Rev. **125**, 1297 (1962).
115. C. Yeretzian, J. Phys. Chem. **99**, 123 (1995).
116. C. Yeretzian, U. Röthlisberger, and E. Schumacher, Chem. Phys. Lett. **237**, 334 (1995).
117. K. Albert, K. M. Neyman, G. Pacchioni and N. Rösch, to be published.
118. V. Bonačić-Koutecký, L. Češpiva, P. Fantucci, C. Fuchs, M. F. Guest, J. Koutecký, and J. Pittner, Chem. Phys. **186**, 275 (1994).
119. P. S. Bagus, K. Herrmann, and C. W. Bauschlicher Jr., J. Chem. Phys. **80**, 4378 (1984).
120. P. Pyykkö and Y. Zhao, Chem. Phys. Lett. **177**, 103 (1991).
121. P. Schwerdtfeger and P. D. W. Boyd, Inorg. Chem. **31**, 327 (1992).
122. A. Haaland, J. Hougen, and H. V. Volden, J. Organomet. Chem. **325**, 311 (1987).
123. G. E. Coates and C. Parkin, J. Chem. Soc. **1963**, 421.
124. G. M. Bancroft, T. Chan, R. J. Puddephatt, and J. S. Tse, Inorg. Chem. **21**, 2946 (1982).
125. P. D. Gavens, J. J. Guy, M. J. Mays, and G. Sheldrick, Acta Cryst. B **33**, 137 (1977).
126. G. Calvin, G. E. Coates, and P. S. Dixon, Chem. Ind. (London) **1959**, 1628.
127. O. D. Häberlen and N. Rösch, J. Phys. Chem. **97**, 4970 (1993).
128. Y. L. Slovokhotov and Y. T. Struchkov, J. Organomet. Chem. **277**, 143 (1984).
129. F. Scherbaum, A. Grohmann, B. Huber, C. Krüger, and H. Schmidbaur, Angew. Chem. Int. Ed. Engl. **27**, 1544 (1988).
130. E. Zeller, H. Beruda, A. Kolb, P. Bissinger, J. Riede, and H. Schmidbaur, Nature **352**,

141 (1991).
131. H. Schmidbaur, Pure Appl. Chem. **65**, 691 (1993).
132. D. M. P. Mingos, J. Chem. Soc. Dalton Trans. **1976**, 1163.
133. A. Brodbeck and J. Strähle, Acta Cryst. A **46**, C-232 (1990).
134. H. Schmidbaur, G. Weidenhiller, and O. Steigelmann, Angew. Chem. Int. Ed. Engl. **29**, 433 (1991).
135. O. D. Häberlen, H. Schmidbaur, and N. Rösch, J. Am. Chem. Soc. **116**, 8241 (1994).
136. H. Schmidbaur, in *Gmelin Handbook of Inorganic Chemistry, Organogold Compounds*, A. Slawisch (ed.), Springer, New York, 1980.
137. F. Scherbaum, B. Huber, G. Müller, and H. Schmidbaur, Angew. Chem. Int. Ed. Engl. **27**, 1542 (1988).
138. F. Scherbaum, A. Grohmann, G. Müller, and H. Schmidbaur, Angew. Chem. Int. Ed. Engl. **28**, 463 (1989).
139. A. Grohmann, J. Riede, and H. Schmidbaur, Nature **345**, 140 (1990).
140. A. N. Nesmeyanov, E. G. Perevalova, Y. T. Struchkov, M. Y. Antipin, K. I. Grandberg, and V. P. Dyadchenko, J. Organomet. Chem. **201**, 343 (1980).
141. Y. Yang, V. Ramamoorthy, and P. R. Sharp, Inorg. Chem. **32**, 1946 (1993).
142. K. Angermaier and H. Schmidbaur, Inorg. Chem. **10**, 2069 (1994).
143. K. Angermaier and H. Schmidbaur, Acta Cryst. C **51**, 1793 (1995).
144. H. Schmidbaur, A. Kolb, E. Zeller, A. Schier, and H. Beruda, Z. Anorg. Allg. Chem. **619**, 1575 (1993).
145. K. P. Hall and D. M. P. Mingos, Prog. Inorg. Chem. **32**, 237 (1984).
146. S.-C. Chung, S. Krüger, H. Schmidbaur, and N. Rösch, Inorg. Chem., in print.
147. F. G. Gabbai, A. Schier, J. Riede, and H. Schmidbaur, Inorg. Chem. **34**, 3855 (1995).
148. D. M. P. Mingos and M. J. Watson, Adv. Inorg. Chem. **39**, 327 (1992).
149. S.-C. Chung, S. Krüger and N. Rösch, unpublished results.
150. F. A. Cotton and G. Wilkinson, *Advanced Inorganic Chemistry*, 3rd ed. (Wiley, New York, 1972).
151. A. Zangwill, *Physics at Surfaces* (Cambridge University Press, Cambridge, 1988).
152. G. A. Somorjai, *Introduction of Surface Chemistry* (Wiley, New York, 1993).
153. M. J. S. Dewar, Bull. Soc. Chem. Fr. **18**, C71 (1951).
154. J. Chatt and L. A. Duncanson, J. Chem. Soc. 2939 (1953).
155. G. Blyholder, J. Phys. Chem. **68**, 2772 (1964).
156. S.-C. Chung, S. Krüger, S. P. Ruzankin, G. Pacchioni, and N. Rösch, Chem. Phys. Lett. **248**, 109 (1996).
157. J. Li, G. Schreckenbach, and T. Ziegler, J. Phys. Chem. **98**, 4838 (1994).
158. P. S. Bagus and W. Muller, Chem. Phys. Lett. **115**, 540 (1984).
159. P. S. Bagus and G. Pacchioni, Surf. Sci. **236**, 233 (1990).
160. J. G. Louderback, A. J. Cox, L. J. Lising, D. C. Douglas, and L. A. Bloomfield, Z. Phys. D **26**, 301 (1993).
161. N. Rösch, L. Ackermann, G. Pacchioni, and B. I. Dunlap, J. Chem. Phys. **95**, 7004 (1991).
162. N. Rösch, L. Ackermann, G. Pacchioni, and B. I. Dunlap, J. Am. Chem. Soc. **114**, 3550

(1992).
163. L. Ackermann, N. Rösch, B. I. Dunlap, and G. Pacchioni, Int. J. Quantum Chem. **26**, 605 (1992).
164. D. A. van Leeuwen, J. M. van Ruitenbeek, L. J. de Jongh, A. Cerotti, G. Pacchioni, O. D. Häberlen, and N. Rösch, Phys. Rev. Lett. **73**, 1432 (1994).
165. C. S. Feigerle, A. Guler, J. L. Peña, K. J. Celotta, and D. T. Pierce, Phys. Rev. Lett. **56**, 2207 (1986).
166. H. Danan, A. Herr, and A. J. P. Meyer, J. Appl. Phys. **39**, 669 (1968).
167. C. E. Briant, R. G. Smith, and D. M. P. Mingos, J. Chem. Soc. Chem. Comm. **1984**, 586.
168. G. Doyle, K. A. Eriksen, and D. V. Engen, J. Am. Chem. Soc. **108**, 445 (1986).
169. V. G. Albano, L. Grossi, G. Longoni, M. Monari, S. Mulley, and A. Sironi, J. Am. Chem. Soc. **114**, 5708 (1992).
170. V. G. Albano, F. Azzaroni, M. C. Iapalucci, G. Longoni, M. Monari, S. Mulley, D. M. Proserpio, and A. Sironi, Inorg. Chem. **33**, 5320 (1994).
171. V. G. Albano, F. Calderoni, M. C. Iapalucci, G. Longoni, and M. Monari, J. Chem. Soc. Chem. Comm. **1995**, 433.
172. P. Schwerdtfeger, M. Dolg, W. H. E. Schwarz, G. A. Bowmaker, and P. D. W. Boyd, J. Chem. Phys. **91**, 1762 (1989).
173. H. W. Kroto, J. R. Heath, S. C. O'Brian, R. F. Curl, and R. E. Smalley, Nature **318**, 529 (1985).
174. W. Krätschmer, L. D. Lamb, K. Fostiropolous, and D. R. Huffman, Nature **347**, 354 (1990).
175. A. Hirsch, *The Chemistry of Fullerenes* (Thieme, Stuttgart, 1994).
176. H. W. Kroto and D. R. M. Walton, Eds., *The Fullerenes* (Cambridge University Press, Cambridge, 1994).
177. T. Guo, M. D. Diener, Y. Chai, M. J. Alford, R. E. Haufler, S. M. McClure, T. Ohno, J. H. Weaver, G. E. Scuseria, and R. E. Smalley, Science **257**, 1661 (1992).
178. B. I. Dunlap, O. D. Häberlen, and N. Rösch, J. Phys. Chem. **96**, 9095 (1992).
179. O. D. Häberlen, N. Rösch, and B. I. Dunlap, Chem. Phys. Lett **200**, 418 (1992).
180. N. Rösch, O. D. Häberlen, and B. I. Dunlap, Angew. Chem. Int. Ed. Eng. **32**, 108 (1993).
181. N. Rösch, Inorg. Chim. Acta **94**, 297 (1984).
182. J. Li, G. Schreckenbach, and T. Ziegler, Inorg. Chem. **34**, 3245 (1995).
183. K. M. Neyman, V. A. Nasluzov, J. Hahn, C. R. Landis, and N. Rösch, to be published.
184. Q. Cui, D. G. Musaev, M. Svensson, S. Sieber, and K. Morokuma, J. Am. Chem. Soc. **117**, 12366 (1995).
185. C. E. Laplaza and C. C. Cummins, Science **268**, 861 (1995).
186. G. J. Leigh, Acc. Chem. Res. **25**, 177 (1992).
187. G. J. Leigh, Science **268**, 827 (1995).
188. M. Hidai and Y. Mizobe, Chem. Rev. **95**, 1115 (1995).
189. C. E. Laplaza, A. L. Odom, W. M. Davis, and C. C. Cummins, J. Am. Chem. Soc. **117**, 4999 (1995).
190. Z. Gebeyehu, F. Weller, B. Neumüller, and K. Dehnicke, Z. Anorg. Allg. Chem. **593**,

99 (1991).
191. R. R. Schrock, R. M. Kolodziej, A. H. Liu, W. M. Davis, and M. G. Vale, J. Am. Chem. Soc. **112**, 4338 (1990).
192. K. Y. Shih, R. R. Schrock, and R. Kempe, J. Am. Chem. Soc. **116**, 8804 (1994).
193. K. A. Jørgensen, Chem. Rev. **89**, 431 (1989).
194. W. A. Herrmann, J. Organomet. Chem., **500,** 149 (1995).
195. A. M. Al-Ajlouni and J. H. Espenson, J. Am. Chem. Soc. **117**, 9243 (1995).
196. R. S. Drago, Coord. Chem. Rev. **117**, 185 (1992).
197. H. Mimoun, Angew. Chem. Int. Ed. Engl. **21**, 734 (1982).
198. K. B. Sharpless, J. M. Townsend, and D. R. Williams, J. Am. Chem. Soc. **94**, 295 (1972).
199. H. Mimoun, Catal. Today **1**, 281 (1987).
200. W. Thiel, M. Angstl, and N. Hansen, J. Mol. Catal. A **103**, 5 (1995).
201. K. A. Jørgensen and R. Hoffmann, Acta Chem. Scand. B **40**, 411 (1986).
202. M. J. Filatov, K. V. Shalyaev, and E. P. Talsi, J. Mol. Catal. **87**, L5 (1994).
203. K. F. Purcell, Organometallics **4**, 509 (1984).
204. Y. D. Wu and D. K. W. Lai, J. Org. Chem. **60**, 673 (1995).
205. W. A. Herrmann, R. Alberto, P. Kiprof, and F. Baumgärtner, Angew. Chem. Int. Ed. Engl. **29**, 189 (1990).
206. W. A. Herrmann, S. J. Eder, P. Kiprof, K. Rypdal, and P. Watzlowik, Angew. Chem. Int. Ed. Engl. **29**, 1445 (1990).
207. A. Veldkamp and G. Frenking, J. Am. Chem. Soc. **116**, 4937 (1994).
208. A. Neuhaus, A. Veldkamp, and G. Frenking, Inorg. Chem. **33**, 5278 (1994).
209. P.-O. Norrby, H. C. Kolb, and K. B. Sharpless, J. Am. Chem. Soc. **116**, 8470 (1994).
210. P.-O. Norrby, H. C. Kolb, and K. B. Sharpless, Organometallics **13**, 344 (1994).
211. S. Köstlmeier, V. A. Nasluzov, W. A. Herrmann, and N. Rösch, to be published.
212. M. Berkowitz, S. K. Ghosh, and R. G. Parr, J. Am. Chem. Soc. **107**, 6811 (1985).
213. M. Berkowitz, J. Am. Chem. Soc. **109**, 4823 (1987).
214. R. G. Pearson, J. Am. Chem. Soc. **85**, 3533 (1963).
215. R. G. Pearson, Inorg. Chem. **27**, 734 (1988).
216. W. A. Herrmann, R. W. Fischer, W. Scherer, and M. U. Rauch, Angew. Chem. Int. Ed. Engl. **32**, 1157 (1993).
217. Z. Zhu, A. M. Al-Ajlouni, and J. H. Espenson, Inorg. Chem **35**, 1408 (1996).

PART IV:

APPLICATIONS

J.M. Seminario (Editor)
Recent Developments and Applications of Modern Density Functional Theory
Theoretical and Computational Chemistry, Vol. 4

Adsorption complexes on oxides: Density functional model cluster studies

K. M. Neyman,[a] G. Pacchioni[b] and N. Rösch[a]

[a]Lehrstuhl für Theoretische Chemie, Technische Universität München, D-85747 Garching, Germany

[b]Dipartimento di Chimica Inorganica, Metallorganica e Analitica, Universitá di Milano, via Venezian 21, 20133 Milano, Italy

An overview of recent density functional studies of adsorption complexes formed with oxide and zeolite materials is given. The investigations carried out with the help of the Munich version of the LCGTO-DF program are exemplified by (i) complexes of a CO molecule on regular, defect and impurity sites of a MgO(001) surface, (ii) Brønsted and Lewis centres of zeolites and alumina interacting with CO and N_2 probes, (iii) metal clusters supported on Al_2O_3 and MgO surfaces, as well as electron-deficient palladium and platinum species in zeolite cavities. A brief account of selected features implemented lately in the LCGTO-DF code is also provided.

1. INTRODUCTION

In this article we shall review a series of first principle studies on oxide surfaces based on cluster models and density functional (DF) theory. From a historical point of view, the first attempts employing accurate quantum mechanical methods to investigate processes at surfaces within the cluster approach were directed to metal surfaces [1]. On the other hand, cluster methods for solid-state studies were introduced long time ago by Hans Bethe in his seminal work on the crystal field splitting of energy levels of impurity ions in ionic crystals [2]. Apart from a few examples reported at the beginning of the eighties [3], where (wave function based) ab initio methods where applied to the description of oxides, it is only in the last decade that this field has attracted considerable interest among theoreticians [4–6]. Experimentalists have always been interested in oxides as these compounds are very important materials in many areas of physics and chemistry [7–9]. However, oxides and oxide surfaces of technological importance are often complex, exhibiting plenty of defects and irregularities, so that their exact structure and composition is not precisely known. This lack of information is also due to experimental difficulties that arise when many surface science techniques are applied to characterize oxide surfaces (like the problem of sample charging in photoemission experiments). In the last few years, however, a number of new techniques for studying oxide surfaces have been developed [9]. This experimental progress turned out to be extremely useful for theoreticians because the existence of "clean" structural information permitted the design of more accurate models of these surfaces. Furthermore, in this way it became possible to complement experimental measurements by theoretical analysis to achieve well grounded explanations of the observed phenomena.

The improvement in experimental techniques coincided with a fast growth of computing power and an impressive development of computational methodologies. It is nowadays possible to perform on a workstation such elaborate calculations which, a few years ago, were not doable even on a supercomputer. At the same time, the accuracy and the reliability of the calculations has improved tremendously. The more classical wave function based methods, like the Hartree-Fock (HF) and post-HF approaches, have been complemented by the DF methods whose popularity has increased enormously during the last few years in the quantum chemistry community. Advances have been achieved not only within the static electronic structure approach, but also in the field of dynamical methods, like the Car-Parrinello method [10]. Each theoretical method exhibits special advantages depending on the problem under investigation, but there is no doubt that the arsenal of theoretical tools to investigate electronic structure problems in field of solid state and surfaces has grown to a point where it is possible to answer a very large number of questions.

Oxides represent a large and complex class of materials. Some are dominantly ionic, with high formal oxidation numbers of the metal atoms, some are largely covalent with little or no charge transfer; some are insulating, some metallic, some feature peculiar superconducting properties; most are diamagnetic, but many manifest a complex ferromagnetic or antiferromagnetic behaviour; some surfaces exhibit features typical for an acid substance, other have a definitely basic character, etc. It is therefore clear that the electronic structure of oxides is very diverse and that the theoretical methods used for their characterization must be flexible enough to cover the wide range of bonding situations encountered and of properties observed in nature. It is the primary advantage of first principle approaches that with their use one is able to address these problems without recourse to empirical parameters.

To some extent, the choice of methodology remains a matter of convenience, but for a given problem one will select a treatment which is able to provide the better answer with the smaller effort. It is obvious that a highly localized process, e. g. the photoionization of core levels, can be better treated with the local approach of a cluster model where the surface is approximated by a small number of atoms [11]. On the other hand, extended phenomena like a surface reconstruction require band structure methods where the periodic nature of the surface is properly taken into account. Chemisorption issues provide examples for problems which can be accurately described with both approaches, local and periodic. The reaction of a carbon dioxide molecule with a surface oxide anion to form a surface carbonate, the interaction of a CO molecule with acid protons of a zeolite cavity, the deposition of a small metal cluster on an insulating substrate are all phenomena which involve very local electronic interactions.

When the characterization of this interaction in terms of classical chemical bonds is the main objective of the investigation, the cluster model approach presents particular advantages. In fact, the use of molecular orbitals to describe bonding at surfaces allows one to compare in a very direct way this kind of interaction with that of molecules and molecular complexes using a language which is very familiar to chemists. Furthermore, as will be illustrated below, special tools and techniques have been developed within the cluster model method to analyse the wave function or the charge density of a system with the particular aim to evaluate the importance of individual bonding mechanisms, like covalent or ionic effects, electrostatic contributions, polarization terms, etc.

A problem connected with the cluster modelling of condensed matter systems is that in some way the effect of the surrounding must be taken into account. This is particularly important when dealing with highly ionic systems where the long-range Coulomb potential, the Madelung potential, plays a central role. In this case simple techniques based on the use of point charges to represent the host crystal have been used with some success. At the other extreme are covalent solids where a simple ionic model breaks down. For this type of systems saturation of the dangling bonds by hydrogen atoms is commonly used, often with satisfactory results. Of course, more elaborate formalisms are also available, like rigorous embedding procedures [12,13], but their use is not always straightforward and therefore still restricted to a few examples.

In the following we shall discuss the status of the field based on our personal experience. Therefore, we make no pretence of completeness or of a unified view. Instead, a discussion is given of the potential of the combined use of the cluster model approach and of DF methods through selected case studies. These examples cover quite a broad range of applications, from the analysis of vibrational modes to the determination of adsorption energies, from adsorption at oxide defects and impurities to the metal/oxide interface.

This review is organized as follows. In the next chapter we shall give an outline of features recently implemented in the DF computing code employed and of some computational details. Chapter 3 is devoted to magnesium oxide, exemplifying ionic metal oxide substrates, and to the bonding and vibrations of a typical probe molecule, CO, adsorbed in various positions of a MgO surface. Interactions of probe molecules with Brønsted and Lewis acid centres in zeolites and on aluminum oxide are discussed in chapter 4. The important issue of transition metal species supported on oxide surfaces or encapsulated in zeolite cavities is addressed in chapter 5. The last chapter summarizes the data presented and gives an outlook for studies of oxide materials and of their chemical interactions within DF methods.

2. METHOD AND COMPUTATIONAL DETAILS

Most of the results discussed below were obtained in the model cluster approach with the help of the Munich version of the linear combination of Gaussian-type orbital density functional (LCGTO-DF) code [14–16]. Selected options recently implemented in the LCGTO-DF computing suite as well as basic computational details are outlined in this chapter. Periodic DF slab model calculations of MgO(001) surface were performed with the FILMS program [17,18], a band structure code which, like the LCGTO-DF cluster method, employs Gaussian basis set techniques for representing crystal orbitals and the electron charge density. The program SCHAKAL92 [19] has been utilized to sketch structures of the cluster models considered.

Generalized gradient approximation (GGA). The computationally economic local density approximation (LDA) to the exchange-correlation functional has been successfully used in chemical applications since the seventies. This functional, e. g. in the parametrisation suggested by Vosko, Wilk and Nusair (VWN) [20], has been demonstrated to accurately describe important properties of adsorption complexes and other molecular species such as bond lengths and vibrational parameters. However, the calculated LDA binding energies were often significantly overestimated [16]. In the past few years it became apparent that density gradient corrections to LDA exchange-correlation functional considerably improve the calculated molecular energetics [21]. Several "nonlocal" exchange-correlation functionals of the GGA form [22–27] have been implemented in the LCGTO-DF program. The GGA expressions suggested by Becke [22] or, alternatively, by Perdew and Wang [26,27] were utilized to calculate density gradient corrections to the exchange energy functional. The corresponding correlation contribution was computed following the work of Perdew and collaborators [25,26]. The pertinent exchange-correlation contributions to the Kohn-Sham matrix are calculated using a direct technique involving scalar products of the first order partial derivatives of the energy functionals and of the wave functions [28]. The following three GGA variants are in use: Becke-Lee-Yang-Parr (BLYP) [22–24], Becke-Perdew (BP) [22,25] and Perdew-Wang (PW) [26,27].

Geometry optimization and vibrational analysis based on analytic gradients of the total energy. Analytically calculated gradients of the total energy with respect to the nuclear coordinates are required for the automatic geometry optimization of molecular systems. They have been implemented in the LCGTO-DF code in a form which ensures a strictly variational definition of the total energy. A high-quality numerical integration scheme [29] was implemented together with the appropriate gradient technique [28] to obtain values of the forces acting on the nuclei which are consistent with the total energy and its variation upon displacement of the nuclei. Options are available to optimize either all degrees of freedom of a given cluster model or a limited number of specified structural parameters. A further extension allows one to automatically carry out a vibrational analysis of molecular moieties and to locate saddle points of the total energy surface using a rational function approximation [30]. For any given symmetry constraint, it is possible to calculate totally symmetric vibrational frequencies and infra-red (IR) intensities with or without geometry optimization of the whole cluster. Very recently the self-consistent scalar-relativistic (SR) version of LCGTO-DF method [31–33] has also been extended for calculating analytic energy gradients. The SR variant employs a unitary second order Douglas-Kroll-Hess (DKH) transformation for decoupling large and small components of the full four-component spinor solutions to the Dirac-Kohn-Sham equation. The approximate DKH transformation most appropriate and most efficient for molecular calculations has been implemented; this variant utilizes nuclear potential-based projectors and it leaves the electron-electron interaction untransformed. This extension of the LCGTO-DF code for the first time furnishes the possibility of optimizing geometries of polyatomic systems containing heavy elements by applying a rigorously variational DF scheme; previous relativistic DF implementations relied on some perturbation schemes [34]. For a more detailed account of the recent SR-DKH extension see [35] and a paper by Rösch et al. in this volume.

Non-empirical effective-core potentials. Despite the high efficiency of the all-electron LCGTO-DF code, it is still desirable to reduce the required computational effort without loss of accuracy in order to treat more realistic models. An attractive strategy for achieving this goal is to limit the number of electrons that are explicitly taken into consideration to those of the valence shells. To follow this line the LCGTO-DF program has been supplemented by a module which allows for the construction and use of "norm-conserving" atomic effective-core potentials (pseudopotentials) [36–38] in the framework of the DF method.

Constraint space orbital variation (CSOV) analysis. The CSOV method [39] is a useful tool to analyse various contributions to the binding energy, to the dipole moment and to other observables in a system AB in which two fragments A and B interact. A CSOV analysis permits a separation of contributions of various interaction channels, like the Pauli repulsion, the polarization of both fragments, and charge transfer A $\rightarrow$ B and B $\rightarrow$ A. The CSOV technique has been recently implemented for the first time in the framework of a DF program [40–42].

Basic computational details. Relatively large orbital Gaussian basis sets of the triple-zeta or better quality have been used in the LCGTO-DF and FILMS all-electron calculations to be discussed below. The two auxiliary bases sets employed in the LCGTO-DF method to represent the electron charge density and, in some cases, the exchange-correlation potential were constructed from the orbital exponents as proposed in ref. 15. For weakly bound molecular systems the "counterpoise" technique [43] was applied to correct interaction energies for the basis set superposition error (BSSE). Vibrational frequencies have been obtained either by diagonalizing the mass-weighted force constant matrix which had been computed from finite differences of total energy gradients or by using a polynomial fit to a pointwise representation of the energy hypersurface. In the latter case both harmonic and anharmonic frequencies can be calculated by assuming a negligible coupling of the vibrational modes under consideration. Absolute intensities of the IR bands (integrated absorption coefficients) have been calculated in the double-harmonic approach [44] according to which the IR intensity is proportional to the square of the dynamic dipole moment of the corresponding normal mode.

3. MgO AS AN EXAMPLE OF IONIC METAL OXIDE SUBSTRATES

Solid magnesium oxide is, on the one hand, a textbook example of typical ionic crystals with a simple rock salt structure. On the other hand, this substance represents an important class of metal oxide catalytic materials [7,9]. Being composed of light Mg and O atoms containing few electrons, MgO is a system computationally tractable at rather high levels of sophistication. These reasons make MgO as well as adsorption complexes formed on its surfaces attractive touchstones for quantum chemical studies. This statement holds in particular for the MgO(001)/CO adsorption complex which recently has been the subject of a series of first principles HF [45–49], post-HF [47] and DF [41,48,50–52] calculations. In spite of the seemingly simple nature of the bonding within the MgO substrate, often referred to as purely ionic, as well as of the interaction between the MgO(001) surface and a CO molecule, in experiments widely used as a probe, several important issues concerning the bonding mechanism still remain open. Some of these questions will be discussed in the following.

3.1. Slab studies of MgO(001) surface

One of the questions frequently addressed when investigating adsorption processes at ionic surfaces concerns the amount of charge separation in the substrate. Information on the charge separation in the vicinity of the surface is especially important because molecular adsorption at ionic materials is to a large extent governed by electrostatic interaction. Furthermore, a quantification of the substrate ionicity is a prerequisite for constructing appropriate models for cluster calculations. In the model cluster approach a finite part of an ionic substrate is commonly embedded in an array of point charges chosen to reproduce the correct Madelung field near the adsorption site. Hence, an analysis of a clean ionic surface is necessary if the point charges are to be determined in a consistent way, rather than by simply following chemical intuition or by assuming a fully ionic model. Before discussing the CO adsorption on MgO in the sections 3.2-3.6, let us first outline the DF slab model results on the charge distribution in (001) surface of MgO [52,53].

LDA calculations of two-dimensional periodic MgO slabs have been carried out with the help of the FILMS code [17,18]. The binding energy and the Kohn-Sham one-particle core levels of slab models of increasing thickness from one to five layers reveals that the influence of the (001) surface on the MgO substrate is essentially limited to the first crystal plane [53]. Slab models of three atomic layers are sufficient for a quantitative description of many relevant aspects of the surface. These three-layer slabs already approach the convergence limit of a semi-infinite half-space filled with the ionic substrate. This finding validates the widely used strategy of constructing "flat" cluster models of the (001) surface of MgO and of other rock-salt crystals. These models are only few atomic layers thick, but cover a rather extended horizontal area. The required large surface area of the clusters can be reached by embedding of the quantum-chemically treated moiety into a suitable array of point charges.

Two schemes based entirely on the electron density have been employed to quantify the ionic charges in the surface region. The first procedure is basically the topological atom approach suggested by Bader [55]. The results of a topological atom analysis of the one-, two- and three-layer MgO(001) slab are displayed in Figure 1. Not unexpectedly, the oxygen anions are found to occupy a significantly larger volume (dotted lines) than the cations. Also, a direct interaction between the nearest anions can clearly be seen. The second scheme is based on a direct Fourier analysis of the electrostatic (Madelung) field above a slab; from the resulting moments a charge value is extracted by comparing them to those of an isostructural slab of point charges. The integrated charge of the topological atoms of the MgO(001) slab as well as the effective point charges derived from the analysis of the long-range behaviour

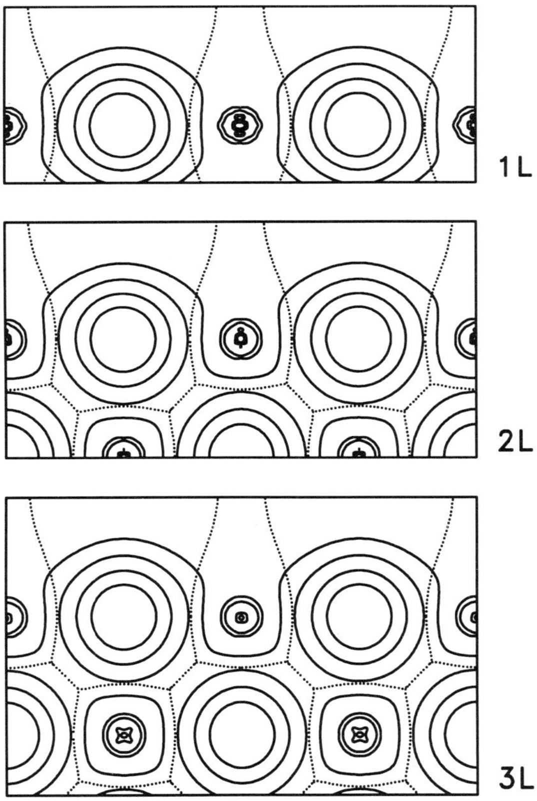

Figure 1. FILMS valence charge density of one- to three-layer MgO(001) slabs in the plane perpendicular to the surface. The dotted lines mark the borders of the topological atoms: oxygen anions – larger regions, magnesium cations – smaller regions. The vacuum region is toward the top of the various panels. The values of the contours (solid lines) correspond to 0.010, 0.032, 0.10 and 0.32 au [54].

Table 1
Atomic charges (in au) in MgO slab models of one to five layers

Layers	Topological atom method						Madelung field
	O(1)	Mg(1)	O(2)	Mg(2)	O(3)	Mg(3)	analysis
1	-1.700	+1.700					±1.603
2	-1.657	+1.657					±1.774
3	-1.658	+1.660	-1.619	+1.615			±1.810
4	-1.658	+1.660	-1.619	+1.617			±1.816
5	-1.658	+1.660	-1.619	+1.617	-1.619	+1.620	±1.810

of the electrostatic potential above the (001) surface (Table 1, for more details see [53]) yield a charge separation of ±1.66 and ±1.81 au, respectively. These values are significantly lower than the formal charges of ±2.0 au corresponding to perfect ionicity of MgO. This result is consistent with indications for a reduced charge separation in the MgO surface derived from a comparison of calculated and measured values of the vibrational frequency shift of CO adsorbed on MgO(001) [50,52]. As we shall see in chapter 5, the conclusion on a partially reduced state of the surface oxygen anions of MgO is also in line with their oxidative ability found with respect to supported metals. The reduced charge separation is related to a noticeable covalent admixture of Mg 2s and 2p orbital character to the Mg-O bonds in magnesia. On the other hand, this admixture was calculated to have only a minor effect (at most 5%) on the substrate cohesive energy [53].

As shown below, the bonding energy of adsorbates on MgO is to a significant extent due to electrostatic interaction. Therefore, the effective charges derived from the Madelung field above the surface will be preferred over the topological atom values when constructing an appropriate point charge array for embedding a DF cluster model of adsorption complexes on MgO.

3.2. Bonding and vibrations of isolated CO probe molecules on MgO(001)

Various aspects of the adsorption system MgO/CO are still under debate, among them energy and nature of the MgO-CO interaction as well as the size and the mechanism of the adsorption-induced alterations of vibrational parameters of a CO probe. The first DF data on MgO(001)-CO complexes were obtained within the LDA approach [48,50–52], therefore a comparison with recent GGA-DF results [41] obtained for the BLYP and BP energy functionals is also in order. To start, we shall discuss how the computed MgO(001)-CO interaction energy depends on the size of the substrate cluster model [41].

The problem of stability of the calculated data with respect to changes in the cluster size is of particular importance for judging the reliability of the model cluster results. Compared to other observables, like structural and vibrational data, the model cluster adsorption energy is

Table 2
Cluster size dependence of the MgO(001)-CO adsorption energy D_e calculated using the BLYP gradient-corrected exchange-correlation potential

Cluster model[a]	$D_e{}^b$, (eV)
Mg_5O_5(5,5)-CO	0.20
Mg_9O_9(9,9)-CO	0.21
$Mg_{17}O_{17}$(25,9)-CO	0.06
$Mg_{21}O_{21}$(21,21)-CO	0.11
$Mg_{25}O_{25}$(25,25)-CO	0.09

[a] The numbers of atoms in the first and second layers of the Mg_nO_n cluster models are given in parentheses. Point charges of ±1.8 au (see previous section) were used for embedding.
[b] Corrected for the BSSE.

known to be significantly more sensitive to the cluster size and geometry (see e. g. [56,57]). Also, this quantity is in general not easy to measure experimentally. For instance, experimental adsorption energies for CO on MgO(001) cover a broad range of values from 0.15 eV [58] to 0.31-0.42 eV [59] and 0.43-0.46 eV [60]. Calculated MgO(001)-CO adsorption energies scatter over an even wider range. The LDA values of 0.37-0.56 eV [50,51] probably provide an upper estimate; they were obtained for extended stoichiometric cluster models. HF calculations yield binding energy values from 0.08 to 0.25 eV, depending on cluster size [45,46,48]. On the other hand, post-HF calculations using large non-stoichiometric cluster models [47,49] resulted in the extremely low binding energy value of 0.01 eV only.

The energy oscillations for the stoichiometric clusters in Table 2 are rather modest. A noticeably lower value of the adsorption energy of CO compared to that calculated for the other models has been obtained for the substrate cluster $Mg_{17}O_{17}$(25,9). This finding may indicate a certain shape dependence since 25 - 9 = 16 peripheral atoms of the upper layer of this substrate cluster are exposed to the influence of nearest-neighbour point charges beneath them. In all other cluster models listed in Table 2 the numbers m_1 and m_2 of the first- and second-layer Mg and O atoms are equal. The electronic configuration of these extra top-layer atoms, if distorted, affects the Madelung field in the adsorption region much stronger than that of atoms in the second layer. Of course, any unphysical exposure to neighbouring point charges should be more dramatic for the highly polarizable oxygen anions affected by positive point charges than for the "hard" magnesium cations in the presence of negative point charges. Results of a CSOV analysis for the system $Mg_{17}O_{17}$-CO (where m_1 = 25 and m_2 = 9) revealed that the most important effect, noticeably decreasing the calculated adsorption energy compared to the values obtained for cluster models with $m_1 = m_2$ is the increased contribution that comprises the sum of Pauli repulsion and electrostatic attraction interactions of the initial frozen charge distributions of the separated subsystems. These two interactions are not to be separated in a CSOV analysis. The repulsive contribution computed for $Mg_{17}O_{17}$(25,9)-CO, 0.35 eV, is significantly larger than that of 0.13 eV, calculated for the cluster model Mg_9O_9(9,9)-CO.

From Table 2 it can be seen that the best DF BLYP estimate of the CO adsorption energy on MgO(001) is about 0.1 eV. The binding energy does not seem to decrease further with cluster size, at least not to a significant degree. To check the stability of the computed adsorption energy with respect to a variation of the generalized gradient-corrected energy functional the model Mg_9O_9(9,9)-CO has also been studied with the BP exchange-correlation potential applied self-consistently. The resulting BSSE corrected MgO-CO interaction energy, 0.30 eV, is 0.09 eV larger compared to CO adsorption energy of this cluster, 0.21 eV, obtained at the BLYP level of theory. Assuming that this energy difference is also valid for the largest MgO-CO models considered with the BLYP functional one estimates the adsorption energy for the cluster model $Mg_{25}O_{25}$(25,25)-CO at the BP level to about 0.18 eV which is close to the lowest experimental value of 0.15 eV [58]. This is in line with the results for a variety of systems where BP binding energy values were calculated to lie closer to experiment than those obtained with the BLYP energy functional which in turn were often calculated to underestimate the measured quantities [35]. Further experimental and computational studies are required to clarify which of the measured values of the MgO(001)-CO adsorption energy reported so far [58–60] is the most reliable one.

DF CSOV results calculated for the embedded Mg_9O_9(9,9)-CO cluster model are given in Table 3. The BSSE in the interaction energies and in other parameters have been taken into account by describing both CSOV fragments in the full basis of the supermolecule.

Inspection of Table 3 shows that quite a few mechanisms, none of them clearly dominating the others, contribute to make up the adsorbate-substrate interaction. Nevertheless, the following interpretation may be gleaned from these data. An estimate of the electrostatic attraction of -0.21 eV supports the generally accepted mechanism of CO adsorption on MgO(001) as being to a large extent of electrostatic nature [45]. A significant MgO(001)-CO bond destabilization, 0.34 eV, due to Pauli repulsion results in an unbound state of the adsorption complex in the first CSOV step which simultaneously accounts for Pauli repulsion and electrostatic attraction. The polarization of CO reduces this repulsive interaction by -0.03 eV only. The MgO(001)-CO system becomes bound after the σ charge transfer CO $\rightarrow$ MgO is taken into account. More importantly, the magnitude of the bond stabilization due to σ charge transfer, -0.20 eV, is comparable to the electrostatic attraction. Wave function based methods invariably estimated the charge transfer component to the bonding as negligibly small compared to the predominant electrostatic factor [45–49], while LDA studies [48,50–52] showed, although at a qualitative level, that the CO $\rightarrow$ MgO σ charge transfer has to be taken into account. The CSOV data confirm the previous graphical analysis of charge density differences [50] which identified a π charge transfer MgO $\rightarrow$ CO contribution as a discernible component of the overall binding. This π backdonation has been rationalized as a charge transfer from the orbitals of the oxygen anions surrounding the Mg^{2+} centre to

Table 3
DF CSOV data for various contributions to the interaction energy, E, to the C-O frequency shift, $\Delta\omega$(C-O), and to the absolute IR intensity of the C-O mode, I(C-O), calculated for the cluster model Mg_9O_9-CO[a]

Interaction channel	ΔE (eV)	$\Delta^2\omega$(C-O) (cm^{-1})	ΔI(C-O) (km/mol)
Electrostatic	-0.21	7	0
Pauli repulsion	0.34	76	-2
Polarization of CO	-0.03	2	5
CO $\rightarrow$ Mg_9O_9 σ charge transfer	-0.20	-19	-31
CO $\rightarrow$ Mg_9O_9 π charge transfer	0.00	-3	0
Polarization of Mg_9O_9	-0.06	-1	13
Mg_9O_9 $\rightarrow$ CO σ charge transfer	0.00	5	2
Mg_9O_9 $\rightarrow$ CO π charge transfer	-0.06	-26	47
Total CSOV	-0.22	34	34
SCF	-0.22	35	41

[a] The distance r_e(Mg-CO) from the closest cation to the carbon atom of the adsorbate equals 2.51 Å as obtained after BSSE correction of the distance. Mg_9O_9 and CO fragments are computed employing the basis of the Mg_9O_9-CO cluster to counterpoise the BSSE. The BSSE corrected reference values for free CO are ω(C-O) = 2125 cm^{-1}, I(C-O) = 61 km/mol.

which the adsorbed CO molecule is attached [51]. Two CSOV charge transfer contributions to the MgO(001)-CO bonding, CO $\rightarrow$ MgO π as well as σ MgO $\rightarrow$ CO charge transfer, are vanishing in line with the known low π donating ability of the CO molecule as well as with the high energetic costs of an electron transfer from a cation like Mg^{2+}.

The CSOV analysis of the adsorption-induced C-O frequency shift reveals that the observed ($\Delta\omega$(C-O) = +14 cm^{-1} [61]) and calculated increase of the frequency is mainly caused by the large contribution of the Pauli repulsion. This result, obtained at the DF level, confirms the previous analyses of the CO vibrational shift based on HF cluster models [45,46]. Other channels produce either small frequency shifts or act to reduce the shift obtained after this first CSOV step. The largest decrease of the shift $\Delta\omega$(C-O) results from the π charge transfer MgO $\rightarrow$ CO as is to be expected from the weakening of the C-O bond caused by a population of the strongly antibonding CO $2\pi^*$ level. Based on similar arguments, the σ charge transfer CO $\rightarrow$ MgO leading to a depopulation of the slightly antibonding CO 5σ orbital might be expected to increase the force constant of the adsorbed CO. However, another consequence of the σ charge transfer seems to be more important as on overall reduction of the C-O force constant (and of the frequency) is computed for this CSOV step: namely, the

σ charge transfer CO $\rightarrow$ MgO causes not only a depopulation of the CO 5σ orbital (resulting in a frequency blue shift), but also a significant relaxation of the Pauli repulsion responsible for the increased C-O frequency and thus a larger red shift. Therefore, an overall negative value of the C-O frequency shift is associated with the σ channel CO $\rightarrow$ MgO. This effect has been identified before at the HF level [45,46].

The nature of the adsorption-induced variation of the IR intensity can also be analysed within the CSOV scheme. Two main CSOV contributions leading to the predicted overall enhancement of the IR intensity of the C-O mode, ΔI(C-O), are the MgO substrate polarization (a modest effect) and the π charge transfer MgO$\rightarrow$ CO (a larger effect); for a discussion of the agreement with the experimentally derived intensity value see ref. 50. π backdonation is usually invoked to rationalize the C-O IR intensity increase which is observed and calculated for CO adsorbed at transition metal surfaces (see e. g. [62]) where the π charge transfer from the metal d orbitals to the CO adsorbate forms an important component of the classical binding mechanism. The σ charge transfer CO $\rightarrow$ MgO causes a partial decrease of the IR intensity of the C-O mode on MgO(001), but the size of this effect is smaller compared to the sum of the contributions mentioned above which act to enhance the integral adsorption coefficient.

To complete the discussion of isolated CO molecules adsorbed on MgO(001) surface a note is in order that inclusion of gradient corrections to the exchange-correlation functional [41] do not change the main conclusions initially reached at the LDA level [48,50,51]. The principal (and expected) effect of the density gradient corrections to the exchange-correlation energy functional is that the MgO-CO interaction energy is reduced by about 50% and a concomitant elongation of the Mg-CO distance.

3.3. Lateral interactions between adsorbed CO molecules on MgO(001)

Modest adsorbate-adsorbate effects may often be noticed for a moderate to high surface coverage on the background of stronger adsorbate-substrate interactions. The investigation of lateral effects in adsorbed overlayers offers an opportunity to thoroughly examine the accuracy of DF calculations in describing intermolecular interactions. Furthermore, clarifying the coverage dependence of various observables helps to gain deeper insight into the electronic properties of the substrate itself. For instance, using IR spectroscopy the propagation of electronic effects through metal oxide materials has been studied by varying the CO adsorption coverage [63]. IR spectroscopy combined with isotopic dilution provides a straightforward way to analyse the lateral effect in an adsorbate overlayer, discriminating between static and dynamic contributions. The latter component accounts for the mutual perturbation of vibrating molecules; it may be well described by existing models (see e. g. references in [51]). On the other hand, the static mechanism which comprises the remaining lateral interactions has not thoroughly been theoretically studied for adsorption complexes on oxides because detailed direct information on the formation of surface chemical bonds was lacking.

On an insulating substrate like MgO, one usually neglects the interaction of a vibrating molecule with its own dipole image and with the images of the neighbouring adsorbates. As a

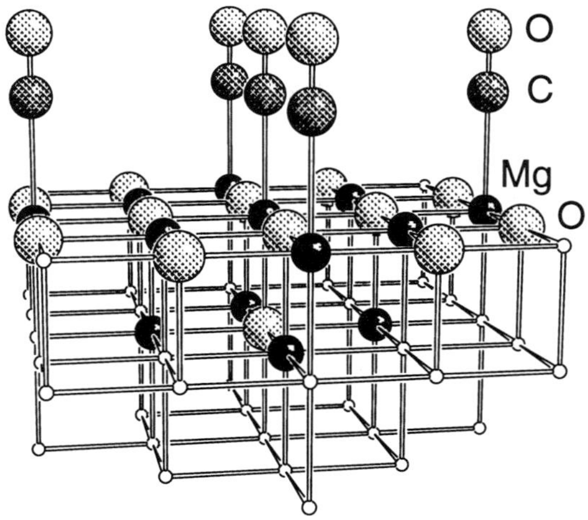

Figure 2. Sketch of the cluster model $Mg_{13}O_{13}(21,5)PC-(CO)_5$. A small fraction of the $(17\times17\times6 - 26)$ embedding point charges (PC) is also displayed.

rule, a vibrational coupling mechanism "through the solid" involving common bonding electrons may also be excluded because of the low adsorption enthalpy [64]. The remaining direct ("through-space") dipole-dipole coupling has been experimentally determined for MgO/CO by the ^{12}CO-^{13}CO isotopic dilution technique. A frequency shift of -12.5 cm^{-1} is attributed exclusively to the static adsorbate-adsorbate interaction in a MgO/CO monolayer [61].

Two mechanisms have been proposed to rationalize this negative static shift: a "solvent" and a "chemical" effect [64]. The solvent effect describes through-space interaction caused by changes in the local environment at a given site which are brought about by filling the other sites. The chemical effect is a through-solid ("through-bond") perturbation among molecules adsorbed at neighbouring sites. For an oxide like MgO, one expects major effects of this type only if an adsorbate-substrate charge transfer takes place. If there is an electron donation from CO to the surface cation, such an electronic effect may, in principle, be transmitted through the substrate. In this case the next adsorbate molecule coming to the surface may be thought of finding its cationic site in a state with lower acceptor ability. Therefore, the effective occupation of the CO antibonding 5σ orbital would be larger and, concomitantly, the corresponding C-O frequency lower than that of a singleton CO on MgO. As demonstrated in the previous section by the CSOV analysis of the DF data for isolated CO on MgO, the assumption about a direct effect of the CO 5σ orbital depletion on the value of a blue shift of the C-O frequency in MgO(001)/CO oversimplifies the more complicated interaction mechanism.

The first DF investigation of the static component of the lateral interaction between CO molecules adsorbed on MgO(001) has been carried out at LDA level [51]. The most

Table 4
Calculated and experimental parameters of the C-O vibrational mode in MgO(001)/CO model adsorption complexes and for a free CO molecule: anharmonic frequency ω(C-O) and absolute IR intensity I(C-O)

System	ω(C-O) (cm^{-1})	ω(C-O)^{BSSE} (cm^{-1})	I(C-O) (km/mol)
MgO(001)/CO, singleton[a]	2157		
PC-CO	2161	2157	40
$Mg_{13}O_{13}$PC-CO	2199	2192	165
MgO(001)/CO, monolayer[a]	2144.5[e]		
$(CO)_5$	2184		53
PC-$(CO)_5$	2172		33
PC(nn-CT)-CO[b]	2156		43
PC(CT)-CO[c]	2149		48
$Mg_{13}O_{13}$PC-$(CO)_5$	2188		148
CO, gas[d]	2143		60.5
CO	2169	2167, 2164[f]	66

[a] Experiment [61].
[b] The four nearest CO neighbours of the central CO are simulated by point charges.
[c] Adsorbates around the central CO at all 80 cationic sites of the PC array which correspond to a c(2×2) structure are simulated by point charges.
[d] Experiment [65,66].
[e] Frequency of ^{12}CO at the monolayer coverage in the ^{12}CO-^{13}CO mixture (15:85) where the dynamic interaction between the ^{12}CO adsorbates may be neglected [61,64]. The frequency of a ^{12}CO monolayer is 2148 cm^{-1}.
[f] Corrected for the BSSE caused by four surrounding CO groups (kept apart at the Mg-Mg distance) and by the substrate atoms, respectively.

elaborate cluster employed to model the MgO(001)/CO monolayer, $Mg_{13}O_{13}$(21,5)PC-$(CO)_5$, is displayed in Figure 2.

According to the LCGTO-DF cluster model data, CO adsorption on MgO(001) is accompanied by a charge transfer (CT) of the order of 0.1 au from the CO 5σ MO to the substrate [51]. The cluster $Mg_{13}O_{13}$PC-$(CO)_5$ models the static effect of the lateral adsorbate interaction with the nearest neighbours in the monolayer (1 CO : 2 Mg^{2+}). For the central adsorbed CO molecule the calculated vibrational frequency of 2188 cm^{-1} is red shifted relative to the frequency determined for the singleton cluster $Mg_{13}O_{13}$PC-CO, ω(C-O) = 2199 cm^{-1} (Table 4). The resulting shift remains negative even after corrections for the BSSE; it agrees qualitatively with the static shift derived from experiment, -12.5 cm^{-1} [61]. The experimental value, unlike the cluster model, includes long-range effects corresponding to a large number of neighbouring adsorbates. It is important to note (Table 4) that the calculated static C-O frequency shift is positive for those models where the substrate is either completely absent ($(CO)_5$ vs. CO) or described exclusively as an array of point charges (PC-$(CO)_5$ vs. PC-CO). This means that the adsorbate interaction with a more realistic model of the substrate, $Mg_{13}O_{13}$PC, induces a pronounced static red shift of the frequency to compensate the blue shift calculated for the clusters $(CO)_5$ and PC-$(CO)_5$.

A chemical mechanism to determine the static contribution to the CO frequency shift on coverage is in principle possible if a redistribution of the charge transferred from CO takes place in the substrate. However, essentially no electron redistribution of this kind was found when the charge density difference was examined [51]. Also, almost the same adsorption energy is calculated for the central CO in $Mg_{13}O_{13}$PC-$(CO)_5$ (0.35 eV) and in $Mg_{13}O_{13}$PC-CO (0.37 eV). Thus, the propagation of an electronic effect through the prototypical insulator substrate MgO has to be ruled out as the dominating mechanism which causes the static red shift of the C-O frequency upon increase of the surface coverage.

Another interpretation of this situation has been proposed [51]: the charge transfer due to the neighbouring adsorbates changes the effective electrostatic field in which a given CO molecule vibrates. (Note that a direct static interaction not involving the substrate would lead to a blue shift of the CO frequency when the coverage increases). The influence of this CT mechanism was studied by modeling the effect of each of the surrounding surface CO molecules by two point charges in a point charge model. One charge was placed at the centre of mass of an adsorbate (q_{PC} = 0.1 au), the other was used to modify the charge representing the corresponding substrate cation (change Δq_{PC} = -0.1 au). Vibrational frequency shifts ω(C-O) were computed for two models of this type (Table 4): (i) PC(nn-CT)-CO, where only the four nearest neighbours (nn) of the central CO are simulated electrostatically and (ii) PC(CT)-CO, where all cationic positions of the upper layer of the PC array (80 model adsorbates corresponding to a c(2×2) overlayer) are modified in the same way. For the cluster PC(CT)-CO a frequency shift of -12 cm^{-1} is calculated relative to the value for the model PC-CO, a value which is more than twice larger than that for the cluster PC(nn-CT)-CO, -5 cm^{-1}. This clearly illustrates the importance of long-range contributions. The agreement of the PC(CT)-CO model result with the measured shift of -12.5 cm^{-1} [61] is certainly fortuitous because important quadrupole-quadrupole and higher order multipole couplings within the CO adsorbed layer [67] are missing in the point charge model. Nevertheless, this simplified model provides a rationalization of the coverage-induced static frequency shift of CO on MgO(001): the surrounding field is modified due to the charge transfer resulting in an electrostatic solvent effect.

3.4. CO adsorption on surface irregularities of MgO

Structural defects of various kinds are usually thought to determine the surface reactivity of MgO(001) [7–9], at variance with the relatively inert regular sites considered so far. In the present and next sections we shall discuss LCGTO-DF results obtained at the LDA level for CO probes interacting with two types of cationic irregularities of MgO: low-coordinate centres located on the edges and corners of crystalline samples [68] and transition metal impurities where Ni^{2+}, Co^{2+}, or Cu^{2+} species are substituted for surface Mg^{2+} cations of MgO(001) [69].

In crystalline MgO, surface ions may occupy the following positions: five-fold coordinate Mg_{5c} and O_{5c} on perfect MgO(001) faces, four-fold coordinate Mg_{4c} and O_{4c} on the edges and three-fold coordinate Mg_{3c} and O_{3c} at the corners of (001) terraces. Complexes of CO on MgO samples with a large surface area exhibit IR absorption maxima at 2157, 2170 and 2203 cm^{-1} which have been attributed to Mg_{5c}-CO, Mg_{4c}-CO and Mg_{3c}-CO species, respectively [61]. To study these surface moieties the stoichiometric clusters Mg_9O_9(9,9), Mg_6O_6(3,6,3), Mg_4O_4(1,3,3,1), representing five-, four- and three-coordinate surface centres, respectively, were calculated at LDA level [68]. The model clusters were embedded in matrices of point charges of ±2.0 or ±1.5 au. Only two of the four possible vertical adsorption situations at on-top sites, namely the C-end and O-end binding of CO to cationic centres, will be discussed because the adsorbate interaction with the oxygen anions was found to be repulsive [3,50].

On ionic materials, CO molecules are usually considered to be adsorbed with the carbon atom directed to a surface cation [4 and references therein]. However, in some special cases at low-coordinate sites where the electrostatic interaction is stronger, theoretical models do not seem to exclude an alternative coordination where CO is coordinated with its oxygen atom toward the adsorption site ("O-down"). Therefore, it is of interest to compare various characteristics calculated for such unusual moieties Mg-OC with those for the more common Mg-CO complexes.

The results summarized in Table 5 indicate a rather weak interaction of CO via carbon with the five-coordinate cationic sites of MgO. The calculated C-O distance is only slightly compressed (by less than 0.01 Å) compared to the bond length of free CO. As measured by the binding energy which is significantly overestimated in LDA models [41], the interaction becomes stronger for low-coordinate centres in agreement with experiment [64] and with previous HF calculations [70]. This trend is mainly due to a stronger electrostatic attraction of the carbonyl in the vicinity of the more exposed cations at edges and corners; point charge models (PC-CO; see Table 5) are able to essentially reproduce the trends of the binding energy. The calculated C-O vibrational frequencies support the assignment of the high-energy features in the carbonyl region of the IR spectra to the adsorption on the coordinatively unsaturated cations located at edge and corner positions; these spectra were obtained for CO on polycrystalline MgO [61,64]. Electrostatic contributions to the blue shift $\Delta\omega$(C-O), as estimated from the PC-CO models, exhibit a strong dependence on the cation coordination: the lower the coordination, the larger the blue shift. It is worth mentioning that the calculated frequency shifts $\Delta\omega$(C-O) are noticeably overestimated when the surrounding point charges of nominal ionicity $q = \pm 2.0$ au are used in the models. As discussed in section 3.1, this model exaggerates the Madelung field on MgO(001). The MgO/C-O frequency for clusters embedded

Table 5
Selected calculated and experimental observables for MgO/CO and MgO/OC complexes[a]

System	r_e(C-O) (Å)	$\Delta\omega$(C-O)[b] (cm^{-1})	I(C-O) (km/mol)	D_e[c] (eV)
Mg_{5c}-CO[d]		14		
Mg_9O_9-CO	1.124	54 (16)	124	0.97
PC-CO	1.132	-8 (0)	40	0.62
Mg_{4c}-CO[d]		27		
Mg_6O_6-CO	1.124	56 (33)	125	0.94
PC-CO	1.131	-6 (7)	28	0.92
Mg_{3c}-CO[d]		60		
Mg_4O_4-CO	1.121	85 (59)	88	1.07
PC-CO	1.128	25 (32)	12	1.32
(MgO_3^{4-})-CO	1.120	99	94	1.25
($Mg_7O_3^{8+}$)-CO	1.125	32	357	0.74
Mg_{5c}-OC				
Mg_9O_9-OC	1.137	-60	201	0.51
PC-OC	1.132	-18	99	0.24
Mg_{4c}-OC				
Mg_6O_6-OC	1.139	-76	196	0.49
PC-OC	1.136	-49	133	0.48
Mg_{3c}-OC				
Mg_4O_4-OC	1.143	-98	269	0.68
PC-OC	1.146	-124	201	0.82

[a] r_e(C-O) is the calculated C-O distance. For an explanation of other notations see Tables 2 and 4.
[b] Adsorption-induced C-O frequency shifts, $\Delta\omega$(C-O), calculated for clusters with surrounding point charges of ±1.5 au are given in parentheses.
[c] No BSSE corrections are applied for the adsorption energy.
[d] Experiment [61].

in arrays of smaller charges, $q = \pm 1.5$ au, agree significantly better with the experiment; the fortuitous agreement of these values will diminish if a more realistic charge separation as discussed in section 3.1 would have been modelled. The calculated IR intensity of the C-O mode for adsorption on MgO(001) [50,51] does not show a decrease relative to the value for gas phase CO. This change might have been expected from the analysis of the dynamic contribution to the coverage dependence of the C-O frequency shift [64] with the help of the modified Hammaker equation [71]. The present results demonstrate that C-down adsorption at edge Mg_{4c} and corner Mg_{3c} cations is also accompanied with an enhancement of the C-O mode intensity.

The LDA adsorption energies computed for the Mg_{nc}-OC models are smaller than those for the Mg_{nc}-CO clusters. The main driving force for the O-down CO adsorption is also the electrostatic field; again, its contribution increases for the low-coordinate cations. Possible reasons for such an unusual orientation of a CO molecule may be connected to its very low static dipole moment [65] and to important polarization contributions of the adsorbate-substrate interaction. The longer C-O distances in the Mg_{nc}-OC models compared to free CO molecule reflect a weakening of the intra-adsorbate bond and are in line with a lowering of the C-O vibrational frequencies (Table 5). According to these results it should, in principle, be possible to detect an oxygen-bound terminal CO molecule on MgO or on similar ionic surfaces by an IR band red shifted relative to the value of free CO. The calculated vibrational shifts of -60 to -98 cm^{-1} are probably overestimated in our models, just as are those for C-end bound CO. Nevertheless, a trend of increasing shifts along the series Mg_{5c}-OC $\rightarrow$ Mg_{4c}-OC $\rightarrow$ Mg_{3c}-OC due to a stronger Coulomb field is obvious. Therefore, an IR band shifted to lower energies by several tens of wave numbers compared to the position for free CO might well be a manifestation of an anomalous carbonyl complex where the CO adsorbate is bound to a cation via its oxygen atom. For CO on MgO with the O-end down the adsorption-induced IR C-O intensity enhancement is mainly caused by the electrostatic field (cf. the results for the PC-OC clusters).

It should be mentioned that the discussion in the present and in the following sections is limited to LDA results because unfortunately the GGA calculations are still to be done. The LDA adsorption energies, in particular those calculated without the BSSE taken into account, are expected to change significantly at the BSSE-corrected GGA level. Therefore, only relative values of the LDA adsorption energies should be considered meaningful. Another remark concerns the finding that the adsorbate-substrate distances in various MgO/CO and MgO/OC complexes are rather sensitive to a variation of computational parameters because of the very flat shape of the MgO-CO potential curves. Instead of analysing this sensitive observable here, we refer the reader to the discussion given in the original papers [41,48,50,52,68,69].

We shall complete this section with a brief account how the cluster composition affects calculated adsorption properties. All models discussed so far contain a *stoichiometric* (and consequently electroneutral) atomic fragment of the MgO substrate. However, charged cluster models of ionic crystals are also quite common [4]. Often they are chosen to contain less atoms than the electroneutral cluster models and therefore they require less computational effort. A tacit assumption underlying the use of a charged cluster model is that uncompensated electronic and nuclear charges have negligible effects on the cluster properties. A rather straightforward and promising way to construct model clusters for metal oxide surfaces has been suggested recently [47]. The idea is to take into account a major missing component of the Pauli repulsion between the model cluster and the surrounding bulk by requesting that all cluster anions at the "bulk" edges have a complete shell of nearest-neighbour cations which are to be treated either at the all-electron or at a pseudopotential level.

Calculations for clusters of a corner Mg_{3c} site, (MgO_3^{4-})-CO and $(Mg_7O_3^{8+})$-CO, have been carried in order to examine, within the DF approach, the aforementioned technique and its efficiency in accounting for the Pauli repulsion effect on the adsorption characteristics and on other cluster properties. In the first cluster $MgO_3(1,3)$, as an artefact, the electrons of the three oxygen atoms in the second layer are strongly polarized by the attraction to the positive point charges beneath them. In order to follow the construction principle outlined above all six third-layer neighbours of the three oxygen anions are taken into account in the second cluster $(Mg_7O_3^{8+})$-CO; the substrate moiety is $Mg_7O_3(1,3,6)$. The overall electroneutrality of these clusters is achieved by means of surrounding arrays of point charges. We can compare the DF charged-cluster results for instance to the known experimental CO stretching frequency shift 60 cm^{-1} [61], caused by the adsorption at Mg_{3c} site of MgO (Table 5). The C-O frequency shift is overestimated in the negatively charged cluster (MgO_3^{4-})-CO (99 cm^{-1}), underestimated in the positively charged cluster $(Mg_7O_3^{8+})$-CO (32 cm^{-1}); the result for the stoichiometric model Mg_4O_4-CO (85 cm^{-1}) falls in-between. The most interesting result is that, for $(Mg_7O_3^{8+})$-CO, the C-O frequency shift is computed too small compared to experiment. Overbinding between carbon and oxygen in an adsorbed carbonyl molecule (or an overestimation of $\Delta\omega$(C-O)) might be easily rationalized either by the peculiarities of LDA or by the exaggerated value of the point charges which had been used to represent nominal ionicity. However, the frequency shift $\Delta\omega$(C-O) is underestimated, a finding not observed so far in LCGTO-DF investigations of MgO/CO [41,50,51]; in some of these DF calculations rather extended stoichiometric models of MgO/CO had been employed. Furthermore, it is worth mentioning that a charged cluster model of CO adsorption at MgO(001), $(Mg_{14}O_5(13,5,1)^{18+})$-CO [47], with all the anionic centres totally surrounded by their nearest-neighbour cations to account for the Pauli repulsion furnishes, at the LDA level of theory, a red shift of the C-O frequency, -20 cm^{-1} [68], at variance with the experimental blue shift of +14 cm^{-1} [61] for MgO(001)/CO. On the other hand, large stoichiometric clusters reproduce the measured MgO(001)/C-O frequency shift reasonably well, apart from the usual slight overestimation [41,50,51]. The present discussion of the merits and demerits of charged cluster models for MgO and related materials shows that the electroneutral models are preferable, for DF as well as for HF studies [72].

3.5. CO molecules adsorbed at transition metal impurity sites on MgO(001)

It is generally believed that the presence of filled d orbitals in transition metal oxides is responsible for a different behaviour of these substrate materials compared, for instance, to alkaline-earth oxides [61,64]. Experimental studies reveal a somewhat stronger interaction of adsorbed CO with Ni_{5c} than with Mg_{5c} cations in the corresponding oxides. This is manifest from a lower frequency and a noticeably larger IR intensity of the C-O vibrational mode for Ni_{5c}-CO species compared to Mg_{5c}-CO [64]. A special IR study [61] of CO adsorbed on the (001) faces of sintered polycrystalline MgO and NiO/MgO and CoO/MgO solid solutions, containing 10% of Ni and Co atoms, has been carried out to eliminate the consequences of a possible difference in the substrate ionicity on the vibrational parameters of M_{5c}-CO complexes (M = Mg, Ni, Co) and to focus on the role played by the d-π overlap in the bonding between the transition metal cations and the CO molecules adsorbed on them. The observed difference in the measured C-O stretching frequency and intensity of these systems has been ascribed to a small d-π overlap contribution which was assumed to be present in addition to the electrostatic adsorbate-substrate interactions [61]. One of the goals of the DF studies [52,69] was to examine if DF cluster models are capable of reproducing and rationalizing the observed differences between the vibrational characteristics of the CO adsorption at pure MgO(001) and on transition metal impurity defects.

Despite the quantitative alteration of some of the observables expected at the GGA-DF level, the LDA picture of adsorbate-substrate interactions is likely to yield a qualitatively adequate description of the surface moieties under consideration [41]. Selected LDA results calculated for embedded cluster models MMg_8O_9-CO and MMg_8O_9-OC (M = Mg, Ni, Co) [69] are collected in Table 6. We start with a discussion of C-end down adsorption, and we shall compare the data for Ni_{5c} and Co_{5c} sites to those obtained for the Mg_{5c} models (models II and III vs. I).

Shorter adsorption distances z_e(Ni-CO), z_e(Co-CO) and higher frequencies ω(Ni-CO), ω(Co-CO) of the corresponding vibrational mode indicate a stronger interaction of CO with the nickel and cobalt sites than with the magnesium one. The charge rearrangement between CO and the transition metal cations is more intense than that in model I for Mg^{2+}. This

Table 6
Selected calculated and experimental[a] (in parentheses) parameters for (M)MgO/CO and (M)MgO/OC complexes[b]

Model		z_e(M-CO) (Å)	ω(M-CO) (cm^{-1})	$\partial\mu/\partial z$ (au)	r_e(C-O) (Å)	$\Delta\omega$(C-O) (cm^{-1})	I(C-O) (km/mol)
I	Mg_9O_9-CO	2.22	183	0.12	1.124	54 (14)	124 (1[c])
II	$NiMg_8O_9$-CO	1.88	327	0.41	1.126	45 (1)	283 (2[c])
III	$CoMg_8O_9$-CO	1.89	333	0.44	1.127	38 (-25)	330 (4[c])
IV	Mg_9O_9-OC	2.19	122	-0.17	1.137	-60	201
V	$NiMg_8O_9$-OC	2.06	202	-0.19	1.136	-42	206
VI	$CoMg_8O_9$-OC	2.10	191	-0.11	1.137	-57 (-60?)	153

[a] Ref. 61.
[b] z_e(M-CO) is the calculated adsorbate-substrate M-C or M-O distance; ω(M-CO) and $\partial\mu/\partial z$ are the vibrational frequency and the dynamic dipole moment computed for the adsorbate-substrate mode, respectively. For an explanation of other notations see Tables 4 and 5.
[c] Experimental IR intensity of the C-O mode relative to that in the MgO/CO complex.

can be seen from the increase of the dynamic dipole moment $\partial\mu/\partial z$, a measure of the total charge moved during a vertical displacement of CO against the substrate. Electron density difference plots for CO on transition metal ions in oxide matrices [69] (not shown here) support a classical charge transfer mechanism of σ and π type where the σ channel dominates in magnitude. The reduction of the C-O frequency by 9 and 16 cm^{-1} in the clusters $NiMg_8O_9$-CO, $CoMg_8O_9$-CO (II, III), respectively, relative to that of Mg_9O_9-CO (I) is mainly due to the π interaction. These differences are in line with the values of 13 and 39 cm^{-1} measured for the solid solutions [61]. A further argument in favour of a non-negligible π contribution to the Ni_{5c}-CO and Co_{5c}-CO bonds, an intrinsic feature of d metal complexes, can be gleaned from the changes of the intensity I(C-O) of the C-O vibrational mode. In the models II and III an intensity enhancement is calculated compared to model I, in agreement with experiment [61]. The DF results are to some extent at variance with those of HF calculations [45]. There, essentially no difference was found for CO adsorbed at Ni_{5c} and Mg_{5c} oxide centres. However, important experimental evidence [61] of specific interactions with the d metal sites are qualitatively reproduced and rationalized at the LDA level.

In the IR spectra of CO adsorbed on the solid solutions NiO/MgO and CoO/MgO [61] one finds several characteristic peaks around 2100-2060 cm^{-1}, some of which have been tentatively assigned to O-down adsorbed CO at Ni_{5c}, Co_{5c} or other Ni and Co sites. Most maxima were assigned to ketenic products of CO reactions with coordinatively unsaturated oxygen anions [61]. The stretching mode of the adsorbed ^{13}C-O isotope also contributes to this spectral region. The most probable candidate to be identified with a vibration of an iso-carbonyl species, bound via the oxygen atom, is a rather intensive peak at 2083 cm^{-1} in the IR spectra of CoO/MgO solid solution samples. The models $NiMg_8O_9$-OC (V) and $CoMg_8O_9$-OC (VI) have been investigated to evaluate vibrational parameters of conceivable iso-carbonyl species, bound to a d metal cation in oxide matrices (Table 6) [69]. One has to point out, however, that the experimental conditions (foremost the quality of the sintered polycrystalline samples) are not very well defined and that the IR spectra exhibit a very complicated structure in the relevant frequency region [61]. Thus, without a thorough computational analysis of alternative structures like adsorbates lying flat, surface ketenic species etc., any conclusions concerning the presence of O-end down CO will only be of preliminary nature. The frequency shift $\Delta\omega$(C-O) = -57 cm^{-1} calculated for model VI fits that of the measured IR peak at 2083 cm^{-1} very well, $\Delta\omega$(C-O) = -60 cm^{-1} [61]. Thus, this spectral feature provides an argument in favour of an iso-carbonyl adsorption complex Co^{2+}-OC. Clearly, more experimental and computational efforts are necessary before this structural hypothesis can be accepted.

3.6. CO on MgO: a comparison of HF and DF results

As was mentioned in the previous sections, CO adsorbed at MgO is one of the most intensively studied systems in the context of quantum mechanical investigations of adsorption at oxide surfaces. Therefore it provides an ideal case for comparing different methodological approaches. Of course, the general description of the interaction between CO and regular MgO surfaces is similar in all methods. In particular, there is consensus on the fact that the surface is rather unreactive and that CO binds on the Mg cations with a largely electrostatic interaction [45,46,50]. However, despite this general broad agreement of the various methods, some differences emerge in a detailed analysis of the results. These differences are not insignificant as they are connected to a more general tendency of the methods to overemphasize (or to downgrade) some aspects of the interaction of molecules at oxide surfaces. One important question which is particularly difficult to answer, as previously discussed, is the binding energy of CO on MgO. We do not want to summarize here the results of the many studies reported on this subject. It may suffice to state that from a series of measures, in particular from the analysis of the CO vibrational frequency and the corresponding dynamic dipole moment, a consistent picture of the differences between DF and wave function based methods emerges. While the latter methods describe the CO adsorption bonding as entirely electrostatic with very little, if any, charge transfer from or to the surface [45,46], DF methods suggest the existence of some covalent interaction (see the previous section and [41,50]). Although this covalent contribution is rather small compared to the electrostatic interactions, its role is important because it shows that the MgO surface is chemically inert, but not entirely unreactive, as predicted by HF-based methods. This is a significant issue, particularly in the context of describing the metal deposition at oxide surfaces, as will be discussed below at some length (see chapter 5).

The partially different response of HF and DF methods in the treatment of the MgO/CO bonding is strictly related to the description of the MgO substrate. It is well known that while HF methods predict too large a band gap for MgO, DF-based methods underestimate this quantity. In other terms, wave function based methods describe MgO as an entirely ionic system, with little or no covalent contributions [73]. On the other hand, in DF methods MgO is described as a largely ionic system with moderate, but non-vanishing covalent contributions to the Mg-O bonding [53] (section 3.1). Deviations from the ideal ionic model for MgO imply that the oxide anions do not have their valence shell completely filled, i. e. there would be no O^{2-} ions in the strict sense; complementary, Mg cations still retain a small formal occupation of their 3s and 3p valence orbitals in line with the experimentally recorded $MgL_{2,3}$ X-ray emission band in the valence energy region of solid MgO [74] which indicates an occupation of the Mg 3s states.

In order to clarify this issue, we have undertaken a direct comparison of HF and LDA results for the case of CO adsorption at a MgO_5 cluster model [48]. This cluster is not stoichiometric and is charged in order to maintain its formal ionic character; it is stabilized by a large array of ideal point charges, $q = \pm 2.0$. To render a direct comparison of the two quantum chemical approaches meaningful, exactly the same clusters, point charges, and basis sets have been used in the calculations. These are not the best basis sets and clusters to model the MgO/CO adsorption system, but the purpose of this study was to achieve, for a case of interest, as direct a comparison of the two theoretical approaches as possible. It was no surprise that LDA gives a shorter MgO-CO bond than HF, 2.28 Å vs. 2.59 Å, and a stronger bonding, 0.75 eV (no BSSE correction) compared to 0.24 eV, respectively. We have seen that the LDA overestimation of the binding is considerably reduced when an energy functional with non-local corrections is applied [41].

More important for our comparison is the analysis of the dynamic dipole moment, $\partial\mu/\partial r$(C-O), of adsorbed compared to free CO. In HF $\partial\mu/\partial r$ of free CO is -1.02 au; when CO is adsorbed on MgO this quantity is calculated to -0.99 au. Thus no change occurs in the dynamic dipole moment of adsorbed CO; this indicates the absence of a charge transfer between CO and the substrate. In LDA, with the present CO basis set, the dynamic dipole moment is calculated to -0.66 au for free CO and to -1.07 au when CO is adsorbed on the MgO_5 cluster. Usually, such a change of the dynamic dipole moment is characteristic for CO adsorbed on metals where this quantity is more negative because of the charge transfer from the metal to the $2\pi^*$ MOs of CO. Thus, a possible interpretation of the adsorption-induced change of $\partial\mu/\partial r$ in LDA is that some charge transfer occurs from the substrate oxygen atoms to CO. Of course, other effects may contribute to change the dynamic dipole moment; in particular, polarization of the CO 1π orbitals has been identified as a possible source for such a change [50]. Nevertheless, it is clear that the response of CO to the substrate bonding is differently described by the two electronic structure methods: all bonding properties point toward a somewhat more covalent picture of the MgO surface in LDA.

To examine the degree of ionic character of the substrate we have determined an electron density difference map for the MgO_5 cluster, computed in HF and LDA, see Figure 3. This plot of the density difference, $\Delta\rho = \rho(\text{LDA}) - \rho(\text{HF})$, clearly shows that the oxygen anions

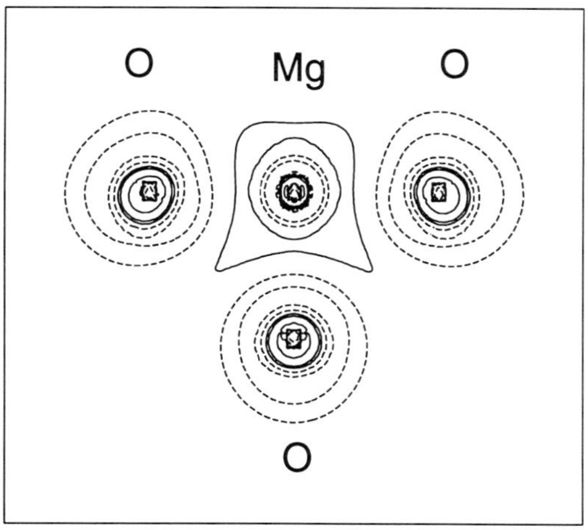

Figure 3. Contour map of the total electron density difference $\Delta\rho = \rho(\text{LDA}) - \rho(\text{HF})$ in the vertical plane containing the Mg and three O atoms of the cluster MgO_5. Dashed lines indicate negative values, or a decrease of the charge density when going from HF to LDA.

feature a higher electron charge density in the HF approach. The positive contour in the Mg-O region indicates some electron charge flow when going from the HF to the LDA method, pointing to some covalent interaction between the Mg and O centres in the LDA description. These results, obtained by using exactly the same numerical "ingredients" for both types of calculations, therefore show that there is a general tendency of the DF method to describe the surface as less ionic than the HF approximation. This finding rationalizes the somewhat different reactivity of a MgO surface observed within the two methods. The slight differences in the description of highly ionic materials by HF and LDA should not be neglected when the bonding of adsorbates on oxide surfaces is discussed.

4. INTERACTION OF PROBE MOLECULES WITH BRØNSTED AND LEWIS ACID CENTRES OF ZEOLITES AND Al_2O_3

The Brønsted and Lewis acidity of zeolite and metal oxide catalysts often determines their reactivity in industrially important chemical processes. However, the acid strength is not an easily measurable quantity. One usually monitors the surface acid strength by adsorbing probe molecules and by measuring IR spectra of the adsorption complexes. In the following three sections we demonstrate how useful DF cluster models can be to rationalize structure, bonding and vibrational observables of complexes formed between diatomic probes CO and N_2 and acid surface sites of different nature [40,75,76]. A description of weak hydrogen bonds in the moieties built up with Brønsted centres [40,75] represents a particularly critical test for the accuracy of DF calculations [6].

4.1. CO and N_2 interactions with acid bridging OH groups in zeolites

In combination with IR spectroscopy homonuclear diatomics, like N_2, may have various advantages as probes of the surface acidity [77–80] over the widely used CO molecules (ref. 81 and references therein). N_2 is a weaker base than CO, and the N-N mode, IR inactive in free N_2, may become dipole allowed in the adsorbed molecule. The interaction of N_2 and CO

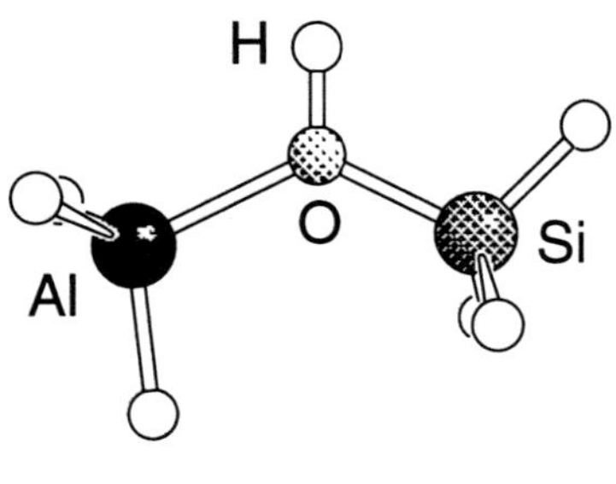

Figure 4. Sketch of the cluster model $H_3Si(OH)AlH_3$.

with Brønsted acid centres of H-ZSM-5 zeolite has been investigated by FTIR spectroscopy and by cluster model calculations based on a GGA-DF method [40]. The goal was to identify those characteristic features of the electronic structure of CO and N_2 that are responsible for the differences in their bonding properties with acid hydroxyl groups. The latter sites were represented by the minimal cluster model $H_3Si(OH)AlH_3$ (Figure 4) which was demonstrated to be adequate for the modeling bridging OH groups in aluminosilicates.

The adsorption-induced frequency shifts $\Delta\omega$(O-H) and $\Delta\omega$(C-O) or $\Delta\omega$(N-N) for carbonyl and dinitrogen complexes are in good agreement with the measured values (Table 7 [40]). It should be emphasized that these are the observables employed to quantify the acidity of surface OH groups. An almost vanishing shift of the N-N frequency for the linear dinitrogen complex compared to free N_2 is measured and calculated. The larger C-O frequency shift reflects stronger OH···CO bonding compared to OH···N_2, in agreement with a higher proton affinity of CO compared to that of N_2 [82]. The adsorption is accompanied by a red shift of the O-H band which is again larger for the CO adsorbate. Note that the agreement with the measured shift $\Delta\omega$(O-H) for carbonyl is much better when the anharmonicity of the O-H mode is taken into account as it displays a significant change upon adsorption (by 63 cm^{-1}). This finding, also obtained in a later DF study [83], may partly explain why the CO-induced

Table 7
Calculated and experimental vibrational parameters for the Brønsted site -Si(OH)Al- and for H-bonded complexes with CO and N_2 adsorbates[a]

System	$\Delta\omega$(O-H) (cm^{-1})	$\Delta\omega_e$(O-H) (cm^{-1})	I(O-H) (km/mol)	$\Delta\omega$(ads) (cm^{-1})	I(ads) (km/mol)	D_e[b] (eV)
$H_3Si(OH)AlH_3$						
calc.	3459[c]	3625[c]	141			
exp.[d]	3615[c]					
$H_3Si(OH)AlH_3$-CO						
calc.	-293	-230	953	43	58	0.19
exp.[d]	-312			32		
$H_3Si(OH)AlH_3$-N_2						
calc.	-109	-107	683	-4	3	0.11
exp.[d]	-109			2		

[a] $\Delta\omega_e$(O-H) is the adsorption-induced shift of the O-H vibrational frequency computed in the harmonic approximation; $\Delta\omega$(ads) and I(ads) are the calculated frequency shift and the absolute IR intensity of the intraadsorbate (C-O or N-N) vibrational mode, respectively.

[b] Adsorption energy corrected for BSSE.

[c] Reference value for the O-H frequency shift.

[d] Ref. 40.

shift $\Delta\omega$(O-H) for the same cluster model $H_3Si(OH)AlH_3$-CO is substantially lower in a HF study [84] where only *harmonic* frequencies have been considered (see also a post-HF investigation [85]). The change in the anharmonicity of the O-H mode due to the formation of the dinitrogen complex is much smaller as a consequence of the weaker interaction with N_2.

Another important aspect of the adsorption of the probe molecules is the change of the integral intensity of the vibrational bands compared to those of the isolated adsorbate and substrate fragments. The most extreme case is the N_2 molecule for which the IR forbidden N-N vibration becomes allowed by the interaction with the substrate and the absorption increases from zero to a measurable value. According to the computed data the absolute IR N-N intensity of adsorbed N_2 is more than an order of magnitude smaller than that of the strongly absorbing C-O mode. Nevertheless, this is sufficient for measuring the adsorption-induced N-N mode by IR spectroscopy. On the other hand, the hydrogen bonding interaction has essentially no effect on the intensity of the C-O mode. Both for CO and N_2 adsorption, DF calculations predict a very significant increase of the IR intensity of shifted O-H bands compared to the intensity of the $H_3Si(OH)AlH_3$ cluster. The experimental spectra are in general, qualitative agreement with the DF results.

According to the results of a CSOV analysis [40], the difference in the adsorbate-to-substrate bond strength for CO and N_2 is mostly due to the fact that N_2 is a weaker σ donor than CO. The stronger σ bonding of CO is consistent with the primary localization of the CO 5σ orbital in the region of the hydrogen-bond. In summary, the CSOV analysis reveals that the adsorption bonds are formed mainly due to σ charge transfer from the adsorbate to the substrate and, less importantly, to the polarization of the adsorbate and of the substrate. The calculated interaction energy for the $OH\cdots N_2$ complex (0.11 eV [40]) is very close to the value of 0.12 eV determined experimentally from temperature dependence of IR intensities for N_2 adsorption on H-mordenite [79]. In contrast, the spectroscopically determined binding energy of CO on bridging OH groups of HY zeolites was reported to be only 0.12 eV [86], while the computed value is 0.19 eV [40]. These coincidences and discrepancies between calculated and spectroscopically determined binding energies should certainly not be overestimated. The experimental values may not be very accurate since the measurement of the temperature of the wafer in the IR beam is extremely critical, thus introducing a significant error in the derived thermodynamic data. Note also that the experimental CO adsorption energy on Brønsted centres of HX zeolites is larger, 0.14-0.18 eV [87]. On the other hand, accuracy limitations of the exchange-correlation potentials and of the small cluster model employed may also affect calculated binding energies to a certain extent.

4.2. Acid strength of the bridging OH groups in [Al]-, [Ga]-, [Fe]-isomorphously substituted zeolites

The modification of the acid strength of zeolites and the quantification of such acidity changes is of special importance when searching for new catalysts with improved performance. One strategy to achieve this goal proceeds via an isomorphous substitution of framework

Al centres by Ga, Fe, B, and other trivalent atoms T [88,89]. LCGTO-DF BLYP cluster calculations on bridging OH groups in isomorphously substituted [Al]-HZSM-5, [Ga]-HZSM-5 and [Fe]-HZSM-5 zeolites and of the complexes with adsorbed CO have been carried out [75] to analyse the reliability of various criteria for the Brønsted acidity and to predict acidic properties of the [Fe]-substituted zeolite. We shall first discuss the results obtained for cluster models of the bridging OH groups and of their deprotonation products; then we shall turn to the results for adsorption complexes with probe CO.

A variety of criteria have been suggested to characterize the acidity of the bridging OH groups statically, i. e. based on the properties of the group itself, leaving aside its interactions with adsorbed probe molecules (e. g. [90–92]). However, some of these static characteristics of acidity seem questionable because the acid strength as a measure of the proton donating power is intrinsically connected to the interactions with a base [81,90]. Probably the most reliable calculated characteristic of acid strength is the energy of deprotonation. Inspection of Table 8 shows this index to yield the expected order of acid strength, Al > Ga > Fe. The frequency ω(O-H) does not follow this trend, probably because the acid strength of the bridging OH groups is particularly affected by "chemical" changes. On the other hand, O-H vibrational frequencies are strongly influenced by structural changes of the -Si(OH)T- moieties, first of all by the alteration of the Si-O-T angle [93] which may not be precisely enough reproduced in small cluster models. The charge q_H of the hydroxyl H atom, computed as the O-H displacement derivative of the dipole moment, $\partial\mu/\partial r$(O-H), is not found to provide an accurate characteristic of acidity, in line with previous investigations [92].

The DF BLYP adsorption energies of CO at the bridging OH groups in [Al]-, [Ga]- and [Fe]-zeolites, 0.19, 0.17 and 0.13 eV, respectively, are in agreement with the proposed acidity trend Al(OH)Si > Ga(OH)Si > Fe(OH)Si [91,94]. According to a CSOV analysis, the OH-CO interactions in [Al]-, [Ga]- and [Fe]-zeolites are of rather similar nature. The CO $\rightarrow$ OH σ charge transfer defines the OH-CO bond strength as well as its decrease in the series of [Al]-, [Ga]- and [Fe]-zeolites: for [Al]-zeolite (no BSSE correction applied), 0.17 eV of the overall

Table 8
Calculated parameters for Brønsted sites -Si(OH)T- [a]

Cluster model	q_H	r_e(O-H)	ω(O-H)	I(O-H)	E_d
	(au)	(Å)	(cm^{-1})	(km/mol)	(eV)
$H_3Si(OH)AlH_3$	0.34	0.979	3459	141	13.17
$H_3Si(OH)GaH_3$	0.34	0.977	3484	133	13.62
$H_3Si(OH)FeH_3$	0.26	0.978	3456	76	13.82

[a] Dynamic dipole moment of the O-H mode $q_H = \partial\mu/\partial r$(O-H), O-H bond length r_e(O-H), frequency and absolute IR intensity of the O-H vibrational mode ω(O-H) and I(O-H), respectively, and deprotonation energy E_d.

binding energy, 0.23 eV, is associated with the OH-CO charge transfer; the corresponding values for Ga and Fe are 0.15 eV out of 0.20 eV (Ga) as well as 0.12 eV out of 0.17 eV binding energy (Fe).

The adsorption of CO molecules on the hydrogen atom of acid OH groups is known to be accompanied by a noticeable red shift of the O-H vibrational frequency, by a small blue shift of the C-O frequency as well as by a significant enhancement of the absolute IR intensity of the O-H mode [81]. Commonly, the change of these parameters is expected to correlate with the acid strength of the OH groups under consideration [81]. The adsorption-induced changes of the vibrational characteristics of the OH group, the decrease of the O-H frequency and the increase of the absolute IR intensity, computed for the models of [Al]-, [Ga]-, and [Fe]-zeolites [75], are in line with a decreasing acidity strength. Also, the calculated frequency shifts and the intensity of the O-H mode are in agreement with the experimental IR data for CO adsorbed on [Al]- and [Ga]-HZSM5 zeolites [95]. These calculations predict smaller effects on the O-H vibrational parameters for the adsorption complex with isostructural [Fe]-zeolites. The calculated adsorption induced C-O frequency shift, while in good agreement with the reported measured IR data [95], does not seem to be a particularly sensitive parameter to distinguish between the acid strength of bridging OH groups in [Al]-, [Ga]- and [Fe]-zeolites: both the experimental and the calculated C-O frequency shifts are localized in a rather narrow interval of less than 5 cm^{-1}. Interestingly, about the same IR intensity of the C-O mode as in free CO has been calculated for the models of [Al]- and [Ga]-zeolites, while for [Fe]-zeolites an enhancement by about 50 % was predicted. However, when comparing these results one should keep in mind that the cluster models employed comprise rather small sections of the zeolite framework. So far, the influence of the model size on various calculated properties of DF cluster models has not yet been as thoroughly investigated as for metal oxide substrates.

In summary, the DF results [75] are in agreement with IR and other measured data for adsorption complexes of CO in [Al]-HZSM-5 and [Ga]-HZSM-5. This finding corroborates the ability of the theoretical models to adequately describe various properties of adsorption complexes with weak hydrogen bonds. The calculated trend in the deprotonation energy for [Al]-, [Ga]- and [Fe]-HZSM-5 correlates with the CO adsorption-induced change of the O-H IR frequencies and intensities. Therefore, the frequency shift and the intensity change can be used to quantify the acidity of O-H groups. Other parameters, such as the charge on the acid H atom or the value of the O-H frequency of samples without the adsorbate, are less reliable criteria for the acid strength. In conclusion, the Brønsted acidity of [Fe]-HZSM-5 is predicted to be significantly weaker compared to that of the isomorphous [Al]-HZSM-5 and [Ga]-HZSM-5 zeolites.

4.3. CO on strong and medium power Al^{3+} Lewis centres in oxides

When considering the Lewis acid strength of alumina (Al_2O_3) and of zeolites it is common to differentiate between strong and medium power Lewis acid sites (LAS). These two kinds of LAS can be identified using vibrational spectra of adsorption complexes with probe CO molecules [96–99]. Based on the spinel structure of γ-Al_2O_3, strong LAS are assigned to three-coordinate Al ions, $Al_{3c}{}^{3+}$; precursors are Al ions located in the tetrahedral spinel positions. The LAS of medium strength are ascribed to five-coordinate Al ions, $Al_{5c}{}^{3+}$, with the prototype Al species placed in the octahedral spinel sites. So far few theoretical efforts (e. g. [84,99]) have been focused on rationalizing the difference in acid strength of $Al_{3c}{}^{3+}$ and $Al_{5c}{}^{3+}$ LAS with the help of calculated vibrational parameters of an adsorbed CO probe. LCGTO-DF BLYP studies on the bonding and the vibration of a CO molecule adsorbed on 3- and 5-coordinate Al centres (Al_{nc}) in oxides [76] have addressed the following topics: (i) the intrinsic uncertainties of the design of cluster models of the $Al_{nc}{}^{3+}$ LAS (n = 3, 5); (ii) the mechanism of CO interaction with $Al_{nc}{}^{3+}$ LAS as a function of the coordination number n; (iii) various contributions to the C-O frequency shift in the $Al_{nc}{}^{3+}$-CO adsorption complexes.

The DF results definitely support the assignment of the $Al_{3c}{}^{3+}$ centres as strong LAS and of the $Al_{5c}{}^{3+}$ sites to LAS of medium strength: this follows not only from the CO adsorption energies and from the frequency shifts $\Delta\omega$(C-O), but also from the Al-CO vibrational frequencies and from the Al-CO bond lengths. Table 9 displays the results of a CSOV analysis of the interaction energy and of the C-O vibrational frequency shift for models of $Al_{3c}{}^{3+}$-CO and $Al_{5c}{}^{3+}$-CO adsorption complexes, $Al(OH)_3$-CO and $Al(OH)_3(OH_2)_2$-CO, respectively (Figure 5). Inspection of Table 9 shows that the σ charge transfer CO $\rightarrow$ $Al_{nc}{}^{3+}$ contributes substantially to the stability of CO at LAS of both types investigated. A similar statement holds for the contribution of the CO polarization which, however, is somewhat less important for the complex of $Al_{5c}{}^{3+}$. A non-negligible charge transfer in the opposite direction, $Al_{nc}{}^{3+}$ $\rightarrow$ CO, taking place in the π channel, is due to the oxygen ions surrounding the cation under consideration. This interaction component is more significant for $Al_{5c}{}^{3+}$ moieties having four oxygen neighbours available for interactions with the adsorbate molecule than for the three anions attached to $Al_{3c}{}^{3+}$. A similar bonding mechanism has been found in LCGTO-DF calculations of CO adsorption on top of Mg^{2+} cations at the MgO(001) surface, carried

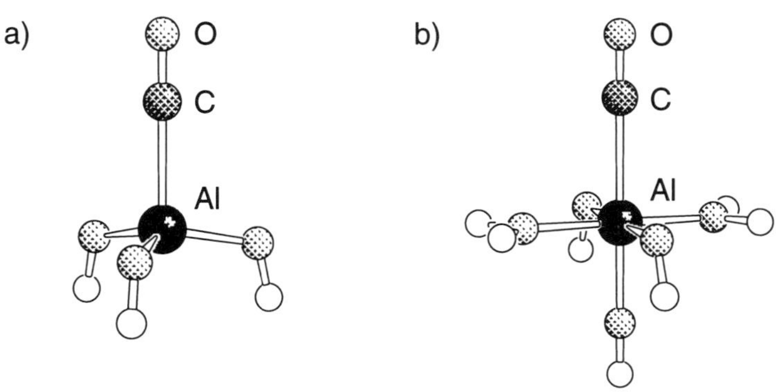

Figure 5. Cluster models a) $Al(OH)_3$-CO and b) $Al(OH)_3(OH_2)_2$-CO.

Table 9
DF CSOV analysis of various contributions to the interaction energy, E, and to the induced C-O frequency shift with respect to free CO[a], $\Delta\omega$(C-O), calculated for adsorption complexes of CO with LAS Al_{3c}^{3+} and Al_{5c}^{3+}.[b]

Cluster model	$Al(OH)_3$-CO		$Al(OH)_3(OH_2)_2$-CO	
Interaction channel	ΔE (eV)	$\Delta^2\omega$(C-O) (cm^{-1})	ΔE (eV)	$\Delta^2\omega$(C-O) (cm^{-1})
Pauli repulsion + electrostatic	0.13	182	0.33	136
Polarization of CO	- 0.39	-45	- 0.19	-25
CO → substrate charge transfer	-0.41	-18	-0.33	-22
Polarization of substrate	-0.08	0	-0.09	0
Substarte → CO charge transfer	-0.12	-36	-0.18	-58
Total CSOV	-0.87	83	-0.46	31
SCF	-0.87	84	- 0.46	30

[a] Calculated for free CO ω(C-O) = 2120 cm^{-1}.
[b] No BSSE corrections applied.

out within LDA [50,51] and gradient-corrected [41] DF approaches (see section 3.2). The adsorption-induced C-O frequency increase of the CO interacting with LAS is due to the Pauli repulsion which contributes 182 and 136 cm^{-1} to the overall calculated blue shift in Al_{3c}^{3+}-CO and Al_{5c}^{3+}-CO, respectively. The remaining CSOV components act to reduce the vibrational frequency in rather similar fashion for both types of LAS, by 99 (Al_{3c}^{3+}-CO) and 105 cm^{-1} (Al_{5c}^{3+}-CO). Therefore, the basic difference between the resulting C-O frequency shifts in Al_{nc}^{3+}-CO complexes is already present in the first CSOV step, which represents the sum of Pauli repulsion and electrostatic attraction. The stronger electrostatic attraction of CO to Al_{3c}^{3+} than to Al_{5c}^{3+} species is accompanied by a stronger Pauli repulsion in the former case. This effect is also revealed by a larger increase of the intramolecular frequency of CO adsorbed on Al_{3c}^{3+}. The aforementioned considerations are by no means in contradiction with the finding that the overall energetic balance computed at the first CSOV step is in favour of interactions with Al_{3c}^{3+} rather than with Al_{5c}^{3+}. A large decrease of the shift $\Delta\omega$(C-O) results from the π charge transfer Al_{nc}^{3+} → CO as is to be expected from the weakening of the C-O bond which is caused by a population of the strongly antibonding CO $2\pi^*$ level. Arguing in similar fashion, one may expect that the σ charge transfer CO → Al_{nc}^{3+} leads to a depopulation of the slightly antibonding CO 5σ orbital and thus to an increase the force constant of adsorbed CO. However, another consequence of the σ charge transfer is more important because a reduction of the C-O force constant (and of the frequency) is computed for this CSOV step: the σ charge transfer CO → Al_{nc}^{3+} causes not only a depopulation of the CO 5σ orbital, which in free CO would result in a frequency blue shift, but also a significant relaxation of the Pauli repulsion responsible for the increased C-O frequency and thus a larger red shift. Therefore, ultimately an overall negative value of the C-O frequency shift is associated with the σ channel CO $\rightarrow$ Al_{nc}^{3+}. This analysis further supports the original suggestion based on MgO/CO HF calculations that the blue shift of the stretching frequency of CO adsorbed on metal oxides has an essentially electrostatically origin and is not due, as often assumed, to the CO σ donation (see section 3.2) [45,46].

5. METAL SPECIES ON OXIDE SURFACES AND IN ZEOLITE CAVITIES

Metal-ceramic interaction is relevant in several areas of modern science such as corrosion, adhesion, microelectronic devices, photovoltaic cells, protective coating for metals, etc. Highly dispersed metal particles supported on oxides make up an important class of heterogeneous catalysts [100]. Despite this considerable technological importance very little is known about the microscopic nature of the interface between the surface of an oxide substrate, like MgO, SiO_2, Al_2O_3 or zeolite, and the contact metal atoms of a supported particle or of a metallic overlayer. This lack of information is both structural and theoretical. Most of the technologically interesting materials are based on amorphous components, but even when one deals with single-crystal experiments there are open structural questions, like the possibility of surface reconstruction, the presence of defects and dislocations, local disorder, etc. For these reasons the theoretical description of the metal-oxide interaction is especially challenging and must be based on a series of structural and electronic assumptions.

Oxide surfaces serve not only as inert support for the active component, but they are actually able to stabilize metal particles of a particular dispersion grade or to alter electronic and chemisorption properties of these species [101,102]. Experimental findings of significant changes in the catalytic performance of supported transition metal species have stimulated a particularly high interest in the question of how the support affects the chemisorption and the catalytic properties of active sites [103]. The complicated phenomenon of metal-support interaction comprises a variety of mechanisms, among them the direct local electronic effect of oxide surfaces on supported metal particles. Metal particles of a few atoms stabilized in zeolite cages may undergo a considerable change in their electronic structure due to interactions with the zeolite framework as well as with exchangeable cations.

In this chapter we shall review the results of DF calculations on four examples of metal/support interaction. The first case is that of Ni clusters supported on alumina [104], while the second is dealing with the interaction of Cu and Ni clusters on MgO [105]. The last two examples address electron-deficient palladium and platinum species in zeolites.

5.1. Ni clusters deposited on Al_2O_3

A number of studies on the electronic structure of the metal/Al_2O_3 interaction have been reported [106–116], based on both band structure calculations and cluster models and on various computational methodologies. The somewhat confusing picture emerging from these different approaches is not so much a question of the level of theory used, but rather is connected to the assumptions introduced in the model chosen to represent the "real" situation. In fact, in the absence of reliable structural information, the conclusions about the nature of the metal/Al_2O_3 interaction vary dramatically depending on the assumptions made on the type of termination and oxidation state of the Al_2O_3 surface. Not surprisingly, an aluminum-terminated Al_2O_3 surface behaves differently from an oxygen-terminated one. Moreover, the chemical nature of the surface oxygen atoms, fully reduced, O^{2-}, partially reduced, O^-, or neutral, O, will result in a completely different interaction with the adsorbed metal.

If the theoretical situation looks rather complicated, the experimental side [117,118] is not more transparent either because the deposition of different metals on the same substrate under similar conditions may result in completely different interfacial bonds. Of particular importance for the present work is the study of Chen et al. [119,120] on the Ni/Al_2O_3/Al(111) interface. These authors found that the Ni atoms interact with the surface Al-O units of Al_2O_3 and that the d electron density on the Ni particles is decreased. This mechanism results in a reduced backdonation from the supported Ni particle to an adsorbed CO probe molecule [120].

The calculations on Ni clusters supported on Al_2O_3 [104] have been performed at the local density level using the LCGTO-DF approach [14–16]. We have used an $Al_6O_{12}H_6$ cluster to simulate an oxygen terminated Al_2O_3 surface and we have deposited one- and two-layer Ni_6 clusters on this substrate. Except for the cluster "embedding", no a priori assumption was made concerning the oxidation state of the surface oxygen atoms; the geometry of the supported clusters has been partially optimized. Carbon monoxide was used as a probe molecule to monitor the modifications induced in the Ni_6 clusters by the interaction with the support.

To represent the Al_2O_3 surface we have used a simplified model where 12 oxygen and 6 aluminum atoms are forming a two-layer cluster with overall C_{3v} symmetry, as shown in Figure 6. The Al-O distance was fixed at 1.89 Å so that the oxygen centres form triangles at the surface with edges of 2.52 Å. This may be considered as a model of an oxygen-covered Al(111) surface rather than as one of a basal (0001) plane of α-Al_2O_3. At the edge, the substrate model cluster was saturated with hydrogen atoms; thus the final substrate model is $Al_6O_{12}H_6$, see Figure 6. In this way, both the stoichiometric ratio, 2 Al : 3 O, and the number of valence electrons in the cluster are as close as possible to the real situation.

A single Ni atom adsorbed on this cluster gives rise to a significant charge transfer from the metal to the substrate, indicating a substantial oxidizing power of the oxide surface. Of course, the character of the bonding may change considerably when a small Ni cluster interacts with the surface; in fact, in this case the formation of strong metal-oxygen bonds occurs at the expense of the metal-metal bonds within the metal cluster. We consider first the interaction of a $Ni_6(6,0)$ one-layer cluster of triangular shape, Figure 6. The optimal Ni-Ni distance for the free cluster, 2.22 Å, is much shorter than the Ni bulk distance, 2.49 Å, but for the adsorption on alumina we assumed an epitaxial growth of the metal film with the Ni atoms occupying

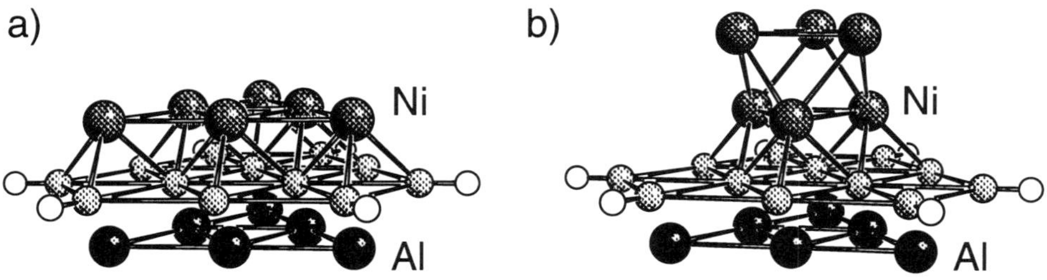

Figure 6. Structure of the Ni_6 clusters supported on a $Al_6O_{12}H_6$ cluster model of an oxygen-terminated alumina surface: a) one-layer Ni_6 species; b) two-layer Ni_6 species.

the three-fold hollow positions, Figure 6; this corresponds to an intra-layer Ni-Ni distance of 2.52 Å, very close to the bulk value of Ni metal. The optimal Ni-O distances between Ni_6 and the oxide surface are about 2 Å.

The Ni_6(6,0) cluster is strongly bound, by 17.7 eV, to the substrate. Of course, this is partially overestimated due to the local density approximation used for the exchange-correlation potential. The calculated interaction energy corresponds to a binding energy of about 3 eV per Ni atom, much lower than that for a single adsorbed Ni atom, about 8 eV, and reflects the fact the Ni atoms are involved not only in the bonding with the oxygen centres, but also in the bonding among each other. Indeed, the energy required to atomize the adsorbed Ni_6 cluster, 31.1 eV, is 13.4 eV higher than the Ni_6-$Al_6O_{12}H_6$ dissociation energy, indicating the existence of considerable metal-metal bonds also in the adsorbed phase. Analysis of the Mulliken charges and of the core level shifts of the Ni 2p levels suggests the occurrence of a charge transfer from the cluster to the oxygen atoms, similarly as for a single Ni atom. The consequences of this moderate charge transfer on the CO adsorption properties are discussed below.

We consider now an adsorbed two-layer Ni cluster, Ni_6(3,3) of D_{3d} symmetry, Figure 6. We assume the same Ni-O contact distance as found for Ni_6(6,0); the inter-layer Ni-Ni distance has been optimized with the constraint that the intra-layer Ni-Ni distances are fixed at 2.52 Å. Ni_6(3,3) is bound to the $Al_6O_{12}H_6$ substrate cluster by 12.5 eV, or 4.2 eV per Ni-O bond. The estimated charge transfer from the first Ni layer to the oxygen centres is higher than in the case of Ni_6(6,0), as shown by both the population analysis and the Ni 2p core level shift. On the other hand, we found a moderate perturbation of the Ni atoms in the second layer of the cluster; essentially, the electronic environment of the Ni atoms in the second layer is not too different in free and supported Ni_6(3,3) moieties.

The vibrational modes of CO chemisorbed on the model catalyst Ni/Al_2O_3/Al(111) [120] may be used to characterize the electronic modifications induced by the support on deposited Ni clusters and to probe the nature of the metal-oxide interface. As a reference, we have

Table 10
Calculated properties of CO chemisorbed on free and supported Ni_6 clusters[a]

	Ni_6(6,0) one-layer		Ni_6(3,3) two-layer	
	Supported	Free	Supported	Free
z_e(Ni-CO), (Å)	1.28	1.16	1.13	1.16
r_e(Ni-CO), (Å)	1.94	1.86	1.84	1.86
ω(Ni-CO), (cm^{-1})	391	470	505	428
r_e(C-O), (Å)	1.19	1.20[b]	1.21	1.20[b]
ω(C-O), (cm^{-1})	1789	1689[b]	1717	1757[b]
D_e, (eV)	3.32	3.54	3.28	3.51

[a] Height of the C atom above the upper plane of the Ni_6 cluster z_e(Ni-CO), distance between the carbon atom and the nearest-neighbour Ni atom r_e(Ni-CO). For an explanation of other notations see Tables 5 and 6.
[b] Computed at the Ni-CO distance optimized for the supported cluster.

also calculated the properties of CO chemisorbed on the three-fold hollow sites of free and supported Ni_6 clusters.

CO is strongly chemisorbed at both free and supported Ni_6 clusters, Table 10; however, for the supported one-layer cluster the Ni-CO distance, vibrational frequency, and dissociation energy indicate a weakening of the adsorption bond, see Table 10. The most important property for the present analysis is the shift in the CO vibrational frequency on the supported cluster compared to that of CO on the free cluster. For the case of the one-layer Ni_6(6,0) cluster, which gives rise to a moderate charge transfer to the surface, the CO vibrational frequency is 100 cm^{-1} higher than on free Ni_6(6,0), consistently with the view of a partially oxidized monolayer. The vibrational blue shift is not found when CO is adsorbed on the two-layer Ni_6(3,3) cluster, Table 10. In fact, the ω(C-O) shift between the supported and the free Ni_6(3,3) cluster is negative, -40 cm^{-1}, thus providing no evidence of a reduced backdonation for this system. Clearly, a qualitatively different behaviour is observed for one-layer and two-layer Ni clusters.

These calculations allow us to draw some general conclusions concerning the nature of the interface between a small Ni particle and an oxidized Al surface. Starting from the assumption that the interaction occurs at an oxygen terminated surface, we found that the Ni cluster adheres quite strongly to the substrate. We found this bond to be 3-4 eV per adsorbed Ni atom, although this value is overestimated in LDA. This strong bond is connected to a partial oxidation of the Ni atoms which are in direct contact with the oxygen centres. Although several measures are consistent with a partial oxidation of the Ni atoms, a quantitative evaluation of the amount of charge transfer is difficult.

The temperature-programmed desorption (TPD) spectrum of CO adsorbed on the Ni/Al_2O_3/Al(111) surface [120] shows a peak at 375 K which has been attributed to the desorption of bridge bonded CO; the corresponding peak on a single crystal Ni(111) surface is found at 410 K, indicating that a weaker interaction occurs between CO and the Ni particle when this is supported on the oxide substrate. Our model calculations, which show a moderate decrease of the binding energy for the supported compared to the free clusters (Table 10), are consistent with these measurements.

The adsorption of CO on a supported cluster provides a rather accurate way to probe the reduced electron density at the metal overlayer. We find that when CO is adsorbed on a one-layer Ni cluster supported on the $Al_6O_{12}H_6$ substrate there is a positive shift, of 100 cm^{-1}, of the CO vibrational frequency. This is in qualitative agreement with the measured electron energy loss (EEL) spectra of Chen et al. [120] for CO adsorbed on Ni/Al_2O_3/Al(111).

The blue shift in the CO vibrational frequency ω(C-O) is not found when a two-layer Ni cluster is considered. In this case, in fact, the calculated vibrational frequency shift is smaller and negative. This indicates that the partial oxidation at the "interface" is quite localized and probably restricted to the first layer of metal atoms.

5.2. Ni and Cu clusters deposition on MgO

The deposition of Cu on MgO surfaces has been studied in detail experimentally by means of low energy electron diffraction, and by various electron spectroscopy methods, like Auger, EEL, and X-ray photoemission [121–124]; the sticking probability of incident Cu atoms has also been measured [125,126]. This system is thus ideal for a theoretical investigation, also because it is less problematic to model the MgO(001) surface than a polar surface like the Al_2O_3 surface just described. The deposition of Al, Ag, Fe, and Ti metal overlayers on MgO(001) has been studied by means of band structure calculations at various level of theoretical sophistication [127–131]. Other investigations have considered the adsorption of single atoms on MgO [132–134].

Here we study the deposition of Ni_4 and Cu_4 clusters on various sites of the MgO(001) surface and we shall discuss the nature of the cluster/substrate interface as well as the modifications induced on the cluster electronic structure. The calculations [105] have been performed with the LCGTO-DF program employing a gradient corrected energy functional (GGA). The total energy has been calculated by taking into account density gradient corrections to the homogeneous electron gas exchange functional [22] and to the correlation energy functional [22–24] (BLYP approach). These gradient corrections were introduced based on the self-consistent VWN density; all results presented in this section are at the BLYP GGA-DF level. Two cluster models have been employed to simulate the MgO(001) surface, namely Mg_9O_9 and O_9Mg_9, Figure 7. These clusters model adsorption on a oxygen anion or a magnesium cation, respectively. The clusters are embedded in an array of ($17\times17\times6$ - 18) point charges.

Before addressing the problem of cluster deposition on MgO, we briefly summarize some aspects of the bonding of single Ni and Cu atoms with the MgO surface. The interaction of

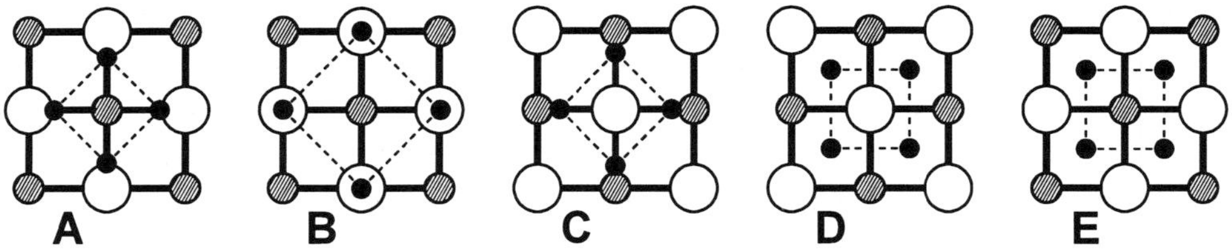

Figure 7. Schematic representation of the sites investigated for the adsorption of Ni_4 and Cu_4 clusters on the (001) surface of MgO. Small hatched circles represent Mg ions, large empty circles O ions, and small full circles Ni or Cu atoms. The cluster models Mg_9O_9 (A, B, E) and O_9Mg_9 (C, D) also contain the corresponding atoms from the "second" crystal layer, not shown here.

a Cu atom with an oxide anion of MgO is rather weak, about 0.3 eV, with an equilibrium distance of 2.18 Å. This is consistent with the low reactivity of the regular MgO surface shown by the measured sticking probability of Cu on MgO [125] and predicted by HF calculations [132]. For Ni, on the other hand, we found a much stronger bonding, 0.88 eV, and an equilibrium distance of 1.87 Å. Both atoms, Ni and Cu, show a moderate charge transfer to the oxide ion on the surface which, however, is much smaller than on Al_2O_3. The different bond strength is largely due to the possibility for Ni to form a covalent polar bond through the partially filled 3d orbitals. The adsorbate-substrate interaction was found to be always very weak when it involves the surface Mg cations [105].

To model the metal/oxide interface we have considered square planar (D_{4h}) clusters which may be viewed as precursors of an epitaxially growing metal film. An unsupported Ni_4 cluster is strongly spin-polarized, with about one unpaired electron per atom. The metal-metal distance r_e(Ni-Ni), is 2.23 Å (2.49 Å is the bulk distance); the corresponding binding energy per atom is 1.76 eV. An unsupported square planar Cu_4 cluster, with a $^3A_2$ ground state, has an optimal Cu-Cu distance of 2.36 Å and a binding energy per atom of 1.3 eV. These two clusters have been deposited on the MgO(001) clusters in various orientations, see Figure 7. A partial optimization of the adsorbed cluster has been carried out. In two of the adsorption geometries, denoted as site A and site B, Figure 7, the interaction is primarily between the cluster metal atoms and the surface oxygen atoms. In site B each metal atom is placed on-top of the surface oxygen centres; this corresponds to epitaxial growth of a metal overlayer. In site C the metal atoms are closer to the surface cations; on-top adsorption has not been considered, given the weak binding of the single metal atoms. Sites D and E represent possible alternative orientations of the M_4 cluster with the metal atoms near the four-fold hollow sites of the (001) surface.

In general, adhesion of the metal clusters to the surface is moderately strong when a direct interaction with the oxygen cations is possible, while it is very weak when the interaction primarily occurs with the substrate cations or in the four-fold hollow sites. As already mentioned, the competition between the metal-metal and the metal-oxygen bonds is an important issue. If the adsorptive metal-oxygen bond strength is larger, the metal cluster may grow epitaxially until completion of the first monolayer (Frank-von der Merwe mechanism [135]). If the intra-cluster metal-metal bond prevails, growth of three-dimensional metal particles is expected (Vollmer-Weber mechanism [135]). When the square planar Ni_4 and Cu_4 clusters are adsorbed on site A, a bond of 0.62 and 0.36 eV/atom is found. The surface-cluster distance is similar for the two metals, about 1.95-2.0 Å. Upon adsorption the metal-metal bond length expands from 2.2 to 2.5 Å for Ni and from 2.4 to 2.65 Å for Cu. This reflects a weakening of the metal-metal bond due to the interaction with the surface oxygen atoms. It is interesting to note that the metal-metal distance in the supported clusters is similar to the bulk metal values (2.49 Å for Ni and 2.56 Å for Cu). If the metal-oxygen bond were strong, the M_4 cluster "epitaxially" grown on-top of the surface oxygen atoms, site B (see Figure 7), should have a lower total energy; however, this is not the case. When the four metal atoms are placed directly above the four peripheral oxygen atoms of the cluster, site B, the bonding with respect to dissociation into free Ni_4 and Cu_4 clusters is weaker than for site A. It should be noted that in site B the metal-metal distance is 2.97 Å; at this distance, there is still considerable metal-metal bonding as shown by the atomization energies of the free clusters Ni_4 and Cu_4 of this geometry which is 0.98 and 0.87 eV/atom, respectively. Thus, the calculations show that the metal-oxygen bonding is not strong enough to prevail over the metal-metal interaction and that the adsorbed clusters maintain to a large extent their identity. For adsorption on site C, with the metal atoms close to the Mg ions, or for sites D and E, where the metal atoms sit on four-fold hollow sites, the bonding is weak as expected.

The electronic interactions at the interface can be better visualized by means of density of states (DOS) plots, Figure 8. As usual, these plots have been generated by broadening the Kohn-Sham one-electron level spectrum using Gaussian functions of 0.2 eV half-width. The "band structure" of our MgO cluster exhibits a gap of about 3.5 eV, i. e. it is too small compared to the experimental value, 7.8 eV; this result, however, is not due to the cluster approximation, but rather is a general characteristic of the DF approach. When a Ni_4 cluster is deposited on the surface the Fermi level moves to the edge of the Ni 3d band. The Ni cluster remains spin-polarized although the interaction with the substrate reduces the magnetization from 4 unpaired electrons in unsupported Ni_4 to 1.75 unpaired electrons in the supported cluster. The formation of some covalent bonding between the O 2p band and the Ni 4s and 3d levels can be seen from the small tails of the Ni DOS into the O 2p band, Figure 8. The electronic structure of Cu_4 on MgO is similar, but here the 3d levels are completely filled and higher in energy than in Ni_4. The bonding involves primarily the 4sp Cu orbitals and not the 3d orbitals. For both Ni_4 and Cu_4, the 2s and 2p core levels are shifted to higher binding energies by 3-4 eV, a fact which is usually interpreted as indication for some charge transfer

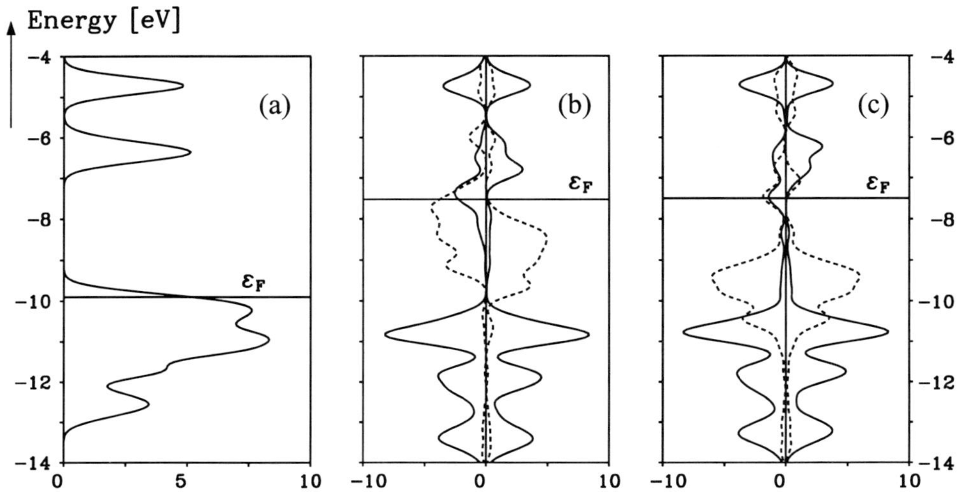

Figure 8. Density of states of (a) Mg_9O_9, (b) Ni_4/Mg_9O_9, (c) Cu_4/Mg_9O_9. Solid line MgO, dotted line M_4; left-hand side minority spin, right-hand side majority spin.

from the cluster to the surface. However, given the existence of an electric field associated with the largely ionic MgO surface, some care is necessary in interpreting these shifts. They are not exclusively caused by charge transfer, but also by the interaction with the field.

To elaborate electronic modifications induced by the oxide surface on the supported cluster, we have adsorbed a CO probe molecule on the centre of the metal cluster. We shall compare the corresponding chemisorption properties with those of free Ni_4 and Cu_4 clusters, following the same procedure used for Ni/Al_2O_3. We first determined the potential energy curve for the interaction of CO with the free and supported clusters, Figure 9. All energy curves exhibit a barrier followed by a deep minimum, indicative of an avoided crossing and of a configuration change as the CO molecule approaches the clusters. For instance, the magnetization of Ni_4 disappears for short Ni-CO distances.

The physical reason for this magnetic quenching and for the appearance of a barrier is that at longer distances a repulsive interaction occurs between the CO 5σ lone pair and the diffuse 4s-derived cluster levels. The depth of the attractive well is much larger on Ni than on Cu, reflecting the different electronic structure of the two metals and the possibility of a configuration change in Ni (from $3d^94s^1$ to $3d^{10}$) to reduce the initial repulsion [136]. As a result, CO is only weakly bound to free Cu_4, by about 0.3 eV, while it is bound by 1.2 eV to an unsupported Ni_4 cluster.

On the supported clusters the height of the barrier is higher than for the free clusters, Figure 9, in particular for Cu. The position of the barrier also moves somewhat. The shape of the attractive well changes as shown by the decrease of the M-C vibrational frequency, by 71 cm^{-1} for Ni and by 47 cm^{-1} for Cu. Similarly, the bond strength is reduced, by 0.1 eV for Ni and by 0.5 eV for Cu. The fact that the height of the barrier increases and that its position moves away from the surface suggests that the incoming CO molecule encounters a larger repulsion on the supported cluster. This is because the 4sp orbitals of the free metal cluster polarize away from the oxide surface, toward the vacuum, in order to reduce the repulsion with the substrate. On the other hand, a polarization away from the incoming probe molecule is prevented through the presence of the rather rigid and unpolarizable MgO surface.

The reduction of the adsorption energy and the decrease of the vibrational frequency ω_e(M-C) suggest a weakening of the metal-CO bond in the supported cluster. To identify the origin of this weakening we have computed the vibrational frequency of the C-O stretching mode. On free Ni_4 the CO vibrational frequency is 1521 cm^{-1}, i. e. 622 cm^{-1} lower than calculated for gas-phase CO; this is consistent with a substantial backdonation of charge from the Ni 3d orbitals to the CO $2\pi^*$ MO. The corresponding C-O vibrational frequency on Cu_4 is 1552 cm^{-1}. On the supported clusters the red shift is reduced, by 104 cm^{-1} on Ni and by 33 cm^{-1} on Cu. Since the surface-CO distance used for the calculations is the same, the higher CO frequency in the supported clusters reflects a reduced back bonding. This is fully consistent with the picture of a weak oxidation of the metal atoms in contact with the surface oxygen

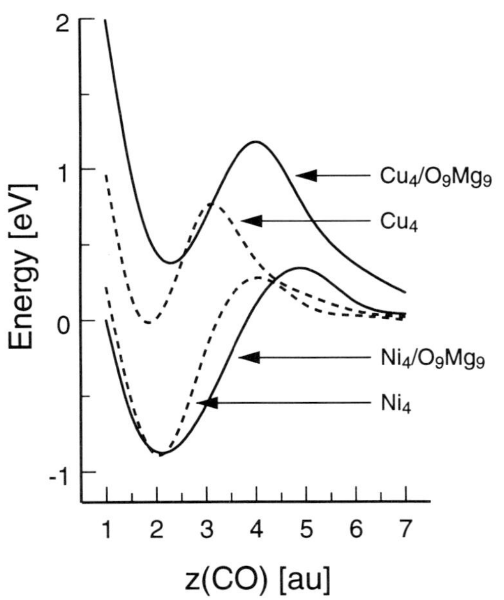

Figure 9. Potential energy curves for the interaction of a CO molecule with various free and supported clusters M_4, M = Cu, Ni.

atoms. The shifts, however, are not very large suggesting that there is only a moderate charge transfer from the cluster to the oxide substrate.

An important conclusion is that on MgO the metal-metal bond prevails over the metal-oxygen bond, in particular for Ni where the intra-cluster metallic bonds are stronger. As a consequence, both Ni_4 and Cu_4 moieties deposited on MgO keep their "cluster" identity. The metal-metal bonds are elongated upon adsorption, but the interaction with the substrate is not sufficient to stabilize a structure where the metal atoms are arranged directly on top of the substrate oxygen atoms. On an extended surface this latter structure would correspond to an epitaxial growth of the metal film. Thus, the calculations suggest that on MgO deposited Cu and, in particular, Ni particles should preferentially grow in a three-dimensional fashion.

5.3. Protonation of palladium clusters in zeolites

The calculations on oxide-supported metal clusters just discussed have indicated a small amount of electron transfer from the metal particle to the supporting oxide. The oxide system studied was an idealized one; real oxide surfaces with structural defects may also feature surface termination by OH groups which may affect the interaction with the metal particle [137]. The situation at such more realistic oxide surfaces is then to some extent reminiscent of that in zeolite cavities where electron-deficient metallic particles were detected. These particles in zeolites are often discussed to be the result of interactions with zeolitic protons and are rationalized in terms of protonated species of the type $[M_nH_x]^{x+}$ [138–140]. An electron-deficient state of small (5-7 Å in diameter [141]) palladium clusters in faujasites has been suggested mainly based on the observation of the unusual vibrational frequency of an adsorbed CO probe, 1860 cm^{-1} [142]. This frequency is blue shifted relative to the value of 1823 cm^{-1} for a CO molecule triply bonded either to small neutral palladium species in zeolites [142] or at a Pd(111) surface [143]. The catalytic activity of these electron-deficient metal species may differ considerably from that of electroneutral moieties of similar size. For example, electron-deficient Pd_n particles have been shown to be responsible for the exceptionally high activity of Pd/NaY catalysts in the hydrogenolysis of neopentane [144].

We examine here the hypothesis that small protonated palladium species are formed in zeolites [145]. For this purpose tetrahedral and octahedral clusters with an entrapped proton, $[Pd_4H]^+$ and $[Pd_6H]^+$, were considered as models of electron-deficient palladium species encaged in a zeolite matrix. GGA-DF calculations including geometry optimization have been carried out on the bare clusters Pd_4 and Pd_6 and on their protonated forms as well as on the corresponding complexes with a CO molecule adsorbed at the three-fold hollow site.

Protonation of the palladium clusters increases the metal-metal distances. The insertion of a proton into the centre of a Pd_4 tetrahedron is accompanied by a bond elongation of 0.18 Å. The structural variation of the larger cluster, Pd_6, due to an interstitial H^+ species is not as strong as that of the smaller cluster Pd_4; the Pd-Pd distance of $[Pd_6H]^+$ is only 0.05 Å larger than in the parent cluster. The adsorption of CO on the Pd_4 cluster causes a sizeable lengthening of the Pd-Pd distances, both between the three Pd atoms closest to the

μ_3-bound CO molecule and from these atoms to the fourth Pd centre, by 0.16 Å and 0.07 Å, respectively. Apparently, the formation of the chemisorption bond with CO occurs, at least partially, at the expense of a weakened metal-metal interaction, by a familiar motif [146,147]. For the protonated $[Pd_4H]^+$ species a slight adsorption-induced bond shortening (by 0.04 Å) is calculated within the Pd_3 triangle which defines the μ_3 adsorption site, while the "lower" Pd atom of the cluster significantly moves away from the atoms of the "upper" Pd_3 unit, by 0.24 Å. The finding that the proton insertion affects the smaller Pd_4 species more strongly than the larger cluster Pd_6 is corroborated by the calculated Pd-CO and C-O interatomic distances of the adsorption complexes. For example, the Pd-CO distance in Pd_4-CO increases upon protonation by 0.09 Å, but by only 0.03 Å in Pd_6-CO. The calculated lengthening of the Pd-CO bond which accompanies the protonation of the metal particle is in agreement with the experimental evidence that adsorbed CO molecules are easier released from electron-deficient (protonated) palladium particles in faujasites than from neutral clusters [142].

The calculated protonation energies of the palladium species under consideration provide a crucial check for the hypothesis on the formation of protonated $[Pd_nH_x]^{x+}$ species in Pd/NaY catalysts [142,144,148]. The results, 9.40 eV and 9.92 eV for Pd_4 and Pd_6, respectively, even exceed the proton affinity of such a strong base as the ammonia molecule. The experimental proton affinity of NH_3, 8.83 eV, is close to the DF result, 9.20 eV (without corrections for the zero-point energy and for the BSSE). Thus, the energetic data support the assumption that electron-deficient metal species are formed by the protonation of Pd_n clusters in zeolites. The error margins of the calculated protonation energy of the palladium clusters may also be estimated by comparison with the experimentally determined bond energy of the species $[Pd\text{-}H^+]$, 7.30 eV [149]. The DF result is 7.15 eV on the non-relativistic and 7.49 eV at the scalar-relativistic level of theory [31–33] outlined in chapter 2. The latter values may be compared to the result obtained from accurate generalized valence bond plus configuration interaction calculations, 7.02 eV [150].

Calculations on the $[Pd_6H]^+$ model with the hydrogen attached to outside of the Pd_6 species at a μ_3-site resulted in a protonation energy close to that of the moiety with the interstitial hydrogen. This means that mobile protons are probably able to change their positions with respect to the Pd_n cluster depending on several factors, including alterations of the zeolite environment and a variation of the cluster nuclearity n. Clearly, a proton situated further away from the adsorption site will affect the adsorbate in a qualitatively similar fashion, but to a lesser degree.

One may expect that the protonation of palladium clusters is accompanied by a rather significant shift of the Pd core levels to higher binding energies, which should be measurable by photoelectron spectroscopy. Indeed, a shift of 5.3 and 4.7 eV to higher binding energies is predicted from the analysis of the Pd 3d and Pd 4s Kohn-Sham eigenvalues for Pd_4 and Pd_6, respectively. These data may be of help to experimentally detect the protonation of palladium moieties in zeolites, independently of the technique of probe molecule adsorption.

The Pd_n-CO binding energy decreases significantly, by 0.39 eV and 0.33 eV, upon protonation of the Pd_4 and Pd_6 substrate species, respectively. This finding is in accord with the observations that the CO adsorption energy on the palladium particles encapsulated in zeolites is lowered by protonation [138,142] .

The measured relative blue shift of the intramolecular vibrational frequency of μ_3-CO molecules adsorbed on palladium clusters in Pd/NaY is one of the most important experimental arguments in favour of the formation of electron-deficient $[Pd_nH_x]^{x+}$ species [142,148]. Depending on the cluster model, the calculated vibrational frequency for μ_3-CO adsorbed on the unprotonated small palladium clusters is underestimated by 50-80 cm^{-1} compared to the experimental value of 1823 cm^{-1}. Protonation of the cluster Pd_4 leads to an increase of the vibrational frequency of adsorbed CO by 99 cm^{-1}. This calculated shift, although in the proper direction, is too large compared to the experimentally determined value, 37 cm^{-1} [142]. For the larger cluster Pd_6 the calculated proton-induced shift of the CO vibrational frequency is somewhat smaller, 82 cm^{-1}, but is still significantly larger the experimental quantity. Nevertheless, even the results for the protonation-induced C-O frequency shift are very encouraging: the effect of the inserted proton on various properties of the host metal species, including the energy of CO adsorption and the C-O vibrational frequency, decreases with cluster size. The only exception to this trend is the protonation energy which increases with cluster size, probably due to a better screening of the proton in the larger cluster. Thus, the present DF data [145], obtained without taking into account any direct interaction between the metal particle and the zeolite matrix, provide a sufficiently adequate description of the protonation phenomenon of transition metal clusters in a zeolite cage. The whole set of relevant observations on palladium species in faujasite cavities is qualitatively reproduced and rationalized by this quantum chemical study: the hypothesis [142,148] that palladium species in faujasite cavities are indeed stabilized in the form of larger Pd clusters seems quite appropriate. However, no arguments may be gleaned from the present calculations supporting the assumed nuclearity, Pd_{13}, or shape of the encaged particles [142,148].

5.4. Atomic platinum in mordenite

The concept of single Pt atoms formed in zeolites has gained strong support through investigations by Sachtler et al. on faujasite (HY) and mordenite (HMor) [151,152]. They proposed that Pt atoms are stabilized either in sodalite cages of HY or in the side pockets of the main channels of HMor, basing their arguments on TPD data for hydrogen which they interpreted in terms of an oxidation of the Pt atoms by zeolite protons: $Pt^0 + 2\,H^+ \rightarrow Pt^{2+} + H_2$. A FTIR spectroscopy study of adsorbed CO [153] provides additional arguments for the formation of stable monoatomic Pt species in mordenite. The IR absorption band at 2123 cm^{-1} was found not to shift with CO coverage, but to correlate with the presence of zeolite protons. If the protons are neutralized or exchanged by Na^+ cations, this IR band disappears. Thus, this band was assigned to CO interacting with Pt atoms that are anchored to zeolitic protons in complexes like $[Pt–H_z]^{z+}$. The C-O vibrational frequency shift of -20 cm^{-1} relative to free CO [65] is significantly smaller (in absolute value) than shifts observed for on-top CO adsorption complexes (-40 to -60 cm^{-1} [154,155]), for terminal carbonyl groups in platinum complexes (-80 to -150 cm^{-1} [156,157]), or for a "free" PtCO species in matrix isolation (-90 cm^{-1} [158]). This unusual magnitude of the C-O frequency shift has been attributed to an electron-deficient state of the Pt atoms [153].

We have considered a series of model complexes where a CO molecule is adsorbed on a Pt atom which in turn is anchored at one or two Brønsted acid groups of a zeolite fragment [159]. Furthermore, we have investigated models where CO interacts with a PtH^+ moiety or with a charged Pt species. The coordination and electronic structure of monoatomic Pt species in mordenite has been investigated by a scalar-relativistic LCGTO-DF model cluster method [31–33] (see chapter 2).

In the case of mordenite, Pt atoms are likely to form complexes with either one bridging hydroxyl group (Figure 10a, Z-Pt) or with H atoms of two neighbouring hydroxyl groups (Figure 10b, Z_2-Pt) situated on opposite walls of a side pocket or on a so-called four-ring (an eight-membered ring) (Figure 11, Z(4r)-Pt).

Taking the calculated C-O stretching frequency shifts as a criterion, the clusters Z-Pt-CO and Z_2-Pt-CO seem to adequately represent how Pt species in mordenite interact with CO probe molecules. For Z-Pt-CO and Z_2-Pt-CO the vibrational frequency shift $\Delta\omega$(C-O) is calculated to -43 cm^{-1} and -40 cm^{-1}, respectively. These values are somewhat larger (in absolute value) than the experimental shift of -20 cm^{-1} [153]. In part, this slight discrepancy may be ascribed to deficiencies of the cluster models, like the compensation of dangling bonds by hydrogen atoms and the neglect of long-range electric field effects due to the zeolite surrounding. The negative C-O frequency shift computed for the model cluster Z(4r)-Pt-CO

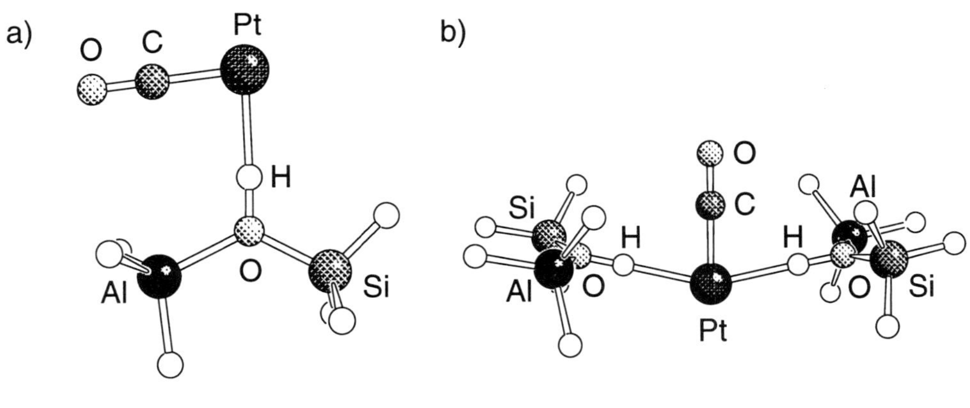

Figure 10. Calculated structures of cluster models of a Pt-CO fragment interacting with a) one (Z-Pt-CO) and b) two (Z_2-Pt-CO) separated bridging hydroxyl groups in mordenite. The hydrogen atoms employed to compensate the dangling bonds in the Si(OH)Al moieties are not marked.

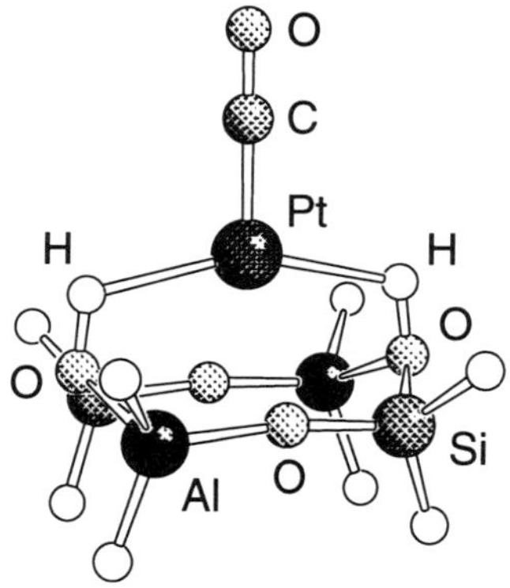

Figure 11. Calculated structure of the cluster model Z(4r)-Pt-CO.

is even larger, -84 cm^{-1}, thus indicating that this kind of Pt atom anchoring is less probable. Also, as manifested by the calculated Pt 4f level shift [159], the electronic state of Pt in the latter complex differs markedly from that of Z-Pt-CO and Z_2-Pt-CO moieties. This difference can be rationalized by the interaction of the Pt-CO species with the two nearest oxygen atoms of the four-ring which is absent in complexes of other structural arrangements. Another model, H^+-Pt-CO, was chosen to represent an extreme case of interaction with a Brønsted site: it describes possible effects if a platinum species, with CO adsorbed, is directly protonated by a "naked" (non-solvated) proton. The value of the C-O vibrational frequency shift calculated for the complex H^+-Pt-CO is positive, +37 cm^{-1}, at variance with the negative experimental value. Therefore, in the light of both the results of DF calculations and of experiment [153] a direct protonation of Pt-CO species in mordenite is unlikely.

The DF model results suggest that the binding energy of the Pt-CO species to the zeolite framework increases if the acid strength and the number of the OH groups interacting with the Pt atom increases. Particularly, the Pt-H binding energy of Z_2-Pt-CO is higher than that of Z-Pt-CO. The calculated difference in the stabilization energy of these two complexes may help to rationalize the following seemingly contradictive experimental data. Zholobenko et al. [153] have found that the adsorption of CO on Pt/HMor samples results both in the appearance of the FTIR band at 2123 cm^{-1}, associated with the formation of carbonyl complexes with electron-deficient Pt species, and in the increase of the O-H stretching band intensity. They attributed the latter observation to the breaking of some zeolite-Pt bonds caused by the interaction between CO molecules and Pt atoms. Using the cluster model results, this experimental finding may be rationalized by assuming that platinum complexes formed with one OH group are destroyed whereas complexes attached to two OH groups remain stable during the adsorption of the CO probe molecules.

6. OUTLOOK

We have presented an overview of recent results of density functional calculations on the general problem of adsorption at oxide materials. This is a relatively new area as the work done in the last two decades on quantum chemical modelling of surfaces has been dedicated mostly to metals and semiconductors rather than to ionic or semi-ionic materials. On the other hand, oxides play a very important role in many technological applications and the amount of scientific computing which will be dedicated to this problem will significantly grow in the near future. The chemistry of oxide surfaces is largely the chemistry of defects and vacancies. This is why local models, like the cluster model, will play a particular role.

Of course, there are several open questions which have to be answered in order to make the quantum chemical description of the reactivity of oxide surfaces an important tool in the study of these phenomena. Some of these questions are of purely theoretical nature. For instance, more experience is needed in order to assess the chemical accuracy which can be obtained with gradient-corrected energy functionals for binding energies and geometries. The inclusion of scalar-relativistic effects will be an important aspect of the treatment of heavy metals interacting with oxide surfaces, and more elaborate embedding schemes will be required to better approximate the effect of the surrounding when using local cluster models, in particular in cases where the substrate exhibits a partially covalent character.

A very important point is the underestimation of the band gap of ionic materials by the majority of the available DF methods. This is related to the "self-interaction" problem of these methods [160,161]. This methodological defect may, to some extent, be reflected in the tendency of these DF methods to provide a more delocalized description of the bonding compared, for instance, to the HF approach. Recent efforts to construct DF methods that exhibit potentials which are free of self-interaction artifacts, may provide a promising alternative [162,163]. A good description of band gaps and of the oxide electronic structure will be an important goal to improve the predictive power of the calculations. This is particularly relevant if one considers how little is still known about the changes in the bonding character at the surface of an oxide material, where the atoms have a completely different coordination than in the bulk and a modified reactivity which is a direct function of their bonding and coordination.

On the other hand, the technical improvement of the quantum chemical treatments must be accompanied by a more detailed and accurate structural information. The design of "good" models of a surface process requires the knowledge of the geometric structure of the substrate material. New experimental techniques are very promising in this respect, but a considerable input to the theoretical calculations will come from dynamical simulations based on pair potentials or on analytical force fields. Also, the comparison of rigorous "first principles" calculations will be essential to validate simulations which, in principle, can be used to predict the complex reconstruction of polar surfaces.

In general, the future of quantum chemical investigations in this area looks pretty bright.

Reasons of optimism are the possibility to use combined approaches (force fields, cluster models, periodic calculations) to solve a given problem from different viewpoints, the increasing amount of reliable experimental data, and, very importantly, the continuous growth of computing power, in particular in connection with massive parallel computers. In the next decade, the combination of all these developments will definitely increase the role of theory as an essential tool to complement experimental results.

Acknowledgment. The authors are indebted to Dr. U. Birkenheuer, Prof. H. Knözinger, Dr. V. A. Nasluzov and Prof. G. M. Zhidomirov for valuable contributions to various joint projects. We thank Dipl. Chem. K. Albert for her help during the preparation of figures. We have also profited much from the contributions made by numerous gifted graduate students of the Munich group. The studies described above have been made possible by grants from Deutsche Forschungsgemeinschaft (SFB 338), Bayerischer Forschungsverbund Katalyse (FORKAT), Volkswagen Foundation (Project I/68 691), INTAS (Project 93-1876), and Fonds der Chemischen Industrie. G. P. thanks the Alexander von Humboldt foundation for supporting his work at the Technische Universität München.

REFERENCES

1. C. W. Bauschlicher Jr., P. S. Bagus and H. F. Schaefer III, IBM J. Res. Develop. 22 (1978) 213.
2. H. Bethe, Ann. Physik 3 (1929) 133.
3. E. A. Colbourn and W. C. Mackrodt, Surf. Sci. 143 (1984) 391.
4. E. A. Colbourn, Surf. Sci. Rep. 15 (1992) 281.
5. J. Sauer, Chem. Rev. 89 (1989) 199.
6. J. Sauer, P. Ugliengo, E. Garrone and V. R. Saunders, Chem. Rev. 94 (1994) 2095.
7. H. H. Kung, Transition Metal Oxides: Surface Chemistry and Catalysis, Elsevier, Amsterdam, 1989.
8. H.-J. Freund and E. Umbach (eds.), Adsorption on Ordered Surfaces of Ionic Solids and Thin Films, Springer-Verlag, Berlin, 1993.
9. V. E. Henrich and P. A. Cox, The Surface Science of Metal Oxides, Cambridge University Press, Cambridge, 1994.
10. R. Car and M. Parrinello, Phys. Rev. Lett. 55 (1985) 2471.
11. G. Pacchioni, P. S. Bagus and F. Parmigiani (eds.), Cluster Models for Surface and Bulk Phenomena, NATO ASI Series, Vol. 283, Plenum Press, New York, 1992.
12. C. Pisani, R. Nada and L. Kantorovich, J. Chem. Phys. 92 (1990) 7448.
13. S. Krüger and N. Rösch, J. Phys. C 6 (1994) 8149.
14. N. Rösch, P. Knappe, P. Sandl, A. Görling and B. I. Dunlap, in: The Challenge of d and f Electrons. Theory and Computation, ACS Symposium Series No. 394, eds. D. R. Salahub and M. C. Zerner, Am. Chem. Soc., Washington, DC, 1989, p. 180.

15. B. I. Dunlap and N. Rösch, Adv. Quantum Chem. 21 (1990) 317.
16. N. Rösch, in: Cluster Models for Surface and Bulk Phenomena, NATO ASI Series B, eds. G. Pacchioni, P. S. Bagus and F. Parmigiani, Plenum Press, New York, 1992, p. 251.
17. J. W. Mintmire, J. R. Sabin and S. B. Trickey, Phys. Rev. B 26 (1982) 1743.
18. J. C. Boettger and S. B. Trickey, Phys. Rev. B 45 (1992) 1363.
19. E. Keller, Program SCHAKAL 92, Kristallographisches Institut der Universität Freiburg 1992, Germany.
20. S. H. Vosko, L. Wilk and M. Nusair, Can. J. Phys. 58 (1980) 1200.
21. T. Ziegler, Chem. Rev. 91 (1991) 651.
22. A. D. Becke, Phys. Rev. A 38 (1988) 3098.
23. C. Lee, W. Yang and R. G. Parr, Phys. Rev. B 37 (1988) 785.
24. B. Miehlich, A. Savin, H. Stoll and H. Preuss, Chem. Phys. Lett. 157 (1989) 200.
25. J. P. Perdew, Phys. Rev. B 33 (1986) 8822, erratum ibid. 34 (1986) 7406.
26. J. P. Perdew, J. A. Chevary, S. H. Vosko, K. A. Jackson, M. R. Pederson, D. J. Singh and C. Fiolhais, Phys. Rev. B 46 (1992) 6671.
27. J. P. Perdew and Y. Wang, Phys. Rev. B 45 (1992) 13244.
28. B. G. Johnson, P. M. W. Gill and J. A. Pople, J. Chem. Phys. 98 (1993) 5612.
29. P. M. W. Gill, B. G. Johnson and J. A. Pople, Chem. Phys. Lett. 209 (1993) 506.
30. J. Baker, J. Comput. Chem. 7 (1986) 385.
31. O. D. Häberlen and N. Rösch, Chem. Phys. Lett. 199 (1992) 491.
32. N. Rösch and O. D. Häberlen, J. Chem. Phys. 96 (1992) 6322.
33. O. D. Häberlen, S.-C. Chung and N. Rösch, Int. J. Quant. Chem.: Quant. Chem. Symp. 28 (1994) 595.
34. T. Ziegler, V. Tschinke, E. J. Baerends, J. Snijder and W. Ravenek, J. Phys. Chem. 93 (1989) 3050.
35. V. A. Nasluzov and N. Rösch, Chem. Phys., in press.
36. N. W. Winter, R. M. Pitzer and D. K. Temple, J. Chem. Phys. 86 (1987) 3549.
37. G. L. Gutsev, V. K. Gryaznov and V. A. Nasluzov, Chem. Phys. 154 (1991) 291.
38. V. A. Nasluzov, G. L. Gutsev, V. V. Rivanenkov, K. M. Neyman and A. G. Anshits, Soviet J. Struct. Chem. 33 (1992) 15.
39. P. S. Bagus, K. Hermann and C. W. Bauschlicher Jr., J. Chem. Phys. 80 (1984) 4378.
40. K. M. Neyman, P. Strodel, S. P. Ruzankin, N. Schlensog, H. Knözinger and N. Rösch, Catal. Lett. 31 (1995) 273.
41. K. M. Neyman, S. P. Ruzankin and N. Rösch, Chem. Phys. Lett. 246 (1995) 546.
42. K. Albert, K. M. Neyman, V. A. Nasluzov, S. P. Ruzankin, C. Yeretzian and N. Rösch, Chem. Phys. Lett. 245 (1995) 671.
43. S. F. Boys and F. Bernardi, Mol. Phys. 19 (1970) 553.
44. R. D. Amos, Adv. Chem. Phys. 67 (1987) 99.
45. G. Pacchioni, G. Cogliandro and P. S. Bagus, Surf. Sci. 255 (1991) 344.

46. G. Pacchioni, G. Cogliandro and P. S. Bagus, Int. J. Quantum Chem. 42 (1992) 1115.
47. M. A. Nygren, L. G. M. Pettersson, Z. Barandiarán and L. Seijo, J. Chem. Phys. 100 (1994) 2010.
48. G. Pacchioni, K. M. Neyman and N. Rösch, J. Electron Spectrosc. Relat. Phenom. 69 (1994) 13.
49. J. A. Mejías, A. M. Márquez, J. F. Sanz, M. Fernández-García, J. M. Ricart, C. Sousa and F. Illas, Surf. Sci. 327 (1995) 59.
50. K. M. Neyman and N. Rösch, Chem. Phys. 168 (1992) 267.
51. K. M. Neyman and N. Rösch, Ber. Bunsenges. Phys. Chem. 96 (1992) 1711.
52. N. Rösch, K. M. Neyman and U. Birkenheuer, in: Adsorption on Ordered Surfaces of Ionic Solids and Thin Films, Springer Series in Surface Sciences, Vol.33, eds. H.-J. Freund and E. Umbach, Springer-Verlag, Berlin, 1993, p. 206.
53. U. Birkenheuer, N. Rösch and J. C. Boettger, J. Chem. Phys. 100 (1994) 6826.
54. U. Birkenheuer, Quantenchemische Beschreibung von Oberflächen und geordneten Adsorbatsystemen. Methoden und Anwendungen, Dissertation, Technische Universität München, München, 1994.
55. R. F. W. Bader, Atoms in Molecules: A Quantum Theory, International Series of Monographs on Chemistry, Vol. 22, Oxford University Press, Oxford, 1990.
56. K. Hermann, P. S. Bagus and C. J. Nelin, Phys. Rev. B 35 (1987) 9467.
57. L. Ackermann and N. Rösch, J. Chem. Phys. 100 (1994) 6578.
58. S. Furuyama, H. Fujii, M. Kawamura and T. Morimoto, J. Phys. Chem. 82 (1978) 1028.
59. C. R. Henry, C. Chapon and C. Duriez, J. Chem. Phys. 95 (1991) 700.
60. J.-W. He, C. A. Estrada, J. S. Corneille, M.-C. Wu and D. W. Goodman, Surf. Sci. 261 (1992) 164.
61. D. Scarano, G. Spoto, S. Bordiga, S. Coluccia and A. Zecchina, J. Chem. Soc., Faraday Trans. 88 (1992) 291.
62. K. Hermann, P. S. Bagus and C. W. Bauschlicher Jr., Phys. Rev. B 30 (1984) 7313.
63. V. Bolis, B. Fubini, E. Garrone, C. Morterra and P. Ugliengo, J. Chem. Soc., Faraday Trans. 88 (1992) 391.
64. E. Escalona Platero, D. Scarano, G. Spoto and A. Zecchina, Faraday Disc. Chem. Soc. 80 (1985) 183.
65. K. P. Huber and G. Herzberg, Molecular Spectra and Molecular Structure. Constants of Diatomic Molecules, vol. 4, Van Nostrand Reinhold, New York, 1979.
66. J.-P. Bouanich and C. Brodbeck, J. Quant. Spectry. Radiat. Transfer. 14 (1974) 1199.
67. R. Disselkamp, H.-C. Chang and G. Ewing, Surf. Sci. 240 (1990) 193.
68. K. M. Neyman and N. Rösch, Surf. Sci. 297 (1993) 223.
69. K. M. Neyman and N. Rösch, Chem. Phys. 177 (1993) 561.
70. G. Pacchioni, T. Minerva and P. S. Bagus, Surf. Sci. 275 (1992) 450.
71. G. D. Mahan and A. A. Lucas, J. Chem. Phys. 68 (1978) 1344.
72. A. M. Ferrari and G. Pacchioni, Int. J. Quant. Chem. 58 (1996) 241.

73. G. Pacchioni, C. Sousa, F. Illas, F. Parmigiani and P. S. Bagus, Phys. Rev. B 48 (1993) 11573.
74. V. A. Fomichev, T. M. Zimkina and I. I. Zhukova, Soviet Solid State Phys. 10 (1969) 2421.
75. P. Strodel, K. M. Neyman, H. Knözinger and N. Rösch, Chem. Phys. Lett. 240 (1995) 547.
76. K. M. Neyman, V. A. Nasluzov and G. M. Zhidomirov, Catal. Lett., in press.
77. S. A. Zubkov, V. Y. Borovkov, S. G. Gagarin and V. B. Kazansky, Chem. Phys. Lett. 107 (1984) 337.
78. S. M. Zverev, K. S. Smirnov and A. A. Tsyganenko, Kinet. Katal. 29 (1988) 1439.
79. F. Wakabayashi, J. Kondo, A. Wada, K. Domen and C. Hirose, J. Phys. Chem. 97 (1993) 10761.
80. F. Wakabayashi, J. Kondo, K. Domen and C. Hirose, Catal. Lett. 21 (1993) 257.
81. H. Knözinger, in: Elementary reaction steps in Heterogenous Catalysis, eds. R. W. Joyner and R. A. van Santen, Plenum Press, New York, 1993, p. 267.
82. S. G. Lias, J. F. Liebman and R. D. Levin, J. Phys. Chem. Ref. Data 13 (1984) 695.
83. K. J. Farnworth and P. J. O'Malley, J. Phys. Chem. 100 (1996) 1814.
84. S. Bates and J. Dwyer, J. Phys. Chem. 97 (1993) 5897.
85. I. N. Senchenya, P. Ugliengo and E. Garrone, J. Mol. Struct. (Theochem), in press.
86. N. Echoufi and P. Gélin, J. Chem. Soc. Faraday Trans. 88 (1992) 1067.
87. N. M. Gupta, V. S. Kamble, K. A. Rao and R. M. Iyer, J. Catal. 120 (1989) 432.
88. R. M. Barrer, Hydrothermal Chemistry of Zeolites, Academic Press, London, 1982, p. 251.
89. R. Szostak, Molecular Sieves. Principles of Synthesis and Identification, Van Nostarnd Reinhold, New York, 1989.
90. P. J. O'Malley and J. Dwyer, Chem. Phys. Lett. 143 (1988) 97.
91. R. Zahradnik, P. Hobza, B. Wichterlová and J. Čejka, Collect. Czech. Chem. Commun. 58 (1993) 2474.
92. W. Langenaeker, N. Coussement, F. DeProft and P. Geerlings, J. Phys. Chem. 98 (1994) 3010.
93. A. G. Pelmenshchikov, E. A. Paukshtis, V. G. Stepanov, V. I. Pavlov, E. N. Yurchenko, K. G. Ione and G. M. Zhidomirov, J. Phys. Chem. 93 (1989) 6725.
94. C. T.-W. Chu and C. D. Chang, J. Phys. Chem. 89 (1985) 1569.
95. I. Mirsojew, S. Ernst, J. Weitkamp and H. Knözinger, Catal. Lett. 24 (1994) 235.
96. H. Knözinger, in: Adsorption on Ordered Surfaces of Ionic Solids and Thin Films, Springer Series in Surface Sciences, Vol. 33, eds. H.-J. Freund and E. Umbach, Springer-Verlag, Berlin, 1993, p. 257.
97. A. Zecchina, E. E. Platero and C. O. Areán, J. Catal. 107 (1987) 244, 244-247.
98. E. A. Paukshtis, Infrared Spectroscopy in Heterogeneous Acid-Basic Catalysis, Nauka, Novosivbirsk, 1992, (in Russian).
99. L. M. Kustov, V. B. Kazansky, S. Beran, L. Kubelková and P. Jirů, J. Phys. Chem. 91 (1987) 5247.

100. J. R. Anderson, Structure of Metallic Catalysts, Academic Press, New York, 1975.
101. S. A. Stevenson, J. A. Dumestic and E. Ruckenstein (eds.), Metal-Support Interaction in Catalysis, Sintering and Redispersion, Van Nostrand Reinhold, New York, 1987.
102. M. Che and C. O. Bennett, Adv. Catal. 36 (1989) 55.
103. S. J. Tauster, S. C. Fung and R. L. Garten, J. Am. Chem. Soc. 100 (1978) 170.
104. G. Pacchioni and N. Rösch, Surf. Sci. 306 (1994) 169.
105. G. Pacchioni and N. Rösch, J. Chem. Phys. 104 (1996) 7329.
106. K. H. Johnson and S. V. Pepper, J. Appl. Phys. 53 (1982) 6634.
107. A. Zdetsis and A. B. Kunz, Phys. Rev. B 32 (1985) 6358.
108. F. Raatz and D. R. Salahub, Surf. Sci. 156 (1985) 982.
109. A. B. Anderson, S. P. Mehandru and J. Smialek, J. Electrochem. Soc. 132 (1985) 1695.
110. A. B. Anderson, C. Ravimohan and S. P. Mehandru, Surf. Sci. 183 (1987) 438.
111. K. Nath and A. B. Anderson, Phys. Rev. B 39 (1989) 1013.
112. F. S. Ohuchi, R. H. French and R. V. Kasowski, J. Appl. Phys. 62 (1987) 2286.
113. R. V. Kasowski, F. S. Ohuchi and R. French, Physica B 150 (1988) 44.
114. M. Kohyama, S. Kose, M. Kinoshita and R. Yamamoto, J. Phys. Chem. Solids 53 (1992) 345.
115. P. Alemany, R. S. Boorse, J. M. Burlitch and R. Hoffmann, J. Phys. Chem. 97 (1993) 8464.
116. T. R. Ward, P. Alemany and R. Hoffmann, J. Phys. Chem. 97 (1993) 7691.
117. T. J. Sarapatka, Chem. Phys. Lett. 212 (1993) 37.
118. Q. Zhong and F. S. Ohuchi, J. Vac. Sci. Technol. A 8 (1990) 2107.
119. J. G. Chen, J. E. Crowell and J. T. Yates Jr., Surf. Sci. 185 (1987) 373.
120. J. G. Chen, J. E. Crowell and J. T. Yates Jr., Surf. Sci. 187 (1987) 243.
121. J.-W. He and P. J. Møller, Surf. Sci. 180 (1987) 411.
122. J.-W. He and P. J. Møller, Surf. Sci. 178 (1986) 934.
123. J.-W. He and P. J. Møller, Chem. Phys. Lett. 129 (1986) 13.
124. I. Alstrup and P. J. Møller, Appl. Surf. Sci. 33/34 (1988) 143.
125. J. B. Zhou, H. C. Lu, T. Gustafsson and E. Garfunkel, Surf. Sci. 293 (1993) L887.
126. M.-C. Wu, W. S. Oh and D. W. Goodman, Surf. Sci. 330 (1995) 61.
127. U. Schunberger, O. K. Andersen and M. Methfessel, Acta Metall. Mater. 40 (1992) S1.
128. C. Li and A. D. Freeman, Phys. Rev. B 43 (1990) 780.
129. D. M. Duffy, J. H. Harding and A. M. Stoneham, Acta Metall. Mater. 40 (1992) S11.
130. C. Li, R. Wu, A. J. Freeman and C. L. Fu, Phys. Rev. B 48 (1993) 8317.
131. J. R. Smith, T. Hong and D. J. Srolovitz, Phys. Rev. Lett. 25 (1994) 4021.
132. N. C. Bacalis and A. B. Kunz, Phys. Rev. B 32 (1985) 4857.
133. A. B. Kunz, Philosoph. Mag. 51 (1985) 209.
134. Y. Li, D. C. Langreth and M. R. Pederson, Phys. Rev. B 52 (1995) 6065.
135. H. Lüth, Surfaces and Interactions of Solids, 2nd ed., Springer-Verlag, Berlin, 1993.

136. G. Pacchioni and N. Rösch, Acc. Chem. Res. 28 (1995) 390.
137. B. C. Gates, Chem. Rev. 95 (1995) 511.
138. L. L. Sheu, H. Knözinger and W. M. H. Sachtler, J. Mol. Catal. 57 (1989) 61.
139. Z. Karpiński, Adv. Catal. 37 (1990) 45.
140. W. M. H. Sachtler and Z. Zhang, Adv. Catal. 39 (1992) 129.
141. G. Bergeret, T. M. Tri and P. Gallezot, J. Phys. Chem. 87 (1983) 1160.
142. L. L. Sheu, H. Knözinger and W. M. H. Sachtler, J. Am. Chem. Soc. 111 (1989) 8125.
143. A. M. Bradshaw and F. Hoffmann, Surf. Sci. 72 (1978) 513.
144. S. T. Homeyer, Z. Karpiński and W. M. H. Sachtler, J. Catal. 123 (1990) 60.
145. A. L. Yakovlev, K. M. Neyman, G. M. Zhidomirov, V. A. Nasluzov and N. Rösch, Ber. Bunsenges. Phys. Chem. 100 (1996) 413.
146. G. Pacchioni and N. Rösch, Inorg. Chem. 29 (1990) 2901.
147. N. Rösch, L. Ackermann and G. Pacchioni, Inorg. Chem. 32 (1993) 2963.
148. L. L. Sheu, H. Knözinger and W. M. H. Sachtler, Catal. Lett. 2 (1989) 129.
149. J. A. Martinho Simões and J. L. Beauchamp, Chem. Rev. 90 (1990) 629.
150. J. B. Schilling, W. A. Goddard III and J. L. Beauchamp, J. Am. Chem. Soc. 109 (1987) 5564.
151. M. S. Tzou, B. K. Teo and W. M. H. Sachtler, J. Catal. 113 (1988) 220.
152. B. A. Lerner, B. T. Carvill and W. M. H. Sachtler, Mol. Catal. 77 (1992) 99.
153. V. L. Zholobenko, G.-D. Lei, B. T. Carvill, B. A. Lerner and W. M. H. Sachtler, J. Chem. Soc. Faraday Trans. 90 (1994) 233.
154. B. E. Hayden, K. Kretzschmar and A. M. Bradshaw, Surf. Sci. 125 (1983) 366.
155. R. Gardner, R. Martin, M. Tüshans and A. M. Bradshaw, J. Electron Spectrosc. Relat. Phenom. 54/55 (1990) 619.
156. J. C. Calabrese, L. F. Dahl, P. Chini, G. Longoni and S. Martinengo, J. Am. Chem. Soc. 96 (1974) 2614.
157. G.-J. Li, T. Fujimoto, A. Fukuoka and M. Ichikawa, J. Catal. 12 (1992) 171.
158. E. P. Kündig, D. McIntosh, E. P. Moskovits and G. A. Ozin, J. Am. Chem. Soc. 95 (1973) 7234.
159. A. L. Yakovlev, K. M. Neyman, G. M. Zhidomirov and N. Rösch, J. Phys. Chem. 100 (1996) 3482.
160. J. P. Perdew and A. Zunger, Phys. Rev. B 23 (1981) 5048.
161. A. Svane and O. Gunnarsson, Phys. Rev. Lett. 65 (1990) 1148.
162. J. B. Kriger, Y. Li and G. J. Iafrate, Phys. Rev. A 46 (1992) 5453.
163. A. Görling, Phys. Rev. B 53 (1996) 7024.

J.M. Seminario (Editor)
Recent Developments and Applications of Modern Density Functional Theory
Theoretical and Computational Chemistry, Vol. 4

Density Functional Theory as a Tool in Studying Catalytic Processes

Ewa Broclawik[a], Rajappan Vetrivel[b] and Akira Miyamoto[c]

[a]Institute of Catalysis and Surface Chemistry, Polish Academy of Sciences, Cracow, Poland

[b]Catalysis Division, National Chemical Laboratory, Pune, India

[c]Department of Molecular Chemistry and Engineering, Faculty of Engineering, Tohoku University, Sendai, Japan

1. INTRODUCTION

The understanding and modelling of heterogeneous catalysis and catalytic behaviour at molecular level can benefit from accurate and significant quantum chemical calculations or molecular statistics and dynamic simulations or all of these procedures. Although theory has had in the past a limited impact on the understanding and predicting of catalysis, the advent of more powerful computing facilities, which permit more complex calculations and simulations representing more realistically the system of interest, makes this area of research one of potential and rapid growth. Advanced computer systems, including large massively-parallel machines, and recent improvements in quantum chemical computer codes allow calculations on considerable large systems, often involving more than 1000 basis functions and more than 100 atoms depending on symmetry. Nevertheless, the quantum chemical evaluation of potential energy surfaces for three-dimensional systems is highly complex. The enormous sizes of unit cells of such materials as zeolites, metal oxides or mixed catalysts make periodic *ab initio* calculations very demanding and only recently such calculations have become feasible for certain structures. Active sites, amorphous regions, or crystal perturbations introduce further limitations due to the broken translatory and space symmetry, because these require larger pseudounit cells.

One issue to solve this problem is a finite cluster approach. Such clusters may consist of atoms or ions, real or hypothetical molecules. The standard "molecular" *ab initio* methods can be applied in calculations involving these models. However, adoption of the finite models means confinement of the explicit treatment to selected atoms and interactions of significance for the effect under study but neglecting or only approximately including influence of the environment by suitably chosen boundary conditions. Hence, this approach is particularly suited to tackle local phenomena, e.g. to describe active sites of catalysts or impurity centers in semiconductors, but it also helps to extend knowledge on structure and bonding of crystals and on their bulk properties. Cutting the interior part out of large organic molecules, macromolecules and solids with covalent or partially covalent bonds, leaves dangling bonds. These have to be saturated with e.g. hydrogen atoms in order to create a realistic model for the quantum chemical description. Such approach will be implemented in discussing the properties of active sites in metal-exchanged zeolites. Sometimes it could prove very profitable to pursue properties of complex catalysts, especially when their structures are not defined or claimed amorphous, by assuming simple molecular models for the active site in belief that the most important factors might be deduced from the properties of ingredient atoms bonded in a proper way. Such approach will be followed in studies on supported oxidized metal catalysts.

The other side of the problem of modelling catalysis is the choice of a suitable quantum chemical method. As the final target is of chemical accuracy - fine enrgetical effects measured in units of a few kcal/mol - the most sophisticated quantum methodologies should be used. It is already well known that the treatment of highly open-shell systems is very difficult, if at all feasible, at the Hartree-Fock level. Moreover, the HF wavefunctions must be viewed, at best, as not always convenient starting point for more elaborate correlated calculations. On the other hand, there exists already growing evidence that the density functional theory (DFT) is capable of meeting such challenges and provide a unified theoretical framework for the study of electronic, geometric, and vibrational properties of solids, surfaces, interfaces, clusters, and molecular systems [1-3] encompassing metallic as well as covalent bonding. Density functional theory deals also with correlation where the use of fully correlated methods is generally limited. DFT is the example of intrinsically many-body approach to electronic structure of the system of interest, in principle avoiding many-electron wave function. Nevertheless, in accord with the tendency to retain the features of one-particle picture of the electronic structure which cannot be overestimated in interpreting chemistry, one-particle equations have been delivered by Kohn and Sham [4]. Even if so called Kohn-Sham orbitals may be interpreted only in approximate way, they bring about great help in understanding mechanisms of chemical processes. Thus in this work we will refer frequently to qualitative properties of Kohn-Sham orbitals to highlight e.g. spatial requirements of some interactions or directional bonding.

DFT methods appear as especially attractive for describing catalytic, frequently weakly interacting systems. The major gain of the DFT based methods is explicitly included electron correlation which compromises with their computational modesty. There exist nowadays numerous implementations of the DFT methodology in a variety of software packages. We are by no means intending to make any benchmark comparisons either with respect to their numerical efficiency or calculational accuracy. We would rather give the reader the impression how the available softwares could be used as tools in formulating and testing hypotheses regarding intricate chemical processes. In this chapter we will describe the results of the calculations obtained by means of the DMOL software of BIOSYM Technologies, Inc. [5]. Numerical spin-polarized density functional calculations were performed at two levels of approximation to the exchange-correlation potential: local density approximation (LDA) and the nonlocal gradient-corrected potential (NLDA). Among variety of functionals offered in numerous implementations of DFT we have selected the local Janak-Moruzzi-Williams (JMW) [6] and the nonlocal exchange of Becke combined with Lee-Yang-Parr correlation (BLYP) [7] functionals as our numerical choice together with a set of double numerical plus polarization (DNP) basis functions of chemical quality. These constitute an advised level of computation in the DMOL software.

All original results presented here follow from the calculations on molecular or cluster models constructed not so much to mimic original structures of active centers in real systems, but rather to simulate the properties of such sites selected as important factors governing their activity. We call such models functional models to distinguish them from the process of direct modelling either the bulk, or the surface, or the active site in a supermolecule.

2. DENSITY FUNCTIONAL DESCRIPTION OF THE ELECTRONIC STRUCTURE OF A CATALYTIC ACTIVE SITE

2.1. Precious metals and their oxides

Theoretical modelling in modern material science and catalysis cannot be accomplished without knowing electronic properties of transition metals and their compounds as they are present in a majority of systems of practical interest. In particular the accurate description of the bonding involving transition metal atoms is of fundamental importance for both theory and experiment. This is, however, a very difficult target for quantum chemistry because the electronic structure of such species is highly complicated and the computation requires explicit treatment of electron correlation. In particular, characterization of the electronic structure and molecular bonding in transition metal oxide diatomic molecules and transition metal dimers is an attractive topic and an extremely challenging task for both experimental spectroscopy and theoretical chemistry. Electronic properties of these compounds are intricate, the difficulties come from many unpaired electrons, which produce large number of electronic states, many with high spin multiplicity. On the other hand, such variety of the availble close-lying energetical states, frequently with substantially different properties, may be the prerequisite factor for high activity and extremely delicate and flexible nature of the interaction of these species with other molecules. The density functional theory has already been presented by us as a very promising tool for investigation of such systems, in particular transition metal oxides [8,9].

The simplest system where the interaction of a hydrocarbon molecule with a metal oxide can be investigated is the metal oxide MeO diatomic molecule. Although it can be hardly taken as a realistic model of a bulk oxide or oxidized metal catalyst, it offers wide possibility of deep studies on the electronic structure of the system composed of a cluster and a hydrocarbon molecule and its evolution along the adsorption and prospective dissociation processes, within high-level quantum chemical methodology. Our studies on the electronic structure of palladium systems and the methane activation [10,11] have clearly indicated that only detailed knowledge of the electronic state of the system, be it a simple diatomic cluster, may bring some insight into the intricate mechanism of a hydrocarbon adsorption and the activation of a carbon - hydrogen bond. On this basis we have succesfully drawn qualitative conclusions regarding some factors governing catalytic combustion of methane on an active site in supported metal catalyst. In the same way the differences in behaviour of oxidized and metallic phases may be studied on the basis of the interaction of methane with a metallic dimer [11]. One may attempt to extend this simple model by extending the cluster size which is the commonly adopted practice for metal surfaces (see e.g. recent publicationd by van Santen *et al.* [12] or Goursot *et al.* [13]). The matter is much more complicated in the case of highly covalent oxides or complex supported transition metal based catalysts. For example the fine structure of the surface in supported metal/oxidized metal catalysts (where Me= Pd, Rh, Pt etc.), which are often used for catalytic combustion of hydrocarbons, is still not well known and any model may be verified only by the validity of the offered predictions. Thus we strongly believe that high-level and high-accuracy description of the dimers aided occasionally with some extended systems modelling strictly selected features of the interaction between different phases, such as the Pd_2O triatomic linear cluster, could be of great imporance for pursuing intrinsic factors which rule the catalysis.

Details of the electronic structure of PdO, Pd_2O and Pd_2 have been already given elsewhere [10,11] and we will only briefly recall our results here. The detailed structure of diatomic clusters has been derived after analysing potential energy curves fitted to a number of points close to the energy minimum. The structure of Pd_2O has been obtained by automatic energy optimization with the constraint of linearity. Table 1 compares dissociation energies (NLDA) and structural parameters (equilibrium bond distances and vibrational frequencies where available, LDA) for the ground state of both diatomic molecules and the Pd_2O cluster. In the case of the PdO the ground state was found to be the triplet $^3\Pi$ coming from $\pi_d^3\sigma_s^1$ configuration with the bond energy of about 59 kcal/mol. It has been shown that the properties of singly occupied SOMO composed mainly of s-d_0 hybrid on palladium was responsible for relatively strong collinear adsorption of a methane molecule. The other SOMO was of π character, composed of the antibonding combination of metal d_π and oxygen p_π orbitals. The ground state of Pd_2 was found here to be the singlet $^1\Sigma_g$ formed in principle by two d^{10} palladium atoms, with small promotion of ca. 0.1 electron to the 5s orbitals and the bond energy of about 18.5 kcal/mol (at nonlocal level). The triplet state formed by promoting an electron out of d_{z2} σ^* into 5s σ orbital was found 0.19 eV above the singlet.

Table 1
Dissociation energies D_e (eV, nonlocal approximation) and structural parameters (equilibrium bond distance, R_e (Å) and vibrational frequencies ω_e (cm^{-1}), local approximation) for ground states of studied clusters; a) D_e for Pd_2O -> Pd + PdO
b) D_e for Pd_2O -> Pd_2 + O

Cluster	R_{Pd-Pd}(Å)	R_{Pd-O}(Å)	D_e (eV)	ω_e (cm^{-1})
Pd_2	2.69	...	0.807	150
PdO	...	1.84	2.56	651
Pd_2O	2.55	1.82	1.02^a) 3.52^b)	...

Palladium dimer is a very weakly bonded molecule formed by two closed-shell palladium atoms. We will see in the following paragraphs that the chemisorptive bonding of methane to Pd_2 will be fairly strong like that to the palladium atom [14], the best position being bridged η^2-bonding of a methane mediated by hydrogen atoms. The activation of methane on Pd_2 will be very easy but no stable product of the C-H bond scission will be formed [11]. The metal-oxygen bonding in the PdO is stronger, nevertheless, it is also relatively weak compared to other transition metal oxides. This open-shell molecule will bind the methane with intermediate adsorption energy, preferably in collinear position on palladium [10]. Therefore high activation energy will be required here for initiating the interaction of a hydrogen with an oxygen and formation of a stable, OH containing, dissociation product.

The simplest model combining both functions, easy activation of the C-H bond on a metallic dimer and effective insertion of oxygen on oxidized cluster, may be composed of a linear Pd_2O cluster. This cluster is not supposed to represent structural features of the surface oxide, neither shows it any real correspondence to the bulk PdO. We have already discussed the differences between the electronic structure of PdO and Pd_2 together with some expectations emerging from their eletron distributions. The ground state of a linear Pd_2O cluster is the triplet $^3\Sigma^-$, by -1.02 eV stable with respect to the PdO + Pd system and by -3.52 eV stable with respect to the Pd_2 + O separate moieties. The optimized bond distances are $R_{Pd\text{-}Pd}$ = 2.55 Å and $R_{Pd\text{-}O}$ = 1.82 Å, both distances being slightly shorter than in corresponding diatomic systems. On such a cluster a methane molecule adsorbs in a way similar to that on Pd_2 while the role of oxygen may be analogous to that in the case of PdO. The ground state upper valence electronic structure of the cluster is formed by doubly occupied σ orbital with an admixture of a diffuse Pd 5s σ^*, and two singly occupied palladium π_d antibonding orbitals with some admixture of O 2p functions. It must be pointed out that there is a qualitative difference of the Pd_2O valence electron density and that of the PdO molecule. In the latter case the highest open-shell orbital was derived from the diffuse Pd 5s function which has been found prerequisite to collinear methane adsorption on palladium. For the linear Pd_2O the d_π derived orbital forms semioccupied HOMO, thus favorable conditions for bridging adsorption of methane on the Pd_2 fragment of Pd_2O may be provided, in contrast to the PdO case. In such a way even a very simple molecular model may give functional information on the interaction of an adsorbate with a site and its influence on the adsorbed moiety.

Electronic structure of the rhodium oxide and rhodium dimer has only recently been studied by us to larger extent [15] and we will report our results here in more detail. Table 2 presents the results of our calculations for the quartet ground state, and selected doublet states of the RhO molecule. In the table the equilibrium bond distance R_e (Å), the dissociation energy D_e (eV), the vibrational energy between the two lowest vibrational levels ω_e (cm^{-1}), the rotational constant B_e (cm^{-1}), and the excitation energy from the ground state (cm^{-1}) are listed. The dissociation energy has been calculated with respect to the triplet oxygen and the ^4F atomic state of rhodium. We can see from the table that the ground state is the $\pi^2\sigma^1$ $^4\Sigma^-$ open shell quartet, in agreement with the prevoius calculations of Siegbahn [16]. It is also clear that our DF calculations within nonlocal approximation give much better dissociation energy than correlated traditional ab initio calculations. The same was found true in the case of other transition metal oxides and homonuclear dimers [8-11]] thus we feel authorized to strongly advertise DFT methodology as a reliable tool for calculation of energetical effects. Among states originating from dublet-coupled $\pi^2\sigma^1$ configuration only the $^2\Delta$ state can be expressed as a single determinant and thus may be the subject of rigorous DF calculations. It is located about 9400 cm^{-1} above the ground state and displays very similar spectroscopic properties. Other two $\pi^2\sigma^1$ states, $^2\Sigma^+$ and $^2\Sigma^-$, may be expressed only by a combination of determinants. Our estimation from the sum rules places these states about 5900 cm^{-1} and 9700 cm^{-1} above the ground state, respectively, with the bond distance by 0.01 Å shorter. The last low-lying doublet is the $^2\Pi$ state coming from π^3 configuration. It is located only 5800 cm^{-1} above the ground state, has the same equilibrium bond distance and similar vibrational frequency. We will see in forthcoming paragraphs that the properties of the ground state and the first excited doublet have substantial influence on the adsorption and dissociation of a methane molecule.

Table 2
Properties of electronic states of RhO: optimum bond distance, dissocation energy, vibrational frequency, rotational constatnt and excitation energy (L-with local potential, NL-with nonlocal corrections)

		R_e (Å)	D_e (eV)	ω_e (cm^{-1})	B_e (cm^{-1})	Δ (cm^{-1})
$^4\Sigma^-$	L	1.75	5.58	805	0.397	
	NL	1.80	4.54	727	0.375	
	ref. 16	1.74	3.32			
	exp		4.2			
$^2\Pi$	L	1.75	4.86	831	0.397	5807
	NL	1.80	3.81	743	0.376	5807
$^2\Delta$	L	1.75	4.41	818	0.400	9436
	NL	1.80	3.38	740	0.378	9275

Rhodium dimer might be described as a relatively strongly bonded molecule compared to the $^1\Sigma_g$ palladium dimer found by us the most stable. The assignement of highly open-shell electronic states is even more difficult for Rh_2 than for Pd_2 [13,17]. We have established its lowest electronic state as the quintet $^5\Delta$ coming from $\delta_d^{*3}\pi_d^{*2}\sigma_s^1$ open-shell electronic structure [15]. The equilibrium bond distance of 2.30 Å and the dissociation energy of 2.53 eV may be compared with the values of R_e=2.70 Å and D_e=0.82 eV for Pd_2 dimer. In the closed-shell palladium dimer bonding and antibonding interactions merely cancel whereas in the rhodium dimer antibonding counterparts to π_d orbitals are only partially occupied. Moreover, 4d -> 5s electron promotion has been calculated as ca. 0.1 e in the former molecule while it is increased to one electron in the latter. Thus repulsive interaction with 5s electrons should make the adsorption of methane weaker in the case of Rh_2.

Table 3 summarizes the results for the lowest electronic states of the studied dimers. The table will serve as a comparative material to show relative trends among the dimers thus only local calculations are shown there. In the case of rhodium systems the ground state is well separated (at least 5000 cm^{-1}) from the next states. For palladium systems this is not the case thus the two states are listed. These states are still under debate regarding their relative positions [16-19].

Table 3
Properties of electronic states of dimers: optimum bond distance, dissocation energy, vibrational frequency, rotational constant and excitation energy (local approximation)

	R_e (Å)	D_e (eV)	ω_e (cm^{-1})	Δ (cm^{-1})
RhO				
$^4\Sigma^-$ $\pi_d^{*2}\sigma_s^1$	1.75	5.58	805	
PdO				
$^3\Pi$ $\pi_d^{*3}\sigma_s^1$	1.84	3.23	651	
$^3\Sigma^-$ $\pi_d^{*2}\sigma_s^2$	1.87	3.21	616	161
Rh_2				
$^5\Delta$ $\delta_d^{*3}\pi_d^{*2}\sigma_s^1$	2.30	2.53	312	
Pd_2				
$^1\Sigma^+$ $\pi_d^{*4}\sigma_d^{*2}$	2.69	0.81	150	
$^3\Sigma^-$ $\pi_d^{*4}\sigma_d^{*1}\sigma_s^1$	2.88	1.00	...	1532

On the basis of the table we may discuss the character of the upper region π and σ molecular orbitals in the dimers, which contribute to the electron distribution responsible for formation of the adsorptive bond. In the oxides qualitative picture is the same for both RhO and PdO molecules: they posses open shells both of d_π and σ_s character, the latter known to be strongly repulsive. Metal dimers differ substantially in electron distribution. Closed shell palladium dimer has in principle no diffuse, repulsive σ_s electrons (the levels composed of d_π and d_σ orbitals are fully occupied while the repulsive one, σ_s, is empty) whereas in quintet Rh_2 all these levels are singly occupied, including the repulsive σ_s orbital. The same expectations with respect to prospective adsorption of a hydrocarbon molecule on various dimers have been already mentioned in preceding paragraphs. Here they are compared and related directly to spatial properties of theoretical electronic structures of the studied dimers. In sections

3 and 4 they will be confronted with actual calculations of appropriate reaction pathways and reaction energetics.

2.2. Metal-exchanged sites in zeolites: Ga- and In-exchanged ZSM-5 zeolite

In this paragraph we discuss entirely different approach to designing the model of an active site for theoretical investigations. The example is taken from a very important field of zeolite catalysis, with the metal site formed in metal-exchanged silicalites. Recently, much attention has been given to Ga- and In- exchanged ZSM-5 which were found to promote the reduction of NO with methane [20]. Despite of extremely challenging character of the topic, the mechanism of this process is basically unknown, for example little is known about the mechanism of mathane activation, of water poisoning, the specific role of the reductant and the catalyst in the reaction [21], also the identification of intermediates has not yet been accomplished. Thus theoretical modelling of this reaction system would be of enormous value for both basic scientific understanding and the practice.

The cluster model for quantum chemical calculations has been based here upon the known crystal structure of ZSM-5 reported by X-ray diffraction study modified according to the equilibrium geometry given by molecular dynamics (MD) simulation [22]. The physical and chemical state of the active metal species in H-ZSM-5 zeolites, however, has not yet been unequivocally established. There exist, however, strong evidence [23] that gallium or indium is present in highly dispersed monomeric form, coordinated to basic oxygens within zeolite channels. In reducing atmosphere it could be a reduced hydride moiety while in the presence of excess oxygen an oxidized $[MeO]^+$ unit may be proposed. Fig. 1 shows the conformation of the latter form of the Ga-exchanged site within zeolitic lattice as obtained from large model MD calculation.

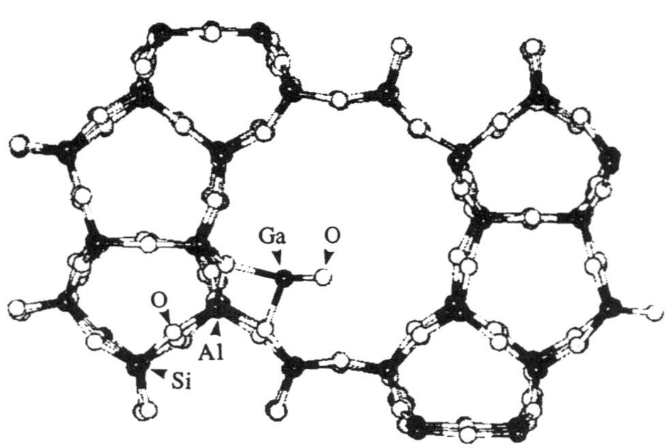

Figure 1. MD-simulated ZSM-5 framework with the exchanged gallium site.

In our quantum chemical calculations, single AlO_4 tetrahedron was considered as the representative of the active site, with adjacent silicones replaced by hydrogens and charge deficiency compensated by $[MeO]^+$ cationic species (see Fig. 2). We believe the cluster model specified above is adequate to reproduce basic features of the zeolitic active site local electronic structure. At the same time, the modest size of the cluster allows for meeting more rigorous demands of quantum chemical methodology and testing various computational parameters. It should be admitted here that also other forms of gallium supported on H-ZSM-5 may coexist, such as the precipitated gallium oxide [20]. Our model explores one of a few possibilities selected in accord with experiment.

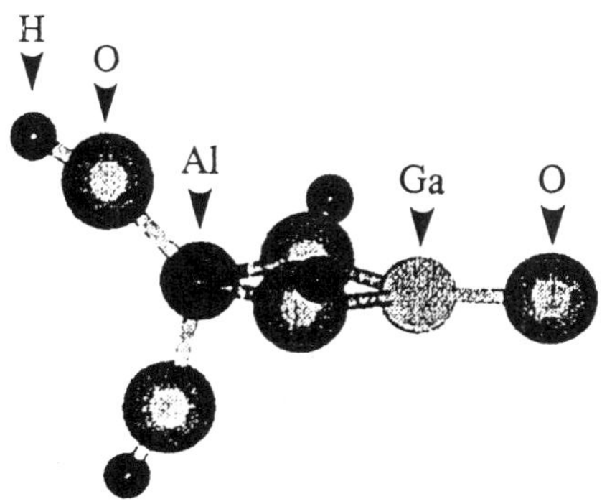

Figure 2. Cluster model for the Ga site with double coordination of the [GaO] unit.

As molecular dynamics, which is very good for qualitative modelling, not always can provide quantitative description of chemical bonding, the final coordination of the gallium unit to the zeolite framework has been verified by quantum chemical DFT calculations. [GaO] can coordinate to $Al(OH)_4$ tetrahedron in three different ways: single coordination, double coordination in a bridging position between two oxygen atoms and triple coordination in a three-fold site between three framework oxygens. The DFT total energy minimization for the three configurations showed [24] that the doubly coordinated bridging site rendered the minimum energy geometry and that this should be the most favorable structure of the ion-exchange center. The final coordination of gallium in this conformation was the planar triple one, with one bond formed with the extraframework oxygen and the other two pointing towards zaolitic framework oxygens. Thus the gallium, usually within fourfold coordination, showed here structural unsaturation and increased reactivity of such center could be expected. The final optimized geometry of the cluster was in full agreement with the MD prediction which indicates that classical molecular dynamics can be used as an efficient tool in determining cluster models for first principle quantum chemical calculations. Our ultimate cluster model for the active site in Ga- or In-ZSM-5 employed in quantum chemical calculations was the one shown in Fig. 2. Thus the cluster $[Al(OH)_4][MeO]$ and the methane molecule constitute the reactants system in our model for actual adsorption and activation process.

3. ADSORPTION OF HYDROCARBONS ON METALLIC CENTERS ON THE CATALYST SURFACES

3.1. Adsorption of a methane on palladium and rhodium sites

In this chapter the interaction of a methane with diatomic clusters: PdO, Pd_2, RhO and Rh_2 is assumed as a first attempt to model the interaction of a hydrocarbon with metallic and oxidized sites in supported metal catalysts [10,11,15]. To construct the model, approximate potential energy curves for the methane adsorption with respect to a distance between carbon and metal atoms or between carbon and midpoint in a cluster bond have been calculated with fixed geometry of the reactants. Next, the geometry of the system has been optimized starting from approximate energy minima. Potential energy surfaces close to minima have been scanned for various electronic and spin states as after the example of bare transition metal atoms [14] it could be expected that the basins for substrates, products and the transition state might belong to different surfaces and the surface crossing might be an important factor in determining energetical properties of the process.

There have been already a number of theoretical studies of the CH_4 oxidative coupling by nontransition metal oxides such as LiO, AlO, and MgO in gas phase [25] and of simple models of perfect and defected lattices [26]. None of them reported a stable methane adsorption prior to dissociation. Transition metal oxides, however, may reveal differnet behaviour and there exists already theoretical evidence for stable methane adsorption on some transition metal atoms and complexes [14]. Also the stability of the so called encounter complexes has been qualitatively predicted to increase the reactivity of transition metal MO^+ species toward hydrocarbons [27].

To the best of our knowledge, little is known about the interaction of a methane with transition metal clusters or surfaces. There were attempts to study consecutive scission of the C-H bonds in methane on cobalt and nickel atoms or small clusters [12]. The one-metal atom models were shown to be useful in describing the insertion of a metal into the C-H bond in saturated hydrocarbons via the oxidative addition mechanism [14]. The methane was claimed to form a stable preadsorbed η^2-complex only with the palladium atom without the addition barrier due to its d^0 ground state and no repulsive 5s electrons. Metal dimers might be regarded as an extension of a model composed of a neutral transition metal atom and a methane. At the same time they might be useful in comparing properties of metal-metal and metal-oxygen bonds with respect to methane adsorption and activation.

The better known, yet relevant topic is the interaction of hydrogen with metals, in particular with palladium, which has recently attracted much attention owing to the importance of this system in catalysis and relative simplicity of the problem. As the dissociative adsorption of H_2 is the first elementary step in many catalytic reactions, a large amount of work has been devoted to hydrogen-palladium systems by both experimental and theoretical approach. Till now, however, most theoretical efforts have been focused on the chemisorption of H on various metal surfaces, i.e., the final structure of already dissociated adsorbates on the surfaces. Only very recently, the whole process of the dissociative adsorption has become the subject of theoretical studies [28]. The approach based on finite temperature density functional theory (Vienna Ab-initio Molecular-dynamics Program) applicable in condensed matter physics showed the existence of spontaneous hydrogen dissociation paths with negligible barrier [29]. The existence of a precursor, preadsorbed state has, however, been and continues to be a hotly debated topic.

It seems clear that similar questions should be raised as to the dissociation of the carbon-hydrogen bond in hydrocarbons and it is obvious that the problem must be much more complex. Thus the first task would be to probe and locate possible stable methane adsorption complexes with metal oxide and metal-metal dimers. To this end we have investigated various assumed approach geometries to the metal, to oxygen, and with methane entering between the two atoms. No energy minimum emerged for the approach toward oxygen while stable adsorption complexes were found on metal or bridging positions.

According to the procedure described in computational details various adsorption modes defined by the choice of the reaction coordinate have been investigated in the present work. Here the investigated modes were the following: i) collinear adsorption (C-Me distance selected as a working coordinate, collinear approach), ii) metal on-top (C-Me distance defining the reaction coordinate, perpendicular approach), iii) bridging adsorption (the distance between carbon atom and the center of a dimeric bond selected) and finally the dissociated mode (iv), with a methyl bonded to a metal atom and a hydrogen bonded to an oxygen in the case of the oxide or moving around the metal dimer after the scission of the C-H bond.

The results of the calculations are shown in Table 4. The table lists the relative energies (in kcal/mol) of various adsorption modes of a methane on MeO and Me_2 (where Me=Pd or Rh). All listed energies were obtained within the local DFT approximation as they are subjected only to comparative analyzis. The energies were calculated with respect to the separate methane molecule and the ground state dimer. Appropriate energies for bare metal atoms taken from ref. 14 are also included in the table for comparison. The entries in the table are arranged according to the strength of the adsorptive bond.

Table 4. Properties of methane preadsorption complexes on MeO and Me_2 dimers and bare metal atoms and cations: adsorption energies, selected bond distances ([a] bridging geometry, [b] collinear adsorption, [c] from ref. 14)

	E_{ads} (kcal/mol)	$R_{Me\text{-}C}$ (Å)	$R_{Me\text{-}H}$ (Å)	$R_{C\text{-}H}$ (Å)
Pd_2[a]	-26	2.20	1.80	1.23
PdO[b]	-20	2.36	2.02	1.13
RhO[b]	-19	2.43	2.06	1.13
Rh_2[b]	-11	2.44	2.02	1.13
Pd (Pd^+)[c]	-4 (-17)	2.50 (2.54)		
Rh (Rh^+)[c]	+6 (-13)	2.50 (2.65)		

The strongest adsorption complex has been found for the palladium dimer, where it occurs in the bridging position. This result agrees perfectly with the expectations inferred from the features of its electronic structure. The low-lying state without repulsive 5s electrons should be apparently responsible for the strong bonding of a hydrocarbon molecule. Geometrical parameters indicate that the bondng is mediated by hydrogen atoms not by the carbon and, what follows, the C-H bond becoms strongly activated by increasing the distance up to 1.23 Å. This is apparently the unique example among the studied systems. Other diatomic species bind the methane molecule in collinear geometry, the bonds between the metal and carbon or hydrogen atoms are ca. 0.2 Å longer and the activation of the C-H bond is much weaker. Other adsorption modes were found also stable, however, with much lower binding energy. This may be ascribed to the presence of the repulsive 5s electrons in the valence electronic structure of the these dimers. Nevertheless, the adsorptive interaction is at most moderate (we must recall here that binding energies are significantly overestimated by the local approximation) which manifests itself also by the same spin state of the reactants and the complex. It is low spin state for the palladium dimer and palladium atom while it is high spin state for other species. Bare neutral metal atoms show similar tendencies towards formation of the chemisorptive bond with methane. In the case of palladium with closed-shell d^{10} electronic structure and no repulsive 5s electrons the adsorption complex is stable while for rhodium atom it is not. In the case of cations where the 5s electron has been removed, both cations form stable complexes, with the adsorption energies similar to those on metal oxides. Thus the electronic state of the metal in the oxide molecule resambles to some extent that of the metal cation.

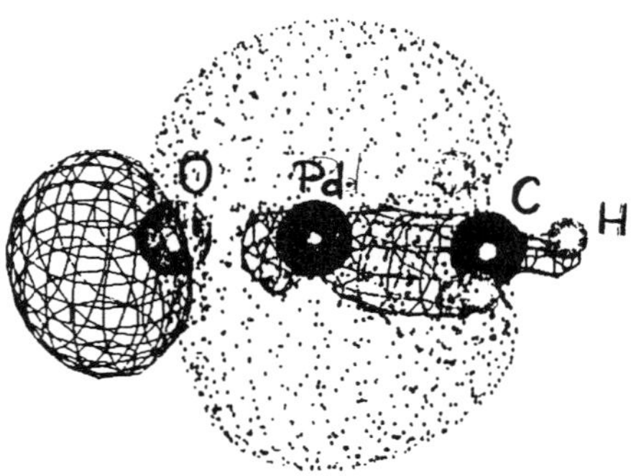

Figure 3. Orbital contour for the highest open-shell orbital for collinear methane adsorption on PdO.

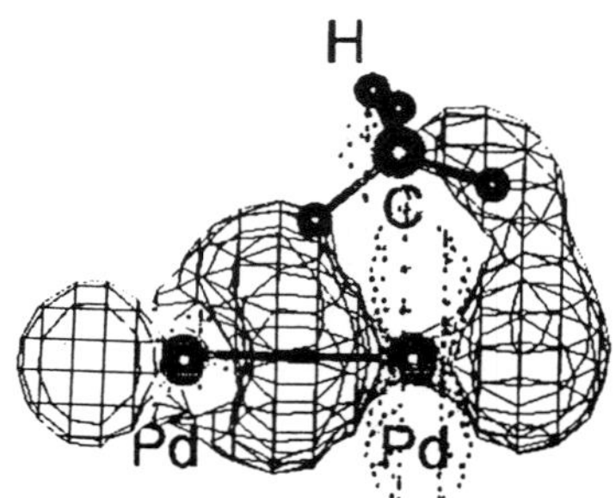

Figure 4. Orbital contour for the valence orbital for bridging methane adsorption on Pd_2.

Figures 3 and 4 show representative optimum geometries for the bridging and collinear adsorption modes, respectively, described in Table 4. In the figure also the contours of the valence region molecular orbital relevant for the metal-methane bonding are given (contour values are +0.01, full lines, and -0.01, dashed lines). It can be seen qualitatively that in the bridging position hydrogens are alraedy "embedded" in the sea of metal d electrons (they bear negative charge of -0.06 e). There exists no experimental evidence as to the adsorption geometry of methane on metal or oxide surfaces. High stability of the bridging adsorption mode on Pd_2 has been, however, predicted theoretically for CO adsorbed on a Pd_4 cluster [13]. Also IR and TDP experiments were reported to indicate that CO adsorbs on palladium at 3-fold and bridging sites at intermediate coverages. The top site is only populated at high covarage on palladium while it is first occupied on Rh(111) surface [13]. Thus we believe that our simple model for the adsorption of methane on the transition metal involved in the bonding with the other metal or oxygen atom may bring a valuable insight into the merits of such interaction.

3.2. Interaction of hydrocarbons with the gallium site in ZSM-5 zeolites

The model for the gallium site in ZSM-5 zeolite described in preceding sections has been constructed on the basis of known zeolitic structure aided by molecular dynamics simulations thus it should be viewed as more realistic one than functional models built of transition metal dimers. On the other hand, however, it suffers from all drawbacks ascribed to cluster models: boundaries to be saturated or the question of limited size and neglected interactions with the lattice. Thus this model is not the appropriate subject for deep analysis of fine details of electronic states and their impact on the interaction with adsorbed molecules. We would rather pay attention to structural factors governing the process of methane adsorption and activation. Nevertheless, this reaction pathway, be it qualitative, should resamble the actual reaction route and reaction energetics should be comparable to experiment.

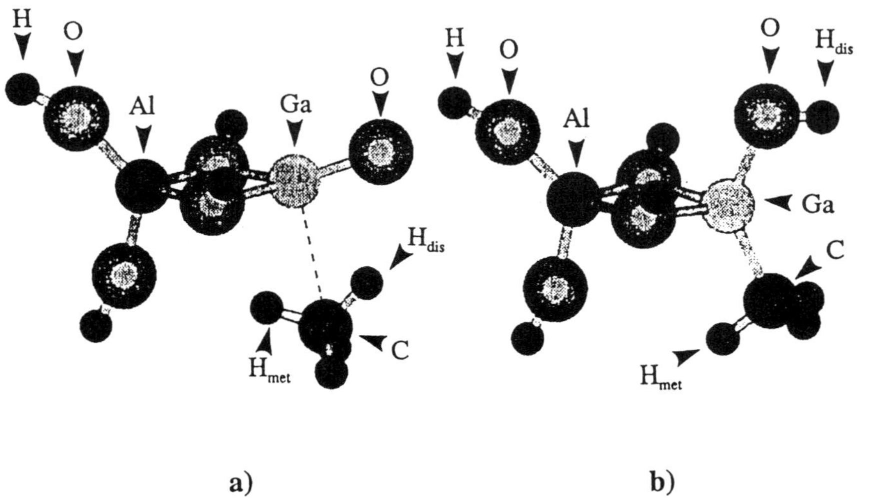

Figure 5. Optimized structures of two adsorption modes of methane: (a) physisorbed methane, (b) dissociated methane.

Two stable structures on approach of the methane molecule towards the active center in the Ga-exchanged catalyst are the weakly bound van der Waals surface complex and chemically bonded product of the C-H bond scission shown in Fig. 5. The former system may be described as the practically unmodified CH_4 molecule interacting weakly with the Ga cation. Its stability with respect to the hydrocarbon molecule separated from the site amounts to only 5 kcal/mol. The optimum distance between the two is equal to 2.7 Å, however, the potential energy surface in this region is very flat and the actual distance is very sensitive to any changes in computational parameters. In the physisorbed state neither the geometry of the methane molecule nor the conformation of the center undergo any significant relaxation, although the hydrogen atom closest to the extraframework oxygen bears slightly increased positive charge. From now on we will call this hydrogen the *dissociative hydrogen*. This weak preadsorption may be viewed as the effect of structural unsaturation of the gallium site, where the gallium takes planar, triple coordination. The direction of the methane molecule approach towards the site is perpendicular to this plane. Usual coordination of the gallium in its compounds is a tetrahedral one thus the change in the gallium hybridization might be regarded as a driving force initializing system reorganization towards fourfold coordination. Indeed, the dissociation process in which the OH group would be formed and CH_3 moiety would remain supported on the Ga cation leads to the structure found to be highly stabilized being composed of chemically bonded hydroxyl and methyl group strongly interacting with the metal. The refined optimization leads to the OH distance of 0.99 Å and the Ga-C separation of 1.92 Å. The latter value may be compared with the Ga-C bond length in methylgallanes, calculated as 1.997 Å for trimethylgallane and 2.054 Å for monomethylgallane [30]. The stability of the dissociated system with respect to the reactants was calculated as 56 kcal/mol in the LDA approximation.

Table 5
Charge distribution on the methane molecule and Ga-exchanged center (in e) for various stages of adsorption

	Ga	O	C	H_{met}	H_{dis}
Cluster and molecule (R)	+0.47	-0.45	-0.19	+0.048	-
Physisorbed methane (S)	+0.45	-0.47	-0.17	+0.045	+0.060
TS	+0.46	-0.45	-0.21	+0.045	+0.061
Dissociated methane (P)	+0.45	-0.36	-0.23	+0.044	+0.131

Table 5 shows the charge distribution in the relevant part of the system in various stages of the methane transformation. The transition state, which will be discussed in detail in the following paragraphs, is included here for clarity. The charges were calculated according to the Hirschfeld definition which has been found more appropriate to describe charge distribution in a hydrocarbon containing system. Although the population analysis from density functional theory in numerical implementation is not unambiguous and may serve rather for qualitative interpretation, it may provide some guidance into the physical mechanism of the process. In a preliminary, weakly bound adsorption complex, the methane molecule bears a positive charge (Q_{CH4}=+0.02e) with the polarization towards the hydrogen closest to the extraframework oxygen, which, in turn, is negatively polarized. In the transition state some flow of charge towards the methyl group is noticeable, which is even bigger in the final, dissociatively adsorbed state, and leads to negatively charged methyl (Q_{CH3}= -0.10e) and hydroxyl (Q_{OH}=-0.26e) groups bonded to the gallium center.

When geometrical rearrangement between the two stable structures is analyzed, the following most important changes should be noted. One obvious element is the transfer of a hydrogen between the carbon and the extraframework oxygen. The others are the relaxation changes: the shortening of the carbon - gallium bond from 2.70 Å to 1.93 Å, accompanied by the readjustment of methyl hydrogens, and the change in the coordination of Ga from tri-coordinated, planar structure, to almost tetrahedrally coordinated one. The extraframework oxygen becomes finally involved in the hydroxyl group and, as a consequence, its bond with the gallium is elongated from 1.67 Å to 1.80 Å. Thus it is clear that the relaxation of the whole fragment will contribute to the description of the process and must be taken into account.

4. PERFORMANCE OF DENSITY FUNCTIONAL APPROACH IN DESCRIBING CATALYTIC REACTION MECHANISMS: C-H BOND ACTIVATION AND DISSOCIATION

The practical aspect of this chapter is devoted entirely to the study of the activation and prospective scission of the hydrogen - carbon bond in hydrocarbons. With this end, we have investigated the interaction of the methane molecule with various prospective active sites, and tried to locate and analyse possible stable structures. In the next step, however, the question must be raised, whether the dissociation process is allowed from the kinetic point of view, in particular what is the probable mechanism and what is the activation energy of the reaction and, finally, what role does the active center play. Answering these questions requires the explicit calculation of the transition state for the reaction. Thus at least partial reaction profile for the scission of the C-H bond in a hydrocarbon should be discussed in the next step. We attempted to search for the transition state of this process by applying the procedure consisting in step-wise increasing the C-H bond length and optimizing other geometrical parameters (an analogue to Minimum Energy Path).

In the case of transition metal oxides or homonuclear dimers the starting point was the bridging adsorption which, even if being not the lowest energy geometry in each case, appeared to be stable and the most promising for the activation of the C-H bond. In this procedure the energy differnce between the most stable complex and this bridging position has been added to the activation barrier. After crossing the barrier the final product of the methane dissociation has been fully optimized with all geometrical constraints lifted up.

The case of palladium dimer is distinct from other systems both regarding strong adsorption and easy scission of the carbon-hydrogen bond. Thus we will disuss it in more detail. Table 6 contains the values of the most important properties of these structures. First column of Table 6 displays important geometrical parameters and Hirschfeld atomic charges in the adsorbed system. It may be easily noticed that one hydrogen atom is actually within the coordination sphere of both metal atoms, whereas the carbon is weakly bonded by Pd_1. The bond between the carbon and active hydrogen atoms becomes substantially stretched up to 1.23 Å. Further stretching of the bond is very cheap energetically and for the C-H distance equal to 1.48 Å the transition state is reached with the expense of only 2 kcal/mol at the local level and 5 kcal/mol for gardient-corrected potential. After crossing this point hydrogen atom is almost free to move around the palladium dimer as slightly negative moiety. We have found weakly bound optimum geometry where both the hydrogen (Q_H=-0.15 e) and the methyl (Q_C= -0.22 e) take the bridging positions between the two palladiums, at the opposite sides of the molecule. The stability of this system with respect to the adsorbed conformation is very low, -11 kcal/mol at the local level and -1.5 kcal/mol at the nonlocal one. This indicates at strong methane adsorption but weak bonds in the dissociated complex and high probability for the reverse reaction, recombination of methyl and hydrogen moieties.

Table 6
Selected optimum bond distances (Å) and atomic charges (in e) for the methane adsorbed on Pd_2 (ADS), transition state for the C-H* bond scission (TS) and final system of CH_3 and H* adsorbed on Pd_2 (DIS)

	ADS	TS	DIS
R_{Pd1C} (Å)	2.20	2.15	2.14
R_{Pd1H^*} (Å)	1.91	1.80	1.70
R_{Pd2H^*} (Å)	1.79	1.70	1.60
R_{CH^*} (Å)	1.23	1.48	...
Q_{Pd1}	+0.01	0.0	+0.20
Q_{Pd2}	+0.13	+0.16	+0.17
Q_{H^*}	-0.06	-0.07	-0.15
Q_C	-0.20	-0.19	-0.22

Very similar behaviour of a methane molecule has been registered on a Pd_2O linear cluster. Among tested conformations the most stable geometry appeared to be the bridging adsorption on the Pd_2 subsystem, very similar to that on palladium dimer. Figure 6 displays optimized structures of critical points along the pathway of the C-H bond scission on the Pd_2 fragment of the cluster. These are the bridging adsorption, the transition state and the dissociated product. Energy relations show that it is indeed strong chemisorption with the methane bonding energy equal to -21.3 kcal/mol at the local level and -4 kcal/mol at the nonlocal one. Methane chemisorptive bonding is mediated by a hydrogen which in turn becomes highly activeted (R_{CH} in adsorbed complex is equal to 1.17 Å) thus easy scission of the bond may be expected. Stretching the C-H bond up to 1.6 Å which costs only 6 kcal/mol (LDA) or 14 kcal/mol (NLDA) brings about actual scission of the bond and almost freely moving hydrogen moiety. After crossing the transition state negative hydrogen species (Q_H= -0.13 e) migrates almost freely in the subspace of the Pd_2 fragment. The charges on palladium atoms are more polarized in this system due to the presence of oxygen (+0.40 and +0.24 e for Pd_1 and Pd_2 in the optimal conformation after the C-H bond scission), other geometrical and electronic properties being almost the same as in the case of Pd_2 cluster. These results indicate that most probably it is the Pd-Pd bond which is responsible for strong adsorption and easy activation of a methane molecule. In addition, the entire rection pathway is located here on the same spin surface, the singlet in the case of Pd_2 and the triplet in the case of Pd_2O.

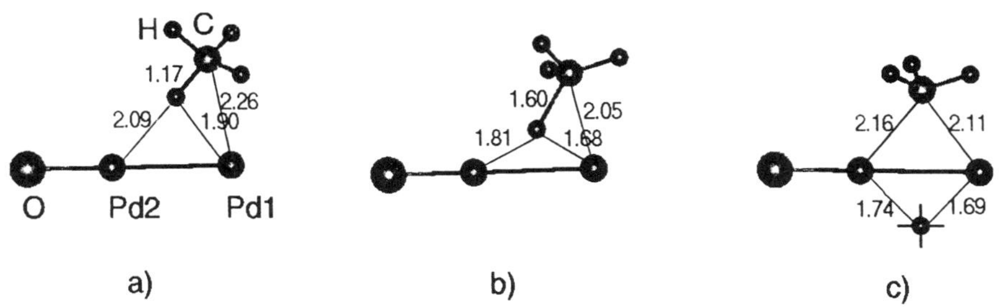

Figure 6. Optimized structures of critical points along the C-H bond scission on Pd_2O: (a) bridging adsorption, (b) transition state and (c) adsorbed CH3 + H (bond distances given in Å).

The preceding discussion shows that the activation of a methane molecule should occur easily on both metallic palladium and the system composed of a metal and partly oxidized phases. Such an activated species should be next the target for oxygen insertion. The first step of such investigation consisted in the calculation of the reaction profile on a PdO diatomic. Partial reaction profile was calculated by stretching the activated hydrogen - carbon bond up to 1.4 Å, where the approximate saddle point has been located. Finally the dissociation product with the hydrogen forming a hydroxyl group and the methyl bonded to the palladium has been fully optimized.

The most probable conformation of the methane + PdO system was found to be the collinear position on the palladium (see Fig. 7a). The adsorption geometry was similar to that found by Blomberg *et al.* for palldium atom [14], namely with with the Pd-H distance equal to 2.05 Å and the carbon by 2.36 Å apart from the metal. The stability of this complex has been found significant, -19.4 kcal/mol at the local level. Though the stability has been strongly decreased by introducing the nonlocal corrctions, the complex remained bound by -3.3 kcal/mol. As B88-LYP potential was found already to agree with experimental values of dissociation energies of metal-carbonyl bonds within few kcal/mol but for elongated bond distances [13], we believe that the methane molecule undergoes fairly strong adsorption on PdO, in variance with nontransition metal oxides.

The other very stable structure appeared to be the C-H bond scission product with the hydrogen forming the OH group and the methyl carbon - palladium bond of 1.99 Å (see Fig. 7c). The stability of this sytem with respect to separate reactants was -54.7 kcal/mol at LDA and -32.8 kcal/mol at NLDA level. Among other possible conformations, the bridging adsorption with R_{PdC}=2.86 Å and R_{OC}=2.89 Å showed also a shallow local minimum (Fig. 7b) which was located, however, ca. 13 kcal/mol above the collinear adsorption level.

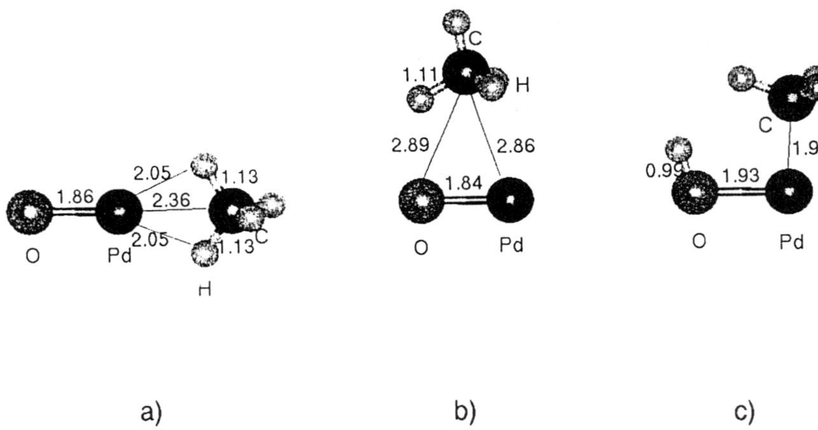

Figure 7. Optimized structures of critical points along the C-H bond scission on PdO: (a) collinear adsorption, (b) Bridging adsorption and (c) dissociation product.

It must be pointed out that the pronounced reorganization occured in the electronic structure of this system. Both adsorption structures form triplet states originating from the $^3\Pi$ ground state of the PdO molecule with the singlet lying 13 kcal/mol and 7 kcal/mol for the on-top and bridging complexes, respectively. The ground state of the activated complex is still the triplet one, although the singlet nears it as close as 0.8 kcal/mol. In this region the crossing of singlet and triplet potential energy surfaces is expected as the dissociated system forms the highly stabilized singlet state whereas no dissociation occurs on the triplet spin-surface. The close vicinity of the singlet-triplet crossing disabled direct calculation of the transition state for the C-H bond scission due to severe mixing and convergence difficulties. Thus only a few points along the reaction pathway have been calculated in this region which allowed for the discussion of the approximate properties of the transition state located at the C-H bond stretch of 1.4 Å.

The calculations have been repeated for the same adsorption and reaction modes on rhodium dimer and rhodium oxide molecules. The results of the calculations are summarized in Table 7. The table lists the relative energies (in kcal/mol) of the transition state and dissociative adsorption mode of a methane on MeO and Me_2 (where Me=Pd or Rh). All listed energies were obtained within the local DFT approximation. The activation energies and stabilities of the dissociated system were calculated with respect to the most stable preadsorbed complex of methane with a selected system. Appropriate energies for bare metal atoms taken from ref. 14 are also included in the table for comparison.

Table 7
Activation energy for methane dissociation and stability of the dissociated complex with selected bond distances for Pd and Rh systems ([a] with respect to bridging adsorption, [b] with respect to collinear adsorption, [c] from ref. 14)

	$\Delta E^{\#}$ (kcal/mol)	ΔE^{d} (kcal/mol)	$R_{Me\text{-}C}$ (Å)	$R_{Me\text{-}H}$ (Å)	$R_{O\text{-}H}$ (Å)
Pd_2[a]	3	-11	2.14	1.60	-
PdO[b]	25	-33	1.99	-	0.99
RhO[b]	29	-14	1.99	-	0.99
Rh_2[b]	>5	-14	1.99	1.55	-
Pd[c]	14	-5	1.95	1.52	-
Rh[c]	16	+9	2.03	...	-

After inspecting the table the following features of the adsorption process may be inferred:

i) In the case of palladium dimer the best adsorption geometry is apparently the bridging one, this is the strongest non-dissociatve adsorption among the studied systems. Here, very strong activation of the C-H bond is already observed and the scission of the bond costs no more than 3 kcal/mol. The final product, however, is only 11 kcal/mol more stable than the adsorption complex while this differnce amounts to 33 kcal/mol for the oxide.

ii) In the case of palladium oxide, strong adsorption occurs in the collinear position. There is also a shallower energy minimum for the bridging position but it requires activation with respect to the most stable one. Both adsorption modes are bound relatively weakly, the spin coupling is the same as in an isolated PdO dimer, that is the

triplet. Further activation of the system is, however, very cheap, small change in geometry leads to the change of the spin state to singlet and costs only a few kcal/mol. This singlet state is located already on the dissociative potential energy surface and thus C-H bond scission occurs spontaneously and the stable final product is obtained.
iii) Rhodium oxide behaves in a very similar way, shows comparable preliminary collinear adsorption and activation towards bridging adsorption and transition state. The difference, however, arises in the vicinity of the transition state. Here the spin transition cost mounts to extra circa 10 kcal/mol which would rise the barrier and indicate lower stability of the final product.
iv) Rhodium dimer has been included in the table although the calculations of the transition state have not yet been fully completed. It does not exhibit any position for strong adsorption of a methane, in accord with the predictions coming from its electronic structure. A methane molecule binds weakly in collinear position, other positions are higher in energy, in addition no significant activation of the C-H bond may be noticed. Also the dissociated complex has the lowest stability among studied dimers.
v) Bare metal atoms show similar important features towards adsorption of a methane and formation of the insertion products into the C-H bond. The discussion of these properties in metal atoms and dimers bears similar qualitative features regarding the inerdependence between the electronic state and reactivity. In the case of palladium atom with closed-shell d^{10} electronic structure no repulsive 5s electrons are present thus the adsorption complex is relatively stable but the insertion product is not. Rhodium atom has the open-shell electronic d^8s^1 configuration and due to repulsion with 5s electrons the adsorption complex is unstable but the insertion product has substantial stability.

Analogous procedure of estimating the reaction pathway has been applied to the case of methane interacting with the gallium site in ZSM-5 zeolite. It is apparent from the previous paragraphs that it is the dissociated form of methane which is the most energetically favorable mode of adsorption. However, following the discussion of the structural changes on methane transformation on adsorption, the geometrical parameters relevant to the description of TS must include the coordinates of constituent atoms in a methane molecule and the positions of a gallium and an extraframework oxygen. Thus the system becomes too complicated to calculate the potential energy surface in the vicinity of TS in explicit grid of coordinates. Therefore we have performed the search for the transition state in two steps. We have selected the distance between the dissociative hydrogen and the oxygen as the working model for the reaction coordinate. In the first step the energy profile has been calculated with respect to this coordinate keeping all other parameters fixed in order to approximately locate the transition state region. The energy maximum on this profile was found at the O-H and C-H distances of 1.6 Å and 1.8 Å, respectively. This geometry was then taken as the starting point to the next step of the calculations in which the optimization to the saddle point was performed within the same conditions as in the search for minima, that is with the geometrical parameters apart from the framework cluster released.

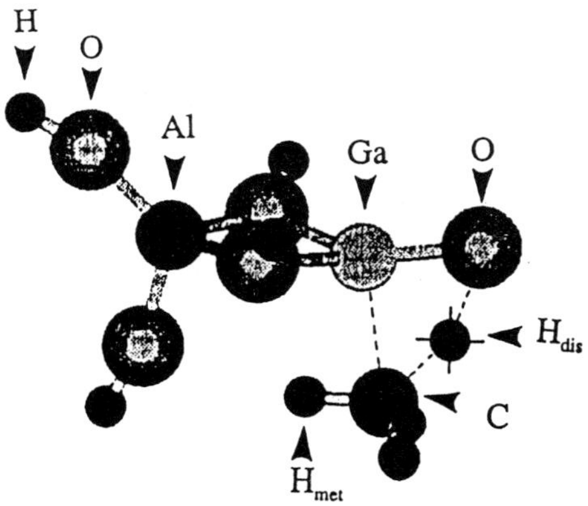

Figure 8. Optimized structure of the transition state complex for methane dissociation reaction on the Ga active site.

The optimized geometry for the transition state is shown in Fig. 8. The saddle point character of the structure was confirmed by the analysis of the Hessian restricted to optimized coordinates. The only negative frequency corresponded almost exclusively to the in-plane movement of the dissociative hydrogen between its neighboring carbon and oxygen atoms, supplemented by the Ga-O stretch, which confirmed our choice of the reaction coordinate.

Table 9
Relative total energies (LDA, in Ry) and selected geometrical parameters for reactants, substrate, transition state and dissociation product for methane and Ga site

	Total energy (Ry)	$R_{Ga\text{-}C}$ (Å)	$R_{C\text{-}Hmet}$ (Å)	$R_{C\text{-}Hdis}$ (Å)	$R_{O\text{-}H}$ (Å)	$R_{Ga\text{-}O}$ (Å)	α_{CGaO} (deg)
Cluster and molecule (R)	0.0	-	1.10	1.10	-	1.67	-
Physisorbed methane (S)	-0.00687	2.77	1.11	1.12	2.46	1.67	97
TS	+0.01378	2.19	1.10	1.35	1.47	1.72	88
Dissociated methane (P)	-0.08866	1.93	1.10	3.54	0.98	1.80	133

Table 9 shows the evolution of the selected relevant geometrical parameters while going along the reaction pathway from the physisorbed methane, through the transition state, towards dissociative adsorption. As we have already pointed out, when small positive charge is present on the physisorbed methane molecule, the carbon-hydrogen bonds become slightly weakened. In the transition structure, the dissociative hydrogen is already half-way between the carbon and oxygen atoms. The carbon involved now in the methyl group, nears the gallium up to 2.2 Å. However, the gallium retains its original planar coordination to oxygens, with the methyl perpendicular to this plane. The interaction between the oxygen and the dissociative hydrogen is apparently set on in TS, which is revealed by the increase in the Ga-O bond length by 0.05 Å and the decrease in the C-Ga-O angle. From this point on a new strong OH bond is being formed and full relaxation of the system towards the final dissociated structure takes place. In this structure methyl becomes the fourth ligand of the gallium cation, which results in tetrahedral coordination of the latter and further weakening of Ga-O bonds.

The energy parameters of the adsorption and dissociation reaction calculated already within the local density approximation appeared to be reasonable for a catalytic reaction. The height of the reaction barrier, however, which is one of the determining factors in reaction kinetics, emerges from the fine interplay of various effects. These involve both energy losses, on breaking stable bonds and changing preferable coordination of the center, and energy gains on forming new strong chemical bonds such as OH. Thus the good description of correlation effects is crucial for determining quantitative reaction energetics and the calculations should employ the most accurate quantum chemical methodology to account for bond breaking and forming. It is known that the local approximation to exchange-correlation potential underestimates correlation and overestimates the bonding. Therefore, for the critical points on the reaction pathway, the total energy has been recalculated including selfconsistently nonlocal gradient corrections to the potential. Although presently available nonlocal potentials are known to overestimate correlation effects, they provide an overall better description of the energy parameters in question.

Table 10
Calculated stabilization energy for the dissociative adsorption and electronic activation barrier for the dissociation of methane on the Ga site (all values in kcal/mol, local and nonlocal approximation)

	$\Delta H(0)$	$\Delta E^{\#}$
LDA	51	13
NLDA	39	31

Table 10 displays the comparison between local and nonlocal estimates of the reaction heat at 0 K, $\Delta H(0)$, calculated as the difference between the adsorption energies of both stable structures, and the electronic activation energy, $\Delta E^{\#}$, calculated as the total energy difference between the substrate and TS. It can be seen that including nonlocal corrections to the LDA changes dramatically the values of energetical parameters. The corrected value of the activation energy is 31 kcal/mol. It compares perfectly with the activation barrier of 31-33 kcal/mol calculated by Blaszkowski *et al.* [12] for the hydrogen exchange reaction between methane and acidic zeolite clusters. This latter value was related to a process different from the ours, namely the hydrogen exchange reaction between the methane molecule and a Brönsted zeolitic proton. The theoretical barrier for the actual dehydrogenation of the methane on such acidic center was estimated at a much higher value of 84 kcal/mol. Our value of ca. 30 kcal/mol indicates clearly how important the influence of the exchanged gallium center is when compared to unmodified H-ZSM-5. Firstly, the Ga cation provides strong stabilization for a methyl moiety and compensates for the broken C-H bond. Furthermore, the interaction of the dissociative hydrogen with the lattice oxygen facilitates C-H bond scission and finally highly stabilizes the system energetically by forming strong OH bond. In this way the activation energy for the transition state for CH_4 dissociation on Ga-ZSM-5 active site may become comparable to the hydrogen exchange process on H-ZSM-5. The gallium cation is a better stabilizing center for the methyl than zeolitic acidic hydrogen, while the extraframework oxygen seems to be even more active towards bonding the dissociative hydrogen than the regular lattice oxygen.

No experimentally measured activation barrier for methane dehydrogenation is available for either H-ZSM-5 or Ga-ZSM-5. It has been shown, however, that while both zeolites can activate the C-H bond in CH_4, on H-ZSM-5 the rate of C-H activation is about 20 times slower for methane than for propane, whereas on Ga-ZSM-5 it is considerably more efficient [23]. Thus it seems clear that not only the C-H bond strength but also the stabilization of the transition state and the final products must be of great importance for the C-H bond dissociation. Kramer *et al.* [31] also gave an estimate of the effective reaction barrier for the methane deutero exchange on zeolites to be 29 kcal/mol. In order to compare with the last value, our result should be corrected for zero point vibrational energy (ZPE).

For further explaining the influence of the Ga-exchanged ZSM-5 on the dissociation of methane, this barrier may be compared with the value of 104 kcal/mol required for gas phase homolytic scission of the C-H bond in methane which makes this process extremely difficult. In addition, the calculations for the direct reaction between the gas phase CH_4 and NO_2 molecules indicated that the height of the barrier in this case should amount to at least 60 kcal/mol [32]. In this view, the lowering of the activation energy for the C-H bond dissociation in the presence of the Ga center to about 29 kcal/mol must be regarded as a very important effect with a significant impact on the reaction dynamics.

5. CONCLUDING REMARKS.

After analysing the details of the geometrical and electronic structure ot the postulated active sites and their interaction with the approaching hydrocarbon molecule the main factors governing the transformation of methane after adsorption on a variety of catalytic centers may be summarized and conluded in a general way.

The discussed results seem to indicate that the properties responsible for efficient adsorption and activation of methane on oxidized transition metals are directly related to fine details of their electronic structure and its rearrangement. Table 11 shows the particulars of charge distribution and its reorganization in studied systems composed of transition metal dimers and oxides for the dissociation product (DIS) and the most

Table 11
Charge on active hydrogen for preadsorption or dissociative adsorption of methane on studied systems and the effective cluster -> methane charge transfer during dissociation

	Pd_2		PdO		RhO		Rh_2	
	ADS	DIS	ADS	DIS	ADS	DIS	ADS	DIS
Q_H (e)	-0.06	-0.15	+0.02	+0.11	+0.04	+0.13	+0.03	-0.06
CT (e) (ads-dis)	+0.23		-0.07		-0.07		+0.15	

stable preadsorption (ADS). The dimers are arranged in the order of increasing stability of the preadsorbed complex. The charge on active hydrogen and the value of the charge transfered from the cluster to the adsorbed species during the dissociation process (positive sign indicates the charge transfer from the cluster) are given in the table. Chemisorptive bonding of a methane molecule with a metal atom is generally mediated by hydrogens with the most pronouncing effect and the shortest metal-hydrogen bond displayed for palladium. The strongest methane adsorption and activation of hydrogen on palladium dimer may be fully confirmed here in accord with high affinity of palladium towards hydrogen. Rhodium dimer occupies the position of the least active agent.The difference between palladium and rhodium with respect to their affinity towards hydrogen might be also related to their metallic properties. Since bonding and antibonding interactions within the palladium dimer are weak thus the splitting of the fully occupied d "band" is small (ca. 2 eV) with the empty s "band" lying 0.8 eV higher. In the case of rhodium dimer partially occupied d and s "bands" overlap and span over energy range wider than 4 eV. Thus we can expect easier bond formation and localization of electrons in the former case. The hydrogen dissociated on a metal bears negative charge and the overall electron transfer occurs from the metal dimer, in the case of rhodium, however, the effect is small. The effect is the opposite on oxides where slightly positive hydrogen is finally strongly covalently bonded within the OH group.

According to our previous findings [10,11] and other authors [14] two main factors influence the hight of the activation barrier and the reaction energetical effect: moderate adsorption strength in the entrance channel and strong bonds formation in the reaction channel. This implies that in the case of oxides we should expect moderate barriers and high exothermicity. Indeed, in the case of oxides the product is very stable and the adsorption is only moderately strong which manifests itself by the change of spin state in the transition complex, from the quartet (RhO) or triplet (PdO) to doublet or singlet, respectively. The same was found true by Siegbahn et.al. [14] in the case of bare rhodium atom. It imposes substantial hight of the barrier increased by prerequisite spin transition from the high spin state to the low spin state of the system. Palladium atom is an exception among late second row transition metals and does not show such spin transition.

For metal dimers the dissociated methyl and hydrogen do not form strong bonds and the reaction exothermicity is smaller than for appropriate oxides. In the case of the palladium dimer it manifests itself also in disappearance of the spin transition and all adsorption modes lie on the same spin surface. As the adsorption of methane is also moderate, the dissociation barrier is rather low (a few kcal/mol) but the bonds in dissociated complex are weak, thus the reverse recombination reaction seems to be of comparable probability. For rhodium dimer the adsorption complex and the product lie on different spin surfaces thus although due to repulsive 5s electrons adsorption is weak, required spin-flip would raise the dissociation energy barrier.

The most promising situation would be invoked when both phases, metallic and oxidized one, were present on the catalyst surface, which is realized in supported oxidized palladium catalyst. We have supported this hypothesis in our previous work concerning the activation of methane on PdO, Pd_2 and mixed Pd_2O clusters [11]. Here from the comparison of the properties of rhodium and palladium small systems we can infer more information on the properties of the catalyst components prerequisite to assure its high activity. In the case of palladium systems the most favourable predissociative adsorption mode is the bridging adsorption on palladium dimer. In this position a methane molecule becomes strongly activated and the scission of the C-H bond occurs with almost no barrier. After dissociation it is the complex of PdO with methyl and hydrogen fragments which is the most stable state of the system. Thus in the presence of oxygen the process in question should have low activation barrier and high efficiency of methane degradation. In the case of rhodium system the preferred adsorption is the collinear mode on RhO while the one on rhodium dimer cannot concurre as it is fairly weak and brings not so distinct methane activation. Thus we may expect higher activation barrier in this case and lower efficiency of the methane combustion on rhodium catalysts.

In the case of the gallium site in Ga-ZSM-5 the geometrical structure, in particular structural unsaturation of the center, seems to provide the driving force for preliminary methane adsorption. The molecular electrostatic potential near the gallium site has also been calculated [24] which shows that there exists a polarizing electrostatic field around GaO: extraframework oxygen has large attractive potential for protons and can enhance the activation of a hydrocarbon over this center. Indeed, it was concluded previously [24] that the interaction between the metal ion and the polarizable methane molecule via the long-range charge-induced dipole force causes a precoordination of methane to the cation as the first step in the reaction. Due to structural unsaturation of the gallium site and strong electrostatic potential extended in its vicinity, the molecule may experience charge polarization leading to the initialization of the dissociation process.

In the physisorbed state a methane molecule becomes slightly positive, whereas in the transition state, some flow of charge towards already forming methyl group is noticeable. The calculated activation barrier is almost 3.5 times lower than that for direct scission of the C-H bond in the gas phase and 2.5 times lower than the theoretical barrier for the dissociation promoted by H-ZSM-5. In the case of Ga-ZSM-5, the transition state for the hydrogen transfer between the carbon and the oxygen is additionally stabilized not only by increased interaction of the gallium cation with the methyl carbon but also by the interaction of the polarized extraframework oxygen and the dissociative hydrogen. Thus our model reveals that both constituents of the postulated GaO^+ active unit play important roles in the C-H bond scission.

The preceding considerations seem to render very clear arguments in discussing the character of the gallium center and the transformation of the system in various stages of the interaction between the center and the adsorbed molecule. Moreover, the quantitative energy parameters and qualitative evolution of the process, inferred from vibrational analysis, reasonably describe basic interactions in the system. This indicates that the density functional quantum chemistry methodology is a very promising tool for studying intricate catalytic reactions in zeolytic systems. In particular the inclusion of nonlocal corrections, and thus better description of electron correlation, gives reliable energy parameters, accurate enough to describe reaction profiles.

The extension of the model towards more realistic image of the zeolite structure would be obviously desirable. The significant enlargement of the cluster size is, however, not likely in the nearest future, as incorporating the next-neighbor coordination sphere would lead to a cluster too big for rigorous reaction pathway calculations. The rapidly developing embedding techniques would hopefully offer the alternative future solution to this problem. Further development of our theoretical model comprises similar studies for various metal-exchanged centers in zeolytic lattices to confirm the proposed mechanism. These would include transition metals such as cobalt, known to be active for the reduction of nitrogen oxides by hydrocarbons.

It must be, finally, stressed out that the surface activity is subjected to many other factors which were not included in reported investigations. Different idealized scenarios for different reaction sites may appear to be useful, depending on which part of the reaction path is to be described. Nevertheless, we attempted to construct models for the catalytic reaction, either based on classical simulation, followed by quantum chemical calculations, or based on molecules representing functions of the site, to search for stable structures and transition states on the reaction pathway. Determining full reaction energetics requires the calculation of the entire reaction pathway with the most accurate quantum mechanical methodology. Density functional theory which was the sole basis for quantitative considerations has, in our opinion, proved to be an extremely efficient and reliable tool in this area.

REFERENCES

1. D.R. Salahub and M. Zerner (eds.) The Challenge of d and f Electrons, ACS Symposium Series, vol. 394, ACS, Washington, 1989.
2. J. Labanowski and J. Andzelm (eds) Theory and Application of Density Functional Methods in Chemistry, Springer Verlag, New York, 1991.
3. D.R. Salahub and N. Russo (eds) Metal-Ligand interactions, from Atoms, to Clusters, to Surfaces, NATO ASI Series, Series C, vol. 378, Kluwer, Dodrecht, 1992.
4. P. Hohenberg and W. Kohn, Phys. Rev., 136 (1964) B864; W.Kohn and L.J. Sham, Phys. Rev. A, 140 (1965) 1133.
5. DMol version 2. 3. 5 San Diego : Biosym Technologies, 1993.
6. V. L. Moruzzi, J. F. Janak and A. R. Williams, Calculated Electronic Properties of Metals, Pergamon, New York (1978).
7. A. Becke, J. Chem. Phys., 88 (1988) 2547; C. Lee, W. Yang and R. G. Parr, Phys. Rev.B, 37 (1988) 786.
8. E. Broclawik and D.R. Salahub, J. Mol. Catal. 82 (1993) 117.
9. E. Broclawik, in: Modern Density Functional Theory: A Tool for Chemistry, J.M. Seminario and P. Politzer (eds), Elsevier, Amsterdam, 1995, p. 349.
10. E. Broclawik, R. Yamauchi, A. Endou, M. Kubo and A. Miyamoto, J. Chem. Phys., 104 (1996) 4098.
11. E. Broclawik, R. Yamauchi, A. Endou, M. Kubo and A. Miyamoto, Int. J. Quant. Chem., in print.

12.H. Burghraef, A.P.J. Jansen and R.A. van Santen, J. Chem. Phys., 101 (1994) 11012; S.R. Blaszkowski, A.P.J. Jansen, M.A.C. Nascimento and R.A. van Santen, J. Phys. Chem., 98 (1994) 12938.
13.A. Goursot, I. Papai and D.R. Salahub, J. Am Chem. Soc. 114 (1992) 7452; A. Goursot, I. Papai and C. Daul, Int. J. Quant. Chem. 52 (1994) 799.
14.M.R.A. Blomberg, P.E.M. Siegbahn, and M.J. Svensson, J. Am. Chem. Soc., 114 (1992) 207; M.R.A. Blomberg, P.E.M. Siegbahn and M.J. Svensson, J. Phys. Chem., 98 (1994) 2062; M. Hada, H. Nakatsuji, S. Gyoba, and S.J. Miki, J. Mol. Struct. THEOCHEM, 281 (1993) 207.
15.E. Broclawik, J. Haber, A. Endou, R. Yamauchi, M. Kubo and A. Miyamoto, J. Mol. Catal., submitted.
16.P. E. M. Siegbahn, Chem. Phys. Lett., 201 (1993) 15.
17.M. Harada and H. Dexpert, J. Phys. Chem. 100 (1996) 565.
18.S.-C. Chung, S. Kruger, G. Pacchioni and N. Rosch, J. Chem. Phys. 102 (1995) 3695; P.E.M. Siegbahn, Chem. Phys. Lett. 201 (1993) 15.
19.D.G. Musaev and K. Morokuma, J. Am. Chem. Soc., 117 (1995) 799; K.G. Dobbs and D.A. Dixon, J. Phys. Chem., 98 (1994) 12584.
20.E. Kikuchi and K. Yogo, Catal. Today, 22 (1994) 73.
21.R.A. Grinsted, H. -W. Jen, C.N. Montreuil, M.J. Rokosz and M. Shelef, Zeolites, 13 (1993) 602.
22.A. Miyamoto and M. Kubo, J. Jpn. Petrol. Inst., 36 (1993 282; A. Miyamoto, H. Himei, Y. Oka E. Maruya, M. Katagiri, R. Vetrivel and M. Kubo, Catal. Today, 22 (1994) 87; A. Miyamoto, M. Kubo, K. Matsubaand T. Inui, in: "Computer aided innovation of new materials II" (M. Doyama *et al.* eds.), Elsevier Science Publishers, Amsterdam, 1993, p. 1025.
23.E. Iglesia and J.E. Baumgartner, Catal. Lett., 21 (1993) 55; E. Iglesia, J.E. Baumgartner, G.L. Price, J. Catal., 134 (1992) 549; G.D. Meitzner, E. Iglesia, J.E. Baumgartner and E.S. Huang, J. Catal., 140 (1993) 209.
24.R. Vetrivel, M. Kubo, H. Himei, E. Maruya, M. Katagiri, E. Broclawik, and A. Miyamoto, in: "Science and Technology in Catalysis 1994", Y. Izumi, H. Arai and M. Iwamoto (eds.) Studies in Surface Science and Catalysis 92; Kodansha, Tokyo, 1995, p. 233.
25.K. Borve and L.G.M. Pettersson, J. Phys. Chem., 95 (1991) 3214; Y. Aray, J. Rodriguez, J. Murgich and F.J. Ruette, J. Phys. Chem.,97 (1993) 8393.
26.K. Borve, J. Chem. Phys., 95 (1991) 4626.
27.A. Fiedler, D. Schroder, S. Shaik and H. Schwartz, J. Am. Chem. Soc., 116 (1994) 10734; P.E.M. Siegbahn and M. Svensson, J. Chem. Phys. 100 (1994) 423.
28.B. Hammer and J.K. Norskov, Surf. Sci., 343 (1995) 211; A. Gross, S. Wilke and M. Schaffler, Phys. Rev. Lett., 75 (1995) 2718; G. Wiesenekker, G.J. Kroes, E.J. Baerends and R.C. Mowrey, J. Chem. Phys., 102 (1995) 3873.
29.W. Dong, G. Kresse and J. Hoffner, J. Mol. Catal., submitted.
30.C.W. Bock and M. Trachtman, J. Phys. Chem., 98 (1994) 95.
31.G.J. Kramer, R.A. van Santen, C.A. Emeis and A.K. Novak, Nature, 363 (1993) 529.
32.E. Broclawik, H. Himei, M. Yamadaya, and A. Miyamoto, A., unpublished results.

J.M. Seminario (Editor)
Recent Developments and Applications of Modern Density Functional Theory
Theoretical and Computational Chemistry, Vol. 4

DFT Study of Nickel: Towards the MD Simulation of the Nickel-Water Interface

Perla B. Balbuena[1] and Jorge M. Seminario[2]

[1]Department of Chemical Engineering, The University of Texas at Austin, Austin, TX 78712, USA
[2]Department of Chemistry, University of New Orleans, New Orleans, LA 70148, USA.

A systematic study has been carried out for the determination and characterization of the lowest states of Ni. This allowed determining a suitable basis set that performs considerably well with nickel atoms, while others of double-ζ and triple-ζ quality were found insufficient to reproduce basic characteristics of the smallest nickel systems. Then we calculated, using DFT, a force field for nickel-water interactions. Using this DFT force field, we constructed Ni_3, Ni_4, Ni_7 and Ni_{22} clusters to perform molecular dynamics (MD) simulations of the clusters immersed in liquid water at 298 K. The evolution of the *cluster to surface* properties was investigated through radial distribution functions, binding energies, average dipole orientations and characteristic reorientation times of the water molecules surrounding the metal atoms.

1. INTRODUCTION

The study of small nickel clusters has motivated a great deal of interest in several areas of chemistry and technology, especially in catalytic processes and materials science. There are several experimental techniques like molecular beams that allow the synthesis and characterization of clusters of a defined size. Several studies [1-7] have been able to determine dissociation energies, ionization potentials, magnetic moments and chemical reactivity for some metallic clusters. The study of nickel clusters is interesting because clusters can be considered a transition between a molecule and a surface combining properties of metallic surfaces and metallic compounds together. Three of the major routes of interest include the study of reduction, nucleation, and transformation of the clusters.

On the other hand, the interfacial properties of the water-metal systems attract much attention due to their implications for electrochemistry, catalysis and corrosion studies. For these complex interfaces, where the experimental information at the molecular level is very scarce, computer simulations can play a significant role. In these methods, a certain number of particles are placed in a box, with interactions described by pair potentials, and the classical equations of motion are integrated numerically for all particles in the box. As a result, all positions and velocities are known as a function of time for each of the particles, allowing a complete description of the structure, dynamics and thermodynamics of the system. The first simulations of aqueous solutions near model metallic surfaces were reported in 1986 by Spohr and Heinzinger [8, 9]. Both uncharged Lennard-Jones and charged walls were studied. In the first case, the metallic character was described by image charges. Further improvements emphasized the development of the water-metal potentials [10] towards a better agreement on the adsorption properties as obtained from experiments [11-14] and molecular orbital (MO) calculations [15, 16]. In some of these simulations the surface is explicitly represented by a set of metal atoms [17], while others use an effective external field [18]. Recent review articles [19, 20] summarize the state of the art of these calculations that have been extended to include adsorption of ionic solutions onto charged electrodes [21] and to confined geometries [22]. Here we present results of MD simulations where the metal surface, represented by clusters of increasing number of atoms, is embedded in the bulk water. The objective is to analyze the structural and dynamical properties of the water molecules surrounding the metal atoms towards the characterization of the minimum number of atoms in the metal cluster that best approximates the surface.

Thus, the goal of the first part of this study is to determine the basic properties of the lowest states of the nickel atom using the best theoretical tools available at present; thus, a proper tool can be determined to study much larger clusters and nickel complexes. Since the present technology in hardware and software allows us to perform high level calculations on small nickel systems, we have been able to obtain precise estimates of the quality of several ab initio and DFT techniques. The techniques used here are very similar to those used with chemical accuracy in small first and second row compounds, such as those used in the Gaussian 1 and the Gaussian 2 models [23-26]. In the next section we provide a short description of the DFT formalism. In section 3, a summary of the methods is given. The results for the monomer and the calculation of the force fields are presented in section 4. In the second part of this work we study the water-nickel interface by the analysis of the trajectories from MD simulations performed using the force fields described in section 4. In section 5 we review the MD methodology. MD calculations using clusters of 3, 4, 7 and 22 nickel atoms are presented in section 6. An analysis of the structure and dynamics of the water molecules in the neighborhood of the metal atoms is done based on radial distribution functions, binding energies and orientation of the dipoles.

2. DENSITY FUNCTIONAL THEORY

Present DFT methods are strictly ab initio in character [27-30] (see also introduction by Levy in this book). They can be directly derived from the electronic time-independent, non-relativistic, Born-Oppenheimer-approximated, Schrödinger equation for a molecular system,

$$\hat{H}\Psi = E\Psi \tag{1}$$

where the Hamiltonian is given by the contributions from the external potential due to the nuclei, the kinetic energy operator, and the electron-electron repulsion operators, i. e.,

$$\hat{H} = \hat{\upsilon}_{ext} + \hat{T} + \hat{V}_{ee} \tag{2}$$

The wave function Ψ depends on the N electrons in the molecule,

$$\Psi = \Psi(\vec{r}_1, \vec{r}_2, \vec{r}_3, \ldots, \vec{r}_N) \tag{3}$$

Instead of solving for the wavefunction in Eq. 1, DFT transforms the Schrödinger equation to

$$\hat{H}_\lambda \Psi_\lambda = E_\lambda \Psi_\lambda \tag{4}$$

with the modified Hamiltonian [31] defined as

$$\hat{H}_\lambda = \hat{T} + \hat{\upsilon}_\lambda + \lambda \hat{V}_{ee} \tag{5}$$

where the potential $\hat{\upsilon}_\lambda$ is tuned in order to reproduce the density ρ of the unmodified Hamiltonian $\hat{H}$ [32, 33],

$$\rho(\vec{r}_1) = N \int d\vec{r}_2 d\vec{r}_3 \ldots d\vec{r}_N \left| \Psi(\vec{r}_1, \vec{r}_2, \vec{r}_3, \ldots, \vec{r}_N \right|^2 \tag{6}$$

for any λ between zero (noninteracting-electron system), and one (real system). The total energy E of the real system is given by

$$E = T_s + \int \upsilon_{ext}(\vec{r})\rho(\vec{r})d\vec{r} + \int_0^1 \langle \Psi_\lambda | V_{ee} | \Psi_\lambda \rangle d\lambda \tag{7}$$

where T_s is the kinetic energy of the noninteracting system of electrons (λ=0). T_s and $\rho(\vec{r})$ can be expressed exactly using the wavefunction of a non interactive system of electrons whose density is identical to the one of the real system. The integral in λ is a functional of the density ρ, but its analytical treatment is still very difficult and only few non-fitted approximations have been suggested, like the local, the gradient-corrected, and the generalized gradient-corrected of the Perdew-Wang approximations. Most approximations start with the following partition,

$$\int_0^1 \langle \Psi_\lambda | V_{ee} | \Psi_\lambda \rangle d\lambda = \frac{1}{2} \int \int d\vec{r}_1 d\vec{r}_2 \frac{\rho(\vec{r}_1)\rho(\vec{r}_2)}{|\vec{r}_1 - \vec{r}_2|} + E_{xc}(\rho) \tag{8}$$

where the classical electron-electron repulsion has been subtracted from the unknown integral, and the remainder is being defined as the exchange-correlation energy, E_{xc}.This partitioning permits finding an exact solution for an ideal uniform electron gas (jellium) which in turn yields the so-called local functionals. These functionals can be further modified into the nonlocal or gradient-corrected functionals. Once a functional of the exchange-correlation energy has been obtained, the problem reduces to a set of one-electron equations of the form,

$$\left[-\tfrac{1}{2}\nabla^2 + \upsilon_{ext} + \upsilon_{class} + \upsilon_{xc}\right]\varphi_i = \varepsilon_i\varphi_i \tag{9}$$

that can be solved in a self-consistent manner. υ_{ext}, υ_{class}, and υ_{xc} are the functional derivatives with respect to the density of the nuclear-electron, classical electron-electron and exchange-correlation energies respectively.

3. METHODS

We analyzed atomic nickel using ab initio and DFT tools available in the program Gaussian-94 [34]. Based on the results obtained by these methods, we chose the B3PW91 functional as described below to calculate the nickel-water force field parameters. In this work, we have used two basis sets to study atomic nickel (with both, the standard ab initio and the DFT methods). One is the so-called "6-311G*" [35-37] shown in Table 1, which is actually a (15s, 11p, 6d, 1f) contracted to (10s, 7p, 4d, 1f). The other basis set is the uncontracted modified Huzinaga (12s, 8p, 4d) shown in Table 2. The modification of the Huzinaga [38] basis set was necessary because the original one was not able to represent the lowest electronic states of the nickel atom as indicated below.

We have used the Becke-3 (B3) [39] exchange functional combined with the Perdew-Wang functionals PW91 for correlation [40-42], or with the Lee-Yang-Parr (LYP) [43] correlation functional. All of them belong to the generalized gradient approximation (GGA) scheme with the addition that in the B3 exchange functional, a portion of the exchange energy is computed in the same fashion as the Hartree-Fock exchange energy is calculated, but using the noninteractive wave function. We have also used the standard ab initio methods: Møller-Plesset (MP) perturbation theory [44] up to fourth order in the energy, including single, double, triple and quadruple excitations (MP4SDTQ), and the coupled cluster single double (CCSD) [45-48] excitations method alone and also with non-iterative triplet excitations, CCSD(T) [49].

4. SMALL Ni SYSTEMS

4.1. Atomic Nickel:

The following states were investigated:

$1s^2\, 2s^2\, 2p^6\, 3s^2\, 3p^6\, 3d^8\, 4s^2$ Singlet $3d^84s^2$,

$1s^2\, 2s^2\, 2p^6\, 3s^2\, 3p^6\, 3d^8\, 4s^2$ Triplet $3d^84s^2$,

$1s^2\ 2s^2\ 2p^6\ 3s^2\ 3p^6\ 3d^8\ 4s^1\ 4p^1$	Quintet $3d^84s^14p^1$, and
$1s^2\ 2s^2\ 2p^6\ 3s^2\ 3p^6\ 3d^{10}$	Singlet $3d^{10}$

These states were chosen because they correspond to the lowest states of their corresponding symmetry and also because of the availability of precise experimental information. The results are shown in Table 3, in which can be observed the good agreement with experiment of DFT results for these lowest states of atomic nickel. It was practically impossible to reach reasonable results with the standard ab initio methods. The tolerance indicated in Table 3 for the range of experimental values was obtained as follow: the ground state of Ni corresponds to the triplet $3d^84s^2$ a^3F; the splitting of the J = 4, 2, 1 states spans an energy range of 4.9 kcal/mol which has been added to the other states in order to compare with our nonrelativistic calculations.

Table 1
Gaussian exponents of the 6-311G* basis set for Ni; coefficients for the contracted Gaussians are shown in parenthesis. This basis set correspond to a (15s, 11p, 6d, 1f) contracted to (10s, 7p, 4d, 1f). Notice that the 6-311G* nomenclature is bogus here.

S-exponent	P-exponent	D-exponent	F-exp.
284878. (0.00032)	1774.18 (0.00295)	48.9403 (0.02706)	1.29
41997.9 (0.00246)	423.403 (0.02337)	13.7169 (0.14598)	
9627.67 (0.01254)	138.311 (0.10406)	4.63951 (0.36418)	
2761.96 (0.04926)	53.1703 (0.28226)	1.57433	
920.488 (0.1495)	22.38742 (0.43486)	0.486409	
341.805 (0.3264)	9.92848	0.1316	
138.023	4.11625		
59.2587	1.71031		
20.3712	0.672528		
8.59400	0.146588		
2.394171	0.044447		
0.918169			
0.130176			
0.046392			
0.013918			

Table 2
Gaussian exponents of the modified Huzinaga uncontracted (12s, 8p, 4d) basis set of Ni.

S	P	D	F
921.28732	53.248593	4.6256646	
342.06529	22.421150	1.5877750	
138.10746	9.9349303	0.47179676	
59.260278	4.0867350	0.1651544	
20.358991	1.7180932		
8.5942749	0.66515353		
2.3574855	0.2947343		
0.93250293	0.1335633		
0.31727122			
0.12890620			
0.04609800			
0.019780135			

The calculation of the singlet $3d^{10}$ state starts at the HF with a pure $3d^{10}$; however with the inclusion of higher order perturbation contributions, this state should tend to the singlet $3d^84s^2$. Unfortunately, at the MP4 level the energy with respect to the triplet $3d^84s^2$ (ground state) is still too large to be considered within the range of chemical accuracy. There is not much difference between the B3PW91 and the B3LYP functionals, therefore we decided to choose the B3PW91 functional for calculations of nickel-containing systems.

Practically the 6-311G* and the modified (Huzinaga) basis sets yielded similar results; however, only the modified Huzinaga was able to yield the excited states of the Ni atom. Since this ability is important when forming nickel-containing systems we decided to use the simpler modified basis set. It is interesting to note that a calculation using DFT having as initial state the singlet $3d^{10}$ will immediately converge to the lower singlet $3d^84s^2$, yielding a result very close to the experimental values. This modified Huzinaga basis set with the B3PW91 functional yielded, even for the quintet state, very good agreement with the experimental values. This is very important when the basis set is going to be used to form complexes. The problem with the original Huzinaga basis set was that their exponents are shifted to higher values which are only to take care of the behavior of the core electrons, leaving a small number of functions to handle the valence electrons. Therefore our solution was to modify the Huzinaga basis, shifting the values of the exponents to smaller numbers. The first four exponents were eliminated and the 0.31727122 and 0.019780135 exponents were added in the S-gaussians. The highest three exponents were eliminated, and the 0.2947343 and 0.1335633 exponents were added to the P-gaussians.

Table 3
Calculations of the lowest multiplicities for Ni atom using several conventional ab initio, and DFT methods. The basis sets used are defined as follows: G is the 6-311G* which contains f-functions, H is the Huzinaga basis set, and M is the modified Huzinaga basis set. (the last two do not contain f-functions)

method	m=1 $3d^84s^2$	m=3 $3d^84s^2$	m=5 $3d^84s^14p^1$	m=1 $3d^{10}$
HF/G	-	-1506.74074	-1506.49733	-1506.37367
		(0.0)	(152.7)	(230.3)
MP2/G*	-	-1506.95974	-1506.71037	-1506.75665
		(0.0)	(156.5)	(127.4)
MP3/G	-	-1506.94669	-1506.69797	-1506.68427
		(0.0)	(156.1)	(164.7)
MP4/G	-	-1506.95774	-1506.70785	-1506.73092
		(0.0)	(156.8)	(142.3)
CCSD/G	-	-	-	-1506.70635
CCSD(T)/G	-	-	-	-1506.71548
B3PW91/H	-1508.20147	-1508.25746	-1507.97581	-
	(35.1)	(0.0)	(176.7)	
B3LYP/H	-1508.25402	-1508.30598	-1508.01838	-
	(32.6)	(0)	(179.0)	
B3PW91/M	-1471.67100	-1471.74829	-1471.59247	-
	(48.5)	(0)	(97.8)	
B3PW91/G	-1508.06148	-	-	-
Experiment	(37.9-42.8)	(0.0-4.9)	(73.7-91.8)	(311.4-316.3)

The highest two exponents were eliminated, and the 0.1651544 exponent was added to the D-functions of Ni. This modified basis set represents a very good compromise between quality and efficiency to deal with complexes of Ni. It performs even better than the 6-311G* which contains f-functions. Probably the reason for the failure of the conventional basis sets is because they have been developed using a HF procedure where no electron correlation was considered.

4.2. Calculation of Force Fields

Using the B3PW91 hybrid functional, the modified Huzinaga basis set for Ni atom and the 6-311G** basis set for H and O, we performed calculations on the singlet, triplet and quintet lowest states of the Ni-water complex in order to calculate the Lennard-Jones parameters for the molecular dynamics simulations. These results are shown in Table 4.

Table 4
Energies (Hartree) and optimized parameters (lengths in Å's, angles and dihedrals in degrees) for the Ni-water, water, and atomic nickel systems.

System	Energy	Ni-O	O-H	H-O-H	Ni-O-H	H-O-Ni-H
Ni-OH_2 Singlet	-1548.15435	1.794	0.975	100.8	106.2	109.1
Ni-OH_2 Triplet	-1548.18030	2.063	0.967	106.9	105.8	113.0
Ni-OH_2 Quintet	-1548.04266	2.086	0.963	104.5	108.2	113.6
OH_2	-76.41888		0.960	103.8		
Ni Singlet	-1471.67100					
Ni Triplet	-1471.74829					
Ni Quintet	-1471.59247					

From the above table, we chose the triplet Ni-water complex with its corresponding Ni-O distance of 2.063 Å and its depth of the potential well of 8.239 kcal/mol to be used in the molecular dynamics simulations. The higher energy singlet complex would make a strong complex (40.5 kcal/mol) with respect to the dissociation of ground state water and singlet atomic nickel; and the quintet would make a smaller but still strong complex (19.6 kcal/mol) with respect to the dissociation to a quintet Ni and singlet water. Since the singlet complex Ni-water is 13.3 kcal/mol above the triplet and the quintet is 86.4 kcal/mol above the triplet, there is not much problem in selecting the triplet complex as the basic system to study Ni-water interactions. This choice is also consistent with the individual ground states of the complex components atomic nickel atoms and the singlet ground state of water. The DFT self-consistent convergence was practically impossible or very slow using the STO-3G, Huzinaga or 6-311G* basis sets of the nickel atom.

5. MD SIMULATIONS

5.1. Overview of the method

In this section we present a general outlook of the main concepts and algorithms used for the integration of Newton's equations in MD simulations. For further details the reader is referred to a more comprehensive discussion[50]. A central box of cubic (or another) geometry with volume V constitutes the basic cell containing N particles that interact with pair potentials $U_{ij}(\mathbf{r}_i, \mathbf{r}_j, \Omega_i, \Omega_j)$ depending on the particle positions ($\mathbf{r}$) and orientations (Ω). In its simplest form the total potential energy $\mathcal{V}$ is the sum over all distinct pairs i and j,

$$\mathcal{V} = \sum_i \sum_{j>i} U_{ij}(\mathbf{r}_i, \mathbf{r}_j, \Omega_i, \Omega_j). \tag{10}$$

Periodic boundary conditions are imposed in order to replicate the central box all over the space, emulating the thermodynamic limit, $N \to \infty$. The classical equations of motion for molecule i, can be written as a system of 3N second order differential equations,

$$m_i \frac{d^2\mathbf{r}_i}{dt^2} = \mathbf{f}_i \tag{11}$$

where m is the mass of molecule i and **f** the total force exerted on it. Several numerical methods have been proposed for the integration of these equations, mostly based on variations of finite difference methods. By far the most popular is the Verlet algorithm [51] that is used by many standard MD packages. Within this method, the positions at $t+\delta t$ are calculated using the accelerations and positions at time t, and the positions at the previous time,

$$\mathbf{r}(t+\delta t) = 2\mathbf{r}(t) - \mathbf{r}(t-\delta t) + \delta t^2 \mathbf{a}(t) \tag{12}$$

The velocities are obtained from

$$\mathbf{v}(t) = \frac{\mathbf{r}(t+\delta t) - \mathbf{r}(t-\delta t)}{2\delta t} \tag{13}$$

The choice of the time step δt is a compromise between accuracy and availability of computer time, it is also dependent on the system and model potentials used. Several modifications have been proposed in order to improve the computational efficiency of the Verlet algorithm; however, as discussed extensively elsewhere [50], the most computationally intensive part is the calculation of the forces and energies, and thus here is where the major effort for improvement has been directed. Special considerations must be taken into account when dealing with systems with long range interactions, which decay no faster than r^{-d} where d is the dimensionality of the system, such as ionic solutions. We refer to one of these techniques (Ewald summation) in the next section. Once the trajectories (positions and velocities) for all particles in the system are obtained by numerical solution of equations (11), this microscopic information can be used to compute macroscopic thermodynamic and transport properties by statistical averages, using the Gibbs concept [52] of substituting time averages by ensemble averages. One last point regarding the ensembles: Since Newton's equations ensure conservation of energy, the MD method is usually defined in the NVE (N molecules in a given Volume and with a given total Energy) microcanonical ensemble. However, it can be adapted to generate trajectories in other thermodynamic ensembles [50]. For example, constant temperature MD is a representation of the canonical NVT ensemble. It can be achieved using different modifications to the standard MD procedure [53].

The actual simulation is started by defining an initial configuration and letting the system evolve in time according to classical dynamics, until proper equilibration is obtained. Some key properties (which ones depends on the thermodynamic ensemble) are monitored periodically during the equilibration period. After equilibration is achieved, it is followed by another set of runs where the trajectories are saved for ulterior analysis. Structural information is usually calculated in the form of radial distribution functions (rdf's). For a simple monoatomic fluid, this function g(r) gives the probability of finding one atom at a given distance r from a central atom. From the simulation results, the rdf's can be obtained from

$$g(r) = \frac{V}{N^2}\left\langle \sum_i \sum_{j\neq i} \delta(\mathbf{r} - \mathbf{r}_{ij}) \right\rangle \tag{14}$$

where the brackets indicate a time average. Much other information is readily obtainable, either directly from the trajectories, as in Eq. 14, or by simple processing of the data. For example, time correlations of velocities and orientations yield valuable information about the microscopic structure and dynamics of the system.

5.2. Details of the simulations.

As described in the introduction, in this work we present results from MD simulations of metallic clusters immersed in water at 298 K. The pair potentials described in section 4 define the interactions among water and metal atoms in the system. The water-water interactions are modeled with the Simple Point Charge/Extended (SPC/E)[54] model. In the SPC/E model, the water molecule is a Lennard Jones (LJ) sphere with the oxygen atom in its center. Each hydrogen atom is separated by 1 Å from the oxygen and the HOH bond angle is 109.47°. The atomic charges (–0.8476*e* for oxygen and 0.4238*e* for hydrogen) are slightly larger than those given by the original SPC model [55] to include a mean field polarization effect. The potential between a pair of molecules w_1 and w_2 is the sum of the site-site(α and β) Coulombic and LJ terms (σ_{OO} = 3.166 Å and ε_{OO} = 0.65 kJ/mol; σ_{HH} and ε_{HH} are taken as zero):

$$u_{w_1w_2} = \sum_{\alpha=1}^{3}\sum_{\beta=1}^{3}\frac{q_1^{\alpha}q_2^{\beta}}{r_{12}^{\alpha\beta}} + 4\varepsilon_{oo}\left[\left(\frac{\sigma_{oo}}{r_{oo}}\right)^{12} - \left(\frac{\sigma_{oo}}{r_{oo}}\right)^{6}\right] \tag{15}$$

The SPC/E is a rigid model, i.e., all the bond lengths are fixed, atomic vibrations are not allowed. For this reason, a modification needs to be introduced into the formulation of the equations of motion, since the dynamics are constrained [56]. The procedure (called SHAKE) is based on the use of a set of undetermined Lagrange multipliers to represent the magnitudes of the forces in the direction of the bonds, which are required to keep the bond lengths constant. Long-range forces are handled by a special technique, the Ewald summation [57, 58]. It consists of a mathematical transformation of the long-range interactions into short-ranged screened interactions by the introduction of Gaussian distributions of charges surrounding each point charge in the basic cell. The total screened potential is then calculated by summing over all molecules in the central cell and all their images.

The simulations were started by placing the Ni_x cluster (with x = 3, 4, 7 and 22 respectively in each simulation) with the experimental geometry in the central box and 500 water molecules conveniently distributed around it. The box length was calculated in each case including the volume of the cluster in order to have the correct liquid water density at 298 K and atmospheric pressure. The time step used was 0.002 ps. The system was initialized at 0 K, and the temperature was gradually increased in intervals of 30 K, allowing 2 ps of stabilization at each temperature, until a temperature of 298 K was obtained. At this temperature, the system was run for 20 ps for equilibration plus 40 ps for the production stage. The water velocities were adjusted every 250 steps (0.5 ps) according to a Maxwell-Boltzmann distribution in order to keep the temperature at the desired setting.

6. ANALYSIS OF THE NICKEL-WATER INTERFACE

In this section we focus on the adsorption of molecular water on nickel clusters of increasing numbers of atoms by an examination of the structural, energetic and dynamical properties of the molecules surrounding the first shell of the metal atoms in the cluster.

6.1. Small clusters: Ni_3 and Ni_4.

The first cluster investigated is a set of Ni atoms arranged in an equilateral triangle, with a bond length of 2.49 Å. This is obviously the elementary representation of a surface containing two of the adsorption sites corresponding to the (111) crystal face of nickel. One is the hollow site located in the center of the triangle with no second layer nickel atom underneath. The other is the top atom site located at each of the vertices. We have calculated site-site pair correlation functions between each of the Ni atoms in the cluster and the oxygen and hydrogen in the water molecule, respectively.

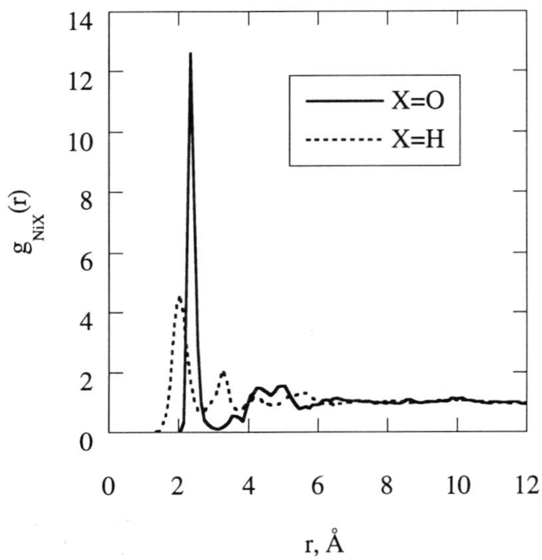

Figure 1. Spherically-averaged pair correlation functions for Ni_3 in water at 298 K.

From Figure 1 it can be seen that on the average, the water molecules in the first shell of this cluster orient with one of the hydrogens pointing to its nearest metal atom. This spherically-averaged first shell includes both the molecules located in the position corresponding to the hollow site, i.e., the center of the triangle, and those on the top site. The separation between the first and second peaks in $g_{NiO}(r)$ is similar to that in pure liquid water where tetrahedral ordering is assumed [59]. A certain degree of similarity was found by MD results for water adsorbed on a Pt (111) surface [18] yielding a density profile where the oxygen and hydrogen peaks in the first layer are coincident, and a second layer is H-bonded to the first.

A second bit of information is obtained from the distribution of binding energies. The number of water molecules interacting with one of the Ni atoms in the cluster is given in Figure 2 as a function of the pair energy interactions. Since all metal atoms in the cluster are located in equivalent positions, the interactions represented in Figure 2 correspond to one of them with all water molecules in the system. The peak centered near 0 kcal/mol is characteristic of the large number of water molecules that are located far away from the cluster. The peak with a negative energy (about -8 kcal/mol) represents the preferential adsorption between Ni and H_2O. Experimental measurements using electron energy loss spectroscopy (EELS) [60] for water adsorbed on various transition metals yield an average energy of -10/-11.5 kcal/mol. MO calculations [16] have also been reported for water adsorbed on clean Ni (111) for the several types of sites characterized in this study.

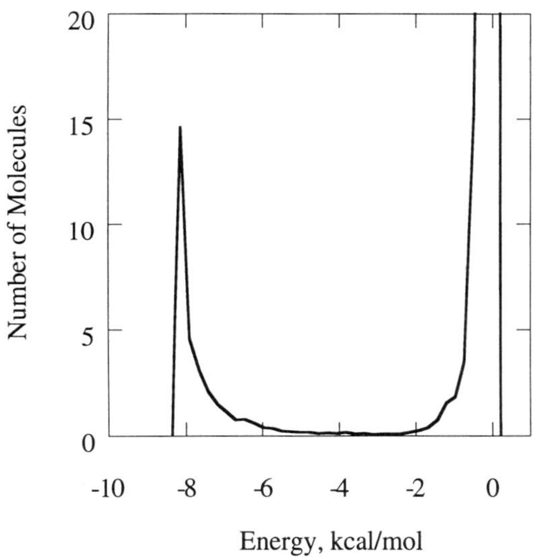

Figure 2. Distribution of pair interaction energies for water molecules with a Ni atom in the Ni_3 cluster.

The values of the adsorption energies from the peak on the left in Figure 2 agree well with the ones corresponding to a hollow three-fold site with no second layer nickel underneath [16], although to a lesser degree there are also contributions from the top site, with a slightly lower energy. A snapshot of the molecules in the first shell of the cluster, shown in Figure 3, gives further information about the orientations and preferred structures. In this and the following snapshots, the numbers on the oxygens and hydrogens correspond to a sequential numbering for all atoms belonging to water molecules found in a first shell of 2.6 Å computed from the center of mass of each atom in the metal cluster.

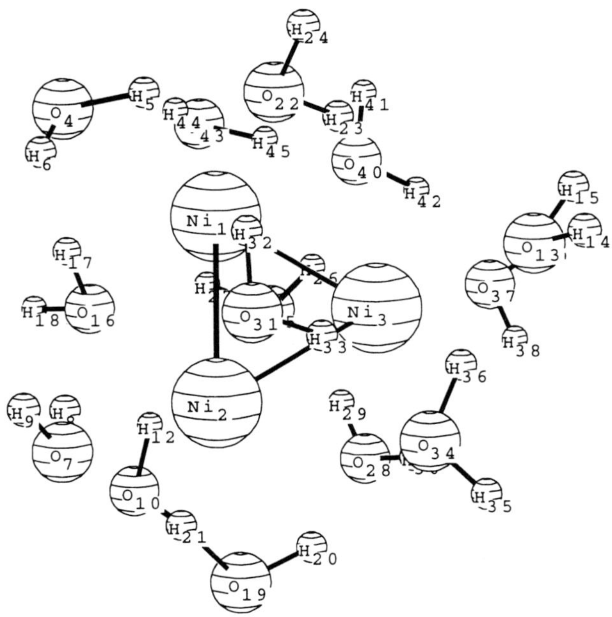

Figure 3. Snapshot of water molecules surrounding the Ni_3 cluster.

The numbering on the Ni atoms is also sequential and should not be confused with the notation used for the clusters. From Figure 3 is clear the presence of one water molecule at each side of the plane of the cluster (hollow sites), in the center of the triangular surface.

Next, we perform a similar analysis on a cluster constituted of four Ni atoms organized in a tetrahedral geometry. In this case, the possible adsorption sites are: a filled site with a second Ni underneath, represented by the center of any of the tetrahedral faces, and again the top atom site located at each of the vertices. With regard to binding energies, no substantial differences were observed between Figure 2 and its equivalent (not shown) for the Ni_4 cluster indicating similar water-Ni interaction patterns for the two clusters. For Ni_4, there is also a peak corresponding to adsorption energies that yields a maximum value of about -8 kcal/mol. According to MO calculations, [16] a filled adsorption site with a second layer atom underneath gives an adsorption energy of -5.7 kcal/mol. This suggests that in this case there are approximately equal contributions from adsorbed molecules on top sites (with energies of about -10/-11 kcal/mol) and those from filled sites (and there is only one of this last type in the Ni_4 cluster). The snapshot in Figure 4 reveals some other aspects of the interactions, such as preferred orientations in the top sites. H-bonds between water molecules are shown in the figure (example O_{26} with H_{45}). Adsorption on the filled sites is also visualized, in particular over the upper face.

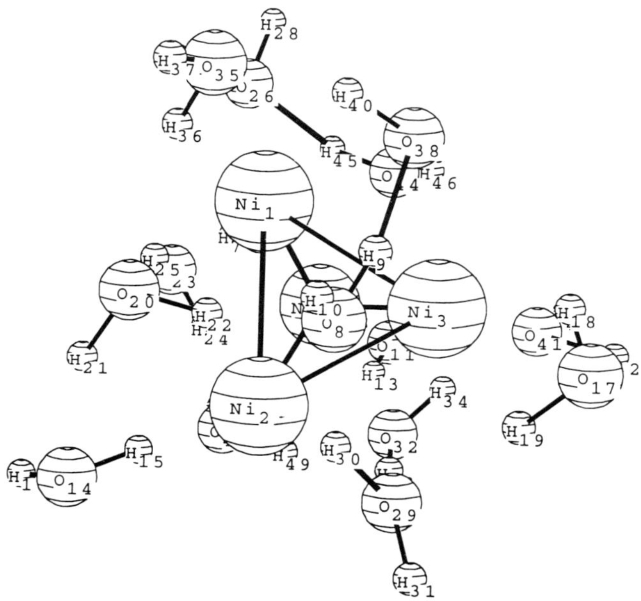

Figure 4. Snapshot of water molecules surrounding the Ni_4 cluster.

6.2 . Medium clusters: Ni_7 and Ni_{22}.

The next cluster we chose for simulations is the seven-atom planar geometry which contains the same adsorption sites as the Ni_3 cluster plus the bridge type which is located on each one of the inside bonds connecting two triangular structures. This cluster only misses the fourth adsorption type, a filled site with a second Ni atom underneath. It is interesting to analyze the structure shown in Figure 5. For this particular configuration, the water molecules with oxygen atoms labeled 23, 35 and 41 lay over the hollow sites over the plane of the figure, while those labeled 44, 57 and 47 are in the remaining hollow sites below the plane of the figure. Thus, they arrange alternatively on each of the hollow sites. The oxygen labeled 62 is an example of a water molecule located on a bridge site, while 56 and 14 are positioned on top sites. The distribution of binding energies looks different for the peripheral sites than for the central atom, as seen in Figure 6. The bridge site yields a binding energy of -7.1 kcal/mol [16], which is intermediate between the values corresponding to the hollow and the filled sites respectively, mentioned in our discussion above.

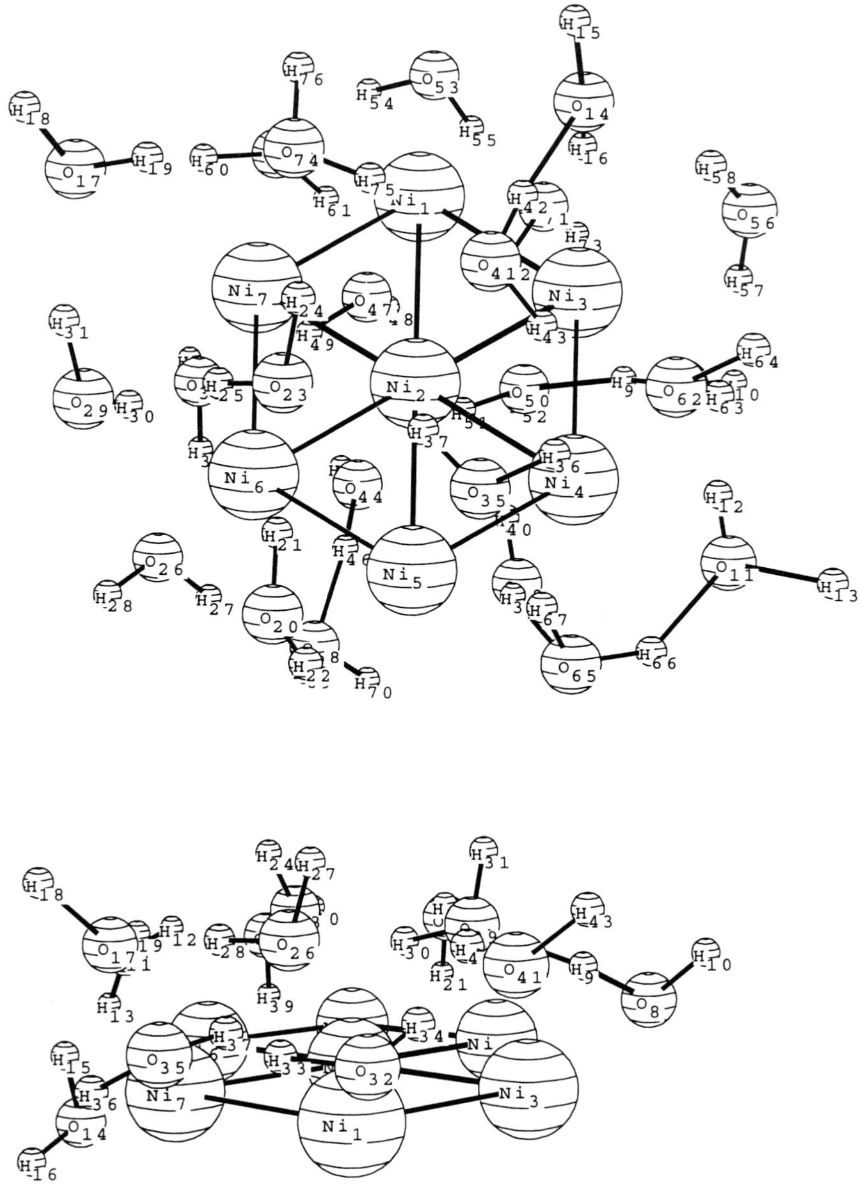

Figure 5. Snapshots of water molecules surrounding the Ni_7 cluster. Bottom: only the molecules laying on top of the surface plane are shown.

Note that the central atoms are more likely to emulate the atoms in a surface, while the peripheral atoms may describe defects or impurities. The broadening of the peak close to 0 kcal/mol indicates the increase in the number of water molecules bonded to the metal atom with less attractive energies, which is more evident for the central site. In addition, as seen in Figure 2, there is a peak of preferential adsorption for both type of sites.

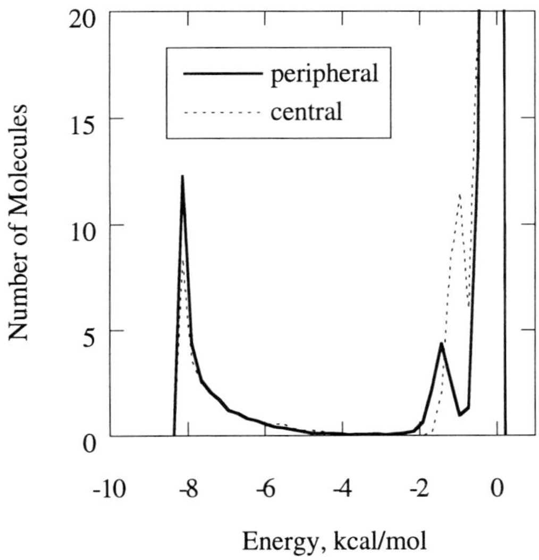

Figure 6. Distribution of pair interaction energies for water molecules with a Ni atom in the Ni_7 cluster.

We have also calculated the average orientation of the water molecules near each site on the cluster, defined as the angle θ (Figure 7) formed by the direction of the water dipole vector μ and the vector joining the centers of the atom and the water oxygen. The angular distribution for molecules located in a spherical region of 2.6 Å from the center of one of the metal atoms in the cluster (central or peripheral) is shown in Figure 8; the total nickel-water binding energy for a given atom in the cluster has its minimum value at a radial distance of 2.32 Å. For the site in the center of the cluster (Figure 8, top) there is a trimodal distribution indicating that the dipoles are either pointing to the surface (peak at 57°), parallel to it (peak at 105°) or pointing away from the surface (peak at 147°). Since the water model is rigid, these orientations can be expressed also in terms of β (Figure 7), the HONi angle, by subtracting half of the fixed HOH angle (109.47 °), yielding β = 3°, 51° and 93° respectively.

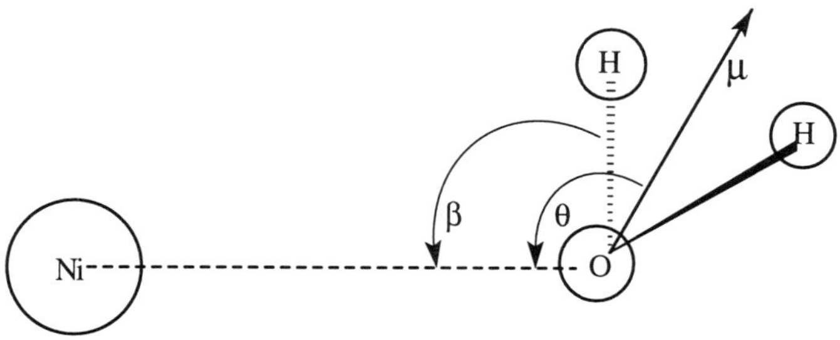

Figure 7. Orientations between a water molecule and an atom in the cluster.

MO calculations [16] have shown that the most stable geometry for water adsorbed on the top site of Ni (111) corresponds to a value of β =127.75°, yielding an energy of -10.6 kcal/mol, and the energy increases when β is reduced compared to the geometry of the minimum energy. This suggests that there is a correspondence of the three peaks in Figure 8 (top) with the ones in Figure 6. Since the force field that we have used takes into account metal-oxygen interactions only, these preferred orientations are a consequence of the water-water interactions in the adsorbed layer, possibly forming H-bonds, as was observed using high resolution electron-energy-loss spectroscopy for water adsorbed on platinum surfaces [61] and by MD simulations [18]. The molecules in the periphery of the cluster show a more random set of orientations (Figure 8, bottom), although a peak is observed at θ = 160° (β = 105°), in correspondence with the peak of attractive energies in Figure 6. It has been concluded [19] that for water near metallic surfaces there is a preferred orientation of the dipoles parallel to the interface (θ = 90°) because in this way water is able to maximize the number of H-bonds with other water molecules.

The distribution in Figure 8 (top) resembles the one found for the first layer of water molecules adsorbed on Pt (111) by MD simulations [18], defined within a radius corresponding to half of the one considered here. Figure 8 (bottom) has exact correspondence with the dipole orientations reported by Raghavan and coworkers for their second layer, which is equivalent to the second half of the radial distance used in this work. Thus, it is possible that the spherical region considered here contains features belonging to both first and second layers of previous MD reports. It is interesting to notice that even though the number of atoms in the cluster is still rather small, some of the characteristics of simulations on "surfaces" are obtained.

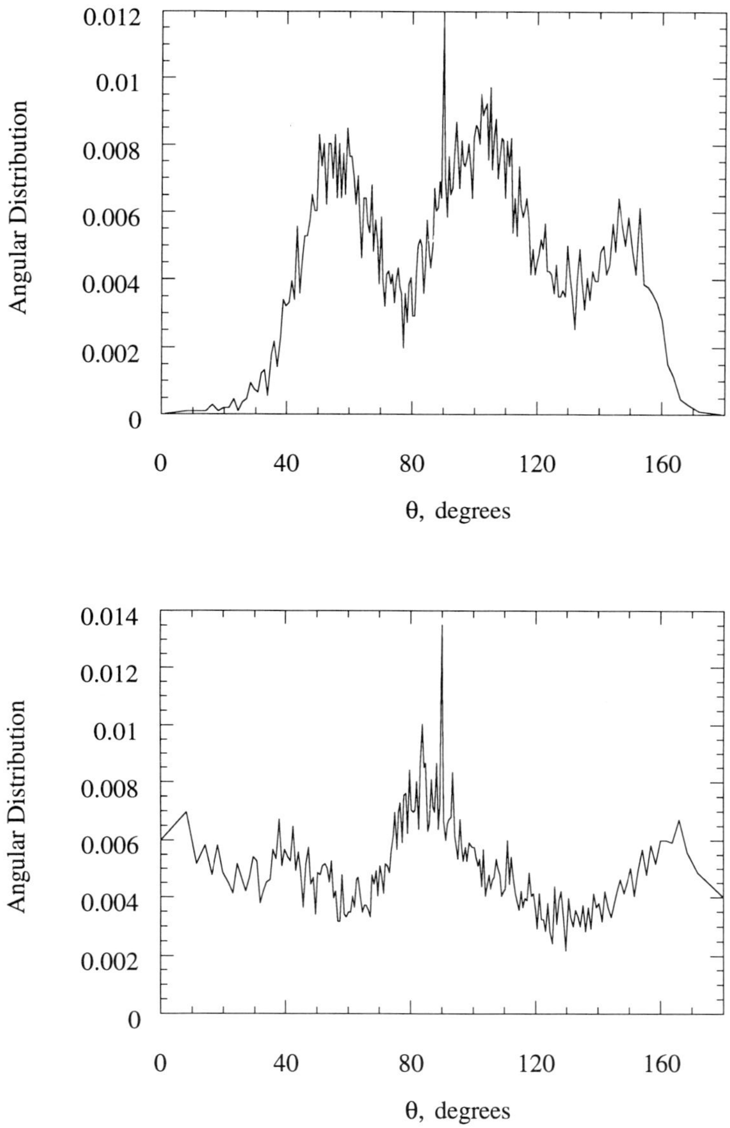

Figure 8. Distribution of orientations for the water molecules in the neighborhood of a central (top) and a peripheral (bottom) atom in the Ni_7 cluster.

Another important point concerns to the dynamics of the adsorbed water molecules. A time autocorrelation function (TACF) is defined as:

$$c_l(t) = \left\langle P_l(\underline{u}_i(t).\underline{u}_i(t_0)) \right\rangle_{t_0} \tag{16}$$

where $\underline{u}_i$ is a unit vector that determines the orientation of molecule i and P_l is the Legendre polynomial of degree l yielding an appropriate characterization of the reorientation dynamics [62]. The brackets indicate average over all water molecules within a given distance of the metal atom and over all time origins t_0. Several choices can be made for $\underline{u}_i$, such as the direction of the water dipole or the direction of the OH bond. It can be shown that these TACF's can be related to the spectral bandshapes measured from infrared absorption (l = 1) and Raman scattering (l = 2) spectra [62]. Information on the TACF's is obtained by Fourier transforming the experimental spectra, although the interpretation is complex. Figure 9 shows $c_1(t)$ given by Eq. 16 using the first Legendre polynomial (cosθ) for the correlation of the average dipole direction of water molecules located in a spherical region of radius 2.6 Å from the center of the Ni atom. Characteristic integral relaxation times are obtained from

$$\tau_1 = \int_0^\infty c_1(t)dt \tag{17}$$

where the upper integration limit is extended to a time where $c_1(t)$ vanishes.

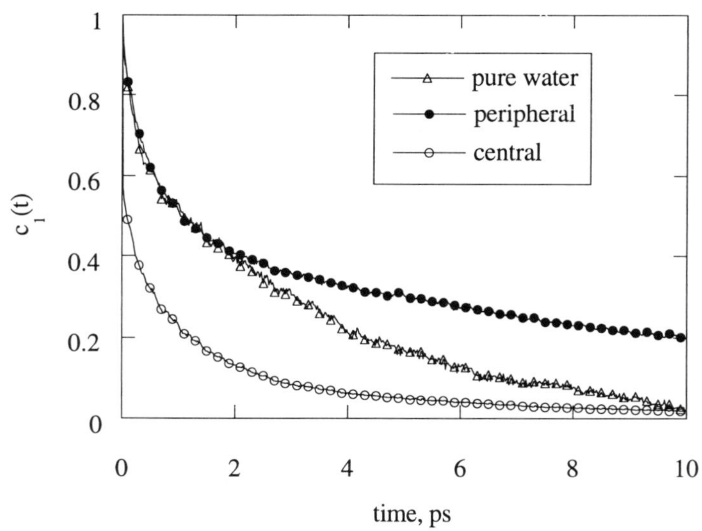

Figure 9. Reorientation times for the water molecules surrounding the Ni_7 cluster.

In the cases where we could not obtain values for such long times from the current data, we have extended the $c_1(t)$ using an exponential function that best describes the decay between 5 and 10 ps. Then the reorientation times are calculated from Eq. (17).

For comparison, Figure 9 displays also the pure water $c_1(t)$ obtained by MD simulations with the same SPC/E model. Reorientation times for water molecules surrounding the atoms in the periphery of the cluster are of the order of 10.3 ps, about seven times longer than those in the vicinity of the central atom (1.41 ps) and four times longer than the ones in the neighborhood of a bulk water molecule (2.3 ps). From the information of Figures 8 and 9 it can be concluded that the molecules interacting with the central atoms in the cluster experience a relatively fast rotation switching among three preferred orientations. The molecules surrounding the peripheral sites, on the other hand, are less perturbed. This difference might also be rationalized in terms of the pair binding energies (Figure 6). A larger percent of molecules are bounded to peripheral rather than to central metal atoms at the positions corresponding to the minimum in energy.

The last type of cluster we have examined using MD in this study is the Ni_{22}

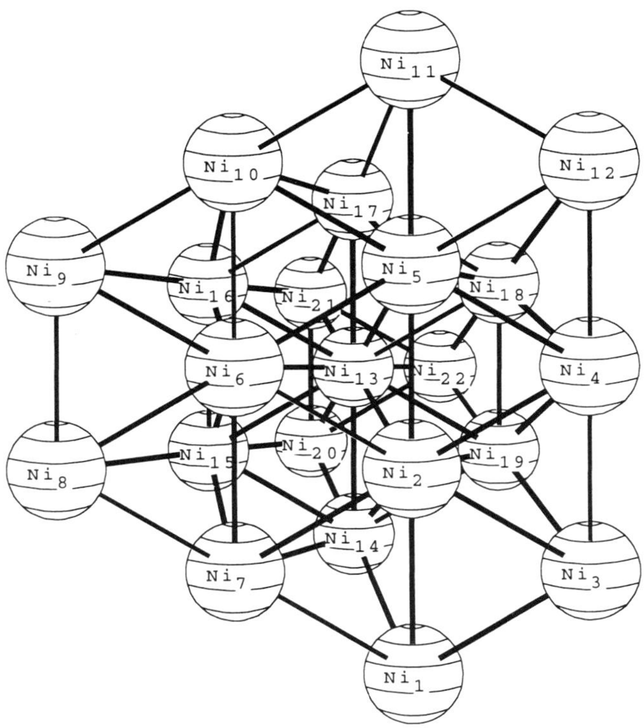

Figure 10. Top view of the Ni_{22} cluster showing the 12-atom surface layer.

with experimental geometry. The 22-atom cluster is constituted of three layers of 12, 7 and 3 atoms respectively. Figure 10 shows a top view of the 12-atom plane that contains all four adsorption types introduced in the above discussion.

A snapshot of the water molecules adsorbed on the surface is displayed in Figure 11, where all the other molecules were omitted for clarity. The distribution of binding energies is depicted in Figure 12 for the central and peripheral atoms respectively. The labeling of the curves corresponds to that in Figure 10. The second distribution is characterized by an adsorption peak centered at about -8 kcal/mol, corresponding to top and bridge adsorption sites. A smaller number of molecules adsorbs with less favorable energies (small peak centered near -2 kcal/mol) coming from the hollow and filled sites. A similar distribution is observed for the peripheral atoms; however the contributions from the top and bridge sites are dominant over the other two types. It has been observed that the nickel-water adsorption energies are in the same range as the water-water H-bond energies [60], i.e. on the order of -4.7 kcal/mol [63]. This fact would obscure the identification of adsorption energies by experimental measurements. However, as seen from Figure 12, we have not found any water molecules bonded to nickel atoms in this particular range.

The angular distribution for water molecules located in a spherical region of radius 2.6 Å centered on a Ni atom from the Ni_{22} cluster is depicted in Figure 13. The case of a central atom (Fig. 13, top) shows a distribution similar (although not identical) to the corresponding function for the Ni_7 cluster (Figure 8, top). In comparison to Figure 8, the first peak is shifted to the left and centered at about $\theta = 40°$ ($\beta = 14°$), while the other two maxima are at $\theta = 110°$ and $150°$ ($\beta = 56°$ and $96°$). As described for the case of Ni_7, these orientations correspond to dipoles either pointing toward the surface plane, parallel to it, or pointing away from the surface. Again in this case, it is interesting to compare these results with the ones reported from MD simulations on a Pt (111) surface modeled with an effective potential that takes into account surface symmetry and corrugation. Raghavan et al. [18] analyzed the orientations in a direction perpendicular to the surface plane at intervals of about 1 Å. Figure 13 (top) is in agreement with the MD results for the first and second sublayers of the first shell, while Figure 13 (bottom) shows a striking similarity with the second sublayer only [18], with a bimodal distribution corresponding to flip-up and flop-down molecules. With respect to the energetic values reported for these configurations [16], the first two peaks correspond to relatively high energy configurations (see Figure 12, top), while the last one conforms to the preferred energetic value. Note that the competition between adsorbate-adsorbate and adsorbate-adsorbent interactions influences the structures particularly for molecules adsorbed on sites located not at the periphery of the cluster. A different distribution is shown for molecules adsorbed on the peripheral atoms (Figure 13, bottom), which somehow resembles its equivalent for the Ni_7 cluster (Figure 8, bottom), although the distribution in Figure 13 shows a more defined bimodal pattern. As expected, when Figures 8 and 13 are compared to results from MD simulations using an effective potential [18], there is observed a tendency of the biggest cluster to improve the emulation of the surface.

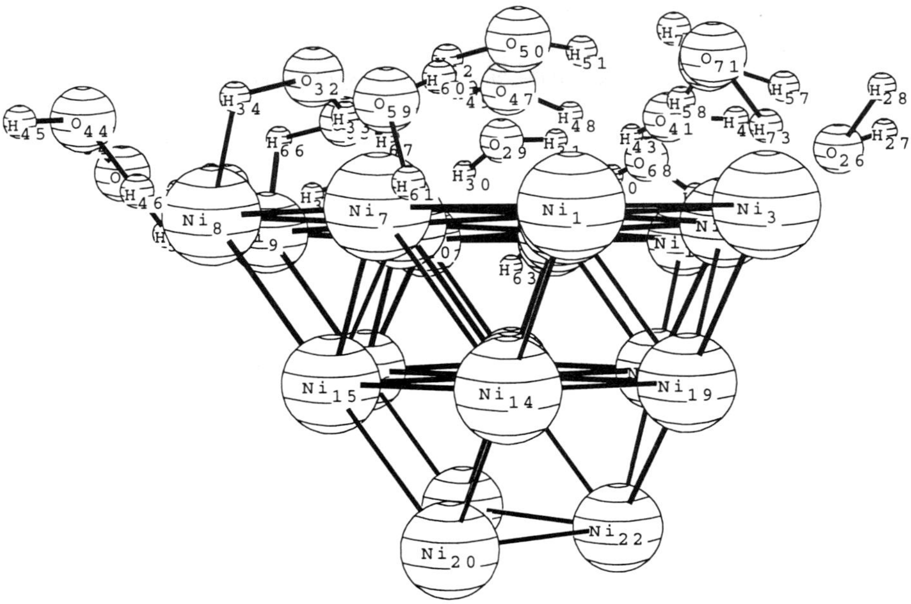

Figure 11. Water molecules adsorbed on the surface layer of Ni_{22}.

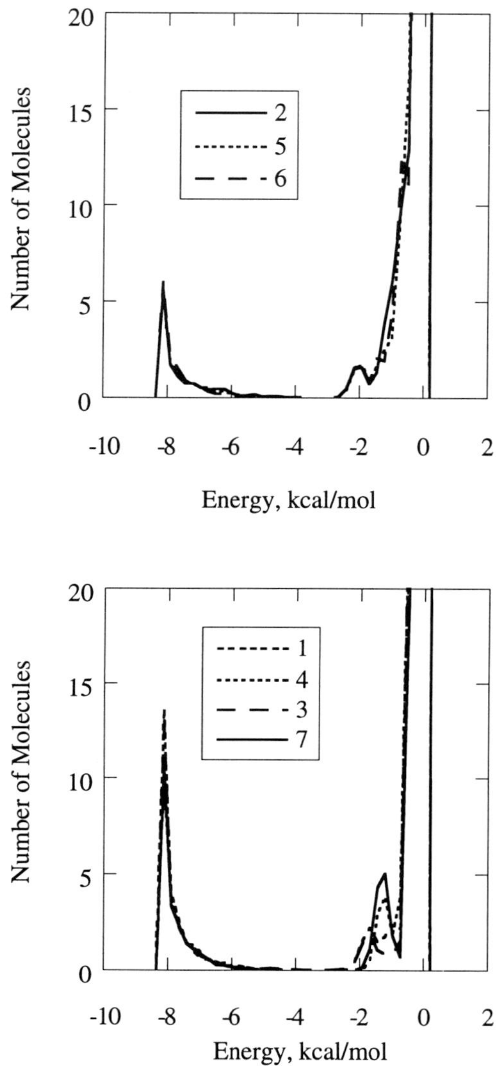

Figure 12. Distribution of binding energies water-nickel for the 12-atom layer in Ni_{22} (top: central atoms; bottom: atoms in the periphery)

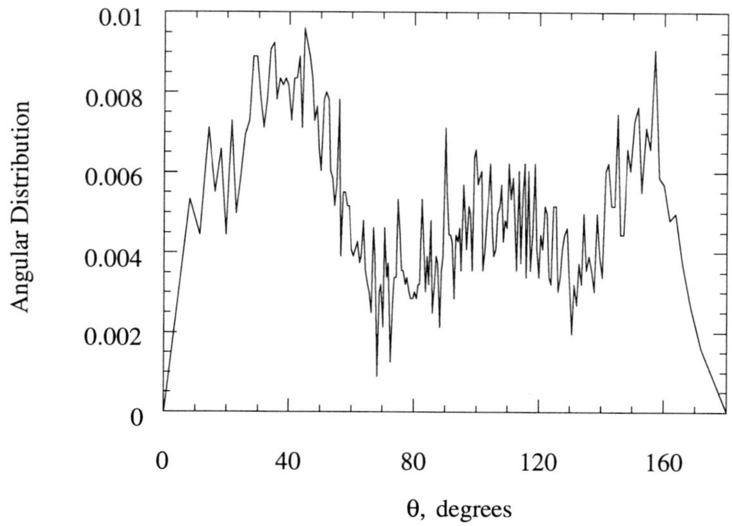

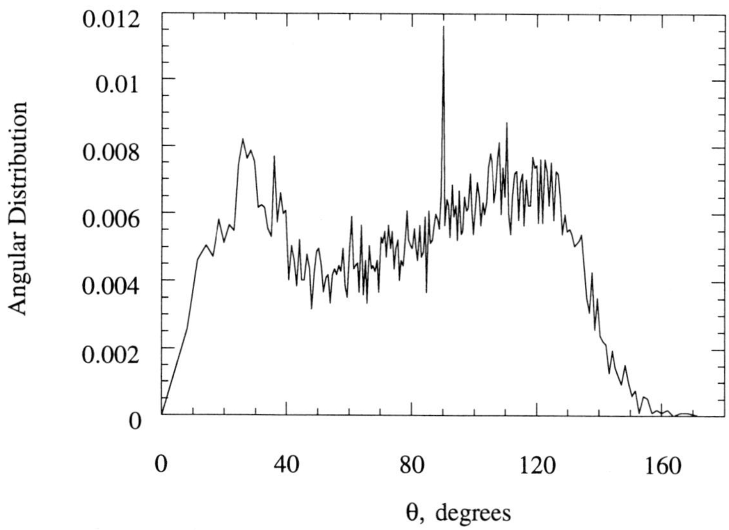

Figure 13. Distribution of orientations for the water molecules in the neighborhood of a central (top) and a peripheral (bottom) atom in the Ni_{22} cluster.

Reorientation times are depicted in Figure 14. As was observed from Figure 9, the water molecules adsorbed on the periphery of the cluster relax much more slowly than the ones on the central sites. But for the Ni_{22} cluster the differences among the corresponding reorientation times are larger. For peripheral molecules, τ_1 is 28 ps compared with 3 ps for the molecules adsorbed on the central atoms. The latter value is slightly longer than the reorientation time found in the first shell of molecules in pure water experimentally [64] and by our MD simulations, and similar to those found for the reorientation times of water molecules in the proximity of hydrophobic and hydrophilic surfaces based on MD simulations [65]. Note (Figure 14) that molecules adsorbed on the central atoms of the metal cluster decay rapidly in the first femtoseconds but are correlated for longer times than those in the neat liquid which have a slower initial decay.

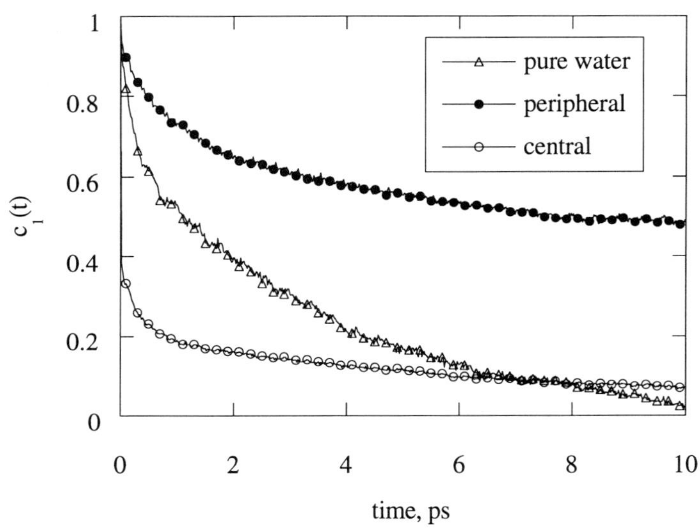

Figure 14. Reorientation times for the water molecules surrounding the Ni_{22} cluster.

The results from the 22-atom cluster determine that the molecules on the peripheral atoms do not change their orientations for intervals three times longer than in the seven-atom cluster. We expect this value to be closer to the surface value; however the values obtained for the central atoms are more likely to change, due to the presence of extra neighbors when more of this type of site become available as it occurs on a surface.

7. CONCLUSIONS

The present investigation of atomic nickel using the best available full-electron ab initio and DFT techniques showed that nonrelativistic calculations on atomic nickel can be performed with DFT methods obtaining results comparable to the experimental information available. This allowed us to compute the nickel-water interaction potential, which in turn was used for the molecular dynamics simulation of the nickel-water interface.

Nickel-water binding energies corresponding to well defined adsorption sites were identified based on pair energy distributions obtained from histograms of MD equilibrium trajectories. For Ni_3 and Ni_4, a single adsorption peak was obtained at about -8 kcal/mol, while Ni_7 and Ni_{22} yielded a bimodal distribution (with maxima at -8 and -2 kcal/mol) with different weights according to the position of the Ni atom in the cluster.

For the largest clusters investigated in this work, Ni_7 and Ni_{22}, the interaction of water molecules at the metal-water interface was further described by characterization of the average orientations of the water dipoles in the neighborhood of the cluster. For atoms in the periphery (edges and/or vertices of the clusters), the average direction of the water dipoles was found to be perpendicular to the direction of the vector joining the centers of the water molecule and the Ni atom. On the other hand, water molecules in the vicinity of metal atoms located in central positions (i.e. surrounded by other Ni atoms) were found to be much more labile, yielding a trimodal distribution of average orientations.

Characteristic times associated with the reorientation of the dipoles were also calculated. These times might give a first indication of the dynamics of the adsorption/desorption process. Moreover, assuming that the desorption process is initiated by rotation followed by translation of the water molecule, the rate of desorption would be inversely proportional to the reorientation times. In both clusters (Ni_7 and Ni_{22}), water molecules surrounding peripheral atoms were found to be much more stable (7 to 9 longer reorientation times) than those near central atoms. Increasing the number of atoms in the cluster from 7 to 22 resulted in a 2 to 3-fold increase in the respective reorientation times.

In summary, the present study gives a fair characterization of equilibrium adsorption properties of molecular (undissociated) water on Ni clusters. Regarding dynamic properties, further work using larger clusters and similar studies on simulated surfaces (instead of clusters) would be required if one wishes to determine asymptotic values where the cluster best approximates the surface. Comparisons to experimental values (rates of desorption, activation energies) would also be very valuable, particularly in considering situations where clusters may be a good representation of the actual process, for example in catalysts where the active metal is incrusted on an inert surface.

REFERENCES

1 K. Balasubramanian, J. Chem. Phys. **89**, 5731 (1988).
2 K. Balasubramanian, J. Chem. Phys. **91**, 307 (1989).
3 A. Goursot, I. Papai and D. R. Salahub, J. Am. Chem. Soc. **114**, 7452 (1992).

4 K. Balasubramanian, P. Y. Feng and M. Z. Liao, J. Chem. Phys. **87**, 3981 (1987).
5 S. Sirois, M. Castro and D. R. Salahub, Int. J. Quantum Chem. (Symposium) **28**, 645 (1994).
6 J. M. Seminario, M. C. Concha and P. Politzer, Int. J. Quantum Chem. **S27**, 263 (1993).
7 T. Nakao, D. A. Dixon and H. Chem., J. Phys. Chem. **97**, 12665 (1993).
8 E. Spohr and K. Heinzinger, J. Chem. Phys. **84**, 2304 (1986).
9 E. Spohr and K. Heinzinger, Chem. Phys. Lett. **123**, 218 (1986).
10 E. Spohr and K. Heinzinger, Ber. Bunsenges. Phys. Chem. **92**, 1358 (1988).
11 T. E. Madey and F. P. Netzer, Surf. Sci. **117**, 549 (1982).
12 C. Nobl and C. Benndorf, Surf. Sci. **182**, 499 (1987).
13 F. T. Wagner and T. E. Moylan, in *Structural effects in electrocatalysis and oxygen electrochemistry*, 92-11,D. Scherson, D. Tryk, M. Daroux and X. Xing Eds, (The Electrochemical Society, Inc., Case Western Reserve University, 1991),
14 G. U. Kulkarni, C. N. R. Rao and M. W. Roberts, Langmuir **11**, 2572 (1995).
15 S. Holloway and K. H. Bennemann, Surf. Sci. **101**, 327 (1980).
16 H. Yang and J. L. Whitten, Surf. Sci. **223**, 131 (1989).
17 E. Spohr and K. Heinzinger, Electrochim. Acta **33**, 1211 (1988).
18 K. Raghavan, K. Foster, K. Motakabbir and M. Berkowitz, J. Chem. Phys. **94**, 2110 (1991).
19 I. Benjamin, Chem. Rev. **96**, 1449 (1996).
20 K. Heinzinger, Fluid Phase Equil. **104**, 277 (1995).
21 J. N. Glosli and M. R. Philpott, J. Chem. Phys. **96**, 6962 (1992).
22 L. Zhang, H. T. Davis, D. M. Kroll and H. S. White, J. Phys. Chem. **99**, 2878 (1995).
23 L. A. Curtiss, C. Jones, G. W. Trucks, K. Raghavachari and J. A. Pople, J. Chem. Phys. **93**, 2537 (1990).
24 L. A. Curtiss, K. Raghavachari, G. W. Trucks and J. A. Pople, J. Chem. Phys. **94**, 7221 (1991).
25 J. A. Pople, M. Head-Gordon, D. J. Fox, K. Raghavachari and L. A. Curtiss, J. Chem. Phys. **90**, 5622 (1989).
26 J. A. Pople, M. Head-Gordon, D. J. Fox, K. Raghavachari and L. A. Curtiss, J. Chem. Phys. **90**, 5622 (1990).
27 R. M. Dreizler and E. K. U. Gross, *Density Functional Theory*, (Springer-Verlag, Berlin, 1990).
28 E. S. Kryachko and E. Ludeña, *Energy Density Functional Theory of Many-electron Systems*, (Academic, New York, 1990).
29 R. G. Parr and W. Yang, *Density Functional Theory of Atoms and Molecules*, (Oxford University Press, Oxford, 1989).
30 J. M. Seminario and P. Politzer, Eds., *Density Functional Theory: A Tool for Chemistry*, (Elsevier, Amsterdam, 1995).
31 J. Harris and R. O. Jones, J. Phys. F **4**, 1170 (1974).
32 D. C. Langreth and J. P. Perdew, Solid State Commun. **17**, 425 (1975).
33 O. Gunnarsson and B. I. Lundqvist, Phys. Rev. B **13**, 4274 (1976).
34 Gaussian 94, Revision B.3, M. J. Frisch, G. W. Trucks, H. B. Schlegel, P. M. W. Gill, B. G. Johnson, M. A. Robb, J. R. Cheeseman, T. Keith, G. A. Petersson, J. A. Montgomery, K. Raghavachari, M. A. Al-Laham, V. G.

Zakrzewski, J. V. Ortiz, J. B. Foresman, C. Y. Peng, P. Y. Ayala, W. Chen, M. W. Wong, J. L. Andres, E. S. Replogle, R. Gomperts, R. L. Martin, D. J. Fox, J. S. Binkley, D. J. Defrees, J. Baker, J. P. Stewart, M. Head-Gordon, C. Gonzalez and J. A. Pople, Gaussian, Inc., Pittsburgh PA., 1995.
35 A. J. H. Wachters, J. Chem. Phys. **52**, 1033 (1970).
36 P. J. Hay, J. Chem. Phys. **66**, 4377 (1977).
37 K. Raghavachari and G. W. Trucks, J. Chem. Phys. **91**, 1062 (1989).
38 S. Huzinaga, J. Chem. Phys. **66**, 4245 (1977).
39 A. D. Becke, J. Chem. Phys. **98**, 5648 (1993).
40 J. P. Perdew, in *Electronic Structure of Solids*, P. Ziesche and H. Eschrig Eds, (Akademie Verlag, Berlin, 1991),
41 J. P. Perdew, J. A. Chevary, S. H. Vosko, K. A. Jackson, M. R. Pederson, D. J. Singh and C. Fiolhais, Phys. Rev. B **46**, 6671 (1992).
42 J. P. Perdew and Y. Wang, Phys. Rev. B **45**, 13244 (1992).
43 C. Lee, W. Yang and R. G. Parr, Phys. Rev. B **37**, 785 (1988).
44 C. Møller and M. S. Plesset, Phys. Rev. **46**, 618 (1934).
45 J. Cizek, Adv. Chem. Phys. **14**, 35 (1969).
46 G. D. Purvis and R. J. Bartlett, J. Chem. Phys. **76**, 7382 (1982).
47 G. E. Scuseria, C. L. Janssen and I. H. F. Schaefer, J. Chem. Phys. **89**, 7382 (1988).
48 G. E. Scuseria and I. H. F. Schaefer, J. Chem. Phys. **90**, 3700 (1989).
49 J. A. Pople, M. Head-Gordon and K. Raghavachari, J. Chem. Phys. **87**, 5968 (1987).
50 M. P. Allen and D. J. Tildesley, *Computer Simulation of Liquids*, (Clarendon Press, Oxford, 1990).
51 L. Verlet, Phys. Rev. **159**, 98 (1967).
52 D. A. McQuarrie, *Statistical Mechanics*, (Harper Collins, New York, 1976).
53 H. C. Andersen, J. Chem. Phys. **72**, 2384 (1980).
54 H. J. C. Berendsen, J. R. Grigera and T. P. Straatsma, J. Phys. Chem. **91**, 6269 (1987).
55 H. J. C. Berendsen, J. P. M. Postma, W. F. van Gunsteren and J. Hermans, *Intermolecular Forces*, (Reidel, Dordrecht, 1981).
56 J. P. Ryckaert, G. Cicotti and H. J. C. Berendsen, J. Comput. Phys. **23**, 327 (1977).
57 S. W. de Leeuw, J. W. Perram and E. R. Smith, Proc. R. Soc. Lond. **A 373**, 27 (1980).
58 M. Belhadj, H. E. Alper and R. M. Levy, Chem. Phys. Lett. **179**, 13 (1991).
59 Y. Guissani and B. Guillot, J. Chem. Phys. **98**, 8221 (1993).
60 P. A. Thiel and T. E. Madey, Surface Sci. Rept. **7**, 211 (1987).
61 H. Ibach and S. Lehwald, Surf. Sci. **91**, 187 (1980).
62 J.-P. Hansen and I. R. McDonald, *Theory of simple liquids*, (Academic Press, San Diego, CA, 1990).
63 F. H. Stillinger and A. Rahman, J. Chem. Phys. **68**, 666 (1972).
64 J. Jonas, T. DeFries and D. J. Wilber, J. Chem. Phys. **65**, 582 (1976).
65 S. H. Lee and P. J. Rossky, J. Chem. Phys. **100**, 3334 (1994).

J.M. Seminario (Editor)
Recent Developments and Applications of Modern Density Functional Theory
Theoretical and Computational Chemistry, Vol. 4

Systematic Model Chemistries Based on Density Functional Theory: Comparison with traditional Models and with Experiment

Michael J. Frisch, Gary W. Trucks and James R. Cheeseman

Lorentzian, Inc., 140 Washington Avenue, North Haven, CT 06472, USA

Model chemistries based on DFT methods are compared to one another and to those based on traditional ab initio methods such as Hartree-Fock and MP2 for a variety of job types including energies, structures, vibrational frequencies and related properties and NMR chemical shifts. In general, model chemistries based on hybrid functionals perform best of all DFT models, many times achieving equal or better accuracy to the more computationally expensive MP2 theory.

1. INTRODUCTION

Kohn-Sham density functional theory (DFT) [1, 2] has been used increasingly frequently in recent years to model molecular systems and chemical reactions. Very good results have been reported by many different researchers on a wide variety of compounds and problems [3-25]. In such cases, various DFT methods have proven to provide an effective tradeoff between computational cost and accuracy. However, as with any computational method, model chemistries based on DFT methods require systematic calibration in order to determine their range of applicability and expected accuracy for previously-unstudied molecular systems. This chapter will examine such results.

Current implementations of DFT methods replace the exchange-correlation energy by a one-electron integral of the electron spin densities and possibly their gradients; functionals involving only the density itself are termed "local," while those involving both the density and its gradient are termed "gradient-corrected." Many different functionals have been proposed, and separate functionals have been devised for the exchange and correlation portions of this term. These pure functionals are the ones which will be considered:

- SVWN5, comprised of the standard local exchange functional [1, 2, 26] and the local correlation functional of Vosko, Wilk and Nusair (VWN5, denoted functional V in the paper) [27].

- SVWN3, comprised of the standard local exchange functional [1, 2, 26] and an alternate local correlation functional of Vosko, Wilk and Nusair (VWN3, denoted functional III in the paper) [27].
- BLYP, which combines Becke's 1988 gradient-corrected exchange functional (B88) [28] with the gradient-corrected correlation functional of Lee, Yang and Parr (LYP) [29, 30].

In addition to these "pure" functionals, Becke has introduced functionals comprised of a mixture of Hartree-Fock and DFT exchange along with a gradient-corrected DFT correlation functional. Becke has proposed several of hybrid functionals [31-33], one of which has the following form [31]:

$$E_{XC} = E_{XC}^{LSDA} + a_0(E_X^{exact} - E_X^{LSDA}) + a_X \Delta E_X^{B88} + a_C \Delta E_C^{PW91} \quad (1)$$

This functional, denoted B3PW91, is a mixture of exact (Hartree-Fock) and DFT exchange (the latter is Becke's gradient-corrected exchange functional [28]) and the 1991 gradient-corrected correlation functional of Perdew and Wang (PW91) [34]. The three constants were determined by fitting to 116 calculations drawn from the G1 molecule set [35] and its components (atomization energies, ionization potentials, proton affinities and first-row atomic energies). Becke arrived at values of a_0=0.20, a_X=0.72, and a_C=0.81 for the three parameters.

Two hybrid functionals are considered in this work:

- B3PW91, as defined in Equation 1.
- B3LYP, a similar hybrid functional constructed using the LYP correlation functional as follows:

$$E_{XC}^{B3LYP} = E_X^{LSDA} + a_0(E_X^{HF} - E_X^{LSDA}) + a_X \Delta E_X^{B88} + E_C^{VWN3} + a_C(E_C^{LYP} - E_C^{VWN3}) \quad (2)$$

The parameters are again those suggested by Becke.

Practical model chemistries based on DFT methods will combine a pure or hybrid functional with a specific basis set, creating a computational procedure which may be generally applied to chemical problems and whose results may be compared for related compounds. However, most systematic studies of density functional methods were completed prior to the widespread use of hybrid functionals. Our goal here is to consider calibration studies which include the whole range of DFT functionals using a variety of basis sets in an effort to determine which are most applicable to the chemical problems of greatest current interest.

2. GEOMETRIES AND ENERGIES

The choice of a model for the optimized molecular structure is a natural starting point for any chemical investigation undertaken with electronic structure theory. Thus, this section begins by considering the performance of several model chemistries involving DFT methods for geometry optimizations. It then goes on to discuss models for predicting high-accuracy energies and thermochemical properties of molecular systems.

2.1 Optimized Geometries

Large scale, systematic studies of optimized geometries using the full range of DFT functionals are still in preparation by several researchers as of this writing. Nevertheless, the performance of a range of functionals on several compounds known to cause difficulty for traditional ab initio methods can be considered.

Murray and coworkers reported results from calculations for this set of several small molecules known to be troublesome at the Hartree-Fock level for a variety of DFT methods, using the 6-31G(d), DZ2P and TZ2P basis sets [14]. Their optimized geometries agree very well with the corresponding calculations in Table 1 (the HF, MP2, SVWN5 and BLYP optimizations were run in order to use identical basis sets for all calculations). In addition, the table provides results using the B3LYP and B3PW91 hybrid functionals.

In general, using the larger, triple-zeta basis set decreases the predicted bond lengths (the sole exception is the bond length in F_2 for the BLYP functional). For these systems, Hartree-Fock bond lengths are too short, sometimes quite significantly so. MP2 theory improves on the Hartree-Fock values, often producing bond lengths in excellent agreement with experiment for the larger basis set. Like MP2, the models employing pure DFT functionals tend to predict bond lengths that are longer than the observed values. The models using hybrid functionals produce errors whose sign is less predictable but generally are smaller in magnitude than those for the pure functionals; their values are comparable to or more accurate than the MP2 predictions for all of the bond lengths not involving fluorine.

In virtually all cases, the models including the hybrid functionals perform significantly better than the models using SVWN5 and BLYP (the one exception is hydrogen molecule where BLYP with the larger basis set performs about as well as the hybrid functionals). For these molecules, the accuracies of B3LYP and B3PW91 are comparable.

Both Hartree-Fock theory and MP2 provide very good agreement with experiment for the bond angle in formaldehyde. The DFT models using the smaller basis set all significantly underestimate the bond angle in this structure. The bond angle computed by the DFT-based models with the larger basis set tends to be slightly too small (and this is especially true for the hybrid functionals).

Table 1
Optimized geometries for several small molecules using traditional ab initio and DFT-based model chemistries (bond lengths in Å; angles in degrees).

	H_2: R_{HH}		HF: R_{HF}	
	6-31G(d,p)	6-311+G(2d,2p)	6-31G(d,p)	6-311+G(2d,2p)
HF	0.732	0.734	0.901	0.897
MP2	0.734	0.737	0.921	0.918
SVWN5	0.765	0.766	0.933	0.932
BLYP	0.747	0.747	0.937	0.933
B3LYP	0.743	0.743	0.925	0.923
B3PW91	0.743	0.744	0.922	0.920
Experiment[a]	0.741		0.917	
	N_2: R_{NN}		F_2: R_{FF}	
	6-31G(d,p)	6-311+G(2d,2p)	6-31G(d,p)	6-311+G(2d,2p)
HF	1.078	1.067	1.345	1.330
MP2	1.131	1.114	1.421	1.410
SVWN5	1.111	1.096	1.389	1.388
BLYP	1.118	1.104	1.434	1.437
B3LYP	1.106	1.092	1.403	1.400
B3PW91	1.105	1.091	1.393	1.388
Experiment[a]	1.098		1.412	
	CO: R_{CO}		Formaldehyde: R_{CO}	
	6-31G(d,p)	6-311+G(2d,2p)	6-31G(d,p)	6-311+G(2d,2p)
HF	1.114	1.104	1.184	1.179
MP2	1.151	1.138	1.220	1.213
SVWN5	1.142	1.128	1.207	1.199
BLYP	1.150	1.138	1.218	1.213
B3LYP	1.138	1.126	1.207	1.201
B3PW91	1.137	1.125	1.205	1.199
Experiment[a]	1.128		1.203	
	Formaldehyde: R_{CH}		Formaldehyde: A_{CHC}	
	6-31G(d,p)	6-311+G(2d,2p)	6-31G(d,p)	6-311+G(2d,2p)
HF	1.093	1.091	115.7	116.4
MP2	1.100	1.098	115.5	116.7
SVWN5	1.126	1.121	115.1	116.3
BLYP	1.120	1.114	114.9	116.1
B3LYP	1.110	1.105	115.2	116.1
B3PW91	1.111	1.107	115.3	116.1
Experiment[a]	1.099		116.5	

[a] Reference [14].

Table 2 gives the absolute average deviation from experiment for the bond lengths in these compounds for the models under consideration as well as the standard deviation of the absolute errors. MP2 with the larger, triple-zeta basis set and the models including the hybrid functionals have the smallest overall error, one third to a quarter that of Hartree-Fock. The models using the pure functionals have overall errors comparable to Hartree-Fock with the smaller basis sets and less than half that of Hartree-Fock for the larger basis set. However, when the standard deviations are examined, which are generally comparable in size to the average errors themselves, it is evident that these differences are not statistically significant.

FOOF is another molecule which has historically proven to be a challenge for successive electronic structure methods. This molecule has an unusually long F-O bond length, indicating a fairly weak interactions. FOOF is a case where MP2 theory performs rather poorly, and CCSD(T) or QCISD(T) with a large basis set is necessary for good agreement with experiment.

Amos and coworkers [3] have published geometries for FOOF optimized with the LDA and BLYP functionals using the TZ2P and TZ2Pf basis sets. These optimizations were run with the 6-311+G(2d) and 6-311+G(2df) basis sets and obtained results in good agreement with Amos and coworkers.

Table 2
Average absolute bond length differences with experiment (Å) and their standard deviations for several small molecules using traditional ab initio and DFT-based model chemistries.

	6-31G(d,p)		6-311+G(2d,2p)	
	$\lvert \Delta R^{exp} \rvert$	Std. Dev.	$\lvert \Delta R^{exp} \rvert$	Std. Dev.
HF	0.022	0.021	0.028	0.025
MP2	0.013	0.011	0.006	0.006
SVWN5	0.017	0.008	0.013	0.011
BLYP	0.018	0.006	0.013	0.007
B3LYP	0.007	0.003	0.005	0.004
B3PW91	0.008	0.006	0.007	0.008

Geometry optimizations were also performed with the B3LYP and B3PW91 hybrid functionals. Table 3 compares our results to Scuseria's CCSD(T) results [36] for the optimization of FOOF; Hartree-Fock and MP2 results with a very large basis set are also given for reference.

Table 3
Optimized structure of FOOF using various model chemistries (bond lengths in Å; bond angles in degrees).

	6-311+G(2d)[a]				6-311+G(2df)[b]			
	R_{OO}	R_{OF}	A_{FOO}	D_{FOOF}	R_{OO}	R_{OF}	A_{FOO}	D_{FOOF}
SVWN5	1.189	1.568	110.7	88.6	1.187	1.559	110.7	88.4
BLYP	1.208	1.643	111.6	89.7	1.205	1.634	111.5	89.6
B3LYP	1.227	1.530	109.4	88.3	1.226	1.522	109.3	88.1
B3PW91	1.221	1.515	109.4	88.3	1.221	1.506	109.4	88.1
CCSD(T)[c]	1.216	1.614	108.9	87.8	1.218	1.589	108.9[d]	87.8[d]
Experiment[c]	1.217	1.575	109.5	87.5	1.217	1.575	109.5	87.5
HF/6-311+G(3df)					1.296	1.352	106.3	85.1
MP2/6-311+G(3df)					1.143	1.672	111.4	89.3

[a] TZ2P for the CCSD(T) calculation.
[b] TZ2Pf for the CCSD(T) calculation.
[c] Reference [36].
[d] These parameters were held fixed during the geometry optimization [36].

All the DFT-based models perform somewhat less accurately than CCSD(T), especially with respect to the O-F bond length, but their results are nevertheless substantially superior to Hartree-Fock and MP2 for this difficult case. Note that the addition of f functions to the basis set does not generally improve the predicted structure of this molecule for these model chemistries. The larger basis set results in a slight lengthening of the O-O bond and a shortening of the O-F bond, so it is of substantial help only when the latter bond length is overestimated at the TZ2P level.

Some additional geometry optimization results are discussed as part of our consideration of accurate energies predictions using DFT-based model chemistries, the subject of the next section.

2.2 Thermochemical Results

This section considers the performance of density functional theory for predicting thermochemical quantities: atomization energies, electron affinities, ionization potentials, and proton affinities. A standard series of reference calculations for which very accurate experimental values are known is used for this purpose [35, 37, 38]. This molecule set is commonly referred to as the G2 set, and it consists of 55 atomization energies, 38 ionization potentials, 25 electron affinities, and 7 proton affinities; it and its component subsets are frequently used to test new methods.

This test set of molecules has certain inherent limitations. For example, all of the component molecules are very small systems containing only first and second row atoms. Only a limited number of bond types are represented (there are no ring systems). Nevertheless, it is useful for comparing the relative accuracies of different model chemistries. The computations in the molecule set are difficult to model very accurately, and thus a model's performance on the set is also very likely to be a lower bound on its true accuracy.

Foresman and coworkers have performed all of the calculations in the molecule set for over 30 different model chemistries [39]. Table 4 presents some of their overall results for various model chemistries. (Seminario performed a related study comparing LDA atomization energies to those computed by G2 theory [18].)

Table 4
Overall results on the 125 calculations in the G2 molecule set for various model chemistries (all values are in kcal/mol).[a]

	Mean Absolute Deviation	Standard Deviation	Maximum Errors	
			Positive	Negative
B3LYP/6-311+G(2d,p) // B3LYP/6-311+G(2d,p)	3.1	3.0	13.6	-19.7
BLYP/6-311+G(2d,p) // BLYP/6-311+G(2d,p)	3.9	3.2	14.3	-15.9
MP2/6-311+G(2d,p) // MP2/6-311+G(2d,p)	8.9	7.8	18.3	-39.2
SVWN5/6-311+G(2d,p) // SVWN5/6-311+G(2d,p)	18.1	19.8	81.0	-10.1
SVWN3/6-311+G(2d,p) // SVWN3/6-311+G(2d,p)	24.9	19.2	89.3	-10.4
BLYP/6-31+G(d,p) // BLYP/6-31+G(d,p)	3.9	3.2	15.2	-15.2
B3LYP/6-31+G(d,p) // B3LYP/6-31+G(d,p)	3.9	4.2	17.6	-33.8
B3LYP/6-31G(d) // B3LYP/6-31G(d)	7.9	9.5	12.2	-54.2
MP2/6-31+G(d,p) // MP2/6-31+G(d,p)	11.4	8.1	15.6	-44.0
HF/6-31+G(d,p) // HF/6-31+G(d,p)	46.7	40.6	10.1	-179.8
HF/6-31G(d) // HF/6-31G(d)	51.0	41.2	11.5	-184.2

[a] Reference [39].

In each case, the geometry of each system was first optimized at the given level, and vibrational frequencies were computed in order to predict the zero-point energy (ZPE), which was scaled using the scaling factors of Wong [25] and added to the predicted total energy. Stability calculations were performed for each system for every model chemistry to ensure that the lowest energy wavefunction of the proper electronic state had been located.

The columns in the table give the mean absolute deviation (MAD) from experiment (the average difference between the calculated and experimental values ignoring sign), the standard deviation of the absolute errors, and the largest positive and negative errors for any calculation in the set. The MAD gives a measure of the average magnitude of error with respect to experiment, the standard deviation gives a sense of how widely the model's performance varied across the various calculations, and the maximum errors indicate each model's worst-case accuracy for this molecule set.

The five models in the top portion of the table all use the large 6-311+G(2d,p) basis set, and accordingly may be used to compare the overall results for these methods somewhat independently of basis set effects. B3LYP performs best of all of them, with a MAD of only 3.1 kcal/mol. It's maximum error is around 20 kcal/mol, however, indicating that it still requires calibration for new classes of systems. The performance of the BLYP functional is worse than that of the hybrid functional with a MAD of 5.0 kcal/mol, and its largest error also greater.

The results for MP2 theory at the same basis set are significantly inferior to those for B3LYP and BLYP. The two local DFT functionals perform very poorly on this set of calculations, having errors at least twice as large as MP2.

The models in the lower section of Table 4 use smaller basis sets, and these results are provided for comparison purposes. The B3LYP method using the 6-31+G(d,p) basis set performs almost as well as B3LYP with the larger basis set, although the maximum error increases substantially. BLYP using the 6-31+G(d,p) basis set has a lower MAD than BLYP with the larger basis set although the other statistics are essentially the same for the two BLYP-based model chemistries. B3LYP using the modest 6-31G(d) basis set is still more accurate than MP2 with the larger basis set. The two Hartree-Fock model chemistries indicate the unsuitability of Hartree-Fock theory for calculations at this desired level of accuracy.

The 6-311+G(2d,p) basis set is rather large, and will be out of reach for geometry optimizations and frequency calculations for many large systems of chemical interest. For such cases, it is often desirable to perform the geometry optimization and frequency calculation at a lower level of theory than the final energy calculation. Overall G2 set results for several models based upon B3LYP/6-31G(d), HF/3-21G(d) and AM1 geometries and frequencies are given in Table 5, along with the result for Gaussian-2 theory for comparison purposes [38].

Table 5
Overall results on the 125 calculations in the G2 molecule set for various model chemistries at several specified geometry optimization and frequency calculation levels (all values are in kcal/mol).[a]

	Mean Absolute Deviation	Standard Deviation	Maximum Errors Positive	Maximum Errors Negative
G2 Theory[b]	1.2	0.9	5.1	-3.5
B3LYP/6-311+G(3df,2df,2p) // B3LYP/6-31G(d)	2.7	2.6	12.5	-9.3
B3LYP/6-311+G(2d,p) // B3LYP/6-31G(d)	3.2	3.0	13.6	-20.1
B3LYP/6-31+G(d,p) // B3LYP/6-31G(d)	4.0	4.2	17.6	-33.9
B3LYP/6-31G(d) // B3LYP/6-31G(d)	7.9	9.5	12.2	-54.2
MP2/6-311+G(2d,p) // B3LYP/6-31G(d)	8.9	7.8	29.7	-39.2
B3LYP/6-311+G(2d,p) // HF/3-21G(d)	3.2	3.0	13.8	-21.2
B3LYP/6-31G(d) // HF/3-21G(d)	8.0	9.4	9.4	-54.2
B3LYP/6-31G(d)//AM1[c]	10.5	11.3	14.7	-54.2
AM1//AM1[c]	18.8	16.9	47.8	-95.5
HF/6-31G+(d,p)//AM1[c]	49.4	43.1	8.0	-206.1
HF/6-31G(d)//AM1[c]	54.2	43.1	8.6	-207.2

[a] Reference [39].
[b] Derived from data in Reference [38].
[c] AM1 zero-point energies were not scaled. The calculations involving Na, Li and Mg were excluded from models using AM1. The non-zero energy for H^+ predicted by AM1 was computed and taken into account in the calculation of proton affinities.

Using B3LYP/6-311+G(3df,2df,2p) energies computed at the B3LYP/6-31G(d) geometry and corrected by the (scaled) ZPE at the optimized geometry produces very good overall results across the computations in the G2 set. In fact, this model chemistry had the lowest MAD of all of those considered in this study. (The basis set notation specifies different sets of polarization functions for second-row and third-row and higher heavy atoms.) This basis set is similar to one recommended by Bauschlicher and Partridge [40].

The B3LYP/6-311+G(2d,p) model chemistry at the B3LYP/6-31G(d) geometry also produces good results, although the maximum error is almost twice that using the larger basis set. Using smaller basis sets for the final energy calculation correspondingly increases the mean absolute deviation and maximum errors. The MP2/6-311+G(2d,p) model performs less accurately than the hybrid functionals for all of the models using the B3LYP/6-31G(d) geometries.

For systems where B3LYP/6-31G(d) is too expensive for a geometry optimization and frequency calculation, Hartree-Fock geometries and frequencies using the 6-31G(d) or 3-21G(d) basis sets may be a viable alternative. Two examples of the latter are given in the middle section of the table.

The final section of the table gives results for three models using AM1 geometries and frequencies. Such models are not very accurate, but they may be all that is practical for some very large molecules. For such systems, using B3LYP for a final single point energy calculation substantially improves the overall accuracy of the thermochemistry.

These data suggest some interesting observations about methods for geometry optimizations and frequency calculations when thermochemical predictions is the final calculation goal. Table 6 summarizes the relevant results.

Table 6
B3LYP-based model chemistries with the best overall results on the 125 calculations in the G2 molecule set (all values are in kcal/mol).[a]

	Mean Absolute Deviation	Standard Deviation	Maximum Errors Positive	Maximum Errors Negative
B3LYP/6-311+G(3df,2df,2p) // B3LYP/6-31G(d)	2.7	2.6	12.5	-9.3
B3LYP/6-311+G(2d,p) // B3LYP/6-311+G(2d,p)	3.1	3.0	13.6	-19.7
B3LYP/6-311+G(2d,p) // B3LYP/6-31G(d)	3.2	3.0	13.6	-20.1
B3LYP/6-311+G(2d,p) // HF/3-21G(d)	3.2	3.0	13.8	-21.2
B3LYP/6-31+G(d,p) // B3LYP/6-31+G(d,p)	3.9	4.2	17.6	-33.8
B3LYP/6-31+G(d,p) // B3LYP/6-31G(d)	4.0	4.2	17.6	-33.9
B3LYP/6-31G(d) // B3LYP/6-31G(d)	7.9	9.5	12.2	-54.2

[a] Reference [39].

Note that the overall results for B3LYP/6-311+G(2d,p) final energies are not very sensitive to the basis set used for the optimized geometry. Similarly, the two models using B3LYP/6-31+G(d,p) total energies are essentially unchanged by the addition of hydrogen polarization and heavy-atom diffuse functions in the geometry optimization and frequency calculation. However, these additional functions do make a significant difference in the final energy calculation (compare the final two models).

Table 7 lists the results for various model chemistries on the atomization energy calculations in the G2 molecule set.

Table 7
Results on the 55 atomization energy calculations in the G2 molecule set for various model chemistries (all values are in kcal/mol).[a]

	Mean Absolute Deviation	Standard Deviation	Maximum Errors Positive	Maximum Errors Negative
B3LYP/6-311+G(3df,2df,2p) // B3LYP/6-31G(d)	2.3	2.1	8.1	-9.3
B3LYP/6-311+G(2d,p) // B3LYP/6-311G+(2d,p)	3.0	3.0	8.0	-19.7
B3LYP/6-311+G(2d,p) // B3LYP/6-31G(d)	3.2	3.1	8.0	-20.1
B3LYP/6-311+G(2d,p) // HF/3-21G(d)	3.3	3.1	8.1	-21.2
BLYP/6-31+G(d,p) // BLYP/6-31+G(d,p)	4.0	3.4	13.5	-15.2
BLYP/6-311+G(2d,p) // BLYP/6-311+G(2d,p)	4.5	3.5	14.3	-15.9
B3LYP/6-31+G(d,p) // B3LYP/6-31+G(d,p)	4.4	5.0	8.1	-33.8
B3LYP/6-31+G(d,p) // B3LYP/6-31G(d)	4.6	5.1	8.3	-33.9
B3LYP/6-31G(d) // B3LYP/6-31G(d)	5.4	5.0	8.1	-31.5
B3LYP/6-31G(d) // HF/3-21G(d)	5.7	5.4	8.3	-35.0
B3LYP/6-31G(d)//AM1[b]	8.6	8.2	8.0	-44.2
SVWN5/6-311+G(2d,p) // SVWN5/6-311+G(2d,p)	34.0	20.8	81.0	-0.6
SVWN3/6-311+G(2d,p) // SVWN3/6-311+G(2d,p)	37.6	22.8	89.3	-0.4

[a] Reference [39].
[b] AM1 zero-point energies were not scaled. The calculations involving Na, Li and Mg were excluded from the model using AM1.

The relative performance of the various model chemistries is quite similar to that of the overall results. The major exception occurs with the two LSDA functionals which perform much more poorly for atomization energies than they do for other calculation types (and hence on the overall set). As was true for the overall results, the 6-311+G(3df,2df,2p)//B3LYP/6-31G(d) is the most accurate.

Table 8 similarly lists the results for the ionization potential calculations. The ordering of the various model chemistries is very similar to that for the atomization energies.

Table 8
Results on the 38 ionization potential calculations in the G2 molecule set for various model chemistries (all values are in kcal/mol).[a]

	Mean Absolute Deviation	Standard Deviation	Maximum Errors Positive	Negative
B3LYP/6-311+G(3df,2df,2p) // B3LYP/6-31G(d)	3.6	3.2	12.5	-4.8
B3LYP/6-311+G(2d,p) // B3LYP/6-31G(d)	3.6	3.2	13.2	-4.8
B3LYP/6-311+G(2d,p) // HF/3-21G(d)	3.6	3.1	12.0	-5.6
B3LYP/6-311+G(2d,p) // B3LYP/6-311+G(2d,p)	3.7	3.2	13.1	-5.2
B3LYP/6-31+G(d,p) // B3LYP/6-31G(d)	4.0	3.3	15.7	-6.0
B3LYP/6-31+G(d,p) // B3LYP/6-31+G(d,p)	4.1	3.3	15.7	-6.0
B3LYP/6-31G(d) // B3LYP/6-31G(d)	4.1	3.5	12.2	-13.8
B3LYP/6-31G(d)//HF/3-21G(d)	4.2	3.3	8.5	-13.8
BLYP/6-311+G(d2,p) // BLYP/6-311+G(1d,p)	4.4	3.7	12.2	-9.0
BLYP/6-31+G(d,p) // BLYP/6-31+G(d,p)	4.7	3.1	12.4	-9.4
SVWN5/6-311+G(2d,p) // SVWN5/6-311+G(2d,p)	4.9	3.4	13.5	-6.9
B3LYP/6-31G(d)//AM1[b]	5.2	6.0	14.7	-31.2
SVWN3/6-311+G(2d,p) // SVWN3/6-311+G(2d,p)	14.9	5.1	26.5	-10.4

[a] Reference [39].
[b] AM1 zero-point energies were not scaled. The calculations involving Na, Li and Mg were excluded from the model using AM1.

Once again the model chemistries based on local functionals perform significantly less well than the gradient-corrected functionals. The hybrid functionals continue to perform better than the pure ones.

Table 9 shows similar trends for electron affinities and proton affinities, although BLYP performs slightly better than B3LYP for electron affinities. For proton affinities, BLYP and B3LYP perform equally well when using the 6-31+G(d,p) basis set, but the hybrid functional has a significant advantage when the larger 6-311+G(2d,p) basis set is used.

2.3 Using DFT Geometries within Compound Model Chemistries

Several groups are investigating incorporating DFT methods into composite methods for computing high accuracy total energies such as Gaussian-2 theory [38]. For example, Bauschlicher and Partridge have examined a modification of the G2(MP2) method which uses B3LYP structures and vibrational frequencies [41]. Morokuma and coworkers have proposed a modified G2 theory incorporating coupled cluster energies and DFT geometries and frequencies [42].

Bauschlicher and Partridge denote their method G2(B3LYP/MP2/CC). Figure 1 summarizes the steps involved in the original G2(MP2) method [43], which simplifies G2 by computing the basis-set-extension corrections at the MP2 level rather than the MP4 level (in the interests of reducing computation time and disk storage requirements), as well as the variations proposed by Bauschlicher and Partridge.

Figure 1
Summary of the G2(MP2) methods and the G2(B3LYP/MP2/CC) modifications to the procedure.

Step	G2(MP2) Method	G2(B3LYP/MP2/CC) Variations
1	HF/6-31G(d) Opt./Freq. for ZPE	B3LYP/6-31G(d) Opt./Freq. for ZPE
2	MP2/6-31G(d) Opt. for final structure	Use B3LYP/6-31G(d) geometry
3	QCISD(T)/6-311G(d,p) energy	CCSD(T)/6-311G(d,p) energy
4	MP2/6-311+G(3df,2p) energy	Same
5	Higher level correction (HLC) computed from # of valence α and β electrons	Same
	$E^{G2(MP2)} = E^{QCISD(T)} + (E^{MP2(Step\ 4)} - E^{MP2(Step\ 3)}) + 0.8929(ZPE^{HF}) + HLC$	$E^{G2(B3LYP/MP2/CC)} = E^{CCSD(T)} + (E^{MP2(Step\ 4)} - E^{MP2(Step\ 3)}) + 0.98(ZPE^{B3LYP}) + HLC$

Table 9
Results on the 25 electron affinity calculations (top section) and 7 proton affinity calculations in the G2 molecule set for various model chemistries (all values are in kcal/mol).[a]

	MAD	Standard Deviation	Maximum Positive	Errors Negative
BLYP/6-311+G(2d,p) // BLYP/6-311+G(2d,p)	2.5	2.5	11.4	-4.5
B3LYP/6-311+G(2d,p) // B3LYP/6-311+G(2d,p)	2.6	2.9	13.6	-1.7
B3LYP/6-311+G(3df,2df,2p) // B3LYP/6-31G(d)	2.6	2.5	10.5	-7.0
BLYP/6-31+G(d,p) // BLYP/6-31+G(d,p)	2.8	3.1	15.2	-4.3
B3LYP/6-311+G(2d,p) // B3LYP/6-31G(d)	2.9	3.0	13.6	-7.0
B3LYP/6-311+G(2d,p) // HF/3-21G(d)	2.9	3.0	13.8	-2.6
B3LYP/6-31+G(d,p) // B3LYP/6-31+G(d,p)	3.0	3.7	17.6	-1.9
SVWN5/6-311+G(2d,p) // SVWN5/6-311+G(2d,p)	6.6	3.3	15.1	0.0
SVWN3/6-311+G(2d,p) // SVWN3/6-311+G(2d,p)	17.3	3.5	26.9	0.0
B3LYP/6-311+G(3df,2df,2p) // B3LYP/6-31G(d)	1.0	0.6	0.6	-2.2
B3LYP/6-311+G(2d,p) // B3LYP/6-311+G(2d,p)	1.2	0.7	1.2	-2.4
B3LYP/6-311+G(2d,p) // B3LYP/6-31G(d)	1.3	0.6	0.7	-2.3
B3LYP/6-311+G(2d,p) // HF/3-21G(d)	1.4	0.9	1.0	-2.9
BLYP/6-311+G(2d,p) // BLYP/6-311+G(2d,p)	1.7	1.5	0.7	-4.3
BLYP/6-31+G(d,p) // BLYP/6-31+G(d,p)	1.9	0.9	2.7	-2.7
B3LYP/6-31+G(d,p) // B3LYP/6-31+G(d,p)	2.0	1.0	3.0	-3.5
SVWN3/6-311+G(2d,p) // SVWN3/6-311+G(2d,p)	6.0	2.0	0.0	-9.4
SVWN5/6-311+G(2d,p) // SVWN5/6-311+G(2d,p)	6.5	2.0	0.0	-10.1

[a] Reference [39].

Bauschlicher and Partridge's variations are designed to better address systems which require an electron correlation treatment to obtain accurate vibrational frequencies and zero-point energy and to reduce the computational requirements of the overall method by replacing the MP2 optimization with the less expensive but comparably accurate B3LYP calculation. The CCSD(T) coupled cluster theory method also replaces QCISD(T).

These researchers tested their approach on the 55 atomization energies in the G2 molecule set. Table 10 summarizes their results for the B3LYP/6-311+G(3df,2p) model chemistry and for the G2(MP2/B3LYP/CC) method, as well as for the Complete Basis Set method known as CBS-Q, another compound model chemistry designed to produce highly accurate thermochemistry results [44, 45].

Table 10
Results on the 55 atomization energy calculations in the G2 molecule set for variations of the G2(MP2) model chemistry (all values are in kcal/mol).

	Mean Absolute Deviation	Standard Deviation	Maximum Absolute Error
CBS-Q[a]	0.8	0.6[b]	2.3
G2(MP2)[a]	1.3	0.9[b]	4.2
G2(B3LYP/MP2)[c]	1.3	1.0	3.7
G2(B3LYP/MP2/CC)[c]	1.3	0.9	3.1
B3LYP/6-311+G(3df,2p)[c]	2.2	1.9	8.1

[a] Reference [45].
[b] Standard deviation calculated from the reported RMS value.
[c] Reference [41]; ZPEs scaled by 0.989.

Bauschlicher and Partridge considered their two changes to the G2(MP2) procedure separately. Using the B3LYP/6-31G(d) geometry and ZPE—denoted G2(B3LYP/MP2)—reduced the maximum absolute error slightly. Replacing QCISD(T) with CCSD(T) further reduced this maximum error, although neither substitution had any significant effect on the MAD or its standard deviation for this series of calculations. Thus, this approach does reduce the computational intensity of G2(MP2) without a decrease in accuracy. None of the G2 variations is as accurate as CBS-Q, and CBS-Q is significantly less computationally expensive than those based on G2 theory.

The atomization energies computed via the B3LYP/6-311+G(3df,2p) model chemistry are not as accurate and the maximum error is about twice as large as those of the compound models incorporating high-level and expensive traditional electronic structure methods (the cost of which scales as N^7). This result is also in line with those for other B3LYP-based model chemistries examined in the previous section of this chapter.

3. MOLECULAR AND VIBRATIONAL PROPERTIES

Molecular properties may also be predicted with DFT-based methods, including both ones derived from the electron density and the vibrational frequencies of the molecule and its related properties.

3.1 Molecular Properties

De Proft and coworkers have computed atomic populations according to several schemes using several different ab initio and DFT methods [46]. Their preliminary examination at the QCISD/cc-pVDZ level indicated that charges computed with the atomic polar tensor (APT) method [47] were the most sensitive to the effects of electron correlation. Accordingly, they computed charges with this scheme for a variety of methods. Their results are summarized in Table 11.

This table compare the results for several theoretical methods (all using the cc-pVDZ basis set) to the QCISD/cc-pVDZ charges. All of the methods involving electron correlation achieve reasonable agreement with the high-level results, and the differences between them are much smaller than the difference between the Hartree-Fock and QCISD results. For this set of molecules, the smallest overall difference from the QCISD results was exhibited by the hybrid functionals.

Table 11
Mean absolute deviation of APT charges from the QCISD results for 15 first-row molecules using a variety of methods and the cc-pVDZ basis set (units of atomic charge).

		MAD(ΔQCISD)[a]			
HF	MP2	LDA	BLYP	B3LYP	B3PW91
0.104	0.046	0.034	0.032	0.013	0.019

[a] Reference [46].

Table 12 summarizes De Proft and coworkers' comparison of the dipole moment predicted by B3LYP and QCISD, using the 6-31G(d) and cc-pVDZ basis sets, for a set of 12 compounds containing first-row atoms to the observed values for this quantity. With the smaller basis set, B3LYP and QCISD perform comparably for this set of compounds. Agreement with experiment improves with the larger basis set for both methods, but the improvement is greater for QCISD, and the B3LYP average absolute deviation is twice as large as that for QCISD.

Table 12
Mean absolute deviation from experiment of the predicted dipole moment and its standard deviation for 12 first-row compounds using the B3LYP and QCISD methods and the cc-pVDZ basis set (values in debye).

	MAD(ΔExp)[a]	Standard Deviation[a]
B3LYP/6-31G(d)	0.197	0.129
QCISD/6-31G(d)	0.196	0.166
B3LYP/cc-pVDZ	0.116	0.112
QCISD/cc-pVDZ[b]	0.059	0.049

[a] Computed from data in Reference [46].
[b] Only 10 of 12 QCISD values are reported at this basis set.

Wiberg and coworkers have investigated spin polarization in heterosubstituted allyl radicals using a range of ab initio and DFT methods [23]. Good experimental values exist only for acetyl radical; Wiberg and coworkers' results for this compound as well as some smaller calculations of our own are given in Table 13.

Table 13
Predicted spin polarizations for acetyl radical using several different model chemistries.

	O		C_1 (away from substituent)		C_2 (near substituent)	
	6-31G(d)	6-311G(d,p)[a]	6-31G(d)	6-311G(d,p)[a]	6-31G(d)	6-311G(d,p)[a]
HF	0.651		1.091		-0.615	
MP2	0.023	0.039	0.994	0.980	0.127	0.117
B3LYP	0.386	0.292	0.864	0.911	-0.154	-0.122
QCISD	0.334	0.275	0.964	1.002	-0.187	-0.168
Experiment[a]	0.2		0.7			

[a] Reference [23].

Based on these calculations, spin densities appear to be another property for which the hybrid B3LYP functional produces excellent agreement with the high-level ab initio method, in this case QCISD. The spin densities in acetyl radical is a problem where MP2 theory performs poorly, predicting a value for the oxygen atom that is much too small and a value for the second carbon atom that has the wrong sign. Spin contamination is the major factor reducing the accuracy of the Hartree-Fock and MP2 results.

3.2 Vibrational Frequencies and Zero-Point Energies

Vibrational frequencies computed with ab initio methods are known to suffer from systematic errors. Accordingly, computed frequencies are typically scaled by standard factors before they are compared with experimental observations [48, 49]. Frequencies used to compute zero-point energies and other thermal corrections to the total energy are also scaled, and the optimal scaling factors for each of these uses differ from one another.

Wong [25] and Scott and Radom [17] have investigated the performance of several DFT models for predicting vibrational frequencies and zero-point energies, and they have independently computed optimal scaling factors for each model chemistry. Where their investigations overlapped, there is very good agreement between the two studies. Their findings are also in agreement with the scale factors for zero-point energies computed by Bauschlicher and Partridge [41].

Table 14 summarizes the computed scaling factors for a variety of ab initio and DFT model chemistries based on the 6-31G(d) basis set, as well as the errors that remain in the computed frequencies and zero-point energies after scaling.

Table 14
Optimal scaling factors and overall RMS errors after scaling for fundamental frequencies and zero-point energies for various model chemistries (RMS errors in cm^{-1} for frequencies and kJ/mol for ZPEs).

	Fundamental Frequencies		Zero-point Energies	
	Scale Factor	RMS Error	Scale Factor	RMS Error
HF/6-31G(d)[b]	0.8953	50	0.9182	0.46
SVWN3/6-31G(d)[a]	0.9833	43	1.0079	0.57
B-LYP/6-31G(d)[d]	0.9940	45	1.0057	0.44
B3LYP/6-31G(d)[d]	0.9613	34	0.9826	0.30
B3P86/6-31G(d)[d]	0.9559	34	0.9760	0.28
B3PW91/6-31G(d)[b]	0.9573	34	0.9774	0.28
MP2(Full)/6-31G(d)[c]	0.9427	61	0.9643	0.50
MP2(FC)/6-31G(d)[b]	0.9434	63	0.9676	0.50
QCISD/6-31G(d)[b]	0.9537	37	0.9805	0.38

[a] Reference [25].
[b] Reference [17].
[c] Reference [50].
[d] Frequency data from Reference [25] and ZPE data from Reference [17].

Both studies used a test set of 1066 frequencies of 122 molecules (known as the Z1 set) to determine frequency scaling factors. Table 14 reports the ZPE scale factors of Scott and Radom, who used a set of 89 frequencies from 39 molecules to compute them.

All of the computed scaling factors are near unity, indicating a correlation of modest remaining errors in the harmonic force field with the (neglected) anharmonic contributions. The DFT-based model chemistries perform very well. Those using pure functionals produce overall RMS errors of about the same magnitude as traditional ab initio methods such as Hartree-Fock and MP2. The hybrid functionals produce significantly smaller overall RMS errors, with values on the same order as QCISD.

These results demonstrate that DFT methods, and especially those based on hybrid functionals, are likely to be reliable for predicting vibrational frequencies and properties derived from them. Indeed, the B3LYP and B3PW91 models produce the best results of all of those considered.

3.3 Vibrational Circular Dichroism

Vibrational circular dichroism (VCD) is exhibited by chiral molecules [51], and can accordingly be used as a direct method for determining the absolute configuration of a chiral molecule. To do so in practice requires the accurate prediction of VCD intensities; these are determined by vibrational rotational strengths, which are themselves in turn determined by vibrational electric and magnetic dipole transition moments. Computation of VCD spectra within the harmonic approximation thus requires the calculation of the Harmonic Force Field (HFF), the Atomic Polar Tensors (APTs), and the Atomic Axial Tensors (AATs) [52, 53].

The development of a method for predicting VCD spectra which employs density functional theory to compute each of the required components of the vibrational rotational strengths has recently been reported [10, 54, 55]. This implementation includes a new, efficient DFT methodology for the calculation of AATs using direct methods and gauge-including atomic orbital (GIAO) basis sets.

Systematic calibration studies on a large set of molecules are still underway. However, preliminary results suggest that hybrid functionals produce excellent agreement with experimental VCD spectra. For example, Table 15 compares the rotational strengths predicted by Hartree-Fock theory and three DFT-based methods to available observed values for 6,8-dioxabicyclo[3.2.1]octane.

Table 15
Calculated and experimental rotational strengths for the (+)–(1R,5S) enantiomer of 6,8-dioxabicyclo[3.2.1]octane for various methods using the TZ2P basis set (values in 10^{-44} esu^2 cm^2).

Fundamental	HF[a]	LSDA[a]	BLYP[a]	B3LYP[a]	Exp.[b]
38	6.0	10.6	9.3	8.5	19.4
33	-16.5	0.5	-9.0	-11.9	-18.8
32	0.8	-5.9	6.8	8.8	24.9
31	13.5	-7.4	-4.9	-13.8	-29.8
30	7.4	32.5	12.3	25.7	52.3
29	-8.0	-13.6	2.7	-0.7	-15.2
28	2.8	12.0	-1.7	-1.4	11.6
26	-13.0	-22.7	-9.9	-11.9	-16.4
23	-13.1	4.8	5.5	7.5	13.6
21	0.2	-12.7	36.6	6.1	-9.6
20	73.7	12.6	6.8	12.6	19.3
19	-44.4	-19.4	7.5	6.8	20.3
18	19.1	-13.2	-82.1	-44.1	-62.3
17	-37.4	-26.9	-63.3	-45.4	-80.0
16	-2.7	38.7	9.0	-2.2	6.4
15	51.2	28.1	57.2	74.8	88.0
14	-12.0	21.2	66.1	47.8	63.3
13	12.8	-15.6	-9.2	-44.0	-74.0
12	12.0	3.1	23.1	30.6	32.5

[a] Reference [55].
[b] Reference [56].

The B3LYP spectra best reproduces the experimental spectra for this molecule. In fact, this calculation yields a nearly unambiguous assignment of fundamentals 10-38. The BLYP spectra is significantly less accurate than the B3LYP results, and the differences are even greater for the LSDA spectra. For this problem, the functionals can be rated LSDA < BLYP < B3LYP in terms of relative accuracy. Finally, the poor Hartree-Fock results illustrate the importance of electron correlation for this property.

3.4 NMR Chemical Shifts

Electronic structure methods have been applied to the prediction of NMR properties for several years [57, 58], and recently DFT methods using GIAOs have also proven effective for predicting nuclear magnetic resonance shielding tensors and calculating NMR chemical shifts [7, 59, 60]. Various methods for computing NMR chemical shifts are compared for a set of 16 molecules in Table 16.

The MP2 results are the most accurate. However, the gradient-corrected DFT-based models represent a significant improvement over the Hartree-Fock results. Interestingly, the raw differences between the predicted values and experiment indicate that the DFT-based methods generally predict shifts which are too deshielded, while Hartree-Fock theory makes errors in both directions.

These shifts were computed using a very large basis set. For general use, we recommend the B3LYP/6-311+G(2d,p) model for computing NMR properties of chemical systems, using a geometry optimized at the B3LYP/6-31G(d) level. For systems where it is computationally practical, this level represents a reasonable trade-off between accuracy and cost, since B3LYP is significantly more cost-effective than MP2. This model includes the effects of electron correlation in the calculation and provides quantitatively good agreement with experiment.

Table 17 summarizes results computed with these models for a set of 32 ^{13}C and 5 ^{15}N chemical shifts from 22 molecules. Even the Hartree-Fock NMR predictions are adequate for assignment of most experimental spectra, and the B3LYP results are quite accurate.

Table 16
Mean absolute deviation, standard deviation of the absolute error, and maximum absolute deviation of ^{13}C, ^{15}N and ^{17}O chemical shifts[a] from experiment for a set of 16 molecules[b] using a variety of methods and the QZ2P basis set (values in ppm).

		HF[c]	LSDA[c]	BLYP[c]	B3LYP[c]	B3PW91[c]	MP2[d]
Overall	MAD	19.0	20.2	12.1	14.1	13.2	4.0
(25 shifts)	Std. Dev.	17.9	11.6	8.0	10.6	10.6	5.2
	Max. Error	65.2	44.8	30.2	39.6	38.8	23.9
^{13}C only	MAD	8.9	14.7	8.1	8.0	7.1	1.6
(17 shifts)	Std. Dev.	7.4	7.3	4.7	5.5	5.2	1.8
	Max. Error	30.8	26.3	18.6	17.3	17.3	6.5
^{15}N only	MAD	47.8	32.8	20.6	28.4	26.6	11.8
(5 shifts)	Std. Dev.	12.1	10.8	6.8	6.5	7.0	6.4
	Max. Error	65.2	44.8	30.2	39.6	38.8	23.9

[a] ^{13}C shifts relative to CH_4, ^{15}N shifts relative to NH_3, and ^{17}O shifts relative to H_2O.
[b] Geometries optimized with MP2/TZ2P.
[c] Reference [7].
[d] MP2 statistics based on data from Reference [58].

Table 17
Mean absolute deviation, standard deviation of the absolute errors, and maximum absolute deviation of ^{13}C and ^{15}N chemical shifts[a] from experiment for a set of 22 molecules[b] using the model chemistries recommended in the text (values in ppm).

		HF/6-31G(d)[c]	B3LYP/6-311+G(2d,p)[c]
^{13}C only	MAD	9.3	3.6
(32 shifts)	Std. Dev.	5.9	2.2
	Max. Error	22.4	8.4
^{15}N only	MAD	12.5	16.9
(5 shifts)	Std. Dev.	7.2	3.0
	Max. Error	19.7	21.6

[a] ^{13}C shifts relative to TMS, and ^{15}N shifts relative to NH_3.
[b] Geometries optimized with B3LYP/6-31G(d).
[c] Reference [7].

Unfortunately, many interesting molecules are substantially too large for their NMR properties to be practically studied with these DFT-based methods. For such systems, Hartree-Fock theory, using the modest 6-31G(d) basis set, may provide a reasonable alternative approach to computing ^{13}C chemical shifts (HF/6-31G(d) ^{13}C shifts are only slightly less accurate than ones computed with the HF/6-311+G(2d,p) model chemistry [7]). Taxol provides an appropriate test case (this molecule's structure is illustrated in Figure 2).

The structure of this molecule was optimized using the HF/STO-3G model, and its NMR properties were computed at the HF/6-31G(d) level. The RMS error of the computed values for the 41 ^{13}C chemical shifts for which experimental data exists was 6.4 ppm, and the mean absolute deviation was 5.0 ppm. The maximum error for carbon shifts is 18.7 ppm. These results appear to be sufficiently accurate to be of aid in experimental peak assignments.

The DFT functionals used here to predict NMR properties do not currently include a specific magnetic-dependent term. It is possible that the inclusion of such a term would improve upon these results.

Figure 2
Structure of the taxol molecule.

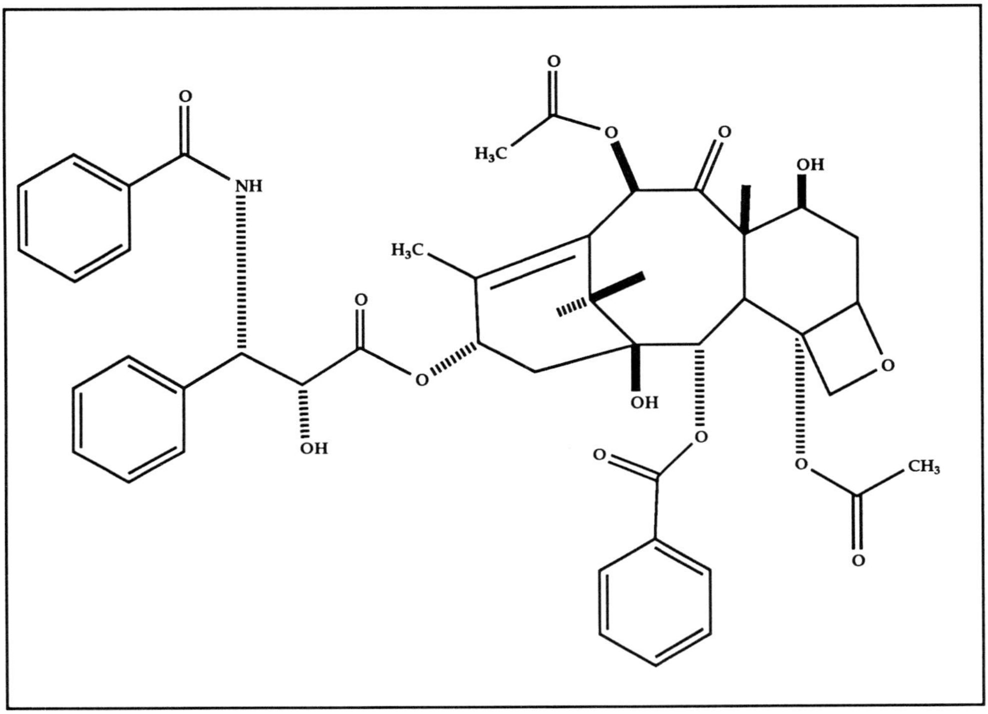

4. IMPORTANCE AND CALIBRATION OF INTEGRATION GRIDS

In contrast to conventional ab initio methods, which rely on analytic algorithms for their principal calculation types, DFT calculations currently proceed by employing three-dimensional numerical quadrature techniques, which involves numerical integration over a three dimensional grid of points in space [61-67]. Such grids are defined as a number of radial shells around each atom, each of which contains a set number of integration points.

There is a strong correlation between the number and choice of grid points and both the accuracy and cost of the quantum chemical calculation. Grids are usually denoted as a parenthesized pair of integers. For example, the designation (75,302) refers to a grid having 75 radial shells around each atom each of which contains 302 points.

Pruning a uniform grid—a grid containing the same number of angular points at each radial distance—to eliminate some of the points in specific regions of space provides a means of reducing the computational cost for a calculation with negligible loss in accuracy. Pruned grids are reduced from the full, uniform grid form so that fewer points are used on the shells near the core, where the density is nearly spherically symmetric, and far from the nucleus, where the density is small. Thus, pruned grids are densest in the region of the atom where properties are changing most rapidly.

Available pruned grids differ in three main ways: their computational accuracy, the degree to which they exhibit rotational invariance and their range of applicability by element. In this section, two widely-used pruned grids are compared [68]: the SG1 grid suggested by Gill and coworkers [69], which is a pruned (50,194) grid, and the (75,302)p grid, a pruned version of (75,302).

4.1 Accuracy

The choice of an integration grid can influence the accuracy of the final DFT calculation. Table 18 gives the overall accuracy differences between several commonly-used grids with respect to the energies computed with the very large (96,32,64) grid of Murray and coworkers [70] (a grid of 96 radial shells, each composed of a 32x64 uniform angular grid).

Table 18
Mean absolute deviation, standard deviation of the absolute error, and maximum absolute errors for the differences in predicted total energy with respect to the (96,32,64) grid for a set of 36 molecules[a] containing first and second-row atoms, computed with the BLYP/6-31G(d) model chemistry using various integration grids (energy differences in kcal/mol).[b]

	(99,302)	(75,302)	(75,302)p	(50,194)	SG1 [(50,194)p]
MAD	0.009	0.009	0.008	0.043	0.044
Standard Deviation	0.017	0.017	0.016	0.059	0.062
Maximum Error	0.079	0.078	0.078	0.265	0.280

[a] Tight-convergence geometry optimizations were performed at the MP2/6-31G(d) level for systems with up to one heavy atom and at the HF/6-31G(d) level for those with two or more heavy atoms.
[b] Reference [68].

Although all of these energy differences are relatively small for these small test systems, there is a significant difference in overall accuracy between the SG1 and (75,302)p grids. Larger differences are to be expected with large molecules. SG1 has a MAD about 5 times as large as (75,302)p, and their maximum errors also differ by a factor of 4. Both grids perform just as well as the larger uniform grids from which they were pruned.

4.2 Rotational Invariance

All finite grids suffer from some degree of rotational invariance: changing the orientation of the molecular system under investigation can alter the predicted energy. (75,302)p exhibits relatively little rotational invariance while for SG1 this effect can be substantial. Table 19 gives the results for an example case which illustrates this difference, Al_4P_4.

Table 19
Difference in computed total energy between two orientations of Al_4P_4 for two integration grids, computed at the BLYP/6-31G(d) level (values in kcal/mol).[a]

	$\Delta E^{orientation}$
SG1 [(50,192)p]	0.354
(75,302)p	0.079

[a] Reference [68].

The change in energy arising solely from the differing orientation of the molecule is over 4 times as large for SG1 as for the (75,302)p grid. Once again, the error produced by SG1 is small but significant.

4.3 Extensibility

Ideally, the same integration grid should be used for all DFT calculations undertaken, regardless of the atom types within the molecular system under investigation. Initially, both SG1 and (75,302)p were defined for the first two rows of the periodic table. (75,302)p has been extended to the third row. However, SG1 cannot be so extended in any straightforward way since its parent grid, the uniform (50,194) grid, performs very poorly on third-row atoms. Table 20 illustrates this effect.

Table 20
Difference in predicted total energy with respect to the (96,32,64) grid for two third-row atoms, using the BLYP/6-31G(d) model chemistry (values in kcal/mol).[a]

	(99,302)	(75,302)p	(50,194)
Mn	0.000	0.008	-0.727
Zn	0.000	0.009	-0.660

[a] Reference [68].

The energies predicted by the (75,302) grid differ from those of the large grid only by microhartrees. In contrast, those predicted by the full, uniform (50,194) grid differ by over 0.6 kcal/mol, indicating that this grid is unsuitable for compounds involving third-row and higher atoms even before pruning.

5. CONCLUSION

Model chemistries based on gradient-corrected DFT methods generally provide good reliability for a wide range of electronic structure calculations. The performance of models based on local DFT methods is both less accurate and more erratic. While some previous studies have claimed that LSDA is better than the gradient-corrected functionals for some properties and the reverse is true for others, we find that DFT methods using hybrid functionals are in general as good as or better than either type of pure functional.

Hybrid functionals perform very well in predicting the energies, geometries, vibrational frequencies of chemical systems, as well as many molecular properties including atomic populations, dipole moments, spin densities, VCD spectra and NMR chemical shifts. They also predict thermochemical properties of systems with very good accuracy when combined with an accurate geometry and a large basis set. In these cases, such model chemistries perform as well as or better than traditional, more expensive electron correlation methods such as MP2. In the case of NMR chemical shifts, DFT methods represent an improvement over Hartree-Fock theory, but they are not as accurate MP2 for this property.

The various hybrid functionals do not differ substantially in accuracy. In particular, B3LYP and B3PW91 appear to be comparable in accuracy for the calculation types considered here. The same parameters, chosen on the basis of thermochemical calculations at fixed geometries for one functional can be used with other functionals and give improved accuracy over pure functionals for a wide range of calculations. This reflects the fact that inclusion of exact exchange is based on sound underlying physics [31, 32] and is not simply an ad-hoc fitting procedure.

Acknowledgments

The authors thank Æleen Frisch for her help in preparing this chapter. All results reported for the first time in this work were computed with the *Gaussian 94* program [71].

References

1. P. Hohenberg and W. Kohn, *Phys. Rev.*, **136**, B864 (1964).
2. W. Kohn and L. J. Sham, *Phys. Rev.*, **140**, A1133 (1965).
3. R. D. Amos, C. W. Murray and N. C. Handy, *Chem. Phys. Lett.*, **202**, 489 (1993).
4. J. Andzelm and E. Wimmer, *J. Chem. Phys.*, **96**, 1280 (1992).
5. J. Andzelm, C. Sosa and R. A. Eades, *J. Phys. Chem.*, **97**, 4664 (1993).
6. J. E. Carpenter and C. P. Sosa, *J. Mol. Struct.*, (1993).
7. J. R. Cheeseman, M. J. Frisch, G. W. Trucks and T. A. Keith, *J. Chem. Phys.*, **104**, 5497 (1996)
8. S. J. Collins and P. J. O'Malley, *Chem. Phys. Lett.*, **228**, 246 (1994).
9. S. M. Colwell, C. W. Murray, N. C. Handy and R. D. Amos, *Chem. Phys. Lett.*, **210**, 261 (1993).
10. F. J. Devlin, P. J. Stephens, J. R. Cheeseman and M. J. Frisch, *J. Am. Chem. Soc.*, in press (1996).
11. N. C. Handy, P. E. Maslen, R. D. Amos, J. S. Andrews, C. W. Murray and G. J. Laming, *Chem. Phys. Lett.*, **197**, 506 (1992).
12. K. Laasonen, M. Parrinello, R. Car, C. Lee and D. Vanderbilt, *Chem. Phys. Lett.*, **207**, 208 (1993).
13. R. Merkle, A. Savin and H. Preuss, *Chem. Phys. Lett.*, **194**, 32 (1992).
14. C. W. Murray, G. J. Laming, N. C. Handy and R. D. Amos, *Chem. Phys. Lett.*, **199**, 551 (1992).
15. T. V. Russo, R. L. Martin and P. J. Hay, *J. Chem. Phys.*, (1993).
16. B. G. Johnson, P. M. W. Gill and J. A. Pople, *J. Chem. Phys.*, **98**, 5612 (1993).
17. A. P. Scott and L. Radom, *J. Phys. Chem.*, submitted (1996).
18. J. M. Seminario, *Chem. Phys. Lett.*, **206**, 547 (1993).
19. C. Sosa, J. Andzelm, B. C. Elkin, E. Wimmer, K. D. Dobbs and D. A. Dixon, *J. Phys. Chem.*, **96**, 6630 (1992).
20. C. Sosa, C. Lee, G. Fitzgerald and R. A. Eades, *Chem. Phys. Lett.*, **211**, 265 (1993).
21. P. J. Stephens, F. J. Devlin, C. F. Chabalowski and M. J. Frisch, *J. Phys. Chem.*, **98**, 11623 (1994).
22. D. J. Tozer and C. P. Sosa, *J. Chem. Phys.*, submitted (1995).
23. K. B. Wiberg, J. R. Cheeseman, J. W. Ochterski and M. J. Frisch, *J. Am. Chem. Soc.*, **117**, 6535 (1995).
24. K. B. Wiberg and J. W. Ochterski, *J. Comp. Chem.*, accepted (1996).
25. M. W. Wong, *Chem. Phys. Lett.*, in press (1996).
26. J. C. Slater, *Quantum Theory of Molecular and Solids. Vol. 4: The Self-Consistent Field for Molecular and Solids*, McGraw-Hill: New York, 1974.
27. S. H. Vosko, L. Wilk, and M. Nusair, *Can. J. Phys.*, **58**, 1200 (1980).

28. A. D. Becke, *Phys. Rev. A*, **38**, 3098 (1988).
29. C. Lee, W. Yang and R. G. Parr, *Physical Review B*, **37**, 785 (1988).
30. B. Miehlich, A. Savin, H. Stoll and H. Preuss, *Chem. Phys. Lett.*, **157**, 200 (1989).
31. A. D. Becke, *J. Chem. Phys.*, **98**, 5648 (1993).
32. A. D. Becke, *J. Chem. Phys.*, **98**, 1372 (1993).
33. A. Becke, *J. Chem. Phys.*, **104**, 1040 (1996).
34. J. P. Perdew and Y. Wang, *Phys. Rev. B*, **45**, 13244 (1992).
35. J. A. Pople, M. Head-Gordon, D. J. Fox, K. Raghavachari and L. A. Curtiss, *J. Chem. Phys.*, **90**, 5622 (1989).
36. G. E. Scuseria, *J. Chem. Phys.*, **94**, 442 (1991).
37. L. A. Curtiss, C. Jones, G. W. Trucks, K. Raghavachari and J. A. Pople, *J. Chem. Phys.*, **93**, 2537 (1990).
38. L. A. Curtiss, K. Raghavachari, G. W. Trucks and J. A. Pople, *J. Chem. Phys.*, **94**, 7221 (1991).
39. J. B. Foresman, Æ. Frisch, J. W. Ochterski and M. J. Frisch, in preparation (1996).
40. C. W. Bauschlicher Jr. and H. Partridge, *Chem. Phys. Lett.*, **240**, 533 (1995).
41. C. W. Bauschlicher Jr. and H. Partridge, *J. Chem. Phys.*, **103**, 1788 (1995).
42. A. M. Mebel, K. Morokuma, and M.C. Lin, *J. Chem. Phys.*, **103**, 7414 (1995).
43. L. A. Curtiss, K. Raghavachari and J. A. Pople, *J. Chem. Phys.*, **98**, 1293 (1993).
44. G. A. Petersson, T. G. Tensfeldt and J. A. Montgomery Jr., *J. Chem. Phys.*, **94**, 6091 (1991).
45. J. W. Ochterski, G.A. Petersson, and J.A. Montgomery, *J. Chem. Phys.*, **104**, 2598 (1996).
46. F. De Proft, J. M. L. Martin and P. Geerlings, *Chem. Phys. Lett.*, **250**, 393 (1996).
47. J. Cioslowski, *J. Am. Chem. Soc.*, **111**, 8333 (1989).
48. J. A. Pople, R. Krishnan, H. B. Schlegel and J. S. Binkley, *Int. J. Quant. Chem., Quant. Chem. Symp.*, **13**, 325 (1979).
49. J. A. Pople, R. Krishnan, H. B. Schlegel, D. DeFrees, J. S. Binkley, M. J. Frisch, R. F. Whiteside, R. F. Hout and W. J. Hehre, *Int. J. Quant. Chem., Quant. Chem. Symp.*, **15**, 269 (1981).
50. J. A. Pople, A. P. Scott, M. W. Wong and L. Radom, *Isr. J. Chem.*, **33**, 345 (1993).
51. P. J. Stephens and M. A. Lowe, *Ann. Rev. Phys. Chem.*, **36**, 213 (1985).
52. P. J. Stephens, *J. Phys. Chem.*, **89**, 748 (1985).
53. P. J. Stephens, *J. Phys. Chem.*, **91**, 1712 (1987).
54. J. R. Cheeseman, C. S. Ashvar, M. J. Frisch, F. J. Devlin and P. J. Stephens, *Chem. Phys. Lett.*, **252**, 211 (1996).
55. P. J. Stephens, C. S. Ashvar, F. J. Devlin, J. R. Cheeseman and M. J. Frisch, *Mol. Phys.*, **88**, in press (1996).
56. T. Eggimann, R. A. Shaw and H. Weiser, *J. Phys. Chem.*, **95**, 591 (1991).

57. K. Wolinski, J. F. Hilton and P. Pulay, *J. Am. Chem. Soc.*, **112**, 8251 (1990).
58. J. Gauss, *J. Chem. Phys.*, **99**, 3629 (1993).
59. G. Schreckenbach and T. Ziegler, *J. Phys. Chem.*, **99**, 606 (1995).
60. A. M. Lee, N. C. Handy and S. M. Colwell, *J. Chem. Phys.*, **103**, 10095 (1995).
61. A. D. Becke, *J. Chem. Phys.*, **88**, 2547 (1988).
62. V. I. Lebedev, *Zh. Vychisl. Mat. Mat. Fiz.*, **15**, 48 (1975).
63. V. I. Lebedev, *Zh. Vychisl. Mat. Mat. Fiz.*, **16**, 293 (1976).
64. V. I. Lebedev, in *Proc. Conf. Diff. Eqn. Numer. Math., Novosibivsk, 1978*, Ed. S. L. Sobolev, Nauka: Novosibivsk, 1980, 110.
65. V. I. Lebedev, *Russian Acad. Sci. Dokl. Math.*, **45**, 587 (1992).
66. N. C. Handy and S. F. Boys, *Theo. Chim. Acta*, **31**, 195 (1973).
67. C. W. Murray, N. C. Handy and G. J. Laming, *Mol. Phys.*, **78**, 997 (1993).
68. G. W. Trucks and M. J. Frisch, in preparation (1996).
69. P. M. W. Gill, B. G. Johnson and J. A. Pople, *Chem. Phys. Lett.*, **209**, 506 (1993).
70. C. W. Murray, N. C. Handy and G. J. Laming, *Molecular Physics*, **78**, 997 (1993).
71. M. J. Frisch, G. W. Trucks, H. B. Schlegel, P. M. W. Gill, B. G. Johnson, M. A. Robb, J. R. Cheeseman, T. A. Keith, G. A. Petersson, J. A. Montgomery, K. Raghavachari, M. A. Al-Laham, V. G. Zakrzewski, J. V. Ortiz, J. B. Foresman, J. Cioslowski, B. B. Stefanov, A. Nanayakkara, M. Challacombe, C. Y. Peng, P. Y. Ayala, W. Chen, M. W. Wong, J. L. Andres, E. S. Replogle, R. Gomperts, R. L. Martin, D. J. Fox, J. S. Binkley, D. J. Defrees, J. Baker, J. P. Stewart, M. Head-Gordon, C. Gonzalez and J. A. Pople, *Gaussian 94, Revision D.3*, Gaussian, Inc.: Pittsburgh, PA, 1995.

J.M. Seminario (Editor)
Recent Developments and Applications of Modern Density Functional Theory
Theoretical and Computational Chemistry, Vol. 4

Computing Transition State Structures with Density Functional Theory Methods

Branko S. Jursic

Department of Chemistry, University of New Orleans, New Orleans, Louisiana 70148, USA

1. INTRODUCTION

A proper description of the potential energy surface for a chosen chemical reaction has a very important place in chemical science. Using a one dimensional potential energy diagram simplifies the matter but it is a fact that location of the reactants, transition state, and the products for many reactions is sufficient enough information. This information tells us more than just reaction rate. For example, the energy barrier will tell us whether the reaction is possible at all. A comparison of the structural features of the reactants and the transition state structure tells us what changes in the reactants' structures must happen to become transition state. Reactants should show a "transition state like" structure. The geometry of the transition state will tell us what product we can obtain. If two parallel reactions are possible, what is the energy difference between the concurrent transition states? Is there a major change in charge distribution, orbital overlap, or steric interactions? Will the influence of acid or base increase or inverse the energy difference between these two concurrent transition state structures? The answer to these questions is invaluable for the experimentalist. By choosing proper reaction conditions, experimentalists can perform chemical transformations in the desired direction; the computation of transition state structures should be a powerful tool for achieving it.

2. COMPUTATIONAL METHODOLOGY

All of the calculations are performed with the GAUSSIAN 92 or 94 computational package [1]. Traditional Hartree-Fock (HF) [2-6], Møller-Plesset energy correction (MP2, MP3, MP4) [7-12] and Quadratic Configuration Interactions, Singlet and Doublet substitution with Triples contribution to the energies [QCISD(T)] [13] *ab initio* methods are used. The gaussian theoretical models (G1, G2, and G2MP2) [14-17] which are known to produce highly accurate energies are also used to compute the energies.

Three hybrid DFT methods are applied. Becke's three parameter functional (B3) [18] in combination with LYP [19,20] (B3LYP), P86 [21] (B3P86), and PW91 [22] (B3PW91) correlation functionals. Three exchange DFT functionals: Xα [23], Slater's exchange functional (HFS or S in combination with the correlation functionals) [23] and Becke's 88 (HFB or B in combination with the correlation functionals) [24] are used in combination with the local (non-gradient corrected) functional of Perdew (PL) [25], with the gradient-corrected Perdew 86 (P86) correlation functional [21], with Perdew and Wang's 1991 (PW91) gradient-corrected correlation functional [22], with local spin density correlation (LSD) of Vosko, Wilk and Nusair (VWN) [26], and with the fifth functional of VWN [26].

For all of the calculations, the gaussian type basis sets [3-21G, 6-31G(d), 6-311G(d,p); 6-311G(2d,2p), etc.] are employed. The explanation and abbreviations of the basis sets are as found in the GAUSSIAN manuals [27, 28]. The finding, optimization, and verification of the transition states are performed as explained above.

3. EVALUATION OF REACTION BARRIERS WITH DFT METHODS

To determine the reliability of density functional theory methods for computing chemical reaction barriers, the computational results for several types of the chemical reactions that are neutral, with spin, and charge will be presented. The computed values will be compared with experimental results, where available and with computational study at higher level of *ab initio* theory. First we will start with pericyclic reactions.

3.1. Computing Transition States for Pericyclic Reactions

Thirty years ago Woodward and Hoffmann [29-34] offered an explanation for a very important class of organic reactions, "pericyclic reactions". It can be easily said that these reactions are the most explored by theoretical chemists. The detailed analysis of the geometries, energies and electronic characteristics of the transition state structures for these reactions were performed with empirical, semiempirical and ab initio methods [35-39]. For many of these reactions experimental energies are available; they are a perfect candidate for evaluating the reliability of DFT methods for computing transition state structures.

3.1.1. Ring Opening-Closing for Cyclobutene-Butadiene Electrocyclization

The reaction has been extensively studied by both semiempirical and ab initio methods [40-44]. The geometric parameters for the corresponding transition state structure **1** are very

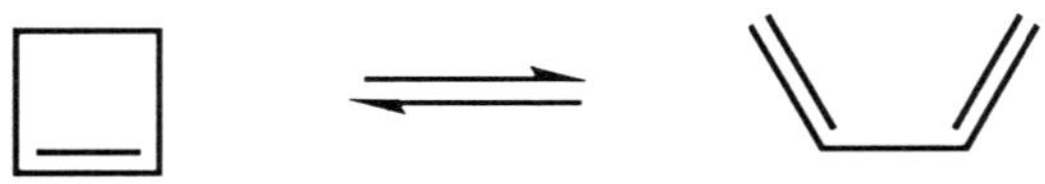

similar regardless of the level of theory employed [40-44]. Our calculations extend this observation to the transition state structure **1** computed with SVWN (or LSDA) and BLYP DFT methods (Table 1) [45]. The largest deviation is in the bond lengths; this noticed in the

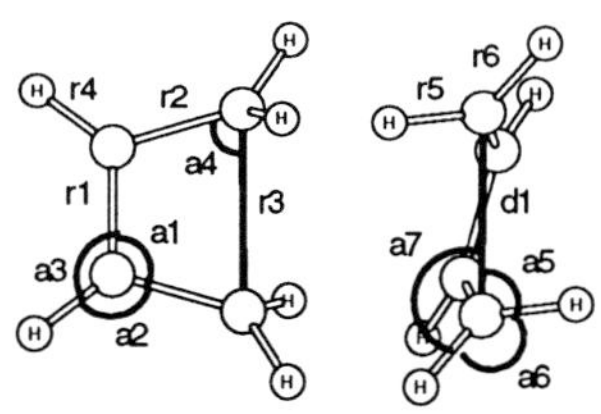

1

Table 1. Computed* geometry characteristics of transition state **1** for cyclobutene ring opening obtained by using 6-311+G(d) gaussian type of basis set [45]

	AM1	RHF	MP2	SVWN	BLYP
r1/Å	1.389	1.367	1.385	1.372	1.380
r2/Å	1.428	1.416	1.431	1.413	1.441
r3/Å	2.120	2.127	2.131	2.126	2.154
r4/Å	1.088	1.077	1.087	1.095	1.093
r5/Å	1.098	1.072	1.084	1.091	1.088
r6/Å	1.096	1.084	1.094	1.102	1.100
a1/Å	103.8	104.2	103.7	104.3	104.5
a2/°	126.6	126.2	126.2	126.0	126.0
a3/°	129.2	129.4	129.9	129.5	129.3
a4/°	74.5	73.6	74.0	73.8	73.7
a5/°	87.5	85.8	83.8	82.9	86.2
a6/°	115.2	113.9	114.6	114.2	113.8
a7/°	127.0	133.0	133.2	133.8	132.3
d1/°	-18.7	-21.8	-22.2	-20.7	-19.7

elongated, cleaving C-C bond (r3), which differ by 0.034 Å. The agreement with the bond angles is also quite good. In all calculations, the nonplanarity of the carbon skeleton is pronounced with the torsion angle d1 ranging from 18.7 to 22.2°, which is also the maximal disagreement in the computed angles (Table 1). The reaction barrier for the ring opening is estimated to be 32.9±0.5 kcal/mol [46, 47]. All zero-point energy corrected (ZPEC) activation barriers are around 2 kcal/mol lower than the uncorrected energies. The MP2 ab initio, the local (SVWN), and the hybrid (B3LYP) DFT methods predict an activation barrier almost identical to the experimental value. As expected, HF, that by definition does not include

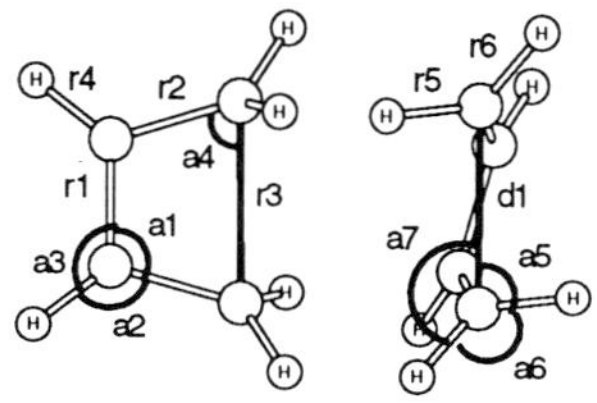

1

Table 2. Computed activation barrier (kcal/mol) for cyclobutene ring opening through **1** [45]

Theory Model	without ZPEC	with ZPEC
RHF/6-311+G(d,p)	44.7	42.8
MP2/6-311+G(d,p)//HF/6-311+G(d,p)	34.2	32.4
MP2/6-311+G(d,p)	34.2	32.5
SVWN/6-311+G(d,p)	34.3	32.6
BLYP/6-311+G(d,p)	29.7	27.4
B3LYP/6-311+G(d,p)[48]	33.9	32.2
HF/6-311+G(d,p)/AM1[48]	44.5	
MP2/6-311+G(d,p)/AM1[48]	34.4	
SVWN/6-311+G(d,p)/AM1[48]	34.4	
BLYP/6-311+G(d,p)/AM1[48]	29.4	
B3LYP/6-311+G(d,p)AM1[48]	33.5	
Experimental [46, 47]		32.9±0.5

electron correlation, considerably overestimates the activation barriers. Surprisingly, the gradient corrected DFT method, BLYP, does not produce an acceptable activation barrier. We will demonstrate later that a local DFT method such as SVWN performs well for activation barriers involving ring opening and rearrangement reactions, but not for many other chemical reactions.

Due to excellent geometric similarities of transition state structures computed with different theory model, one may assume that only energy evaluations with a suitable level of theory on geometries obtained with less costly methods will generate a viable activation barrier. That is clearly demonstrated in the example of ring opening with cyclobutene. The computed activation barrier with MP2/6-311+G(2d,2p) on HF/6-311+G geometry is almost identical to full MP2/6-311+G(d,p) computed energies. The same is true for full MP2, SVWN, BLYP, and B3LYP computed activation barriers and for their single point energy evaluation on AM1 computed geometry (Table 2).

To further explore the reliability of DFT methods for computing the activation barrier for ring opening we have chosen the same system, although now the bonds that are breaking are between two heteroatoms. The computed transition state structure **2** with different methods does not agree as well. The deviation is substantial; all have an atom bond distance (r1, r2, and r3) with a maximal discrepancy between AM1 and BLYP computed **2** (Table 3).

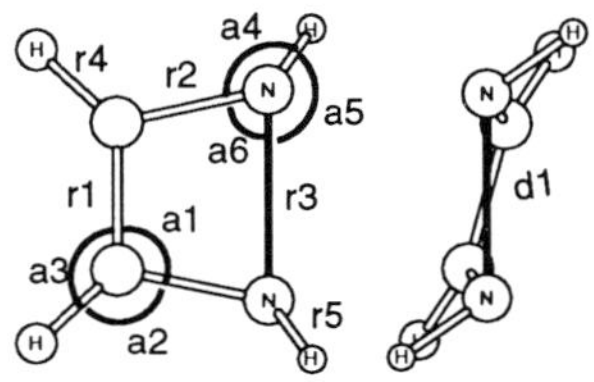

2

Table 3. Calculated geometry characteristics of transition state **2** for *trans*-1,2-dihydro-1,2-diazete ring opening [45]

	AM1 [48]	RHF	MP2	SVWN	BLYP
r1/Å	1.431	1.365	1.380	1.375	1.379
r2/Å	1.351	1.354	1.388	1.362	1.397
r3/Å	1.791	1.889	1.896	1.900	1.925
r4/Å	1.088	1.073	1.084	1.092	1.089
r5/Å	1.003	1.016	1.038	1.045	1.045
a1/Å	95.8	99.8	99.4	99.9	100.0
a2/°	131.5	125.7	125.5	125.5	125.0
a3/°	132.2	133.8	134.7	133.5	134.0
a4/°	116.4	113.8	111.8	113.4	112.7
a5/°	127.8	124.4	122.0	125.0	121.7
d1/°	23.9	21.0	20.5	19.6	19.0

* the calculation was carried out with 6-311+G(d,p) gaussian type of basis set

Unfortunately, we do not have experimental data available for this system, however we can demonstrate that when the energy evaluation with different methods is performed, the structures obtained are not acceptable when a heteroatom is involved in bond breaking and bond formation (Table 3). If we assume MP2 produces the most reliable activation barriers, then SVWN is not accurate, but B3LYP continues to perform well generating an energy almost identical to the MP2 computed value (Table 4). The single point MP2 energies computed on the HF and the MP2 are almost identical while the energy estimated on AM1 geometries are considerably higher. Wile this approach consumes less computational time, it also produces too high of activation energies; consequently, we do not recommend it when ring opening of heterocycles is studied. Furthermore, B3LYP hybrid DFT methods perform well in both cases and it is a recommended method.

Table 4. Computed activation barrier (kcal/mol) for *trans*-1,2-dihydro-1,2-diazete ring opening through **2** [45]

Theory Model	without ZPEC	with ZPEC
RHF/6-311+G(d,p)	38.7	36.3
MP2/6-311+G(d,p)//HF/6-311+G(d,p)	22.7	20.3
MP2/6-311+G(d,p)	22.9	20.8
SVWN/6-311+G(d,p)	27.7	25.6
BLYP/6-311+G(d,p)	18.5	16.5
B3LYP/6-311+G(d,p)	23.5	21.6
HF/6-311+G(d,p)/AM1[48]	41.2	
MP2/6-311+G(d,p)/AM1[48]	27.8	
SVWN/6-311+G(d,p)/AM1[48]	31.4	
BLYP/6-311+G(d,p)/AM1[48]	24.2	
B3LYP/6-311+G(d,p)/AM1[48]]	28.9	

3.1.2. [3,3] Sigmatropic Shifts

There are at least two possible reaction mechanisms, one through a concerted and the other through a two-step mechanism (fragmentation to two allyl fragments and than recombination). It has been concluded that for an unsubstituted system the reaction goes through an aromatic

transition state with synchronous breaking and making of carbon-carbon bonds [49-51]. There are also two possible transition states structures, a chair-like transition state **CTS** and a boat-like transition state **BTS**.(Figure 1). The experimental activation barriers through three

Figure 1. The B3LYP/6-31+G(d) optimized chair-like and boat-like transition state structures for the Cope rearrangement

possible transition states are available [52-55]. Again, the local spin density approximation does not produce activation barriers that agree well with experiment. All computed energies are considerably lower. Of all the studied DFT methods, the hybrid B3LYP produces the best agreement with experiment and high level ab initio calculations. Interestingly, the MP2/6-31G(d) computes the **CTS** with a bond distance of 1.783 Å, which is much closer to the 1.753 Å computed by SVWN than to 2.007 Å as generated with B3LY. AM1 generates an even shorter carbon -carbon bond (1.68 Å). The energies computed with high level theory on AM1 geometry dos not produce results agreeable with the experiment.

Table 5. Computed activation barrier (kcal/mol) for the Cope rearrangement through transition states **CTS**, **BTS**, and allyl radical recombination

Theory Model	CTS	BTS	allyl
HF/6-31G(d)[56]		56.6	
MP2/6-31G(d)[56]		31.4	
MP4/6-31G(d)//MP2/6-31G(d)[51]	44.9	32.7	
QCISD(T)/6-31G(d)//MP2/6-31G(d)[51]	47.2	35.3	
QCISD(T)/6-31G(d)//CISD/6-31G(d)[51]	45.0	34.8	57.9
SVWN/6-31G(d)[57]	33.2	19.8	
BLYP/6-31G(d)[57]	36.0	29.3	
B3LYP/6-31G(d)[57]	42.0	34.2	
B3LYP/6-31+G(d)	41.7	33.7	54.0
SVWN/6-31G(d)/AM1	34.5	21.3	
BLYP/6-31G(d)/AM1	37.1	31.2	
B3LYP/6-31G(d)/AM1	43.3	35.9	
Experimental	44.7±2.0	33.3±0.5	59.7

We will now examine the case of the Cope rearrangement with a heteroatom involved in bond breaking and bond making such as oxygen, the new rearrangement is called the Claisen

Rearrangement. This reaction is also well studied by experimentalists [58, 59] and theoreticians [57]. This system represents a difficult computational problem and it is quite difficalt to reproduce the experimental value of 36 kcal/mol [58, 59]. HF/6-31G(d) overestimates the activation barrier (47.7 kcal/mol) with slightly better energy (42.5 kcal/mol) obtained by CASSCF/6-31G*(6,6). The best energy (26.8 kcal/mol) was obtained with B3LYP/6-31G(d) calculation. Both BLYP and SVWN produce even lower activation energies (~20 kcal/mol). Although lower than experimental, B3LYP again computes that best activation barrier.

3.1.3. [1.5] Sigmatropic Shifts

We will present the results for a 1,5-hydrogen shift in an open chain molecule, 1,3-pentadiene and the 1,5-hydrogen shift in bicyclo molecule on the example of the isomerization of Bicyclo[6.1.0]nona-3,5-diene to Bicyclo[6.1.0]nona-2,4-diene. For both of these systems experimental isomerization energies are available. Let us first examine of the activation energy

Table 6. Computed activation barrier (kcal/mol) for [1,5] proton shift in 1,3-pentadiene [60]

Theory Model	E_RI	E_RII	E_{TS}	ΔE_I	ΔE_{II}
B3LYP	-195.312249	-195.309806	-195.248126	40.2	38.7
B3LYP(0 K)	-195.198451	-195.195765	-195.137492	38.2	36.5
BLYP	-195.196720	-195.194232	-195.137649	37.1	35.5
BLYP(0 K)	-195.086202	-195.083488	-195.030327	35.1	33.4
SVWN	-194.185873	-194.184212	-194.143354	26.7	25.6
SVWN(0 K)	-194.075067	-194.073148	-194.035307	24.9	23.7
B3LYP/6-31G(d)/AM1	-195.309752	-195.307126	-195.242996	41.9	40.2
BLYP/6-31G(d)/AM1	-195.194099	-195.191410	-195.132270	38.8	37.1
SVWN/6-31G(d)/AM1	-194.184646	-194.183145	-194.139529	28.3	27.4
Experimental [61-62]					36.1(2.2)

E_RI = total energy (a.u.) of (*E*)-1,3-pentadiene; E_RII = total energy (a.u.) of (*Z*)-1,3-pentadiene; E_{TS} = total energy (a.u.) of the transition state; ΔE_I = activation barrier (kcal/mol) in regard to (*E*)-1,3-pentadiene; ΔE_{II} = activation barrier (kcal/mol) in regard to (*Z*)-1,3-pentadiene

for 1,3-pentadiene isomerization. It was experimentally determined that the reaction threshold is 36.1±2.2 kcal/mol [61-62]. The computed transition state structures with various DFT methods are very similar with a C-H bond distance of C-H-C hydrogen transfer part of around 1.41 Å. This suggest that a single point energy calculation on AM1 geometries might also be a correct approach to calculate the activation energies. The best agreement between computed and experimental values are again obtained with hybrid B3LYP and gradient-corrected BLYP DFT methods. The B3LYP computed energies differ by 0.4 kcal/mol which is in the margin of experimental error. The local DFT method, SVWN, computes energies that are too low (Table 6)

Let us now investigate [1,5] sigmatropic hydrogen shift if vinylallene. This reaction was studied by Skattebol [63] with determined experimental barrier of 28.9 kcal/mol. All three hybrid DFT methods (B3LYP, B3P86, and B3PW91) compute activation barriers that are between 28 and 30 kcal/mol [64].

The next example for the [1,5] sigmatropic shift is bicyclo[6.1.0]nona-3,5-diene isomerization to bicyclo[6.1.0]nona-2,4-diene. There are two possible transition state structures as presented in Figure 2, depending of the mutual orientation of the rings. It was

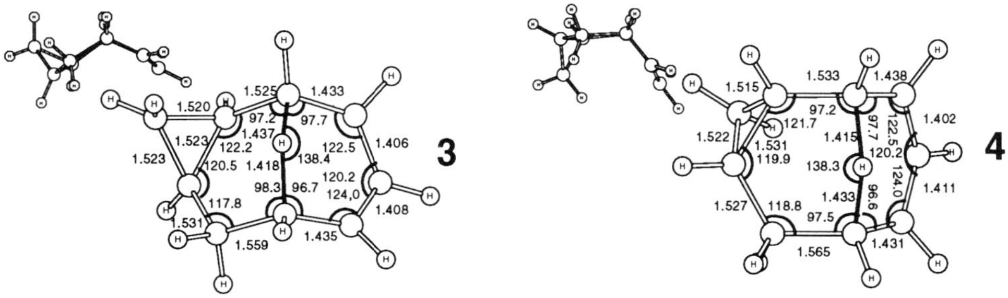

Figure 2. Geometries of the transition state structures **3** and **4** calculated with BLYP/6-31G(d)

determined experimentally that the isomerization barrier through transition state **3** is 30.5 kcal/mol, while the same barrier through **4** is 32.5 kcal/mol [65]. Our computed reaction threshold [66] is extremely close to the experimental values (Table 7). Unfortunately, the computational study was not done with the B3LYP method, nevertheless it demonstrates that gradient-corrected BLYP produces energies that are far more reliable than nonlocal SVWN and are closer to the experimental results that one obtains by MP2 ab initio study.

Table 7. Activation and relative energies (kcal/mol) for [1,5]-hydrogen shift in bicyclo[6.1.0]-nona-3,5-diene [66]

Theory Model	through **3**	through **4**	E_4-E_3
HF/3-21G	53.2	55.3	2.1
HF/6-31G(d)	57.2	59.9	2.7
MP2/6-31G(d)//HF/6-31G(d)	34.1	35.7	1.8
SVWN/6-31G(d)	24.1	24.8	0.7
BLYP/6-31G(d)	31.3	33.1	1.8
Experimental [65]	30.5	32.5	2.0

3.2. Cycloaddition Reactions

Cycloaddition reactions are reactions involving addition of two or more unsaturated molecules to form a new ring. Their nomenclature is based on the number of π electrons involved in bonding changes when the two reactants are used. Thus [2 + 2] cycloaddition is the formation of cyclobutane from two molecules of ethylene and a [4 + 2] cycloaddition is the formation of cyclohexene from butadiene and ethylene.

3.2.1. [4 + 2] Cycloaddition (Diels-Alder) Reactions

The parent Diels-Alder reaction of 1,3-butadiene with ethylene to form cyclohexene is the prototype of a thermally allowed cycloaddition. It has been studied extensively both by semiempirical [67-69] and ab initio techniques [70-72]. The Diels-Alder reaction is the core of pericyclic reactions. The mechanism of the reaction has been the subject of much extensive study and disagreement. After a decade of controversy, involving a heated debate on the usefulness and quality of semiempirical vs. ab initio methods, the concerted mechanism of this reaction has been generally accepted. It has been shown, however, that the energy barriers calculated by using Hartree-Fock theory are grossly overestimated, while the second order

Møller-Plesset correlation energy correction level are underestimated. Consequently, the very expensive MP3 or superior MP4 or QCID levels are necessary for this reaction. Considering the above, an evaluation of the performance of DFT methods in predicting the reaction barriers would be beneficial for the investigation of other all-carbon reactions.

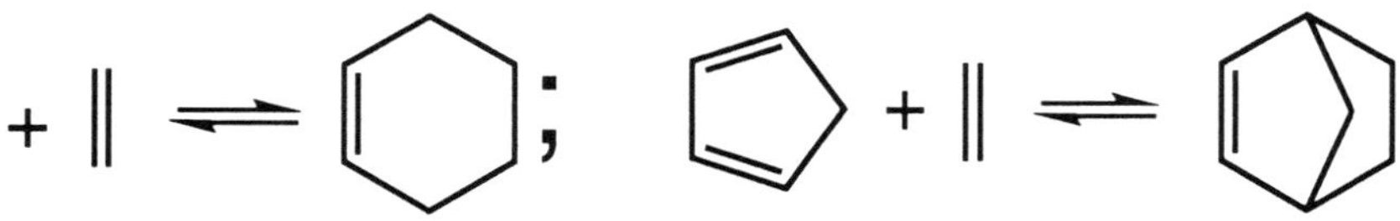

Some of the geometric features of the transition state obtained by various DFT methods for ethylene addition to 1,3-butadiene are presented in Table 8. The major disagreement is

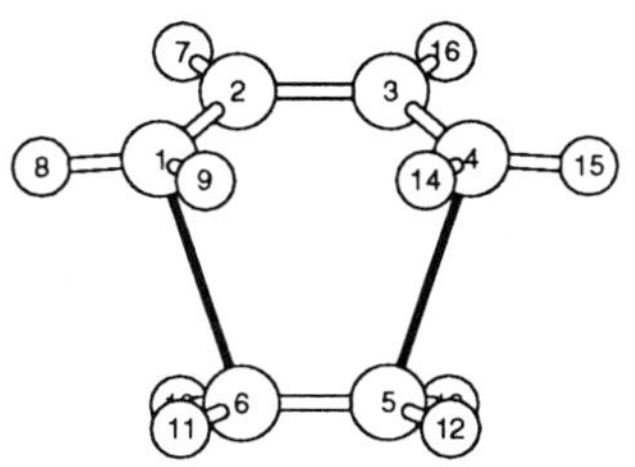

5

Table 8. Structural parameters for the transition state of the ethylene addition to 1,3-butadiene

Theory Model	r12/Å	r16/Å	r23/Å	r56/Å	a216/°	a123/°	d6123/°
HF/6-31G(d)[73]	1.377	2.201	1.393	1.383		121.5	
CASSCF/6-31G(d)[74]	1.398	2.223	1.397	1.404			
SVWN/6-31G(d,p)[75]	1.363	2.400	1.415	1.361	100.8	122.5	59.7
BLYP/6-31G(d,p)[75]	1.395	2.286	1.414	1.398	102.5	122.2	57.6
B3LYP/6-31G(d,p)[75]	1.383	2.268	1.406	1.386	102.3	122.0	57.9

for computed bond distance formation. The highest level ab initio calculations reported here estimate the bond distance to be 2.223 Å. Both gradient corrected BLYP and hybrid B3LYP generates transitions state structures with newforming bond distances that deviate 0.04-0.06 Å. SVWN predicts the transition state structure with considerably longer newforming carbon-carbon bonds. Recently Goldstein, Beno and Houk [76] have studied ethylene addition to butadiene. Energies of reaction and geometries of the reactants and products are in good agreement with available experimental results.

Similar structural differences are obtained for the transition state of ethylene addition to cyclopentadiene (Table 9). The BLYP and B3LYP computed transition state **6** are in close agreement, while SVWN generates too long of carbon-carbon bond distances for the forming bonds. Geometries are product of energy optimization for the studied chemical system, thus it is more appropriate to compare the computed energies with experimental data (Table 10). The acceptable activation barrier for the Diels-Alder reaction was obtained with the QCISD(T) theory model. It is interesting to point out that both MP2 as well as MP4 underestimates the energies for [4+2] cycloaddition reactions. In regard to DFT methods, SVWN, as well as exchange functionals, only (Xα and HFB) produces the activation energies that are quite different from experimental values. In the case of ethylene addition to cyclopentadiene, SVWN even predicts a negative reaction barrier (Table 10). The best values are obtained with the B3LYP hybrid DFT method. The computed and experimental energies differ by 1 kcal/mol (Table 10).

The Diels-Alder reaction is one of the key reactions in the preparation of large organic compounds that have important biological activities and/or are isolated from natural resources. Although hybrid and gradient-corrected DFT methods require less computational resources, they are applicable to large molecular systems and are equally accurate ab initio methods, they are still not applicable to large organic molecules. Due to the close similarity of the Diels-Alder transition state computed with the AM1 and B3LYP methods, we have explored the possibility to use DFT-AM1 approach for the evaluation of the Diels-Alder activation barrier [79].

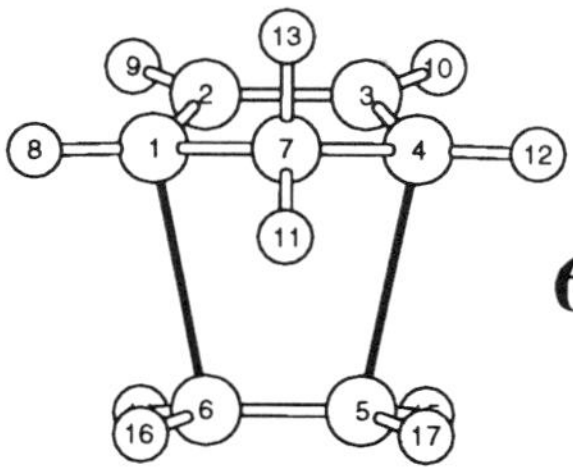

6

Table 9. Structural parameters for the transition state of ethylene addition to cyclopentadiene

Theory Model	r12/Å	r16/Å	r23/Å	r56/Å	a216/°	a123/°	d6123/°
HF/6-31G(d)[73]	1.389	2.193	1.392	1.382			
SVWN/6-31G(d,p)[75]	1.379	2.367	1.412	1.361	97.8	109.0	74.7
BLYP/6-31G(d,p)[75]	1.412	2.260	1.413	1.400	100.3	109.0	73.4
B3LYP/6-31G(d,p)[75]	1.398	2.244	1.405	1.387	100.1	108.9	73.5

Table 10. Computed activation barrier (kcal/mol) for ethylene addition to 1,3-butadiene and cyclopentadiene

	transition state **5**		transition state **6**	
Theory Model	without ZPEC	with ZPEC	without ZPEC	with ZPEC
RHF/6-31G(d) [73]	45.0		39.7	
MP2/6-31G(d)/HF/6-31G(d)[73]	16.6		10.6	
QCISD(T)/CASSCF/6-31G(d)[74]	25.5			
Xα/6-31G(d,p)[75]	4.7	6.7	2.3	4.4
HFB/6-31G(d,p)[75]	31.4	33.8	29.7	32.0
SVWN/6-31G(d,p)[75]	1.7	3.6	-0.6	1.4
BLYP/6-31G(d,p)[75]	20.7	23.1	18.8	21.2
B3LYP/6-31G(d,p)[75]	22.4	24.9	20.1	22.5
Experimental[77,78]		25.1		22.5

The computed AM1 and B3LYP/6-31G(d) transition state structures for cyanoethene addition to cyclopentadiene are not so similar as one would expect based on the parent cycloaddition reaction. The addition of 1,1-dicyanoethylene to cyclopentadiene has an asynchronous transition state. There is disagreement on the level of asynchronicity (Δr is 0.384 Å for AM1 and 0.789 Å for B3LYP/6-31G(d) computed transition states). Furthermore, AM1 computes a more compact transition state structure. Therefore, it is not surprising that the computed activation barrier does not agree well for the 1,3-dicyanoethylene addition. On the contrary hand, B3LYP/6-31G(d) energy on BLYP/6-31G(d) geometries closely follows experimental activation barriers (Table 11) [79]. It is conceivable that B3LYP/6-31G(d)/AM1 will produce acurate activation barriers for chemical systems without heteroatoms in close proximity of the reaction centers.

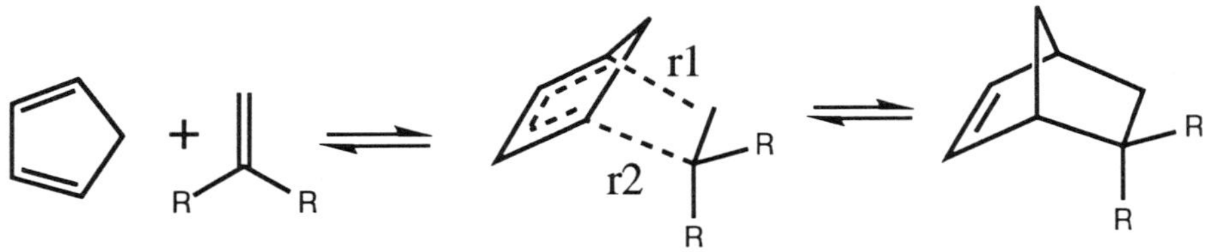

Table 11. Energies (total energies in a.u., activation energies in kcal/mol) and geometries for cyanoalkenes addition to cyclopentadiene

Theory Model	transition state r1(Å)	r2(Å)	ΔE(kcal/mol) (calculated)	ΔE(kcal/mol) (exp.)
dienophile=ethylene				
AM1	2.106	2.106	28.0[69]	
B3LYP/6-31G(d)/AM1			20.9	
B3LYP/6-31G*	2.244	2.244	19.9[69]	22.5[78]
dienophile=*endo*-acrylonitrile				
AM1	2.059	2.189	30.4[69]	
B3LYP/6-31G(d)/AM1			18.0	
B3LYP/6-31G(d)	2.076	2.469	16.4	15.2[80]
dienophile=*exo*-acrylonitrile				
AM1	2.052	2.156	29.0[69]	
B3LYP/6-31G(d)/AM1			17.6	
B3LYP/6-31G(d)	2.078	2.469	16.0	
dienophile=1,1-dicyanoethylene				
AM1	1.953	2.337	28.7[69]	
B3LYP/6-31G(d)/AM1			12.3	
B3LYP/6-31G(d)	1.990	2.779	9.31	~8.0[70]

There are some examples where B3LYP/6-31G(d) drastically fails to predict the activationbarrier for highly exothermic reactions having low activation barrier. It was estimated that the cycloaddition of benzene to N-methylpyrrole proceeds with an activation barrier between 3 to 5 kcal/mol [81]. The computed activation energy by the Becke3LYP/6-31G* and Becke3LYP//AM1 and are -2.71 and -0.33 kcal/mol, respectively. It seems that all DFT methods evaluate negative activation barriers for highly exothermic low barrier chemical reactions.

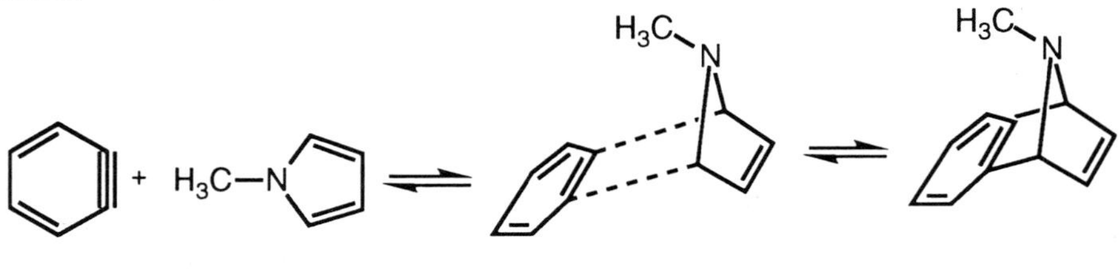

3.3. Proton Abstraction and Addition Reactions

The proton transfer between neutral and radical molecules is an important organic reaction [82, 83]. It is the only reaction that has any practical application for nonfunctionalized alkanes. A large amount of information is available regarding the relative reactivity of different kinds of hydrocarbons toward free radicals and the variations of behavior observed among different abstracting radicals [84, 85]. It is generally accepted that for alkanes, hydrogen becomes more easily abstracted on proceeding along the series primary, secondary and tertiary. The magnitudes of the differences depend on the radical removing the hydrogen; the more reactive ones are less selective and those of lower reactivity are more selective. Radical selectivity can be modified by the conditions; aromatic solvents, for example, apparently complex with the chlorine radical and increase its selectivity [86]. The selectivity is not very high and there is considerable interest in the chemical community to have a reliable computational method that shows the details of this kind of chemical transformation.

3.3.1. Proton Transfer between Methane and Methyl Radical

For this reaction there is an experimentally determined activation barrier of 14.1 kcal/mol [87]. We have performed an extensive study of this system with hybrid DFT methods [88]. In this particular case B3LYP/6-31G(d) theory model predicts an activation barrier of 14.4

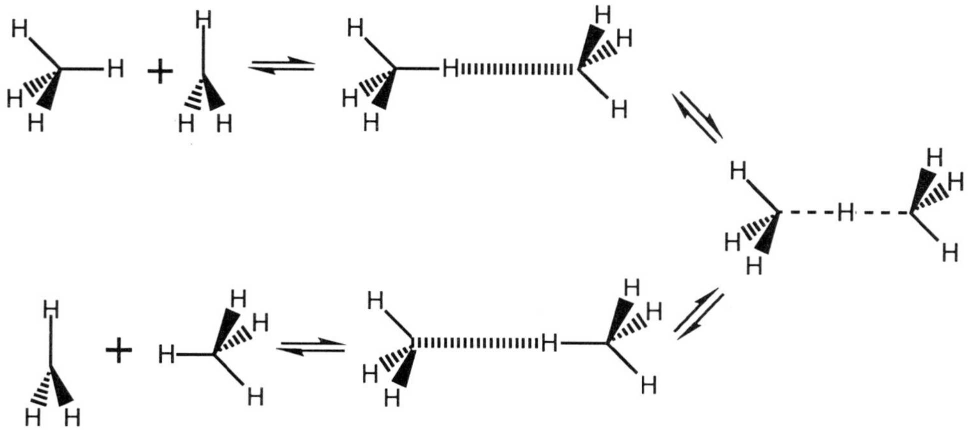

kcal/mol, however with a larger basis set, 6-311++G(3df,3pd), the computed activation barrier is 15.6 kcal/mol. Other DFT methods such as B3P86/6-311+G(2d,2p) also generate an accurate activation barrier of 13.8 kcal/mol [138]. Both ROHF/6-311+G(2df,2pd) and MP2/6-311+G(2df,2pd) overestimate the activation barriers (35.2 kcal/mol, and 19.2 kcal/mol), while local DFT methods generate a low activation barrier (~ 3 kcal/mol). The double-well potential energy surface of the proton transfer is favored by ab initio methods and by low level hybrid DFT methods. When an extended basis set, such as 6-31+G(d,p) or higher, was used, the energy of the complex and reactants was indistinguishable. Thus, the activation barriers computed by ab initio methods were much higher than the experimental estimates. Therefore, one has reason to suspect that the shallow minimum predicted by ab initio methods might be artifact.

3.3.2. Trichloromethyl Radical Proton Abstraction Reactions

The trichloromethyl radical has the highest selectivity for the hydrogen abstraction reaction [89, 90]. There is considerable interest in computational chemistry toward free radical behavior [91, 92]. Unfortunately, inexpensive ab initio methods [93] do not produce satisfactory results. Our results demonstrate that hybrid and gradient-correlated DFT methods can be methods of choice for analyzing these systems [94]. We successfully demonstrated the superiority of hybrid and non-local DFT methods, with LYP correlation over both UHF and MP2 ab initio methods (Table 12) for the computation of acurate activation energies. One

Table 12. Activation energies (kcal/mol) for hydrogen abstraction with trichloromethyl radical from alkanes calculated by using 6-31G(d) basis set

	UHF	MP2	B3LYP	BLYP	Exp.[95]
Methane	31.5	21.2	19.3	16.9	17.9±1.2
Ethane	29.2	17.2	15.7	12.9	14.2±0.6
Propane	27.4	13.7	13.1	10.1	10.6±1.6
2-methylpropane	26.0	10.6	11.0	8.0	7.7±0.5

method, BLYP, should be considered for the calculation of qualitative values for activation energies, because the computed activation energy should be 1-2 kcal/mol in difference from the experimental values. It could also be possible that the MP2 *ab initio* method, with a large basis set such as 6-311++G(3df,3pd), would also produce reliable values. Because this theory model is fairly computationally expensive and can only be applied to small chemical systems, we believe that using Becke3LYP or better BLYP with a modest basis set such as 6-31G(d) will have broad application for similar chemical systems.

3.3.3. Hydrogen Abstraction with the Fluorine Radical from Hydrogen Molecules

This is of "difficult" computational problems. Low, and middle level ab initio methods do not even approximate the experimentally determined activation barrier [96]. Even the heat of reaction [97] is not easy to compute. We have examined this reaction with both ab initio and DFT methods [98]. Neither ROHF nor MP2 compute satisfactory energies for this system. It is necessary to perform complete QCISD(T)/6-311++G(3df,3pd) to obtain excellent agreement with experiment. The ab initio and DFT methods compute totally different transition state structures. With some of the DFT methods, such as LSDA, the transition state was not possible to locate (Table 13). Although hybrid DFT methods show lower basis set sensitivity,

Table 13. Classical barrier height (ΔE_I) and exothermicity (ΔE_{II}) in kcal/mol for $H_2 + F \mapsto HF + H$ calculated with the assumption that the transition state structure is linear

Theory model	r_{HH}(Å)	r_{HF}(Å)	ΔE_I	ΔE_{II}
ROHF/6-311++G(3df,3pd)	0.829	1.179	27.60	17.43
SVWN/6-311++G(3df,3pd)				51.26
B3LYP/6-311++G(3df,3pd)	0.743	2.908	0.07	29.82
B3P86/6-311++G(3df,3pd)	0.744	2.880	0.21	31.83
BLYP/6-311++G(3df,3pd)	0.748	3.009	0.11	32.50
BP86/6-311++G(3df,3pd)	0.752	2.941	0.09	34.09
MP2/6-311++G(3df,3pd)	0.769	1.424	4.56	41.15
QCISD(T)/6-311++G(3df,3pd)	0.764	1.550	1.86	32.09
Experimental[96, 97]			~2.0	33.7

it was demonstrated that the calculation with lower basis sets, such as 6-31+G(d,p), produce satisfactory results that are in excellent agreement with 6-311+G(2d,2p) , and with 6-311++G(3df,3pd) computed values. The hybrid DFT methods prefer a slightly bent transition state structure (angle is ~175°). Hybrid density functional methods suggest that the reaction occurs without a very small reaction barrier. That observation is generally correct, if we assume that hybrid DFT methods underestimate the activation energies by around 2 kcal/mol. The computed heat of the reaction with Hybrid DFT methods agree with the experimental results quite well.

3.3.4. Hydrogen Abstraction with the Hydrogen and Hydroxyl Radicals from Hydrogen Molecules.

These are quite difficult computational problems. Even high level computational methods cannot reproduce experimental barriers. Let us first consider the reaction of hydrogen radical abstraction. A similar problem was attempted by Johnson and Pople [99]; they found that some DFT methods do not correctly predict reaction barriers. In light of Johnson and Pople's work we conducted a systematic study of the simplest hydrogen abstraction reaction calculated at full optimized geometry with extended 6-311+G(2d,2d).

All computational methods produced symmetric linear transition state structures with H-H bond distances between 0.917-0.973 Å (Table 14). Our attempt to locate the complex between the hydrogen molecule and hydrogen radical was not successful. After optimization, two reactants were calculated to be separated by a distance of at least 5 Å, with their total energy exactly the same as the sum of the individual energies of the hydrogen molecule and the hydrogen radical. Both ab initio and DFT methods suggest the reaction proceeds without

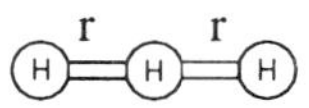

Table 14. Bond Distance in the Transition State Structure, Total Energies, and Barrier Heights of $H_2 + H \mapsto H + H_2$ Computed by Using 6-311+G(2d,2p) [100]

Method	r/Å	ΔE/kcalmol^{-1}
HF	0.932	17.67
MP2	0.917	13.77
G2		8.94
B3LYP	0.929	4.32
B3P86	0.928	3.27
B3PW91	0.929	5.23
HFS	0.973	5.62
HFB	0.950	11.90
XALPHA	0.960	5.76
BLYP	0.935	2.96
BP86	0.936	1.65
XALYP	0.945	-3.29
XAP86	0.946	-1.59
XAVWN	0.932	-1.74
SLYP	0.958	-3.33
SP86	0.960	-4.79
Experimental [101]		9.7

formation of the reactant complex. This is an astonishing finding because according to hybrid DFT methods, proton transfer between the methyl radical and methane should occur via a methane-methyl radical complex formation While HF and MP2 computational studies produce activation energies that are considerably higher than experimental value, almost all DFT methods drastically underestimate the barrier, with the exception of the gradient-corrected exchange-only method of HFB (Table 14). The qualitative performance is seen to depend simply on the type of exchange functional used. The barriers lie within a range of 3 kcal/mol. While hybrid and Becke exchange functional based DFT methods produce positive reaction barriers, the Slater and Xα based DFT methods compute negative barriers. Of the four ab initio methods, HF and MP2 both overestimate the barrier significantly, while G1 and G2 perform well, predicting a barrier of less than 1 kcal/mol from the experimental value. A remaining question is the performance of HFB that calculates a barrier which is much closer to experiment than any other DFT method. Direct examination of the features of the potential surface reveals that the agreement with experiment is indeed fortuitous. It seems obvious that hybrid B3PW91 performs the best of all DFT methods, although the activation barrier is underestimated by more than 4 kcal/mol. We have to be aware that this reaction is a special case due to the large tunneling effect.

Next we will examine the performance of DFT methods for the $H_2 + OH \mapsto H + HOH$ reaction (Table 15). This reaction was previously studied by Andzelm and coworkers by local and non-local DFT methods [103]. They were not able to determine the transition state, but walked the reactants through the transition state with a constant change in the H-O forming bond distance. They determined that all of the DFT methods, except Becke's three exchange functional (acm), would generate a negative reaction and suggest that an exact exchange should be incorporated in the DFT methods. Our extensive ab initio and hybrid DFT studies of this system are presented in Table 15 [102]. The best agreement between computational and experimental data was obtained with G2. That is expected, since this method was proven to produce highly reliable energies. Unfortunately, we have to state that the currently available DFT methods are not capable of correctly computing the energies for the $H_2 + OH \mapsto H + H_2O$ reaction. Only the hybrid DFT produce energies that are 2-4 kcal/mol below the experimental

Table 15. Computed activation barriers (kcal/mol) and heat (kcal/mol) for $H_2 + OH \mapsto H + H_2O$ reaction [102]

Theory Model	ΔE_I	ΔE_{II}	ΔE_{III}
UHF/6-31G(d)	26.7	26.6	0.1
UHF/6-31G(d)(0 K)	26.5	24.4	2.1
MP2/6-31G(d)	12.8	31.7	-18.9
MP2/6-311G(d,p)	9.9	29.0	-20.0
MP2/6-311+G(d,p)	9.2	30.8	-21.6
MP2/6-311G(2df,p)	9.3	30.0	-20.7
MP2/6-311+G(3df,2p)	8.0	28.8	-18.8
MP4/6-311G(d,p)	10.0	22.5	-12.5
MP4/6-311+G(d,p)	9.2	24.0	-14.8
MP4/6-311G(2df,p)	9.4	23.2	-13.8
QCISD/6-311G(d,p)	8.8	20.3	-11.5
G1(0 K)	7.2	20.5	-13.3
G1	5.6	19.2	-13.6
G2(0 K)	6.7	20.4	-13.7
G2	5.1	19.1	-14.0
G2MP2(0 K)	6.8	21.0	-14.2
G2MP2	5.2	19.7	-14.5
UB3LYP/6-31G(d)	3.2	9.7	-6.5
UB3LYP/6-311++G(3df,3pd)	0.9	13.7	-12.9
UB3LYP/6-311++G(3df,3pd)(0 K)	1.9	13.0	-11.1
UB3P86/6-311++G(3df,3pd)	-0.1	14.9	-15.0
UB3P86/6-311++G(3df,3pd)(0 K)	1.1	14.2	-13.1
UB3PW91/6-311++G(3df,3pd)	1.3	14.3	-14.3
UB3PW91/6-311++G(3df,3pd)(0 K)	2.4	13.6	-11.2
experimental[103]	4.0±0.7	18.6±3	-14.6±3

ΔE_I = activation barrier for forward reaction; ΔE_{II} = activation barrier for reverse reaction; ΔE_{III} = heat of the reaction

values. Based on these and our previous results, it is likely that the DFT methods will fail to produce a correct activation barrier for the reaction which includes the hydrogen radical as a reactant or a product. To improve the hybrid DFT methods, mixing with an HF portion should increase the energy, as HF estimates activation energies too high. Preliminary calculations confirmed that almost any barrier could be obtained with fitting the to resemble the semiempirical methods. Besides, the heat of reaction cannot be calculated correctly, even with this approach, because both the DFT and HF energies are too positive.

This clearly indicates that the majority of DFT methods are not capable of handling hydrogen abstraction reactions with small radicals. Thus, it is not surprising that there is much skepticism present in the scientific community regarding the DFT performance on radical abstraction reactions. To clarify this problem, the HFB computational results for five hydrogen abstraction reactions were compared with G1, G2, and experimental values [104]. It is well established that the latter two methods produce accurate energies.

It is well known that the *ab initio* computational methods which include an electron correlation will produce transition state structures that have shorter forming or breaking bond distances than the HF computed transition state structure. Generally, HFB would compute transition state structures for hydrogen radical abstraction reactions which are much closer to HF computed structures than to MP2 computed structures, but the computed activation barriers would be quite different. Both HF and MP2 considerably overestimate activation barriers.

The computed activation barriers for five radical proton abstraction reactions are presented in Table 16. As mentioned before, for these reactions, the majority of the DFT methods will

Table 16. Activation Barriers (kcal/mol) Computed with Becke's 88 (HFB) DFT and G1 and G2 *Ab Initio* Methods for Some Hydrogen Abstraction Reactions [105]

Theory Model	ΔE_I	ΔE_{II}	ΔE_{III}	ΔE_{IV}	ΔE_V
HFB6-31G(d)	11.9	7.8	16.5	15.4	8.3
HFB/6-31G(d) (0 K)	11.5	7.6	15.1	11.1	6.3
HFB/6-31G(d,p)	11.7	6.2	15.6	13.5	6.6
HFB/6-31G(d,p)(0 K)	11.0	6.3	14.2	9.4	4.7
HFB/6-311G(2d,2p)	11.9	5.3	16.3	13.4	6.2
HFB/6-311G(2d,2p)(0 K)	11.2	5.6	14.9	9.7	4.6
HFB/6-311++G(3df,3pd)	11.7	5.3	15.8	11.2	5.8
HFB/6-311++G(3df,3pd) (0 K)	11.0	5.8	14.4	7.8	4.4
G1	9.2	5.6	13.7	10.3	5.0
G1(0 K)	10.6	7.2	14.9	11.3	5.9
G2	8.9	5.1	13.1	10.2	4.9
G2(0 K)	10.4	6.7	14.3	11.2	5.8
Experimental	9.7	4.0±0.7	11-12	9.0-11.4	3.6

ΔE_I = Activation barriers for the $H_2 + H \mapsto H + H_2$ reaction; ΔE_{II} = Activation barriers for the $H_2 + OH \mapsto H + H_2O$ reaction; ΔE_{III} = Activation barriers for the $CH_4 + H \mapsto CH_3 + H_2$ reaction ; ΔE_{IV} = Activation barriers for the $CH_4 + {}^3O \mapsto CH_3 + HO$; ΔE_V = Activation barriers for the $CH_4 + OH \mapsto CH_3 + H_2O$ reaction; (0 K) is for the energies at 0 K.

compute negative activation barriers. HFB computed activation barriers which are all positive and close to experimental values (Table 16). There was also a slight sensitivity to the size of the basis set used in the calculation. Middle size basis sets, such as 6-311G(d,p) or 6-311G(2d,2p), are sufficient for the chosen reactions because of the *d* functional on heavy atoms and the *p* functional on hydrogens. The computed activation barriers with HFB/6-311G(d,p) or HFB/6-311G(2d,2p) DFT theory models have a 1-2 kcal/mol difference from the experimental values. If the HF/6-311G(2d,2p) activation barriers computed with a zero-point energy correction (0 K) are compared with the G2 (0 K) energies, then the DFT method performs even better. Only for the $H_2 + H \mapsto H + H_2$ reaction is the G2 (0 K) computation closer to the experimental value. It should also be noted that the HFB/6-311G(2d,2p) theory model overestimated activation barriers for pericyclic reactions and it should only be applied for hydrogen radical abstraction reactions.

Based on the five difficult computational systems, the HFB/6-311G(2d,2p) DFT theory model should compute accurate reaction barriers which are even closer to the experimental values than the energies generated by the G2 *ab initio* approach. Thus, we recommend this theory model for the computation of activation barriers of hydrogen abstraction reactions with small radicals.

3.3.5. Addition of the Hydrogen Radical to Ethylene

There are extensive experimental [106, 107] and theoretical [108] studies of the $C_2H_4 + H \mapsto C_2H_5$ reaction system. Thus it is a classical case for determining the accuracy of DFT methods for computing the energy profile for hydrogen radical addition to alkenes. Our ab initio and hybrid computational results are presented in Table 17 [109]. It is well known that accurate heats of reaction and barrier heights are more difficult to calculate directly than the geometries. It is surprising that even higher level of ab initio methods have a problem with computing the $C_2H_4 + H$ reaction threshold, contrary to the threshold computed with the UHF (Table 17). Usually UHF suffers from spin contamination and cannot reproduce the experimental reaction barriers. However, here it almost matches perfectly with the experimental value of 2.5 kcal/mol (Table 17). We believe that the agreement was coincidental. For the transition state, the spin unrestricted MP2 also suffers from serious spin contamination problems; the barrier was about 8-10 kcal/mol higher than it should have been. This problem

Table 17. Computed forward and reverse activation barriers (kcal/mol) and enthalpy (kcal/mol) for $C_2H_4 + H \mapsto C_2H_5$ reaction [109]

Theory model	ΔE_I	ΔE_{II}	ΔE_{III}
UHF/6-311G(2d,2p)	3.8	44.6	-40.8
UHF/6-311G(2d,2p)(0 K)	4.1	39.6	-35.5
MP2/6-311+G(2df,p)	10.3	45.5	-35.1
MP2/6-311+G(2df,p)(0 K)	12.1	41.6	-29.5
QCISD/6-311G(d,p)	3.6	44.7	-41.1
QCISD/6-311G(d,p)(0 K)	4.6	40.1	-35.5
G1(0 K)	3.5	36.4	-32.9
G1	2.4	36.2	-33.8
G2(0K)	2.9	37.0	-34.0
G2	1.9	36.8	-34.9
G2MP2(0 K)	3.3	37.0	-33.7
G2MP2	2.2	36.9	-34.7
B3LYP/6-311G(2d,2p)	0.1	42.6	-42.5
B3LYP/6-311G(2d,2p)(0 K)	0.8	38.1	-37.3
B3PW91/6-311G(2d,2p)		44.5	-59.1
B3PW91/6-311G(2d,2p) (0 K)		40.0	-54.0
Experimental [104, 105]	2.5	38.0	-35.5

ΔE_I = forward reaction barrier in kcal/mol; ΔE_{II} = reverse reaction barrier in kcal/mol; ΔE_{III} = enthalpy of the reaction in kcal/mol

was partially solved by using the less spin contamination affected QCISD ab initio method; with use of this method, the computed barriers are 1-3 kcal higher than they should be. The problem of spin contamination was avoided by using G1, G2 and G2MP2 computational approaches. Indeed, these computational methods produced an activation barrier that was around 2.5 kcal/mol.

There was a considerable problem in computing the transition state structure with the DFT methods. Our attempt to compute the transition state with BLYP, BP86 and BPW91 was unsuccessful. With Becke's three exchange functional, we had limited success. Two of these DFT methods were able to find the transition state structures, after continuous changes in the newforming C-H bond distances. The procedure was tedious and rather time consuming. When obtained, the structures with the maximal energy were further optimized as previously described. With many of the DFT methods (B3P86, BLYP, BPW91 and BP86), the optimization does not lead to a transition state structure, but rather to the separated reactants. To find and optimize the transition state structure with ab initio methods was a straightforward process. We have experienced the DFT optimization convergence problem in many previous cases, but it was very computationally expensive to (keyword scf=qc) bring the optimization towards a minimum. In the optimization of these radical transition state structures, even this approach was not successful.

Thus we were able to compute the reaction threshold only with the B3LYP hybrid over all other DFT methods. The B3LYP with a modest basis set, such as 6-31G(d), estimates the hydrogen radical addition barrier to ethylene to be negative. By putting a *p* orbital on the hydrogen, the computed activation barrier was now in better agreement with the experimental value. The zero point energy correction does not substantially change the computed activation barrier (Table 17).

Due to the lack of optimization of the transition state structure with other DFT methods, we cannot firmly state that they would also fail to correctly reproduce an experimental value for the activation barrier. If we compared the activation energies computed from the transition state structure obtained by the C-H bond distance variation, using BPW91 and BP86, the computed activation barrier should be around -16 kcal/mol.

The reaction barrier for the hydrogen elimination from the ethyl radical computed with the ab initio methods was too high. An excellent activation barrier of 38.1 kcal/mol was computed with B3LYP/6-311G(2d,2p). In addition, an acceptable heat of reaction of -37.3 kcal/mol was also computed. The other hybrid DFT method predicted a slightly worse activation barrier (40 kcal/mol), but the heat of reaction differed by 18.5 kcal/mol from the experimental values. These results are clearly not acceptable. We believe that with the proper corrected hybrid DFT methods, the total energies should be computed more accurately. Our results suggested, that the major cause for an incorrect behavior lies in the Slater exchange term in the exchange functional of the DFT methods. The Becke's three exchange functional and BLYP correlation functional included a portion of the exact exchange, and the B3LYP provided a much better description of the reaction profile than any other DFT method. The properly corrected DFT methods should be able to compute the potential energy surface for the reaction that involves the hydrogen radical.

3.4. Elimination reactions

The hybrid DFT methods compute the elimination reaction threshold generally well, if in the reaction combination carbon and hydrogen bonds are breaking. The accuracy is similar to the previously discussed cycloaddition reactions. For example, the computed activation barrier for the elimination of HONH2 from the N-oxyethylamine computed with BLYP and B3LYP is comparable to high level ab initio calculations [110].

3.4.1. Elimination of Hydrogen from Formaldehyde

The formaldehyde photodissociation has been the subject of experimental [111-113] and theoretical [114-117] computational studies. Two dissociation products, $H_2(^1\Sigma_g^+)$ and $CO(^1\Sigma^+)$, close to the formaldehyde first singlet exited state (80.6 kcal/mol) , were observed. Because the dissociation process is almost thermoneutral, it is believed that the activation barrier should be near the first singlet excitation threshold of 80.6 kcal/mol [111-113]. It was determined computationally that with increasing the basis set and the electron correlation by the selection of higher level ab initio methods, the estimated reaction barrier decreases considerably. Electronic correlation is crucial during the dissociation process because the ground state of formaldehyde is more related to $CO(^1\Pi)$ than to $CO\ (^1\Sigma^+)$.

The structural parameters for the transition state are given in Table 18 [118]. It is believed that a computational study of this reaction performed by Dupes and coworkers [119] is the most accurate up today. The MC10(DZDP) method is very close to full CI. It involves CI calculations including single and double excitations from the valence orbitals of the MCSCF configurations that are obtained by ten-electron, ten-orbital MCSCF calculations. The computed geometry is included in Table 18. All DFT methods generate a transition state structure quite similar to the MC10(DZDP) computed geometry; the best agreement was obtained by Slater's exchange functional DFT method (HFS). Relatively good agreement was obtained also with Becke's 88 exchange based methods. With these methods the computed transition state structures differ by less than 0.02Å in bond length and less than 2° in bond angle from the MC10(DZDP) computed transition state structure (Table 18). Other DFT methods compute a transition state structure that differs by a slight amount.

The MC10(DZDP) method estimates the reaction barrier to be 84.4 kcal/mol. Dupius and coworkers [119] concluded that the basis set error should be ~4.0 kcal/mol and the final barrier should be 80.9±3.0 kcal/mol. Frisch and coworkers [120] performed other high level ab initio calculations. Their MP4(SDTQ)/6-31G**//MP2/6-31G* estimates the reaction barrier to be 79.6 kcal/mol. We demonstrated that a HF calculation of the reaction barrier for formaldehyde dissociation reaction is not reliable. The estimated energy is too high. All three hybrid DFT methods (B3LYP, B3P86, and B3PW91) compute almost identical activation barrier that is around 84.5 kcal/mol. Excellent agreement is with the MC10(DZP) computed activation barrier [168]. If a computational study is performed with exchange functional only (HFS, HFB, Xα), considerably lower activation barriers are obtained (Table 18). The non-local Becke's 88 exchange functional based DFT methods (BPL, BP86, BPW91, BLYP, and BVWN) estimate the activation barrier to be around 79-80 kcal/mol. These values agree well with the MP4

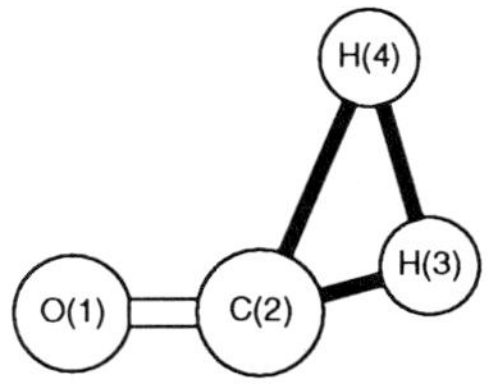

Table 18. Computed Geometric Parameters Reaction Barriers (ΔE_I, kcal/mol) and Heat of Reaction (ΔE_{II}, kcal/mol) for Dissociation of Formaldehyde To Hydrogen and Carbon Monoxide Computed by Using 6-311++G(3df,3pd) Basis Set

Method	r21/Å	r32/Å	r43/Å	a321/°	a432/°	d432/°	ΔE_I	ΔE_{II}
HF	1.126	1.099	1.286	164.5	89.9	0.1	100.	0.7
B3LYP	1.157	1.095	1.310	163.7	88.7	0.0	84.3	-8.2
B3P86	1.157	1.097	1.286	162.9	87.9	0.0	84.7	-12.2
B3PW91	1.157	1.098	1.291	163.1	88.0	0.2	84.5	-10.8
HFS	1.172	1.120	1.371	164.5	87.4	0.1	72.9	-14.9
HFB	1.178	1.102	1.425	166.4	89.4	0.1	74.4	-3.9
Xα	1.161	1.109	1.333	163.7	87.4	0.1	75.9	-18.8
BPL	1.168	1.090	1.361	164.9	89.3	0.0	79.2	-2.6
BP86	1.170	1.100	1.333	164.0	88.0	0.0	78.2	-10.6
BPW91	1.169	1.099	1.327	1.63.8	88.1	0.1	78.8	-10.2
BLYP	1.169	1.095	1.356	164.6	88.8	0.1	78.2	-6.8
BVWN	1.167	1.088	1.352	164.7	89.3	0.0	80.0	-3.7
XAPL	1.152	1.097	1.275	162.1	87.3	0.0	80.5	-25.8
XAP86	1.154	1.108	1.256	161.1	86.0	0.0	79.0	-34.0
XAPW91	1.153	1.106	1.250	161.0	86.1	0.0	79.6	-33.7
XALYP	1.153	1.203	1.272	161.8	86.8	0.1	79.3	-30.0
XAVWN	1.151	1.095	1.266	161.8	87.3	0.0	81.2	-27.0
SPL	1.163	1.108	1.309	162.9	87.2	0.1	77.5	-21.9
SP86	1.165	1.119	1.290	162.0	85.9	0.0	76.1	-30.0
SPW91	1.164	1.117	1.283	161.8	86.0	0.0	76.6	-29.7
SLYP	1.164	1.113	1.307	162.6	86.8	0.0	76.3	-26.0
SPW91	1.162	1.106	1.300	162.7	87.2	0.0	78.2	-23.1
MC10(DZDP)	1.190	1.100	1.371	163.7	87.4	0.0	84.4	-12.0

computed reaction barrier of 80.9±3.0 kcal/mol. To out surprise Xα exchange DFT methods produce excellent activation barriers as well. They are all around 80 kcal/mol (Table 18). In this case Slater's exchange functional based DFT methods (SPL, SP86, SPW91, SLYP, and SVWN) behave as expected and underestimate the activation barrier (Table 18).

3.4.2. Elimination of Hydrogen from Protonated Formaldehyde

The elimination of hydrogen from protonated formaldehyde is a computationally difficult problem. The experimental energy profile for the $CH_2OH^+ \mapsto CHO^+ + H_2$ reactions are well established [121-123] and allow us to properly estimate the accuracy of the DFT methods. Our computational results are presented in Table 19 [124]. Previously, the computed activation energies were not in satisfactory agreement with experiment. For example Lee and coworkers computed the activation energies with HF/6-31G** theory model to be 30-40% higher than the experimental values [121]. Slightly better agreement with experiment for the forward reaction was obtained, but not for the reverse activation barrier; complete active space self-consistent field (CASSCF) calculations were used [122, 123]. These two independent computational

Table 19. The 6-311++G(3df,3pd) computed reaction barriers (kcal/mol) and heat of $CH_2OH^+ \mapsto CHO^+ + H_2$ reaction (kcal/mol)

Method	ΔE_I	ΔE_{II}	ΔE_{III}
HF	104.5	64.9	39.6
QCISD(T)	89.1	61.2	27.9
G1	81.3	54.5	26.8
G2	81.3	54.3	27.0
G2MP2	81.3	54.5	26.8
B3LYP	87.7	49.2	38.5
B3P86	86.0	43.7	44.3
B3PW91	85.8	45.2	40.6
HFS	75.4	32.9	42.5
HFB	82.5	58.9	23.6
Xα	77.5	30.9	46.6
BPL	84.8	55.0	29.8
BP86	80.5	42.4	38.1
BPW91	80.7	43.5	37.2
BVWN	85.2	54.3	30.9
BVWN5	85.0	55.0	30.0
XAPL	79.7	26.2	53.5
XAP86	75.6	13.6	62.0
XAPW91	75.7	14.5	61.2
XAVWN	80.0	25.5	54.5
XAVWN5	79.8	26.2	53.6
SPL	77.6	28.3	49.3
SP86	73.6	15.8	57.8
SPW91	73.7	16.7	57.0
SVWN	80.0	27.5	52.5
SVWN5	77.8	28.2	49.6
HF/6-31G**[121]	108.4	66.4	42.0
CASSCF [122, 123]	93.5	73.4	20.1
Experimental [121, 122]	83.7	47.9	35.8

ΔE_I = activation barrier for H_2 elimination from CH_2OH^+ in kcal/mol; ΔE_{II}= the reverse activation barrier in kcal/mol; ΔE_{III} = enthalpy of the reaction in kcal/mol

studies suggest that *ab initio* computational methods may have a problem correctly evaluating the energy profile for a $CH_2OH^+ \mapsto OCH^+ + H_2$ type reaction. To confirm this finding, we evaluated energies for this reaction with the HF and G2 *ab initio* methods by using the large 6-311++G(3df,3pd) basis set [124]. Our HF/6-311++G(3df,3pd) computed energies are in close agreement with the results of Lee and coworkers HF/6-31G** computed values [121], indicating little basis set sensitivity after 6-31G**. The quadratic configuration interaction calculation, including single and double substituents with a triple contribution added to the energy [QCISD(T)], was selected as one of very reliable *ab initio* computational methods. Indeed, the computed activation barriers are now closer to the experimental values. The improvement is very good for the forward activation barrier, but the computed value for the reverse reaction is only slightly improved Next we turn to one of the most accurate *ab initio* methods, G2; it was designed for accurate energy evaluation. The computed energies are usually close to experimental results. In our case the computed forward activation barrier with G1, G2, and G2 are ~ 2 kcal/mol lower than the experimental values, which is considered as excellent agreement between experiment and theory. Unfortunately, the reverse reaction barrier is ~15% higher. With this overestimation of the reverse reaction barrier, the enthalpy of the reaction is underestimated by around 10 kcal/mol. These results clearly demonstrate that *ab* *initio* methods have a problem correctly describing the energy profile for the $CH_2OH^+ \mapsto OCH^+ + H_2$ reaction.

Hybrid DFT methods have proven in many of our studies, on neutral chemical systems, to produce energies that agree with experimental values quite well. The B3LYP/6-311++G(3df,3pd) theory model indeed produces a much better energy profile for the studied elimination. All energies are ~5% higher than the experimental values (Table 19), with zero-point energy correction the energy usually decreases by around 2-3 kcal/mol. The computed energy profile is almost identical to the experimental energy profile. In this case we have again demonstrated the high accuracy of the hybrid B3LYP method for the computation of the reaction energy profile.

3.4.3. *Syn* Elimination Reaction of Organic Amine Oxide and Sulfoxide

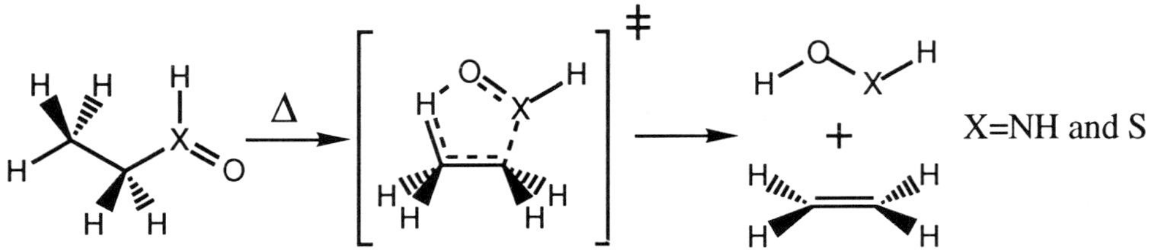

Many organic molecules that contain neighboring hydrogen and a proton abstracting group upon heating are known to undergo decomposition with the formation of anolefinic product. The mechanism for those reactions are generally believed to be cyclic. A transition state in which an intramolecular proton is transferred, accompanies the elimination and produces a new carbon-carbon double bond. These reactions normally do not require acidic or basic catalysts; however, there is a wide variation in the temperature at which the elimination proceeds at a convenient rate [125].

The cyclic transition state dictates that elimination will occur with *syn* stereochemistry [126, 127]. This mechanism, known as the Ei mechanism or as the internal elimination mechanism, was first proposed by Hurd and Blunck [128]. In our case, a five-membered Ei mechanism operates. Almost all evidence suggests that the transition state must be planar. For example, the deviation from planarity is not found in the elimination of hydroxylamine from amine oxides. Indeed, this is why the six-membered heterocyclic nitrogen compounds do not

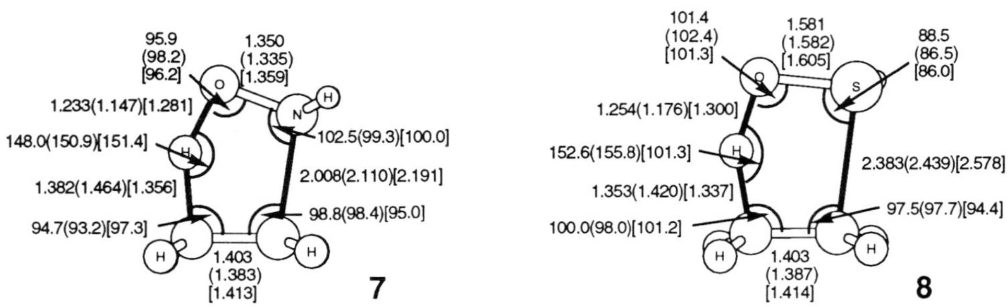

Figure 3. The transition structures for *syn* elimination of the oxides calculated with MP2/6-31G(d), SVWN/6-31G(d) (in parentheses) and BLYP/6-31G(d) (in brackets)

react [127]. In addition, only moderate reaction conditions are necessary to perform this reaction; usually, the amine oxides are eliminated at 100-150°C [125], selenoxides and sulfoxides are eliminated below 100°C [129-133]. Because of the stereoselectivity of this reaction and the lack of rearrangement of the products, it is useful for the formation of trans cycloolefins with eight-membered rings or higher. Recently, Erickson and Kahn performed the *ab initio* study (geometries optimized at RHF/6-31G(d) and the energy was evaluated at the MP2/6-31G(d)) for ethyl formate, xanthate, and phosphinate eliminations [134].

The DFT computed geometric parameters for transition states **7** and **8** are quite different from the MP2/6-31G(d) generated structures (Figure 3) [135]. As usual, the SVWN/6-31G(d) predicts a more compact structure than the BLYP/6-31G(d), while the MP2/6-31G(d) parameters are somewhere in the middle. The bond distance disagreement is maximal for bonds in formation or while breaking for associated bond angles (up to 5%). Nevertheless, all applied methods predicted for all transition states tend to be planar.

The activation energies were calculated as the energy difference between the transition structures and the reactants in eclipsed and the reactant in global minima (Table 20). All levels of theory predict the correct order of reactivity, as known from experimental data.

Table 20. Estimated activation energies (kcal/mol) for the elimination reactions [135]

Structure	RHF/6-31G(d)	MP2/6-31G(d)	SVWN/6-31G(d)	BLYP/6-31G(d)
7[a]	37.1	20.3	12.9	12.9
7[b]	45.0	26.4	15.6	16.3
8[a]	42.0	28.4	18.5	18.3
8[b]	47.2	32.7	21.1	21.3

[a]calculated with the reactant in eclipsed conformation
[b]calculated with the reactant in global minima

The nitrogen oxides are widely used for the preparation of olefins (Cope reaction); the sulfoxide requires a slightly elevated temperature. The predicted activation energies at the MP2/6-31G(d) theory level are predicted to be 26.4 and 32.7 kcal/mol respectively. The MP2/6-31G(d) estimation tends to be too high as compared with experimental observations. It is well known that the majority of elimination reactions with nitroxide can be performed under 100^oC and the sulfoxide under 150^oC [125, 133]. Thus the activation energy should be around 20 kcal/mol. It is somewhat unusual for MP2 calculations to actually overestimate the activation energy, that is unusual because the MP2 is known to underestimates the activation energies [136]. The DFT predicted activation energies are considerably lower than one estimated with MP2/6-31G(d) and they are in the case of sulfoxide **8** around 10 kcal/mol lower than MP2/6-31G(d). Furthermore, one unusual phenomena was observed; both DFT methods predict very similar activation energies that would be expected considering the experimental observations.

3.5. Isomerization Reactions

Hybrid and gradient-corrected DFT methods in general compute isomerization reaction energy profiles with a high degree of accuracy for molecules that contain only carbon and hydrogen atoms. A slightly better result can be obtained with hybrid DFT methods, particularly with B3LYP and B3PW91. We would like to demonstrate that some DFT methods are reliable for the computation of isomerization reaction thresholds for chemical systems that include heteroatoms.

3.5.1. Isomerization of Isocyanide and methylisocyanide

The isomerization of $CH_3NC \mapsto NCCH_3$ and $HNC \mapsto NCH$ has attracted much theoretical and experimental interest. Most of the interest for $CH_3NC \mapsto NCCH_3$ isomerization is related to whether it follows the RRKM mechanism [137]. On the other hand, $HNC \mapsto NCH$ has served as a model system for new computational methods [138, 139]. Ziegler and coworkers investigated both of these isomerizations with the HFS [140], LDA, and NL [141] DFT methods. The transition state structure and the potential energy surface for $HNC \mapsto HCN$ has been studied by a number of authors [139-142]. All three DFT methods produce transition state structures that are much closer to the structure computed with QCISD(T), than with the RHF and MP2 ab initio methods. (Table 21). In this way, we have demonstrated that these three DFT methods produce geometries that are QCISD(T) ab initio quality.

The QCISD(T)/6-311+G(2d,2p) computed activation energy with zero-point correction is 30.1 kcal/mol. The barriers calculated with RHF, LSDA, B3LYP, BLYP, and MP2 by using the 6-311+G(2d,2p) basis set are 34.9, 30.7, 30.1, 31.7, and 32.0 kcal/mol, respectively. Thus B3LYP produces the identical reaction barrier as QCISD(T) ab initio study [142].

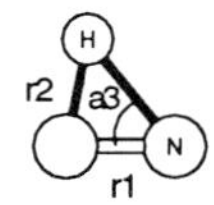

Table 21. Geometric parameters for The transition State Involved in HNC $\mapsto$ HCN computed with ab initio and DFT methods [143]

Theory model	r_1/Å	r_2/Å	a_3/Å
RHF/6-311+G(2d,2p)	1.161	1.157	78.2
LSDA/6-311+G(2d,2p)	1.182	1.203	69.9
B3LYP/6-311+G(2d,2p)	1.181	1.193	71.4
BLYP/6-311+G(2d,2p)	1.193	1.214	69.0
MP2/6-311+G(2d,2p)	1.188	1.175	73.4
QCISD(T)/6-311+G(2d,2p)	1.195	1.192	70.9

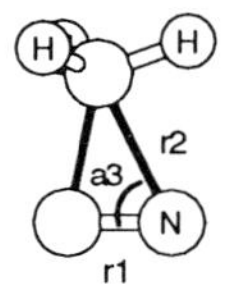

Table 22. Geometric parameters for The transition State Involved in $CH_3NC \mapsto CH_3CN$ computed with *ab initio* and DFT methods [143]

Theory model	r_1/Å	r_2/Å	a_3/°
RHF/6-311+G(2d,2p)	1.165	1.755	79.0
LSDA/6-311+G(2d,2p)	1.193	1.751	70.1
B3LYP/6-311+G(2d,2p)	1.187	1.805	73.1
BLYP/6-311+G(2d,2p)	1.199	1.850	71.5
MP2/6-311+G(2d,2p)	1.197	1.777	75.5

The computed transition state geometry for methylisocyanide (Table 22) shows the same behavior, as it was observed for the isocyanide transition state structure (Table 21) . Nevertheless, their structural features are very similar; the C-N multiple bond distance is lightly below 1.20 Å and the C-C is around 1.8 Å. The transition structure computed with B3LYP and BLYP is slightly different than that computed with both RHF, LSDA, and MP2 computational methods; it is more symmetrical [143].

The energy barrier for the CH_3NC isomerization is determined experimentally [144] as 38.4 kcal/mol. The best computed value with RHF was obtained with a 6-311+G(2d,2p) basis set and zero-point correction (Table 23). It is 41.8 kcal/mol which is 3.4 kcal higher than the experimental value. The MP2 calculation with 6-311+G(2d,2p) affords a barrier of 40.6 kcal/mol, which is still too high compared with the experimental value. To the best of our knowledge, the barrier calculated by the NL method of 39.5 kcal/mol [145] is the best computed value obtained with DFT methods up to now. If corrected for zero-point energy, the barrier becomes 37.5 kcal/mol. Our calculations with LSDA using 6-311+G(2d,2p) produced a reaction barrier of 40.2 kcal/mol; with the zero-point correction (~2.0 kcal/mol), the activation barrier becomes 38.2 kcal/mol. As stated above, we have performed many computational studies of ab initio hard to handle systems; in these computational studies we have found that hybrid, B3LYP, DFT methods produce remarkable agreement with experimentally determined values. Here, we have some astonishing results. The B3LYP/6-311+G(2d,2p) computes the activation energy to be 39.9 kcal/mol; with zero-point correction it is 38.2 kcal/mol; this differs by only 0.2 kcal/mol from the experimental value (Table 23). B3LYP is close in performance to the gradient-corrected BLYP, with 38.0 kcal/mol for the activation barrier (Table 23).

Table 23. Total Energies (Hartrees), Activation Barrier (kcal/mol), and Exothermicity (kcal/mol) for $CH_3NC \mapsto CH_3CN$ Reaction [143]

Theory model	$\Delta E_{barrier}$(kcal/mol)	$\Delta E_{exo.}$(kcal/mol)
RHF/6-311+G(3df,3pd)/ZPC	42.8	-21.1
LSDA/6-311+G(2d,2p)	40.2	-25.3
B3LYP/3-21G	50.1	-26.4
B3LYP/3-21G/ZPC	48.0	
B3LYP/6-31G(d)	41.0	-24.1
B3LYP/6-31G(d)/ZPC	39.3	
B3LYP/6-31+G(d)	41.5	-22.2
B3LYP/6-31+G(d)/ZPC	39.8	
B3LYP/6-311+G(d,p)	40.4	-24.0
B3LYP/6-311+G(d,p)/ZPC	38.7	
B3LYP/6-311+G(2d,2p)	39.9	-23.9
B3LYP/6-311+G(2d,2p)/ZPC	38.2	-23.8
BLYP/3-21G	46.8	-27.1
BLYP/6-31G(d)	39.2	-24.5
BLYP/6-31+G(d)	39.5	-22.5
BLYP/6-311+G(d,p)	38.5	-24.5
BLYP/6-311+G(2d,2p)	38.0	-24.4
MP2/6-31G(d)	42.6	-28.5
MP2/6-31G(d)/ZPC	41.1	-28.4
MP2/6-31+G(d)	42.5	-26.3
MP2/6-31+G(d)/ZPC	41.5	-26.0
MP26-311+G(d,p)	40.6	-27.4
MP2/6-311+G(2d,2p)	41.2	-27.0
Experimental [144]	38.4	-23.7

The DFT computed heat of isomerization is also very close to the experimental results; both RHF and MP2 *ab initio* methods have problems to reproduce the experimental values. The RHF underestimates, while MP2 overestimates the energy by about 3 kcal/mol. Fore example B3LYP/6-311+G(2d,2p) with zero-point energy correction estimates the heat of the isomerization to be -23.8 kcal/mol, this differs by 0.1 kcal from experiment (Table 23). This is remarkable performance of the B3LYP hybrid DFT for computing both the reaction threshold and heat of reaction for $CH_3NC \mapsto NCCH_3$ isomerization.

3.5.2 Protonated Carbon Monoxide Isomerization

Now we will demonstrate the accuracy of DFT methods for the computation of activation barriers of charged molecular system. Unfortunately, there is no experimental data available. There is Nobes and Radom transition state structures computed with MP3/6-31G** ab initio theory model [138]. The best agreement is with the B3LYP hybrid DFT method. The bond distances are 0.008Å away from the reported values (Table 24). Becke's 88 exchange functional based DFT methods predict O-C bond distances that are acceptable, while the C-H bond distance is ~ 0.04 Å shorter. This deviation is quite similar for transition state structures computed with all other DFT methods (Table 24).

Nobes and Radom studied the basis set and CI effect on the theoretical energy barriers for the $HOC^+ \mapsto OCH^+$ transformation [138]. The result obtained by different ab initio methods ranged from 77.0 kcal/mol at HF/4-31G to 39.4 kcal/mol for MP3/6-311G***//MP3/6-31G**. The study revealed that a large basis set and higher level calculation always decreased the reaction barrier. In our calculation, 51.3 kcal/mol was obtained with HF calculations by using 6-311+G(2d,2p), quite a decrease from 100 kcal/mol obtained with 4-31G calculations previously [190]. Hybrid DFT methods (B3LYP, B3LYP, and B3PW91) compute the reaction barrier to be between 37 and 38 kcal/mol (Table 24) [146], which is quit close to the MP3/6-311G*** 39.4 kcal/mol. The other DFT methods estimate the reaction barrier even lower than 37 kcal/mol. They are in range from 30-36 kcal/mol. The difference between Slater's and Becke's exchange functional based DFT method is ~ 3 kcal/mol indicating that the influence of nonlocal correlation's to the energy barrier is not substantial. This wide variety of computed reaction barrier (30-38 kcal/mol) might suggest that with more extensive CI and larger basis sets the evaluated reaction barrier might be even lower.

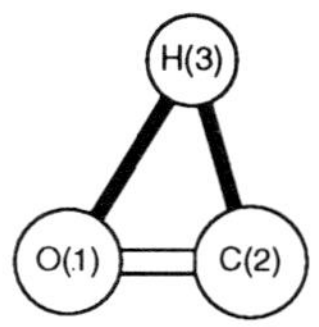

Table 24. Geometric Parameters Reaction Barrier (ΔE, kcal/mol) for Isomerization of Formyl Cation into Protonated Carbon Monoxide Computed by Using 6-311++G(3df,3pd) Basis Set

Method	r21/Å	r32/Å	a321	ΔE
HF	1.112	1.270	73.7	51.3
B3LYP	1.142	1.371	67.3	38.0
B3P86	1.141	1.308	67.3	37.3
B3PW91	1.142	1.309	67.2	37.2
HFS	1.159	1.367	65.0	31.5
HFB	1.170	1.359	65.6	34.4
Xα	1.147	1.349	65.0	32.9
BP86	1.158	1.334	65.5	33.9
BPW91	1.157	1.330	65.3	33.9
BLYP	1.157	1.343	65.4	35.0
XAP86	1.136	1.326	64.8	32.0
XAPW91	1.136	1.322	64.6	32.1
XALYP	1.135	1.225	64.8	33.2
SP86	1.148	1.343	64.8	30.6
SPW91	1.148	1.340	64.7	30.7
SLYP	1.147	1.352	64.8	31.8
SPW91	1.146	1.341	64.8	33.1
MP3/6-31G**	1.150	1.379		39.4

3.5.3. Methylperoxide Isomerization

The oxidation of the methyl radical with molecular oxygen involved many different competition reactions. It is believed that the first step of the reaction, combining the oxygen and methyl radical together, proceeds without barriers. The crucial process in the formation of the formaldehyde and hydroxyl radical is the isomerization reaction $CH_3OO \mapsto CH_2OOH$ isomerization which occurs [147]. Because of the well known difficulties in the theoretical characterization of radical reactions and the importance of the role that these reactions play in the combustion process, we have selected a systematic DFT study of this radical transformation.

There is no reliable information about the threshold energy for the $CH3OO \mapsto CH_2OOH$ transformation, since the reverse reaction has a barrier and the threshold energy is not related to the heat of reaction directly. Therefore, the calculations strongly depend on the assumption for the activation barrier for the methylperoxyl radical rearrangement. It is interesting that the barrier height for the peroxide radical rearrangement was assumed to be 36 kcal/mol by Reitel'boim and coworkers [148], while Hsu and coworkers [149] estimate that this barrier should be 60 kcal/mol. Saito and coworkers [150] believe that the rearrangement activation barrier should be around 50 kcal/mol. The DFT computed reaction barrier and threshold are presented in Table 25 [151]. The HF computed values are considerably higher than expected.

MP2, with electron correlation, considerably decreases both barriers. The agreement between the computed and experimental barrier for the peroxy radical isomerization is excellent. All of the hybrid DFT methods estimate energies which are in good agreement with both experimental and MP2/6-311+G(2d,2p) computed energies.

Table 25. Computed geometric parameters and activation energy for $CH_3OO \mapsto CH_2OOH$ isomerization obtained by using 6-311+G(2d,2p) basis set [151]

Method	r21/Å	r32/Å	r41/Å	r53/Å	a321/°	a432/°	a532/°	d5321/°	ΔE
HF	1.439	1.363	1.372	1.076	90.4	84.1	115.2	108.1	64.8
MP2	1.434	1.391	1.286	1.082	87.6	87.2	114.7	108.1	51.0
B3LYP	1.506	1.382	1.332	1.087	88.7	86.4	115.3	107.4	50.1
B3P86	1.484	1.377	1.322	1.087	88.7	86.3	115.3	107.4	48.9
B3PW91	1.486	1.378	1.324	1.088	88.7	86.1	115.3	107.4	49.3
HFS	1.509	1.390	1.328	1.118	88.8	88.2	115.3	107.0	42.2
HFB	1.610	1.414	1.353	1.107	87.7	87.7	115.3	107.7	47.2
Xα	1.482	1.374	1.310	1.104	88.9	88.0	115.4	107.0	43.3
BVWN	1.561	1.396	1.325	1.088	88.0	87.0	115.3	107.6	48.6
BVWN5	1.565	1.398	1.328	1.090	88.0	87.0	115.3	107.6	45.5
BL	1.566	1.399	1.329	1.091	88.0	87.1	115.3	107.6	46.0
BLYP	1.544	1.396	1.329	1.095	88.1	87.3	115.3	107.4	46.9
BP86	1.526	1.392	1.320	1.098	88.2	87.2	115.3	107.4	48.8
BPW91	1.524	1.391	1.319	1.096	88.2	87.1	115.3	107.4	48.7
XAVWN	1.450	1.356	1.283	1.086	88.8	87.9	115.4	107.0	44.4
XAVWN5	1.453	1.358	1.286	1.087	88.9	87.8	115.4	107.0	41.2
XAPL	1.454	1.359	1.286	1.088	88.9	87.8	115.4	107.0	41.6
XALYP	1.446	1.356	1.286	1.092	88.9	88.2	115.4	106.8	42.9
XAP86	1.427	1.352	1.276	1.095	88.9	88.6	115.3	106.8	44.5
XAPW91	1.427	1.351	1.275	1.093	88.9	88.4	113.4	106.7	44.5
SVWN	1.475	1.372	1.301	1.099	88.8	87.9	115.4	107.0	43.3
SVWN5	1.478	1.373	1.303	1.101	88.8	88.0	115.3	107.0	40.2
SPL	1.479	1.375	1.304	1.102	88.7	88.0	115.3	107.0	40.6
SLYP	1.471	1.372	1.304	1.106	88.8	88.3	115.3	107.8	41.8
SP86	1.450	1.368	1.293	1.109	88.8	88.7	115.2	106.8	43.4
SPW91	1.449	1.367	1.293	1.107	88.8	88.5	115.3	106.7	43.4
Experimental									50.0

The gradient corrected DFT methods (BPL, BP86, BPW91, BLYP, BVWN, and BVWN5), which are Becke's 88 exchange functional based DFT methods, slightly underestimate the reaction barrier. In many of our previous studies we have been discouraged from using Xα and Slater's exchange functional because the computed energies are usually much lower than the experimental values. Once again, we have demonstrated this phenomena. The computed isomerization barrier is too low. This suggests once again that these methods should be excluded when the reaction barrier for the organic free radical reactions are computed.

The isomerization barrier was previously computed by Abashkin and coworkers [145] at MP2/MP2, MP4/MP2, QCISD(T)/MP2, and NLSD level of theory. The barriers are 51.7, 52.4, 52.6, and 49.1 kcal/mol, respectively; they are in excellent agreement with our DFT computed values (Table 25).

3.5.4. Oxywater Isomerization

There are no available experimental results for oxywater isomerization. The most accurate *ab initio* calculation was performed by Schaefer, III and coworkers [152] with the CCSD(T)/TZ2Pf theory model. According to the CCSD(T)/TZ2Pf the activation barrier is 3.3 kcal/mol and the heat of reaction is -6.8 kcal/mol (Table 26). All hybrid DFT methods [151] find the oxywater molecule to be a stationary point on the PES of the $H_2OO \mapsto HOOH$ isomerization, thus supporting its existence and kinetic stability. In this way our DFT methods fully agree with Schaefer's [152] conclusion, which was based on a high level of *ab initio* calculations, that oxywater can be made. The G2 and hybrid DFT methods predict that the activation barrier for the [1,2]-hydrogen shift in oxywater is around 4 kcal/mol. The oxywater molecule is 46 kcal/mol less stable than hydrogen peroxide. These results are again in excellent agreement with the high level of *ab initio* calculations [152]. And, finally, we have demonstrated that hybrid DFT methods are reliable techniques to compute activation energies for low barrier, highly exothermic reactions if the reaction does not include a hydrogen radical in any of its steps.

Table 26. The Computed Isomerization Barrier (ΔE_I) and Exothermicity (ΔE_{II}) in kcal/mol for $H_2O \mapsto$ HOOH isomerization [151]

Theory	ΔE_I	ΔE_{II}
HF	20.9	-39.4
HF(0 K)	17.6	-38.4
CCSD(T)/TZ2P+f	5.7	-46.8
CCSD(T)/TZ2P+f (0 K)	3.3	-46.8
G2	3.9	-45.8
G2 (0 K)	4.0	-45.8
B3LYP	7.6	-45.2
B3LYP (0 K)	4.8	-45.1
B3P86	7.0	-46.6
B3P86 (0 K)	4.4	-46.5
B3PW91	7.1	-46.6
B3PW91 (0 K)	4.4	-46.5

3.5.5. Oxywater Radical Cation Isomerization

Recently, Schaefer, III and coworkers [153] explored the stability of the oxywater radical cation with a high level of *ab initio* methods, in particularly with the CCSD(T)/TZ2Pf theory model. Thus, fairly accurate energies are available for this system, making it a perfect system for the evaluation of DFT accuracy in the computation of radical cations. The relative energies in regard to the oxywater radical cation of these four stationary points (oxywater, transition state, cis-hidrogen peroxide, and trans-hydrogen peroxide radical cations) on the potential energy surface are presented in Table 27 [151]. It was demonstrated that the geometries for

Table 27. Relative energies (kcal/mol) for species involved in $H_2OO^+ \mapsto HOOH^+$ isomerization relative to H_2OO^+ [151]

Theory	$\Delta E(H_2OO^+)$	ΔE(TS)	$\Delta E(cHOOH^+)$	$\Delta E(tHOOH^+)$
ROHF/TZ2Pf	0.0	50.3	4.6	-3.5
ROHF/TZpf (0 K)	0.0	54.4	5.3	-2.5
CCSD/TZ2Pf	0.0	42.6	-15.4	-23.6
CCSD/TZ2Pf (0 K)	0.0	32.6	-14.8	-22.6
G2	0.0	32.5		-22.3
G2 (0 K)	0.0	32.6		-22.4
B3LYP	0.0	37.7	-15.8	-23.9
B3LYP (0 K)	0.0	34.2	-15.2	-23.0
B3P86	0.0	37.1	-16.6	-24.7
B3P86 (0 K)	0.0	33.6	-16.0	-23.8
B3PW91	0.0	37.3	-16.5	-24.6
B3PW91 (0 K)	0.0	33.8	-15.9	-23.6

four thes four stationary points on the potential energy surface of $H_2OO^+ \mapsto HOOH^+$ are computed accurately with the B3LYP hybrid DFT computational method. It was further shown that three hybrid DFT methods (B3LYP, B3P86, and B3PW91) will generate accurate energies for this species, at least as accurate as the CCSD and G2 *ab initio* computational methods will produce. Contrary to neutral hydrogen peroxide, which was shown to have an open book structure according to the experimental data and our DFT computational studies, hydrogen peroxide is planar with a *trans* isomer preference by 7.8 kcal/mol when computed with the hybrid DFT method. The same energy difference was obtained with both the CCSD and G2 *ab initio* methods. The reaction barrier of the oxywater radical cation isomerization is considerably higher than that of neutral oxywater. The hybrid DFT barrier is between 33-34 kcal/mol, which is in excellent agreement with the 32-33 kcal/mol computed by CCSD and G2.

In general, it can be concluded that hybrid DFT methods are powerful computational tools for computing the geometries, energies and vibrational spectra of "distonic" open shell, charged isomers. Other DFT methods usually do not produce such reliable results.

3.6. Multiple Proton Transfer Reactions

A multiple proton transfer is one of the most important reactions in nature, so it has been studied extensively [154, 155]. Nevertheless, the majority of studies were performed on systems where a single proton transfer occurred. Therefore, it is of particular interest to perform theoretical studies of these systems.

3.6.1. Multiple Proton Transfer in the Formic Acid Dimer

In formic acid dimer the proton transfer might occur in a stepwise or concerted mechanism. To the best of our knowledge, there is no decisive experimental evidence for either of these two mechanisms. For example, Limbach and coworkers have studied the double proton transfer in prototropic tautomerisms for many of the formamidine systems and porphyrins by using the NMR technique [156-160]. Their results support both a stepwise and concerted mechanism for the proton transfer. Recently, Kim [161] has shown by using a high level of *ab initio* computational study on the example of formic acid that the two protons are transferred synchronous with a transition state structure of D_{2h} symmetry. Actually, the formic acid dimer is one of the most extensively studied systems. There are plenty of experimental and theoretical data. The formic acid dimer has strong hydrogen bonding and it is very easily determined by both IR and Raman spectroscopies [162-164]. Most of the earlier studies were focused on the geometrical changes in the course of dimerization and the energy stabilization due to dimerization [165-167]. There are many *ab initio* quantum chemical studies of dimerization and proton transfer. For example, Scheiner and coworkers [168] have studied the proton transfer and dimerization energy in hydrogen bonded systems. Most of the studies for the double proton transfer have been based on a one-dimensional double well potential. The most accurate computational studies predict an energy dimerization of -14.2 kcal/mol and an activation barrier for the synchronous two proton transfer of 8.9 kcal/mol [161].

The computed transition state geometries for the synchronous double proton transfer vary with different DFT methods. As expected, the hybrid DFT methods compute the transition state geometry that is very close to the MP2 geometry. The Xα and Slater's exchange based DFT methods generate the transition state geometry which is almost identical to the formic acid dimer, indicating the downfall of these DFT methods.

The estimated relative energies follow the pattern observed in the computation of geometries. The energetic profile computed with the G2 computational method is in excellent agreement with the experimental values (Table 28). The MP2, MP4, and QCISD(T) ab initio methods overestimate the dimerization stabilization. Consequently, the computed activation barrier is too high. To obtain accurate energies for this system a high level of ab initio study is necessary. The G2 produces very good energies because many corrections are included (corrections for diffuse functional on non-hydrogen atoms, correction for higher polarization functions on non-hydrogen atoms, correction for non-additivity, correction for basis set enhancement, correction effects beyond fourth-order perturbation, and zero-point energy correction). Only two of the DFT methods come close to the experimental values, and they are B3 and B exchange functional incorporated DFT methods. Their uncorrected energies deviate by ~ 4 kcal/mol from the experimental values. If corrected for the zero point vibrational energies, the computed activation barrier becomes closer to zero.

The results suggest that both of the geometries for the formic acid dimer, and energy profile formic acid dimer proton transfer cannot be described adequately by MP2, MP4, QCISD(T) and current density functional theory methods. It was demonstrated that the computed data are less sensitive due to the inherent nature of the correlation functional incorporated in the DFT method than on the exchange functional. With a proper correction of the exchange portion of the gradient-corrected DFT methods this system should be described adequately. The results that are obtained with the current DFT methods for similar systems should therefore be treated with extreme caution and verified at higher levels of theory where possible.

Table 27. Computed Dimerization Stabilization (ΔE_I), Relative Energy of Transition State (ΔE_{II}), and Activation Barrier for Proton Transfer in Formic Acid Dimer (ΔE_{III}) [169]

Theory Model	ΔE_I/kcalmol^{-1}	ΔE_{II}/kcalmol^{-1}	ΔE_{III}/kcalmol^{-1}
HF/6-311G(2d,2p)	-14.0	3.1	17.1
MP2/6-311G(d,p)	-16.5	-7.9	8.7
MP2/6-311+G(3df,2p)	-16.3	-8.4	8.0
MP4/6-311G(2df,p)	-17.8	-10.5	7.3
QCISD(T)/6-311G(d,p)	-16.5	-6.9	9.6
G2	-14.2	-9.0	5.2
B3LYP/6-311G(2d,2p)	-17.9	-11.5	6.4
B3LYP/6-311G(2d,2p)(0 K)	-16.0	-13.5	2.5
B3P86/6-311G(2d,2p)	-19.2	-14.7	4.4
B3PW91/6-311G(2d,2p)	-17.2	-12.2	5.0
HFS/6-311G(2d,2p)	-28.1	-27.4	0.8
HFB/6-311G(2d,2p)	-10.4	-0.6	9.8
Xα/6-311G(2d,2p)	-29.9	-29.4	0.6
BP86/6-311G(2d,2p)	-18.6	-15.4	3.2
BPW91/6-311G(2d,2p)	-16.7	-13.1	3.6
BLYP/6-311G(2d,2p)	-17.2	-12.0	5.3
BVWN/6-311G(2d,2p)	-12.8	-5.0	7.8
XAP86/6-311G(2d,2p)	-45.0	-45.0	0.0
XAPW91/6-311G(2d,2p)	-42.7	-42.7	0.0
XALYP/6-311G(2d,2p)	-41.3	-41.3	0.0
XAVWN/6-311G(2d,2p)	-34.3	-34.3	0.0
SP86/6-311G(2d,2p)	-42.8	-42.8	0.0
SPW91/6-311G(2d,2p)	-40.6	-40.6	0.0
SLYP/6-311G(2d,2p)	-39.1	-39.1	0.0
SVWN/6-311G(2d,2p)	-32.3	-32.3	0.0
Experimental	-14.4	~-9	

4. CONCLUSION

There are twenty-six density theory computational methods that are presented in this chapter. As demonstrated, only a handful of them should be used for the study of reaction potential surfaces. Their performance strongly depends on the exchange functional of the DFT method. Thus, hybrid or Becke's three exchange functional based DFT methods perform the best in evaluating the activation energies. Particularly reliable agreement between experimental or high level of ab initio data and DFT computed data were obtained with B3LYP. It is recommended that the highest possible basis set should be used, since DFT methods are basis set sensitive. The DFT methods based on Xα and Slater's (S) exchange functionals are not recommended in computing reaction activation barriers. One in particular, SVWN, widely known as local spin density approximation (LSDA) might compute good energies for ring opening and some isomerizations, however, it is not in general reliable. This statement is in regard to the activation barrier, for the geometries of the reactants and products SVWN may perform well.

Although in many of our examples B3LYP performs exceptionally well, there are cases when even this DFT methods produces unacceptable energies. The highly exothermic reactions with a small activation barrier (2-5 kcal/mol) which involve reactions with the hydrogen radical are particularly difficult to reproduce. In some of our examples all DFT methods, including B3LYP, produce a negative activation barrier. Careful examination of the DFT methods suggest that the exchange functional part of the DFT method is responsible. Our preliminary results indicate that by increasing the Slater term in the exchange DFT functional, the problem can be corrected and better agreement with experimental data for low barrier reaction is achievable. We should be cautious with this approach because by changing the Slater portion in the exchange functional we can obtain any reaction barrier.

In using currently available DFT methods for computing the activation barriers, we suggest B3LYP or B3PW91 with an appropriate basis set. The computed activation energies above 10 kcal/mol should be reliable. Lower activation energies should be underestimated by 3-4 kcal/mol.

REFERENCES

1. Gaussian 94, Revision B.3, M. J. Frisch, G. W. Trucks, H. B. Schlegel, P. M. W. Gill, B. G. Johnson, M. A. Robb, J. R. Cheeseman, T. Keith, G. A. Petersson, J. A. Montgomery, K. Raghavachari, M. A. Al-Laham, V. G. Zakrzewski, J. V. Ortiz, J. B. Foresman, C. Y. Peng, P. Y. Ayala, W. Chen, M. W. Wong, J. L. Andres, E. S. Replogle, R. Gomperts, R. L. Martin, D. J. Fox, J. S. Binkley, D. J. Defrees, J. Baker, J. P. Stewart, M. Head-Gordon, C. Gonzalez and J. A. Pople, Gaussian, Inc., Pittsburgh PA, 1995.
2. D. R. Hartree, Proc. Cambridge Phil. Soc. 24 (1928) 89.
3. V. Fock, Z. Physik, 61 (1930) 126.
4. C. C. J. Roothan, Rev. Mod. Phys. 23 (1951) 69.
5. J. A. Pople and R. K. Nesbet, J. Chem. Phys. 22 (1959) 571.
6. R. McWeeny and G. Dirksen, J. Chem. Phys. 49 (1968) 4852.
7. C. Møller and M. S. Plesset, Phys. Rev. 46 (1934) 618.
8. M. J. Frisch, M. Head-Gordon and J. A. Pople, Chem. Phys. Lett. 166 (1990) 281.
9. J. A. Pople, R. Seeger and R. Krishnan, Int. J. Quantum Chem. Symp. 11 (1977) 149.
10. R. Krishnan and J. Pople, Int. J. Quantum Chem. 14 (1978) 91.
11. R. Krishnan, M. J. Frisch and J. A. Pople, J. Chem. Phys. 72 (1980) 4244.
12. K. Raghavachari, J. A. Pople, E. S. Replogle and M. Head-Gordon, J. Phys. Chem. 94 (1990) 5579.
13. J. A. Pople, M. Head-Gordon and K. Raghavachari, J. Chem. Phys., 87 (1987) 5968.
14. J. A. Pople, M. Head-Gordon, D. J. Fox, K. Raghavachari and L. A. Curtiss, J. Phys. Chem. 90 (1989) 5622.
15. L. A. Curtiss, C. Jones, G. W. Trucks, K. Raghavachari and J. A. Pople, J. Chem. Phys. 93 (1990) 2537.
16. L. A. Curtiss, K. Raghavachari, G. W. Trucks and J. A. Pople, J. Chem. Phys. 94 (1991).
17. L. A. Curtiss, K. Raghavachari and J. A. Pople, J. Chem. Phys. 98 (1993) 1293.
18. A. D. Becke, J. Chem. Phys. 98 (1993) 5648.
19. C. Lee, W. Yang and R. G. Parr, Phys. Rev. B, 37 (1988) 785.
20. B. Miehlich, A. Savin, H, Stoll and H. Preuss, Chem. Phys. Lett 157 (1989) 200.
21. J. P. Perdew, Phys. Rev. B 33 (1986) 8822.
22. J. P. Perdew and Y. Wang, Y. Phys. Rev. B 45 (1992) 13244.
23. J. C. Slater, Quantum Theory of Molecular and Solids. Vol. 4: Self-Consistent Field for Molecular and Solids, McGraw-Hill, New York, 1974.
24. A. D. Becke, Phys. Rev. A, 38 (1988) 3098.
25. J. P. Perdew and A. Zunger, Phys. Rev. B, 23 (1981) 5048.
26. S. H. Vosko, L. Wilk and M. Nusair, Can. J. Phys., 58 (1980) 1200.

27. M. J. Frisch, Æ. Frisch and J. B. Foresman, Gaussian 94 User's Reference, Gaussian, Inc. Pittsburgh, PA, 1995.
28. J. B Foresman and Æ. Frisch, Exploring Chemistry with Electronic Structure Methods:
29. R. B. Woodward and R. Hoffmann, J. Am. Chem. Soc. 87 (1965) 2046.
30. R. B. Woodward and R. Hoffmann, J. Am. Chem. Soc. 87 (1965) 4389.
31. R. Hoffmann and R. B. Woodward, J. Am. Chem. Soc. 87 (1965) 395.
32. R. Hoffmann and R. B. Woodward, J. Am. Chem. Soc. 87 (1965) 2511.
33. R. B. Woodward and R. Hoffmann, Angew. Chim. Int. Ed. Engl. 8 (1969) 781.
34. R. B. Woodward and R. Hoffmann, The Cnservation of Orbital Symmetry, Verlag Chemie, Weinheim, 1970.
35. J. E. Eksterowicz and K. N. Houk, Chem. Rev. 93 (1993) 2439.
36. M. J. S. Dewar and C. Jie, Acc. Chem. Res. 25 (1992) 537.
37. M. J. S. Dewar, A Semiempirical Life Profiles, in J. L. Seeman (Ed.0, Profiles, Pathways and Dreams, 1992
38. K. N. Houk, Y. Li and J. D. Evanseck, Angew. Chem. Int. Ed. Engl. 31 (1992) 682.
39. K. N. Houk, J. Gonzalez and Y. Li, Acc. Chem. Res. 28 (1995) 81.
40. M. J. S. Dewar, E. G. Zoebisch, E. F. Healy and J. J. P. Stewart, J. Am. Chem. Soc. 107 (1985) 3902.
41. M. J. S. Dewar and Kirchner, J. Am. Chem. Soc. 93 (1971) 4290, 4291, 4292.
42. J. Breulet and H. F. Schaefer III, J. Am. Chem. Soc. 106 (1984) 1221.
43. F. Bernardi, S. De, M. Olivucci and M. A. Robb, J. Am. Chem. Soc. 112 (1990) 1737
44. E. A. Kallel, Y. Wang and K. N. Houk, J. Org. Chem. 54 (1989) 6006.
45. B. S. Jursic and Z. Zdravkovski, Int. J. Quant. Chem. 56 (1995) 115.
46. R. W. Carr, Jr. and W. D. Walter, J. Phys. Chem. 69 (1965) 1073.
47. W. Cooper and W. D. Walters, J. Am. Chem. Soc. 80 (1958) 4220.
48. B. S. Jursic, unpublished results.
49. J. J. Gajewski, Hydrocarbon Thermal Isomerization, Academic Press, New York, 1981.
50. W. T. Borden, R. J. Loncharich and K. N. Houk, Ann. Rev. Phys. Chem. 39 (1988) 213.
51. H. Jiao and P. von R. Schleyer, Angev. Chem. Int. Ed. Engl. 43 (1995) 334 and references therein.
52. W. von E. Doering, V. G. Toscano, G. H. Beasley, Tetrahedron, 27 (1971) 5299.
53. M. J. Goldstein, M. S. Benzon, J. Am. Chem. Soc. 94 (1972) 5119.
54. M. J. Goldstein, M. S. Benzon, J. Am. Chem. Soc. 94 (1972) 7147.
55. W. R. Roth, F. Bauer, A. Beitat, T. Ebbrecht and M Wüstefeld, Chem. Ber. 124 (1991) 1453.
56. K. N. Houk, S. M. Gustafson and K. A. Black, J. Am. Chem. Soc. 114 (1992) 8565.
57. O. Wiest, K. A. Black and K. N. Houk, J. Am. Chem. Soc. 116 (1994) 10336 and references therein.
58. J. J. Gajewski and N. D. Conrad, J. Am. Chem. Soc. 101 (1979) 6693.
59. L. Kupczyk-Subotkowska, W. H. Saunders, Jr, H. J. Shine and W. Sobotkowski, J. Am. Chem. Soc. 115 (1993) 5957.
60. B. S. Jursic. *J. Chem. Soc. Perkin Trans 2* in press.
61. W. R. Roth and J. König, Ann. Chem. 699 (1966) 24.
62. W. R. Roth, Chimia 229 (1966) 20.
63. T. Skattebol, Tetrahedron, 25 (1969) 4933.
64. B. S. Jursic, unpublished results.
65. D. S. Glass, R. S. Boikess and S. Winstein, Tetrahedron Lett. (1966) 999.
66. B. S. Jursic, J. Mol. Struct. (Theochem) in press.
67. M. J. S. Dewar, S. Olivella and H. S. Rezepa, J. Am. Chem. Soc. 100 (1978) 5650.
68. M. V. Basilevsky, A. G. Shamov and V. A. Tikhomirov, J. Am. Chem. Soc. 23 (1971) 75.
69. B. S. Jursic and Z. Zdravkovski, J. Mol. Struc. (Theochem), 309 (1994) 249 and references therein.
70. K. N. Houk, R. J. Loncharich, J. F. Blake and W. L. Jorgensen, J. Am. Chem. Soc. 111 (1989) 9172. and references therein.
71. K. N. Houk, Y.-T. Lin and F. K. Brown, J. Am. Chem. Soc. 108 (1986) 554.

72. W. L. Jorgensen, D. Lim and J. F. Blake, J. Am. Chem. Soc. 115 (1993) 2936. and references therein.
73. R. D. Bach, J. W. Mc.Douall and H. B. Schlegel, J. Org. Chem. 54 (1989) 2931.
74. Y. Li and K. N. Houk, J. Am. Chem. Soc. 115 (1993) 7478.
75. B. S. Jursic and Z. Zdravkovski, J. Chem. Soc. Perkin Trans. 2 (1995) 1223.
76. E. Goldstein, B. Beno and K. N. Houk, J. Am. Chem. Soc. 118 (1996) 6036.
77. The experimental value for the ethylene addition to 1,3-butadiene is 27.5 kcal/mol. For the estimated activation barrier of 25.1±2 kcal/mol at O K, see 72.
78. R. Walsh and J.M. Wells, J. Chem. Soc., Perkin Trans. 2, (1976) 72.
79. B. S. Jursic, J. Mol. Struct. (Theochem) 358 (1995) 139.
80. J. Sauer, H. Wiest, and A. Mielert, Chem. Ber. 97 (1964) 3183.
81. B. S. Jursic, J. Mol. Struct. (Theochem) in press.
82. M. L. Poutsman, in "Free Radicals"; Vol. II; Kochi, J. K. ; Ed.; Wiley: New York, 1973.
83. C. Rüchardt,. Angew. Chem. Int. Ed. Engl. 97 (1970) 2757.
84. A. F. Trotman-Dickenson, Adv. Free Radical Chem. 1 (1965) 1.
85. J. M. Tedder, Angew. Chem. Int. Ed. Engl.21 (1982) 401.
86. J. C. Martin, in "Free Radicals"; Vol II; Kochi, J. K., Ed.; Wiley: New York, 1973, p.493.
87. J. A. Kerr and S. J. Moss, CRC Handbook of biomolecular and thermolecular gas reactions, Vol. 1 (CRC Press, Boca Raton, 1981)
88. B. S. Jursic, Chem. Phys. Lett. 244 (1995) 263.
89. W. H. S. Yu and M. H. Wijnen, J. Chem. Phys. 52 (1970) 4166.
90. G. A. Russell and C. De Boer, J. Am. Chem. Soc. 85 (1963) 3136 and references therein.
91. M. L. McKee, J. Am. Chem. Soc. 112 (1990) 7957
92. D. J. Pasto, R. Krasnansky and C. Zercher, J. Org. Chem. 52 (1987) 3062 and references therein.
93. W. J. Hehre, "Practical Strategies for Electronic Structure Calculations" Wavefunction, Inc.: Irvine, 1995.
94. B. S. Jursic, J. Mol. Struct. (Theochem) 366 (1996) 75.
95. I. Matheson and J. Tedder, Int. J. Chem. Kint. 14 (1982) 1033.
96. P. J. Knowles, K. Stark and H. J. Werner, Chem. Phys. Lett. 185 (1991) 555.
97. K. P. Huber and Herzberg, Constants of Diatomic Molecules , Van Nostrand Reinhold, New York, 1979.
98. B. S. Jursic, J. Chem. Phys. 104 (1996) 4151.
99. B. G. Johnson, C. A. Gonzalez, P. M. W. Gill and J. A. Pople, Chem. Phys. Lett. 221 (1994) 100 and reference theirein.
100. B. S. Jursic, J. Chem. Soc. Faraday Trans. submitted.
101. W. R. Schultz and D. J. LeRoy, J. Chem. Phys. 42 (1995) 3869.
102. B. S. Jursic, Int. J. Quant. Chem. in press.
103. J. Baker, J. Andzelm, M. Muir and P. R. Taylor, Chem. Phys. Lett. 237 (1995) 53. and references thereuin.
104. J. W. Ochterski, G. A. Petersson and K. B. Wiberg, J. Am. Chem. Soc. 117 (1995) 11299.
105. B. S. Jursic, J. Am. Chem. Soc. submitted.
106. Y. Feng, J. T. Niiranen, A. Benosura, V. D. Knyazev, D. Gutman and W. Tsang, J. Phys. Chem. 97 (1993) 871.
107. P. D. Pacey and J. H. Wimalasen, J. Phys. Chem., 88 (1984) 5657.
108. W. L. Hase, H. B. Schlegel, V. Balbyshev and M. Page, J. Phys. Chem. 100 (1996) 5354.
109. B. S. Jursic, J. Chem. Soc. Prkin Trans. 2 in press.
110. B. S. Jursic, J. Mol. Struct. (Theochem) in press.
111. C. B. Moore and J. C. Weisshaar, Ann. Rev. Phys. Chem. 34 (1983) 525.
112. W. M. Gelbart, M. L. Elert and D. F. Heller, Chem. Rev. 80 (1980) 403.
113. P. L. Houston and C. B. Moore, J. Chem. Phys. 65 (1976) 757.
114 L. Fan and T. Ziegler, J. Am. Chem. Soc. 114 (1992) 10890 (1992).

115. J. D. Goddard, Y. Yamaguchi and H. F. Schaefer III, J. Chem. Phys.75 (1981) 3459.
116. L. B. Harding, H. B. Schlegel, R. Krishnan and J. A. Pople, J. Phys. Chem. 84 (1980) 3394.
117. G. F. Adams, G. D. Bent, R. J. Bartlett and G. D. Purvis, J. Chem. Phys. 75 (1981), 834.
118. B. S. Jursic, Chem. Phys. Lett. submitted.
119. M. Dupuis, W. A. Lester Jr., B. H. Lengsfield III and B. Liu, J. Chem. Phys. 79 (1983) 6167.
120. M. J. Frisch, R. Krishnaa and J. A. Pople, J. Phys. Chem. 79 (1983) 6167.
121. T. G. Lee, S. C. Park, and M. S. Kim, J. Chem. Phys. 104 (1996) 4517.
122. G. Hvistendal and E. Uggerud, Org. Mass Spectrum. 26 (1991) 67.
123. T. Helgaker, E. Uggerud, and H. J. Aa. Jensen, Chem. Phys. Lett. 173 (1990) 145.
124. B. S. Jursic, Int. J. Quantum Chem. in press.
125. A. C. Cope and E. R. Trumbull, Org. React. 11 (1960) 317.
126. A. C. Cope and N. A. LeBel, J. Am. Chem. Soc. 82 (1960) 4656.
127. A. C. Cope, E. Ciganek, C. F. Howell and E. E. Schweizer, J. Am. Chem. Soc. 82 (1960) 4663.
128. C. D. Hurd and F. H. Blunck, J. Am. Chem. Soc. 60 (1938) 2419.
129. D. J. L. Clive, Tetrahedron 34 (1978) 1049.
130. C. A. Kingsbury. and D. J. Cram J. Am. Chem. Soc. 84 (1960) 1810.
131. C. Walling and L. Bollyky, J. Org. Chem. 29 (1964) 2699.
132. I. D. Entwistle amd R. A. W. Johnstone, J. Chem. Soc. Chem. Comm. (1965) 29.
133. T. Yoshimura, E. Tsukurimichi, Y. Iizuka, H. Mizuno, H. Isaji and C. Shimasaki, Bull. Chem. Soc. Jpn. 62 (1989) 1891.
134. J. A. Erickson and S. D. Kahn, J. Am. Chem. Soc. 116 (1994) 6271.
135. B. S. Jursic, J. Mol. Struct. (Theochem) in press.
136. W. J. Hehre, R. F. Stewart and J. A. Pople, *J. Chem. Phys.*, 1969, **51**, 2657.
137. P. Saxe, Y. Yamaguchi, P. Pulay, H. F. Schaefer III, J. Am. Chem. Soc. 102 (1980) 3718.
138. R. H. Nobes and L. Radom, Chem. Phys. 60 (1981) 1.
139. P. K. Pearson, H. F. Schaefer III, J. Chem. Phys. 62 (1975) 350.
140. L. Fan and T. Zigler, J. Chem. Phys., 90 (1990) 3645.
141. L. Fan and T. Zigler, J. Am. Chem. Soc., 114 (1992) 10890 and references 46-51
142. K. Saito, R. Ito; T. Kakumeto and A. Imamura, J. Phys. Chem. 90 (1986) 1422.
143. B. S. Jursic, Chem. Phys. Lett. 256 (1996) 213.
144. M. A. Reitel'boim, L. B. Romanovich and B. I. Vedeneev, Kinet. Katal. 19 (1978) 1399.
145. Y. Abashkin, N. Russo, E. Sicilia and M. Toscano, in "Modern density functional theory a tool for chemistry" J. M. Seminario and P. Politzer Eds., Elsevier, Amsterdam, 1995 p.255
146. B. S. Jursic, Int. J. Quantum Chem. submitted.
147. D. S. Y. Hsu, W. M. Shaub, T. Creamer, D. Gutman and M. C. Lin, Ber. Bunsenges. Phys. Chem. 87 (1983) 909.
148. M. A. Reitel'boim, L. B. Romanovich and B. I. Vedeneev, Kinet. Katal. 19 (1978) 1399.
149. D. S. Y. Hsu, W. M. Shaub, T. Creamer, D. Gutman and M. C. Lin, Ber. Bunsenges. Phys. Chem. 87 (1983) 909.
150. K. Saito, R. Ito; T. Kakumeto and A. Imamura, J. Phys. Chem. 90 (1986) 1422.
151. B. S. Jursic, J. Chem. Phys. Submitted.
152. H. H. Huang, Y. Xie and H. F. Schaefer III, J. Phys. Chem. 100 (1996) 6076.
153. Y. Xie, W. D. Allen, Y. Yamaguchi and H. F. Schaefer III, J. Chem. Phys. 104 (1996) 7615.
154 L. Melander and W. H. J. Saunders, Reaction Rates of Isotopic Molecules; Wiley: New York, 1980.
155. M. L. Bender, Mechanisms of Homogeneous Catalysis from Protons to Proteins, Wiley, New York, 1972.
156. G. Scherer and H.-H. Limbach, J. Am. Chem. Soc. 116 (1994) 1230.

157. M. Schlabach, H. -H. Limbach, E. Bunnenberg, A. Y. L Shu, B. -R. Tolf, B.-R. C. J. Djerassi, J. Am. Chem. Soc. 115 (1993) 4554.
158. L. Meschede and H. -H. Limbach, J. Phys. Chem. 95 (1991) 10267
159. G. Scherer and H. -H. Limbach, J. Am. Chem. Soc. 111 (1989) 5946.
160. D. Gerritzen and H. -H. Limbach, J. Am. Chem. Soc. 106 (1984) 869.
161. Y. J. Kim, J. Am. Chem. Soc. 118 (1996) 1522 and references therein.
162. J. E. Bertie, K. H. Michaelian and H. H. Eysel, H. H. J. Chem. Phys. 85 (1986) 4779.
163. J. E. Bertie and K. H. Michaelian, J. Chem. Phys. 76 (1982) 886.
164. R. C. Millikan and K. S. Pitzer, J. Am. Chem. Soc. 107 (1985) 3769.
165. K. I. Lazaar and S. H. Bauer, J. Am. Chem. Soc. 1985, 107, 3769.
166. A. Karpfen, Chem. Phys. 1984, 88, 415.
167. A. Almenningen, O. Bastiansen and T. Motzfeldt, Acta. Chem. Scand. 24 (1970) 747.
168. S. Scheiner, In *Proton Transfer in Hydrogen Bonded Systems*; Bountis, T.; Ed., NATO ASI Series B291; Plenum: New York, 1992; p 29.
169. B. S. Jursic, J. Mol. Struct. (Theochem) in press.

J.M. Seminario (Editor)
Recent Developments and Applications of Modern Density Functional Theory
Theoretical and Computational Chemistry, Vol. 4

Density Functional Theory as a Tool for the Prediction of the Properties in Molecules with Biological and Pharmacological Significance

Marcella Belcastro, Tiziana Marino, Tzonka Mineva, Nino Russo, Emilia Sicilia and Marirosa Toscano

Dipartimento di Chimica, Università della Calabria, I-87030 Arcavacata di Rende (CS), Italy

The performance of Gaussian Density Functional methods in the prediction of physico-chemical properties of a series of molecules with biological and pharmacological importance is briefly reviewed. The examples reported here cover a large range of systems going from nucleic bases, to neurotransmitters, to enzymes, to amino acids. Computed properties include structural parameters, vibrational frequencies, ionization potentials, proton affinities, isomerization processes, potential energy surfaces and solvent effects.
The agreement with the available experimental and high level theoretical data suggests that the density functional based methods are powerful challenges for the studies of bio-macromolecules and drug design. They can, in a near future, be of significant computational aid for predicting the properties of compounds relevant to the biotechnological and pharmaceutical industries.

1. INTRODUCTION

Computational molecular modelling of biomolecular systems has enjoyed rapid development in the last decade and is becoming a significant part of industrial research. Today many chemical and pharmaceutical industries (i.e. Monsanto, Dupont, Exxon, Merck, Bayer, Ciba, Rhone-Poulenc, etc.) have integrated preliminary computational strategies into their research activity for proposing and characterizing new drugs or materials. Starting from the basic idea that the properties and the behaviour of such systems can be described in terms of interactions between atoms, different theoretical procedures have been applied to macromolecules with biological and pharmacological significance.

For large biomolecules the high number of degrees of freedom restricts us to choose empirical methods which require reliable and transferable molecular force fields and are parametrized for well defined classes of compounds. Static and dynamical properties of systems with more than 10000 atoms have been investigated using modern computer architectures.

The elucidation of the nature of the chemical bonds as well as the energetic paths on potential energy surface (i.e. in chemical reactions catalyzed by enzymes), the electronic properties (i.e. electron transfers, charge distributions) and the nature of the substrate-receptor interactions are often of crucial importance for understanding the elementary mechanisms of fundamental biochemical processes. This information cannot be obtained from empirical or semiempirical calculations, but requires a more sophisticated quantum mechanical treatment that only the so-called ab-initio methods can provide.

Notwithstanding the availability of powerful supercomputers, the use of ab-initio methods is limited, however, to small and medium sized systems (less than 100 atoms).

For this reason the development of quantum chemistry in the near future will be based on Quantum Molecular Modelling, in which the performance of quantum chemical methods will be integrated by the effect of the surrounding environment treated at appropriate levels of approximation. Many problems in the study of biomolecular systems can be partially solved or understood by studying smaller molecules with similar chemical groups and physical characteristics or by considering model systems physically consistent with the real system.

This kind of investigations can be performed by using the available first principles based methods such as the traditional ab-initio Hartree-Fock [1] and the correlated post-HF methods or density functional ones [2-4].

The evolution of Hartree-Fock and Density Functional theories are depicted in Scheme 1.

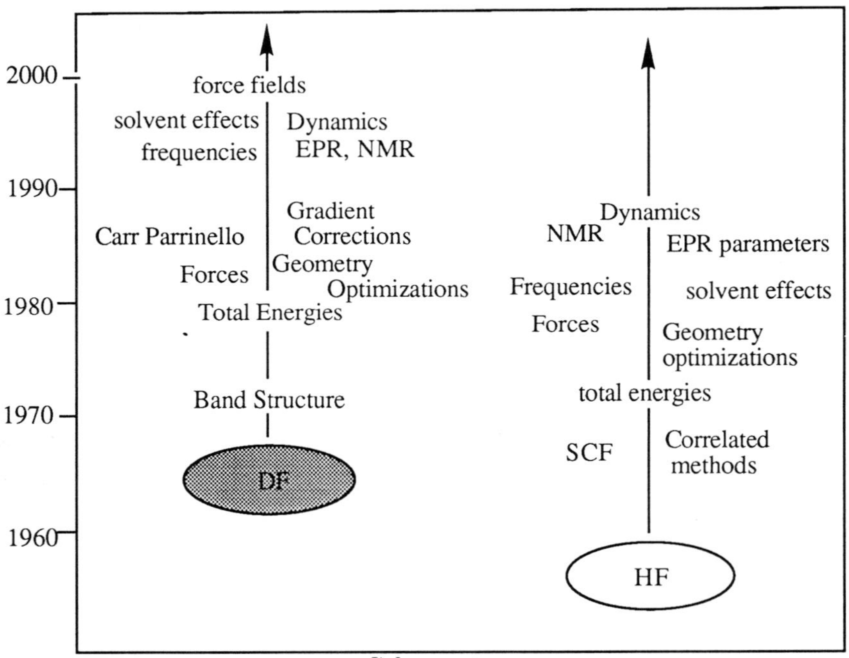

Scheme 1

Concerning the HF methods the following points can be made:

i. They generally work better for stable organic molecules and fail drastically for metallic compounds;

ii. bond lengths, valence and dihedral angles are predicted within 0.03 Å and a few degrees respectively;

iii. binding energies are strongly underestimated (about 50% with respect to experimental data) and can be corrected by the introduction of correlation effects by post-Hartree-Fock calculations;
iv. the energy difference between isomers or conformations are usually well reproduced if medium-large basis sets are used;
v. the harmonic vibrational frequencies are higher by about 10% compared with the harmonic experimental data. Better agreement with experiment can be reached if further computations, taking into account the correlation energy, are performed (i.e. vibrational frequencies computed at the MP2 level of theory reproduce well the experimental data), but, also in this case, the computations are too expensive even for medium-size systems (i.e. 10-20 atoms);
vi. modern codes based on the HF method coupled with the present high-performance workstations are able to carry out computations for systems of about 50 atoms in the time-range of 20-100 hours. Generally the computational effort increases with the fourth power in the number of basis functions (a large portion of the computational time is devoted to the evaluation of four-centre integrals). For large systems many of these integrals vanish or are less significant so they can be neglected a priori by using cut-offs. In these cases the computational time grows with the third power in the number of basis functions.

The inclusion of electron correlation into the HF procedure (i.e. via perturbation theory, MPn, or Coupled-Cluster theory) increases drastically the computational effort (i.e. it scales approximately with the fifth and the seventh power in the case of MP2 and MP4 calculations respectively) while the accuracy of the results is reached slowly.

In this situation only small systems, and certainly not biological macromolecules, can be treated even with the use of modern supercomputers.

Alternative tools for obtaining accurate physico-chemical properties with relatively low computational requirements are those based on the density functional methods (DF) that solve the Kohn-Sham equations.

Starting from the work of Thomas [5] and Fermi [6], Hohenberg and Kohn [7] and Kohn and Sham [8] demonstrated that the total energy of a non degenerate system can be expressed as a functional of the electron density. This means that one can use the electron density rather than the many-body wave function as the fundamental quantity for a rigorous quantum mechanical approach.

During the 1960's [9] this method has been largely used in solid state physics.

The introduction of muffin-tin and Xα approximations, in the 1970's, has allowed the extension of the use of DF based methods also to molecular physics and quantum chemistry. The development of computational strategies in HF methods, during the 1980's, was used also in the density functional related codes. In addition the reliability of local [10] and gradient-corrected functionals [11-14] for the calculation of exchange-correlation energy has improved dramatically the efficiency of the DF methods [15-24].

The main advantages of modern DF methods with respect to the traditional and correlated HF approaches are:

- the calculated energies and spectroscopic parameters are not very sensitive to the choice of the basis set;
- the computational cost is smaller;
- they include a significant amount of correlation in a single set of calculations;
- reliable results are obtained also for transition metal containing systems;

- optimized geometries and vibrational frequencies are reproduced well using the local density approximation;
- with the availability of the modern gradient-corrected exchange-correlation functionals, the binding energies are close to the experimental values;

Many computers codes based on the DF procedure [25-30] are used currently for solving a great variety of problems in chemistry and physics but few applications have been devoted to their use in the biomolecular and pharmacological fields.

In this work we will show the potential of the Gaussian Density Functional method- Linear Combination of Gaussian Type Orbitals-Local (Nonlocal) Density Functional, LCGTO-(N)LSD- in the prediction of several properties of systems with biochemical and pharmacological interest.

2. METHOD AND COMPUTATIONAL DETAILS

Density functional based deMon [28] and DGauss [27] codes have been used for solving the Kohn-Sham equations [8]

$$\left[-\frac{1}{2}\nabla^2 + V(r) + \int \frac{\rho(r)}{|r-r'|}dr' + \frac{\delta E_{XC}}{\delta\rho}\right]\Phi_i = \varepsilon_i\Phi_i \tag{1}$$

$$\rho(\mathbf{r}) = \sum_i n_i \, \Phi_i^*(\mathbf{r})\Phi_i(\mathbf{r}) \qquad n_i\text{=occupation numbers.}$$

in the framework of the spin polarized version (LSD) of the local approximation (LDA)

$$E_{XC}^{LSD}[\rho(r)] = \int \varepsilon_{XC}[\rho(\mathbf{r}_s,\xi)]d\rho \tag{2}$$

(ε_{XC} is the exchange-correlation energy per particle of an interacting spin-polarized homogeneous electron gas, and $\mathbf{r}_s,\xi$ are electron gas parameters) and using the Vosko, Wilk and Nusair VWN [10] analytic form of the correlation energy for the homogeneous electron gas and the $X\alpha(\alpha=2/3)$ approximation for the exchange potential. Both the codes use the Linear Combination of Gaussian-type Orbitals (LCGTO) approach [31,32]. The electron density, $\rho(\mathbf{r})$, and the exchange-correlation potential, $\mu(\mathbf{r})$, are obtained by Gaussian auxiliary basis sets, $g(\mathbf{r})$:

$$\rho(\mathbf{r}) = \sum_r \rho_r g_r(\mathbf{r})$$
$$\mu(\mathbf{r}) = \sum_s \mu_s g_s(\mathbf{r}) \tag{3}$$

ρ_r and μ_s are the fitting coefficients of the density and of the exchange-correlation potential, respectively. In this approach, the Hamiltonian matrix can be expressed as follows:

$$H_{pq} = h_{pq} + \sum_r \rho_r [pq|r] + \sum_s \mu_s [pq|s] \tag{4}$$

the first term is the one-electron Hamiltonian and those in square brackets are the two- and one-electron integrals respectively [33].

The LSD gradients in the LCGTO method are obtained through the following formula [34]

$$\partial E_{LSD} / \partial_x = F_{HBF} + F_D \tag{5}$$

where F_{HBF} and F_D are the Hellman-Feynmann forces plus the corrections for the orbital basis set dependence on the nuclear coordinates x and the correction due to the incompleteness of the density fit basis set respectively.

The non local corrections are taken into account *via* the Generalized Gradient Approximation (GGA). The total energy E^{NLSD}, considering the gradient exchange (E_X^G) and correlation (E_C^G) contributions, is written as follows:

$$E^{NLSD} = E^{LSD} + E_X^G + E_C^G \tag{6}$$

In the framework of the GGA, different functionals for exchange and correlation energy have been proposed. In the computations presented here, we have used those of Perdew [11] (P), Becke [13] (B) for exchange and those of Perdew and Wang [12] (PW) for correlation.

The solvent effect has been computed in the framework of the Self Consistent Reaction Field (SCRF) model [35]. In our implementation the SCRF can be obtained by using only the dipole term or multipole expansion of solute-solvent interaction containing the dipole and the quadrupole contributions [36]

$$E_{int} = -\frac{1}{2}\sum_{l=0}^{\infty} \sum_{m=-l}^{l} g_l \langle M_l^m \rangle \langle M_l^m \rangle \tag{7}$$

where the medium response functions g_l, for l=1,2, are expressed as a function of the dielectric constant ε and the cavity radii a_0 :

$$g_l = \frac{(l+1)(\varepsilon-1)}{(l+1)\varepsilon+1} \frac{1}{a_0^{2l+1}} \tag{8}$$

$$M_l^m = \left(\frac{4\pi}{2l+1}\right)^{\frac{1}{2}} \int \rho r^l Y_l^m(\delta\varphi) d^3r \tag{9}$$

where $Y_l^m(\delta\varphi)$ are the spherical harmonics and $\langle M_l^m \rangle$ are the values of the charge (l=0), dipole (l=1), quadrupole (l=2) and higher l-poles.

The dipole and quadrupole moments in the deMon package are obtained as the expectation values of the solute Hamiltonian.

Very recently [37] we have implemented in the deMon code the polarizable continuum model (PCM).

In the original version, proposed by Miertus, Scrocco and Tomasi [38,39], the cavity is defined in terms of spheres with radii Ra proportional to the appropriate van der Waals radii.

The polarization of the solvent by the solute charge distribution induces a surface charge distribution, $s(s_i)$, on the solute-solvent interface.

The solute-solvent interface is devided into N elements (S_i) small enough to consider the charge distribution constant inside them.

Thus the continuum charge distribution is represented by a discrete distribution of apparent point charges and the reaction potential of the solvent also assumes a discrete form:

$$\Psi \approx \sum_{i=1}^{N} \frac{\sigma(s_i)S_i}{|r_o - r|} = \sum_{i=1}^{N} \frac{q_i}{|r_o - r|} \tag{10}$$

The evaluation of the charge density in the different surface elements, $\sigma(s_i)$, is performed by solving the Laplace equation

$$\sigma(s_i) = \frac{\varepsilon - 1}{4\pi}\left(\frac{\partial \Psi}{\partial n}\right)_n \tag{11}$$

Because of the linearity of the electrostatic equation (eq. 10) the potential Ψ may be conveniently divided into two contributions, deriving from the nuclear (Ψ_{nuc}) and electronic (Ψ_{elec}) charge distribution of the solute:

$$\begin{aligned} \Psi &= \Psi_{nuc} + \Psi_{elec} \\ \sigma &= \sigma_{nuc} + \sigma_{elec} \end{aligned} \tag{12}$$

The interaction energy thus is decomposed into the four terms:

$$W = W_{nuc,nuc} + W_{nuc,elec} + W_{elec,nuc} + W_{elec,elec} \tag{13}$$

where

$$
\begin{aligned}
W_{nuc,nuc} &= \int \rho_{nuc} \Psi_{nuc} dv \\
W_{nuc,elec} &= \int \rho_{nuc} \Psi_{elec} dv \\
W_{elec,nuc} &= \int \rho_{elec} \Psi_{nuc} dv \\
W_{elec,elec} &= \int \rho_{elec} \Psi_{elec} dv
\end{aligned}
\tag{14}
$$

For a cavity of general shape it is difficult to get a direct solution of the total solute energy in presence of the solvent, and the equation for the electrostatic potential [39] must be solved numerically. On the basis of the original iterative formalism, alternative computational methods for the evaluation of the induced surface point charges are proposed: the closure solution (CLS) [40] and the matrix inversion boundary element method (BEM) [41]. In our implementation we used the BEM procedure.

Full geometry optimization by Broyden-Fletcher-Goldfarb-Shanno algorithm [42] has been performed for all the reported systems. The defaults DZVP and TZVP orbital and the corresponding auxiliary basis sets have been used [43].

3. APPLICATIONS

3.1. Structure and Oxo-Hydroxy Tautomerism of 5-Fluorouracil

5-fluorouracil shows interesting biological properties (i.e. physiological activity against tumours and viruses) and has been extensively studied both from theoretical [44-47] and experimental [48] points of view.

In the computations reported here DZVP and TZVP gaussian basis sets have been employed and the effects of the non local corrections have been considered using different functionals for the exchange and correlation energy contributions. In addition the effect of the solvent has been considered in the framework of the Self-Consistent Reaction Field model (SCRF) taking into account only the dipolar contribution [36].

The calculations have been performed in water (ε=80) and using cavity radii of 7.740, 7.912 and 7.720 bohr respectively for FU1, FU2 and FU3 tautomers (see Figure 1) obtained from the corresponding larger internuclear distances (optimized in vacuo at the PP level) by adding the proper van der Waals radii of the two atoms involved. The optimized structures for the three lowest energy tautomers of 5-fluorouracil calculated employing both the BP and PP exchange-correlation functionals are reported in Table 1. The obtained bond lengths and valence angles are almost the same at the two levels of theory and are in reasonable agreement with those obtained with the Hartree-Fock method [44]. The effect of the substitution of fluorine in the geometry of uracil is very small as confirmed by the comparison of our data with experiment for uracil [49].

As expected the change in the molecular geometry going from oxo- to hydroxy- tautomers concerns mainly the C-O and C-N bond lengths and the valence angles around the carbonyl carbon because of the bond order changes.

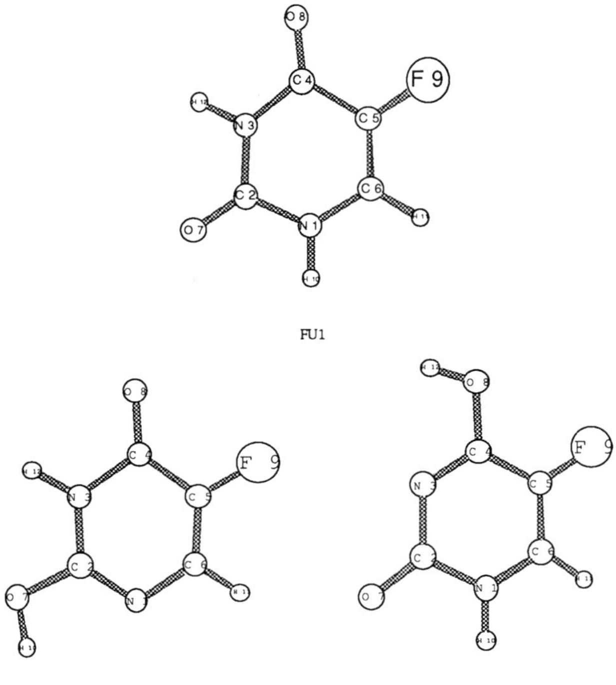

FU1

FU2 FU3

Figure 1. 5-fluorouracil tautomers.

The computed total energies of the three tautomers reveal that the dioxo form is always the most stable followed by the 2-hydroxy-4-oxo (FU2) and the 2-oxo-4-hydroxy (FU3) forms in agreement with experimental observations [50].

The relative stabilities, ΔE, are given in Table 2 and Figure 2.

The BP-double zeta energy difference between FU1 and FU2 forms is 39.2 kJ/mol and becomes 34.5 if the larger triple zeta bases are used. The same effect is observed in the relative stability of FU3 tautomer. This order of stability has been previously found at different levels of theory [44-46] in uracil and thymine, so we can confirm that the substitution of a fluorine at the 5-position has no effect.

The observation that FU2 is more stable than FU3 implies that the proton at N1 is more acid than that at N3. Table 2 shows also the tautomeric behaviour in aqueous solution. The in vacuo order is kept in water solvent but the energy gaps are now different. In agreement with the dipole moment values (see Table 2) the FU2 tautomer appears slightly stabilized with respect to the lactam (FU1) form, while FU3 is destabilized by 3.6 kJ/mol.

These results are consistent with a previous ab-initio study [44].

Although the fact the FU2 tautomer is more stable than FU3, from a biological point of view the latter is particularly interesting because it can form the characteristic N1-glycosidic bond in the nucleic acids. For this reason we have calculated the FU1-FU3 equilibrium constant using the following expression:

$$K = \exp(-\Delta H + T\Delta S) / kT$$

where ΔH is the tautomeric energy difference between FU1 and FU3 corrected by the Zero Point Energy (ZPE), TΔS is the entropic contribution estimated to be 1.3 kJ/mol [51] at 298 K and k is equal to 0.008314 kJ/mol K.

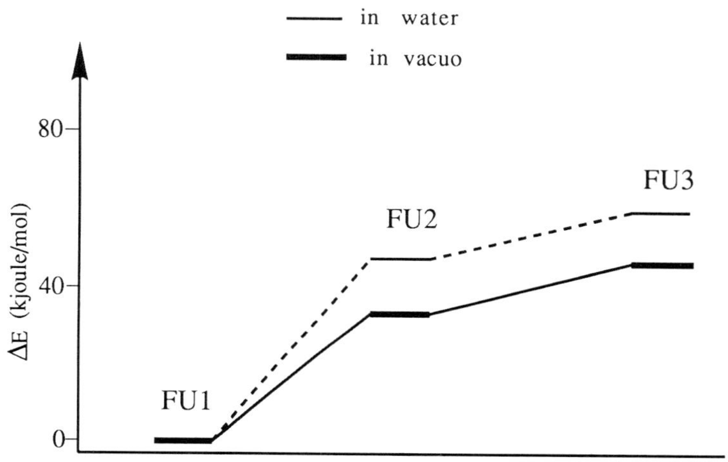

Figure 2. Relative stabilities of 5- Fluorouracil tautomers

Table 1
Geometrical parameters of 5-fluorouracil (FU1).
Distances are in Å and angles in degrees.

Parameter	BP opt	PP opt	HF[a]
N1-C2	1.400	1.399	1.375
C2-N3	1.399	1.397	1.378
N3-C4	1.418	1.415	1.388
C4-C5	1.470	1.468	1.456
C5-C6	1.359	1.358	1.316
C6-N1	1.385	1.385	1.377
C2-O7	1.231	1.231	1.210
C4-O8	1.231	1.231	1.208
C5-F	1.348	1.359	1.352
N1-C2-N3	112.7	113.5	113.1
C2-N3-C4	128.9	128.2	128.8
N3-C4-C5	112.0	112.3	112.3
C4-C5-C6	121.5	122.0	121.2
C5-C6-N1	120.7	120.3	121.6
C6-N1-C2	124.0	123.8	123.1
O7-C2-N3	124.2	123.8	123.5
N3-C4-O8	121.7	121.7	122.0
C6-C5-F	121.0	121.0	117.2

a) from reference 44

Table 2
Relative energies (kJ/mol) for the FU1, FU2 and FU3 tautomers. Ionization potential (eV) is reported only for FU1.

Methods	ΔEFU1-FU2	ΔEFU1-FU3	IPvert(FU1)
BP opt	39.2	53.9	9.28
PP opt	34.9	46.8	9.35
BP TZ	34.5	47.6	9.30
PP solv	33.9	50.4	-
HF[a]	33.1	57.5	-
MBPT(2)[b]	29.0	51.6	-

a) from ref 44; b) from ref. 45.

Equilibrium constants (PP values in vacuo and in solution) for the FU1-FU3 tautomerization process together with the PP values obtained for the corresponding process in uracil are reported in Table 3.

Table 3
Equilibrium costants for the tautomerization processes FU1-FU3 and U1-U3

System	PP	PPsolv
FU1-FU3	$1.0\ 10^{-9}$	$5.0\ 10^{-9}$
U1-U3	$2.0\ 10^{-9}$	$2.9\ 10^{-8}$

The values suggest that the FU3 form, although less probable than FU1, occurs with a frequency of about 10^{-9} which falls in the range of the observed frequency of spontaneous mutations (10^{-8} - 10^{-11}) [48]. Taking into account the corresponding equilibrium constant values for uracil we can see that the 5-fluorouracil, tautomerizing less frequently than uracil, could cause a decrease of the mutations.

Finally we have calculated the vertical ionization potential for the most stable isomer considering different exchange correlation potentials (see Table 2). The obtained values for BP and PP are similar (9.28, 9.35 eV respectively).

3.2. Relative stabilities of uracil tautomers

Due to its presence in the RNA structure, the properties of uracil are the subject of current interest. This base can exist in different tautomeric forms differing by the position of the protons which can be bound to either ring nitrogens or to oxygen atoms.

Full geometry optimization at the local level has been performed for the six tautomers of uracil shown in Scheme 2 and previously studied at semiempirical [52], ab-initio [44,45,53,54] and DF levels [46].

The obtained geometrical parameters for the most stable structure (U1 of Scheme 2) are reported in Table 4 together with other selected theoretical [46,53] and experimental [55] data.

The local results compare well with previous HF ones, and with those obtained by a recent density functional study [46] performed at the non local level (Becke exchange and Perdew correlation functionals).

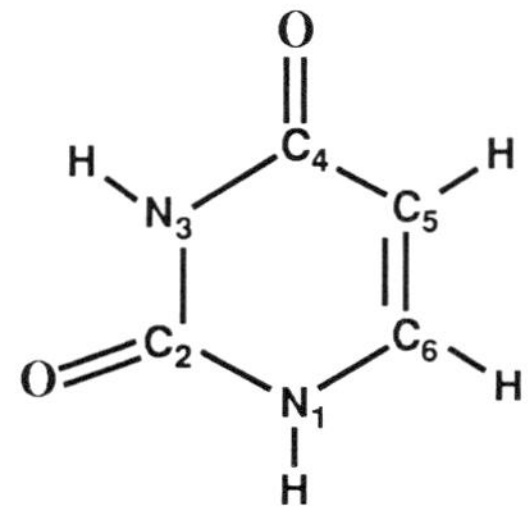

U 1

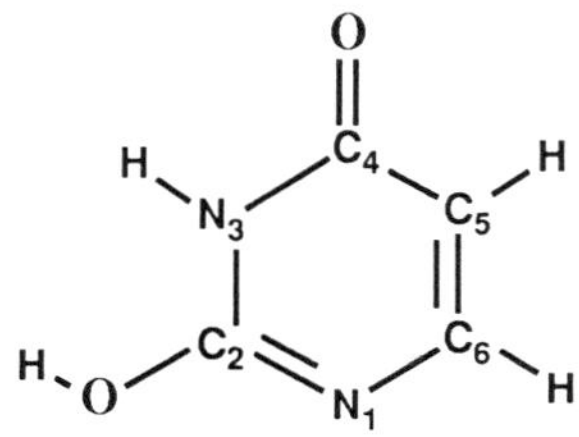

U 2

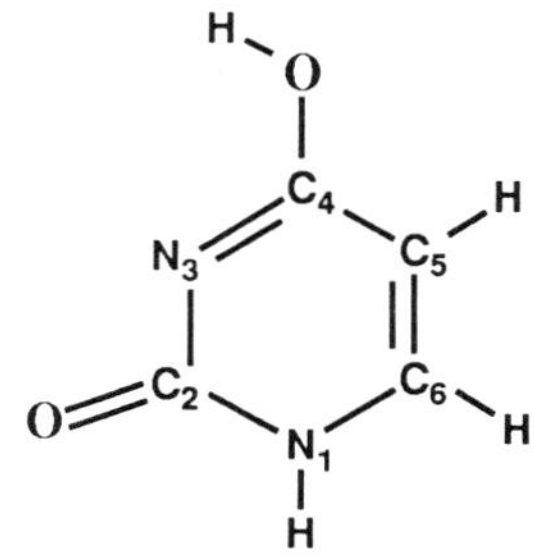

U 3

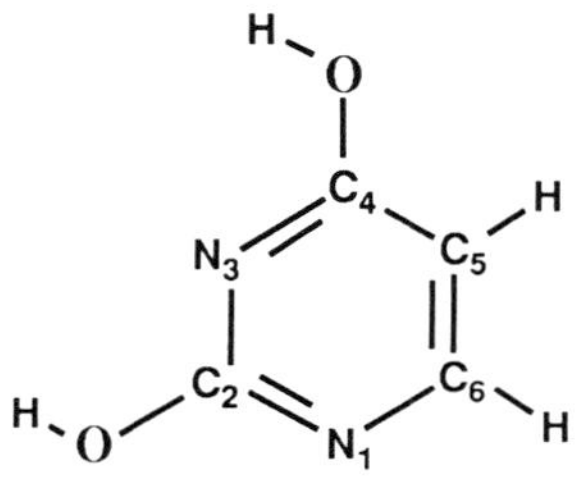

U 4

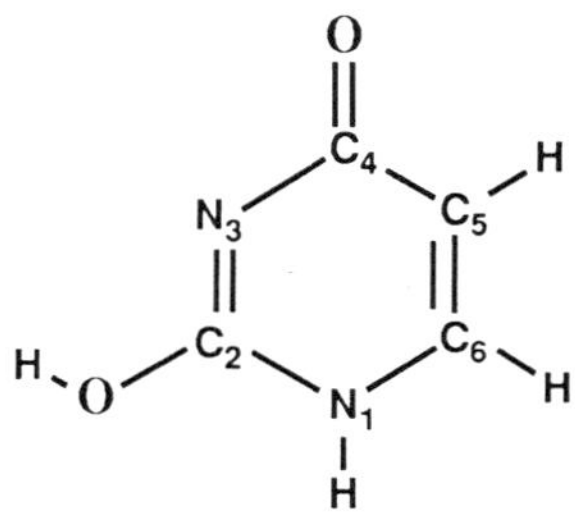

U 5

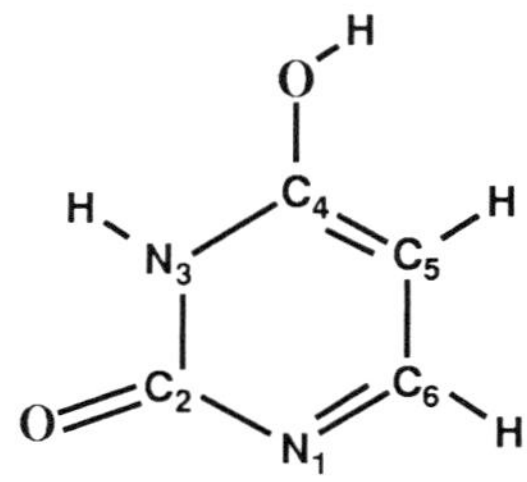

U 6

Scheme 2.

Our structure is in general good agreement also with the X-ray structure. The major deviation concerns the C4-N5 bond length for which both the DF results are overestimated by about 0.03 Å.

In Table 5 we report the energy differences between the most stable dioxo tautomer and the other five considered forms. As it is evident from the Table 5 all levels of theory and experiment give the 2-oxo-4-oxo tautomer as favoured.

The values of the relative energies strongly depend on the level of the correlation treatment. MP2 and MP3 give the U4 form more stable than U3, while the higher perturbation order (MP4) reverses the relative stability.

The simple HF method gives the tautomer U5 very stable and the values of the relative energies for all other forms too high. The relative energies of U2 and U3 are similar in both the gradient corrected DF computations. In particular the PP functional gives U3 slightly more stable than U2, while the BP one prefers U2 over U3. Our PP computations are performed without thermal corrections and this can affect the results, as demonstrated in a previous work [45], just in the case of uracil tautomerization energies . The PP and BP computations yield similar trends for the other tautomers. In general the DF results agree better with the high level ab-initio correlated methods. From a biological point of view the data on U2 and U3 can be significant. In fact, although the U2 tautomer cannot appear in the normal nucleic acids, it can be found in those modified nucleic acids which contain pseudouridine, which is an unusual nucleoside with the sugar moiety bound to the C5 atom of the pyrimidine ring. As mentioned before, the importance of the U3 tautomer is related to its ability to form the N(1)-glycosidic bond characteristic of nucleic acids.

Table 4
Bond lengths (Å) and valence angles (degrees) for the most stable 2-Oxo-4-oxo (U1 of Scheme 2) uracil tautomers.

Parameter	VWN	BP[a]	HF[b]	EXP[c]
N1-C2	1.381	1.378	1.382	1.370
C2-C3	1.355	1.361	1.348	1.343
C3-C4	1.458	1.463	1.457	1.444
C4-N5	1.422	1.422	1.403	1.380
N5-C6	1.392	1.392	1.378	1.381
C6-O7	1.232	1.228	1.214	1.219
C4-O8	1.232	1.233	1.215	1.233
N1-H	1.021	1.027	1.004	/
C2-H	1.089	1.100	1.079	/
C3-H	1.087	1.097	1.076	/
N5-H	1.026	1.030	1.007	/
N1-C2-C3	121.8	122.2	120.0	121.2
C2-C3-C4	120.3	119.4	119.2	120.7
C3-C4-N5	113.7	114.0	114.3	114.1
C4-N5-C6	127.6	127.8	127.6	126.4
N1-C6-O7	122.6	122.6	122.6	122.9
N5-C4-O8	120.1	120.1	120.1	/
H-N1-C2	121.1	121.0	121.3	/
H-C2-N1	121.7	122.3	122.8	/
H-C3-C4	117.8	118.2	118.5	/
H-N5-C6	115.7	116.0	115.8	/

a) from ref. 46; b) from ref. 53; c) from ref. 55.

Table 5
Relative energies (kJ/mol) for uracil tautomers.

Method	**U1**	**U2**	**U3**	**U4**	**U5**	**U6**
HF[a]	0.0	81.8	119.6	109.3	72.1	101.1
PP	0.0	52.7	52.1	64.6	84.3	90.3
BP[b]	0.0	46.3	48.5	53.7	82.9	85.6
MP2[c]	0.0	46.0	52.3	50.4	82.9	97.2
MP3[c]	0.0	44.1	51.6	47.7	77.7	93.2
MP4[c]	0.0	54.7	61.5	66.9	88.4	104.3
MP4[d]	0.0	50.4	/	59.6	/	/
CISD[d]	0.0	56.8	/	71.5	/	/
MBPT[e]	0.0	43.9	50.2	/	/	/
EXP[f]	0.0	53	62.7	79+25	92+42	/

a) from ref. 44; b) from ref. 46; c) from ref. 53; d) from ref. 54; e) from ref. 45; f) from ref. 56.

3.3. Structure and conformational behaviour of acetylcholine

The fundamental neurotransmitter acetylcholine (ACh^{+}) has been the subject of many theoretical and experimental studies [57-81] (see Table 6) devoted to the determination of its conformational flexibility. The reason for this interest is related to the high biological activity that this substrate, exhibits in both the nicotinic and muscarinic sites. So, the knowledge of its conformational behaviour is fundamental to propose structure-activity relationships. From different conformational energy maps [57,58,65,66] it is clear that ACh^+ has a high conformational mobility.

Table 6
Conformations of acetylcholine from previous theoretical and experimental studies. Torsional angles are in degrees.

Conf.	τ1	τ2	Method	Ref.
TG	180	60	PCILO	57
GG*	120	-80	PCILO	57
TT	180	180	VdW	58
G*T	-76	180	"	58
TG*	178	-74	"	58
G*G*	-75	-72	"	58
TG	193	85	X-ray Cl^-	67
GG	79	78	X-ray Br^-	67
GG	83	89	X-ray I^-	68
TG	180	78	X-ray ClO_4^-	68
TG	165	80	X-ray resor.	68
GG	105	75	X-ray bitart.	68
TG	172	90	X-ray Br_4O_2U	68
GG	118	85	X-ray Br_4O_2U	68
TT	180	180	NMR	69
TG	171	65	HF-4-21G	70
G	70	225	Lennard-Jones	71
TT	180	180	FSGO	59
TG	180	60.0	HF- STO-3G	60
	150.0	60.0	HF/STO-3G	61
TG	180	80.0	EHT	62
	100	300	INDO	63
TG	179	274	Theor. vacuo	64
TT	177	172	Theor. aqueous	64
G*G	-102	60	Theor. vacuo	65
TG	160	100	Theor. aqueous	65
G*T	-75	150	Emp. En. Funct.	66
G			NMR	72,73, 74
TG	180	60	^{13}C NMR	75
TG	180	90	X-ray	76
TG	180	75	X-ray	77
TG	160	300	Raman solution	78
GG	260	280	Raman solution	78
TG	169	78	MNDO	79
TT			NMR	80

The relative stability of the minima that lie in a narrow range of energy, depends on the employed level of theory (see Table 6). A variety of minima is also obtained from experimental measurements performed with different techniques and physico-chemical conditions.

Results of LSD full geometry optimization with double zeta basis sets are reported in Table 7 together with three different sets of X-ray data [67,81,82].

The experimental geometrical parameters are quite different expecially for the bond lengths. The agreement with our results is satisfactory if the comparison is made with an average of the experimental values. Our C1-C4 distance is 1.526 Å while the corresponding one in the perchlorate crystal [82] is 1.47 Å and that in two different bromide crystals are 1.50 [67] and 1.52 Å [81] respectively. Similar differencies are observed also for other distances (see Table 7). The variation of valence angles is smaller in going from perchlorate to bromide crystal and comparisons with our results are more reliable (the average deviation with respect to all the three experimental data is less than 3°).

Table 7
Theoretical and experimental geometrical parameters for ACh^+ conformer TG. a) from ref. 82, b) from ref. 67, c) from ref. 81.

Distance/Å	TG	X-ray[a]	X-ray[b]	X-ray[c]
C1-C4	1.526	1.47	1.50	1.52
C1-N7	1.540	1.52	1.53	1.54
C4-O20	1.444	1.43	1.42	1.50
O20-C25	1.401	1.38	1.38	1.32
C25-C21	1.510	1.51	1.47	1.55
C25-O26	1.217	1.20	1.12	1.25
N7-C8	1.515	1.49	1.48	1.52
N7-C12	1.514	1.45	1.48	1.52
N7-C16	1.510	1.51	1.52	1.50
Angles/ °				
C4-C1-N7	118.05	119.0	116.4	118.0
C1-C4-O20	108.62	109.0	111.6	113.0
C4-O20-C25	115.12	115.0	115.7	115.0
O20-C25-C21	110.26	111.0	111.3	110.0
O20-C25-O26	121.64	121.0	122.8	124.0
C1-N7-C8	111.99	109.0	112.2	111.0
C1-N7-C12	107.49	108.0	107.1	109.0
C1-N7-C16	110.81	111.0	110.7	111.0
C8-N7-C12	108.16	107.0	108.3	109.0
C8-N7-C16	109.19	110.0	108.6	-
C12-N7-C16	109.10	111.0	109.8	106.0

From our nonlocal computations performed employing the Becke [13] and Perdew [11] functionals for the exchange and correlation energy respectively, it results that the absolute minumum is that with the angle τ_1 in trans and τ_2 in gauche positions (TG).

The other three considered conformers lie in a very narrow range of energy, above the absolute one. In particular, the GT structure is practically isoenergetic with the absolute minimum, the GG and TT conformers lie at only 0.27 and 1.12 kcal/mole above the TG one respectively. The ΔE and the relative values of the torsional angles are reported in Table 8, while the structures of the minima are shown in Figure 3 .

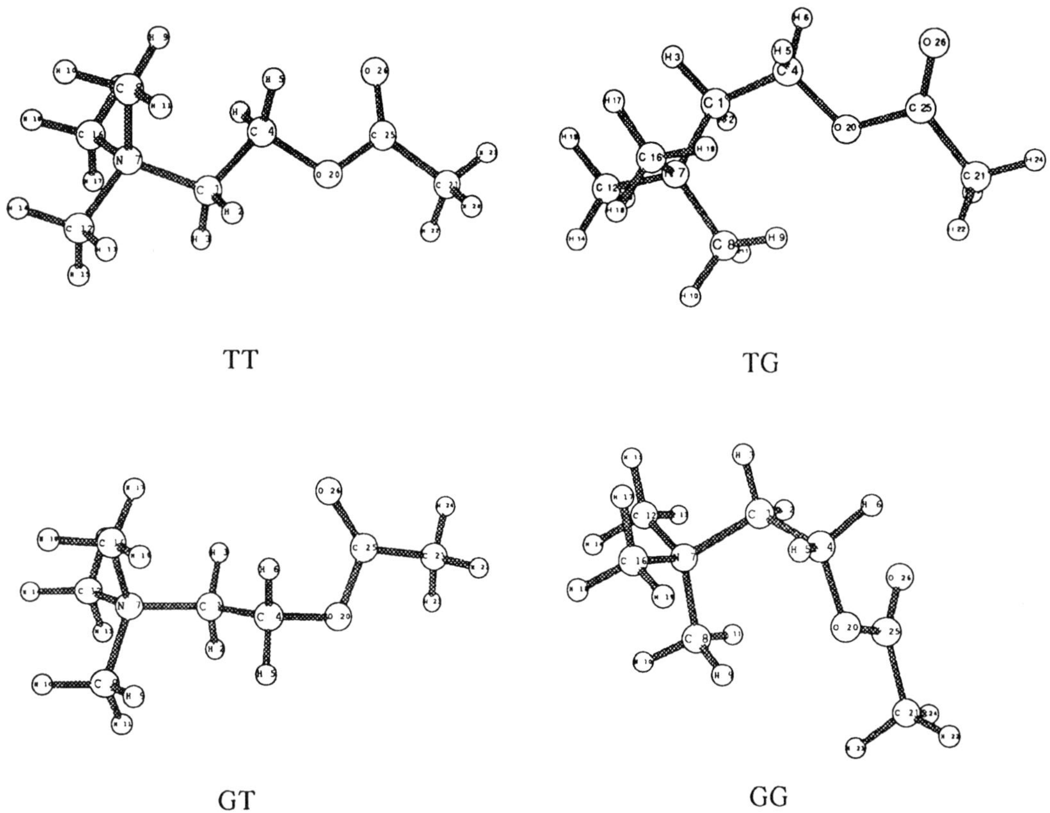

Figure 3. NLSD optimized low energy conformers of acetylcholine.

Table 8
Relative stability (ΔE) (in kcal/mol) and main torsional angles (in °) of the different conformers of acetylcholine from DZ-BP density functional computations.

Conformers	ΔE	τ1	τ2
TG	0.0	-178.48	67.55
GT	0.01	78.20	177.39
GG	0.27	68.80	58.98
TT	1.12	175.68	177.43

Our results confirm the possibility, for the ACh^+, to assume different conformations in agreement with previous theoretical and experimental works. The small energetic differences between the various minima explain the strong dependence of the conformations on the experimental conditions and are the reason that acetylcholine can assume readily or rearrange its conformation in interaction with receptors (i.e. nicotinic, muscarinic and enzyme) to give different biological activity.

3.4. Proton affinities of glycine and cytosine

The prediction of proton affinities (PA) for amino acids and nucleic bases is of considerable importance to rationalize their biological functions. In particular the protonation process has a relevant effect on the hydrogen bonding ability of these compounds and consequently on the variety of pairing schemes to which nucleic bases participate other than on the stabilization of particular conformations of the peptide backbone in which the protonated amino acids are present.

From an experimental point of view, the gas-phase proton affinities can be obtained from mass spectrometric measurements.

Different experimental values for both amino acids and nucleic bases are reported in the recent literature [83-87].

The possibility to compute theoretically the PA contributes significant ly to solution of many problems related to the protonation processes (i.e. the protonation sites, the influence of low energy isomers) that cannot be solved experimentally.

The full optimization of the geometry of the neutral and protonated species is mandatory to obtain reliable theoretical PA values.

The optimized structure of glycine performed at the PP level is reported in Table 9 together with previous theoretical (MP2 [88], CCSD[89], B3LYP [90]) and experimental data [91].

The agreement between all reported high level theoretical studies and experiment is good both for the bond lengths and valence angles.

For both glycine and cytosine systems we have considered the possibility that H^+ can protonate the nitrogen as well as the carbonyl oxygen (see scheme 3 and 4). However the protonation of the nitrogen is strongly favoured with respect to that of oxygen.

The proton affinity is calculated as

$$PA = -[E_{SCF}(BH^+) - E_{SCF}(B) + (E_{vib}(BH^+) - E_{vib}(B))] + 5/2\ RT$$

Table 9
Theoretical geometrical parameters (in Å and degrees) for the most stable conformer of glycine

	PP	MP2/ TZP[b]	CCSD/ DZP[a]	B3LYP/ TZ2P[c]	Expt.[d]
NH	1.024	1.014	1.011	1.012	(1.001)
NC	1.468	1.447	1.458	1.451	1.467
CH	1.098	1.094	1.098	1.091	(1.081)
CC	1.528	1.519	1.525	1.522	1.526
C-O	1.372	1.356	1.359	1.356	1.355
OH	0.984	0.968	0.972	0.968	(0.966)
C=O	1.221	1.209	1.216	1.204	1.205
HNH	103.1	106.2	104.7	105.7	(110.3)
HNC	109.3	110.2	108.8	109.9	(113.3)
NCC	115.1	115.6	115.4	115.5	112.1
HCH	105.8	106.1	106.3	105.6	(107.0)
HCC	109.0	107.4	107.5	107.7	
CC-O	113.0	110.9	111.5	111.5	111.6
COH	106.3	106.6	105.9	107.1	(112.3)
CC=O	125.0	125.7	125.7	125.7	125.1
C=O--N	2.87		2.782	2.867	

[a]From ref. 89,[b] From ref. 88,[c]From ref. 90, [d]From ref. 91, the values in parentheses were assumed from low-level theoretical predictions (HF/4-21G, Ref. 92)

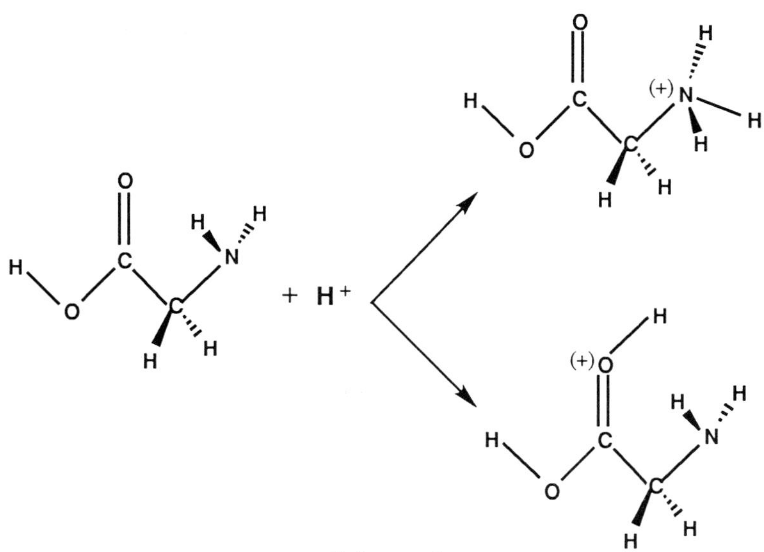

Scheme 3.

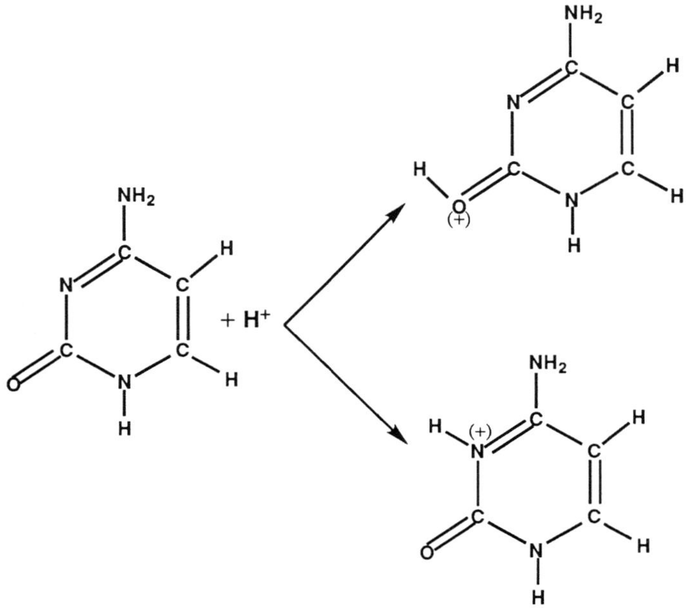

Scheme 4

where E_{SCF} are obtained from the SCF calculations, E_{vib} includes the zero point energy and temperature corrections to the vibrational enthalpy and 5/2RT includes the translational energy of proton and the $\Delta(PV)$ term.

For the protonation of the nitrogen site, the obtained PA at 298 K are 214.2 and 226.8 kcal/mol for glycine and cytosine respectively. For cytosine, the most recent experimental PA value is 225.9 kcal/mol [87]. For glycine three different experimental values of 213.0 [83], 211.7 [84] and 216.1 [85] kcal/mol are available. Our value falls in the range of the proposed values and is closer to the value obtained at the MP4 level of theory (213.6 kcal/mol) [93].

3.5. N-Methylacetamide

N-Methylacetamide (NMA) continues to attract theoretical and experimental attention since it is considered to be a simple but realistic model system for the peptide group in proteins [94-106]. The main problem is related to the conformation of this molecule. In fact the orientation of the two backbone torsional angles and the energetics of the corresponding rotations are of great importance in the stabilization of peptide and protein structures. NMA trans and cis conformers are close in energy, so high level theoretical methods are necessary for the reliable description of its conformational properties. In the gas phase the trans form dominates [102], while NMR measurements, in aqueous solution [100,103], shows the cis form to be the absolute minimum. In peptides and proteins, both conformers have been observed, but the trans form appears to be much more common (it is favoured 10^3-fold over the cis configuration). Solvation might be one of the most significant factors that contributes to the difference of stability between trans and cis conformers in peptides, so, it is very useful to perform studies taking into account the solvent effects.

Full geometry optimized structures of both cis and trans isomers of NMA, performed with double zeta basis sets and taking into account the non local corrections with the Perdew and Wang and Perdew functionals for the exchange-correlation energy, are shown in Figure 4. Also for this system, the obtained parameters agree well with those coming from previous theoretical ab-initio [96] and experimental [102] studies. We found that in the gas-phase the trans isomer is 1.9 kcal/mol more stable than the cis one. The corresponding HF, MP2 and B3LYP ΔE values are 2.5 [99], 2.2 [99] and 2.4 [95] kcal/mol respectively. A "rough" experimental estimate from IR spectroscopy gives the trans form favoured by 2.3 kcal/mol with respect to the cis one [105].

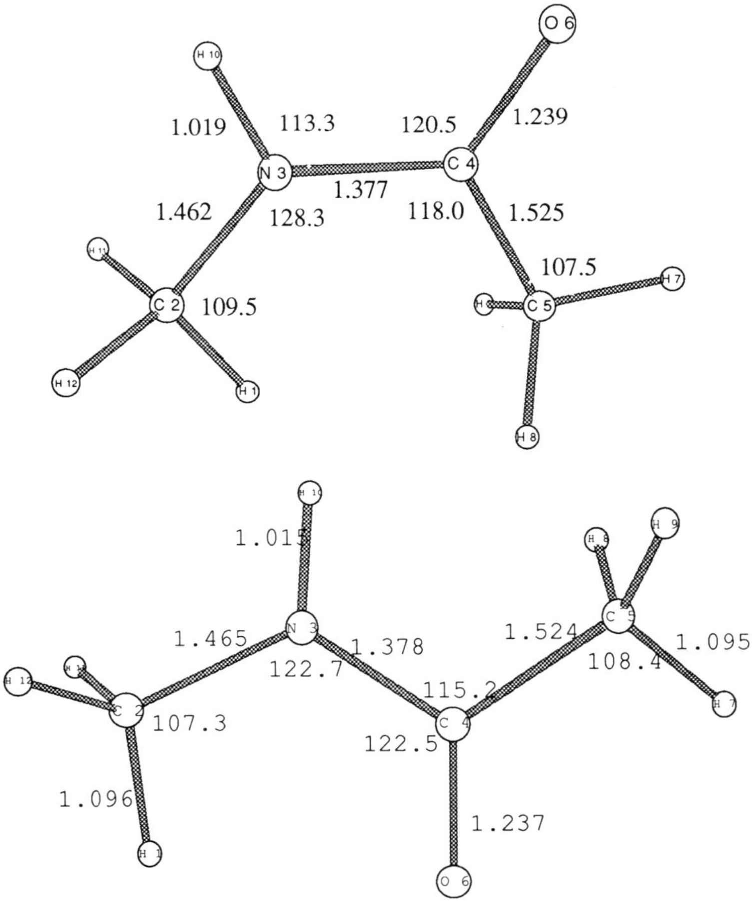

Figure 4. Non local PP optimized geometrical parameters for trans and cis conformers of N-Methylacetamide. Distances are in Å and angles in degrees.

In aqueous solution, a ^{1}H and ^{13}C NMR study [100], suggests that the trans isomer is again more stable (the population of the cis form is only of 1.5%). Our computation in water, considering both the dipole and quadrupole contributes to the reaction field, gives the trans conformer more stable by 4.3 kcal/mol. Monte Carlo simulations [99] of cis and trans NMA, with a modified OPLS charge distribution for the amides, reveals no statistically significant difference in the free energies of hydratation of the two conformers.

This means that the trans form is more stable.

Our results agree with another NMR study [103] that indicates the trans form of NMA is more stable than the cis one by 2.5 kcal/mol in water. The computed heats of solvation for trans- and cis-NMA are -7.7 kcal/mol and -6.4 kcal/mol respectively.

Calculated dipole moments are 3.90 and 4.22 Debye in the gas-phase. The value for the trans isomer, determined by electron diffraction, is 3.70 Debye. The agreement is also good with previous theoretical determinations (see Table 10). In aqueous solution the dipole moment of trans-NMA becomes 4.41 at PP level, while the corresponding experimental values in benzene (ε=2.27) and in 1,4-dioxane (ε=2.27) are 3.85 and 4.22 Debye respectively.

Table 10
Dipole moments (Debye) for NMA from theory and experiment.

Method	trans-NMA		cis-NMA	
	ε=1	ε=80	ε=1	ε=80
PP	3.90	4.41	4.22	4.99
HF-6-31G(d)[a]	4.04	/	4.21	/
B3-LYP[b]	3.92	/	4.32	/
Exp.[c]	3.70			
Exp.(ε=2.3)[d]	3.85			
Exp.(ε=2.2)	4.22			

a) from ref. 99; b) from ref. 95; c) from ref. 102; d) from ref. 106

For the trans-NMA we have calculated the harmonic vibrational frequencies taking into account the non local corrections. Results are collected in Table 11 together with the HF [94] and experimental [105] ones.

The agreement with experiment and with scaled HF data is very satisfactory and shows that the DF method is able to reproduce correctly the harmonic vibrational frequencies without any scaling factor.

It is well known that for the elucidation of the structural properties and biological functions of proteins it is important to know how they interact with the aqueous environment through hydrogen bonding.

The NMA molecule can be used also as the simplest model to study the interaction of the chain of proteins with water molecules. Different theoretical studies have been performed on the NMA-water supermolecular system [95-99].

We have investigated the interaction of two water molecules with trans-NMA considering different sites and topological possibilities (see Scheme 5): in the structure A the water molecules can act as acceptors and bind to the amide proton, in B they can act as donors and bind to the carbonyl oxygen, in C one water molecule acts as acceptor and the other one as donor.

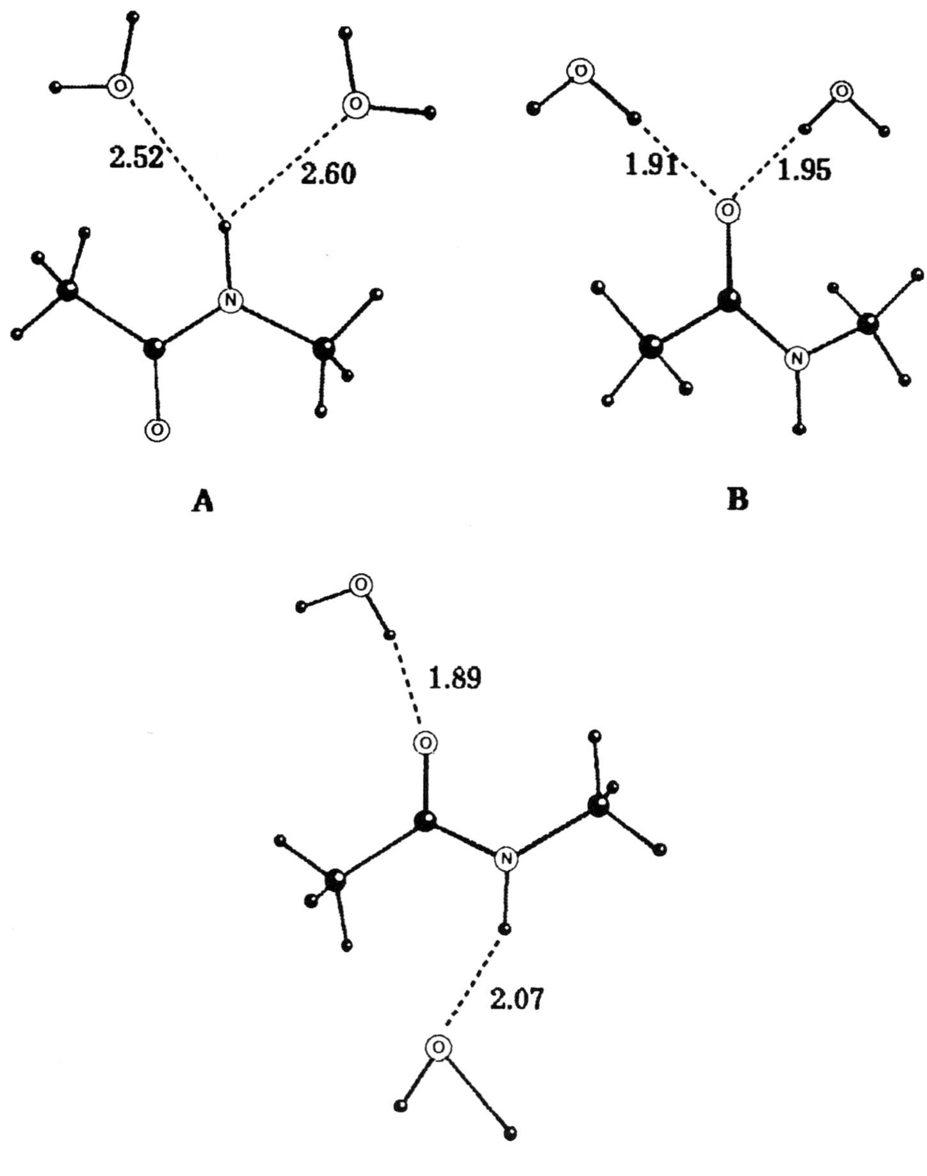

A **B** **C**

Scheme 5

Table 11
Harmonic vibrational frequencies (cm^{-1}) for the trans-N-methylacetamide in the gas-phase.

Parameters[a]	NLSD	HF[b]	EXP[c]
N_3-H_{10} s	3557	3498	3498
C_4-C_5 as	3088	3028	3008
N_3-C_2 as	3080	3001	2973
C_4-C_5 as	3039	2968	3008
N_3-C_2 as	3035	2988	3008
C_4-C_5 ss	2981	2918	2915
C_4-C_5 ss	2976	2918	2915
C_4-O_6 s	1690	1708	1706
N_3-H_{10} ib	1492	1469	1472
N_3-C_2 ab	1466	_	1446
C_4-C_5 ab	1446	1428	1432
" "	1436	1422	1419
" "	1428	_	1400
" "	1399	_	_
C_4-C_5 sb	1345	1375	1370
N_3-H_{10} ib; C_4-O_6 ib; N_3-C_4 s	1234	1265	1265
N_3-C_2 r	1150	1183	1168
" "	1120	1099	1089
C_4-C_5 r	1050	1049	1037
C_4-N_3 s; N_3-C_2 r	1018	_	1037
C_2-N_3-C_4 d	943	961	980
C_4-C_5 r	846	859	857
C_4-C_5 r; C_4-O_6 ob	614	631	619
C_5-O_6 ib	602	_	_
C_4-O_6 ib; C_4-C_5 r	416	444	429

a) s=stretch; ss=symmetric stretch; as=antisymmetric stretch; sb=simmetric bend; ab=antisymmetric bend; ib=in-plane bend; ob=out of plane bend; r=rock; d=deformation. b) from ref. 94; c) from ref. 105

In the gas-phase the complex C appears to be the most stable followed by B and A at 0.8 and 7.0 kcal/mol respectively. Previous HF/6-31G [99] and B3LYP [95] computations give the B structure more stable than C by 0.2 and 0.6 kcal/mol respectively. Both structures are practically isoenergetic and the discrepancy with our PP computations can be explained by considering the different conformations of the methyl groups. Our H bond interaction energy for structure C and B is -11.3 kcal/mol and -10.5 kcal/mol respectively that is of the same order of magnitude as that obtained by B3LYP computations (-14.6 kcal/mol for B and -13.8 kcal/mol for C) [95] and there is also agreement with previous HF values (-17.8 kcal/mol for B and -17.5 kcal/mol for C) [96].

We have redone the computations for structure B and C considering the solvent effect computed with our extension of the Onsager method. In the aqueous solution the structure B becomes more stable than C. Structure A lies at the very high energy.

A recent experimental IR study [101] has measured the shift of the C=O stretch in going from the gas- to the liquid-phase. Our computations on the complexes B and C show that the value of $\nu_{C=O}$ decreases in presence of the two water molecules. In particular we found that, in the case of structure B, the C=O bond stretch is 43 cm^{-1} less than the corresponding value in isolated NMA. This result agrees well with the experimental result [101].

3.6. Potential energy surface of peptide bond cleavage by Carboxypeptidase A

The study of the mechanisms of the reactions catalyzed by enzymes, is of great interest in biochemistry.

Insight to these processes comes from a combination of theoretical and experimental investigations.

Experiment can give precious information on the initial and final state of the reactions, on the kinetic constants, but is unable to predict and to explain the reaction pathways.

On the other hand reliable theoretical methods offer the possibility to compute the potential energy hypersurfaces including the localization of the mimima and maxima.

The BP non local functional, double-zeta basis set and our recent algorithm that allows the possibility to locate the transition states by using the first derivatives [107], have been used to explore the potential energy hypersurface of the peptide bond cleavage catalyzed by Carboxypeptidase A enzyme [108].

In our active site model, constructed on the basis of a 1.54 Å crystal structure from the Protein Data Bank [109], the zinc atom is coordinated by Glu-72, Glu-270, His-196, His-69 residues and a water molecule, simulated with an appropriate functional group, occupies another coordination site. A formamide molecule has been used to represent the peptide bond.

The obtained potential energy hypersurface is depicted in Figure 5.

The first step of the reaction is the nucleophilic attack of a Zn-coordinated anion on the carbonyl carbon of the formamide substrate.

Through the transition state TS1 a tetrahedral intermediate (see Figures 6, 7 and 8) is formed.

The second step of the reaction is the cleavage of the hydrated peptide bond.

The intermediate **1** leads, *via* the transition state TS2, to another intermediate (structure **3** of Figure 7) in which the amino group of the peptide bond is protonated by a proton coming from the Glu-270 residue.

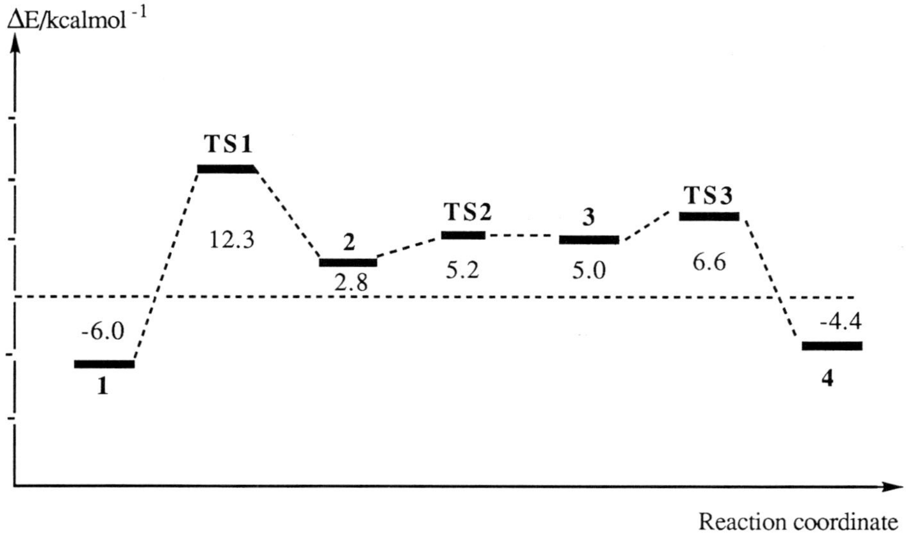

Figure 5. Potential energy surface of peptide bond cleavage catalized by carboxypeptidase A computed at BP/DZVP2 level of theory. For the structures of the reactants, products, intermediates and transition states see Figures 6, 7 and 8.

Then, the transfer of the hydroxyl hydrogen of the substrate to the Glu-270 residue, through the transition state TS3, causes the cleavage of the peptide bond yelding the final products of the reaction (structure **4** of Figure 6).

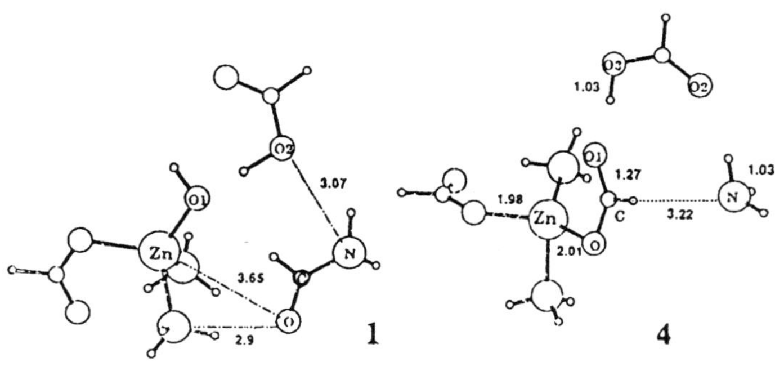

Figure 6. BP optimized structure of (**1**) the initial enzyme-substrate complex and of (**4**) the product-enzyme complex. The peptide bond is cleaved (CH---NH_3 bond distance is 3.22 Å), but the products are still coordinated to the active site. All distances are in Å.

From the computed reaction path we note that: i. the rate-determining step is the formation of the intermediate **2**; ii. two proton transfers are required for the cleavage of the peptide bond; iii. the barriers for these proton transfers are

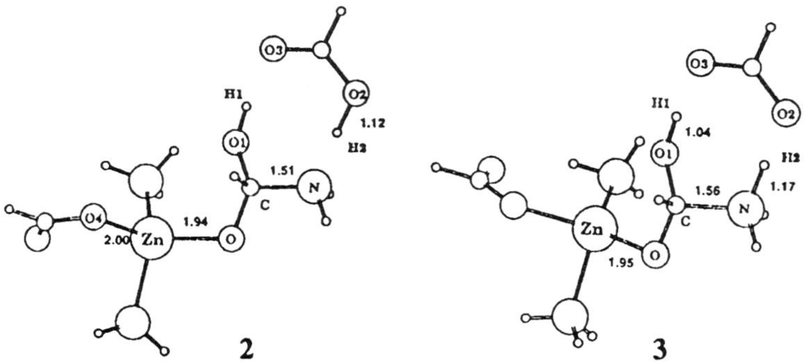

Figure 7. BP optimized structures of the tetrahedral intermediate (**2**) and of the intermediate (**3**) in which the amino group of the formamide has been protonated. All distances are in Å.

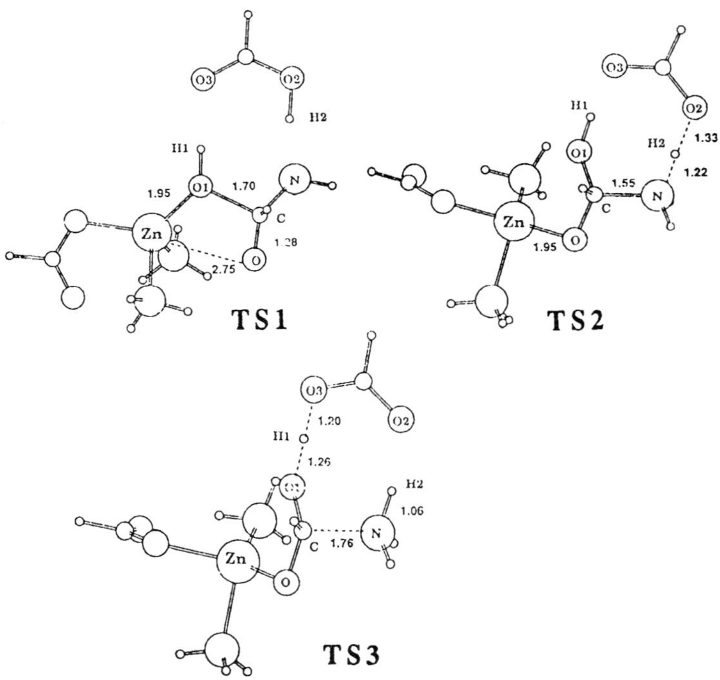

Figure 8. BP optimized structures of the transition state (TS1) for the nucleophilic attack of the hydroxide group to the peptide bond and those (TS2, TS3) for the two sequential proton transfers. All distances are in Å.

small with respect to that to generate the tetrahedral intermediate; iv. the catalytic role of the Zn atom is to generate the base (OH^-) for the initial attack on the amide bond (it serves essentially to polarize the water molecule involved in the initial step).

3.7. Zwitterion-neutral equilibrium of glycine

The neutral form of the glycine molecule is found in isolated gas-phase or matrix conditions [92,110,111], whereas the zwitterionic isomer is present in solid phase and aqueous medium [112,113,114]. In water, at room temperature and in a pH range of 2-10, the zwitterionic form is fovoured by free energy and enthalpy values of 7.2 and 10.3 kcal/mol respectively [113].

The experimental heat of solvation is measured to be -19.2 kcal/mol [115].

Because glycine is the smallest amino acid it has been the subject of many theoretical studies employing different methods [88,89,90,116-120]. The hydratation process has been studied theoretically by the HF (using minimal and 4-31G basis sets in the framework of polarizable continuum model (PCM) [119]) and by PM3 [118] methods. Very recently, the stability of glycine zwitterion, has been investigated by an high level correlated method considering a glycine-water supermolecule [117]. It is found that two water molecules stabilize the glycine zwitterion. We have studied the stabilities of neutral and zwitterionic forms both in gas-phase and aqueous environments by using PP non local functionals for exchange and correlation potentials, DZVP2 basis set and, in the case of computation in water solvent, the PCM procedure recently implemented [37] in the deMon code. Results show that in gas-phase the zwitterionic form lies at very high energy above the neutral one in agreement with previous theoretical and experimental studies. On the other hand, in water, the neutral form does not exist (its electrostatic solvation energy is positive) and the glycine zwitterion is strongly stabilized.

The hydratation energy is found to be -20.4 kcal/mol (by using the cavity, given by spheres centered on the solute nuclei with radii equal to 1.4 time the Van der Waals values) in satisfactory agreement with the experimental (-19.2 kcal/mol) [105] and HF-PCM (-19.5 kcal/mol) [109] values.

4. CONCLUSIONS

In this contribution we have reported a series of density functional computations for some systems of biological and pharmacological significance.

The calculated properties range from geometrical structures, harmonic vibrational frequencies, ionization potentials, proton affinities, conformational equilibria and potential energy surfaces of metal containing systems.

On the basis of our and other already published results, we believe that density functional methods can become rapidly promising tools for the theoretical study of biomolecular systems. Their relatively low computational cost and the inclusion of correlation effects allows the study of complex phenomena also for medium-large systems that are difficult to investigate with the more traditional high level correlated ab-initio methods.

The extension of the use of DF methods to other fundamental biological processes can provide further elements to give a final judgement on their reliability in this field.

ACKNOWLEDGEMENT

This work has been supported by MURST and CNR (Comitato Scienze Chimiche and Comitato Tecnologico). We thank the CINECA for the generous gift of computer time on Cray supercomputers. T. Mineva thanks the Fondazione Catizone, Cosenza, for a fellowship.

REFERENCES

1. A. Szabo and N. Ostlund, *Modern Quantum Chemistry: Introduction to Advanced Electronic Structure Theory* (MacMillan New York, 1982).
2. R. Parr, and W. Yang , *Density-Functional Theory of Atoms and Molecules* (Oxford University Press, NY, 1989).
3. E. Kryatchko and E. Ludena, *Energy Density Functional Theory of Many Electron Systems* (Kluwer, Dordrecht, 1990).
4. N. March, *Electron Density Theory of Atoms and Molecules*; (Academic Press, New York, 1991).
5. L. H. Thomas, Proc. Cambridge Phil. Soc. 23, (1926) 542.
6. E. Fermi, Z. Phys. 48 (1926), 542.
7. P. Hohenberg and W. Kohn, Phys. Rev. 136, (1964) B864.
8. W. Kohn and L. J. Sham, Phys. Rev. A140, (1965) 1133.
9. J. C. Slater, *The Self-Consistent Field for Molecules and Solids* (Mc Graw Hill, New York, 1974).
10. S. H. Vosko, L. Wilk and M. Nusair, Can. J. Phys. 58, (1980) 1200.
11. J. P. Perdew, Phys. Rev. B33, (1986) 8822.
12. J. P. Perdew and Y. Wang, Phys. Rev. B33, (1986) 8800.
13. a) A. D. Becke, Phys. Rev. A38, (1988) 3098 ; b) J. Chem. Phys. 88, (1988) 2547.
14. J. P. Perdew, Phys. Rev. B45, (1992) 13244; ibid. B44, (1991) 13298
15. R. O. Jones and O. Gunnarson, Rev. Mod. Physics 61, (1990) 1280.
16. T. Ziegler, Chem. Rev. 91, (1990) 651.
17. J. Labanowski and J. Andzelm (eds.) *Density Functional Methods n Chemistry* (Springer Verlag: NY, 1991).
18. D. R. Salahub and N. Russo (eds.) *Metal Ligand Interaction. From Atoms, to Clusters, to Surfaces* (Kluwer: Dordrecht, 1992).
19. J. M. Seminario and P. Politzer (eds.) *Density Functional Theory: a Tool for Chemistry* (Elsevier, New York, 1995).
20. B. G. Johnson, P. M. W. Gill and J. A. Pople, J. Chem. Phys. 98, (1993) 5612
21. A. M. Lee and N. C. Handy, J. Chem. Soc. Faraday Disc. 2, (1993) 1
22. Y. Abashkin, N. Russo and M. Toscano, Int. J. Quantum Chem. 52, (1994) 695.
23. N. Russo and D. R. Salahub (eds.) *Metal Ligand Interaction: Structure and Reactivity* (Kluwer: Dordrecht, 1995).
24. D. P. Chong (ed) *Recent Advances in Density Functional Methods*. Vol 1 (World Scientific, Singapore, 1995).

25. Amsterdam Density Functional System (ADF), Department of Theoretical Chemistry, Vrije Universiteit, Amsterdam, The Netherlands; E.J. Baerends and P. Ros, Int. J. Quantum Chem. S12, (1978) 69.
26. B. Delley, J. Chem. Phys. 92 (1990) 508.
27. J. Andzelm and E. Wimmer, J. Chem. Phys. 96 (1992) 508.
28. A. St. Amant, PhD Thesis, Universite' de Montreal, 1992.
29. GAUSSIAN 92/DFT, Gaussian Inc., Pittsburgh, PA, 1994.
30. CADPAC5: The Cambridge Analityc Derivatives Package issue 5, Cambridge, 1992.
31. H. Sambe and R. H. Felton, J. Chem. Phys. 62. (1975) 1122.
32. B. I. Dunlap, J. W. D. Connolly and J. R. Sabin, J. Chem. Phys. 71, (1979) 3396.
33. S. Obara and A. Saika, J. Chem. Phys. 84 (1986) 3963.
34. R. Fournier, J. Andzelm and D.R. Salahub, J. Chem. Phys. 90, (1989) 6371.
35. L. Onsager, J. Am. Chem. Soc. 58, (1936) 1486.
36. T. Mineva, N. Russo and M. Toscano, Int. J. Quantum Chem. 56, (1995) 663.
37. T. Mineva, E. Sicilia and N. Russo, to be published.
38. S. Miertus, E. Scrocco and J. Tomasi, Chem Phys. **55** (1981) 117.
39. S. Miertus and J. Tomasi, Chem Phys. **65** (1982) 239.
40. E. L. Coitiño, J. Tomasi and R. Cammi, J. Comput. Chem. **16** (1995) 120.
41. H. Hoshi, M Sakurai, Y. Inoue and Chûjo, J. Chem Phys. **87** (1987) 1107.
42. C.C. Broyden, J. Inst. Math. Appl. 6 (1970) 76; R. Fletcher, Comput. J. 13, (1970) 317; D. Goldfarb, Math. Comput. 24, (1970) 23; D.F. Shanno, Math. Comput. 24, (1970) 647.
43. N. Godbout, D. R. Salahub, J. Andzelm and E. Wimmer, Can. J. Chem. 70, (1992) 560.
44. M. J. Scanlan and H. Hillier, J. Am. Chem. Soc. 106, (1984) 3737.
45. A. Les and L. Adamowicz, J. Phys. Chem. 93, (1989) 7078.
46. D. A. Estrin, L. Paglieri and G. Corongiu, J. Phys. Chem. 98 (1994) 5653; ibidem Int. J. Quantum Chem. 56, (1995) 615.
47. R. J. Hall, N. A. Burton, I. Hillier and P. E. Young, Chem. Phys. Lett. 220, (1994) 129.
48. N. Bodor and J. Kaminski, J. Annu. Rep. Med. Chem. 22, (1987) 303.
49. D. Voet, and A. Rich, Prog. Nucleic Acid Res. Mol. Biol. 10 (1970) 196.
50. A. R. Katritzky, P. Linda Eds., *The Tautomerism of Heterocycles, Advances in Heterocyclic Chemistry, Suppl. 1* (Academic Press, New York 1976).
51. M. Shibata, T. J. Zielinski and R. Rein, Int. J. Quantum Chem. 18, (1980) 323.
52. A. R. Katritzky and M. Karelson, J. Am. Chem. Soc. 113 (1991) 1561.
53. J. W. Boughton and P. Pulay, Int. J. Quant. Chem. 47 (1993) 49.
54. J. Leszcynski, J. Phys. Chem. 96 (1992) 1649.
55. D. Voet and A. Rich, Prog. Nucleic Acid Res. Mol. Biol. 10 (1970)196.
56. P. Beak and J. M. White, J. Am. Chem. Soc. 104 (1982) 7073.
57. B. Pullman, P. Courriere and J.L. Coubeils, Mol. Pharm. 7, (1971) 397.
58. A. M. Liquori, A. Damiani, and J. L. De Coen, J. Mol. Biol., 33, (1968) 445.
59. D. W. Genson and R. E. Christoffersen, J. Am. Chem. Soc. , 95, (1973) 362.
60. G. N. J. Port, A. Pullman, J. Am. Chem. Soc. , 95, (1973) 4059.
61. A. Pullman, G. N. Port, J., Theor. Chem., 32, (1973) 77.
62. L. B. Kier, Mol. Pharmacol., 3, (1967) 487.

63. R. J. Radna, D. L. Beveridge and A. L. Bender, J. Am. Chem. Soc., 95, (1973) 3831.
64. H. J. Weintraub and A. H. Hopfinger, Jerusalem Symp. Quantum Chem. Biochem. 7, (1974) 131.
65. D. L. Beveridge, R. J. Radna, G. W. Schnuelle and M. M. Kelly, Jerusalem Symp. Quantum Chem. Biochem. 7, (1974) 153.
66. B. R. Gelin, M. Karplus, J. Am. Chem. Soc. , 97, (1975) 6996.
67. T. Svinning, H. Sorum, Acta Crystallogr., Sect. B: Struct. Crystallogr. Cryst. Chem. B31, (1975) 1581.
68. S. Jagner, B. Jensen, Acta Cryst. B33, (1977) 2757.
69. R. J. Cushley, H. G. Mautner, Tetrahedron 26, (1970) 2151.
70. V. J. Klimkowski, L. Schafer, J. N. Scarsdale and C.U. Alsenoy, J. Mol Struct (Theochem) 109, (1984) 311.
71. M. Froimowitz and P. J. Gans, J. Am. Chem. Soc. 94, (1972) 8020.
72. P. Partington, J. Feeney and A. S. V. Burgen, Mol. Pharm. , 8, (1972) 269.
73. H. G. Mautner, D. D. Dexter and B. Low, Nature (London) New Biol. 238, (1972) 87.
74. D. Lichtenberg, P. Kroon and S. Chan, J. Am. Chem. Soc. 96, (1974) 5934.
75. L. Cassidei, O. Sciacovelli, J. Am. Chem. Soc. , 103, (1981) 933.
76. C. Chothia, Nature (London) 225, (1970) 36.
77. C. Chothia, and P. Pauling, Proc. Nat. Acad. Sci., U. S. 65, (1970) 447.
78. K. J. Wilson, P. Derremaux, G. Vergoten, and W. L. Peticolas, J. Phys. Chem. 93, (1989) 1351.
79. M. A Ergan, . and R. W. Zoeller, J. Org. Chem. 58, (1993) 1719.
80. J-P Behr, and J. M. Lehn, Biochem. Biophys. Res. Commun. 40, (1970) 583.
81. F. G. Canepa, P. Pauling and M. Sorum, Nature (London) 210,(1966) 907.
82. V. Mahjan and R. L. Sass, J. Cryst. Mol. Struct. 4, (1974) 15.
83. M. J. Locke and R. T. Mc Iver, J. Am. Chem. Soc. 105 (1983) 4226.
84. G. S. Gorman, J. P. Speir, C. A. Turner and I. J. Amster, J. Am. Chem. Soc. 114 (1992) 3986.
85. X. Li and A. G. Harrison, Org. Mass Spectrometry 28 (1993) 366.
86. S. G. Lias, J. F. Liebman and R. D. Levine, J. Phys. Chem. Ref. Data 13 (1984) 695.
87. A. Liguori, F. Greco, G. Sindona and N. Uccella, J. Am. Chem. Soc. 112 (1990) 9092.
88. A. Csazar, J. Am. Chem. Soc. 114, (1992) 9568.
89. C. H. Hu, M. Shan and H. F. Schaefer III, J. Am. Chem. Soc. 115, 2923 (1993).
90. V. Barone, C. Adamo and F. Lelj, J. Chem. Phys. 102, (1995) 364.
91. K. Iijima, K. Tanaka and S. Onuma, J.Mol. Struct. 246, (1991) 257.
92. L. Schafer, H. L. Sellers, F. J. Lovas and R. D. Suenram, J. Am. Chem. Soc. 102, (1980) 6566.
93. K. Zhang, D. M. Zimmerman, A. Chung-Phillips and C. J. Cassady, J. Am. Chem. Soc. 115 (1993) 10812.
94. N. G. Mirkin and S. Krimm, J. Mol. Struct. 242, (1991)143.
95. W. G. Han and S. Suhai, J. Phys. Chem. 100, (1996) 3942.
96. H. Guo and M. Karplus, J. Phys. Chem. 96, 7273 (1992).
97. D. A. Dixon, D. K. Dobbs and J. J. Valentini, J. Phys. Chem. 98, (1994) 13435.
98. W. L. Jorgensen and C. J. Swenson, J. Am. Chem. Soc. 107, (1985) 1489.
99. W. L. Jorgensen and J. Gao, J. Am. Chem. Soc. 110, (1988) 4212 .

100. A. Radzicka, L. Pedersen and R. Wolfenden, Biochemistry 27, (1988) 4538.
101. I. P. Gerothanassis and C. Vakka, J. Org. Chem. 59, (1994) 2341.
102. M. Kitano, T. Fukuyama and K. Kuchitsu, Bull. Chem. Soc. Japn. 46, (1973) 384.
103. T. Drakenbey and S. Forsen, J. Chem. Soc. Chem. Commun. 1404 (1971)
104. J. E. Del Bene, J. Am. Chem. Soc. 100 (1978) 1387.
105. S. Ataka, H. Takouchi and M. Tasumi, J. Mol. Struct. 113, (1984) 147.
106. K. Pralat, J. Jadzyn and S. Balamicka, J. Phys. Chem. 87, (1983) 1385.
107. Y. Abashkin and N. Russo, J. Chem. Phys. 100, (1994) 4477.
108. Y. G. Abashkin, S. K. Burt, J. R. Collins, R. E. Cachau, N. Russo and J. W. Erickson in *Metal-Ligand Interactions Structure and Reactivity*, (eds) N. Russo and D. R. Salahub (Kluwer Dordrecht 1995), p. 1-22).
109. E. E. Abola, F. C. Bernstein, S. H. Bryant, T. F. Koetzle and J. Weng Protein Data Bank. In *Crystallographic Data Bases-Information Content; Software Systems Scientific Applications*; F. H. Allen, G. Bergerhoff and R. Sievers Eds.; Data Commission of the International Union of Crystallographic; Bonn/ Cambridge/Chester, pp 107-132 (1987).
110. R. D. Brown, P. D. Godfrey, J. W. V. Storey and M. P. Bassez, J. Chem. Soc. Chem. Commun., 547 (1978).
111. R. D. Suenram and F. J. Lovas, J. Am. Chem. Soc., 102, (1980) 7180 .
112. J. Almlof, A. Knich and J. O. Thomas, J. Chem. Phys. 59, (1973) 3901.
113. G. Wada, E. Tamura, M. Okima and M. Nakamura, Bull. Chem. Soc. Jpn. 55, (1992).
114. M. A. Slifkin and S. M. Ali, J. Mol. Liq. 28, (1984) 215.
115. J. S. Goffrey, R. C. Pierce and L. Friedman, J. Am. Chem. Soc. 99, (1977) 4293.
116. Y. Ding and K. Krogh-jespersen, Chem. Phys. Lett. 199, (1992) 261.
117. J. H. Jensen and M. S. Gordon, J. Am. Chem. Soc. 117, (1995) 8159.
118. M. S. Rzepa and M. Y. Yi, J. Chem. Soc. Perkin Trans. 2 (1991) 531.
119. R. Bonaccorsi, P. Palla and T. Tomasi, J. Am. Chem. Soc. 106, (1984) 1945.
120. L. R. Wright and R. S. Borkman, J. Am. Chem. Soc. 102 (1980) 6207.

J.M. Seminario (Editor)
Recent Developments and Applications of Modern Density Functional Theory
Theoretical and Computational Chemistry, Vol. 4

Density-Functional Theory Concepts and Techniques for Studying Molecular Charge Distributions and Related Properties

Paul Geerlings [a,$] , Frank De Proft [a] and Jan M. L. Martin [b]

[a] Eenheid Algemene Chemie (ALGC), Free University of Brussels, Faculty of Sciences, Pleinlaan 2, B-1050 Brussels, Belgium
[b] The Institute of Chemistry, The Hebrew University of Jerusalem, Givat-Ram, Jerusalem 91904, Israel* .

1. INTRODUCTION

The impact of density-functional theory (DFT) in the development of quantum chemistry in the last 10 years can hardly be overestimated [1]. Based on the famous Hohenberg-Kohn theorems [2], density-functional theory focuses on the electron density function $\rho(\underline{r})$ as the carrier of all information in the atomic or molecular ground state. It thereby tremendously reduces the complexity of the more conventional wave function information carrier $\Psi(x^{4N})$, depending on the spatial and spin coordinates of all N electrons of the system.
Although Nature permits this conceptual simplification, it does not lose its grip on the situation. While in the time-independent Schrödinger equation (or its pseudo one-particle analogue, the Hartree-Fock equations [3]) the potential energy term in the Hamiltonian is known, in the celebrated Kohn-Sham equations [4] (which yield the electron density) part of the potential is unknown. This prompted both physicists and chemists to look for adequate functionals, at ever increasing rate in recent years (for a review, see [5]).
The avalanche of papers presented in order to select an appropriate exchange-correlation functional did not concentrate, in the first place, on the quality of the electron density as such, but rather on the performance for energies and related properties.

In this contribution, we concentrate (Chapter 3) on the molecular charge distribution obtained from DFT calculations (3.1) as represented via atomic populations (3.2), thereby trying to establish standards, well known in wave function quantum chemical methods [6-8]. In 3.3., highly sensitive probes for molecular charge distributions, the molecular dipole moment [9] and its geometrical derivatives (which are related to the infrared intensities [10, 11]) are used to check the validity of the conclusions put forward in 3.1. Finally, in 3.4, the link with molecular reactivity is established by evaluating the performance of various exchange-correlation functionals for calculating two well-known reactivity descriptors: one of the "traditional" type, i.e. the molecular electrostatic potential, already introduced by Bonaccorsi et al. [12-14] in the early seventies; another a DFT based reactivity

---

$ Corresponding author
* On sabbatical from: Limburgs Universitair Centrum, Institute for Materials Science (IMO), Department SBG, Universitaire Campus, B-3590 Diepenbeek, Belgium and University of Antwerp (UIA), Institute for Materials Science, Department of Chemistry, Universiteitsplein 1, B-2610 Wilrijk, Belgium

descriptor (vide infra), the Fukui function introduced by Parr and Yang [15] as a generalization of Fukui's HOMO-LUMO treatment [16, 17].

However, we start this contribution with a chapter (Chapter 2) which concentrates on the ideas emerging from DFT rather than on the computational techniques. This kind of topic has attracted a lot of interest in recent years, and, in our opinion, it is at least as important an aspect of the DFT framework as its computational advantages. Sanderson's electronegativity equalization principle was already formulated in the early fifties [18, 19], but mathematically proven in the late seventies by Parr et al. [20]. Together with the identification of the first and second energy derivative with respect to the number of electrons of a system as the electronegativity [20] and hardness [21] respectively, this in priciple created the possibility to obtain and interpret molecular charge distributions in a fundamentally different way from the "brute force" techniques mentioned above. Already in the late seventies and early eighties, various techniques have been put forward, all involving one or more adjustable parameters [22]. After a general introduction (2.1) and a description of Sanderson's principle (2.2), we show in 2.3 how such a scheme can be developed (up to second order) without adjustable parameters, thereby fully exploiting the fact that those parts of (the molecular) space that should equalize in electronegativity should not be restricted to "atomic" spaces (i.e. working in atomic resolution (2.3.1)), but can be extended to functional groups (2.3.2) or amino acid residue (2.3.3) space, the latter possibility permitting charge calculations on peptides and proteins.

Chapter 2 is closed with a simple illustration of how DFT concepts can be used to interpret charge distributions, obtained either by DFT or conventional wave function techniques. It concentrates on the subtle interplay of electronegativity and hardness in the determination of atomic charges.

The results discussed in Chapter 2 and 3 will be mainly, but not exclusively, taken from our own research in recent years, which can be summarized as the calculation and use of DFT based properties in the study of molecular charge distributions and reactivity [23, 24].

2. DFT CONCEPTS AND THEIR USE IN THE STUDY OF MOLECULAR CHARGE DISTRIBUTIONS

2.1 Density-functional theory based molecular descriptors

The chemical reactivity of a molecule is often interpreted in terms of its charge distribution. Although the chemist has an intuitive feeling for the qualitative nature of these charge distributions, assigning atomic charges quantitatively is far from trivial.

Charge distributions are accessible through quantum chemical calculations. However, different methods are available and results seem to be heavily dependent on the quantum mechanical method chosen. Moreover, the charge on an atom in a molecule is not a directly observable experimental quantity, so that a direct comparison of the calculated atomic charges to experiment is impossible. In addition, computing times increase superlinearly (typically quadratically to quartically) with the number of electrons in the molecular system of interest, thus precluding very accurate calculations. Moreover, the practising chemist likes to hold on to intuitive and easy to use chemical concepts in order to interpret the charge distribution.

One of these concepts is electronegativity. Being by far one of the oldest chemical concepts, electronegativity gives a measure for the unequal sharing of electrons in a chemical bond, resulting in bond polarity. Introduced in 1932 by Pauling [25] as "the power of an atom in a molecule to attract electrons to itself", the concept has undergone great changes since then. Many scales were introduced, the most important due to Mulliken [26], Gordy [27], Allred and Rochow [28] and Sanderson [18]. Despite all the proposed electronegativity scales, a rigorous quantum mechanical definition of the electronegativity concept was clearly lacking. This changed in 1978 when Parr et al. [20], after preliminary work of Iczkowski and Margrave [29], identified the electronegativity with the negative of the Lagrange multiplier from density-functional theory, which appears when the energy functional of the electron density is minimized with the constraint of constant number of electrons N:

$$\delta\left(E-\mu\int\rho(\underline{r})d\underline{r}\right) \tag{1}$$

This variational principle leads to the famous equation:

$$\mu = v(\underline{r}) + \frac{\delta F[\rho(\underline{r})]}{\delta\rho(\underline{r})} \tag{2}$$

with $v(\underline{r})$ the external potential (i.e. due to the nuclei) and $F[\rho(\underline{r})]$ the functional containing the electron-electron repulsion, the kinetic energy and exchange-correlation energy.

The physical significance of the Lagrange multiplier μ can be appreciated by considering some equations governing the change from one ground state to another. Since the atomic or molecular energy is a functional of the number of electrons N and the external potential $v(\underline{r})$ (i.e. $E = E[N, v(\underline{r})]$), the total differential of E for the change of one ground state to another will take the form:

$$dE = \left(\frac{\partial E}{\partial N}\right)_v dN + \int\left(\frac{\delta E}{\delta v(\underline{r})}\right)_N \delta v(\underline{r})d\underline{r} \tag{3}$$

This expression must be the same as the total differential for E using $\rho(\underline{r})$ and $v(\underline{r})$ as the basic variables:

$$dE = \int\left(\frac{\delta E}{\delta\rho(\underline{r})}\right)_v \delta\rho(\underline{r})d\underline{r} + \int\left(\frac{\delta E}{\delta v(\underline{r})}\right)_\rho \delta v(\underline{r})d\underline{r} \tag{4}$$

Since, using the Euler equation (2)

$$\left(\frac{\delta E}{\delta\rho(\underline{r})}\right)_v = \mu \tag{5}$$

and

$$\int\delta\rho(\underline{r})d\underline{r} = N \tag{6}$$

equation (4) becomes

$$dE = \mu dN + \int\rho(\underline{r})\delta v(\underline{r})d\underline{r} \tag{7}$$

Comparing (3) and (7), one thus obtains:

$$\mu = \left(\frac{\partial E}{\partial N}\right)_v \tag{8}$$

which can be identified with the negative of the electronegativity (i.e. $\mu = -\chi$): the electronegativity is the negative partial derivative of the system's energy E with respect to its number of electrons N, taken at a constant external potential (i.e. constant nuclear positions and charges).

Using a second-order polynomial for the energy versus the number of electrons and approximating the loss or gain of charge by the ionization energy I and electron affinity A respectively, one obtains for the electronegativity:

$$\chi = \frac{I+A}{2} \qquad (9)$$

which is the well-known Mulliken expression [26] for this quantity.

In 1983, Parr and Pearson introduced the chemical hardness as a companion parameter to electronegativity [21]. This concept, empirically introduced by Pearson in the 1960s in the rationalization of certain acid-base reactions [30], was now quantified as:

$$\eta = \frac{1}{2}\left(\frac{\partial^2 E}{\partial N^2}\right)_v \qquad (10)$$

Again using a quadratic *E=E(N)* curve, one obtains:

$$\eta = \frac{I-A}{2} \qquad (11)$$

Finally, the softness *S* can be introduced as [31]:

$$S = \frac{1}{2\eta} \qquad (12)$$

which, using a finite difference approximation and a quadratic *E=E(N)* curve, reduces to:

$$S = \frac{1}{I-A} \qquad (13)$$

Both quantities, electronegativity and hardness(softness), have proven to be very important tools in the qualitative description of molecular charge distributions. The quantitative link is provided by density-functional theory, by means of the electronegativity equalization principle.

2.2. Sanderson's electronegativity equalization principle: the role of hardness/softness

The principle of electronegativity equalization has led to a series of formalisms involving the calculation of atomic charges. This principle, proposed in the early fifties by Sanderson [18], states that:

> *When two or more atoms, initially different in electronegativity, unite to form a compound, their electronegativities become adjusted to the same intermediate value within the compound.*

This principle was proven quantum mechanically by Parr later on in the seventies [20]. It was also confirmed by Politzer and Weinstein [32], who showed that for a diatomic molecule in the ground state, the electronegativities of all arbitrary portions of the total number of electrons are the same.

According to Sanderson, the "intermediate value within the compound" is the geometric average S_m of the Sanderson electronegativities S_A of the constituting atoms A of the molecule. Moreover, he derived a relatively simple scheme to obtain partial charges from this principle. The charge q_A of an atom A in a molecule was defined as the ratio of change in electronegativity $S_m - S_A$ the atom undergoes upon molecule formation, to its electronegativity change when acquiring a positive or negative unit charge δS_A:

$$q_A = \frac{S_m - S_A}{\delta S_A} \tag{14}$$

with

$$\delta S_A = k S_A^{1/2} \tag{15}$$

where it was assumed that the electronegativity of an atom changes linearly with the partial charge of the atom and that the bond in an isolated molecule of NaF is 75 % ionic (later revised to 80 %). Taking these approximations into consideration, the constant k equals 1.56.

Parr and Bartolloti [33] proved this geometric mean principle to be valid when the valence state energy E of an atom was taken to be an exponentially decreasing function of the number of electrons N:

$$E(N) = A e^{-\gamma(N-Z)} + B \tag{16}$$

where Z is the nuclear charge of the corresponding atom. The principle was found to hold when γ was the same for each atom that was bound within the molecule. An average value of (2.15 ± 0.59) was found for γ.

Using Sanderson's method however, all atoms of a given element obtain the same atomic charge within a molecule, which is a serious drawback. All protons in acetic acid for example would obtain the same charges, and so will all carbons and oxygens.

Huheey proposed that the electronegativity of an atom in a molecule could be expressed as a function of its charge q [19, 34]:

$$\chi = a + bq \tag{17}$$

with a the inherent electronegativity and b the charge coefficient. This formula was used to calculate group electronegativities for a large number of substituents [35,36] ; moreover, the equation was succesfully applied to discuss the alkyl group substituent effect in alkoxide ions [34]. Unfortunately, this approach suffers the same drawback as Sanderson's method, as will any method which neglects the electrostatic interaction between the atoms.

The partial equalization of orbital electronegativity (PEOE), introduced by Gasteiger and Marsili, performs a partial equalization of electronegativity [37]. It consists of an iterative scheme, in which the charge shifts in each bond are evaluated separately. This yields connectivity dependent charges and residual electronegativities for an atom in a molecule. The atomic charges obtained in this way showed a good correlation with experimental measures for this quantity, such as core electron binding energies (ESCA shifts) [37], acidity constants [37] and NMR chemical shifts [38]. Moreover, the residual electronegativity can be used as a measure of the inductive effect: in conjunction with the effective polarizability, it was used in a quantitative study of the energetics of gas phase proton transfers of amines, alcohols and ethers [39, 40].

EEM, the Electronegativity Equalization Method, is due to Mortier et al [22, 41, 42]. The electronegativity χ_α of an atom α in a molecule is expressed as:

$$\chi_\alpha = \overset{*}{\chi}_\alpha + \overset{*}{\eta}_\alpha q_\alpha + \sum_{\beta \neq \alpha} \frac{q_\beta}{R_{\alpha\beta}} \tag{18}$$

with q_α and q_β the charges on the atoms α and β respectively and $R_{\alpha\beta}$ their interatomic distance within the molecule. Use of this equation for every atom, together with the electronegativity equalization principle and the closure relation (for a neutral molecule)

$$\sum_\alpha q_\alpha = 0 \tag{19}$$

yields a very powerful method for the calculation of molecular charge distributions. The parameters $\overset{*}{\chi}_\alpha$ and $\overset{*}{\eta}_\alpha$ were calibrated on STO-3G Mulliken charge distributions. The methods was mainly used for problems in inorganic chemistry, all related to subjects involving zeolite catalysis.

As will be shown now, the Huheey and Mortier equations can both be considered as special, approximate cases of (the "density-functional") equation (20).

From equation (8) and the functional relationship $E=E[N,v]$, it follows that $\mu=\mu[N,v]$. This total differential becomes:

$$d\mu = \left(\frac{\partial \mu}{\partial N}\right)_v dN + \int \left(\frac{\delta \mu}{\delta v(\underline{r})}\right)_N \delta v(\underline{r}) d\underline{r} \tag{20}$$

with

$$\left(\frac{\partial \mu}{\partial N}\right)_v = 2\eta \tag{21}$$

the chemical hardness and

$$\left(\frac{\delta \mu}{\delta v(\underline{r})}\right)_N = \left(\frac{\partial \rho(\underline{r})}{\partial N}\right)_v = f(\underline{r}) \tag{22}$$

the Fukui function. This function, on which attention will be focussed in Chapter 3, is very important in describing chemical reactivity of molecules and in elucidating site-selectivity upon electrophilic, nucleophilic or radical attack. In a finite difference approximation, the differential (20) can be written as:

$$\Delta\mu = 2\eta \Delta N + \int f(\underline{r}) \Delta v(r) d\underline{r} \tag{23}$$

The change in chemical potential from one ground state to another for a chemical species can thus be accomplished in two manners: changing its number of electrons or the external potential.

Consider an atom A with a chemical potential μ_A^0 and hardness η_A^0, which is surrounded by n atoms B_1, B_2, ..., B_i, ...,B_n upon molecule formation. To first order, and evaluating all derivates in the point $N_A^0, v_A^0(\underline{r})$ (i.e. referring to the isolated atom state), the change in chemical potential $\Delta\mu_A$, associated with the given ΔN_A and $\Delta v_A(\underline{r})$ can be written as:

$$\Delta\mu_A = 2\eta_A^0 \Delta N_A + \int f_A^0(\underline{r}) \Delta v_A(\underline{r}) d\underline{r} \tag{24}$$

Neglecting the external potential perturbation, i.e. $\Delta v_A(\underline{r}) = 0$, one obtains

$$\Delta\mu_A = 2\eta_A^0 \Delta N_A \tag{25}$$

Using the fact that $\mu_A^0 = -\chi_A^0$ and

$$\Delta N_A = N_A - N_A^0 = -q_A \tag{26}$$

one obtains

$$\chi_A = \chi_A^0 + 2\eta_A^0 q_A \tag{27}$$

which is exactly Huheey's expression. As can be seen, the fact that all atoms acquire the same charge within the molecule with this approach is due to the neglect of the external potential perturbation. The charge coefficient b from Huheey's work can thus be identified with the hardness η. Its importance can be judged when considering a diatomic molecule A-B. The charge transfer ΔN_A to A within the molecule upon equalizing χ_A and χ_B can be expressed as:

$$\Delta N_A = \frac{\chi_A^0 - \chi_B^0}{2\left(\eta_A^0 + \eta_A^0\right)} \tag{28}$$

The charge transfer to A is proportional to the difference in electronegativity of the two atoms: the electronegativity drives the electron transfer while the hardness prevents it. The hardness will thus be a very important parameter, next to electronegativity. Atoms or functional groups with smaller electronegativities can sometimes bear more negative charge within a molecule than atoms or groups with higher electronegativities if the former do have smaller η values, i.e. are softer. Despite the fact that the more electronegative atom or substituent has a stronger initial tendency to attract negative charge, it is not able to accumulate it completely, i.e. it has a lower charge capacity [43,44].

In a second step, the external potential perturbation on A can be approximated as

$$\Delta v_A(\underline{r}) = -\sum_{i=1}^{n} \frac{q_{B_i}}{\left|\underline{r} - \underline{R}_{B_i}\right|} \tag{29}$$

i.e. using an atomic point charge model for the Coulomb interaction.
Inserting (29) in (24) yields

$$\Delta\mu_A = 2\eta_A^0 \Delta N_A - \sum_{i=1}^{n} q_{B_i} \frac{\partial}{\partial N_A} \int \frac{\rho_A^0(\underline{r})}{|\underline{r} - \underline{R}_{B_i}|} d\underline{r} \tag{30}$$

which after some manipulation leads to:

$$\chi_A = \chi_A^0 + 2\eta_A^0 q_A + \sum_{i=1}^{n} q_{B_i} \left(\frac{\partial V_A^{el}(R_{B_iA})}{\partial N_A} \right)_{N_A^0} \tag{31}$$

involving the derivative of the electronic part of the electrostatic potential V^{el} of an isolated atom A at a distance R_{B_iA}. At large distance:

$$\left(\frac{\partial V_A^{el}(R_{B_iA})}{\partial N_A} \right)_{N_A^0} \approx \frac{1}{R_{B_iA}} \tag{32}$$

yielding

$$\chi_A = \chi_A^0 + 2\eta_A^0 q_A + \sum_{i=1}^{n} \frac{q_{B_i}}{R_{B_iA}} \tag{33}$$

which is the EEM equation (18), in which the χ^* and η^* parameters are identified as χ_A^0 and η_A^0.

Nalewajski et al. derived the following expression for the chemical potential $\mu_A[N_A, Z_A]$ of an atom in a molecule [45, 46]:

$$\mu_A[N_A, Z_A] = \mu_A^0 + 2\left(\eta_A^0 \Delta N_A + \alpha_A \Delta Z_A\right) \tag{34}$$

starting from a full second order expansion of the energy $E_A[N_A, Z_A]$ with:

$$\alpha_A = \frac{1}{2}\left(\frac{\partial \mu_A}{\partial Z_A} \right)_{N_A} \tag{35}$$

In this formula, the increase in core charge $\Delta Z_A (>0)$ models the external potential perturbation. Finally, it should be mentioned that recent and important work on electronegativity equalization has been presented by Cioslowski [47, 48] and by Rappé and Goddard [49] (the charge equilibration approach (QEq)).

2.3. A parameter-free implementation of Sanderson's electronegativity equalization principle

2.3.1. Atomic resolution: theory and application to diatomics

Consider again equation (24). Supposing that the external potential perturbation on an atom A upon molecule formation has the following form [50]:

$$\Delta v_A(\underline{r}) = -\sum_{i=1}^{n} \frac{Z_{B_i,A}^{eff}}{\left|\underline{r} - \underline{R}_{B_i}\right|} \tag{36}$$

with $Z_{B_i,A}^{eff}$ the effective nuclear charge of the atom B_i as it is experienced by A.

In analogy with the derivation of equation (31) in section 2.2, the electronegativity of an atom A in the molecule can then be shown to be:

$$\chi_A = \chi_A^0 + 2\eta_A^0 q_A + \sum_{i=1}^{n} Z_{B_i,A}^{eff} \left(\frac{\partial V_A^{el}(R_{B_i,A})}{\partial N_A} \right)_{N_A^0} \tag{37}$$

When this equation is extended to second order, three additional terms appear in the expression for χ_A:

$$\frac{1}{2}\left(\frac{\partial^2 \mu_A}{\partial N_A^2}\right)_{v_A} \Delta N_A^2 + \int \left(\frac{\partial}{\partial N_A} \left(\frac{\delta \mu_A}{\delta v_A(\underline{r})} \right)_{N_A} \right)_{v_A} \Delta N_A \Delta v_A(\underline{r}) d\underline{r} + \tag{38}$$

$$\frac{1}{2}\iint \left(\frac{\delta^2 \mu_A}{\delta v_A(\underline{r}) \delta v_A(\underline{r}')} \right)_{N_A} \Delta v_A(\underline{r}) \Delta v_A(\underline{r}') d\underline{r} d\underline{r}'$$

where

$$\left(\frac{\partial^2 \mu_A}{\partial N_A^2}\right)_{v_A} = \left(\frac{\partial \eta_A}{\partial N_A}\right)_{v_A} \tag{39}$$

This hardness derivative will of course vanish when a quadratic relationship between the energy E and the number of electrons N is assumed. Fuentalba and Parr showed this derivative to be small, starting from the following form of the energy curve [51]:

$$E(N) = \frac{aN^2 + bN}{1 + cN} \tag{40}$$

which reduces to the quadratic expression when c is small, which indeed turned out to be the case. The second term in equation (38) can easily be rearranged to

$$\int \left(\frac{\partial f_A}{\partial N_A}\right)_{v_A} \Delta N_A \Delta v_A(\underline{r}) d\underline{r} \tag{41}$$

Using (36) and a reasoning analogous to the above in the derivation of the first order perturbation formula, this finally becomes:

$$q_A \sum_{i=1}^{n} Z_{B_i,A}^{eff} \left(\frac{\partial^2 V_A^{el}(R_{B_i,A})}{\partial N_A^2} \right)_{N_A^0} \tag{42}$$

which vanishes under assumption (32). This means that N_A and $v_A(\underline{r})$, within a reasonable approximation, show no coupling at second order. The third term and most difficult to evaluate term of (38) can be rewritten as:

$$\frac{1}{2} \iint \left(\frac{\delta f_A(\underline{r})}{\delta v_A(\underline{r}')} \right)_{N_A} \Delta v_A(\underline{r}) \Delta v_A(\underline{r}') d\underline{r} d\underline{r}' \tag{43}$$

where

$$\left(\frac{\delta f_A(\underline{r})}{\delta v_A(\underline{r}')} \right)_{N_A} = f_A(\underline{r},\underline{r}') \tag{44}$$

can be identified with the linear Fukui response function $f_A(\underline{r},\underline{r}')$, also introduced in the work of Fuentalba and Parr [51]. This response function can be calculated, although this is far from trivial, from the function $t_A(\underline{r})$ and its kernel $t_A(\underline{r},\underline{r}')$, where $t_A(\underline{r})$ has the following form:

$$t_A(\underline{r}) = \left(\frac{\partial s_A(\underline{r})}{\partial N_A} \right)_{v_A} \tag{45}$$

with $s_A(\underline{r})$ the local softness [52] (see Chapter 3). Introducing this $t_A(\underline{r})$ and its kernel, $f_A(\underline{r},\underline{r}')$ can be written as:

$$\left(\frac{\delta f_A(\underline{r})}{\delta v_A(\underline{r}')} \right)_{N_A} = -t_A(\underline{r},\underline{r}') + t_A(\underline{r}) f_A(\underline{r}') \tag{46}$$

In view of the smallness of the contributions of the two previous second order terms, this term, originating from a third order derivative of the energy of the system, might also be expected to be small.

The main problem for the calculation of the electronegativity of an atom in a molecule involves the evaluation of the effective nuclear charge Z^{eff} of a certain atom as it is experienced by another atom within a molecule.

In order to achieve this, an approach similar to the one adopted by Nalewajski et al. was used [46, 53]. In their work, it was suggested that a screened nuclear charge of the bond partner should be used ; the bare nuclear charge Z_B^0 is multiplied by an appropriate R-dependent (i.e. dependent on the bond distance) factor $f_B(\underline{r})$:

$$\tilde{Z}_B = Z_B^0 f_B(\underline{r}) \tag{47}$$

It was thereby suggested that the screening factor $f_B(\underline{r})$ shows an exponential decrease with R:

$$f_B(\underline{r}) = e^{-c_B R} \tag{48}$$

In our work, we adopted the following methodology. Since the nuclear charge of an atom is shielded by its electrons, i.e. the electron density $\rho(\underline{r})$ surrounding it, the nuclear charge can be expected to show the same radial behavior as the electron density. It is generally recognized that the spherically averaged electron density of an atom is an exponentially decreasing function of the distance r from the nucleus, which shows one unique maximum, located at the nucleus [54]. The rate of this exponential decrease changes at certain intervals or points, which were shown to coincide with the minima in the radial density distribution function $D(r)$ [55], defined as:

$$D(r) = 4\pi r^2 \rho(r) \tag{49}$$

Parr and Politzer have argued that the outermost minimum in $D(r)$ could be used as a boundary surface, separating the core and valence regions [56].

The integration of the radial density distribution function between two successive minima r_{n-1} and r_n (with $n>1$), yields the number of electrons contained in that shell:

$$N_n = \int_{r_{n-1}}^{r_n} D(r)dr \tag{50}$$

Taking all these facts in consideration, we finally propose the following expression for the effective nuclear charge of an atom B, sensed by an atom A at a distance R_{AB}:

$$Z_{B,A}^{eff} = \left(Z_B^0 - \int_0^{r_{\min,out}} D(r)dr \right) e^{-\lambda_{out}\left(R_{AB} - r_{\min,out}\right)} \tag{51}$$

This expression, together with equation (37), the closure relation and the electronegativity equalization principle enables us to calculate charge distributions in molecules. As an example, calculations on a selected number of diatomics are presented. The charge distribution was calculated for CO, ClF, NaF, NaCl, LiF, LiCl and MgO. The effective nuclear charges were calculated using the HF/6-31G* atomic electron density data $D(r)$, λ_{out} and $r_{\min,out}$ taken from [57]. The derivative of the electronic part of the electrostatic potential was calculated using a finite difference approximation: V_A^{el} was fitted through the points $N_A = N_A^0, N_A^0 - 1, N_A^0 + 1$ (a linear dependence was encountered in all cases), the derivative being taken at the point $N_A = N_A^0$. For χ^0 and η^0, the experimental Parr and Pearson values from [1] were used.

The resulting charge distributions (q, i.e. the charge on the electropositive atom) for these small molecular systems calculated using equation (37) and the equalized molecular electronegativities χ are given in Table 1. The same table lists charge distributions obtained using Sanderson's method (q_S) (see chapter 2.2 for a short introduction to this methodology) and Cioslowski's Atomic Polar Tensor (APT) method (q_{APT}) [58] - which will be the focus of more attention in Chapter 3 - from density functional calculations using the B3LYP exchange-correlation functional and the 6-31G* basis set (for a detailed account on these type of basis sets, see [6]). Moreover, we list some other estimates of the molecular electronegativity for these diatomic molecules ; these include the geometric mean of the Sanderson (S_m) and Parr and Pearson (χ_S) electronegativities of the constituting atoms, the electronegativity (χ_{Huheey}) obtained using Huheey's approximate equation (27) and the electronic chemical potentials *m* (with $\mu = -\chi$) calculated by Ohwada [59]. We also calculated *c* using equation (52) where I and A are calculated using Koopmans' theorem [60] with the Hartree-Fock 6-31G* orbital energies. The electronegativity thus becomes:

$$\chi = -\frac{\varepsilon_{HOMO} + \varepsilon_{LUMO}}{2} \qquad (52)$$

Table 1: Charge distributions and global electronegativities for some selected diatomics, calculated via equation (37), together with B3LYP/APT and Sanderson charges and global electronegativities, calculated using other methods (see text). All values are in au, except the molecular electronegativities which are in eV.

Molecule	q	χ	q_{APT}	q_S	χ_S	χ_{Huheey}	μ	$(I+A)/2$	S_m
ClF	0.118	9.42	0.210	0.086	9.30	9.14	-9.15	5.61	3.728
LiF	0.602	5.90	0.739	0.753	5.60	4.89	-4.88	6.14	1.882
LiCl	0.493	5.48	0.744	0.666	5.00	4.80	-4.79	5.22	1.755
NaF	0.684	5.99	0.732	0.797	5.45	4.72	-4.71	5.53	1.828
NaCl	0.562	5.46	0.757	0.711	4.86	4.65	-4.64	4.96	1.703
CO	0.126	8.16	0.223	0.162	6.88	6.84	-6.84	5.25	3.168
MgO	0.447	7.25	0.547	0.486	5.31	5.23	-	4.75	2.194
r					0.916	0.944	0.979	0.027	0.977

Table 1 shows that the density-functional charges roughly correlate with the charges obtained from the electronegativity equalization principle. However, there are some differences, more specifically in the series LiF, LiCl, NaF and NaCl. Comparing LiF and LiCl for example, the electronegativity equalization principle predicts a decrease of the positive charge on Li, in accordance with the decrease of electronegativity when going from F to Cl. The APT charges however show an increase, but the difference is very small. The same effect is noticable in the case of NaF and NaCl, and also when LiF and NaF and LiCl and NaCl are compared. Comparing with the Sanderson charges however yields exactly the same trend.

This example clearly illustrates the usefulness of electronegativity equalization schemes in the determination of charge distributions as compared to ab initio methods. The DFT charges are in agreement with electronegativity arguments and thus appeal to practising chemists, whereas the ab initio charges can not be rationalized using simple chemical arguments in this particular case.

Finally, we list the correlation coefficients r of the average molecular electronegativities resulting from our approach with the other estimates at the bottom of the table. The correlation is very satisfactory, except for the values obtained using Koopmans' theorem.

In this chapter, a numerical study of the electronegativity equalization principle was performed. An expression was derived for the electronegativity of an atom in a molecule, which could be identified with other formalisms proposed in the past, and which was successfully applied to the calculation of charge distribution of diatomics. In this approach, the explicit calculation of the external potential perturbation was achieved introducing effective nuclear charges, obtainable for atomic electron density data. Moreover, the expression was extended to second order, and it could be shown that second order terms are probably small in magnitude.

Since electronegativity equalization has to take place in all arbitrary parts of the molecule, the principle can also be applied using parts of the molecule larger than atoms, e.g. functional groups or even amino-acid residues in a peptide chain or protein.

In the case of functional groups, this idea permits to calculate charge transfer to or from functional groups in a molecule. It thus appears that density functional theory not only provides the framework to determine these functional group properties in a nonempirical manner, but also provides the framework to calculate the effect of the functional group on the molecular charge distribution and thus the reactivity, which is one of the major interest of physical organic chemistry.

2.3.2. Functional group resolution (first order treatment): acidity of alkyl alcohols

In the past, a number of studies (theoretical and experimental) were devoted to the apparent anomaly between the behaviour of alkyl groups in solution and the gas phase [23, 34, 61, 62]. In solution, the acidity of alkyl alcohols increases upon decreasing alkyl group size. Alkyl groups thus behave as electron donors, in accordance with the traditional view from organic chemistry. In the gas phase however, the acidity sequence is exactly the opposite, thus assigning electron withdrawing properties to these functional groups. In order to quantitatively study the charge transfer occurring to or from the alkyl group within the alcohol and the conjugate base, electronegativities and hardnesses of a selected number of alkyl groups were calculated and used in an electronegativity equalization scheme.

These group properties, listed in Table 2 were calculated at the MP4 level using the 6-31G* basis set at a geometry they adopt in a molecular environments [63].

Table 2: MP4 electronegativities χ (eV) and hardnesses η (eV) for some selected alkyl groups, calculated with the 6-31G* basis set.

Group	χ	η
Me	4.35	5.98
Et	3.76	5.45
n-Pr	3.68	5.17
i-Pr	3.44	5.04
n-Bu	3.57	5.05
t-Bu	3.30	4.69

Neglecting the external potential perturbation, the electronegativity of an alkyl group R in the acidic or the conjugate base forms ROH and RO^- respectively can be expressed as:

$$\chi_R = \chi_R^0 - 2\eta_R^0 \Delta N_R \qquad (53)$$

with χ_R^0 and η_R^0 again the electronegativity and hardness of the isolated functional group R and ΔN_R the charge transfer to or away from R.

Writing this relationship for the group R and the OH group in ROH and for R and O in RO⁻ and using charge conservation relations (i.e. $\Delta N_R + \Delta N_{OH} = 0$ and $\Delta N_R + \Delta N_O = 1$) for acidic and conjugate base form respectively, the charge transfer to ΔN_{R_I} to R in ROH and $\Delta N_{R_{II}}$ in RO⁻ can be calculated as:

$$\Delta N_{R_I} = \frac{\chi_R^0 - \chi_{OH}^0}{2\left(\eta_R^0 + \eta_{OH}^0\right)} \tag{54}$$

and

$$\Delta N_{R_{II}} = \frac{\chi_R^0 - \chi_O^0}{2\left(\eta_R^0 + \eta_O^0\right)} + \frac{\eta_O^0}{\eta_R^0 + \eta_O^0} = \Delta N_{R_{II,a}} + \Delta N_{R_{II,b}} \tag{55}$$

The sign of the charge transfer to the alkyl groups in the alcohols ROH is determined by the electronegativity difference of the hydroxyl group and the alkyl groups. Since the alkyl groups are less electronegative than the hydroxyl group for all R, the resulting charge transfers to the alkyl groups are all negative, so that the R groups all loose electrons.

Since, in addition, the group electronegativity and hardness decrease with increasing alkyl group size, ΔN_{R_I} becomes more negative with increasing alkyl group size, so that, in the case of ROH, the alkyl group behaviour is in accordance with the traditional viewpoint from organic chemistry. The expression for the charge transfer in the conjugate base, $\Delta N_{R_{II}}$, contains two terms: one term, indicated as $\Delta N_{R_{II,a}}$, which is of the same form as ΔN_{R_I}, and an additional term $\Delta N_{R_{II,b}}$, which appears due to the fact that the charge transfer is measured in a charged system, and which contains only group hardnesses. The first term in this expression is again always negative and becomes more negative as the alkyl group size increases ; the second term however is always positive, and will become more positive as the alkyl group size increases, indicating that an electron-withdrawing effect is exhibited by alkyl groups. It can thus be concluded that, besides an electronegativity dominated term with negative values for the alkyl groups, an additional and hardness dominated term appears due to which they globally act as electron acceptors.

Finally, the charge transfer $\Delta(\Delta N_R)$ undergone by the alkyl groups upon deprotonation can be calculated as:

$$\Delta(\Delta N_R) = \Delta N_{R_{II}} - \Delta N_{R_I} \tag{56}$$

The values for $\Delta(\Delta N_R)$ can be found in Table 3, together with the alkyl alcohol acidities in aqueous phase [64] and the gas phase [65].

When the approximation is made that $\chi_{OH}^0 \approx \chi_O^0$ and $\eta_{OH}^0 \approx \eta_O^0$ (i.e. the reasonable approximation that the electronegativity and hardness of the functional group is mainly determined by the electronegativity and hardness of the central atom in the case of a group consisting of a central heavy atom saturated with hydrogens), this expression can be rewritten as:

$$\Delta(\Delta N_R) \approx \frac{\eta_O^0}{\eta_R^0 + \eta_O^0} \tag{57}$$

Table 3: Experimental acidities of the alkyl alcohols ROH in the gas phase ΔG_{acid} (kcal/mol) and in aqueous solution $\Delta G_{acid}(aq)$ (kcal/mol) together with the calculated charge transfer $\Delta(\Delta N_R)$ (a.u.) to the alkyl group upon deprotonation of the acid.

ROH	ΔG_{acid}	$\Delta G_{acid}(aq)$	$\Delta(\Delta N_R)$
MeOH	375.0	20.7	0.507
EtOH	370.7	21.8	0.532
n-PrOH	369.5	22.1	0.545
i-PrOH	368.8	23.4	0.552
n-BuOH	368.8	22.1	0.551
t-BuOH	368.1	26.3	0.570

This term only contains group hardnesses. It can be considered as the ability of alkyl groups to stabilize negative charge in the process of deprotonation. This ability increases with increasing alkyl group size, resulting in a larger gas phase acidity. The opposite trend in the solution phase acidities is due to steric effects and to the fact that in solution, the stabilization of the conjugate base is due to the solvent and not the intrinsic electronic properties of the conjugate base itself.

In combination with calculated substituent electronegativities and hardnesses, electronegativity equalization offers a relatively simple and efficient scheme to elucidate charge distributions in alcohols and alkoxide ions. The calculated substituent effects are completely in accordance with trends for the acidity of these systems.

2.3.3. Amino acid residue resolution (first order treatment): charge distributions in peptides and proteins

Using the same ansatz as in the functional group case, and in order to be able to calculate charge distributions in large biomolecules such as peptides and proteins, we recently calculated electronegativity and hardness values for amino acid residues, considered as biradicals:

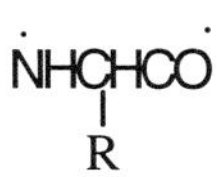

taken out of their protein environment. Both a-helix and b-sheet geometries were considered [66]. These values [67] (given in Table 4) were used to interpret the experimental gas phase proton affinity (PA) sequence of the amino acids [68]. In the regression equations

$$\text{PA(kcal/mol)}=-0.33\,\chi^0_{\alpha-helix}(\text{eV}) - 25.54\,\eta^0_{\alpha-helix}(\text{eV}) + 349.79 \tag{58}$$

(r=0.941 ; N=13)

$$PA(kcal/mol)=2.21\,\chi^0_{\beta-sheet}(eV) - 20.40\,\eta^0_{\beta-sheet}(eV) + 308.77 \quad (59)$$

(r=0.907 ; N=14)

it was found that group hardness plays the dominant role whereas group electronegativity only had a minor influence on the sequence, thereby stressing the importance of the charged form in the acid base equilibrium [23].

Using the χ^0 and η^0 values of Table 4, we calculated the charge distribution in some dipeptides, terminated by methyl groups:

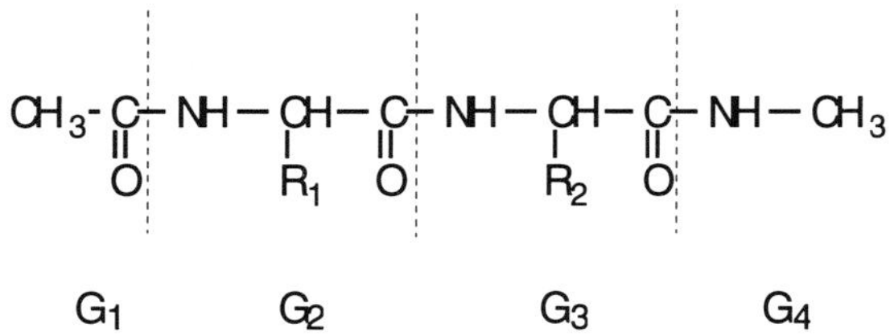

A few preliminary results are given below. Using a Huheey type of approach (eqn (27)), neglecting $\Delta\nu$ perturbations, we obtained the charges in the regions G_2 and G_3 for the following dipeptides AV, YA, IN, TV, DK and AA (Code, see Table 4),using the b-sheet data. As can be seen, the correlation is satisfying (r=0.909).

Table 4: Calculated group electronegativities χ (eV) and hardnesses η (eV) for the a-helix and b-sheet conformations of the amino acid residues, together with the experimental proton affinities (kcal/mol), taken from reference [68].

a-helix		b-sheet			
Amino Acid	χ	η	χ	η	PA
Glycine (G)	4.28	5.36	4.40	5.24	212.0
Alanine (A)	4.20	5.22	4.26	5.05	214.8
Valine (V)	4.09	5.09	4.18	4.88	217.4
Leucine (L)	4.09	5.05	4.19	4.86	218.5
Isoleucine (I)	4.08	5.06	4.15	4.85	219.3
Phenylalanine (F)	4.10	4.98	4.20	4.81	220.5
Serine (S)	4.59	5.11	4.70	5.05	217.2
Asparagine (N)	-	-	4.57	4.96	220.2
Glutamine (Q)	4.44	5.11	4.52	4.90	218.8
Aspartic acid (D)	4.57	5.15	4.68	5.00	217.1
Glutamic acid (E)	4.38	5.10	-	-	216.9
Tyrosine (Y)	4.11	4.98	4.21	4.75	222.7
Threonine (T)	4.32	5.26	4.63	4.83	218.8
Cysteine (C)	4.42	5.22	4.56	5.03	214.7
Methionine (M)	-	-	4.36	4.92	217.8
Lysine (K)	4.17	5.07	4.21	4.97	220.7

It turned out that using the a-helix data, the correlation deteriorates appreciably (r=0.627)as compared to the b-sheet case, which is in line with the smaller electrostatic interactions of neighbouring amino acids in the b-sheet as compared to the a-helix. We therefore tested the influence of the introduction of a Δv perturbation (cf. Chapter 2.3.1). Using a analogous formalism as in Chapter 2.3.1., on easily arrives at our expression:

$$\chi_{G_i} = \chi_{G_i}^0 + 2\eta_{G_i}^0 q_{G_i} + \sum_{j \neq i} \sum_{A_j \in G_j} q_{A_j} \left(\frac{\partial V_{G_i}^{el}(R_{A_j})}{\partial N_{G_i}} \right)_{N_{G_i}^0} \quad (60)$$

where the summation runs over all atoms of amino acid residue G_j, different from the one considered G_i and the remaining symbols do have their usual meaning. As shown in Figure 2, a considerable improvement was found for the a-helix case (r=0.899), confirming the above hypothesis.

Further work in this direction is in progress [69].

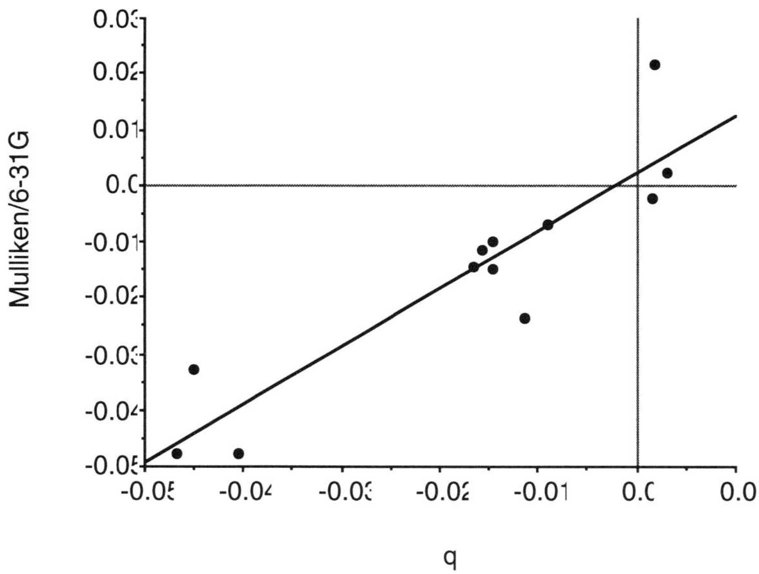

Figure 1: Correlation of the charges q in dipeptides AV, YA, IN, AA, TV and DK (b-sheet data) obtained using the simple Huheey equation with the ab initio Mulliken 6-31G charges.

2.4. Interpretation of charge distributions with DFT based functional group properties: a case study on halogenated alcohols and silanols

In several of our previous studies, it was found that charge distributions in acids and their conjugated bases are very often good descriptors for the observed experimental acidities.

As a case study and as an example how electronegativity and hardness of substituents can be used to qualitatively interpret molecular charge distributions, we summarize the results of a study on halogenated alcohols (methanols) $X_nCH_{3-n}OH$ and silanols $X_nSiH_{3-n}OH$, with X=F, Cl and Br [70]. These systems can be expected to give beautiful examples of the interplay of electronegativity and hardness due to the opposite behaviour of electronegativity and softness when going down in the periodic Table (FÆBr ; CÆSi). In this chapter, we will not explicitly discuss the acidity of these systems as such, but we will concentrate mainly on the effect of the halogen substituents on the charge distribution within the "reaction centre" of these molecules (i.e. the proton in the OH group of the alcohol or silanol and the O atom in the alkoxide and siloxide ions).

Table 5 lists the charges on the acidic proton, calculated at the Hartree Fock level with the 3-21+G and 6-31+G* basis sets. As can be seen, substitution of hydrogen by a halogen atom, i.e. increasing electronegativity, always results in a larger charge on the acidic proton, in accordance with chemical intuition. However, the trends obtained, $q_H(Cl_n) > q_H(Br_n) > q_H(F_n)$ cannot be explained without considering only the electronegativity or charge capacity [43,44] of the substituent. If only electronegativity is at work, the sequence F > Cl > Br should be found ; if only charge capacity would be of importance, the trend should be inverted, being Br > Cl > F. With the 6-31+G* basis set, in the case of the substituted alcohols, the obtained sequences $q_H(Cl_n) > q_H(F_n)$ for n=1,2 and $q_H(F_3) > q_H(Cl_3)$ elucidate clearly the interplay of the two above-mentioned effects. The slightly greater $q_H(F_n)$ compared to $q_H(Cl_n)$ in the silanol derivatives shows that the effect of the greater electronegativity of fluorine vs. chlorine is weakened by the high polarizability of silicon. As with the 3-21G basis set, this charge increases with increasing n for a given halogen.

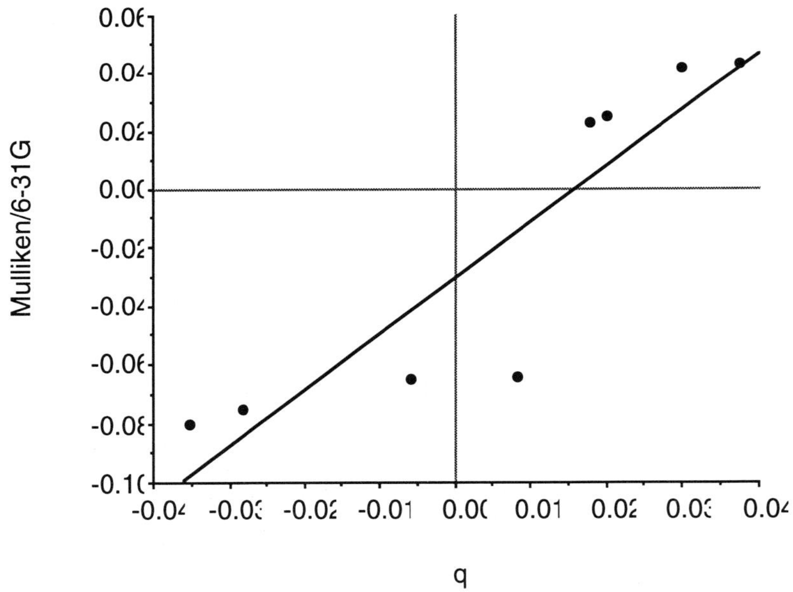

Figure 2: Correlation of the charges q in the dipeptides AV, YA, IN, and AA obtained using equation (60)with the ab initio Mulliken 6-31G charges (r=0.899).

Table 5: Charges on the acidic protons in halogenated alcohols and silanols, calculated at the Hartree Fock level using the 3-21+G and 6-31+G* basis sets.

3-21+G			
Alcohol	q_H	Silanol	q_H
CH_3OH	0.3731	SiH_3OH	0.4113
FCH_2OH	0.3949	$FSiH_2OH$	0.4284
F_2CHOH	0.4155	F_2SiHOH	0.4489
F_3COH	0.4373	F_3SiOH	0.4758
$ClCH_2OH$	0.4150	$ClSiH_2OH$	0.4403
Cl_2CHOH	0.4379	Cl_2SiHOH	0.4601
Cl_3COH	0.4465	Cl_3SiOH	0.4742
$BrCH_2OH$	0.4049	$BrSiH_2OH$	0.4351
Br_2CHOH	0.4213	Br_2SiHOH	0.4504
Br_3COH	0.4265	Br_3SiOH	0.4591

6-31+G*			
Alcohol	q_H	Silanol	q_H
CH_3OH	0.4862	SiH_3OH	0.5112
FCH_2OH	0.5071	$FSiH_2OH$	0.5273
F_2CHOH	0.5255	F_2SiHOH	0.5415
F_3COH	0.5480	F_3SiOH	0.5525
$ClCH_2OH$	0.5112	$ClSiH_2OH$	0.5263
Cl_2CHOH	0.5226	Cl_2SiHOH	0.5355
Cl_3COH	0.5390	Cl_3SiOH	0.5438

Table 6 gives the charge on the oxygen atom in the conjugate base for the different systems with the 3-21+G and 6-31+G* basis set respectively.

The trend in our results on q_O for H_3CO^- and F_3CO^- corresponds to the sequence found by Olsen and Howell [71] using both minimal STO-3G and the extended 4-31G basis and by Wiberg [72] (including the FH_2CO^- case) using a large basis set. In the silanols, the charge on the oxygen decreases with increasing chlorine and bromine substitution and increases with increasing fluorine substitution. The results for the fluoro compounds clearly do not highlight the electronegativity effect but can only be explained by the decreasing softness or charge capacity when a hydrogen atom is substitued by fluorine. In the case of chlorine and bromine, their decreasing electronegativity, in comparison with fluorine, is more than compensated by their increasing softness or charge capacity, resulting in a larger withdrawal of electrons from the oxygen atom. Obviously, for Cl and Br, the combination of electronegaivity and charge capacity seems to be more effective.

The comparison of the charge on the acidic proton with the charge on the oxygen atom in the conjugated base when F is replaced by Cl or Br confirms that polarizability or charge capacity effects are more important in the charged form in agreement with our earlier findings.

Table 6: Charges on the oxygen atoms in the conjugate bases of the halogenated alcohols and silanols, calculated at the Hartree Fock level using the 3-21+G and 6-31+G* basis set

3-21+G			
Alkoxide ion	q_O	Siloxide ion	q_O
CH_3O^-	-0.8656	SiH_3O^-	-0.9179
FCH_2O^-	-0.8458	$FSiH_2O^-$	-0.9364
F_2CHO^-	-0.8325	F_2SiHO^-	-0.9585
F_3CO^-	-0.8255	F_3SiO^-	-0.9778
$ClCH_2O^-$	-0.5668	$ClSiH_2O^-$	-0.8454
Cl_2CHO^-	-0.4160	Cl_2SiHO^-	-0.7959
Cl_3CO^-	-0.5416	Cl_3SiO^-	-0.7492
$BrCH_2O^-$	-0.5625	$BrSiH_2O^-$	-0.8490
Br_2CHO^-	-0.5349	Br_2SiHO^-	-0.8044
Br_3CO^-	-0.5520	Br_3SiO^-	-0.7673
6-31+G*			
Alkoxide ion	q_O	Siloxide ion	q_O
CH_3O^-	-0.9355	SiH_3O^-	-0.9749
FCH_2O^-	-0.8319	$FSiH_2O^-$	-0.9796
F_2CHO^-	-0.8184	F_2SiHO^-	-0.9853
F_3CO^-	-0.8269	F_3SiO^-	-1.0288
$ClCH_2O^-$	-0.4575	$ClSiH_2O^-$	-0.9137
Cl_2CHO^-	-0.3430	Cl_2SiHO^-	-0.8558
Cl_3CO^-	-0.4133	Cl_3SiO^-	-0.8081

This study on halogenated alcohols and silanols clearly illustrates the usefulness of a combined use of electronegativity and hardness in the qualitative interpretation of molecular charge distributions. Traditionally, electronegativity was used exclusively for this purpose; in this case however, hardness plays a very important role, enabling us to interpret apparently anomalous charge distributions.

3. DFT CALCULATIONS OF MOLECULAR CHARGE DISTRIBUTIONS AND PROPERTIES DERIVED FROM IT

3.1. Methodology

Until now, we have been focusing on concepts, such as electronegativity and hardness emerging in a natural way from density-functional theory. These concepts were used both qualitatively and quantitatively to determine and interpret molecular charge distributions. However, this is not the only advantage associated to working with DFT, as will be seen in this chapter.

Hohenberg and Kohn proved that the ground state energy E of a many-electron system can be obtained as a minimum of the energy functional:

$$E[\rho(\underline{r})] = T[\rho(\underline{r})] + V_{ee}[\rho(\underline{r})] + \int \rho(\underline{r}) v(\underline{r}) d\underline{r} \tag{61}$$

with $T[\rho(\underline{r})]$ and $V_{ee}[\rho(\underline{r})]$ the kinetic and electron-electron repulsion functionals and the integral being the nuclear attraction energy. The functional $V_{ee}[\rho(\underline{r})]$ consists of a classical part $J[\rho(\underline{r})]$, i.e.

$$J[\rho(\underline{r})] = \frac{1}{2} \int \frac{\rho(\underline{r})\rho(\underline{r}')}{|\underline{r} - \underline{r}'|} d\underline{r} d\underline{r}' \tag{62}$$

and a non-classical part, the unknown exchange-correlation energy functional $E_{XC}[\rho(\underline{r})]$.

Minimization of (61) with respect to r leads to the Euler equation (2). The transformation of DFT into a tool for rigorous calculations was established by Kohn and Sham [4]. They proposed to write the unknown kinetic energy functional as:

$$T[\rho(\underline{r})] = \sum_i n_i \left\langle \psi_i \middle| -\tfrac{1}{2}\nabla_i^2 \middle| \psi_i \right\rangle \tag{63}$$

with ψ_i and n_i the natural spin orbitals and occupation numbers respectively. In this way, orbitals are introduced in the problem. Putting n_i=1 for N natural spin orbitals (N being the number of electrons) and n_i=0 for all others, (63) can be approximated as:

$$T_s[\rho(\underline{r})] = \sum_i^N \left\langle \psi_i \middle| -\tfrac{1}{2}\nabla_i^2 \middle| \psi_i \right\rangle \tag{64}$$

In terms of these orbitals, the electron density becomes:

$$\rho(\underline{r}) = \int \sum_i^N |\psi_i(\underline{r}, \sigma)|^2 d\sigma \tag{65}$$

Introducing these formulas in the Euler equation (2) yields the Kohn-Sham orbital equations:

$$\left[-\tfrac{1}{2}\nabla^2 + v_{eff}(\underline{r})\right]\psi_i = \varepsilon_i \psi_i$$

where the effective Kohn-Sham potential $v_{eff}(\underline{r})$ has the form:

$$v_{eff}(\underline{r}) = v(\underline{r}) + \int \frac{\rho(\underline{r}')}{|\underline{r} - \underline{r}'|} d\underline{r}' + v_{XC}[\rho(\underline{r})] \tag{66}$$

Here, $v_{XC}[\rho(\underline{r})]$ is the exchange-correlation potential:

$$v_{XC}[\rho(\underline{r})] = \frac{\delta E_{XC}[\rho(\underline{r})]}{\delta \rho(\underline{r})} \tag{67}$$

Unfortunately, one needs an explicit form of this exchange-correlation functional. Many proposals as to its form have been put forward.

In all of the functionals considered in this work, exchange and correlation are separated in the exchange-correlation functional. These exchange and correlation functionals can then again be divided into two classes, i.e. the local and non-local functionals. Examples of local functionals are the Slater functional for exchange [73] and the VWN (Vosko, Wilk, Nusair) functional [74] for correlation. We will therefore call the S-VWN method the "local density approximation" (LDA). Among the non-local functionals we mention Becke's 1988 gradient corrected functional for exchange [75], the LYP (Lee-Yang-Parr) [76], P86 (Perdew 1986) [77] and PW91 (Perdew-Wang 1991) [78] correlation functionals. Finally, we also used Becke's three parameter hybrid exchange functional, which includes the exact Hartree-Fock exchange based on Kohn-Sham orbitals as one of its components (designated as B3) [79]. This thus brings the used functionals in this work to five, being SVWN (or LDA), BP86, BLYP, B3LYP and B3PW91.

Among the wavefunction methods, we considered the Hartree-Fock level (SCF), second order perturbation theory [80] (MP2) and Quadratic Configuration Interaction with all Single and Double Substitutions [81] (QCISD), which is a simplified form of coupled cluster theory with all single and double substitutions (CCSD).

Although some DFT calculations of charge distributions and reactivity indices have been published [82-85], the performance of DFT methods in their determination has not been investigated in a comprehensive way: this will be the subject of the following paragraphs.

All calculations have been performed using the Gaussian 94 program [86].

The following basis sets have been considered: the Pople group 6-31G* (for a detailed account, see [6]), because of its widespread use in applied quantum chemistry ; Dunning's correlation consistent basis sets [87], limited in our work to the cc-pVDZ (correlation consistent polarized valence double zeta, i.e. [3s2p1d/2s1p]) and cc-pVTZ (correlation consistent polarized valence triple zeta, i.e. [4s3p2d1f/3s2p1d]) basis sets.

3.2. Application to atomic populations: testing population analysis techniques

Atomic populations were calculated for some molecules containing first and second row atoms [88]. The following population analysis techniques were considered (for a review, see [89]): the Mulliken method [90], the electrostatic potential based ChelpG method [91], the Natural population analysis (NPA) [92, 93], the atomic polar tensor method (APT) of Cioslowski [58], and Bader's topological atoms-in-molecules charges [94], calculated using Cioslowski's implementation [95]. Geometries were optimized at the respective levels of theory.

In Table 7, we depict the dependence of the different population analysis methods on electron correlation for the cc-pVDZ basis set. Since evidently, the APT and Bader charges are the most sensitive to electron correlation, we will consider both of these methods for the remainder of the discussion.

Bader and APT charges computed at the HF, MP2 and QCISD levels, as well as charges obtained using the five different exchange-correlation functionals, are listed in Tables 8 and 9, using the cc-pVDZ basis set. For all methods, the mean absolute deviations from quadratic CI is listed at the bottom of the Table.

Focussing first on the Bader results, it can immediately be seen that all DFT methods perform better than Hartree-Fock. The exact exchange-correlation functionals B3LYP and B3PW91 perform approximately the same as MP2. The two gradient-corrected functionals BP86 and BLYP perform comparably; essentially, there is really not much to choose between them. The worst performance among the density-functional methods is, not surprisingly, put in by the local density approximation (LDA), since it constitutes the most approximate method.

Table 7: Correlation contribution to different charge definitions at QCISD/cc-pVDZ level.

		q(QCISD)-q(SCF)				
Molecule		Mulliken	ChelpG	NPA	APT	Bader
BH	H	0.000	-0.068	0.018	-0.048	0.053
	B	0.000	0.068	-0.018	0.048	-0.053
C_2H_2	H	-0.005	-0.031	-0.008	-0.021	-0.013
	C	0.005	0.031	0.008	0.021	0.013
CH_4	H	0.005	-0.006	0.003	0.007	0.036
	C	-0.021	0.025	-0.011	-0.028	-0.145
CO	C	-0.084	-0.055	-0.121	-0.151	-0.184
	O	0.084	0.055	0.121	0.151	0.184
CO_2	O	0.087	0.085	0.109	0.202	0.186
	C	-0.173	-0.170	-0.217	-0.406	-0.373
H_2CO	O	0.081	0.074	0.097	-0.169	0.204
	C	-0.079	-0.051	-0.100	-0.127	-0.261
	H	-0.001	-0.011	0.002	-0.021	0.028
H_2O	O	0.019	0.057	0.042	0.095	0.103
	H	-0.009	-0.028	-0.021	-0.048	-0.051
H_2S	S	0.016	0.029	0.010	0.030	-0.221
	H	-0.008	-0.015	-0.005	-0.015	0.110
HCl	H	-0.018	-0.017	-0.016	-0.037	-0.043
	Cl	0.018	0.017	0.016	0.037	0.043
HCN	H	-0.016	-0.034	-0.010	-0.031	-0.033
	C	-0.032	-0.002	-0.039	-0.068	-0.250
	N	0.048	0.037	0.049	0.098	0.283
HF	H	-0.012	-0.028	-0.026	-0.053	-0.064
	F	0.012	0.028	0.026	0.053	0.064
NH_3	N	0.021	0.079	0.039	0.096	0.114
	H	-0.007	-0.026	-0.013	-0.032	-0.038
NNO	N	0.008	0.024	0.003	0.107	-0.181
	N	-0.135	-0.123	-0.121	-0.345	0.111
	O	0.127	0.099	0.118	0.238	0.070
PH_3	P	-0.018	0.033	-0.032	-0.035	-0.181
	H	0.006	-0.011	0.010	0.012	0.060
SO_2	S	-0.203	-0.122	-0.293	-0.405	-0.420
	O	0.092	0.061	0.146	0.203	0.210

Focussing first on the Bader results, it can immediately be seen that all DFT methods perform better than Hartree-Fock. The exact exchange-correlation functionals B3LYP and B3PW91 perform approximately the same as MP2. The two gradient-corrected functionals BP86 and BLYP perform comparably; essentially, there is really not much to choose between them. The worst performance among the density-functional methods is, not surprisingly, put in by the local density approximation (LDA), since it constitutes the most approximate method.

An interesting point immediately stands out: all density-functional methods perform extremely bad for H_2S and PH_3. Eliminating the results for these systems from the statistical treatment, the mean absolute deviations become 0.131, 0.042, 0.082, 0.055, 0.054, 0.026 and 0.028 for HF, MP2, LDA, BP86, BLYP, B3LYP and B3PW91 respectively. Thus, it becomes clear that B3LYP and B3PW91 perform considerably better than MP2.

Table 8: Charge distributions at the different levels using Bader's population analysis method and the cc-pVDZ basis set.

Molecule		SCF	MP2	QCISD	LDA	BP86	BLYP	B3LYP	B3PW91
BH	H	-0.811	-0.782	-0.758	-0.625	-0.699	-0.716	-0.745	-0.736
	B	0.811	0.782	0.758	0.625	0.699	0.716	0.745	0.736
C_2H_2	H	0.153	0.158	0.140	0.210	0.165	0.156	0.163	0.165
	C	-0.153	-0.158	-0.140	-0.210	-0.165	-0.156	-0.163	-0.165
CH_4	H	-0.077	-0.028	-0.040	0.049	-0.002	-0.010	-0.015	-0.012
	C	0.306	0.113	0.161	-0.196	0.007	0.041	0.062	0.046
CO	C	1.382	1.151	1.198	1.161	1.156	1.151	1.210	1.212
	O	-1.382	-1.151	-1.198	-1.161	-1.156	-1.151	-1.210	-1.212
CO_2	O	-1.353	-1.116	-1.166	-1.080	-1.072	-1.066	-1.142	-1.145
	C	2.706	2.232	2.333	2.161	2.144	2.132	2.284	2.290
H_2CO	O	-1.265	-1.083	-1.061	-1.053	-1.043	-1.035	-1.093	-1.099
	C	1.376	1.148	1.115	0.967	1.033	1.035	1.100	1.104
	H	-0.055	-0.032	-0.027	0.043	0.005	0.000	-0.004	-0.003
H_2O	O	-1.164	-1.187	-1.162	-1.190	-1.136	-1.122	-1.168	-1.177
	H	0.632	0.593	0.581	0.595	0.568	0.561	0.584	0.588
H_2S	S	0.677	0.464	0.456	0.042	0.166	0.205	0.281	0.264
	H	-0.338	-0.232	-0.228	-0.021	-0.083	-0.102	-0.141	-0.132
HCl	H	0.269	0.254	0.227	0.314	0.266	0.251	0.263	0.272
	Cl	-0.269	-0.254	-0.227	-0.314	-0.266	-0.251	-0.263	-0.272
HCN	H	0.227	0.212	0.194	0.264	0.216	0.212	0.223	0.221
	C	1.257	0.919	1.007	0.918	0.959	0.961	1.025	1.026
	N	-1.484	-1.131	-1.201	-1.182	-1.175	-1.172	-1.248	-1.248
HF	H	0.781	0.726	0.717	0.714	0.692	0.687	0.712	0.716
	F	-0.781	-0.726	-0.717	-0.714	-0.692	-0.687	-0.712	-0.716
NH_3	N	-1.231	-1.176	-1.116	-1.253	-1.134	-1.110	-1.164	-1.177
	H	0.410	0.392	0.372	0.418	0.378	0.370	0.388	0.392
NNO	N	0.332	0.083	0.151	0.095	0.104	0.104	0.139	0.138
	N	0.099	0.261	0.211	0.247	0.222	0.216	0.214	0.220
	O	-0.432	-0.344	-0.362	-0.342	-0.326	-0.320	-0.353	-0.358
PH_3	P	1.908	1.749	1.727	1.376	1.492	1.533	1.620	1.592
	H	-0.636	-0.583	-0.576	-0.459	-0.497	-0.511	-0.540	-0.531
SO_2	S	2.640	2.044	2.220	2.039	1.978	1.951	2.151	2.168
	O	-1.320	-1.022	-1.110	-1.016	-0.989	-0.975	-1.076	-1.084
		Mean absolute deviation from QCISD							
		0.133	0.036	0.000	0.105	0.071	0.067	0.035	0.038

Some other interesting observations can be made from Table 8. All methods predict a C^{d+}-H^{d-} polarity in CH_4. In H_2CO, B3LYP and B3PW91 are the only density-functionals which are able to reproduce the negative charges on the hydrogen atoms. In all other cases, there is no change in polarity between the different levels.

From Table 9, it can again be noticed that all density-functional show a much better performance than the Hartree-Fock level. Moreover, all (density functional) methods now even perform better the MP2, the overall best performance exhibited by the two exact exchange functionals B3LYP and B3PW91. B3LYP appears to do slightly better than

Table 9: Charge distributions at the different levels using Cioslowski's Atomic Polar Tensor population analysis method and the cc-pVDZ basis set.

Molecule		SCF	MP2	QCISD	LDA	BP86	BLYP	B3LYP	B3PW91
BH	H	-0.020	-0.039	-0.068	-0.055	-0.066	-0.060	-0.047	-0.052
	B	0.020	0.039	0.068	0.055	0.066	0.060	0.047	0.052
C_2H_2	H	0.216	0.201	0.195	0.209	0.196	0.189	0.199	0.203
	C	-0.216	-0.201	-0.195	-0.209	-0.196	-0.189	-0.199	-0.203
CH_4	H	-0.015	0.002	-0.008	0.030	0.004	-0.005	-0.001	0.007
	C	0.061	-0.009	0.032	-0.122	-0.018	0.022	0.002	-0.029
CO	C	0.354	0.105	0.203	0.177	0.185	0.178	0.215	0.221
	O	-0.354	-0.105	-0.203	-0.177	-0.185	-0.178	-0.215	-0.221
CO_2	O	-0.764	-0.495	-0.562	-0.491	-0.482	-0.475	-0.548	-0.555
	C	1.529	0.990	1.123	0.982	0.965	0.951	1.096	1.110
H_2CO	O	-0.684	-0.467	-0.514	-0.513	-0.507	-0.501	-0.543	-0.549
	C	0.797	0.629	0.670	0.669	0.688	0.691	0.701	0.698
	H	-0.057	-0.081	-0.078	-0.078	-0.091	-0.095	-0.079	-0.075
H_2O	O	-0.562	-0.486	-0.467	-0.474	-0.428	-0.403	-0.455	-0.474
	H	0.281	0.243	0.234	0.237	0.214	0.201	0.227	0.237
H_2S	S	-0.101	-0.104	-0.071	-0.133	-0.097	-0.069	-0.087	-0.111
	H	0.051	0.052	0.036	0.066	0.048	0.034	0.044	0.055
HCl	H	0.229	0.212	0.192	0.218	0.199	0.185	0.200	0.212
	Cl	-0.229	-0.212	-0.192	-0.218	-0.199	-0.185	-0.200	-0.212
HCN	H	0.272	0.245	0.241	0.248	0.237	0.231	0.246	0.249
	C	-0.003	-0.113	-0.071	-0.083	-0.081	-0.076	-0.061	-0.062
	N	-0.269	-0.132	-0.170	-0.166	-0.156	-0.155	-0.185	-0.187
HF	H	0.434	0.391	0.382	0.377	0.356	0.343	0.371	0.379
	F	-0.434	-0.391	-0.382	-0.377	-0.356	-0.343	-0.371	-0.379
NH_3	N	-0.422	-0.363	-0.326	-0.377	-0.298	-0.262	-0.326	-0.356
	H	0.141	0.121	0.109	0.126	0.099	0.087	0.109	0.119
NNO	N	-0.390	-0.259	-0.283	-0.288	-0.275	-0.274	-0.312	-0.316
	N	1.136	0.673	0.791	0.728	0.703	0.695	0.814	0.825
	O	-0.747	-0.414	-0.509	-0.440	-0.428	-0.421	-0.502	-0.510
PH_3	P	0.365	0.305	0.330	0.228	0.286	0.324	0.316	0.284
	H	-0.122	-0.102	-0.110	-0.076	-0.095	-0.108	-0.105	-0.095
SO_2	S	1.409	0.769	1.004	0.943	0.920	0.903	1.022	1.038
	O	-0.705	-0.384	-0.502	-0.471	-0.460	-0.451	-0.511	-0.519
			Mean absolute deviation from QCISD						
		0.104	0.046	0.000	0.034	0.029	0.032	0.013	0.019

B3PW91. In this case, the difference between LDA and the two gradient corrected exchange-correlation functionals (BP86 and BLYP) is not very significant.

The fact that the exact-exchange contributions are definitely more important than gradient corrections, even for very correlation-sensitive properties such as APT charges, is a consequence of the fact that the exchange energy is 1 to 2 orders of magnitude more important than the correlation energy (e.g. compare Tables 8.3 and 8.4 in reference [1]).

3.3. Dipole moments and infrared intensities as sensitive probes for molecular charge distributions

In a next step, we investigated the performance of the B3LYP functional in the calculation of molecular dipole moments and infrared intensities, two sensitive probes for the molecular charge distribution.

The performance of B3LYP in the calculation of these quantities is depicted in Tables 10 and 11. The generally excellent performance of this functional for frequencies was already discussed extensively elsewhere (e.g. [97]). As can be seen from Table 10, the B3LYP computed dipole moments are usually in very good agreement with high level ab initio calculations and with experiment as long as a basis set of spdf quality is used. When a basis set of only split-valence plus polarization quality is used, the overall good agreement rapidly deteriorates.

Table 10: Computed and experimental dipole moments (Debye) for some selected small molecules.

Molecule	6-31G*	B3LYP cc-pVDZ	cc-pVTZ	B3PW91 cc-pVTZ	CCSD(T) TZ2Pf [96]	Experiment [a]
HCN	2.905	2.795	2.981	2.986	2.990	2.984
H_2O	2.095	1.938	1.919	1.919	1.920	1.854
H_2CO	2.186	2.071	2.280	2.270	2.341	2.332
NH_3	1.912	1.684	1.586	1.595	1.603	1.471
CO [b]	0.060	0.180	0.125	0.148		0.110
HF	1.860	1.818	1.831	1.825		1.826
HCl	1.468	1.338	1.206	1.220		1.109
SO_2	1.776	1.899	1.812	1.751		1.633
PH_3	0.962	0.710	0.626	0.689		0.574
H_2S	1.427	1.219	1.088	1.126		0.97
BH	1.285	1.447	1.531	1.528		1.270 [c]
NNO	0.016	0.020	0.037	0.009		0.161

(a) All experimental data taken from Handbook of Chemistry and Physics, 75th Edition (CRC Press, Boca Raton, FL, 1995)
(b) All calculated values show a $C^{\delta-}O^{\delta+}$ orientation in agreement with experiment
(c) K. P. Huber and G. Herzberg, *Constants of Diatomic Molecules* (Van Nostrand Reinhold, New York, 1979)

In the next Table, the performance of the B3LYP functional in the calculation of harmonic frequencies and infrared intensities is depicted. This table is self-explanatory, indicating very good agreement between B3LYP/cc-pVTZ intensities and high-level ab initio calculations. Two exceptions however are the CO stretch and the antisymmetric CH stretch in formaldehyde. The smaller basis set results also appear to be satisfactory if only a semiquantitative description of the intensity pattern is required.

How does B3PW91 perform for these properties, compared to B3LYP? We will restrict our comparison to the cc-pVTZ basis set. As seen in Table 10, performance for dipole moments is essentially the same, and in Table 11 we hardly see any differences in the computed IR intensities. For some molecules, however, the computed frequencies differ markedly. This is particularly the case for the A-H stretching frequencies in the strongly polar hydrides NH_3, H_2O, and HF, where the B3PW91 harmonic frequencies are in excellent agreement with experiment while their B3LYP counterparts still have appreciable errors.

Table 11: Computed and observed harmonic frequencies w (cm^{-1}) and infrared intensities I (km/mol)

Experiment [a]		CCSD(T)		B3LYP				B3PW91					
TZ2Pf [96]				6-31G*		cc-pVDZ		cc-pVTZ		ccpVTZ			
		w	I	w	I	w	I	w	I	w	I	w	I
HCN	w_1	3444	54	3470	64	3481	52.5	3466	63.2	3450	60.7	3453	60.7
	w_2	725	46	791	71	766	71.0	772	68.2	762	72.6	763	73.1
	w_3	2130	0.1	2128	0.06	2214	2.0	2200	2.0	2202	1.3	2205	1.3
CO_2	w_2	673	48	672	59	640	61.4	655	28.7	672	63.7	676	62.7
	w_3	2397	548	2391	634	2436	545.8	2422	577.6	2417	629.2	2441	630.7
CH_4	w_3	3157	69±3 72±11 70±3	3147	64	3163	79.2	3145	63.0	3132	73.1	3151	63.9
	w_4	1367	29±1 41±6	1351	32	1374	46.2	1311	35.9	1343	37.5	1330	41.4
C_2H_2	w_3	3415	71±2	3434	81	3442	76.5	3428	95.3	3416	89.1	3417	89.7
	w_5	747	175±5	794	182	775	162.2	773	153.3	767	192.4	768	194.5
H_2O	w_1	3832	2.2 2.2 2.2 2.5	3835	4.7	3726	1.7	3751	2.9	3801	3.2	3834	3.9
	w_2	1649	53.6 63.9 66.6 71.9	1650	69.5	1713	75.8	1659	55.8	1640	69.5	1640	68.8
	w_3	3943	44.6 48.2 39.8	3944	48.4	3848	19.4	3852	20.0	3901	40.8	3937	44.0
H_2CO	w_1	2923 [b]	75.5±7.1	2926	59.4	2916	55.9	2866	59.5	2876	69.2	2881	69.1
	w_2	1778 [b]	74.0±5.3	1776	74.5	1850	98.7	1831	108.8	1825	107.1	1839	107.7
	w_3	1539 [b]	11.2±1.0	1541	10.6	1563	6.4	1515	5.8	1537	9.9	1531	9.3
	w_4	1194 [b]	6.5±0.6	1197	4.4	1198	1.4	1186	1.6	1203	3.2	1200	3.6
	w_5	2999 [b]	87.6±8.0	2996	108.4	2967	164.7	2919	177.3	2929	145.1	2936	140.6
	w_6	1276 [b]	9.9±1.0	1272	12.0	1280	12.6	1253	16.0	1268	12.8	1265	12.5
HF	w_1	4138	100.6 [c]	—	—	3978	62.2	4022	89.3	4097	97.7	4132	100.3
NH_3	w_1	3478	7.6±0.9	3471	2.3	3437	0.8	3418	2.0	3461	2.4	3482	2.0
	w_2	[1022]	138±6	1076	147	1132	156.1	1122	105.5	1068	146.7	1061	146.7
	w_3	3597	3.8±0.8	3600	3.8	3568	1.0	3532	1.9	3577	1.1	3605	2.1
	w_4	1684	28.2±0.5	1679	31	1727	30.6	1673	17.9	1677	33.6	1671	34.5

(a) For experimental references, see [96] except for HF: K.P. Huber and G. Herzberg, *Constants of diatomic molecules* (Van Nostrand Reinhold, New York, 1979) and [98].
(b) Reference [99]
(c) Intensity obtained from reference [99], from the vibrational dipole moment matrix element (0-1) and the experimental ground vibrational frequency, considered as band origin.

For less polar molecules, the frequencies either hardly differ at all, or (as in CO) there is a difference, but both values deviate substantially from experiment.

These results suggest that the more sophisticated Perdew-Wang (1991) correlation functional in B3PW91 does represent an improvement in situations with strong polarity, that is, a large density gradient. It will be seen below that B3PW91 also outperforms B3LYP for the prediction of molecular electrostatic potential minima of anions and, to a lesser extent, neutral molecules, which probably has a related explanation. For multiply-bonded systems or systems with very high electron density (such as F_2), B3LYP performs less well than for single bonds [97], and B3PW91 appears to follow the same pattern.

3.4. The Molecular electrostatic potential and the Fukui function: reactivity descriptors related to the molecular charge distribution

3.4.1. Introduction

In this section, two descriptors of the molecular reactivity, the molecular electrostatic potential (MEP) and the Fukui function will be studied. The MEP, a traditional reactivity descriptor, giving the interaction energy of a molecule with a unit positive charge at a position $\underline{r}$ is calculated as [12-14]:

$$V(\underline{r}) = \sum_A \frac{Z_A}{|\underline{r} - \underline{R}_A|} - \int \frac{\rho(\underline{r}')}{|\underline{r} - \underline{r}'|} d\underline{r}' \tag{68}$$

Following the definition, it can be seen that the MEP is used in the interpretation of site selectivity of electrophilic reactions. The extension to nucleophilic processes is possible, but has to be performed with great care [100, 101]. Moreover, the MEP can be expected to describe essentially the hard-hard interactions between the reacting molecules.

The Fukui function finds its origin within density-functional theory and is introduced (see also section 2.2) as [15]:

$$f(\underline{r}) = \left(\frac{\delta\mu}{\delta v(\underline{r})}\right)_N = \left(\frac{\partial\rho(\underline{r})}{\partial N}\right)_v \tag{69}$$

Because of the discontinuity of this derivative, Parr and Yang associated a backward ($f^-(\underline{r})$) and forward ($f^+(\underline{r})$) derivative with the Fukui function, corresponding to an electrophilic and nucleophilic attack respectively [15]. Using a finite difference approximation, both functions can be written as:

$$f^-(\underline{r}) = \rho_N(\underline{r}) - \rho_{N-1}(\underline{r}) \tag{70}$$

and

$$f^+(\underline{r}) = \rho_{N+1}(\underline{r}) - \rho_N(\underline{r}) \tag{71}$$

where $\rho_N(\underline{r})$, $\rho_{N+1}(\underline{r})$ and $\rho_{N-1}(\underline{r})$ are the electron densities of the N, N+1 and N-1 electron system respectively, all calculated at the same external potential $v(\underline{r})$ of the N electron system (cf. the constraint of a constant external potential). In addition, one can introduce a Fukui function $f^0(\underline{r})$, being the average of $f^+(\underline{r})$ and $f^-(\underline{r})$, which can be considered to be a reactivity index for a radical attack:

$$f^0(\underline{r}) = \frac{f^+(\underline{r}) + f^-(\underline{r})}{2} = \frac{\rho_{N+1}(\underline{r}) + \rho_{N-1}(\underline{r})}{2} \tag{72}$$

Yang and Mortier [102] introduced a condensed version of these functions, condensing the function to one value per atom *k* in the molecule using a population analysis method:

$$f_k^- = q_k(N) - q_k(N-1) \tag{73}$$

$$f_k^+ = q_k(N+1) - q_k(N) \tag{74}$$

and

$$f_k^0 = \frac{q_k(N+1) - q_k(N-1)}{2} \tag{75}$$

with $q_k(N)$, $q_k(N+1)$ and $q_k(N-1)$ the populations on atom k in the N, N+1 and N-1 electron system. In the work of Yang and Mortier and in most of the other applications hitherto, use was made of the Mulliken populations.

These Fukui functions were already calculated in many papers and were indeed shown to be very valuable in the description of molecular reactivity [103]. Finally, we mention that the Fukui function can be expected to describe the soft-soft interactions between molecules.

In analogy with (10) and (12), local softness, which, upon integration, should yield the global molecular softness can written as:

$$s(\underline{r}) = \left(\frac{\delta \rho(\underline{r})}{\delta \mu} \right)_v \tag{76}$$

with

$$S = \int s(\underline{r}) d\underline{r} \tag{77}$$

Decomposing equation (76) yields:

$$s(\underline{r}) = \left(\frac{\partial \rho(\underline{r})}{\partial N} \right)_v \left(\frac{\partial N}{\partial \mu} \right)_v = f(\underline{r}) S \tag{78}$$

One sees that the Fukui function distributes softness among the various parts of a molecule; within a molecule, regions with high values for the Fukui function thus correspond to soft regions. From (77), it also follows that:

$$\int f(\underline{r}) d\underline{r} = 1 \tag{79}$$

In this chapter, we will investigate the performance of the different density-functionals in the calculation of the above discussed reactivity descriptors. Again, comparison will be made with Hartree-Fock and two correlated methods, MP2 and QCISD.

3.4.2. The molecular electrostatic potential

In a first part, we compare the energy of MEP minima for some selected neutral molecules and some closed shell anions. Those were calculated with both the cc-pVDZ and cc-pVTZ basis sets (for the anionic systems, an augmented function was added to the heavy atom [104]).

Table 12: Minima in the molecular electrostatic potential (kcal/mol) at the different levels with the cc-pVDZ basis set (augmented functions were placed on the heavy atoms in the anions).

Molecule		SCF	MP2	QCISD	LDA	BP86	BLYP	B3LYP	B3PW91
NH_3		-81.10	-78.03	-77.02	-78.12	-75.58	-74.37	-76.95	-77.86
H_2O		-61.33	-57.61	-57.15	-57.82	-55.48	-54.70	-56.97	-57.46
HF		-33.77	-31.83	-31.62	-32.04	-30.84	-30.59	-31.63	-31.73
HCN		-47.82	-43.59	-43.94	-41.86	-40.97	-40.89	-42.89	-43.10
CO	C	-17.32	-20.33	-17.99	-19.14	-18.66	-19.22	-18.83	-18.52
	O	-12.55	-5.07	-8.09	-5.18	-5.72	-5.83	-7.21	-7.09
H_2CO		-47.35	-38.13	-39.48	-35.47	-35.56	-34.98	-38.25	-38.81
N_2		-12.87	-14.02	-13.79	-12.98	-12.60	-13.07	-13.06	-12.71
PH_3		-27.56	-23.66	-22.13	-21.29	-21.05	-20.03	-22.05	-22.95
H_2S		-27.05	-25.62	-24.91	-26.21	-25.66	-25.01	-25.61	-26.13
HCl		-13.62	-12.99	-12.68	-13.27	-12.92	-12.64	-12.94	-13.14
Mean absolute deviation from QCISD									
		3.42	1.01	0.00	1.30	1.59	1.88	0.73	0.96
CH_3^-		-230.17	-214.80	-218.64	-211.97	-211.18	-209.15	-215.50	-218.93
NH_2^-		-242.40	-225.25	-229.09	-223.43	-222.41	-220.86	-227.18	-229.88
OH^-		-247.69	-229.36	-233.05	-227.67	-226.93	-225.29	-231.55	-234.15
F^-		-245.78	-230.71	-233.10	-227.24	-227.28	-225.40	-230.95	-233.42
SiH_3^-		-169.84	-160.55	-160.79	-153.49	-155.13	-153.03	-157.92	-160.83
PH_2^-		-174.86	-168.47	-169.23	-167.29	-166.78	-164.36	-168.00	-171.21
SH^-		-174.72	-169.23	-169.92	-167.52	-167.38	-165.12	-168.48	-171.32
Cl^-		-172.10	-168.19	-168.62	-166.15	-166.33	-164.29	-167.11	-169.48
Mean absolute deviation from QCISD									
		9.39	1.99	0.00	4.71	4.88	6.87	1.97	0.85

For the neutral species in the study and using the cc-pVTZ basis set, all DFT methods clearly surpass Hartree-Fock. B3LYP and B3PW91, the exact-exchange functionals globally perform the best in comparison with the high level correlated QCISD results.

However, MP2 puts in a surprisingly good performance. All the other functionals perform poorer, BLYP and BP86 even giving worse results than the local density approximation. Concerning this point, it can be remarked that MEP minima occur in regions of considerable pile-up of negative charge, so that the gradient correction to non-locality could perhaps be too simple, the LDA agreement with experiment being fortuitous.

Table 13: Minima in the molecular electrostatic potential (kcal/mol) at the different levels with the cc-pVTZ basis set (augmented functions were placed on the heavy atoms in the anions).

Molecule		SCF	MP2	QCISD	LDA	BP86	BLYP	B3LYP	B3PW91
NH_3		-73.62	-69.55	-69.71	-68.48	-67.39	-66.85	-68.88	-69.48
H_2O		-55.47	-50.82	-51.12	-50.65	-49.14	-48.89	-50.87	-51.12
HF		-29.99	-27.43	-27.57	-27.54	-26.69	-26.65	-27.61	-27.65
HCN		-47.37	-42.75	-43.51	-41.93	-40.94	-40.89	-42.76	-42.97
CO	C	-17.39	-19.02	-17.20	-17.37	-17.30	-17.15	-17.24	-17.73
	O	-10.94	-4.27	-6.93	-5.02	-5.36	-5.83	-6.83	-6.40
H_2CO		-46.29	-38.45	-39.97	-36.13	-36.10	-36.06	-38.71	-38.76
N_2		-11.52	-11.97	-11.79	-11.29	-11.14	-11.44	-11.47	-11.35
PH_3		-26.12	-20.49	-20.43	-17.53	-17.88	-17.41	-19.31	-19.88
H_2S		-24.06	-22.01	-21.95	-21.52	-21.47	-21.01	-21.69	-22.15
HCl		-11.70	-10.83	-10.73	-10.73	-10.60	-10.41	-10.72	-10.90
Mean absolute deviation from QCISD									
		3.44	0.61	0.00	2.07	2.31	2.37	0.80	0.50
CH_3^-		-226.82	-211.97	-217.86	-208.89	-207.47	-205.29	-212.31	-216.10
NH_2^-		-240.79	-222.82	-228.63	-221.01	-219.71	-218.27	-225.06	-227.74
OH^-		-246.99	-227.83	-233.27	-226.26	-225.13	-223.50	-230.25	-232.84
F^-		-246.03	-230.84	-234.41	-227.28	-226.87	-224.91	-230.85	-233.32
SiH_3^-		-169.26	-160.61	-162.26	-153.87	-155.47	-158.05	-158.05	-160.91
PH_2^-		-171.46	-165.63	-167.90	-163.47	-163.32	-161.03	-164.67	-165.42
SH^-		-173.21	-168.15	-169.95	-166.05	-165.87	-163.62	-167.04	-167.43
Cl^-		-172.35	-169.04	-170.13	-166.74	-166.74	-164.66	-167.50	-169.80
Mean absolute deviation from QCISD									
		7.81	3.44	0.00	6.34	6.73	8.14	3.59	1.36

As expected, this effect is even more pronounced for the anions, where BLYP actually performs worse that Hartree-Fock. For these molecules, B3PW91 performs considerably better than B3LYP and BP86 than BLYP; this effect was not encountered in the previous chapter on atomic charges. In addition, we can remark that all results become somewhat poorer in absolute value as compared to the neutral closed shell molecules, but in relative terms, the quality remains comparable due to the somewhat larger magnitudes of the MEP minima for the anions.

For the smaller cc-pVDZ basis set, largely the same conclusions hold as for the cc-pVTZ basis set. For the anions, the B3PW91 results are again remarably improved over the B3LYP results. In general, B3PW91 seems to be the DFT method of choice for the calculation of molecular electrostatic potentials, followed by B3LYP.

3.4.3. The Fukui function

In order to avoid the issue of metastable anions [105, 106], we restrict ourselves in this discussion to the (condensed) Fukui function f^-. In Table 14, this function is listed for a number of small molecules, using Bader atoms-in-molecules charges.

Our conclusions drawn from the study of charge distributions and the molecular electrostatic potential are largely confirmed. B3LYP and B3PW91 show a very good and comparable qualtity ; the same can be said about BP86 and BLYP, although performance has deteriorated a bit.

Surprisingly, the MP2 method performs worse than Hartree-Fock. This is largely due to the bad performance of the methods for the CO molecule, which results from the unusually large spin contamination in the cation of this system. It is known that the density-functional methods are very much less affected by this contamination, as is witnessed by their results for this molecule [107]. Removing this species from the statistical treatment yields for the mean absolute deviations 0.055, 0.026, 0.026, 0.020, 0.021, 0.013 and 0.011. As can be seen, the gradient and exact exchange functionals again all surpass MP2.

Table 14: Condensed Fukui functions f^- for some selected molecules at the different levels with the cc-pVDZ basis set and using Bader atoms-in-molecules charges.

Molecule		SCF	MP2	QCISD	LDA	BP86	BLYP	B3LYP	B3PW91
CO	C	0.602	0.912	0.576	0.605	0.594	0.591	0.612	0.616
	O	0.398	0.088	0.425	0.395	0.406	0.408	0.388	0.384
H_2CO	H	0.189	0.353	0.343	0.340	0.332	0.336	0.346	0.342
	C	-0.188	-0.225	-0.222	-0.145	-0.148	-0.153	-0.200	-0.196
	O	0.311	0.518	0.537	0.466	0.484	0.481	0.508	0.512
NH_3	N	0.445	0.390	0.374	0.424	0.383	0.364	0.379	0.394
	H	0.205	0.203	0.209	0.192	0.206	0.212	0.225	0.202
H_2O	O	0.676	0.676	0.676	0.684	0.669	0.658	0.667	0.676
	H	0.162	0.162	0.162	0.158	0.165	0.171	0.166	0.162
HCN	H	0.227	0.233	0.234	0.209	0.223	0.227	0.225	0.223
	C	0.144	0.300	0.211	0.236	0.227	0.225	0.213	0.214
	N	0.628	0.467	0.554	0.555	0.549	0.548	0.562	0.563
NNO	N	0.368	0.456	0.401	0.418	0.416	0.414	0.411	0.412
	N	0.121	0.107	0.128	0.150	0.150	0.153	0.141	0.138
	O	0.511	0.436	0.471	0.432	0.434	0.434	0.449	0.449
		Mean absolute deviation from QCISD							
		0.051	0.068	0.000	0.028	0.020	0.020	0.016	0.015

Table 15 lists the Fukui function f^- for the ambident nucleophiles SCN^-, CH_2CHO^- and NO_2^- with the aug-cc-pVDZ basis set, which are ideal test cases to assess the performance of the Fukui function [1, 49, 108]. Again, B3LYP and B3PW91 surpass the other methods. The S atom in SCN^- emerges as the soft center, in agreement with experimental observations [109]. For NO_2^-, the O atom is found to be the "softer" center at all levels of theory, in contradiction with Klopman's analysis [109].

In the case of the enolate ion, O is again the softest. Only at the MP2 level, the local softness at the a-carbon is the highest. It has to be mentioned however that using NPA charges, it is the carbon which is the soft center [110]. With this method, the mean absolute deviations are 0.055, 0.043, 0.037, 0.032, 0.030, 0.020 and 0.022 for HF, MP2, LDA, BP86, BLYP, B3LYP and B3PW91 respectively. Although performance is comparable as with the Bader charges, there is an important difference in the conclusion of site selectivity for the enolate species. The relative softness of the atoms in this enolate ion is however still a matter of current debate [111].

It can be concluded from this example that, as far as the condensed Fukui function is concerned, the choice of the population analysis method will become decisive in the correct description of the reactivity of the molecule considered.

Table 15: Condensed Fukui function f^- for the three ambident nucleophiles using the Bader population analysis, together with the aug-cc-pVDZ basis set

Molecule		HF	MP2	QCISD	LDA	BP86	BLYP	B3LYP	B3PW91
SCN^-	S	0.711	0.664	0.677	0.614	0.613	0.611	0.637	0.634
	C	-0.115	0.041	-0.019	0.037	0.043	0.049	0.015	0.019
	N	0.404	0.295	0.342	0.350	0.344	0.340	0.348	0.346
NO_2^-	N	0.175	0.191	0.189	0.219	0.209	0.208	0.205	0.205
	O	0.413	0.405	0.406	0.390	0.396	0.396	0.398	0.397
CH_2CHO^-	C	0.207	0.235	0.225	0.229	0.237	0.233	0.225	0.230
	C	-0.075	0.115	0.023	0.051	0.051	0.050	0.034	0.035
	O	0.419	0.202	0.334	0.286	0.288	0.291	0.306	0.304
		Mean absolute deviation from QCISD							
		0.052	0.045	0.000	0.032	0.031	0.030	0.018	0.020

4. CONCLUSIONS

In this contribution, molecular charge distributions were studied from a density-functional theory viewpoint. In a first part, our attention was focussed on the electronegativity equalization principle. Postulated by Sanderson and proven by Parr within the framework of DFT, this principle provides the quantitative link between molecular charge distributions and readily used chemical concepts such as electronegativity and hardness. A numerical study of this principle was performed at the atomic level. An expression was derived for the electronegativity of an atom in a molecule ; this expression was used to calculate charge distributions in diatomics. At the functional group level, the principle was invoked to rationalize the dual alkyl substituent effect as witnessed by the comparision of gas phase and solution phase alkyl alcohol acidities. Finally, the extension of the principle to amino acid resolution was presented. Some preliminary results indicated the validity of the principle to calculate charge distribution in peptides and proteins.

Moreover, we showed the validity of concepts such as electronegativity, hardness and softness in the qualitative study of molecular charge distributions. These concepts, obtainable in a non-empirical way, either in a DFT or in a wave function context, were used to interpret charges in a series of halogenated alcohols and silanols, illustrating the competition between electronegativity and softness in the periodic Table.

In the second part of this contribution, the performance of different density-functionals (i.e. exchange-correlation functionals) in the calculation of charge distribution (i.e. atomic populations, dipole moments and infrared intensities) and reactivity (i.e. the molecular electrostatic potential and the Fukui function) descriptors was studied. In all cases, it was found that the two exact exchange correlation functionals B3LYP and B3PW91 yield results close to high level correlated calculations, and substantially better that MP2. B3PW91 appears to outperform B3LYP for harmonic frequencies of highly polar molecules and for molecular electrostatic potential minima: otherwise the two methods perform comparably.

ACKNOWLEDGEMENTS

P. Geerlings is indebted to the Free University of Brussels (VUB) and to the D.P.W.B. (National Science Programmation of Scientific Policy, Impulse Program on Information Technology, Contract IT/SC/36) for generous computer grants in support of this work and to the Belgian National Science Foundation (NFWO) for extensive financial support in the set up of the computational infrastructure in the ALGC unit. Dr. P. Raeymaekers, Director of the Computer Center of the Free Universities of Brussels (VUB and ULB) is gratefully acknowledged for continuous advice and help. Finally, Prof. R.G. Parr (Chapel Hill), Prof. R. Nalewajski (Cracow), Prof. K.D. Sen (Hyderabad), Prof. W. Mortier, Prof. R. Schoonheydt and Dr. B. Baekelandt (Leuven) are acknowledged for stimulating discussions and continuous exchange of ideas.

F. De Proft thanks the NFWO for pre- and postdoctoral fellowships. J.M.L. Martin is an NFWO Senior Reseach Associate ("Onderzoeksleider") and a Golda Meir Fellow at the Hebrew University of Jerusalem.

REFERENCES

1. R. G. Parr and W. Yang, Density-Functional Theory of Atoms and Molecules, Oxford University Press, Oxford, 1989.
2. P. Hohenberg and W. Kohn, Phys Rev B136 (1964) 864.
3. For a comprehensive account, see A. Szabo and N. S. Ostlund, Modern Quantum Chemisty, MacMillan, New York, 1982.
4. W. Kohn and L. Sham, Phys. Rev. 140 (1965) 1133.
5. R. Neumann, R. H. Nobes and N. C. Handy, Mol. Phys. 87 (1996) 1.
6. W. J. Hehre, L. Radom, P. von Ragué Schleyer and J. A. Pople, Ab Initio Molecular Orbital Theory, Wiley, New York, 1986.
7. P. R. Taylor in "Lecture Notes in Chemistry", Vol. 88, B. O. Roos Editor, Springer, Berlin, 1992.
8. T. J. Lee and G. E. Scuseria in "Quantum Mechanical Electronic Structure Calculations with Chemical Accuracy", S. R. Langhoff Editor, Kluwer, Dordrecht, The Netherlands, 1995.
9. C. E. Dykstra, Ab Initio Calculation of the Structures and Properties of Molecules, in Studies in Physical and Theoretical Chemistry, Vol. 58, Elsevier, Amsterdam, 1988.
10. J. Overend in "Infrared Spectroscopy and Molecular Structure", M. Davies Editor, Elsevier, Amsterdam, 1963.
11. H. P. Figeys and P. Geerlings in "Theoretical Models of Chemical Bonding", Vol. 3, Z. B. Maksić Editor, Springer-Verlag, Berlin Heidelberg, 1991.
12. R. Bonaccorsi, E. Scrocco and J. Tomasi, J. Chem. Phys. 52 (1970) 5270.
13. E. Scrocco and J. Tomasi, Adv. Quant. Chem. 11 (1978) 115.
14. E. Scrocco and J. Tomasi, Top. Curr. Chem. 42 (1973) 95.
15. R. G. Parr and W. Yang, J. Am. Chem. Soc. 106 (1984) 4049.
16. K. Fukui, T. Yonezawa and H. Shingu, J. Chem. Phys. 20 (1952) 722.
17. H. Fujimoto and K. Fukui in "Chemical Reactivity and Reaction Maps", G. Klopman Editor, Wiley, New York, 1974.
18. R. T. Sanderson, Polar Covalence, Academic Press, New York, 1983.
19. J. A. Huheey, E. A. Keiter and R. L. Keiter, Inorganic Chemistry, Harper Collins College Publishers, New York, 1993.
20. R. G. Parr, R. A. Donnelly, M. Levy and W. E. Palke, J. Chem. Phys. 68 (1978) 3801.
21. R. G. Parr and R. G. Pearson, J. Am. Chem. Soc. 105 (1983) 1503.

22. For a review see W. J. Mortier in "Electronegativity", K. D. Sen and C. K. Jørgenson Editors, Structure and Bonding, Vol. 66, Springer-Verlag, Berlin Heidelberg, 1987, p. 125.
23. P. Geerlings, F. De Proft and W. Langenaeker: "Acidity and Basicity: the Role of Electronegativity, Hardness and Softness" in "Density Functional Methods: Applications in Chemistry and Materials Science", M. Springborg Editor, Wiley, 1996, in press.
24. P. Geerlings, W. Langenaeker, F. De Proft and A. Baeten: "Molecular Electrostatic Potentials vs. DFT descriptors of Reactivity" in "Theoretical and Computational Chemistry", K. D. Sen and P. Politzer Editors, Vol. 3, Elsevier, 1996, in press.
25. L. Pauling, The Nature of the Chemical Bond, Cornell University Press, Ithaca, New York, 1960.
26. R. S. Mulliken, J. Chem. Phys. 2 (1934) 782.
27. W. Gordy, Phys. Rev. 69 (1946) 604.
28. A. L. Allred and E. G. Rochow, J. Inorg. Nucl. Chem. 5 (1958) 264.
29. R. P. Iczkowski and J. L. Margrave, J. Am. Chem. Soc. 83 (1961) 3547.
30. R. G. Pearson, J. Am. Chem. Soc. 85 (1963) 3533.
31. W. Yang and R. G. Parr, Proc. Natl. Acad. Sci. USA 82 (1985) 6723.
32. P. Politzer and H. Weinstein, J. Chem. Phys. 71 (1979) 4218.
33. R. G. Parr and L. J. Bartolotti, J. Am. Chem. Soc. 104 (1982) 3801.
34. J. E. Huheey, J. Org. Chem. 36 (1971) 204.
35. J. E. Huheey, J. Phys. Chem. 69 (1965) 3284.
36. J. E. Huheey, J. Phys. Chem. 70 (1966) 2086.
37. J. Gasteiger and M. Marsili, Tetrahedron 36 (1980) 3219.
38. J. Gasteiger and M. Marsili, Org. Magn. Res. 15 (1981) 353.
39. M. G. Hutchings and J. Gasteiger, Tetrahedron Letters (1983) 2541.
40. J. Gasteiger and M. G. Hutchings, J. Am. Chem. Soc. 106 (1984) 6489.
41. W. J. Mortier, S. K. Ghosh and S. Shankar, J. Am. Chem. Soc. 108 (1986) 4315.
42. K. A. Van Genechten, W. J. Mortier and P. Geerlings, J. Chem. Phys. 86 (1987) 5063.
43. P. Politzer, J. E. Huheey, J. S. Murray and M. Grodzicki, J. Mol. Struct. (Theochem) 259 (1992) 99.
44. P. Politzer, J. S. Murray and M. E. Grice in "Chemical Hardness", K. D. Sen Editor, Structure and Bonding, Vol. 80, Springer Verlag, Berlin Heidelberg, 1993, p. 101.
45. R. F. Nalewajksi, J. Am. Chem. Soc. 106 (1984) 944.
46. R. F. Nalewajski and M. Koninski, J.Phys. Chem. 88 (1984) 6234.
47. J. Cioslowski and S. T. Mixton, J. Am. Chem. Soc. 115 (1993) 1084.
48. J. Cioslowski and B. B. Stefanov, J. Chem. Phys. 99 (1993) 5151.
49. A. K. Rappé and W. A. Goddard III, J. Phys. Chem. 95 (1991) 3358.
50. F. De Proft, W. Langenaeker and P. Geerlings, J. Mol. Struct. (Theochem) 339 (1995) 45.
51. P. Fuentalba and R. G. Parr, J. Chem. Phys. 94 (1991) 5559.
52. C. Lee, W. Yang and R. G. Parr., J. Mol. Struct. (Theochem) 163 (1987) 305.
53. R. F. Nalewajski, J. Phys. Chem. 89 (1985) 2831.
54. H. Weinstein, P. Politzer and S. Srebrenik, Theoret. Chim. Acta 38 (1975) 159.
55. R. J. Boyd, J. Phys. B 9 (1976) L69.
56. P. Politzer and R. G. Parr, J. Chem. Phys. 64 (1976) 4634.
57. F. De Proft and P. Geerlings, Chem. Phys. Lett. 220 (1994) 405.
58. J. Cioslowski, J. Am. Chem. Soc. 111 (1989) 8333.
59. K. Ohwada, J. Phys. Chem. 96 (1992) 5825.
60. T. Koopmans, Physica 1 (1934) 104.
61. F. De Proft, W. Langenaeker and P. Geerlings, Tetrahedron 51 (1995) 4021.
62. see e.g. I. Tuñón, E. Silla and J-L Pascual-Ahuir, J. Am. Chem. Soc. 115 (1993) 2226 and references therein.
63. F. De Proft, W. Langenaeker and P. Geerlings, J. Phys. Chem. 97 (1993) 1826.

64. J. E. Bartmess, J. A. Scott and R. T. McIver, Jr., J. Am. Chem. Soc. 101 (1979) 6056.
65. S. G. Lias, J. E. Bartmess, J. F. Liebman, J. L. Holmes, R. D. Levin, W. G. Mallard, J. Phys. Chem. Ref. Data 17 (1988), Supplement 1.
66. C. K. Mathews and K. E. Holde, Biochemistry, Benjamin/Cunnings, Redwood City, 1990.
67. A. Baeten, F. De Proft and P. Geerlings, Int. J. Quant. Chem., in press.
68. S. Campbell, J. L. Beauchamp, M. Rempe and D. L. Lichtenberg, Int. J. Mass Spectrom. Ion. Process. 117 (1992) 83.
69. A. Baeten, Ph. D. Thesis, Free University of Brussels (VUB), in preparation.
70. S. Damoun, W. Langenaeker, G. Van De Woude and P. Geerlings, J. Phys. Chem. 99 (1995) 12151.
71. J. F. Olsen and J. M. Howell, Theor. Chim. Acta 47 (1978) 39.
72. K. B. Wiberg, J. Am. Chem. Soc. 112 (1993) 3379.
73. J. C. Slater, Phys. Rev. 81 (1951) 385.
74. S. H. Vosko, L. Wilk and M. Nusair, Can. J. Phys. 58 (1980) 1200.
75. A. D. Becke, Phys. Rev. A 38 (1988) 3098.
76. C. Lee, W. Yang and R. G. Parr, Phys. Rev. B 37 (1988) 785.
77. J. P. Perdew, Phys. Rev. B 33 (1986) 8822.
78. J. P. Perdew and Y. Wang, Phys. Rev. B 45 (1992) 13244.
79. A. D. Becke, J. Chem. Phys. 98 (1993) 5648.
80. C. Møller and M. S. Plesset Phys. Rev. 46 (1934) 618.
81. J. A. Pople, M. Head-Gordon and K. Raghavachari, J. Chem. Phys. 87 (1987) 5968.
82. J. Wang, Z. Shi, R. J. Boyd and C. A. Gonzalez, J. Phys. Chem. 98 (1994) 6988.
83. K. E. Laidig, Chem. Phys. Lett. 225 (1994) 285.
84. A. St-Amant, W. D. Cornell, P. A. Kollman and T. A. Halgren, J. Comput. Chem. 16 (1995) 1483.
85. J. Wang, L. A. Eriksson, B. G. Johnson and R. J. Boyd, J. Phys. Chem. 100 (1996) 5274.
86. (a) Gaussian 94, Revision B.3, M. J. Frisch, G. W. Trucks, H. B. Schlegel, P. M. W. Gill, B. G. Johnson, M. A. Robb, J. R. Cheeseman, T. Keith, G. A. Petersson, J. A. Montgomery, K. Raghavachari, M. A. Al-Laham, V. G. Zakrzewski, J. V. Ortiz, J. B. Foresman, C. Y. Peng, P. Y. Ayala, W. Chen, M. W. Wong, J. L. Andres, E. S. Replogle, R. Gomperts, R. L. Martin, D. J. Fox, J. S. Binkley, D. J. Defrees, J. Baker, J. P. Stewart, M. Head-Gordon, C. Gonzalez, and J. A. Pople, Gaussian, Inc., Pittsburgh PA, 1995.
(b) Gaussian 94, Revision C.3, M. J. Frisch, G. W. Trucks, H. B. Schlegel, P. M. W. Gill, B. G. Johnson, M. A. Robb, J. R. Cheeseman, T. Keith, G. A. Petersson, J. A. Montgomery, K. Raghavachari, M. A. Al-Laham, V. G. Zakrzewski, J. V. Ortiz, J. B. Foresman, J. Cioslowski, B. B. Stefanov, A. Nanayakkara, M. Challacombe, C. Y. Peng, P. Y. Ayala, W. Chen, M. W. Wong, J. L. Andres, E. S. Replogle, R. Gomperts, R. L. Martin, D. J. Fox, J. S. Binkley, D. J. Defrees, J. Baker, J. P. Stewart, M. Head-Gordon, C. Gonzalez, and J. A. Pople, Gaussian, Inc., Pittsburgh PA, 1995.
87. T. H. Dunning Jr., J. Chem. Phys. 90 (1989) 1007.
88. F. De Proft, J. M. L. Martin and P. Geerlings, Chem. Phys. Lett. 250 (1996) 393.
89. S. M. Bachrach in "Reviews in Computational Chemistry", Vol. V, K. B. Lipkowitz and D. B. Boyd Editors, VCH, New York, 1995.
90. R. S. Mulliken, J. Chem. Phys. 23 (1955) 1833, 1841, 2338, 2343.
91. C. M. Breneman and K. B. Wiberg, J. Comput. Chem. 11 (1990) 361.
92. A. E. Reed, L. A. Curtiss and F. Weinhold, Chem. Rev. 88 (1988) 899.
93. F. Weinhold and J. E. Carpenter in The Structure of Small Molecules and Ions, R. Na'aman and Z. Vager, Plenum Press, New York, 1988.
94. R. F. W. Bader, Atoms in Molecules: A Quantum Theory, Oxford University Press, Oxford, 1990.
95. J. Cioslowski, Chem. Phys. Lett. 194 (1992) 73.

96. J. R. Thomas, B. J. DeLeeuw, G. Vacek, T. D. Crawford, Y. Yamaguchi and H. F. Schaefer III, J. Chem. Phys. 99 (1993) 403.
97. J. M. L. Martin, J. El-Yazal and J.P. François, Mol. Phys. 86 (1995) 1437.
98. R. N. Sileo and T. A. Cool, J. Chem. Phys. 65 (1976) 117.
99. J. M. L. Martin, T. J. Lee and P. R. Taylor, J. Mol. Spectry 160 (1993) 105.
100. P. Politzer, S. J. Landry and T. Wärnhelm, J. Phys. Chem. 86 (1982) 4767.
101. P. Sjoberg and P. Politzer, J. Phys. Chem. 94 (1990) 3959.
102. W. Yang and W. J. Mortier, J. Am. Chem. Soc. 108 (1986) 5708.
103. R. G. Parr and W. Yang, Ann. Rev. Phys. Chem. 46 (1995) 701.
104. R. A. Kendall, T. H. Dunning, Jr. and R. J. Harrison, J. Chem. Phys. 96 (1992) 6796.
105. J. Simons and K. D. Jordan, Chem. Rev. 87 (1989) 535.
106. W. Langenaeker, M. De Decker, P. Geerlings and P. Rayemaekers, J. Mol. Struct. (Theochem) 115 (1990) 207.
107. J. Baker, A. Scheiner and J. Andzelm, Chem. Phys. Lett. 216 (1993) 380.
108. W. Langenaeker, F. De Proft and P. Geerlings, J. Mol. Struct. (Theochem) 363 (1996) 175.
109. G. Klopman, Chemical Reactivity and Reaction Paths, Wiley, New York, 1974.
110. F. De Proft, J. M. L. Martin and P. Geerlings, Chem. Phys. Lett., in press.
111. F. Méndez and J. L. Gázquez, J. Am. Chem. Soc. 116 (1994) 9298.

J.M. Seminario (Editor)
Recent Developments and Applications of Modern Density Functional Theory
Theoretical and Computational Chemistry, Vol. 4

Density Functional Calculations of Heats of Reaction

Peter Politzer, John J. M. Wiener and Jorge M. Seminario

Department of Chemistry, University of New Orleans
New Orleans, Louisiana, 70148, USA

1. BACKGROUND

In the context of our continuing interest in the decomposition processes of energetic materials—explosives, propellants, fuels, etc.—we need to compute as accurately as possible the energetics of the numerous intermediate steps that may occur. These involve fragments (possibly rearranged) of the original energetic molecules, and typically contain large proportions of the "electron–rich" atoms N, O, F and/or Cl, which have high outer–shell electron densities [1]. For such systems, it is particularly important that computational analyses take account of electronic correlation [2-6]. Since the original molecules as well as some of the fragments are relatively large, a density functional technique is the most practical approach.

It is known, however, that various combinations of exchange and correlation functionals can give significantly different results [7-12]. We have accordingly tested extensively the effectiveness for computing bond dissociation energies of several exchange/correlation combinations [13]: the Becke [14] and Becke–3 [15] for exchange, and the Lee/Yang/Parr [16], Perdew 86 [17] and Perdew/Wang 91 [18] for correlation. The code used was Gaussian 94 [19], and the basis sets were the 6–31G**, the 6–31+G*, and the 6–31+G**. Overall, we obtained the best results, in terms of agreement with experimental values, with the Becke–3 (B3) and Perdew 86 (P86) functionals at the 6–31+G** level [13].

2. METHOD AND RESULTS

We have used the DF/B3P86/6–31+G** procedure to compute the energies (and zero–point corrections) at 0 K and the enthalpies and heat capacities at both 298 K and 800 K for 92 atoms, molecules and ions. Many of these are intermediates or products obtained in the decompositions of energetic molecules.

The results are in Table 1. E (0 K) is the energy minimum at 0 K; it does not include the zero–point energy. ΔH (T) is the enthalpy change in going from 0 K to T; it includes the zero–point contribution. Since E (0 K) = H (0 K), it follows that the enthalpy at T is given by H (T) = E (0 K) + ΔH (T). The heat capacities are at constant volume; they can be converted to constant pressure values by adding R = 1.987 cal·deg^{-1}·mole^{-1} (assuming ideal gas behavior).

Table 1. Computed data (DF/B3P86/6–31+G**).

System	E (0 K), hartrees	Zero–pt. energy, $kcal \cdot mole^{-1}$	ΔH (298 K), $kcal \cdot mole^{-1}$	C_v (298 K), $cal \cdot deg^{-1} \cdot mole^{-1}$	ΔH (800 K), $kcal \cdot mole^{-1}$	C_v (800 K), $cal \cdot deg^{-1} \cdot mole^{-1}$
Atoms and monatomic ions:						
H	–0.51682	0.00	1.48	2.98	3.97	2.98
N	–54.70396	0.00	1.48	2.98	3.97	2.98
O	–75.19989	0.00	1.48	2.98	3.97	2.98
O^-	–75.27650	0.00	1.48	2.98	3.97	2.98
F	–99.88129	0.00	1.48	2.98	3.97	2.98
Cl	–460.45700	0.00	1.48	2.98	3.97	2.98
Homonuclear diatomic molecules:						
H_2	–1.21417	6.36	8.44	4.97	11.94	5.01
N_2	–109.76926	3.53	5.61	4.97	9.18	5.44
O_2	–150.60213	2.41	4.49	5.01	8.22	5.94
F_2	–199.82392	1.52	3.62	5.28	7.62	6.44
Cl_2	–920.99771	0.77	2.97	6.13	7.28	6.81

C,H molecules and molecular–ions:						
CH	–38.60155	4.06	6.13	4.97	9.67	5.29
CH_2	–39.29940	10.91	13.30	6.28	17.85	7.82
CH_3	–40.00720	18.76	21.27	7.35	26.85	10.78
CH_3^+	–39.62184	19.73	22.12	6.26	27.22	10.17
CH_3^-	–40.02430	18.07	20.51	6.92	25.99	10.82
CH_4	–40.70666	28.20	30.56	6.47	36.40	12.75
C_2H_2	–77.57956	16.91	19.28	8.40	25.87	12.95
C_2H_6	–80.16827	46.99	49.77	10.09	59.48	23.70
C_6H_6	–233.02088	63.27	66.63	17.19	83.67	42.69
N,H molecules and molecular–ions:						
NH	–55.36366	4.72	6.79	4.97	10.31	5.15
NH_2	–56.04295	12.00	14.38	6.03	18.70	7.38
NH_3	–56.74653	21.71	24.10	6.42	29.18	9.84
NH_3^+	–56.35428	20.67	23.08	6.62	28.33	10.20
HN_2	–110.29961	8.48	10.87	6.25	15.56	8.50
NH_4^+	–57.08856	31.34	33.72	6.26	39.17	11.80
N_2H_4	–112.20533	33.77	36.40	8.95	44.26	17.51

(continued)

Table 1. Computed data (DF/B3P86/6–31+G**). (continued)

System	E (0 K), hartrees	Zero–pt. energy, kcal·mole^{-1}	ΔH (298 K), kcal·mole^{-1}	C_V (298 K), cal·deg^{-1}·mole^{-1}	ΔH (800 K), kcal·mole^{-1}	C_V (800 K), cal·deg^{-1}·mole^{-1}
O,H molecules:						
OH	–75.89290	5.35	7.43	4.97	10.93	5.08
H_2O	–76.61055	13.49	15.86	6.01	20.14	7.17
HO_2	–151.21000	8.97	11.36	6.25	16.12	8.62
H_2O_2	–151.87357	16.82	19.42	8.08	25.59	12.03
C,O and N,O molecules and molecular–ions:						
CO	–113.55453	3.17	5.23	4.97	8.85	5.58
CO_2	–188.97159	7.34	9.58	6.91	15.02	10.22
NO	–130.15256	2.88	4.95	4.98	8.60	5.70
NO_2	–205.48464	5.65	8.08	6.81	13.31	9.76
NO_2^+	–205.09982	7.52	9.77	6.94	15.17	10.13
NO_2^-	–205.58670	5.06	7.49	6.85	12.86	10.10
NO_3	–280.77536	7.10	10.08	10.61	17.94	15.64
N_2O	–185.05242	7.08	9.36	7.16	14.88	10.32
N_2O_4	–410.99638	14.96	18.87	16.35	30.55	24.72

C,N and N,F molecules:						
CN	–92.93718	3.11	5.18	4.97	8.80	5.60
C_2N_2	–186.10841	10.42	13.36	11.15	21.14	15.17
NF_2	–254.71552	3.81	6.33	7.72	12.16	10.76
NF_3	–354.69139	6.52	9.34	10.65	17.45	16.10
N_2F_4	–509.46053	11.13	15.10	17.88	27.99	26.88
H,C,O and H,N,O molecules:						
HCO	–114.11830	8.28	10.66	6.25	15.36	8.54
H_2CO	–114.79028	16.80	19.19	6.43	24.60	11.15
H_3CO	–115.36200	22.88	25.38	7.80	32.28	15.07
H_2COH	–115.37271	23.53	26.23	9.18	33.58	15.37
H_3COH	–116.05579	32.32	35.00	8.78	42.78	17.64
$HO(CH_2)_2OH$	–230.87957	53.58	57.55	17.20	71.94	34.18
HNO	–130.75773	8.85	11.23	6.06	15.78	8.27
HONO	–206.13464	12.91	15.51	8.62	22.26	13.42
$HONO_2$	–281.46468	16.79	19.58	10.59	28.15	18.12

(continued)

Table 1. Computed data (DF/B3P86/6–31+G**). (continued)

System	E (0 K), hartrees	Zero–pt. energy, kcal·mole^{-1}	ΔH (298 K), kcal·mole^{-1}	C_v (298 K), cal·deg^{-1}·mole^{-1}	ΔH (800 K), kcal·mole^{-1}	C_v (800 K), cal·deg^{-1}·mole^{-1}
H,C,N and H,N,F molecules:						
HCN	–93.67108	10.30	12.49	6.42	17.49	9.14
HNC	–93.64976	9.93	12.29	7.50	17.61	9.46
H_2CN	–94.25282	15.91	18.34	6.95	24.11	11.73
H_2CNH_2	–95.51991	31.63	34.39	10.12	42.79	18.36
H_2CNH_3	–96.09028	39.73	42.50	10.12	51.53	20.90
H_3CNH_2	–96.19691	40.40	43.14	9.54	51.89	20.57
H_3CCN	–133.15282	28.52	31.37	10.34	39.94	19.01
$(H_3C)_2N$	–134.97322	48.45	52.09	13.82	64.20	29.19
HNF_2	–255.36245	12.35	14.92	8.25	21.59	13.49
H,C,F and H,C,N,F molecules:						
H_3CF	–140.07132	24.72	27.14	6.92	33.55	14.26
H_3CNF_2	–294.83189	29.57	32.82	13.68	44.01	25.15

C or F,N,O and H,C,N,O molecules:						
NCO	–168.37037	6.27	8.61	7.51	14.29	10.67
FNO	–230.13762	4.74	7.28	7.77	12.91	10.24
HNCO	–169.07558	13.42	16.06	8.92	22.79	13.17
HOCN	–169.03058	13.58	16.28	9.02	22.99	13.09
H_2C=N–NO	–224.45129	23.09	26.36	12.47	36.15	21.36
H_2C=N–NO_2	–299.79583	27.03	30.56	14.40	42.16	26.04
H_3CNO_2	–245.59583	31.50	34.81	12.48	45.33	24.06
H_3CONO_2	–320.91789	34.40	38.09	15.57	50.78	29.17
$(H_3C)_2N$–NO_2	–340.54001	59.93	64.77	21.86	82.74	43.35
Molecules and molecular–ions containing second–row atoms:						
SiH_3	–291.54659	13.38	15.87	7.62	22.06	12.61
SiH_4	–292.21732	19.65	22.17	8.23	29.50	16.17
H_3CSiH_3	–331.69600	38.45	41.68	13.70	53.43	27.34
SH	–399.06153	3.87	5.94	4.97	9.49	5.34
H_2S	–399.73164	9.54	11.92	6.16	16.46	8.03
H_3CSH	–439.19250	29.08	31.92	10.10	40.41	18.84
ClCN	–553.56183	5.53	8.07	8.68	14.14	11.03
ClO	–535.75877	1.24	3.35	5.52	7.46	6.60

(continued)

Table 1. Computed data (DF/B3P86/6–31+G**). (continued)

System	E (0 K), hartrees	Zero–pt. energy, kcal·mole^{-1}	ΔH (298 K), kcal·mole^{-1}	C_v (298 K), cal·deg^{-1}·mole^{-1}	ΔH (800 K), kcal·mole^{-1}	C_v (800 K), cal·deg^{-1}·mole^{-1}
ClO^-	–535.85983	0.92	3.08	5.91	7.33	6.75
HOCl	–536.43789	8.30	10.74	6.87	15.84	9.10
ClO_2	–611.03334	3.46	6.06	8.15	12.02	10.91
ClO_2^-	–611.13844	2.76	5.46	8.87	11.70	11.25
HOClO	–611.68446	10.34	13.29	10.71	20.90	14.64
ClO_3	–686.27294	6.11	9.05	11.24	17.31	16.26
ClO_3^-	–686.44843	5.99	8.92	11.33	17.24	16.33
$HOClO_2$	–686.96513	12.76	16.39	14.10	26.27	19.87
ClO_4^-	–761.74139	9.33	12.52	13.72	22.86	21.33
$HOClO_3$	–762.22886	16.51	20.21	16.33	32.05	24.76

3. DISCUSSION

In order to assess the accuracies of the energies and enthalpies in Table 1, we have computed ΔH (298 K) for a variety of randomly–selected bond dissociation processes and general chemical reactions. The latter are a particularly stringent test, since they involve rearrangements of atoms and hence there is less likelihood of cancellation of errors within groups that are maintaining their identities; indeed the general reactions present opportunities for accumulation of errors, since they involve as many as eight molecules.

The results are given in Table 2. The computed dissociation energies are, in general, in excellent agreement with the experimental values, the one exception being O_2. It is particularly pleasing, however, to note the remarkable overall accuracy obtained for the general reactions; in most instances, the error is ≤5 kcal·mole^{-1}. In view of the fact that these ΔH are very small differences between extremely large numbers, these results are indeed gratifying.

In Table 3, we compare some of our calculated heat capacities to the experimental values. In general, the agreement is very good, both at 298 K and at 800 K. Since the change in C_v over this 502 K temperature range is often relatively small, it is reasonable to investigate whether linear interpolation can be used to find its magnitudes at intermediate temperatures. Table 4 presents the heat capacities at 600 K obtained by interpolating between those computed for 298 K and 800 K. As is to be expected, the errors are greater than in Table 3, but many of the results are quite satisfactory, indicating that these heat capacities can be approximated as linear functions of the temperature (between 298 K and 800 K).

4. SUMMARY

The data in Table 1 can be used to calculate the heats of reaction of a wide variety of gaseous phase chemical processes over a range of temperatures. The most important point, however, is that density functional computational methods are now capable of giving meaningful results for chemical reactions of practical interest and relevance.

ACKNOWLEDGMENTS

We greatly appreciate the assistance of Dr. Jane S. Murray, and the financial support provided by the Ballistic Missile Defense Organization and the Office of Naval Research through contract N00014–95–1–1339, Program Officers Dr. Leonard H. Caveny (BMDO) and Dr. Richard S. Miller (ONR).

Table 2. Some calculated and experimental heats of reaction.[a]

Process	ΔH (298 K), kcal·mole^{-1}	
	calc.	exp.[a]
Bond–dissociation reactions:		
$H_2O \longrightarrow H + OH$	119.1	119.2
$F_2 \longrightarrow F + F$	37.8	38.0
$N_2O_4 \longrightarrow NO_2 + NO_2$	14.3	14
$H_3C{-}SiH_3 \longrightarrow H_3C + SiH_3$	84.7	88.2[b]
$H_3C{-}NO_2 \longrightarrow H_3C + NO_2$	59.8	60.6
$HNF_2 \longrightarrow H + NF_2$	74.5	75.7[b]
$H_2S \longrightarrow H + SH$	91.7	90.3
$Cl{-}CN \longrightarrow Cl + CN$	103.8	100.0
$O_2 \longrightarrow O + O$	125.4	119.2
General reactions:		
$2HCN \longrightarrow C_2N_2 + H_2$	9.1	8.7
$H_3COH + HONO_2 \longrightarrow H_3CONO_2 + H_2O$	–5.6	–7
$Cl_2 + 2H_2O_2 \longrightarrow 2ClO + 2H_2O$	0.51	–1.6
$H_2CO + NH_3 + H_2 \longrightarrow H_3CNH_2 + H_2O$	–28.2	–26.3

(continued)

Table 2. Some calculated and experimental heats of reaction.[a] (continued)

Process	ΔH (298 K), kcal·mole^{-1} calc.	exp.[a]
$H_2CO + H_2 \longrightarrow H_3COH$	–24.8	–22.2
$NH_3 + F_2 \longrightarrow HNF_2 + H_2$	–8.2	–5
$2CO + 2NO \longrightarrow 2CO_2 + N_2$	–182.8	–178.90
$C_2H_6 + F_2 \longrightarrow 2H_3CF$	–93.5	–98
$H_3CNO_2 + 3H_2 \longrightarrow H_3CNH_2 + 2H_2O$	–98.0	–103.2
$CO + H_2O \longrightarrow H_2 + CO_2$	–16.1	–9.83
$2N_2O + O_2 \longrightarrow 4NO$	57.3	48.1
$C_2H_2 + 2H_2 \longrightarrow C_2H_6$	–87.0	–74.6
$2NH_3 + 2O_2 \longrightarrow N_2O + 3H_2O$	–117.4	–131.8

[a]Unless otherwise indicated, experimental values are obtained from heats of formation given by S. G. Lias, J. E. Bartmess, J. F. Liebman, J. L. Holmes, R. D. Levin and W. G. Mallard, J. Chem. Phys. Ref. Data, 17 (1988), suppl. 1.

[b]D. F. McMillen and D. M. Golden, Ann. Rev. Phys. Chem. 33 (1982) 493.

Table 3. Some calculated and experimental heat capacities.[a]

Compound	C_v (298 K), cal·deg^{-1}·mole^{-1}		C_v (800 K), cal·deg^{-1}·mole^{-1}	
	calc.	exp.[a]	calc.	exp.[a]
H_2O	6.01	6.04	7.17	7.27
Cl_2	6.13	6.13	6.81	6.88
HCN	6.42	6.58	9.14	9.46
NH_3	6.42	6.53	9.84	10.26
H_2CO	6.43	6.47	11.15	11.4
H_2O_2	8.08	8.32	12.03	12.31
HOCl	6.87	6.89	9.10	9.14
CO_2	6.91	6.89	10.22	10.31
H_3CF	6.92	6.98	14.26	14.5
HONO	8.62	9.02	13.42	13.70
ClCN	8.68	8.83	11.03	11.32
H_3CNH_2	9.54	9.99	20.57	20.5
C_2H_6	10.09	10.6	23.70	23.8
H_3CNO_2	12.48	11.7	24.06	23.6

[a]Experimental C_v were determined by subtracting R = 1.987 cal·deg^{-1}·mole^{-1} from C_p values given in Lange's Handbook of Chemistry, J. A. Dean (Ed.), 14th ed., McGraw–Hill, New York, 1992.

Table 4. Some interpolated and experimental heat capacities.[a]

Compound	C_v (600 K), cal•deg^{-1}•mole^{-1}	
	interp.[b]	exp.[a]
H_2O	6.71	6.70
Cl_2	6.54	6.75
HCN	8.06	8.57
NH_3	8.48	8.84
H_2CO	9.27	9.53
H_2O_2	10.46	11.32
HOCl	8.21	8.52
CO_2	8.90	9.32
H_3CF	11.34	11.9
HONO	11.51	12.44
ClCN	10.09	10.64
H_3CNH_2	16.18	16.9
C_2H_6	18.28	19.4
H_3CNO_2	19.45	19.9

[a]Experimental C_v were determined by subtracting R = 1.987 cal•deg^{-1}•mole^{-1} from C_p values given in Lange's Handbook of Chemistry, J. A. Dean (Ed.), 14th ed., McGraw–Hill, New York, 1992.

[b]These values were obtained from the calculated data in Table 3 by linear interpolation.

REFERENCES

1. P. Politzer, J. S. Murray and M. E. Grice, in K. D. Sen (Ed.) Chemical Hardness, Structure and Bonding No. 80, Springer-Verlag, Berlin, 1993, 101.
2. D. J. DeFrees, B. A. Levi, S. K. Pollack, W. J. Hehre, J. S. Binkley and J. A. Pople, J. Am. Chem. Soc., 101 (1979) 4085.
3. D. A. Clabo and H. F. Schaefer III, Int. J. Quant. Chem., 31 (1987) 429.
4. J. M. Coffin and P. Pulay, J. Phys. Chem., 95 (1991) 118.
5. J. M. Seminario, M. C. Concha and P. Politzer, J. Comp. Chem., 13 (1992) 177.
6. D. H. Phillips and G. E. Quelch, J. Phys. Chem., 100 (1996) 11270.
7. B. S. Jursic, Chem. Phys. Lett., 244 (1995) 263.
8. H. Chermette, A. Lembarki, P. Gulbinat and J. Weber, Int. J. Quant. Chem., 56 (1995) 753.
9. B. S. Jursic, J. Mol. Struct. (Theochem), 358 (1995) 145.
10. B. S. Jursic, J. Chem. Phys., 104 (1996) 4151.
11. J. Wang, B. G. Johnson, R. J. Boyd and L. A. Eriksson, J. Phys. Chem., 100 (1996) 6317.
12. P. Politzer and P. Lane, J. Mol. Struct. (Theochem), (in press) .
13. J. J. M. Wiener and P. Politzer, (to be published) .
14. A. D. Becke, Phys. Rev. A, 38 (1988) 3098.
15. A. D. Becke, J. Chem. Phys., 98 (1993) 5648.
16. C. Lee, W. Yang and R. G. Parr, Phys. Rev. B, 37 (1988) 785.
17. J. P. Perdew, Phys. Rev. B, 33 (1986) 8822.
18. J. P. Perdew and Y. Wang, Phys. Rev. B, 45 (1992) 13244.
19. M. J. Frisch, G. W. Trucks, H. B. Schlegel, P. M. W. Gill, B. G. Johnson, M. A. Robb, J. R. Cheeseman, T. A. Keith, G. A. Petersson, J. A. Montgomery, K. Raghavachari, M. A. Al-Laham, V. G. Zakrezewski, J. V. Ortiz, J. B. Foresman, J. Cioslowski, B. B. Stefanov, A. Nanayakkara, M. Challacombe, C. Y. Peng, P. Y. Ayala, W. Chen, M. W. Wong, J. L. Andres, E. S. Replogle, R. Gomperts, R. L. Martin, D. J. Fox, J. S. Binkley, D. J. Defrees, J. Baker, J. P. Stewart, M. Head-Gordon, C. Gonzalez and J. A. Pople, Gaussian 94 (Revision B.3), Gaussian, Inc., Pittsburgh, PA, 1995.

INDEX